Occupational Hearing Loss

Third Edition

Occupational Hearing Loss

Third Edition

Robert Thayer Sataloff

Drexel University College of Medicine
Philadelphia, Pennsylvania

Joseph Sataloff

Jefferson Medical College
Thomas Jefferson University
Philadelphia, Pennsylvania

CRC is an imprint of the Taylor & Francis Group,
an informa business

Published in 2006 by
CRC Press
Taylor & Francis Group
6000 Broken Sound Parkway NW, Suite 300
Boca Raton, FL 33487-2742

Printed in the United States of America on acid-free paper
10 9 8 7 6 5 4 3 2 1

International Standard Book Number-10: 0-8247-5383-6 (Hardcover)
International Standard Book Number-13: 978-0-8247-5383-2 (Hardcover)
Library of Congress Card Number 2005046734

Library of Congress Cataloging-in-Publication Data

Occupational hearing loss / edited by Robert Thayer Sataloff, Joseph Sataloff. -- 3rd ed.
p. cm.
Rev. ed. of: Occupational hearing loss / Robert Thayer Sataloff. 2nd ed.
Includes bibliographical references and index.
ISBN 0-8247-5383-6 (alk. paper)
1. Deafness, Noise induced. 2. Occupational diseases. I. Sataloff, Robert Thayer. II. Sataloff, Joseph. III. Sataloff, Robert Thayer. Occupational hearing loss.

RF293.5.S28 1993
617.8--dc22 2005046734

Visit the Taylor & Francis Web site at
http://www.taylorandfrancis.com

and the CRC Press Web site at
http://www.crcpress.com

To our wives, Dahlia and Ruth

Foreword

On the occasion of the publication of the third edition of this volume of *Occupational Hearing Loss*, it is indeed a pleasure to express my sincere admiration of the authors, Drs. Robert Thayer Sataloff and Joseph Sataloff, for their tireless and career-long dedication to the prevention and reduction of noise-induced hearing loss.

Following World War II, more than 60 years ago, the ravaging effects of noise upon the inner ear became more widely recognized. Soon documentation increased, resulting in the passage of the major legislative standard in the United States, the Occupational Safety and Health Act (OSHA) in 1971. While this act has had a major impact upon industry's control of noise levels and indeed society's interest in compensation in this country, the adverse effects of noise upon hearing remain a major problem, largely due to the increasing popularity of amplified music, the continuing use of firearms in hunting, the lack of use by individuals of ear protection, and the often direct coupling of headphones to the individual's ear when listening to music.

The world's population at this time is approximately 6 billion. Of this rapidly increasing number, 4 to 5 billion live in the developing world. Many of these nations have rapidly expanding economies with occupational noise exposure increasing in the manufacturing, construction, and mining industries; and many of the workers in these nations are neither aware of the need for, nor do they employ, hearing protection.

Thus, this increasing worldwide hazard of occupational hearing loss places the greatest value upon this volume and emphasizes the very great need for increasing interest and involvement by those individuals, organizations, and governments motivated toward, and/or charged with, the prevention of this major global health impairment.

Malcolm D. Graham, M.D.
Department of Otolaryngology—Head and Neck Surgery
Emory University School of Medicine
Atlanta, Georgia

Preface to the Third Edition

Otologists assume a major responsibility for diagnosing and preventing occupational hearing loss and resolving its many medical and legal ramifications. Responsibility for managing this important problem is shared closely by audiologists, industrial physicians, nurses, occupational safety and health personnel, legislators, attorneys, and others. *Occupational Hearing Loss, Third Edition*, is intended to serve as a handy reference volume for practitioners in all these fields, as well as a resource for medical students and paramedical personnel. In practical and comprehensive form, this book presents principles and procedures for determining the causes of hearing loss, including information on the otologic history, otologic examination, and hearing tests, stressing criteria for distinguishing occupational hearing loss from other causes. It also reviews noise control, hearing conservation, legislation, and other relevant subjects.

For more than six decades, the hazards of occupational noise exposure have been recognized and documented. However, it took many years to raise sufficient interest in addressing the problem on a large scale. In 1971, the Occupational Safety and Health Act (OSHA) included a noise standard for all American industries, making it mandatory for them to prevent occupational hearing loss. This regulation, the subsequent hearing conservation amendment, and growing interest in compensation for hearing loss have made occupational hearing loss a major concern for industry, government, medicine, and law.

When the first edition of *Occupational Hearing Loss* was published in 1987, hearing loss was the leading occupational disease in the United States in both prevalence and potential cost. Regrettably, this is still the case, despite considerable progress. Since the first edition was published, there have been improvements in our understanding of the physiology of the ear, the ability to differentiate occupational hearing loss from other etiologies, diagnosis and therapy of hearing loss from many causes, hearing protection, hearing aids, and many other facets of the field.

The second edition updated these and other subjects. It included new material on cochlear biology, criteria for diagnosis of occupational hearing loss, techniques for establishing hearing conservation programs, numerous "new" causes of hearing loss such as Lyme disease and AIDS, advances in audiometry and aural rehabilitation, otologic symptoms associated with head trauma, and many other areas. In addition, it contained new chapters on tinnitus, vertigo, facial paralysis, hearing loss in musicians, hearing conservation in divers, problems associated with taped simulation of hearing loss, and the attorney's approach to cases concerning hearing loss. It also provided updated information on legislation in the United States, Canada, and Great Britain, and a vastly expanded chapter on hearing loss in the railroad industry. The first and second editions of the book were rich in illustrative case reports, and even more have been added to highlight important concepts introduced in the third edition.

The third edition of *Occupational Hearing Loss* has been rewritten and updated extensively. In addition to expanding the discussions of topics included previously (cochlear biology, causes of hearing loss, and other subjects), much new information has been added; and the book has grown from 35 chapters and 2 appendices to 42 chapters and 6 appendices. In addition to expanded discussions on topics such as autoimmune inner ear disease, diagnosing occupational hearing loss, and other issues, entire new chapters have been added on auditory evoked potentials, sudden sensorineural hearing loss, malignancies of the ear, otologic complications of scuba diving, and hearing in dogs. There is also a personal perspective on the history of occupational hearing loss by Joseph Sataloff, who has been involved in the field since its inception.

This book has been written with attention to the needs of industrial physicians, internists, audiologists, audiometric technicians, nurses, health and safety personnel, attorneys, legislators, and others whose practices intersect only incidentally with problems of hearing loss. Every effort has been made to present the material in language that will be accessible to an interdisciplinary readership. The book includes introductory and more advanced reference material of value to otolaryngologists, as well as to readers in other disciplines who have an interest in the ear and hearing. Unavoidably, much of what appears in this book must seem elementary to specialists in otology. Nevertheless, it is hoped that they will find this book useful because it helps establish a bridge between otology and medicine as a whole, because it fosters greater awareness on the part of non-otologists of the vital importance of hearing and early recognition of hearing loss, and because it includes specialized information on hearing conservation, noise control, principles of compensation and legislation that are not included routinely in otolaryngology training.

Since it is hoped that many physicians will use this volume as a reference work, an effort has been made to include most of the known causes of hearing impairment, not only those that may be confused with occupational hearing loss, but also others that may be detected through clinical examination or a hearing conservation program. Both typical and atypical clinical case reports are presented with emphasis on the criteria used to establish the diagnosis. All case reports are from the authors' personal practices or from the practices of our colleagues as specified. Treatment of otologic disorders receives only incidental mention in this book, since this subject is covered widely in medical journals and other books. Nevertheless, forceful references are made to common errors and misconceptions in treatment that have an unhappy tendency to survive the passage of time. It is essential for anyone concerned with occupational hearing loss to be familiar with the existence of other causes of hearing loss and to appreciate the need for comprehensive otological evaluation in order to establish with reasonable certainly whether any individual hearing impairment is causally related to occupational noise exposure.

Chapter 1 provides a succinct overview of hearing loss, emphasizing both medical and societal considerations. Chapter 2 is an accessible review of aspects of the physics of sound pertinent to clinical measurement of hearing and noise. It provides the reader with an understanding of the decibel, hertz, weighting networks, and techniques for calculating the effect of multiple noise sources. Chapter 3 introduces the anatomy and physiology of hearing, and a basic classification of hearing loss. Chapter 4 introduces clinical elements of the history and physical examination for patients with otologic complaints. The third edition also includes an expanded list of questions to be asked when taking a history.

Chapter 5 reviews the principles and techniques for measuring hearing loss, and it includes a discussion of who should do audiometry. It also includes suggested subjects to be covered in training programs for audiometric technicians and hearing conservationists. In addition, it provides updated information on computerized audiometry. Chapter 6 explains the audiogram and its interpretation, including basic information about masking. This is followed in Chapter 7 by a discussion of special hearing tests, which includes material updated from the second edition. Chapter 8 has been added to the third edition and includes information on auditory-evoked response testing, one of the newer methodologies. Chapter 9 offers an in-depth overview of causes of conductive hearing loss and their management. Before providing a similar overview of over 100 pages for sensorineural hearing loss, Chapter 10 includes a new summary of the latest concepts in cochlear biology essential for understanding sensorineural hearing loss. Chapter 11 discusses the complex and controversial problem of sudden sensorineural hearing loss and its treatment. Chapter 12 presents the problems of mixed, central, and functional hearing loss, including new information on auditory processing disorders.

Chapter 13 provides updates to the unique chapter on systemic causes of hearing loss. This chapter summarizes not only the most common and important hereditary causes of hearing loss, but also many of the systemic causes of sensorineural hearing loss, including hypertension, diabetes, syphilis, and many other conditions. In Chapter 14, the author (Joseph Sataloff) provides a personal reflection of the history of occupational hearing loss. This chapter offers insight into many of the events and people responsible for the development of this field. Chapter 15 reviews the difficulties and complexities associated with establishing an accurate diagnosis of occupational hearing loss. It summarizes pertinent literature on this subject and includes a review of the criteria established by the American Occupational Medicine Association. It also includes new information on many co-factors associated with occupational hearing loss as well as non-occupational factors that may affect or be confused with occupational hearing impairment.

Chapter 16 has been revised substantially, particularly regarding hearing aids. Information about the latest concepts in amplification has been added, as well as information about cochlear implants. Chapter 17 discusses hearing protection devices for industry, and for special applications such as use by musicians. Chapter 18 reviews the problems of tinnitus, a condition often associated with hearing loss. Chapter 19, a newly expanded chapter, provides a much more comprehensive treatment of the problems associated with balance disorders, as well as review of modern techniques of evaluation, including posturography. Chapter 20 covers the problem of facial paralysis. This condition often accompanies sensorineural hearing loss, especially following trauma or tumor. Neoplasms of the ear and related structures are among the more serious causes of hearing loss. In the previous editions, only acoustic neuromas have been discussed in any detail. Chapters 21 and 22 have been added to provide the reader with overviews of our current understanding of squamous cell carcinoma (Chapter 21) and sarcoma (Chapter 22) of the temporal bone. Because malignancies of the temporal bone may be associated with actinic exposure, exposure to radiation and other factors that may be encountered in the work place, these chapters should prove valuable to those interested in occupational otologic disorders. Chapter 23 is presented for general information and fascination, as well as for reference for people working in industries where dogs may be employed as guards or to perform work in noisy areas. It describes what is known about hearing in dogs. This information is included in the book not only because of its intrinsic interest, but also because, understanding the species differences described so beautifully in this chapter, all of us must give considerable thought to how we interpret research data based upon experimentation in animals, many of which are even further distant from humans than are canines. Chapter 24 is composed of tables summarizing the differential diagnosis of hearing conditions and is especially valuable for a quick review. Many of the first 24 chapters also appear in our companion book *Hearing Loss, Fourth Edition*, which also includes a chapter on pediatric hearing loss and the potential benefits of nutraceutical modification upon hearing. Although these chapters are excluded from *Occupational Hearing Loss, Third Edition*, the authors consider it likely that neutraceutical modification, particularly substances such as antioxidants, may prove important in occupational hearing loss in the future.

Chapters 25–42 present a unique compendium of information about specific problems of occupational hearing loss and hearing conservation. In Chapters 25 and 26, Drs. Paul and Kevin Michael provide a comprehensive and practical introduction to the principles and problems of noise measurement and noise control. Chapter 27 explores principles of establishing of damage risk criteria and provides invaluable insights based on Terrence Dear's extensive experience with hearing and noise problems at E.I. du Pont de Nemours & Company. Chapter 28 is a new chapter on hearing loss and other complications of scuba diving that should be of value to anyone working in industries that involve underwater activities. Chapter 29 by Dr. Harry Hollien summarizes the problems of hearing conservation in the diving industry. This is a relatively new field, and this chapter introduces important concepts in establishing damage risk criteria and how to approach the problem. Chapter 30 offers a comprehensive review of the problem of hearing loss in musicians, a unique subject in the field of occupational hearing loss. The chapter includes a comprehensive, updated review of the available literature and an explanation of special aspects of the problem. Chapter 31 defines the components of a hearing conservation program in industry, including a review of the physician's responsibilities. Chapter 32 provides invaluable, practical information based on more than 30 years of experience in establishing hearing conservation programs. These two chapters are supplemented by a useful collection of tables and forms, developed over many years for managing day-to-day hearing conservation operations, and published in two appendices.

Chapter 33 discusses legislation and compensation aspects of occupational hearing loss including a historical overview, and in-depth information about calculating hearing impairment. Chapter 34 reiterates the important aspects of the Occupational Safety and Health Act noise regulation, including useful tables (such as age correction values). Chapter 35 is an updated review of the formulae used in different jurisdictions for calculating compensation for hearing loss. It provides an extremely convenient overview for physicians, judges, and attorneys, highlighting the marked variability among jurisdictions. Chapter 36 reviews hearing loss problems in the railroad industry. Written by one of the most experienced attorneys in the country, its insights are invaluable. Chapter 37 is updated to included the latest developments in hearing loss compensation under the Longshore and Harbor Workers' Compensation Act. Chapter 38 provides a similarly updated review of the current situation in Canada.

Chapter 39 reviews the latest developments in occupational hearing loss in the United Kingdom. Following this is a unique chapter (Chapter 40) on taped simulation of hearing loss. Recent attempted use of simulation tapes as evidence in court prompted scientific assessment of the validity of such tapes and a search for standards and criteria that could render them useful and acceptable. This chapter reviews some of the work and explains why such tapes are not valid, reliable, or scientifically justified for medical or legal use. Chapter 41 was written largely for attorneys but should be valuable for all readers of the book. Written by an experienced plaintiffs' attorney, this chapter explains the considerations involved in assessing occupational hearing loss claims, estimating their value, and deciding how to manage them. Irvin Stander's Chapter 42 on presenting medical evidence in hearing loss cases is classic, written by one of the country's most experienced attorneys and jurists. The book also contains six appendices, four of which are new. Appendices 1–4 provide additional information on anatomy of the ear, pathology of the ear, temporal bone dyscrasias (otosclerosis, Paget's disease, and osteogenesis imperfecta) and neurofibromatosis, respectively. Appendix 5 provides examples of forms used in hearing conservation programs, and Appendix 6 illustrates material describing hearing conservation program procedures and reports.

We are deeply indebted to Mary Hawkshaw, R.N., B.S.N, our indefatigable editorial assistant; our audiologists, especially Caren Sokolow, John Luckhurst, and Tracey Virag; and Helen Caputo and Beth V. Luby, our expert and tireless proofreaders/manuscript typists. Their contributions were invaluable.

We hope that our readers will find this book useful and easy to read; and we hope that it inspires in them an increasing passion to answer many of the remaining questions about hearing, hearing loss, our ability to restore hearing to the many patients whom we are still unable to help, and practical methods for protecting the hearing of every industrial worker worldwide.

Robert Thayer Sataloff
Joseph Sataloff

Authors

Robert T. Sataloff, M.D., D.M.A., F.A.C.S., is Professor and Chairman, Department of Otolaryngology—Head and Neck Surgery and Associate Dean for Clinical Academic Specialties at Drexel University College of Medicine, Philadelphia; Professor of Otolaryngology—Head and Neck Surgery at Jefferson Medical College, Thomas Jefferson University; Chairman of the Department of Otolaryngology—Head and Neck Surgery at Graduate Hospital; Adjunct Professor of Otorhinolaryngology—Head and Neck Surgery, The University of Pennsylvania; on the faculty of the Academy of Vocal Arts; Conductor of the Thomas Jefferson University Choir and Orchestra; Director of the Jefferson Arts-Medicine Center; and Director of the Voice Foundation's Annual Symposium on Care of the Professional Voice. Dr. Sataloff is also a professional singer and singing teacher.

He holds an undergraduate degree in Music Composition from Haverford College, graduated from Jefferson Medical College, Thomas Jefferson University, received a Doctor of Musical Arts in Voice Performance from Combs College of Music, and completed his Residency in Otolaryngology—Head and Neck Surgery at the University of Michigan. He also completed a fellowship in Otology, Neurotology, and Skull Base Surgery at the University of Michigan. Dr. Sataloff is Chairman of the Boards of Directors of the Voice Foundation and of the American Institute for Voice and Ear Research. He has also served as Chairman of the Board of Governors of Graduate Hospital, President of the International Association of Phonosurgery, President of the American Laryngological Association, and President of the Pennsylvania Academy of Otolaryngology—Head and Neck Surgery. He is Editor-in-Chief of the *Journal of Voice*, Editor-in-Chief of *Ear, Nose and Throat Journal*, and on the Editorial Boards of the *Journal of Singing*, *Medical Problems of Performing Artists*, and numerous major otolaryngology journals in the United States. Dr. Sataloff has written more than 600 publications, including 35 books. Dr. Sataloff's medical practice is limited to care of the professional voice and to otology/neurotology/skull base surgery.

Joseph Sataloff, M.D., D.Sc., is Professor of Otolaryngology—Head and Neck Surgery, Jefferson Medical College, Thomas Jefferson University, Philadelphia, Pennsylvania. He was the founder of Hearing Conservation Noise Control, Inc. and served as its President for close to five decades. For more than 40 years, he also co-directed an annual institute on occupational hearing loss. Dr. Sataloff has served as a scientific member of the Secretary of Labor's Noise Advisory Committee and is a Consultant to the Surgeon General, the U.S. Navy, and numerous industries and labor groups. He has written or co-authored more than 150 journal articles, many book chapters, and 11 books. Dr. Sataloff has served as Chapter Chair for the American Medical Association's *Guides to the Evaluation of Impairment and Disability*. He is a member of the American Academy of Otolaryngology—Head and Neck Surgery, the American Otological Society, the Otosclerosis Study Group, and numerous other organizations and societies.

Contributors

Mark T. Agrama *River Region Health System, Vicksburg, Mississippi, USA*

D. Caroline Coile *Georgia Pines Station, Ochlocknee, Georgia, USA*

Terrence A. Dear *E. I. du Pont de Nemours and Company, Wilmington, Delaware, USA*

Ann F. Dingle *Department of Otolaryngology, Leighton Hospital, Crewe, England, UK*

Liam M. Flood *Department of Otolaryngology, James Cook University Hospital, Middlesbrough, UK*

Gregory J. Hannon *Hannon and Joyce, Philadelphia, Pennsylvania, USA*

Mary Hawkshaw *American Institute for Voice and Ear Research, Philadelphia, Pennsylvania, USA*

David J. Hickton *Burns, White and Hickton, Pittsburgh, Pennsylvania, USA*

Harry Hollien *University of Florida, Gainesville, Florida, USA*

Christine Johnston *Workplace Safety and Insurance Board of Ontario, Toronto, Ontario, Canada*

John Luckhurst *American Institute for Voice and Ear Research, Philadelphia, Pennsylvania, USA*

Heidi Mandel *Philadelphia College of Podiatric Medicine, Philadelphia, Pennsylvania, USA*

Steven Mandel *Thomas Jefferson University, Philadelphia, Pennsylvania, USA*

Sandra Markowitz *University of Pennsylvania, Philadelphia, Pennsylvania, USA*

Kevin L. Michael *Michael and Associates, Inc., State College, Pennsylvania, USA*

Paul L. Michael *Paul L. Michael and Associates, Inc., State College, Pennsylvania State University, USA (Deceased)*

Brian A. Neff *Department of Otorhinolaryngology, Mayo Clinic, Southwest Rochester, Minnesota, USA*

Michael C. Neuenschwander *Director, Southeastern Voice Disorders Center, Riverdale, Georgia, USA*

Lawrence P. Postol *Seyfarth Shaw, LLP, Washington, DC, USA*

John T. Rutka *Division of Neurosurgery, University of Toronto, and University Health Network, Toronto, Ontario, Canada*

Robert T. Sataloff *Drexel University College of Medicine, American Institute for Voice and Ear Research, and Graduate Hospital, Philadelphia, Pennsylvania, USA*

Joseph Sataloff *Thomas Jefferson University, Philadelphia, Pennsylvania, USA*

Jeremy F. Sharp *Department of Otolaryngology, Derbyshire Royal Infirmary, England, UK*

Caren J. Sokolow *American Institute for Voice and Ear Research, Philadelphia, Pennsylvania, USA*

Irvin Stander *Bureau of Workers' Compensation of Pennsylvania and Pennsylvania Bar Institute, Harrisburg, Pennsylvania, and Temple University, Philadelphia, Pennsylvania, USA (Deceased)*

F. Javed Uddin *Department of Otolaryngology, Derbyshire Royal Infirmary, UK*

Mikhail Vaysberg *Philadelphia College of Osteopathic Medicine, Philadelphia, Pennsylvania, USA*

Tracy M. Virag *American Institute for Voice and Ear Research, Philadelphia, Pennsylvania, USA*

Thomas O. Willcox *Thomas Jefferson University, Philadelphia, Pennsylvania, USA*

Jeffrey M. Zimmerman *Ear, Nose and Throat Specialists of Southern New Hampshire, Manchester, New Hampshire, USA*

Contents

1
Occupational Hearing Loss: An Overview

Robert T. Sataloff
Drexel University College of Medicine, American Institute for Voice and Ear Research, and Graduate Hospital, Philadelphia, Pennsylvania, USA

Joseph Sataloff
Thomas Jefferson University, Philadelphia, Pennsylvania, USA

Hearing loss due to occupational noise exposure is our most prevalent industrial malady and has been recognized since the Industrial Revolution. There are millions of employees with occupational hearing loss in industry. Our neglect of hearing loss, especially occupational hearing loss, has resulted in human and economic consequences that affect virtually every household in industrialized countries. This is especially regrettable since noise-induced hearing loss is almost always preventable at relatively little cost.

Although the importance of good hearing can hardly be overestimated, it has not been appreciated by the public, or even by the medical community. Over 40 million Americans have hearing loss, and there is still a stigma attached to deafness. Little has changed from the days when society had to be admonished: "Thou shalt not curse the deaf." Although hearing loss may not be regarded widely as a punishment from God, it is seen still as an embarrassing infirmity, or a sign of aging and senility, and it is associated with a loss of sexual attractiveness. Too often patients do not seek medical attention of their own accord. Many deny and tolerate hearing loss for a considerable period of time before being coerced by a family member to seek medical care. Patients accept the need for eyeglasses easily, but it is unusual to tell someone that he/she needs a hearing aid without causing distress. This is every bit as common in 70-year-old patients as it is in teenagers.

When hearing loss occurs early in childhood, its devastating consequences are more obvious than when it occurs insidiously in adult life. Normal psychological maturation involves progression from oneness with a child's mother to self-image definition. In this process, the child develops patterns of human interrelationship and modes of emotional expression. A substantial hearing deficit in infancy interferes with this process. It delays self-image development, impairs the child's expression of needs, and often results in alienation from the child's family, sometimes to a permanent deficit in his ability to establish relationships. Severe hearing loss makes learning a mammoth task for the child, and he/she frequently reacts with frustration or isolation. The personality distortion that results from this sequence affects the person and his family throughout their lives.

Even more mild forms of hearing loss early in life can cause great difficulty. We frequently see a child who developed within normal limits but is not doing well in school, is inattentive, and is frequently considered “not too bright.” It is not uncommon to discover a moderate hearing loss in such a child. When the hearing loss is corrected, the parents invariably report that the child is “like a different person.” Fortunately, many of the hearing impairments that lead to these and other consequences are preventable.

When hearing loss occurs in an adult, more subtle manifestations of many of the same problems may be found. Most people with age- or noise-induced hearing loss lose hearing in the high frequencies first, making it difficult for them to distinguish consonants, especially s, f, t, and z. This makes a person strain to understand what is being said in everyday conversation. The person knows that there is speech because he/she can hear the vowels, but he/she cannot easily distinguish the difference between “yes” and “get.” This makes talking to one’s spouse, going to the movies, going to church, and other pleasures that most of us take for granted stressful chores. It is also the unrecognized source of considerable marital discord. For example, a man who has worked hard for many years in a weaving mill or as a boilermaker usually has a substantial hearing loss, especially if he has not worn ear protectors. When he comes home and sits down to read a newspaper, if his wife starts talking to him from another room (especially if there is competing noise such as running water or air conditioners), he will be able to hear her talking but not understand her words. Before long, it becomes so difficult to say “what” all the time that he stops listening. Soon she thinks he does not pay any attention to her or love her anymore, and neither of them realizes that he has a hearing loss underlying their friction. Busy otologists see this scenario daily in the office. Although each of these patients can be helped through counseling and rehabilitation, we still have no cure for most cases of sensorineural deafness. Despite that, there is relatively little support for research. We undergo a constant barrage of requests for funds for sight, cancer, muscular dystrophy, multiple sclerosis, and numerous other entities, but it is hard to remember the last call for help for the deaf.

Although otologic advances have made almost all forms of conductive hearing loss surgically curable, sensorineural hearing loss can be treated and potentially cured in only a few conditions (Meniere’s disease, syphilis, hypothyroidism, and a few others). Despite advances in our understanding of hearing loss and in hearing aid technology, which make it possible to improve the lives of almost every patient with hearing loss, deafness prevention is still our best cure. A thorough understanding of the function of the ear and related structures reveals many possibilities for prevention of injury and restoration of function. Avoidance of damaging noise, of ototoxic drugs, of insufficient treatment of diseases such as syphilis, and other measures often prevent hearing loss. Even when sensorineural hearing loss occurs, systematic, comprehensive assessment may reveal a treatable underlying cause. Fortunately, this is true in an increasing number of patients, as medical knowledge increases.

Today, although not all hearing loss can be cured, virtually every patient with hearing impairment can be helped through accurate diagnosis, understanding, education, medication, aural rehabilitation, amplification, and/or cochlear implantation. Major advances have been made in the last few decades, and even more may be anticipated as exploration continues into the mysteries of the inner ear and ear–brain interface.

2
The Physics of Sound

Robert T. Sataloff
Drexel University College of Medicine, American Institute for Voice and Ear Research, and Graduate Hospital, Philadelphia, Pennsylvania, USA

Joseph Sataloff
Thomas Jefferson University, Philadelphia, Pennsylvania, USA

Fortunately, one need not be a physicist in order to function well in professions involved with hearing and sound. However, a fundamental understanding of the nature of sound and terms used to describe it is essential to comprehend the language of otologists, audiologists, and engineers. Moreover, studying basic physics of sound helps one recognize complexities and potential pitfalls in measuring and describing sound and helps clarify the special difficulties encountered in trying to modify sources of noise.

1. SOUND

Sound is a form of motion. Consequently, the laws of physics that govern actions of all moving bodies apply to sound. Because sound and all acoustic conditions behave consistently as described by the laws of physics, we are able to predict and analyze the nature of a sound and its interactions. Sound measurement is not particularly simple. The study of physics helps us to understand many practical aspects of our daily encounters with sound. For example, why does an audiologist or otologist use a different baseline for decibels in his office from that used by an engineer or industrial physician who measures noise in a factory? Why is it that when hearing at high frequencies is tested, a patient may hear nothing and then suddenly hear a loud tone? Yet, all the examiner did was move the earphone a fraction of an inch. Why is it when two machines are placed close together, each making 60 dB of noise, the total noise is not 120 dB? Everyone concerned with noise or hearing loss must have at least sufficient familiarity with the physics of sound to understand the answers to these basic questions.

2. SOUND WAVES

Sound is the propagation of pressure waves radiating from a vibrating body through an elastic medium. A *vibrating body* is essential to cause particle displacement in the propagating medium. An *elastic medium* is any substance whose particles return to their point of origin as soon as possible after they have been displaced. *Propagation* occurs because displaced particles in the medium displace neighboring particles. Therefore, sound travels over linear distance. *Pressure waves* are composed of areas of slightly greater than ambient air pressure (compression) and slightly less than ambient air pressure (rarefaction). These are associated with the bunching together or spreading apart of the particles in the propagating medium. The pressure wave makes receiving structures such as the eardrum move back and forth with the alternating pressure. For example, when a sound wave is generated by striking a tuning fork, by vocalizing, or by others means, the vibrating object moves molecules in air, causing them alternately to be compressed and rarefied in a rhythmical pattern. This sets up a chain reaction with adjacent air molecules and spreads at a rate of $\sim$1100 ft/s (the speed of sound). This is *propagation* of the pressure waves.

Sound requires energy. Energy is used to set a body into vibration. The energy is imparted to particles in the propagating medium and then is distributed over the surface of the receiver (eardrum or microphone) in the form of sound pressure. Energy is equal to the square of pressure ($E = P^2$). However, we are unable to measure sound energy directly. Only the pressure exerted on the surface of a microphone can be quantified by sound-measuring equipment.

2.1. Characteristics of Sound Waves

Sound waves travel in straight lines in all directions from the source, decreasing in intensity at a rate inversely proportional to the square of the distance from their source. This is called the inverse-square law. This means that if a person shortens his distance from the source of a sound and moves from a position 4 ft away to only 2 ft from the source, the sound will be four times as intense rather than merely twice as intense. In practical application, this inverse-square law applies only in instances in which there are no walls or ceiling. It is not strictly valid in a room where sound waves encounter obstruction or reflection, and increasing the distance of a whisper or a ticking watch from the subject rarely can provide a truly accurate or reliable assessment of hearing.

Sound waves travel through air more rapidly than through water. They are conducted by solids also at different speeds. An ear placed close to the iron rail of a train track will detect the approach of the train before the airborne sounds can reach the observer. Thus, sounds travel through different media at different speeds; the speed also varies when the medium is not uniform. However, sound waves are not transmitted through a vacuum. This can be demonstrated by the classic experiment of placing a ringing alarm clock inside a bell jar and then exhausting the air through an outlet. The ringing will no longer be heard when the air is exhausted, but it will be heard again immediately when air is readmitted. This experiment emphasizes the importance of the medium through which sound waves travel.

The bones of the head also conduct sounds, but ordinarily the ear is much more sensitive to sounds that are airborne. Under certain abnormal conditions, as in cases of conductive hearing loss, a patient may hear better by bone conduction than by air conduction. Such an individual can hear the vibrations of a tuning fork much better when it is held directly touching the skull than when it is held next to the ear but without touching the head.

Distortion of sound waves by wind is common. The effect also varies according to whether the wind blows faster near the ground or above it. When sound travels through air and encounters an obstruction such as a wall, the sound waves can bend around the obstacle almost like water passing around a rock in a stream. The behavior of sound waves striking an object depends upon several factors, including wavelength. Sound waves may pass through an object unaffected, be reflected off the object, or may be partially reflected and partially passed through or around the object (shadow effect). Low-frequency sounds of long wavelength tend to bend (diffraction) when encountering objects, while diffraction is less prominent with sounds >2000 Hz. The behavior of sound waves encountering an object also depends upon the nature of the object. The resistance of an object or system to the transmission of sound is called impedance. This depends upon a variety of factors, such as mass reactants, stiffness reactants, and friction. The ability of an object to allow transmission of sound is called its admittance, which may be thought of as the opposite of impedance.

2.1.1. Components of Sound

A simple type of sound wave, called a *pure tone*, is pictured in Fig. 2.1. This is a graphic representation of one and one-half complete vibrations or cycles, or periods, with the area of compression represented by the top curve and the area of rarefaction by the bottom curve. Although pure tones do not occur in nature, the more complicated sounds that we actually encounter are composed of combinations of pure tones. Understanding the makeup of this relatively simple sound helps us analyze more complex sounds. Fourier analysis is used to separate complex signals into their simple tonal components.

A pure tone has several important characteristics: One complete vibration consists of one compression and one rarefaction (Fig. 2.2). The number of times such a cycle occurs

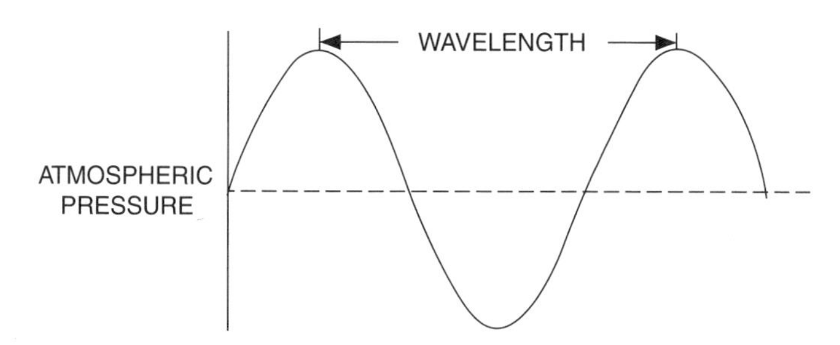

Figure 2.1 Diagram of a pure tone (sine wave).

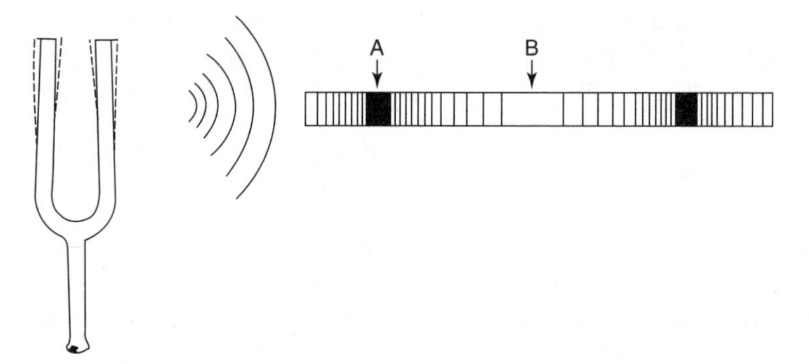

Figure 2.2 Areas of compression (A) and rarefaction (B) produced by a vibrating tuning fork.

in a given period of time (usually 1 s) is called *frequency*. Frequency is usually recorded in cycles per sound, or hertz. The psychological correlate of frequency is *pitch*. In general, the greater the frequency the higher the pitch, and the greater the intensity the louder the sound. However, there is a difference between actual physical phenomena (such as frequency or intensity) and peoples' perceptions of them (pitch and loudness). A tuning fork is constructed so that it vibrates at a fixed frequency no matter how hard it is struck. However, although it will vibrate the same number of times per second, the prongs of the tuning fork will cover a greater distance when the fork is struck hard than when it is struck softly. This increased *intensity* we perceive as increased *loudness*. In the sine wave diagram of a pure tone, a more intense sound will have a higher peak and lower valley than a softer sound. Greater intensity also means that the particles in the propagating medium are more compressed. The height or depth of the sine wave is called its amplitude. Amplitude is measured in decibels (dB). It reflects the amount of pressure (or energy) existing in the sound wave.

Wavelength is the linear distance between any point in one cycle and the same point on the next cycle (peak-to-peak, for example). It may be calculated as the speed of sound divided by the frequency. This is also one *period*. Wavelength is symbolized by the Greek letter lambda (λ) and is inversely proportional to frequency (Fig. 2.3). This is easy to understand. If it is recalled that sound waves travel at ~1100 ft/s, simple division tells us that a 1000-Hz frequency will have a wavelength of 1.1 ft/cycle. A 2000-Hz tone has a wavelength of ~6.5 in. A 100-Hz tone has a wavelength of ~11 ft. The wavelength of a frequency of 8000 Hz would be 1100 divided by 8000, or 0.013 ft (~1 in.). Wavelength has a great deal to do with sound penetration. For example, if someone is playing a stereo too loudly several rooms away, the bass notes will be heard clearly, but the high notes of violins or trumpets will be attenuated by intervening walls. Low-frequency sounds (long wavelengths) are extremely difficult to attenuate or to absorb, and they require very different acoustic treatment from high-frequency sounds of short wavelengths. Fortunately, they are also less damaging to hearing.

Any point along the cycle of the wave is its *phase*. Because a sine wave is a cyclical event, it can be described in degrees like a circle. The halfway point of the sine wave is the 180° phase point. The first peak occurs at 90°, etc. The interaction of two pure tones depends on their phase relationship. For example, if the two sound sources are identical and are perfectly in phase, the resulting sound will be considerably more intense than either one alone (constructive interference). If they are 180° out of phase, they will theoretically nullify each other and no sound will be heard (destructive interference) (Fig. 2.4). Interaction of sound forces also depends upon other complicated factors, such as

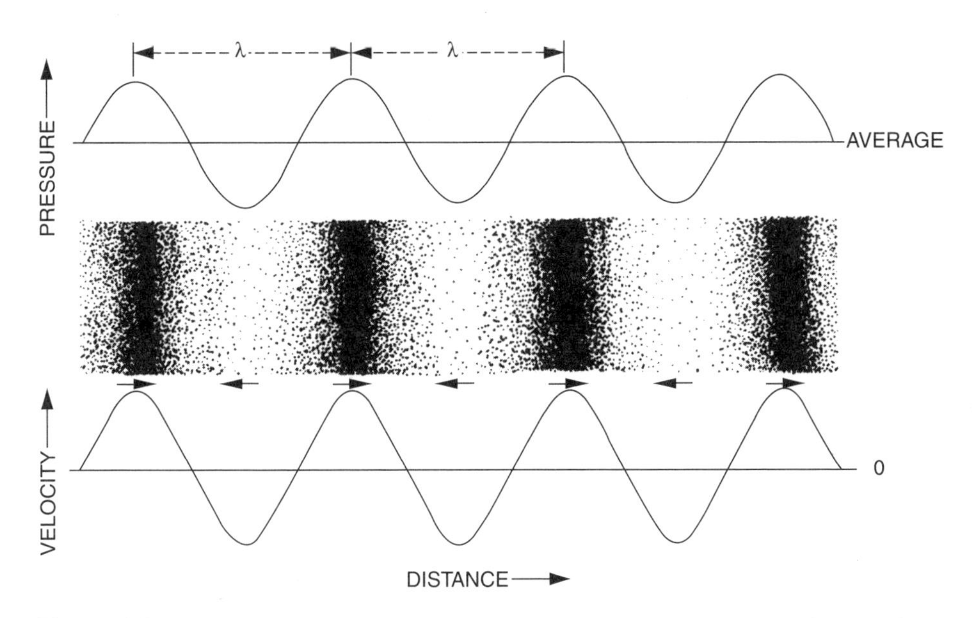

Figure 2.3 Diagram showing wavelength in relation to other components of a sound wave. [Adapted from Van Bergeijk et al. (1).]

resonance, which is affected by the environment and the characteristics of the receiver (such as the ear canal and ear).

Speech, music, and noise are *complex sounds*, rather than pure tones. Most sounds are very complex, with many different wave forms superimposed on each other. Musical tones usually are related to one another and show a regular pattern (complex periodic sound), whereas street noise shows a random pattern (complex aperiodic sound) (Fig. 2.5).

It is somewhat difficult to define noise accurately, because so much of its meaning depends on its effect at any specific time and place rather than on its physical characteristics. Sound can in one instance or by one individual be considered as very annoying noise, whereas on another occasion or to another observer the same sound may seem pleasant and undeserving of being designated "noise." For the purpose of this book, the term *noise* is used broadly to designate any unwanted sound.

An interesting aspect of sound waves related to hearing testing is a phenomenon called the *standing wave*. Under certain circumstances, two wave trains of equal amplitude and frequency traveling in opposite directions can cancel out at certain points called "nodes." Figure 2.6 is a diagram of such a situation. It will be noted that when a violin string is plucked in a certain manner, at point "n" (node) there is no displacement. If this point falls at the eardrum, the listener will not be aware of any sound because the point has no amplitude and cannot excite the ear. This phenomenon occasionally occurs in hearing tests, particularly in testing at $\geq$8000 Hz. These higher frequencies are likely to be involved, because the ear canal is $\sim$2.5 cm long, and the wavelength of sound at such high frequencies is of the same order of magnitude. The point of maximum displacement is called the antinode.

Furthermore, when sound waves are produced within small enclosures, as when an earphone is placed over the ear, the sound waves encounter many reflections, and much of the sound at high frequencies is likely to be in the form of standing waves. Such waves often do not serve as exciting stimuli to the inner ear, and no sensation of hearing is produced because of the absence of transmission of sound energy.

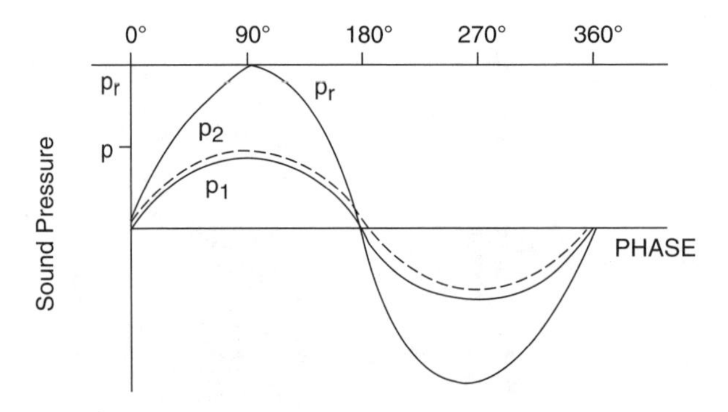

(A) 0° PHASE DIFFERENCE $p_r = 2_p$ ($p_r = p + 6$ dB)

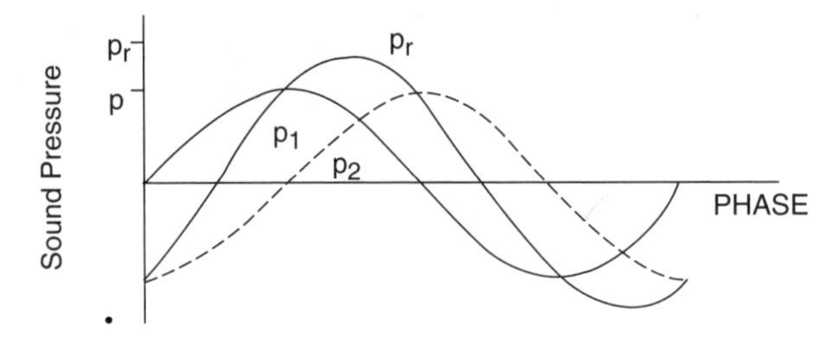

(B) 90° PHASE DIFFERENCE $p_r = 1.4_p$ ($p_r = p + 3$ dB)

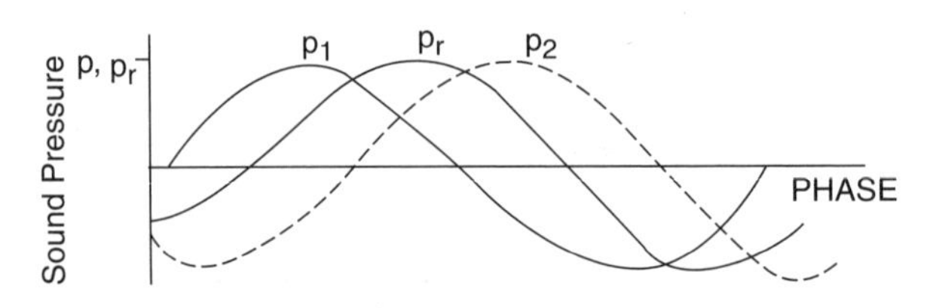

(C) 120° PHASE DIFFERENCE $p_r = p$ ($p_r = p + 0$ dB)

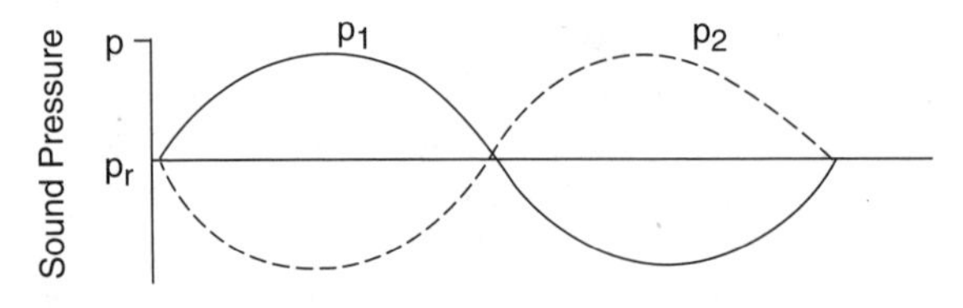

(D) 180° PHASE DIFFERENCE $p_r = 0$

Figure 2.4 Combination of two pure tone noises (p_1 and p_2) with various phase differences.

Sometimes, by simply holding the earphone a little more tightly or loosely to the ear in testing the higher frequencies, suddenly no sound may be produced at all when it should be loud, or a loud sound may be heard when a moment before there seemed to be no sound. This phenomenon occurs because of the presence of standing waves. During hearing

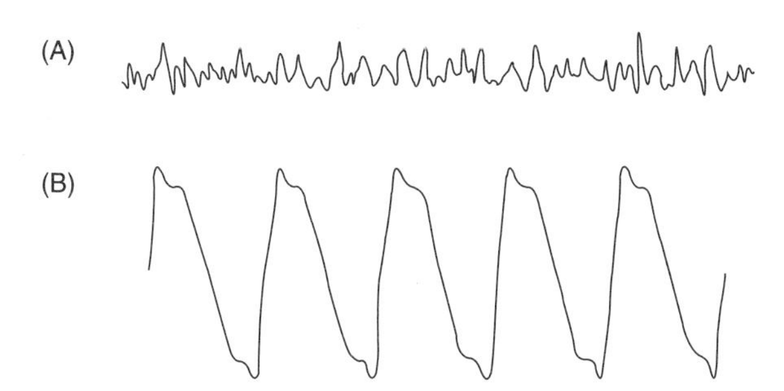

Figure 2.5 (A) Typical street noise and (B) C on a piano.

testing, one often uses modulated or "warbled" tones to help eliminate standing wave problems that might result in misleading test results.

In addition, resonant characteristics of the ear canal play a role in audition. Just like organ pipes and soda bottles, the ear may be thought of as a pipe. It is closed at one end and has a length of ~2.5 cm. Its calculated resonant frequency is ~3400 Hz (actually 3430 Hz if the length is exactly 2.5 cm, and if the ear were really a straight pipe). At such a resonant frequency, a node occurs at the external auditory meatus, and an antinode is present at the tympanic membrane, resulting in sound pressure amplification at the closed end of the pipe (eardrum). This phenomenon may cause sound amplification of up to 20 dB between 2000 and 5000 Hz. The resonance characteristics of the ear canal change if the open end is occluded, such as with an ear insert or muff used for hearing testing, and such factors must be taken into account during equipment design and calibration, and when interpreting hearing tests.

The form of a complex sound is determined by the interaction of each of its pure tones at a particular time. This aspect of a sound is called *complexity*, and the psychological counterpart is *timbre*. This is the quality of sound that allows us to distinguish among a piano, oboe, violin, or voice all producing a middle "C" (256 Hz). These sound sources combine frequencies differently and consequently have different qualities.

3. MEASURING SOUND

The principal components of sound that we need to measure are frequency and intensity. Both are measured with a technique called scaling. The frequency scale is generally familiar because it is based on the musical scale, or *octave*. This is a logarithmic scale with a base of 2. This means that each octave increase corresponds to a doubling of frequency (Fig. 2.7). Linear increases (octaves) correspond with progressively increasing frequency

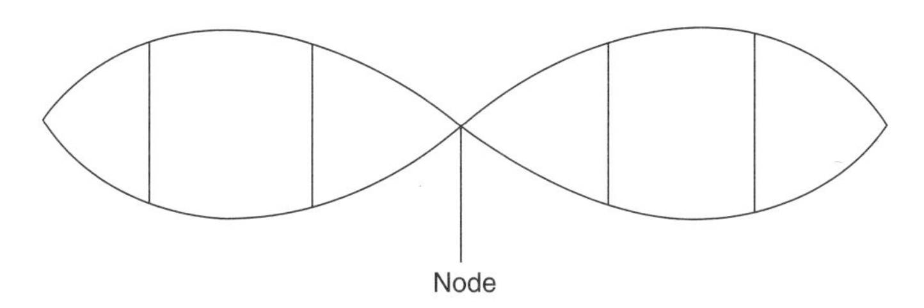

Figure 2.6 Diagram of a standing wave, showing the nodal point at which there is no amplitude.

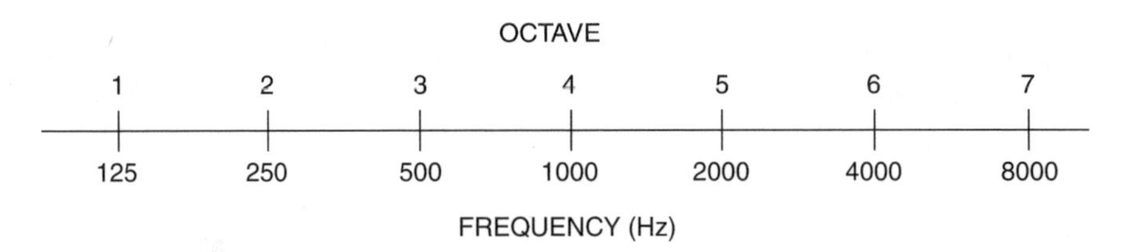

Figure 2.7 Scaling for octave notation of frequency levels.

units. For example, the octave between 4000 and 8000 Hz contains 4000 frequency units, but the same octave space between 125 and 250 Hz contains only 125 frequency units. This makes it much easier to deal with progressively larger numbers and helps show relationships that might not be obvious if absolute numbers were used (Fig. 2.8).

Another reason for using octave scaling was pointed out in the 19th century by psychophysicist Gustav Fechner. He noted that sensation increases as the log of the stimulus. This means that ever-increasing amounts of sound pressure are needed to produce equal increments in sensation. For example, loudness is measured in units called SONES. Other psychoacoustic measures include the PHON scale of loudness level and the MEL scale for pitch. The sone scale was developed by asking trained listeners to judge when a sound level had doubled in loudness relative to a 1000-Hz reference at 40 dB. Each doubling was called one sone. (This is similar to doubling in pitch being referred to as one octave). One-sone increments correspond to ~10-dB increases in sound pressure, or about a 10-fold energy increase. So, in addition to being arithmetically convenient, logarithmic scaling helps describe sound more as we hear it.

In the kind of noise measurement done in industry, the chief concern is with very intense noise. In the testing of hearing, the primary concern is with very weak sounds, because the purpose is to determine the individual's thresholds of hearing. Accurate intensity measurement and a scale that covers a very large range are necessary to measure and compare the many intensities with which we have to work.

The weakest sound pressure that the keen, young human ear can detect under very quiet conditions is ~0.0002 μbar, and this very small amount of pressure is used as the basis or the reference level for sound measurements. This base usually is determined by using a 1000-Hz tone (a frequency in the range of the maximum sensitivity of the ear) and reducing the pressure to the weakest measurable sound pressure to which the young ear will respond. In some instances, the keen ear under ideal conditions will respond to a pressure even weaker that 0.0002 μbar, but it is the 0.0002-μbar pressure that is used as a base.

Of course, sound pressures can be increased tremendously above the weakest tone. The usual range of audible sound pressures extends upward to ~2000 μbar, a point at which the pressure causes discomfort and pain in the ears. Higher pressures can damage or even destroy the inner ear. Because this range (0.0002–2000 μbar) is so great, the use of the microbar as a measurement of sound is too cumbersome.

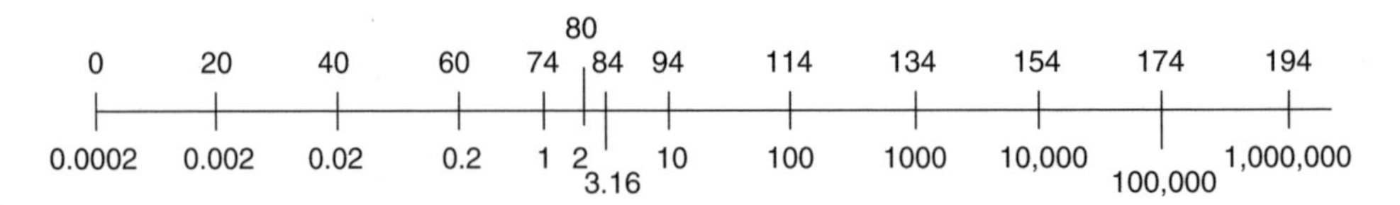

Figure 2.8 Decibel scaling (SPL). [After Lipscomb (2).]

3.1. Intensity

Measuring intensity or amplitude is considerably more complex than measuring frequency. Intensity is also measured on a logarithmic ratio scale. All such scales require an arbitrarily established zero point and a statement of the phenomenon being measured. Usually, sound is measured in decibels. However, many other phenomena (such as heat and light) also are measured in decibels.

3.2. Decibel

The term "decibel" has been borrowed from the field of communication engineering, and it is this term that most generally is used to describe sound intensity. The detailed manner in which this unit was derived and the manner in which it is converted to other units are somewhat complicated and not within the scope of this book. However, a very clear understanding of the nature of the decibel and the proper use of the term is most valuable in understanding how hearing is tested and sound is measured.

3.2.1. A Unit of Comparison

The decibel is simply a unit of comparison, a ratio, between two sound pressures. In general, it is not a unit of measurement with an absolute value, such as an inch or a pound. The concept of the decibel is based on the pressure of one sound or reference level, with which the pressure of another sound is compared. Thus, a sound of 60 dB is a sound that is 60 dB more intense than a sound that has been standardized as the reference level. The reference level must be either implied or specifically stated in all sound measurement, for without the reference level, the expression of intensity in terms of decibels is meaningless. It would be the same as saying that something is "twice," without either implying or referring specifically to the other object with which it is being compared.

3.2.2. Two Reference Levels

For the purpose of this book, two important reference levels are used. In making physical sound measurements, as in a noisy industry, the base used is the sound pressure of 0.0002 μbar (one millionth of one barometric pressure or of one atmosphere), which is known as *acoustical zero* decibels. Sound-measuring instruments such as sound-level meters and noise analyzers are calibrated with this reference level. Several other terms have been used to describe acoustical zero. They include 0.0002 dyne/cm^2, 20 $\mu\text{N/m}^2$, and 20 μPa. Now, 0.0002 μbar has been accepted. When a reading is made in a room and the meter reads so many decibels, the reading means that the sound–pressure level (SPL) in that room is so many decibels greater than acoustical zero. The designation SPL means that the measurement is sound–pressure level relative to 0.0002 μbar. When SPL is written, it tells us both the reference level and the phenomenon being measured.

The other important reference level is used in audiometry and is known as *zero decibels* (0 dB) *of hearing loss or average normal hearing*. This level is not the same as that used as a base for noise measurement. Rather, it is known as hearing threshold level, or HTL. In the middle-frequency range (~3000 Hz), it is 10 dB above the reference level known as acoustical zero. In testing hearing with an audiometer, 40-dB loss in hearing on the audiogram means that the individual requires 40 dB more sound pressure than the average normal person to be able to hear the tone presented.

As the baseline or reference level is different for the audiometer than it is for noise-measuring devices, it should be clear now that a noise of, say, 60 dB in a room is not the same intensity as the 60-dB tone on the audiometer. The noise will sound less loud because it is measured from a weaker reference level.

3.2.3. Formula for the Decibel

With these reference levels established, the formula for the decibel is worked out. To compare the two pressures, we have designated them as Pressure 1 (P_1) and Pressure 2 (P_2), with P_2 being the reference level. The ratio can be expressed as P_1/P_2.

Another factor that must be taken into account is that in computing this ratio in terms of decibels, the computation must be logarithmic. A logarithm is the exponent or the power to which a fixed number or base (usually 10) must be raised in order to produce a given number. For instance, if the base is 10, the log of 100 is 2, because $10 \times 10 = 100$. In such a case, 10 is written with the exponent 2 as 10^2. Similarly, if 10 is raised to the fourth power and written as 10^4, the result is $10 \times 10 \times 10 \times 10$, or 10,000; the logarithm of 10,000 is, therefore, 4. If only this logarithmic function is considered, the formula has evolved as far as $\text{dB} = \log P_1/P_2$. But it is not yet complete.

When the decibel was borrowed from the engineering field, it was a comparison of sound powers and not pressures, and it was expressed in bels and not decibels. The decibel is 1/10 of a bel, and the sound pressure is proportional to the square root of the corresponding sound power. It is necessary, therefore, to multiply the logarithm of the ratio of pressures by 2 (for the square root relationship) and by 10 (for the bel–decibel relationship). When this is done, the decibel formula is complete, and the decibel in terms of sound–pressure levels is defined thus:

$$\text{dB} = \frac{20 \log P_1}{P_2}$$

For instance, if the pressure designated as P_1 is 100× greater than the reference level of P_2, substitution in the formula gives $\text{dB} = 20 \times \log 100/1$. As it is known that the log of 100 is 2 (as $10^2 = 100$), it can be seen that the formula reduces to $\text{dB} = 20 \times 2$, or 40 dB. Therefore, whenever the pressure of one sound is 100× greater than that of the reference level, the first sound can be referred to as 40 dB. Likewise, if P_1 is 1000× greater, then the number of decibels would be 60, and if it is 10,000× greater, the number of decibels is 80. A few other relationships are convenient to remember. If sound intensity is multiplied by 2, sound pressure increases by 6 dB. If intensity is multiplied by 3.16 (the square root of 10), sound pressure increases by 10 dB. When intensity is multiplied by 10, sound pressure increases by 20 dB. These relationships can be seen clearly in Fig. 2.8.

In actual sound measurement, if P_1 is 1 μbar—being a pressure of 1 dyne/cm^2—then the ratio is 1/0.0002 or 5000. By the use of a logarithmic table or a special table prepared to convert pressure ratios to decibels, the pressure level in such a case is found to be 74 dB, based on a reference level of 0.0002 μbar (Fig. 2.8). Figure 2.9 shows where a number of common sounds fall on this decibel scale in relation to a 0.0002-μbar reference level. This base level is used for calibrating standard sound-measuring instruments.

In *audiometric testing*, which uses a higher reference level than that for sound measurement, the tester does not need to concern himself with additional mathematical formulas, because the audiometer used in testing is calibrated to take into account the increase above acoustical zero to provide the necessary reference level for audiometry of average normal hearing (0 dB of hearing loss) at each frequency tested.

AT A GIVEN DISTANCE FROM NOISE SOURCE	decibels re 0.0002 microbar	ENVIRONMENTAL
	-140-	
F-84 AT TAKE-OFF (80′ FROM TAIL)		
HYDRAULIC PRESS (3′)	-130-	
LARGE PNEUMATIC RIVETER (4′)		BOILER SHOP (MAXIMUM LEVEL)
PNEUMATIC CHIPPER (5′)		
	-120-	
MULTIPLE SAND BLAST UNIT (4′)		JET ENGINE TEST CONTROL ROOM
TRUMPET AUTO HORN (3′)		
AUTOMATIC PUNCH PRESS (3′)	-110-	
CHIPPING HAMMER (3′)		WOODWORKING SHOP
CUT-OFF SAW (2′)		INSIDE DC-6 AIRLINER
		WEAVING ROOM
ANNEALING FURNACE (4′)	-100-	
AUTOMATIC LATHE (3′)		CAN MANUFACTURING PLANT
		POWER LAWN MOWER (OPERATOR'S EAR)
SUBWAY TRAIN (20′)		INSIDE SUBWAY CAR
HEAVY TRUCKS (20′)		
TRAIN WHISTLES (500′)		INSIDE COMMERCIAL JET
	-90-	
10-HP OUTBOARD (50′)		
		INSIDE SEDAN IN CITY TRAFFIC
SMALL TRUCKS ACCELERATING (30′)		
	-80-	
LIGHT TRUCKS IN CITY (20′)		GABAGE DISPOSAL (3′)
		HEAVY TRAFFIC (25′ TO 50′)
AUTOS (20′)		
	-70-	VACUUM CLEANER
		AVERAGE TRAFFIC (100′)
		ACCOUNTING OFFICE
CONVERSATIONAL SPEECH (6′)		CHICAGO INDUSTRIAL AREAS
	-60-	WINDOW AIR CONDITIONER (25′)
15,000 KVA, 115 KV TRANSFORMER 3 (200′)		
	-50-	PRIVATE BUSINESS OFFICE
		LIGHT TRAFFIC (100′)
		AVERAGE RESIDENCE
	-40-	QUIET ROOM
		MINIMUM LEVELS FOR RESIDENTIAL AREAS AT NIGHT
	-30-	BROADCASTING STUDIO (SPEECH)
		BROADCASTING STUDIO (MUSIC)
	-20-	STUDIO FOR SOUND PICTURES
	-10-	
	-0-	

Figure 2.9 Typical overall sound levels measured with a sound-level meter.

3.2.4. Important Points

The important thing to remember is that the decibel is a logarithmic ratio. It is a convenient unit, because 1 dB approaches the smallest change in intensity between two sounds that the human ear can distinguish.

An important aspect of the logarithmic ratio is that as the ratio of the pressures becomes larger, because the sound becomes more intense, the rate of increase in decibels becomes smaller. Even if the ratio of the pressures is enormous, such as one pressure being 10,000,000× that of another, the number of decibels by which this ratio is expressed does not become inordinately large, being only 140 dB sound–pressure level (SPL). This is the principal reason for using the decibel scale. From the psychoacoustic aspect, it takes comparatively little increase in sound pressure to go from 0 to 1 dB, and the normal ear can detect this. However, when an attempt is made to increase the sound pressure from 140 to 141 dB—also an increase of 1 dB, which the ear can barely detect—it takes an increase of $\sim$10,000,000$\times$ as much in absolute pressure.

A point to be remembered is that the effect of adding decibels together is quite different from that of adding ordinary numbers. For example, if one machine whose noise has been measured as 70 dB is turned on next to another machine producing 70 dB, the resulting level is 73 dB and not 140 dB. This is obtained as follows: When combining decibels, it is necessary to use an equation that takes into account the energy or power exerted by the sound sources, rather than the sound pressure exerted by this energy. The equation is:

$$\mathrm{dB}_{\mathrm{power}} = 10 \log_{10} \frac{E_1}{E_0}$$

where E_1 is a known power (energy) and E_0 is the reference quantity.

$$\mathrm{dB} = 10 \log_{10} \frac{2}{1}$$

(there were two machines operating rather than one, resulting in a 2:1 ratio)

$$= (10)(0.3010) \quad \text{(the logarithm of 2 is 0.3010)}$$
$$= 3.01$$

Figure 2.10 is a chart showing the results obtained from adding noise levels. It may be used instead of the formulas. On this chart, it will be seen that if 70 and 76 dB are being added, the difference of 6 dB is located on the graph, and this difference is found to produce an increase of 1 dB, which is added to the higher number. Therefore, the combined level of noise produced by the two machines is 77 dB above the reference level.

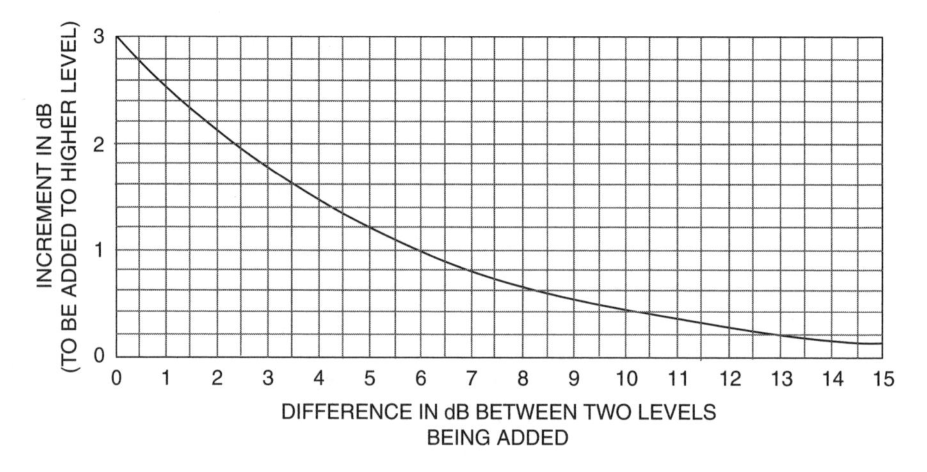

Figure 2.10 Results obtained from adding noise levels.

4. dBA MEASUREMENT

Most sound level meters that are used to measure noise levels do not simply record sound pressure level relative to 0.0002 μbar (dB SPL). Rather, they are generally equipped with three filtering networks: A, B, and C. Use of these filters allows one to approximate the frequency distribution of a given noise over the audible spectrum (Figs. 2.11 and 2.12). In practice, the frequency distribution of a noise can be approximated by comparing the levels measured with each of the frequency ratings. For example, if the noise level is measured with the A and C networks and they are almost equal, then most of the noise energy is >1000 Hz, because this is the only portion of the spectrum where the networks are similar. If there is a large difference between A and C measurements, most of the energy is likely to be <1000 Hz. The use of these filters and other capabilities of sound level meters are discussed in Chapter 14.

The A network is now used when measuring sound to estimate the risk of noise-induced hearing loss, because it represents more accurately the ear's response to loud noise. It is not possible to describe a noise's damaging effect on hearing simply by stating its intensity. For instance, if one noise has a spectrum similar to that shown in curve A in Fig. 2.11, with most of its energy in the low frequencies, it may have little or no effect on hearing. Another noise of the same overall intensity, having most of its sound energy in the higher frequencies (curve C), could produce substantial hearing damage after years of exposure. Examples of low-frequency noises are motors, fans, and trains. High-frequency noises are produced by sheet metal work, boiler making, and air pressure hoses. Although the human ear is more sensitive in the frequency range 1000–3000 Hz than it is in the range <500 and >4000 Hz (Fig. 2.13), this frequency-specific differential sensitivity does not explain fully the ear's vulnerability to high-frequency sounds. Although various explanations have been proposed involving everything from teleology to redundancy of low-frequency loci on the cochlea to cochlear shearing mechanics, the phenomenon is not completely understood. Mechanisms of noise-induced hearing loss are discussed in Chapter 14.

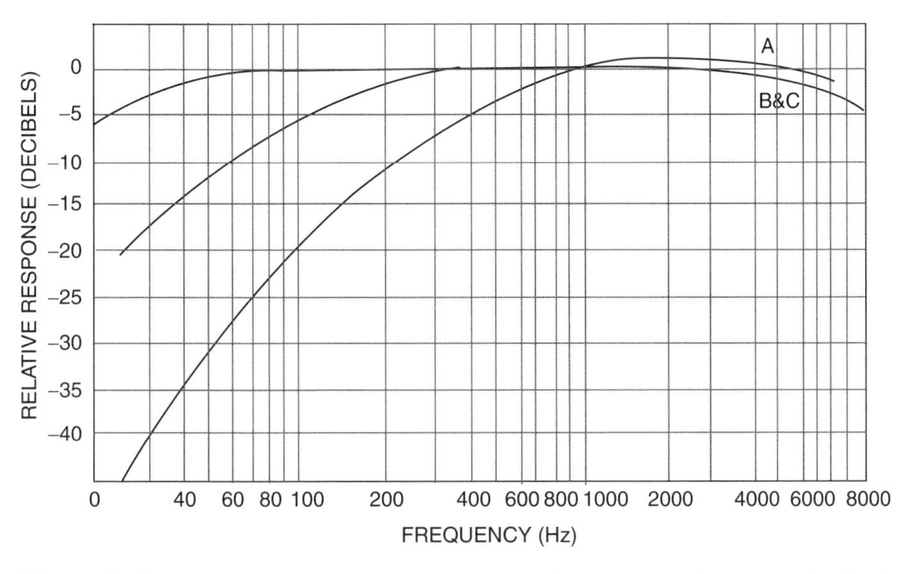

Figure 2.11 Frequency–response characteristics of a sound level meter with A, B, and C weighting.

Center frequency (Hz)	A-weighting adjustment (dB)	Center frequency (Hz)	A-weighting adjustment (dB)
10	–70.4	500	–3.2
12.5	–63.4	630	–1.9
16	–56.7	800	–0.8
20	–50.5	1000	0.0
25	–44.7	1250	+0.6
31.5	–39.4	1600	+1.0
40	–34.6	2000	+1.2
50	–30.2	2500	+1.3
63	–26.2	3150	+1.2
80	–22.5	4000	+1.0
100	–19.1	5000	+0.5
125	–16.1	6300	–0.1
160	–13.4	8000	–1.1
200	–10.9	10000	–2.5
250	–8.6	12500	–4.3
315	–6.6	16000	–6.6
400	–4.8	20000	–9.3

Figure 2.12 Adjustments by frequency for A-weighting scale.

There are other important and interesting aspects of the physics of sound that might be discussed; the subject is a complex and fascinating one. The average physician concerned with the problems of hearing loss will find that a reasonable comprehension of the material thus far presented will be helpful—especially the fact that the term "decibel" expresses a *logarithmic ratio to an established reference level.*

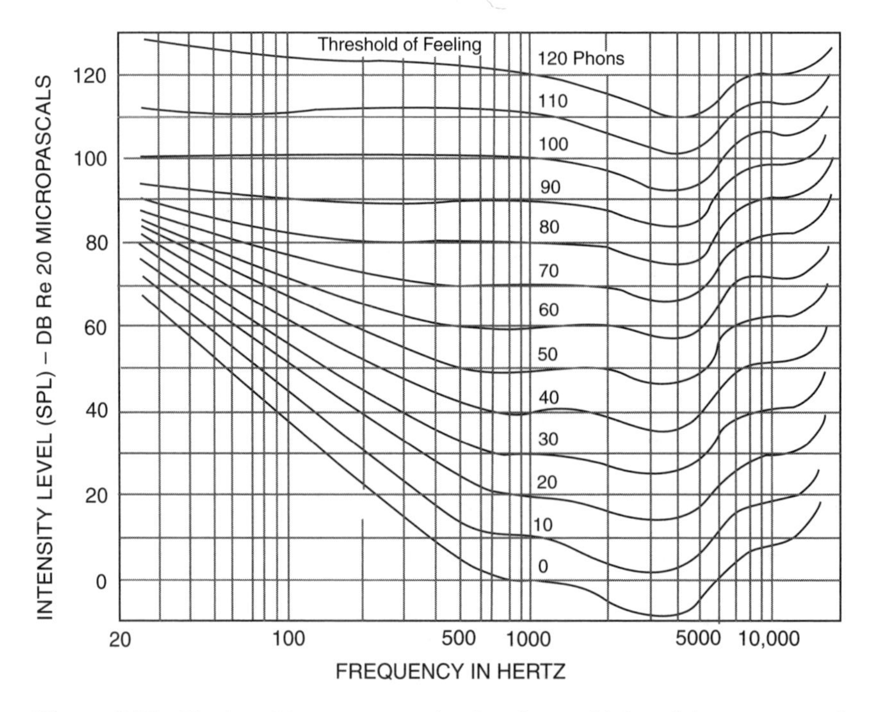

Figure 2.13 Fletcher–Munson curves showing the sensitivity of the ear to sounds of various frequencies.

Understanding the basic physics of sound is invaluable in medical and legal interactions involving hearing loss. Deeper knowledge is required in many instances, and the consultation services of good physicists and engineers with practical understanding of the problems of medicine and industry are indispensable.

REFERENCES

1. Van Bergeijk WA, Pierce JR, David EE. Waves and the Ears. New York: Doubleday and Co., 1960:44.
2. Lipscomb DM. Noise and occupational hearing impairment. Ear Nose Throat 1980; 59:13–23.

3
The Nature of Hearing Loss

Robert T. Sataloff
Drexel University College of Medicine, American Institute for Voice and Ear Research, and Graduate Hospital, Philadelphia, Pennsylvania, USA

Joseph Sataloff
Thomas Jefferson University, Philadelphia, Pennsylvania, USA

Hearing loss is one of the most challenging problems confronting medicine, not only because there are some 40 million Americans with hearing loss, but especially because it can affect personality so adversely. A mild hearing loss sometimes may produce more psychological disturbance than a greater hearing deficit in conditions such as Meniere's disease. It is this effect of hearing loss on the patient's emotions, rather than the actual deafness, that may persuade the patient to seek the help of a physician. The hearing loss may even bother the people around him/her more than it does the patient. Deafness is a rather strange symptom, for it is not accompanied by pain, discomfort, itch, or fear, as is true of cancer and other diseases that compel patients to seek medical aid. Hearing loss is really a symptom rather than a disease.

Consider, for a moment, what motivates a patient to visit a doctor and complain of problems with hearing. Perhaps a number of embarrassing situations begin to occur with greater frequency in everyday life. For example, it may be a failure to hear or to understand an employer when given directions, especially amid much noise. A secretary may fail to take dictation correctly, and the resulting mistakes may cause a great deal of tension in the office. A young lady may try to hide from herself and her friends that her hearing is impaired, but when she goes out on a date, she repeatedly gives the wrong answers, especially when it is dark in the car and she is unable to read the lips of her companion. A husband may sit and read his newspaper and fail to understand what his wife is saying while she is running water in the kitchen; this may lead to constant friction between husband and wife, with complete lack of communication and, eventually, serious marital stress.

These situations are typical of the embarrassing circumstances that produce feelings of inadequacy and insecurity. Yet the patient is unable to face this problem and seek the help of a physician. When he/she does visit a physician to complain of deafness, it is rarely of his/her own free will. The patient usually is nagged into going by a spouse, friend, or boss, who has been trying for years to get the person to do something about his/her hearing difficulty.

Not infrequently, when a husband and a wife walk into the otologist's office, the dialogue follows a familiar pattern. The physician asks the man what his difficulty is, and before he has a chance to answer, his wife blurts out that he is deaf and does not pay attention to her. The husband generally looks meek and bewildered, as if he is not sure what is going on, but he certainly does not want to assume the blame for all of his difficulty. It soon becomes apparent that bickering and strife are his key problems and that they were brought on by his hearing loss. One of the most heartbreaking episodes the authors have encountered in otologic practice occurred with a 21-year-old man who pleaded for some cure for his bilateral nerve deafness secondary to meningitis contracted 6 years earlier. The patient offered to turn over all of the small amount of money he had in the bank if he could be given even a moderate cure for his hearing loss. When asked what prompted him to seek help now, 6 years after the onset of deafness, he replied tearfully that it was important to hear his new baby when she cried during the night.

1. EARLY STAGES OF A HEARING LOSS

People do not really notice any hearing deficit until their hearing level has dropped rather markedly or suddenly. In the early stages of a high-frequency hearing loss, for example, there actually are no symptoms, except that the patient may say he/she cannot hear the watch tick in one ear as well as in the other. If the patient notices this, he/she may seek help early, but most people are not this fortunate. Often, the hearing loss is usually rather substantial before one seeks medical attention.

2. EVERY PATIENT CAN BE HELPED

Because hearing deficiency can cause emotional trauma, it can be stated categorically that every patient with a hearing loss who visits a physician can be helped in some way. It may not always be possible to restore hearing to normal capacity or to improve it to a nonhandicapping level, but it is always possible to mitigate the psychological impact of a hearing loss on the patient. He/she can be taught to hear better with the hearing that is left; correct the pessimistic and antagonistic attitude toward the hearing problem, and in many ways, he/she can be helped to communicate better. It is the physician's responsibility to improve the patient's quality of life.

3. REFERRAL OF PATIENTS BY VARIOUS SPECIALISTS

Interestingly enough, many patients are referred to an otolaryngologist, not by the general practitioner, internist, or pediatrician, but by a surprising variety of other physicians such as obstetricians, dermatologists, psychiatrists, and even proctologists. Patients seem to have a strange inclination to reveal their hearing loss symptoms under the most unusual circumstances. For example, astute obstetricians have referred to us many patients who complained of buzzing in their ears during the last month of pregnancy or early post-partum. The usual finding in these cases is otosclerosis, which accounts for the buzzing tinnitus. Obstetricians should be alert for a family history of hearing loss, because this is associated with otosclerosis, a condition that is often aggravated by pregnancy and may present initially after delivery.

Hearing loss commonly is reported in dermatologists' offices by patients whose ear canals repeatedly collect debris from an exfoliative dermatitis. Unless the debris is removed carefully, the canal walls can be injured and the dermatitis aggravated. Psychiatrists should be cognizant of the relationship between hearing loss and emotional disturbances. Hard-of-hearing patients often have been under a psychiatrist's care for a long time before being referred to an otologist for a hearing evaluation. Many psychological and emotional disturbances can be corrected or mitigated by early attention to the patient's hearing. Lamentably, some deaf but otherwise normal children have been found in mental institutions.

Curiously, some patients are most disturbed by their hearing loss when they fail to hear the little sounds, such as the passing of urine or flatus, that eventually may cause them serious embarrassment. Several patients have been referred to us by proctologists and urologists for complaints that the patients never had discussed with their general practitioners.

With the advent of the Occupational Safety and Health Act (OSHA) and routine audiometry in industry, occupational physicians and nurses have become chief sources of referral of large numbers of employees for otologic examination and diagnosis. This is important not only to detect noise-induced hearing loss, but also to find patients with correctable or serious causes of hearing impairment. Industry is in a unique position to help improve the hearing health of the working force.

When hard-of-hearing patients are not under a physician's responsible guidance, they may go directly to hearing aid dealers without a diagnosis or even a medical examination. Although many "hearing centers" are operated in an ethical manner, there are some that will sell any patient a hearing aid without investigating whether the deafness could have been cured. Some people use their hearing aids incorrectly because they fail to receive proper instruction at the time of purchase.

In many types of hearing loss, a hearing aid provides the only possible improvement. It is necessary for the patient to realize that no hearing aid can overcome the distortion produced by sensorineural deafness and that his/her hearing never will be "normal." Nevertheless, it is painful for a patient to accept these disagreeable facts. It takes skill to select the right hearing aid and time to learn to use it. Often, the patient's only chance to receive a clear explanation of the hearing trouble—what he/she has to look forward to and what can be done about it—is to go to a physician and audiologist who can communicate the facts and advise him/her properly.

The patient's personality and financial need to hear, as well as his/her willingness to wear a hearing aid and perhaps to learn speech reading, also enter into the way he/she adapts to a hearing impairment. A person who must earn a living may put up with these inconveniences more readily than an elderly person, who may be more willing to retire into the comfortable silence of a restricted existence.

4. THE VALUE OF UNDERSTANDING AND CONDUCTING HEARING TESTS

Detecting hearing loss is often such a simple procedure that physicians of every specialty should have some understanding of hearing tests and be able to perform basic screening in their offices whenever the need arises. In most instances, a simple test can be done with a 512-Hz tuning fork. This instrument can prove to be invaluable and very reliable in discovering substantial hearing impairment. It is used to help determine whether a hearing loss is caused by damage to the outer and middle ear or to the sensorineural mechanism. However, a tuning fork (even tuning forks of different frequencies) may not supply sufficient information in cases involving minimal high-frequency losses >1000–2000 Hz and other mild hearing deficiencies. Because the tuning fork will not provide a quantitative determination of hearing loss, it is necessary to use an audiometer, particularly in cases in which the loss is limited to the high-frequency range.

A good audiometer is well worth the price and effort needed to acquire skill in its use. The audiometer should be equipped to do air conduction and bone conduction tests with masking. Directions for using an audiometer and for avoiding some of the pitfalls in its use are discussed in Chapter 5.

Every community should have facilities to test hearing. An otolaryngologist can supply such services. In small communities, the pediatrician, general practitioner, or school nurse may be doing hearing tests. Many industries and small hospitals throughout the country have established hearing centers directed by well-trained audiologists and technicians. It is advisable for all physicians to have a clear understanding of audiometrics, so they may interpret the reports sent to them from the ear specialist. For these reasons a basic discussion of audiometry is included in this book, and the results of various types of hearing tests are interpreted.

4.1. Anatomy and Physiology of the Human Ear

The ear is divided into three major anatomical divisions: (a) the outer ear, (b) the middle ear, and (c) the inner ear (Fig. 3.1).

The outer ear has two parts: (a) the "trumpet-shaped" apparatus on the side of the head called the auricle or pinna and (b) the tube leading from the auricle into the temporal bone called the external auditory canal. This opening is called the meatus and external auditory canal.

The tympanic membrane, or "eardrum," stretches across the inner end of the external ear canal separating the outer ear from the middle ear.

The middle ear is a tiny cavity in the temporal bone. The three auditory ossicles, malleus (hammer), incus (anvil), and stapes (stirrup), form a bony bridge from the external ear to the inner ear. The bony bridge is held in place by muscles and ligaments. The middle-ear chamber is filled with air and opens into the throat through the eustachian tube. The eustachian tube helps to equalize pressure on both sides of the eardrum.

The inner ear is a fluid-filled chamber divided into two parts: (a) the vestibular labyrinth, which functions as part of the body's balance mechanism, and (b) the cochlea, which contains the hearing–sensing nerve. Within the cochlea is the organ of Corti, which contains thousands of minute, sensory, hairlike cells (Fig. 3.2). The organ of Corti functions as the switchboard of the auditory system. The eighth cranial or acoustic nerve leads from the inner ear to the brain, serving as the pathway for the impulses the brain will interpret as sound. Additional information on ear anatomy may be found in Appendix I and in Chapter 10.

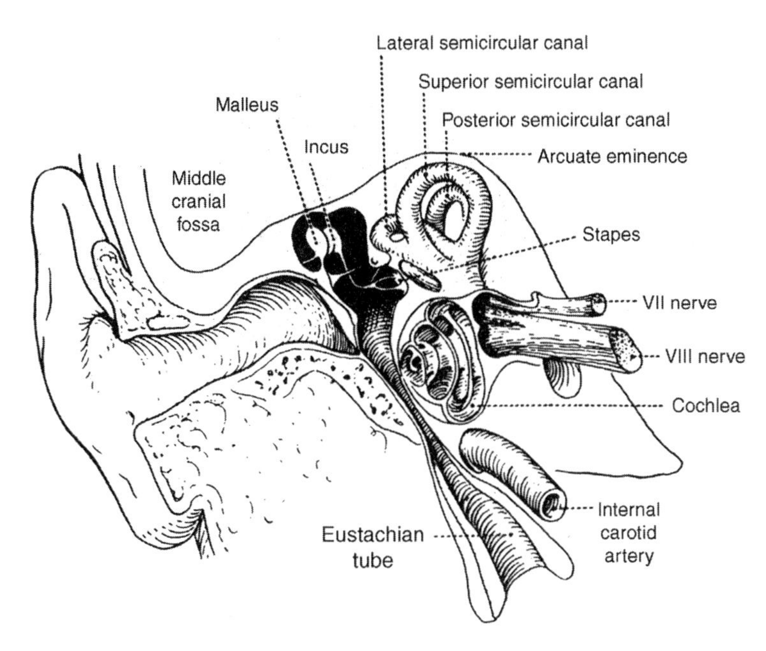

Figure 3.1 Diagrammatic cross-section of the ear. The semicircular canals are connected with maintaining balance.

Sound creates vibrations in the air somewhat similar to the "waves" created when a stone is thrown into a pond. The outer-ear "trumpet" collects these sound waves, and they are funneled down the external ear canal to the eardrum. As the sound waves strike the eardrum, they cause it to vibrate. The vibrations are transmitted by mechanical action through the middle ear over the bony bridge (ossicular chain) formed by the malleus, incus, and stapes. These vibrations, in turn, cause the membranes over the openings to the inner ear to vibrate, causing the fluid in the inner ear to be set in motion. The motion of the fluid in the inner ear excites the nerve cells in the organ of Corti, producing electrochemical impulses that are gathered together and transmitted to the brain along the acoustic nerve. As the impulses reach the brain, we experience the sensation of hearing.

The sensitivity of the hearing mechanism is most extraordinary. Near threshold, the eardrum only moves approximately one 1,000,000th of an inch. Our intensity range spans extremes from the softest sounds, to sounds of jet-engine intensity, covering an intensity range of ~100,000,000–1. Over this range, we are able to detect tiny changes in intensity and in frequency. Young, healthy humans can hear frequencies from ~20 to 20,000 Hz and can detect frequency differences as small as 0.2%. That is, we can tell the difference between a sound of 1000 Hz and one of 1002 Hz. Consequently, it is no surprise that such a remarkably complex system can be damaged by various illnesses and injuries.

4.1.1. Causes of Hearing Loss

There are two basic types of hearing loss: *conductive* and *sensorineural.* They are discussed in detail throughout this book.

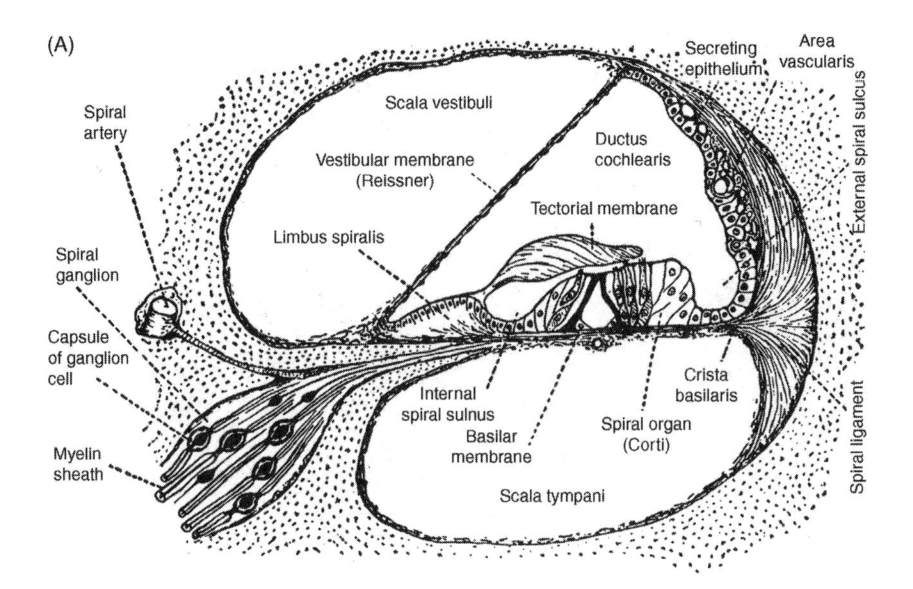

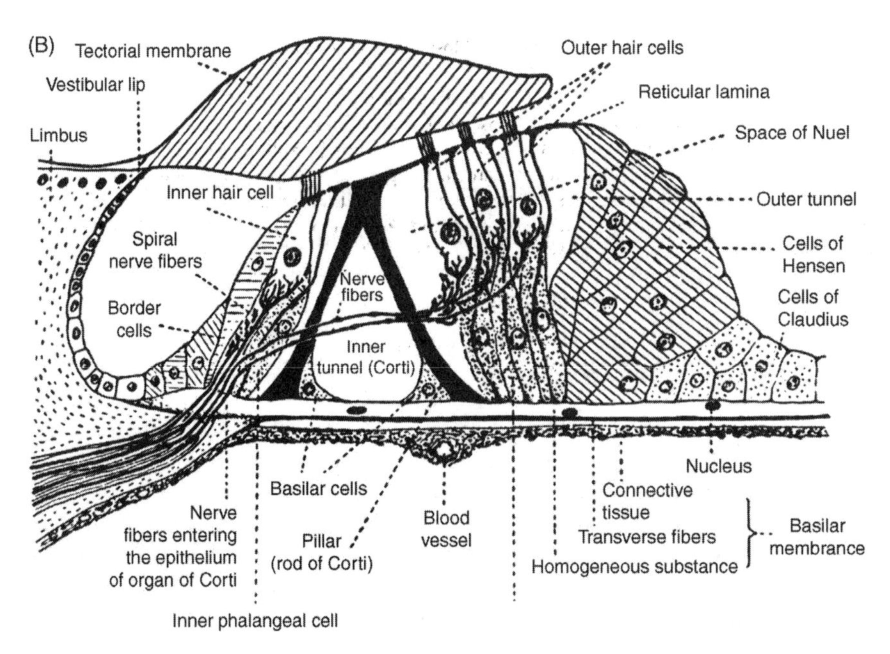

Figure 3.2 A cross-section of the organ of Corti. (A) Low magnification. (B) Higher magnification. [After Rasmussen (1).]

5. ESTABLISHING THE SITE OF DAMAGE IN THE AUDITORY SYSTEM

The cause of a hearing loss, like that of any other medical condition, is determined by carefully obtaining a meaningful history, performing a physical examination, and obtaining certain laboratory tests. In otology, hearing tests parallel the function of clinical laboratory tests in general medicine.

Despite recent advances in otology, we still lack certain information about the ear and, as a result, cannot always determine the cause of hearing impairment. Fortunately, if the site of damage in the auditory system can be established, it is possible to decide on the best available treatment and the prognosis. When a hearing loss is classified, the point at which the auditory pathway has broken down is localized, and it is determined whether the patient's hearing loss is conductive, sensorineural, central, functional, or a combination of these.

Conductive hearing loss is due to any condition that interferes with the transmission of sound through the external and middle to the inner ear. If it is in the middle ear, the damage may involve the footplate of the stapes, as in otosclerosis, or the mobility of the drum and ossicles caused by fluid. Conductive hearing losses generally are correctable.

In *sensorineural hearing loss* the damage lies medial to the stapedial footplate—in the inner ear, the auditory nerve, or both. Many physicians call this condition "nerve deafness." In the majority of cases, it is not curable. The cochlea has ~30,000 hair cells that connect with nerve endings. Those hair cells in the large end of the cochlea respond to very high-pitched sounds, and those in the small end (and throughout much of the rest of the cochlea) respond to low-pitched sounds. These hair cells, and the nerve that connects them to the brain, are susceptible to damage from a variety of causes.

In *central hearing loss* the damage is situated in the central nervous system at some point from the auditory nuclei (in the medulla oblongata) to the cortex. Formerly, central hearing loss was described as a type of "perceptive deafness," a term now obsolete. Knowledge about the subject still is limited.

In *functional hearing loss* there is no detectable organic damage to the auditory pathways, but some underlying psychological or emotional problem is at fault.

Frequently, a patient experiences two or more types of hearing impairment, a problem called *mixed hearing loss*. However, for practical purposes this term is used only when both conductive and sensorineural hearing losses are present in the same ear.

Each type of hearing loss has specific, distinctive characteristics, which make it possible to classify the vast majority of cases seen in clinical practice. When certain basic features are found, the classification usually can be made with confidence.

6. CONDUCTIVE HEARING LOSS

In cases of conductive hearing loss, sound waves are not transmitted effectively to the inner ear because of some interference in the external ear canal, the eardrum, the ossicular chain, the middle-ear cavity, the oval window, the round window, or the eustachian tube. For example, damage to either the middle ear, which transmits sound energy efficiently, or the eustachian tube, which maintains equal air pressure between the middle ear cavity and the external canal, could result in a mechanical defect in sound transmission. In pure conductive hearing loss, there is no damage to the inner ear or the neural pathway.

Patients diagnosed as having conductive hearing loss receive a much better prognosis than those with sensorineural loss because modern techniques make it possible to cure or at least improve the vast majority of cases in which the damage occurs in the outer or middle ear. Even if they are not improved medically or surgically, these patients benefit greatly from a hearing aid, because what they need most is amplification. They are not bothered by distortion and other hearing abnormalities that may occur in sensorineural loss.

7. SENSORINEURAL HEARING LOSS

The word "sensorineural" was introduced to replace the ambiguous terms "perceptive deafness" and "nerve deafness." It is a more descriptive and more accurate anatomical term. Its dual character suggests that two separate areas may be affected, and actually, this is the case. The term "sensory" hearing loss is applied when the damage is localized in the inner ear. Useful synonyms are "cochlear" or "inner-ear" hearing loss. "Neural" hearing loss is the correct term to use when the damage is in the auditory nerve, anywhere between its fibers at the base of the hair cells and the auditory nuclei. This area includes the bipolar ganglion of the eighth cranial nerve. Other common names for this type of loss are "nerve deafness" and "retrocochlear hearing loss." These names are useful if applied appropriately and meaningfully, but too often they are used improperly.

Although at present it is common practice to group together both sensory and neural components, it has become possible in many cases to attribute a predominant part of the damage, if not all of it, to either the inner ear or the nerve. Because of some success in this area and the likelihood that ongoing research will allow us to differentiate between even more cases of sensory and neural hearing loss, we shall divide the terms and describe the distinctive features of each type. This separation is advisable because the prognosis and the treatment of the two kinds of impairment differ. For example, in all cases of unilateral sensorineural hearing loss, it is important to distinguish between a sensory and neural hearing impairment, because the neural type may be due to an acoustic neuroma which could be serious or even life-threatening. Those cases which we cannot identify as either sensory or neural and those cases in which there is damage in both regions we shall classify as sensorineural.

There are various and complex causes of sensorineural hearing loss, but certain features are characteristic and basic to all of them. Because the histories obtained from patients are so diverse, they may contribute more insight into the etiology than into the classification of a case.

Sensorineural hearing loss is one of the most challenging problems in medicine. A large variety of hearing impairments fall under this category. The prognosis for restoring a sensorineural hearing loss with presently available therapy is poor in most cases. Although some spontaneous remissions and hearing improvements have occurred with therapy, particularly in cases involving sensory loss, a great need for further research still exists.

8. MIXED HEARING LOSS

For practical purposes, in this book a "mixed hearing loss" should be understood to mean a conductive hearing loss accompanied by a sensory or a neural (or a sensorineural) loss in the same ear. However, the emphasis is on the conductive hearing loss because available therapy is so much more effective for this group. Consequently, the otologic surgeon has a special interest in cases of mixed hearing loss in which there is primarily a conductive loss complicated by some sensorineural damage.

9. FUNCTIONAL HEARING LOSS

Functional hearing loss occurs in clinical practice more frequently than many physicians realize. This is the type of condition in which the patient does not seem to hear or respond, yet the handicap may be caused not by any organic pathology in the peripheral or the central auditory pathways, but rather by psychological factors.

The hearing difficulty may have an entirely emotional or psychological etiology, or it may be superimposed on some mild organic hearing loss, in which case it is called a functional or a psychogenic overlay. Often, the patient really has normal hearing underlying the functional hearing loss. A carefully recorded history usually will reveal some hearing impairment in the patient's family or some reference to deafness which served as the nucleus for the patient's functional hearing loss.

The most important challenge in such a case is to classify the condition properly. It may be quite difficult to determine the specific emotional cause, but if the classification is made accurately, the proper therapy can be instituted. Too often, the emotional origin of a functional hearing loss is not recognized, and patients receive useless otologic treatments for prolonged periods. In turn, this process may aggravate the emotional element and cause the condition to become more resistant. Therefore, early and accurate classification is imperative.

10. CENTRAL HEARING LOSS (CENTRAL DYSACUSIS)

Although information about central hearing loss is accumulating, it remains somewhat a mystery in otology. Physicians know that some patients cannot interpret or understand what is being said and that the cause of the difficulty is not in the peripheral mechanism but somewhere in the central nervous system. In central hearing loss the problem is not a lowered pure-tone threshold but rather the patient's ability to interpret what he/she hears. Obviously, it is a more complex task to interpret speech than to respond to a pure-tone threshold; consequently, the tests necessary to diagnose central hearing impairment must be designed to assess a patient's ability to handle complex information. Most of the tests now available were not created specifically for this purpose, and so, it still requires a very experienced audiologist and almost intuitive judgment on the physician's part to make an accurate diagnosis. (Although aphasia sometimes is considered to be related to central hearing loss, it is outside the realm of otology.)

Central auditory processing disorders of lesser severity actually are much more common than recognized previously. Patients with this problem have difficulty filtering out competing auditory signals. Typically, they have trouble "hearing" in crowds, concentrating or reading with a radio on, and having conversations in noisy restaurants. The cause is often unknown, but the condition can result from head trauma. Thresholds often are normal, but impaired thresholds and increasing age may aggravate the symptoms.

REFERENCE

1. Rasmussen AT. Outlines of Neuro-Anatomy. Dubuque, IA: WC Brown, 1947.

4

The Otologic History and Physical Examination

Robert T. Sataloff
Drexel University College of Medicine, American Institute for Voice and Ear Research, and Graduate Hospital, Philadelphia, Pennsylvania, USA

Joseph Sataloff
Thomas Jefferson University, Philadelphia, Pennsylvania, USA

The first concern of a physician who is consulted by a patient with a hearing problem should be to put him at ease. The patient is likely to be anxious because he/she already has suffered much embarrassment from failure to understand other people who have not always been patient with his handicap. The first step is to face the patient and to speak in a distinct and moderate tone. If the patient is wearing a hearing aid, there usually is no need to use a loud voice, but it helps to speak slowly and distinctly.

A hearing difficulty is quite different from the usual complaints presented to a physician. Other patients may be concerned about discomfort, itching, or pain. Perhaps they are worried that they have cancer. In comparison, the patient with a hearing loss is likely to be in good health.

A person who experiences a hearing loss usually sees a physician because of an inability to communicate successfully in social and vocational situations. The hearing loss itself is not the main issue. Therefore, when he/she first tells the doctor about the hearing trouble, the patient probably will have a great deal to unburden about psychological, social, and business problems.

Otolaryngologists (ear, nose, and throat doctors) are specialists in ear problems, among other things. Otology is a subspecialty of otolaryngology, practiced by physicians with special interests and concentration on ear problems. Neurotology is a subspecialty of otolaryngology, and really a subspecialty of otology. Although the field is more than 40 years old, there are still few practitioners who have the experience or fellowship training beyond otolaryngology residency to qualify them as neurotologists. Otolaryngologists subspecializing in this area are specially trained in the diseases of the ear and ear–brain interface, and in skull base surgery for problems such as acoustic neuroma, glomus jugulare, intractable vertigo, total deafness, and traditionally "unresectable" neoplasms. They are distinct from otoneurologists, whose background is in neurology but who have special interest in disorders afflicting the hearing and balance system. Consultation with an otologist or neurotologist often is advisable during evaluation of ear and hearing problems, which are often more complex than they appear to be at first.

1. ESSENTIAL QUESTIONS

The physician often can save much time and be more helpful to the patient by asking certain meaningful questions that have a direct bearing on the nature of the patient's problem. The answers to these questions may help to establish a differential diagnosis of the hearing impairment.

These are among the helpful questions:

1. In which ear do you think you have hearing loss?
2. How long have you had a hearing loss?
3. If you first noticed it in relation to a head injury, exactly when did you become aware of it?
4. Who noticed it: You, family members, or others?
5. Did your hearing decrease slowly, rapidly, or suddenly?
6. Is your hearing now stable?
7. Does your hearing fluctuate?
8. Do you have distortion of pitch?
9. Do you have distortion of loudness (bothered by loud noises)?
10. Can you use both ears on the telephone?
11. Do you have a feeling of fullness in your ears?
12. Are you aware of anything (foods, weather, sounds) that makes your hearing loss better or worse?
13. Does your hearing change with straining, bending, nose blowing, or lifting?
14. Did you have ear problems as a child?
15. Have you ever had ear drainage?
16. Have you had recent or frequent ear infections?
17. Have you ever had ear surgery?
18. Have you ever had ear surgery recommended, but not performed?
19. Have you ever had a direct injury to your ears?
20. Have you ever had problems similar to your current complaints prior to your current injury?
21. Do you have ear pain?
22. Have you had recent dental work?
23. Do you have any medical problems (diabetes, blood pressure, others)?
24. Have you ever had syphilis or gonorrhea? Do you have AIDS?
25. Does anyone in your family have a hearing loss?
26. Has anyone in your family undergone surgery for hearing?
27. Do you have parents, brothers, or sisters with syphilis?
28. Have you ever worked at a job noisy enough to require you to speak loudly in order to be heard?
29. Do your ears ring?
30. Do you have temporary hearing loss when you leave your noisy work environment?
31. Do you have any noisy recreational activities, such as rifle shooting, listening to rock and roll music, snowmobiling, motor cycling, wood working, etc.?
32. Do you wear ear protectors when exposed to loud noise?
33. Do you frequently scuba dive?
34. Do you fly private aircraft or skydive?
35. Do you have ear noises or dizziness?

When the patient is a child, parents should be asked to supply information about any possible difficulties at birth and early childhood problems such as anoxia, severe jaundice, blood dyscrasias, or a hemorrhagic tendency. They should also be asked to provide information regarding any history and causes of convulsions and high fevers.

2. THE IMPORTANCE OF GETTING ACCURATE ANSWERS

The patient often, inadvertently, gives inaccurate answers to some of the questions in the history. This is particularly true when he/she is asked to specify how long he/she has had hearing loss. Usually, the patient underestimates the duration of the impairment.

It is an advantage to have the patient's husband or wife present at the taking of the history because he or she often supplies more accurate information. Many hearing losses develop insidiously, and the patient may not be aware of any trouble until it is quite pronounced, long after the problem has become obvious to everyone else. Some patients refuse to recognize or to admit that their hearing is defective, even though others have suggested the possibility to them. This common observation illustrates the psychological overtones that frequently complicate certain forms of hearing impairment.

The exact time of onset of hearing loss may be crucial, particularly when it is rather sudden. Hearing may be lost instantly when a patient puts a finger in his/her ear and thereby blocks the ear canal with a plug of wax. Sudden onset may be caused by Meniere's disease, mumps, viruses, rupture of the round window membrane or a blood vessel, or by other causes, including acoustic neuroma. In practically all such cases, only one of the ears is involved. Both ears may be affected as a result of meningitis, autoimmune inner ear disorders, or a severe head injury. Most frequently, hearing loss develops slowly over many years, especially in presbycusis, otosclerosis, deafness following exposure to intense noise, and hereditary nerve deafness.

Because otosclerosis, presbycusis, and hereditary nerve deafness are determined genetically, or at least have a tendency to recur in families, the question of familial occurrence is of some importance. Frequently, the patient is certain there is no history of deafness in the family, and yet, when the studies are completed, the diagnosis may point to a hereditary or familial condition. After further questioning, the patient sometimes recalls that one or several members of his family were afflicted with a hearing impairment. More often the patient insists that there has been no deafness in the family. This statement can be true, for in many cases of hereditary deafness, the hearing loss may not manifest itself for many generations, may be recessive, or the patient may be the first person in the family with a hereditary condition. In otosclerosis, for example, there may be no recent evidence of deafness in any living member of the family, and yet it is known that otosclerosis tends to be inherited.

3. PROGNOSIS AND DIAGNOSIS

A vital question in the minds of both the physician and the patient is: "Will the hearing loss get worse?" The answer depends on the diagnosis. For example, some cases of hereditary nerve deafness are likely to get worse and may become quite profound. Otosclerosis, on the other hand, may progress but level off and not become very severe. Some otosclerotics, however, tend to develop sensorineural loss and suffer hearing deterioration early in life. Congenital hearing loss progresses infrequently.

The patient often can help the otologist arrive at a more definite diagnosis, even on the first visit, by indicating with some certainty whether the hearing loss has remained constant or has been getting worse over a period of months or years. It helps the otologist to distinguish between deafness caused by noise and that caused by hereditary or advancing age. A diagnosis that reveals an inherently nonprogressive condition is a source of great comfort and satisfaction to both the otologist and the patient.

4. DIFFERENTIATING SYMPTOMS

4.1. Sensorineural or Conductive Hearing Loss

Asking a patient whether he/she hears better in a quiet or a noisy environment usually provokes an expression of bewilderment. "I wonder what that means?" seems to be written on the patient's face. Actually, the answer to this question provides a valuable preliminary clue as to whether the patient has a sensorineural or a conductive hearing loss. In many cases of conductive hearing loss, especially in otosclerosis, there is a tendency to hear better in noisy places, whereas in sensorineural hearing loss there often is a tendency to hear much more poorly in a noisy environment. The ability to hear better in the presence of noise is called paracusis of Willis, or paracusis Willisiana, and is named after Thomas Willis, the physician who first described this phenomenon.

4.2. Tinnitus

One of the least understood and therefore most frustrating conditions encountered by the otologist is tinnitus, or "ear noises." Because various types of tinnitus are associated so often with specific types of hearing impairment, a complaint of tinnitus and a description of its characteristics can be helpful. This subject is discussed throughout this book, and especially in Chapter 18.

4.3. Vertigo

Vertigo is also a frequent companion of hearing loss. Because the hearing and the balance mechanisms are related so intimately and bathed in the same labyrinthine fluid, vertigo often accompanies hearing difficulties. Some disturbances in the labyrinthine fluid such as Meniere's syndrome produce not only hearing loss but also interference with balance. To the otologist, vertigo does not mean lightheadedness, fainting, or seeing spots before the eyes. It does not mean merely a slight sensation of loss of balance. Rather, it conveys a sensation of turning, a feeling that the room or the patient is revolving. A sick feeling in the stomach, or nausea, often accompanies the sensation of rotary vertigo, which is similar to that felt by an inexperienced sailor on a storm-tossed ship. Labyrinthine vertigo can cause a loss of balance during walking; the patient may find it difficult to walk in a straight line because of a sensation of swaying from side to side. Full discussions of tinnitus and vertigo are found in Chapters 18 and 19.

4.4. Fluctuation in Hearing

Almost all people with hearing trouble are aware of fluctuation in their hearing. Many patients seem to hear better in the morning than at night. Some claim they hear better after they inflate their ears (by pinching the nose, closing the lips, and blowing—the so-called Valsalva maneuver). Several factors are involved in a fluctuating hearing level. For example, most people seem to hear much better when they are rested and relaxed than when they are tired and upset, as at the end of a hard day. Alertness may sharpen, whereas inattention may dull auditory efficiency. Sharp fluctuations in hearing are inherent in some types of hearing loss, such as Meniere's disease.

4.5. Self-Inflation

Although it is true that some people actually can improve their hearing temporarily by inflating their ears, the vast majority experience only a clear feeling in their ears, which leads them to feel subjectively better without really hearing better. This subjective improvement usually is short-lived and provides little benefit. Moreover, indiscreet self-inflation can lead to ear infections and abnormal eardrums.

5. FACTS IN THE HISTORY THAT MAY BE IMPORTANT

5.1. Ototoxic Drugs

Hearing loss can be caused by taking large doses of certain drugs such as gentamycin, neomycin, kanamycin, aspirin, and many others. The physician should learn whether the patient has used any of these drugs extensively because the information may have an important bearing on the diagnosis. Knowledge of previous ear or systemic infections also may yield important clues.

5.2. Speech Defect

Because the development of speech depends on hearing, deafness or defective hearing in infancy or early childhood can result in speech problems. This is the reason that for centuries people who were deaf from birth also were considered to be mentally impaired, since they could neither hear nor speak. One of the important clues the otologist uses in diagnosis is the patient's speech. Features such as loudness, strain, and poor articulation all help to indicate the type of hearing loss and its prognosis.

For example, a hard-of-hearing patient who speaks in a loud and strained voice probably has a sensorineural type of hearing loss. If his voice is unusually soft, the loss probably is conductive. If a child has a speech deficit, particularly involving consonants, he most likely has a high-frequency sensorineural defect.

5.3. Noise-Induced Loss

Questions that determine the patient's line of work help to establish whether he/she has been exposed to very intense noise. A history of military service or other exposure to gunfire also is important. The large number of people who lose their hearing because of exposure to intense industrial noise is a subject of serious concern.

The importance of the patient's history cannot be emphasized enough. Often it suggests a diagnosis that then may require only a few special tests for confirmation. In such instances, many needless studies can be eliminated, and much time and energy can be saved.

5.4. Previous Ear Surgery

At one time it could be established readily that the patient had had ear surgery by looking for the postauricular mastoidectomy scar. Today, modern otologic surgery leaves little or no scarring. Surgery for correction of otosclerosis and ossicular defects leaves no detectable scar. Even the most observant otologist cannot know whether a patient has had previous surgery, in many cases. Not uncommonly, a patient who has had stapes surgery is embarrassed or unwilling to admit that he/she is undergoing a revision rather than an initial procedure. Therefore, it is always important to ask directly whether a patient has had any surgery to correct deafness.

Every patient who complains of a hearing loss, tinnitus, vertigo, or any other aural symptom requires a complete examination of the head and the neck. It is not sufficient to examine only the ears because the source of some otologic symptoms lies in the nasopharynx, the posterior choanal region, the temporomandibular joint, or even the throat. Ear pain, for example, may be a presenting symptom of cancer of the larynx.

6. A STANDARD PATTERN FOR A COMPLETE EXAMINATION

It is advisable to develop a standard pattern for a complete examination so that nothing is overlooked. If this pattern is followed consistently, it becomes routine to examine the opposite ear and the nasopharynx even when the presumptive cause of the patient's symptoms is located immediately and cleared up, as in the case of removing impacted cerumen from one ear.

Physical examination of the patient with otological complaints should begin with a general assessment as the patient enters the physician's office. While physical examination outside the head and neck is deferred to other specialists, initial observations of skin color and turgor, gait, affect, and other characteristics frequently provide valuable information that may trigger the appropriate referral.

Some physicians begin their examinations by first inspecting the ear about which the patient complained. The authors prefer to examine first the nose, then the neck and the throat, then the presumably normal ear, and finally the so-called bad ear. We suggest this sequence because it is possible to overlook the opposite ear, nasopharynx, and other structures while concentrating on the symptomatic ear. Occasionally, the patient asks pointedly, "Doctor, aren't you going to look at my other ear or my throat?", but we should not have to be reminded.

6.1. Endoscopic Examination of the Nose and Nasopharynx

The time required to perform a complete examination can be shortened by first inspecting the nose with a nasal speculum and at the same time spraying the nasal cavities with a vasoconstrictor and 1% cocaine or Pontocaine to prepare them to receive the nasal endoscope or nasopharyngoscope.

It is important to check the condition of the turbinates and osteomeatal complex and also look for possible discharge and obstruction caused by polyps. If pus is found in the nose, its source should be identified. Does it originate in the middle meatus or further back toward the nasopharynx? Any discharge should be evacuated with a fine nasal suction before passing the nasopharyngoscope. The appearance and consistency of the mucosa over the turbinates also should be noted. Is it pale and boggy, or is it red and tense? Does it shrink markedly after the cocaine is applied? What about the nasal airway? Is it adequate before shrinkage? What happens after shrinking?

The authors avoid passing the nasopharyngoscope until the last part of the nasal examination. A small amount of 4% cocaine should be applied to the floor of the nose with a fine probe tipped with cotton. In only a few minutes, the medication anesthetizes the floor sufficiently to pass the nasopharyngoscope. Many patients tolerate this procedure well even without topical anaesthesia. Extreme gentleness should be practiced when passing probes, especially a nasopharyngoscope, into the nasal passage. If a direct passage still is not clearly visible after shrinking the mucosa, use the other naris. In many cases, it is also possible to visualize the nasopharynx by placing a small mirror at the back of the patient's throat (Fig. 4.1). Whichever method is used, it is important to obtain a good view. The nasopharyngeal examination enables the physician to see

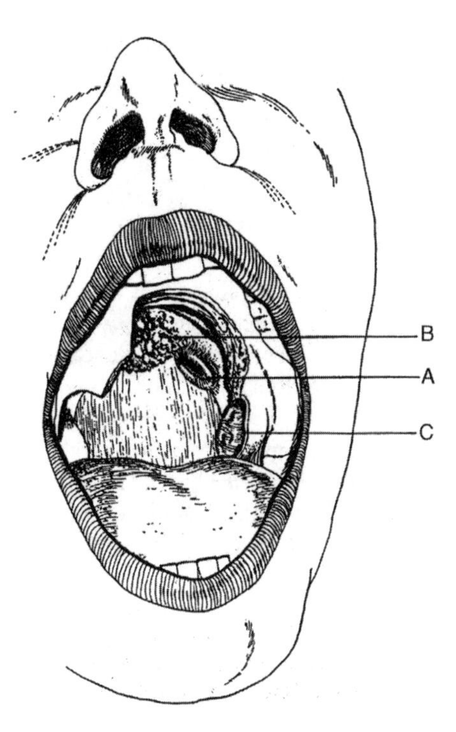

Figure 4.1 (A) The eustachian tube opening in the back of the throat behind the soft palate that has been partially removed; (B) adenoids; (C) tonsils.

the roof of the nose and the posterior aspects of the turbinates. Finally, the nasopharynx, eustachian tube, and fossae of Rosenmüller are scrutinized carefully on both sides.

6.2. The Eustachian Tube

The functional efficiency of the eustachian tube can be estimated in part by looking at the prominent cartilaginous lips (tori) of the tubal orifice through the nasopharyngoscope while the patient swallows. The tori should move freely. If a bubble covers the tubal orifice and the tube is normal, the bubble should break during swallowing. It is most important to look for thick bands of adhesions or growths of adenoid tissue in Rosenmüller's fossae behind the tubal opening. Sometimes these can be seen best by placing a moderate-sized (1 cm) mirror on the depressed tongue and looking up into the nasopharynx.

6.3. The Mouth and Throat

The oral cavity should be examined carefully with attention not only to all mucosal surfaces but also to the teeth and gingiva. Wear facets on teeth and malocclusion may indicate that temporomandibular joint problems are causing ear pain. Palpation is an essential part of oral cavity examination because the examining finger can detect early tumors in the oral cavity, the base of the tongue, and the nasopharynx. The physician should always look specifically for a submucous cleft palate, which may be associated with ear disease. This is particularly important if a bifid uvula is found.

Examination of the larynx with a mirror should provide a good view of the epiglottis, true and false vocal folds, pyriform sinuses, and the base of the tongue at rest and during phonation. Better visualization can be obtained with a flexible fiberoptic nasopharyngeal laryngoscope (NPL) and through strobovideolaryngoscopy.

Systematic examination of the neck includes bimanual palpation of the temporomandibular joints and parotid and submandibular glands. In addition, various triangles of the neck, the thyroid gland, and the carotid arteries must be palpated. Auscultation of the carotids for bruits should be routine, especially when the patient complains of dizziness or pulsatile tinnitus. A test for laryngeal crepitus also should be performed, because absence of crepitus may be the only clue to a postcricoid cancer.

A basic neurological examination with special attention to the cranial nerves also is a routine part of a good otologic assessment and is especially important in the evaluation of unilateral or asymmetrical hearing loss, where neurological findings may lead to the diagnosis of acoustic neuroma, multiple sclerosis, or other important health problems.

6.4. Ears

A head mirror, rather than an otoscope, provides a better view of the auricle and the entrance to the external canal. The shape of the external ear and its position on the head should be noted. The examiner also must look behind the ear for scars, cysts, or other abnormalities. Sometimes a small cyst or furuncle situated just at the entrance to the canal may be overlooked and cause pain when the otoscope is inserted. Choose an otoscope with as large a tip as will fit comfortably. A large tip affords broader vision and fits more snugly in the canal, so that alternative positive and negative pressure can be applied to the eardrum with a small rubber bulb to test the mobility of the eardrum. This procedure is useful also in determining whether there is a perforation in the eardrum. Some patients may experience dizziness and eye movement as a result of this test if the eardrum is perforated. If there is an eardrum perforation or a mastoid cavity, this is called a positive fistula test and may signify an erosion of a semicircular canal. If the eardrum is intact, it is a positive Hennebert's sign, which may suggest Meniere's syndrome, or a perilymph fistula. Actually, a positive fistula test requires conjugate deviation of the eyes, but in practice, subjective dizziness is accepted as a positive fistula test or Hennebert's test.

6.5. Eardrum

6.5.1. Removal of Wax or Debris

If wax or debris is present, it should be removed carefully, so that the entire drum can be seen. Whenever possible, the wax should be picked out gently with a dull ear curette or an alligator forceps. Irrigation should be reserved for those cases in which there is no likelihood of a perforation, and the wax is impacted and difficult to pick or wipe out. When the drum has a perforation, irrigation may result in middle-ear infection. Any debris, such as that caused by external otitis or otitis media, should be wiped out carefully with a thin cotton-tipped applicator or removed with a fine suction tip. Generally, a culture should be obtained if infection is suspected. If the physician notes bony protrusions (exostoses) in the canal, he/she should be especially gentle because injury to the thin skin covering them could result in bleeding and infection. If a large tip on the otoscope makes it difficult to see the drum, a smaller tip is used; care should be taken when inserting it deeply in order to avoid pain or injury. Findings can be documented easily using a videootoscope.

6.5.2. Cone of Light

We can derive much information from scrutiny of the eardrum. In a normal eardrum, a cone of light is seen coming from the end of the umbo or handle of the malleus because of the way in which the sloping drum reflects the otoscopic light. In some eardrums, the cone of light may not be seen, but this does not necessarily mean that an important abnormality exists. Absence of the cone of light may be due to abnormal slope of the drum or the angle of the external canal, a thickening of the eardrum, or senile changes that perhaps do not allow the light to be reflected. It also may be caused by middle ear effusion and other abnormalities.

6.5.3. Intact or Perforated Eardrum

It is essential to find out whether the drum is intact. Most of the time a hole (perforation) in the eardrum is readily visible (Fig. 4.2). Sometimes, however, it is difficult to see a perforation (Fig. 4.3), and occasionally what appears to be one really is an old perforation that has healed over completely with a thin, transparent film of epithelium. If a patient has a discharge that does not come from the external canal, or if the discharge is mucoid, the physician always must look carefully for the perforation through which the discharge issues. A pinpoint perforation should be suspected when a patient complains that air whistles through the ear whenever he/she blows his/her nose or sneezes.

There are several ways to detect a perforation in the eardrum. One method is to move the drum back and forth with air pressure in the external canal (pneumatic otoscopy). This is done with a special otoscope or the rubber bulb attached to some otoscopes. If the drum moves back and forth freely, it probably is intact. If it does not move or moves only slightly, the perforation may become visible because the perforated area moves more sluggishly than the rest of the drum. Decreased eardrum mobility also may be caused by middle-ear fluid or adhesions. If a perforation is high in the area of Shrapnell's membrane, the eardrum still may move fairly well. Another technique (politzerization) to detect

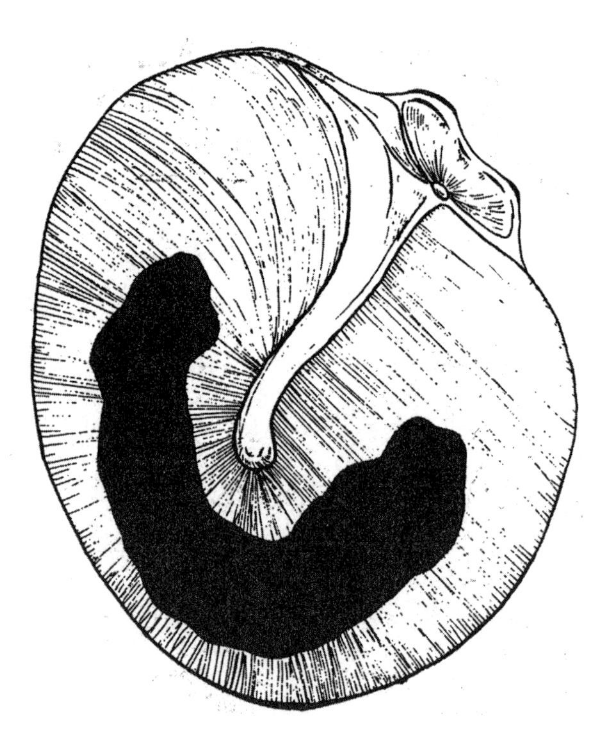

Figure 4.2 Tympanic membrane with a large central perforation.

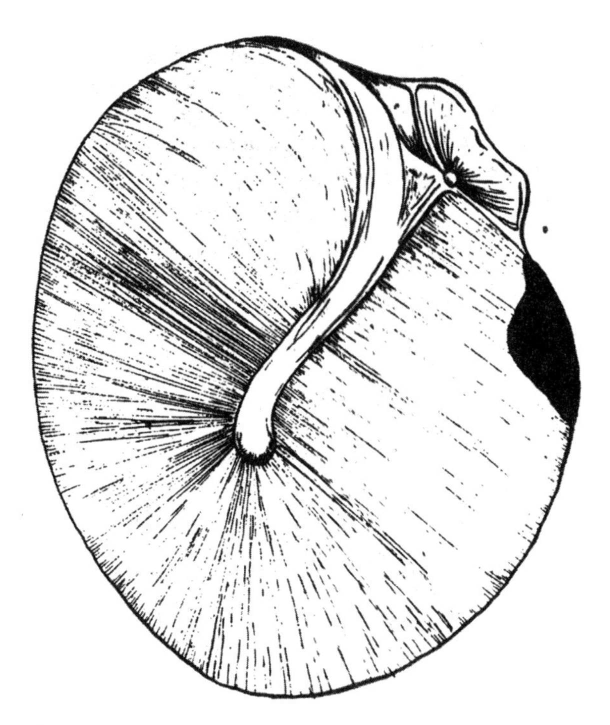

Figure 4.3 Tympanic membrane with a small marginal perforation.

a perforation is to have the patient swallow while a camphorated mist is forced into one nostril and the other nostril is pinched shut. If the eustachian tube is patent and there is a perforation in the eardrum, the examiner, looking into the external auditory canal, will see the mist coming out through the small perforation. Sometimes spraying a film of powder, such as boric acid powder, on the drum delineates the edges of the perforation. This can also be done with the patient lying down, with the eardrum covered with water or a clear eardrop. When air passes through the eustachian tube, the examiner sees bubbles at the site of the perforation. All these procedures can be used also to see whether there is a transparent film over a healed perforation; gentleness and care are essential to avoid breaking the film.

The use of an otologic microscope enhances accurate assessment of the eardrum and middle ear.

6.5.4. *Shadow Formations*

It is important to look for shadow formations behind the drum, particularly those caused by fluid in the middle ear. To accomplish this, the otologist should try to look through the drum rather than merely at it. In this way, what seemed to be a simple surface becomes a map with a dark shadow for the round window niche, a lighter area for the promontory, a pink area for the incus, and many other features.

6.5.5. *Fluid in the Middle Ear*

Fluid in the middle ear often eludes detection, even though it causes hearing loss. Failure to discover the fluid could result in a wrong diagnosis. For example, a patient may have a 30-dB conductive hearing loss with an eardrum that appears to be practically normal.

The diagnosis naturally would be otosclerosis, and stapes surgery would be indicated. When the eardrum is reflected during surgery, however, a thick mucoid, gelatinous mass is found, especially around the oval and the round windows, and the correct diagnosis is not otosclerosis but secretory otitis media. The fluid was simply not detected preoperatively.

A diligent search should be made for fluid in the middle ear if bone conduction is reduced slightly in an otherwise classic picture of conductive hearing loss. There are several ways to detect fluid in the middle ear. If a well-defined fluid level is seen through the eardrum, the diagnosis is simple. It should be borne in mind, however, that strands of scar tissue in the drum and bands in the middle ear can stimulate a fluid level. It helps to see whether the apparent fluid level stays in position while the patient's head is bent forward and backward. With air pressure in the external canal, it is difficult to get free to-and-fro motion of the drum if there is much fluid behind it in the middle ear. By contrast, a normal drum moves easily. Occasionally, bubbles can be seen in the fluid; they assure the diagnosis. Impedance audiometry is useful in patients who might have an abnormality in the eardrum and middle ear, and computerized tomography (CT) scans can help to clarify the status of the middle ear in selected patients.

Politzerization is of great help in detecting fluid but should not be performed in the presence of an upper respiratory infection, particularly one affecting the nose. During politzerization, the fluid and the bubbles can be seen briefly through the drum; then they usually disappear. The patient may profess to suddenly hear better. Politzerization also should be avoided in patients who have undergone stapedectomy.

Whenever there is any suspicion of the pressure of middle-ear fluid, a *myringotomy* should be performed for diagnostic and therapeutic reasons. In an adult, this can be done without local or general anesthesia by using a sharp knife to puncture the inferior portion of the drum. If fluid is present, some usually will ooze out spontaneously, or it can be forced out by politzerization or suction through the myringotomy. A ventilation tube can be placed, if appropriate.

6.5.6. Scars and Plaques; Color; Tumor

The eardrum may reveal still other findings such as scars and plaques. These reflect previous infections and tissue changes in the eardrum. They rarely in themselves cause any significant degree of hearing loss. Occasionally, an eardrum appears to be blue or purple. This may be due to a blockage in the middle ear or to entrapped fluid, or it may be merely a peculiar type of retracted eardrum. A reddish color sometimes is caused by a tumor (glomus jugulare) extending into the middle ear. If there is any possibility of the presence of such a tumor, manipulation should be avoided until further studies have clarified the diagnosis.

6.5.7. Retracted Eardrum

This is another abnormal finding. It is easy to understand why one physician will look at a drum and consider it to be normal, whereas another will say it is retracted. Eardrums vary in their appearance, and the concept of retraction is subject to comparable variations. Even a moderate amount of retraction *per se* may not cause any significant hearing loss. Only when the drum is retracted markedly, and especially when it is pulled onto the promontory, is there a correlation between retraction and hearing. In such instances, politzerization can restore hearing by returning the drum to its original position. Occasionally, the drum is overdistended during politzerization, and then it appears to be flaccid and relaxed. It recovers spontaneously in most cases, often after a swallow or two.

In all cascs of rctractcd cardrum (cspccially unilatcral), thc causc should bc sought in the nasopharynx, the sinuses, and the eustachian tube. Allergies and enlarged adenoids are the most common causes, but neoplasms also must be ruled out, especially in unilateral cases. Aerotitis media may be another cause of a retracted eardrum. In some patients politzerization is not possible, and a small eustachian catheter has to be introduced gently into the mouth of the eustachian tube, generally guided by a nasopharyngoscope positioned through the other naris. The air can be forced in carefully until the tube is opened. By placing one end of a listening tube (Toynbee tube) in the patient's ear and the other end in the physician's ear, the sound of air can be heard as it enters the middle ear.

6.5.8. Erosion; Previous Surgery; Drainage

Perhaps the most confusing otologic picture presents itself when the eardrum is largely eroded and the middle ear is discharging; a similar problem arises when some kind of mastoid surgery has deformed the normal landmarks. In such cases it is necessary to appraise the condition of the middle ear in order to decide on the proper treatment and to evaluate the chances of restoring hearing. In view of the extensive amount of otologic surgery that has been performed in recent years, it is always wise to look at scars of previous operations, both postauricular and endaural, the latter being situated just above the tragus. A postauricular scar usually indicates mastoid surgery. If the eardrum is practically normal, most likely a simple mastoidectomy was done, and the hearing may be within normal limits. If the eardrum is gone and the malleus and the incus also are absent, the operation may have been a radical mastoidectomy, and the conductive hearing loss should be $\sim$50–60 dB. Intermediate between the simple and the radical mastoidectomies are various surgical procedures aimed at both preserving as much hearing as possible and eradicating the infection or cholesteatoma. These procedures usually are called modified radical mastoidectomies or tympanoplasties. Most of the time, the eardrum or part of it is visible, and some form of ossicular chain is present. In modern techniques, a prosthesis may have been inserted to restore ossicular continuity; also, a graft may have been applied to replace the eardrum that had been removed or damaged previously. The endaural scar also could indicate a fenestration operation; an eardrum will be visible, but it will seem to be out of place, and at least some part of the mastoid bone will have been exenterated. Quite often these cavities are covered with debris and require gentle cleaning to permit a clear view. Caution is necessary in cleaning such a cavity around the fenestrated area to avoid inducing vertigo and nystagmus.

Many stapes operations are performed. Because this procedure rarely leaves an evident scar, previous stapes surgery must be uncovered in the history. In stapes surgery, the incision is made inside the external auditory canal on its posterior wall, and the eardrum is reflected forward upon itself so that the surgeon can work in the middle ear. Healing is almost free of visible scars in the canal.

It is becoming much more common to see eardrums of a very peculiar appearance in which infection has played no part. In most cases, the unusual features are the result of myringoplasties (surgical repair of the eardrum) with skin, fascia, or other grafts. The drum may appear to be thick and flaccid or whitish, and it may show few landmarks. Ideally, the patient or previous medical records can supply the pertinent information in such instances. Sometimes one will see an eardrum with something that looks like a small tube sticking out of it. A tiny piece of polyethylene, silastic, metal, or Teflon tubing has been inserted through a small perforation to prevent closure and to allow ventilation of the middle ear. This usually is done in cases of persistent secretory otitis media.

7. NEUROTOLOGICAL EXAMINATION

In addition to complete otoscopic examination, a pneumatic otoscope is used to move each eardrum back and forth to determine whether this maneuver causes dizziness and/or nystagmus. If there is a hole in the eardrum, this is called a fistula test. If the eardrum is intact, it is called Hennebert's test. Although technically conjugate deviation of the eyes is required for the test to be positive, in general practice a clear subjective response of dizziness is considered a positive test, especially if nystagmus is present. A positive Hennebert's test may occur with endolympatic hydrops or a fistula. Hitselberger's sign is sought by testing sensation of the lateral posterior/superior aspect of the external auditory canal. This is the area that receives sensory supply from the facial nerve. Lesions putting pressure on the nerve such as acoustic neuromas or anterior/inferior cerebellar auditory vascular loops often cause a sensory deficit in this area, or positive Hitselberger's sign. Prior ear surgery may also cause decreased sensation in this area. The eyes are examined for extraocular muscle function and spontaneous nystagmus. This examination is aided by Frenzel glasses, which prevent visual fixation. It is important to note that the examiner's eye is an order of magnitude more sensitive in detecting nystagmus than an electronystagmograph. So, direct observation of the eyes should not be omitted. Other cranial nerves should be examined as well. The olfactory nerve may be tested by asking the patient to inhale vapors from a collection of different scents. The optic nerve is tested at least by visual confrontation, if not by referral to an ophthalmologist. Fundoscopic examination is added in some cases, especially if increased intracranial pressure or significant microvascular disease is suspected.

Trigeminal nerve sensation is tested by assessing sensation in all three divisions on both sides. The trigeminal nerve also supplies motor fibers to muscles of mastication, which can be evaluated by assessing jaw movement and muscle strength. In addition to Hitselberger's sign, the facial nerve is assessed through observations of facial movement and tone. Tear flow, stapedius muscle reflex, salivary flow, and taste can also be tested as part of a facial nerve evaluation. The glossopharyngeal nerve is evaluated by testing gag reflex in the posterior third of the tongue or oropharynx and sensation along the posterior portion of the palate, uvula, and tonsil ossopharyngeal sensory function also can be tested and quantified using functional endoscopic sensory testing with air pulses that trigger the laryngeal reflex. Abnormal vocal fold or palatal motion is often the most obvious sign of tenth nerve dysfunction. Eleventh cranial nerve abnormality is diagnosed in the presence of sternocleidomastoid or trapezius muscle weakness, and twelfth nerve dysfunction causes unilateral tongue paralysis. Examination of the nose and oral cavity is performed routinely. Special attention is paid to nasal obstruction when taste and smell disorders have been identified and to clear rhinorrhea, which may indicate a cerebrospinal fluid leak following head injury. Examination of the larynx should include special attention to symmetry of vocal fold motion, and to any signs of direct laryngeal trauma. In addition, hoarseness or any other voice change should be noted and investigated with strobovideolaryngoscopy, objective voice measures, and other state-of-the-art evaluations. Examination of the neck includes palpation not only of the anterior neck, but also of the posterior aspect looking for muscle spasm and tenderness of the cervical vertebrae. These findings are associated often with limitation of motion, especially in patients who have dizziness associated with changes in head and neck position. Attention should be paid to the regions of C1 and C2, especially in patients with posttraumatic dizziness. Neck examination should also include auscultation of the carotid arteries and palpation of the superficial temporal arteries. If there is any question of vascular insufficiency, ultrasound of the carotid and vertebral arteries, MRA, MRV, or arteriography should be considered. In addition, Romberg testing, gait assessment, cerebellar function testing, and other neurologic evaluation should be carried out.

8. OTHER CONDITIONS TO CONSIDER

In addition to asking questions directed specifically to otologic problems, the physician must obtain a complete general medical history. Many systemic conditions are associated with otologic symptoms such as hearing loss, tinnitus, and dizziness. Such conditions include diabetes, hypoglycemia, thyroid dysfunction, cardiac arrythmia, hypertension, hypotension, renal disease, collagen vascular disease, previous meningitis, multiple sclerosis, herpes infection, previous syphilis infection (even from decades ago), glaucoma, seizure disorders, and many other conditions. Psychiatric conditions also are relevant because many of the medications used to treat them and to treat various systemic diseases may cause otologic symptoms as side effects. So can a variety of antibiotics and toxic chemicals, such as lead and mercury. Previous radiation treatment to the head and neck may result in microvascular changes that cause hearing loss, tinnitus, or dizziness. Even excess consumption of alcohol or caffeine may produce symptoms that could be confused with other etiologies. Chickenpox occasionally leaves a small pockmark on the eardrum that may persist for many years and be confused with a clinically relevant problem. Blood in the middle ear following head injury generally indicates a fracture in the temporal bone and may result in eardrum scar or hearing loss. Hearing tests help to determine the extent of involvement, especially if the inner ear is damaged. Whenever there is a possibility of an acoustic neuroma, additional tests, MRI, and neurological evaluation are indicated. The interpretation of many of these tests is not included in this book.

Objective tinnitus is a noise that can be heard by the examiner as well as the patient. To detect these cases, the physician should put his ear to the patient's ear or use a listening tube (Toynbee Tube) to find out whether there is a bruit indicating some vascular disorder or a click from the nasopharynx or the middle ear. Another cause of objective tinnitus is an intermittent spasm of the soft palate that produces a clicking sensation heard in the ear. The cause is unknown, but the condition is recognized when it is encountered. There are also many other conditions that should be considered in otologic patients. In addition to those discussed later in this book, a great many more may be reviewed in medical textbooks of otology/neurotology.

9. TESTS

9.1. Tests of Hearing and Balance

Testing of hearing and balance function are fundamental in patients with otological complaints. Specific appropriate tests will be discussed in detail in subsequent chapters.

9.2. Metabolic Tests

Metabolic tests must be selected on the basis of clinical need in each individual case, of course. However, certain conditions have such profound importance in otologic symptoms that they are sought with nearly routine frequency. Most of these conditions are discussed in greater detail in later chapters.

Luetic labyrinthitis is a highly specific syphilis infection of the inner ear. Luetic labyrinthitis can cause hearing loss, tinnitus, and vertigo. Untreated, it may eventually cause total deafness. Routine serologic testing (RPR and VDRL) is normal. In order to detect luetic labyrinthitis, an FTA absorption test, MHA-TP, or other sophisticated syphilis antibody test must be obtained.

Diabetes and especially reactive hypoglycemia may produce symptoms of dizziness. In some cases, hypoglycemia may provoke symptoms similar to endolympathic hydrops (Meniere's syndrome). A 5-h glucose tolerance test is often necessary in dizzy patients to rule out hypoglycemia, although other tests are available to detect diabetes.

Even mild hypothyroidism may produce fluctuating hearing loss, tinnitus, and disequilibrium in some patients. It is necessary to obtain T3, T4, and TSH to establish this diagnosis.

Hyperlipoproteinemia has been associated with sensorineural hearing loss, as well. When sensorineural hearing loss of unknown etiology is under investigation, measurement of cholesterol and triglyceride levels should be included in most cases.

Diabetes mellitus and collagen vascular disease produce vascular changes which compromise perfusion and may cause otologic symptoms. In addition to routine screening for diabetes, tests for collagen vascular disease including rheumatoid factor, antinuclear antibody, and sedimentation rate may be indicated.

Autoimmune inner ear pathology has been well documented. When suspected, a variety of tests of immune function is required, as discussed later in this book.

In at least a small number of patients, allergies may cause otologic symptoms. In the authors' experience, this association is less common than some literature would suggest. However, in the appropriate clinical setting, allergy evaluation and treatment may be required for otologic symptoms including dizziness, tinnitus, and hearing loss.

A great many other tests may be appropriate depending upon clinical presentation. Many viruses, Lyme disease, sickle cell disease, and numerous other problems may cause neurotologic symptoms that may be difficult to differentiate from symptoms caused by noise or other etiologies without appropriate studies.

9.3. Radiologic Tests

Neurotologic diagnosis has been revolutionized by modern radiologic technology. In many patients with otologic symptoms, radiologic investigation is essential.

9.4. Magnetic Resonance Imaging

Magnetic resonance imaging (MRI) is the mainstay of radiologic evaluation of the neurotologic patient. MRI of the brain and internal auditory canals is required for complete assessment. In the neurotologic patient, it is essential to rule out demyelinating disease, neoplasms, subdural hematomas, and other conditions that may be responsible for the patient's neurotologic complaints. High-resolution gadolinium-enhanced MRI of the internal auditory canal is required to rule out acoustic neuroma. Very high quality studies are necessary, and MRI should be performed on a magnet of at least 1.5-Tesla strength.

9.5. Computerized Tomography

CT scan has been performed much less frequently in the recent years because of improvements in MRI. However, CT testing may still be extremely valuable. MRI does not show bony detail. A high-resolution CT of the ears may show birth defects or even hairline fractures or other bony abnormalities of great clinical importance that may be invisible on MRI. One should not hesitate to order both studies.

9.6. Air Contrast CT

Air contrast CT involves infusion of 3–5 cc of air through a lumbar puncture. The procedure should be performed with a small-gauge spinal needle and is done routinely on an out-patient basis. Air is allowed to rise into the cerebellopontine angle (CPA), and the internal auditory canal and neurovascular bundle can be visualized well. This test was standard for detection of small acoustic neuromas before MRI was developed. Now it is performed much less commonly, but it still has use. It shows the anatomy much more clearly than MRI and often allows detection of abnormalities such as arachnoid cysts and anterior inferior cerebellar artery loop compression of the eighth cranial nerve, a condition that cannot be assessed usefully by MRI.

9.7. Ultrasound and Arteriography

When there is a question regarding the adequacy of carotid or vertebral blood flow, ultrasound provides a noninvasive, painless, expeditious method for assessing blood flow. In patients with positional vertigo, vertebral ultrasound with the neck in neutral, flexed, extended, and turned positions may reveal intermittent vertebrobasilar artery system occlusion. If the results are equivocal, if significant vascular compromise is identified, or if there is very strong clinical suspicion of vascular occlusion despite an unimpressive ultrasound, arteriography may be required. While this test is more definitive, it may be associated with serious complications and is ordered only when truly necessary. MR angiography (MRA) is less invasive and may provide the needed information in many cases. Angiography also provides information about intracranial vascular anatomy. In some cases, additional information about intracranial vascular flow may be necessary, and new techniques of intracranial doppler study are available for this purpose.

9.8. Dynamic Imaging

Single Photon Emission Computed Tomography (SPECT) or Positron Emission Tomography (PET) also have proven useful for neurotologic evaluation. In some patients, they may show blood flow abnormalities in the microcirculation of the inner ears that may be the only objective abnormalities identified for some patients. These studies should be considered in many patients with hearing loss, tinnitus, or vertigo, and they are especially useful following head trauma.

5

Classification and Measurement of Hearing Loss

Robert T. Sataloff
Drexel University College of Medicine, American Institute for Voice and Ear Research, and Graduate Hospital, Philadelphia, Pennsylvania, USA

Joseph Sataloff
Thomas Jefferson University, Philadelphia, Pennsylvania, USA

1. TWO BASIC TYPES OF TESTING

Two basic types of testing are done with a tuning fork routinely: air conduction and bone conduction. The best all-purpose tuning fork to use is a 512 Hz steel tuning fork. We use a 512 Hz tone because forks of lower frequencies produce a greater tactile sensation that sometimes can be felt rather than heard or that can be felt before the tone is heard. However, tuning forks of various frequencies from 125 to 8000 Hz may be useful in some situations. Although forks of frequencies higher than 512 Hz are attenuated readily, they can supply useful information. With the 512 Hz tuning fork, it is possible for the examiner to obtain a rough estimate of the extent of the hearing loss and to speculate whether the cause is conductive or sensorineural. In some cases it is even possible to establish whether the damage is in the inner ear or in the nerve.

A tuning fork should not be struck very hard; a blow that is too forceful produces overtones that might give false information. Furthermore, a very loud tone may startle some patients who are especially sensitive to noise because of hyperrecruitment, a condition often present in Meniere's disease. Tuning forks should be struck on something that is firm but not too hard. The knuckle, the elbow, and the neurological rubber hammer are all satisfactory for this purpose. Tabletops and wooden chairs generally should not be used to activate tuning forks.

In testing for air conduction, the tuning fork should be held close to, but should not touch, the ear, and the broad side of one of the prongs must face the ear (Fig. 5.1). It is wrong to hold the two prongs parallel to the side of the ear. Such an application often produces a dead spot, and, consequently, the listener may hear no tones, even though his hearing may be normal. To verify this fact, the examiner should place the tuning fork to his own ear and rotate it. The tone will appear to go off and on as the fork turns.

Air conduction measures the ability of airborne sound waves to be transmitted to the inner ear along the external auditory canal, the eardrum, and the ossicular chain. This is done by holding the vibrating tuning fork near, but not touching, the external auditory

Figure 5.1 Position of the tuning fork for air conduction testing.

canal. Bone conduction measures, to some degree, the ability of the inner ear and the nerve to receive and to utilize sound stimuli. In this test, the external auditory canal and the middle-ear areas are bypassed. The base or the handle of the vibrating tuning fork is held directly on the skull so that the vibrations can reach the inner ear directly (Fig. 5.2). The fork may be held on the mastoid bone, the forehead, the nasal dorsum, the closed mandible, or the upper teeth. Gentle application to the upper incisors or even to dentures provides the best clinical measurement of bone conduction. These sites are preferable to the mastoid area or forehead with regard to sensitivity.

2. FACTORS IN EVALUATION

2.1. Approximate Results with the Tuning Fork

Despite efforts, no reliable method to calibrate tuning forks quantitatively has been found. Therefore, the tuning fork makes it possible to obtain only a rough quantitative approximation of a patient's ability to hear. For example, the physician can strike the fork gently,

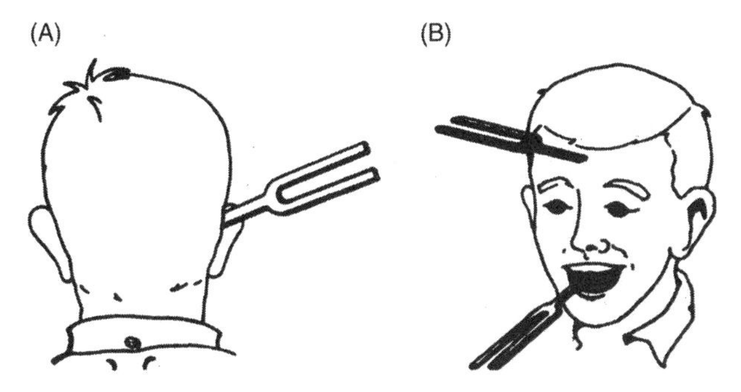

Figure 5.2 Placement of the handle of the fork on the mastoid (A) and on the forehead and the upper incisors (B) for bone conduction testing.

apply it successively to the left and to the right ear, and ask the patient to specify in which ear the tone sounds louder; or the examiner can compare the patient's ability to hear a tuning fork with his own, presumably normal (or, at least, known) hearing. However, because it is not possible to express the results of tuning fork tests in quantitative terms such as decibels, the use of an audiometer is required.

2.2. Transmission of Sound to the Opposite Ear

When testing hearing with a tuning fork or an audiometer by either air or bone conduction, it is important to remember that sounds of sufficient intensity, when applied to or held near one ear, are transmitted around the head or through the bones of the skull and are heard by the opposite ear.

In air conduction testing, the tone near one ear has to be quite loud to be carried around the head and heard by the opposite ear. There is roughly a 40-dB attenuation between the ears; in other words, if a tone near one ear is 40 dB or louder, it can be heard by the opposite ear. In bone conduction testing the problem is far more complex because there is little or no attenuation by the skull of low-frequency bone-conducted sound. Although the examiner thinks he/she is testing the *left* ear when he/she holds a tuning fork to the *left* mastoid bone, he/she actually is testing *both* ears, because the right ear receives the sound at almost the same intensity as the left ear.

2.3. Masking the Opposite Ear

For these reasons it is necessary, especially in bone conduction testing, to mask the opposite ear so that responses only from the ear being tested are received. This also applies to air conduction testing, particularly when there is a difference of $\sim$40 dB or more in hearing acuity between the two ears.

It might seem that the opposite ear could be masked by inserting a plug or covering it with the patient's hand. Actually, these measures would not only fail to mask the ear, but they would cause the tuning fork to sound even louder. To prove this, you need merely strike a tuning fork and hold the handle to your upper incisors. Generally, if your hearing is normal, the fork will be heard throughout the head. Plugging an ear with a finger does not mask it; instead, it produces a conductive hearing loss and misleading results. When testing one ear with a tuning fork, a good way to mask the opposite ear is to have an assistant or the patient rub a page of stiff writing paper over the opposite ear. The noise made by this paper occupies the nerve pathway of that ear, and then you can be sure of getting a response only from the ear to which you apply the tuning fork. The air-pressure hose that is used to spray noses also can serve as a masking device. It should be used cautiously in order to avoid injuring the eardrum. The air nozzle should be applied somewhat sideways in the ear, so that the noise goes into the canal without too much air pressure. A special noisemaker called a Bárány noise apparatus also is available. This device is very inexpensive and small; it is activated by winding it up and pressing a button, and it is our preferred device for washing in this scenario.

2.4. Using the Tuning Fork in Diagnosis

Now that we know the basics of tuning fork testing, let us proceed to use the fork in helping to make a diagnosis in routine office practice. Bear in mind that if only a mild hearing loss is present (i.e., $<$25 dB for the frequency being tested by the fork), the tuning fork may not provide accurate information. It is reliable only when the hearing loss is $\geq$25 dB.

2.5. Systematic Evaluation

Let us suppose a patient complains of hearing trouble in his left ear, and the otoscopic examination shows normal external canals and eardrums in both ears. What steps are taken to find exactly where in the auditory pathways the damage has occurred, and what is the most likely cause? First, it is necessary to determine whether there really is a hearing loss in the patient's left ear. To do this, strike the tuning fork gently, hold it to your own normal ear until the tone gets weak, then quickly put the fork near the patient's left ear and ask if he/she hears it. If not, put it to the right and presumably normal ear to be sure the fork is still vibrating. Finally, put it back to your own ear to be certain the tone is still present.

Obviously, if the fork is heard either in your ear or in the patient's good right ear but not in his left, or bad, ear, he/she probably has some hearing loss in the left one. Then, by striking the tuning fork a little harder each time and listening with your own normal ear before and after you place it to the patient's left ear, you can determine how loud the fork's vibrations must be for the patient to hear them. In this way, a rough idea of the degree of his hearing loss can be gained.

Converting this finding to decibels by an "educated guess" is surprisingly difficult. It should not be attempted even by so-called experts. If the patient seems to hear what you believe to be an extremely weak tone (almost as weak as you can hear), his hearing loss, if it exists, probably is too mild to be studied with a tuning fork, or perhaps it is present only in the higher frequencies. In either case, audiometric studies are required.

2.6. Rinne Test

Now let us presume that the patient does have a moderate hearing loss in the left ear and does not respond to a vibrating tuning fork that you hear. The next step is to determine whether the damage is in the conductive area (the outer or the middle ear) or in the sensorineural pathways (the inner ear or the auditory nerve). The patient now is asked to tell you whether the vibrating tuning fork seems to sound louder when it is held beside his left ear (by air) or behind his left ear directly on the mastoid bone (by bone). The tuning fork is struck hard enough so that the patient should be able to hear it fairly well, and it is held beside his left ear for about a second. Then, it is quickly moved until its handle touches the left mastoid bone, where it is held for another second. Move the fork back and forth between these two points, striking the fork again, if necessary, until the patient can tell whether it is louder by air or behind the ear. This is called the Rinne test. Applying the handle to the upper incisors is another good method of testing and will detect smaller air–bone gaps (20–25 dB) than will mastoid testing (30–35 dB).

If the fork is louder behind the ear on the patient's mastoid bone or on the teeth, bone conduction is considered to be better than air conduction, and therefore he/she has a conductive hearing loss. In other words, the sensorineural pathway is working, but something is blocking the sound waves from reaching the inner ear.

Furthermore, because the outer ear had been examined and was found to be normal, the probable diagnosis is some defect in the ossicular chain, such as otosclerosis.

2.7. Weber Test

To confirm these findings, the tuning fork now is struck again, and the handle is placed on the patient's forehead nasal dorsum or gently touching his upper incisor teeth. The patient is then asked to indicate in which ear the fork sounds louder (the Weber test).

In a conductive hearing loss, the tone will sound louder in the bad ear, the left one in this case. Something like this happens when you plug your own ear while the fork is on your teeth; you probably will be surprised that it sounds louder in the plugged ear. Plugging the ear produces a conductive hearing loss just as otosclerosis does. In sensorineural hearing loss, the tuning fork sounds louder in the good ear.

2.8. Schwabach Test

Occasionally, a patient will find it difficult to lateralize the fork to either ear, that is, to tell in which ear the fork sounds louder. This reaction does not rule out conductive deafness but, rather, suggests the need for further studies, particularly with an audiometer. To complete the tuning fork tests on this particular patient, who has a presumptive diagnosis of otosclerosis, strike the fork gently, press it against the patient's left mastoid area until he/she barely hears it, and then move the instrument quickly to your own mastoid (the Schwabach test). The patient will hear the tuning fork much better and longer by bone conduction on his/her mastoid than even you yourself can hear it. This is called prolonged bone conduction and substantiates a diagnosis of conductive hearing loss.

2.9. Further Tests

Now let us examine a different patient with normal otoscopic findings who also is complaining of left-sided hearing loss. When we compare his/her ability to hear by air and bone conduction (Rinne test), we find he/she hears much better by air conduction than by bone, in contrast to the previous patient. Furthermore, when we put the tuning fork to his/her teeth, it sounds louder in his/her good ear than in his/her bad ear (Weber test). Then, when you compare his/her bone conduction with your own (Schwabach test), you find that you can hear a tone much longer and that it sounds louder to you. It is important in this last test to mask the good ear with noise so that the sound of the fork pressed against the mastoid of the bad ear may not be heard in the good ear. The site of damage in this patient is not in his/her middle or outer ear, as in the previous case, but in his/her inner ear or auditory nerve. The patient has sensorineural hearing loss, and the most likely cause will have to be determined by exploring the history and performing many more tests, some of which require special equipment. Chapter 10 includes an explanation of how it is possible with a tuning fork to decide in some patients whether the damage is located specifically in the inner ear proper or auditory nerve by testing for displacusis (pitch distortion) or recruitment (loudness distortion).

2.10. Difficult Cases

The two patients mentioned earlier were comparatively easy to classify. Actually, most patients are just as easy to test, but, occasionally, more difficult cases are encountered, as when the same patient has a severe or even total loss of hearing in one ear and a partial conductive hearing loss in the other. A great deal of masking and careful interpretation of hearing tests are necessary in such cases; yet the tuning fork can provide the essential information, and each ear often can be classified properly.

Determining the site of damage usually can be done readily with a tuning fork. In some instances, however, this may become difficult, and it may be especially hard to decide which of many possible causes applies. More sophisticated tests with an audiometer and other equipment are helpful in such cases, but the tuning fork should be used routinely to confirm or to challenge the results obtained with the more discriminating and complex equipment.

2.11. The Audiometer and the Tuning Fork

Few general practitioners have audiometers today, but many more would find it rewarding to purchase one and learn to perform audiometric hearing tests. It would enable them to render better service to some of their patients, just as electrocardiography refines their service to others. The technique for performing good audiometry and the pitfalls to be avoided are described later in this chapter. If a practitioner prefers not to use an audiometer and refers his patients to a local otolaryngologist, audiologist, or hearing center, he/she should make it a routine policy to confirm all studies done by consultants with his/her own tuning fork tests. Although a consultant may have more elaborate equipment and testing experience, the general practitioner should not underestimate the importance of the simple tuning fork in the diagnosis of hearing loss, or the possibility for errors in audiometric test results.

3. ASPECTS OF TESTING

3.1. Cooperation of the Patient

At present physicians have at their disposal only a few reliable *objective* methods of measuring hearing (Chapter 7). In routine testing, some voluntary response of the patient is necessary as an indication that he/she hears the sound used to test hearing. The sound may be word, a sentence, a pure tone, a noise, or even the blast of a loud horn. The patient's response may consist of raising his/her finger or hand, pressing a button, answering a question, repeating a sentence, turning in the direction of the sound, or merely blinking his/her eyes. The test sound is reduced in intensity until the patient hears it ~50% of the times it is presented. In such an instance, the intensity level at which the patient just hears the sound is called his *threshold of hearing*. Speech may be used at a reasonably loud level, and the patient is asked to repeat words or combinations of words to determine how well he/she distinguishes certain speech sounds. This is called *discrimination testing*.

4. DEVELOPMENT OF THE AUDIOMETER

The ideal method of testing hearing would be to measure and to control everyday speech accurately and to present it to the patient in such a manner that, without requiring any voluntary response from the patient, the physician could determine whether the patient had received and understood it clearly through both ears and with the participation of his/her brain. Unfortunately, many unsolved complex problems have prevented the development of such a method. Even the very first step of controlling and measuring the intensity of speech itself has not yet been perfected.

The old method of testing hearing—one that too many physicians still use—is to stand 15 ft away and to whisper a number of words to the patient, who is plugging his/her distal ear with a finger. The examiner gradually comes closer until the patient just begins to repeat the whispered words correctly. If the patient responds correctly to every word at 15 ft, the examiner gives him/her a score of 15/15, or normal hearing, but if the physician must approach to within 5 ft to obtain a correct response, the patient receives a 5/15 score for his/her hearing impairment. Then the patient faces the opposite way, with a finger plugging the other ear, and the procedure is repeated.

It is extremely difficult to duplicate this test under identical conditions. Furthermore, it is virtually impossible to compare the results of different patients or to maintain an accurate sound-intensity level. Factors such as the acoustics of the test room, the choice of words, the examiner's accent, enunciation, and ability to control and to project his/her voice as well as the degree of hearing loss in either or both of the patient's ears render this testing procedure highly inaccurate.

4.1. The Pure-Tone Audiometer

These shortcomings and the impossibility of controlling and reproducing speech sounds accurately with the early forms of electronic equipment led to the development of the pure-tone audiometer. Its designers recognized that their chief objective was to determine to what degree speech must be amplified to be just heard by the patient. They therefore analyzed speech and found that it encompassed frequencies from $\sim$128 to 8192 Hz.

At that time electronic equipment was readily available to measure pure tones with a great deal of accuracy, though no equipment was available to measure speech. Because it was not practical to test all frequencies from 128 to 8192 Hz, the developers of the audiometer decided to sample certain pure-tone frequencies within the speech range. They selected a series of doubles: 128, 256, 512, and so on to 8192. Frequencies bearing this double relationship to each other represent octaves on the musical scale.

4.2. Frequency Range

Later, the numbers were rounded off, so that today audiometers are calibrated for frequencies 250, 500, 1000, 2000, 3000, 4000, 6000, and 8000 Hz. This frequency range does not cover the entire gamut of normal hearing; it covers only the speech range. The young human ear is sensitive to sound waves from frequencies as low as 16 Hz to as high as 20,000 Hz. Its most sensitive area is in the range between 1000 and 3000 Hz because of ear canal resonance, as discussed in Chapter 2. In this so-called middle-frequency range, it takes less sound energy to reach the threshold of hearing than it does for tones $>$3000 and $<$1000 Hz. Figure 5.3 shows that the ear is more sensitive at the middle frequencies than at the higher and lower frequencies, and so it is necessary to make higher and lower tones louder to permit the normal ear to hear them.

Because the audiogram has a 0-dB reference level for normal hearing, which is depicted as a straight line across the audiogram, it was necessary to introduce a *correction factor* to the lower and the higher frequencies to adjust the reference level to a straight, rather than a curved, line. The straight-line reference level makes it easier to read and to interpret an audiogram.

Above the range of normal human hearing is the ultrasonic domain. Dogs and bats can hear those high frequencies that are inaudible to the human ear. Bats have such sensitive ears that they use these frequencies to guide their flight in a manner that parallels closely the principle of the modern sonar system. Because the human ear is not sensitive to these frequencies, it does not suffer damage from ultrasonics, even at reasonably high intensities.

5. REFERENCE HEARING THRESHOLD LEVELS

The American Standards Association (ASA) 1951 reference level of 0 dB was derived from studying "normal" individuals in a national hearing survey. A large number of hearing tests were performed on young people with presumably normal hearing to find

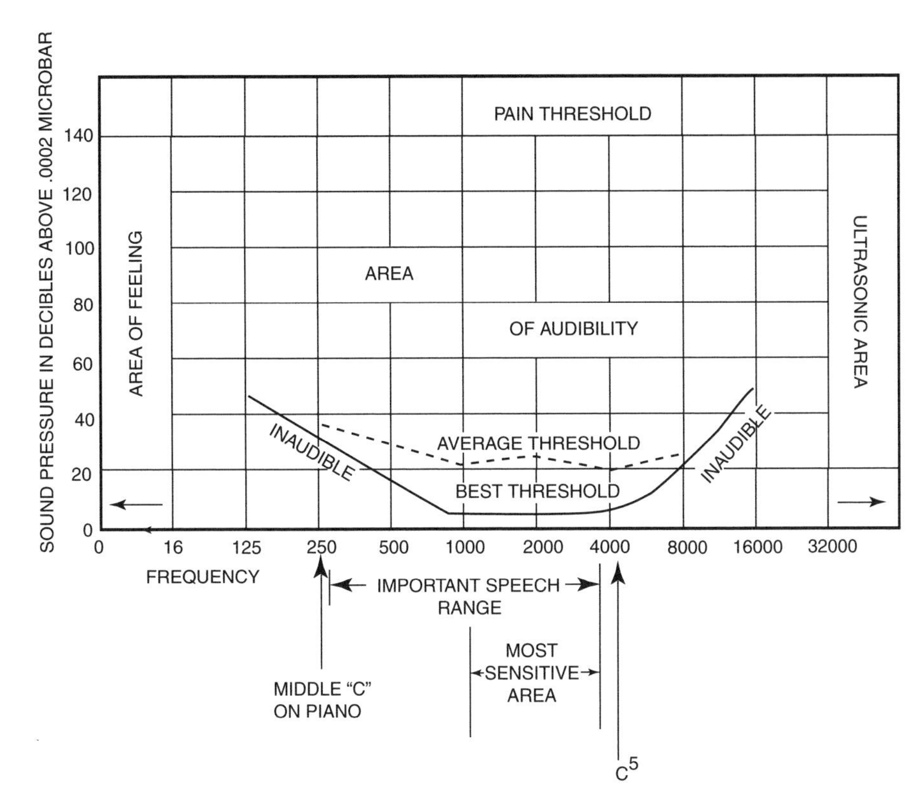

Figure 5.3 Graph showing area of audibility and sensitivity of the human ear, known as the minimum audibility curve. The best threshold (solid line) separates the audible from the inaudible sounds. This level generally is the reference level for sound-level meters. The average threshold of hearing (dashed line) lies considerably above the best threshold and is the reference level used in audiometers. The ear is most sensitive between 1000 and 3000 Hz. The sound pressures are measured in the ear under the receiver of an audiometer. [Modified from Davis and Silverman (1).]

the intensity that could be considered normal at each frequency. Because this was an average value, some subjects had better-than-average normal hearing, causing some otologists to complain that it was awkward to express the results of a hearing test on a person with above-normal hearing in minus figures (e.g., as −5 dB). It also was cumbersome to have to speak in terms of "hearing loss" rather than "residual hearing." Because the reference hearing level used in some European countries was lower, or better, than that of the American Standards Association, the International Standards Organization (ISO), of which the American National Standards Institute (ANSI) is the United States' member body, changed the standard to conform to the European standard.

Table 5.1 shows the difference in decibel readings between the ASA and ISO/ANSI standards.

6. THE NEED FOR EXAMINERS WITH SPECIALIZED TRAINING

Presently the most reliable and accepted way to test hearing is to use a standard pure-tone audiometer. The ability to use this instrument satisfactorily requires specialized training, because the testing requires the voluntary cooperation of the subject.

Table 5.1 A Comparison of ASA-1951 and ISO-1963/ANSI-1969 Reference Hearing Threshold Levels

	Reference threshold level (dB)		
Frequency (Hz)	ASA-1951[a]	ISO/ANSI[b]	Difference
125	54.5	45.5	9.0
250	39.5	24.5	15.0
500	25.0	11.0	14.0
1000	16.5	6.5	10.0
1500	(16.5)	6.5	10.0
2000	17.0	8.5	8.5
3000	(16.0)	7.5	8.5
4000	15.0	9.0	6.0
6000	(17.5)	8.0	9.5
8000	21.0	9.5	11.5

[a]The figures in parentheses are interpolations.
[b]It is common practice to add 10 dB at 500, 1000, and 2000 Hz when converting hearing thresholds from ASA to ANSI. ISO/ANSI values are from W.E. 705A earphone data.

One of the most important functions of every training program is to teach the tester how to make the responses of the subject a reliable indication of whether or not he/she is hearing the test tone. This can be accomplished best by following these principles: (a) The method of testing should be explained to the listener in a simple and positive manner, and a practical demonstration should be given if the person ever had a hearing test before. (b) The method of response should be as simple as possible—for example, raising a finger or a hand, or pressing a button is simpler than writing down an answer. (c) The subject should be conditioned to give a positive response and encouraged to give reliable answers quickly and concisely. (d) The subject should be given just enough time to respond after the presentation of each sound signal.

One of the responsibilities of a trained tester is *to be certain* that the responses obtained are reliable and are an accurate indication of the subject's hearing. Frequently, the experienced tester develops an intuitive feeling as to the reliability of the test and can change the technique when there is any question about the cooperation of the subject. For example, if a tester notes that the subject seems to be indecisive and does not give precise answers, he/she may ask the subject to raise his/her entire hand or to say "yes" instead of using the finger response.

6.1. Answers to Questions

Why is special training necessary to perform hearing tests? Why cannot one become proficient just by following the directions supplied with the audiometer or be trained by the salesman who sells the audiometer? The answers already were suggested in part when it was emphasized that audiometry is a subjective test and that subjects are not always anxious or able to give reliable responses. Accurate audiometry requires a very carefully trained tester to determine when these instances occur.

Experience has shown that though they have performed several hundreds or even thousands of audiograms and consider themselves to be authorities on audiometry, testers without adequate audiometric training do in fact make serious mistakes of which they are unaware and thus produce hearing tests that are neither reliable nor valid. This kind of circumstance has been verified in every phase of audiometry in otologic practice, industry, and school systems. It was demonstrated dramatically in a report by a subcommittee of the ASA in which only a few hundred audiograms were found to be reliable out of several thousands performed in industry by presumably trained people.

To perform satisfactory audiometry, a tester must be trained thoroughly to understand the importance of his/her responsibility and to take pride and interest in his/her work. Without this training, unsatisfactory test results may be obtained that may prove to be more a liability than an asset. Such training in hearing testing is available in numerous institutions throughout the country, or it can be supplied by well-trained audiologists and otologists.

7. WHO SHOULD PERFORM AUDIOMETRY?

Ideally, the physician should do his/her own audiometry, because in this way he can make a good appraisal of the hearing level of his patient. Unfortunately, this is not usually possible because of the time factor in the busy schedule of the otologist, general practitioner, pediatrician, industrial physician, or the school physician. Partly as a consequence, the profession of audiology has developed. Audiologists are trained professionals, usually with a Masters degree, Au.D. or Ph.D., and with certification. They are generally the most fully trained personnel at performing routine and specialized hearing tests. However, it is neither necessary nor practical to use a fully trained audiologist for screening audiometry in every noisy work place, school, or clinical setting. Nurses, medical assistants, and other personnel available in industry can be trained to perform excellent audiometry. The training may take several days or weeks, depending on individual aptitude and the manner in which the program is organized. The principal purposes of the training program are to teach the tester to utilize the best available technique, to be completely aware of the potential pitfalls in hearing testing, to understand the limitations of his/her training and the situations in which testing should be referred to a certified audiologist, and to understand the serious consequences of an incorrect report. The responsibility of the tester, unless he/she is a physician, is not to interpret results but, rather, to produce valid, reliable test results. People trained to do hearing tests must also have a firm basic understanding of the hearing mechanism.

7.1. Training of Audiometric Technicians

Audiometric technicians are used commonly in industry and in some physicians' offices. No certifying agency has been accepted universally, although there are now two recognized organizations supervising technician training (one for office practice, one for industry), as discussed below. So, many such technicians have no certification credential as audiometric technicians or occupational hearing conservationists. Consequently, the content and quality of training programs for technicians is not consistently good. While this problem is being addressed, a few principles should be kept in mind. Any training course should consider the following objectives:

1. Introduction to basic anatomy and physiology of the ear.
2. Introduction to basic physics of sound and hearing.
3. Understanding of the types of hearing loss, audiometric patterns, and variations.
4. Basic understanding of audiometric techniques and the ability to perform basic audiometry, for example, pure-tone air and bone conduction and speech audiometry, and tympanometry.
5. Awareness of other audiometric techniques, such as evoked response, electrocochleography (ECoG), and central testing.

6. Basic examination of the external ear, including evaluation for impacted cerumen, and the use of tuning forks.
7. Basic introduction to causes of hearing loss and audiometric patterns.
8. Recognition of the limits of knowledge and understanding the need for supervision by a physician or audiologist. Understanding of the need for diagnosis by a physician.

The following course outline details the kind of information the authors feel would be necessary and useful for fully trained audiometric technicians. A few of the subjects listed are omitted routinely in training programs specifically designed for hearing conservationists whose practice is limited to the industrial setting.

Programs such as the one outlined here should include ample "hands-on training" under supervision. This kind of training program is useful for audiometric technicians who will be working in a physician's office, preferably under the supervision of an

---

I. Basic Science of the Ear
- A. Anatomy
- B. Physiology
- C. Examination
- D. Hearing loss
  1. Types
  2. Causes

II. Basic Physics of Sound
- A. Sound waves
- B. Measurement of sound
- C. The decibel
- D. Frequency and pitch
- E. Intensity and loudness
- F. Complex sounds and speech

Laboratory
- A. Study of ear models
- B. Physics of sound
- C. Physical examination of ear
- D. View relevant video or films

Audiometry

Equipment
- A. Development and history
- B. Types of audiometers
  1. Manual
  2. Self-recording
  3. Computerized
- C. Terminology
- D. Reference hearing thresholds
- E. Audiometer performance check
- F. Calibration
- G. Record keeping
- H. Testing environment
- I. Ambient noise levels

III. The Audiogram
- A. Definition and terms
- B. Reference hearing levels
- C. Calibration
- D. Graphic representation and symbols
- E. Numerical representation
- F. Audiometric forms

IV. Audiometric Technique
- A. Basic concepts
- B. Subject instructions
- C. Demonstration of test procedures
- D. Routine audiometry
- E. Special situations
- F. Screening audiometry
- G. Errors of audiometry
- H. Pure-tone air conduction
- I. Pure-tone bone conduction
- J. Basic concepts of masking
- K. Speech reception threshold
- L. Speech discrimination

V. Impedance Technique
- A. Impedance audiometry
- B. Tympanometry
- C. Compliance
- D. Acoustic reflex
- E. Demonstration of test procedure

Laboratory

Practicum
- A. Pure-tone air and bone
- B. Speech reception threshold
- C. Speech discrimination
- D. Impedance and tympanometry
- E. Acoustic reflex
- F. Equipment check

Audiometric Interpretation

Interpretation
- A. Sensorineural hearing loss
- B. Conductive hearing loss
- C. Mixed hearing loss
- D. Differential diagnosis
- E. Functional vs. malingering
- F. Inability to diagnose from audiogram alone
- G. Technician's responsibilities and limitations

H. Need for physician's diagnosis

Special Hearing Tests

A. Recruitment
B. SISI
C. Békésy
D. Alternate binaural loudness balance
E. Tone decay testing
F. Central auditory testing
G. Pseudo-hypoacusis tests ("functional" hearing loss)
H. Electrocochleography
I. Evoked response audiometry
J. Acoustic emissions

Hearing Aids and Devices

Evaluation

A. Candidates
B. Types of aids available
C. Making an ear mold
D. Auditory rehabilitation
E. Auditory devices
F. Cochlear and hearing aid implantation
G. Hearing protectors

Other Considerations

Diagnostic problems

A. Tinnitus
B. Vertigo
C. Conductive hearing loss
D. Sensorineural hearing loss
E. Anatomical problems
   1. Ventilation tubes
   2. Collapsing canals
   3. Draining ears
   4. Others
F. Uncooperative patient

VI. Hearing Conservation Programs
   A. Federal regulations
   B. State and local regulations
   C. Worker's compensation
   D. Impairment vs. disability
   E. Reporting and record keeping

VII. Limitations of Training
   A. Masking
   B. Problem patients
   C. Unusual patterns of audiograms
   D. Referrals necessary
   E. Interpretation of results
   F. Hearing aids
   G. Special tests

otologist and audiologist. It is especially important that such technicians understand the great difference between their training and that of a certified audiologist. Technicians should have specific guidelines regarding appropriate patients to test, those who require a referral, and exactly which patients they are qualified to test with masking. The American Academy of Otolaryngology—Head and Neck Surgery has established a certificate program recently (but not a certification program). It is excellent and includes the important features listed above.

Training for occupational hearing conservationists whose duties are limited to the industrial setting can be slightly different. Industrial audiometric technicians and hearing conservationists must acquire considerable knowledge in order to generate reliable, valid results, and participate in an effective hearing conservation program. The required information can usually be obtained during a 2- or 3-day course, supplemented with reading materials. The Council for Accreditation in Occupational Hearing Conservation (CAOHC) has an excellent certification program for hearing conservationists in industry, which we endorse. If CAOHC training is not available, a similar curriculum should be utilized.

At present, as a minimum we recommend the following training curriculum for hearing conservationists:

Training Program for Occupational Hearing Conservationists	
Day One	
8:00	Introduction: Hearing Conservation in Noise
9:00	Physics of Hearing
9:15	The Audiometer and Audiometry—OSHA Requirements
10:30	Supervised Audiometric Testing

(Continued)

Training Program for Occupational Hearing Conservationists	
12:15	Review of Technique and Pitfalls
12:45	Supervised Audiometric Testing
1:45	Record Keeping and OSHA Requirements
3:00	Physics of Sound/Noise Analysis
4:00	Review—Questions and Answers
Day Two	
8:00	Physiology and Pathology of the Ear
9:00	The Audiometer/Calibration of OSHA Requirements
10:15	Audiogram Review
10:45	Supervised Audiometric Testing
12:30	Hearing Protection and OSHA Requirements
1:30	Use of Tuning Fork and Otoscope, Fitting of Hearing Protection
2:15	Federal and State Noise Regulations Worker's Compensation
3:15	Review—Questions and Answers
Day Three	
8:00	Self-Recorders, Microprocessors
9:00	Noise Analysis Continued
9:45	Written Examination
11:00	Review of Written Examination
11:30	Summary of Hearing Conservation Program Adjourn

8. METHODS OF TESTING HEARING

There are several methods of using pure tones to test hearing. The best technique and the one that we recommend for all physicians is the procedure for performing an individual audiogram described in Chapter 6. Some other methods also are useful, as discussed subsequently.

8.1. Screening Method

If a large number of people, such as in the armed forces and school systems, have to be tested in a short time and routine threshold audiometry would be too time-consuming and impractical, it may be necessary to use a screening method. In a screening test, the intensity of each frequency is set at a specific level, for example, 20 dB, and the individual is asked whether he/she hears each tone—yes or no. If the tones are heard, the test is passed; if not, the person may be scheduled for threshold testing. Such screening audiometry, using above-threshold levels, does not meet the proper requirements of medical practice or industrial medicine, nor do testing environments with excessive ambient noise that may alter test results.

8.2. Group Pure-Tone Testing

Another method of testing hearing is group pure-tone testing, in which many subjects are given the same tones at the same time through multiple earphones; then the subjects are asked to fill out certain coded forms establishing whether or not they hear the levels tested. Although this method may have some usefulness in the military and in certain school systems, it is recommended that all personnel tested in this manner be rechecked by threshold audiometry when the opportunity arises.

8.3. Self-Recording Audiometry

Self-recording audiometry with either individual pure tones or continuously changing tones is a modification of the standard manual technique. Self-recording audiometers are being used in increasing numbers, and often a hearing tester is not required to be in constant attendance. The sounds are presented to the subject in a standardized manner using electronic controls, and the subject records a written record of his auditory acuity by pressing and releasing the button in response to the presence or absence of the tone. In industries where compensation is an issue, the tester should be in attendance during most of the test to be certain the responses are accurate.

9. HOW TO SELECT THE TEST ENVIRONMENT

9.1. Criteria

Because it is difficult to hear very soft sounds in a noisy room and audiometers are calibrated on subjects tested in a quiet room, all testing of hearing should be done in a quiet room. If there is excessive ambient noise in the test room when hearing tests are done, the subject invariably has difficulty hearing the weaker threshold tones and gives invalid responses. Prefabricated "soundproof" rooms are desirable but expensive and generally are not purchased by general practitioners or even by all otologists. Some schools and industries have purchased such rooms, but often efforts are made to find a satisfactory test room without spending a great deal of money.

When selecting the location for testing hearing, the physician should bear in mind that the room should be close to his regular examining office. The test room should be as far as possible from ringing telephones, elevators, air-conditioning systems, air ducts, water pipes, drain pipes, and other sources of disturbing and extraneous noise. The ceiling should not transmit the noise of people walking on the floor above. The room should have as few windows as possible and only one door. In spite of these apparent limitations, a satisfactory test room can be found in almost every physician's office. If a physician does a great amount of hearing testing and is interested in optimal results, he/she should purchase a prefabricated hearing test booth.

9.2. Tests for Ambient Noise

9.2.1. Sound-Level Meter

Several methods can be used to determine whether or not a room is satisfactory for testing hearing. One of the best methods is to use a sound-level meter/octave band analyzer. The physician can obtain the services of a local engineer, particularly if there are large industries in the area, to do the measurements for him. Most major industries have properly trained personnel and equipment and are glad to cooperate with physicians. If the sound study shows that the ambient noise falls at or below the levels in Table 5.2, the room can be considered satisfactory for the test frequencies listed and for using earphones to test hearing.

Table 5.2 Maximum Allowable Sound–Pressure Levels—ANSI-S3.1, 1977 and 1991

Test tone frequency (Hz)	500	1000	2000	3000	4000	6000	8000
Octave band level	21.5	29.5	34.5	39.0	42.0	41.0	45.0
One-third octave band level	16.5	24.5	29.5	34.5	37.0	36.0	40.0

The readings on the sound-level meter should be taken several times during the day, especially when the outside environment is noisiest. In this way the physician can determine what is probably the best time to do hearing tests.

9.2.2. *A Comparison of Hearing Tests*

If a sound-level meter is not available, another good method can be used. In a very quiet test room, perhaps available at a local institution or university, the examiner should use his/her audiometer to test several subjects who have already been found to possess normal hearing ability. These same subjects should then be retested by the examiner in the room that he/she wants to use for hearing tests. If it can be determined that the hearing of these individuals is similar in both rooms, it may be assumed that the physician's room is suitable for obtaining satisfactory audiograms.

9.3. Reducing Ambient Noise

If there is a significant difference between the thresholds of the tests performed in the room in question and those in the very quiet room, and if other variables have been ruled out, the ambient noise level of the test room is probably *too high.* Some measures have to be taken to reduce it. Most of the time, the interference occurs in the low frequencies at 250 and 500 Hz. When comparing the thresholds, the physician should keep in mind that a 5-dB difference is within expected variation in audiograms and is not caused necessarily by the masking effect of the ambient noise. However, if it is >5 dB, something will have to be done to quiet the room before it can be used. The following steps can be taken. Make sure there is a tight-fitting, solid door or, better yet, a double door at the entrance to the room. Put a soft rug on the floor and perhaps acoustic tile on the ceiling and walls. Hang drapes on the windows. Signs reading "Quiet, please" also should be posted.

9.4. Furniture and Regulations

The test room should be furnished as simply as possible with a table, a chair, and lighting fixtures that do not produce a loud hum. If air-conditioning units, electric fans, or telephones are in the room, shutoff switches should be easily accessible, so that they may be silenced during the testing period. Smoking should not be permitted in the test room because of ventilation problems.

9.5. The Audiometric Booth

All audiometric testing in industry should be done in an audiometric booth that meets standards specified in the Occupational, Safety and Health Act (OSHA) Hearing Conservation Amendment. (These levels are less restrictive than those shown in Table 5.2.)

There are several good commercially manufactured booths available. Major differences between them include the placement of the doors, view-ports, ventilation systems, and, of course, sound reduction capabilities.

They should be noiseproof enough to bring the ambient noise down to acceptable testing levels in accordance with OSHA specifications.

Windows and earphone jack plugs are situated so that the instrument can be placed and the tester can be seated so as to see the patient being tested without the patient being able to see the operation of the audiometer.

The patient should not be required to sit in the closed booth any longer than necessary to minimize claustrophobia. Showing the subject how to open the door from the inside should help to ease apprehension. Instructions should be given to the seated subject with the door wide open. Only when the actual testing is under way should the door be closed.

If it becomes necessary to interrupt the test or leave in the middle of a test to attend to a more urgent matter, open the door, remove the earphones, and invite the subject to leave the booth until you are able to give him/her your attention again.

Although the use of a booth may be unfamiliar, usually the patient will not become nervous unless he/she senses insecurity or hesitancy on the part of the tester. Practice is necessary to ensure confidence of the patient and accuracy of the procedure.

Loud sounds and talking in the immediate area of the booth will disturb the subject. Stop the test during any such disturbance. Sometimes low-flying aircraft or street traffic cause this problem, and the examiner should be familiar with the attenuation qualities of the booth.

9.6. Summary

1. Booths will be used in all industrial audiometric testing.
2. The booth must lessen ambient noise but does not completely deaden the spoken voice or factory rumble.
3. Once inside the booth, the patient's test should proceed with dispatch.
4. Practice and familiarity of the instrument and booth will result in very accurate test results.
5. Noise levels inside the booth should be within ANSI specifications.
6. Periodic checks on booth noise levels should be made, especially if repair work has been done on the booth (leaky seals), or if production noise levels increase near the test area.
7. All locations will prepare a designated area to conduct audiometric tests that must meet prevailing standards. If existing facilities with or without modifications are not available, an approved sound booth must be installed.

10. THE KIND OF AUDIOMETER TO PURCHASE

An audiometer is a precision instrument that produces pure tones of known intensity. The instrument is very delicate and must be handled carefully, with particular attention to the earphones. Although standard specifications for approved audiometers are established by ANSI, audiometers differ in many features. These differences help to determine which audiometer is best suited for a particular use. For the general practitioner, the pediatrician, or the director of a school testing program, it is advisable that the audiometer be as simple as possible, as it is likely that only air and perhaps bone conduction tests are to be done. If bone conduction audiometry is to be performed, it is essential that masking of known intensity be available. In industry, bone conduction rarely is done, and only air conduction aspects of the audiometer are of importance. Most commercial audiometers are now of reasonably high quality. Certain features are essential to all audiometers:

1. They should meet ANSI S3.6–1989 standards.
2. They should be easy to operate and should have as few complicating and intricate extras as possible.

3. They should be able to test at least the following frequencies: 500, 1000, 2000, 3000, 4000, 6000, and 8000 Hz.
4. Unless the physician intends to use more complicated tests, accessories such as speech testing apparatus, dual channels, and various others should not influence the purchase of the equipment.
5. The tone presenter switch should not click when it is operated.
6. The audiometer should be purchased from a supplier who will guarantee prompt attention in case of malfunction and supply a substitute calibrated audiometer if the instrument has to be removed for repairs. This last feature is of utmost importance; otherwise a physician or facility may be without an audiometer for long periods of time.

11. THE MANUAL AUDIOMETER

A manual audiometer has a series of switches and controls to direct its operation (Fig. 5.4). The on–off switch controls the power on the audiometer. This should be turned on at least 15 min before use, especially in older audiometers, and, if possible, it should be left on all day rather than turned on and off as the need dictates. A frequency selector dial designates the tone that is produced in the earphones.

The attenuator determines the intensity of the tone produced. Attenuators usually are calibrated in 5-dB steps from 0 to 110 dB, with the exception of the very low and high frequencies, for which the maximum usually is ~90 dB. Readings should not be made between the 5-dB steps. As the maximum output at 250 and 8000 Hz generally is not

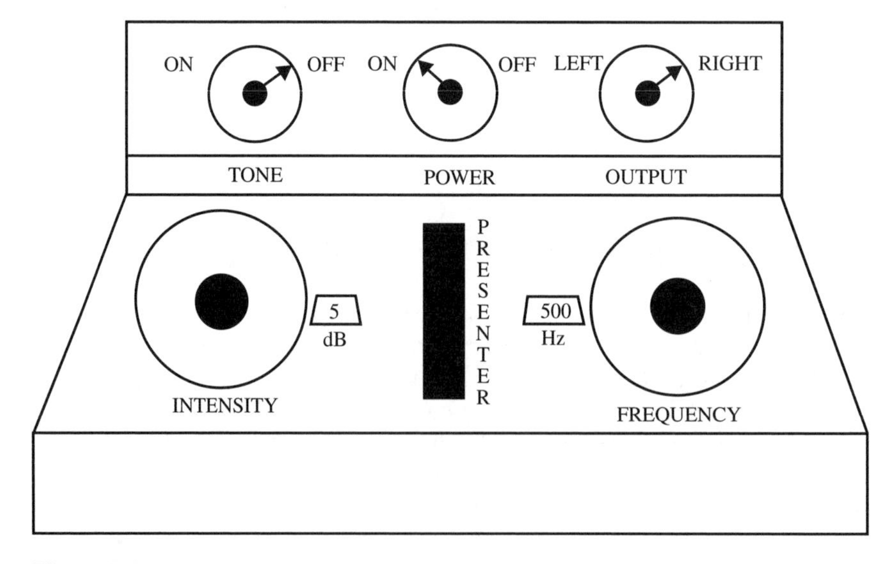

Figure 5.4 Main switches and dials of a typical clinical screening audiometer. Included are the frequency dial, which controls the tonal output, in this case set at 500 Hz; the attenuator, which controls the volume output in decibels, here set at 5 dB; the earphone selector switch; and the tone presenter, which turns the signal on and off. The presenter is left in the "off" position and pressed to the "on" position when the tone is to be presented. Modern sophisticated diagnostic audiometers have many additional switches, functions, and designs, but the basic principles remain unchanged.

>90 dB, one should bear in mind while testing that exceeding 90 dB on the dial does not increase the intensity of the tone produced. The tone presenter switch is used to turn the tone on and off. The tone always should be off, unless the tester wishes to turn it on for testing purposes.

Two wires leading from the audiometer are attached to earphones connected by a spring headband. The earphones should be handled with extreme caution because they are very delicate and can easily be thrown out of adjustment. They are part of a mechanism that converts electric current into sound. The earphones are equipped with a rubber cushion that is of considerable importance. It must be of a specified size so that the volume of air it encloses is precisely the same as that provided when the instrument was calibrated (~6 cc). The cushion cannot be replaced by a larger or smaller one for reasons of comfort without disturbing the calibration of the instrument. If a type of cushion is purchased to provide better attenuation, the audiometer then must be calibrated with the cushion in place.

Some audiometers are accompanied by a push-button cord that allows the patient to signal his response by pressing the button. When the subject hears the tone, he/she presses the button, causing a light to appear on the instrument panelboard. This is one way of getting a response from the subject. Many experienced testers prefer to have the patient raise a finger or a hand.

12. THE MICROPROCESSOR AUDIOMETER

The millions of audiograms being performed in industry because of OSHA and worker's compensation are stimulating development to make audiometry more reliable and data handling more efficient. Noteworthy is the microprocessor audiometer, which is useful especially for larger plants and corporations.

It would take a great deal of time, effort, and money to record and recall by hand the information that can be programmed into a microprocessor for ready use at little cost. The instrument is extremely helpful for plants performing large numbers of audiograms. It is also an efficient way of obtaining audiometric data and otologic histories and sending them in print or on a disk to a consultant otologist for diagnosis.

The microprocessor generally consists of a control panel with a table of functions, self-explanatory operation keys, backlit LEDs to read out complete test status as it occurs, a quiet, high-speed electrostatic printer, and a computer programmed for data management. The audiogram can be performed manually, semiautomatically, or automatically. Information questions can be inserted, deleted, or modified. Automatic calculation of standard or significant threshold shift with or without age correction can be programmed along with any other useful information. Calibration is performed readily and documented. Automatic validity tests are generally standard in each testing sequence. In the best instruments, the computer can be used for other purposes when it is not used for hearing testing. Additional copies of printed data can be reprinted from stored data. The data can also be transferred to a central computer or an attached microcomputer. Figure 5.5 shows a sample printout obtained with a microprocessor.

The efficiency of data management with a microcomputer depends on the software. Many of the most commonly used instruments employ a software program developed by HCNC (Hearing Conservation Noise Control, Inc.) in conjunction with the E. I. DuPont de Nemours & Company, Inc., whose hearing conservation program has been in operation for >50 years. The HCNC Computerized Audiometric Testing Program is a combined data storage and evaluation program which runs on a personal computer. This program

```
TRACOR INSTRUMENTS
  AUSTIN, TEXAS
DATE  00 00 00
TIME     00:00

SUBJECT:

X.................

SS#/ID#   000000000
JOB#   000000000000
NOISE EXP.   000000
TEST TYPE   0
TYPE PROTECTOR   0
BIRTH DATE 00 00 00
SEX   F

CURRENT AUDIOGRAM
FREQ.    L/DB  R/DB
1KHZ TEST 50   50
 500HZ     60   50
1000HZ     50   50
2000HZ     30   30
3000HZ     30   30
4000HZ     50   30
6000HZ     30   30
8000HZ     30   30
AV 234     36   30

MODE       PULSED

RA400  SER#.
VERSION   3.4
CAL. ANSI 1969 STD

CAL. DATE  00/00
EXAMINER ID#
000000000

X.................

LOCATION CODE
000000
```

Figure 5.5 Typical computerized audiogram.

interfaces with audiometers from MONITOR™, TREMETRICS™, and MAICO™ and provides:

A printed copy of the audiogram for permanent records.
An immediate evaluation of the hearing test for the occurrence of OSHA Standard Threshold Shift (STS).
Data storage of employee information, hearing thresholds, and otoscopic/otologic examination histories.

The HCNC Computerized Audiometric Testing Program is a state-of-the-art, menu-driven program that can save time and eliminate tedious filling out of audiogram forms. Data are transmitted for professional evaluation by the periodic mailing of a diskette. The information on the diskette is read and evaluated, with results and recommendations returned on printed forms for permanent records. This program is now installed in chemical, electronics, steel, pharmaceutical, transportation, pulp and paper, and other manufacturing industries. The continued success of the program, first installed in 1984, stems from its constant evaluation and enhancement with regard to the current needs of industry.

12.1. Calibration Procedure

If a change in an audiometer's operating characteristics is sudden and extensive, it is obvious that the instrument should be serviced. However, a slow change may not be obvious. If changes in instrument accuracy go undetected, poor measurements may be made over many weeks or months before a calibration check discloses the inaccuracy. Wasted measurements can be prevented by simple daily checks designed to detect changes in an instrument's operating characteristics and potential trouble spots.

The following tests and inspection should be made by the technician at the beginning of each day:

1. All control knobs on the audiometer should be checked to be sure that they are tight on their shafts and not misaligned.
2. Earphone cords should be straightened so that there are no sharp bends or knots. If the cords are worn or cracked, they should be replaced. A recalibration is not necessary when earphone cords are replaced.
3. Earphone cushions should be replaced if they are not resilient or if cracks, bubbles, or crevices develop. A recalibration should not be necessary when earphone cushions are replaced.
4. The audiometer calibration should be checked by measuring the hearing threshold at each test frequency of a person who has normal hearing and whose hearing levels are well known (biological calibration). If a persistent change of >10 dB occurs from day to day in this person's hearing threshold at any test frequency, and it cannot be explained by temporary threshold shifts caused by colds, noise exposure, or other factors, an instrument recalibration is indicated. A technician who has normal hearing may serve as the test subject. At the beginning and end of each day of testing, these threshold levels should be recorded serially in ink, with no erasures, if the records are to have legal significance. Any mistakes made in these record entries should be crossed once with ink, initialed, and dated.
5. The linearity of the hearing level control should be checked, with the tone control set on 2000 Hz, by listening to the earphones while slowly increasing

the hearing level from threshold. Each 5-dB step should produce a small, but noticeable increase in level without changes in tone quality or audible extraneous noise.

6. Test the earphone cords electrically, with the dials set at 2000 Hz and 60 dB, by listening to the earphones while bending the cords along their length. Any scratching noise, intermittency, or change in test tone indicates a need for new cords.
7. Test the operation of the tone presented, with dials set at 2000 Hz and 60 dB, by listening to the earphones and operating the presenter several times. Neither audible noise, such as clicks or scratches, nor changes in test tone quality should be heard when the tone presenter is used.
8. Check extraneous noises from the case and the earphone not in use, with the hearing level control set at 60 dB and the test earphone jack disconnected from the amplifier. No audible noise should be heard while wearing the earphones when the tone control is switched to each test tone.
9. Check the headband tension by observing the distance between the inner surfaces of the earphone cushions when the headset is held in a free, unmounted condition. At the center of its adjustment range, the distance between cushions should be ~0.5 in. The band may be bent to reach this adjustment.

When a tester observes that audiograms indicate a persistent, unexplainable hearing impairment at either all or specific frequencies, he/she should stop testing patients and test several ears known to be normal, in order to confirm that the instrument is calibrated properly.

After testing for the day has been completed, testers should recheck the instrument on themselves and on several normal ears and then record all results. This final check is necessary because the tester may not be aware that at some time during the day the audiometer had gone out of adjustment. By establishing the calibration of the audiometer at the end of the day, the tester will know whether there is a need to retest any subjects, and he/she also will have an important confirmation for medico-legal purposes if ever required.

When a tester uses his/her own or other normal ears to verify the calibration of an audiometer, he/she is testing only the calibration at threshold and also is assuming that the attenuator is working properly and producing accurate readings at above-threshold levels. Electronic instruments are necessary to perform above-threshold and other calibration measurements.

When an instrument's accuracy is suspect, the possibility that the tester may not be using the instrument properly should not be overlooked. For this reason, it is always wise to discuss the problem with some person who is very familiar with the operation of the instrument to determine whether a simple solution is available.

12.2. Record of Calibration Testing

It is important for testers to keep a daily record of their biological calibration testing, indicating the exact time and manner in which the audiometer in use was checked for calibration. Electronic calibration records also should be maintained. This is an important precaution, particularly in medico-legal cases.

12.3. Audiometer Repairs

When an audiometer's accuracy is suspect, it should be serviced and calibrated by a qualified laboratory. Preferably, the instrument should be hand-carried to and from the laboratory because rough handling encountered in normal shipping procedures may change its operating characteristics after calibration.

Because there are no certified facilities for the calibration of audiometers, it may be difficult to find competent laboratories nearby. Even the audiometer distributor may not be competent in repair or calibration procedures. When a service and calibration facility is located, some understanding should be reached about the kind of calibration to be performed. Determine that hearing level accuracy will be checked for each test tone at each 5-dB interval throughout the operating range. For industrial applications the range should cover hearing levels from 10 to 70 dB re ANSI S.6–1989 and at least test tones of 500, 1000, 2000, 3000, 4000, 6000, and 8000 Hz. Calibration specifications include tolerance limits on attenuator linearity, test tone accuracy and purity, tone presenter operation, masking noise, and the effects of power supply variations. It is always good practice to require a written report that includes all measurement data. A simple statement that the audiometer meets ANSI specifications should not be accepted without some evidence that all tests have been made. Electronic-acoustical calibration should be conducted routinely on an annual basis.

If the instrument must be shipped away for service or calibration, it may be advisable to have another instrument on hand. Another solution may be to borrow an instrument during the repair period. If a second instrument is used, its calibration must be established carefully and a note must be made on each audiogram stating the change of instruments.

Accuracy of the instrument should not be taken for granted following a factory adjustment, particularly after shipment. The instrument always should be checked subjectively as described earlier.

REFERENCE

1. Davis H, Silverman SR, eds. Hearing and Deafness, revised edition. New York: Holt, 1960.

6
The Audiogram

Robert T. Sataloff
Drexel University College of Medicine, American Institute for Voice and Ear Research, and Graduate Hospital, Philadelphia, Pennsylvania, USA

Joseph Sataloff
Thomas Jefferson University, Philadelphia, Pennsylvania, USA

In this chapter, a more detailed discussion of audiometry and the basic information needed to interpret an audiogram are presented. What do the numbers mean? What is 0-dB, or normal, hearing, and what is a −5 dB-hearing level? What is a high-tone or a low-tone loss?

1. DEFINITION

An audiogram is a written record of a person's hearing level measured with specific pure tones. The pure tones generally used are the frequencies 250, 500, 1000, 2000, 3000, 4000, 6000, and 8000 Hz; these tones are generated electronically by the audiometer. This frequency range includes the speech spectrum.

2. TERMS

If 0 dB represents ideal normal hearing, a 60-dB hearing threshold level also can be called a 60-dB "hearing loss." Although both terms describe the same condition, the term "hearing level" is currently more popular in otology because it emphasizes the hearing that the patient still has left rather than the hearing that he has lost. Also, because a −5-dB level is a gain rather than a loss among persons with normal hearing, the "level" avoids the confusion of "negative losses" and contributes to a more positive approach in helping patients with deficient hearing.

However, because the term "hearing loss" still is in common use and this book is concerned with the diagnosis of hearing loss, the words "loss" and "level" will be used interchangeably to indicate the threshold of a person's hearing.

3. REFERENCE HEARING LEVEL

The original American Standards Association (ASA 1951) 0-dB reference level was established from data obtained between 1935 and 1936. Newer data obtained some 30 years later indicated that the human ear was ~10 dB more sensitive. This information led to the International Standards Organization (ISO 1964) reference level, which was later adopted by the American National Standards Institute (formerly ASA), now known as ANSI 1989. The actual differences are shown in Chapter 5. Because many otologists may still be using ASA-calibrated audiometers and thousands of old audiograms are based on the old reference level, graphic audiograms throughout this book will show both references—ASA on the left and ANSI on the right. Most hearing levels recorded in numerals and serial form will be on the ANSI reference level.

4. HORIZONTAL AND VERTICAL VARIABLES IN A GRAPHIC AUDIOGRAM

Ideally, one should measure a patient's ability to hear speech, but because of several difficulties, pure tones are still used. For simplicity, only specific frequencies were selected for routine use. They are called *octave frequencies* because each successive tone is an octave above the one immediately below it, and the number of cycles per second from one tone to the next is doubled. Octave frequencies constitute the horizontal variable in an audiogram: how well the patient hears at each frequency. Does he/she hear the tone as well as a person with normal hearing, or does it have to be made louder to be heard; if so, how much louder? To make this comparison, one must have a normal baseline for each frequency. There are certain shortcomings inherent in any system that relies on sampling of this sort. In order to improve the tests, frequencies between the usual octaves are also often tested, particularly 3000 and 6000 Hz. Frequencies above the usual test range may also be useful, including 10,000, 12,000 Hz, and sometimes higher frequencies.

5. WHAT IS NORMAL HEARING?

The threshold at the various frequencies of a person with normal hearing ability originally was obtained by testing a large number of young people between the ages of 20 and 29 and determining the intensities of the thresholds at the various frequencies. It was found that the thresholds fluctuated over an ~25-dB range even in subjects with assumably normal hearing. An average was reached at each frequency tested and defined as 0 dB of hearing. Some subjects heard better than 0; many of them heard tones as weak as −10; others did not hear a sound until it was amplified to a level between 0 and 15 dB. This variation indicated that the range between −10 and +15 dB could be considered normal for the average young person (ASA). The range for the ANSI scale is 0–25 dB. In clinical practice, however, a patient with a 15-dB hearing level in most frequencies may be considered as having a hearing loss.

Average normal hearing, or 0 dB, is the reference level on the audiometer. A hearing loss at some specific frequency is expressed and recorded as the number of decibels by which a tone must be amplified for a patient to hear it.

6. WHAT THE AUDIOMETER MEASURES

Commercial audiometers are calibrated and recording methods are standardized in such a way that what is recorded is not a patient's ability to hear but, rather, his/her hearing *loss* in the frequencies tested. If he/she can hear 0 dB, he/she has no hearing loss, but if he/she cannot hear until the tone is 30 dB louder than 0 dB, he/she has a hearing loss of 30 dB. The pure-tone audiometer offers the best means yet devised for routine hearing measurement.

7. FORMS OF THE AUDIOGRAM

The audiogram, which shows a patient's hearing threshold for the standard range of frequencies, can be recorded in several ways. The most common is a *graph* on which the frequencies are marked off from left to right, and the tone intensities range up and down. A statement that O is for the right ear and X for the left should appear at the bottom of each graph. Figure 6.1 shows such a graph and indicates the conventional marks by which the curve for the left ear can be distinguished from that for the right ear. A series of X marks connected by a dashed line denotes the left; circles connected by a solid line represent the right ear. The short arrows found on some audiograms hereafter indicate that there was no response to the test tone at the output limits of the audiometer.

One of the chief disadvantages of using this type of graph in industry and otology is that if 8 or 10 audiograms are done in a year on a certain subject, the record becomes bulky, and it becomes difficult to compare one curve with another performed on a different date.

Figure 6.2 shows another form of recording these thresholds in which the intensities are recorded numerically and serially instead of being plotted. This form, which is more acceptable and practical, is recommended for industry, schools, and otologic practice. It is called a serial form. The authors routinely use this form in their practice, but because the graph form may be familiar to some readers, it is used often in this book. In a serial audiogram, instead of using a symbol for the right and left ears, the number of decibels designating the threshold is recorded at each frequency. The notation NR on the serial form denotes that there was "no response" to the test tone at the output limits of the audiometer. The letters WN indicate that "white noise" was used for masking. Furthermore, a place for comments and a brief history is available on this type of serial audiogram.

Serial audiograms make it easier to record all thresholds obtained independently. For example, in the routine retesting of 1000 Hz, both thresholds, even if alike, should be recorded in the space provided. In legal situations, these multiple numbers will confirm that the threshold was rechecked several times. It is necessary to record *every*

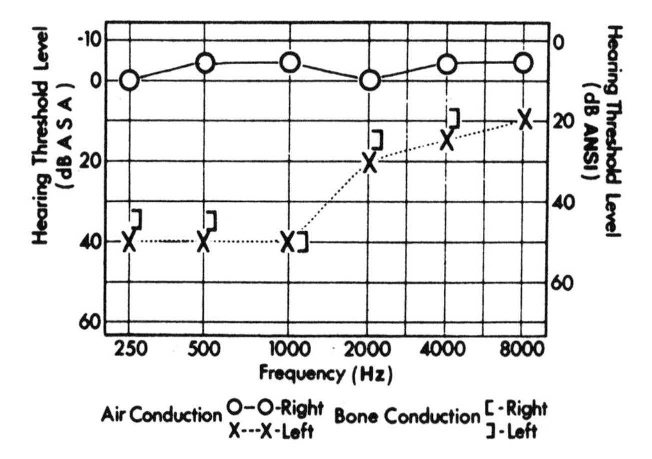

Figure 6.1 Audiometric findings: Right-ear thresholds for air and bone conduction pure tones are normal. Left ear has reduced air and bone thresholds of about the same magnitude (no air–bone gap). Right ear was masked during all testing of the left ear. Speech reception threshold: right, 5 dB; left, 45 dB. Discrimination score: right, 98%; left, 62%. Tuning fork lateralizes to right ear. Tuning fork on left shows air is better than bone conduction (A > B), and bone conduction is reduced on the left mastoid. In general, brackets ([and]) are used to indicate masked bone conduction. Unmasked right ear is symbolized with a O, unmasked left ear with an ×.

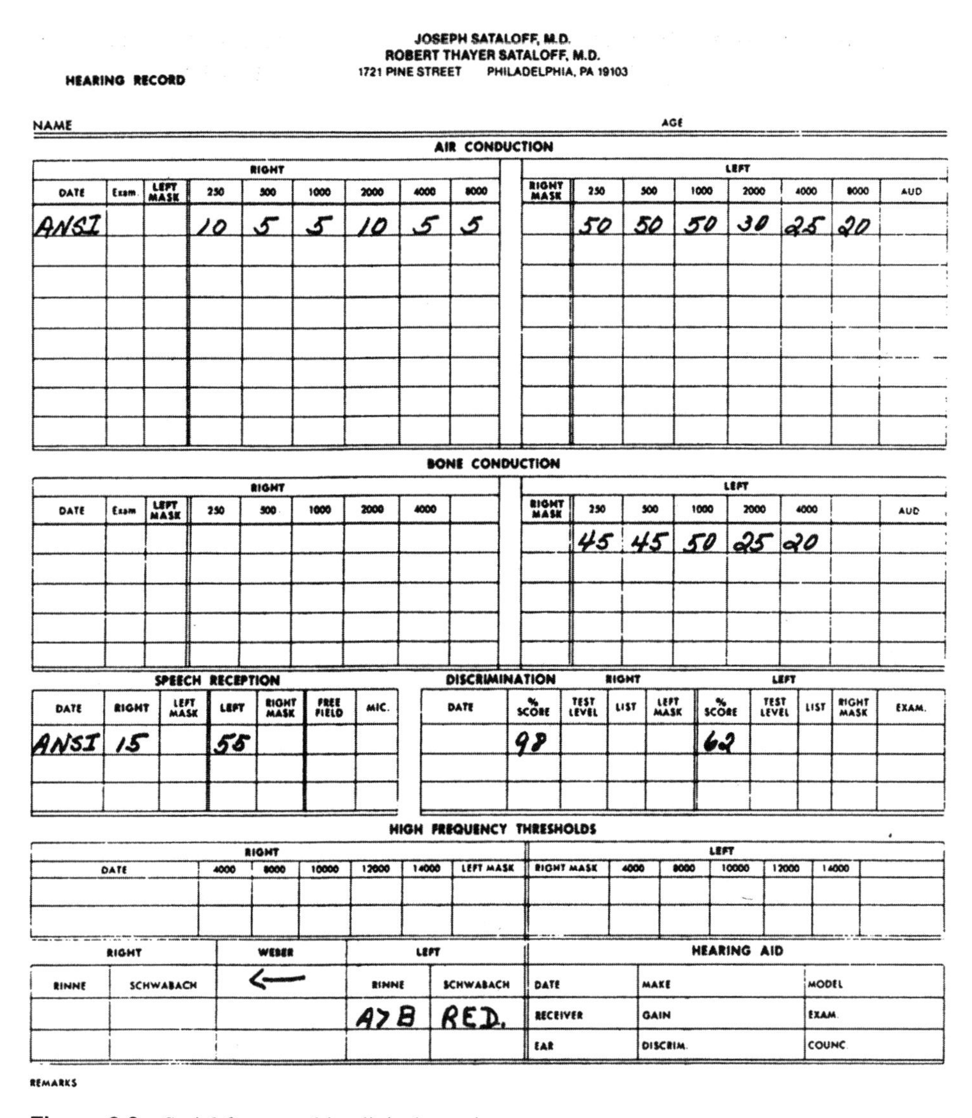

JOSEPH SATALOFF, M.D.
ROBERT THAYER SATALOFF, M.D.
1721 PINE STREET PHILADELPHIA, PA 19103

HEARING RECORD

NAME AGE

AIR CONDUCTION

			RIGHT							LEFT						
DATE	Exam	LEFT MASK	250	500	1000	2000	4000	8000	RIGHT MASK	250	500	1000	2000	4000	8000	AUD
ANSI			10	5	5	10	5	5		50	50	50	30	25	20	

BONE CONDUCTION

			RIGHT							LEFT						
DATE	Exam	LEFT MASK	250	500	1000	2000	4000		RIGHT MASK	250	500	1000	2000	4000		AUD
										45	45	50	25	20		

SPEECH RECEPTION

DATE	RIGHT	LEFT MASK	LEFT	RIGHT MASK	FREE FIELD	MIC.
ANSI	15		55			

DISCRIMINATION

	RIGHT				LEFT				
DATE	% SCORE	TEST LEVEL	LIST	LEFT MASK	% SCORE	TEST LEVEL	LIST	RIGHT MASK	EXAM.
	98				62				

HIGH FREQUENCY THRESHOLDS

	RIGHT							LEFT					
DATE	4000	8000	10000	12000	14000	LEFT MASK	RIGHT MASK	4000	8000	10000	12000	14000	

RIGHT		WEBER	LEFT	
RINNE	SCHWABACH	←	RINNE	SCHWABACH
			A>B	RED.

HEARING AID

DATE	MAKE	MODEL
RECEIVER	GAIN	EXAM.
EAR	DISCRIM.	COUNC.

REMARKS

Figure 6.2 Serial form used in clinical practice.

threshold that is derived independently, even if there is marked variation, for this may have considerable significance. It is important that every serial audiogram include the date and the signature of the tester as well as other information that can be placed under Comments.

8. INTERPRETING THE TYPICAL AUDIOGRAM

Now let us look at a typical audiogram and interpret it.

In Fig. 6.1 the left ear (X's connected with dashed lines) shows a level of ~40 dB (ASA) up to 1000 Hz; then the curve approaches more normal hearing (the 0 line on the graph). We would say, then, that this patient hears high tones better than low tones. This is an ascending curve, and the patient is said to have a low-tone or low-frequency hearing loss. Hearing in the right ear is normal.

9. SOME BASIC CONCEPTS

The reader is referred to Chapter 7 for information on special studies, including bone conduction, adaptation, recruitment, and speech discrimination. However, because these concepts are discussed in the earlier chapters on classification of hearing loss, it is necessary to define them briefly here.

9.1. Air Conduction

This denotes the ability of the ear to receive and conduct sound waves entering the external ear canal. Normally, these waves cause the eardrum to vibrate, and the vibrations are transmitted through the chain of ossicles to the oval window. When air conduction is impaired as a result of damage to the outer or middle ear and the sensorineural mechanism of the inner ear is intact, the maximum difference between air and bone conduction thresholds is ~60–70 dB. This is so because when the sound is >60–70 dB, it will be conducted by the bones of the skull directly to the cochlea.

9.2. Bone Conduction

To some extent this is a measure of the patient's ability to hear sound vibrations that are transmitted directly to the cochlea through the bones of the skull, bypassing the outer and the middle ear. Bone conduction is unimpaired in simple conductive hearing loss. Thus, conductive hearing loss can be distinguished from sensorineural hearing loss by tuning forks or bone conduction audiometry.

Tuning-fork tests, reviewed in Chapter 5, always should be done to confirm the audiometric findings. When inconsistencies occur, the tuning-fork test often turns out to be correct.

Although sounds up to ~50 dB directed to one ear by air conduction through an earphone usually are heard by the ear alone, this is not the case with bone conduction. Bone-conducted sounds are heard almost equally well by both ears no matter where the vibrations are impressed upon the skull. This holds true of both the tuning fork and the audiometer vibrator. The proper way to minimize confusion is to mask the opposite ear by introducing enough neutral sound into it to occupy its auditory pathway and prevent the test from reaching it.

A common method of recording bone conduction by a graph type of audiogram is shown in Fig. 6.1. The open cusps, or brackets, are used to symbolize the ear as it faces the examiner. A caricature of a face showing that "]" is the left ear and "[" the right should clarify this concept (Fig. 6.3). Bone conduction also can be recorded numerically, as in the serial type of audiogram (Fig. 6.2).

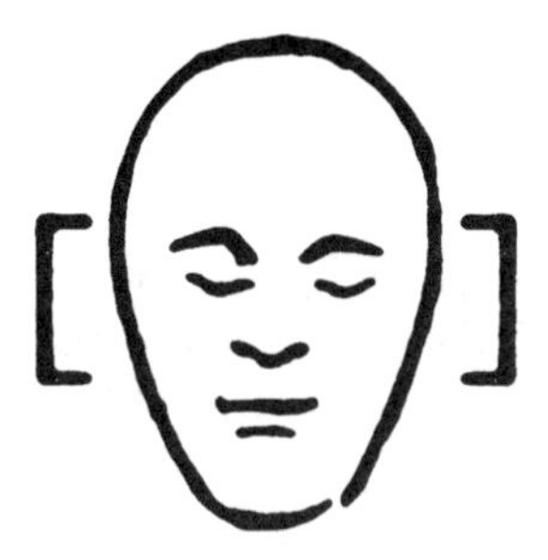

Figure 6.3 Brackets are used in graphic audiograms to indicate left and right ears. The correct side may be remembered if they are thought of as earmuffs.

When the bone conduction and the air conduction curves or levels are of the same magnitude, there is *no air–bone gap*. But if the bone conduction level is better (i.e., shows less hearing loss and is closer to the normal hearing level), an *air–bone gap* is said to exist.

9.3. Speech Reception Threshold

This is a measure of a person's ability to hear speech, not pure tones, using a speech audiometer that controls the intensity of the speech output. One can test the speech reception threshold (SRT) by means of simple two-syllable words or sentences to determine the weakest intensity at which the subject can hear well enough to repeat the spoken words or the sentences. A person who hears normally can hear and repeat these words at a level of ~15 dB. For hard-of-hearing individuals, the SRT is higher (i.e., the speech has to be louder to enable them to repeat it). The higher the number of decibels, the greater the hearing loss.

9.4. Discrimination Score

This does not measure the weakest intensity at which the patient hears speech sounds but, rather, how well he/she can repeat correctly certain representative words delivered to the ear at ~30 or 40 dB above the individual's SRT. The person with normal hearing discriminates between 90% and 100% of the words. Patients with sensorineural losses may have moderate and sometimes severe discrimination losses, although they also may be normal.

9.5. Recruitment

To a patient with recruitment, compared with someone without it, a tone that sounds soft becomes loud much more suddenly and rapidly when its intensity is increased. This abnormally great and abrupt increase in the sensation of loudness, especially marked in patients with sensory hearing loss, generally is absent in patients with conductive and neural hearing loss.

9.6. Abnormal Tone Decay or Pathological Adaptation

This finding occurs predominantly in neural hearing loss. A patient who exhibits abnormal tone decay is unable to continue hearing a tone at threshold when it is prolonged at a uniform level of intensity; his hearing fatigues rapidly. The phenomenon is called pathological fatigue or abnormal tone decay.

Someone who has normal hearing continues to hear a very weak threshold tone for several minutes, but an individual with abnormal tone decay may hear the sound only for several seconds. Then, the sound is made louder. When this is done, the patient will hear the sound again for a few seconds, only to lose it again quickly and request that the volume be increased and so on.

10. HOW TO PERFORM A ROUTINE AUDIOGRAM

It has been demonstrated repeatedly that, although hearing testers without adequate training may have performed many hundreds of audiograms and may consider themselves to be experts, many of them are making mistakes of which they are unaware, and they are producing test results that often are inaccurate.

Although audiometric testing may appear to be disarmingly simple, by no means is it easy to obtain accurate thresholds consistently. Because reliable and valid audiograms are of such great diagnostic importance, such valuable guides to therapy, and so decisive in medico-legal cases, it is essential that they be performed well.

11. PREPARING THE SUBJECT

Before starting the audiogram the tester should consider the following preliminary steps:

1. Seating the subject
2. Instructing the subject
3. Placing the earphones on the subject

11.1. Seating the Subject

If the tester and the subject are in the same room, the subject should be seated in a comfortable, squeakless chair so that his/her hands can rest on the arms of the chair or on the far end of the tester's table. The subject's profile should be turned toward the tester because the tester must be able to observe the subject's hands, face, and head, but the subject must not be able to observe the tester's hands and arms or the control panel of the audiometer.

Having the subject close his/her eyes while listening may help the patient to concentrate more on threshold sounds and prevents him/her from receiving visual clues from the examiner's movements.

Although some technicians advocate seating the subject with his/her back to the tester, the authors feel this is unsatisfactory, even if the push button is used to register the response. This arrangement hinders the tester from numerous indications that help establish an opinion as to the validity of the thresholds and the cooperation of the subject, including the subject's possible malingering.

11.2. Instructing the Subject

Proper, concise instructions directed to the subject are critical for obtaining reliable responses. A successful method of instruction is as follows:

> "Have you had your hearing checked before?"
> "Yes."
> "That's fine—let me remind you of what we're going to do. You will be listening for some tones. Each time you hear a tone, raise your finger. [Demonstrate how you want this done.] When the tone goes away, lower your finger. [Demonstrate how you want this done.] No matter how faint the tone, raise your finger when you hear it. [Demonstrate how you want this done.] And lower it when the tone goes away. [Demonstrate how you want this done.] Do you hear better in one ear than the other?"
> "No."
> "Then we will check your right ear first." [If there is a difference in hearing, check the better ear first.]

11.3. Placing the Earphones

It is very important to place the receivers snugly on the subject's ears so that no leakage exists between them and the sides of the head. The headband should be adjusted to head size so that the phones are comfortable. The tester should ask the subject to remove earrings and push away hair covering the ears. Items such as glasses, hearing aids, and cotton also should be removed, because they can prevent the receiver from fitting snugly or can block out the test tones. The phone should not bend the ear over; the center of the phone should be directly aligned with the opening of the ear canal.

Make sure that the cord leading from the earphones is not draped over the front of the subject. The subject's movements possibly could rub the cord and introduce distracting noises into the earphones.

When the tester places the earphones, the red phone must be placed on the right ear.

The earphones should not be placed on the subject until directions have been given and the tester is ready to proceed with the production of the audiogram. Also, it is preferable to remove eyeglasses of hearing aids after the instructions have been given. When the earphones are in place, start the test immediately. Do not make the subject wait while you fill in the date, signature, serial number, and other information.

12. THE AUDIOGRAM

The tester is to determine the subject's threshold at the specific frequencies in the following order: 1000, 2000, 3000, 4000, 6000, 8000, 1000 (repeat), 500, and 250 Hz. The 1000-Hz tone is tested first because usually this is the easiest one for which to establish a definitive threshold. The threshold at 1000 Hz is confirmed by repeating it, to help confirm test accuracy, and because the subject who has not previously had an audiogram may not have recognized the tone as such the first time. Both readings are recorded on the form. Tests at higher frequencies such as 10,000, 12,000, and 14,000 Hz also may be performed. Findings at these frequencies may be helpful in differentiating occupational hearing loss from presbycusis, and in early damage detection, for example, from ototoxic drugs.

With the earphone selector switch properly set and the frequency dial at 1000 Hz, turn on the tone with the tone presenter, and roll the hearing level dial slowly upward from 0 dB until the subject responds. Stop the tone, allowing the subject to lower his/her finger. Present the tone once again at this response level to confirm the initial response. If the patient responds, turn off the tone, and *decrease* the intensity by 10 dB. Present the tone. Generally, there should be no response to this 10-dB reduction in intensity. If there is no response, turn the tone off, *increase* intensity 5 dB, and then present the tone. If there is a response to this 5-dB increase, turn off the tone and decrease by 5 dB. Present the tone. If no response occurs, turn the tone off, *increase* by 5 dB and present the tone. If there is a response, this is the threshold; record that number of last response. The objective is to get at least two "no" and two "yes" responses. For example: "yes," "no," "yes," "no," "yes," or "no," "yes," "no," "yes." Always end on a "yes" response.

Most young subjects will have extremely good hearing and will respond to the tone while the hearing level dial is still at 0 dB. It is good practice to obtain at least two or three "yes" responses to ascertain the threshold. Do not neglect to get a confirmation of the initial response, which was obtained during the rollup of the volume control dial.

If there is no confirmation of the initial response, turn on the tone again and continue to roll the volume control dial up until there is a response and a *confirmation.* Then, make the 10-dB reduction and proceed from that point. A "yes" response requires the tone to be made softer until a "no" response is obtained. A "no" response requires the tone to be made louder until a "yes" response is obtained.

All tones presented to the subject should be brief bursts of sound and should be held for no longer than 1 or 2 s.

When the threshold for 1000 Hz has been determined and recorded, the threshold for each succeeding frequency is determined in the same manner. After rechecking 1000 Hz, test 500 Hz and then 250 Hz. For good test–retest reliability at 1000 Hz, the threshold should not deviate more than (plus or minus) 10 dB when rechecked. However, a (plus or minus) 10-dB difference indicates to the technician that the first threshold may be invalid, and the same may be true for the other thresholds. When this difference is noted, several of the succeeding frequencies should be rechecked until reliability is obtained. Then, recheck and record the 1000-Hz threshold before testing 500 Hz. When the tester has completed the recording of thresholds for one ear, the earphone selector is then switched to the opposite ear, and the identical procedure is repeated, except that it is not necessary to retest 1000 Hz on the second ear. Recheck the thresholds for those frequencies indicating a loss >25 dB. The tester always should record all independently obtained thresholds on the audiogram.

12.1. Order of Test

First ear: 1000, 2000, 3000, 4000, 6000, 8000, 1000, 500, and 250 Hz. Second ear: 1000, 2000, 3000, 4000, 6000, 8000, 500, and 250 Hz.

13. RECORDING AUDIOGRAMS

The relative merits of *graph* and *serial* audiograms were discussed earlier. For industrial applications, we strongly recommend the use of serial audiograms.

13.1. Securing Objectivity

Because of the possibility that the tester, in an attempt to complete the test quickly, may be influenced by a previously obtained threshold, it is advisable to avoid looking at the preceding audiogram on the serial chart. For this reason, it is helpful to have an assistant record while the tester calls off the threshold obtained at each frequency. If a tester cannot find a satisfactory assistant for this purpose, he/she should place a card over every previously obtained audiogram so that he/she is not influenced in any way. A special mask that allows only the blank spaces on the audiogram to be seen also can be prepared. This type of self-restriction will assure the tester that he/she is performing as objective a test as possible.

13.2. Preserving Records

Original audiograms should not be destroyed, even if they are transcribed from one form to another. They are important written records of a subject's hearing. Recording of thresholds never should be erased when a repeat check finds the original threshold to differ from the new. Instead, all thresholds should be recorded. The difference may be significant, and this may have an important bearing on the interpretation of any hearing loss.

14. AVOIDING ERRORS IN AUDIOMETRY

In the performance of audiometry it is important to avoid certain errors encountered by testers who are not adequately acquainted with the limitations of this testing procedure.

1. When depressing the tone presenter switch, the tone should not be presented for >1–3 s. The tones should be *short bursts* rather than prolonged. Each tone should be presented for about the same length of time, except in situations where the tester may want to check the subject's response to the cessation as well as the onset of the tone.
2. Every audiogram should be done as *rapidly as possible* without sacrificing the reliability and the validity of the threshold testing. Taking too long to do an audiogram will fatigue the subject and result in inaccurate responses.
3. Rushing through the test too rapidly is as bad as wasting time. The tester should appreciate that some subjects take longer to respond than others. It is essential that the tester *allow sufficient time for the response.* Faster and more definitive responses often can be obtained from the subject if he/she is given concise and explicit directions prior to testing.
4. The tester should *always be certain that the subject is not directly or indirectly watching* the control panel of the audiometer and/or the tester.
5. Some testers tend to present the signal and then look up at the subject as if to ask if he/she has heard the tone; others may move a hand away from the audiometer after a dial change. Both actions constitute visual, rather than aural, clues and can elicit false–positive responses. *It is poor audiometric testing technique to signal your move.*
6. Another possible error is to place the wrong phone on the ear. Repeated checks should be made to see that the *phones are placed on the proper ear*, that they correspond with the switch on the control panel, and that the threshold is recorded for the ear being tested.
7. The tester should remember to recheck the threshold at 1000 Hz after the ear is tested for other frequencies, as the initial determination may not have been completely accurate.
8. If, during the testing of many subjects, significant hearing losses are found repeatedly in the same frequencies, it is wise for the tester to *recheck the earphones* on himself/herself to be sure that nothing has gone wrong during the testing procedures. Conversely, if all subjects seem to have normal hearing, the audiometer may be generating tones louder than the hearing level dial indicates.
9. At all times the tester must *avoid rhythmical presentation* of the signals.
10. Some subjects will complain, particularly after listening to very loud tones, that the tones continue to linger even after the signal itself has stopped. This so-called *aftertone* happens occasionally in certain ears and *must be taken into consideration.* More time and more careful determination of threshold are indicated for such subjects.
11. Occasionally, a tester will encounter a subject who has tinnitus, such as a ringing noise in his/her ears. When the threshold on certain frequencies is being obtained, such a subject may state that the "head noise" is confusing his/her responses. If a *threshold cannot be determined in a routine manner to the tester's satisfaction, several other methods are available.* One of these is to use several short, interrupted bursts of tone two or three times,

instead of the single tone generally presented in the routine audiogram, or to use a marble tone that can be produced by some audiometers. Sometimes this will enable a subject to respond more accurately. This change in technique and the fact that the subject complains of tinnitus should be noted on his/her record.

12. Sometimes a tester may encounter a subject whose responses are so varied that an accurate threshold cannot be obtained at the particular time. In this case, the test should be terminated and repeated on another day. It is *unsatisfactory for a tester to report vague general thresholds* on such a subject, especially because accurate responses may be attainable.
13. When recording the hearing losses at 250 and 8000 Hz, *one should not record higher levels than the maximum output of the audiometer.* The tester should be familiar with the limitations of the audiometer. These output limits are generally printed on the frequency dial.
14. When depressing the *tone presenter switch*, the tester must be particularly *careful not to press it down too hard or let it spring back too quickly*, otherwise it will make a click and may result in a subjective response to the click rather than to the pure tone presented.
15. *If a hearing loss is present*, particularly at frequencies of 1000 Hz or higher, *each threshold should be rechecked and recorded.*
16. All threshold readings should be recorded in *5-dB multiples.*

15. MASKING

When an individual wishes to test vision in only one eye, he/she merely closes the other eye or covers it with a patch to exclude it from the test. It is not as simple to test hearing in only one ear because sound waves, unlike light waves, travel in all directions and are not stopped easily by merely plugging the opposite ear. When a person hears normally in both ears, it is easy to test each ear separately, because he hears the very weak threshold tones in one ear long before they become loud enough to be heard in the opposite ear. However, when there is a difference in threshold between the two ears, the test tone intended for one ear could be heard by the other ear instead. This phenomenon that prevents correct measurement of the ear being tested is called shadow hearing, shadow response, crossover, and cross hearing. The only way to keep the other ear from participating in the test is to keep it busy with a masking noise. The noise produces a temporary artificial loss in the nontest ear. We want to distract the nontest ear sufficiently so that tones presented to the test ear are heard by the test ear only. However, we do not want to use so much masking noise in the nontest ear that it also raises the threshold of the test ear—an error called overmasking. Ideally, proper masking isolates the two ears from one another—the test signal is heard in the test ear, and the masking noise is heard in the masked ear—and neither is heard in both.

16. WHAT STARTING LEVEL OF MASKING SHOULD BE USED?

16.1. Air Conduction Testing

The starting masking level should be based on the air conduction thresholds for the frequency being tested in the masked ear plus ~20–25 dB.

16.2. Bone Conduction Testing

The starting masking level should be the air conduction threshold plus 25 dB. If the masked ear is normal or has a sensorineural involvement, another 15 dB should be added to the above sum at 250 and 500 Hz, and 10 dB should be added at 1000 Hz. This will account for the occlusion effect or the enhancement of threshold when the ear is covered, in this case with the masking earphone. There is a minimal occlusion effect at frequencies >1000 Hz. In conductive losses, the occlusion effect has already compensated and should not be counted twice.

17. WHAT TYPE OF MASKING DOES YOUR AUDIOMETER HAVE?

The type of masking noise produced by many audiometers is broad-band or white noise. A more efficient masking noise is called narrow-band noise. The temporary hearing loss produced by masking noise is not the same for all frequencies at a particular dial setting of the masking control. It follows, then, that the numbers appearing on the masking control should be regarded as practically meaningless unless appropriate correction tables are compiled and formulas applied.

There is a way, however, of determining proper masking without going through the complicated process of establishing masking tables and formulas. The procedure is referred to variously as the plateau method, the threshold shift method, or the shadow method.

18. WHAT IS THE PLATEAU METHOD OF MASKING?

The following procedure applies to both air and bone conduction threshold determinations:

1. Obtain air and bone conduction thresholds in both ears without masking.
2. If there is a possibility that shadow responses were obtained for the poor ear at some or all test frequencies, those frequencies will need to be retested with masking in the nontest, or better, ear.
3. Inform the subject that steamlike noise will be heard in the ear to be masked. The subject also will be listening to the test tones heard previously and is to respond only to the test tones, not the noise.
4. With the selector switch turned to "bone conduction," the masking noise automatically will be directed to one of the earphones. Determine which one is the masking earphone when doing bone conduction testing. When masking is used in air conduction testing, the noise automatically is directed to the earphone opposite the one in which the earphone selected is placed. Earphone select in "right" means the test tone goes to the right ear and masking noise to the left ear.
5. If doing air conduction with masking, be sure to place the headset snugly against the ears to prevent leakage of the noise to the opposite side.
6. If doing bone conduction with masking, first place the vibrator on the prominent portion of the mastoid area behind the ear. Be certain that the vibrator does not touch the ear. Then, put the masking earphone on the opposite ear and the deadened earphone on the temple of the side of the head where the vibrator was placed. Do not permit the headbands of the vibrators and the earphones to touch. Recheck the bone vibrator for proper position. Be sure that the vibrator is flat against the mastoid.
7. Introduce masking at a 30- to 40-dB effective level above the patient's threshold. Obtain a threshold (do not record it yet). If this produces a threshold

shift from the unmasked level, increase the masking by another 5 dB and obtain another threshold. If the threshold shifts 5 dB, you are still undermasking. Continue this procedure until the threshold stabilizes, even though masking is increased by two more 5-dB steps. At this point you probably have found the true threshold of hearing (the plateau). At some point, further increases in masking will start to shift thresholds again. This is the point of overmasking.

8. Procedure no. 7 is followed for each test frequency for both air and bone conduction testing.

If the initial introduction of masking is not sufficient, and with two or more successive increases in masking you do not shift threshold correspondingly, you still may not have reached the plateau because the threshold in the masked ear was not and has not been shifted successfully.

19. WHEN IS MASKING USED?

19.1. Air Conduction Testing

Generally, masking should be used whenever there is a difference of ≥40 dB between the air conduction reading in the poorer ear and the *bone* conduction threshold in the better ear. For example, if the air conduction thresholds in the better ear are normal, it follows that the bone conduction thresholds are also normal. Then, when testing air conduction of the poorer ear, masking will be necessary if the difference between ears is ≥40 dB. (Each frequency is compared individually.)

If you have not performed bone conduction testing and there is a difference of ≥40 dB between the left and right ears, masking should be used in the better ear while testing the poorer.

Example 1 Masking Needed in Right Ear When Testing the Left

Frequency (k/Hz)	0.5	1.0	2.0	3.0	4.0	6.0	8.0
Right ear	5	5/5	10	10	15	15	5
Left ear	50	55	50	55	55	55	50

One can assume that the right ear has normal bone conduction.

Example 2 Masking Needed in Right Ear at 2000–8000 Hz When Testing the Left

Frequency (k/Hz)	0.5	1.0	2.0	3.0	4.0	6.0	8.0
Right ear	40	35	45	40	40	45	40
Left ear	40	40	90	85	90	85	NR

As yet, bone conduction of the right ear is unknown, but about a 40-dB difference exists at frequencies 2000–8000 Hz.

19.2. Bone Conduction Testing

The purpose of the mastoid vibrator in bone conduction testing is to present the test tone directly to the inner ear, using the cranial bones instead of the ossicles.

Bone conduction testing is performed to determine whether an elevated air conduction threshold of hearing is caused by problems in the outer or middle ear. Bone conduction testing should be performed whenever the air conduction hearing threshold in either ear exceeds 15 dB.

Unlike the 40-dB criteria in air conduction testing, there is little or no transmission loss between ears when tones are presented via the bone vibrator. In most cases, therefore, masking should be used routinely when doing bone conduction testing.

There are some occasions, however, when masking is not necessary during bone conduction testing. These include:

1. When bone conduction thresholds are equal to the air conduction threshold for that ear (no air–bone gap).
2. When bone conduction thresholds for the ear being tested are better than those of the opposite ear.
3. When there is no response at the upper bone conduction testing limits.

20. WHAT FREQUENCIES SHOULD BE TESTED BY BONE CONDUCTION?

It is necessary to test only the frequencies 250–4000 Hz for bone conduction thresholds. It is not necessary to do bone conduction tests at frequencies where the air conduction thresholds are 15 dB or better.

21. HOW IS RESIDUAL HEARING TESTED CLINICALLY?

Too frequently, a patient with unilateral deafness is tested with inadequate masking or no masking at all, and he/she is told he/she has residual hearing in an ear that is deaf. In some instances, the patient even may be misdiagnosed as having otosclerosis and be operated on with no chance of success. Such errors can be avoided through proper testing technique.

21.1. Simple Methods of Masking

In clinical practice, several simple methods of masking are available to determine grossly whether an ear has some useful hearing or is profoundly deaf, as is commonly the case in unilateral hearing loss caused by mumps and in some ears postoperatively. These methods employ readily available sources of noise: (a) the air hose available in most powered nose spray equipment, (b) douching the ear with water at body temperature, (c) a piece of onionskin paper crackled against the nontest ear, and (d) a Bárány noise apparatus. A Bárány noise apparatus is inexpensive and easy to operate. Merely wind it up, insert it into the ear with a small tip, and press a button. The noise produced is of a high level and can mask most conductively impaired ears.

Frequently, patients are seen who have conductive hearing loss in both ears, usually as a result of otosclerosis, but occasionally due to bilateral chronic otitis media. Some of these patients have had mastoid, fenestration, or stapes surgery in one ear only, and they seek the physician's help to determine whether it is possible to restore hearing in either ear. If the hearing level in both ears is ~50–60 dB, it is very difficult to mask either ear with commercially available masking devices. The important fact to be determined is whether the ear that had the surgery now is totally deaf or actually does have a good amount of residual hearing. The most effective way of ascertaining this in everyday practice is to strike a 500-Hz tuning fork, apply it to the operated ear by both air and bone conduction, and ask the patient whether he/she still hears it while a loud masking noise is placed in the other ear. If, while the fork is still vibrating, the patient's hand comes down when the masking is turned on and then reappears when the masking is turned off again, the operated ear probably has little or no residual hearing, and further surgery is not warranted (Figs. 6.4 and 6.5). Incidentally, under such circumstances it would be injudicious to operate on the opposite ear, for in the event of a surgical complication, the patient might lose the hearing in his one useful ear and become totally deaf.

21.2. Sensorineural Acuity Level: A Supplementary Procedure

In order to overcome some of the difficulties in testing conductively impaired ears by bone conduction, Rainville, Jerger, and others developed a test called sensorineural acuity level. In this technique, the bone conduction vibrator must be connected to a noise generator and applied to the patient's forehead. First, the earphones of the audiometer are placed on both ears, and the air conduction threshold is obtained for each ear with no noise produced by the bone vibrator. Then, the air conduction thresholds are obtained with the noise vibrator turned on to a specific level: a signal strength of 2 V. By comparing these two air conduction thresholds with presumably normal levels, it is possible to determine the level of sensorineural hearing loss. This technique avoids masking in a bone conduction test, but it really is not a substitute for standard bone conduction testing. It does not assess the bone conduction level in some patients any better than the standard technique, but it may be a useful supplementary procedure.

22. HOW IS THE BONE CONDUCTION VIBRATOR CALIBRATED?

Major problems still exist in bone conduction testing. One involves calibration of the bone conduction vibrator; another concerns the relation of bone conduction threshold levels to the actual function of the sensorineural mechanism.

One method of calibrating the bone conduction vibrator is through the use of an artificial mastoid. The equipment necessary for this calibration procedure is fairly expensive and therefore usually is available only from the audiometer manufacturer or a well-equipped instrument laboratory. The authors prefer the biological method of checking vibrator responses.

Two biological techniques for calibration are used commonly. In one, using people with normal ears in a very quiet room, the manufacturer sets the bone conduction reference thresholds level at 0 dB, so that the subjects with a 0 air threshold level get a 0 bone threshold level. In the other technique, the same reasoning is applied to patients who have true sensorineural hearing loss, so that the air and the bone conduction thresholds match. None of these techniques is really quite satisfactory, but for the time being they must be used until a more sophisticated and reliable method is developed.

JOSEPH SATALOFF, M.D.
ROBERT THAYER SATALOFF, M.D.
1721 PINE STREET PHILADELPHIA, PA 19103

HEARING RECORD

NAME AGE

AIR CONDUCTION

RIGHT									LEFT							
DATE	Exam	LEFT MASK	250	500	1000	2000	4000	8000	RIGHT MASK	250	500	1000	2000	4000	8000	AUD
			-5	-5	-10	-10	-10	-10	-	55	60	60	50	65	60	
									60	75	NR	NR	NR	NR	NR	
									80	NR	NR	NR	NR	NR	NR	
									100	NR	NR	NR	NR	NR	NR	

BONE CONDUCTION

RIGHT									LEFT							
DATE	Exam	LEFT MASK	250	500	1000	2000	4000		RIGHT MASK	250	500	1000	2000	4000		AUD
									-	5	5	10	15	15		
									60	25	35	50	60	NR		
									100	NR	NR	NR	NR	NR		

SPEECH RECEPTION

DATE	RIGHT	LEFT MASK	LEFT	RIGHT MASK	FREE FIELD	MIC.

DISCRIMINATION

	RIGHT				LEFT				
DATE	% SCORE	TEST LEVEL	LIST	LEFT MASK	% SCORE	TEST LEVEL	LIST	RIGHT MASK	EXAM.

HIGH FREQUENCY THRESHOLDS

	RIGHT						LEFT					
DATE	4000	8000	10000	12000	14000	LEFT MASK	RIGHT MASK	4000	8000	10000	12000	14000

RIGHT		WEBER	LEFT	
RINNE	SCHWABACH		RINNE	SCHWABACH

HEARING AID

DATE	MAKE	MODEL
RECEIVER	GAIN	EXAM
EAR	DISCRIM	COUNC

REMARKS

Figure 6.4 Hearing record of patient with left total deafness caused by mumps. Right ear is normal. Without masking the right ear in testing the left, a threshold of ~60 dB seems to be present by air conduction, and bone conduction appears to be almost normal. These are shadow curves. With 60-dB white noise (WN) masking in the right ear, the shadow curve for air almost disappears, but it does not for bone conduction. With 80-dB WN in the right ear, the shadow curve for air is gone. With 100-dB WN in the right ear, there is no hearing for air or bone in the left ear. If, however, the 100-dB WN is put into the left ear during testing of the right normal ear, there is overmasking, and the right ear will show a reduced threshold. WN in this case was measured above the zero average on the audiometer.

The problems of bone conduction audiometry are numerous and complex. They include factors such as placement and pressure of the oscillator, thickness of the skin and underlying tissue, density of the petrous bone, the frequency being tested, noise conditions of the test room, condition of the ear not under testing, masking occlusion effect, and the vibrations of the skull and the bony otic capsule. Another important consideration

JOSEPH SATALOFF, M.D.
ROBERT THAYER SATALOFF, M.D.
1721 PINE STREET PHILADELPHIA, PA 19103

HEARING RECORD

NAME AGE

AIR CONDUCTION

DATE	Exam.	RIGHT: LEFT MASK	250	500	1000	2000	4000	8000	LEFT: RIGHT MASK	250	500	1000	2000	4000	8000	AUD
		100	70	80	70	80	80	NR		45	55	50	35	50	50	
AIR HOSE NOISE			NR	NR	NR	NR	NR	NR								

BONE CONDUCTION

DATE	Exam	RIGHT: LEFT MASK	250	500	1000	2000	4000		LEFT: RIGHT MASK	250	500	1000	2000	4000		AUD
		-	5	10	10	30	30		-	-5	-5	5	30	30		
		80	5	10	10	35	30									
		100	20	30	20	40	35									
AIR HOSE			25	NR	NR	NR	NR									

SPEECH RECEPTION

DATE	RIGHT	LEFT MASK	LEFT	RIGHT MASK	FREE FIELD	MIC.

DISCRIMINATION

DATE	RIGHT: % SCORE	TEST LEVEL	LIST	LEFT MASK	LEFT: % SCORE	TEST LEVEL	LIST	RIGHT MASK	EXAM.

HIGH FREQUENCY THRESHOLDS

DATE	RIGHT: 4000	8000	10000	12000	14000	LEFT MASK	LEFT: RIGHT MASK	4000	8000	10000	12000	14000

RIGHT: RINNE	SCHWABACH	WEBER	LEFT: RINNE	SCHWABACH

HEARING AID

DATE	MAKE	MODEL
RECEIVER	GAIN	EXAM
EAR	DISCRIM.	COUNC.

REMARKS

Figure 6.5 The right ear in this patient actually has no hearing, and it gives no caloric response; yet it gives a measurable threshold by air and bone conduction despite large amounts of masking. The reason is that the masking noise is not enough to overcome the conductive hearing loss in the left ear. Discrimination tests with masking are more helpful. The use of compressed-air noise, a Bárány noisemaker, or narrow-band noise through an insert receiver effectively masks the left ear. With the present findings, a misdiagnosis of right conductive hearing loss with an air–bone gap could be made, and the patient could be operated on with no chance of success.

is the air arising from the bone vibrator, especially when testing is done in the higher frequencies and at higher intensities. A strong air signal emanating from the bone vibrator may be detected by air conduction before actual bone vibration takes place to elicit responses.

A strong air signal emanates from the bone vibrator. If a bone vibrator is calibrated on ears with pure sensorineural hearing loss and is used on ears with conductive hearing loss, it may not be of proper calibration at the higher frequencies and intensities. In these ears, the air signal may be suppressed, and this might produce poorer bone thresholds than the actual sensorineural losses of these patients would warrant.

23. WHAT IS THE RELATION OF THE BONE CONDUCTION THRESHOLD LEVELS TO THE FUNCTION OF THE SENSORINEURAL MECHANISM?

In patients whose audiograms have isolated dips, as at 3000 and 4000 Hz, there is no need to test bone conduction, for in these cases the classification is almost invariably sensorineural hearing loss. It is not likely that a conductive or a mixed hearing loss would produce such an isolated dip. However, this statement does not hold for all high-tone hearing losses. One may find that in a patient with a high-tone hearing loss, whose bone conduction even may be somewhat reduced, the cause is not in the sensorineural pathway but in the middle ear. Fluid in the middle ear is the most common cause, although often it produces a low-frequency hearing loss. Otosclerosis may also produce a dip in bone conduction that is not really due to sensorineural impairment (Carhart's notch).

When interpreting bone conduction audiograms, it is essential to bear in mind that the threshold obtained by bone conduction testing provides only a rough approximation of the function of the sensorineural mechanism. In certain cases of conductive hearing loss, especially in adhesive otitis and in the presence of fluid in the middle ears, the lower bone conduction values that frequently are obtained may not indicate true sensorineural loss but rather impaired mobility of the oval and the round windows.

Bone conduction audiometry on the mastoid bone often shows a reduction that is not borne out by testing on the upper incisor tooth with a 512-Hz tuning fork. In such instances, the better dental bone conduction is more likely to reflect the true sensorineural hearing than the poorer bone conduction via the mastoid bone. Of course, it is important to rule out tactile sensation in evaluating these conflicting findings. Another important factor in assessing bone conduction threshold is that the maximum intensity obtainable on commercial audiometers for bone vibrators is ~55 or 60 dB. Failure to obtain any bone conduction in routine testing does not indicate necessarily that the sensorineural mechanism is dead. It indicates only that the mechanism does not respond at the maximum intensity of the tone.

As bone conduction testing really consists of a vibratory stimulus, it has created a problem in interpretation. Many patients with very severe air conduction hearing losses give responses to bone conduction at 250 and 512 Hz as low as 20 or 25 dB. This gives the impression that an air–bone gap exists in these low frequencies, an impression that is not justified. Actually, in most patients with severe nerve deafness, the false thresholds at these low frequencies are probably a response to tactile sensation and not to auditory stimuli. On the basis of these findings alone, middle-ear surgery is not warranted.

24. HIGH-FREQUENCY AUDIOMETRY

High-frequency audiometry, also called ultrahigh-frequency audiometry and very-high-frequency audiometry, refers to threshold testing of frequencies >8000 Hz. Several commercially available audiometers include capabilities for testing at 10,000 and 12,000 Hz, and high-frequency audiometers are available to test hearing up to 20,000 Hz. High-frequency audiograms can help in early detection of hearing loss from ototoxicity and other conditions, revealing hearing damage before it is detectable at the frequencies measured routinely. In selected cases, high-frequency audiometry is also helpful in differentiating between hearing loss due to noise and that due to presbycusis. For example, in most cases of presbycusis, hearing levels continue to deteriorate progressively at higher frequencies. In noise-induced hearing loss, improvement in hearing levels may be seen at 10,000, 12,000, or 14,000 Hz, revealing an audiometric "dip" similar to that usually centered around 4000 or 6000 Hz.

25. CONTINUOUS-FREQUENCY AUDIOMETRY

It is important to remember that routine audiometry samples hearing at only eight frequency regions out of thousands used by the ear. Equipment is available to test the *in between* frequencies, and such testing can be invaluable. For example, in a patient claiming to have tinnitus in one ear, after exposure to a blast, but who has normal hearing at 3, 4, 6, and 8 Hz, finding a 40 dB dip between 5240 and 5628 Hz can provide useful confirmation that an organic problem is present.

7

Special Hearing Tests

Robert T. Sataloff
Drexel University College of Medicine, American Institute for Voice and Ear Research, and Graduate Hospital, Philadelphia, Pennsylvania, USA

Joseph Sataloff
Thomas Jefferson University, Philadelphia, Pennsylvania, USA

For the majority of patients who complain of hearing loss, the history, ear examination, tuning-fork tests, and air and bone conduction audiometry provide sufficient information to make a preliminary differential diagnosis. This information tells the physician how much hearing is lost and what frequencies are affected and even helps to determine whether the loss is conductive or sensorineural in nature.

These tests are adequate in most cases of conductive hearing loss. However, there are some cases of conductive hearing loss and many cases of sensorineural loss in which these tests do not disclose enough information to make an accurate diagnosis possible, and additional hearing studies become necessary.

Because air and bone conduction audiometry measures only thresholds for pure tones, it provides limited information. For instance, it does not help the physician to detect certain phenomena that aid in localizing the site of a lesion in sensorineural involvement or to assess a patient's capacity to use his/her residual hearing. Phenomena such as recruitment, tone decay, tone distortion, and the patient's ability to discriminate speech, to localize sound, or to understand speech in a noisy environment give clues to the site of a lesion. To obtain this information, special hearing tests have been devised. These rarely are done by general practitioners or pediatricians, but many are carried out in an otologist's office, and most are performed in hospital or university hearing centers.

Although family doctors and occupational physicians may not perform all these tests, they should know when they are indicated, and they should be able to interpret the results.

1. THE DIFFERENCE BETWEEN AUDIOMETRIC THRESHOLD LEVELS AND DISCRIMINATION

It is common practice to interpret a patient's ability to hear speech by averaging his/her pure-tone thresholds in the four speech frequencies in the audiogram. For example, if his/her hearing loss is 30 dB at 500 Hz, 40 dB at 1000 Hz, 50 dB at 2000 Hz, and 40 dB at 3000 Hz, his/her average hearing loss would come to 40 dB for the speech frequencies, and this would be called his/her hearing loss for pure tones and for speech. It is common practice to express this as an average and to tell the patient that he/she has a 40% hearing loss. This procedure has serious shortcomings and should be avoided in clinical practice. Percentages have some application in determining hearing impairment in compensation problems, but this is not the same as using them to describe a patient's hearing capacity, especially based directly on decibels measured, rather than calculated impairment percentages.

If a patient is told that he/she has a 40% hearing loss in the involved ear, he/she naturally infers that the ear has only 60% of its hearing left. This is not true, especially because maximum hearing does not stop at 100 dB (maximum output of some audiometers), but rather, the ear will respond to much higher intensities. Percentages are particularly deceiving in cases of sensory hearing loss with poor discrimination. When the physician tells a patient with Meniere's disease that he/she has a 40% hearing loss, the patient may reply that, as far as he/she is concerned, his/her hearing loss is nearly 100% because he/she cannot use that ear in telephone conversations and the patient gets little or no use of it in daily communication. The problem exists because the threshold determination shows a 40-dB hearing level for pure tones, whereas the patient is referring to his ability to distinguish or to discriminate what he/she hears. In some patients with Meniere's disease or with an acoustic neuroma, the ability to discriminate may be impaired so severely that the ear is useless even with only a 40-dB threshold level in the speech frequencies.

Still another shortcoming of expressing hearing loss on the basis of speech frequency averaging is demonstrated in Fig. 7.1. Both patients whose audiograms are shown in A and B have an average hearing loss of 40 dB; but patient A can get along fairly well without the use of a hearing aid, whereas patient B needs a hearing aid to get along reasonably well.

The physician should explain to the patient the difference between audiometric threshold levels and discriminating ability. The patient also should be told the facts about his/her other difficulties, such as sound localization in unilateral hearing losses, recruitment, hearing in the presence of noise, and so on.

JOSEPH SATALOFF, M.D.
ROBERT THAYER SATALOFF, M.D.
1721 PINE STREET PHILADELPHIA, PA 19103

HEARING RECORD

NAME AGE

AIR CONDUCTION

RIGHT									LEFT							
DATE	Exam.	LEFT MASK	250	500	1000	2000	4000	8000	RIGHT MASK	250	500	1000	2000	4000	8000	AUD
(A)			15	20	40	60	65	NR								
(B)			50	50	40	30	30	35								

BONE CONDUCTION

RIGHT									LEFT							
DATE	Exam	LEFT MASK	250	500	1000	2000	4000		RIGHT MASK	250	500	1000	2000	4000		AUD

SPEECH RECEPTION

DATE	RIGHT	LEFT MASK	LEFT	RIGHT MASK	FREE FIELD	MIC.

DISCRIMINATION

	RIGHT				LEFT				
DATE	% SCORE	TEST LEVEL	LIST	LEFT MASK	% SCORE	TEST LEVEL	LIST	RIGHT MASK	EXAM.

HIGH FREQUENCY THRESHOLDS

RIGHT							LEFT						
DATE	4000	8000	10000	12000	14000	LEFT MASK	RIGHT MASK	4000	8000	10000	12000	14000	

RIGHT		WEBER	LEFT	
RINNE	SCHWABACH		RINNE	SCHWABACH

HEARING AID

DATE	MAKE	MODEL
RECEIVER	GAIN	EXAM.
EAR	DISCRIM.	COUNC.

REMARKS

Figure 7.1 Patients A and B have an average pure-tone loss of 40 dB, but patient A can get along well without the use of a hearing aid, whereas patient B will need amplification.

2. TESTING HEARING WITH SPEECH

2.1. Speech Reception Test

In addition to pure tones for testing threshold and for calculating hearing acuity, the physician can use speech of electronically controlled intensity. What is measured by this test is called a speech reception threshold (SRT). This represents the faintest level at which speech is heard and repeated. The SRT is recorded in dBHL. The results should be in good agreement with the average hearing levels obtained at 500, 1000, 2000, and 3000 Hz, the range that comprises the so-called speech frequencies, known as the pure-tone average. The pure-tone average and SRT normally vary from each other by no more than $\sim$10 dB.

2.1.1. *Materials*

Several types of speech material can be used to determine SRT. These include isolated words, individual sentences, and connected discourse. The most commonly used material is the standardized word list composed of spondaic words (spondees). These are two-syllable words equally stressed on both syllables, prepared in several lists (Fig. 7.2).

2.1.2. *Administration*

The test can be administered by recordings or by monitored live voice through a microphone attached to the speech audiometer. Earphones are placed on the patient's ears to test the hearing in each ear separately, or the patient may be tested through a loudspeaker, in which case binaural hearing is tested. The patient is instructed to repeat the spondaic words or sentences as they are presented. The patient should be told also that the words or the sentences will become fainter as the test proceeds but that he/she should repeat them until they are no longer audible.

List A	List B	List C	List D	List E	List F
Greyhound	Playground	Birthday	Hothouse	Northwest	Padlock
Schoolboy	Grandson	Hothouse	Padlock	Doormat	Daybreak
Inkwell	Doormat	Toothbrush	Eardrum	Railroad	Sunset
Whitewash	Woodwork	Horseshoe	Sidewalk	Woodwork	Farewell
Pancake	Armchair	Airplane	Cowboy	Hardware	Northwest
Mousetrap	Stairway	Northwest	Mushroom	Stairway	Airplane
Eardrum	Cowboy	Whitewash	Farewell	Sidewalk	Playground
Headlight	Oatmeal	Hotdog	Horseshoe	Birthday	Iceberg
Birthday	Railroad	Hardware	Workshop	Farewell	Drawbridge
Duckpond	Baseball	Woodwork	Duckpond	Greyhound	Baseball
Sidewalk	Padlock	Stairway	Baseball	Cowboy	Woodwork
Hotdog	Hardware	Daybreak	Railroad	Daybreak	Inkwell
Padlock	Whitewash	Sidewalk	Hardware	Drawbridge	Pancake
Mushroom	Hotdog	Railroad	Toothbrush	Duckpond	Toothbrush
Hardware	Sunset	Oatmeal	Airplane	Horseshoe	Hardware
Workshop	Headlight	Headlight	Iceberg	Armchair	Railroad
Horseshoe	Drawbridge	Pancake	Armchair	Padlock	Oatmeal
Armchair	Toothbrush	Doormat	Grandson	Mousetrap	Grandson
Baseball	Mushroom	Farewell	Playground	Headlight	Mousetrap
Stairway	Farewell	Mousetrap	Oatmeal	Airplane	Workshop
Cowboy	Horseshoe	Armchair	Northwest	Inkwell	Eardrum
Iceberg	Pancake	Drawbridge	Woodwork	Grandson	Greyhound
Northwest	Inkwell	Mushroom	Stairway	Workshop	Doormat
Railroad	Mousetrap	Baseball	Hotdog	Hotdog	Horseshoe
Playground	Airplane	Grandson	Headlight	Oatmeal	Stairway
Airplane	Sidewalk	Padlock	Pancake	Sunset	Cowboy
Woodwork	Eardrum	Greyhound	Birthday	Pancake	Sidewalk
Oatmeal	Birthday	Sunset	Greyhound	Eardrum	Mushroom
Toothbrush	Hothouse	Cowboy	Mousetrap	Mushroom	Armchair
Farewell	Iceberg	Duckpond	Schoolboy	Whitewash	Whitewash
Grandson	Schoolboy	Playground	Whitewash	Hothouse	Hotdog
Drawbridge	Duckpond	Inkwell	Inkwell	Toothbrush	Schoolboy
Doormat	Workshop	Eardrum	Doormat	Playground	Headlight
Hothouse	Northwest	Workshop	Daybreak	Baseball	Duckpond
Daybreak	Greyhound	Schoolboy	Drawbridge	Iceberg	Birthday
Sunset	Daybreak	Iceberg	Sunset	Schoolboy	Hothouse

Figure 7.2 Lists of spondees. (Auditory Test W-1, Spondee Word Lists, courtesy of Central Institute for the Deaf.)

2.1.3. The Significance of Various SRTs

The point at which 50% of the items are heard correctly as the intensity is reduced is considered to be the SRT. In clinical practice 5 dB steps of attenuation for every three words is a satisfactory rate. A person with normal hearing has an SRT of between 0 and 15 dB. An SRT up to ~25 dB usually presents no important handicapping hearing impairment. However, when the loss exceeds this level, the patient may experience difficulties in everyday communication, and a hearing aid usually is recommended.

In general, thresholds for speech tests and thresholds obtained by averaging the pure tones in the speech frequency should differ by no more than 6 dB. Discrepancies may be found between the two in cases in which there is a sharp dropoff of pure-tone thresholds across the 500–3000 Hz frequency range. Discrimination difficulty may result in a wide disparity between SRT and pure-tone average losses. There also may be a wide disparity in cases in which the loss is produced by emotional rather than organic causes.

2.2. Discrimination Testing

Determining the SRT is a rather imperfect measure of a person's ability to hear speech. Generally, it does not tell whether the patient is able to distinguish sounds that he/she hears, particularly difficult sounds. A special test to measure speech discrimination, devised to satisfy this need, differs from tests previously described in that it does not measure a minimal sound level but the *ability to understand speech* when it is amplified to a comfortable level.

Discrimination testing usually is done at 30 or 40 dB above the SRT. This level has been found to yield maximum discrimination scores. In some patients with a severe hearing loss, a level of 40 dB above the SRT is very difficult to obtain, because most instruments cannot go >100 dB. In other subjects, such a level might be painful or uncomfortable, or it might cause distortion in the instrument or the ear itself, resulting in invalid scores. In such cases the level of presentation of the test material should be adjusted to a comfortable listening level for the particular subject.

2.2.1. Materials

The speech materials used to test discrimination ability are known as phonetically balanced (PB) word lists. These are lists of monosyllabic words balanced for their phonetic content so that they have about the same distribution of vowels and consonants as that found in everyday speech. The lists are made up of 50 words each, and because each list is balanced in a particular way, it is necessary to administer the full list of 50 words when the test is performed (Fig. 7.3). Occasionally, the list is cut in half, but this leaves the resulting scores open to question.

2.2.2. Administration

The test usually is administered through each earphone in order to test the discrimination ability of each ear individually. Testing can be done by means of a recording on which the PB word lists have been recorded or monitored live-voice testing through a microphone. Each word is preceded by an introductory phrase, such as, "Say the word ________," and the subject is asked to repeat the word that he/she thinks he/she hears.

2.2.3. Rating

Only those words understood perfectly are counted as correct. Each word has a value of two points, so that a perfect score would be 100 points. If 10 of the 50 words are repeated

	List 4E	List 4F	List 3E	List 3F
1.	Ought (aught)	Out (hour)	Add (ad)	West
2.	Wood (would)	Art	We	Start
3.	Through (thru)	Darn	Ears	Farm
4.	Ear	Ought (aught)	Start	Out
5.	Men	Stiff	Is	Book
6.	Darn	Am	On	When
7.	Can	Go	Jar	This
8.	Shoe	Few	Oil	Oil
9.	Tin	Arm	Smooth	Lie (lye)
10.	So (sew)	Yet	End	Owes
11.	My	Jump	Use (yews)	Glove
12.	Am	Pale (pail)	Book	Cute
13.	Few	Yes	Aim	Three
14.	All (awl)	Bee (be)	Wool	Chair
15.	Clothes	Eyes (ayes)	Do	Hand
16.	Save	Than	This	Knit
17.	Near	Save	Have	Pie
18.	Yet	Toy	Pie	Ten
19.	Toy	My	May	Wool
20.	Eyes (ayes)	Chin	Lie (lye)	Camp
21.	Bread (bred)	Shoe	Raw	End
22.	Pale (pail)	His	Hand	King
23.	Leaves	Ear	Through	On
24.	Yes	Tea (tee)	Cute	Tan
25.	They	At	Year	We
26.	Be (bee)	Wood (would)	Three	Ears
27.	Dolls	In (inn)	Bill	Ate (eight)
28.	Jump	Men	Chair	Jar
29.	Of	Cook	Say	If
30.	Than	Tin	Glove	Use (yews)
31.	Why	Where	Nest	Shove
32.	Arm	All (awl)	Farm	Do
33.	Hang	Hang	He	Are
34.	Nuts	Near	Owes	May
35.	Aid	Why	Done (dun)	He
36.	Net	Bread (bred)	Ten	Through
37.	Who	Dolls	Are	Say
38.	Chin	They	When	Bill
39.	Where	Leave	Tie	Year
40.	Still	Of	Camp	Nest
41.	Go	Aid	Shove	Raw
42.	His	Nuts	Knit	Done (dun)
43.	Cook	Clothes	No (know)	Have
44.	Art	Who	King	Tie
45.	Will	So (sew)	If	Aim
46.	Tea (tee)	Net	Out	No (know)
47.	In (inn)	Can	Dull	Smooth
48.	Our (hour)	Will	Tan	Dull
49.	Dust	Through (thru)	Ate (eight)	Is
50.	At	Dust	West	Add (ad)

Figure 7.3 Phonetically balanced word lists (PB lists).

incorrectly, the patient has an 80% discrimination score. It is obvious, then, that an individual can suffer not only from a quantitative reduction in the ability to hear sounds but also from a reduction in discrimination ability, so that even when the speech is well above his threshold, he/she still cannot understand what is being said.

2.2.4. Arrangement of Equipment

Ideally, pure-tone and speech tests should be conducted via a two-room arrangement, which provides a soundproof room with the necessary earphones, microphone, and loudspeakers, in which the patient is seated, and an adjacent control room housing the pure-tone audiometer, the speech audiometer, and the examiner's microphone. The examiner conducts the tests from this room. All electrical connections are accomplished through wall plugs, and the addition of an observation window between the two rooms permits the examiner to watch the patient's reaction during the testing period. The two-room arrangement is essential when monitored live-voice testing is done through a microphone; otherwise, the subject might very well hear the speech directly from the tester's voice rather than through the equipment that electronically controls the intensity of the speech. One other precaution is to have the subject seated so that he/she cannot observe the tester's lips during the procedure. Even a fair lip reader can render the speech tests inaccurate if the patient is allowed to complement the auditory signals with visual clues.

2.2.5. *Evaluation of Discrimination Score*

It is essential to remember that the discrimination score cannot be translated directly into the percentage of difficulty that the patient will have in everyday life. A discrimination score provides only a rough idea of the patient's ability to distinguish certain sounds in a quiet environment. With experience, a broader interpretation can be made, but when this is done, other factors, such as personal auditory needs and adaptability and the effects of environmental noise, must be taken into consideration.

Figure 7.4 compares the discrimination scores in a series of patients with about the same SRT but with different diagnoses. Discrimination scores between 90% and 100% are considered to be normal. Scores between 90% and 100% are obtained usually by patients who have conductive hearing losses or those who have dips at high frequencies, such as a 40-dB dip at 3000 Hz; these scores are common also in cases in which the losses encompass only the higher frequencies of 4000, 6000, and 8000 Hz. Discrimination usually is not affected adversely when the loss is in this higher frequency range, because most speech sounds are in the area <4000 Hz. A slight reduction in the discrimination score may occur in patients with sensorineural hearing loss involving ≥2000 Hz. Generally, patients with these high-frequency losses experience more difficulty in daily conversation than their discrimination score would suggest.

In the interpretation of a discrimination score, it should be recalled that the test was done in a quiet room and therefore does not measure the patient's ability to discriminate against a noisy background. In such an environment, discrimination probably would be

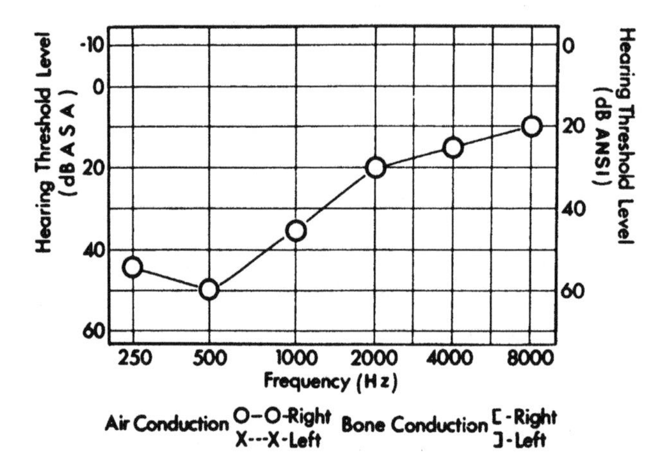

Figure 7.4 One patient had this audiogram as a result of otosclerosis. The discrimination score was 98%. Another patient had this same audiogram as a result of acoustic neuroma. The discrimination score was 42%. Still another patient had this same audiogram as a result of Meniere's disease. The discrimination score was 62%.

worse, because consonants are masked by the noise. At present, there is no completely reliable method of measuring an individual's ability to get along in everyday conversation under varied circumstances.

2.2.6. Causes of Reduced Discrimination

Discrimination scores are moderately lower in patients with hearing loss due to presbycusis or most congenital causes. Severe reductions in discrimination are associated particularly with two principal causes, Meniere's syndrome and acoustic neuroma.

In Meniere's syndrome, the discrimination difficulty generally is attributed to the distortion produced in the cochlea, that is, distortion in loudness, pitch, and clarity of speech. The characteristic finding in severe Meniere's syndrome is that the discrimination score becomes even worse as speech is made louder (Fig. 7.5).

Discrimination scores generally remain good in hearing loss caused by noise alone. They are rarely much <85%, although they may be slightly lower in far advanced cases.

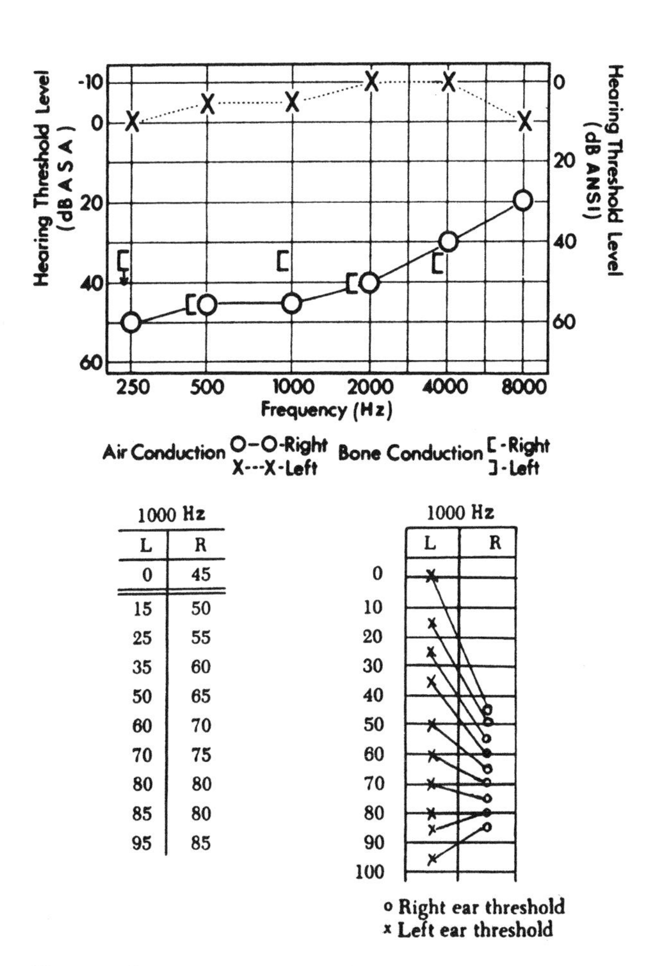

1000 Hz

L	R
0	45
15	50
25	55
35	60
50	65
60	70
70	75
80	80
85	80
95	85

Figure 7.5 Audiogram of a patient with Meniere's disease in the right ear. Discrimination score: 62% at 70 dB, 42% at 80 dB, and 30% at 90 dB. Note that not only does the 1000-Hz tone sound equally loud in both ears at 80 dB, but an 85-dB tone in the bad right ear sounds as loud as a 95-dB tone in the normal left ear. This is called hyperrecruitment. Diplacusis is marked, with the higher-sounding and distorted tone in the right ear.

When discrimination is significantly worse, other causes should be considered. Noise damages primarily outer hair cells, as discussed in Chapter 14.

In acoustic neuroma, the damage to the auditory nerve fibers may produce very little reduction in the hearing threshold level for pure tones but a disproportionately large reduction in the discrimination score. This disproportion is an important clue to the presence of acoustic neuroma, one that should be watched for in every case of sensorineural hearing loss, especially unilateral cases.

2.2.7. *The Glycerol Test*

This test is used by some physicians to help establish a diagnosis of Meniere's syndrome and to help predict response to diuretic therapy or endolymphatic sac operation. Pure-tone and speech audiometry is performed immediately before and 3 h after a single oral dose of glycerol, 2.3 mL/kg of body weight of 50% glycerol solution. Pure-tone threshold improvement of at least 10 dB in three adjacent octave bands and/or a speech discrimination improvement >12% constitutes a positive test and suggests reversibility of the inner-ear hydrops. Agents other than glycerol also may be used.

2.2.8. *Masking*

Because masking often is overlooked in discrimination testing and speech reception testing, its importance must be reiterated. If the nontest ear is not masked properly, serious errors in diagnosis can be made (Fig. 7.6).

2.2.9. *Etiology*

From the examples in Figs. 7.4–7.6, it is obvious that it is not always possible to predict the discrimination score from the pure-tone audiogram. The etiology is of the utmost importance in assessing the discrimination.

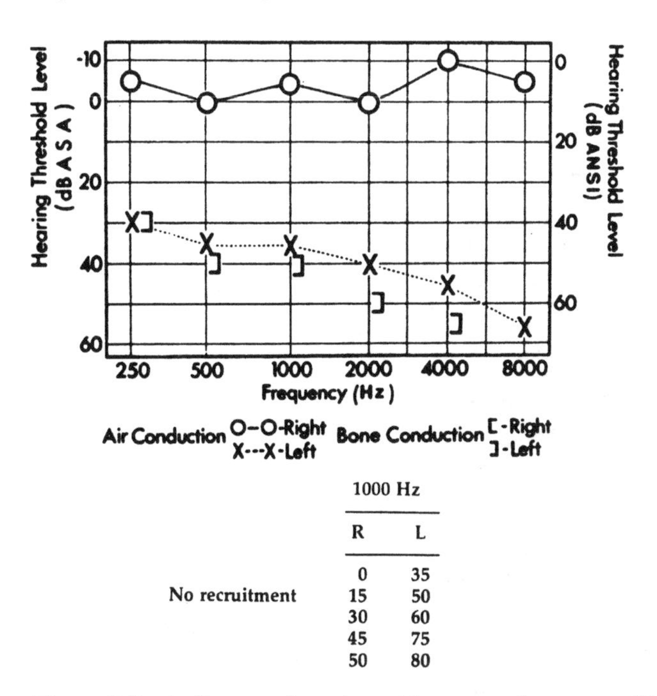

	1000 Hz	
	R	L
	0	35
No recruitment	15	50
	30	60
	45	75
	50	80

Figure 7.6 Audiogram of a patient with an acoustic neuroma. There was abnormal tone decay in the left ear during the 60-s test (right ear masked).

2.2.10. Clues to Discrimination

These tests require equipment that may not be available in the offices of many practitioners. Nevertheless, even a careful history and examination will give a rough estimate of a patient's ability to hear and to distinguish speech. For example, the patient should be asked how well he/she does with each ear when using the telephone. The patient may volunteer that, although voices sound loud enough, he/she cannot understand conversation on the telephone. This usually indicates a reduction in discrimination ability.

Another important clue to discrimination ability is whether the patient experiences more difficulty in noisy environments. If he/she does, it is probably because he/she has reduced discrimination, which is aggravated further by environmental noise that masks out the normally weak consonants. He/she also may have trouble with higher-pitched female voices. The physician who has no hearing testing equipment in his/her office may obtain a fair idea of a patient's discrimination score by asking the patient to repeat words in the PB lists when the physician presents them without letting the patient see the examiner's face. The words should be spoken at a normal level of intensity and enunciated normally. If the loss is unilateral, the better ear must be masked with noise intense enough to prevent its participation in the test. The same precaution is necessary during testing with audiometric equipment. Scores obtained without masking the opposite ear can result in a wrong diagnosis.

3. TESTS FOR RECRUITMENT

The phenomenon of recruitment has been defined elsewhere in this book as a disproportionate increase in the sensation of loudness of a sound when its intensity is raised. The principal value of detecting recruitment is that it helps to trace the site of the lesion in the auditory pathway to the hair cells of the cochlea. The patient often provides clues to the presence of cochlear damage when he/she is questioned about hearing difficulty. The patient may say that loud noises are very bothersome in his/her bad ear or that the sound seems to be tinny, harsh, and very unclear. He/she may volunteer that music sounds distorted or flat. These complaints should not be confused with the annoyance voiced by a neurotic patient who hears well but is bothered by such noises as the shouting of children. A well-defined sensorineural hearing loss is a prerequisite to utilize recruitment as a basis for localizing an auditory deficit to the cochlea.

3.1. Tuning Fork

Recruitment can be detected in some cases with the aid of a 512 Hz tuning fork if a hearing loss affects the speech frequencies, as often occurs in Meniere's disease. The test is done by comparing the growth of loudness in the good ear with that in the bad ear. The fork is struck once gently and held up to each of the patient's ears, and he/she is asked in which ear the tone sounds louder. Naturally, he/she will say the tone is louder in the ear that has better or normal hearing. Then, the intensity of the tone is increased by striking the fork once again quite hard (but not too hard, or the tone may be distorted). The patient then is asked again to indicate in which ear the tone now sounds louder, with the fork held first near the good ear and then quickly moved to about the same distance from the bad ear. If complete or hyperrecruitment is present, the patient will now exclaim in surprise that the tone is as loud or louder in the bad ear. This means that there has been a larger growth in the sensation of loudness in the bad ear, in spite of a hearing loss; that is, recruitment is present in the bad ear.

3.2. Alternate Binaural Loudness Balance Test

Testing for recruitment with a tuning fork is a rather rough technique, but it may help in diagnosing recruitment. More precise tests have been devised to test for recruitment, but most of these are suitable for use only when one ear is impaired and the other is normal. The technique in common use is called the alternate binaural loudness balance (ABLB) test, which matches the loudness of a given tone in each ear.

This is done with an audiometer and involves presenting a tone of a certain intensity to the good ear and then alternately applying it to the bad ear at various intensities; the patient is asked to report when the tone is equally loud in both ears. Initially, a brief tone 15 dB above threshold is applied to the good ear. Then the tone is presented briefly to the bad ear 15 dB above its threshold, and the patient is asked whether the tone was louder or softer than that heard in the good ear. According to the response, necessary adjustments are made to the intensity going to the bad ear until a loudness balance is obtained with the good ear. Then, the intensity to the good ear is increased by another 15 dB, and another balance is obtained with the bad ear. Loudness balancing is continued in 15 dB steps until sufficient information is obtained about the growth of loudness in the bad ear. This technique requires that the same frequency be balanced in the two ears and that the tone be presented alternately to the good ear, which serves as the reference. Also, the difference in threshold between the two ears should be at least 20 dB for this test to be valid.

If the difference in loudness level between the two ears is unchanged at higher intensities, recruitment is absent. If the loudness difference gradually decreased at higher intensities, recruitment is present. If the loudness difference completely disappears between the two ears at higher intensities, the condition is called complete recruitment and is indicative of damage to the inner ear. There may be hyperrecruitment, in which the tone in the bad ear sounds even louder than the tone in the good ear at some point above threshold. Recruitment may occur at varying speeds. If it continues regularly with each increase of intensity, it is called continuous recruitment, and this is indicative of inner ear damage. Recruitment that is found only at or near-threshold levels is not necessarily characteristic of inner ear damage but occurs often in patients with sensorineural hearing impairment. Figures 7.5 and 7.7 show two cases that exhibit recruitment, with methods of recording the result.

3.3. Detection of Small Changes in Intensity

Another method of demonstrating recruitment involves the patient's ability to detect small changes in intensity. A recruiting ear detects smaller changes in intensity than normal ears or those with conductive hearing loss. At levels near threshold, a normal ear is likely to require a change of ~2 dB to recognize a difference in loudness, but in an ear that recruits, only a 0.5-dB increase may be necessary to detect the loudness change. As the intensity of the tone then increases in normal ears, the change necessary for detecting the difference in loudness becomes smaller, whereas the recruiting ear requires about the same change as it did near-threshold level.

3.4. Short-Increment Sensitivity Index Test

Another test for localizing the site of damage to the cochlea is the short increment sensitivity index (SISI) test. It measures the patient's ability to hear small, short changes of sound intensity. The test is done monaurally by fixing the level of a steady tone at 20 dB above the patient's threshold at each frequency to be tested and superimposing on this steady tone 1 dB increments of ~200 ms duration, interspersing the increments at 5-s intervals.

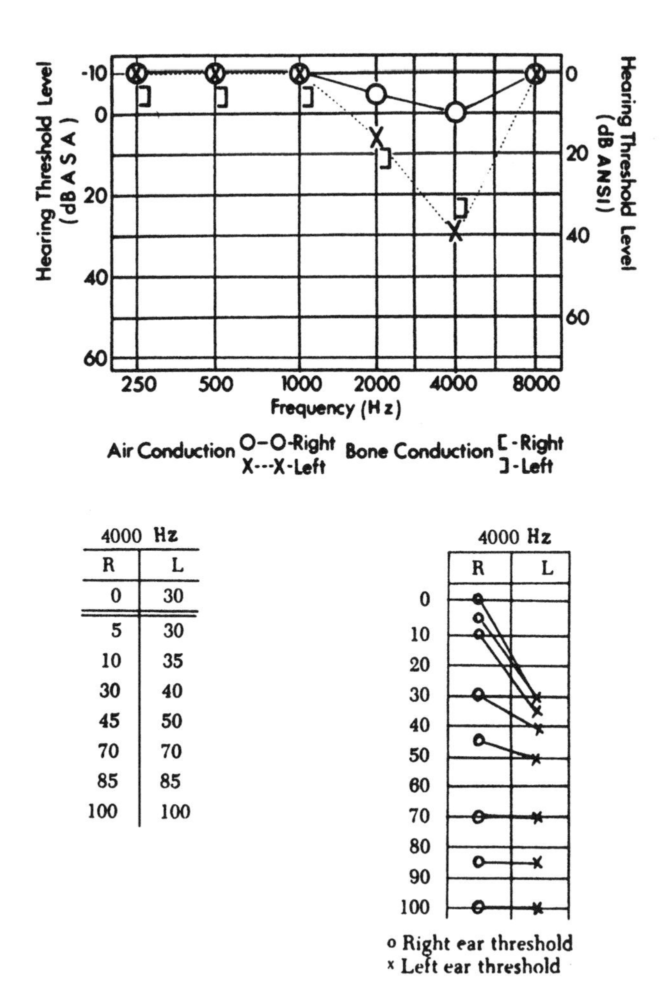

4000 Hz	
R	L
0	30
5	30
10	35
30	40
45	50
70	70
85	85
100	100

Figure 7.7 Audiogram of a 21-year-old male with acoustic trauma of the left ear caused by explosion of a firecracker. Two methods of recording results of loudness balance tests are shown.

The patient is to respond each time he hears any "jumps in loudness." If he/she hears five of the 1-dB increments, his/her sensitivity index is 25%. A score of between 60% and 100% at frequencies >1000 Hz is positive for cochlear disorders, whereas a score <20% is considered to be negative. Scores between 20% and 60% are inconclusive.

A revised form of the test called the high-intensity SISI uses a similar technique, but with very loud tones rather than near-threshold tones. In high-intensity SISI testing, patients with cochlear hearing loss and those with normal hearing will exhibit a high percentage of responses to the short-increment increases. However, patients with retrocochlear disease, such as an acoustic neuroma, will continue to have low percentage scores. Thus, both the classical SISI and the high-intensity tests help to differentiate not only between cochlear and noncochlear but also cochlear and retrocochlear pathology. However, the SISI test has many shortcomings and is of limited value.

3.5. Other Tests

There are more tests using speech discrimination and Békésy audiometry that also help to determine the presence of cochlear damage; these are supplementary to the basic tests described here. They are especially helpful when both ears have a hearing loss, because a "control" ear is not essential to the test procedure. Stapedius reflex testing also is useful in these cases (see *Impedance Audiometry* section).

4. TESTING FOR DIPLACUSIS—DISTORTION OF PITCH

Another simple and fairly reliable office test can be done with a tuning fork to help to localize the site of auditory damage in the cochlea. This test explores not distortion of loudness (recruitment) but distortion of pitch, which is called diplacusis. Distortion is the hallmark of hair cell damage.

A 512-Hz tuning fork is struck and held near the normal ear and then near the opposite ear. If the damage is localized in the cochlea, the patient may report that the same tuning fork has a different pitch when it is heard in the bad ear. Usually, he/she will say that the pitch is higher or lower and not as clear but rather fuzzy. It is important to clarify to the patient when this test is performed that he/she is being asked to evaluate pitch, not loudness, otherwise inaccurate results may be obtained. This test may be performed with tuning forks of various frequencies. It is not rare for pitch distortion to occur primarily at frequencies that are affected by hearing loss.

5. HEARING TESTS USING SPEECH TO DETECT CENTRAL HEARING LOSS

Special tests using modified speech are becoming very useful in deciding whether a hearing loss is caused by damage in the central nervous system. Lesions in the cortex do not result in any reduction in pure-tone thresholds, but brainstem lesions may cause some high-frequency hearing loss. Routine speech audiometry is almost always normal in cortical lesions. Sometimes it is impaired in brainstem lesions but without a characteristic pattern. As neither pure-tone nor routine speech tests help to localize damage in the central nervous system, more complex tests have been developed to help to provide this information.

A chief function of the cortex is to convert neural impulses into meaningful information. Words and sentences acquire their significance at the cortical level. Because quality, space, and time are factors governing the cortical identification of a verbal pattern, the tests are designed so that they explore the synthesizing ability of the cortex when one or more of these factors is purposely changed.

5.1. Binaural Test of Auditory Fusion

One such test of central auditory dysfunction is the binaural test of auditory fusion. Speech signals are transmitted through two different narrow-band filters. Each band by itself is too narrow to allow recognition of test words. Subjects who have normal hearing show excellent integration of test words when they receive the signals from one filter in one ear and the other filter in the other ear. Poorer scores are made by patients with brain lesions and are indications of a functional failure within the cortex.

5.2. Sound Localization Tests

Sound localization tests also are being used in the diagnosis of central lesions. Deviation of the localization band to one side points to a cerebral lesion on the contralateral side or to a brainstem lesion on the ipsilateral side.

5.3. Other Tests

Distorted-voice, *interrupted-voice*, and *accelerated-voice tests* likewise are used in detecting central lesions. In the distorted-voice test, PB words are administered ~50 dB above threshold through a low-pass filter that is able to reduce the discrimination to ~70% or 80% in normal subjects. Patients with temporal lobe tumors present an average discrimination score that is poorer in the ear contralateral to the tumor.

The interrupted-voice test presents PB words at ~50 dB above thresholds, interrupting them periodically 10 times/s. Subjects with normal hearing obtain ~80% discrimination; those with temporal lobe tumors have reduced discrimination in the ear contralateral to the tumor.

In the accelerated-voice test, when the number of words per minute is increased from ~150 to ~350 words, the discrimination approaches 100% in subjects who have normal hearing, but the threshold is raised by 10–15 dB. In patients with tumors of the temporal lobe, there is a normal threshold shift, but the discrimination never attains 100% in the contralateral ear. In cortical lesions the impairment always seems to be in the ear contralateral to the neoplasma and moderate in extent. Brainstem lesions exhibit ipsilateral or bilateral impairments.

Ipsilateral and contralateral stapedius reflex tests also provide useful information.

5.4. Neuropsychological Testing

When central auditory processing disorders are suspected, it is often helpful to refer the patient for formal neuropsychological testing. This approach helps map the site of brain dysfunction, confirm the auditory processing ability, and detect any other cognitive deficits. Neuropsychologists' treatment recommendations also may be extremely useful.

6. TESTING FOR FUNCTIONAL HEARING LOSS

Whenever a patient claims to have a hearing loss that does not seem to be based on organic damage to the auditory pathway, or whenever the test responses and the general behavior of the patient appear to be questionable, a variety of tests can be performed to help determine whether the loss is functional rather than organic.

6.1. Suggestive Clues

The most suggestive findings are inconsistencies in the hearing tests. For instance, a patient has a hearing threshold level of 70 dB in one test and a 40 dB threshold when the test is repeated several minutes later; or the audiogram of a patient shows a 60 dB average hearing loss bilaterally, but the patient inadvertently replies to soft speech behind his/her back; or the patient has an SRT of 20 dB in contrast to a 60 dB puretone average; or the patient gives poor or no responses in bone conduction tests, indicating severe sensorineural involvement, but has suspiciously good discrimination ability for the alleged degree and type of loss. However, care must be exercised. Certain organic conditions such as Meniere's disease, multiple sclerosis, and severe tinnitus also may cause inconsistent responses.

In addition, the patient's behavior may not be consistent with the degree of loss claimed, especially in cases of bilateral functional hearing loss. Usually, a patient with severe bilateral deafness is very attentive to the speaker's face and mouth in order to benefit from lipreading. A functionally deaf person may not show this attentiveness. He/she also may have unusually good voice control, which is not consistent with the degree of loss. Occasionally, a functionally deaf person will assume a moronic attitude or repeat part of a test word correctly and labor excessively over the last half of the word. These and other clues should alert the examiner to the possibility of the presence of a purely functional hearing loss or a functional overlay on an organic hearing loss.

6.2. Lombard or Voice-Reflex Test

When a patient claims deafness in one ear, but it is suspected of being functional, several simple tests are available to determine the validity of the loss. The patient is given a newspaper or a magazine article to read aloud without stopping. While he/she is reading, the tester presents noise to the good ear. This may be done by rubbing a piece of typing paper such as onionskin paper over the patient's good ear. If the patient's voice does not get significantly louder, it is highly suggestive that he can hear in his supposedly "bad" ear. Because hearing is partly a feedback mechanism that informs the speaker how loud he is speaking, a person with normal hearing will speak more loudly in a very noisy area in order to be heard above the noise. If the patient does not raise his/her voice when noise is applied to one ear, it means that he/she is hearing himself speak in the other ear, and consequently that ear does not have the marked hearing loss indicated on the pure-tone or speech audiogram. Instead of rubbing paper against the patient's ears as the source of noise, a Bárány noise apparatus or the noise from an audiometer noise generator is extremely effective in this test, because the level of the noise can be controlled. This type of test is called the Lombard or voice-reflex test, and although it does not help to establish thresholds, it does give the examiner some idea of whether or not the loss is exaggerated.

6.3. Stenger Test

The Stenger test also is used for detecting unilateral functional hearing loss and evaluating the approximate amount of residual hearing (1). This test can be done with tuning forks or with an audiometer, the latter being the more reliable.

The Stenger test depends on a given pure tone presented to both ears simultaneously. The tone will be perceived in the ear where it is louder. If the sound in one ear is made louder, then the listener will hear it in that ear and will not realize that a weaker sound exists in the other ear.

A tone is presented to the good ear ~10 dB above threshold and, at the same time, 10 dB below the admitted threshold in the bad ear. If the patient responds, the test is a negative Stenger because he/she heard the tone in the good ear without realizing there was a weaker tone in the bad ear. If the patient does not respond, it is a positive Stenger because he/she heard the sound in the assumed bad ear without realizing there was a tone of weaker intensity presented to the good ear.

This test can be done with speech as well as pure tones. There must be a difference of at least 30 dB between thresholds of the good and allegedly bad ear for the test to be effective. Also, a two-channel audiometer is needed to administer the test.

The Stenger test also enables the examiner to obtain an approximation of the patient's true thresholds in the bad ear. This is done by presenting the pure tone 10 dB above threshold in the good ear and presenting a pure tone at 0 Hz in the bad ear. The tone in the bad ear is increased in 5 dB steps until the patients stops responding.

(Remember, he/she is hearing the tone is his/her good ear at first.) The Stenger pure-tone threshold of the bad ear is $\sim$15 dB above his or her true threshold.

6.4. Repetition of Audiogram without Masking

Still another test to indicate whether a patient really has a severe or total unilateral hearing loss or may be malingering is to repeat the audiogram but this time without masking the good ear. As a pure tone presented to the test ear can be heard also in the nontest ear when the loudness of the tone is 50–55 dB above the threshold of the nontest ear, at least some shadow curve should be present in the absence of masking. If the patient does not respond when the intensity levels reach this point, then the chances are that he/she has a functional deafness in the test ear or is malingering. If the patient does report hearing the tone, he/she should be questioned carefully about its location. Again, total lack of response is an indication of the dilemma that the functional patient faces when he/she feels that his/her claim is threatened with exposure.

6.5. Delayed Auditory-Feedback Test

The monitoring effect of an ear also can be disrupted if a person listens to himself/herself speak through earphones while the return voice is delayed in time. A delay of 0.1–0.2 s causes symptoms similar to stuttering. If this occurs when the feedback level is lower than the admitted threshold, functional loss or malingering is present. In the delayed talk-back test (also called the delayed auditory-feedback test), which is done through a modified tape recorder, it is possible to detect hearing losses of sizable degree but not the minor exaggerations that occur occasionally in medico-legal situations. This is so because delayed feedback affects the rhythm and the rate of the patient's speech at levels averaging 20–40 dB above threshold.

6.6. Psychogalvanic Skin Response Test

A great deal of testing has been done with the psychogalvanic skin response (PGSR) test, which is close to being an objective test of hearing, though it still has many shortcomings. This test is done with special equipment and is based on the conditioned response mechanism. The patient is conditioned so that each time he/she hears a tone, it is followed about a second later by a definite electric shock in the leg, to which is strapped an electrode. Through electrodes placed on the patient's fingers or palms, it is possible to measure the change in skin resistance or the so-called electrodermal response excited by the electric shock in the patient's leg. Each time that the patient receives a shock, the skin resistance is altered and can be read on a meter or recorded on a moving graph. After the patient is conditioned well, the electric shock is stopped, and only the sound is given. In a well-conditioned patient, about a second after the sound is applied, he/she will "expect" the shock again and show a typical change in the electrodermal responses. It can be concluded, then, that each time the patient gives a positive reading on the recording equipment after a sound is given, he/she has heard the sound. By lowering the intensity of the stimulus, a threshold level can be obtained. At certain intervals it is necessary to reinforce the conditioned response mechanism by reapplying the electric shock.

This is an excellent technique based to some extent on the traditional lie detector method, but many complicating factors make it far from a completely reliable method of measuring a hearing threshold level objectively and reliably. It may be helpful, however, if it is used in conjunction with a battery of other tests, in helping to establish the organic or the nonorganic basis of any hearing threshold. PGSR is rarely used anymore.

6.7. Brainstem Evoked Response Audiometry in Malingering

Brainstem evoked response audiometry (BERA) is discussed in greater detail later in this chapter. However, because BERA testing is objective, the technique may provide valuable information in malingerers and patients with functional hearing loss. However, in its present state of development, there are shortcomings to BERA threshold testing. It is often helpful in assessing auditory function in patients who are unable or unwilling to cooperate.

6.8. Use of Excellent Audiometric Technique

One of the most effective methods of obviating intentional functional hearing loss, particularly in industry and in school hearing testing programs, is to use excellent audiometric technique. Malingering and inaccurate responses are discouraged by a tester who uses excellent technique. Malingering normal hearing also is possible. If a patient is given a sound and is asked repeatedly, "Do you hear it?" he will be tempted to say, "yes," even if he does not hear it, whenever some advantage or remuneration is at stake, such as obtaining employment.

6.8.1. Responsibility of the Tester

The question as to what a tester in industry or in a school should do when he/she suspects or detects a malingerer or someone with functional hearing loss is important. It is not the tester's responsibility to accuse or even to imply to the subject the he/she is suspected of giving inaccurate responses. Quite often, inaccurate responses are the result of disturbances in the auditory tract or nervous pathway, or the loss may be a true hysterical deafness. The tester may not be qualified to express so sophisticated a judgment. The tester's only responsibility is to suspect that the audiogram does not represent the accurate threshold of hearing of the individual tested.

The subjects should be handled in a routine manner, and subsequently the findings should be brought to the attention of the physician in charge of the hearing testing program. If the physician is suitably trained to study the patient on a more comprehensive basis, he should proceed to do so. If not, the patient should be referred to an otologist or a hearing center that is equipped to study the problem.

7. TESTING FOR AUDITORY TONE DECAY

Just as marked recruitment usually is indicative of damage in the inner ear, abnormal tone decay (abnormal auditory fatigue) usually is a sign of pressure on or damage to the auditory nerve fibers. This phenomenon may be of particular importance in that it can be an early sign of an acoustic neuroma or some other neoplasm invading the posterior fossa.

7.1. Administration of Test

The test for abnormal tone decay is very simple to perform and should be done routinely in every case of unilateral sensorineural hearing loss, especially when no recruitment is found. The test is based on the fact that, whereas a person with normal hearing can continue to hear a steady threshold tone for at least 1 min, the patient who has a tumor pressing on his auditory nerve is unable to keep hearing a threshold tone for this length of time. The test is performed monaurally with an audiometer. A frequency that shows reduced threshold is selected, and the patient is instructed to raise a finger as long as he/she can hear the tone. The tone then is presented at threshold or 5 dB above threshold, and a stopwatch is started with the presentation of the tone. Each time the patient lowers his/her finger, the intensity is increased 5 dB, and the time is noted for that period of hearing. The tone interrupter switch never is released from the "on" position during any of the intensity changes. The test is 1 min in duration.

7.2. Findings

A person with normal tone decay usually will continue to hold up his/her finger during the entire 60 s. Occasionally, he/she may require a 5- or 10-dB increase during the first part of the test, but he/she then maintains the tone for the remainder of the time. A patient who has abnormal tone decay may lower the finger after only ~10 s, and when the tone is raised 5 dB, he/she may lower his/her finger again after another 10 s and continue to indicate that the tone fades out repeatedly, until after 60 s there may be an increase of ≥25 dB above the original threshold. Some patients may even fail to hear the tone at the maximum intensity of the audiometer after 1 min, whereas originally they may have heard the threshold tone at 25 dB. Masking may be required. This finding of abnormal tone decay is highly suggestive of pressure on the auditory nerve fibers. Figure 7.6 describes a typical case. A tuning fork can also be used to detect abnormal tone decay by testing for threshold, then fatiguing the ear and retesting for threshold. Modifications of this method can also be used.

8. DIAGNOSTIC SELF-RECORDING AUDIOMETRY

8.1. Békésy Audiometry

Another method of measuring abnormal tone decay is with a Békésy audiometer. This is a type of self-recording audiometer that is being used with increased frequency for threshold and special testing. Physicians should be acquainted with its operating principles and the information that it can supply.

8.2. Procedure and Mechanism

This method of establishing pure-tone thresholds permits the patient to trace his/her own audiogram as the tone or tones are presented automatically. Each ear is tested separately. The patient holds a hand switch and has on a set of earphones through which he hears the tone. As soon as he/she hears the tone, he/she presses the switch, which causes the sound to decrease in loudness, and holds it down until the tone is gone. This procedure continues until the full range of frequencies has been tested.

The switch controls the attenuator of the audiometer that decreases or increases the intensity of the tone. A pen geared to the attenuator makes a continuous record on an audiogram blank of the patient's intensity adjustments. The audiogram is placed on a table that moves in relation to the frequency being presented. Several methods of frequency selection are available. The audiometer can be set up to produce the frequency range continuously from 100 to 10,000 Hz, or it can be arranged to test hearing in a two-octave range or, if desired, to test threshold for a single frequency for several minutes.

The test signal can be continuous, that is, without interruption, or pulsed at a rate of ~2.5 times/s. Operation of the patient's hand switch attenuates the signal at a rate of ~2.5–5 dB/s according to the speed selection made by the examiner. Thus, a test routinely performed with Békésy self-recording audiometry can determine thresholds with both pulsed- and continuous-tone presentations. If the pulsed-tone is used first, a pen with a specific colored ink is placed in the penholder, and the thresholds are recorded. When the pulsed-tone testing is completed, a pen with a different color is placed in the penholder, the frequency is reset to the original point, and the switching is changed to provide a continuous tone. The patient traces another audiogram, as he/she did for the pulsed tones. It is important that the patient not see the equipment in operation, because awareness of the movements of the pen and the action of the hand switch may affect his/her responses and result in an invalid audiogram.

8.3. Value

With proper instruction to the patient, Békésy audiometry not only provides an accurate picture of thresholds but also supplies other valuable information. By comparing the thresholds for the pulsed and the continuous tone, the physician can get a reasonably good indication of the site of the lesion within the auditory system.

8.4. Types of Békésy Audiograms

In normal or conductively impaired ears, the pulsed and the continuous tracings overlap for the entire frequency range tested. This is a type I audiogram. In the type II audiogram, the pulsed and the continuous tones overlap in the low frequencies, but between 500 and 1000 Hz the continuous-frequency tracing drops ~15 dB below the pulsed tracing and then remains parallel with the high frequencies. Type II audiograms occur in cases of cochlear involvement. Sometimes the pen excursions narrow down to ~5 dB in the higher frequencies. Cases of cochlear involvement sometimes also show type I tracings.

In the type III audiogram, the continuous tracing drops suddenly away from the pulsed tracing and usually continues down to the intensity limits of the audiometer. Eighth-nerve disorders usually show type III audiograms. Another type of audiogram found in eighth-nerve disorders is the type IV tracing, in which the continuous-tone tracing stays well below the pulsed tracing at all frequencies.

8.5. Interpretation of the Békésy Audiogram

There is some feeling that the amplitude of the Békésy audiogram provides considerable information about the presence or the absence of recruitment. For example, tracings of very small amplitude might lead one to believe that the patient can detect changes in intensity much smaller than the average subject and that he/she therefore has recruitment. Unfortunately, this is not the case. It is more likely that tracings of small amplitude are suggestive of abnormal tone decay rather than of recruitment. A great deal of the interpretation depends on the on-and-off period of the tone presented. Figures 7.8 and 7.9 show examples of Békésy audiograms.

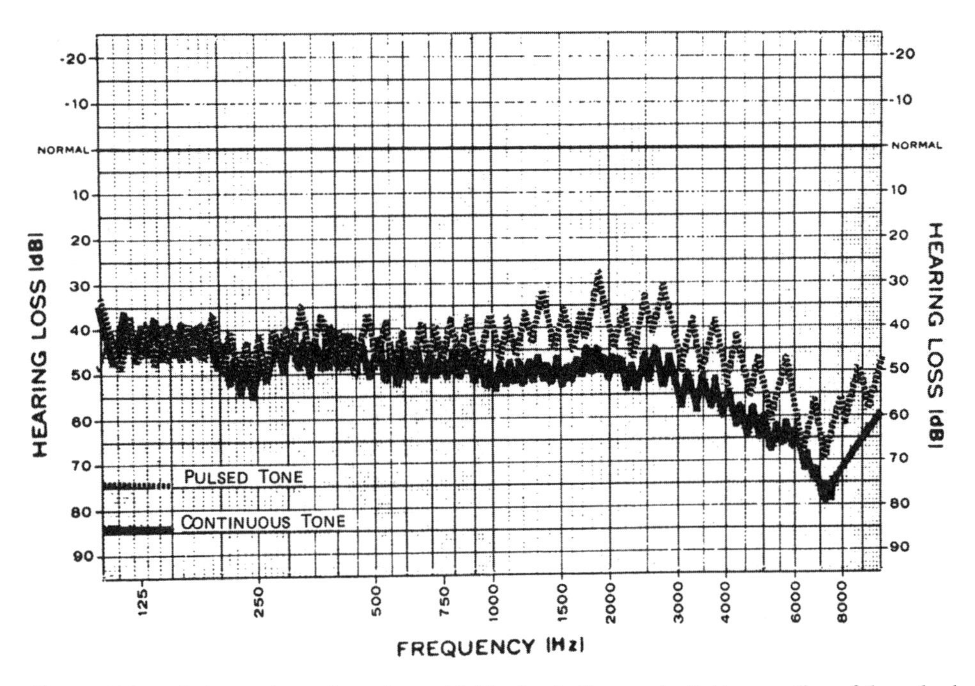

Figure 7.8 Békésy tracings of a patient with Meniere's disease. A slight separation of the pulsed and the continuous-tone tracings appears only in the higher frequencies.

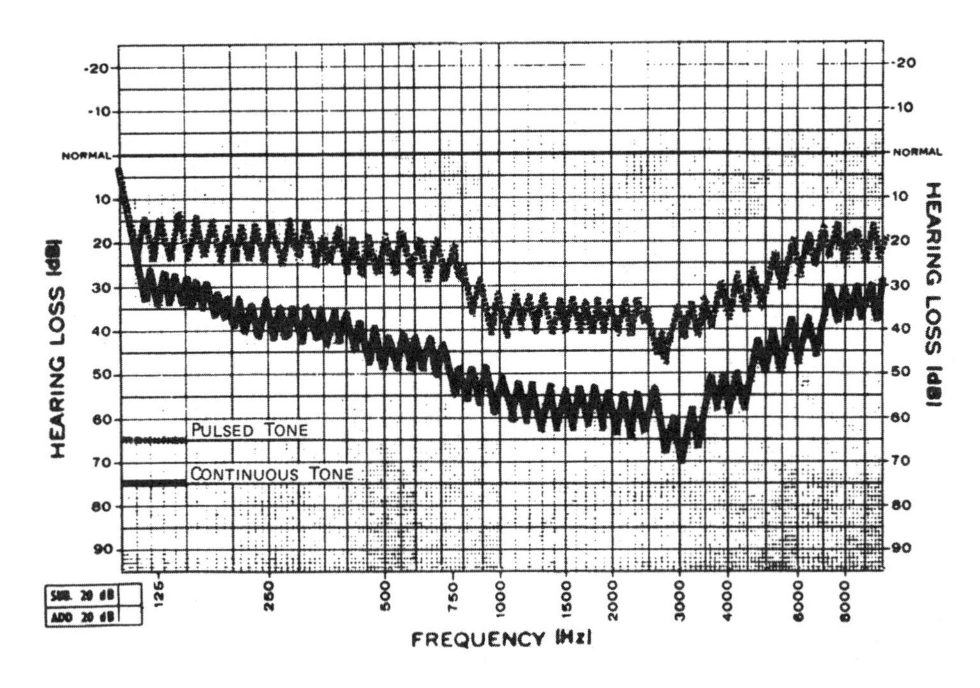

Figure 7.9 Békésy tracings in a case of acoustic neuroma. Tone decay is demonstrated by a large gap between the pulsed and the continuous-tone tracings. In some cases the continuous-tone tracing may break completely away at ~500 Hz, and the tone may not be heard even at the maximum output of the audiometer.

9. NONDIAGNOSTIC SELF-RECORDING AUDIOMETRY

Whereas the Békésy audiometer is used in the battery of diagnostic tests, other self-recording audiometers are used for establishing pure-tone thresholds only. These self-recording audiometers present discrete individual frequencies that are changed automatically at 30-s intervals and can present both pulsed or continuous tones at the option of the operator. Figure 7.10 shows an example of an audiogram obtained with a self-recording audiometer.

There is some sentiment that hearing thresholds obtained with self-recording audiometers in the industrial setting are more legally acceptable because the operator does not participate in the test. There are, however, specific procedures that must be adhered to if validity and reliability are to be assured. These include proper instructions to the subject and monitoring the test in the beginning and at intervals as the test proceeds (a common misconception is that the operator can "walk away" from the area while the audiogram table and recording pen cannot be observed by the subject). When the test is completed, the tracings should be studied by the operator to ascertain that all are acceptable.

A trained and experienced technician generally can recognize an unreliable or invalid self-recorded audiogram or a reticent subject by using certain clues: (a) Barring a language problem, an employee may refuse to follow the directions of the operator or claim he/she does not understand the instructions about his/her role in performing the hearing test. Repeated attempts to instruct the subject do not result in improved operation. (b) On repeat tests the employee may be unable to give reasonably similar responses, and this will result in a wide disparity between threshold tracings. (c) The subject may not respond at all to the tones, as though totally deaf, yet will be able to carry on a normal conversation with the examiner. (d) The audiogram may show extremely wide tracing sweeps, which makes it impossible to ascertain actual threshold. There should be at least six crossings of threshold at each frequency, and a line drawn through the midpoints should be parallel to the time axis on the audiogram (horizontal). (e) Occasionally, an employee may show a moderate to severe, mostly flat, loss, which ordinarily would cause little doubt as to its validity in the mind of the operator, except that such a pattern is uncommon and should raise the suspicion that perhaps the subject was pressing and releasing his hand

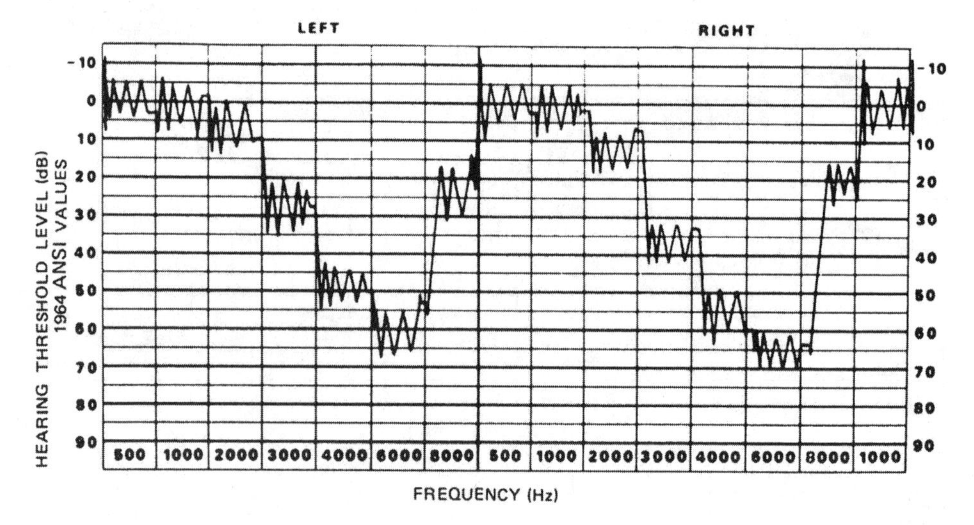

Figure 7.10 Self-recorded audiogram showing normal hearing in the low frequencies and at 8000 Hz and impairment at 3000, 4000, and 6000 Hz.

switch in a perfectly timed sequence. (f) In some types of self-recording audiometers, it is possible to feign normal hearing by keeping the button depressed during the entire test sequence. If the operator has difficulty in communicating with the subject (again barring a language problem) and yet the audiogram indicates extremely keen hearing, there is reason to doubt the accuracy of the audiogram. Should any pattern appear suspicious, the questionable frequencies or the entire audiogram should be repeated. If responses remain unsatisfactory, the test should be repeated in the manual mode if the audiometer is so equipped, or it should be repeated on a backup manual audiometer.

Self-recording audiometers do allow some mobility of the technician (although this aspect tends to be abused) and, in addition, permit one technician to test several persons at the same time. But, as in all modes of hearing testing, proper audiometer calibration and satisfactory ambient-noise levels in the test area must be maintained. All technicians, no matter what test method is employed, should be properly trained and certified.

10. IMPEDANCE AUDIOMETRY

Impedance audiometry is a comparatively new evaluation tool. It supplements otoscopic and audiometric findings and adds new capabilities to hearing evaluation.

Impedance audiometry is an objective method for evaluating the integrity and function of the auditory mechanism. It includes four separate types of measurement and has the potential for a much wider role as research in its use continues. The procedures most often used are: (a) tympanometry, (b) static compliance, (c) acoustic reflex thresholds, and (d) acoustic reflex decay test.

10.1. Tympanometry

The eardrum and connecting ossicles comprise a mechanism that should transfer vibrating energy efficiently. Tympanometry measures the mobility of this system. It is analogous to pneumatic otoscopy. If the system becomes stiffer and more resistant because some condition impedes its free movement, we are able to measure the abnormal impedance (or its reciprocal "compliance"). The compliance or impedance of the middle-ear system is measured by its response to variations in air pressure on the eardrum. The entrance of the ear canal is sealed with a probe tip containing three holes: one for supply of air pressure, one for a low-frequency probe tone (usually of 220 Hz), and the third opening connected to a pick-up microphone. As controlled degrees of positive and negative air pressure are introduced into the sealed ear canal, the resulting movement (or reduced movement) of the mechanism is plotted or automatically graphed on a chart called a tympanogram (Fig. 7.11).

10.2. Static Compliance

Static compliance is a measure of middle-ear mobility. It is measured in terms of equivalent volume in cm^3, based on two volume measurements. (a) C1 is made with the tympanic membrane in a position of poor compliance with $+200$ mm H_2O in the external canal. (b) C2 is made with the tympanic membrane at maximal compliance. $C1 - C2 =$ static compliance, which cancels out the compliance due to the column of air in the external canal. The remainder is the compliance due to the middle-ear mechanisms.

Static compliance is low when the value is <0.28 cm^3 and high when >2.5 cm^3. Its major contribution is to differentiate between fixed middle ear and a middle-ear discontinuity.

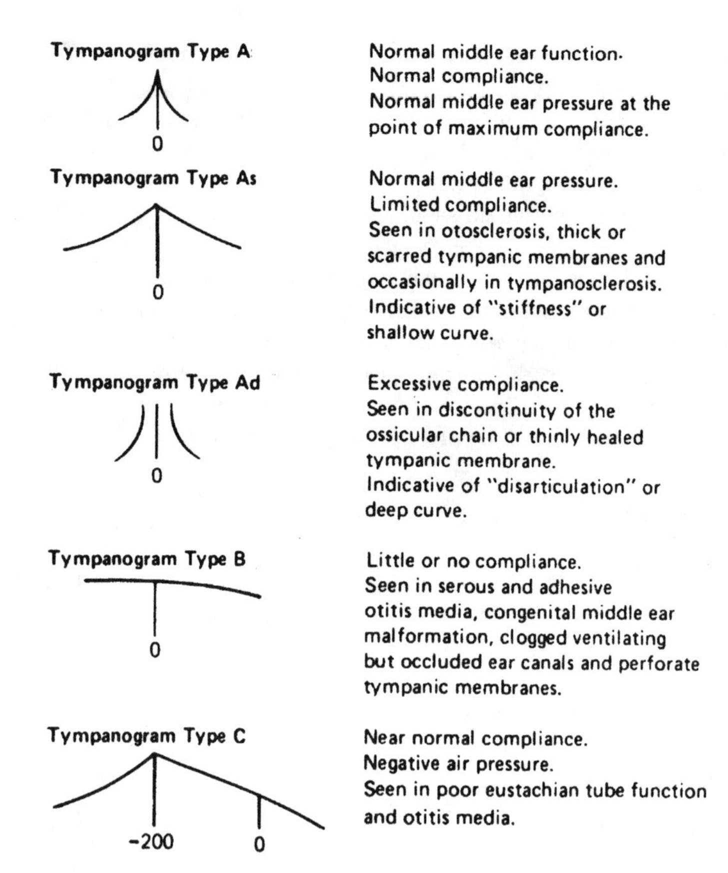

Figure 7.11 Characteristic tympanograms.

10.3. Physical Volume

The physical volume (PV) test uses a signal of fixed intensities in the ear canal. With an intact tympanic membrane the meter will balance at a cm^3 value of 1.0–1.5 in an adult and 0.7–1.0 in a child. If the eardrum is not totally intact, the PV measures will be large, often exceeding 5.0 cm^3. It is helpful in ruling out a nonobservable perforation, or it can also help to identify obstruction of a ventilating tube.

10.4. Acoustic Reflex Thresholds

This test determines the level in dB at which the stapedius muscle contracts. Normally, the reflex for pure tones is elicited at ~90 dB above the hearing threshold. For broad-band noise, it occurs at ~70 dB above threshold. The contraction occurs bilaterally, even when only one ear is stimulated. In patients with cochlear damage and associated recruitment, the reflex may occur at sensation levels <60 dB above the auditory pure-tone threshold (Metz recruitment). In bilateral conductive losses and in some unilateral losses, the acoustic reflex may be absent bilaterally. In a unilateral cochlear loss not >80 dB, acoustic reflexes may appear unilaterally when the stimulation earphone is on the "dead-ear" side. These factors then can be diagnostically important, especially when masking is impractical.

10.5. Acoustic Reflex Decay Test

In the normal ear, contraction of the middle-ear muscles to a sound 10 dB above the acoustic reflex threshold can be maintained for at least 45 s without detectable decay or adaptation. In the presence of an acoustic neuroma or other retrocochlear lesion, however, the middle-ear muscle contraction may show fatigue or decay in <10 s. In some cases it may be entirely absent.

Other impedance tests include the *ipsilateral reflex test* for the differential diagnosis of brainstem lesions, *facial nerve test* for localizing the site of a lesion in facial paralysis, *eustachian-tube tests* for determining eustachian-tube function, *fistula test* in which positive air pressure will cause dizziness and deviation of the eyes if a fistula into the inner ear exists, and a test for presence of a *glomus tumor* in which meter variations in synchrony with the pulsebeat can be observed.

Impedance audiometry is especially useful in difficult-to-test patients such as very young children, the mentally impaired, the physically disabled, and the malingerer. Like all other tests, it is not 100% accurate and must be interpreted with expertise.

10.6. Continuous-Frequency Audiometry

It must be remembered that audiograms merely sample hearing at selected frequencies, leaving many frequencies between them untested. In some cases, it is helpful to test these frequencies. This can be done with a Békésy audiometer, or with several commercially available audiometers that permit either continuous-frequency testing, or testing at ~60-Hz intervals. This kind of test may be useful, for example, for a person who complains of ringing tinnitus and fullness in one ear, but whose routine audiogram is normal. Continuous-frequency testing may reveal a dip at an in-between frequency, say 6450 Hz, that helps establish the cause of the symptom.

10.7. Tinnitus Matching

Several devices are available to help quantify tinnitus, and some newer audiometers include tinnitus-matching capabilities. These tests allow reasonably good quantifications of tinnitus pitch and loudness. Interestingly, even very loud tinnitus is rarely $>5-10$ dB above threshold.

10.8. High-Frequency Audiometry

It is often useful to test frequencies >8000 Hz, as discussed in Chapter 6. Testing to 12,000 or 14,000 Hz provides the desired information in most cases, and testing at frequencies $>14{,}000$ Hz presents special difficulties. High-frequency testing is especially valuable for differentiating presbycusis from occupational hearing loss, detecting early effects of ototoxic drugs, and potentially in selected trauma cases (such as those with tinnitus and normal routine audiograms).

11. ELECTROCOCHLEOGRAPHY

This method of assessing difficult-to-test patients involves placing an electrode in the ear and measuring directly the ear's electrical response to a sound stimulus. Most commonly, the electrode is placed through the eardrum into the promontory. In children, this may require a short general anesthetic. Newer electrodes permit high-quality testing with the electrode placed in the ear canal rather than through the tympanic membrane. Electrocochleography has proven clinically valuable particularly for confirming wave I in the brainstem response if the BERA results are equivocal, and for confirming endolympathic hydrops in patients with Meniere's syndrome. This topic is covered in greater detail in the following chapter. The ratio of summating potential (SP) to action potential (AP), the SP/AP ratio, is used to help determine whether hydrops is present. If the electrode is placed through the eardrum in contact with a promontory, an SP/AP ratio of greater than 30% is considered abnormal. With electrodes placed in the ear canal, normal values may vary depending upon electrode type. An ear canal electrode that makes contact with the tympanic membrane usually yields an abnormal SP/AP ratio of greater than 35%. For electrodes placed in the ear canal more laterally, an abnormal SP/AP ratio greater than 50% is typical.

11.1. Promontory Stimulation

The promontory stimulation test is suitable for patients with profound deafness and is rarely appropriate in patients with occupational hearing loss. The test involves placement of an electrode through the tympanic membrane. The electrode is placed against the promontory, and the cochlea is stimulated electrically. The test is used most commonly when assessing patients for possible cochlear implant candidacy.

12. EVOKED RESPONSE AUDIOMETRY

Accurate hearing testing in infants, mentally deficient patients, neurologically disabled patients, and others who cannot or will not volunteer accurate responses is a special problem. A few objective tests (those requiring no patient cooperation) are now available. Impedance audiometry is objective, but it is difficult to determine hearing thresholds from it in some cases.

Evoked-response audiometry is similar to electroencephalography or brain wave testing. Painless electrodes are attached to the patient. A darkened, "soundproof" room is used. A computer is required to isolate the auditory response from the rest of the electrical activity from the brain. Pure-tone or broad-band stimuli can be used. There are two types of evoked-response audiometry. More information on evoked responses appears in the next chapter.

12.1. Cortical Evoked Response Audiometry

This method focuses on electrical activity at the cerebral–cortex level. It allows measurement not only of auditory signals but also of other brain wave variations that are associated with the perception of sound. Therefore, it may prove a valuable tool in evaluating not only thresholds but also whether or not a sound actually reaches a level of perception in the brain. Cortical evoked responses occur at ~200 ms after the stimulus. Unfortunately, they are of limited value for threshold testing because they can be effected volitionally. For example, responses are better if a patient concentrates on an auditory signal than if he/she attempts to ignore it. Cortical evoked responses may also be altered substantially by drugs and state of consciousness.

12.2. Middle Latency Responses

Middle latency responses (MLRs) are electrical potentials whose origins are still uncertain. They are thought to be generated by sites central to the brainstem generators, such as the primary auditory cortex, association cortex, and thalamus. Although responses ~40 ms are considered most common, MLRs may be observed between 8 and 50 ms following stimulus onset. MLRs appear to be somewhat more amenable to use for special testing than brainstem responses, but they are also more subject to extraneous influence. Although MLR testing still appears to hold promise for special populations who are difficult to test by traditional means, shortcomings of this procedure have limited its routine clinical applicability.

12.3. Brainstem Evoked Response Audiometry

Brainstem evoked responses occur within the first 10 ms, and they are unaffected by behavior, attention, drugs, or level of consciousness. In fact, they can be measured under general anesthesia or during deep coma. They give information about the ear and central auditory pathways at the brainstem level, although not about cortical perception of hearing. BERA has become very popular because it is objective, consistent, and provides a great deal of valuable information. The test measures electrical peaks generated in the brainstem along the auditory pathways. The sites of origin of the waves are still controversial. Traditionally, the following scheme has been believed: Wave I is actually generated at the junction of the hair cell and VIII nerve, but the measured peak occurs in the distal auditory nerve where it leaves the bone and enters the CSF and the internal auditory canal; Wave II comes from the proximal portion of the acoustic nerve, although previously it was believed to originate at the cochlear nucleus; Wave III, at the superior olive; Wave IV, probably at the level of the lateral lemniscus in the pons; Wave V, at the inferior calyculus; Wave VI, probably at the thalamus; and Wave VII, possibly at the cortical level. Although there is good research to support these beliefs, other opinions remain possible. The most common alternate schema is as follows: Wave I, as above; Wave II, proximal portion of the auditory nerve; Wave III, cochlear nucleus; Wave IV, contralateral superior olivary complex; Wave V, lateral lemniscus. At present, only Waves I–V are used clinically for audiological purposes. Absence or distortion of peaks, or delay between peaks, can help localize lesions in the auditory pathway. For example, difference in latency between a patient's two ears of >0.2 ms currently appears to be the most sensitive audiological test for detecting acoustic neuromas (2). Nevertheless, it must be stressed that BERA is not sensitive enough to be relied on or to rule out acoustic neuroma and may miss ~40% of tumors ≤1 cm. However, BERA can have other localizing value. The presence of Wave I with absence of later waves may occur following a brainstem vascular accident with normal peripheral hearing and damaged central pathways. Increasing interwave latencies with clear separation of Waves IV and V (which usually overlap) occurs in conditions that cause conduction delay, classically multiple sclerosis. In demyelinating disease, one also sees degradation of later waves aggravated by higher rates of stimulation. Testing can be done with pure tones, broad-band noise, logons, or clicks. Approximate threshold levels can be determined. The tests can be used on infants and children and have even been advocated for routine screening in newborn nurseries.

12.4. Otoacoustic Emissions

The study of otoacoustic emission (OAE) is a growing area of interest in the scientific community. In 1977, Kemp discovered that the cochlea was capable of producing sound emissions (3). Specifically, Kemp proposed that a biomechanical amplifier within the organ of Corti underlies these properties. This amplifier is the origin of OAEs that are apparently generated as a byproduct of the traveling wave-initiated amplitude enhancement of basilar membrane vibration. There are four categories of OAEs: spontaneous, evoked, stimulus frequency, and distortion product. Two groups of OAEs appear to hold promise for future clinical use. Evoked OAEs may be conceptualized as an echo in response to a sound stimuli. They may produce consistent patterns in cochlear pathology with involvement of the outer hair cells such as noise-induced hearing loss, ototoxicity, and hereditary hearing loss. These emissions are also generally absent in hearing loss >30 dB and thus may be a good hearing screening tool for infants.

Distortion product otoacoustic emissions (DPOAE) are generated in response to paired pure tones and are felt to be more frequency specific (4). Some researchers feel that DPOAE's can accurately assess boundaries between normal and abnormal hearing with losses up to 50 dB. This category of OAEs may be clinically useful in monitoring changes in the cochlea due to hereditary hearing loss, progressive disease, and ototoxic agents. Research in the area of OAEs is still young, and many of the proposed theories have yet to be widely accepted and utilized in a clinical setting. The following chapter provides more detail on this subject.

REFERENCES

1. Rintelmann W, ed. Hearing Assessment. Baltimore, MD: University Park Press, 1979: 404–406.
2. Schmidt RJ, Sataloff RT, Newman J, Spiegel JR, Myers DL. The sensitivity of auditory brainstem response testing for the diagnosis of acoustic neuromas. Arch Otolaryngol Head Neck Surg 2001; 127:19–22.
3. Kemp OT. Stimulated acoustic emissions from within the human auditory system. J Acoust Soc Am 1978; 64:1386–1391.
4. Lonsbury-Martin BL, Harris FP, Stanger BB, Hawkins MD, Martin GK. Distortion product emissions in humans. I. Basic properties in normally hearing subjects. Ann Otol Rhinol Laryngol 1990; 99:3–14.

8

Auditory-Evoked Phenomena: Theory and Clinical Applications

Mark T. Agrama
River Region Health System, Vicksburg, Mississippi, USA

Robert T. Sataloff
Drexel University College of Medicine, American Institute for Voice and Ear Research, and Graduate Hospital, Philadelphia, Pennsylvania, USA

Thomas O. Willcox
Thomas Jefferson University, Philadelphia, Pennsylvania, USA

The diagnosis and evaluation of hearing disorders is one of the foundations of modern otology. Much of our evaluation of hearing consists of a variety of audiometric tests (1). Audiometry helps to identify, characterize, and quantify hearing loss with great accuracy. Traditional audiometry, however, requires the patient to be an active participant in the process. These behavioral tests have only limited utility when the patient cannot or will not provide a subjective response. Examples of these patients include infants, the unconscious, and malingerers (2,3).

As we have expanded our understanding of the cochlea and its neural pathways, we have developed various objective measures that reflect hearing function. These tests do not rely on the behavioral response of the patient. Instead, they record various electrical and sound phenomena produced by the ear in response to sound stimulus. Electrocochleography (ECOG) provides a measurement of the electrical potentials in the cochlea and the auditory nerve (4). Auditory brainstem response (ABR) measures the electrical potentials in the auditory nerve and higher brainstem structures (5). The testing of otoacoustic emissions (OAEs) consists of recording the sounds produced in the cochlea either spontaneously or in response to sound stimuli (6). Each of these physiologic tests has specific limitations, yet they provide a powerful complement to traditional audiometry.

This chapter will begin with a brief overview of the history of auditory theory. Each of the three tests of auditory-evoked phenomena, ECOG, ABR, and OAEs, will be reviewed in detail. The description will include their basis in anatomy and physiology and the methods for their recording and analysis. Current and potential applications of auditory-evoked phenomena will then be discussed in the context of the clinical practice of otolaryngology.

1. HISTORY OF AUDITORY THEORY

Helmholz described the resonance theory of cochlear function in 1862 (7). He studied acoustics using a series of sound resonators. He theorized that, similar to the strings on a harp, each location on the basilar membrane of the cochlea was tuned to a specific frequency.

In 1944, von Békésy presented his traveling wave theory (8). Using cadaveric cochleas, he showed that the basilar membrane vibrates maximally at different places along its length in response to pure-tone stimulation. The resulting pattern of vibration appears as a wave traveling from the base to the apex. The stiffer and less massive basal end is more responsive to higher frequencies, whereas the more flexible and massive apex vibrates more in response to lower frequencies. The peak of the amplitude of the traveling wave corresponds with the point on the basilar membrane most tuned to that frequency.

Although Békésy's traveling wave theory describes an overall frequency organization of the cochlea, it fails to account for its fine tuning mechanism. The cadaveric cochleas produce waves that distribute energy passively over a broad section of the basilar membrane. Yet the amount of vibration at the peaks of the traveling waves seen in live animals is much greater than mechanically possible for a passively vibrating system.

In 1948, Gold (9) argued that the cochlea must use active mechanical processes to augment its sensitivity to fine differences in pitch and sound intensity. He theorized that the cochlea expends energy to actively vibrate a narrow segment of the basilar membrane in response to a specific frequency. Gold essentially described a cochlear amplifier which enhances the sensitivity of the cochlea to specific frequencies at low intensity. In addition, he predicted that these vibrations would transmit in reverse to the middle ear, where they would resonate as sound. Gold never proved the existence of these vibrations, however, because of the limitations of his testing equipment.

It was not until 1978 that OAEs were reported by Kemp (6). He delivered click stimuli to the ears of both normal hearing and cochlear deaf patients. He found "cochlear echoes" that are present only in normal hearing subjects. Using loop diuretics and noise trauma, he was able to suppress what he called "acoustic emissions." Kemp demonstrated what Gold had hypothesized: the cochlea not only transduces acoustic energy into nervous impulses, but produces its own detectable sounds.

In 1983, Brownell (10) showed that the outer hair cells will reversibly change their length in response to electrical stimulus. The outer hair cells have high actin content, numerous mitochondria, and smooth endoplasmic reticulum. Resembling skeletal muscle cells, they act as cellular motors. The outer hair cells appear to be the anatomic basis of the cochlear amplifier. They enhance both the fine frequency selectivity of the cochlea as well as its sensitivity and ability to detect very faint sounds.

The cochlea receives both afferent and efferent innervation (11,12). Most of the afferent neurons travel from the inner hair cells and carry sensory potentials to the brainstem. Efferent nerve innervation travels from the olivocochlear bundle and reaches the outer hair cells near the inferior vestibular nerve (13). Of the efferent fibers, 90% synapse directly onto outer hair cells, with the remaining 10% synapsing on the afferent fibers arising from the inner hair cells. Presumably, the efferent innervation regulates outer hair cell function and generates the active tuning mechanism of the cochlear amplifier (14).

2. ELECTROCOCHLEOGRAPHY

ECOG is the measurement of electrical potentials arising in the cochlea and the auditory nerve (4,15,16). The cochlea is transiently stimulated either with a wide-band click or a specific frequency tone-burst sound. The wide-band click produces more robust potentials by stimulating the entire cochlea. Tone-burst stimulation produces potentials that are more accurate indicators of hearing levels at specific frequencies. Recordings are made by placing electrodes either through the tympanic membrane directly onto the promontory or noninvasively within the external auditory meatus (16,17). The accuracy of these recordings improves as the electrodes are placed closer to the cochlea. The three potentials recorded are the cochlear microphonic, the summating potential, and the eighth nerve compound action potential (4,15). The fourth cochlear potential, the endolymphatic potential, represents the baseline direct current battery of the endolymph produced by the stria vascularis. It is not recorded clinically.

The cochlear microphonic is an alternating current potential (16). It represents the changing potassium currents that flow through the outer hair cells. The alternating current shifts mimic the waveform of the auditory stimulus. As the basilar membrane vibrates, the outer hair cell resistance is altered, generating a current.

The summating potential is a direct current potential generated not only by the outer hair cells, but also by the inner hair cells (16). The current is possibly reflective of basilar membrane distention. The summating potential can be observed as the baseline shift in the cochlear microphonic.

The compound action potential represents the average response of the discharging neurons in the auditory nerve (16). Each auditory nerve fiber produces a very small amplitude biphasic pulse. Continuous stimuli tend to cause a random firing sequence of neurons. Because they are biphasic, they tend to cancel one another. Transient stimuli, however, produce a synchronized response of action potentials in neurons that tend to sum with one another. This effect creates the measurable compound action potential.

One of the most important applications of ECOG is in the diagnosis of Meniere's disease (17–20). Meniere's is a fluctuating, progressive process that manifests with variable hearing loss, tinnitus, aural fullness, and vertigo. In addition to the distention of Reissner's membrane by endolymphatic hydrops, there is a significant loss of hair cells and cochlear neurons. Early in the disease course, the stretching of Reissner's membrane is thought to increase the summating potential. Conversely, hydrops is thought to progressively kill sensory neurons, leading to an absolute decline in the compound action potential. An SP/AP ratio of $>0.45–0.5$ is considered to have a 95% specificity for Meniere's disease, but its sensitivity is reported to be only 56–68%. Other clinical indicators include increases in the absolute summating potential and increases in the latency of the compound action potential.

Monitoring of ECOG potentials can also be performed intraoperatively (21,22). Their presence or absence reflects the status of the cochlea. They have been used during operations on acoustic neuromas to monitor hearing. ECOG has exquisite sensitivity to disruption of cochlear blood supply. It is limited, however, to monitoring only the peripheral auditory function of the cochlea. It cannot detect changes to the intracranial portion of the auditory nerve.

3. AUDITORY BRAINSTEM RESPONSE

Jewett et al. (5) first described ABR in 1970. The ABR is generated by transient acoustic stimuli, either clicks or tone bursts, similar to ECOG. Surface electrodes are placed on the forehead and near the ears, with the contralateral mastoid acting as a ground. These electrodes record the electrical activity of the auditory pathways that occurs in the first 10–15 ms after sound stimulus. The signal is filtered to eliminate ECG and EEG interference. Multiple stimuli are presented to the ear each second, and a computer is used to average the electrical responses. Reliable waveforms are produced from the average of hundreds of responses.

A series of five distinct waveforms is typically identified (23). Each succeeding waveform is thought to represent neural activity at progressively higher auditory structures. ECOLI is a mnemonic developed for the waveforms, roughly corresponding with the distal and proximal **E**ighth nerve, **C**ochlear nucleus, superior **O**livary body, **L**ateral lemniscus, and **I**nferior colliculus.

In normal hearing ears, wave V is the largest and most consistent component of the ABR. Normative data also has been collected to describe the absolute latencies for waves I, III, and V. The wave I latency is 1.5 ms, the interval between wave I and wave III is 2 ms, and the interval between wave III and wave V is also 2 ms.

The applications of ABR include threshold testing (24), neonatal screening (25), site of lesion diagnosis (26), and intraoperative monitoring (27).

Threshold testing with ABR relies upon the first appearance and latency of wave V (24). As stimulus intensity decreases, wave V latency increases, while the amplitude decreases until it disappears below threshold. Threshold corresponds with the lowest intensity level that produces a wave V. ABR correlates well with the behavioral audiogram at frequencies between 1000 and 4000 Hz, but is still deficient in accurately predicting hearing loss in the lower frequencies.

Objective threshold testing is valuable for testing infant hearing. Because of its reliability, ABR is considered by many to be the most objective method of evaluating neonates (24,25). It remains, however, a relatively expensive and time-consuming method that requires an otolaryngologist or trained audiologist to interpret the results. A 1993 National Institutes of Health consensus panel recommended ABR screening of infants who fail an initial test, such as OAEs (28). Automated machines also have been developed that extract wave V data and provide a "pass" or "refer" response based on comparison with normative data. This enables untrained professionals to screen large populations of infants with a reported sensitivity between 93% and 100% for potential hearing disorders. Patients who are referred on initial screening undergo a full ABR test.

Site of lesion testing with ABR relies upon waves I, III, and V (26). Wave I serves as the benchmark for peripheral auditory function. A prolonged wave I latency with normal waveforms and interwave latencies correlates with conductive hearing loss. A prolonged wave I latency with abnormal waveforms with normal latencies suggests sensory hearing loss. A normal wave I latency with delayed wave I to wave III interval suggests retrocochlear pathology. And, an isolated prolongation of the wave III to wave V interval suggests isolated axial brainstem pathology. While acoustic neuroma is the tumor best identified by ABR, other cerebellopontine angle tumors such as meningioma, cholesteatoma, and facial neuromas are detected with 75% sensitivity.

Several findings are particularly suggestive of acoustic neuroma. The most common is an interaural difference in the wave V latency of 0.2 ms. The absence of all waves with the presence of only mild to moderate hearing loss (60 dB levels) in the high frequencies or the presence of wave I only suggests acoustic neuroma. Earlier estimates of the sensitivity of ABR for acoustic neuromas were ~95%. With the introduction of high-resolution MRI, however, ABR has a high false-negative rate for small tumors, with a reported sensitivity of only 85–90% overall (29). ABR is presently advocated as a first-line screening test only for patients in whom the possibility of acoustic neuroma is low.

ABR also has found use during neuro-otologic procedures for acoustic neuromas and other lesions (27,30–32). The presence of robust waves correlates with the physiologic integrity of both the cochlea and auditory nerve. A preoperative baseline is obtained from which to compare intraoperative recordings. Changes in the recordable potentials may correspond with injury caused by surgical manipulation. With sufficient warning, it is assumed that reversal of the specific surgical manipulation can reverse or at least halt the progression of the injury.

Most investigators have followed the recommendations of Grundy et al. (30), who advocate, for reasons of expediency, the measurement of only wave V. There is a practical trade-off between generating a more complete picture vs. faster real-time results. Wave V ABR has been shown to increase postoperative hearing preservation in surgery for acoustic neuromas ≤3 cc. Slavit et al. (32) conducted a case-control study of intraoperative ABR monitoring. A total of 60 consecutive patients who underwent acoustic neuroma resection with ABR were compared with a matched group of 60 controls. Overall, 42% of those monitored vs. 25% of the controls had preserved hearing. Monitoring during operations for tumors <2 cm resulted in an even higher relative percentage of hearing preservation. Intraoperative monitoring has significantly improved the preservation of hearing during operations for tumors <3 cm.

4. OTOACOUSTIC EMISSIONS

OAEs are sounds produced by the active vibration of the basilar membrane generated by the outer hair cells (6,10,33). Unlike the electrical potentials of ECOG and ABR, OAEs are mechanical auditory phenomena. In response to acoustic stimulus, the olivocochlear bundle activates the efferent nerves to the outer hair cells. Stimulation of the outer hair cells causes them to vibrate the basilar membrane. These vibrations are driven through the cochlea, through the ossicular chain, to produce sound at the tympanic membrane. In theory, the outer hair cells act as the cochlear amplifier as their vibrations enhance the tuning of the traveling wave. OAEs are thought to be a byproduct of this cochlear amplifier.

OAEs are either spontaneously produced or evoked by external stimuli (34,35). Spontaneous OAEs can be recorded intermittently in the ears of subjects in the absence of external stimuli. A 50–60% of normal-hearing ears are reported to produce spontaneous OAEs. Their incidence in women is twice that of men. They are not associated with pathological conditions and are probably indicative of normal hearing function.

Evoked OAEs come in two forms (35). Transiently evoked OAEs were the first type recorded by Kemp (6). These emissions occur in response to brief acoustic stimuli such as a click or a tone burst. Because of the broad spectrum of the stimulus, most of the basilar membrane is involved. The responding emissions consist of a broad spectrum of frequencies at varying intensity. Conditions known to cause sensorineural hearing loss have been shown to reduce or eliminate transiently evoked OAEs at thresholds of 30 dB or worse. In contrast, the presence of these emissions implies gross hearing thresholds of 30 dB or better.

Distortion product OAEs are single-frequency emissions produced in response to two simultaneous stimulus tones, F1 and F2 (36). The emission is called a distortion product because it originates from the cochlea at a frequency not present in either of the stimulus tones. Several distortion products are produced in response F1 and F2, but the largest is found at their cubic-difference of 2F1 − F2. As the primary frequencies are incremented from 500 to 8000 Hz, a distortion product audiogram can be generated similar to a pure-tone audiogram. Distortion product OAEs are suppressed at hearing thresholds of 50 dB or worse (37).

To generate robust emissions, multiple samples of transiently evoked and distortion product OAEs are obtained by repeatedly stimulating the external ear canal with sound (36). Transiently evoked involves one sound stimulus, whereas distortion product requires two sustained stimuli. The evoked responses are recorded serially by a probe microphone in the external canal. The recorded signals are then amplified, averaged, and processed by fast Fourier transform. A frequency spectrum of the OAEs is produced. The quality of the OAEs can be affected by poor stimulus delivery as well as internal and ambient noise. These factors can be minimized with proper equipment controls, noise floor thresholds, and patient cooperation.

OAE testing can also be affected by pathologic conditions involving the external and middle ear (35). Inflammation of the external ear canal can prevent comfortable placement of the stimulus and microphone probes. Middle ear effusion or ossicular discontinuity can prevent passage of the stimulus or production of the emission. OAEs are measurable in only a small percentage of children with middle ear effusion and flat tympanograms (38). After myringotomy, OAEs usually become measurable or show larger amplitudes than preoperatively.

Numerous clinical applications exist for OAEs in everyday otologic practice (35,39,40). These include the screening of newborns (41–43), assessment of sensorineural hearing loss (44,45), ototoxicity monitoring (35,39), pseudohypacusis testing (39), site of lesion diagnosis (35,39), and intraoperative monitoring (46–48).

The screening of newborns has perhaps the highest volume of clinical application of OAEs (41–43). As discussed previously, ABR testing is considered the most objective current screening method. Rhodes et al. (41) conducted a study comparing the utility of transiently evoked, distortion product OAEs, ABR, tympanometry, and acoustic stapedial reflex for initial screening, followed by ABR testing for confirmation of those who failed. Distortion product OAE testing had a 89% initial pass rate and a 92% pass rate with ABR. Infants who failed the initial distortion product test also had the highest proportion of failures on subsequent ABR. Validation of these studies remains a difficult problem, however, because there is no acceptable gold standard.

OAEs are sensitive to sensorineural hearing loss and can augment traditional audiometric diagnosis (40,44,45). The presence of transient OAEs indicates a hearing threshold of 30 dB or better for the frequency range in their spectrum. Distortion product OAEs reveal amplitude decreases at hearing thresholds >15 dB, and are generally absent at thresholds of ≥ 50 dB. Kimberley (40) used multivariate statistical analysis of distortion product OAEs amplitude to demonstrate up to 85–90% accuracy of predicting pure-tone thresholds as normal or abnormal, where normal is a 30 dB pure-tone threshold or better. The association between thresholds and distortion product OAEs, however, is not strong enough to extrapolate audiometric thresholds directly from a distortion product audiogram.

The presence or absence of OAEs can also help to distinguish pseudohypacusis from true hearing loss (35). Because behavioral audiometry relies upon accurate responses from patients, malingerers can provide false responses in an effort to receive compensation or attention. In addition to the Stenger test and ABR, OAE can provide objective results to confirm or reject the patient's claim of hearing loss.

Perhaps a more useful role for OAEs is monitoring for dynamic changes in cochlear function and ototoxicity (35,39). Noise trauma and ototoxic agents have been shown to affect outer hair cell function in animals. Distortion product OAEs have been shown to correlate well with the typical 4000-Hz notch pattern associated with noise-induced hearing loss. OAEs also have been shown to be an early and sensitive indicator of the ototoxic effects of *cis*-platinum and aminoglycosides. Serial monitoring of OAEs in workers in noisy environments or patients receiving ototoxic agents may provide a simple and objective measure of hearing loss.

OAEs may have some utility in site of lesion diagnosis of cochlear and retrocochlear pathology. Bonfils and Uziel (49) studied OAEs and conventional audiometry in 24 patients with acoustic neuromas. They discovered several emission patterns in which the OAE activity associated with the tumor was purely cochlear involvement (54%), solely retrocochlear (33%), or associated with mixed cochlear and retrocochlear symptoms (3%). For initial diagnosis, however, OAEs are not as sensitive as either ABR testing or MRI. Their role in the initial diagnosis of cerebellopontine tumors is adjunctive.

Distortion product OAEs may also have a role in the management of these tumors (39). Preoperative testing of smaller tumors (<3 cm) may assist in predicting the success of hearing preservation surgery. Robust emissions would indicate a high likelihood of cochlear viability.

Perhaps even more promising is the potential for intraoperative monitoring of OAEs during cerebellopontine surgery (46–48). Because distortion product OAEs can be continuously stimulated and monitored, they can provide up-to-date feedback on the state of the cochlea similar to ABR. Telischi (48) studied the effects of compression of the internal auditory artery in rabbits by measuring ABR and distortion product OAEs. Wave I latency was prolonged after a mean compression of 18.3 s which was comparable to the decrease in OAE amplitudes after a mean compression of 14.8 s. These results suggest that ABR wave I latency and OAE amplitude shifts occur by a similar mechanism.

Significant phase shifts in the distortion product OAE waveform, however, occurred much sooner—within a mean 5 s of compression (48). These results suggest a separate mechanism by which cochlear ischemia affects outer hair cell function. These phase changes are empirical observations of what may be changes in the outer hair cell microenvironment or their internal biochemical state. Whatever their origin, the practical importance is their more immediate feedback, which may improve the ability of surgeons to alter their technique to prevent disruption of cochlear vascular supply. Certain technical limitations still exist that interfere with intraoperative distortion product monitoring. These include the ambient noise conditions in the operating room that interfere with OAE stimulation and collection as well as the lack of OAE sensitivity to retrocochlear function. Techniques are now being developed for concurrent monitoring of OAEs and ABR that should enhance the intraoperative monitoring during cerebellopontine surgery.

5. CONCLUSIONS

Auditory-evoked phenomena provide a valuable adjunct to audiometric testing because they rely upon physiologic rather than behavioral measurements. One of their primary uses is evaluating hearing in patients who are unable or unwilling to provide behavioral responses. These phenomena also can be used in conjunction with traditional audiometric tests to provide objective confirmation.

In addition, auditory-evoked phenomena can provide information about subclinical hearing loss, occult tumors, and the real-time status of the cochlea and auditory pathways. Because ECOG, ABR, and OAEs provide a window to cochlear mechanics and auditory pathways, future research may lead to more precise understanding of sensorineural hearing loss.

REFERENCES

1. Hall JW III, Mueller HG III. Audiologists' desk reference volume I. San Diego: Singular Publishing Group, 1997.
2. Rhodes MC, Margolis RH, Hirsch JE et al. Hearing screening in the newborn intensive care nursery: comparison of methods. Otolaryngol Head Neck Surg 1999; 120:799–808.
3. Sataloff RT, Sataloff J. Hearing Loss. 3rd ed. New York, NY: Marcel Dekker, 1993.
4. Davis H. Principle of electric response audiometry. Ann Otol Rhinol Laryngol 1976; 85(suppl 28):5–96.
5. Jewett DL, Romano MN, Williston JS. Human auditory evoked potentials: possible brain stem components detected on the scalp. Science 1970; 167:1517–1518.
6. Kemp DT. Stimulated acoustic emissions from within the human auditory system. J Acoust Soc Am 1978; 64:1386–1391.
7. Helmholtz HLF. Die Lehre von den Tonempfindungen als Physiologische Grundlage fur die Theories der Musik. In: Ellis AJ, ed. On the Sensations of Tone. 3rd ed. London, UK: Longmans, 1863.
8. von Békésy G. Experiments in Hearing. New York, NY: McGraw-Hill, 1960.
9. Gold T. Hearing II: the physical basis of the action of the cochlea. Proc R Soc Lond B Biol Sci 1948; 135:492–498.
10. Brownell WE. Observations on a motile response in isolated outer hair cells. In: Webster WR, Aitken LM, eds. Mechanisms of Hearing. Melbourne: Monash University Press, 1983:5–10.
11. Kiang NY-S et al. Discharge patterns of single fibers in the cat's auditory nerve. Res Monog 1965; 65.
12. Warr WB. Organization of olivocochlear efferent systems in mammals. In: Webster DB, Popper AN, Fay RR, eds. Mammalian Auditory Pathway: Neuroanatomy. New York, NY: Springer-Verlag, 1992.
13. Spoendlin H. Sensory neural organization of the cochlea. J Laryngol Otol 1979; 93:853–877.
14. Kemp DT. Otoacoustic emissions, traveling waves and cochlear mechanics. Hear Res 1986; 22:95–104.
15. Brackmann DE, Don M, Selters WA. Electric response audiometry. In: Paparella MM, Shumrick DA, eds. Otolaryngology. 3rd ed. Philadelphia, PA: WB Saunders, 1991.
16. Eggermont JJ. Electrocochleography. In: Keidel WD, Neff WD, eds. Handbook of Sensory Physiology. Vol. V. Part 3. Berlin, Germany: Springer-Verlag, 1976.
17. Margolis RH, Rieks D, Fournier EM et al. Tympanic electrocochleography for diagnosis of Meniere's disease. Arch Otolaryngol Head Neck Surg 1995; 121:44–55.
18. Coats AC, Jenkins HA, Monroe B. Auditory evoked potentials—the cochlear summating potential in detection of endolymphatic hydrops. Am J Otol 1984; 5:443–446.
19. Dauman R, Aran JM, Charlet de Sauvage R et al. Clinical significance of the summating potential in Meniere's disease. Am J Otol 1988; 9:31–38.
20. Ferraro J, Best LG, Arenberg IK. The use of electrocochleography in the diagnosis, assessment, and monitoring of endolymphatic hydrops. Otolaryngol Clin N Am 1983; 16:69–82.
21. Levine RA. Short-latency auditory evoked potentials: intraoperative applications. Int Anesthesiol Clin 1990; 28:147–153.

22. Sabin HL, Bentivoglio P, Symon L et al. Intraoperative electrocochleography to monitor cochlear potentials during acoustic neuroma excision. Acta Neurochir 1987; 85:110–116.
23. Moller AR, Jannetta PJ. Neural generators of the auditory brainstem response. In: Jacobson JT, ed. The Auditory Brainstem Response. Boston, MA: College-Hill Press, 1985.
24. Mahoney TM. Auditory brainstem response hearing aid applications. In: Jacobsen J, ed. The Auditory Brainstem Response. San Diego, CA: College-Hill Press, 1985.
25. Bluestone CD. Universal newborn screening for hearing loss: ideal vs. reality and the role of otolaryngologists. Otolaryngol Head Neck Surg 1996; 115:89–93.
26. Selters WA, Brackmann DE. Acoustic tumor detection with brain stem electric response audiometry. Arch Otolaryngol 1997; 103:181–187.
27. Moller AR. Intraoperative neurophysiologic monitoring. In: Brackman DE, Shelton C, Arriaga MA, eds. Otologic Surgery. Philadelphia, PA: WB Saunders, 2001.
28. National Institutes of Health. Early identification of hearing impairment in infants and young children. Int J Pediatr Otorhinolaryngol 1993; 27:215–217.
29. Wilson DF, Hodgson RS, Gustafson MF et al. The sensitivity of auditory brainstem response testing in small acoustic neuromas. Laryngoscope 1992; 102:961–964.
30. Grundy BL, Jannetta PJ, Procopio PT et al. Intraoperative monitoring of brainstem auditory evoked potentials. J Neurosurg 1982; 52:674–681.
31. Telischi FF, Mom T, Agrama M et al. Comparison of the auditory-evoked brainstem response wave I to distortion-product otoacoustic emissions resulting from changes to inner ear blood flow. Laryngoscope 1999; 109:186–191.
32. Slavit DH, Harner SG, Harper M et al. Auditory monitoring during acoustic neuroma removal. Arch Otolaryngol Head Neck Surg 1991; 117:1153–1157.
33. Brownell WE. Outer hair cell electromotility and otoacoustic emissions. Ear Hear 1990; 11:82–92.
34. Martin GK, Probst R, Lonsbury-Martin BL. Otoacoustic emissions in human ears: normative findings. Ear Hear 1990; 11:47–61.
35. Lonsbury-Martin BL, Martin GK, McCoy MJ et al. New approaches to the evaluation of the auditory system and a current analysis of otoacoustic emissions. Otolaryngol Head Neck Surg 1995; 112:50–63.
36. Lonsbury-Martin BL, Harris FP, Stagner BB et al. Distortion product emissions in humans I. Basic properties in normally hearing subjects. Ann Otol Rhinol Laryngol 1990; 99:3–14.
37. Martin GK, Ohlms LA, Franklin DJ et al. Distortion product emissions in humans III. Influence of sensorineural hearing loss. Ann Otol Rhinol Laryngol 1990; 99:30–42.
38. Lonsbury-Martin BL, Martin GK, McCoy MR et al. Otoacoustic emissions testing in young children: middle ear influences. Am J Otol 1994; 15(suppl 1):13–20.
39. Balkany T, Telischi FF, McCoy MJ et al. Otoacoustic emissions in otologic practice. Am J Otol 1994; 15(suppl 1):29–38.
40. Kimberly BP. Applications of distortion-product emissions to an otological practice. Laryngoscope 1999; 109:1908–1918.
41. Rhodes MC, Margolis RH, Hirsch JE et al. Hearing screening in the newborn intensive care nursery: comparison of methods. Otolaryngol Head Neck Surg 1999; 120:799–808.
42. Choi SS, Pafitis IA, Zalzal GH et al. Clinical applications of transiently evoked otoacoustic emissions in the pediatric population. Ann Otol Rhinol Laryngol 1999; 108:132–138.
43. Norton SJ. Emerging role of evoked otoacoustic emissions in neonatal hearing screening. Am J Otol 1994; 15(suppl 1):4–12.
44. Ohlms LA, Lonsbury-Martin BL, Martin GK. Acoustic-distortion products: separation of sensory from neural dysfunction in sensorineural hearing loss in human beings and rabbits. Otolaryngol Head Neck Surg 1991; 104:159–174.
45. Schweinfurth JM, Cacace AT, Parnes SM. Clinical applications of otoacoustic emissions in sudden hearing loss. Laryngoscope 1997; 107:1457–1463.
46. Telischi FF, Widick MP, Lonsbury-Martin BL et al. Monitoring cochlear function intraoperatively using distortion product otoacoustic emissions. Am J Otol 1995; 15 (suppl 1):597–608.

47. Widick MP, Telischi FF, Lonsbury-Martin BL et al. Early effects of cerebellopontine angle compression on rabbit distortion-product otoacoustic emissions: a model for monitoring cochlear function during acoustic neuroma surgery. Otolaryngol Head Neck Surg 1994; 111:407–416.
48. Telischi FF, Mom T, Agrama M et al. Comparison of the auditory-evoked brainstem response wave I to distortion-product otoacoustic emissions resulting from changes to inner ear blood flow. Laryngoscope 1999; 109:186–191.
49. Bonfils P, Uziel A. Evoked otoacoustic emissions in patients with acoustic neuromas. Am J Otol 1988; 9:412–417.

9
Conductive Hearing Loss

Robert T. Sataloff
Drexel University College of Medicine, American Institute for Voice and Ear Research, and Graduate Hospital, Philadelphia, Pennsylvania, USA

Joseph Sataloff
Thomas Jefferson University, Philadelphia, Pennsylvania, USA

1. DIAGNOSTIC CRITERIA

Certain findings are characteristic of conductive hearing loss. The most essential ones are that the patient hears better by bone than by air conduction, and that the bone conduction is approximately normal. It would seem that these observations should suffice for classifying a case as conductive. Unfortunately, they are not always reliable, because some patients are encountered who have conductive hearing loss and yet also have reduced bone conduction, as it now is measured. The difficulty is that bone conduction tests alone do not always provide an accurate assessment of the sensorineural mechanism. Other tests often are necessary. It is therefore essential to have a clear understanding of the symptoms and the features that characterize a conductive hearing loss. They should also be placed in context with the many other conditions that can affect the ear discussed throughout this book, some of which are summarized in Appendix II.

2. CHARACTERISTIC FEATURES

These features are provided by the history, the otologic examination, and the hearing tests:

1. The history may reveal a discharging ear or a previous ear infection. A feeling of fullness, as if fluid were trapped in the ear, may accompany the hearing loss, or a sudden unilateral hearing loss may follow an effort to clean out wax with a fingertip. There may be a history of a ruptured or perforated eardrum or of head trauma. Often the hearing loss is of gradual onset and is aggravated by pregnancy. The hearing loss may even have been present at birth, or it may have been noted in early childhood.
2. Tinnitus may be present, and most frequently it is described as low-pitched or buzzing.
3. If the hearing loss is bilateral, the patient generally speaks with a soft voice, especially if the etiology is otosclerosis.
4. The patient hears better in noisy areas (paracusis of Willis).
5. Occasionally, the patient claims he/she does not hear well when eating foods that make loud noises when chewed, such as celery or carrots (because these noises are easily transmitted to the ears by bone conduction and produce a masking effect, with the result that he/she cannot hear airborne speech as well).
6. The hearing loss by air conduction is generally greater in the lower frequencies.
7. The bone conduction threshold is normal or almost normal.
8. An air–bone gap is present.
9. Otologic examination may reveal abnormality in the external auditory canal, the eardrum, or the middle ear. Sometimes bubbles or a fluid level may be seen behind the eardrum. When only the ossicular chain is involved, the visible findings through an otoscope may be normal.
10. There is no difficulty in discriminating speech if it is loud enough.
11. Recruitment and abnormal tone decay are absent.
12. If the two ears have different hearing levels, the tuning fork lateralizes to the ear with the worse hearing.
13. The maximum hearing loss possible in pure conductive hearing impairment is $\sim$70 dB.
14. The patient's hearing responses often are indecisive when they are tested at threshold during audiometry. This is in contrast to sharp end points in testing for sensorineural hearing losses.
15. Impedance audiometry may show abnormal findings.

It is helpful to understand the reasons for these characteristics so that they may be logically interpreted rather than merely memorized for classifying a clinical case.

When the patient's history reveals external or middle-ear infections associated with hearing impairment, conductive hearing loss may be suspected. Complaints may include a discharging ear, an infected external or middle ear, impacted wax, and a feeling of fluid in the ear accompanied by fullness. Often the fluid seems to move, and hearing improves with a change in position of the head. These symptoms justify a suspicion of conductive hearing loss and suggest the need for confirmatory tests. Too hasty a classification may lead to an error. It is wise to recall that fluid in the middle ear, as in secretory otitis media, produces not only a reduced air conduction threshold but may affect bone conduction as well—especially at high frequencies—even though the sensorineural mechanism is intact. When the fluid is removed surgically, both air and bone conduction levels quickly return to normal.

The hard-of-hearing patient's answers to the physician's questions invariably provide essential clues to both classification and etiology, as discussed elsewhere in this book.

Among the distinguishing features of conductive hearing loss elicited by careful questioning are the following (in addition to those listed earlier):

1. The patient has no difficulty in understanding what he/she hears as long as people speak loudly enough (because in uncomplicated conductive hearing loss only the threshold is affected, not the discrimination).

2. The patient often hears better in noisy areas such as on a bus or at a cocktail party. The reason for this is that people speak louder in noisy places, and the patient can hear the speaker's voice better above the background noises.

3. Another related finding in conductive hearing loss, but most prominent in otosclerosis, is the patient's remarkably soft voice. A soft voice in the presence of insidious hearing loss immediately should suggest otosclerosis, particularly if a low-pitched tinnitus is present. The voice is soft because the patient's excellent bone conduction gives the impression that his/her voice is louder than it actually is. Consequently, he/she lowers his/her voice to such a soft level that often it is difficult to hear the patient. Tinnitus rarely is present in conductive hearing loss, except in otosclerosis. When otosclerosis is present, there is frequently a family history of similar hearing loss. Otosclerotic hearing impairment may be aggravated by pregnancy.

When inspection of the external ear canal or middle ear reveals any obstructive abnormality, a conductive hearing loss should be suspected, but before deciding on this classification, one should first rule out a possible sensorineural loss, which also may be present. Bone conduction studies can help resolve this question. If otologic findings are normal, and there is a significant air–bone gap (i.e., if the bone conduction is better than the air conduction), there is a conductive loss—probably caused by some defect in the ossicular chain. Occasionally, fluid causing a conductive hearing loss escapes otoscopic detection because of its position, and the hearing loss is attributed to a defect in the ossicular chain rather than to the fluid. Tympanometry may be helpful in such cases.

Whenever a hearing loss is substantially in excess of 70 dB (ANSI), some other type of damage is almost certain to be superimposed on that which produces the conductive hearing loss. The reason for this is that even total disruption of the sound transmission mechanism of the middle ear produces a loss of only ~70 dB (ANSI). For example, a complete break of this kind may occur after surgery or a fracture of the temporal bone, or in the presence of congenital aplasia.

3. FINDINGS IN VARIOUS TESTS

In a pure conductive hearing loss the bone conduction is normal or almost normal, because there is no damage to the inner ear or the auditory nerve. We say "almost normal" for good reason. In some cases of pure conductive hearing loss, there is a mild reduction in bone conduction, especially in the higher frequencies, even though the sensorineural mechanism is normal. This observation emphasizes a significant blind spot in bone conduction tests; they really are not a completely valid measure of the function of the sensorineural mechanism or cochlear reserve, and reduced bone conduction sometimes may be due to purely mechanical reasons. In otosclerosis a dip often occurs at $\sim$2000 Hz in bone conduction ("Carhart's notch"). This is called a *stiffness curve.*

An air–bone gap is present in uncomplicated conductive hearing loss because the air conduction threshold is reduced, but the bone conduction threshold is essentially normal. A simple and excellent way to demonstrate this is to take a 512-Hz tuning fork and strike it gently. To a patient with a conductive hearing loss, the fork will sound weak when it is held close to his ear, but he/she will hear it better when the shaft is placed so that it presses on the mastoid bone or the teeth. This test is so useful that it should be used in every case of hearing loss, no matter how many other tests an audiologist or a technician may perform. In the hands of the experienced otologist the tuning fork is an essential diagnostic instrument. This is the basis of the Rinne test.

If there is a conductive loss in only one ear or if the conductive loss is considerably greater in one ear than in the other, a vibrating tuning fork pressed against the skull will be heard in the ear with the greater hearing loss (Weber test). Tests depending on *lateralization*, as this phenomenon is called, are not as reliable as some other tuning-fork tests, such as those comparing the efficiency of air and bone conduction.

Finally, in every case, carefully done hearing tests are analyzed to confirm the classification. Air and bone conduction audiometry are fundamental but should be corroborated by tuning-fork tests. If there is still any doubt as to the classification, impedance audiometry, recruitment studies, discrimination, and special tests are performed. In this type of loss there will be no evidence of recruitment or abnormal tone decay. The patient's ability to discriminate is always excellent in conductive hearing loss. Impedance changes, as described in Chapter 7, may be found. It is helpful to do all these tests routinely in every case, but this is not always possible in private practice. There are some audiologic centers in which an entire battery of hearing tests is carried out on all patients, even before they are examined and questioned. The experienced clinical physician performs only sufficient tests to make the classification or diagnosis reasonably certain. This selectivity saves the patient money and saves the patient and clinician considerable time.

4. PROGNOSIS

In the vast majority of clinical cases of conductive hearing loss, the prognosis is excellent. With available medical and surgical treatment, most conductive hearing losses now can be corrected.

5. AUDIOMETRIC PATTERNS

There is a popular impression that the different classifications of hearing loss are characterized by distinctly shaped audiometric patterns. One cannot be sure whether the loss is conductive or sensorineural merely by inspecting an air conduction audiogram. However, it is true that many conductive hearing losses have audiometric patterns of a distinctive appearance. One characteristic audiogram generally is described as an ascending curve, indicating that the hearing loss is greater in the lower than in the higher frequencies. This classic audiogram, illustrated in Fig. 9.1, demonstrates almost all of the findings characteristically present in conductive hearing loss resulting from otosclerosis.

Another audiometric pattern commonly found in conductive defects is shown in Fig. 9.2, which illustrates the audiologic findings in chronic otitis media.

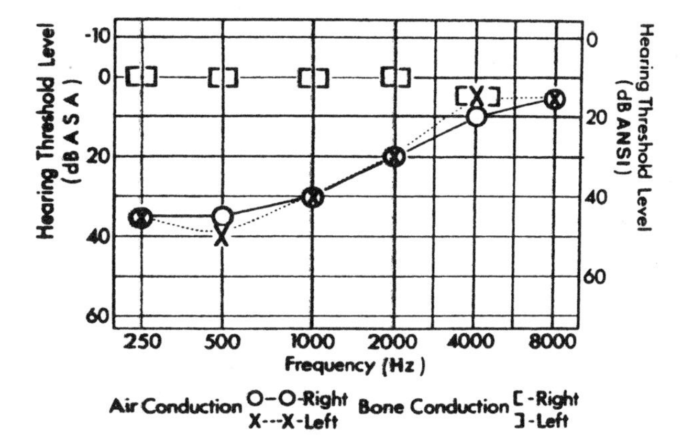

Figure 9.1 Ascending audiogram typical of conductive hearing loss. Bone conduction is better than air conduction and is normal. Discrimination is excellent. *History*: A 24-year-old woman complaining of insidious hearing loss and buzzing tinnitus over 10 years. One aunt uses hearing aid. Has trouble hearing soft voices but understands clearly when voice is loud. Hears better at cocktail parties and in noisy places. Voice is soft and barely audible. *Otologic*: Normal. *Audiologic*: Vague responses during audiometry, and fluctuating hearing levels. Bilateral air–bone gap present. Opposite ear masked during bone conduction tests. *Speech reception threshold*: Right 30 dB. Left 35 dB. *Discrimination score*: Right 100%. Left 100%. Bone conduction prolonged and better than air conduction by tuning fork. Vague lateralization to right ear. *Classification*: Conductive. *Diagnosis*: Otosclerosis. Confirmed at surgery and hearing restored. *Aids to diagnosis*: The combination of the patient's ability to hear better in noisy places and soft voice indicates a conductive lesion, which was confirmed by audiometric testing (air–bone gap). Normal otologic findings suggest stapes fixation (otosclerosis).

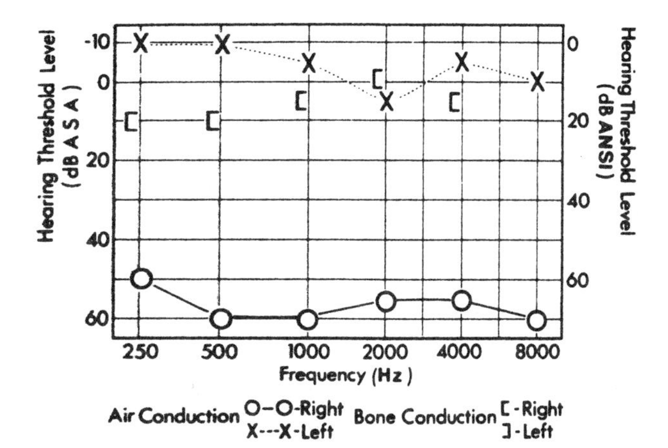

Figure 9.2 Flat audiogram found in conductive hearing loss caused by chronic otitis media. Bone conduction and discrimination are normal. *History*: An 8-year-old girl with discharge from right ear, since age of 6 months. *Otologic*: Right tympanic membrane eroded with putrid discharge in middle ear. Ossicles absent at surgery. *Audiologic*: Reduced air and normal bone conduction responses in right ear with left ear masked. Tuning fork lateralizes to right ear. *Speech reception threshold*: Right 60 dB. *Discrimination score*: Right 98%. *Expected impedance findings*: Type B curve with bilateral, absent stapedius reflexes. *Classification*: Conductive. *Diagnosis*: Chronic otitis media.

Not every case of conductive hearing loss presents all distinctive characteristics. For example, Fig. 9.3 illustrates a case of secretory otitis media in which there is a high-frequency loss and reduced bone conduction, even though the sensorineural mechanism is normal.

JOSEPH SATALOFF, M.D.
ROBERT THAYER SATALOFF, M.D.
1721 PINE STREET PHILADELPHIA, PA 19103

HEARING RECORD

NAME AGE

AIR CONDUCTION

RIGHT									LEFT							
DATE	Exam.	LEFT MASK	250	500	1000	2000	4000	8000	RIGHT MASK	250	500	1000	2000	4000	8000	AUD
4-10-78		85	-5	-5	-10	-10	5			30	30	40	40	50	FLUID	
4-18-78		80								0	10	5	15	20	FLUID REMOVED	
4-24-78										0	-5	0	0	5		

BONE CONDUCTION

RIGHT									LEFT							
DATE	Exam	LEFT MASK	250	500	1000	2000	4000	TYPE	RIGHT MASK	250	500	1000	2000	4000		AUD
4-10-78		85	5	-5	-10	15	20	WN	85	15	15	5	10	30		
4-24-78									85	0	0	0	0	5		

SPEECH RECEPTION

DATE	RIGHT	LEFT MASK	LEFT	RIGHT MASK	FREE FIELD	MIC.

DISCRIMINATION

	RIGHT				LEFT				
DATE	% SCORE	TEST LEVEL	LIST	LEFT MASK	% SCORE	TEST LEVEL	LIST	RIGHT MASK	EXAM.

HIGH FREQUENCY THRESHOLDS

	RIGHT						LEFT					
DATE	4000	8000	10000	12000	14000	LEFT MASK	RIGHT MASK	4000	8000	10000	12000	14000

RIGHT		WEBER	LEFT	
RINNE	SCHWABACH		RINNE	SCHWABACH

HEARING AID

DATE	MAKE	MODEL
RECEIVER	GAIN	EXAM.
EAR	DISCRIM	COUNC.

REMARKS

Figure 9.3 A typical audiogram in conductive hearing loss caused by thick fluid in the middle ear. The bone conduction is reduced but returns to normal after fluid is removed. *History*: Recurrent fullness in left ear with hearing loss for past 6 months. *Otologic*: Eardrum not freely mobile with air pressure. No fluid level seen. Thick, clear jelly in middle ear. Removed with myringotomy and suction. *Audiologic*: Reduced air and somewhat reduced bone conduction thresholds in left ear. Normal discrimination. Bone thresholds returned to normal after removal of fluid. Tuning fork lateralizes to left ear. *Expected impedance findings*: Type B tympanogram on the left. There is stapedius reflex with high-intensity stimulus in the left ear and none when stimulus tone is in right ear. *Classification*: Conductive. Initial testing indicated some sensorineural involvement because of the reduced bone thresholds. Removal of the fluid allowed bone responses to return to normal. *Diagnosis*: Secretory otitis. *Aids to diagnosis*: Immobile eardrum; fluctuating hearing loss with slightly reduced bone conduction; normal discrimination and lateralization to left ear suggest fluid in middle ear. Tympanogram will show abnormal findings.

6. THE DANGER OF INCOMPLETE STUDIES

These cases illustrate the inadvisability of making a classification of conductive or any other type of hearing loss solely on the basis of the air conduction alone or even on the air and the bone conduction combined. Without a careful otologic history and examination, some of these patients may be classified inaccurately and thus may be treated incorrectly or deprived of effective therapy and relief from their hearing handicap.

7. ESSENTIAL CRITERIA

To qualify as an uncomplicated conductive hearing loss, a case must have these features:

1. The bone conduction must be better than the air conduction.
2. The air–bone gap must be ~15 dB, especially in the lower frequencies.
3. The bone conduction must be normal or near normal.
4. The discrimination must be good.
5. The hearing threshold should not exceed ~70 dB (ANSI).
6. Although tests for recruitment and abnormal tone decay rarely are performed in the presence of the above features, both these phenomena should be absent.
7. In many instances, impedance audiometry helps to confirm the site and the type of damage in the middle ear.

8. VISIBLE OBSTRUCTION OF THE EXTERNAL AUDITORY CANAL

If a patient's audiologic findings show a conductive hearing loss and an obstruction is seen in the external canal, the cause of his hearing loss may be one of the following: congenital aplasia, Treacher Collins syndrome, stenosis, exostosis, impacted cerumen, fluid in external canal, collapse of ear canal, external otitis, foreign body, carcinoma, granuloma, cysts, or other causes.

9. CONGENITAL APLASIA

When the external auditory canal is absent at birth, the condition is called congenital aplasia. It is the result of a defect in fetal development. When the physician is examining an aplasia in an infant, or young child, he/she may wonder whether the neural mechanism also is defective. It is helpful to recall that the outer ear and the sensorineural apparatus are of different embryonic origins. The sound-conducting mechanism of the middle ear develops from the branchial system, whereas the sensorineural mechanism is derived from the ectodermal cyst. Therefore, it is rare to find embryonic defects in both the conductive and the sensorineural systems in the same ear. Malformations associated with syndromes are discussed in Chapter 13.

9.1. Embryonic Development

Another interesting question concerns the ability to predict from the appearance of the outer ear the likely presence of ossicles in the middle ear. The auricle starts forming at 6 weeks of embryonic life and is almost complete after 12 weeks. In some patients there is a small pit just in front and above the meatus of the canal. This shows the point at which the hillocks from the first and the second branchial arches fused to from the auricle. The embryonic structure occasionally may persist as a fistula or as a congenital preauricular cyst. The eardrum and the external auditory canal start forming about the end of 2 months of fetal life and are complete by ~7 months. First, the eardrum is formed, and a short time later, the meatus. It is possible, but uncommon, to find an aplasia of the meatus with a normal eardrum and ossicles. The ossicles begin to form from cartilage at 8 weeks of fetal life and are fully developed by 4 months, although not ossified until shortly before birth.

9.2. Congenital Abnormalities

All the causes of congenital abnormalities are not yet known, and the variety is great. There may be only a membranous layer closing the canal, while everything else is normal, or there may be no canal, no eardrum, no ossicles, and even very little middle ear. Often the aplasia is unilateral, but even in such cases evidence of a slight congenital defect may be found in the other ear.

It is not always possible to predict preoperatively whether a functioning eardrum and normal ossicles are present in congenital aplasia. Imaging techniques, although extremely useful, are not yet sophisticated enough to provide complete information. This can be obtained only at surgery. However, the shape of the auricle does provide some indication of the condition of the deeper structures in the ear. As the auricle is formed completely by the third month of fetal life, a deformed auricle suggests probable deformities of the eardrum and the ossicles. The chances of improving hearing are better when the auricle is well formed.

9.3. The Question of Surgery

The question of what can be accomplished by surgery is of decisive importance in congenital aplasia and other malformations. Parents always are anxious to know how much hearing their children lost as a result of these conditions and whether hearing can be restored.

If the aplasia is unilateral and the other ear is normal, surgery becomes an elective procedure. In bilateral aplasia, early surgery is desirable to restore hearing in at least one ear, so that the child can learn to speak at the normal stage in his development. If for any reason surgery is inadvisable or should be postponed for several years, a bone conduction hearing aid should be fitted on children as young as 3 months of age, or even younger.

In the past, the results of surgical procedures were often disappointing, and, consequently, physicians find many cases of unilateral congenital aplasia in adults. With new surgical approaches, the results are much more rewarding. The great variety of abnormal conditions that may be found at surgery and the distortion or the absence of landmarks make this surgery very difficult. It should be undertaken only by trained and experienced otologists.

9.4. Studying the Extent of Involvement

Although the diagnosis of aplasia may be made by simple inspection, the extent of involvement must be determined by careful study. CT scans or other imaging studies can be helpful in demonstrating the presence of an external auditory canal, ossicles, and semicircular canals. Tomography can show the shape of the ossicles but not their function (Fig. 9.4). The absence of semicircular canals makes the prognosis for restoring hearing

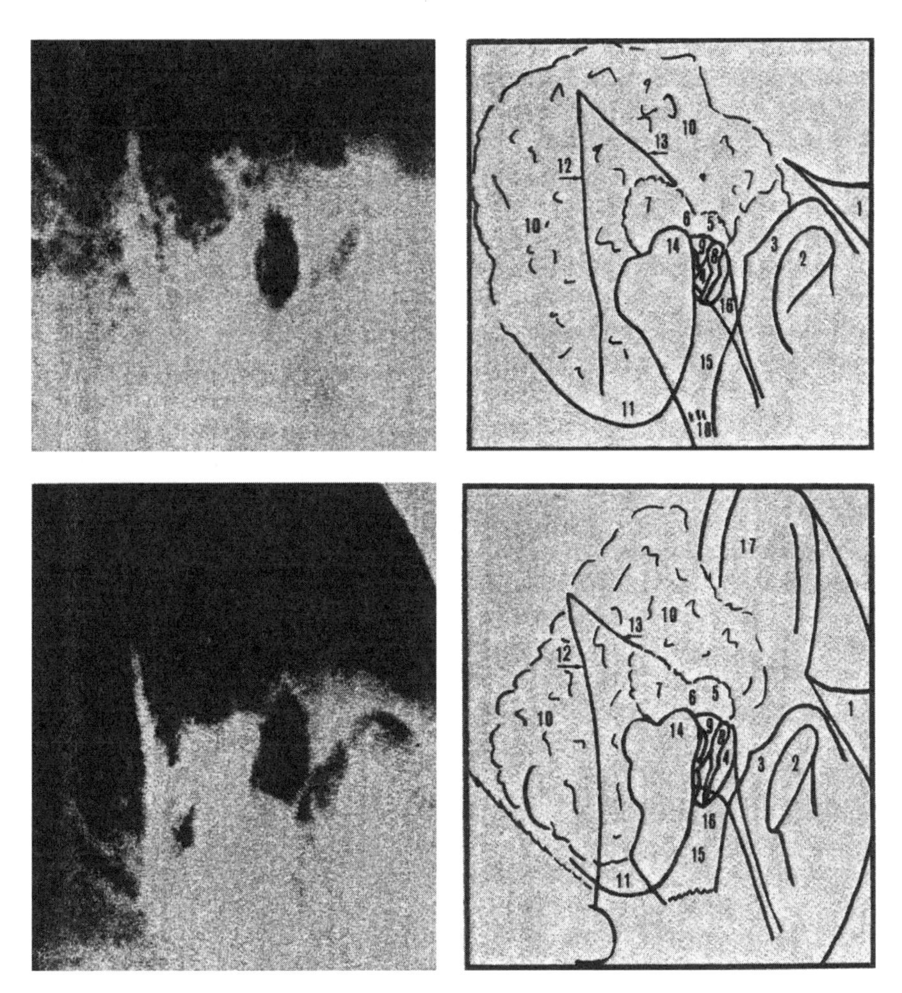

Figure 9.4 Tomograms of ossicles, Mayer's position. (1) Root of Zygoma. (2) Condyle of mandible. (3) Temporomandibular joint. (4) Tympanic activity. (5) Epitympanic space. (6) Area of aditus. (7) Area of antrum. (8) Malleus. (9) Incus. (10) Mastoid air cells. (11) Mastoid tip. (12) Anterior plate of lateral sinus. (13) Dural plate. (14) Labyrinth. (15) Petrous pyramid. (16) Eustachian tube. (17) Auricle. (18) Styloid process. (Tomograms courtesy of Dr. WE Compere Jr.)

very poor because this indicates a defect in the labyrinth. Hearing studies are essential, though these are difficult to perform in an infant. Nevertheless, a reasonable estimate of the child's hearing level can be obtained by a comprehensive study. For example, if a child consistently turns his head in the direction of a vibrating tuning fork that was not struck too forcefully, there is a reasonably good basis for optimism about restoring hearing. If there is no response to this simple test, it does not necessarily mean that hearing is absent. More elaborate tests are then indicated.

10. TREACHER COLLINS SYNDROME

One type of congenital aplasia is so distinctive that it warrants separate consideration. In Treacher Collins syndrome both auricles are deformed, and there is complete absence of both external auditory canals and eardrums; the malleus and the incus are also deformed. In addition, the patient's eyes are slanted downward at the lateral corners in the so-called antimongoloid fashion. Congenital defects, known as colobomas, are observed in ocular structures such as the iris, the retina, or the choroid. The mandible is small, and the lower jaw markedly recedes in the "Andy Gump" manner. The cheeks are pulled in, causing the lower eyelids and the face to droop.

Although no mental retardation accompanies this syndrome, children afflicted with it present such a strange appearance that they generally are taken to be "backward." A contributory factor to this impression is the marked conductive hearing loss resulting from the aplasia. Because these children hear so poorly, they are slow in developing speech, and this in turn often is attributed unjustly to lack of mental acuity. These children usually have normal sensorineural mechanisms, and their hearing can be improved by successful surgery or by the early use of a bone conduction hearing aid. Preferably, this help should be provided in infancy to avoid retardation of speech development and thus to obviate many psychological problems. The hearing level usually is ~70 dB for all frequencies. Preoperatively, it is very difficult to ascertain even by CT scan the condition of the ossicles or the presence or absence of an eardrum.

11. STENOSIS OF THE EAR CANAL

Stenosis is diagnosed readily when otoscopic examination reveals complete obstruction of the external auditory canal leading to the eardrum. This obstruction may occur anywhere along the length of the canal. Occasionally, a skin layer is the only block present; this causes a hearing loss of ~40 dB (sometimes more) in the speech frequency range (Fig. 9.5). More often, however, there is a bony wall behind the skin, and the hearing loss then may range from 50 to 60 dB in all frequencies.

Stenosis of the ear canal generally is detected on routine otologic examination in infancy. Sometimes, however, it is not picked up until a hearing loss is revealed in a school hearing test, after which physical examination shows closure of the ear canal.

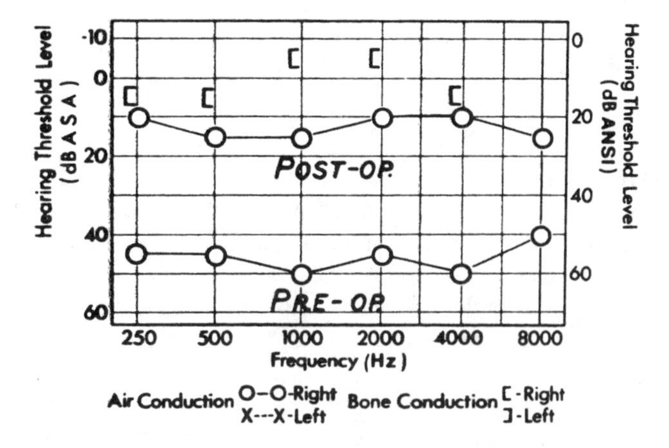

Figure 9.5 *History*: A 6-year-old boy with right congenital aplasia. *Otologic*: Left ear normal. Mild microtia of right auricle. Meatus of right external auditory canal completely occluded by firm, thick skin. X-ray showed normal ossicles. At surgery the aplasia was corrected, and the eardrum was found to be almost normal.

Stenosis is not always congenital. It may be a sequel to infections, or it may result from complications from surgery on the ears as well as from burns. In these instances, the obstruction usually is fibrous.

In some instances stenosis is not complete, but it narrows the opening of the ear canal to such an extent that any small accumulation of wax or debris causes impaction and hearing loss. In such cases the canal can be enlarged and the hearing problem resolved. However, it is essential to enlarge the canal in such a manner that it does not close again.

Correction of a stenotic ear canal generally is advisable, so that in the event of a subsequent middle-ear infection necessitating myringotomy or ear treatment, the eardrum can be visualized adequately for diagnosis and treatment.

12. EXOSTOSIS OF THE EAR CANAL

In exostosis of the ear canal, bony projections can be seen arising from the walls of the canal. They are not uncommon in adults, but they are rare in children. Although the cause of this condition is unknown, it seems to occur more frequently in individuals whose ears are exposed excessively to cold water, such as swimmers.

When it is recalled that the outer portion of the external ear is cartilaginous, it seems logical that bony exostoses are found only in the bony or inner portion of the canal. Generally, these growths are small and do not of themselves occlude the lumen completely. However, they do narrow it to such a degree that any slight accumulations of water, wax, or dead skin or any infection may cause complete blockage and immediate hearing loss. These types of episodes may be so frequent in some patients that surgery is necessary to prevent recurrent hearing loss and infections.

The hearing loss is generally ~30–40 dB when the lumen of the canal becomes occluded. The loss is not as great as in complete atresia, possibly because some of the sound waves apparently traverse the flexible material that completes the closure of the ear canal. The loss is predominantly in the lower frequencies. Upon removal of the wax or the debris, hearing returns to normal. Utmost care is necessary in examining ear canals with exostoses, because trauma to the very thin skin covering them can readily produce bleeding, swelling, infection, and further hearing loss.

In such cases, if water enters, it is very difficult for the patient to remove it by ordinary means. The water accumulates in the pockets between the exostoses and the eardrum. A diagnosis of hearing loss caused by exostoses must be based on the findings in the ear canal and the audiologic testing.

With local anesthesia in the ear canal skin, exostoses can be removed easily by elevating the skin off the bony projections, removing them with a surgical drill, and replacing the skin.

13. IMPACTED CERUMEN

13.1. Occurrence

As wax glands are situated only in the skin covering the cartilaginous or outer part of the ear canal, wax is formed exclusively in this outer area. If it is found impacted more deeply in the bony portion or against the eardrum, it probably was pushed in there, usually with a cotton swab or hairpin. Because of the structure of some ear canals and the consistency of the wax, the excess cerumen, instead of falling out of the ear, accumulates and plugs up the canal, thus causing hearing loss. This is common in infants because they have very narrow ear canals, and because mothers are likely to use large, cotton-tripped applicators that push the wax into a baby's ear canal instead of removing it.

Obstruction caused by excessive cerumen occurs frequently among workers in industrial areas because dirt gets into their ear canals. Individuals with an abundance of hairs in the ear canal easily accumulate cerumen because it becomes enmeshed in the hairs and thus is prevented from falling out by itself.

Interestingly enough, the patient with impacted wax often gives a history of sudden rather than gradual hearing loss. The person may say that while chewing or poking a finger or a probe of some kind into the ear in an effort to clean it, he/she suddenly went "deaf" in that ear. The patient may have experienced some itching or fullness in the ear, and by probing into it with a large object, pushed the cerumen into the narrower portion of the ear canal until it caused an impaction. If the canal closes while the patient is chewing, it may be explained by the proximity of the temporomandibular joint to the cartilaginous portion of the ear canal. In such instances, pressure of the joint on the soft ear canal may displace wax from its normal position and block the narrow lumen of the canal.

13.2. Tinnitus

If cerumen becomes lodged against the eardrum, the patient sometimes reports a rushing type of tinnitus or complains of hearing his/her own heartbeat. The noise stops at once when the cerumen is removed. The hearing loss invariably is accompanied by a feeling of fullness, and the loss usually is greater in the lower frequencies. Rarely is the loss $>$40 dB, and most often it is $\sim$30 dB.

13.3. Ruling Out an Organic Defect

Patients with hearing loss due to other causes commonly tell their physicians that wax in their ears probably is causing the loss. To help rule out another organic defect, it is essential to inquire whether the patient had impaired hearing and tinnitus before the present episode.

13.4. Testing

A common mistake is to look in an ear that has a large amount of cerumen and then to assure the patient that wax in the ear is his only trouble. Such a hasty diagnosis may necessitate an embarrassing retraction if the hearing loss persists after the wax is removed. Bear in mind that the severity of the loss cannot be estimated merely from the presence of a large amount of cerumen in the ear canal. Even if there is only a small pinpoint opening through the cerumen, the patient can hear almost normally, provided there is no organic defect. It is only when the ear canal is blocked completely that the hearing loss occurs. Therefore, it is highly advisable to do at least air and bone conduction audiometry before venturing to establish the cause and the prognosis of any hearing impairment. It also is important to perform air conduction audiograms after removal of the wax to be certain that hearing has been restored fully.

13.5. Removal of Cerumen

If irrigation is used to remove impacted cerumen, the ear canal should be dried afterward; otherwise, some water may remain in the deep pit at the anterior–inferior portion of the ear canal, which might cause a feeling of fullness as well as a slight interference with hearing. Figure 9.6 illustrates the type of hearing loss that frequently results from impacted

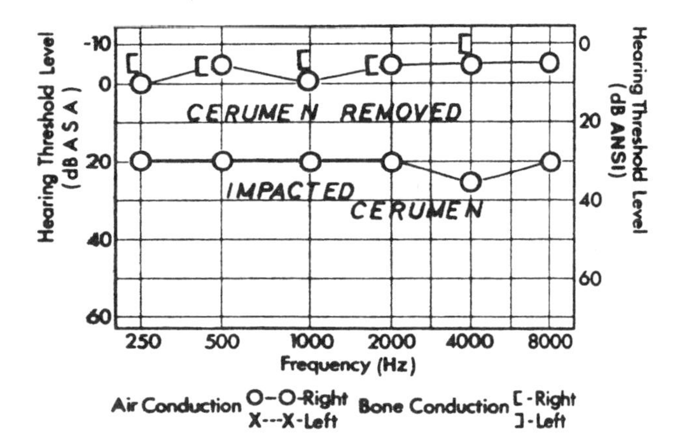

Figure 9.6 *History*: Fullness and hearing loss in right ear for several weeks after trying to clean ear with a cotton probe. *Otologic*: Right external auditory canal impacted with cerumen. Removal of cerumen revealed normal eardrum. *Audiologic*: Mild, flat hearing loss in right ear with normal bone conduction. Left ear masked. Removal of cerumen resulted in restoration of hearing to normal (upper curve). *Classification*: Conductive. *Diagnosis*: Impacted cerumen.

cerumen. Figure 9.7 illustrates an important reason for taking a careful history and doing audiologic testing before assuring a patient that his/her hearing loss can be cured merely by removing impacted cerumen. Note that in this case some hearing loss still was present even though the ear canal was cleared entirely of cerumen.

The removal of cerumen requires gentleness and patience and always should be performed with good illumination. The simplest method that suits the situation should be used. Firm plugs of wax can be removed best *en masse* by gently teasing them out with fine forceps. The forceps or any other instrument should come in contact with the wax only and not with the skin of the canal, which is thin and tender. Soft wax can be wiped out with a curette or a very thin, cotton-tipped probe. Using cotton swabs to remove wax from a narrow canal only pushes it in further and impacts it. Sometimes, it may be necessary to irrigate wax from an ear canal, but this should be avoided if the canal already is inflamed. Irrigation should not be performed in the presence of a known perforation of the eardrum, because this may cause a dry ear to flare up and result in otitis media. When irrigation is performed, the water used should be at body temperature to avoid stimulating the labyrinth and producing vertigo. The stream of water is most effective when it is directed forcefully toward one edge of the wax so that the water can get behind the plug and force it out. The ear canal should be dried carefully at the end of the procedure.

Harsh chemicals that are supposed to soften cerumen when introduced into the ear often irritate the tender skin of the canal and cause external otitis. They should be avoided or used with great caution.

14. FLUID IN THE EXTERNAL AUDITORY CANAL

The external auditory canals in some people are angled such that when water gets in, it is very difficult to remove. This may be a problem after swimming, showering, or bathing, and it is common after people spray their hair with certain lotions and after they use shampoos. The reader may recall seeing a bather after a swim tilt his/her head, slap it on one side,

NAME ____________________

DATE	RIGHT EAR AIR CONDUCTION 250	500	1000	2000	4000	8000		LEFT EAR AIR CONDUCTION 250	500	1000	2000	4000	8000
	45	50	40	50	65	60		50	50	55	65	70	60
AFTER REMOVAL OF WAX	20	20	20	25	40	45		20	25	35	35	45	50

RIGHT EAR BONE CONDUCTION							LEFT EAR BONE CONDUCTION					
10	10	10	10	40			5	10	10	20	50	

SPEECH RECEPTION: Right ______ Left ______ DISCRIMINATION: Right ______ Left ______

(Each ear is tested separately with pure tones for air conduction and bone conduction, if necessary. The tones increase in pitch in octave steps from 250 to 8,000 Hz. Normal hearing in each frequency lies between 0 and 25 decibels. The larger the number above 25 decibels in each frequency the greater the hearing loss. When the two ears differ greatly in threshold, one ear is masked with noise to test the other ear. Speech reception is the patient's threshold for everyday speech, rather than pure tones. A speech reception threshold of over 30 decibels is handicapping in many situations. The discrimination score indicates the ability to understand selected test words at a comfortable level above the speech reception threshold.)

Figure 9.7 *History*: Patient claims deafness started several months ago after attempt to remove wax from both ears. Occasional buzzing tinnitus. No vertigo. Denies family history of deafness. *Otologic*: Bilateral impacted cerumen. Removed. Eardrums normal. *Audiologic*: Bilateral reduced air conduction with near-normal bone conduction, except at 4000 Hz. Removal of cerumen closed the air–bone gap somewhat, but a residual conductive loss remained. *Classification*: Conductive. *Diagnosis*: Conductive hearing loss caused by impacted cerumen and an underlying condition of otosclerosis. *Aids to diagnosis*: Impacted cerumen does not cause such a severe conductive loss. Stapes fixation was found at surgery, and hearing improved in left ear.

then jump up and down—all this merely to get a little water out of an ear. People like this swimmer may have deformed ear canals or excess ear wax that prevents water from coming out readily. Exostoses in the ear canal also may account for this difficulty. An example of hearing loss caused by fluid in the external canal is seen in Fig. 9.8. Note the high-frequency drop in air conduction that is so suggestive of sensorineural hearing loss, but also note that the bone conduction is normal. This example should be compared with a case showing fluid in the middle ear; in such a case, a drop in bone conduction usually accompanies the reduction in air conduction. When fluid in the external auditory canal is the only cause, removing it will restore normal hearing.

Occasionally, oily medicine dropped into a child's canal for treating an ear infection may be trapped there for a long time and cause hearing loss. It is well worth noting that if only air conduction testing is done, as is customary in most industrial and school hearing test programs, one might conclude erroneously that the two cases presented in Figs. 9.7 and 9.8 should be classified as sensorineural, because the hearing loss was most pronounced in the higher frequencies.

15. COLLAPSE OF THE EAR CANAL DURING AUDIOMETRY

In rare instances, auditory canals may be shaped in a way that when pressure is directed on the pinna, the canal wall completely collapses, and conductive hearing loss results.

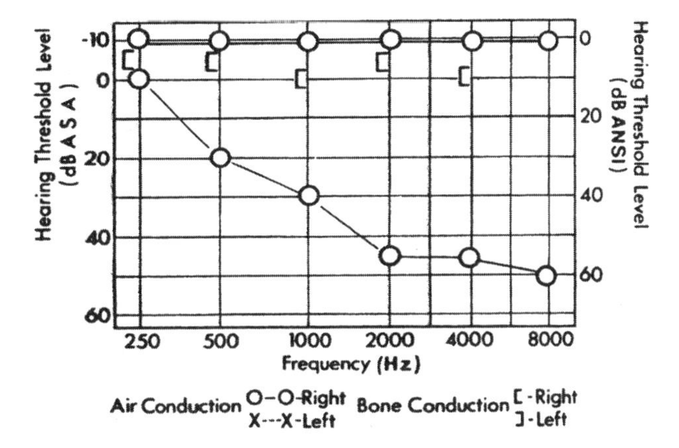

Figure 9.8 Conductive hearing loss induced in right ear by filling the external auditory canal with mineral oil. Note that the shape of the curve in this conductive loss is atypical because the greater loss is in the high frequencies. No recruitment is present with the ear filled with mineral oil. Hearing returned to normal after the mineral oil was removed.

This condition may be produced when earphones are placed over the ears during a routine hearing test. Therefore, the examiner should adjust earphones to the ears carefully to avoid collapse of the canal and a spurious hearing level. Usually, the patient will complain that his/her ears feel full and he/she unable to hear well as soon as the earphones are placed over the ears. The patient also may make some effort to readjust the earphones. The examiner should be alert to this type of situation and correct it. In some cases, it is necessary to place small tubes in the ear canal to prevent collapse.

Figure 9.9 gives an example of such a circumstance in a patient who already had a sensorineural hearing loss. To demonstrate that there was a conductive overlay produced by the collapse of the ear canal, a plastic tube was inserted to keep the ear canal open, and the hearing level improved instantly.

If there is no collapse of the canal, and yet air conduction is reduced despite apparently normal speech reception (judged by the patient's response to conversation), functional hearing loss should be suspected.

16. EXTERNAL OTITIS

16.1. Cause

Occlusion of the auditory canal with debris or swelling due to inflammation of the surrounding skin is a common cause of hearing loss, particularly in summer and in tropical climates. The most common cause is prolonged exposure of the skin to water, especially during swimming, but inflammation may result also from excessive washing or irrigation of the ear. Trauma to the skin of the canal during the removal of cerumen or foreign bodies may be another cause of external otitis, or it may result from dermatitis, infections, allergies, and systemic diseases.

16.2. Diagnosis

In diagnosing external otitis, the first consideration is to distinguish it from otitis media; this is difficult unless the eardrum is visible and appears to be normal. Occasionally,

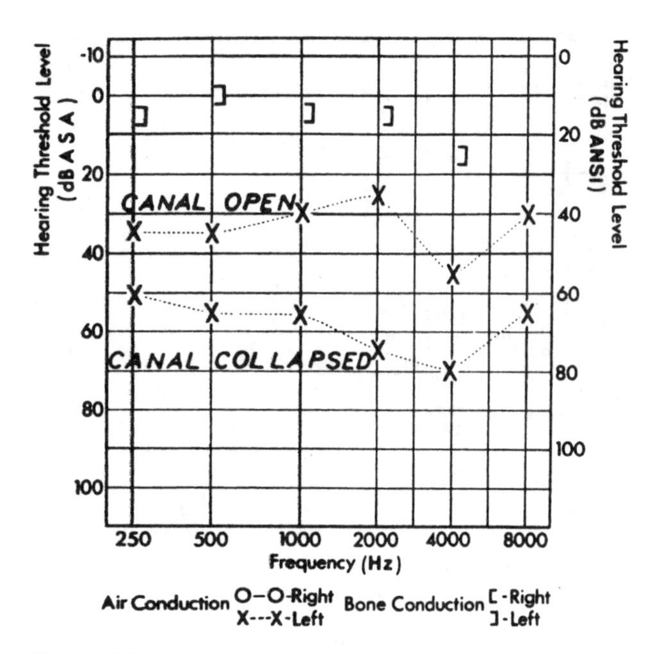

Figure 9.9 *History*: A 37-year-old man with gradual progressive hearing loss. Occasional ringing tinnitus. Maternal aunt hard-of-hearing. *Otologic*: Ears normal. Stapes fixation confirmed at surgery. *Audiologic*: Right and left ears showed a moderately severe air conduction loss. The patient's responses to conversational voice did not seem to be in keeping with the pure-tone responses, and a functional hearing loss was suspected. After removal of the earphones the patient reported that his hearing seemed to be blocked when the earphones were in place. A stock ear mold used for hearing aid evaluations was placed in the left ear canal, the earphones were put on again, and the hearing was retested. Significant improvements in thresholds were obtained when the ear canal was held open with the ear mold. Bone conduction thresholds were normal with the opposite ear masked. *Classification*: Conductive. *Diagnosis*: Otosclerosis. Inconsistencies in initial pure-tone thresholds and subjective responses to normal voice indicated a functional aspect to the problem presented. The patient's report pointed to the possibility of canal closure when earphones were worn. This was confirmed, and the original moderately severe loss actually was found to be mild in degree. *Comment*: In applying the earphones, care must be taken not to compress the ear canal.

both external otitis and otitis media occur at the same time. When the eardrum is not visible, certain features aid in establishing the diagnosis. In external otitis the skin of the auditory canal generally is edematous or excoriated. Tenderness around the entire ear is pronounced, and pain in the ear is aggravated by chewing or pressing on the ear. However, in most cases of otitis media, there is comparatively little swelling in the external canal unless mastoiditis is present or there is a profuse, irritating discharge from the middle ear. The pain in otitis media usually is very deep in the ear and is not aggravated by movements of the jaw during eating. However, sneezing and coughing often produce severe, sharp pain because of the increased pressure in the inflamed middle ear. If the discharge in an ear canal has a stringy mucoid appearance, as is found in the nose during rhinitis, it almost invariably indicates otitis media with a perforated eardrum.

Whenever a clear-cut diagnosis is not possible, therapy should be directed to both the external and middle ears. For external otitis, medication principally is applied locally to the outer ear; for otitis media, it is directed to the middle ear, the nasopharynx, and primarily systemically.

Overly strenuous efforts to introduce an otoscope into a tender, inflamed ear canal should be avoided. In some instances, a preliminary diagnosis must be based on the history and the superficial examination, as well as the clinical experience, until the infection subsides and visualization of the eardrum becomes possible.

16.3. Treatment

Of the many successful methods of treating "swimmers' external otitis," one of the best is to insert snugly into the swollen ear a large wick soaked in ear drops. The same wick should be kept in place and wetted with the same solution for at least 24–48 h. The wick should not be allowed to dry. The drops may contain antibiotics with or without steroids, or may contain acetic acid. This treatment (acetic acid) merely changes the pH in the ear canal and thus inhibits the growth of certain pathogenic organisms while the ear is healing. More specific medication is indicated if this mild therapy does not resolve the infection. Too often, a resistant external otitis is misdiagnosed as a fungus infection; the fungus usually is a secondary invader, and more careful study will reveal a bacterial or an allergic problem. Steroid and antibiotic eardrops are helpful in treating severe external otitis.

The use of strong chemicals and overtreatment should be avoided, as should excessive manipulation in a swollen ear canal. Strong medications frequently will aggravate an external otitis, and prolonged use of medication will cause an infection to persist when it would have cleared up otherwise.

After the infection has subsided in the auditory canal and after the eardrum is visualized, it is advisable to perform a hearing test to be certain that there is no underlying hearing impairment from some other condition in the middle or the inner ear.

17. FOREIGN BODY IN THE EAR CANAL

Hearing loss and fullness in the ear are often the only symptoms produced by a foreign body in the ear canal. It is surprising how long a patient can remain unaware of a piece of absorbent cotton or other foreign matter in the ear canal. Only when this foreign body becomes impacted with wax or swollen with moisture do fullness and hearing loss ensue, and then medical attention is sought. The hearing loss is caused by the occlusion of the canal; it is usually mild and greater in the lower frequencies.

The variety of foreign bodies removed from ear canals, especially in children, ranges from rubber erasers to peas. Most of these cause enough ear discomfort to attract attention before hearing loss becomes prominent, but not always. Caution always must be observed when attempting to remove a foreign body from the ear canal. Usually, special grasping instruments are essential, depending on the nature of the foreign body. General anesthesia may be advisable for a child unless the foreign body is obviously simple to grasp and to remove in one painless maneuver. It is easy to underestimate the difficulty in removing a foreign body and to run into unexpected problems; excessive preparation is better than too little. Irrigation is contraindicated if the foreign body swells in water.

Figure 9.10 illustrates the hearing loss and the findings in a man who was unaware that he had left a piece of cotton in his ear 3 months before. Only when shower water seemed to get trapped in his ear and cause fullness did he seek medical attention.

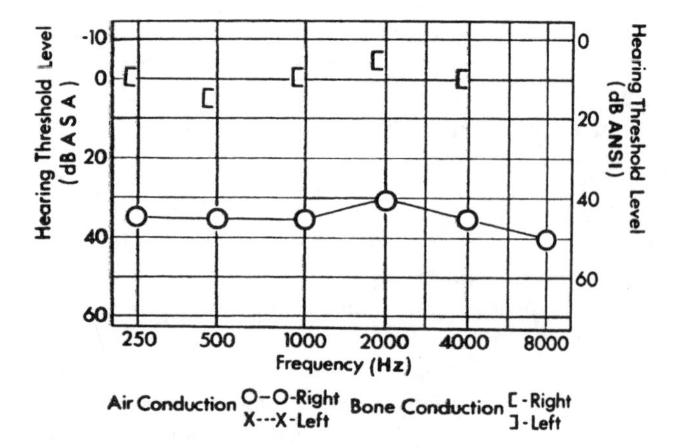

Figure 9.10 *History*: A 27-year-old man who complained of hearing loss in the right ear, which had begun 3 months before. It started with itching and fullness in the canal. He did not seek medical attention until the ear had started to discharge 2 days before. No tinnitus. *Otologic*: Left ear clear. Right ear had a putrid external otitis, and behind the discharge was a thick plug of absorbent cotton and debris. The patient recalled putting the cotton in the ear ~3 months previously. The foreign matter was removed. *Audiologic*: Right-ear air conduction thresholds revealed a flat, moderate loss. Bone conduction was normal with the left ear masked. Hearing returned to normal with removal of plug. *Classification*: Conductive. *Diagnosis*: Foreign matter in ear canal. *Note*: For an accurate diagnosis, the eardrums should be made visible by clearing out any debris that prevents inspection.

18. CARCINOMA OF THE EXTERNAL CANAL

Whenever a granuloma or a similar mass is seen in the external canal, carcinoma should be suspected. Although carcinoma in this area is not common, the possibility is serious enough to warrant constant alertness. The most common complaints associated with carcinoma are fullness in the ear, hearing loss, pain, and bleeding from the canal. The mass does not have to be large to occlude the ear canal. Often the symptoms have existed only for a short time, so that it may be possible to diagnose the malignancy comparatively early, even though the prognosis in such cases is not always good. Hearing loss is an almost insignificant aspect of this important entity, but it is frequently the presenting symptom. Early attention to a complaint of hearing impairment may be essential to prompt diagnosis of carcinoma and early surgical intervention.

19. GRANULOMA

Although granuloma in the external auditory canal is comparatively rare, it warrants special comment, because hearing loss generally is the only presenting symptom. Occasionally, there may be some drainage from the ear caused by secondary infection, but more often, the patient complains of a gradual hearing loss for no apparent reason, or possibly may attribute it to wax in the ear. A granuloma of the external auditory canal is seen readily by examination with an otoscope. This condition should not be mistaken for the fragile polypoid soft granulation tissue that arises in chronically diseased middle ears and sometimes extends into the canal. Granulomata are usually firm or hard masses that resemble neoplasms and regenerate when they are removed. Occasionally, their cause and diagnosis are most difficult to determine; for example, in the case cited in Fig. 9.11 the patient's chief complaint was an insidious hearing loss in the left ear, present for about a year, with no related symptoms or obvious cause. The right ear was normal, but the left showed an obvious complete atresia of the external canal a short distance from the opening.

The appearance was the picture of a stenosis with firm, thick, normal skin covering a bony undersurface. The major diagnostic feature was that the patient had had a normal audiogram and normal hearing in this same ear 2 years prior to his latest visit to the physician's office. This case illustrates the importance of considering the possibility of a granuloma as a cause of conductive hearing loss due to visible damage in the external canal.

The causes of granuloma include tuberculosis, eosinophilic granuloma, fungus infection, carcinoma, and others. Biopsies and special tests help to establish a definitive etiology.

20. CYSTS IN THE EAR CANAL

A large variety of cysts can occur in the ear canal and may cause hearing loss by obstructing the lumen. The common types found in the canal are sebaceous and dermoid cysts, but others also are found.

Hematomas in the auricle may become large enough to extend into the external canal and completely occlude it, producing hearing loss. In a chronic type of hematoma

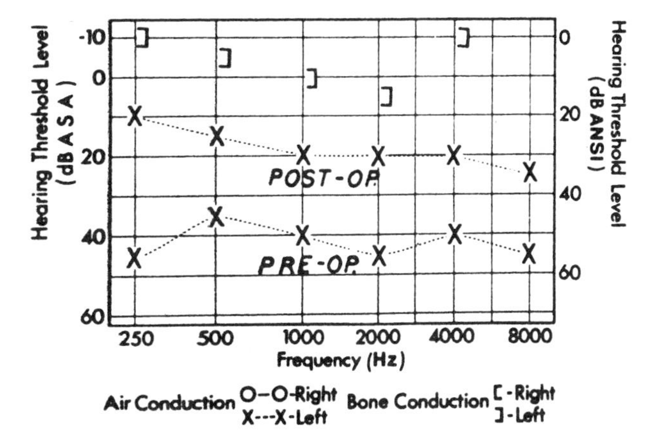

Figure 9.11 *History*: Prior otologic and audiologic examinations revealed normal findings. For the past year patient noted first intermittent, then gradual and constant fullness and hearing loss in left ear. No pain or tenderness in ear. *Otologic*: Complete atresia of left canal just inside opening. Under the thick skin was a firm bony layer that could not be penetrated with a needle. At surgery, a large, fibrous granuloma was removed from under the skin. The eardrum was thick and white but intact. A pathological diagnosis of foreign-body granuloma was established. The patient did not recall any event or symptom that might explain the diagnosis. X-ray films showed normal mastoid but bony occlusion in the left external canal. *Audiologic*: Reduced air and normal bone conduction thresholds in the left ear with masking in the right ear. Tuning fork lateralized to the left ear, and bone was better than air conduction. Postoperative air conduction responses were improved, but the air–bone gap was not closed completely. The eardrum was thick, opaque, immobile, and suggestive of a long-standing external otitis. *Classification*: Conductive. *Diagnosis*: Atresia with foreign-body granuloma.

that occurs in wrestlers' or boxers' ears, the opening of the canal can be so constricted by an old accumulation of blood and scar tissue that the canal is closed entirely, and hearing loss results.

21. OTHER CAUSES

Other causes of conductive hearing loss with abnormal findings visible in the canal include furuncles, keloids, angiomas, papillomas, osteomas, acute infectious diseases, and malignancy. None of these is a very common cause, but hearing impairment and fullness in the ear may be the chief or even the only symptoms to direct attention to the condition. The audiometric findings usually show hearing losses of <30 dB, with the lower frequencies involved to a greater extent.

22. ABNORMALITIES VISIBLE IN THE EARDRUM

An abnormal appearance of the eardrum detected by otoscopic examination of patients with a conductive hearing loss may indicate a condition largely restricted to the drum itself. Conditions of this type are reviewed in this chapter. More often the abnormality visible in the eardrum is produced by injury or disease of the middle ear and communicating structures; such conditions are considered in Chapter 9. Careful inspection of the drum and the external ear also may reveal evidence of surgical procedures previously performed on the patient.

23. MYRINGITIS

Occasionally, the eardrum is attacked by diseases without involvement of the rest of the external or the middle ear. This is called myringitis. In most cases this condition is the result of a little-understood viral infection. The most common type is described as myringitis bullosa, in which blebs appear on the drum owing to pouching out of its outer layer with fluid. These blebs appear to be clear, and when punctured, they discharge a thin, clear or slightly blood-tinged fluid. Sometimes the blebs may extend to the skin of the external canal. Myringitis starts rather abruptly and causes pain and fullness in the ear along with mild hearing loss. Usually, only one ear is involved. When the blister is punctured, and the eardrum is freed of its burden of fluid, the hearing promptly improves, and the feeling of fullness and pain diminishes.

The diagnosis occasionally is difficult because myringitis may be confused with a bulging drum caused by acute otitis media. The absence of an immobile eardrum or of any upper respiratory infection and the normal appearance of the portions of the drum not affected by belbs help to distinguish myringitis from a middle-ear infection.

Furthermore, when a bleb is punctured carefully, no hole is made through the drum, and only the outer layer is incised, so that with air pressure in the canal or politzerization through the nose the intact eardrum moves. By contrast, after myringotomy, because of the perforation, the drum does not move in response to a small difference in air pressure.

Bullous myringitis is commonly associated with mycoplasma infection. Herpes zoster, which can produce a picture somewhat similar to that of myringitis, is described in Chapter 10, as the sensorineural mechanism also is affected in many instances.

24. RUPTURED EARDRUM

24.1. Definition

An eardrum is said to be ruptured when it is penetrated suddenly by a foreign body (like a hairpin) or when it tears from the force of a slap across the ear. Otherwise, a hole in an eardrum is called a perforation rather than a rupture. Usually, the edge of a rupture is more irregular, and it is not accompanied immediately by signs of inflammation.

24.2. Causes

Most ruptures caused by penetrating objects are situated in the posterior portion of the drum because of the curve of the external canal and the slope of the drum. Ruptures caused by a sudden and intense pressure change, as by a blow to the side of the head or by an explosion, are more frequently in the anterior–inferior quadrant and occasionally in the pars flaccida. Another common cause of eardrum rupture is a slap or other impact on the ear while swimming under water. Because water gets into the middle ear, infections are more likely to ensue. In all cases of ruptured eardrum, the history is most pertinent because marked pain, fullness, and ringing in the ear often are experienced.

24.3. Treatment

Systemic antibiotics and eardrops may be useful in treating contaminated ruptures such as underwater injuries. Eardrops are not indicated for dry ruptures. Nose blowing should be avoided. In most cases the small rupture heals spontaneously if infection is prevented. If healing does not take place, it may be necessary to encourage it by cauterization or by doing a myringoplasty at some future time. The hearing loss may be as large as 60 dB if the rupture was caused by a force severe enough to impair the ossicular chain, but usually the loss is >30 dB and involves almost all frequencies. Figure 9.12 shows an example of hearing loss caused by a ruptured eardrum.

24.4. A Tear in the Eardrum

This may be considered a special type of rupture of the eardrum. It may result from a direct blow to the head that causes a longitudinal fracture of the temporal bone extending into the roof of the middle ear. Usually, the top of the eardrum is torn, blood gets into the middle and the outer ear, and occasionally the ossicular chain also is disrupted. Sometimes, there also may be a facial paralysis and cerebrospinal otorrhea. In longitudinal fractures, the roentgenograms may show a fracture line extending into the middle cranial fossa from the outer ear inward and parallel with the superior petrosal sinus. Because there is blood in the middle ear, bone conduction also is reduced but returns to normal after resolution occurs. If the fracture involves the sensorineural mechanism, bone conduction usually is affected more seriously and permanently.

25. SPARK IN THE EARDRUM

In industry a spark may hit the eardrum and have a severe effect on hearing. Figure 9.13 points out what happens in such a case. Unfortunately, this is not a rare experience when workers are welding, grinding, chipping, or burning, and an occasional spark may find its way into the canal. It hits the drum with a devastating effect. Usually, the entire drum is

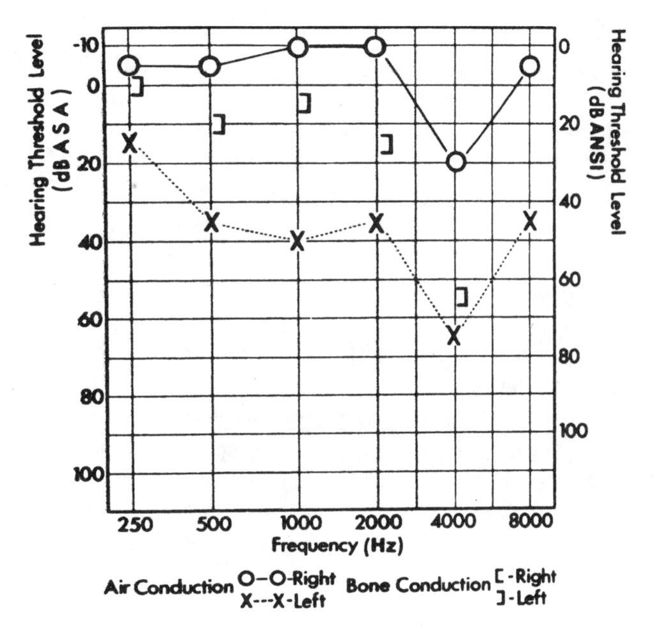

Figure 9.12 *History*: A 35-year-old man exposed to a firecracker explosion next to left ear. Intermittent ringing tinnitus. Had some previous exposure to gunfire in service. *Otologic*: Right, normal. Left ear had a large rupture of the eardrum without visible infection. *Audiologic*: The right ear had normal thresholds, except for a 20-dB dip at 4000 Hz (C-5 dip). The left ear had a moderate air conduction loss and a 65-dB C-5 dip. Masked bone conduction thresholds on the left ear showed a good degree of air–bone gap. *Classification*: Conductive. *Diagnosis*: Acoustic trauma with ruptured eardrum. The hearing returned to a level similar to that in the right ear after a myringoplasty. The ossicles were found to be functioning.

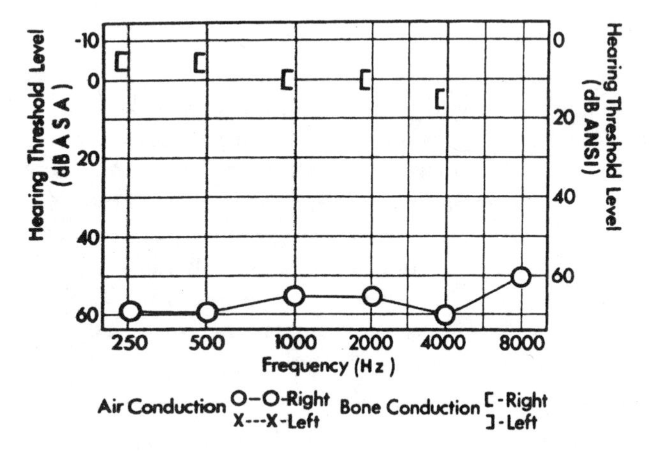

Figure 9.13 *History*: Patient works as a welder, and a spark flew into his right ear, causing severe burning pain and subsequent hearing loss. *Otologic*: Patient was seen 24 h after accident. Right eardrum almost cleanly and completely destroyed without evidence of inflammation. Handle of malleus visible as well as incudostapedial joint. External canal not affected visibly. *Audiologic*: Flat 60-dB loss with good bone conduction (left ear masked) and lateralization to right. *Classification*: Conductive. *Diagnosis*: Eardrum destroyed by hot spark.

destroyed cleanly, leaving only the handle of the malleus hanging down. Little or no infection accompanies this trauma. The pain is severe but of short duration. The hearing loss is usually ~50 or 60 dB and affects all frequencies. As in cases of ruptured eardrum, forceful nose blowing should be avoided after such an accident. Applying eardrops is not advisable. The spark cauterizes the area, and infection does not ensue if antibiotics are administered orally and local probing is avoided. Myringoplasty usually is indicated, and the hearing usually can be improved or restored.

Wherever free sparks are produced in industry, steps should be taken to shield personnel from them. In places where this is not practical, as in some forges or foundries, it is essential to protect the ears with ear protection and the eyes with safety glasses.

26. PERFORATED EARDRUM

The role of the eardrum in hearing often is misunderstood, especially by laypeople. Many believe that without an eardrum, it is not possible to hear at all. Some parents understandably become very concerned when a myringotomy is suggested for their child, because they are fearful of having the hearing destroyed by a hole in the drum. There even are physicians who attribute marked hearing defects to such conditions as "hardening of the eardrum" and "too small an eardrum."

26.1. Extent of Hearing Loss

Actually, it is possible to have very little loss in hearing though a large hole may be visible in the drum. In a nationwide study of school children, 60% of ears with dry perforations went undetected by hearing screening tests because the hearing loss of these children was $<$20 dB. The hearing level in another 40% of the ears showing perforations was better than 5 dB on the audiometer. It really is not possible to predict the degree of hearing loss from the appearance of the tympanic membrane. In some cases in which the membrane is perforated and even scarred, the hearing is nearly normal. There may be a severe hearing loss with only a pinpoint perforation. The cause of the pathology and its effect on the middle ear are more important criteria than the appearance of the eardrum.

However, the *location of the perforation* does have some *diagnostic meaning*. A persistent posterior perforation suggests associated mastoid infection, whereas an anterior perforation is not quite as serious. A superior perforation in Shrapnell's area also suggests that a fairly serious infection has been present.

A dry perforation in the tympanic membrane indicates that at some time infection probably was present in the middle ear. Only in rare instances does a perforation result from an infection in the external ear. Although the character and the location of the perforation are not reliable indications of the degree of hearing loss, they do play a certain role. For example, large perforations usually cause a greater loss. It appears that the principal effect of a perforation is a reduction of the surface on which sound pressure is exerted, the effect being proportional only to the area of the perforation. However, the relationship is not quite so simple.

Because the eardrum is effective only if it is in contact with the handle of the malleus, perforations affecting this area are especially damaging to the hearing. Perforations in the tense part of the eardrum, such as the anterior–inferior portion, affect the hearing more than those in the flaccid and superior portions because the tense part is largely responsible for the stiffness of the eardrum.

26.2. Testing for Hearing Loss

An audiogram is the best way to determine the extent of the hearing loss when a perforation is found.

Caution should be exercised in correlating a perforation with the audiogram. It is natural for a physician to blame a hearing loss on a small perforation found in the eardrum, even if the loss is as severe as 60 dB; yet a perforation alone rarely produces such a marked degree of hearing loss. However, if the drum is entirely eroded and the ossicular chain has become ineffective, a hearing loss of 70 dB ANSI is to be expected. It is more than likely that there is some break in the ossicular chain when a severe degree of conductive hearing loss is found. This can be determined very simply by placing a small patch over the perforation, using Gelfoam or some other artificial material. If the hearing loss is a result of the perforation, then the hearing should be restored immediately.

This therapeutic test may be used when myringoplastic surgery is contemplated. If a hearing test shows that a temporary patch on the perforation in the drum fails to restore hearing, the otologist will know that further exploration of the middle ear is indicated to discover the major cause of the hearing loss, because a simple myringoplasty would yield disappointing results.

26.3. Perforations

26.3.1. Demonstration

It is often difficult to be certain that a perforation actually is present in the tympanic membrane. Sometimes a large hole seems to be visible through the otoscope, but more careful scrutiny may reveal a very fine, thin, transparent film of epithelium covering the area. Such a film can be demonstrated best by gentle pressure on the ear canal with a pneumatic otoscope; by moving the eardrum, the light reflected from it will be visible on the covered perforation. Unless pressure is very gentle, the healed area may be broken readily. Spraying a fine, sterile powder into the ear also can outline a perforation or show an intact drum. It is also possible to fill an ear canal with water or ear drops and "pop" or politzerize the ear, looking for bubbles escaping through a tiny perforation.

26.4. Healing

Perforations in the eardrum heal in several ways. Most heal with a small scar that barely is visible. Others leave a somewhat thick, white scar which can be seen for many years. These scars in themselves produce no measurable hearing loss.

26.5. Retracted Drum with Perforation

In some ears the drum is retracted so much that it envelops the promontory like a sheet of cellophane, and yet it remains intact. In some cases, the retracted drum may have a perforation in it that is scarcely visible. The aperture can be visualized best by blowing camphor mist under pressure into one nostril while the other nostril is compressed, and the patient is asked to swallow. Through an otoscope, the mist can be seen coming out of the perforation. If the eardrum is intact, often it will push out a little bit, or it even may suddenly snap out quite sharply. In the latter case the patient's hearing will improve suddenly.

26.5.1. Methods of Closing

Excellent strides have been made in closing perforations in eardrums. Tympanic perforations cannot and generally should not be closed until the middle ear has been completely clear of infection and dry for at least several weeks; otherwise, the closure often breaks down. A small perforation that is not marginal (i.e., does not have an edge on the annulus) generally can be closed by repeated cauterization of its rim with trichloroacetic acid. This is done to destroy the edge of outer epithelium that has grown over the rim, thus preventing the middle (fibrous) layer from closing the hole spontaneously. The cauterization is repeated every few weeks until the closure is complete. Sometimes a long-standing perforation closes by itself after an acute attack of otitis media. In such cases the edge of the perforation probably has been irritated and traumatized by the acute infection, and the trauma has stimulated growth of tissue around the edges of the perforation.

For large perforations and marginal ones, myringoplasty is necessary. This is done under local anesthesia in adults. The epithelium around the edge of the perforation is removed, and a thin piece of skin, vein, fascia, heart valve tissue, or other graft is applied. This technic produces excellent results. Sometimes, in marginal perforations a sliding skin graft from the external canal is most effective. In cases in which the preliminary closure of the perforation with Gelfoam or artificial membrane does not restore hearing, the middle ear and the ossicular chain should be explored during surgery, especially in cases in which the hearing loss exceeds 30 dB.

27. RETRACTED EARDRUM

Physicians frequently attribute a marked conductive hearing loss, such as 50 dB, to a retracted eardrum, to adhesions in the middle ear, or to so-called catarrh. In most cases these conclusions are unjustified. Rarely do any of these conditions in themselves cause a hearing loss $\gtrsim$40 dB. If the loss is substantially greater, there probably is some accompanying defect in the ossicular chain or the middle ear.

To understand why a retracted drum usually causes only a mild loss, it should be recalled that sound waves can be transmitted through the ossicles even if only part of the eardrum is present, particularly the portion around the malleus; therefore, even some badly retracted drums cause comparatively little hearing loss. In a few exceptional cases a hearing loss much $>$40 dB is caused by retraction, and in these cases hearing returns to normal with successful inflation of the middle ear. This is demonstrated in Fig. 9.14.

Actually, a retracted eardrum is not pulled in; it is pushed in because the atmospheric pressure in the external canal is greater than that in the middle ear. Reduced middle-ear pressure is produced most frequently by dysfunction in the eustachian tube. In children, hypertrophied adenoids and allergies are the chief causes. In adults, the principal causes are infections and allergies.

In any case, the eustachian tube becomes congested and blocks access to air needed to balance middle-ear pressure with that in the external canal. The air already in the middle ear slowly is absorbed, and as a result the eardrum gradually is pushed in toward the promontory. Sometimes the drum is so thin that when it is retracted, it envelops the promontory and is hardly discernible as a drum. By forced politzerization it occasionally can be distended. If it does distend, it usually is flaccid and redundant, and later it retracts again unless the cause of the original retraction has been corrected.

In retracted drums the reflected cone of light disappears, and the handle of the malleus appears to be prominent and shorter. Tinnitus rarely is caused by a retracted drum. If tinnitus is present, another cause should be suspected. Because a retracted eardrum often is associated with a perforation, it is important to use forced air pressure (pneumatic otoscopy) to be sure that the drum is intact.

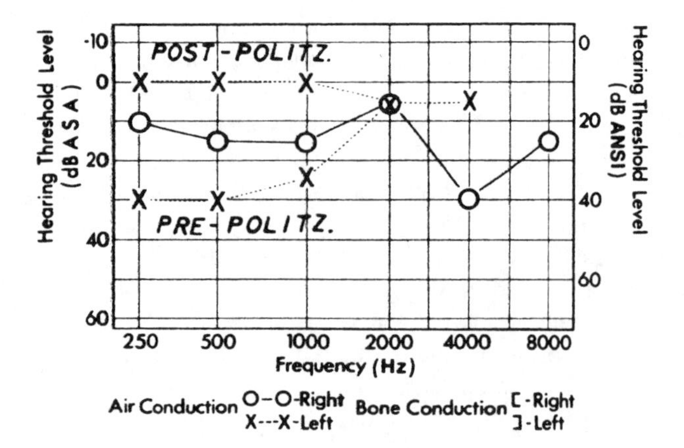

Figure 9.14 *History*: A 40-year-old woman with insidious hearing loss for past 3 years following an upper respiratory infection. Hears better intermittently. Has occasional pulsebeat tinnitus. *Otologic*: Scarred eardrums with healed perforations. Drums retracted. With politzerization the left eardrums ballooned out, and hearing immediately returned to normal. *Audiologic*: Air conduction thresholds revealed a mild right hearing loss and a moderate left loss with greater loss in the low frequencies. After politzerization, the left ear thresholds returned to normal. *Impedance*: Type C tympanogram before politzerization. *Diagnosis*: Retracted and scarred eardrums caused by previous infections.

28. FLACCID EARDRUM

A flaccid eardrum rarely causes hearing loss. The cause of the wrinkled and redundant eardrum is not always clear, but frequently it follows some long-standing malfunction of the eustachian tube, with alternate retraction and bulging of the eardrum resulting in the wrinkled appearance. Associated with a flaccid eardrum is a sensation of fullness, and occasionally the patient may report hearing his/her own breathing sounds. There is no general agreement on the best course of treatment, but in any case politzerization and inflations should be avoided.

29. THE SENILE EARDRUM

The appearance of a normal eardrum described in textbooks is always that of a young person. As human beings age, changes occur in the eardrum as they do in the eye and in the skin. The aged drum no longer reflects a cone of light, and it is no longer shiny. Its outer layer now appears to be thick and white or gray. It also loses its elasticity and becomes difficult to move with air pressure. White plaques and strands of fibrous tissue are evident in the middle fibrous layer of the drum.

It is difficult to ascertain how much, if any, hearing loss is caused by senile drum changes, but it probably is negligible. The high-frequency hearing loss associated with aging is caused by sensorineural changes. Usually, low-tone losses are negligible; if present, they too are likely to be sensorineural. Many aged patients whose eardrums show marked senile changes have normal hearing. However, it is important to expect senile changes in the drums of aged persons and not to attribute to them hearing losses which, in fact, have different etiologies.

30. CAUSES IN MIDDLE EAR AND COMMUNICATING STRUCTURES

Although the causes of conductive hearing loss enumerated in this chapter originate in the middle ear, the eustachian tube, or the nasopharynx, they frequently can be detected by transilluminating the eardrum and looking not necessarily *at* the drum, but *through* it for telltale evidence of fluid, air bubbles, reflections, shadows, and the ossicles. The appearance of the drum itself also may be changed by pathological conditions behind it. What the otoscope reveals must be interpreted using otologic experience.

31. CATARRHAL DEAFNESS AND ADHESIONS

One still commonly hears of diagnosis of "catarrhal deafness." The pathology implied by this term is indefinite. Occasionally, it refers to a cloudy eardrum, which is blamed for a marked conductive hearing loss. Many times the term is applied to any conductive hearing loss in which the drum appears to be slightly opaque or retracted. Actually, usually this is not catarrhal deafness but otosclerosis or tympanosclerosis with some incidental minor changes in the eardrum owing to malfunction of the eustachian tube. In all likelihood, the term catarrhal deafness was intended to apply to any one of the several conditions which today are identified separately as otitis media, secretory otitis media, and slight thickening and retraction of the eardrum suggesting middle-ear adhesions. A better understanding of ear physiology and pathology now makes it possible to apply more specific and meaningful terms to these conditions.

There also is some doubt concerning the role of adhesions in the middle ear as a major cause of conductive hearing loss. Experience with stapes surgery has convinced otologists that occasional adhesions about the ossicles produce negligible hearing loss. Even when many adhesions have to be cut during stapes surgery, very little of the hearing impairment seems to be attributable justly to this cause. Apparently, the ossicles transmit sound waves quite readily in spite of being tied down with adhesions. However, in those instances in which the adhesions add weight and mass to the ossicular chain and drum, hearing is likely to be impaired measurably.

Figure 9.15 illustrates the mild type of conductive hearing loss that probably is caused solely by adhesions in the middle ear. Figure 9.16 shows a case that was diagnosed as catarrhal deafness by several otologists and proved on surgical exploration to be tympanosclerosis. Schuknecht reported a case of hearing loss caused by a bridge of bone binding the neck of the stapes and preventing mobility (Fig. 9.17).

32. AEROTITIS MEDIA (MIDDLE EAR BAROTRAUMA)

Hearing loss in aerotitis media usually is rather mild and temporary unless complications develop. The immediate cause of hearing impairment is retraction of the eardrum, but if the pressure disparity persists, fluid may accumulate in the middle ear and further aggravate the hearing loss.

The pressure disparity between the outer and middle ears almost invariably occurs during rapid descent in an airplane, when the atmospheric pressure increases rapidly as the plane comes down. If congestion prevents air from passing up the eustachian tube to equalize the increased pressure in the external canal, this pressure pushes in the eardrum toward the middle ear, causing sudden pain, fullness, and hearing loss. It occurs most

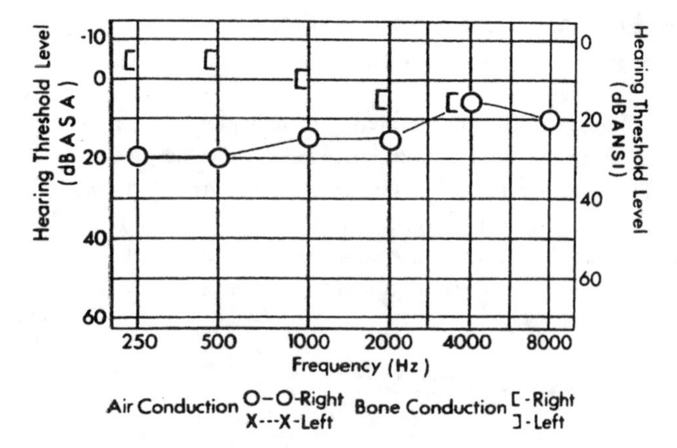

Figure 9.15 *History*: A 42-year-old man with a history of recurrent otitis media in the right ear as a child. Nonprogressive. No tinnitus. No familial deafness. *Otologic*: Right eardrum was scarred and reflected previous infection, but it was intact and moved normally. No fluid visible. Hearing not improved after politzerization. Exploratory surgery on the right ear revealed muliple bands of scar tissue around incus and crura, binding ossicles to promontory. Stapes footplate was mobile. The ossicles were freed of the adhesions, and hearing in the right ear returned to normal. *Audiologic*: Mildly reduced low- and midfrequency thresholds with normal bone conduction thresholds in the right ear. Right ear bone thresholds were obtained with left ear masked. *Impedance*: Shallow Type A tympanogram. *Classification*: Conductive. *Diagnosis*: Adhesive deafness.

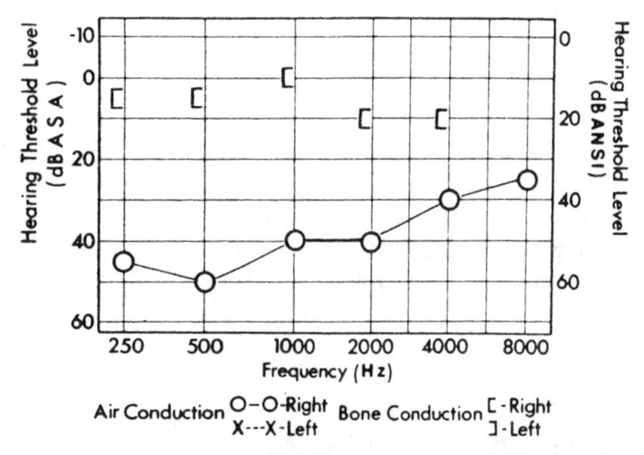

Figure 9.16 *History*: For a period of 6 years patient had had chronic otitis media, which then had cleared up and remained so for the last 10 years. The original hearing loss had not progressed. Condition originally had been diagnosed as catarrhal deafness. *Otologic*: Right eardrum was intact but was thick, scarred, and had white calcific plaques. Exploratory surgery revealed no fluid in the middle ear but layers of shalelike tympanosclerotic bone around the footplate stapes. The plate was mobilized. *Audiologic*: Moderately reduced air conduction thresholds with normal bone conduction threshold in the right ear. Masking used in the left ear in testing for right-ear air and bone thresholds. Hearing improved with footplate mobilization. *Impedance*: Type A_S tympanogram, stapedius reflex absent. *Classification*: Conductive hearing loss. *Diagnosis*: Deafness caused by tympanosclerosis rather than catarrhal deafness, as originally diagnosed. *Aids to diagnosis*: The severe, nonprogressive conductive hearing loss with white calcific deposits is suggestive of tympanosclerosis.

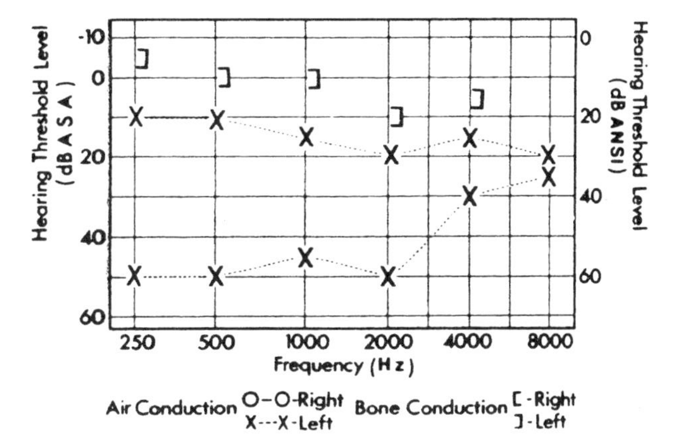

Figure 9.17 *History*: Progressive hearing loss for 12 years. Patient has worn a hearing aid in the left ear for 5 years. *Otologic*: Normal eardrums. *Audiologic*: Air conduction thresholds revealed a moderately severe loss in the left ear and a moderate loss in the right ear. Tuning-fork tests indicated better hearing by bone conduction than by air conduction in the left ear. *Impedance*: Would have revealed type A_S tympanogram with absent stapedius reflex. *Classification*: Conductive. *Diagnosis*: Bridge of bone binding stapes visualized and corrected at surgery. [From Schuknecht (1).]

frequently in people whose eustachian tubes are congested because of infection or allergy. To prevent aerotitis media, people with upper respiratory infections and acute allergies should be cautioned against flying. In adequately pressurized airplanes, aerotitis media is minimized.

If a patient is seen soon after the onset of aerotitis media, the symptoms can be relieved by politzerization followed by steroids, or by decongestants and myringotomy, followed by administration of an oral decongestant and politzerization. When the air pressure in the middle ear is made equal to that in the outer ear, hearing is restored, and the feeling of fullness gradually disappears. Figure 9.18 describes a classic example of aerotitis media with recovery. If a patient is seen several days after the onset of aerotitis media, it may be necessary to use antibiotics and steroids. In all cases, hearing should return to normal when the eardrum is restored to its normal position and the middle ear and the eustachian tube are clear.

33. HEMOTYMPANUM

Blood in the middle ear with an intact drum produces a dark-red or bluish hue. This is a common finding after head injury with fracture of the middle cranial fossa. Sometimes a fluid level is visible. If the sensorineural mechanism has not been injured, the conductive hearing loss generally is ~40 dB, usually involving all frequencies. Occasionally, the high frequencies are involved to a greater degree. Interestingly, the bone conduction shows a high-frequency drop (Fig. 9.19) and prompts the erroneous belief that sensorineural damage has occurred. Both air conduction and bone conduction may return to normal as the blood is absorbed. Blood also may be seen in the middle ear for a short time following stapes surgery, and when this occurs, hearing improvement is delayed until absorption takes place.

The normal middle ear has a remarkable ability to remove fluid and debris by absorption, ciliary action, and phagocytosis. For this reason, it is usually unnecessary

JOSEPH SATALOFF, M.D.
ROBERT THAYER SATALOFF, M.D.
1721 PINE STREET PHILADELPHIA, PA 19103

HEARING RECORD

NAME AGE

AIR CONDUCTION

			RIGHT						LEFT							
DATE	Exam	LEFT MASK	250	500	1000	2000	4000	8000	RIGHT MASK	250	500	1000	2000	4000	8000	AUD
			20	35	35	35	40	25								
			0	10	20	10	10	10	AFTER POLITZERIZATION							
			-5	0	-10	0	5	5	AFTER ASPIRATION TO CLEAR PASSAGE							

BONE CONDUCTION

			RIGHT						LEFT						
DATE	Exam	LEFT MASK	250	500	1000	2000	4000	RIGHT MASK	250	500	1000	2000	4000		AUD
		95dB	-5	-10	0	5	10								
		95dB	0	0	-5	10	10	AFTER POLITZERIZATION							
		95dB	5	5	-10	10	5	AFTER ASPIRATION TO CLEAR PASSAGE							

SPEECH RECEPTION

DATE	RIGHT	LEFT MASK	LEFT	RIGHT MASK	FREE FIELD	MIC.

DISCRIMINATION

		RIGHT				LEFT			
DATE	% SCORE	TEST LEVEL	LIST	LEFT MASK	% SCORE	TEST LEVEL	LIST	RIGHT MASK	EXAM

HIGH FREQUENCY THRESHOLDS

	RIGHT						LEFT				
DATE	4000	8000	10000	12000	14000	LEFT MASK	RIGHT MASK	4000	8000	10000	12000	14000

RIGHT		WEBER	LEFT	
RINNE	SCHWABACH		RINNE	SCHWABACH

HEARING AID

DATE	MAKE	MODEL
RECEIVER	GAIN	EXAM
EAR	DISCRIM	COUNC

REMARKS

Figure 9.18 *History*: Patient developed fullness, pain, and hearing loss in right ear during descent in airplane. *Otologic*: Examination 2 days following the incident showed a slightly vascular handle of the malleus and bubbles in the right ear. The drum was retracted slightly and did not move freely. The patient had received an injection of penicillin 24 h previous to the examination, and the eustachian tube seemed to be patent. The ear was politzerized, followed by a myringotomy and aspiration. *Audiologic*: Air conduction thresholds in the right ear revealed a mild loss and normal bone condution. Hearing improved with politzerization and returned to normal after a myringotomy with aspiration. *Impedance*: Type B tympanogram before politzerization; Type C after politzerization. *Classification*: Conductive. *Diagnosis*: Aerotitis media. *Note*: Politzerization alone did not restore the hearing to normal, because some fluid still remained in the middle ear.

and often unwise to remove blood from the middle ear cavity by myringotomy and suction. This may introduce infection and cause complications. Whenever a red eardrum is seen and blood is suspected of being in the middle ear, a glomus jugulare tumor should be ruled out. Imaging studies and a good history are essential.

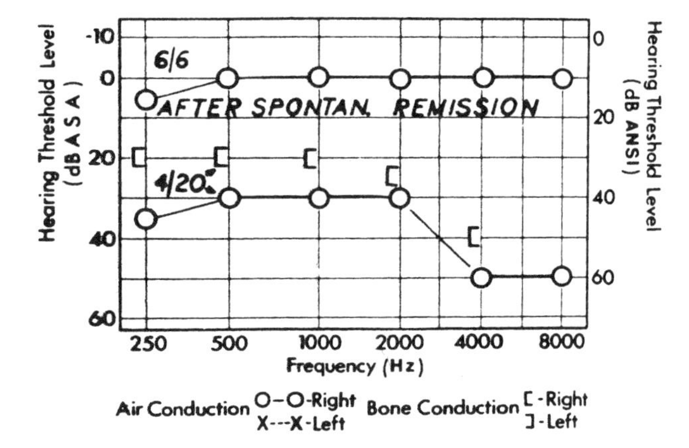

Figure 9.19 *History*: A 24-year-old patient who sustained a head injury causing right middle ear to be filled with blood. *Otologic*: Tympanic membrane was intact but was deep red in color because of blood in the middle ear. The eardrum did not move well with air pressure. Resolution occurred spontaneously without myringotomy. *Audiologic*: Pure-tone thresholds in the right ear revealed a mild air conduction loss with reduced bone conduction (left ear masked). Note the greater loss in the high frequencies and the reduced bone thresholds because of the fluid in the middle ear. Hearing returned to normal at all frequencies after spontaneous resolution. *Classification*: Conductive; damage in middle ear. *Diagnosis*: Posttraumatic hemotympanum. *Aids to diagnosis*: A glomus tumor must be ruled out. Usually, the eardrum moves freely in a glomus tumor, and blanching may occur with positive pressure. Tympanogram may show pulsations. X-ray films are helpful if erosion is present. If a glomus tumor is suspected strongly, retrograde jugular venography may be performed.

34. SECRETORY OTITIS MEDIA

In spite of considerable literature about secretory otitis media, neither the cause nor a specific cure is known. Secretory otitis media seems to have increased in incidence—a fact that has been related to the use of antibiotics.

34.1. Characteristic Features

The major feature of the condition is accumulation of fluid in the middle ear, usually straw-colored and sometimes mucoid or gel-like in consistency. In many cases, the eustachian tube is patent, and the fluid can be removed readily by myringotomy and politzerization. However, the fluid continues to accumulate in spite of numerous myringotomies and various treatments, and hearing loss occurs concomitantly, usually greater in the higher frequencies (Fig. 9.20). This can occur even with myringotomy takes in place. Secretory otitis may be present in one or both ears, and it is found in babies as well as in adults.

Often the condition suddenly stops spontaneously, and the treatment used at that particular time is likely to get the credit. In some cases the secretion continues to form and causes a perforation in the eardrum; the findings resemble those seen in chronic otitis media, but the discharge is free of infectious elements.

When the mucosa of the middle ear in secretory otitis media is examined by biopsy, it is found in most cases to be thickened and hydropic. The thickened mucosa extends to the mouth of the eustachian tube in the middle ear, and perhaps this is a factor in the blockage. In many patients with secretory otitis, one can observe increased secretion in the nasal mucosa and in the nasopharynx as well; this suggests the possibility that secretory otitis may be more than a local middle ear phenomenon. Some ears fill with secretory fluid

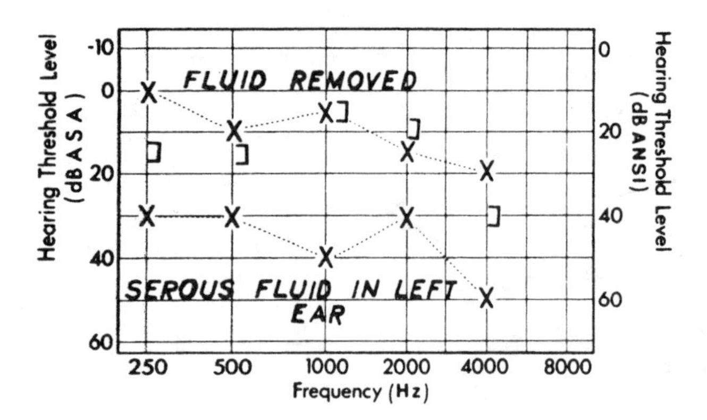

Figure 9.20 *History*: A 12-year-old boy with recurrent episodes of painless hearing loss in left ear. These occurred in the absence of upper respiratory infections. He had a T&A and an adenoid revision, allergy studies, desensitization, autogenous vaccines, and many courses of antibiotics and nose treatments. *Otologic*: The left eardrum was slightly scarred, but a fluid level was seen. The eustachian tube seemed to be patent. The eardrum did not move freely with air pressure. Exploration of the middle ear showed a thick, clear, gelatinous fluid, which was aspirated. The mucosa over the promontory was thick and hydropic. A tube was used to keep open the lower edge of the replaced eardrum and was left in place for several months. The patient has had no recurrence since his surgery, but not all patients respond this well after the tube is removed. *Audiologic*: Pure-tone thresholds in the left ear revealed a mild air conduction loss with reduced bone conduction (right ear masked). After surgery, the air conduction thresholds returned to better levels than the original bone conduction thresholds at most frequencies. Reduced bone thresholds were caused by the thick fluid in the middle ear. *Impedance*: Type B tympanogram, absent stapedius reflex. *Classification*: Conductive. *Diagnosis*: Recurrent secretory otitis. *Aids to diagnosis*: Reduced bone conduction does not mean necessarily that a sensorineural hearing loss is present. An immobile drum, fluctuating hearing loss, and a feeling of fluid in an ear should suggest that reduced bone conduction may be due to middle-ear involvement rather than to sensorineural causes.

after removal of the soft palate for neoplasm. There seems little doubt that in such instances the condition is related to eustachian tube malfunction. Recently, gastric reflux has been identified as a cause of middle ear fluid, and pepsin has actually been recovered from the middle ear.

It often is difficult to distinguish secretory otitis from serous otitis. Caution in treatment is of upmost importance, and unwarranted surgical procedures should be avoided. For example, it is injudicious to perform a tonsillectomy and adenoidectomy (T&A) on an infant who has secretory otitis media without positive evidence that the adenoids are the principal cause of the difficulty, in which case it would be a serous otitis. Too frequently, secretory otitis will continue to recur after a T&A that was unjustified, thereby placing the surgeon in an embarrassing situation. When adenoids are really the major cause of secretion in the middle ears (serous otitis), other symptoms usually are present, including recurrent otitis media and mouth breathing; furthermore, the hypertrophied adenoids can be felt and visualized in the lateral area of the nasopharynx. These findings often are not present if the cause is secretory otitis media.

34.2. Appearance and Treatment

The appearance of the eardrum in secretory otitis often is characteristic, and yet sometimes it may be deceiving. Generally, it is easy to observe air bubbles and a yellowish fluid behind the drum. Politzerization, which shows the eustachian tube to be patent, causes the fluid level to shift or even to disappear, and hearing suddenly improves (though it does not always return to normal).

Conductive hearing loss is usually present, and the bone conduction may be reduced somewhat in the higher frequencies. Myringotomy generally produces a gush of fluid, and the hearing improves markedly. The length of time that the hearing loss has been present is not a factor and may even be misleading, because fluid can remain in the middle ear for many months or years. In many cases, it is advisable to insert a tube through an inferior myringotomy to maintain air in the middle ear.

34.3. Another Type of Secretory Otitis and Its Differentiation from Otosclerosis

Another type of secretory otitis media can lead to the erroneous diagnosis of otosclerosis, because it is so difficult to observe abnormalities in the eardrum or the middle ear. In this type, a gel-like mass of clear secretion or a very thick collection of mucoid secretion is located in the middle ear so that it causes hearing loss without any abnormality being visible through the drum. This leads to a diagnosis of otosclerosis. When the middle ear is exposed with the intention of doing stapes surgery, a thick secretion is found instead. Its removal causes the hearing to return to normal. Such a case is shown in Fig. 9.21. A preliminary myringotomy and the introduction of a suction tip would have revealed the mass of gel-like material.

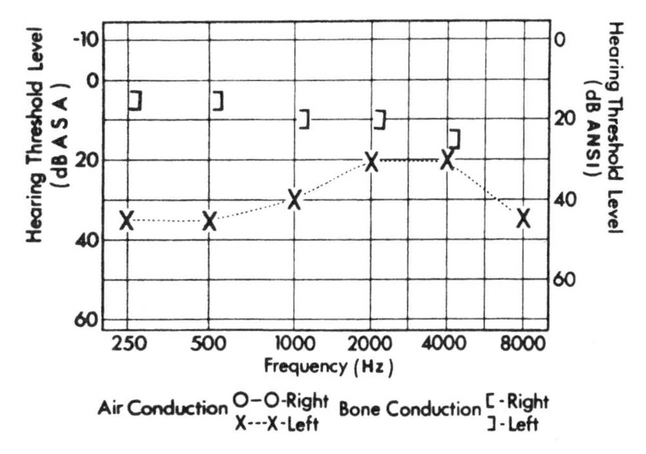

Figure 9.21 *History*: A 27-year-old woman with gradual onset of hearing loss in the left ear for several years. Feeling of fullness and occasional heartbeat tinnitus in the left ear. Several members in family have hearing loss. No history of ear infections. The patient had been diagnosed as having unilateral otosclerosis. *Otologic*: Eardrums appeared to be normal (but did not move well with politzerization). When the eardrum was elevated, a large, gelatinous mass caused by secretory otitis was removed from the oval and the round windows. The stapes was found to be mobile. *Audiologic*: Air conduction thresholds in the left ear were reduced mildly in the lower frequencies. Bone conduction was almost normal with the right ear masked. Tuning fork lateralized to the left ear. Bone conduction was prolonged and better than air. Hearing returned to normal after removal of gelatinous mass. *Impedance*: Type B tympanogram, absent stapedius reflex. *Classification*: Conductive. *Diagnosis*: Secretory otitis. *Aids to diagnosis*: Original misdiagnosis of otosclerosis might have been avoided if the mobilization of the drum had been tested with politzerization and if a diagnostic myringotomy had been performed. Not all cases of fluid in the middle ear cause reduced bone conduction.

35. SEROUS OTITIS MEDIA AND ADENOIDS

There is understandable confusion between serous otitis and secretory otitis media. In the first place, the etiology of secretory otitis is not known, and in the second place, the two conditions sometimes are hard to distinguish clinically. Even outstanding otologists have differences of opinion concerning the differential diagnosis and the distinguishing characteristics.

35.1. Definition

For present purposes, let us consider serous otitis media as a condition in which serous fluid accumulates in the middle ear because of obstruction or infection of the eustachian tube or the nasopharynx. If the middle ear becomes infected, the condition is called acute otitis media. The fluid is a secondary phenomenon and results from pathology external to the middle ear. This is in contrast to secretory otitis media, in which the pathology seems to originate in the middle ear.

35.2. Causes

The chief cause of serous otitis in children is hypertrophied adenoids in the fossae behind the eustachian tube openings (Rosemüller's fossae). The adenoids do not grow over the mouth of the eustachian tube but cause obstruction in its lumen by submucosal compression or congestion. For this reason, it is essential to resect this area of adenoid tissue carefully under direct visualization in doing an adenoidectomy for recurrent otitis media or hearing impairment. Merely removing the central adenoid in such cases leads to recurrent symptoms and the need for adenoid revisions.

Adhesions in the Rosenmüller's fossae usually are caused by previous surgery and may lead to serous otitis by restricting the normal function of the mouth of the eustachian tube; the adhesions may result in hearing impairment. In such cases, the adhesions must be removed meticulously under direct vision, and the tubal ends must be mobilized with care not to injure the submucosal layers or to cause further adhesions.

Nasopharyngeal neoplasms, nasopharyngeal radiation, and allergies also may produce serous otitis.

35.3. Difficult Detection

The diagnosis of serous otitis as a cause of hearing loss in children often is overlooked because the hearing loss rarely exceeds 40 dB and children generally are addressed in a loud voice. Thus, hearing difficulty is not detected until the school audiogram is performed or until the symptom has persisted a long time. In a nationwide study of school children, 85% with visible serous otitis media had hearing losses of $\leq$25 dB, and 50% had hearing losses of $<$15 dB.

A fluid level and air bubbles often are readily apparent through the drum. The serous fluid level may show no movement when the child's head is bent forward or backward. Sometimes the fluid is hidden or fills the ear so completely that it goes undetected even on otoscopic examination. In such a case, the eardrum is hypomobile and may not move even with air pressure. Fullness and dullness in the ear are common complaints. Figure 9.22 shows a typical case, in a child, of hearing loss caused by serous otitis that went undetected for many months until a routine school audiogram was performed.

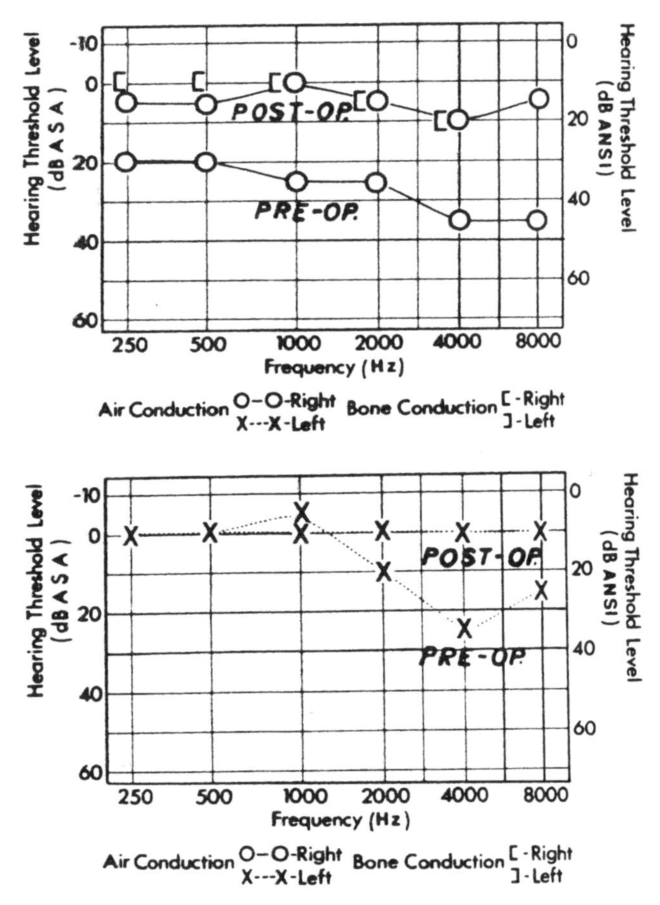

Figure 9.22 *History*: An 8-year-old child who failed to pass a school hearing test. No history of ear trouble. T&A performed at age 4. *Otologic*: Amber fluid in right middle ear. Slight amount of fluid in left middle ear, also. Eustachian tube not clearly patent. Large mass of regrown adenoid tissue, especially in lateral pharynx. No response to a year of conservative therapy. Adenoids revised and thin, clear fluid aspirated through myringotomy. *Audiologic*: Pure-tone thresholds in the right ear revealed a mild air conduction loss with normal bone conduction. The left ear had a mild loss in the higher frequencies. Hearing returned to normal after surgery. *Impedance*: Shallow type C tympanogram on the right; type A on the left. *Classification*: Conductive. *Diagnosis*: Hypertrophied adenoids with serous otitis media.

35.4. Aim of Therapy

Simple myringotomy performed in such cases may clear up the hearing only temporarily, for the cause still is present in the nasopharynx. Therefore, therapy should be directed to the causes as well as to the immediate relief of symptoms in the ear.

36. ACUTE OTITIS MEDIA

36.1. Extent of Hearing Loss

In acute otitis media hearing loss is temporary. It clears up when the inflammation subsides and the debris in the middle ear is absorbed. Depending on the stage of the infection, the hearing loss may be as great as 60 dB if the middle ear is filled with pus. Usually, all frequencies are involved if fluid forms in the ear. If there is no fluid in the ear, sometimes only the lower frequencies are involved. An interesting paradox characterizes some cases of severe otitis media in which inflammation extends to the postauricular area. A tuning fork held to the mastoid region shows considerably reduced bone conduction. The same finding can be obtained by bone conduction audiometry on the mastoid region. It seems that sound waves are conducted poorly through the inflamed mastoid bone, but the sensorineural mechanism is not affected. When bone conduction is tested on the teeth, it is found to be normal. The tuning fork should be cleaned well when this technique is used.

The hearing loss in otitis media is caused by impeded sound transmission across the eardrum and the ossicular chain as a result of the added mass in the middle ear. Tinnitus rarely is present; when it is reported, the patient reports hearing his/her own pulse-beat. Although hearing loss during an attack of acute otitis media may be a cause for concern after the pain has subsided, the immediate and urgent problem is the relief of pain. However, what is most important in the long run is to treat otitis media adequately so that it resolves without leaving any permanent hearing damage.

36.2. The Question of Myringotomy

The question arises whether myringotomy should be done in all cases of acute otitis media to prevent hearing impairment. There are avid supporters for myringotomy in all cases and equally enthusiastic supporters of the concept that myringotomy rarely should be done. Probably, the best approach lies somewhere in between. Whenever a middle ear contains pus and is causing the drum to bulge, myringotomy certainly is indicated to relieve pain and to prevent hearing loss. This is in keeping with the proven surgical principle that incision and drainage are advisable whenever pus is under pressure. In most bulging drums, there is an area of anesthesia in the eardrum created by the pressure in the middle ear, and if the myringotomy is done quickly in this area without pressing deeply into the drum, very little pain is experienced.

36.3. Antibiotics

In spite of the excellent results that have been achieved in the prevention and treatment of acute otitis media by nonsurgical means, a large number of cases do progress to chronic otitis media. One of the causes for some of these failures is the use of inadequate doses or kinds of antibiotics. In many cases of otitis media, a much higher blood level of antibiotic is needed than is generally recognized, because the infection has become walled off or is encased in bone and can be reached only by very high blood levels of the drug.

36.4. Difference between Acute and Chronic Otitis Media

Because acute otitis media frequently still leads to chronic otitis media and hearing loss, a more extensive discussion is in order. First, we should clarify the general difference between acute and chronic otitis media. Not infrequently, a patient is referred to an ear specialist with a diagnosis of "chronic otitis media," and the otologist finds the drum to be practically normal. The history may reveal that the patient has had repeated earaches and infections almost every 3 months. This is recurrent acute otitis media, not what is meant by chronic otitis media. Instead, otologists mean an ear that has been infected continuously for at least many months. An acute otitis media is an ear infection of comparatively short duration. If the acute otitis media does not respond satisfactorily to therapy and the infection persists for many months, it then becomes a chronic otitis media.

If a patient has an acute otitis media that results in a persistent perforation in the eardrum and the infection clears up only to return again in a month or so, this should be considered to be a recurrent acute, not a chronic otitis media. As a matter of fact, this very situation is most common in otologic practice. Many patients with perforations whose ears have been dry either get water in them or blow their noses improperly during an upper respiratory infection, and the ear becomes reinfected. In such cases the otorrhea usually is stringy and mucoid and comes from the area of the eustachian tube. Hearing loss generally is minimal, and closure of the perforation often restores the hearing.

36.5. Common Causes and Prevention

The common causes of acute otitis media are upper respiratory infections and sinusitis, hypertrophied adenoids, allergies, and improper nose blowing and sneezing, as well as eustachian tube blockage and possible exposure to secondhand smoke in children. It is notable that all these causes are external to the ear itself, so that in most instances acute otitis media is a secondary infection, and its prevention must be directed to its causes.

To prevent otitis media and hearing loss, patients should be cautioned to refrain from indiscreet nose blowing and sneezing. Forceful blowing or sneezing while pinching both nares causes a buildup of pressure in the nasopharynx; this pressure may force small amounts of infected mucus through the eustachian tube into the middle ear, with resulting otitis media. These facts should be impressed on patients with upper respiratory infections. In spite of the social stigma attached to the practice, sniffing is far safer than nose blowing. If the nose is blown, both nostrils should be left unpinched.

36.6. Final Checking of Hearing

In all cases of acute otitis media, it is advisable to perform hearing tests after the infection has subsided to be certain that there is no residual hearing impairment that might be permanent or that might require further treatment.

37. CHRONIC OTITIS MEDIA

When infection persists continuously in the middle ear for long periods of time, it is called chronic otitis media, and this is a very frequent cause of hearing impairment. The mechanism varies. There is invariably a perforation in the eardrum.

Occasionally, the entire drum is eroded, and much of the middle ear is visible through the otoscope. The more severe hearing losses are caused by erosion of some part of the ossicular chain. The most common ossicular defect is erosion of the long end of the incus, so that it does not contact the head of the stapes. Occasionally, the handle of the malleus and even the stapedial crura are eroded. In some ears, the entire incus has been destroyed. Scarlet fever and measles are notorious causes of severe erosion of the ossicles and the eardrum.

Another cause of hearing loss in chronic otitis media is discharge in the middle ear. This naturally adds a mass which impedes transmission of sound waves. Strangely enough, a patient may say he/she hears much worse after the ear is cleared of discharge, and the audiogram often will substantiate this complaint. This result may be explained in several ways, but one of the more reasonable theories is that the discharge blocks sound waves bound for the round window niche and thereby permits some semblance of normal phase difference for sound waves hitting the inner ear. To make this clear, it may be pointed out that sound waves in normal ears selectively enter the oval window rather than the round window because of their direct transmission through the ossicular chain. When the eardrum is missing and the ossicular chain is not functioning properly, sound waves occasionally strike both the round and the oval windows almost simultaneously, so that the waves may in part cancel each other before they can reach the inner ear. This causes hearing loss. A discharge in the middle ear sometimes may prevent this effect by blocking sound waves that otherwise would reach the round window niche. Thus, the patient may hear better when his/her ear is moist. The same mechanism sometimes is used purposely to improve hearing in ears that are free of infection.

One type of prosthesis covers the round window opening so that sound waves cannot strike both the round and the oval windows simultaneously, thereby nullifying their effects in the inner ear and causing a hearing loss. Ointments can be used for the same purpose in some of these cases. Aquaphor ointment is an example. If the patient's eardrum is gone so that the round window niche is visible, a patient's hearing sometimes may be improved markedly by carefully applying a plug of Aquaphor ointment over the round window niche. In appropriate cases this ointment maintains the hearing improvement for several weeks; occasionally, the patient can be taught to replace the ointment.

However, these considerations should not lead a physician to disregard discharge in the hope of achieving improved hearing, because infection often can result in serious complications. There are more satisfactory methods to restore hearing after an infection has been controlled. When a perforation (especially a posterior superior perforation) is found in the eardrum with a putrid discharge, it generally means that mastoiditis is present. If the infection is allowed to continue, it may cause a number of complications, including erosion of the ossicles and severe hearing loss. A cholesteatoma may form that could erode the semicircular canals and the facial nerve and even produce a brain abscess. Sensorineural hearing loss often occurs in patients who have long-standing chronic ear infections.

It is important, therefore, to cure a chronic middle-ear infection as rapidly as possible. Unfortunately, systemic antibiotics do not often succeed in clearing up cases of chronic otitis media, because the chronically diseased mastoid cells in the middle ear have such a poor blood supply. Consequently, it is necessary also to treat the ear locally, and occasionally surgery must be performed when local therapy has been unsuccessful and complications threaten.

Figures 9.2, 9.23, and 9.26 illustrate several audiometric patterns that may occur in chronic otitis media.

38. TYMPANOSCLEROSIS

As a result of as yet undetermined causes, sclerotic changes occur in some chronically diseased middle ears. After infection has subsided, a shalelike layer of bony deposit remains over the promontory of the cochlea and around the oval window region.

Occasionally, the stapes and the incus are enveloped by stratified bone that can be peeled off in layers. This condition is called tympanosclerosis, and it produces hearing loss by fixing the stapes and the incus in a manner similar to that of otosclerosis. It differs markedly from otosclerosis in that its onset follows an infection, without a familial history of hearing loss. Furthermore, it is generally unilateral. In addition, pathological changes are visible in the eardrum in most cases of tympanosclerosis. The drums may

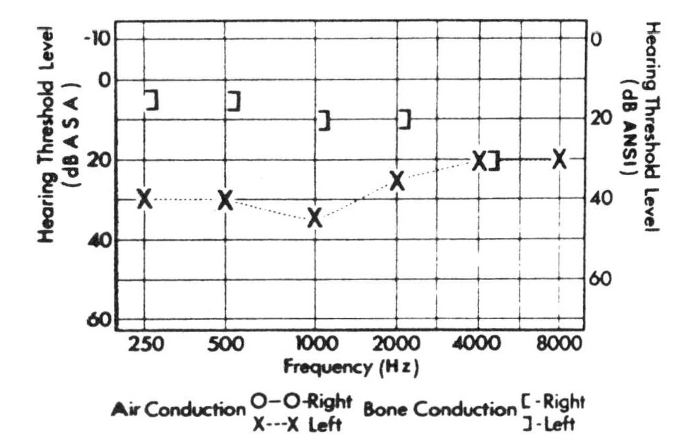

Figure 9.23 *History*: This patient had a discharging left ear for many years. X-ray study showed a cholesteatoma. *Otologic*: Large posterior marginal perforation. The ossicular chain was not disrupted; otherwise, the hearing loss would have exceeded 50 dB. Ossicular continuity was confirmed at surgery. *Audiologic*: Pure-tone thresholds in the left ear revealed a mild air conduction loss with a gradually decreasing air–bone gap at the higher frequencies. Right ear masked. *Impedance*: Type B tympanogram; pressure seal for probe maintained without difficulty at usual test pressures. *Classification*: Conductive. *Diagnosis*: Cholesteatoma with marginal perforation.

show healed scars and are thick and yellowish-white with white areas that suggest sclerosis.

Figure 9.16 describes a typical case of tympanosclerosis in which a stapes mobilization was performed. It is wise to point out that stapedectomies should be undertaken with great caution in cases of tympanosclerosis, because for some unknown reason the incidence of "dead ears" and severe sensorineural loss may be substantial even when the surgery seems to be successful in the operating room. It has even been suggested that an operation on an ear of this type should be done in two stages, in the first removing the tympanosclerosis and in the second mobilizing the stapes or doing a stapedectomy. The hearing loss may be mild or as severe as 65 dB when the fixation is complete. All frequencies usually are affected. In many long-standing cases, a sensorineural hearing loss develops.

39. CARCINOMA

Carcinoma of the middle ear is rare, and frequently it is not recognized at its inception. Its onset resembles chronic otitis media as it causes at first a mild conductive hearing loss and later a loss of up to 60 dB as invasion or restriction of the ossicles ensues. Sometimes carcinoma occurs in an ear with chronic otitis that has been present for many years. This is even more difficult to detect because the change takes place below the typical surface of granulation and polypoid tissue and is not visible to the examiner. Even roentgenograms may be of little help. Biopsy of all granulation tissue is advisable but not always practical in office practice. However, it should be done in all cases of long-standing granulation with recurrent polypoid formation that resists conservative therapy or continues even after surgical intervention, especially when severe pain is present. If the tissue is harder than usual or bleeds readily, a biopsy is indicated. Usually, the eardrum perforates early, and chronic otitis media develops, so that the appearance is not very distinctive. Figure 9.24 shows an unusual case of carcinoma of the middle ear. Hearing is of

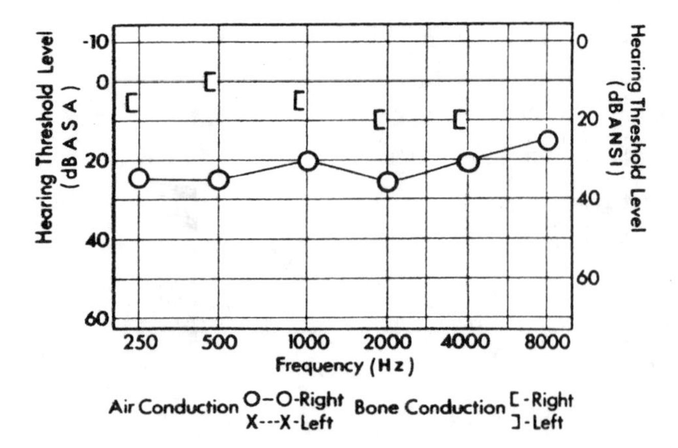

Figure 9.24 *History*: A 42-year-old woman complained of fullness and severe pain in right ear for several months. No vertigo or tinnitus. She had received antibiotics and myringotomies without satisfactory help. X-ray films showed only a mild haziness in the right mastoid but not in the left. *Otologic*: The eardrum was slightly opaque but moved satisfactorily with air pressure. *Audiologic*: Mild conductive ascending hearing loss in right ear. Left normal. *Diagnosis*: After continued unsuccessful conservative therapy, the right ear was explored, and a carcinomatous invasion of the middle ear and the mastoid was found. The patient did not respond to cobalt irradiation, and, unfortunately, radical surgery was not done.

secondary consideration in carcinoma of the middle ear, but it may be important as a presenting symptom to alert the physician to the presence of a serious condition. Malignancies of the ear are discussed in detail in Chapters 21 and 22.

40. NASOPHARYNGEAL TUMORS

Unilateral conductive hearing loss frequently is commonly the first presenting symptom of a tumor in the nasopharynx. The mechanical cause of the resulting hearing loss resembles that described in serous otitis media: gradual obstruction of the eustachian tube. The possibility that such a serious condition may exist adds emphasis to the importance of doing a nasopharyngoscopic examination in all cases of conductive hearing loss, especially when they are unilateral and associated with middle ear fluid. The hearing loss by air conduction is comparatively mild in the beginning and may progress to ~40 dB, and bone conduction usually is normal. The nasopharyngeal examination shows a fullness that is sometimes clear-cut but in other cases uncertain with poor delineation, depending on the nature of the tumor. A better perspective of the tumor generally is obtained through a nasopharyngeal mirror below the soft palate or using fiber optic endoscopes.

In the early stages, when the hearing loss is mild, the eardrum appears to be normal. Later, retraction of the drum or serous fluid becomes visible. When there has been recurrent fluid formation in one ear and multiple myringotomies are necessary, it is essential to rule out a nasopharyngeal mass.

41. ALLERGY

Allergy plays a somewhat vague role in hearing loss of middle ear origin. Undoubtedly, allergic conditions can lead to congestion of the eustachian tube and the middle ear, with resultant serous otitis media. A few cases of hearing loss have been reported that appear to result primarily from allergy in the middle ear. When the allergy is treated, hearing improves. If judgment is based on experience and a review of the literature, this type of case is not very common unless fluid is present in the middle ear.

Certainly, allergic swelling of the eardrum or the mucosa of the middle ear does occur and may produce hearing loss, but this does not happen as frequently as one might expect from the known frequency of allergies of the upper respiratory tract. Medical treatment of the allergy is of major help in clearing up many cases of serous otitis media.

Hearing loss resulting from allergy of the middle ear is mild. If it exceeds ~20 dB, fluid probably is present in the middle ear, or some other causative factor is present.

42. X-RAY TREATMENTS

Irradiation to the region of the nasopharynx frequently produces malfunction of the eustachian tube. Hearing loss along with serous otitis media often is present. The radiation may have been directed to the thyroid, the face, the skull, or the nasopharynx. The hearing loss produced is mild and rarely exceeds 30–35 dB. After some months of conservative therapy, the malfunction may subside, and the hearing may return to normal (Fig. 9.25).

An extraordinary incidence of conductive hearing loss caused by irradiation for a brain tumor was encountered by the author. Notable was an aseptic mastoid cell necrosis that occurred >20 years after the treatment. The severe conductive loss was caused by two factors: a large erosive defect in the posterior external canal wall and profuse serous discharge from the middle ear. The hearing could be improved slightly by occluding the perforation in the canal wall.

43. SYSTEMIC DISEASES

Certain systemic diseases are known to affect the middle ear and to cause conductive hearing loss. The most common of these are measles and scarlet fever. Both are notorious for the marked otitis media that they can cause, with erosion of the eardrum and the ossicles. This complication is not nearly so common now as it was years ago, but the hearing losses resulting from these conditions still are encountered in practice. Figure 9.26 shows an example of conductive hearing loss following scarlet fever in childhood. The eardrum and the ossicles in such cases generally are eroded, and the hearing level is ~60 dB.

Letterer–Siwe's disease, xanthomatosis, eosinophilic granulomatoma, and other granulomata are additional causes of conductive hearing loss. Although not very common, they can damage the middle ear and cause handicapping hearing loss, as discussed in Chapter 13.

44. GLOMUS JUGULARE AND GLOMUS TYMPANICUM

Glomus jugulare tumors are rare, but when they are present, hearing loss and tinnitus frequently are the only symptoms. This neoplasm arises from cells around the jugulare bulb and expands to involve neighboring structures. In doing so, it most frequently extends to the floor of the middle ear, causing conductive hearing loss and pulsating tinnitus (Fig. 9.27). As the disease progresses, it may appear as chronic otitis media and may

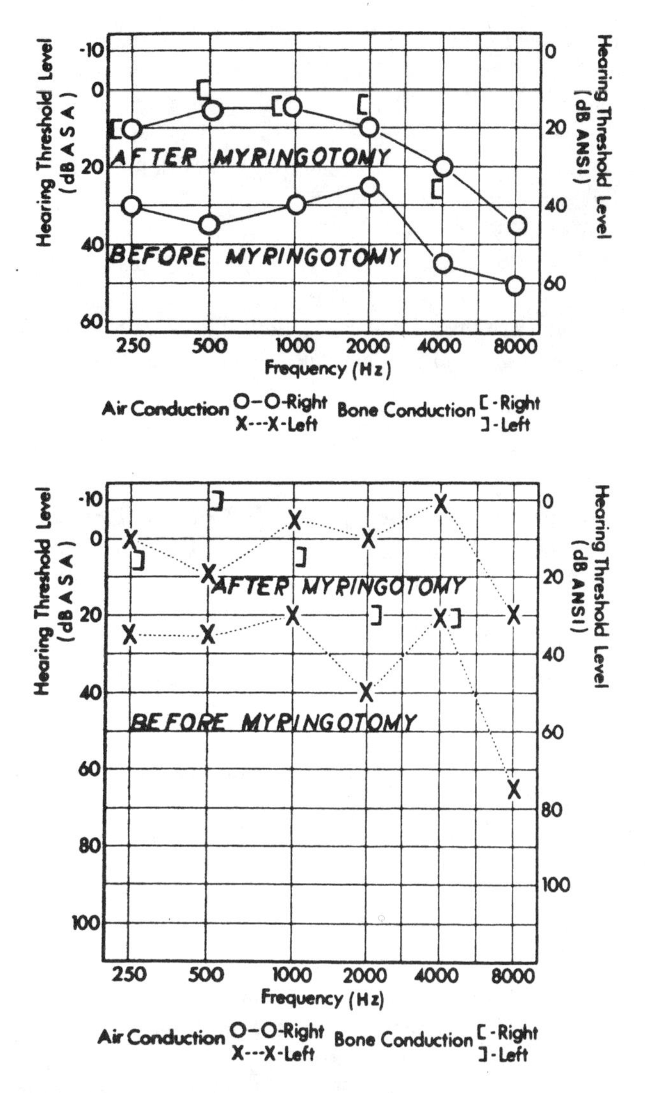

Figure 9.25 *History*: A 51-year-old woman with acromegaly who received radiation to the pituitary gland. Several weeks later, she noted insidious hearing loss, and her own heartbeat sounded loud in both ears. She was treated with tranquilizers with little relief. *Otologic*: Normal eardrums, but highly congested nasopharynx and eustachian tubes. Eardrums mobile but difficult to politzerize. Amber fluid evacuated from both ears through myringotomy, with restoration of hearing. Tinnitus disappeared. *Audiologic*: Right and left ears showed mild conductive losses with normal bone conduction in the low- and midfrequency range. After both middle ears were aspirated, air conduction thresholds returned to preoperative bone levels in the right ear and bettered these levels in the left ear. Reduced bone conduction thresholds were caused by fluid in middle ears. *Impedance*: Type C tympanogram. *Classification*: Conductive. *Diagnosis*: Otitis media caused by X-ray treatment to pituitary for acromegaly.

even extend through the eardrum and appear to be granulation tissue in the ear canal. Unsuspecting biopsy of this apparent granulation tissue may result in profuse bleeding because of the striking vascularity of glomus tumors. As the disease extends, it may destroy portions of the temporal bone and jugular bulb and may spread intracranially.

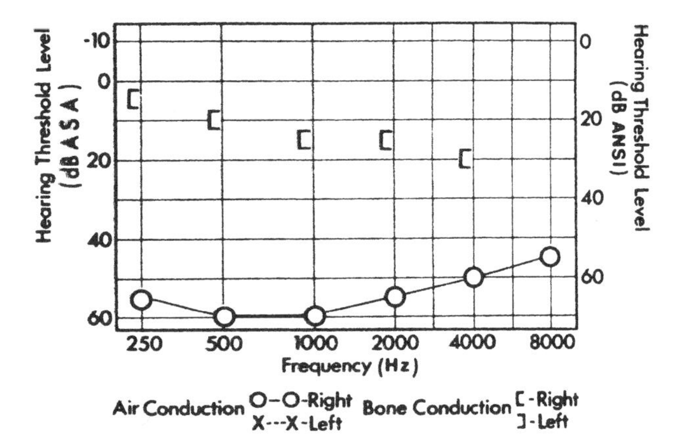

Figure 9.26 *History*: A 20-year-old man who had scarlet fever as a child with right chronic otitis for 1 year. Since that time the hearing has been reduced, but the ear drained only after swimming or a bad cold. *Otologic*: The entire eardrum was eroded, and the handle of the malleus was gone, except for a small nubbin. The incus and the stapedial crura also were eroded. *Audiologic*: Air conduction thresholds in the right ear revealed a moderately severe loss and slightly reduced bone conduction thresholds. The mild reduced bone conduction does not mean necessarily a sensorineural involvement in such a case. *Classification*: Conductive. *Diagnosis*: Right chronic otitis media caused by scarlet fever.

Glomus tumors also may arise from cells along the promontory or medial wall of the middle ear. These are called glomus tympanicum tumors and are generally somewhat easier to manage surgically. It is essential to distinguish between glomus tympanicum and glomus jugulare before attempting surgical intervention.

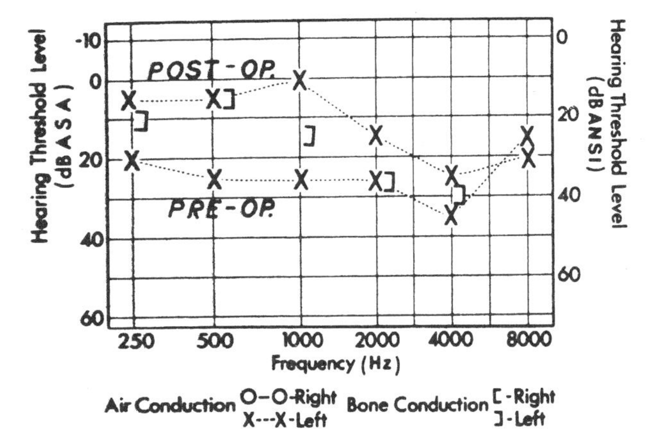

Figure 9.27 *History*: A 57-year-old woman with sudden discomfort in the left ear 6 months before. Thumping tinnitus for past 6 months. No facial paralysis. Some hearing loss. No pain. *Otologic*: Right ear clear. Left middle ear was red and inflamed, but eardrum moved freely. Needle aspiration produced bleeding, which was controlled quickly. *Audiologic*: Left ear air conduction thresholds revealed a mild, flat hearing loss. Bone conduction thresholds showed an air–bone gap at 500, 1000, and 2000 Hz only. Left ear discrimination was 96%. *Classification*: Conductive with high-frequency sensorineural involvement. *Diagnosis*: Glomus jugulare tumor, removed surgically; tympanogram showing pulsations, radiologic evidence of bone erosion, and obstruction or retrograde jugular venogram generally confirm the diagnosis preoperatively.

As in any expanding neoplasm, early diagnosis of a glomus tumor facilitates surgical cure. As conductive hearing loss may be the only symptom for a long time, the physician is obligated to establish the cause for every case of unilateral conductive hearing loss.

Physical examination may disclose a pinkish mass in the middle ear. Positive pressure on the eardrum may reveal blanching of the mass. Pulsating tinnitus may be audible to the examiner by using a Toynbee tube or a stethoscope placed over the ear. The finding of objective tinnitus may occur not only with glomus tumors but also with carotid artery aneurysms, intracranial arteriovenous malformations, carotid artery stenosis, and other conditions. Glomus tumors must be distinguished from other masses that may be present in the middle ear, such as carotid artery aneurysms, high jugular bulbs, meningiomas, and adenomas.

Radiological evaluation is now the mainstay of glomus tumor diagnosis. Biopsy rarely is indicated. CT and MRI of the temporal bone are used to assess bone erosion and soft tissue extent and arteriography and retrograde jugular venography are used to define the extent of the neoplasm. Four-vessel arteriograms are now being recommended by some otologists because of the high incidence of associated tumors. Up to 10% of patients with glomus tumors will have associated bilateral glomus tumors, glomus vagale, carotid body tumor, or thyroid carcinoma. The vast majority of glomus patients are female, and the tumor is extremely rare in children. If the diagnosis is entertained seriously in a child, biopsy then may be appropriate to rule out other lesions. Biopsy also is used in patients who are not surgical candidates prior to instituting palliative radiation therapy.

45. CAUSES WITH APPARENTLY NORMAL EARDRUMS AND MIDDLE EARS

The following are some of the causes of conductive hearing loss in which the external auditory canals, the eardrums, and the middle ears appear to be normal on otoscopic examination:

1. Congenital ossicular defects
2. Acquired ossicular defects
3. Otosclerosis
4. Tympanosclerosis
5. Paget's disease
6. van der Hoeve'e syndrome
7. "Invisible" fluid
8. Malfunction of the eustachian tube
9. Superior canal dehiscence syndrome

45.1. Congenital Ossicular Defects

A variety of defects may arise in the ossicular chain during fetal development and produce a conductive hearing loss of 60–70 dB even though the eardrum and middle ear appear to be normal during physical examination. The impairment, if bilateral, usually is discovered by the time the child is 3 years old, but the discovery may be delayed if the defect affects only one ear.

A congenital ossicular defect should be suspected when a young patient with an apparently normal eardrum "has had a hearing loss all his life," when it is conductive and in the range of 50 dB or more, and particularly when it is unilateral and nonprogressive. Complaints of tinnitus are rare.

If the hearing loss is bilateral, careful questioning usually will disclose that speech development was delayed and that the patient reacted to the impairment in one of two ways: poor schoolwork or some behavior problem related to the child's inability to hear and to respond. Although the nonprogressive character of the hearing loss is of considerable diagnostic importance, the hearing may fluctuate during upper respiratory infections, allergies, and similar conditions, as is true also of hearing losses of other etiologies.

Figure 9.28 shows an audiogram of a patient whose hearing loss is highly suggestive of congenital ossicular defect. Reference to this audiogram at this point serves to emphasize the importance of a careful and critical appraisal of the otologic findings in conjunction with audiology, especially a meticulous inspection through the drum.

In cases in which an ossicular defect is suspected, the procedure of choice is to elevate the eardrum and to examine the ossicular chain. A defect can be found in almost any part of the chain. One of the most common is congenital fixation of the stapes. Another is congenital absence of the long process of the incus and no connection between the incus and the head of the stapes. General malformation of the stapedial crura also has been found to be responsible for some hearing loss. There are other congenital defects that challenge the ingenuity of the surgeon who attempts to establish a functional continuity of the ossicular chain.

45.2. Acquired Ossicular Defects

It is possible for the ossicular chain to be disrupted or damaged without causing visible changes in the tympanic membrane. Figure 9.29 shows an interesting example of ossicular damage following head injury in an industrial accident. A similar defect can be produced

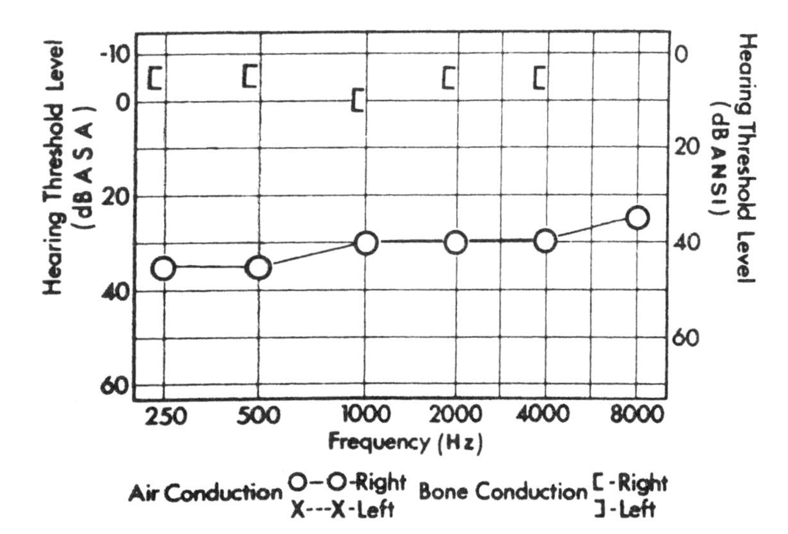

Figure 9.28 *History*: A 16-year-old boy with right ear hearing loss following an ear infection at age 7. Ear was lanced at that time. There have been no further infections, and the hearing loss has not progressed. No tinnitus. *Otologic*: Eardrums almost normal with only slight evidence of scarring. At surgery the end of the incus was found to have been fractured and to have healed with fibrous union. This damage probably was caused by a myringotomy knife. A plastic tube was used to improve the incudostapedial connection. *Audiologic*: Right ear air conduction thresholds revealed a mild, flat hearing loss, with normal bone thresholds. *Impedance*: Deep type A tympanogram; probably not type A_D because of the fibrous union. *Classification*: Conductive. *Diagnosis*: Ossicular disruption. *Aids to diagnosis*: Normal bone conduction thresholds are an important criterion in the diagnosis of ossicular defect (e.g., not depressed as the result of fluid or stapes fixation).

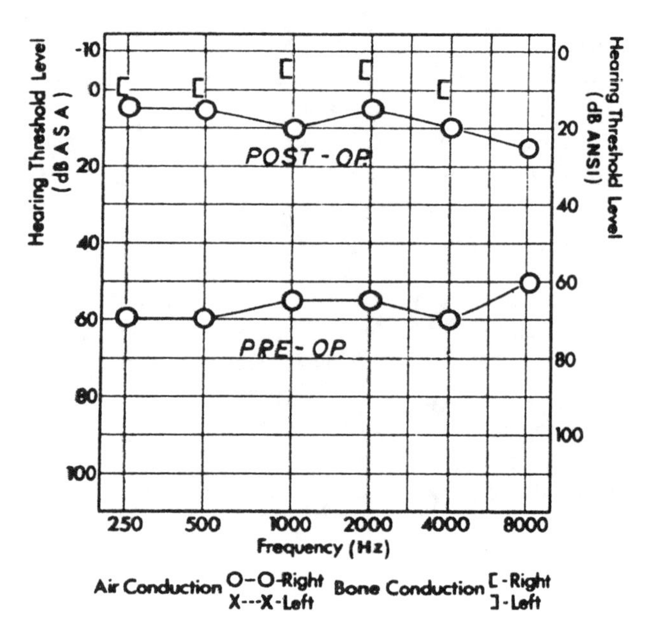

Figure 9.29 *History*: A 50-year-old man who had normal hearing until an industrial head injury. He lost consciousness, but there was no fracture of the skull. No tinnitus or vertigo, but a marked right-ear hearing loss resulted from the injury. *Otologic*: Normal eardrums and middle ears. Exploration revealed a complete dislocation of the end of the incus from the head of the stapes. Continuity was restored and hearing improved. *Audiologic*: Air conduction thresholds in the right ear revealed a moderately severe hearing loss. Bone conduction thresholds were normal. The air–bone gap was maximal for a conduction lesion. Tuning fork lateralized to the right ear, and air was better than bone conduction. *Impedance*: Type A_D tympanogram. *Classification*: Conductive. *Diagnosis*: Ossicular disruption caused by head trauma.

during myringotomy in childhood if the myringotomy is too high and the knife penetrates too deeply and damages the incus. After the operation, the eardrum may heal without scaring, but the damage to the incus remains.

Figure 9.30 shows another case in which a patient heard well following a stapedectomy but suffered a sudden hearing loss when she received a sharp blow to the operated side of her head.

Examination of the middle ear showed that the prosthesis had been dislodged from its proper position by the blow. In a contrary experience, another patient failed to gain better hearing from a stapedectomy and vein graft. A sudden and hard stop in a descending elevator several months later caused her hearing to return immediately. Apparently, the prosthesis was moved into a better position, and the hearing improved. In all these cases, the change in hearing was sudden in comparison with the long history of deafness in congential defects and otosclerosis.

Occasionally, there is fixation of the malleoincudal joint or of the malleus itself. This may occur after a mild infection without evidence in the eardrum, may be congenital, or there may be no obvious cause. Two ossicles may become so fixed that they do not transmit sound waves, and a marked hearing loss may result.

In some instances, the incus or the stapedial crura may be broken by a blow to the head, and healing takes place by fibrous rather than bony union (Fig. 9.28). In such cases, the hearing loss may be comparatively mild. The hearing loss resulting from an ossicular defect generally involves all frequencies, and tinnitus is absent. The precise diagnosis

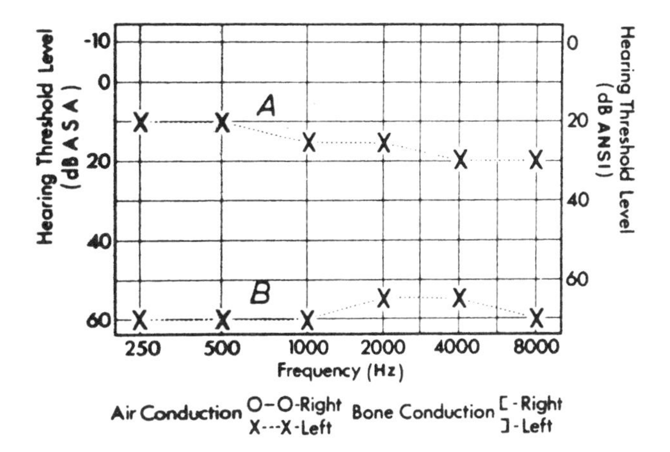

Figure 9.30 *History*: Patient had good stapedectomy result (A). Fifteen months postoperative she received a sharp blow to the left ear and the side of the face. Sudden deafness followed (B). No vertigo was experienced. *Otologic*: Exploration revealed the polyethylene-type prosthesis was dislodged and lay free in the middle ear. This was removed and replaced with a steep piston and wire around the incus. *Audiologic*: Posttraumatic air conduction thresholds represented a 35- to 50-dB reduction in hearing from those obtained after a successful stapedectomy. Hearing was restored with a new prosthesis. *Classification*: Conductive. *Diagnosis*: Dislocated prosthesis.

seldom is made until the ossicular chain is examined surgically. The vast majority of these defects can be repaired and the hearing restored.

One other type of ossicular defect with normal otoscopic findings is the damage that may be done either to the end of the incus or to the stapes during stapes surgery. In some cases in which stapes mobilization is attempted by pushing down on the end of the incus or in which a wire prosthesis is too tight, the incus end is weakened or split off. This impairs the efficiency of the incudostapedial joint and may cancel the benefits of successful mobilization or even make hearing worse. Fracture of the head of the stapes or its crura is another cause of acquired ossicular defect. In some such instances, an initial hearing loss of 30 dB may be increased to 50 dB. The bones rarely heal sufficiently for the hearing to improve noticeably, and stapedectomy or replacement of the crura and the end of the incus with a stapes prosthesis then becomes necessary.

Stapedectomies also may be unsuccessful because, after removal of the stapes and replacement with an artificial stapes, the prosthesis may slip or make poor contact, or the oval window may close over, and the hearing loss may become worse than it was before surgery.

45.3. Otosclerosis

When conductive hearing loss is present in an adult, and the eardrum and the middle ear appear to be normal through the otoscope, the most likely diagnosis is otosclerosis. There are other less common causes. The cause of otosclerosis is hereditary but otherwise unknown, but in some people a gradual fixation of the stapedial footplate in the oval window occurs because of abnormal changes in the cochlear bone. Actually, the term "otosclerosis" is misleading, for it is not initially a sclerotic process; it is more like a vascularization in the bone with formation of spongy-appearing bone. This stage of the disease is called otospongioses, and the mature, less active, later stage is otosclerosis. In common practice, the term otosclerosis is used to include both. Such changes have not been reported anywhere else in the body, only in the bony labyrinth. In view of the high incidence of otosclerosis and the confusion surrounding its diagnosis and because of the very large number of surgical operations being performed for it, a comparatively elaborate description is devoted to the subject.

45.4. Clinical Otosclerosis

Actually, there are many more people with otosclerosis than there are patients who seek medical advice for their hearing loss and receive the benefit of diagnosis. For instance, otosclerosis has been found at autopsy in the ears of many individuals who gave no evidence of hearing impairment while they were alive. It is present in up to 10% of caucasian Americans but symptomatic in only ~1%. Conductive hearing loss is absent when otosclerotic changes affect areas of the bony labyrinth other than the oval window. It is only when the oval window and the stapes are involved that conductive hearing loss develops, and then it is called clinical otosclerosis. For our purposes, we shall call it simply otosclerosis, because the condition does not concern us clinically until hearing impairment develops. Fixation of the stapes may occur over a period of many months or even years. The hearing gradually diminishes as the footplate becomes more fixed. Figure 9.31 shows a serial audiogram of a patient with otosclerosis whose hearing level has been followed repeatedly for years. Otosclerosis can be confused with Paget's disease and osteogenesis imperfecta. Distinguishing features are summarized in Appendix III.

45.4.1. Extent of Hearing Loss

In some cases the hearing loss stops progressing after reaching only a very mild level. We do not know why this occurs. More frequently, the hearing deteriorates until it stabilizes at ~70 dB. The hearing deficit classically starts in the lower frequencies and gradually progresses to the high ones. Eventually, all frequencies are involved. In numerous patients, the high frequencies continue to deteriorate even more than the low frequencies because of superimposed sensorineural hearing loss. Profound loss can occur in *far advanced otosclerosis*.

45.4.2. Characteristic Features

There are many intriguing and challenging features about otosclerosis. For example, it is far more common in females than in males, and it is generally noted first around the age of 18. Often the hearing loss is accompanied by a very annoying buzzing tinnitus; both symptoms usually are aggravated by pregnancy. The tinnitus commonly subsides over a period of many years.

Psychological Aspects. Of particular interest is the psychological complex present commonly in otosclerotic patients. Some of them become suspicious and feel that people are talking about them. Many become introspective and have marked personality changes. They retreat into their own "shells." Some try to appear to be light-hearted and even humorous, but the attempt lacks conviction. Perhaps these symptoms are related to the gradual onset of otosclerosis; yet comparable reactions rarely are observed in patients with hearing loss from other causes, even though these also may have a gradual onset. One of the vital reasons for trying to restore hearing early with surgery or amplification is to prevent adverse psychological changes.

Familial Aspects. Otosclerosis is familial to some extent. It often is found in several people in the same family. In contrast, many patients with otosclerosis deny any hearing loss in their families for as many generations as they may recall. It is not possible

JOSEPH SATALOFF, M.D.
ROBERT THAYER SATALOFF, M.D.
1721 PINE STREET PHILADELPHIA, PA 19103

HEARING RECORD

NAME AGE

AIR CONDUCTION

DATE	Exam	LEFT MASK	RIGHT 250	RIGHT 500	RIGHT 1000	RIGHT 2000	RIGHT 4000	RIGHT 8000	RIGHT MASK	LEFT 250	LEFT 500	LEFT 1000	LEFT 2000	LEFT 4000	LEFT 8000	AUD
1st YR.			35	35	30	20	10	5		35	40	30	20	5	5	
2nd YR.			45	40	40	20	10	10		40	40	30	20	10	10	
3rd YR.			50	50	50	25	15	15		40	45	35	20	15	15	
4th YR.			55	55	50	40	20	25		45	45	40	35	30	25	
7th YR.			60	55	50	50	40	35		50	55	50	45	45	40	
9th YR.			60	55	55	60	65	50		55	55	55	55	60	45	

BONE CONDUCTION

DATE	Exam	LEFT MASK	RIGHT 250	RIGHT 500	RIGHT 1000	RIGHT 2000	RIGHT 4000	RIGHT MASK	LEFT 250	LEFT 500	LEFT 1000	LEFT 2000	LEFT 4000	AUD
1st YR.			0	0	0	5	5		0	0	0	5	5	
4th YR.			0	5	5	10	15		0	5	10	10	15	
9th YR.			0	10	20	25	30		0	10	25	20	35	

SPEECH RECEPTION

DATE	RIGHT	LEFT MASK	LEFT	RIGHT MASK	FREE FIELD	MIC
1st YR.	30		35			

DISCRIMINATION

DATE	RIGHT % SCORE	RIGHT TEST LEVEL	RIGHT LIST	LEFT MASK	LEFT % SCORE	LEFT TEST LEVEL	LEFT LIST	RIGHT MASK	EXAM
	100	70			100	75			

HIGH FREQUENCY THRESHOLDS

DATE	RIGHT 4000	RIGHT 8000	RIGHT 10000	RIGHT 12000	RIGHT 14000	LEFT MASK	RIGHT MASK	LEFT 4000	LEFT 8000	LEFT 10000	LEFT 12000	LEFT 14000

RIGHT RINNE	RIGHT SCHWABACH	WEBER	LEFT RINNE	LEFT SCHWABACH

HEARING AID

DATE	MAKE	MODEL
RECEIVER	GAIN	EXAM
EAR	DISCRIM	COUNC

REMARKS

Figure 9.31 Progressive hearing loss in otosclerosis. Patient refuses surgical intervention and gets along well with a hearing aid. Note that bone conduction is becoming reduced, even though this patient was 38 years old at the time of the last audiogram.

to predict on genetic principles who in a family will exhibit clinical otosclerosis, but on theoretical grounds marriages between members of two families both of which have clear-cut cases of otosclerosis seem to be inadvisable. Nevertheless, the risk does not warrant an extreme position, and one should not try to discourage a man and a woman who wish to get married after they have been made aware of the facts. It is possible that none of their offspring will develop a hearing loss, although generally it follows an autosomal dominant inheritance pattern with variable penetrance. Certainly, there is no justification for a therapeutic abortion in a pregnant woman with otosclerosis merely because her hearing loss might be aggravated during pregnancy. Some older textbooks took a contrary view, but it is unwarranted in the light of what we now know about otosclerosis.

Effect of Excellent Bone Conduction. As noted previously, patients with otosclerosis have excellent bone conduction. This explains why these patients speak in very soft and modulated voices, a prominent feature in otosclerosis. A patient may complain that his/her hearing is worse when chewing crunchy foods. The reason is that excellent bone conduction causes the crunching noises to interfere with the ability to hear conversation.

45.4.3. *Differentiation*

During pure-tone audiometry, patients with otosclerosis often are uncertain whether or not they really hear the tone when they are being tested at threshold. In contrast, patients with sensorineural hearing loss generally are sure when they hear and when they do not.

In spite of the classic symptoms that otosclerotic patients present, many cases still are misdiagnosed as catarrhal deafness, allergic deafness, adhesive deafness, or deafness due to eustachian tube blockage. Because of these errors in diagnosis, the authors deliberately have emphasized that these nonotosclerotic conditions produce only mild hearing loss and that they rarely are accompanied by tinnitus or a family history of deafness. Whenever a patient with normal otoscopic findings has conductive hearing loss exceeding 45 dB in the speech frequencies, the cause in all probability is otosclerosis or ossicular defect, even though there may be a marked allergic history and changes in the nose, or slight retraction of the eardrum or even a demonstrably blocked eustachian tube. It is important to remember that few, if any, causes other than otosclerosis produce progressive conductive hearing losses of >45 dB accompanied by tinnitus and a familial history.

45.4.4. *Variable Progress in the Hearing Loss of Each Ear*

Otosclerosis sometimes does occur unilaterally, with normal hearing in the other ear. The progress of the hearing loss in each ear, when otosclerosis is present in both ears, may differ widely, so that most of the time a patient complains of hearing better with one ear than with the other. This difference may change, however, and the patient often says, "My left ear used to be the better ear, but it has gotten so bad that the right ear is now the better one." This subjective experience, corroborated by hearing tests, may determine which ear should be operated on when an attempt is made to restore hearing by stapes surgery.

45.4.5. *Other Types*

There are many other types of otosclerosis that the otologist can classify. One is the so-called malignant type, which is most disturbing. It may occur in young patients and is noted for a rapidly progressing hearing loss which, often in 1 or 2 years, reaches handicapping proportions and is accompanied by diminished bone conduction as a result of sensorineural involvement (Fig. 9.32). Frequently, these cases are not seen until the sensorineural damage is already so pronounced that the otosclerotic origin is largely obscured, and surgery is of more limited value.

Although a connection has not yet been proven, the frequency with which sensorineural hearing loss accompanies otosclerosis suggests a relationship between them. Otosclerotic changes found in the sensory area of the cochlea, particularly in the stria vascularis, are believed to be responsible for the sensorineural impairment.

Another type of otosclerosis is illustrated in Fig. 9.33. This is hardly recognizable as otosclerosis, and yet the patient has all the classic history and symptoms, with the exception that the hearing loss is >60 dB, and sensorineural loss also is present. Many features of this type of otosclerosis are discussed in Chapter 10. This is definitely otosclerosis, and the hearing can be improved by surgery, as seen in the audiogram. The absence of

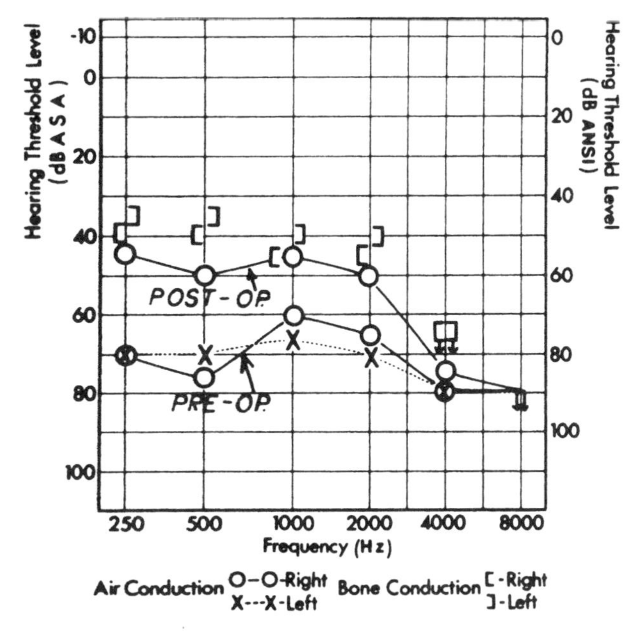

Figure 9.32 *History*: A 24-year-old woman with insidious hearing loss only 3 years from onset. Marked roaring tinnitus and some imbalance. No familial deafness. Voice normal. *Otologic*: Ears normal. Right stapedectomy improved the hearing after removing a thick, white otosclerotic fixed footplate. *Audiologic*: Pure-tone air conduction thresholds revealed a severe bilateral loss. Bone conduction also was reduced. Preoperative tuning-fork tests showed bone better than air conduction in both ears and no lateralization. Postoperatively, air equaled bone conduction in the right ear. Pre- and postoperative discrimination scores in the right ear were in the low 40% range. The air–bone gap in the right ear was closed with surgery. (A similar result was obtained later in the left ear.) *Impedance*: Type A tympanogram. *Classification*: Mixed hearing loss. *Diagnosis*: Otosclerosis with sensorineural involvement. The presence of mild imbalance is not uncommon in this type of otosclerosis with sensorineural involvement.

responses to audiometric testing does not mean necessarily that the ear is "dead." It means merely that the threshold is beyond the limits of the audiometer. Another unusual, but severe type of otosclerosis occurs when both the oval window and the round window are obstructed. Here again, the bone conduction is reduced.

The above cases of otosclerosis that have a sensorineural component are not considered to be purely conductive but, rather, cases of mixed hearing loss. They are included in this chapter because the conductive element predominates.

Occasionally, there will be minimal visible abnormalities in the appearance of the eardrum, yet marked tympanosclerotic changes will be found in the middle ear. This condition, called tympanosclerosis, can cause ossicular damage or fixation and hearing loss.

The hearing loss in Paget's disease may be either conductive or sensorineural. Conductive hearing loss occurs as a result of stapes fixation when calcium deposits prevent normal movement of the stapes footplate. These deposits form in the annular ligament of the stapes at its attachment to the oval window and present a hearing loss pattern similar to that which occurs in otosclerosis (Fig. 9.34).

Sensorineural hearing loss develops because of changes in the labyrinth or narrowing of the internal auditory meatus and pressure on the auditory nerve. Other established findings in patients with Paget's disease are discussed in Chapter 13 and Appendix III.

JOSEPH SATALOFF, M.D.
ROBERT THAYER SATALOFF, M.D.
1721 PINE STREET PHILADELPHIA, PA 19103

HEARING RECORD

NAME AGE

AIR CONDUCTION

			RIGHT							LEFT						
DATE	Exam	LEFT MASK	250	500	1000	2000	4000	8000	RIGHT MASK	250	500	1000	2000	4000	8000	AUD
1st TEST			90	95	85	90	NR	NR		95	NR	NR	NR	NR	NR	
1 mo.			RIGHT STAPES MOBILIZATION													
4 mos.			45	50	45	75	65	65								
5 mos.										LEFT STAPES MOBILIZATION						
6 mos.										60	70	70	70	75	NR	

BONE CONDUCTION

			RIGHT						LEFT						
DATE	Exam	LEFT MASK	250	500	1000	2000	4000	RIGHT MASK	250	500	1000	2000	4000		AUD
1st TEST			40	30	40	50	55		15	35	35	50	45		

SPEECH RECEPTION

DATE	RIGHT	LEFT MASK	LEFT	RIGHT MASK	FREE FIELD	MIC.
5 mos.	58		86			

DISCRIMINATION

	RIGHT				LEFT				
DATE	% SCORE	TEST LEVEL	LIST	LEFT MASK	% SCORE	TEST LEVEL	LIST	RIGHT MASK	EXAM
	44	88			68	98			

HIGH FREQUENCY THRESHOLDS

	RIGHT						LEFT					
DATE	4000	8000	10000	12000	14000	LEFT MASK	RIGHT MASK	4000	8000	10000	12000	14000

RIGHT		WEBER	LEFT	
RINNE	SCHWABACH		RINNE	SCHWABACH

HEARING AID

DATE	MAKE	MODEL
RECEIVER	GAIN	EXAM
EAR	DISCRIM	COUNC

REMARKS

Figure 9.33 *History*: A 67-year-old woman with insidious hearing loss for >30 years. Refuses to wear hearing aid and communicates only in writing. No familial history of deafness. No tinnitus or vertigo. *Otologic*: Normal. *Audiologic*: There is severe sensorineural hearing loss, but patient has an air–bone gap. Reduced discrimination. Hears tuning fork well by teeth. *Impedance*: Type A tympanogram. *Aid to diagnosis*: The excellent results obtained by stapes mobilization emphasize that these ears are not "dead" but have hearing beyond the audiometer's limits.

In van der Hoeve's syndrome, the patient exhibits a combination of a generalized bone affliction, osteogenesis imperfecta (Lobstein's disease), with blue sclera and hearing loss. In this condition, cartilages throughout the body are softened, and the teeth become transparent in a few of these patients. It is common to find several members of the same family having hearing loss and blue sclera. Although the audiologic findings closely resemble those of otosclerosis, the impression during stapes mobilization surgery suggests that the stapes fixation is of a different consistency than that found in otosclerosis (see Chapter 13 and Appendix III).

JOSEPH SATALOFF, M.D.
ROBERT THAYER SATALOFF, M.D.
1721 PINE STREET PHILADELPHIA, PA 19103

HEARING RECORD

NAME AGE

AIR CONDUCTION

RIGHT									LEFT							
DATE	Exam	LEFT MASK	250	500	1000	2000	4000	8000	RIGHT MASK	250	500	1000	2000	4000	8000	AUD
1st TEST			25	15	25	20	15	15		45	35	35	45	60	65	
1 YR.			25	15	20	20	40	25		40	35	45	45	60	65	
3 YRS.			25	25	20	20	45	25		40	40	50	50	65	65	
4 YRS.			35	40	30	50	65	30		45	50	50	60	75	60	
13 YRS.			70	65	75	85	95	NR		65	55	50	70	90	NR	

BONE CONDUCTION

RIGHT									LEFT							
DATE	Exam	LEFT MASK	250	500	1000	2000	4000		RIGHT MASK	250	500	1000	2000	4000		AUD
1st TEST			0	5	10	25	20			5	0	30	55	NR		
1 YR.			10	10	10	30	30			5	10	35	NR	NR		
13 YRS.			10	10	25	NR	NR			5	10	35	NR	NR		

SPEECH RECEPTION

DATE	RIGHT	LEFT MASK	LEFT	RIGHT MASK	FREE FIELD	MIC.

DISCRIMINATION

	RIGHT				LEFT				
DATE	% SCORE	TEST LEVEL	LIST	LEFT MASK	% SCORE	TEST LEVEL	LIST	RIGHT MASK	EXAM.

HIGH FREQUENCY THRESHOLDS

	RIGHT						LEFT					
DATE	4000	8000	10000	12000	14000	LEFT MASK	RIGHT MASK	4000	8000	10000	12000	14000

RIGHT		WEBER	LEFT	
RINNE	SCHWABACH		RINNE	SCHWABACH

HEARING AID

DATE	MAKE	MODEL
RECEIVER	GAIN	EXAM
EAR	DISCRIM.	COUNC

REMARKS

Figure 9.34 *History*: A 59-year-old man first noticed beginning impairment in hearing in the left ear at the age of 44. Hearing loss gradually progressed with increasing discrimination difficulty and occasional tinnitus. Diagnosed as having far-advanced Paget's disease at the age of 46. Tinnitus started in the right ear 6 months after diagnosis was made. Wearing hearing aid in left ear since age 52. Had periods of fleeting imbalance for the past few years. Speech normal. Skull roentgenograms revealed a narrowing of the internal auditory canal and no visualization of the right canal. Neither cochlea could be made out clearly. *Otologic*: Both eardrums were slightly opaque. *Audiologic*: Air and bone conduction thresholds measured over a period of 13 years show a slowly progressive hearing loss with greater loss in the higher frequencies. An air–bone gap was always present in the lower frequencies but not in the higher frequencies. No recruitment was found. *Classification*: Low-frequency conductive loss and subsequent sensorineural involvement in the middle and higher frequencies. *Diagnosis*: Paget's disease.

45.5. Eustachian Tube and Its Relationship to Conductive Hearing Loss

45.5.1. Action of the Eustachian Tube

The eustachian tube generally is straight, but sometimes it has a slight curve and a little twist where the bony portion from the middle ear joins the cartilaginous portion in the nasopharynx. In addition, the opening of the tube is wide in the nasopharynx and in the middle ear, but the lumen narrows down considerably at the bony cartilaginous junction. Normally, the eustachian tube is not open, because the mucous membrane lining the lumen is in contact, except during swallowing, sneezing, yawning, or forceful nose blowing. Behind the opening of the eustachian tube in the nasopharynx is a deep fossa called the fossa of Rosenmüller. Excessive growth of adenoid tissue in this area often compresses the tube and prevents normal aeration. Congestion and infection in the nasopharynx and the adenoid region also can cause closing of the tube by upward progression along the tubal mucosa and the submucosal areas.

Functions. The eustachian tube normally has two important functions. One is to allow drainage from the middle ear into the nasopharynx, and the other is to maintain equal air pressure on both sides of the eardrum. The mucous membrane of the middle ear absorbs air, though slowly. As the eustachian tube normally is closed, it is essential that the eustachian tube be opened at intervals to maintain the pressure equilibrium. This is done during swallowing, not only when eating, but continually throughout the day and night as saliva and mucus collect in the pharynx and stimulate the swallowing reflex.

Methods of Evaluating Function. One simple way to demonstrate that the tube opens during these acts is to hold a vibrating tuning fork in front of the nostrils. The sound received by the patient will be weak until swallowing, and then it will be much louder as the sound waves travel up the eustachian tube into the middle ear.

Another way of determining whether the tube is functioning and opening properly is to use Politzer's method. A large nasal tip on a pressure bottle containing camphor mist (or only air) is inserted snugly into one nostril while the other is occluded, and the patient is told to say "kick" or "cake." Pronouncing the "k" sound causes closure of the nasopharynx by lifting the soft palate, and the air or the mist then is forced up the eustachian tube, causing the eardrum to push out slightly. It is helpful to watch the eardrum through an otoscope during this procedure. The normal eardrum will move out very slightly and then return to its original position. Abnormal results may include a slow return of the eardrum to its initial position, persistent pouching out of large blebs, or failure of the drum to move at all. These irregularities suggest abnormal conditions in the tube and the middle ear. Caution must be exercised not to use too much air pressure, as this may rupture a weakened eardrum or cause dizziness. The whole procedure is contraindicated if infection is present in the nose or the nasopharynx.

Other methods of evaluating the function of the eustachian tube include direct examination through a nasopharyngoscope and catheterization. Politzerization also is especially helpful if a small perforation is suspected in the eardrum but cannot be visualized through the otoscope. The forced-up camphor mist can be seen coming out of the small perforation, thereby pinpointing its site. If air is used, water or ear drops can be placed in the ear, and the perforation can be located by watching for bubbles.

Impedance audiometry can demonstrate eustachian tube function by recording changes in middle ear pressure while swallowing. When a perforation is present, tubal function can be measured directly.

45.6. Extent of Hearing Loss with Eustachian Tube Involvement

No part of the auditory system has been incriminated more wrongly and mistreated more than the eustachian tube. This was understandable before there was clear understanding of the physiology of hearing, of the function of the eustachian tube, and of otosclerosis. At the present time, however, there should be no reasonable excuse for blaming or mistreating the eustachian tube when the cause of the hearing loss actually lies elsewhere.

Malfunction of the eustachian tube causes only conductive hearing loss. If sensory hearing loss is present, the eustachian tube generally should not be the target for treatment, even though politzerization or tubal inflation may give the patient a subjective sense of well-being and an apparent hearing improvement for a few moments. Such treatment often delays the patient's actual aural rehabilitation.

In general, simple blockage of the eustachian tube causes only a comparatively mild hearing loss not exceeding ~35 dB; most of the time it is much less than that. The loss is greater in the lower frequencies than in the higher. The most common causes are acute upper respiratory infections and allergies in which the eustachian tube becomes boggy and obstructed owing to congestion and inflammation. This condition makes the ears feel full, and the individual appears to be slightly hard-of-hearing. If the obstruction in the tube persists, the air in the middle ear is absorbed by the mucosal lining, and the eardrum becomes slightly retracted; thus, the hearing loss is aggravated, so that it may reach a measurable level of ~35 dB. If fluid forms in the middle ear, there may be an even greater hearing loss, but in this case the emphasis is in the higher frequencies. This frequency change is caused by the addition of the fluid mass to the contents of the middle ear. When there is fluid in the middle ear, the bone conduction also may be slightly reduced, and this reduction may suggest a false diagnosis of sensorineural hearing loss. However, removal of the fluid restores both air and bone conduction to normal.

There are exceptions to these generalizations. When the tube has been closed for many months or years, the drum may become retracted so completely that it becomes "plastered" to the promontory, a condition that causes a hearing loss of ~50 dB, or, if the fluid is thick and gelatinous, the loss may be slightly greater. In general, however, the eustachian tube obstruction *per se* causes only a mild loss in hearing. The loss increases only when complications arise in the middle ear, such as the presence of a fluid mass and retraction of the drum, or infection. Therefore, it is safe to conclude that, with rare exceptions, if the hearing loss exceeds 35 dB and the eardrum and the middle ear appear to be practically normal, the fault does not lie in the eustachian tube. The cause is more likely to be found in the ossicular chain and especially in the stapes. Hearing loss commonly attributed to eustachian tube blockage most frequently has other causes, most of them not related to the tube itself.

45.7. Injuries to the Eustachian Tube

Trauma of the eustachian tube is not a frequent cause of hearing loss, but during war-time a number of patients were seen whose eustachian tubes were injured by gunshot or shrapnel wounds to the face. The damage generally is unilateral and can be visualized through the nasopharyngoscope.

Following trauma to the eustachian tube, the scar tissue that forms during healing narrows the tube and causes recurrent or persistent attacks of serous otitis media. In such instances, it is important to avoid harsh manipulation inside the tube, which eventually would aggravate the constriction. Repeated myringotomies, or a temporary tube in the eardrum, and the use of oral and local decongestants are the best alternatives in most cases. Patients usually complain of fullness in the affected ear and of some hearing loss, which clears up when the fluid is removed from the middle ear.

45.8. Patulous Eustachian Tube

In rare instances, a patient complains of a loud swishing sound in the ear whenever he/she inhales through the nose, and the ears also may feel full and slightly hard-of-hearing.

The examination of the mouth of the eustachian tube through a nasopharyngoscope may reveal the opening to appear to be much larger and far more patent than normal. Such a condition is called a patent, or patulous, eustachian tube. The cause is not always clear, but it appears to result from a depletion of fat around the eustachian tube. Anything that causes fat redistribution may produce this condition: weight loss, use of steroids, pregnancy, birth control pills, and other conditions.

Patients with a patulous eustachian tube are far more disturbed by their symptoms than are those with an obstructed tube. It can be most annoying to a patient to hear a swish in the ear with each breath. The condition is aggravated by exercise. The patient eventually may become obsessed with this symptom. The symptoms stop when breathing is done through the mouth. Another complaint is hearing one's own voice as if in a resonant chamber. This is called autophonia. Another unusual observation is to see the eardrum move in and out during respiration. This may be confirmed by tympanometry. Although the patient may believe that there is difficulty hearing in the involved ear, actually there is no hearing loss, and this condition basically is not a hearing defect. Although many treatments have been suggested for this abnormality, there is no one specific cure for these annoying symptoms. Weight gain, myringotomy and tube placement, and, rarely, eustachian tube injection may be helpful.

46. HYPERTROPHIED ADENOIDS

Serous otitis media is the most common cause of mild conductive hearing loss in children and can be caused by hypertrophied adenoids, allergy, and other conditions. School surveys in several states agree that ~3% of the children have significant hearing losses in at least several frequencies. Well over 80% of these are conductive and may be associated with hypertrophied adenoids (Fig. 9.35). Other possible causes of this condition are secretory otitis, cleft palate, allergies, nasopharyngitis, sinusitis, gastric reflux, and nasopharyngeal tumors, but even in some of these conditions hypertrophied adenoids may be a contributing factor.

As audiograms are not always performed routinely in children, hearing loss is probably much more prevalent than commonly believed. Children usually are spoken to in a

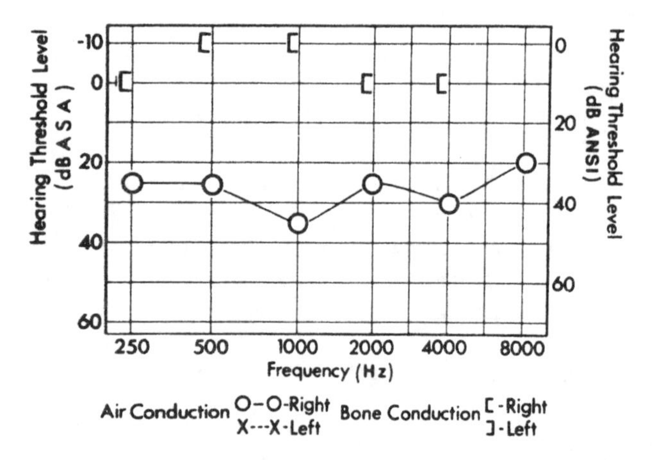

Figure 9.35 *History*: A 6-year-old child with recurrent earaches and hearing loss for the past year. Mouth breather. Patient had had six myringotomies, allergic desensitization, and a T&A before being seen by the author.

raised voice, and as a result, hearing losses seldom are noted until they approach 40 dB. These losses develop insidiously in many children and may be present for many months or even years before they are distinguished from childhood inattention. Hearing loss as mild as 15 dB can have a measurable effect on a child's behavior. Adenoidectomies often are performed because of mouth breathing or chronic ear infection without a preoperative audiogram; it seems reasonable to speculate that routine audiograms would reveal a substantial number of hearing losses in such children. Hence, universal screening has been advocated. In uncomplicated cases, hearing loss may subside under either conservative or surgical treatment without its existence ever having been noted by either the parent or the physician. Routine audiograms clearly are indicated in all children with recurrent attacks of otitis media, chronic inattention, history of mouth breathing, and earache, especially in the presence of hypertrophied adenoids.

Opinions differ widely as to the indication for adenoidectomy in children. These differences range from the routine performance of adenoidectomies in the almost all children having upper respiratory infections to the opposite extreme of never removing the adenoids unless there is complete obstruction in the nasopharynx. Such marked differences of opinion concerning the effectiveness of adenoidectomy in dealing with this problem indicate the need for continuing reappraisal based on objective studies.

Untreated childhood ear disease may result in hearing loss later in life. The sequelae of serous otitis are primarily conductive hearing losses. However, chronic, recurrent infectious otitis media appears to occur more often in people with serous otitis and may result in sensorineural hearing loss in selected cases, as well. It is important to remember that the implications of serous otitis, especially unilateral serous otitis, are different in the adult. Unilateral serous otitis with no apparent etiology should be considered carcinoma of the nasopharynx until proven otherwise. Serous otitis is often the presenting symptom of this disease because of unilateral malignant eustachian tube obstructions.

47. EFFECTS OF EAR SURGERY

Procedures of middle-ear surgery are beyond the scope of this book, but the physician who examines the ear of a patient with conductive hearing loss may discover evidence of previous ear surgery. He/she should be prepared to evaluate the relation of the patient's hearing deficit to past otologic surgical procedures. A brief review of these interventions and the visible traces they are likely to leave may be helpful.

47.1. Microscopes

Otologic microsurgery utilizes magnification, usually provided by an operating microscope. Many surgeons are not familiar with formulas that determine accurately the amount of magnification used, and it is often recorded incorrectly in operative reports. It is not unusual for surgeons to assume that the number on the indicator on the zoom control correlates with the number of times the image is magnified, but accurate determination is more complex than that. This author usually works with a Zeiss operating microscope (Oberkochen, Germany), and the information in this discussion refers specifically to Zeiss instruments. However, the principles are the same for microscopes by other companies. In order to determine the amount of magnification, the focal length of the binocular tube is divided by the focal length of the objective lens, and then multiplied by the magnification of the eye pieces (2). The result is then multiplied by the indicator on the magnification (zoom) knob of the microscope, on a modern microscope. The focal length of the binocular tube is usually a number such as F125, F160, or F170. For example, the old Zeiss OPMI-6 microscope has a binocular tube with a focal length of F160, and the newer design with the wider angle of view has a focal length of F170. The focal length of the objective lens varies depending upon the surgeon's preference. For ear surgery, it is usually 250 or 300 mm. For laryngeal surgery, a 400-mm lens is used most commonly. The usual eye piece magnification is either 10× or 12.5×. The indicator number on a modern Zeiss operating microscope can be read through a small window next to the zoom control knob, and the number ranges from 0.4 to 2.4. The OPMI-6-S, for example, provides a continuous magnification range of 1:4. Older Zeiss operating microscopes (such as the OPMI-1) have magnification changes that are step-like rather than continuous, and have numbers that range from 6 to 40 next to the dial. These provide five magnification steps in a range of 1:6. These numbers should not be used in the formula noted earlier, but can be converted as follows: 40 corresponds to 2.5; 25 corresponds to 1.6; 16 corresponds to 1.0; 10 corresponds to 0.6; 6 corresponds to 0.4. So, for example, if a surgeon is using an OPMI-1 microscope with 10× eyepieces, a 400-mm objective lens, and the magnification set at 40 (maximum), image magnification is 7.8× [i.e., $(125/400)\ 2.5 \times 10 = 7.8\times$], not 40× as misstated commonly. Simply changing the eyepieces from 10× to 12.5× increases the magnification from 7.8× to 9.8×, and using 20× eyepieces increases the magnification to 15.6×. Utilizing an objective lens with a shorter focal length also increases magnification but brings the microscope closer to the operating field. While this approach is used during ear surgery, it is not suitable for laryngeal surgery because the decreased space between the microscope and direct laryngoscope is not sufficient to permit unimpeded manipulation of long-handled laryngeal instruments. It is important for surgeons to be familiar with these principles in order to optimize surgical conditions for each specific case, and in order to document surgery accurately.

47.2. Myringotomy

Incision and drainage through the eardrum is called myringotomy. This incision is made in the bulging part of the eardrum or in the posterior inferior or an anterior quadrant to avoid injuring any middle- or inner-ear structure in case the incision goes too deeply. This might occur if the patient moved inadvertently during the procedure.

In unusual instances, the end of the incus and, even more rarely, the stapedial crura may be injured during myringotomy. Figure 9.28 is a case report in which such an injury occurred in childhood and remained undiagnosed until surgery was performed many years later. These injuries may be avoided if the myringotomy is performed in the anterior inferior quadrant.

In most instances, after myringotomy, the eardrum returns to normal, but sometimes permanent scar tissue is left. The hearing is not damaged by a myringotomy itself or by the scar tissue it occasionally leaves in the eardrum. A perforation may persist, especially when the myringotomy incision has been kept open with a synthetic tube. These perforations can be closed surgically.

47.3. Hearing Loss Associated with Ear Surgery

When many kinds of ear surgery are performed, prime consideration is given to preservation or restoration of hearing. Of course, in surgery for mastoiditis or chronic otitis media, the principal objective is to remove the infection, but the type of surgery done is determined to a great degree by efforts to preserve hearing. A number of surgical procedures that potentially are associated with hearing impairment produce visible changes in the eardrum, over the mastoid, or in the external ear. Some procedures, such as stapes surgery, leave little or no recognizable scarring, and even a skilled otologist may be unable to tell from looking that the ear has been operated on. Therefore, it is important to obtain a careful history of any previous surgery by questioning the patient.

47.4. Simple Mastoidectomy

Extremely common several years ago, simple mastoidectomy for acute infection now is rare. Better management of otitis media by the general practitioner and the pediatrician has practically eliminated the need for this procedure. It was done most commonly in children who had severe otitis media that extended to the mastoid bone and caused postauricular swelling, pain, and tenderness as well as hearing loss. Hearing loss was present because of the middle ear infection, not the mastoiditis.

A simple mastoidectomy consists of making a postauricular incision, removing the mastoid cell, and creating an opening into the tympanic antrum that leads to the middle ear. This opening allows aeration and drainage of pus. Essentially, this procedure is a complex type of incision and drainage for pus under pressure. As the middle ear is not disturbed deliberately, and the eardrum usually remains intact (after the infection subsides), hearing usually returns to normal. It is common to find patients with large postauricular scars who have retained normal hearing after having had several simple mastoidectomies.

Unfortunately, not all patients who have had simple mastoidectomies or myringotomies have normal hearing. Some have persistent hearing loss because of surgical mishaps and others because of persistent perforation in the eardrum or other damage from the infection; still others have mild hearing losses caused by retraction of the eardrum, by middle ear adhesions, or as consequences of the infection.

The chief evidence of a previous simple mastoidectomy is the presence of a postauricular scar with a normal or almost normal eardrum and good hearing. Even if the eardrum is perforated, but the hearing level is better than ~30 dB, the surgery probably was a simple mastoidectomy that left an intact ossicular chain.

Simple mastoidectomy also is done for access to the endolymphatic sac, the facial nerve, or the nerves of the internal auditory canal and cerebellopontine angle, and it is now used widely for chronic otitis media and mastoiditis, with slight expansion of the surgery.

47.5. Myringoplasty

When a perforation is found in the eardrum without any active infection, it is nearly always possible to close the perforation. If the hearing loss is due entirely to the perforation, the closure should restore the hearing. The improvement in hearing that can be expected by closing the perforation can be determined preoperatively by patching the hole with a small piece of cigarette paper, Gelfoam, or silastic film. Audiograms obtained prior to and after the patching show the amount of hearing improvement that can be expected from permanent closure with a graft. If no improvement occurs, it means the perforation is not the sole cause of the hearing loss. In either event, it is usually advisable to close the perforation, but the patient should be advised of the prognosis prior to the surgery. In cases in which the hearing loss is not caused by the perforation alone, a simple myringoplasty is inadequate as far as the hearing is concerned. Then it is necessary to reflect the eardrum and explore the middle ear for ossicular problems or adhesions in addition to closing the perforation in the drum.

Figure 9.36 describes a case of simple myringoplasty.

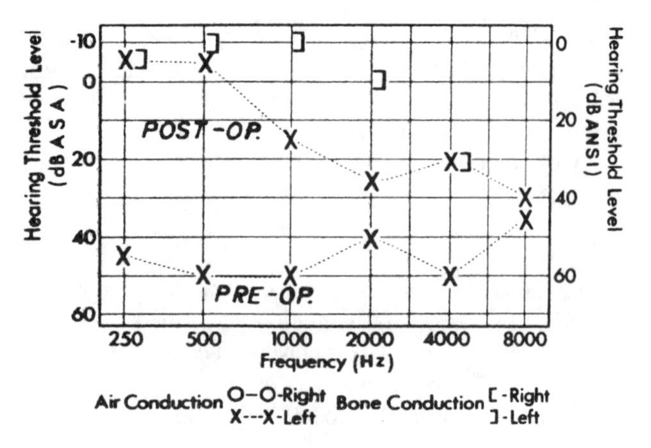

Figure 9.36 *History*: Intermittent discharge from left ear. *Otologic*: Large perforation in eardrum. Discharge was cleared up with conservative therapy, and later the large perforation was closed with a sliding flap from the external auditory canal. *Audiologic*: Left-ear air conduction thresholds revealed a moderate loss. Bone conduction thresholds were normal, except at 4000 Hz. After the myringoplasty, an air–bone gap remained at the two middle frequencies. *Classification*: Conductive. *Diagnosis*: Perforated drum corrected with myringoplasty.

The appearance of the eardrum following myringoplasty varies greatly. It is important to obtain a history of this type of surgical procedure from the patient; otherwise, a physician might be startled by the appearance of the eardrum and the hearing loss present and advise unjustified treatment. If a skin graft was used to close the perforation, the drum may appear to be thick and white, and redundant skin may be present so that the drum is flaccid and sometimes hardly delineated. If a vein graft was used, the drum may appear to be stranded and scarred and sometimes thickened and discolored. The appearance of the drum often depends on the graft used, its thickness, and the manner in which it "took." When a grafted eardrum appears flaccid and retracted, and the patient complains of a fluctuating hearing loss that becomes more noticeable every time there is a feeling of fullness in the ear, gentle politzerization often can restore hearing temporarily to an improved level. Sometimes a myringoplasty heals so well that it leaves the eardrum looking almost normal, and the examining physician is scarcely able to detect that any surgery has been done.

47.6. Ossiculoplasty

The term "ossiculoplasty" denotes repair or restoration of the continuity of the ossicular chain. For example, in the case presented in Fig. 9.37, closure of the perforation did not restore hearing. Therefore, at a later date the then intact eardrum was elevated, and the ossicles were examined. Instead of a normal incus, the connection of the incus with the head of the sapes was fibrous, thin, and very weak, making poor contact. This condition probably resulted from the same infection that had caused the perforation. When the weak connection was replaced with an artificial joint, hearing improved markedly. The only reason that two surgical procedures were required in this case was that prior to the simple myringoplasty no therapeutic test of closure with an artificial membrane had been done and the ossicular defects was not recognized initially.

The many types of deformities and defects sometimes found in the ossicular chain may test the ingenuity of the surgeon. The defects may be congenital or acquired and may

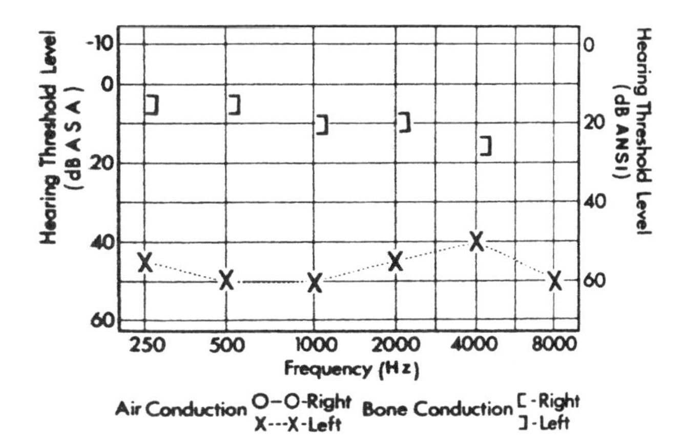

Figure 9.37 *History*: A 44-year-old woman with a 2-year episode of left ear discharge when in her early twenties. No tinnitus. Had had a myringoplasty which did not restore hearing. *Otologic*: Large healed perforation. Further surgery revealed that the incus end was eroded and replaced with a thin band of fibrous tissue. The continuity of the chain was corrected with a plastic prosthesis. Hearing was restored with restoration of ossicular continuity. *Impedance*: Type A tympanogram; absent stapedius reflex. *Classification*: Conductive. *Diagnosis*: Ossicular disruption.

involve any or all of the ossicles. The hearing loss can involve practically all of the frequencies, and when there is complete disruption of the chain with an intact drum, the hearing loss is ~60–70 dB ANSI.

When the eardrum is normal and a nonotosclerotic ossicular defect is found, the etiology is often congenital. Exceptions do occur.

More commonly, however, there is a history of trauma or otitis media, and some abnormality is present in the drum. The diagnosis is not hard to make, but it is difficult to predict just which ossicle is involved.

Whenever a patient has a 60–70 dB hearing loss involving all of the frequencies and patching of the perforation does not improve the hearing, there is a good possibility that the ossicular chain is disrupted (if otosclerosis is excluded). If the hearing loss is only ~40 dB and other circumstances are the same, it is more likely that the ossicular chain is intact but that some fracture or joint damage has healed with fibrous or weakened union, or that one or more ossicles is partially fixed and not moving freely. The middle ear should be examined at least whenever there is any doubt that simple myringoplasty can restore hearing, and many surgeons recommend exploring the middle ear when repairing any perforation.

47.7. Stapedectomy

When the stapes is fixed by otosclerosis or congenital deformity, the resultant conductive hearing loss may be repaired by mobilizing the stapes or by removing it. When removed, it is replaced by a prosthesis (or "artificial bone") made out of stainless steel, Teflon, or a similar material. The prosthesis usually is connected to the incus, and it may be visible as a metallic reflection through the eardrum. This not a problem and may be the only visible evidence of stapedectomy. Occasionally, if bone has been curetted for visibility, the posterior, superior edge of the ear canal may be jagged, and the annulus may be displaced. However, most often there are no visible sequelae of stapes surgery.

47.8. Radical Mastoidectomy

Radical mastoidectomy is performed in selected cases of chronic otitis media and mastoiditis in which it is not advisable or possible to clear up the infection by more limited operations that preserve the ossicular chain and the eardrum, as well as hearing. Usually, in such cases there is extensive cholesteatoma and erosion of the eardrum and the ossicles. Surgery can be done through an endaural or a postauricular incision. The mastoid cells are exenterated, and the malleus and the incus are removed. The stapes (or at least its footplate) is carefully left in place. Because there is no ossicular chain or drum, the hearing level in ears after such surgery usually is between 35 and 70 dB at all frequencies. The appearance of the middle ear varies, but usually skin covers a large mastoid cavity, no eardrum is present, and one can see into the anterior part of the middle ear where the eustachian tube opening lies. These ears have a tendency to collect debris and cerumen and require frequent cleaning.

In many patients with radical mastoid cavities, it is now possible to improve hearing by tympanoplastic surgery that partly reconstructs the middle ear. The ear canal also can be rebuilt to eliminate the cavity.

47.9. Modified Radical Mastoidectomy, Tympanoplasty, and Intact Canal Wall Tympanomastoidectomy

There is now a choice of procedures intermediate between simple and radical mastoidectomies. Aimed at the eradication of infection in the mastoid bone and the middle ear together with the preservation of hearing, these are called modified radical mastoidectomies, tympanomastoidectomies, or tympanoplasties. The latter term has broader coverage because it applies also to procedures on previously operated ears, now dry, for the purpose of restoring hearing. An entire field of reconstructive middle ear surgery has evolved with the new, improved knowledge of hearing and middle ear function. For this reason, in the presence of hearing loss, more and more tympanoplasties are being performed for hearing improvement on ears previously operated upon. Among the most common procedures used are those involving the positioning of a strut or wire connecting the stapes or oval window with either a newly grafted eardrum or the malleus, or even a small remaining section of the malleus.

In modified radical mastoidectomies, the posterior ear canal wall is removed, but some portion of tympanic membrane, malleus, and/or incus is preserved. Some modified radical mastoidectomies and tympanoplasties are done through an endaural incision rather than postauricularly. Very little scar is left following endaural surgery, and this is seen as a fine line slightly above the tragus and directed upward toward the temple. The posterior position of the ear canal may have been removed, and thus a canal larger than normal is left. The eardrum, usually present, generally is not normal because it has been scarred by infection and in most cases has been grafted to close a perforation.

Intact canal wall tympanomastoidectomy is now performed commonly and is the procedure of choice in many cases. This is similar to a simple mastoidectomy, but more extensive. A postauricular incision is used, and the mastoid cells are removed. Bone removal is carried anteriorly into the zygomatic root as far forward as the anterior ear canal wall. An extension from the mastoid into the middle ear may be created between the facial nerve and the bony external auditory canal. This is called a facial recess approach. The middle ear also is entered through a separate incision in the ear canal. When done properly, this somewhat more complicated technique permits eradication of disease, with preservation of the external auditory canal wall, eardrum (with or without graft), and any healthy ossicles. It permits the greatest chance for reconstruction of normal hearing. The disadvantage is that the ear is not exteriorized. So, when intact canal wall tympanomastoidectomy is performed for cholesteatoma, a "second look procedure" is generally required 1 year later to remove any residual cholesteatoma while it is still small enough to be resected completely.

47.10. Fenestration

Many ears were fenestrated in the past, but this complicated procedure has been superceded by stapes surgery. It is important to recognize a fenestrated ear, to be able to care for it, and to know the procedures available to improve hearing in an unsuccessful fenestration or in one in which the hearing has regressed.

Fenestrations were done in cases of otosclerosis with excellent bone conduction, and they usually were performed endaurally. The operation still is performed occasionally for patients with severe congenital malformations. The object was to circumvent the fixed stapes and to allow sound waves to impinge directly on a new window, bypassing the fixed oval window. The new window was made in the horizontal semicircular canal and was covered with a skin flap from the posterior canal wall continuous with the eardrum. Because of the depth and the difficulty of access to the operative site, the eardrum was relocated, and the incus and the head of the malleus were removed. On otoscopic examination, the external auditory canal is found to be enlarged, and the eardrum is seen to be pushed back into the middle ear. Part of the mastoid bone also has been removed, so that a cavity is visible. Because of the disruption of the ossicular chain and other factors, it was not common to obtain consistent hearing levels better than 15 or 20 dB in most fenestrated ears. Figure 9.38 shows a typical successful result. Bone conducion had to be very good with a large air–bone gap to warrant fenestration. Occasionally, hearing regressed owing to closure of the fenestra.

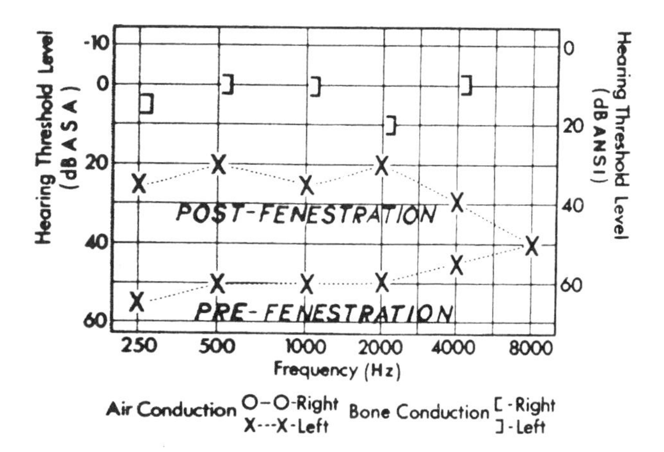

Figure 9.38 *History*: Gradual hearing loss and buzzing tinnitus >10 years. Sister also hard-of-hearing. Soft voice and hears better in noisy room. *Otologic*: Normal. *Audiologic*: Left-ear air conduction thresholds revealed a moderate, flat loss. Bone thresholds were normal. Fenestration surgery reduced the air–bone gap, but because the ossicular continuity had been disrupted during this operation and the hearing pathway altered to go through a fenestration in the lateral semicircular canal, hearing rarely improved beyond the 15 dB level. *Classification*: Conductive. *Diagnosis*: Otosclerosis corrected with fenestration surgery.

In some of these cases it is now possible to improve hearing by stapes surgery. In such cases, the stapes may be mobilized and connected to the eardrum or malleus remnant with a strut or wire. Another good procedure is to remove the stapes and run a wire from the malleus to the oval window.

When cleaning debris from fenestrated ears, the physician should be careful while working deep in the ear. Vertigo is induced readily by pressing near the fenestrated area.

REFERENCES

1. Schuknecht HF. Some interesting middle ear problems. Laryngoscope 1957; 67:395–409.
2. Hoerenz P. The operating microscope: optical principles, illumination systems, and support systems. J Microsurg 1980; 1:364–369.

10

Sensorineural Hearing Loss: Diagnostic Criteria

Robert T. Sataloff
Drexel University College of Medicine, American Institute for Voice and Ear Research, and Graduate Hospital, Philadelphia, Pennsylvania, USA

Joseph Sataloff
Thomas Jefferson University, Philadelphia, Pennsylvania, USA

Sensorineural hearing loss is a challenging problem confronting physicians. Millions of industrial workers and older citizens have this type of impairment. It generally is irreversible and often affects daily communication. The potential psychological implications place sensorineural hearing loss in the foreground of medical importance.

1. CLASSIFICATION

The damage to the auditory pathway may take place both in the inner ear (sensory loss) and in the auditory nerve (neural loss). We exphasize that the damage may be in both areas (as the name sensorineural indicates), but it is possible in many cases to pinpoint the diagnosis specifically as being either a sensory or a neural type. If it can be established that the major damage to the auditory pathway is in the inner ear, the case is classified as a sensory hearing loss. If the chief damage is in the fibers of the eighth nerve proper rather than in the inner ear, it is classified as a neural hearing loss.

2. LIMITED INFORMATION

The precise cause or the detailed pathology of some cases is not known. It is difficult for investigators to explore the very small and intricate cochlea, deeply embedded in the temporal bone. Temporal bones from patients whose hearing characteristics have been studied are available only infrequently—hence the problem of correlating clinical findings with pathological observations. Animal experiments, although they are helpful, provide limited information and must be interpreted with caution.

3. COROLLARIES TO CLASSIFICATION

Although it is not possible to specify the causes of all clinical cases of sensorineural hearing loss, merely classifying them as sensorineural provides important information. In the first place, such classification is a recognition that the site of the damage is not in the middle ear. This contraindicates surgical therapy in the middle ear in most cases. There is no logical ground for eustachian tube inflation, stapes surgery, or removal of adenoids and tonsils when the diagnosis is sensorineural hearing loss. Second, the prognosis is not nearly as optimistic as it is when there is a conductive type of loss. Many times the hearing deficit is irreversible, and it behooves the physician to tell the patient in a forthright manner that recovery of hearing is unlikely. The physician also is in a position to tell the patient whether the hearing loss is likely to be progressive or nonprogressive and what therapy is worthwhile. For example, the patient can be assured that the deafness will not progress if it can be established that it is of some congenital type or possibly a result of intense noise exposure. On the other hand, it is likely to grow progressively worse if it is due to presbycusis.

4. RELATION OF PATHOLOGY TO NEURAL AND ANATOMIC MECHANISMS

Prominent investigators, such as Guild, Rasmussen, Fernandez, Lawrence, and Schuknecht, have broadened our understanding of the mechanisms underlying such phenomena of abnormal hearing as poor discrimination, recruitment, and pitch distortion (diplacusis). Because these symptoms are prominent in sensorineural hearing loss, it seems fitting to crystallize the available information for the purpose of visualizing the pathological conditions likely to be present when these findings are encountered in practice.

For many years, the endolymphatic sac has been considered important in maintaining various aspects of cochlear fluid; and it has been postulated as the site of pathology in patients with endolymphatic hydrops (Meniere's syndrome). Recently, the endolymphatic sac has been found to function as an endocrine gland. Its chief cells produce Sacin, an inhibitor of sodium resorption in the kidneys. This discovery has led to a new theory of endolymphatic hydrops. This theory postulates that the endolymphatic duct becomes obstructed by debris. An increased volume of endolymph within the cochlea results from production of Sacin. This, combined with the secretion of hydrophillic proteins within the sac, is believed to clear the obstruction, suddenly restoring longitudinal flow and precipitating an attack of vertigo. This interesting speculation would explain clinical observations; but further research is required to determine whether it is accurate.

The organ of Corti in the inner ear contains ~25,000 hair cells resting on a basilar membrane. These hair cells are arranged in long rows conforming to the spiral shape of the organ of Corti. There are ~3500 to 5000 inner hair cells arranged in a single row, and >20,000 outer hair cells which run in three to five parallel rows. There is a tunnel between the inner and outer hair cells [Fig. 10.1(A) and (B)] (1). There are also various types of supporting cells in the inner ear that relate to the nerve fibers as well as to the inner ear. About 95% of the auditory nerve fibers terminate on the inner hair cells, and only 5% go to the many outer hair cells (2). As long as there are adequate supporting cells in the inner ear, the nerve fibers do not seem to show much degeneration. However, if the hair cells and the supporting cells are damaged, the nerve fibers supplying them degenerate, so that many cases of sensory hearing loss progress to the sensorineural type.

The hair cells change mechanical vibrations into electrochemical impulses that can be interpreted by the nervous system. As the hairs or cilia covering the tops of the inner and outer hair cells are deflected, an electrical current flows across the top of the cell, leading to a nerve impulse. Outer hair cells are in contact with the tectorial membrane, a gel-like structure that appears to assist in deflection and restoration of position. Contact between the cilia of the inner hair cells and the tectorial membrane is slight, or possibly nonexistent (3). The hair cells are contained in the organ or Corti which rests on the basilar membrane in the cochlear partition. The partition itself also moves because of complex fluid phenomena. Acoustic vibrations result in a wave that travels through the cochlea (4). Motion of the cochlear partition is important to hearing, and impairment of motion may be responsible for some kinds of hearing loss, as discussed later in the section on presbycusis. At present, it is believed that the frequency selective properties of the auditory system are due to mechanical processes within the cochlea, rather than to complex peripheral neural interactions as previously believed. However, many of the mysteries of auditory physiology remain unsolved. For example, there is a suggestion that an active, energy-producing system exists in the cochlear partition and is responsible for the sharp mechanical tuning abilities of the ear (5). However, this structure has not really been identified or explained. In fact, even the roles of the inner and outer hair cells are not well understood. Several histological pecularities of the outer hair cells remain unexplained. For example, there are many subsurface layers in the outer hair cells that appear to be specialized for calcium storage; and the cylindrical external surface is surrounded by the fluid of the organ of Corti (unlike inner hair cells which are closely juxtaposed to supporting cells). These outer hair cell findings are not commonly associated with sensory receptor cells, and they suggest that the outer hair cells may be serving functions as yet undiscovered. Interestingly, the outer hair cells appear to contain the ingredients associated with active nonmuscle contraction. The structure of the lateral wall of the outer hair cells is unusual. It is trilaminate, including a plasma membrane, cortical lattice, and subsurface cisternae. Outer hair cells can alter

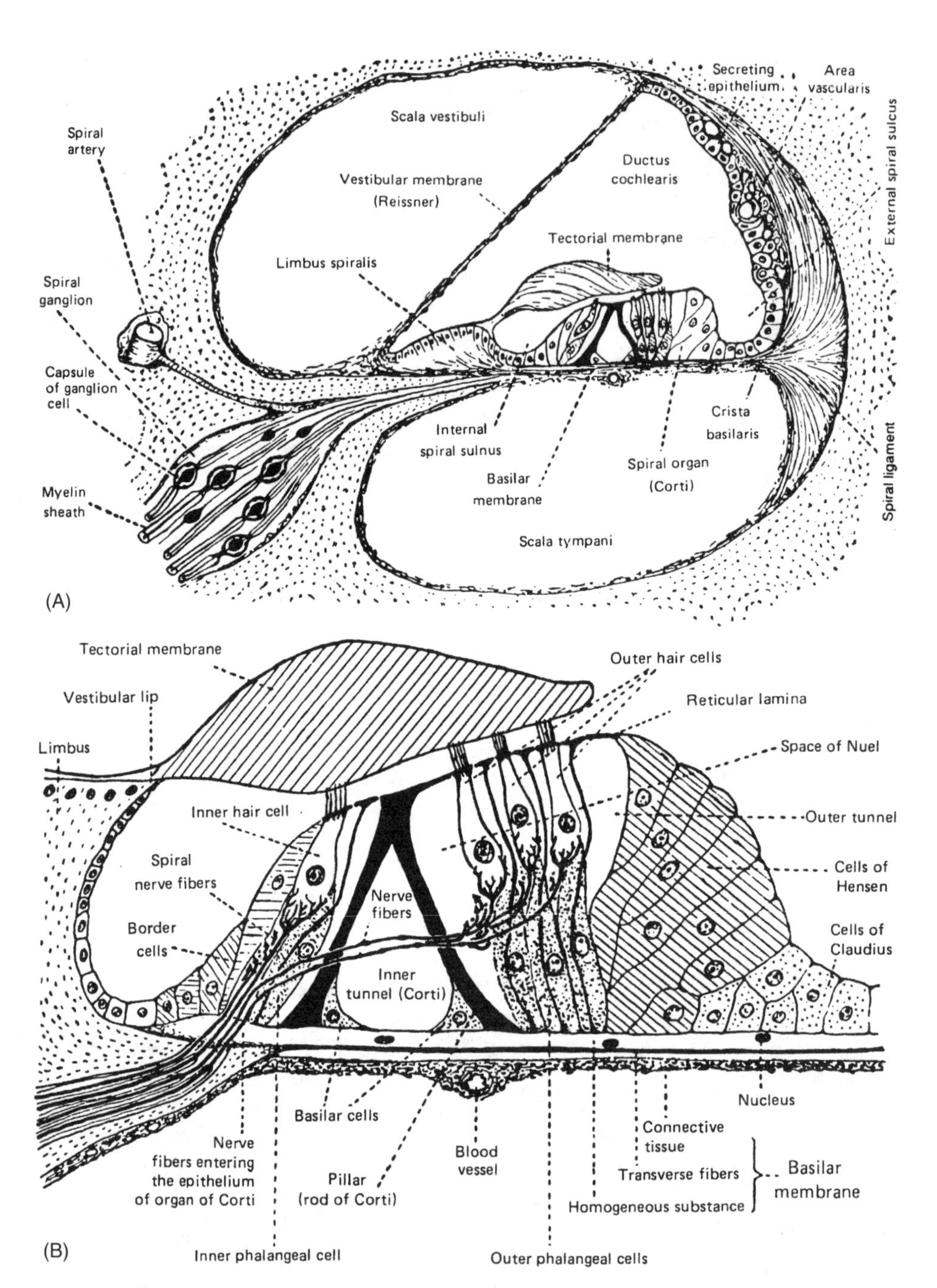

Figure 10.1 A cross-section of the organ of Corti. (A) Low magnification. (B) Higher magnification. [After Rasmussen (1).]

their length in response to voltage changes across the cell membrane. This phenomenon is called an electromotile response. It is believed to be due to conformational changes of Prestin, a membrane-bound protein molecule. When cell membrane potential is changed, axial stiffness and cell motility result, and intracellular anions act as extrinsic voltage sensors. Somatic motility involves both cochlear and vestibular hair cells. There is speculation that somatic motility plays a role in mechanical sound amplification. However, this remains to be proven or disproven. Nevertheless, it is clear that when outer hair cells are damaged, hearing loss occurs, and the frequency of the hearing loss is directly related to the area of outer hair cell damage. Outer hair cell loss also damages or eliminates fine tuning capabilities. In addition, absence of the outer hair cells changes markedly the neural output from the cochlea, even though this outflow originates primarily from the inner hair cells.

Despite the many unanswered questions of cochlear physiology, it is clear that, despite the small neural distributions of the outer hair cells, they are extremely important to hearing. They are also very fragile. Damage to the auditory system usually is first seen as outer hair cell injury, with noise and direct head trauma typically destroying the outer row of outer hair cells first. Interestingly, many ototoxic drugs tend to injure the inner row of outer hair cells. The maximum hearing loss (threshold shift) that occurs when the outer hair cells are lost is ~50 dB. The hearing loss exceeds 60 dB only when the inner hair cells also become damaged. When all the hair cells are lost, no stimuli are available to excite the nerve endings, and consequently there is no sensation of hearing, although the nerve itself may be entirely intact. Considerably more research is necessary to clarify the functions of even the hair cells, let alone the entire auditory system. The reader is encouraged to consult an excellent summary by Dallos for a review of other concepts in cochlear physiology, as well as other sources (6).

In addition to the afferent bundles from the ear to the brain, there is also an efferent bundle of Rasmussen that carries impulses to the ear from the brain. This efferent tract appears to have a role in the inhibition of impulses, and its fibers appear to stimulate the outer hair cells causing a change in the mechanical properties of the organ of Corti and cochlear partition.

The relation of the hair cells to the nerve is the basis for explaining the phenomenon of recruitment which is most evident in two particular causes of hearing loss: noise-induced hearing loss and Meniere's syndrome. It is found also in patients whose hearing has been damaged by certain drugs. The precise explanation for recruitment of loudness in its varying forms still is not clear. The essential element for recruitment is damage to the hair cells that is disproportionately large compared with the nerve fiber supply. The mechanism seems to be that enough hair cells are damaged to reduce the threshold, but enough remain so that when the sound gets loud enough, a normal number of nerve fibers are excited as if all the hair cells were present. Although this explains most types of recruitment, it does not provide a satisfactory explanation for the phenomenon of hyperrecruitment, in which the sound in the damaged ear is not only as loud as that in the normal ear but is perceived by the patient as even louder. This suggests that the number of nerve impulses ascending the auditory nerve is even greater per unit of time than in the normal ear. Patients with hyperrecruitment are those who complain habitually that noises are very bothersome and exceptionally loud.

Some patients with damage to the inner ear and marked recruitment can detect very small changes in sound intensity, smaller than those which even the normal ear can detect. The ear seems to become ultrasensitive to loudness. This phenomenon technically is called reduced intensity difference-limen or, briefly, reduced difference-limen. Its occurrence has a logical explanation similar to that of recruitment of loudness.

Certain facts should be clarified about the ability of present-day hearing tests to detect damage to the sensorineural pathway. Although an audiogram may show a 0-dB hearing level, which is considered to be normal, it does not necessarily indicate that the sensorineural mechanism is undamaged. Crowe, Schuknecht, and others have shown that many nerve fibers can be destroyed without affecting threshold hearing for pure tones. As a matter of fact, as many as 75% of the auditory nerve fibers supplying a certain cochlear area can be sectioned without creating a substantial change in hearing threshold level. This may be hard to believe, but it must be considered in the interpretation of hearing tests and in visualizing auditory pathway damage. It also should be borne in mind that when octave bands are measured, acuity in the many frequencies between the octave points measured (especially in the large area between 4000 and 8000 Hz) is unknown.

5. CHARACTERISTIC FEATURES

5.1. Reduced Bone Conduction: Need for Complete Examination

Because the damage is sustained by the areas that analyze sound waves and transmit nerve impulses, certain features result that are characteristic of sensorineural hearing loss. Also characteristic is difficulty in hearing a vibrating tuning fork by bone conduction. It would seem reasonable to assume that this reduced bone conduction alone would sufficiently warrant classifying hearing loss as sensorineural, but in practice this assumption does not prove to be consistently reliable. There are equivocal cases of conductive hearing loss in which bone conduction is reduced; these make it essential to do a complete otologic and audiologic examination in all cases of hearing loss to be certain of their classification.

5.2. Impaired Discrimination

In comparing the histories of patients with conductive hearing losses and those with sensorineural hearing losses, one clinical difference immediately becomes evident. In addition to inadequate hearing ability for soft sounds, patients with sensorineural hearing loss can have the problem of impaired discrimination. Though they may hear people speak, they may be unable to distinguish words with similar vowel sounds but different consonants. This is predominant when the speech frequency range is involved, but it may be observed also when only the higher-frequencies are affected. In a conductive hearing loss the patient has no difficulty understanding someone if the speech is loud enough. By contrast, the patient with sensorineural hearing loss may mistake one word for another, though it is said in a loud voice. One of the most prominent complaints in these cases is the inability to understand speech, especially on the telephone: voices do not seem to be clear, and people sound as if they are always talking with loose dentures or cigarettes in their mouths. Foreign and unfamiliar accents are a special problem to these patients. Music and voices that compete with conversation make understanding especially difficult. The severity of hearing loss and degree of discrimination impairment are not always directly proportional. Figure 10.2 shows the audiogram of a patient with a severe unilateral hearing loss, but his ability to discriminate is well preserved so long as sounds are loud enough. Figure 10.58 shows the audiogram of a patient with much better hearing threshold but poor discrimination caused by an acoustic neuroma.

Many different etiologies produce sensorineural hearing loss, and the characteristics vary accordingly. The most common etiology is presbycusis, which produces the prominent features associated with sensorineural loss.

JOSEPH SATALOFF, M.D.
ROBERT THAYER SATALOFF, M.D.
1721 PINE STREET PHILADELPHIA, PA 19103

DATE	RIGHT EAR AIR CONDUCTION							LEFT EAR AIR CONDUCTION					
	250	500	1000	2000	4000	8000		250	500	1000	2000	4000	8000
3/91	5	10	10	10	15	20		90	90	85	85	100	100

RIGHT EAR BONE CONDUCTION							LEFT EAR BONE CONDUCTION					
							NR	NR	NR	NR	NR	NR

SPEECH RECEPTION: Right ______ Left ______ DISCRIMINATION: Right 100% Left 96%

Figure 10.2 This 39-year-old man was involved in a motor vehicle accident in which he suffered a head injury with unconsciousness for 1–2 h. He had a left oval window fistula which caused hearing loss and intractable dizziness. The dizziness improved following surgery, but the hearing loss persisted. Notice that despite severe left sensorineural hearing loss, discrimination was normal.

6. GENERAL CHARACTERISTICS

General characteristics of sensorineural hearing loss are the following:

1. If the hearing loss is marked, bilateral, and of long duration, and the patient's voice generally is louder and more strained than normal. This is particularly notable when the voice is compared with the soft voice in otosclerosis.
2. If tinnitus is present, it usually will be described as high-pitched hissing or ringing.
3. The air conduction threshold is reduced.
4. The bone conduction threshold is reduced to about the same level as the air conduction, so that there is no substantial air–bone gap. Often the tuning fork held to the skull is not heard at all, even when it is struck hard.
5. The discrimination score is reduced decidedly when the speech frequencies are involved and to a lesser degree when only the high frequencies are involved.
6. The patient finds it more difficult to understand speech in a noisy environment than in a quiet one.
7. When the discrimination score is reduced, the patient will discriminate as well or slightly better when the intensity of the speech is increased (in contrast to some cases of sensory hearing loss caused by Meniere's disease).
8. In most cases there is little, if any, evidence of abnormal tone decay (adaptation).
9. Recruitment generally is absent; if present, it is not marked and is found only in the region just above threshold (noncontinuous and incomplete).
10. The vibrating tuning fork lateralizes to the ear that has better hearing in patients who have a substantial difference in hearing level between the two ears.
11. Responses to audiometry are usually sharp and clear-cut.

12. Otologic findings are normal.
13. Little or no separation is found between the interrupted and the continuous Békésy audiograms.
14. If the hearing loss originally was sensory but has progressed to a sensorineural loss, some characteristics of sensory hearing loss may persist, such as mild recruitment and diplacusis.
15. The prognosis for recovery with treatment is poor, with rare exceptions.

7. ASPECTS OF CRITERIA

7.1. Loudness of Voice

Many of these criteria have interesting backgrounds. For instance, the loudness of voices depends on many factors such as personality, the distance from a listener, the eagerness to be heard, the amount of background noise, what the individual has been taught to regard as socially acceptable, and so on.

Individuals control the loudness of their voices mainly by hearing themselves speak. They do this by both air and bone conduction to varying degrees. It is a feedback system whereby the individual hears his/her own voice, and if it seems too loud, he/she lowers it, or if too soft, he/she raises it. In otosclerosis the patient hears his/her own speech chiefly by bone conduction because it is far better than the air conduction, and so he/she has a tendency to keep the voice soft; otherwise, the patient thinks he/she is shouting. Furthermore, in otosclerosis the patient does not have the urge to raise his/her voice above the background level because the background noise is hardly heard.

In marked sensorineural hearing loss of long duration involving both ears, the patient cannot hear his/her own voice by either air or bone conduction. When continued efforts fail to hear ones own voice and thus to obtain the feedback control to which one has become accustomed, the voice often becomes louder and more strained.

However, not all patients with severe, long-standing sensorineural hearing impairment develop loud and strained voices. We have seen patients who, despite becoming progressively deaf as adults until no residual hearing remains, have been able for many years to maintain almost normal speech levels. Speech and voice therapy, their own conscious efforts, and tactile feedback have enabled these patients to fix a certain loudness reference level for their own voices similar to what they used when they had normal hearing. However, it becomes strikingly evident that such patients do not lower or raise their voices in different noise environments as most people with normal hearing do. Some of them will inadvertently speak loudly in a quiet room and softly in a noisy one.

7.2. Speech Discrimination

Another intriguing aspect of sensorineural hearing loss concerns discrimination of speech. In otologic practice the physician frequently sees patients whose findings are like those described in Fig. 10.3. This important industralist's only complaint was that he could not discern what was said at meetings or in groups in which several people spoke at the same time. Many people who have losses chiefly >2000 Hz complain of "hearing trouble" when they actually mean discrimination trouble. In a routine audiologic evaluation, tests usually do not duplicate the circumstances under which the patient noted this handicap, such as in noisy areas and in multiple conversations or on the telephone. Some of these difficulties can be attributed to psychological causes as well as

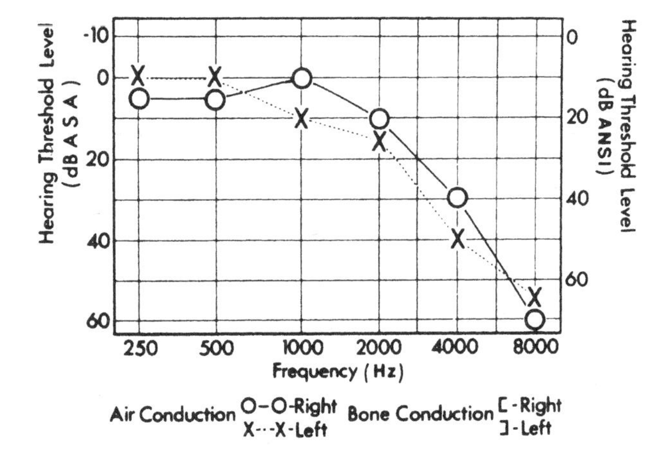

Figure 10.3 *History*: A 64-year-old industrialist complaining of difficulty in understanding people who speak, especially at important meetings. He found it necessary to resign as board chairman of a large corporation because of his hearing handicap and personal stress. *Otologic*: Normal. *Audiologic*: Patient has typical high-frequency hearing loss of presbycusis. Discrimination score is 86%, but he has more trouble in a noisy background. *Classification*: Sensorineural hearing loss. *Diagnosis*: Presbycusis. *Aid to diagnosis*: Two years later the pure-tone threshold dropped slightly, but the discrimination score fell to 65%, indicating more neural damage.

hearing losses. Actually, such cases are merely mild examples of the common type of sensorineural hearing loss involving a bilateral high-frequency loss and measurable reduction in discrimination. One reason for reduced discrimination is a patient's inability to distinguish certain consonants that fall within the frequency range of the hearing loss. As consonants help give meaning to words, and the patient has trouble distinguishing certain consonants, the patient misunderstands certain key words and thus at times misinterprets the speaker. If the speaker talks fast and fails to enunciate well, or if noise masks already weak consonants, the patient with high-frequency loss could have even greater difficulty understanding conversation. This type of discrimination difficulty is different from that encountered in the sensory type produced by Meniere's syndrome, in which there is marked distortion and fuzziness of speech sounds that worsens when the speech becomes louder.

7.3. Tests and Their Shortcomings

In sensorineural hearing loss, another shortcoming of hearing tests occurs in interpreting discrimination testing. In some cases of mild high-frequency loss, discrimination testing shows an excellent score of 90% despite the patient's complaint of difficulty understanding speech. It must not be concluded that the patient's complaint is always unwarranted. Rather, it should be realized that the discrimination test was done under very quiet conditions in a sound-treated room and with a tester who enunciated very clearly and spoke slowly through an expensive amplification system. Under such conditions the patient did well. However, in everyday experiences such favorable conditions are rare. Usually, there is some loud distracting noise to mask what the speaker says. Furthermore, few people speak slowly and enunciate clearly. If the speech is coming through an amplification system (e.g., at a train station), it often is distorted and even muffled. It is no wonder that the patient has complaints under such circumstances.

When a physician talks to a patient, it is common to underestimate the patient's discrimination difficulty because he/she communicates so well in an office setting. A tendency to underrate the patient's difficulty should be avoided, for most patients visit a physician only after such symptoms have persisted and become disturbing. To brush aside the patient's complaint on the strength of a brief face-to-face conversation is to show a lack of understanding of hearing and psychological problems that gives the patient ample motive for discouragement.

Testing bone conduction with a tuning fork in a case of sensorineural hearing loss sometimes may confuse the physician. Occasionally, the physician sees a patient with about a 70-dB hearing loss who absolutely denies hearing a tuning fork by bone conduction regardless of how hard it is struck. Does this mean the nerve is dead? It cannot be completely dead, or there would be no hearing by air conduction. It must be borne in mind that bone conduction is not always a valid test for sensorineural function, and also that the ear is ~60–70 dB more sensitive to air conduction than to bone conduction, so that the fork would have to produce a very loud tone to be heard by the deafened patient. Some forks are unable to reach this intensity even when they are struck very hard.

In patients with unilateral sensorineural hearing loss, the tuning fork placed on the skull or the teeth sounds louder in the good ear. This is a result of a reduced bone conduction in the ear with sensorineural hearing loss. The physician also will observe that he/she can hear the fork on his/her own skull (if the doctor's hearing is normal) much better than the patient with sensorineural hearing loss can hear it. When testing for this, it is necessary to mask the patient's good ear with noise so that the patient will not report erroneously that hearing was present in the bad ear when the vibrations were actually conducted through the skull to the good ear.

7.4. Recruitment and Perception of Intensity

Recruitment is the outstanding feature of sensory rather than neural hearing loss. However, there may be some degree of recruitment in neural hearing loss, particularly if the loss originally started as sensory and progressed to involve the nerve endings. In such cases the recruitment is mild, and the abnormal increase in loudness occurs near threshold. It rarely continues into the high intensities to equal the loudness in the ear that has normal hearing; when it does, this is referred to as "complete recruitment." When the loudness in the recruiting ear exceeds that of the normal ear for the same tone, the condition is called "hyperrecruitment."

It has been pointed out previously that during audiometry the patient with conductive hearing loss waivers in his/her responses as if he/she is uncertain whether or not sounds are heard near threshold, whereas in sensorineural cases the responses are more decisive and sharp. The reason is that the patient with sensorineural loss often has a keener ability to detect small differences in intensity than either a patient who has normal hearing or one with conductive hearing loss.

7.5. Hyperacusis

Hyperacusis may be related to recruitment, but the phenomenon is not well understood. The term describes uncomfortable or painful hypersensitivity to sounds. It may occur in patients with little measurable hearing loss, and no measurable recruitment. Patients with hyperacusis are excessively disturbed by common sounds of intensities routinely encountered in daily life. The term hyperacusis does not imply "super hearing," or the ability to detect sounds softer than normal thresholds, although it is occasionally defined this way in some sources. It is not clear whether hyperacusis is caused by peripheral, central, or psychological factors or a combination; and problems involving inhibatory pathways have been proposed as causal, as well.

7.6. Almost Symptomless Types

Some types of sensorineural hearing impairment are almost symptomless and are detected only by audiometric examinations. The early noise-induced dips and the early high-frequency hearing losses are typical.

8. MEDICAL TREATMENT

An often-repeated maxim is that sensorineural hearing loss is irreversible, and damage to the auditory nerve fibers cannot be cured. Although the generalization may be true in many cases, the catch lies in the term "damage." In numerous cases of sensorineural hearing impairment the patient gets better, mostly spontaneously, but medication may be of some help. What the maxim really means is that there is no cure for permanent damage to the auditory nerve. It is not true that all sensorineural damage is permanent. The simplest example of the reversibility of sensorineural damage is auditory fatigue following exposure to gunfire. There can be a marked sensorineural hearing loss which gradually reverts to normal. There are many other cases of sensorineural hearing loss that have all the characteristics of being permanent, and yet they improve dramatically, often with medication or even without. Sensorineural hearing loss can be very severe and yet return to an excellent level, as demonstrated in Fig. 10.4. The chief lesson to be learned from such examples is that it is possible for some cases of sensorineural hearing loss to improve, but at present there is not sufficient knowledge to forecast consistently when this will happen, nor is there any specific therapy known to promote the return of hearing consistently in many cases. One generalization need to be considered of value. The cases more likely to be reversible are those of sudden onset rather than those which develop slowly over a period of months. As we have gained experience with conditions such as luetic labyrinthitis and autoimmune sensorineural hearing loss, we have realized that this generalization is not a good one. The onset of hearing loss of the sensorineural type is often helpful in establishing its possible cause. Included in this chapter is a chart showing the known and likely causes of sensorineural hearing loss of sudden and insidious onset. Examples of each cause in the list are found in the following chapters.

The classic example of sensorineural hearing loss is presbycusis, because it invariably occurs bilaterally and almost symmetrically. Occupational hearing loss is another common example. The unilateral losses are always interesting and challenging, and it is occasionally difficult to be certain of their cause.

Sensorineural hearing impairment frequently is attributed to factors such as viruses, vascular spasm, vascular embolus, thrombosis, or even severe emotional trauma, although the validity of these attributions is questionable.

A sensorineural hearing loss may occur unilaterally or more commonly bilaterally, either simulataneously or at different rates in each ear. A sudden onset of hearing loss almost always motivates the patient to seek a physician's advice as quickly as possible. The patient with hearing loss of gradual onset infrequently seeks help until he/she has experienced many difficulties in communication or has developed disturbing tinnitus or vertigo.

JOSEPH SATALOFF, M.D.
ROBERT THAYER SATALOFF, M.D.
1721 PINE STREET PHILADELPHIA, PA 19103

HEARING RECORD

NAME AGE

AIR CONDUCTION

			RIGHT							LEFT						
DATE	Exam	LEFT MASK	250	500	1000	2000	4000	8000	RIGHT MASK	250	500	1000	2000	4000	8000	AUD
7th day post-op.			60	55	45	50	50	70		60	55	40	50	45	NR	
9th day post-op.			15	10	20	25	20	30		25	15	15	25	10	20	
16th day post-op.			10	5	5	0	5	25		5	0	5	5	5	10	

BONE CONDUCTION

			RIGHT							LEFT						
DATE	Exam	LEFT MASK	250	500	1000	2000	4000		RIGHT MASK	250	500	1000	2000	4000		AUD
7th day post-op.			35	45	40	40	50			35	40	35	45	55		
9th day post-op.			10	10	15	15	20			10	15	15	10	15		
14th day post-op.			0	0	0	0	5			5	0	5	0	0		

SPEECH RECEPTION

DATE	RIGHT	LEFT MASK	LEFT	RIGHT MASK	FREE FIELD	MIC.

DISCRIMINATION

		RIGHT				LEFT				
DATE	% SCORE	TEST LEVEL	LIST	LEFT MASK	% SCORE	TEST LEVEL	LIST	RIGHT MASK	EXAM.

HIGH FREQUENCY THRESHOLDS

	RIGHT						LEFT					
DATE	4000	8000	10000	12000	14000	LEFT MASK	RIGHT MASK	4000	8000	10000	12000	14000

RIGHT		WEBER	LEFT	
RINNE	SCHWABACH		RINNE	SCHWABACH

HEARING AID

DATE	MAKE	MODEL
RECEIVER	GAIN	EXAM
EAR	DISCRIM.	COUNC

REMARKS

Figure 10.4 A 28-year-old woman had undergone a tonsillectomy. During the first 6 days following surgery she took 200 tablets of aspirin for what she called "terrific pains." On the seventh postoperative day she complained of hearing loss, recurrent tinnitus, and a recurrent sensation of unsteadiness. Examination showed normal eardrums. Audiometric studies indicated bilateral, symmetrical, severe hearing loss by both air conduction and bone conduction and evidence of recruitment. The aspirin was stopped, and the hearing showed considerable improvement 2 days later. The patient was aware of the improved hearing, and the tinnitus ceased altogether. Vestibular tests performed the same day showed no evidence of vestibular disturbance. Audiometric studies revealed normal hearing 9 days after the first hearing test. [From Walters (7).]

9. CHARACTERISTIC AUDIOMETRIC PATTERNS

The audiogram considered to be the classic example of sensorineural hearing loss is the high-frequency loss shown in Fig. 10.3. An audiogram of this shape has come to be called a "nerve-type audiogram," because it is found so often in presbycusis.

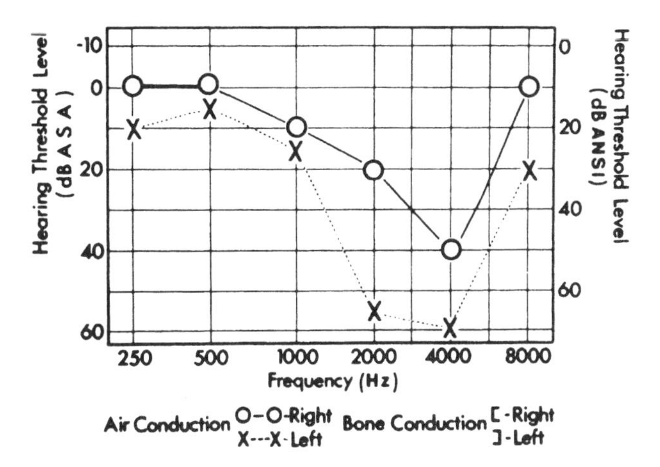

Figure 10.5 Exposure to small-arms gunfire. Complete recruitment present at 2000 Hz.

R	L
60	70
80	80

Classification: Sensorineural. *Diagnosis*: Repeated acoustic trauma.

Another audiometric pattern characteristic of sensorineural hearing impairment is the high-frequency dip shown in Fig. 10.5. This is caused frequently by exposure to intense noise, particularly rifle fire. Initially, this type of loss is sensory rather than sensorineural. Then, as the damage grows more extensive, and the supporting cells and nerve fibers become affected by various factors, the classification becomes sensorineural, but the classic subjective symptoms of sensorineural hearing loss are not so well defined. The same explanation applies to still other audiometric patterns found in sensorineural loss. For instance, Fig. 10.6 shows an audiogram of sensorineural hearing loss associated with otosclerosis.

Although these patterns, especially the first, are highly suggestive of damage in the sensorineural mechanism, it must be remembered that other possibilities exist. For instance, an audiometric pattern similar to that shown in Fig. 10.3 can be obtained after merely dropping oil into an ear canal, and this scarcely affects the sensorineural mechanism. The ascending curve and the flat audiogram certainly are more common in conductive hearing loss. So it becomes even more evident that a diagnosis cannot be made reliably solely on the basis of the air conduction audiogram.

The paramount audiometric characteristic of sensorineutal hearing loss is that the bone conduction curve is not normal and usually assumes the same shape and level as the air conduction curve. The Békésy audiograms show no singular features. Continuous and interrupted Békésy tracings usually reveal little or no separation. Neither marked recruitment nor abnormal tone decay is present.

10. DIFFERENTIATING SENSORY HEARING LOSS FROM NEURAL HEARING LOSS

Until recently, when bone conduction was found to be reduced, a case could be classified only as sensorineural or, as it was more commonly called, perceptive or nerve deafness.

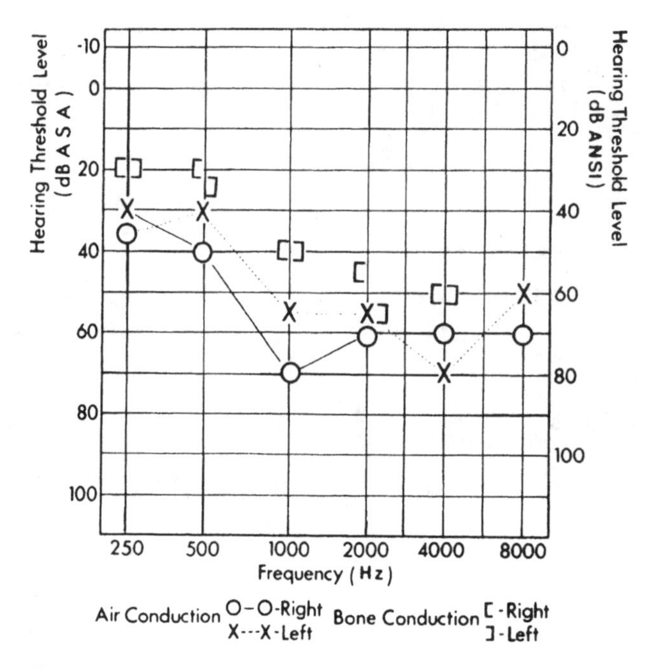

Figure 10.6 *History*: A 37-year-old man with insidious hearing loss for 10 years. No tinnitus or vertigo. A brother and a sister had successful stapes surgery. Patient uses hearing aid successfully. *Otologic*: Normal. *Audiologic*: Bilateral mixed hearing loss. Hears tuning fork only by teeth. *Classification*: Sensorineural hearing loss. *Diagnosis*: Otosclerosis with sensorineural involvement.

With the development of improved tests based on a clearer understanding of auditory pathology, it is now possible in some cases to determine whether the damage is primarily in the sensory or in the neural mechanism. The designations "sensory" and "neural" have become more meaningful as knowledge of ear pathology has improved. The distinction is not possible in every case, because many patients actually have damage both in the inner ear and in the nerve and thus have a sensorineural loss. In some patients the hearing loss has its origin in the inner ear and satisfies all the criteria for sensory hearing loss, but later the damage extends to the nerve, and the classification then becomes sensorineural. Certain criteria, when available, are helpful in differentiating between sensory and neural types. These include recruitment, abnormal tone decay and discrimination. If a patient demonstrates marked recruitment and diplacusis, the site of his auditory damage is most likely in the inner ear, and the hearing loss is classified as sensory. Patients with nerve deafness rarely have this marked degree of recruitment. If a patient demonstrates clear-cut evidence of abnormal tone decay, the damage is in the fibers of the auditory nerve, and the classification then is neural hearing loss. Cases of sensory hearing loss usually do not show abnormal tone decay. This does not necessarily mean that every case of sensory hearing loss must show marked recruitment, but when these phenomena are present, they are highly suggestive. Except for Meniere's syndrome, labyrinthitis, and a few other conditions, low discrimination scores are more indicative of damage to the nerve rather than the inner ear, especially the outer hair cells.

The availability of these criteria is limited. Recruitment tends to disappear when neural factors complicate sensory hearing loss. Abnormal tone decay can be elicited only when damage to the nerve fibers is only partial, whereas it cannot be elicited in those cases in which nerve fibers have presumably lost all function, as in congenital nerve deafness and most cases of advanced presbycusis.

The tests used to establish the presence of recruitment and abnormal tone decay are described in Chapter 7. One testing procedure of increasing importance is continuous- and fixed-frequency audiometry, which was done traditionally with a Békésy-type audiometer. This instrument is rather expensive and at present is used very little even in large hearing centers. However, other instruments are available now to provide similar information. For this reason, the emphasis in this book is on the results of the tests and their interpretation rather than the details of technique.

11. SENSORY HEARING LOSS

The classification of sensory hearing loss includes all cases in which, according to the best information available, the damage to the auditory pathway is in the inner ear. The damage usually is in the hair cells and sometimes in labyrinthine fluid central systems.

11.1. Characteristic Features of Sensory Hearing Loss

1. Some patients give a history of recurrent attacks of labyrinthine vertigo associated with fullness in the ear, ocean-roaring tinnitus, and intermittent hearing loss. This history is highly suggestive of the syndrome variously called Meniere's syndrome or disease, cochlear hypertension, labyrinthosis, or labyrinthine hydrops.
2. In Meniere's syndrome the hearing loss is more likely to be unilateral.
3. In Meniere's syndrome, if tinnitus is present, usually it is described as an ocean roar; it may be likened to the sound of a seashell held against the ear. In cases of hearing loss caused by exposure to loud noises, the tinnitus is said to sound like a high-pitched ring. Occasionally, the history will reveal exposure to intense noise accompanied by ringing tinnitus.
4. In Meniere's syndrome there may be a reduction in the threshold of hearing during an attack and a return of hearing when the attack has subsided. The hearing loss eventually becomes permanent.
5. The patient's voice usually is of normal loudness.
6. The otologic examination is normal.
7. There is a hearing loss by air conduction.
8. There is a hearing loss by bone conduction.
9. There is no air–bone gap.
10. If the hearing loss is moderate or marked in the speech frequencies, speech discrimination is reduced.
11. Frequently, discrimination gets worse when the patient is addressed in a loud voice.
12. Marked recruitment is present; it may be continuous and complete, and occasionally there even may be hyperrecruitment.
13. Diplacusis is present.
14. The threshold of discomfort for loud sounds is lowered.
15. There is no abnormal tone decay or stapedius reflex decay.
16. With some exceptions, the tuning fork lateralizes to the ear that has better hearing when struck softly. When struck forcefully, the tuning fork may lateralize to the bad ear because of recruitment.
17. There is a type II Békésy tracing.

The reasons for some of these distinguishing features will be described briefly because almost all are discussed in the section on hearing testing (see Chapter 7).

The history is helpful because it is distinctive in two notable causes of sensory hearing loss: noise-induced hearing loss and Meniere's disease. The patient with noise-induced hearing loss generally will volunteer the information that the hearing loss and ringing tinnitus started after exposure to gunfire, exploding firecrackers, or industrial noise. In Meniere's disease, the hearing impairment usually will be unilateral and accompanied by ocean-roaring, or seashell, tinnitus and a feeling of fullness in the ear. Vertigo may or may not be a feature, and this as well as the hearing loss will be intermittent at first, though later it may become persistent. Many patients with Meniere's disease, even when it is unilateral, will not say that they do not hear, but they will complain that they are unable to distinguish the exact words they hear. *Bath* sounds like *path, bomb* like *palm.* This indicates a reduction in discrimination. Along with this, they will say that speech and sounds are distorted and irritating, especially if loud. A baby's cry may sound unbearably loud. (This indicates a lowered threshold of discomfort due to recruitment, although hyperacusis may produce a similar complaint.)

According to most authorities on Meniere's disease, this disorder is characterized by a pathological change in the endolymph, which then affects the hair cells. This causes damage which at first is temporary and reversible but later results in permanent damage inside the cochlea, and, ultimately, degeneration of the auditory nerve fibers takes place, making the final result a sensorineural hearing loss.

If both ears have been involved markedly for many years, the patient's voice then becomes loud because bone conduction is affected. However, it usually is not as loud as in some cases of neural deafness of long standing, because many sensory hearing losses are unilateral.

There is no air–bone gap, because the bone conduction hearing level usually is about the same as the air conduction hearing level. In Meniere's disease the patient's ability to distinguish or to discriminate between words that sound somewhat alike often gets worse instead of better when the voice is raised. Much of this is caused by recruitment and increased distorition with higher intensities. The otologic findings are normal because the damage is restricted to the inner ear.

The comment regarding lateralization with a tuning fork is of special interest. If the tuning fork is struck gently and held to the forehead or the teeth in a case of unilateral sensory hearing loss, it will sound louder in the normal ear. However, if marked recruitment is present in the bad ear, and the tuning fork is struck hard, there is a good chance the tone will sound even louder in the bad ear than in the normal ear. For this reason, testing for lateralization in cases of unilateral hearing loss should be performed with deliberate control of the sound intensity of the tuning fork while caution is observed in technic and interpretation of results.

Distortion is an outstanding feature of sensory hearing loss. Pure tones, such as those produced by a tuning fork or a piano, sound raspy and of different pitch in the bad ear than in the unaffected ear—hence the phenomenon of diplacusis. This sensory distortion affects not only pure sounds but also voices and noises generally, and it represents a source of much irritation, frustration, and emotional strain to the patient, a problem insufficiently understood by many physicians.

Patients with bilateral sensory hearing loss are less able to use hearing aids in a satisfactory manner. The amplification provided by the aid many add to speech distortion and thus makes voices even less intelligible, especially in noisy areas.

The prognosis for sensory hearing impairment is better than it is for the neural type but not nearly as favorable as for conductive hearing loss.

In the early stages, many cases of sensory hearing loss seem to be reversible. For example, there are recurrent intermittent attacks of hearing loss with Meniere's disease alternating with periods of better hearing such optimism appears warranted in most cases of neural deafness. There is, however, great need for more facts and information concerning this point.

11.2. Audiometric Patterns in Sensory Hearing Loss

As in conductive hearing loss, the shape of the air conduction audiogram in sensory impairment depends to a great extent on the cause and the pathology. However, two audiometric patterns are typically associated with the two common causes of sensory hearing loss.

Figure 10.7 shows the audiometric pattern found in a classic early case of Meniere's disease. This is described as an ascending air conduction hearing level with a reduced

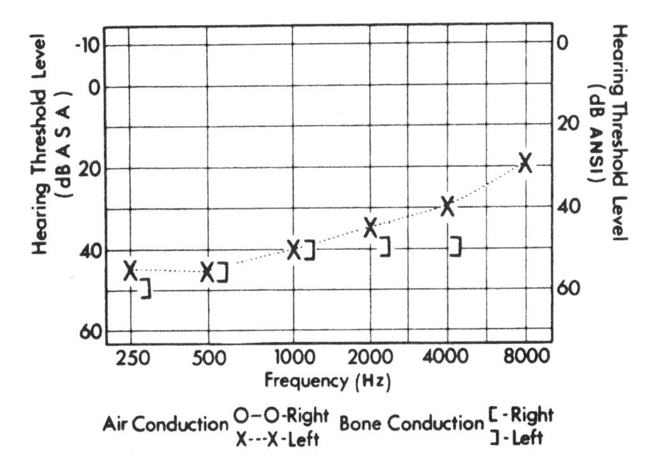

Figure 10.7 *History*: A 37-year-old man with recurrent sudden attacks of rotary vertigo accompanied by nausea, vomiting, and an ocean-roar tinnitus in the left ear. Between attacks the patient reports fullness and hearing loss, aggravated during attacks. The patient is annoyed by loud noise in the left ear, and voices sound fuzzy and unclear. *Otologic*: Normal. *Audiologic*: Moderate hearing loss with no air–bone gap. *Left ear discrimination*: 60%. Binaural loudness balance studies show complete recruitment in the left ear:

1000 Hz

R	L
0	40
10	45
20	50
30	35
40	55
50	60
60	65
70	70

During the loudness balance test the patient reported that the tone in the left ear was not as clear as the tone in the good ear. This is diplacusis and is an important symptom in inner-ear pathology. The patient also has a lowered threshold of discomfort. There was no abnormal tone decay. *Impedance Audiometry*: Normal tympanogram. Stapedius reflex would show Metz recruitment. *Diagnosis*: Meniere's disease.

bone conduction curve following almost exactly the same pattern. This patient presented almost all the symptoms classically associated with sensory hearing loss secondary to Meniere's disease.

Another characteristic type of audiogram found in sensory hearing loss is shown in Fig. 10.8. This is sometimes described as a C-5 dip, and in this instance the cause was exposure to intense noise. It is the serial audiogram of a man who was exposed to extensive small-arms gunfire, with the result that he suffered a permanent high-frequency loss. It is appropriate to point out that the term "C-5 dip" is not very satisfactory. It is commonly designated this way because of the manner in which audiograms are done. If the tests were to be made at several hundred Hertz on either side of 4000 Hz, the so-called C-5, the chances are that the dip would be found there as well. When testing is done with continuous-frequency audiometry, the dips can occur at 3000 Hz, at 5000 or 6000 Hz, or anywhere in between, without involving 4000 Hz. Consequently, the term "C-5 dip" should be replaced by the term "high-frequency dip," which really means that the hearing is relatively normal on either side of a sharp depression in the hearing level.

Although the range of frequencies involved in this high-frequency dip is comparatively small, recruitment studies often show complete and continuous recruitment. However, no significant reduction in speech discrimination is detectable with presently available tests. This is so because speech frequencies have not become involved. A high-frequency dip is not the only type of threshold shift that can be produced in the early stages of noise exposure, nor is noise the only etiology of dips in the hearing level, as discussed in Chapter 12.

Obviously, in sensory as in conductive hearing loss and, as will be noted, in neural hearing loss as well, certain characteristic patterns apparently are associated with each; this does not mean that other patterns do not occur, or that one may not find a so-called characteristic pattern associated with nonclassic types of hearing loss.

11.2.1. The Basic Audiologic Features

In making a classification of sensory hearing loss it is necessary to look for: (a) reduced threshold of hearing by air conduction and bone conduction; (b) the absence of any air–bone gap; (c) recruitment; (d) reduced discrimination if the speech frequencies are

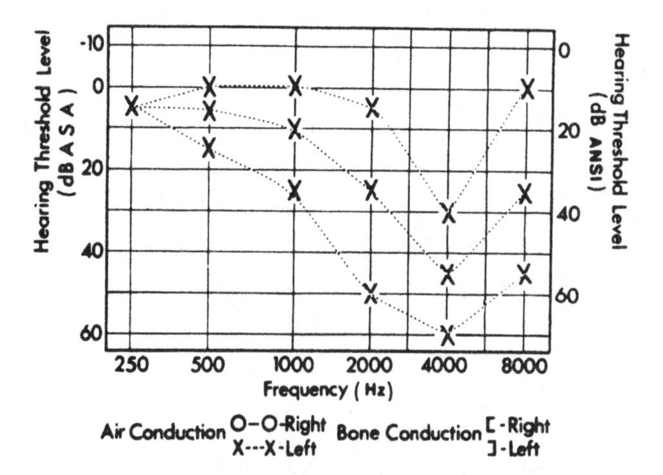

Figure 10.8 Classic audiometric curves showing progressive hearing loss caused by extensive repeated exposure to gunfire. The hearing loss can continue beyond that shown in the upper threshold curve if the exposure continues and if the noise is very intense, as indicated in the two lower threshold curves.

involved and still further reduction as the intensity of speech is increased; (e) lateralization of the sound of the tuning fork to the ear that hears better, with certain exceptions; and (f) an absence of pathological adaptation. When these features are present along with a corroborating history and otologic examination, a case can be classified accurately as sensory hearing loss.

12. NEURAL HEARING LOSS

When hearing impairment results from damage to the fibers of the auditory nerve *per se*, the hearing loss is classified as a nerve or neural type of hearing loss.

12.1. Characteristic Features of Neural Hearing Loss

Certain features are characteristic of neural hearing loss.

1. The history is variable. The deafness may be sudden in onset and practically complete in one ear, as may occur in fracture of the skull involving the internal auditory meatus; or it may be gradual and bilateral, as in progressive hereditary nerve deafness. It may follow many other patterns, as well. A history of familial deafness often is helpful, but it should be borne in mind that a family history of hearing loss also is to be expected in otosclerosis. The patient's age is of little help, because nerve deafness may occur in any age group. Vertigo is an important symptom. If it is present, especially in the presence of a unilateral sensory hearing loss, its cause must be established: at least the presence of an acoustic nerve tumor (acoustic neuroma) must be ruled out. Tinnitus does not aid the differential diagnosis of nerve deafness *per se*; but if present, it is likely to be high-pitched.
2. Air conduction and bone conduction are both reduced.
3. There is no air–bone gap.
4. If the hearing loss is unilateral or more severe in one ear than in the other, the vibrating tuning fork is lateralized to the better-hearing ear.
5. Recruitment usually is absent, but if it is present, it is minimal and not complete.
6. There generally is a striking disparity between the hearing threshold level and the patient's ability to discriminate speech.
7. Abnormal tone decay is present except in cases of congenital nerve deafness and presbycusis. If this feature can be elicited, it localizes the damage to the auditory nerve fibers, and this is of great diagnostic value.
8. Stapedius reflex may be absent, or decay may be present.
9. Békésy audiometry in the presence of abnormal tone decay generally shows a separation between the continuous-tone and interrupted-tone tracings, and the continuous-tone tracings are of small amplitude (type III or IV).

The basic causes underlying some of these criteria have been traced by a study of proven cases of auditory nerve tumors. Of paramount importance is the finding of abnormal tone decay. This phenomenon strongly suggests acoustic neuroma. Any injury that causes what one might call "partial damage" to the auditory nerve is likely to produce abnormal tone decay. A classic example is shown in Fig. 10.9. Here, after the nerve was damaged by direct injection, all features of nerve damage were evident. It is notable also that temporary damage to the auditory nerve can occur; recovery from nerve damage is, indeed, not an uncommon clinical experience. Another example of the reversibility of nerve damage is auditory fatigue or temporary threshold shift (TTS).

HEARING RECORD

NAME AGE

AIR CONDUCTION

DATE	Exam.	LEFT MASK	RIGHT 250	500	1000	2000	4000	8000	RIGHT MASK	LEFT 250	500	1000	2000	4000	8000	AUD
1st test			15	10	0	0	20	15								
17 days		85	90	95	NR	NR	NR	NR	After injury to 8th nerve							
1 yr.		95	65	50	40	30	30	NR								

BONE CONDUCTION

DATE	Exam.	LEFT MASK	RIGHT 250	500	1000	2000	4000		RIGHT MASK	LEFT 250	500	1000	2000	4000		AUD
1st test			20	15	0	15	15									
17 days		85	NR	NR	NR	NR	NR		After injury to 8th nerve							

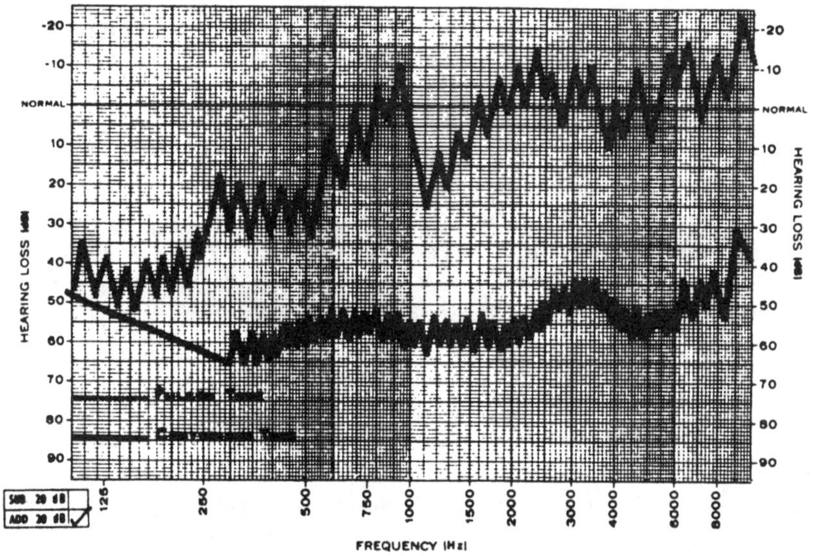

Figure 10.9 *History*: A 43-year-old woman with recurrent attacks of severe pain in the right side of the face for years. No other complaints. Hearing was normal. *Otologic*: Normal. *Audiologic*: Normal hearing. After accidental injection of the eighth nerve with boiling water for relief of the facial pain, there was total deafness, facial palsy, and absence of caloric response due to direct injury to the eighth nerve. *Classification*: Nerve deafness. *Diagnosis*: Injury to auditory nerve. Severe deafness persisted for several months, and then hearing gradually returned and showed no recruitment but marked tone decay in all frequencies. Return of hearing after such a severe injury and profound loss is remarkable. The marked tone decay is evident in the large separation between the pulsed (bottom, upper curve) and continuous (bottom, lower curve) Békésy tracings. [From Harbert and Young (8).]

It is generally assumed that TTS, like permanent threshold shift (PTS), results from damage to the hair cells of the inner ear. Although there is extensive proof that this is true for PTS, there is no conclusive evidence that it is true for TTS. Animal experiments have demonstrated that hair cells do not fatigue readily, and experiments with humans seem to indicate that TTS may be a reversible neural type of hearing loss.

12.1.1. Comment on Criteria

The onset of neural hearing loss is very variable. It may occur slowly, as in a typical case of acoustic neuroma, or suddenly, as a result of herpes virus.

From what little is known about tinnitus, one would expect it to be an important feature accompanying neural hearing loss, but this is not borne out in practice. Many patients with acoustic neuromas complain little, or not at all, of noise in their ears. Even elderly patients who give evidence of a marked neural type of hearing loss do not complain of tinnitus very often. This is in contract to patients having sensorineural hearing loss, who sometimes complain bitterly of hissing or ringing tinnitus.

The reduced bone conduction in nerve deafness is particularly marked. The vibrating tuning fork might not be heard even when struck very hard. This is especially true in older patients. The fork would be expected to lateralize to the ear that has better hearing.

One of the earliest findings in auditory nerve damage caused by pressure from a neoplasm is disturbance of the vestibular pathways. Although there may be no subjective vertigo, vestibular studies often show abnormal findings; such studies should be considered routinely in all cases of unilateral sensorineural hearing loss, especially when marked recruitment is absent.

Another singular feature in nerve deafness is the great discrepancy often found between a hearing threshold level and the patient's ability to discriminate. Figure 10.10 illustrates an excellent example of a patient with a comparatively good hearing level

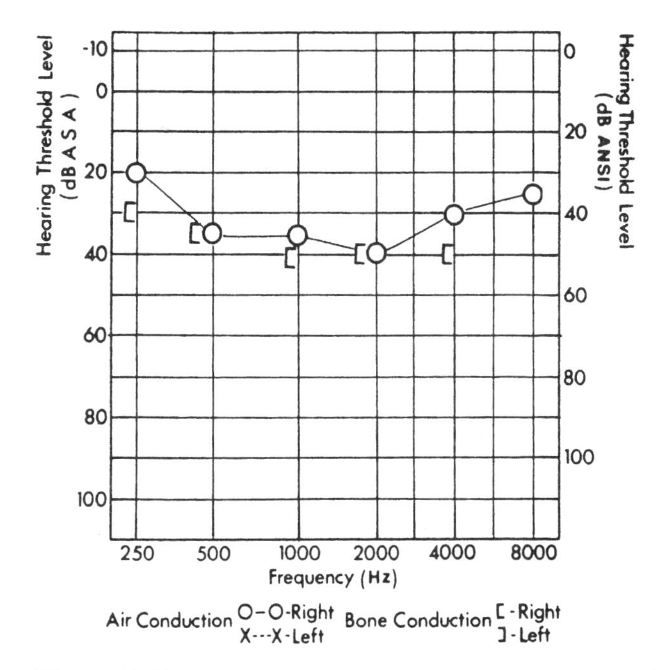

Figure 10.10 *History*: A 40-year-old woman with occasional imbalance and feeling of fatigue. No tinnitus and no rotary vertigo. Noted progressive hearing loss in right ear for 2 years. *Otologic*: Normal. Vestibular studies showed no response in the right ear and perverted response of the vertical canals on the left side. *Audiologic*: Békésy audiometry—wide separation between pulsed- and continuous-tone tracings and small amplitude of continuous-tone tracings. No recruitment but marked tone decay. Moderate pure-tone loss with no air–bone gap. Right ear discrimination was 42%. Left ear masked during all testing of right ear. *Classification*: Neural. Damage in auditory nerve fibers. *Etiology*: Acoustic neuroma right side, removed at surgery.

but a disproportionately poor discrimination score. At surgery, an acoustic neuroma was removed. The reason for the good hearing threshold level in spite of so much auditory nerve damage must be borne in mind: almost three-quarters of the auditory nerve can be severed before the threshold for pure tones is affected markedly. The discrimination score and the patient's ability to understand what is heard, however, are affected, even though the hearing thresholds are not. It seems that the neural patterns carrying information to the auditory cortex are disturbed by the nerve damage so that the patient has great difficulty in understanding what is said. This problem is aggravated further by anything that interferes with understanding, such as ambient noise, distortion, and distraction.

As a rule, recruitment is not present in nerve damage, but it may be found occasionally, probably when there is also some cochlear damage not recognizable by available testing methods.

12.2. Audiometric Patterns in Nerve Deafness

It is difficult to describe definite characteristic patterns associated with pure nerve deafness, except for a cause such as acoustic neuroma. When actual nerve deafness develops in presbycusis, this often is preceded by degeneration in the inner ear, causing the typical high-frequency loss. This makes it difficult to determine precisely what the nerve deafness alone would cause. For our purposes, we can assume that the typical picture of nerve deafness is that of high-frequency hearing loss.

However, it is important to keep in mind that other types of audiometric patterns besides these two are possible in nerve deafness. For example, Fig. 10.11 shows an ascending type of nerve deafness in a very elderly patient. The cause was never established with certainty, but the findings were highly suggestive of nerve deafness.

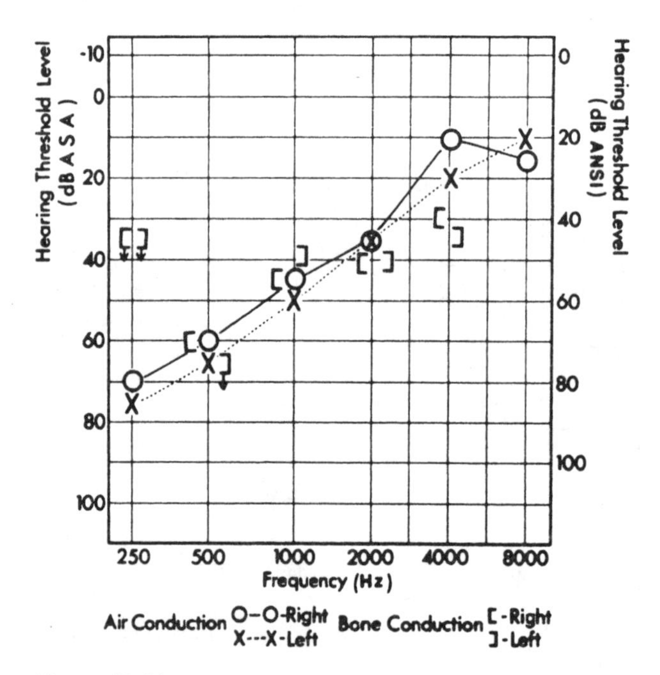

Figure 10.11 *History*: A 74-year-old woman with insidious hearing loss for 30 years. No tinnitus or vertigo. Soft speaking voice. *Otologic*: Normal. *Audiologic*: Moderate to severe loss in the low frequencies. No air–bone gap. *Discrimination score*: Right, 82%; left, 78%. *Classification*: Neural hearing loss. *Diagnosis*: Unknown.

13. CAUSES OF SENSORINEURAL HEARING LOSS

The clinical pictures and causes of sensorineural hearing loss are numerous and varied. Different causes often produce the same clinical profile, and a single cause may produce a variety of clinical pictures. In some cases of sensorineural hearing loss, the cause is speculative or completely unknown. To complicate the problem further, physicians now are beginning to realize that in some patients with histories and audiologic findings highly suggestive of peripheral sensorineural impairment, the temporal bone findings at autopsy do not account for the hearing loss on the basis of any visible pathology.

To provide the reader with a comprehensive perspective of sensorineural hearing loss, we have included as many as possible of the characteristics of the conditions commonly encountered in practice. The cases are classified by causes and on the basis of whether the hearing loss is insidious, sudden, unilateral, bilateral, or congenital. However, patients are not always certain or able to give valid information about the onset of their deafness.

14. CLASSIFICATION

14.1. Causes of Sensorineural Hearing Loss of Gradual Onset

1. Presbycusis
2. Occupational hearing loss
3. Sensorineural aspects of otosclerosis and chronic otitis media
4. Sensorineural aspects of Paget's and van der Hoeve's diseases
5. Effects of hearing aid amplification
6. Auditory nerve damage from chronic systemic diseases (diabetes, etc.)
7. Unknown causes

14.2. Causes of Sudden Bilateral Sensorineural Hearing Loss

1. Meningitis
2. Infections
3. Functional hearing loss
4. Ototoxic drugs
5. Multiple sclerosis
6. Syphilis
7. Autoimmune diseases
8. Unknown causes

14.3. Causes of Sudden Unilateral Sensorineural Hearing Loss

1. Mumps
2. Head trauma and acoustic trauma
3. Meniere's disease
4. Viral infections
5. Rupture of round window membrane or inner-ear membrane
6. Vascular disorders
7. Following ear surgery
8. Fistula of oval window
9. Following general surgery and anesthesia
10. Syphilis

11. Some acoustic neuromas
12. Unknown causes

14.4. Causes of Sensorineural Congenital Hearing Loss

1. Heredity
2. Rh incompatibility with kernicterus
3. Anoxia
4. Viruses
5. Unknown causes

15. CAUSES OF SENSORINEURAL HEARING LOSS OF GRADUAL ONSET

15.1. Presbycusis

The most common cause of sensorineural hearing loss is presbycusis, the gradual reduction in hearing with advancing age. The hearing loss takes place in a well-defined manner. It occurs gradually over a period of years, usually with the very highest frequencies affected first and the lower ones gradually following. Both ears are affected at about the same rate, but sometimes the loss in one ear may progress faster than the other. Whenever an older patient is encountered who has a hearing loss much greater in one ear than in the other, a diagnosis of presbycusis should not be made without reservation, for it is likely that some other cause is also present.

15.1.1. Development

Actually, the process of presbycusis starts quite early in life. Children can hear up to ~20,000 Hz. But many adults can hear only up to 14,000, 12,000, or 10,000 Hz. With advancing years the ability to hear the higher frequencies becomes less acute, and by the age of 70 most people do not hear frequencies >6000 Hz.

Because human beings make very little use of frequencies normally >8000 Hz, they do not become aware of any loss in hearing until the frequencies <8000 Hz are affected. These frequencies start to become affected typically around the age of 50. The rate at which presbycusis advances and the degree to which the individual becomes affected vary widely. To some extent, heredity plays a role, for early or premature presbycusis often is found in several members of the same family.

There is a form of early presbycusis in which the inner ear structures may be affected in such a manner that only the frequencies between 2000 and 6000 Hz become involved before there are any changes in the higher ranges. When this occurs, hearing loss does not progress to a marked degree before the higher frequencies also become involved. This audiometric picture closely resembles that of the hearing loss caused by intense noise exposure.

A Typical Case and Average Values. Figure 10.12 shows the typical manner in which presbycusis develops in one individual. This is a longitudinal study, as the same man has been examined for ~20 years. In most presbycusis cases in which 6000 and 8000 Hz are impaired, there is a greater loss in the higher frequencies, such as at 10,000 and 12,000 Hz. It is possible to find middle-aged people with marked presbycusis, whereas quite elderly people may be found who have very little hearing loss for pure-tone thresholds. Presbycusis should not be confused with hereditary progressive

JOSEPH SATALOFF, M.D.
ROBERT THAYER SATALOFF, M.D.
1721 PINE STREET PHILADELPHIA, PA 19103

HEARING RECORD

NAME AGE

AIR CONDUCTION

AGE	Exam	LEFT MASK	RIGHT 250	RIGHT 500	RIGHT 1000	RIGHT 2000	RIGHT 4000	RIGHT 8000	RIGHT MASK	LEFT 250	LEFT 500	LEFT 1000	LEFT 2000	LEFT 4000	LEFT 8000	AUD
53			0	0	5	15	25	40		0	0	0	10	20	35	
58			0	0	10	15	30	45		0	0	5	10	25	40	
63			5	0	10	15	55	60		0	0	10	15	40	55	
67			5	0	10	20	60	75		5	0	10	20	55	65	
74			10	10	25	35	65	75		10	5	15	30	65	70	

BONE CONDUCTION

DATE	Exam	LEFT MASK	RIGHT 250	RIGHT 500	RIGHT 1000	RIGHT 2000	RIGHT 4000		RIGHT MASK	LEFT 250	LEFT 500	LEFT 1000	LEFT 2000	LEFT 4000		AUD

SPEECH RECEPTION

DATE	RIGHT	LEFT MASK	LEFT	RIGHT MASK	FREE FIELD	MIC.

DISCRIMINATION

DATE	RIGHT % SCORE	RIGHT TEST LEVEL	RIGHT LIST	LEFT MASK	LEFT % SCORE	LEFT TEST LEVEL	LEFT LIST	RIGHT MASK	EXAM

HIGH FREQUENCY THRESHOLDS

DATE	RIGHT 4000	RIGHT 8000	RIGHT 10000	RIGHT 12000	RIGHT 14000	LEFT MASK	RIGHT MASK	LEFT 4000	LEFT 8000	LEFT 10000	LEFT 12000	LEFT 14000

RIGHT RINNE	RIGHT SCHWABACH	WEBER	LEFT RINNE	LEFT SCHWABACH

HEARING AID

DATE	MAKE	MODEL
RECEIVER	GAIN	EXAM
EAR	DISCRIM	COUNC

REMARKS

Figure 10.12 *History*: Routine audiograms were done on this patient while he was being treated for allergic nasal discomfort. At age 63 he started to complain of discrimination difficulty. No tinnitus or vertigo. *Otologic*: Ears clear. *Audiologic*: This is a longitudinal study of gradual progressive deafness due to aging. Important here are the comparatively small change in puretone thresholds at 500, 1000, and 2000 Hz between the ages of 63 and 74 and the dramatic change in discrimination ability during that same period.

Discrimination score:

Age 63	Right 92%	Left 90%
Age 74	Right 60%	Left 56%

Bone conduction thresholds were at the same level as the air thresholds at all frequencies for the ears covered. *Classification*: Sensorineural hearing loss. *Diagnosis*: Presbycusis.

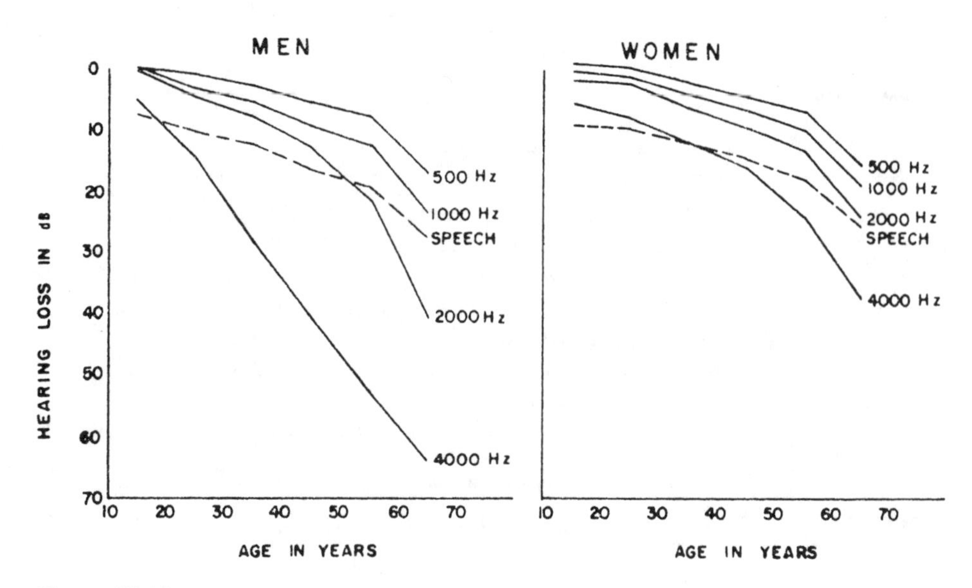

Figure 10.13 Average hearing loss by decades for men and women from age 10 to 70. Median loss in right ear and median loss in left ear were averaged. (Adapted from data collected by Research Center, Subcommittee on Noise in Industry, American Academy of Ophthalmology and Otolaryngology, Wisconsin State Fair Survey, 1954.)

nerve deafness, which, usually starting at a much earlier age and becoming quite severe, is undoubtedly inherited. Figure 10.13 shows the average hearing loss in each frequency that can be expected with aging in the general population. These are average values to be used for statistical purposes; they do not hold necessarily for specific individuals.

15.2. A Complex Phenomenon

Despite its simple definition and predictable development, presbycusis actually is a complex phenomenon. Even its classification is complicated. In the early stages of some cases of presbycusis, only the epithelial elements in the cochlea may be affected (sensory hearing loss); later, the nerve elements also are involved (sensorineural), and eventually, in the last decade of life, the cortex and the central pathways become involved, taking on the aspects of central hearing loss or central dysacusis. In still other cases of presbycusis the nerve fibers seem to be damaged, making it chiefly neural hearing loss. So a single etiology, presbycusis, can produce hearing loss that may fall into one of four different classifications. Schucknecht's classification of presbycusis has shed light on this complex subject (9). **Sensory presbycusis** characteristically produces bilateral, symmetrical high-frequency hearing loss with an abruptly sloping pattern. Discrimination depends upon the frequencies involved. The damage is primarily in the basal end of the cochlea, and sensory presbycusis appears to be genetically modulated. **Neural presbycusis** involves loss of neurons in all turns of the cochlea. It may begin early in life, but does not usually become symptomatic until later in life. Normally, an ear contains ~37,000 neurons during the first decade of life, and only ~19,000 neurons in the ninth decade of life, losing ~2000 per ear per decade. Discrimination correlates with the extent of neuron loss in the 15- to 22-mm region of the cochlea where the speech frequencies are represented. In neural presbycusis, loss of discrimination ability tends to be severe. Loss is usually diffuse throughout the turns, and the audiometric pattern shows a gradually sloping loss involving all frequencies. **Strial presbycusis** is associated with atrophy of the stria vascularis causing a flat, pure-tone hearing loss, with good discrimination. The degree of strial atrophy correslates with the degree of hearing loss (10). **Cochlear conductive prebycusis** is associated with gradual, sloping high frequency hearing loss that generally is noticed first during middle age. Discrimination depends upon the steepness of the audiometric pattern. The histologic correlate of cochlear conductive presbycusis has not been established with certainty, but Schucknecht believed it is associated with thickening of the basilar membrane and stiffening of the cochlear partition that "are related to the physical-anatomic gradients that determine the resonance characteristics of the cochlear duct"(9).

The cause of presbycusis also is complex and not well understood. The simplest hypothesis is based on arteriosclerosis. However, this theory is not borne out by studies, most of which how no pertinent vascular changes in the ear to explain the clinical findings of sensorineural impairment. Still under debate are the factors that predispose to presbycusis or that produce the damage to the hearing mechanism. A study in Africa showed that elderly natives who had spent their lives in an atmosphere free of the everyday noises to which most human beings are exposed in modern society show little clinical evidence of presbycusis. This led a few investigators to propose that presbycusis may be the damage produced by exposure to the everyday noises of modern civilization, but there are many other incidental circumstances to be taken into account. On the basis of comparable physiological studies of the eyes, the skin, the brain, and so on, one would expect a natural wearing down of the hearing mechanism function with age. An interesting epidemiological study using the Framingham Heart Study Cohort looked at hearing loss in an elderly population trying to determine which variables had a significant impact upon hearing loss (11). Age was the most critical risk factor by far, although sex, illness, family history of hearing loss, Meniere's syndrome and noise exposure were also significant population risk factors. Further study of this complex issue is needed.

15.3. The Clinical Picture

The sensorineural type of presbycusis is rather characteristic. The patient generally is >50 and usually >60 years of age. The major complaint is not difficulty in hearing, but, rather, a gradual difficulty in understanding what has been said. At first this happens only in noisy gatherings such as cocktail parties or when several people are speaking at the same time, particularly if the high-pitched voices of women are prominent. As the loss continues to progress, the symptoms become more apparent. Now the patient may miss what is being said on the radio and television or at church or meetings. If the patient is a business person, there will be many difficult moments in cross-conversation, and generally he/she will have to ask with embarrassing frequency, "What did you say?" This difficulty may be even worse when the speaker talks softly, quickly, or has dentures that do not fit well. It is even more marked when the speaker has a foreign accent and poor enunciation. The audiogram by this same time shows that all the high frequencies have become affected, and the loss has extended into the speech range at 2000 Hz. The hearing threshold level at 2000 Hz usually has reached ~30 dB, and the higher frequencies are worse.

The patient's difficulty in understanding speech is similar to that in advanced cases of occupational hearing loss. The vowels are heard, but certain consonants that fall into the high-frequency range cannot be differentiated. For example, there is a problem in determining whether the speaker said, "yes," "yet," "jet," "get," or "guess." This inability to understand what is being said generally is mistaken for inattention or, in the case of an aging parent or a boss who is ready to retire, for "brain deterioration" or "signs of old age."

Ways of Helping. As a result of hearing difficulty, the senior citizen faces major psychological problems that demand better understanding from physicians and all people.

Though physicians have no cure for presybcusis, this does not mean that we cannot help the patient who has discrimination difficulty. It reassures the elderly patient with presbycusis when we explain that the chief difficulty is not in the brain but merely in the hearing mechanism of the ears. The microphones are not working properly, and this is the reason that conversations at the dining-room table or in the living room are not clear.

If members of the patient's family can be made to understand this problem, they will obviate countless strained situations that breed insecurity, depression, and introversion in the senior citizen. Families deliberately should speak not-too-loudly, slowly, and clearly to parents or grandparents with handicapping degrees of presbycusis. Every effort should be made for the speaker and the listener to be in the same room, so that the older person can see the face of the speaker. In this way consonants that cannot be heard are seen on the lips, and conversation can be carried on much more readily and easily. It is difficult to adjust to the loneliness of old age after a successful career, and each family and physician must strive to help the senior citizen to communicate better.

If the loss in the speech frequencies is greater than ~35 dB, a hearing aid sometimes can be of considerable benefit to the elderly patient. Although the hearing aid does not improve the patient's discrimination, it helps the patient to hear with greater ease and less strain. Unfortunately, some elderly patients cannot adjust readily to using an aid and have to be helped by repeated educational talks and reminders, in order to understand that their compliance is essential for improved communication.

Tinnitus. About 25% of healthy people >65 have tinnitus. Occasionally, a patient will complain of ringing or hissing head noises, and sometimes these even may become a severe complaint. In the absence of specific therapy, symptomatic treatment may be necessary to reduce excessive reaction to head noises.

Bone Conduction. A common audiologic finding in presbycusis is inordinately poor bone condition. The bone conduction threshold usually is much worse than the air conduction, and often the bone conduction cannot even be determined with a tuning fork or an audiometer. This does not necessarily mean that the patient's sensorineural mechanism is incapable of responding. It may mean that there is something in the skin or in the bone of the head that prevents the vibrating tuning fork or the audiometric oscillator from being heard. Frequently, if the tuning fork is placed on the patient's teeth or dentures, it will be heard more easily than when it is pressed against the skull. Bone conduction testing in older patients is a particularly poor indication of their sensorineural potential.

The Eardrum. In elderly patients a change in the appearance of the eardrum also is seen often. The cause of sensorineural hearing loss must not be attributed to a thick, scarred, or white eardrum. This does not solve the difficulty and may lead the patient only to seek treatment of the eardrum while the real problem is disregarded.

A Striking Feature. A complaint of difficult hearing without a sufficient corresponding reduction in audiometic threshold is prominent in cases of presbycusis. Even the discrimination score sometimes is not reduced as much as the patient's complaints would lead one to expect. This is characteristic of elderly patients with presbycusis. In the central type of presbycusis and pure nerve damage, the results of tests are even more inconsistent with the patient's complaints. These patients may have a concurrent auditory processing problem, as well.

15.4. Occupational Deafness

Exposure to hazardous noise in industry can result in hearing loss. For details see Chapter 14.

15.5. Otosclerosis (Sensorineural Sequelae)

Otosclerosis is basically a disease of cochlear bone. When it reaches the oval window and the stapedial footplate, it can cause the classic clinical picture of conductive hearing loss known as clinical otosclerosis.

15.6. Bone Conduction

Unfortunately, the clinical findings are not always quite as classic or so simple. Many cases of otosclerosis, even in their incipient stage, do not have normal bone conduction. One feature of the reduced bone conduction associated with clinical otosclerosis is called the "Carhart notch." This is a 20- or 15-dB reduction in bone conduction at 2000 Hz. Its cause has been attributed to some defect in the transmission of bone-conducted sound of this frequency through the patient's skull because of the fixed footplate. When hearing is restored by surgical techniques, the reduced bone conduction often improves to ~5 dB.

In most cases of otosclerosis in the authors' experience, bone conduction rarely is normal for all frequencies. Usually there is some reduction in bone conduction at 2000 and 4000 Hz and even at some other frequencies. The frequent association of otosclerosis with reduced bone conduction has created a suspicion that there must be some relationship between the two. As yet, no experimental or valid clinical relationship has been established. Histopathological studies support this impression.

15.7. Sensorineural Damage

There is little doubt that otosclerosis often leads to sensorineural damage. Cases of otosclerosis in which bone conduction and the sensorineural mechanism remain normal for years are in the minority. High-frequency hearing loss, associated with reduced bone conduction, and recruitment in the early stages are definite manifestations of this sensorineural damage. In many cases the early perceptive effects seem to be only in the cochlea. However, in others the picture is one of sensorineural hearing loss associated with otosclerosis. In unilateral cases with sensorineural damage it often is possible to demonstrate complete recruitment (Fig. 10.14). Even the so-called Carhart notch may in some instances represent a sensorineural loss rather than a mechanical defect. This is especially true in cases in which the hearing and the bone conduction do not improve markedly after surgery or in which the improvement is only to 5 or 10 dB. The normal level for such an individual probably was −10 dB, and actually he/she has a 15-dB sensorineural hearing impairment in that frequency. Although there is impressive evidence that otosclerosis can cause sensorineural hearing loss, there still is no proof that surgical correction of otosclerosis prevents the development of sensorineural hearing impairment. A toxic effect on the sensory mechanism has been speculated but not proven and sensorineural impairment seems to be associated with otosclerotic involvement of bone adjacent to the stria vascularis.

Diagnostic Clues. The onset of sensorineural hearing loss associated with otosclerosis generally is insidious. It may appear long after the conductive hearing loss has been recognized. Occassionally, otosclerosis and sensorineural impairment be diagnosed simultaneously (Fig. 10.15). In a few such cases the diagnosis of otosclerosis may seem to be far-fetched, but it is established definitely at surgery, and hearing sometimes can be improved. The important clues in such cases are that bone conduction is definitely present and better thatn air conduction (especially with the tuning fork in contact with

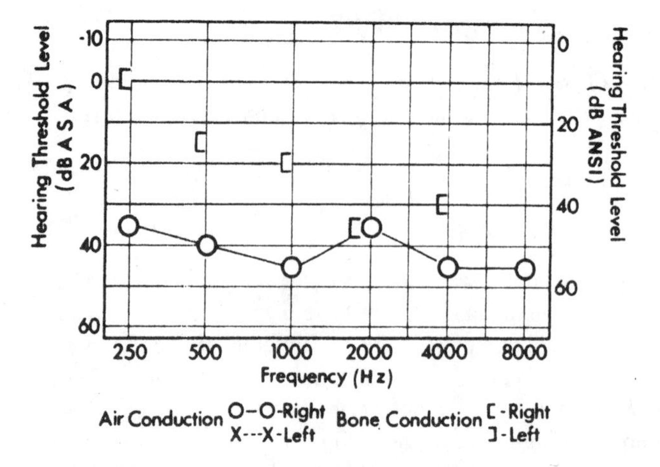

Figure 10.14 *History*: A 22-year-old woman with right-ear deafness for 2 years. Occasional ringing tinnitus. No vertigo. *Otologic*: Normal. *Audiologic*: Left ear normal. Right ear air conduction thresholds revealed a moderate and essentially flat loss. Bone thresholds were reduced, but an air–bone gap existed at all frequencies except 2000 Hz. The right ear had complete recruitment with diplacusis. There was no abnormal adaptation. Right-ear discrimination was 86%. Masking used in the left ear for all right-ear tests. *Classification*: Mixed hearing loss. *Diagnosis*: Unilaterial otosclerosis with secondary sensorineural involvement. *Aids to diagnosis*: The presence of an air–bone gap, good discrimination, and negative otoscopic findings indicates middle-ear damage, specifically, stapes fixation. The presence of recruitment, diplacusis, and reduced bone conduction indicates sensorineural involvement. Otosclerotic stapes fixation was confirmed at surgery, and hearing improved.

Loudness Balance Tests for Recruitment (Preoperative):

1000 Hz		4000 Hz	
L	R	L	R
−5	45	10	45
10	40	25	45
25	45	40	55
40	50	55	60
55	65	70	65
70	80		
85	90		

Interpretation: The columns of figures indicate that, on pure-tone threshold testing, the left ear at 1000 Hz had a threshold of −5 dB, and the right ear had a threshold of 45 dB. When subsequently the intensity of the tone presented to the left ear was raised, and the patient was asked to match the loudness of the tone in the right ear with that in the left, it was noticed that the threshold difference tended to disappear. At 85 dB for the tone in the left ear, the intensity required to produce equal loudness in the right ear was 90 dB. Thus, the threshold difference, which was 50 dB, compares with a difference of only 5 dB when the tone intensity was increased to 85 dB (left) and 90 dB (right), respectively. This is a measure of recruitment. If the tone loudness in the two ears matched exactly at the same intensity, this condition would be called *complete recruitment*. If the tone sounded even louder in the bad ear than in the good ear at the same intensity, it would be called *hyperrecruitment*.

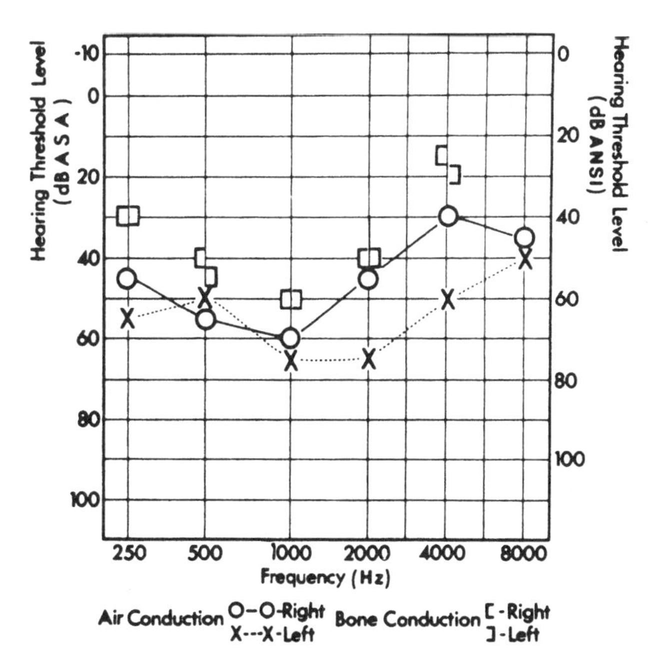

Figure 10.15 *History*: A 50-year-old woman who has a soft speaking voice and uses a hearing aid in the left ear. Insidious hearing loss for several years. Initially, she had buzzing tinnitus, which gradually has disappeared. Two sisters have hearing loss. No vertigo or ear infections. Patient says she is getting satisfactory results with her hearing aid. *Otologic*: Normal. *Audiologic*: Very little air–bone gap is present, but the patient hears better with the tuning fork on the teeth than by air conduction. Dental bone conduction is also better on the teeth than on the mastoid, where there is little or no response. *Discrimination score*: Right, 88%; left, 84%. *Classification*: Mixed hearing loss. *Diagnosis*: Otosclerosis associated with sensorineural hearing loss. *Aids to diagnosis*: Soft voice, satisfactory hearing aid usage, familial history of hearing loss, fairly good discrimination, and bone conduction better than air conduction by tuning fork all point to the possibility of initial conductive involvement. An exploration was done on the left middle ear, and the footplate was found to be markedly fixed, with otosclerosis in the oval window. Mobilization was achieved, and the patient said that she heard better. There was a 10-dB improvement at most frequencies postoperative.

the teeth), and there is reasonably good discrimination. The typical history of a soft voice also is helpful, particularly if the patient gets better results with a hearing aid than would be expected in a case of sensorineural loss. The fact that many cases of this type have been diagnosed successfully and operated on should in no way be misconstrued as encouragement for patients with sensorineural hearing impairment to demand middle-ear surgery for possible otosclerosis. A diagnosis of otosclerosis must be made positively by hearing tests and history, not by surgical exploration, in nearly all cases.

Surgical Complications. So many patients with otosclerosis have had stapedectomies, and so many of these have had some postoperative high-frequency hearing loss and reduced bone conduction, that it is now common to see cases like that in Fig. 10.16. This patient probably had good bone conduction pre-operatively and reduced bone conduction postoperatively. Seeing this patient for the first time postoperatively, a physician may wonder why surgery was done in the first place, as there is little or no air–bone gap, and the diagnosis now is sensorineural hearing loss. The new examiner always should bear in mind that the bone conduction may not be the same

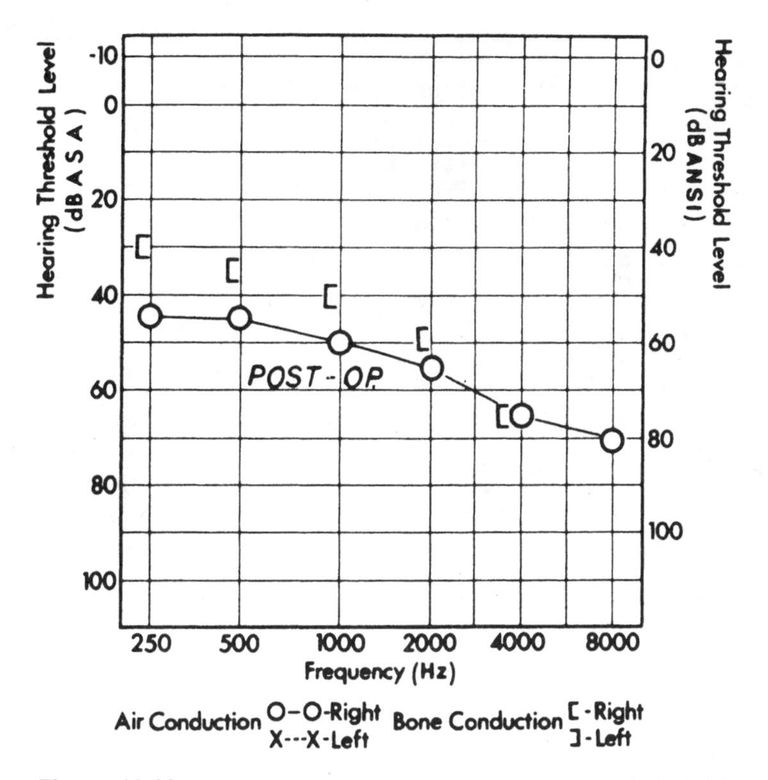

Figure 10.16 *History*: Patient had a right stapedectomy followed by vertigo, fullness, and tinnitus. Hearing worsened after surgery. There is a family history of otosclerosis. *Otologic*: Normal. *Audiologic*: Left ear normal. Right-ear air condition thresholds are reduced moderately, as are the bone thresholds. Right-ear discrimination is 66%. There is complete recruitment with diplacusis. *Classification*: Mixed hearing loss. *Diagnosis*: Otosclerosis with postoperative labyrinthosis. *Aids to diagnosis*: Although this picture is highly suggestive of Meniere's disease, the history points to postoperative labyrinthosis as the etiology.

now as it was preoperatively and may have been decreased by some complication. One such complication is a perilymph fistula, or leak between the inner ear and middle ear. This is particularly important because it may be repaired surgically. It is essential to recognize treatable and reversible causes of sensorineural hearing loss.

15.8. Chronic Otitis Media

Ear infection may produce sensorineural hearing loss. The causal relationship is obvious when the hearing loss follows an acute infection that extends to the inner-ear and/or auditory nerve. However, the relationship may be less apparent when associated with chronic ear disease. Sensorineural hearing loss is a well recognized concomitant of chronic otitis media, particularly in cases with recurrent otorrhea or cholesteatoma. Hearing loss is also not uncommon in the presence of tympanosclerosis involving the middle ear, another condition generally associated with chronic otitis media.

15.9. Paget's and van der Hoeve's Diseases

Both Paget's and van der Hoeve's diseases can affect the auditory nerve pathway. In the early stages, usually only conductive hearing loss is produced, but as the disease

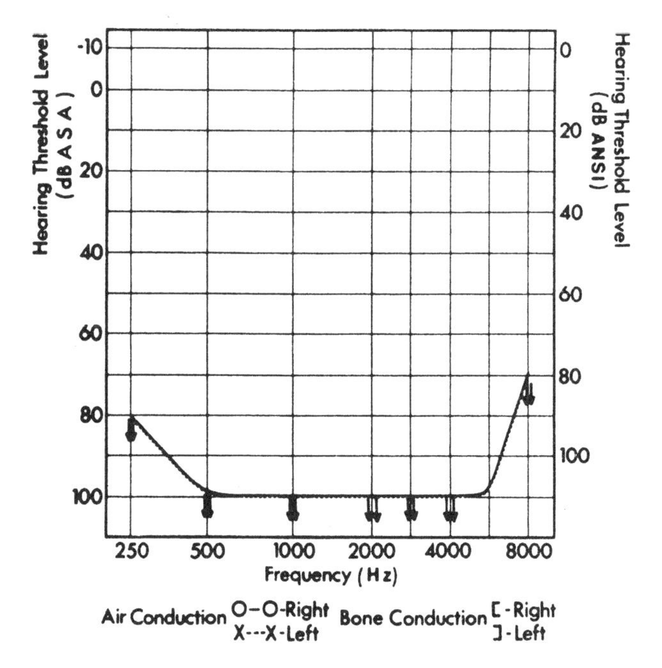

Figure 10.17 *History*: A 70-year-old woman with hearing loss for >30 years. Wears a hearing aid in the right ear and seems to get along with some residual hearing. Other members of her family have long-term deafness. Both this patient and other members of the family have a history of having blue sclera and repeated fractures of the long bones. *Otologic*: Ears normal. *Audiologic*: No response to air conduction audiometry and only tactile response to bone conduction at the lower frequencies. *Classification*: Mixed hearing loss. *Diagnosis*: van der Hoeve's disease. *Aids to diagnosis*: Deafness, blue sclerae, fragile bones, and the hereditary features established the diagnosis.

progresses, the sensorineural mechanism becomes affected, often quite severely. The specific reason for this involvement is not completely clear, but generally there are degenerative processes in the inner ear and the nerve fibers. Sometimes the auditory nerve may be constricted by narrowing of the internal auditory meatus caused by Paget's disease. A toxic effect on the inner ear, such as that which probably exists in some cases of otosclerosis, also warrants consideration in Paget's and van der Hoeve's diseases. In some cases, the conductive and the sensorineural loss seem to start simultaneously and to progress in parallel. An interesting case is described in Fig. 10.17. Note the characteristics of van der Hoeve's disease as a cause of hereditary deafness—blue sclera and fragile bones that sustain repeated fractures (see Chapter 13 and Appendix 3).

15.10. Other Genetic and Systemic Causes

Many other hereditary and nonhereditary conditions are associated with sensorineural hearing loss. Many of these are discussed in Chapter 13.

15.11. Hearing Aid Amplification

In a few cases, a person who has used a hearing aid has noted marked aggravation of hearing loss only in the aided ear. There seems to be no doubt about this causal

JOSEPH SATALOFF, M.D.
ROBERT THAYER SATALOFF, M.D.
1721 PINE STREET PHILADELPHIA, PA 19103

HEARING RECORD

NAME AGE

AIR CONDUCTION

			RIGHT							LEFT						
DATE	Exam	LEFT MASK	250	500	1000	2000	4000	8000	RIGHT MASK	250	500	1000	2000	4000	8000	AUD
5/10/52			25	55	60	70	90	AID	ON	15	40	50	50	45		
6/7/54										70	NR	NR	NR	NR		
6/21/54								AID	OFF	50	70	85	85	75	RING.	TIN.
9/5/54										30	55	60	55	50		
10/2/54								AID	ON	25	50	60	55	50		
11/20/54										60	85	95	85	70		
12/4/54								AID	OFF	30	50	60	60	50		
4/2/55								AID	ON	25	60	85	80	90	SUDDEN	
4/8/55								AID	OFF	15	50	60	60	50	RING. PERSIST.	
9/4/56			25	50	65	70	86	AID	ON	55	NR	NR	NR	NR	RING.	
9/14/56	AID ON			50	65	70	85			60	90	NR	NR	NR		
10/13/56			15	45	60	65	85	AID	OFF	20	50	60	70	60		
1/20/57			20	45	65	70	90			25	45	65	70	70		
4/6/57			15	40	60	75	NR	AID	ON	55	NR	NR	NR	NR		
9/9/57			20	45	70	70	NR	AID	OFF	NR	45	40	75	NR		

Figure 10.18 *History*: This boy was first seen at age 5 for a hearing loss that parents had noted since he was 3 years old. The child had a mild speech defect and had developed speech slowly. The diagnosis was congenital deafness of unknown origin. A hearing aid was placed on his left ear, and he did very well until he noted worse hearing in the left ear several years later. After numerous studies were done to rule out other causes, the hearing aid was removed, and his hearing improved. Each time the aid was used, the hearing got worse but then improved when the aid was removed. Finally it was recommended that the aid be worn on the right ear, and the patient noted no adverse effect. *Diagnosis*: Sensorineural hearing loss caused by noise. *Note*: Even the low frequencies were affected in this case.

relationship, but because only a few ultrasensitive ears are affected, advice against the use of hearing aids is unjustified. The hearing loss, usually associated with the use of a powerful hearing aid, is of a peculiar type. Ordinarily, in a typical picture there would be a temporary or even a permanent threshold change as the result of amplification by the aid. This is not what we generally find, as is shown in the interesting case described in Fig. 10.18. Although the hearing returned to its previous level after the hearing aid was removed, it took much longer to return than it would have taken in a TTS, and it did not have other classic characteristics of TTS.

Whenever a patient who uses a hearing aid complains that hearing is getting much worse in the aided ear, amplification should be suspected as a possible cause if the loss is sensorineural. In such instances, the hearing aid should be removed, and the hearing should be watched to see whether it returns to its original level. It may be necessary for the patient to use the aid in the other ear, and if this ear shows the same sensitivity to amplification, a less powerful type of hearing aid might be indicated.

As in all cases of unilaterial sensorineural hearing loss, the possibility of an acoustic neuroma always should be considered, and all necessary studies should be done.

15.12. Neuritis of the Auditory Nerve

Auditory neuritis is an inflammatory condition of the auditory division of the eighth cranial nerve. It causes hearing loss without dizziness, although tinnitus may be present. It is distinguished from vestibular neuritis which causes dizziness, without hearing loss. When hearing loss and balance disorders are involved in the presence of an inflammatory process, the diagnosis is usually either eighth nerve neuritis or labyrinthitis.

Auditory neuritis may follow systemic infections such as scarlet fever, influenza, typhoid fever, syphilis, and many other infectious diseases that produce high fevers. It is seen more frequently following less severe viral illnesses such as upper respiratory infections, or herpetic infections such as those associated with cold sores or "fever blisters." The hearing loss may be noted immediately in conjunction with the infection, but the onset is often progressive over a period of days or weeks. It may be unilateral, or bilateral; but unilaterality or asymmetry is common. The condition often is associated with a feeling of fullness in the ear. In the early stages, hearing may improve following therapy with corticosteroids. Severity of the hearing loss varies from mild to profound. Figure 10.19 shows the findings in a patient with a presumptive diagnosis of neuritis of the auditory nerve. Figure 10.20 describes a case in which the diagnosis is more certain.

15.13. Vascular Insufficiency

The labyrinthine artery which supplies the inner ear is a nonanastomotic end artery. That is, its branches do not intermingle with collateral vascular supply from other sources. Consequently, the inner ear appears to be more sensitive to vascular degeneration than most other organs. This problem is associated with generalized atherosclerosis, diabetes, and other conditions that alter blood flow to the ear. It may also occur without obvious systemic concomitants. More research is needed to clarify the process of hearing deterioration related to vascular degeneration.

JOSEPH SATALOFF, M.D.
ROBERT THAYER SATALOFF, M.D.
1721 PINE STREET PHILADELPHIA, PA 19103

NAME J.F.

DATE	RIGHT EAR AIR CONDUCTION							LEFT EAR AIR CONDUCTION					
	250	500	1000	2000	4000	8000		250	500	1000	2000	4000	8000
	45	80	80	85	90	70		25	55	55	60	60	55

RIGHT EAR BONE CONDUCTION							LEFT EAR BONE CONDUCTION					
250	500	1000	2000	3000	4000		250	500	1000	2000	3000	4000
35	60	75	NR	NR	NR		25	40	35	55	55	55

SPEECH RECEPTION: Right 70 Left 35 DISCRIMINATION: Right 52% Left 80%

Figure 10.19 The findings in a patient with a presumptive diagnosis of neuritis of the auditory nerve.

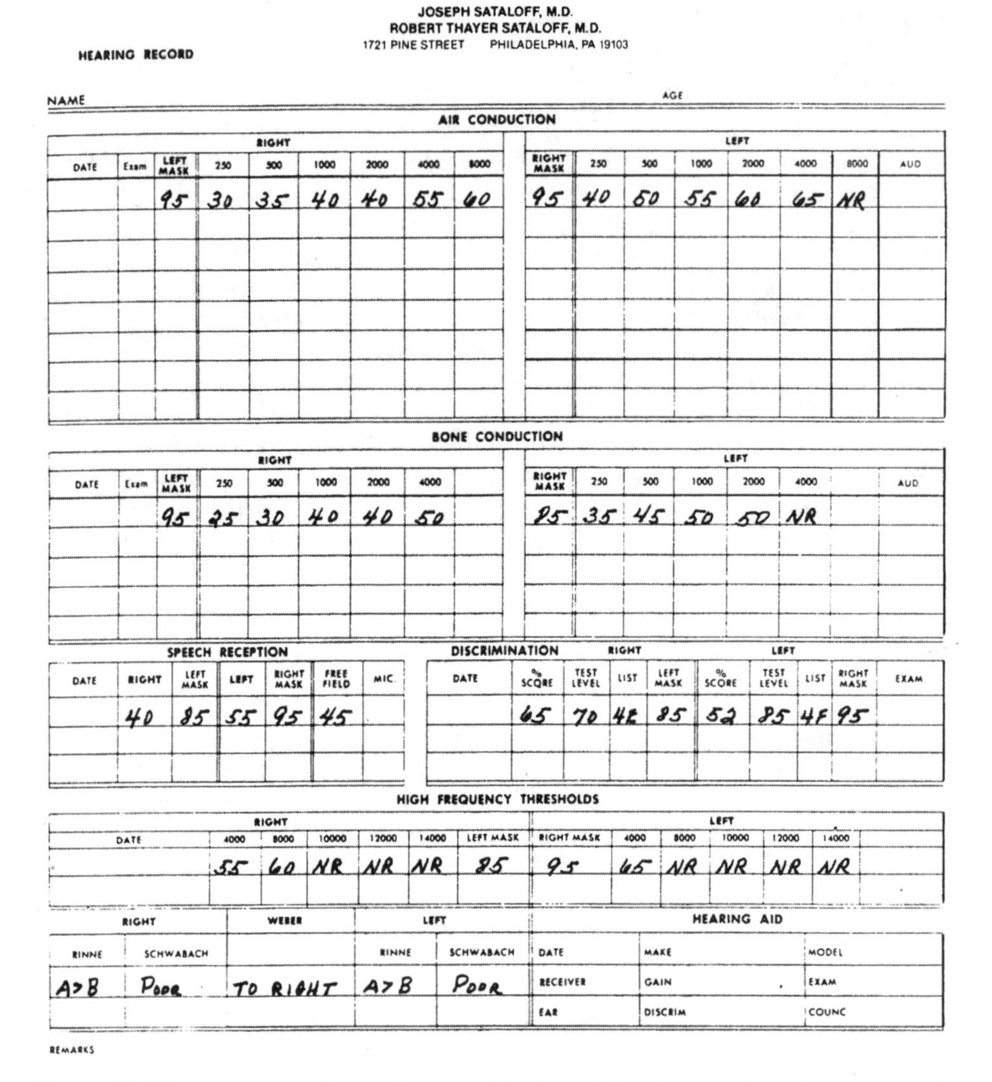

JOSEPH SATALOFF, M.D.
ROBERT THAYER SATALOFF, M.D.
1721 PINE STREET PHILADELPHIA, PA 19103

HEARING RECORD

NAME AGE

AIR CONDUCTION

	RIGHT								LEFT							
DATE	Exam	LEFT MASK	250	500	1000	2000	4000	8000	RIGHT MASK	250	500	1000	2000	4000	8000	AUD
		95	30	35	40	40	55	60	95	40	50	55	60	65	NR	

BONE CONDUCTION

	RIGHT								LEFT							
DATE	Exam	LEFT MASK	250	500	1000	2000	4000		RIGHT MASK	250	500	1000	2000	4000		AUD
		95	25	30	40	40	50		85	35	45	50	50	NR		

SPEECH RECEPTION

DATE	RIGHT	LEFT MASK	LEFT	RIGHT MASK	FREE FIELD	MIC.
	40	85	55	95	45	

DISCRIMINATION

	RIGHT				LEFT				
DATE	% SCORE	TEST LEVEL	LIST	LEFT MASK	% SCORE	TEST LEVEL	LIST	RIGHT MASK	EXAM
	65	70	4E	85	52	85	4F	95	

HIGH FREQUENCY THRESHOLDS

	RIGHT						LEFT					
DATE	4000	8000	10000	12000	14000	LEFT MASK	RIGHT MASK	4000	8000	10000	12000	14000
	55	60	NR	NR	NR	85	95	65	NR	NR	NR	NR

RIGHT		WEBER	LEFT	
RINNE	SCHWABACH		RINNE	SCHWABACH
A>B	Poor	To right	A>B	Poor

HEARING AID

DATE	MAKE	MODEL
RECEIVER	GAIN	EXAM
EAR	DISCRIM	COUNC

REMARKS

Figure 10.20 *History*: A 55-year-old man with bilateral hearing loss following attack of flu during an epidemic. Loss has not progressed. No tinnitus or vertigo. *Otologic*: Normal. Caloric examination was normal. *Audiologic*: Bilateral hearing loss with no air–bone gap, reduced discrimination and poor tuning fork responses. No recruitment or abnormal tone decay. *Classification*: Sensorineural hearing loss. *Diagnosis*: Neuritis of auditory nerve.

15.14. Unknown Causes of Bilateral and Unilateral Sensorineural Hearing Loss of Gradual Onset

There are cases of sensorineural hearing loss for which it is not possible to establish a specific cause. The accompanying series of examples, each of which emphasizes a different factor associated with the deafness, illustrates cases of this type (Figs. 10.21–10.23). Vascular problems such as hypotension and hypertension are most prominent. In addition, Chapter 13 contains other examples of sensory and neural hearing loss. As has been pointed out, causes such as Meniere's disease, noise exposure, and occupational deafness may start as sensory hearing impairment and later progress to sensorineural loss, especially in the presence of other superimposed conditions.

JOSEPH SATALOFF, M.D., D.Sc
ROBERT T. SATALOFF, M.D., D.M.A.
1721 PINE STREET • PHILADELPHIA, PA. 19103

HEARING RECORD

NAME AGE

AIR CONDUCTION

RIGHT											LEFT									
DATE	Exam.	LEFT MASK	250	500	1000	2000	3000	4000	6000	8000	RIGHT MASK	250	500	1000	2000	3000	4000	6000	8000	AUD
1st test			10	20	30	85		90		NR		20	30	60	90		NR		NR	
4 years			80	80	100	NR		NR		NR		80	95	100	NR		NR		NR	
9 years			75	85	NR	NR		NR		NR		NR	NR	NR	NR		NR		NR	
10 years			75	NR	NR	NR		NR		NR		NR	NR	NR	NR		NR		NR	

BONE CONDUCTION

RIGHT									LEFT							
DATE	Exam.	LEFT MASK	250	500	1000	2000	3000	4000	RIGHT MASK	250	500	1000	2000	3000	4000	AUD
1st test			10	20	35	NR		NR		20	40	NR	NR		NR	
9 years			NR	NR	NR	NR		NR		NR	NR	NR	NR		NR	

SPEECH RECEPTION THRESHOLD

DATE	RIGHT	LEFT MASK	LEFT	RIGHT MASK	FREE FIELD	DATE	RIGHT	LEFT MASK	LEFT	RIGHT MASK	FREE FIELD

SPEECH DISCRIMINATION

RIGHT					LEFT						
DATE	% SCORE	TEST LEVEL	LIST	LEFT MASK	% SCORE	TEST LEVEL	LIST	RIGHT MASK	FREE FIELD	AIDED	EXAM

COMMENTS:

Figure 10.21 *History*: Since the age of 18, this healthy 36-year-old man had known that he had a very mild high-frequency hearing loss. It was not progressing, and he participated in active military service. After the age of ~26 he became aware that his hearing was becoming worse, and in spite of extensive studies and a great variety of treatments, he continued to lose his hearing. No tinnitus or vertigo. There is no history of familial deafness. *Otologic*: Normal. Normal caloric responses. *Audiologic*: Rapid progressive bilateral hearing loss. *Classification*: Sensorineural hearing loss. *Diagnosis*: Unknown.

16. CAUSES OF SUDDEN BILATERAL SENSORINEURAL HEARING LOSS

Sensorineural hearing loss of sudden onset is of special interest to the otologist for at least two reasons. First, the cause is hard to establish, and second, it is not

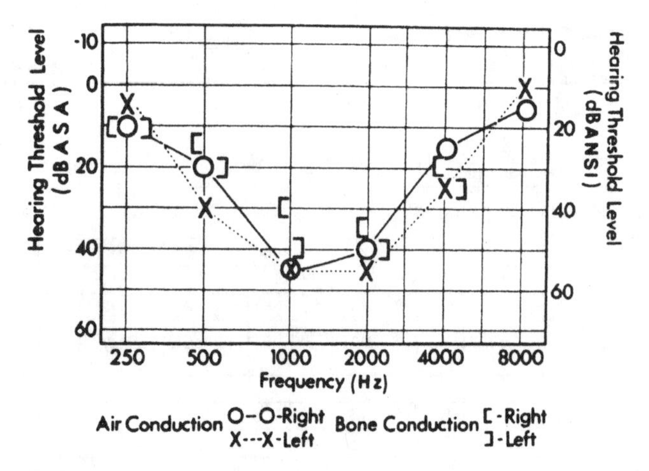

Figure 10.22 *History*: A 50-year-old woman first noticed gradual hearing loss about a year ago and reports that it is progressing. Has had no ear infections or infectious diseases. No tinnitus or vertigo. No history of familial deafness. *Otologic*: Normal. *Audiologic*: Bilateral basin-shaped loss with no significant air–bone gap. Monaural two-frequency loudness balance tests indicate some recruitment in both ears. There was no abnormal tone decay. *Speech reception threshold*: Right, 38 dB; left, 42 dB. *Discrimination score*: Right, 100%; left, 94%. *Classification*: Sensorineural hearing loss. *Diagnosis*: Unknown.

uncommon for the hearing to return. Certain causes more often affect both ears, whereas other causes affect only one ear. The severity of the hearing loss varies, but quite often it can be very profound. The common causes of sudden bilateral sensorineural hearing loss are: (a) meningitis, (b) infections, (c) emotionally induced illness, (d) drugs, (e) multiple sclerosis, (f) autoimmune disease (g) syphilis, and (h) unknown causes.

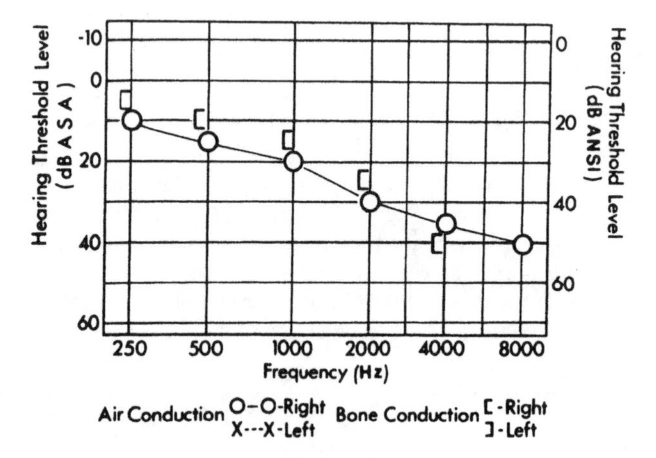

Figure 10.23 *History*: A 45-year-old woman noted gradual onset of hearing loss in the left ear with slight ringing tinnitus. No vertigo. Blood pressure normal. *Otologic*: Normal. *Audiologic*: Left ear normal. Right ear had mild to moderate loss, which is greater in the high frequencies. No air–bone gap. No recruitment or abnormal tone decay. *Classification*: Unilateral sensorineural hearing loss, progressive. *Diagnosis*: No clear-cut cause for this progressive unilateral hearing loss.

16.1. Meningitis

The sudden, profound irreversible bilateral deafness caused by meningitis makes this disease of great and singular concern to all physicians. Figure 10.24 shows the typical, practically total loss of hearing that results when meningitis affects the sensorineural areas. Because the damage is irreversible, every effort must be exerted to prevent meningitis or to treat it vigorously and early to obviate such complications. Occasionally, a small amount of hearing remains, and a powerful hearing aid then is of some value, but only to hear sounds, not to recognize speech. Tinnitus rarely is present, and rotary vertigo usually is of short duration.

Imbalance, particularly in a dark room, is common because of labyrinthine damage. Caloric studies, except in rare cases, reveal absent or poor vestibular function. In selected cases, cochlear implants may be particularly helpful.

16.2. Acute Infections

Systemic infections such as scarlet fever, typhoid fever, measles, and tuberculosis occasionally may cause bilateral sensorineural hearing loss. In the past, syphilis has been blamed for many cases of deafness that actually were because of other causes. At present deafness is caused by syphilis more commonly than one might suspect, and in some cases it can cause sudden bilateral hearing loss. Inner-ear syphilis is a particularly important entity because it is treatable (see Chapter 11). Although a patient may have positive serological findings for syphilis along with sensorineural impairment, there

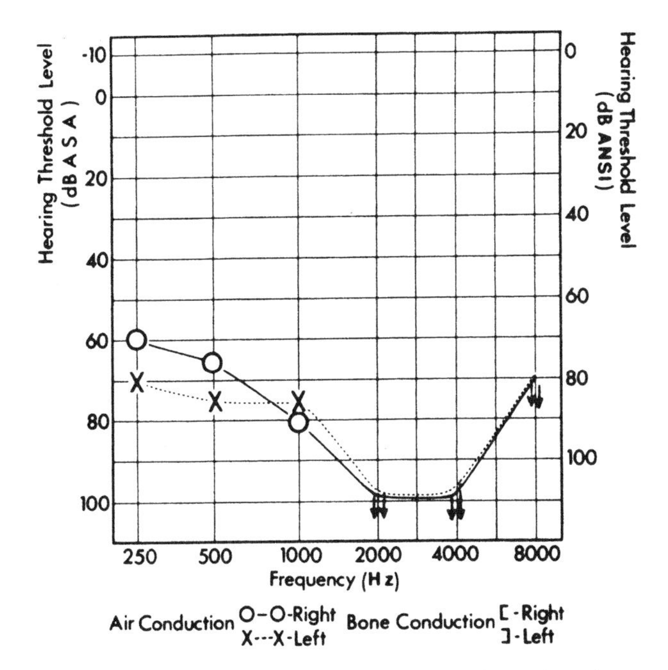

Figure 10.24 *History*: A 16-year-old boy who had meningococcus meningitis at age 6. Speech shows marked evidence of deterioration. He goes to school for the deaf but does not use a hearing aid. *Otologic*: Normal, but caloric responses are absent. *Audiologic*: Patient has some residual hearing in low tones. *Classification*: Sensorineural hearing loss. *Diagnosis*: Effect of meningitis on the cochlea. *Comment*: This boy could benefit from a powerful hearing aid and was urged to use one in his right ear.

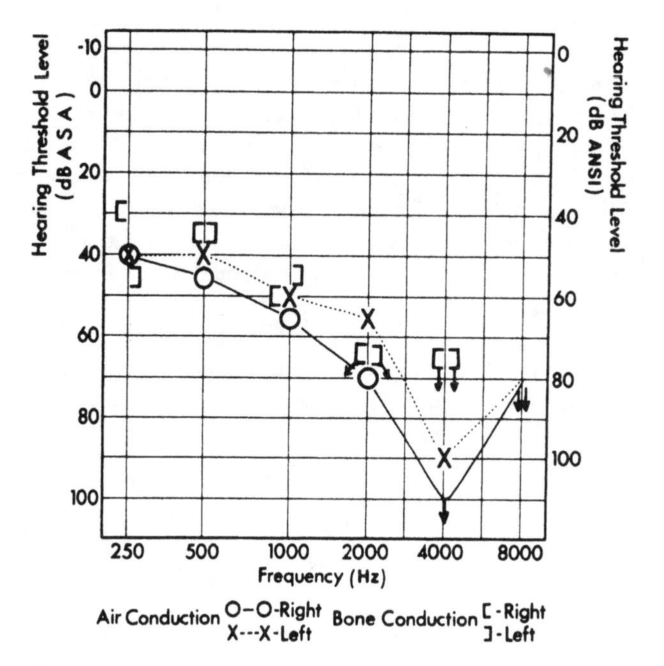

Figure 10.25 *History*: A 48-year-old man with bilateral otitis media and draining ears since age 6 following scarlet fever. Has had some ringing tinnitus but no vertigo. Uses a hearing aid but has trouble because of ear discharge. *Otologic*: Both eardrums eroded and no evidence of ossicles. A putrid discharge is present in both middle ears. X-ray films show sclerotic mastoids but no cholesteatoma. *Audiologic*: Bilateral sloping audiogram shows moderate to severe hearing loss with no air–bone gap. Poor response to tuning fork even on the teeth. *Classification*: Sensorineural hearing loss. *Diagnosis*: Chronic otitis media and labyrinthitis involving the cochlea following scarlet fever. *Comment*: A bone conduction hearing aid usually is not recommended in losses of this type. However, the chronic otitis precludes the use of an air conduction aid, which requires an ear mold in the ear. Use of a bone conduction aid when the ear is discharging might eliminate long periods of lack of amplification.

may not be a causal relationship. The incidence of sensorineural hearing loss is so high that undoubtedly some of the people have syphilis without any relation between the two. Scarlet fever still causes a moderate degree of sensorineural hearing loss, usually accompanied by bilateral acute and later chronic otitis media. Figure 10.25 shows a type that was much more prevalent prior to the use of antibiotics. Usually, the eardrum and the ossicles also were eroded.

16.3. Functional Hearing Loss

Some cases of bilateral sudden hearing loss are caused by emotional disturbances, and although there is no real damage to the sensorineural mechanism, the clinical findings strongly resemble this diagnostic entity. Hysteria is the outstanding feature of such cases, which are common during periods of marked emotional stress, as in wartime. Episodes of severe stress and tension can result in sudden hearing loss, with audiologic findings very similar to those found in sensorineural impairment attributable to other causes. The main distinguishing features are the history and the inconsistent audiologic findings. Because such cases sometimes do occur in civilian life, it is important to rule out functional hearing loss by obtaining consistent audiologic findings and an adequate history. Special audiometric tests are helful (see Chapter 7). Figures 10.26 and 10.27 illustrate examples of function hearing loss encountered in otologic practice.

JOSEPH SATALOFF, M.D.
ROBERT THAYER SATALOFF, M.D.
1721 PINE STREET PHILADELPHIA, PA 19103

HEARING RECORD

NAME AGE

AIR CONDUCTION

			RIGHT							LEFT						
DATE	Exam	LEFT MASK	250	500	1000	2000	4000	8000	RIGHT MASK	250	500	1000	2000	4000	8000	AUD
1ST TEST			40	45	50	50	50	55		45	50	55	50	45	50	
			40	45	50	45	45	45		40	45	50	45	50	50	
1 MO.			45	50	40	50	40	55		45	50	50	55	40	55	
4 MOS.			0	-5	0	0	-5	0		0	-5	-5	-5	-5	0	

BONE CONDUCTION

			RIGHT							LEFT						
DATE	Exam	LEFT MASK	250	500	1000	2000	4000		RIGHT MASK	250	500	1000	2000	4000		AUD
1ST TEST			30	35	40	40	40			35	35	40	40	40		
1 MO.			35	30	40	45	40			30	35	40	45	45		

SPEECH RECEPTION

DATE	RIGHT	LEFT MASK	LEFT	RIGHT MASK	FREE FIELD	MIC.
1ST TEST	10		10			
1 MO.	10		10			

DISCRIMINATION

	RIGHT				LEFT				
DATE	% SCORE	TEST LEVEL	LIST	LEFT MASK	% SCORE	TEST LEVEL	LIST	RIGHT MASK	EXAM
	98	40			100	40			
	96	40			100	40			

HIGH FREQUENCY THRESHOLDS

	RIGHT						LEFT					
DATE	4000	8000	10000	12000	14000	LEFT MASK	RIGHT MASK	4000	8000	10000	12000	14000

RIGHT		WEBER	LEFT	
RINNE	SCHWABACH		RINNE	SCHWABACH

HEARING AID

DATE	MAKE	MODEL
RECEIVER	GAIN	EXAM
EAR	DISCRIM	COUNC

REMARKS

Figure 10.26 *History*: A 14-year-old girl whose hearing loss was suspected by teacher and confirmed by school nurse after audiometry. School work was deteriorating past year. No ear infections or related symptoms. *Otologic*: Normal. *Audiologic*: Note consistent hearing loss in air and bone conduction tests taken days apart, though the speech reception thresholds are normal (10 dB). The girl's normal hearing was confirmed by later testing. She had an emotional problem at school which was rectified by psychotherapy. *Classification*: Functional. *Etiology*: Emotional conflict.

16.4. Ototoxic Drugs

These are discussed elsewhere in this book. The severity and rapidity of onset of ototoxic hearing loss depend upon the patient's medical condtion (renal, etc.), ototoxic potency of the drug, and method of administration. For example, certain diuretics such as Furosemide administered in high doses through rapid intravenous push can cause sudden severe bilateral hearing loss associated with a high peak blood level of the drug. The same drug given over a longer period of time is less likely to result in similar otologic problems.

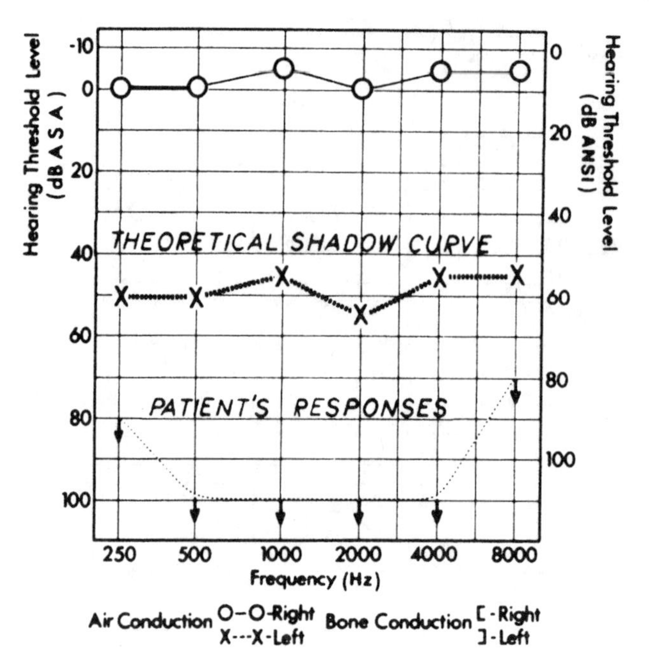

Figure 10.27 *History*: A 36-year-old man received a sharp blow behind the ear, lacerating the scalp but causing no unconsciousness. He claims sudden deafness in left ear following the industrial injury. No tinnitus or vertigo. *Otologic*: Normal. *Audiologic*: Patient denied hearing anything in his left ear when the right ear was unmasked. This led to a suspicion of nonorganic hearing loss. All malingering tests showed good hearing in left ear and did not substantiate a sudden hearing loss following head injury, as the man claimed. *Classification*: Functional hearing loss. *Diagnosis*: Malingering. *Aids to diagnosis*: When one ear is normal and the other is severely impaired, a shadow curve from the good ear (if unmasked) should appear when testing the impaired ear. This did not happen, and after much testing and discussion the patient admitted he was fabricating his deafness, and eventually he gave a normal hearing audiogram in the left ear.

Selected chemotherapeutic agents and other drugs also may cause bilateral hearing loss, in some cases sudden and severe.

16.5. Multiple Sclerosis

Multiple sclerosis is a rare cause of deafness, but it has been reported as a cause of sudden bilateral hearing loss and other patterns of sensorineural impairment. Usually, the deafness fluctuates, and hearing may return to normal even after a very severe depression. Susac's syndrome is misdiagnosed commonly as multiple sclerosis. Susac's syndrome includes the triad of micro-angiopathy of the brain and retina with sensorineural hearing loss in young women. Numerous cases have been recognized since the first report in 1979 (12).

16.6. Autoimmune Hearing Loss

Autoimmune hearing loss is discussed briefly in Chapter 11 and in greater detail here. Figure 10.28 illustrates a case of autoimmune sensorineural hearing loss that developed following surgical manipulation of one ear.

NAME CASE 2 AGE

AIR CONDUCTION																			
			RIGHT								LEFT								
DATE	Exam	LEFT MASK	250	500	1000	2000	3000	4000	6000	8000	RIGHT MASK	250	500	1000	2000	3000	4000	6000	8000	AUD
Pre-Op.			40	35	45	25		35		50		45	40	45	45		65		NR	
Post-Op.			NR	NR	NR	NR		NR		NR		NR	NR	NR	NR		NR		NR	

Figure 10.28 *History*: This is a 26-year-old female who had been born with choanal atresia, bilateral malformed ears and poor vision. Right middle-ear surgery was performed under local anesthesia. Her middle ear was shallow, and her facial nerve was dehiscent in the horizontal portion. The malleus and incus were slightly malformed, but the stapes and oval window were absent. A small fenestra was drilled through the promontory, and a stapes prosthesis was placed through the fenestra and crimped around the incus without difficulty. The patient's hearing improved substantially in the operating room, and there was no dizziness during the procedure. On the second postoperative day she developed sudden deafness in her right ear. Four days later, she developed sudden deafness in her left ear. There was no evidence of infection. Comprehensive testing revealed no abnormalities. Immunophoresis was normal. Other immune system evaluation discussed below had not yet been described for autoimmune hearing loss. She did not respond to high-dose Prednisone. Five years following surgery she has no useful hearing in either ear. Cochlear implant surgery was offered, but the patient declined. This is a rare case of sympathetic cochleitis, an autoimmune phenomenon.

Autoimmune sensorineural hearing loss was first described by McCabe (13) in 1979. In addition to hearing loss, he noted aural fullness and tinnitus and, rarely, destruction of external or middle ear structures. He defined this new entity on the basis of characteristic laboratory findings, as well as its responsiveness to treatment with medications used classically for treatment of autoimmune disorders. McCabe's discussion stressed the importance of accurate diagnosis because of the availability of treatment options for this cause of sensorineural hearing loss.

The characteristics McCabe (13) described continue to be relevant even today, including asymmetric sensorineural hearing loss, occasionally involvement of balance organs, worsening over time, and responsiveness to steroids and other immunosuppressants such as cyclophosphamide. There is a spectrum of disease encompassed by the current terms autoimmune inner ear disease (AIED) and immune-mediated cochleovestibular disorders. The presentations include rapidly progressive sensorineural hearing loss, sudden deafness, and Meniere's syndrome (14). One of the authors (R.T.S.) has treated cases with virtually all audiometric patterns, including total deafness and normal hearing with vestibular symptoms only.

To evaluate suspected autoimmune disease, it is most favorable to use organ biopsy as a diagnostic aid. Clearly, in the case of inner ear disease, the impossibility of biopsy results in a diagnostic dilemma (15). The initial laboratory tests discussed by McCabe (13) was the lymphocyte migration inhibition test, subsequently shown to have inconsistent results. The lymphocyte transformation test was introduced later by Hughes et al. (16) and showed high rates of specificity (93%) and sensitivity (between 50% and 80%). In addition, the indirect immunofluorescence test has been used to screen patients for inner-ear-specific and -nonspecific autoantibodies. Veldman (17) demonstrated immunohistochemical proof of the immune-mediated reaction between spiral ganglia and patients' antibodies. The most definitive diagnostic test appears to be the western blot assay for autoantibodies against cochlear antigens, but the sensitivity and specificity vary depending upon the method used.

Although the treatment proposed by McCabe (13) in his initial research consisted of the classic immunosuppressants corticosteroids and cyclophosphamide (Cytoxan), later research has confirmed the value of methotrexate (Rheumatrex) (18). Some of the most marked benefits of methotrexate therapy include the improved side-effect profile compared with that of steroids and cyclophosphamide, and its utility in patients who require long-term treatment.

When McCabe (13) described autoimmune hearing loss, initially, he defined it on the basis of laboratory findings and clinical characteristics, such as its responsiveness to steroid therapy. The "typical" picture that has emerged since is that of a middle-aged female patient with bilateral progressive, asymmetric sensorineural hearing loss, with or without vertigo, and occasional systemic immune disease (16). However, in our experience, this description is far too limited.

Since the initial description of this entity, autoimmune sensorineural hearing loss has been found to occur in association with numerous immunologic and systemic autoimmune disorders, such as ulcerative colitis (19). As mentioned earlier, ulcerative colitis has been found to occur concurrently. Other diseases of a systemic nature that have been associated with otologic manifestations include connective tissue diseases (rheumatoid arthritis, polyarteritis, polymyositis, Scleroderma, Sjögren's syndrome, amyloidosis), endocrine and associated organ diseases (Hashimoto's thyroiditis, Graves' disease, pernicious anemia), and nonendocrine organ diseases (myasthenia gravis, glomerulonephritis, Wegener's granulomatosis) (20). As a matter of fact, it has been demonstrated that otologic symptoms may be the harbinger of a systemic autoimmune disease that has yet to develop (21). However, the inner ear manifestations of these various diseases are identical. Although the exact mechanism of autoimmunity within the ear has not been elucidated fully, histologic studies have demonstrated the occurrence of an inflammatory reaction throughout the membranous labyrinth (19).

Previously, the presence of the blood–labyrinth barrier led to the notion of the inner ear as an immuno-privileged site. However, it has since been understood that under certain circumstances, immune cells may be present in the inner ear, particularly in the scala tympani (22). Occasionally, their presence is protective as against a viral infection; however, immune cells also may be destructive via a theoretical sensitization mechanism. The immune response likely occurs in the cochlea by the same mechanism as elsewhere throughout the body, with increased vascular permeability allowing the extravasation of leukocytes. The mechanism of this process was detailed by Ryan et al. (23). The endolymphatic sac appears to be important in the development of this reaction, although its precise role is undefined. It may represent as area for initial antigen presentation and processing, resulting in the production of the Cytokines that trigger the increased vascular permeability and lymphocyte tracking that marks the remainder of the immune response (23). One of the authors (R.T.S.) has begun investigating Cytokine system intervention (change from Th2 to Th1 control) as a possible treatment strategy for AIED.

The absence of HLA-DR4 is suggestive of AIED, as 80.6% of AIED patients lack this antigen (24). The other HLA findings that, when present, may indicate a propensity to develop AIED include B35, Cw4, and Cw7.

As described earlier, the lymphocyte transformation test had been used initially in the diagnosis of autoimmune sensorineural hearing loss. However, this test was determined to be fairly insensitive, particularly after empiric corticosteroid treatment in patients in whom AIED was suspected (25). This test has since become an adjunct to the Western blot, which tests serum for antibodies against an inner ear antigen. The sensitivity of the commercially available test (Immco Laboratories, Buffalo, NY) is only 42%; however, its specificity is 90% and its positive predictive value is 91% (25). Attempts to delineate the specific components responsible for the autoimmune reaction have consistently identified several proteins of similar weight. Most of these proteins are of the IgG type, with the rare IgA finding. They have been partially identified as antibodies to myelin protein PO (found in the spiral ganglion in the organ of Corti) and β-actin (in supporting cells and mechanosensory hair cells) (14).

Several antibodies still await identification, such as a 68 kDa protein that had previously been identified as heat shock protein 70 (hsp70) but recently has come into question (14). In order to attain a more accurate diagnosis, investigators have attempted to isolate the inner-ear antigen that would be most closely linked to AIED (26). Although it has been identified repeatedly as the antigenic source, hsp70 has been questioned because of its ubiquity in human tissues. An autoimmune response related to this protein would likely have effects beyond simply the inner ear. The study produced the KHRI-3 antibody, which reacts strongly with supporting cells in the guinea pig organ of Corti. This protein migrates with a relative molecular mass of 65–68 kDa. Additionally, this protein, while abundant in inner ear extracts, is notably absent from extracts of other tissues. Testing of mice inoculated with KHRI-3 hybridoma revealed that mice carrying this hybridoma developed high antibody titres and exhibited a threshold shift on auditory brain stem response testing. High-frequency sensorineural hearing loss was also documented in most of these KHRI-3-bearing mice, and they had the highest degree of hearing loss and the highest circulating immunoglobulin levels. Further, *in vivo* testing of the guinea pig with an intracochlear mini-osmotic pump showed binding of antibody to organ of Corti supporting cells (including inner and outer pillar cells, phalangeal processes of out pillar cells, inner phalangeal cells, and phalangeal process of Deiters' cells). Outer hair cells were lost, as well as, in a few cases, inner hair cells. When the sera of patients suspected to have AIED were tested, the results were strikingly similar to those obtained with the KHRI-3 monoclonal antibody. There was also an observed relationship between antibody to supporting cells, and improvement in hearing after steroid therapy. However, the fact that this same relationship was not determined for antibody to the 68–708 kD protein on Western blot led to the suggestion that Western blot assay may be less sensitive than immunofluorescence and that it may produce some false positives, but 89% of those patients who improved had antibody to supporting cells (positive immunofluorescence). Only 56% of Western blot positive patients improved (26).

A positive Western blot also has a therapeutic implication, in that it has predictive value for the degree of corticosteroid responsiveness (25). Because of the low sensitivity of the Western blot, however, empiric corticosteroid treatment may be used in cases in which a high clinical suspicion exists. One must exercise caution, however, in over-interpreting responsiveness to steroid treatment, as several nonautoimmune processes similarly respond. For example, infectious processes such as syphilitic labyrinthitis and other inflammatory processes show improvement with steroid therapy (27). Another caution regarding steroid treatment is the plethora of side effects that accompany long-term use of high-dose corticosteroids. In addition, the response is rarely sustained. Moreover, failure to respond to steroids does not mean that the patient will not response to cytotoxic drugs, and they should not be withheld on this criterion (24).

One other option that has been explored, cyclophosphamide, has proved to be moderately successful, but again, carries a high price in adverse effects, including susceptibility to infection, malignancy, and death (28). These risks prompted the search for an alternative therapy. Methotrexate had been used in the management of other autoimmune conditions, such as rheumatoid arthritis, with good efficacy and low toxicity (29). In a 1-year prospective study by Matteson et al. (30), methotrexate was started at 7.5 mg/week and increased to 17.5 mg/week over 4–8 weeks. Folic acid was given at 1 mg/day to minimize toxic effects. Of the 11 patients, 82% experienced a response to treatment (defined as improvement in pure tone average of >10 dB or increased speech discrimination at >15% in at least one ear) at 1-year of follow-up (30). A similar study performed by Salley et al. (31) described the response of 53 patients in Methotrexate over an 8-year course. A response rate of 70% was found, according to response criteria similar to those of Matteson et al. (30) with the inclusion of a subjective reduction in vertigo or disequilibrium. Similar results have been observed in other studies (18,32). A study conducted by Lasak et al. (24) involved cytotoxic therapy of Western blot-positive patients with unresponsiveness to steroids or deterioration after steroid taper. These patients received 7.5–15 mg/week of oral methotrexate for at least 6 months or oral cyclophosphamide at 100 mg twice a day. Nonresponders were offered the other medication or azathioprine (Immuran) as an alternative. In total, 63% of patients who receive cytotoxic treatment responded, with modest improvement in mean pure tone average (4.5 dB) and marked improvement in mean speech discrimination (26.2%). Conversely, those patients treated solely with steroids experienced a marked improvement in mean pure tone average (4.5 dB) but only a small improvement in mean speech discrimination (6.9%). On average, those who responded to steroids had a lesser degree of hearing loss before therapy than those who responded to cytotoxic therapy (24). Currently, various regimens are used in our center. Cytoxan is now usually given as intravenous pulsed therapy, but also Embril, CellCept, glutathione precursors, and rarely, plasmapheresis.

Many of the mechanisms and signs of autoimmune inner ear disease remain unknown. For a brief period of time, it appeared that autoimmune sensorineural hearing loss might be associated with cochlear enhancement on the MRI. Zavod et al. (33) investigated this possibility and found that there was no correlation between the presence of antibodies to inner-ear antigens in patients with hearing loss, and cochlear enhancement on MRI scans.

Gupta and Sataloff (34) described a particularly interesting case of noise-induced autoimmune sensorineural hearing loss. This was the first reported case associating autoimmune sensorineural hearing loss with a environmental occurrence such as noise exposure. The patient was a 46-year-old women who developed left-sided hearing loss after exposure to loud music at a party. Over the following decade, she had repeated episodes of temporary hearing loss, usually on the left side, precipitated by noise. Comprehensive neurotological evaluation was negative except for absence of HLA DR4. Initially, Western blot for cochlear antibodies was negative, but a repeat Western blot was positive. Her hearing loss was steroid responsive, and her temporary threshold shifts have been prevented for more than a decade by use of low-dose Imuran (50 mg daily). When she stopped the Imuran and was exposed to noise, she developed hearing loss. As long as she is taking the Imuran, noise exposure has no apparent effect on her hearing. This raises the question as to whether other cases of noise-induced hearing loss may involve an autoimmune mechanism. This question appears to warrant further research.

16.7. Congenital Syphilis

Congenital syphilis can cause sudden bilateral hearing loss (see Chapter 11). More often, it causes unilateral sudden hearing loss or bilateral, asymmetric progressive loss. However, if the patient is unaware that he/she is already deaf in one ear (as is often the case with deafness due to mumps in childhood), syphilitic sudden hearing loss in the other ear will make it appear as if the patient has experienced bilateral sudden hearing loss.

16.8. Unknown Causes

Bilateral sudden hearing loss occurs much less frequently than unilateral sudden hearing loss. Many cases of bilateral sudden hearing loss still remain unexplained. Their causes have been attributed to viruses, vascular rupture or spasm, or toxins, but as yet there is no certain knowledge of the precise mechanism. Cases like those illustrated in Figs. 10.29 and 10.30 occasionally are seen in clinical practice.

17. CAUSES OF SUDDEN UNILATERAL SENSORINEURAL HEARING LOSS

Far more common than bilateral is unilateral sudden hearing loss. There is no satisfactory evidence to establish the specific cause for a large number of such cases. The degree of hearing impairment may range from a slight drop at 4000–8000 Hz to total unilateral deafness. The hearing loss in some instances may disappear completely just as spontaneously as it appeared (and the particular medication used at the time generally gets the credit). Usually, if the hearing loss is going to improve, it does so within 2 or 3 weeks. About 65% of people with sudden unilateral hearing loss improve regardless of treatment. The outlook is not as good if the hearing loss is associated with vertigo. The cause of most of these cases is unknown, but viral and vascular etiologies are believed to be important. Syphilis always must be ruled out, not only because it is one of the

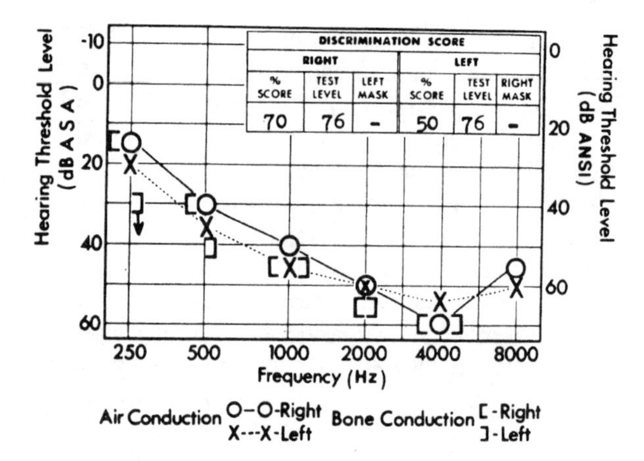

Figure 10.29 *History*: This 40-year-old man noticed sudden deafness after getting chilled and sneezing repeatedly in an air-conditioned restaurant. He had rushing tinnitus but no vertigo. Sounds became fuzzy and distorted. The hearing loss did not clear up with time, nor did it get worse. *Otologic*: Normal. *Audiologic*: Air and bone conduction were reduced. No abnormal tone decay. Recruitment could not be measured accurately. *Classification*: Unknown. *Cause*: Unknown. *Comment*: Sudden hearing loss of this kind with chills and vasospasm is not uncommon but rarely occurs bilaterally and permanently. The patient did not respond to vasodilators or histamine desensitization. A viral cause is unlikely here.

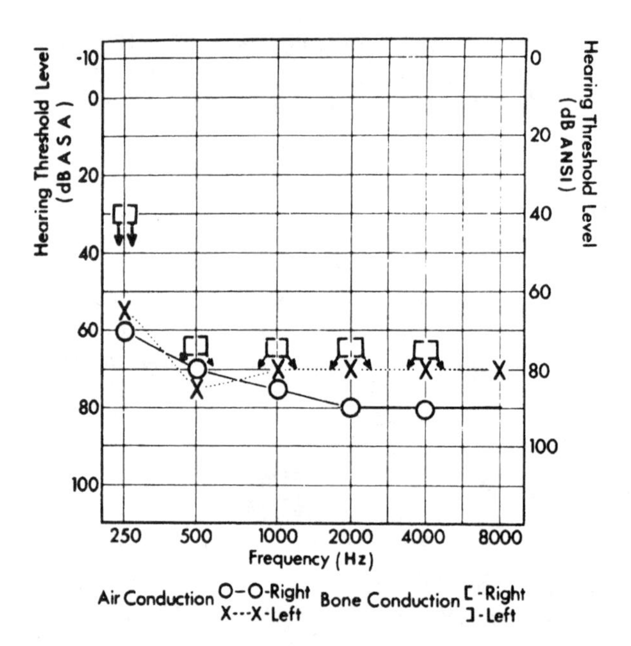

Figure 10.30 *History*: This 8-year-old boy developed speech normally until the age of 5, when he awoke one morning and was severely deaf. This occurred 1 week after a severe virus infection with high fever that might have been encephalitis. He had no other adverse effects. *Otologic*: Normal. No caloric responses. *Audiologic*: Severe air and bone conduction loss. *Classification*: Profound sensorineural hearing loss. *Diagnosis*: Unknown. *Comment*: Delayed deafness caused by virus infection occasionally occurs and may have been the cause here following encephalitis. No treatment is available, but speech therapy is essential to preserve good speech. A hearing aid is of some use.

most readily treatable causes, but also because, if left untreated, it may produce deafness in the other ear. Unilateral sudden hearing loss is, indeed, a most perplexing clinical picture.

17.1. Mumps

Sudden hearing loss may occur with mumps (see Chapter 11).

17.2. Direct Head Trauma and Acoustic Trauma

They are discussed elsewhere in this chapter.

17.3. Meniere's Syndrome

Meniere's syndrome may cause sudden hearing loss.

17.4. Viral Infections

Comparatively little proof is available to show that viral infections produce partial sensory or sensorineural hearing impairment. It is known that mumps and herpes zoster can produce severe sudden hearing loss by attacking the cochlea and the auditory nerve. However, the mechanism is not known with certainty, nor the reason for the loss being almost invariably in one ear. As far as partial sensory or sensorineural hearing loss is concerned, there is a strong feeling among clinicians that patients may sustain this type of damage after such typical viral infections as head colds, mouth ulcers, and influenza. For this reason many clinicians tell patients that their deafness probably was caused by some viral infection.

The authors have seen cases that seems to provide reasonsable clinical proof that viruses can cause certain types of sensory hearing deficiency. For example, we were able to follow very closely an instance of sensory hearing loss associated with a viral upper respiratory infection in an audiologist. The case history is described in Fig. 10.31. This is an interesting example of a cause of hearing loss that is undoubtedly more common than generally is realized. The symptoms of ringing tinnitus, fullness in the ear, and the measurable hearing loss associated with recruitment and diplacusis seem to be related definitely to the typical upper repiratory viral infection that occurred in the patient. It appears that viruses can affect the inner ear and produce temporary partial deafness. The associated ringing tinnitus was found to match the frequency of the hearing loss and confirmed a common clinical impression that tinnitus can be produced by viral infections. The mechanism of these phenomena undoubtedly will be investigated much more extensively in the future. There seems to be justification, however, for considering a viral infection as the possible etiological factor in certain cases of partial sensory hearing impairment.

The clinical impression that viruses may cause sudden deafness in selected cases seems to be convincing (Fig. 10.31). Although no viral studies were performed on this patient, the clinical picture was that of aphthous ulcers with stomatitis, a syndrome often associated with viral infection. The patient has had several recurrences of stomatitis because of the serious complication.

Herpes zoster also is known to cause sudden unilateral deafness of a severe degree. The authors have seen one instance of total loss of hearing that returned to normal in a few days. The patient had a typical case of shingles on his face.

JOSEPH SATALOFF, M.D.
ROBERT THAYER SATALOFF, M.D.
1721 PINE STREET PHILADELPHIA, PA 19103

HEARING RECORD

NAME AGE

AIR CONDUCTION

RIGHT									LEFT							
DATE	Exam	LEFT MASK	250	500	1000	2000	4000	8000	RIGHT MASK	250	500	1000	2000	4000	8000	AUD
1ST TEST			55	60	80	70	80	65								
2 MOS.			45	40	55	60	NR	NR	←	RIGHT STAPES MOBILIZATION.						
6 MOS.			80	95	100	95	NR	NR	←	AFTER ATTACK OF APHTHOUS ULCERS OR HERPES.						

BONE CONDUCTION

RIGHT									LEFT							
DATE	Exam	LEFT MASK	250	500	1000	2000	4000		RIGHT MASK	250	500	1000	2000	4000		AUD
1ST TEST		90WN	5	10	20	25	35									
6 MOS.		90WN	NR	55	NR	NR	NR									

SPEECH RECEPTION

DATE	RIGHT	LEFT MASK	LEFT	RIGHT MASK	FREE FIELD	MIC

DISCRIMINATION

DATE	RIGHT % SCORE	TEST LEVEL	LIST	LEFT MASK	LEFT % SCORE	TEST LEVEL	LIST	RIGHT MASK	EXAM

HIGH FREQUENCY THRESHOLDS

DATE	RIGHT 4000	8000	10000	12000	14000	LEFT MASK	RIGHT MASK	LEFT 4000	8000	10000	12000	14000

RIGHT RINNE	RIGHT SCHWABACH	WEBER	LEFT RINNE	LEFT SCHWABACH

HEARING AID

DATE	MAKE	MODEL
RECEIVER	GAIN	EXAM
EAR	DISCRIM	COUNC

REMARKS

Figure 10.31 *History*: Patient with bilateral otosclerosis had a right stapes mobilization followed by mild vertigo for a few days postoperatively. Patient had a fair hearing improvement, but apparently mobilization was not adequate. About 4 months later he developed either aphthous ulcers or herpes inside his buccal mucosa and on his tongue. During this period he experienced sudden vertigo and roaring tinnitus in his right ear and deafness that has not improved. *Otologic*: Normal, and normal caloric response. *Audiologic*: There was marked depression of air conduction thresholds with no measurable bone conduction responses ("threshold" of 55 dB at 500 Hz is a tactile response). *Classification*: Sensorineural hearing loss. *Diagnosis*: Viral cochleitis and labyrinthitis.

Many cases of sudden unilateral hearing loss that now are attributed to blood vessel spasm or rupture may prove to be caused by viral infections. The peculiar reversibility of some cases of unilateral sudden hearing loss, even when they are very severe, is intriguing. Although now it is customary to consider sensorineural hearing loss to be permanent and incurable, many cases shed doubt on this maxim. Examples are shown in Figs. 10.4 and 10.32. It is hard to believe that such a long-standing hearing loss as that in the subject

JOSEPH SATALOFF, M.D.
ROBERT THAYER SATALOFF, M.D.
1721 PINE STREET PHILADELPHIA, PA 19103

HEARING RECORD

NAME AGE

AIR CONDUCTION

RIGHT										LEFT							
DATE	Exam	LEFT MASK	250	500	1000	2000	4000	8000	RIGHT MASK	250	500	1000	2000	4000	8000	AUD
1st TEST			5	0	0	0	20	30		0	0	0	0	10	25	
46 mos.		80WN	45	40	40	15	25	45		5	-10	-5	-5	15	20	
48 mos.		80WN	60	50	45	15	30	40								
48 mos.		80WN	30	15	10	10	10	35								
55 mos.			20	5	15	15	5	15								

BONE CONDUCTION

RIGHT									LEFT							
DATE	Exam	LEFT MASK	250	500	1000	2000	4000	TYPE	RIGHT MASK	250	500	1000	2000	4000		AUD
46 mos.		80	45	45	35	15	20	WN								
49 mos.		80	45	50	35	20	45	WN								

SPEECH RECEPTION

DATE	RIGHT	LEFT MASK	LEFT	RIGHT MASK	FREE FIELD	MIC.
46 mos.	35	80				
48 mos.	40	80				

DISCRIMINATION

	RIGHT				LEFT				
DATE	% SCORE	TEST LEVEL	LIST	LEFT MASK	% SCORE	TEST LEVEL	LIST	RIGHT MASK	EXAM.
46 mos.	65	65		80					
48 mos.	60	70		80					

HIGH FREQUENCY THRESHOLDS

	RIGHT						LEFT					
DATE	4000	8000	10000	12000	14000	LEFT MASK	RIGHT MASK	4000	8000	10000	12000	14000

RIGHT		WEBER	LEFT	
RINNE	SCHWABACH		RINNE	SCHWABACH
T.F. 512 Hz A>B		→	A>B	

B.C. Comparison to Normal – Poor

HEARING AID		
DATE	MAKE	MODEL
RECEIVER	GAIN	EXAM
EAR	DISCRIM	COUNC

REMARKS

Figure 10.32 *History*: A 50-year-old woman with recurrent attacks of hearing loss, rotary vertigo, and tinnitus in right ear for 2 years. Occasional vomiting with attacks. Marked diplacusis and distortion in right ear during attacks. *Otologic*: Normal. Normal caloric responses. *Audiologic*: Fluctuating hearing loss. Continuous and complete recruitment at 1000 and 2000 Hz during attacks. Normal discrimination when hearing returns to normal. *Classification*: Sensory hearing loss. *Diagnosis*: Meniere's disease. *Comment*: This sensory hearing loss persisted for over 2 months, and then hearing returned.

of Fig. 10.32 can reverse itself. One is always suspicious in such examples and inclined to attribute the cause to psychological factors. In these and many other cases reported in the literature, this attitude merely diverts attention from finding the real cause and explanation for this phenomenon. Viruses play an important role in sensorineural hearing impairment.

The influenza virus has been blamed for many cases of deafness, especially during major epidemics. Here, again, it is hard to find laboratory proof, but the clinical evidence is impressive. In all likelihood, many cases described as auditory neuritis could have been caused by viruses with a predilection for sensorineural tissue, as is true of mumps.

There also is evidence that viral infections during the first trimester of pregnancy can seriously affect the sensorineural mechanism of the fetus and result in congenital hearing loss.

17.5. Rupture of Round Window Membrane, Oval Window Membrane, or Inner Ear Membrane

The membranes of the inner ear or round oval window may rupture, causing hearing loss accompanied at times by vertigo and tinnitus. This occurs most commonly in cases of barotrauma associated with activities such as flying, diving, or severe straining. An interesting case of sudden sensorineural unilateral hearing loss occurred in a patient who blamed it on an attack of severe sneezing (see Fig. 10.33).

17.6. Vascular Disorders

The role of blood vessel spasm, thrombosis, embolism, and rupture as causes of hearing loss still is not clear (see Chapter 11). It is common and logical to attribute a progressive sensorineural hearing loss in an older person to arteriosclerosis or thrombosis. Yet such an explanation rarely is confirmed by histopathological studies. Although physicians continue to blame terminal blood vessels in the ear for many causes of hearing loss, we do not have proof that this is so. Other causes also should be considered.

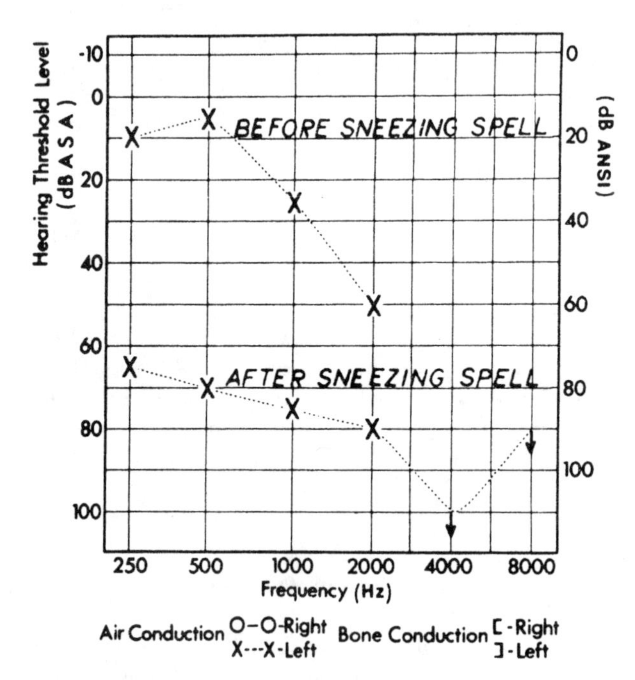

Figure 10.33 *History*: During a severe sneezing spell this 56-year-old man noted ringing tinnius in his left ear, and 2 h later his hearing became bad. The hearing continued to get worse for many months until the ear was practically deaf. He experienced no vertigo. *Otologic*: Normal, and normal caloric responses. *Audiologic*: No abnormal tone decay. Slight separation of Békésy fixed-frequency tracings. Left ear discrimination, 40%. Bone conduction poor; no air–bone gap. *Classification*: Sensorineural. *Diagnosis*: Ruptured round-window membrane observed at surgery.

Sudden unilateral hearing loss may be explained most reasonably on the basis of blood vessel spasm and rupture. In clinical practice one sees quite a few patients with unexplained attacks of sudden unilateral deafness of short duration—or even of longer duration, an attack that may last for several weeks—folowed by a spontaneous return of hearing to normal or improved levels.

Often the hearing losses are accompanied by rotary vertigo, imbalance, and high-pitched tinnitus. In other cases, hearing loss is the only symptom other than a sensation of fullness in the ear. In still other patients, both deafness and tinnitus are permanent, but the imbalance disappears.

Cases of permanent sudden hearing loss in one ear often are blamed on vascular rupture of vascular occlusion, whereas those that recover are attributed to reversible vessel spasm. There are, however, some pitfalls in this explanation. For want of a better understanding at this time, most otologists continue to attribute such cases of deafness to circulatory difficulties, viral diseases, or membrane ruptures.

The role of hypertension and hypotension in deafness also are unknown. However, hypertension associated with hyperlipoproteinemia may be associated with hearing loss, as discussed in Chapter 13. Hypotension seems in some cases to be vaguely associated with progressive high-frequency sensorineural hearing loss. Usually, recurrent attacks of imbalance also are experienced by such patients. Hypotension may predispose to sudden losses in hearing, but here again, convincing proof is lacking. Patients with certain vascular disorders, such as endarteritis, Buerger's disease, diabetes, and others, should have a high incidence of sensorineural hearing impairment, but this has not been borne out by investigations.

The degree of hearing impairment present in patients with sudden unilateral hearing loss varies and may be only 15 or 20 dB and last only a few moments, as most individuals probably have experienced at one time or another when they have felt a sudden fullness and ringing in one ear; or the hearing loss may be up to 70 dB or even total, as described previously. Criteria have been defined to help predict the reversibility of sudden hearing loss. New medical therapies are being developed, and it is necessary for anyone treating this condition to be familiar with the latest literature.

17.7. Following Ear Surgery

The increase in stapes surgery has brought with it a marked increase in the incidence of unilateral sudden hearing loss. Physicians are familiar with the sudden severe deafness that may result as a complication of mastoid and fenestration surgery. The loss is sudden and occurs either at the time of the operation owing to disturbance of the inner ear or shortly after surgery owing to infection. With the advent of stapedectomies, otologists have encountered sudden hearing loss in patients weeks, months, or even years after surgery. Figure 10.34 is an example of a patient who became deaf in his operated ear 4 months after a stapedectomy. Surgical exploration showed invasion of the vestibule with white scar tissue from the middle ear. The hearing in another patient suddenly became poor several months after stapedectomy during a bad infection described as a virus. No unsual findings were encountered in the middle ear or the oval window during exploration, and the perilymphatic fluid in the inner ear appeared to be normal. It is important, therefore, in all cases to inquire about previous ear surgery, but one should not assume that the previous surgery is casually related to the hearing loss. Surgery for tympanosclerosis in which the oval window is opened is a common cause of sudden hearing loss, usually noted within 1 or 2 days postoperatively.

JOSEPH SATALOFF, M.D.
ROBERT THAYER SATALOFF, M.D.
1721 PINE STREET PHILADELPHIA, PA 19103

HEARING RECORD

NAME AGE

AIR CONDUCTION

		RIGHT							LEFT							
DATE	Exam	LEFT MASK	250	500	1000	2000	4000	8000	RIGHT MASK	250	500	1000	2000	4000	8000	AUD
1st TEST		100	55	55	55	35	50	55		40	40	35	40	30	40	
1 YR.		85	10	15	25	15	NR	NR	AFTER RIGHT STAPEDECTOMY							
2 YRS.		100	75	75	80	80	NR	NR	BUZZING TINNITUS, SOUNDS ARE GARBLED.							
2 YRS.		100	75	85	95	90	NR	NR	EXPLORATION: FIBROUS GROWTH INTO VESTIBULE							
3 YRS.		100	NR	85	95	90	NR	NR								

BONE CONDUCTION

		RIGHT							LEFT							
DATE	Exam	LEFT MASK	250	500	1000	2000	4000		RIGHT MASK	250	500	1000	2000	4000		AUD
1 ST TEST		100	0	5	15	10	20			0	5	15	10	20		
2 YRS.		100	NR	NR	NR	NR	NR									

SPEECH RECEPTION

DATE	RIGHT	LEFT MASK	LEFT	RIGHT MASK	FREE FIELD	MIC.

DISCRIMINATION

	RIGHT				LEFT				
DATE	% SCORE	TEST LEVEL	LIST	LEFT MASK	% SCORE	TEST LEVEL	LIST	RIGHT MASK	EXAM

HIGH FREQUENCY THRESHOLDS

	RIGHT						LEFT					
DATE	4000	8000	10000	12000	14000	LEFT MASK	RIGHT MASK	4000	8000	10000	12000	14000

RIGHT		WEBER	LEFT	
RINNE	SCHWABACH		RINNE	SCHWABACH

HEARING AID

DATE	MAKE	MODEL
RECEIVER	GAIN	EXAM
EAR	DISCRIM	COUNC

REMARKS

Figure 10.34 *History*: Four months after a successful right stapedectomy, using a Teflon piston over Gelfoam, this 23-year-old man suddenly noticed a pop in his right ear, and his hearing improved. Several hours later he noted a buzzing tinnitus and distortion. The next day his hearing gradually diminished, became garbled, and was completely gone that evening. Has constant hissing tinnitus but no vertigo. Exploratory surgery showed that the prosthesis was in good position, but a fibrous tissue mass had invaded the labyrinthine vestibule. Removal of much of the mass failed to improve the hearing.

The sensorineural deafness that can follow ear surgery need not be severe or total. Most often it consists of a high-frequency hearing loss that did not exist preoperatively (Fig. 10.35). We generally attribute this to injury to the inner ear during surgery or infection postoperatively. The patient's chief complaint in such situations is reduced ability to understand speech. This is a justified complaint, one that can be demonstrated with discrimination tests. Sometimes a patient with a very successful return of threshold hearing following stapes surgery will appear to be most unhappy about the result and bewilder the surgeon. Figure 10.36 shows such a case, and the discrimination scores show that in spite of the improved threshold level, this patient is far worse postoperatively than she was preoperatively.

JOSEPH SATALOFF, M.D.
ROBERT THAYER SATALOFF, M.D.
1721 PINE STREET PHILADELPHIA, PA 19103

HEARING RECORD

NAME AGE

AIR CONDUCTION

RIGHT									LEFT							
DATE	Exam.	LEFT MASK	250	500	1000	2000	4000	8000	RIGHT MASK	250	500	1000	2000	4000	8000	AUD
STAPEDECTOMY			30	35	50	40	45	60								
			55	60	75	80	NR	NR								

BONE CONDUCTION

RIGHT									LEFT							
DATE	Exam	LEFT MASK	250	500	1000	2000	4000		RIGHT MASK	250	500	1000	2000	4000		AUD
STAPEDECTOMY			-10	0	25	30	30									
			10	30	50	NR	NR									

SPEECH RECEPTION

DATE	RIGHT	LEFT MASK	LEFT	RIGHT MASK	FREE FIELD	MIC.

DISCRIMINATION

DATE	RIGHT % SCORE	TEST LEVEL	LIST	LEFT MASK	LEFT % SCORE	TEST LEVEL	LIST	RIGHT MASK	EXAM.

HIGH FREQUENCY THRESHOLDS

RIGHT							LEFT					
DATE	4000	8000	10000	12000	14000	LEFT MASK	RIGHT MASK	4000	8000	10000	12000	14000

RIGHT		WEBER	LEFT	
RINNE	SCHWABACH		RINNE	SCHWABACH

HEARING AID

DATE	MAKE	MODEL
RECEIVER	GAIN	EXAM.
EAR	DISCRIM.	COUNC.

REMARKS

Figure 10.35 *History*: A 62-year-old woman who had a stapedectomy in right ear and heard well for 2 days. She then developed vertigo and roaring tinnitus, and hearing deteriorated. She complained of distortion and being bothered by loud noise. *Otologic*: Normal. *Audiologic*: Reduced bone conduction postoperatively with recruitment, diplacusis, lowered threshold of discomfort, and reduced discrimination. *Classification*: Sensory hearing loss. *Diagnosis*: Postoperative labyrinthitis.

17.8. Fistula of Oval Window

A fistula in the oval window postoperatively can cause fluctuating hearing loss, recurrent vertigo, fullness, and tinnitus in the ear. This may occur even many years following the surgery. Spontaneous oval or round window fistulas also can occur even in people who have never had ear surgery.

17.9. Following Anesthesia and General Surgery

Hearing impairment following anesthesia and general surgery has been encountered by some otologists, but evidence is unavailable as to which, if either, of these interventions

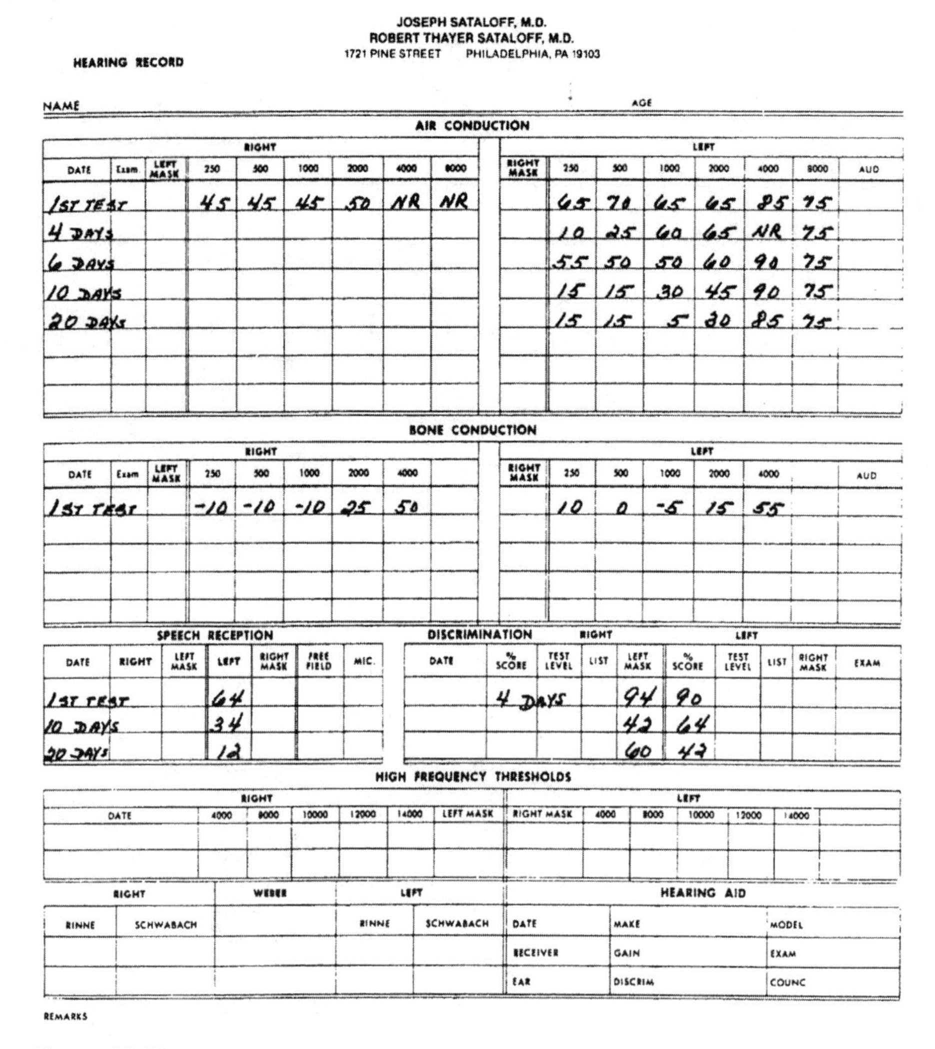

JOSEPH SATALOFF, M.D.
ROBERT THAYER SATALOFF, M.D.
1721 PINE STREET PHILADELPHIA, PA 19103

HEARING RECORD

NAME AGE

AIR CONDUCTION

			RIGHT							LEFT						
DATE	Exam	LEFT MASK	250	500	1000	2000	4000	8000	RIGHT MASK	250	500	1000	2000	4000	8000	AUD
1ST TEST			45	45	45	50	NR	NR		65	70	65	65	85	75	
4 DAYS										10	25	60	65	NR	75	
6 DAYS										55	50	50	60	90	75	
10 DAYS										15	15	30	45	90	75	
20 DAYS										15	15	5	20	85	75	

BONE CONDUCTION

			RIGHT						LEFT					
DATE	Exam	LEFT MASK	250	500	1000	2000	4000	RIGHT MASK	250	500	1000	2000	4000	AUD
1ST TEST			-10	-10	-10	25	50		10	0	-5	15	55	

SPEECH RECEPTION

DATE	RIGHT	LEFT MASK	LEFT	RIGHT MASK	FREE FIELD	MIC.
1ST TEST			64			
10 DAYS			34			
20 DAYS			12			

DISCRIMINATION

	RIGHT				LEFT				
DATE	% SCORE	TEST LEVEL	LIST	LEFT MASK	% SCORE	TEST LEVEL	LIST	RIGHT MASK	EXAM
	4 DAYS			94	90				
				42	64				
				60	42				

HIGH FREQUENCY THRESHOLDS

	RIGHT						LEFT					
DATE	4000	8000	10000	12000	14000	LEFT MASK	RIGHT MASK	4000	8000	10000	12000	14000

RIGHT		WEBER	LEFT	
RINNE	SCHWABACH		RINNE	SCHWABACH

HEARING AID		
DATE	MAKE	MODEL
RECEIVER	GAIN	EXAM
EAR	DISCRIM	COUNC

REMARKS

Figure 10.36 *History*: This patient had an excellent improvement in pure-tone threshold after a left stapedectomy, but she was very unhappy with the result because she could not use the left ear satisfactorily. Speech testing postoperatively showed a real drop in discrimination, which probably was caused by postoperative labyrinthosis. No satisfactory treatment for this is known to the author at present. Note that before surgery she heard 94% of the test words, but after surgery, only 60%. When the test material was presented at a higher level, the discrimination score dropped to 42% owing to distortion.

is the immediate cause. It is not uncommon in otology to find patients who claim that they suffered hearing loss or that their hearing loss was aggravated after some surgical procedure. A causal relationship may be possible, or it may be coincidental.

17.10. Acoustic Neuromas

Approximately 10% of acoustic neuromas present as sudden deafness. It is therefore essential that a complete evaluation be carried out on every patient with sudden deafness searching for the cause that predisposed the deafened ear to injury.

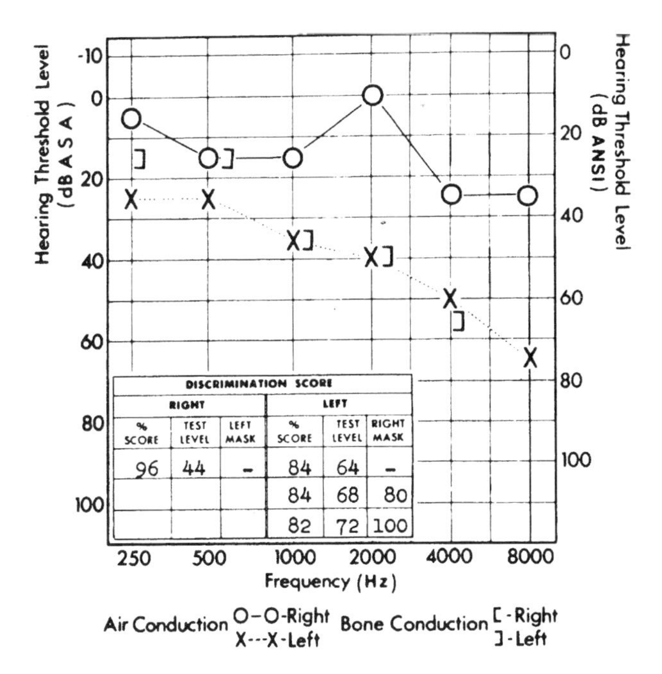

DISCRIMINATION SCORE					
RIGHT			LEFT		
% SCORE	TEST LEVEL	LEFT MASK	% SCORE	TEST LEVEL	RIGHT MASK
96	44	-	84	64	-
			84	68	80
			82	72	100

Figure 10.37 *History*: A 50-year-old patient who developed sudden hearing loss in left ear 3 months before, accompanied by distortion and buzzing but no vertigo of any sort. All symptoms except deafness gradually subsided. *Otologic*: Normal. Only minimal response from both ears after caloric stimulation. No spontaneous nystagmus. *Audiologic*: Left ear hearing loss greater in higher frequencies with reduced bone conduction. Complete recruitment in all frequencies and diplacusis. No abnormal tone decay. *Classification*: Sensory hearing loss. *Etilogy*: Unknown. *Comment*: Findings of complete recruitment and diplacusis and no abnormal tone decay localize the lesion to the inner ear. The good discrimination score, which did not worsen with increased sound intensity, and the cessation of tinnitus and distortion are not indicative of Meniere's disease. Possible causes include vasospasm and viral cochleitis.

17.11. Unknown Causes of Sudden Unilateral Sensorineural Hearing Loss

Because unilateral sudden sensorineural hearing loss is so common in clinical practice, a number of additional examples are presented to help the physician make a better classification and diagnosis (Figs. 10.37–10.39).

18. CAUSES OF SENSORY HEARING LOSS

Certain causes of hearing impairment are known to affect the inner ear primarily and to produce findings characteristic of sensory hearing loss. Because inner-ear damage frequently progresses into the nerve elements, many cases originally starting as sensory progress to the sensorineural type.

The following causes are known to produce sensory hearing impairment. All are characterized, at least in their early stages, by some degree of recruitment and hearing distortion.

1. Meniere's disease
2. Prolonged exposure to intense noise (occupational deafness)

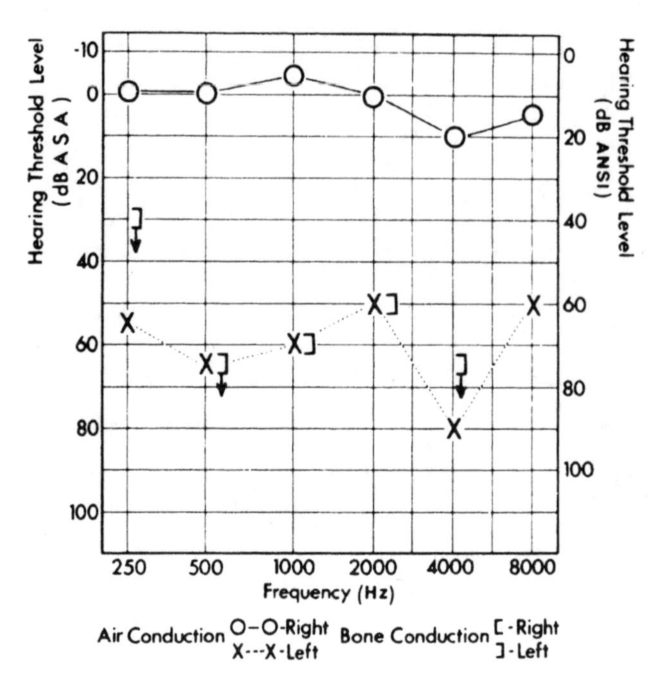

Figure 10.38 *History*: A 45-year-old man with sudden hearing loss in left ear on awakening one morning. Some ringing tinnitus but no vertigo. *Otologic*: Normal. *Audiologic*: Very little recruitment present and no abnormal tone decay. Discrimination about 68% and no diplacusis. *Classification*: Sensorineural hearing loss. *Diagnosis*: Unknown. This damage possibly could be of vascular origin.

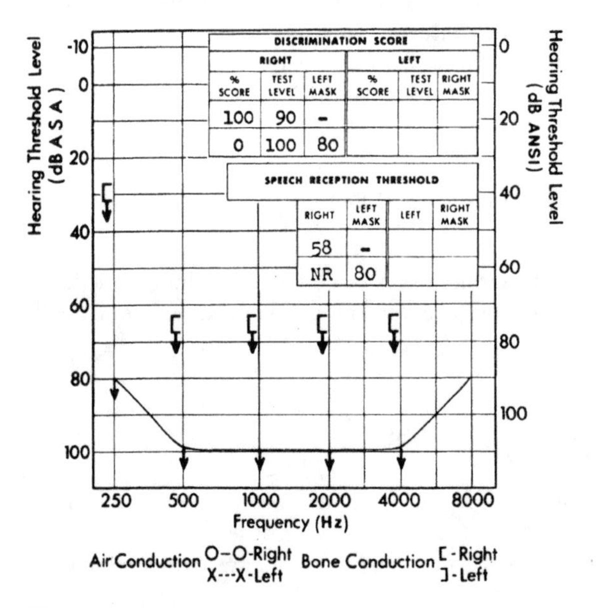

Figure 10.39 *History*: A 45-year-old woman who for several years had experienced a pounding sensation in the right ear whenever she became nervous. Two weeks ago she developed complete deafness, accompanied by momentary severe dizziness and positional vertigo lasting 1 day. Since then she has buzzing tinnitus in the right ear. *Otologic*: Normal. Normal caloric response in right ear. *Audiologic*: No response to pure tones or speech testing with the left ear masked. *Classification*: Sensorineural hearing loss. *Diagnosis*: Unknown, but probably a vascular accident.

3. Acoustic trauma
4. Direct head trauma
5. Ototoxic drugs
6. Virus infections
7. Labyrinthosis following ear surgery
8. Congenital cochlear disease

18.1. Meniere's Disease

Meniere's disease presents all the classic symptoms and findings associated with a sensory hearing loss. There still is confusion concerning the criteria for diagnosing Meniere's disease. Part of the confusion centers about the original findings of the 19th century otologist Prosper Meniere, who reported the autopsy findings on a patient who had suffered from dizziness and deafness. Apparently, what he described and what is now considered to be Meniere's disease are not the same. However, the confusion lies mostly in terminology and the tendency of some physicians to call all cases of vertigo of undetermined origin Meniere's disease.

18.1.1. Terms and Definition

Research studies on patients with vertigo, tinnitus, and deafness have revealed histopathological findings suggesting hydrops of the labyrinth caused by distention with endolymphatic fluid. The cause of the presence of excessive fluid is unknown, but it has now come to be associated with subjective symptoms of Meniere's disease. Because of this hydrops, new names have been applied to the condition. It is variously called labyrinthosis (in contrast with labyrinthitis, which implies an inflammatory involvement) and even labyrinthine hypertension. Common clinical usage now restricts the condition named Meniere's disease to patients who have recurrent attacks of vertigo, hearing loss, and tinnitus. Most often, the hearing loss fluctuates, especially in the lower frequencies and especially early in the disease. Usually the hearing loss is not total, "Seashell" tinnitus is often present.

Atypical Meniere's disease occurs most commonly in four forms. Cochlear Meniere's disease lacks the vertiginous episodes but is typical otherwise. Vestibular Meniere's occurs without the hearing loss. Sometimes these forms are present early but progress to typical Meniere's disease later. Lermoyez's syndrome is Meniere's disease in which vertigo and hearing loss fluctuate in reverse relationship to each other (one condition improves when the other becomes bad) rather than in the usual parallel fashion. Tumerkin's variant, or otolithic catastrophe, in which the patient has abrupt falling attacks of short duration, also is considered a form of Meniere's disease by some clinicians.

Meniere's disease is idiopathic. The diagnosis cannot be made until comprehensive evaluation has ruled out the many specific and often treatable causes of hydrops that produce the same symptoms (Meniere's syndrome).

18.1.2. Symptoms and Their Effects on the Patient

The characteristic findings in Meniere's disease are generally clear-cut, and all are present in most cases. The typical history is marked by a sudden onset of fullness, hearing loss, and a seashell-like tinnitus in one ear. This may last for a period of minutes or several days and then disappear. The same symptoms then may recur at varying time intervals. After several attacks, the hearing loss and the tinnitus may persist. The loss may not be severe, but voices begin to sound tinny and muffled on the telephone, and it becomes difficult for the patient to follow a conversation because of the inability to distinguish between different words that have related sounds. In addition, there is distortion of sound, and the threshold of discomfort for loud noises is reduced. Fullness in the ear is common.

18.2. Vertigo and Tinnitus

Generally, the patient complains of recurrent attacks of vertigo, during which either the room seems to spin, or the patient feels as if he/she is spinning. Everything goes around, and the tinnitus is aggravated severely during the attack. Sometimes nausea and even vomiting may occur.

The vertigo in Meniere's disease is of a specific type that involves some sort of motion. It is "subjective" when the patient has a sensation of moving or "objective" when things seem to move about the patient. Usually, the motion is described as rotary, especially during the acute attack. Occasionally, between acute attacks patients have a mild feeling of motion whereby they seem to fall to one side and have trouble in keeping their balance. Less often, there is a strange up-and-down or to-and-fro motion. Other types of sensation, such as a feeling of faintness or weakness, or seeing spots before the eyes, or just vague "dizziness," should not be attributed to Meniere's disease. The possible causes for these subjective symptoms are numerous, but Meniere's disease is not one of them. There must be some evidence of hearing loss being, or having been, present for the diagnosis to be typical Meniere's disease. It is true that in the very early phases hearing may be lost only during the acute attack and that it may be normal between attacks. But even in such cases the patient will recall having experienced fullness and roaring in one ear. If these subjective symptoms are absent and no hearing loss is present, a diagnosis of Meniere's disease should be made only with the utmost caution, and further studies should explore the possibility of some other cause of the dizziness.

Patients are concerned more frequently about the vertigo and tinnitus than about their hearing loss. For an individual who works on scaffolding or drives a car or a truck, a sudden unexpected attack of vertigo beyond his/her control is indeed a serious problem. These patients are invariably apprehensive and gravely concerned. Nervous tension is a prominent symptom in patients with Meniere's disease. Many of them seem to be much improved and less subject to attacks when they are relaxed and free of tense situations.

The ocean-roaring or seashell-like tinnitus also is a matter of great concern. When patients are in quiet surroundings or when they are under tension, the tinnitus may become so alarming to them that they often are willing to undergo any type of surgery, even if it means the loss of all their hearing in affected ear, provided that the noise can be made to disappear. Under such circumstances the otologist should not let the patient influence him/her to perform irrational and sometimes unsuccessful surgery for the tinnitus. As yet, physicians have no specific, reliable procedure to control tinnitus.

18.2.1. Hearing Loss

Generally, the hearing impairment is not as disturbing to the patient as the other two symptoms, but when it happens to a business person who uses the telephone to conduct business, it becomes an especially important issue.

18.2.2. Audiologic Findings

An audiogram of a typical case of Meniere's disease is shown in Fig. 10.7. Note the low-frequency hearing impairment with reduced bone conduction. If the bone conduction were not reduced, this audiometric pattern would be typical of conductive instead of sensory hearing loss. There is no air–bone gap, because both air conduction and bone conduction are reduced to the same degree. The patient's ability to discriminate in the bad ear is reduced so much that he/she can distinguish only 60% of the speech heard at ordinary levels of loudness. Furthermore, if speech is made louder (from 50 to 70 dB), the patient distinguishes even less (contrary to expectations). This brings out one of the most important features of Meniere's disease: distortion. Distortion explains many of the symtoms, not only the reduced discrimination but also the tinny character that speech assumes in the patient's ears, the inability to discriminate words on the telephone, and diplacusis.

18.2.3. Recruitment

Another interesting phenomenon in patients with Meniere's disease is called a lower threshold of discomfort, and it is manifested by the patient's complaint that loud noises bother him/her. This difficulty is related partially to distortion but principally to the phenomenon of recruitment. Recruitment is a telltale audiologic finding in the diagnosis of inner-ear hearing loss and particularly Meniere's syndrome. It stems from an abnormally rapid increase in the sensation of loudness, and when it is present to a marked degree, it permits the physician to classify a case with reduced bone conduction as being sensory and to localize the damage to the inner ear. In the absence of recruitment, the localization is uncertain, and the condition must be considered to be sensorineural.

Continuous and complete recruitment and sometimes hyperrecruitment occur in Meniere's disease, and according to present-day thinking, they have their origin in disturbances in the hair cells. A simple test for recruitment by means of a tuning fork should be done routinely in cases of reduced bone conduction. Note in Fig. 10.7 that this patient evidences all the phenomena associated with Meniere's disease, including marked recruitment. Abnormal tone decay is absent. The Békésy audiogram (see Chapter 7) also is often characteristic, without a gap between the continuous and the interrupted tone (type II).

18.3. Clinical Studies

The case in Fig. 10.7 is typical Meniere's disease, but there are many cases in which the diagnosis is not quite so certain. For example, Fig. 10.40 shows a patient whose history is suggestive of Meniere's disease. In this patient, however, only the high frequencies were impaired. The audiologic findings were equivocal, except for complete recruitment, and for this reason such a case may be classified as Meniere's disease. the diagnosis was established definitely as the disease progressed. Figure 10.41 shows a difficult and frustrating type of case in which there is bilateral Meniere's disease. Note the comparatively mild reduction in auditory threshold but the severe discrimination. A hearing aid is of no value to this patient because it generally aggravates the distortion, and he sometimes hears more poorly with it, but rarely better. The psychological effects of not being able to understand what his customers were saying and the severe tinnitus have driven this pharmacist to seek psychiatric attention.

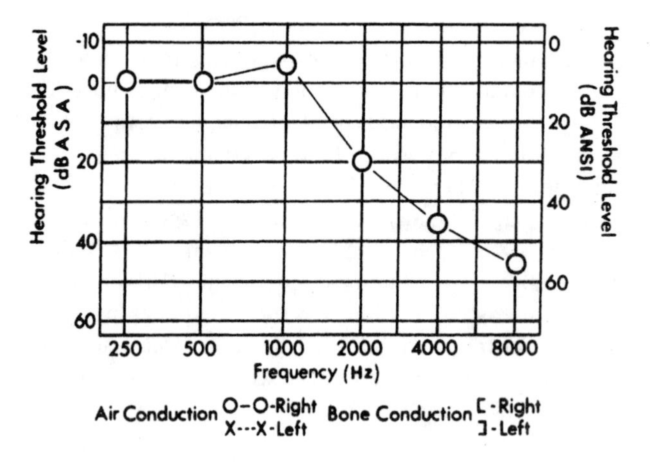

Figure 10.40 *History*: A 42-year-old man with sudden hearing loss and rushing tinnitus in the right ear. Symptoms present for 6 months when first seen for examination. Has slight difficulty distinguishing certain sounds. Feels occasional fullness in the right ear. No vertigo. *Otologic*: Normal with normal caloric responses. *Audiologic*: Left ear normal. Right has a high-frequency loss with complete recruitment at 2000 Hz. No abnormal tone decay. Diplacusis with 2000-Hz tone. *Discrimination score*: Right, 4%. *Classification*: Sensory hearing loss. *Diagnosis*: Meniere's disease. *Aids to diagnosis*: This man worked in a noisy environment, and his unilateral hearing loss originally was misdiagnosed as occupationally related deafness. The presence of complete recruitment, diplacusis, and discrimination difficulty point to sensory hearing loss and probably Meniere's disease without vertigo. In this case the diagnosis was confirmed months later when the patient developed attacks of rotary vertigo. His hearing then changed to levels similar to those in Fig.10.6. Both the unilateral impairment with depression at 8000 Hz and the relatively low noise level (90 dB overall) in the allegedly "noisy plant" mitigated against a diagnosis of noise-induced hearing loss.

18.3.1. Laboratory Tests

Recent work has shown that a number of treatable diseases may be associated with symptoms like those found in Meniere's disease. Inner-ear syphilis, hyperlipoproteinemia, diabetes mellitus, and autoimmune sensorineural hearing loss, lyme disease and hypothyroidism are among the most prominent. These should be investigated, at least by history. Many physicians recommend an FTA-absorption test (not RPR or VDRL), thyroid profile, and screening tests for diabetes and hyperlipoproteinemia in this clinical picture. Collagen vascular disease workup my also be appropriate.

18.3.2. Labyrinthine Tests

Rotary vertigo may occur in both sensory and neural deafness, and when it is essential to differentiate between unilateral inner-ear hearing loss and unilateral nerve hearing loss (possibly due to a tumor), labyrinthine tests are necessary. Not infrequently a patient will experience the vertigo during examination in a physician's office. The patient will complain that the room, or he/she, is starting to turn and that there is a severe noise in his/her ear. At that time, examination of the eyes shows marked nystagmus with a slow and fast component. The nystagmus may be almost in any plane; it may even be oblique, and the direction also may vary. Such a strange type of nystagmus often suggests

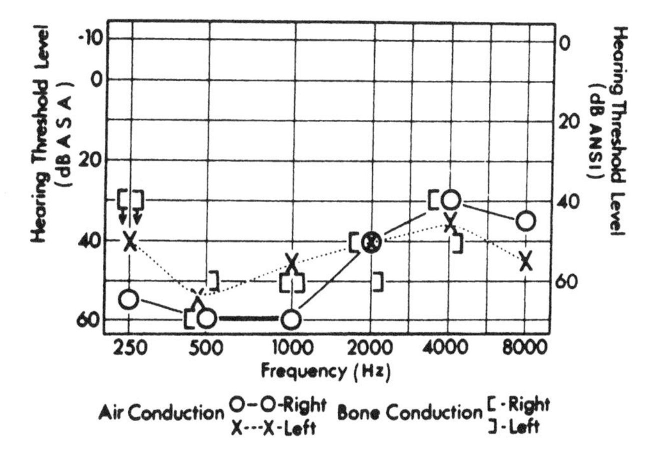

Figure 10.41 *History*: A 31-year-old man with insidious hearing loss for many years with ocean-roar tinnitus and rare mild vertigo. Complains of severe discrimination problem have great difficulty with telephone conversation, get no help from a hearing aid in either ear because of distortion. *Otologic*: Normal. *Audiologic*: Bilateral loss with no air–bone gap. No abnormal tone decay. *Discrimination score*: Right, 32%; left, 36%. Discrimination is reduced when the intensity of the test material is >80 dB. Both ears exhibit lower threshold of discomfortness. *Classification*: Bilateral sensory hearing loss. *Diagnosis*: Bilateral Meniere's disease.

the presence of an intracranial tumor. However, careful watching usually reveals that the nystagmus soon subsides, and the vertigo stops (in contrast with most cases of intracranial involvement). Caloric studies, which should not be performed during the attack, show either diminished or exaggerated labyrinthine responses with nausea and vomiting in Meniere's disease, but the direction and the type of nystagmus are normal with respect to amplitude and direction, unlike the findings in some posterior intracranial fossa neoplasms. Electronystagmograms may be helpful in diagnosing and documenting the vestibular disturbance.

18.4. Treatment

It is beyond the intent of this book to describe treatment of Meniere's disease. However, no specific therapy is available. Many remedies have been suggested as curative, but the spontaneous remissions typical of this disease make it difficult to evaluate the effectiveness of suggested cures. The authors believe that destructive surgery of an ear with Meniere's disease should be performed only as a last resort and, at that, very rarely. Even though discrimination is poor with the residual hearing in the diseased ear, there is always the chance that the other ear may become involved, and thus destructive surgery in the ear first affected could produce a serious handicap. Also, the chance that a specific cure might be found at some future time should be borne in mind and should deter indiscriminate destructive surgery. Vestibular nerve section and endolymptrolic sac surgery may be a better option when surgery is warranted, unless the ear is profoundly deaf.

18.5. Prolonged Exposures to Intense Noise (Occupational Deafness)

Intense noise can produce hearing loss. If the exposure is brief, such as the effect on the ear of a single pistol shot, explosion, or firecracker, the sudden hearing loss produced is called acoustic trauma. If the exposure is prolonged, over many months and years, and the hearing loss develops gradually, the condition is called occupational industrial hearing loss. Acoustic trauma and occupational hearing loss are specific types of noise-induced hearing loss.

18.5.1. Two Components

Both acoustic trauma and occupational hearing loss have two components: one is the temporary hearing loss (auditory fatigue or temporary threshold shift [TTS]) that is of brief duration and clears up, and the other is the permanent hearing loss (or permanent threshold shift [PTS]) that remains. When we speak of occupational hearing loss, we really refer to the permanent loss of hearing from prolonged exposure, not the temporary loss.

18.5.2. Clinical History and Findings

The broad aspects of occupational deafness are discussed in detail in Chapter 14 because of the growing importance of the subject and because the physician will be called upon more and more to express an opinion in cases of this type. In this section, the clinical history and findings that the physician encounters in practice are presented.

The pattern of occupational hearing loss is such that physicians generally do not see patients with this complaint until the impairment has become somewhat disturbing in daily communication. This means that the high-frequency hearing loss is advanced and irreversible. These two points indicate the importance of preventing occupational deafness.

Occupational hearing loss generally develops in a well-defined manner. In the very early stage only the frequencies around 3000, 4000, and 6000 Hz are affected. This is the C-5dip (so called because 4000 Hz corresponds to C-5, or the fifth C, in the normal audiometric testing range seen in the audiogram) (Fig. 10.8). Note that the 8000 Hz frequency is normal. The fact that hearing at the higher frequencies remains normal is an important distinction from presbycusis, which classically progresses from the higher frequencies to the lower ones. This does not necessarily mean that all cases of dips are due to intense noise exposure; they are not (Figs. 10.42 and 10.43). Nor does it mean that the hearing loss is not due to noise exposure if the highest frequencies are damaged, because as exposure to intense noise continues, the frequencies on either side of 3000, 4000, and 6000 Hz also become affected. The classic course of progressive sensorineural hearing loss caused by noise exposure is illustrated in Fig. 10.8. This takes place generally over a period of many years. Susceptible subjects in rare instances may develop some hearing loss after a few months of exposure if the noise is exceptionally intense. There is no valid evidence that noises <90 dB are responsible for clinically significant hearing losses, even after many years of exposure.

The C-5 dip pattern is the most common, but not the only early finding in noise-induced hearing loss. Exposure to certain types of noise produces the most damage principally in the speech frequencies (Fig. 10.44). Communication handicaps manifest themselves much earlier in such a case.

Actually, the degree and the type of hearing loss depends on numerous factors such as the intensity and the spectral characteristics of the noise and the time relation

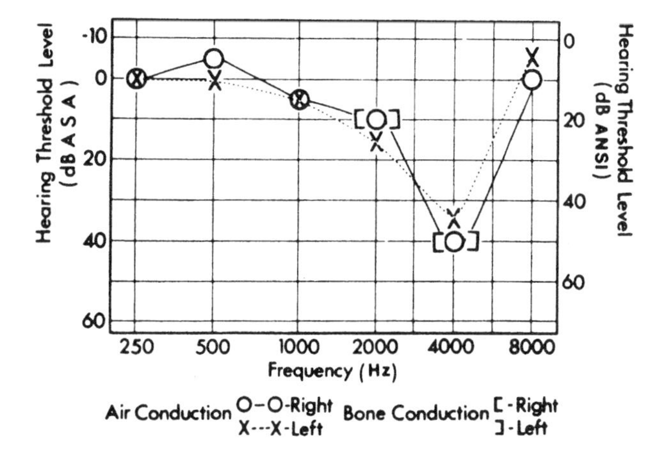

Figure 10.42 *History*: A 40-year-old woman with ringing tinnitus for the past year. No trauma. No noise exposure, and patient denies having had any serious illness or infection. Hearing loss is nonprogressive. *Otologic*: Normal. *Audiologic*: Normal thresholds, except for bilateral 4,000 Hz dip. *Classification*: Sensory hearing loss. *Diagnosis*: Unknown but probably viral cochleitis.

of the noise, that is, its suddenness, its intermittent or continuous character, the duration of exposure, and the little-understood factor individual susceptibility to intense noise.

Extensive studies show conclusively that the early damage caused by intense noise takes place in the outer hair cells of the basilar turn of the cochlea. The site of damage makes it a sensory hearing loss, and this diagnosis in a given patient is often confirmed by showing the presence of recruitment and diplacusis. As the damage progresses, and the loss exceeds ~50–60 dB, the inner hair cells also become involved, and then the supporting cells become damaged. In some cases there also appears to be a decrease in the number of nerve fibers supplying the region of hair cell injury. However, clarification of the reason for and clinical significance of this observation requires additional research.

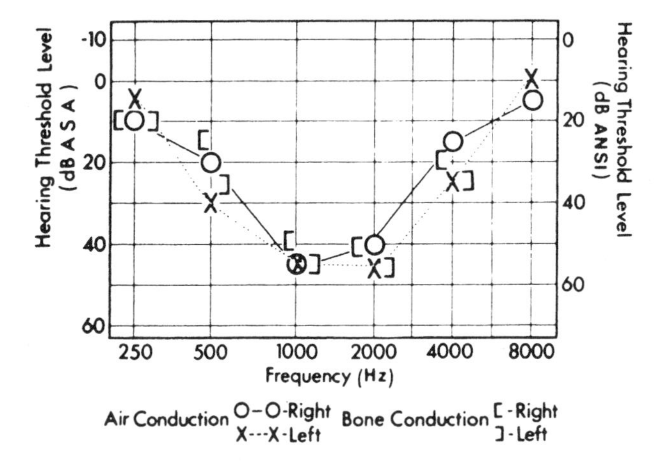

Figure 10.43 *History*: A 50-year-old woman noted gradual hearing loss 1 year ago. No exposure to intense noise. No vertigo or tinnitus. Hearing loss is progressive. *Otologic*: Normal. *Audiologic*: Hearing loss is chiefly in the speech frequencies, and discrimination also is reduced. *Classification*: Sensorineural hearing loss. *Diagnosis*: Unknown.

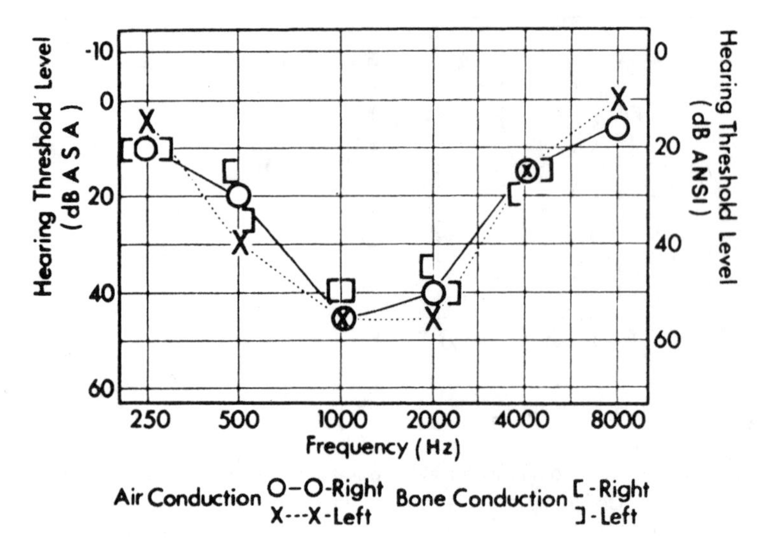

Figure 10.44 *History*: A 48-year-old man employed in a large mill with high noise levels. The noise spectrum is such that the speech frequencies are affected more than the higher frequencies. Fellow employees have similar hearing losses. *Otologic*: Normal. *Audiologic*: Bilateral middle-frequency dip with no air–bone gap. There is some evidence of recruitment but no abnormal tone decay. *Classification*: Bilateral sensory hearing loss. *Diagnosis*: Noise-induced hearing loss.

In advanced cases of occupational hearing loss, it may be difficult to distinguish the portion of the loss caused by noise from the portion that is caused by presbycusis. If, in a specific case, a reasonably good hearing level can be recorded in the frequencies >8000 Hz, the diagnosis could be noise-induced hearing loss. However, if the threshold level at 8000 Hz is about the same or worse than it is at 4000 Hz, then causes other than noise exposure should be considered, unless the employee was exposed for many years to high-frequency, high-intensity noise such as that from chipping and high-pressure air. Frequencies >8000 Hz also should be tested.

Because auditory fatigue can be a factor, it is advisable to evaluate the amount of occupational hearing loss present after the employee has been free of exposure to intense noise for at least 14 h.

Ringing tinnitus is found commonly in acoustic trauma but not very frequently in occupational hearing loss. Slight tinnitus may be present each night but disappears when the auditory fatigue subsides. Vertigo never is caused by long-term occupational noise exposure. If vertigo is present, its true cause should be sought.

Sudden hearing loss is not caused by continuous exposure to occupational noise. If deafness is sudden, and especially if it affects only one ear, another cause should be sought. Military as well as civilian personnel on gunfiring duty, where the noise is intense, may experience much TTS after a day's exposure and perhaps some mild PTS, but more serious PTS involving the speech-frequency range occurs only after repeated exposure. When a person is taken out of a noisy environment, his/her hearing improves at least partially. If, after removal from noise, hearing continues to get worse over a period of months or years, other causes should be investigated.

The term "occupational hearing loss" is somewhat misleading. It implies the presence of obvious difficulties in the ability to hear speech. Actually, the difficulty more often lies in *understanding* speech rather than in *hearing* it. This results from loss of the hearing in the high frequencies, which is the characteristic finding in occupational hearing loss. As many of the consonants that give meaning to words occur in the higher frequencies, it is natural that people who do not hear these frequencies or hear them feebly should be handicapped in understanding certain speech sounds, especially when they are poorly enunciated or masked by a noisy environment. Some of these speech sounds are s, f, t, z, ch, and k. Therefore, unless one specifically looks for this lack of consonant discrimination, the presence of hearing loss is likely to be overlooked, because intense efforts generally are made by the employee to compensate for his/her handicap of reduced discrimination.

18.6. Temporary Hearing Loss

Although temporary hearing loss (auditory fatigue), or TTS, is included in this section on sensory hearing loss, its true position is not certain. Animal experiments and even some observations in humans indicate that hair cells may not be involved in TTS at non-damaging intensities but that nerve fibers are involved. In human beings, TTS differs in many respects from PTS.

There still is no proof that the relationship is simple. For instance, the hearing of many workers who have sustained some degree of temporary threshold shift daily for many years nevertheless returns to normal and remains normal. It is also known definitely that it is not possible to predict the sensitivity of an individual to intense noise by determining TTS characteristics, such as its degree or the rapidity of its return to normal. Until more definite and valid information is available, it is safer to consider the relation between TTS and PTS as still unresolved.

The degree of TTS depends to a great extent on the intensity and the duration of the stimulus in addition to the spectral configuration. The rate of recovery from TTS varies greatly in individuals but seems to be about the same in both ears in the same individual. There is some difference of opinion as to how long it takes for TTS to disappear after long exposure to intense noise. The estimates vary from several days to several months, but there is as yet to well-controlled study to establish the facts. Figure 10.45

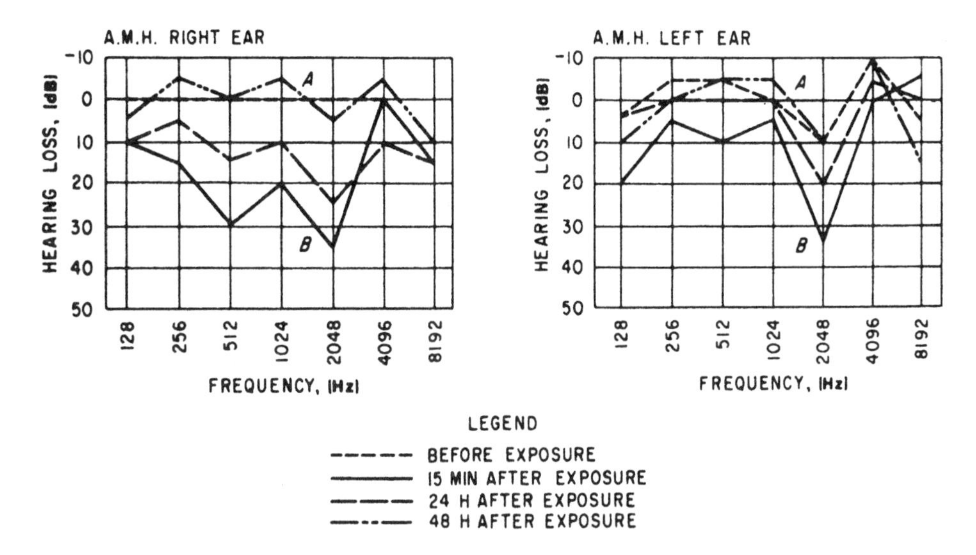

Figure 10.45 Subject was exposed to intense noise of a certain type of jet engine for a long period. The hearing loss (B) returned to its original level (A) after 2 days of rest. During this period the subject experienced ringing tinnitus and much distortion with reduced discrimination.

describes the recovery studies in an interesting case of TTS occurring principally in the speech frequencies.

18.7. Acoustic Trauma

Acoustic trauma, that is, hearing loss caused by a sudden loud noise in one ear, may be produced by firecrackers, cap guns, and firearms, and it is commonly seen in practice. Unless the noise is directed very close to only one ear, the opposite ear also shows a slight hearing loss. In most cases of acoustic trauma the hearing loss is temporary, lasting only a few hours or days, and the hearing returns to normal. Generally, such patients do not reach the otologist, but if they do, it is interesting to watch the return of hearing by taking follow-up audiograms. In some cases of hearing impairment caused by head trauma and acoustic trauma, two types of hearing loss are present: a temporary loss and a permanent loss. As the hearing levels are observed over several days, it is noted that the temporary hearing loss disappears, and a considerably reduced hearing loss that is permanent remains.

Usually, when the hearing loss has persisted for many weeks, it can be considered to be permanent. Some of these cases have a medicolegal aspect. When the hearing loss exceeds 70 dB and involves the speech frequencies, the auditory nerve fibers as well as the inner-ear mechanism undoubtedly are involved in the permanent hearing loss.

In order to differentiate sudden hearing loss due to brief exposure to intense noise from the gradual loss caused by prolonged exposure over many months and years, the term "acoustic trauma" is restricted to the former, and the latter is called industrial or occupational hearing loss.

In acoustic trauma the patient usually is exposed to a very intense noise of short duration like a rifle shot. This causes immediate hearing loss accompanied by fullness and ringing in the ear. If the cause is an explosion, there may also be a rupture of the eardrum and disruption of the ossicular chain. If this occurs, a conductive hearing loss is caused immediately without much serious sensory damage, because the middle-ear defect now serves as a protection for the inner ear.

Following acoustic trauma to the inner ear, the patient usually notes that the fullness and the ringing tinnitus subside, and hearing improves. Generally, the hearing returns to normal. Many human beings have been exposed to gunfire at one time or another and have experienced some temporary hearing loss, only to have their hearing return to normal. In some cases, however, a degree of permanent hearing loss remains. The amount of loss depends on the intensity and the duration of the noise and the sensitivity of the ear. Usually, the permanent loss is very mild and consists only of a high-frequency dip. If the noise is very intense and the ear is particularly sensitive, the loss may be greater and involve a broader range of frequencies. The milder cases of hearing loss involve only one ear, usually the one closer to the gun or the source of the noise. If the noise is very intense and the hearing loss is moderate, then usually both ears are affected to an almost identical degree, or perhaps one ear slightly more than the other. It is hardly possible (as a result of exposure to intense noise) to have a sensory hearing loss in one ear greater than $\sim$50 dB in all frequencies with normal hearing in the other ear. This has important medicolegal aspects.

Because there is almost always some degree of temporary hearing loss or fatigue in acoustic trauma, the amount of permanent damage cannot be established until several months after exposure. In the interim, the individual must be free of other exposure to intense noise that might aggravate the hearing loss. The audiometric patterns in acoustic

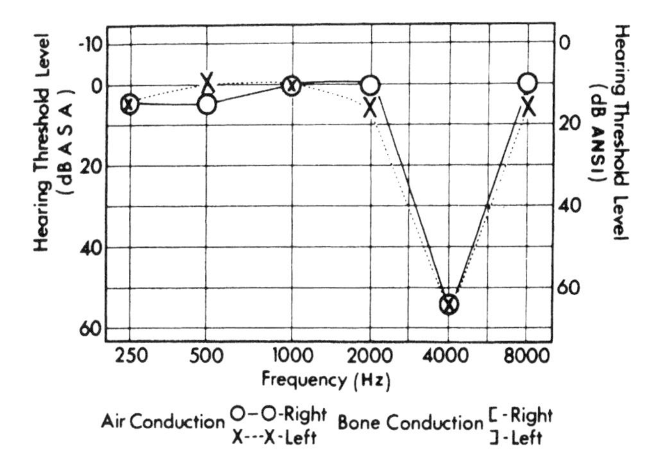

Figure 10.46 *History*: A 9-year-old boy with no clinical symptoms. The hearing defect was detected by the school nurse during routine screening. A year ago this child had been exposed to very loud cap pistol fire, immediately producing deafness and ringing tinnitus in both ears. Shortly thereafter the ringing and the deafness subsided. *Otologic*: Normal. *Audiologic*: The audiogram shows a typical bilateral C-5 dip, characteristic of acoustic trauma. There was no abnormal tone decay. Monaural loudness balance tests show almost complete recruitment:

Right		Left	
2000 Hz	4000 Hz	2000 Hz	4000 Hz
0	55	0	55
15	65	15	70
30	70	30	75
45	75	45	80
60	75	60	85
75	80	75	85

Classification: Sensory hearing loss. *Diagnosis*: Acoustic trauma.

trauma are similar to those in occupational hearing loss, but the history is different, and probably the manner in which the permanent hearing loss is produced also is different. Figure 10.46 shows a typical case of acoustic trauma. Note that recruitment is present, it is complete and continuous, and there is no evidence of abnormal tone decay.

After the temporary hearing loss subsides and only permanent loss remains, the hearing level stabilizes, and according to most investigators, there is no further progression in the hearing loss, although recent research has raised questions about this long-held belief (35).

In children, comparatively little information is available concerning the sensitivity of the inner ears of children to loud noises. The infrequency of exposure to noises loud enough to produce acoustic trauma and the difficulty in testing the hearing of young children accurately may account for this paucity of information.

The example of three patients referred by a school nurse within several weeks dramatizes that more attention should be given to the effect of loud noises on the hearing acuity in children. In all three children, permanent hearing defects were produced by inadvertent exposure to cap pistols or firecrackers. The hearing defects were of the inner ear and resembled those encountered among military personnel who have been exposed to gunfire. The loss was in higher frequencies and occurred predominantly in one frequency, the so-called C-5 dip at 4000 Hz. the hearing loss in such cases is not progressive, but one established, it is irreversible. Middle-ear damage rarely accompanies this type of inner-ear involvement, unless the source of the noise is very close to the ear and is of a very intense low frequency (See Figs. 10.46–10.48). Hearing loss also can be caused by toys (36).

All three patients have been studied closely, and numerous audiograms have been recorded. Repeated hearing tests demonstrated that there was no evidence of progressive hearing loss in any of the youngsters. As might be expected, only the child with substantial bilateral hearing loss presented symptoms of clinical hearing impairment. It is not uncommon to find adults, as well as children, who are completely unaware of even profound unilateral hearing losses if the defects have been present since early childhood. The hearing defects in these three patients were found during the routine testing of 800 schoolchildren.

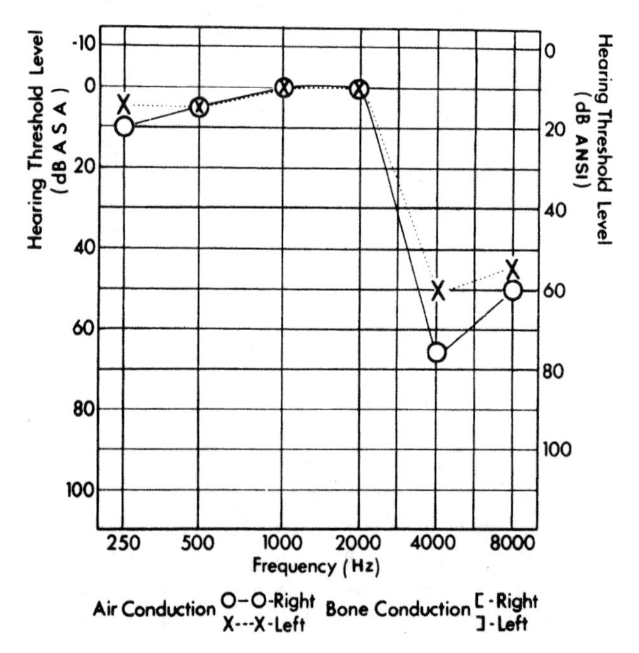

Figure 10.47 *History*: A 16-year-old boy referred because of a slight ringing tinnitus and difficulty in understanding speech in a noisy environment. This difficulty, so typical with high-frequency loss, was aggravated when several people were speaking simultaneously. He also had difficulty in hearing on the telephone. At the age of 6 he had experienced a severe exposure to a cap pistol fire quite close to his ears. From that time he was aware of constant ringing tinnitus and hearing impairment. At the age of 10 his symptoms were aggravated by close exposure to the firing of a 0.22-caliber pistol. *Otologic*: Normal. *Audiologic*: Bilateral C-5 dip with reduced thresholds at 8000 Hz also. This loss is sufficient to produce difficulty in discriminating certain consonants, particularly the sibilants. The difficulty is more pronounced in the presence of ambient noise or speech, caused by the masking effect that these produce on the speech of nearby speakers. *Classification*: Sensory hearing loss. *Diagnosis*: Acoustic trauma.

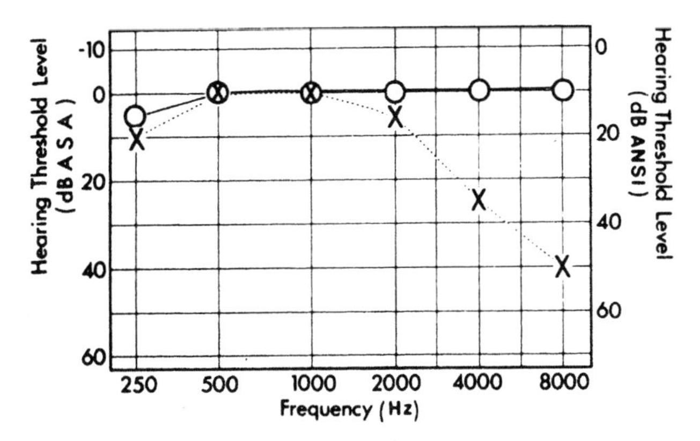

Figure 10.48 *History*: A 12-year-old boy with no clinical symptoms referable to the ears. His hearing defect was detected during the school screening program. Several years ago he experienced sudden hearing loss and ringing tinnitus as a result of the loud report of a cap pistol fired close to his left ear. Both symptoms subsided the same day without recurrence. *Otologic*: Normal. *Audiologic*: Right ear normal. Left ear has a high-frequency loss with recruitment. *Classification*: Sensory hearing loss. *Diagnosis*: Acoustic trauma.

18.8. Direct Head Trauma

A direct blow to the head can produce hearing loss that may belong to almost any classification. To some extent, the type as well as the degree of hearing impairment depends on the severity and the location of the head blow. In general, the harder the blow and the more directly it hits the temporal bone, the more severe the damage and the more likely it is to involve the sensorineural mechanism. Hearing loss may be present with or without evidence of any fracture in the skull. Dizziness, tinnitus, hearing loss, and even facial paralysis may occur following trauma, often in association with headache, memory loss, lethargy, irritability, and other neurologic complaints. For many years, auditory and vestibular symptoms following trauma were considered psychogenic. However, organic causes for these complaints have been established and accepted for at least the last two decades (37,38). Following severe head injury, the mechanisms are clear. They include localized middle- or inner-ear injury from direct trauma to the ear and temporal bone, labyrinthine concussion, injury to the seventh and eighth neurovascular bundles (ipsilateral or contra coup), and injury to the brain stem or higher pathways. In cases of severe head injury, isolated auditory and vestibular symptoms are unusual, but severe dizziness and ataxia, inability to discriminate or process speech signals, tinnitus, and other otologic problems occur frequently in association with other signs and symptoms of neurologic injury. Vestibular symptoms associated with mild head trauma and "inner-ear concussion" are also well recognized (39). When neurotologic symptoms occur following minor head trauma, they may be the most prominent posttraumatic complaints, or they may be subtle. Consequently, all patients with symptoms of hearing loss, tinnitus, dizziness, or facial nerve dysfunction following head trauma should undergo comprehensive neurotological evaluation.

Neurotological examination begins with a comprehensive history. The history must include not only information about the ear and otologic complaints, but also a complete description of the injury, general medical history, and any other information that might help elucidate true causation.

The Head Injury. Teleologically, the ear is an extremely important structure. It is deeply embedded in the head and protected by the otic-capsule, the hardest bone in the body. Hearing and balance were critical to the survival of animals and prehistoric humans. Consequently, they are well protected. In general, if the head injury is not severe enough to cause loss of consciousness, it is unlikely to cause significant, measurable hearing loss, although dips in the 3000–6000 Hz range may occur. However, other otologic complaints such as difficulty processing sentences, tinnitus, and dizziness occasionally occur with lesser injuries. The ear is more readily injured by temporal and parietal blows than by those directed at the frontal or occipital bone. Consequently, it is important to determine the exact nature of the injury, the point at which the head was struck, the force of the injury, whether there was rebound or whiplash injury, whether loss of consciousness occurred, and whether there had been any previous episodes of head trauma or otologic symptoms. As hearing loss, dizziness, and tinnitus are often not noticed until a day or more following injury, the time of onset of symptoms seems less helpful than one might expect. However, it is important to establish, especially if symptoms are noticed immediately at the time of the accident (in which case the trauma is the most likely cause), or many weeks after the accident (in which case a post-traumatic etiology is somewhat less likely).

Longitudinal Temporal Bone Fracture. A moderate sensorineural type of hearing loss due to direct trauma can occur even without evidence of temporal bone fracture. However, it usually accompanies a longitudinal temporal bone fracture. Interestingly enough, a sensory hearing impairment may be found in the ear on the side of the head opposite to the side that sustained the injury (contre coup). The hearing loss has almost the same characteristics as that produced by exposure to intense noise. If the head injury is comparatively mild, only the hair cells of the basilar end of the cochlea are affected, and the effect on the hearing is like that shown in Fig. 10.49. Note the complete recruitment, which localizes the principal site of the damage to the inner ear. If the injury is more severe, a greater area in the cochlea is affected, and even the nerve itself also may be damaged, with the classification then becoming sensorineural.

Transverse Fracture of the Temporal Bone. Another cause of sensorineural hearing loss is transverse fracture of the temporal bone. The fracture line on X-ray examination is perpendicular to the superior petrosal sinus and the long axis of the temporal bone. A severe blow to the back of the head can produce this type of fracture. The accompanying hearing loss is very severe, and often it is total. It is caused by fracture into the vestibule of the inner ear and destruction of the cochlea. Blood fills these areas and can be seen through an intact eardrum (in contrast with the torn eardrum usually caused by longitudinal fracture).

In many cases, damage to the facial nerve causes a complete facial paralysis, including the forehead on one side. Cerebrospinal fluid may fill the middle ear and drain out of the eustachian tube, especially if the eardrum is intact. Vertigo and nystagmus invariably are present after a transverse fracture of the temporal bone, and the hearing loss generally is permanent.

A purely neural type of hearing impairment can occur occasionally with transverse fracture of the temporal bone when the fracture includes the internal meatus and crushes the auditory nerve. Generally, facial paralysis also is present but may clear up (see Fig. 10.50).

Appraising Loss in Claims. The inner ear is so well protected that a blow to the head must be quite severe to produce hearing loss. When there is evidence of fracture, there is little question about the severity of the blow. However, when there is no visible fracture and hearing loss is present, it may be difficult to determine in some cases that

JOSEPH SATALOFF, M.D., D.Sc
ROBERT T. SATALOFF, M.D., D.M.A.
JOSEPH R. SPIEGEL, M.D.
1721 PINE STREET • PHILADELPHIA, PA. 19103

HEARING RECORD

NAME AGE

AIR CONDUCTION

		RIGHT								LEFT									
DATE	Exam	LEFT MASK	250	500	1000	2000	3000	4000	6000	8000	RIGHT MASK	250	500	1000	2000	3000	4000	6000	8000	AUD
1st test			-5	-10	-5	0		-10		0	85	20	20	15	20		35		40	
8 days											85	30	20	30	30		30		35	
38 days											85	10	35	15	30		20		50	
2 mos.											85	-5	10	10	-10		0		10	

BONE CONDUCTION

		RIGHT							LEFT							
DATE	Exam	LEFT MASK	250	500	1000	2000	3000	4000	RIGHT MASK	250	500	1000	2000	3000	4000	AUD
1st test									85	25	20	15	15		30	
38 days									85	15	5	10	10		40	
6 mos.									85	0	0	5	0		10	

SPEECH RECEPTION THRESHOLD

DATE	RIGHT	LEFT MASK	LEFT	RIGHT MASK	FREE FIELD	DATE	RIGHT	LEFT MASK	LEFT	RIGHT MASK	FREE FIELD

SPEECH DISCRIMINATION

	RIGHT				LEFT						
DATE	% SCORE	TEST LEVEL	LIST	LEFT MASK	% SCORE	TEST LEVEL	LIST	RIGHT MASK	FREE FIELD	AIDED	EXAM
					94	50		85			
					92	42		85			

COMMENTS:

Figure 10.49 *History*: A 30-year-old man fell and struck his head 1 week previously. No unconsciousness or vertigo. Fullness and roaring tinnitus in left ear. *Otologic*: Normal. *Audiologic*: Temporary hearing loss with complete recruitment and no tone decay.

2000 Hz

R	L
0	20
15	30
30	40
45	45

Classification: Temporary sensory hearing loss. *Diagnosis*: Direct trauma to head. It is defficult to predict whether such a hearing loss will return to normal levels, as in this case, but it may do so if the damage is not severe.

JOSEPH SATALOFF, M.D.
ROBERT THAYER SATALOFF, M.D.
1721 PINE STREET PHILADELPHIA, PA 19103

HEARING RECORD

NAME AGE

AIR CONDUCTION

		RIGHT								LEFT						
DATE	Exam	LEFT MASK	250	500	1000	2000	4000	8000	RIGHT MASK	250	500	1000	2000	4000	8000	AUD
			0	0	-5	0	10	5	-	55	65	60	50	80	50	
									85	NR	NR	NR	NR	NR	NR	

BONE CONDUCTION

		RIGHT								LEFT						
DATE	Exam	LEFT MASK	250	500	1000	2000	4000		RIGHT MASK	250	500	1000	2000	4000		AUD
			0	0	-5	0	5		-	5	10	0	5	15		
									85	NR	NR	NR	NR	NR		

SPEECH RECEPTION

DATE	RIGHT	LEFT MASK	LEFT	RIGHT MASK	FREE FIELD	MIC.
			58	-		
			NR	85		

DISCRIMINATION

	RIGHT				LEFT				
DATE	% SCORE	TEST LEVEL	LIST	LEFT MASK	% SCORE	TEST LEVEL	LIST	RIGHT MASK	EXAM
					100	90		-	
					NR	100		85	

HIGH FREQUENCY THRESHOLDS

	RIGHT						LEFT					
DATE	4000	8000	10000	12000	14000	LEFT MASK	RIGHT MASK	4000	8000	10000	12000	14000

RIGHT		WEBER	LEFT		HEARING AID		
RINNE	SCHWABACH		RINNE	SCHWABACH	DATE	MAKE	MODEL
					RECEIVER	GAIN	EXAM
					EAR	DISCRIM	COUNC

REMARKS

Figure 10.50 *History*: Sudden onset of hearing loss in the left ear following a fractured skull with unconsciousness. Left facial paralysis. X-ray film shows a fracture through the left internal auditory meatus. *Otologic*: Normal with no caloric responses. *Audiologic*: Without masking the right ear, there seems to be residual hearing in the left ear by air and bone as well as by speech. When masking is used, the absence of residual hearing in the left ear becomes apparent. *Classification*: Sensorineural hearing loss. *Diagnosis*: Injury to the left auditory nerve.

come to litigation whether the hearing loss was caused by the injury or was present previously. In such cases it is well to recall that if the patient did not exhibit any period of unconsciousness, the chances are that the blow was not severe enough to produce cochlear concussion with significant permanent hearing loss. Vertigo and tinnitus usually are associated with such damage, and both are likely to persist for extended periods after the injury.

It also must be borne in mind that a portion of the conductive and the sensorineural hearing loss resulting from trauma may be reversible and that it is necessary to wait at least several months before appraising the degree of permanent hearing loss.

Cases have been reported in which patients did not become aware of their hearing difficulty until several weeks after the injury, and the impairment seemed to progress. This type of loss is unusual; although it is possible, it is difficult to be certain that the injury is the sole etiology.

Audiologic Findings. In the sensory type of hearing defect produced by trauma, audiologic findings reveal continuous and complete recruitment in the frequencies involved. If the speech frequencies also are involved, the discrimination score is reduced, and generally diplacusis is noted. If, in addition to the inner ear, the nerve endings become damaged, recruitment is not quite so prominent and may appear slightly above threshold.

There is no justification for assuming that merely because a patient sustained a head injury, any hearing loss present was caused by the injury. If the deficit is conductive, either there must be evidence of a lesion of the eardrum, or he audiologic findings must indicate some ossicular chain damage. The latter can be confirmed very readily by elevating the eardrum and examining the ossicular chain. If the hearing defect is sensory or sensorineural, then the audiologic findings must fit the characteristic patterns that have been established for these types of hearing impairment.

Figure 10.51 shows a case in which an individual claimed a hearing loss as the result of a head injury. Because this was a total unilateral loss with normal hearing in the other ear, every effort was made to try to explore this case very carefully. After much investigation it was found that the patient had mumps many years ago and that his hearing difficulty was caused by mumps labyrinthitis and was not due to the injury.

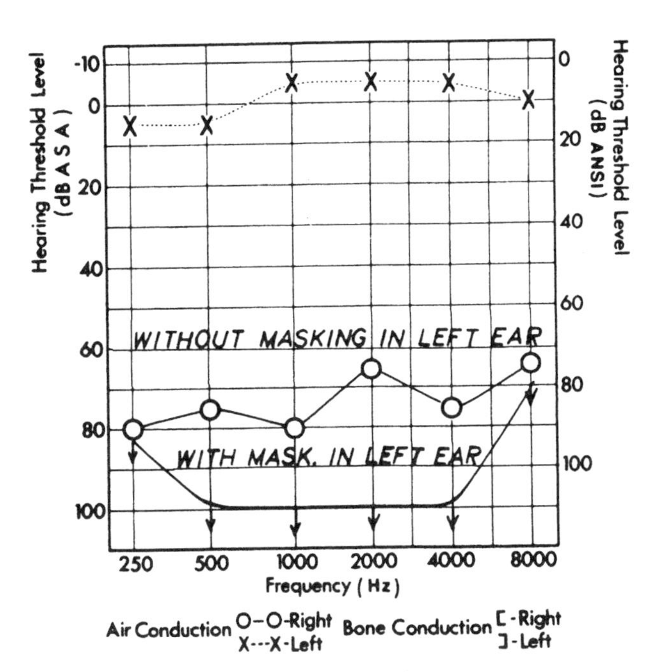

Figure 10.51 *History*: A 33-year-old man who claimed his right ear went deaf after he hit his head on a pole protruding from a building. There was no head wound or unconsciousness. He denied having vertigo or tinnitus. *Otologic*: Normal with normal caloric responses. *Audiologic*: No response in the right ear with the left ear masked. *Classification*: Unilateral sensorineural hearing loss. *Diagnosis*: Deafness caused by mumps. Established by further detailed history. *Comment*: Such a severe unilateral hearing loss with normal labyrinthine responses is not produced by head injury of this type.

Table 10.1 Differential Clinical Findings in Hearing Loss

	Labyrinthine concussion	Longitudinal fracture	Transverse fracture
Bleeding from ear	Never	Very common	Rare
Injury to external auditory canal	Never	Occasional	Never
Rupture of drum	Never	Very common	Rare—commonly a hemotympanum
Presence of cerebrospinal fluid	Never	Occasional	Occasional
Hearing loss	All degrees—partial to complete recovery	All degrees—combined type, partial to complete recovery	Profound nerve type—no recovery
Vertigo	Occasional—mild and transient	Occasional—mild and transient	Severe—subsides; nystagmus to opposite side
Depressed vestibular function	Occasional—mild	Occasional—mild	No response
Facial nerve injury	Never	25%—usually temporary 25%—in squamous and mastoid area	60%—often permanent 60%—in occiput, in pyramid

Source: Ref. (40).

After a better understanding was established with the patient, he freely admitted the situation.

Schuknecht crystallized the differential diagnosis of hearing loss due to head injuries (Table 10.1) (40). Since then, improved *X*-ray technics have greatly increased the percentage of fractures that can be detected. However, some fractures still cannot be visualized.

18.9. Ototoxic Drugs

With the increasing use of ototoxic drugs, it is important to ask every patient with sensorineural hearing loss what kind of drugs he/she has taken, particularly in relation to the onset of hearing loss. Certain drugs can produce sensorineural hearing loss when they are used over a long period of time. Among those drugs reported as ototoxic are streptomycin, neomycin, furosemide, gentamycin, quinine, kanamycin, and dozens of others. Sudden hearing loss from these drugs occurs principally in the presence of impaired kidney function. It may also occur from overdosage. Figures 10.52 and 10.53 provide examples of the severe hearing damage that can result from the improper use of ototoxic drugs. These drugs are very valuable when they are properly used, and they seldom cause hearing damage under such circumstances.

It should be emphasized that the hearing damage usually occurs only after taking the drug systemically for a long period of time. In exceptional cases, such as in patients with renal failure, deafness can result after only a few doses of medication. However, generally the loss is gradual and can be followed audiometrically so that the drug can be stopped before it involves the speech ranges. Ringing tinnitus is a frequently associated symptom and sometimes precedes the hearing loss.

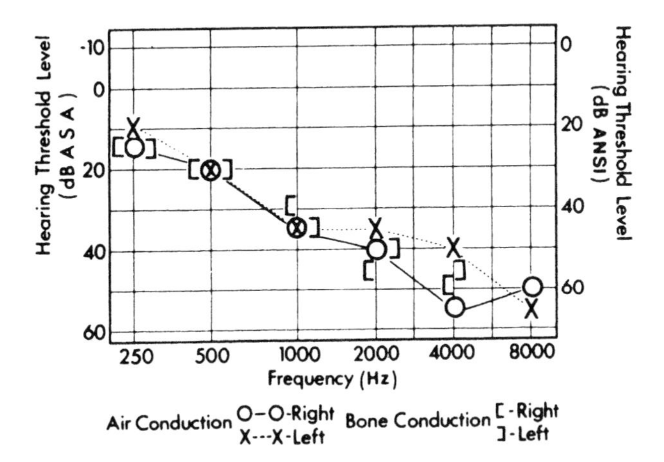

Figure 10.52 *History*: A 10-year-old boy with normal speech and otologic findings. Had a kidney infection and received kanamycin. No progressive hearing loss but has constant ringing tinnitus. *Otologic*: Normal. *Audiologic*: Bilateral downward sloping audiogram with reduced bone conduction. *Diagnosis*: Injudicious use of an ototoxic drug in the presence of kidney dysfunction.

Streptomycin, now used mostly for tuberculosis or for medical labyrinthectomy, is primarily "vestibulotoxic." It is particularly important to keep this drug in mind, however, because the toxicity is related to the total dose of the drug accumulated over a lifetime rather than over a short course. The early impairment results from damage to the hair

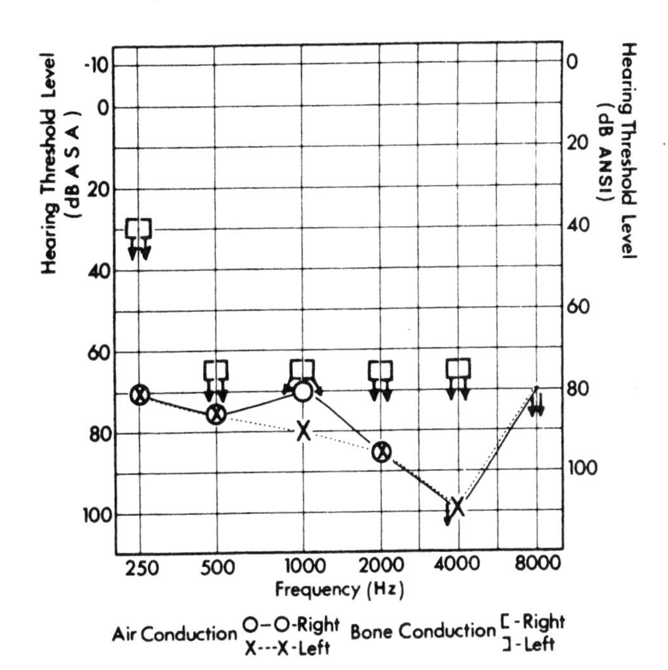

Figure 10.53 *History*: A 49-year-old man who received kanamycin injections daily for 4 weeks because of acute renal failure. Patient claims normal hearing before these injections. *Otologic*: Normal caloric responses. *Audiologic*: Severe to profound bilateral loss with no air–bone gap. No response to speech tests, except for experiencing great discomfort when speech was presented at levels >100 dB (low threshold of discomfort, an indication of recruitment). *Classification*: Sensorineural hearing loss. *Diagnosis*: Deafness caused by ototoxic effects of kanamycin.

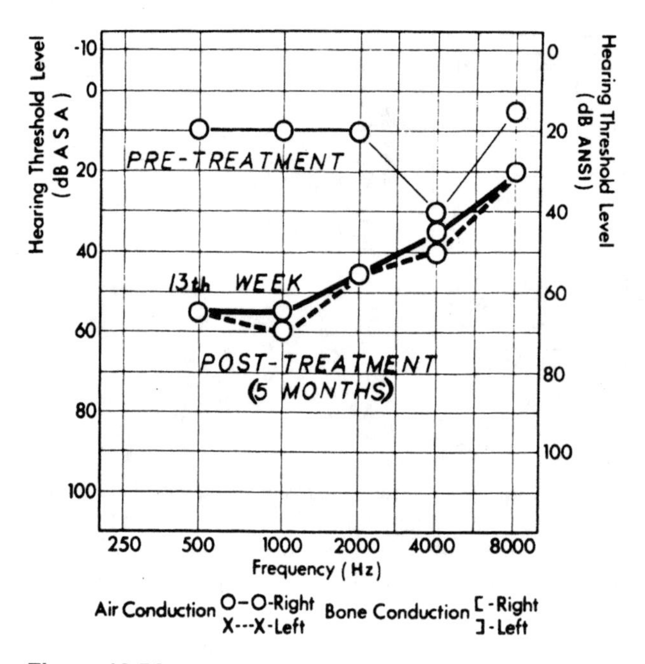

Figure 10.54 *History*: Patient received daily injections of an ototoxic drug for 3 months. His hearing was followed weekly, and in the thirteenth week he developed a ringing tinnitus in the right ear, fullness, and hearing loss. The drug was stopped immediately. No vertigo. *Otologic*: Normal. *Audiologic*: Unilateral ascending hearing loss with recruitment and diplacusis. *Classification*: Sensory hearing loss. *Diagnosis*: Cochlear hearing loss from ototoxic drug.

cells in the inner ear. A classic example of this is described in Fig. 10.54. Most of the time, hearing loss caused by drug ototoxicity is permanent, but it is not progressive. The hearing loss can occur without any vertigo, ringing tinnitus, or other symptoms.

Numerous other drugs and agents may be ototoxic. Roughly 200 such agents have been reported. Some of them are listed in Table 10.2.

18.10. Sensory Hearing Loss after Ear Surgery

The increase in stapedectomies for otosclerosis has focused attention on the sensorineural damage and hearing loss that may be associated with opening the vestibule of the labyrinth, as discussed earlier in this chapter.

18.10.1. Mastoidectomy

Prior to the advent of stapedectomy, the oval-window area and the vestibule of the inner ear were considered to be inviolable, and extreme caution was exercised to avoid disturbing the footplate during mastoid surgery. In spite of this caution, surgical accidents did occur, and hearing often was lost totally in the operated ear during mastoidectomy procedures in which the oval window inadvertently was penetrated. Sometimes the deafness occurred long after surgery as a result of ear infection and cholesteatoma. Partial deafness from toxic labyrinthitis still is observed in chronic mastoiditis even when no surgery has been performed.

Table 10.2 Ototoxic Agents. All medications listed have been reported as associated with hearing impairment except those with an asterisk (*). In the table D denotes dizziness, V denotes vertigo, and T denotes tinnitus The list was compiled from numerous sources.

Accutane capsules (D, T)
Acetazolamide
Actifed w/codeine cough syrup (D, T, V)
Adipex-P tablets and caplets (D)
AerofBid inhaler syrup (D, V)
Alcohol
Aldactazide* (D, T)
Aldoclor tablets (D, T)
Aldomet ester HCI injection (D)
Aldomet oral (D)
Aldoril tablets (D, T)
Alfenta injection* (D)
Alferon-N injection (D, T)
Altace caplets (D, T, V)
Alurate elixir (D)
Amicar syrup, tablets and injection* (D, T)
Amikacin sulfate
Amikin injection
Amocil* (D)
Amphotericin B
Ampicillin
Anafranil caplets (D, V)
Ana-Kit Anaphylaxis Emergency Treatment Kit (D)
Anaprox/Anaprox DST (D, T, V)
Anaproxen sodium
Ancobon capsules
Anestacon solution (D, T)
Anexsia 5/500 tablets (D)
Anexsia 7.5/650 tabs (D)
Aniline dyes
Apresazide capsules (D, V)
Apresoline–Esidrix tablets (D, V)
Apresoline HCI parenteral (D)
Apresoline HCI tablets (D)
Arthritis strength bufferin analgesic capsules
Arthritis strength BC powder (T)
Aristocort suspension (V)
Aristospan suspension* (V)
Arsenic
Asendin tablets (D, T)
Ascriptin A/D capsules (T)
Aspirin (T)
Atgam sterile solution (D)
Ativan injection (D)
Azactam for injection (T, V)
Azo Gantanol tablets* (T, V)
Azo Gantrisin tablets Jenest-28 tabs (D, V, T)
Azulfidine tablets, EN-tabs, oral suspension (T, V)
Bactrim DS tablets* (T, V)
Bactrim intravenous infusion* (T, V)
Bactrim* (T, V)
Bayer Children's chewable aspirin (T)
Bayer Aspirin tablets and caplets (T)
Bayer Plus aspirin tablets (T)
Therapy Bayer aspirin caplets (T)
8-h Bayer timed-release aspirin, BC powder (T)
Bleomycin
Blocarden tablets (T, V)
Brethaire inhaler (V)
Brerylol injection (V)
Bumetanide
Bumex (V)
BuSpar (T)
Butazolidin capsules and tablets (T)
Cama arthritis pain reliever (T)
Capastat sulfate vials (~11%) (T, V)
Capozide (V)
Capreomycin sulfate
Carafate tablets (V)
Carbocaine (T)
Carbon monoxide
Cardene capsules (T, V)
Cardizem SR capsules* 60 mg, 90 mg, 120 mg (T)
Cardura tablets (T, V)
Cartrol tablets (T)
Celestone soluspan suspension* (V)
Children's Advil suspension (T)
Chloramphenicol sodium succinate
Chlordiazepoxide
Chloroform
Chloroquine phosphate
Cholybar* (T, V)
Cibalith-S (T, V)
Cipro intravenous (T)
Cipro tablets (T)
cis-Platinum
Clinoril tablets (T, V)
Colestid granules* (V)
Colistin sulfate
Coly-Mycin* (V)
Coly-Mycin M Parenteral colyte* (V)
Combipres Tablets (V)
Corgard tablets (T)
Cortenema* (V)

(Continued)

Table 10.2 *Continued*

Cortisporin cream
Cortisporin ointment
Cortisporin otic suspension sterile
Cortone acetate sterile suspension (V)
Cortone acetaete tablets (V)
Corzide tablets (T, V)
Cuprimine capsules (T)
Cyclosporine
Cytotec (T)
Dalaone D.P. injectable (V)
Dalgan injection (T, V)
Dapsone USP* (T, V)
Decadron elixir (V)
Decadron-LA sterile solution (V)
Decadron phosphate injection (V)
Decadron phosphate respihaler (V)
Decadron phosphate turbinaire (V)
Decadron phosphate w/Xylocaine injection, sterile (V)
Decadron tablets (V)
Deferoxamine mesylate
Deltasone tablets* (V)
Depen Titratable tablets (T)
Depo-Medrol sterile aqueous Suspension (V)
Deprol tablets (V)
Desferal vials (T)
Desyrel and Desyrel Dividose (T, V)
Diamox parenteral (T)
Diamox sequels (T)
Diamox tablets (T)
Dibekacin
Dihydrostreptomycin
Dilantin w/Phenobarbital kapseals (V)
Dipentum capsules* (V)
Diptheria and tetanus toxoids and pertussis vaccine adsorbed USP (pediatric use)
Diprivan injection (T)
Disalcid (T, V)
Ditropan
Diucardin tablets* (V)
Diulo* (V)
Diunces tablets (V)
Diuril oral suspension* (V)
Diuril sodium intravenous* (V)
Diuril tablets* (V)
Dolobid tablets (V, T)
Duragesic transdermal system (V)
Duranest injections (T)
Dyazide capsules* (V)
Dyclone 0.5% and 1% topical solutions, USP (T)
Easpirin (T, V)
Ecotrin (T)
Edecrin (T, V)
E.E.S. (V)
Elavil (T)
Eldepryl (T, V)
Emcyt capsules* (T)
Eminase* (V)
Empirin with codeine phophate (T)
Emycin tablets
Enalapril
Endep tablets (T)
Enduron tablets* (V)
Enduronyl (V)
Engerix-B Unit-dose vials (V)
Enovid
Equagesic tablets (T, V)
EryPed (V)
Ery-tab tablets (V)
Erythrocin stearate filmtab (V)
Erythromycin
Erythromycin base filmtab (V)
Erythromycin delayed-release capsules, USP (V)
Esidrix tablets* (V)
Esimil tablets* (V)
Eskalith (ethmozin tablets, <2%) (T, V)
Ethracrynic acid
Etretinate
Etrafon (T)
Fansidar tablets* (T, V)
Feldene capsules* (V, T)
Fenoprofen calcium
Fiorinal w/codeine capsules (T, V)
Flagyl intravenous (V)
Flagyl tablets (V)
Flexeril tablets (T, V)
Floxin tablets (V)
Flucytosine
Fluorocitrate
Fungizone intravenous (T, V)
Furosemide
Ganite injection (T)
Gantanol* (T, V)
Gantrisin ophthalmic ointment/solution* (T, V)
Ganrisin* (T, V)
Garamycin injectable (T, V)
Gentamycin
Gold
Halcion tablets (T)

(*Continued*)

Table 10.2 *Continued*

Haldol deconate (V)
Haldol injection, tablets and concentrate (V)
Habistate
Harmonyl tablets
Hexadimethrine bromide
Hedadin
Hexalen capsules (V)
Hibiclens antimicrobial skin cleanser
Hydeltrasol injection, sterile (V)
Hydeltra-T.B.A. sterile suspension (V)
Hydrocortone acetate sterile suspension (V)
Hydrocortone phosphate inj., sterile (V)
Hydrocortone tablets (V)
Hydrodiuril tablets* (V)
Hydromox tablets (V)
Hydromox R tablets (V)
Hydropres tablets (V)
Hylorel tablets
Hyperstat intravenous injection (T, V)
Hytrin tablets (T, V)
Ibuprofen
Ilosone (T)
Inderide tablets (V)
Interide LA long acting capsules (V)
Indocin (T, V)
Intal capsules* (V)
Intal inhaler* (V)
Interferon alfa-26
Intron A (T, V)
Iodoform chemicals
Isomotic (V)
Isoptin injectable (V)
Jenest-28 tablets (D)
Kanamycin
Keflex Pulvules/oral suspension
Keftab tablets (D)
Kemadrin* (D)
Kefurox vials, faspak and ADD-vantage (D)
Kenalog-10 injections (V)
Kenalog-40 injections (V)
Kerlone tables (>2%) (D, T)
Ketalar
Klonopin (V)
Klorves effervescent tablets
Klorves effervescent granules
Klorves effervescent 10% liquid
K-Phos M.F. tablets (D)
K-Phos neutral tablets (D)
K-Phos Original Formula 'Sodium Free' tablets (D)
K-Phos No. 2 tablets (D)
Kwell cream (D)
Kwell lotion (D)
Kwell shampoo (D)
Lanoxicaps (D)
Lanoxin injection pediatric (D)
Lanoxin tablets (D)
Lariam tablets (D, T, V)
Larodopa tablets (D)
Lasix oral solution and intravenous
Lead
Legain (T)
Leuprolide acetate
Levatol (D)
Levien/Tri-Levlen (D)
Levsin/Levsinex* (D)
Levo-Dromoran* (D)
Limbitrol (D)
Lincocin* (T, V)
Lioresal tablets (D, T)
Lithane tablets (D, T, V)
Lithium Carbonate capsules and tablets (D, T, V)
Lithobid tablets (D, T, V)
Lodine capsules (D, T)
Loestrin (D)
Lomotil (D)
Lopid capsules and tablets (D, V)
Lopressor Ampuls (D, T, V)
Lopressor HCT tablets (D, T, V)
Lopressor tablets (D, T, V)
Lorazepam
Lorelco tablets* (D, T)
Lortab ASA tablets (D)
Lortab (D)
Lotensin tablets (D)
Loxitane (D)
Lozol tablets (D, V)
Ludiomil tablets (D, T)
Lupron Depot 7.5 mg (D)
Lupron injection (<5%) (D)
Lysodren (D, V)
Macrodantin capsules (D, V)
Mandelamine
Mannitol
Marax tablets and DF syrup (V)
Marcaine Hydrochloride 0.5% w/epinephrine (D, T)
Marcaine Hydrochloride w/epinephrine (D, T)
Marcaine spinal (D, T)
Marinol (Dronabinol) capsules (D)
Marplan tablets (D, V)

(Continued)

Table 10.2 *Continued*

Matulane capsules (D)
Maxair inhaler (D)
Maxzide* (D, V)
Mebaral tablets (D)
Meclomen capsules* (D, T)
Medrol* (D, V)
Mefenamic acid
Menrium tablets (D)
Mepergan injection (D)
Mephobamate
Mercury
Mesantonin tablets (D)
Methadone hydrochloride diskets (D)
Methergine* (T)
Methodone Hydrochloride oral solution/tablets (D, T)
Methotrexate tablets, parenteral, LPF (D)
Mevacor tablets (D)
Mexitil capsules (D, T)
Micronor tablets (D)
Midamor tablets (D, T, V)
Miltown tablets (D, V)
Minipress capsules* (D, T, V)
Minizide capsules* (D, T, V)
Minocin intravenous* (D, V)
Minocin oral suspension* (D, V)
Minocin pellet-filled capsules* (D, V)
Minocyline hydrochloride
Mintezol (D, T)
Misonidazole
M-M-R (D) vaccine
M-R-V-A-X (D) vaccine
Mobran tablets and concentrate
Modicon (D)
Moduretic tablets (D, T, V)
Mono-Gesic tablets (T, V)
Monopril tablets (D)
Motofen tablets (D)
Motrin tablets (D, T)
Morphine sulfate Contin tablets (D)
MSIR (D)
Mumpsvax
Mustargen (T, V)
Myambutol tablets (D)
Mykrox 2 mg tablets (D, T)
Myochrysine injection (D)
Mysoline (occasional) (V)
Naldecon syrup, tablets, pediatric drops/syrup (D)
Nalfon pulvules and tablets (D, T)
Naphcon A Ophthalmic solution (D)
Naprosyn (D, T, V)
Naprosyn suspension (D)
Naproxen
Nardil (D)
Naturetin tablets* (D, V)
NebuPent for inhalation solution (D, V)
NegGram* (D, V)
Nembutal sodium capsules (D, V)
Nembutal sodium solution (D, V)
Nembutal sodium suppositories (D, V)
NeoDecadron topical cream (D)
Neomycin
Neosporin cream
Neosporin G.U. irrigant sterile
Neptazane tablets (V)
Nescaine/Nescaine MPF* (D, T)
Netilmicin sulfate
Netromycin injection 100 mg/mL (D, T, V)
Nicorette (D, T)
Nipride I.V. infusion (D, T)
Nitro-Bid ointment (D)
Nitrogen mustard
Nitrostat tablets* (D, V)
Nizoral tablets* (D, V)
Norethin (D)
Norgesic (D)
Norinyl (D)
Norisodrine aerotrol (V)
Normodyne injection (D, V)
Normodyne tablets (D, V)
Normozide tablets (D, V)
Noroxin tablets (D)
Norpramin tablets (D, T)
Norzine (D, T)
Novafed A capsules (D)
Novafed capsules (D)
Novahistine DH (D)
Novahistine DMX (D)
Novahistine Elixir (D)
Novahistine expectorant (D)
Novocain hydrochloride for spinal anesthesia (D)
Nubain injection (D, V)
Ogen* (D)
OMIU
Omnipaque (D, T, V)
Optimine tablets (D, T, V)
OptiPranolol sterile ophthalmic solution (D)

(Continued)

Table 10.2 *Continued*

Oramorph SR (morphine sulfate sustained release tabs) (D)
Orap tablets (D)
Oretic tablets* (D, V)
Oreticyl* (D, V)
Ornade spansule capsules (D, T, V)
Ornex* (D)
Opiates
Orthoclone OKT3 sterile solution
Ortho novum (D)
Orudis capsules (D, T, V)
OSM GIYN
Ovcon (D)
P-A-C analgesic tablets* (T)
Pamelor (D, T)
Papaverine HCI vials/ampoules* (V)
Paradione capsules* (V)
Paraplatin for injection
Parlodel (D, V)
Parmade tablets* (D, T)
Pavabid capsules* (V)
Pavabid HP capsulets* (V)
PBZ tablets and elixir (D, T, V)
PBZ SR tablets (D, T, V)
PCE dispertab tablets (D, V)
Pediapred oral liquid* (D, V)
PediaProfen suspension (D, T)
Pediazole (T, V)
PediOptic suspension sterile
Pedvax HIB ampoules
Peganone tablets (D)
Pentam 300 injection (D)
Pepcid (D, T, V)
Pepto-Bismol liquid and tablets* (T)
Percodan* (D)
Periactin (D, T, V)
Permax tablets (D, T, V)
Pfizerpen-AS aqueous suspension (V)
Phenergan w/codeine (D, T)
Phenergan VC w/codeine (D, T)
Phenergan w/dextromethorphan (D, T)
Phenergan suppositories (D)
Phenergan syrup (D)
Phenergan VC (D, T)
Phenergan VC w/codeine (D, T)
Phenobarbital
Phenobarbital elixir and tablets (D, V)
Phenylbutazone
Phenurone tablets* (D, V)
Pipracil* (D)
Pitressin synthetic ampoules* (V)
Placidyl capsules (D)
Plaquenil sulfate tablets (D, T, V)
Platinol (T)
Platinol-AQ injection (T)
Plendil extended-release tablets PMB 200 and PMB 400 (D, T)
Pilocarpine
Polaramine (D, T, V)
Polymyxin B Bulfatel, aerosporin brand sterile powder (D)
Polymyxin B Sulfate
Pondimin tablets (D)
Ponstel (D)
Pontocaine hydrochloride for spinal anesthesia (D, T)
Potassium bromate
Practolol
Prilu* (D)
Prilosec delayed-release capsules (D, T, V)
Primaxin I.M. (D, T, V)
Primaxin I.V. (D, T, V)
Prinivil tablets (D, V)
Prinzide tablets (D, T, V)
Probanthine tablets (D)
Procan SR tablets (D)
Procarbazine hydrochloride
Procardia XL tablets (D, T, V)
Procit for injection (D)
Pronestyl capsules/tablets* (D)
Pronestyl injections* (D)
Propagest* (D)
Propylthrouracil
ProSom tablets (D, T)
Protopam* (D)
Protostat tablets (D, V)
Proventil inhalation aerosol (D, T, V)
Proventil repetables tablets (D, T, V)
Proventil syrup (D, V)
Proventil tablets (D, V)
Prozac pulvules (D, T, V)
Quadrinal tablets (D)
Quarzan capsules (D)
Questran light* (D, T, V)
Questran powder* (D, T, V)
Quibron capsules (D)
Quibron-T (D)
Quibron-T/SR
Quinaglute dura-tabs tablets (T, V)
Quinamm tablets (T, V)
Quinidex extentabls (D, T, V)
Quinidine

(*Continued*)

Table 10.2 *Continued*

Quinine
Q-vel muscle relaxant pain reliever (T)
Raudixin tablets (D)
Rauzide tablets (D, V)
Recombivax HB (T)
Reglan (D)
Renese (D, V)
Renese-R tablets (D, V)
Restoril capsules (D)
Retrovir capsules (D, V)
Retrovir I.V. infusion (D, V)
Retrovir syrup (D, V)
Rheumatrex methotrexate dose pack (D, T)
Rifadin (D)
Rifamate capsules (D)
Rimactane capsules (D)
Robaxin injectable (D, V)
Robinul Forte tablets (D)
Robinul injectable (D)
Robinul tablets (D)
Roferon-A injection (D, V)
Rogaine topical solution (D, V)
Rondec (D)
Rondec-DM (D)
Rondec-TR tablets (D)
Roxanol (D)
Roxicodone tablets, oral solution and intensol (D)
Rufen tablets (D, T)
Ru-Tuss II capsules (D, T, V)
Ru-Tuss DE tablets (T)
Ru-Tuss w/hydrocone (D, T)
Rhythmol tablets (T, V)
Salfex (D, T, V)
Saluron* (D, V)
Salutensin/Salutensin-Demi (D, V)
Sandimmune (T, V)
Sandostatin injection (<1%) (D, V)
Scopolamine
Seconal sodium pulvules (D, V)
Sectral capsules (D)
Sedapap tablets 50 mg/650 mg (D)
Salicylates
Salsalate
Sensorcaine (D, T)
Sensorcaine-MPF spinal (D, T)
Septra* (D, T, V)
Septra I.V. infusion* (D, T, V)
Septra I.V. infusion ADD-Vantage vials* (D, T, V)
Serapes tablets (D, V)
Serax capsules (V)
Serax tablets (V)
Seromycin pulvules (V)
Serpasil tablets (D, V)
Serapsil apresoline tablets (D, V)
Serpasil esidrix tablets (D, V)
Silvadine cream 1%
Sinemet tablets (D)
Sinemet CR tablets (D)
Sinequan (T)
Slo-Bid gyrocaps (D)
Solu-Cortef sterile powder* (V)
Solu-Medrol sterile power* (V)
Soma compound w/codeine tablets (D, T, V)
Soma compound (D, T, V)
Soma tablets (D, V)
Stadol (V)
Streptomycin
Strichnine
Sulfasalazine
Sulfisoxagole/phenazopgridine
Sulindac
Surital ampoules steri-vials (D)
Surmontil capsules (D, T)
Symmetrel capsules and syrup (D)
Syntocinon nasal spray (D)
T-PHYL (Uniphyl) 200 mg tablets (D)
Tacaryl* (D, T)
Talacen (T)
Talwin compound (D, T)
Talwin injection (D, T)
Talwin Nx (D, T)
Tambocor tablets (D, T, V)
Tapazole tablets (D, V)
Tavist syrup (D, T, V)
Tavist-D tablets (D, T, V)
Tavist tablets (D, T, V)
Tegison capsules (D)
Tegral-D (D)
Tegretol chewable tablets (D, T)
Tegretol suspension (D, T)
Tegretol suspension (D, T)
Tegretol tablets (D, T)
Temaril tablets, syrup, spansule sustained release capsules* (D, T)
Tenex tablets (D, T, V)
Tenoretic tablets (D, V)
Tenormin tablets and infusion (D, V)
Tetanus antitoxin
Thalidomide
Thalitone tablets* (D, V)

(*Continued*)

Table 10.2 *Continued*

Thera-Gesic* (T)
Thiosulfil forte tablets* (T, V)
Thorazine (D)
Tigan (D)
Timolide tablets (D, T, V)
Timoptic in ocudose (D, T, V)
Timoptic sterile ophthalmic solution (D, T, V)
Tobramycin
Tocainide
Tofranil ampuls (D, T)
Tofranil tablets (D, T)
Tofranil-PM capsules (D, T)
Tolectin (200, 400, 600 mg)* (D, T)
Tolinase tablets* (D, V)
Tonocard tablets (D, T, V)
Toradol IM injection (D, V)
Torecan (D, T)
Trancopal capsules (D)
Trandate tablets (D, V)
Transdern scop transdermal therapeutic system (D)
Tranxene (D)
Trecator-SC tablets* (D, V)
Trental (D)
Trexan tablets (D, T)
Triaminic cold tablets (D, V)
Triaminic-DM syrup (D, V)
Triaminic expectorant DH (D, T)
Triaminic oral infant drops (D, T, V)
Triaminic repetabs tablets (D)
Triaminic syrup (D)
Triaminicol multi-symptom relief (D)
Triavil tablets (D, T)
Tridione* (V)
Trilifon (D)
Trilisate (D, T)
Tri-Norinyl 28-days tablets (D)
Tussionex extended-release suspension (D)
Tylox capsules (D)
Uricholine* (D)
Urised* (D)
Uroqid-acid (D)
Ursinus inlay-tabls* (D, T)
Valium injectable (V)
Valium tablets (V)
Valpin-50 tablets Valrelease capsules (D)
Vancocin HCI, oral solution and pulvules (D, T, V)
Vancocin HCI, vials & ADD-Vantage (D, T, V)
Vancomycin hydrochloride
Vascor tablets (D, T, V)
Vaseretic tablets (D, T, V)
Vasotec I.V. (D, T, V)
Vasotec tablets (D, T)
Velban vials (D)
Ventolin (D, V)
Verelan capsules (D)
Versed injection (D)
Vicodin tablets (D)
Vicodin ES tablets (D)
Vira-A for injection (D)
Visken tablets (D)
Vivacil tablets (D, T)
Voltaren tablets (D, T)
Vontrol tablets (D)
Wellbutrin tablets (D, T, V)
Wygesic tablets (D)
Xanax tablets (D, T)
Xanax injections
Xylocaine 2% jelly (D, T)
Xylocaine 5% ointment (D, T)
Xylocaine 10% oral spray (D, T)
Xylocaine injections for ventricular arrhythmias (D, T)
4% Xylocaine-MPF sterile solution (D, T)
Xylocaine 2% viscous solution (D)
Yodoxin* (V)
Yohimex tablets (D)
Zantac (D, V)
Zantac injection/Zantac injection Premixed (D, V)
Zarontin syrup (D, V)
Zaroxolyn tablets* (D, V)
Zestril tablets (D, V)
Zestroetic capsules (D)
Zidone capsules* (D)
Zidovudine
Zinacef
Zorontin capsules
Zorprin tablets (T)
Zovirax (D)
Zovirax steril powder (D)
Zydone capsules (D)
Zyloprim tablets (D, T, V)

18.11. Simple Stapes Mobilization Surgery

Sensorineural hearing loss is a rare complication of the procedure, but such instances did and do occur in ~0.1% of cases (Fig. 10.55). The cause still is not established, but the symptoms associated with the deafness typically are highly suggestive of a viral labyrinthitis. Such complications are more common after stapedectomy, and they have occurred also after fenestration surgery.

18.12. Stapedectomy

With more frequent surgical entry into the vestibule during stapedectomy, the incidence of sensory and sensorineural hearing loss has increased markedly as a complication. It has been estimated that after stapedectomy ~1 or 2% of patients will experience either immediate or delayed severe sensorineural hearing loss in the operated ear. Some cautious observers feel that in almost every case in which the footplate is removed or fractured, some degree of temporary sensorineural damage occurs. In most cases in which the surgeon is meticulous and avoids getting blood into the vestibule or suctierning out perilymph, there are minimal sensorineural damage and few clinical symptoms. Yet even some of these patients are known to complain of mild fullness in the ear and very slight imbalance along with a minimal ringing tinnitus.

Occasionally, after stapectomy there is measurable damage to the inner ear. Generally, the effect is temporary, but sometimes there is permanent high-frequency hearing loss associated with recruitment, impaired discrimination, and lowered threshold of discomfort for intense noise. Fig. 10.56 shows these findings.

Another disturbing complication of stapedectomy is seen in patients who had had excellent results and then suddenly, many months after the original surgery, lost practically all their hearing in the operated ear. In most cases this loss is permanent. The causes for this and other sensorineural complications of stapedectomy surgery are not yet entirely understood. Occasionally, there is a fibrous invasion filling the entire vestibule and completely blocking off the cochlea. In other cases of this type of delayed permanent, severe deafness, the vestibule is found to have a normal appearance, and the perilymph also is normal. The cause for the cochlear damage in such cases is still unknown. In other cases, oval window granuloma or perilymph fistula may be the cause.

18.12.1. Discrimination

It is apparent, however, that one should not measure the success of stapes surgery solely by threshold hearing tests. Some patients have excellent pure-tone thresholds, but their discrimination is reduced by the surgery, and their distortion can be distracting. With improving technics and better training for stapes surgery, the incidence and the severity of sensorineural complications can be reduced steadily.

18.12.2. Cases of Tympanosclerosis

The dangers of surgery in these cases also are better recognized. When patients with this disease are operated on to restore hearing, and the stapes footplate is removed in the presence of tympanosclerotic changes in the middle ear, the incidence of severe sensorineural hearing impairment is very high. Many surgeons complete this procedure in two stages or even avoid doing it altogether because of the frequency of complications that occur when the oval window is opened in the presence of this disease.

JOSEPH SATALOFF, M.D.
ROBERT THAYER SATALOFF, M.D.
1721 PINE STREET PHILADELPHIA PA 19103

HEARING RECORD

NAME AGE

AIR CONDUCTION

		RIGHT								LEFT							
DATE	Exam	LEFT MASK	250	500	1000	2000	4000	8000	RIGHT MASK	250	500	1000	2000	4000	8000	AUD
			5	0	5	10	15	25	90	50	45	55	60	70	NR	
LEFT STAPES SIMPLY MOBILIZED.																
			ONE WEEK POST-OP.						90	70	80	85	NR	NR	NR	
			ONE MONTH POST-OP.						90	50	50	40	40	60	65	

BONE CONDUCTION

		RIGHT							LEFT						
DATE	Exam	LEFT MASK	250	500	1000	2000	4000	RIGHT MASK	250	500	1000	2000	4000	AUD
			0	-5	0	5	10	90	0	10	20	25	30	
			ONE WEEK POST-OP.					90	35	40	50	NR	NR	
			ONE MONTH POST-OP.					90	30	30	35	40	NR	

SPEECH RECEPTION

DATE	RIGHT	LEFT MASK	LEFT	RIGHT MASK	FREE FIELD	MIC.

DISCRIMINATION

	RIGHT				LEFT				
DATE	% SCORE	TEST LEVEL	LIST	LEFT MASK	% SCORE	TEST LEVEL	LIST	RIGHT MASK	EXAM
PRE-OP.					82	90		90	
1ST WK. POST-OP.					10	100		90	
1 MTH. POST-OP.					42	85		90	

HIGH FREQUENCY THRESHOLDS

	RIGHT						LEFT					
DATE	4000	8000	10000	12000	14000	LEFT MASK	RIGHT MASK	4000	8000	10000	12000	14000

RIGHT		WEBER	LEFT		HEARING AID		
RINNE	SCHWABACH		RINNE	SCHWABACH	DATE	MAKE	MODEL
					RECEIVER	GAIN	EXAM
					EAR	DISCRIM	COUNC

REMARKS

1000 Hz	
R	L
5	35
20	65
35	70
50	75
65	80
80	80

Figure 10.55 *History*: A 26-year-old woman with a diagnosis of ototsclerosis. Operative notes indicate that the left stapes was mobilized merely by pressure on the end of the incus. No footplate manipulation was performed. Several days postoperatively the patient developed vertigo, roaring tinnitus, fullness, and deafness in the left ear. Loud noises bothered her left ear. *Otologic*: Healed incision. *Audiologic*: Preoperative left ear thresholds revealed a moderate to severe air conduction loss. Bone thresholds were normal in the low frequencies and dropped of somewhat between 1000 and 4000 Hz. The discrimination score was 82%. Postoperative thresholds revealed a great reduction in air and bone thresholds as well as reduced discrimination. For many months postoperatively there were hyperrecruitment, diplacusis, and lowered thresholds of discomfort. Some of these symptoms have gradually subsided. *Classification*: Conductive followed by sensorineural hearing loss. *Diagnosis*: Postoperative labyrinthosis following stapes mobilization.

JOSEPH SATALOFF, M.D.
ROBERT THAYER SATALOFF, M.D.
1721 PINE STREET PHILADELPHIA, PA 19103

HEARING RECORD

NAME AGE

AIR CONDUCTION

RIGHT										LEFT							
DATE	Exam.	LEFT MASK	250	500	1000	2000	4000	8000	RIGHT MASK	250	500	1000	2000	4000	8000	AUD
PRE-OP.		95	45	55	60	60	65	NR	–	10	20	35	45	55	NR	
		95	75	85	NR	NR	NR	NR	← RIGHT STAPEDECTOMY							
		95	50	55	55	85	NR	NR								

BONE CONDUCTION

RIGHT									LEFT							
DATE	Exam	LEFT MASK	250	500	1000	2000	4000		RIGHT MASK	250	500	1000	2000	4000		AUD
PRE-OP.		100	0	15	25	NR	NR		–	10	10	25	45	55		
		115	30	40	50	NR	NR	← RIGHT STAPEDECTOMY								
		100	NR	40	50	NR	NR									

SPEECH RECEPTION

DATE	RIGHT	LEFT MASK	LEFT	RIGHT MASK	FREE FIELD	MIC.
PRE-OP.	66	85	32	–		
	70	85				

DISCRIMINATION

DATE	RIGHT % SCORE	RIGHT TEST LEVEL	RIGHT LIST	LEFT MASK	LEFT % SCORE	LEFT TEST LEVEL	LEFT LIST	RIGHT MASK	EXAM.
	70	96		85	70	62		–	
	40	96		85	80	62		–	
	20	104		85					

HIGH FREQUENCY THRESHOLDS

RIGHT							LEFT					
DATE	4000	8000	10000	12000	14000	LEFT MASK	RIGHT MASK	4000	8000	10000	12000	14000

RIGHT		WEBER	LEFT		HEARING AID		
RINNE	SCHWABACH		RINNE	SCHWABACH	DATE	MAKE	MODEL
					RECEIVER	GAIN	EXAM
					EAR	DISCRIM.	COUNC.

REMARKS

TUNING FORK TESTS (500 Hz)		
AIR-BONE COMPARISON	R	L
LATERALIZATION		
B.C. COMPARISON TO NORMAL		

Figure 10.56 *History:* A 45-year-old woman who several days postoperatively developed hearing loss, vertigo, and nausea. There was a roaring tinnitus in the right ear. *Otologic*: Normal. *Audiologic*: Preoperative low-frequency air–bone gap in the right ear with fair discrimination. Thresholds immediately postoperatively revealed a severe reduction in air and bone conduction thresholds. Three months later thresholds improved, but there were complete recruitment, diplacusis, distortion, and reduced discrimination. There was a lowered threshold of discomfort (loud noises were very annoying). *Classification*: Sensory hearing loss. *Diagnosis*: Postoperative labyrinthosis.

18.13. Early Presbycusis and Genetic Hearing loss

For want of a better explanation, a high-frequency hearing loss occasionally encountered is called early or premature presbycusis if it occurs between the ages of ~30 and 50. Clinically, this condition is easy to overlook, but the audiogram shows a progressive hearing impairment in the high tones, which may be associated with a highpitched tinnitus and no other symptoms. The pathological explanation generally subscribed to today is that the hair cells degenerate because of some hereditary tendency and that the syndrome is not related to metabolism or infection. Little is known about this condition, but the picture is seen. In general, the loss is slowly progressive and eventually causes difficulty in discrimination as the speech frequencies become involved.

When the high-frequency losses are accompanied by recurrent attacks of rotary vertigo, especially when there is tinnitus of some sort, the diagnosis may be atypical Meniere's disease. Actually, it is not certain that this atypical picture truly is related to Meniere's disease. In some cases, especially if the hearing loss is bilateral, it may be due to so-called early presbycusis, whereas the accompanying vertigo is caused by labyrinthitis or some other labyrinthine disorder. Autoimmune inner ear conditions have been recognized recently as important, treatable causes of these symptoms.

This type of case should be distinguished from hereditary progressive sensorineural hearing loss, which generally starts early, progresses faster, and does not exhibit marked recruitment.

18.14. Congenital Defects in the Cochlea

Deafness present at birth can be caused in primarily two ways: (a) malformation of the organ of Corti and (b) toxic effects on the inner ear in utero. Increasing evidence suggests that toxic degeneration is far more common that congenital malformation. Even some cases previously described as hereditary congenital nerve deafness now are being recognized as having been caused actually by toxic effects in the first trimester of pregnancy. German measles in the mother is one such cause, and Rh incompatibility may be another.

Usually, the organ of Corti is affected, and the result commonly is subtotal deafness or a moderately severe hearing loss that is greater in the higher than in the lower frequencies. In both instances, the congenital defect is associated with a speech problem. If the hearing loss is severe, speech may not develop without much special training. When the hearing loss is partial, speech may be defective.

Although complete proof is not available, many otologists are convinced that viral infections during the first trimester of pregnancy do cause cochlear damage and deafness in the fetus. Anoxia shortly after birth also can cause damage to the cochlea, with resultant high-frequency hearing loss.

19. CAUSES OF NEURAL HEARING LOSS

Certain other causes are known to damage the auditory nerve *per se*. Abnormal tone decay becomes an important finding in some cases when there is "partial damage" to the nerve fibers, according to some investigators.

Causes of neural hearing loss include the following (among many other causes):

1. Acoustic neuroma
2. Skull fracture and nerve injury

3. Section of the auditory nerve
4. Virus infections
5. Toxicity and other neural injury

19.1. Acoustic Neuroma

The most urgent reason for differentiating damage to the auditory nerve fibers from damage to the hair cells or the inner ear is that this distinction permits the physician to detect a tumor of the auditory nerve at the earliest possible moment. As the result of advances in hearing tests, it is now possible to diagnose most auditory nerve tumors long before other neurological symptoms or signs become apparent. Figure 10.57 gives an example of such a case and emphasizes the need to perform discrimination, and selected special studies in all cases of sensorineural hearing loss, especially if they are unilateral. A patient who comes in to have wax removed from an ear to correct a hearing loss sometimes actually has a neoplasm of the auditory nerve (Fig. 10.58).

19.1.1. Early Symptoms

The earliest symptom of many acoustic neuromas is a mild unilateral neural hearing loss. Tinnitus is common, and vertigo may or may not be present. The vertigo may be more of a constant imbalance, in contrast to Meniere's syndrome, in which rotary vertigo usually is intermittent and accompanied by a seashell-like tinnitus.

19.1.2. Diagnostic Criteria

In the last few decades, surgery for acoustic neuromas has improved so much that early detection is even more important than it was previously. It is now possible for the otologist to remove relatively small tumors by a middle-cranial fossa or translabyrinthine approach with far less morbidity and mortality than encountered in the suboccipital craniotomy employed by neurosurgeons for larger tumors.

Unexplained dizziness, nystagmus, tinnitus, or hearing loss, especially unilateral progressive sensorineural hearing loss, warrants full evaluation. Even when other ear diseases are present, an undetected acoustic neuroma must be considered. The physical examination should include a complete ear, nose, and throat evaluation, assessment of the cranial nerves, cerebellar testing, and the Romberg test. Corneal sensation and ear canal sensation can be checked quickly with a wisp of cotton, and the gap reflex, with a cotton swab. When the corneal reflex, gag reflex, or facial nerve is involved, the tumor already is fairly large.

Thanks to the pioneering work of William House and other neurologists, a great deal of new information about acoustic tumor diagnosis is available. Unfortunately, there is no routine test that will establish the diagnosis in all cases. Unexpectedly low discrimination scores, type III and type IV Békésy audiograms, low SISI scores, and pathological tone decay each occur in only about two-thirds of patients with proven acoustic tumors. Stapedius reflex decay testing is somewhat more reliable but not as much as initially thought. Brain stem evoked-response audiometry is the only noninvasive test that appeared to have $>90\%$ diagnostic accuracy. However, electronystagmography (ENG) will show reduced vestibular response $\sim 70\%$ of the time. It must be remembered that this test evaluates only the lateral semicircular canal (superior vestibular nerve). Nearly fifteen percent of acoustic neuromas arise from the inferior vestibular nerve and may show a normal ENG even in the presence of vertigo.

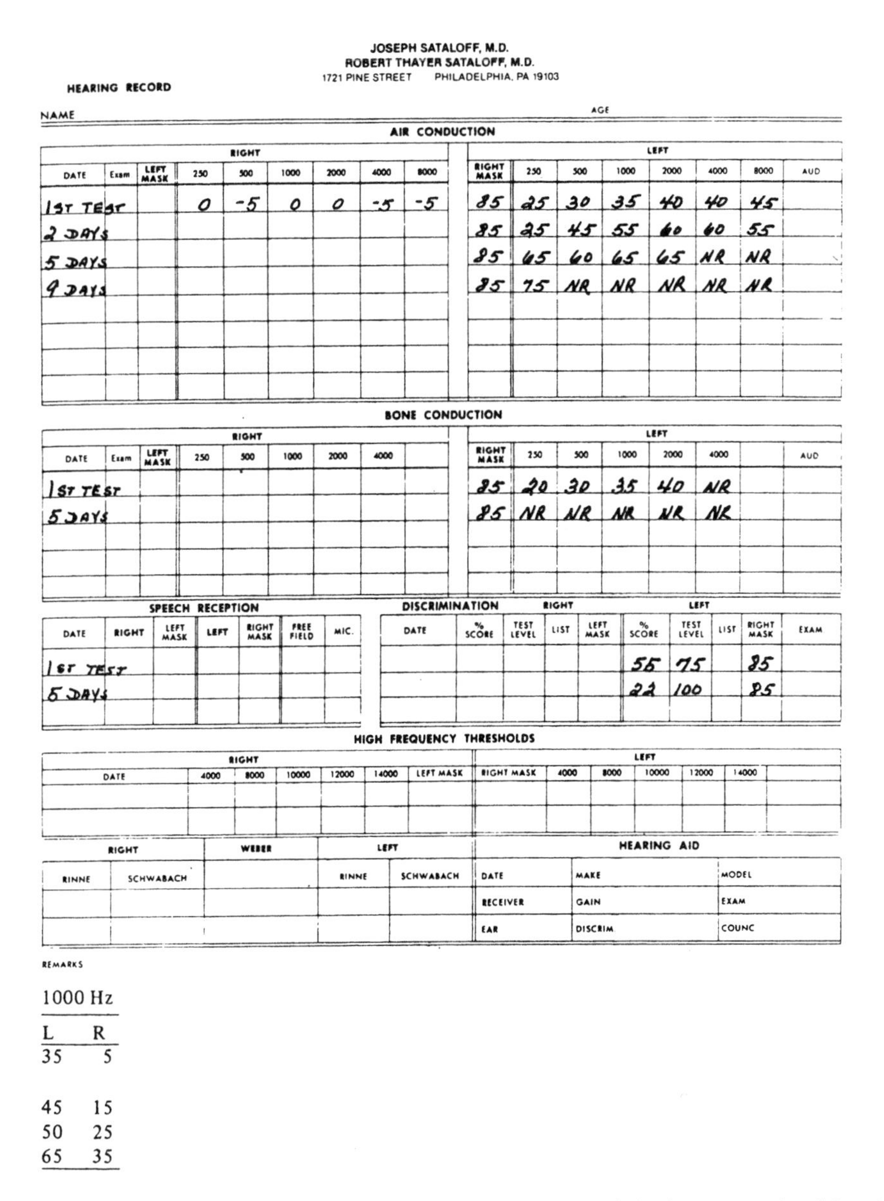

JOSEPH SATALOFF, M.D.
ROBERT THAYER SATALOFF, M.D.
1721 PINE STREET PHILADELPHIA, PA 19103

HEARING RECORD

NAME AGE

AIR CONDUCTION

RIGHT										LEFT								
DATE	Exam	LEFT MASK	250	500	1000	2000	4000	8000	RIGHT MASK	250	500	1000	2000	4000	8000	AUD
1ST TEST			0	-5	0	0	-5	-5	85	25	30	35	40	40	45	
2 DAYS									85	25	45	55	60	60	55	
5 DAYS									85	65	60	65	65	NR	NR	
9 DAYS									85	75	NR	NR	NR	NR	NR	

BONE CONDUCTION

RIGHT									LEFT							
DATE	Exam	LEFT MASK	250	500	1000	2000	4000		RIGHT MASK	250	500	1000	2000	4000		AUD
1ST TEST									85	20	30	35	40	NR		
5 DAYS									85	NR	NR	NR	NR	NR		

SPEECH RECEPTION

DATE	RIGHT	LEFT MASK	LEFT	RIGHT MASK	FREE FIELD	MIC.
1ST TEST						
5 DAYS						

DISCRIMINATION

	RIGHT				LEFT				
DATE	% SCORE	TEST LEVEL	LIST	LEFT MASK	% SCORE	TEST LEVEL	LIST	RIGHT MASK	EXAM
					55	75		85	
					22	100		85	

HIGH FREQUENCY THRESHOLDS

RIGHT							LEFT					
DATE	4000	8000	10000	12000	14000	LEFT MASK	RIGHT MASK	4000	8000	10000	12000	14000

RIGHT		WEBER	LEFT	
RINNE	SCHWABACH		RINNE	SCHWABACH

HEARING AID

DATE	MAKE	MODEL
RECEIVER	GAIN	EXAM
EAR	DISCRIM	COUNC

REMARKS

1000 Hz

L	R
35	5
45	15
50	25
65	35

Figure 10.57 *History*: A 24-year-old man complaining of deafness, occasional buzzing, and slight vertigo. *Otologic*: Normal. No spontaneous nystagmus. *Audiologic*: Unilateral hearing loss with reduced bone conduction. Tuning-fork tests showed lateralization to the right; A > B on left. No recruitment. Marked tone decay and poor discrimination. *Classification*: Neural hearing loss. *Diagnosis*: Acoustic neuroma. *Aids to diagnosis*: Corneal anesthesia on left, no caloric response on left. The patient refused surgery initially because of the mildness of the symptoms. Later his hearing loss increased, and he developed much vertigo. A large neuroma was removed.

Although an acoustic nerve tumor usually occurs on one side, its presence must not be ruled out merely because a patient has bilateral sensorineural hearing loss. In multiple neurofibromatosis, tumors can occur on both auditory nerves and may affect only the cochlear or only the vestibular portions. Characteristics of neurofibromatosis are summarized in Appendix 4. An instance of an auditory nerve tumor causing a comparatively

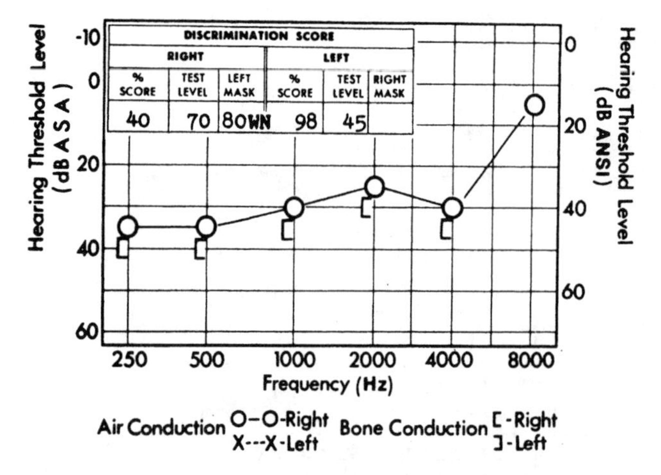

Figure 10.58 *History*: A 36-year-old man complained of having wax in his right ear, causing stuffiness for several weeks. No tinnitus or vertigo. *Otologic*: Normal ear canals and eardrums and no excess wax. No spontaneous nystagmus. *Audiologic*: Mild ascending hearing loss in the right ear with reduced bone conduction. Tuning fork lateralized to the good ear, and air was better than bone conduction in the bad ear. No diplacusis and no recruitment were present. Discrimination was remarkably poor in the right ear, especially in view of the mild hearing loss. Abnormal tone decay was present with the threshold going to 75 dB at 1000 Hz after 1 min. *Classification*: Neural hearing loss. *Diagnosis*: Acoustic neuroma. *Aids to diagnosis*: The presence of unilateral nerve deafness with absent recruitment but abnormal tone decay and reduced discrimination indicates some pressure on the auditory nerve. In addition, there were corneal anesthesia, and no caloric responses in the right ear. An acoustic neuroma was removed surgically.

mild hearing loss and very marked reduction in discrimination threshold, using phonetically balanced word lists, is cited in Fig. 10.59. In such instances the patient may volunteer that he/she simply cannot understand anything in that ear even though he/she hears voices. This may be particularly bothersome when trying to use the telephone. This complaint should always be followed through with a thorough evaluation.

It is essential to bear in mind that the mere presence of abnormal tone decay does not necessarily mean that the patient has a tumor of the acoustic nerve. It indicates merely that nerve fibers have been damaged. Other causes also can produce damage to acoustic nerve fibers.

Acoustic neuroma is not the only retrocochlear tumor that the otologist is likely to see. Meningiomas, cholesteatomas, facial nerve neuromas, and other lesions may present a similar clinical picture.

19.2. Fractured Skull and Acoustic Nerve Injury

A transverse fracture of the temporal bone can go through the internal auditory meatus and compress or sever the acoustic nerve. Occasionally, the seventh nerve also is damaged causing facial paralysis. Since the deafness usually is total, findings such as abnormal tone decay and disproportionately poor discrimination cannot be detected, but CT scans and absent caloric responses on the involved side with normal responses on the other side help to establish the diagnosis.

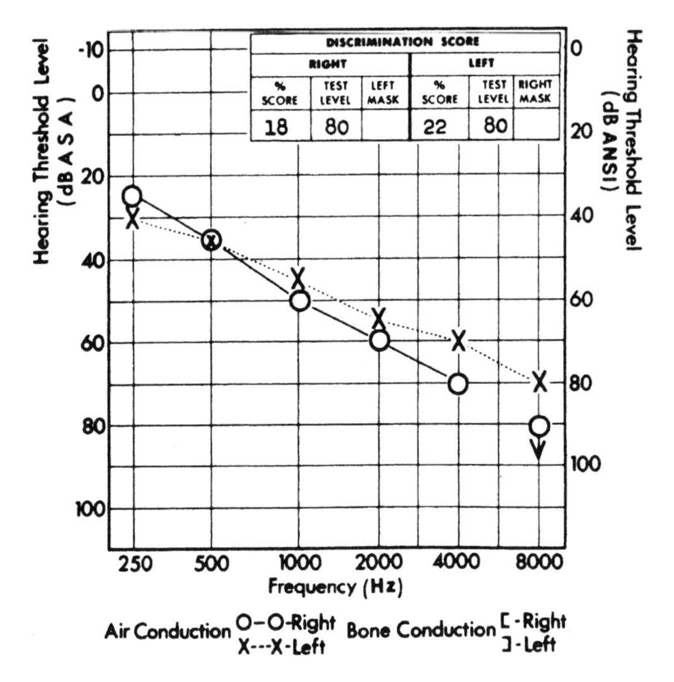

Figure 10.59 *History*: A 52-year-old man had Meniere's disease in both ears and had undergone a bilateral vestibular nerve section. According to the patient, discrimination was sharply reduced postoperatively. *Otologic*: Normal. *Audiologic*: Bilateral mild to moderate hearing loss with no air–bone gap. Very poor discrimination ability, which worsened in the presence of environmental noise. *Classification*: Neural hearing loss. *Diagnosis*: Surgical lesion of auditory nerves resulting from section of vestibular nerves.

19.3. Partial Section of the Acoustic Nerve

Some neurosurgeons in the past have cut the vestibular portion of the auditory nerve to control severe persistent vertigo in patients with Meniere's disease. During the operation it was difficult for the surgeon to avoid severing at least some portion of the adjacent hearing fibers. Although the vertigo was controlled in most instances, many of the patients were left with additional hearing loss owing to section of the auditory nerve fibers. Almost invariably, the high tones were chiefly affected by the surgery. Figure 10.59 describes a case in which both vestibular nerves were sectioned by a neurosurgeon. In this case the low frequencies were affected also, but part or all of this may have been attributable to the preexisting Meniere's disease. Some of the high-frequency loss was present preoperatively, but this was greatly aggravated by the surgery. In addition to the change in threshold, this patient's ability to understand speech was reduced seriously after the nereve section. A hearing aid was practicaly useless for the individual. Neurotologists have applied microsurgical techniques with a middle-cranial fossa, retrolabyrinthine or retrosigmoid approach and have refined vestibular nerve surgery so that hearing generally can be preserved.

19.3. Virus Infection

Certain viral infections, notably herpes, are supposed to cause hearing loss by affecting the acoustic nerve proper. In this instance the site of injury is supposed to be in the

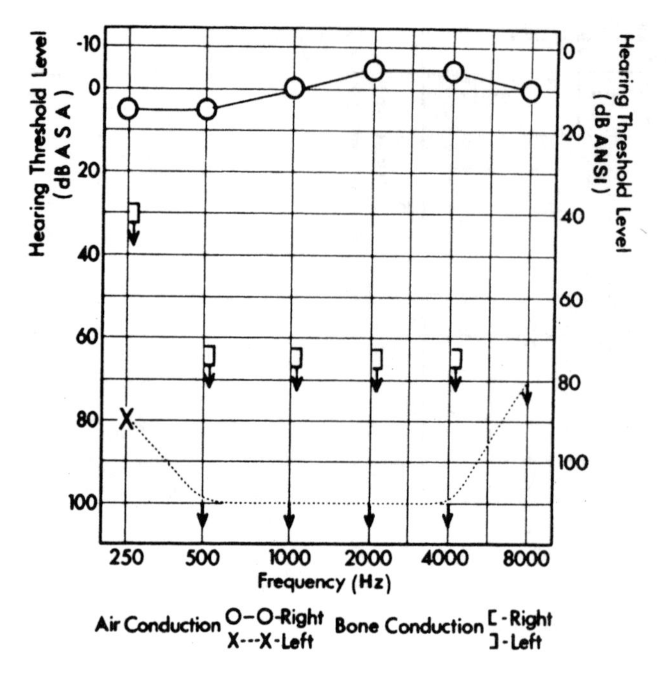

Figure 10.60 *History*: A 58-year-old man who developed severe herpes with pain in the left ear. He had a left facial palsy for several weeks, and this resolved. He also had a buzzing tinnitus that cleared up. No vertigo. *Otologic*: Normal at time of examination. Caloric tests normal. *Classification*: Neural hearing loss. *Diagnosis*: Herpetic acoustic neuritis.

ganglion. The hearing loss produced often is quite severe. Figure 10.60 shows an interesting case.

19.4. Congenital Nerve Deafness

It is common practice to call all cases of deafness present at birth "congenital nerve deafness" or "hereditary nerve deafness," especially if hearing loss is present in other members of the family. It is not certain that the congenital defect always is in the nerve itself, but in many cases it does appear to be in this site, especially if the hearing impairment is severe. Because some cases of congenital hearing loss have their sites of damage in the cochlea, a more nearly accurate term would be cochlear hearing loss, but the distinction is at present primarily of academic interest. However, as neurotherapeutic methods become better developed, the differentiation may become crucial.

19.5. Toxicity and Other Nerve Injury

The acoustic nerve can be injured by any toxin or disorder capable of injuring other nerves. Lead and other heavy metals can cause nerve deafness, for example. Many neurodegenerative diseases also are capable of causing neural hearing loss. For a more comprehensive discussion of these subjects, the reader is referred to neurology literature.

20. SURGERY AND SENSORINEURAL HEARING LOSS

Until very recently, surgical otology has concentrated on disorders of the outer ear and middle ear. The inner ear is the new frontier for the neurotologist and skull base surgeon. Although much remains to be learned, significant advancement has been made in the treatment of both vertigo and sensorineural hearing loss. In some cases, medical management can improve or cure symptoms, or stop progression of inner-ear disease. In selected instances, surgical therapy is appropriate in patients with sensorineural hearing loss either to improve hearing or to treat underlying disease.

A patient with sensorineural hearing loss needs thorough evaluation leading to a specific diagnosis before being considered a surgical candidate. Depending on the history and hearing pattern, this evaluation usually includes auditory, vestibular, neurological, metabolic, and radiological testing. In addition to routine audiometry and impedance studies, brain stem evoked-response audiometry is often extremely helpful even in some cases in which the hearing loss is >80 dB at 4000 Hz. In some such patients, an unexpectedly good brain stem evoked-response audiogram is obtained, and these patients rarely have retrocohlear pathology. In other patients, we are able to obtain useful information with experimental testing using a 500-Hz stimulus rather than a click. Electronystagmogram and computerized dynamic posturography are often helpful, as well. Neurological examination concentrating on the cranial nerves and cerebellum should be performed in all patients. Substantial sensorinerual hearing loss may be associated with more generalized neurological disorders such as multiple sclerosis, neurofibromatosis, or toxic neuropathy. Blood tests are recommended in all cases. An FTA-absorption test to rule out luetic labyrinthitis is most important. When appropriate, tests for diabetes, hypoglycemia, lyme disease, HIV, thyroid dysfunction, hyperlipoproteinemia, collagen vascular disease, polycythemia, and other disorders should also be performed. If there is any question of cerebellopontine angle (CPA) tumor, radiological examination is essential. We have stopped using plane films or polytomograms because they are insufficient to rule out CPA neoplasm with certainty. If radiological evaluation is indicated at all, MRI without and with contrast is obtained on a state-of-the-art MR scanner. The films are always reviewed by the neurotologist as well as by the radiologist. It is not rare to receive a report that says "normal MR scan" and to find that the contents of the internal auditory canals are not visualized on the study. Naturally, this problem can be avoided if the studies are performed regularly by a radiologist who is expert in MR scanning of the ear. High-quality enhanced magnetic resonance imaging is very effective at detecting even small tumors. CT scans and SPECT or PET scans are also often essential in a neurotologic evaluation.

In most patients, it is possible to separate sensory hearing loss from neural hearing loss. Surgery may be indicated despite the presence of substantial sensory or neural hearing loss in such conditions as fistula, infection, far-advanced otosclerosis, CPA tumors, endolymphatic hydrops, profound deafness, and other entities. It is worthwhile reviewing selected situations in which the advisability of surgery is not always immediately apparent. The seven cases discussed below are examples of such situations.

20.1. Infection/Fistula

Surgical management of patients with infection or with perilymph fistula is widely recognized and will not be discussed in detail. When sensorienerual hearing loss occurs in association with acute otitis media or mastoiditis, surgical drainage usually is indicated.

When sudden hearing loss occurs in association with barotrauma, or when fistula is suspected because of fluctuation in symptoms associated with variations of middle-ear pressure, surgery for fistula repair should be considered. Although treatment with bed rest and medication may be helpful, operative repair should be considered early, particularly if the patient has previously undergone surgery in the affected ear.

20.2. Far-Advanced Otosclerosis

Otosclerosis presents infrequently in an advanced state because good therapy has been widely available for so many years. Hence, we may neglect to think of it in a patient with severe sensorineural deafness or a “blank audiogram” (no response by air conduction or bone conduction). Nevertheless, it still occurs. The authors have performed stapes surgery on many such cases in the last decades, representing ~2% of stapes operations. This is similar to the experience of House and Glorig (41) and Sheehy (42), who found slightly >1% of their stapes surgery cases were for far-advanced disease. Stapes surgery may be extremely helpful in patients with severe or profound hearing loss, especially in people who have difficulty wearing hearing aids, as illustrated by the following cases.

Case 1 is an 88-year-old woman who is extremely vital and active, but socially incapacitated by inability to communicate (Fig. 10.61). She could not wear a hearing aid effectively because of distortion and narrow dynamic range. Following surgery she has been able to communicate well using a behind-the-ear aid.

Case 2 is a 58-year-old lady who had undergone prior stapedectomy in one ear and fenestration in the other ear with good results (Fig. 10.62). She had awakened deaf after cardiac bypass surgery and had undergone tow unsuccessful attempts at stapes operations before being referred. She was unable to use powerful body hearing aids effectively. Following surgery, she is able to communicate satisfactorily using a body aid in her left ear.

Frattali and Sataloff (43) reported nine patients and reviewed far advanced otosclerosis in 1993. Postoperatively, seven of the nine patients (78%) used hearing aids successfully, where as none had been aidable preoperatively. Far-advanced otosclerosis generally involves air conduction levels worse than 85 dB HTL, and bone conduction levels beyond the limits of the audiometer. If air conduction levels exceed 85 dB HTL, but bone conduction levels are measurable at some frequencies but worse than 30 dB,

NAME D.G.

DATE	RIGHT EAR AIR CONDUCTION							LEFT EAR AIR CONDUCTION					
	250	500	1000	2000	4000	8000		250	500	1000	2000	4000	8000
8/81	NR	100	100	NR	NR	NR	PRE	90	100	100	100	95	NR
5/82	50	55	65	80	85	90	POST						

RIGHT EAR BONE CONDUCTION							LEFT EAR BONE CONDUCTION					
		65	NR	NR	NR							

SPEECH RECEPTION: Right 95 Left 90 PRE DISCRIMINATION: Right ____ Left ____

Figure 10.61 Preoperative and postoperative audiogram of Case 1, far-advanced otosclerosis.

NAME M.B.

DATE	RIGHT EAR AIR CONDUCTION							LEFT EAR AIR CONDUCTION					
	250	500	1000	2000	4000	8000		250	500	1000	2000	4000	8000
8/82	NR	NR	NR	NR	NR	NR	PRE	NR	NR	NR	NR	NR	NR
							POST	90	110	105	85	NR	NR

RIGHT EAR BONE CONDUCTION							LEFT EAR BONE CONDUCTION					
							30	55	NR	70	55	

SPEECH RECEPTION: Right XXX Left XXX DISCRIMINATION: Right ____ Left ____

Figure 10.62 Preoperative and postoperative audiogram of Case 2, far-advanced otosclerosis.

the condition is called advanced otosclerosis. Far-advanced otosclerosis may be suspected because of clues in the history and physical examination, but definitive diagnosis requires middle-ear surgery. It is not feasible to use the usual criterion (closure of air–bone gap) for assessing the results of surgery; other objective and subjective criteria should be used.

Despite 125 cases of FAO discussed in reports written over >30 years (44–50) (prior to the report by Frattali and Sataloff), the relationship of severe or profound cochlear hearing loss and otosclerosis is poorly understood. Histologically, the correlation between bone disease and perceptive loss remains obscure in most temporal bones described to date. Lindsay and Hemenway (51) described two types of otosclerosis. Type 1 is usually a conductive disturbance limited to specific areas of the oval and round windows. The second type of otosclerosis is a more aggressive form with multiple foci that form early in life. These foci involve multiple areas in the otic capsule.

Wiet et al. (49) studied the temporal bone of a 65-year-old patient with FAO who died of natural causes. There were multiple foci of otosclerosis, diffuse loss of hair cells, and loss of cochlear neurons in the basal turn. There were also areas of strial atrophy near the foci of otosclerosis. Myers and Myers (52) examined the temporal bone of a 66-year-old man who died suddenly several hours after his second stapes operation for FAO. They found atrophy of the spiral ligament with basilar membrane rupture and severe degeneration of the organ of Corti, spiral ganglion cells, and cochlea.

A convincing histologic explanation for the increased bone conduction threshold in FAO remains an issue for continued investigations. In a total of 164 temporal bones examined by Schuknecht and Barber (53), a statistical analysis failed to show any correlation between bone conduction thresholds and size of lesion, activity of lesion, involvement of the endosteum, or presence of a round window lesion.

It is presumed that otosclerotic involvement of the cochlear endosteum causes atrophy of the spiral ligament. Basilar membrane rupture appears related to this atrophy, as two of the layers of the basilar membrane insert into the endosteum. How this is related to sensorineural hearing loss remains uncertain. Otosclerosis may cause atrophic changes leading to alterations of the motion mechanics in the scala media; mechanical and metabolic alterations because of strial atrophy may be causal; or some other mechanism may be response for the sensorineural loss.

Clinically, it is extremely important to recognize FAO. Wiet et al. (49) diagnosed three cases of FAO among 175 patients who were being considered for cochlear implantation. Two of these patients underwent stapedectomy and received substantial benefit from hearing aids following surgery. Stapedectomy involves much less risk than cochlear implantation, and less rehabilitation. Moreover, the hearing achieved with amplification following stapedectomy is generally better than the hearing expected with a cochlear implant. When satisfactory results are not achieved, the patient may still be considered for cochlear implant surgery.

When a patient presents with acquired severe or profound hearing loss, a comprehensive neurotologic evaluation is mandatory to diagnose causes such as neoplasms, luetic labyrinthitis, autoimmune sensorineural hearing loss, ototoxicity, meningitis, and genetic hearing loss. However, because FAO is one of the few conditions that permits treatment capable of converting the patient's hearing from nonserviceable to serviceable with conventional hearing aids, it is especially important to be alert for this entity, and to look for residual, nonmeasurable hearing and an air–bone gap. The definitive diagnosis requires middle ear exploration. However, Sheehy (47) summarized the symptoms and signs that are generally regard as most typical of FAO, and that should alert the otologist to the possible presence of this entity. They include the following:

History. Gradually progressive hearing impairment in early adult life is suggestive, especially with a positive family history of otosclerosis. Eighty-six percent of Sheehy's (47) patients who had an onset of hearing impairment before age 30 years and a positive family history had oval window otosclerosis.

Paracusis. The ability to hear better in noisy surroundings during an early stage of hearing impairment is highly suggestive of otosclerosis.

Hearing Aid Performance. Despite the extent of the hearing loss, patients report some benefit from amplification. Word discrimination ability without lipreading is better than would be predicted on the basis of pure tone thresholds. Eighty percent of patients with FAO wore hearing aids, as opposed to only 30% of patients with nonotosclerotic ears who had similar audiograms in Sheehy's series (47).

Hearing Aid History. The patient may be wearing or in the past was wearing a hearing aid or bone conduction aid.

Previous Audiograms. A prior audiogram may demonstrate an air–bone gap.

Voice Patterning. The voice may not be suggestive of sensorineural hearing loss (modulation, pronunciation, and intensity).

Otoscopic Examination. Schwartze's sign may be present.

Radiographic Confirmation. Although polytomography may further support the diagnosis of FAO, preoperative confirmation with CT (with coronal views) is especially helpful. Von Glass and Philipp (54) investigated 31 patients suspected of having cochlear otosclerosis with high-resolution CT. Fourteen patients had confirmed otosclerosis associated with advanced hearing loss. In four of these patients, free otospongiotic foci were detected in the bone of the cochlear capsule.

Tuning Fork Test. A Weber test lateralizing to the poor ear or a negative Rinne test in presumed sensorineural hearing loss may be suggestive. Because of tactile perception, tuning fork tests may sometimes be misleading. We recommend standard use of a 512-Hz tuning fork. Many patients with FAO do not hear the tuning fork well when it is placed on the forehead or nasal dorsum, but hear clearly by dental bone conduction. In order to help confirm the validity of tuning fork testing, in 1980 one of the authors (R.T.S.) began asking patients to hum the pitch of the tuning fork. This is a simple task for people even with extremely modest musical ability. All of the patients with FAO were able to perform this task. None of the patients with profound sensorineural hearing loss who subsequently went on to cochlear implant surgery was able to perform the task.

No Other Apparent Cause for Hearing Impairment. When hearing loss has been of insidious onset, and when all other known causes have been ruled out, the diagnosis of FAO must be considered, especially if any of the criteria listed earlier are present.

The value of surgery for this disease entity was first discussed in 1960 by House and Glorig (55). Surgical improvements are much more important to a patient with FAO than to a patient with a mild-to-moderate hearing loss, because surgery can convert FAO patients from functionally deaf to easily aidable. House and Glorig were two of many investigators to conclude that opening the oval window in patients with FAO was potentially worthwhile regardless of the level of preoperative air and bone conduction thresholds. This philosophy still prevails some 30 years later.

The reported surgical experience varies (44–50). Our results support the findings of the three previous reports with the best results (44,45,48). The type of prosthesis and surgical technique used, and involvement of the round window by otosclerosis, appear to have no bearing on the success of surgery. House and Glorig (55) opened both windows in a two-stage operation. In three of five patients, the hearing was made worse. The round window should generally not be opened for this reason, and because closure of the round window probably accounts for only 10 dB of bone conduction loss at most (55).

For routine cases of otosclerosis, we generally define surgical success as closure of the air–bone gap to within 10 dB. In experienced hands, we expect a success rate >90%. On the basis of these conventional criteria for stapedectomy surgery, results in FAO would be disappointing. However, the majority of patients with FAO clearly benefit from stapedectomy subjectively, and many have improvements in pure tone air conduction thresholds. Some also have improvement in discrimination ability under earphones; and marked improvement in hearing aid performance is common. Patients are often able to use their amplified hearing more effectively, even when the traditional measures of discrimination do not show improvement. Careful preoperative counseling and informed consent are extremely important in patients with FAO. The patients must be aware not only of the risks of the procedure but also of the relatively limited goals. Normal hearing is not expected. Reported complications in these patients do not appear to differ statistically from those in other patients undergoing stapedectomy surgery; and most patients feel justified in accepting those risks of the potential benefit of advancing from unserviceable hearing to aidable hearing.

Controversy exists as to whether a second stapedectomy for FAO should be performed on the opposite ear. Many otologists believe that there is little justification for this in the event of a successful stapedectomy. We believe, however, that ear cases should be addressed individually, that each patient should be offered the potential benefit of binaural amplification even for a hearing loss of this severity, and that the final decision should be left to the patient after the risks have been explained fully.

If an operation for FAO has failed, many investigators suggest that an attempt at surgery on the opposite ear should also probably fail, and that further surgery is not advisable. We question this anecdotal judgment and believe that the decision regarding surgery on the opposite ear should be individualized in this instance, as well. When hearing loss is profound bilaterally, the patient assumes very little risk from undergoing stapes surgery; and results may be substantially different on the two ears.

In summary, the most important diagnostic aid in recognizing far-advanced otosclerosis is the tuning fork. Despite the audiogram, these patients usually can hear well by dental bone conduction. To help assure that the response is not tactile, it is sometimes useful to ask the patient to hum the tone of the tuning fork. This diagnosis should also be suspected in people who are using powerful hearing aids more effectively than expected, or in people who understand loud spoken voice without lip reading.

In general, discrimination remains better with far-advanced otosclerosis than with other diseases that cause this degree of hearing loss. Although most of us prefer restoring hearing to normal with stapes surgery, converting a patient from unaidable to aidable may be an equally gratifying and valuable intervention.

20.3. Cranial Irradiation

The effects of cranial irradiation on hearing were reviewed by Sataloff and Rosen (56) in 1994. Radiotherapy has been shown to be effective in treatment regimens for a variety of tumors of the head and neck. Both clinical practice and published literature reflect contradictory opinions on the incidence, type, severity, and time of onset of hearing loss as a treatment complication. Review of relevant animal and human studies highlights the available information. References describing cranial irradiation used in conjunction with chemotherapy have been excluded because of the known ototoxic effects of the chemotherapeutic agents themselves.

20.4. Animal Models

Animal models have been used to study this question since 1905 when Ewald (57) placed beads containing radium in the area of the middle ear in pigeons and noted labyrinthine symptoms.

Girden and Culler (58) subjected dogs to various doses of X-rays to evaluate the effects of radiation on hearing. His subjects demonstrated a 5.5-dB gain in acuity after a 7–11 day latent period but returned to baseline within a period of days to weeks. No hearing loss was noted, but the authors could not explain the mechanisms involved. Novotny (59), Kozlow (60), and Gamble et al. (61) studied the effects of ionizing radiation in guinea pigs. In Novotny's research a hearing impairment of 8.4 dB at 4000 Hz was noted. There were no histologic changes in the inner ear. Kozlov's subjects demonstrated 3.9–9.1 dB diminution in the frequency ranges of 500–8000 Hz.

Gamble's study animals were radiated with doses of 500, 1000, 2000, 3000, 4000, 5000, or 6000 rads. The opposite ear was shielded and received a maximum of 500 rads of scatter radiation. Cochlear microphonics were correlated with histologic findings at 24 h, 2 weeks, and 2 months. At 24 h both cochlear microphonic sensitivity and histologic examinations were normal. At 2 weeks there was little inflammatory response in the group receiving lower doses. Animals receiving 5000 rads had mild erythema, and those with a 6000 rads dose had more erythema, some edema and vascular changes that interfered with meaningful measures of cochlear microphonics. Inflammatory response resulted in loss of sensitivity of between 20 and 40 dB, especially at frequencies >7000 Hz. Examination at 2 months revealed resolution of inflammatory response in all animals. The cochlear microphonics demonstrated loss of sensitivity of 30–40 dB at high frequencies. Three animals also demonstrated vestibular signs.

Using light microscopy, histologic changes were noted. There was little detectable difference between the radiated and control ears at doses of ≤2000 rads. At ≥3000 rads the changes in cellular structure were directly dose-related. The earliest changes were seen in the stria vascularis. As the dose increased, changes in hair cells were noted, with distortion and cell boundary dissolution seen at 5000 rads. Doses of 6000 rads led to shrivelling of hair cells and elevation of the tectorial membrane. Additionally, at ≥3000 rads, 50% of temporal bones revealed distention of the endolymphatic sac and displacement of Reissner's membrane.

Temporal bones of rats examined by Keleman (62) after doses of 100–3000 rads demonstrated engorgement of middle ear membranes with extravasation. In the inner ear, damage to the cochlear duct and organ of Corti was noted.

Bohne et al. (63) report late findings in chinchillas treated with 4000–9000 rads and examined 2 years after exposure. There was dose dependent degeneration of sensory and supporting cells and loss of eighth nerve fibers in the organ of Corti.

20.5. Research Formats

Results reported in human subjects are fraught with complications of research design that mandate caution in interpretation. Older reports are affected by the magnitude of radiation voltage (kilovoltage rather than megavoltage dosing). Kilovoltage is more likely to cause radionecrosis of the temporal bone than the megavoltage equipment in current use. Megavoltage X-rays are uniformly absorbed by soft tissues and bone, and kilovoltage has increased bone absorption (64). Radiation doses administered in early studies were much higher than in current standards of care (up to 24,000 rads). Many studies were retrospective, and pretreatment audiometric measures were not available. Some studies reflect sample bias in that patients complaining of subjective hearing loss were recruited as subjects. Finally, the site of the tumor is widely variable and includes carcinoma of the ear and nasopharynx, which may have direct effects on hearing. The treatment portals often irradiated both temporal bones so intrasubject controls were impossible.

20.6. Human Subjects

In 1962, Borsanyi and Blanchard (65) reported on a series of 14 patients receiving 4000–6000 rads in the region of the inner ear for treatment of a variety of head and neck cancers. Pretreatment and immediate posttreatment audiograms were obtained. "Small" shifts in threshold were noted, with the greatest change at 4000 Hz and the mildest at 200 Hz. Discrimination remained within normal limits. Recruitment was present during the treatment phase but resolved spontaneously when it had not been present prior to treatment. No long-term audiologic follow-up was performed. The authors attribute the threshold shifts to a conductive loss secondary to "radiation otitis media." This is known to occur when mucosal edema impairs eustachian tube patency and produces a middle ear fluid collection. They hypothesized that intratreatment recruitment resulted from "temporary vasculitis of stria vascularis and arachnoid mesh in the perilymphatic space." They further observed that the late microscopic changes sometimes occurred in the temporal bone. Obliterating endarteritis resulting from edema and degeneration of collagen and smooth muscle of vessel walls impaired blood supply to the cochlea, labyrinth, and ossicles.

Leach's (66) series of 56 patients received 3000–12,000 rads for treatment of head and neck carcinoma in eight different regions. In the study, 36% of patients experienced sensorineural hearing loss (SNHL) assumed to be secondary to conductive hearing impairment. Eleven patients were evaluated prior to treatment; of these, eight patients showed immediate changes, with recovery in only two. The remaining three showed late changes. Nine additional patients were evaluated 18 months to 10 years after their treatment. Although hearing loss was evident, there were no comparative data available. By patient reports this was noted 9 months to several years after treatment and was gradually progressive. Two of these patients had total deafness and absent ice-water caloric response in one ear. In one patient, this was a 7-year gradual progression. No correlation was found between radiation dose and the amount of hearing loss, nor were treatment duration or patient age significant. Leach also reported radiation otitis media in several cases. Significant tinnitus and vertigo were described anecdotally and preceded subject awareness of hearing loss.

One temporal bone was harvested from a patient with nasopharyngeal carcinoma who became vertiginous and profoundly deaf in a gradually progressive fashion following a 4000 rads dose. He died 1 year later. The pathology report notes:

> . . . exudate present with polymorphs and macrophages in the middle ear and mastoid bilaterally. The mucosa is congested and there is marked bone absorption by granulation tissue resulting in extensive osteoporosis. The organ of Corti is absent. A layer of red fluid covers the basilar membrane. The round window niche contains exudate. The spiral ganglion and nerve is atrophic. No maculae or cristae are seen in the vestibular apparatus.

Dias (67) reports on his series of 29 patients who presented with a variety of head and neck tumor sites. They received a range of kilovoltage radiation from 1000 to 18,000 rads. Using a location 2–3 cm deep to the external meatus, isodose curve calculations proposed a dose to the auditory apparatus of 700–10,000 rads. In this study, 19 patients had pre- and posttreatment audiograms (group 1) and 10 additional patients had evaluation after treatment only (group 2). In group 1, mixed hearing loss predominated. The average decrease in acuity was 9.9 dB, and no particular frequency was more affected. Group 2 also demonstrated mixed hearing loss, with two patients evaluated at 11 and 13 years posttreatment. Their hearing remained "serviceable." One patient had undergone craniotomy for a tumor of the fourth ventricle and received 10,000 rads of radiation. He had ossicular reconstruction for dislocation necrosis of the incudostapedial joint 8 years status postcraniotomy. Postoperative hearing level was 20 dB for the worse side and 13 dB for the operated ear. Dias notes that nervous tissues are among the most radiosensitive in the body and attributes most of the hearing loss to chronic tubal dysfunction secondary to the tumor and/or radiation-induced changes in soft tissue.

Schuknecht and Karmody (68) sectioned a temporal bone in a man who had received 5200 rads over 15 days for squamous cell carcinoma of the ear canal. Hearing loss developed 8 years later. There was radionecrosis of the temporal bone; histologically, "slight atrophy of the organ of Corti was noted with atrophy of the basilar membrane, spiral ligament, and stria vascularis."

In a restrospective study by Moretti (69), 137 patients with carcinoma of the nasopharynx were reviewed. Of these, 13 had received pre- and posttreatment audiograms. Seven of the 13 had SNHL and were self-selected, owing to hearing complaints. Patients ranged in age from 17 to 74 years and received tumor radiation doses between 6000 and 24,000 rads. The onset of severe SNHL was gradual and occurred 3–6 years following radiation therapy. Profound losses had descending audiometric patterns; lesser losses had flat patterns. Although formal statistical analyses were not performed, age appeared to be correlated with hearing loss: affected patients were 42–72 years old (average 60 years); unaffected patients were 17–62 years old (average 40 years). It should be noted that the "unaffected" group included three patients who had only 1 year of audiologic follow-up.

Thibadoux et al. (70) serially assessed the hearing sensitivity of 61 children with acute lymphocytic leukemia. It was postulated that children might be more susceptible to lower radiation doses because of their high growth rate and structural immaturity. Their treatment included combined chemotherapy, 2400 rads of cranial irradiation and intrathecal methotrexate. Pure-tone audiometry at 500, 1000, 2000, 4000, 6000, and 8000 Hz was performed prior to irradiation and at 6, 12, and 36 months thereafter.

Radiation dose was divided into 15 fractions given >18 days to a field including the retro-orbital space, C-2, both temporal bones, and all structures of the auditory system bilaterally. No statistically significant reduction in auditory sensitivity was noted for any test frequency at any test interval. This is a properly controlled prospective study with a reasonable period of follow-up that is continuing on cohort survivors. However, in spite of a theoretic increase in radiation sensitivity, the 2400-rads dose was not found to be correlated with hearing loss in this or other studies. A 3000-rads dose is the minimum with cited hearing effects.

Two case reports of delayed SNHL are provided by Coplan et al. (71) and Talmi et al. (72). Coplan et al. describe a 12-year old girl who underwent sub-total resection of an optic glioma and received postoperative radiation: 5000 rads, total tumor radiation dose, in 25 fractions >5 weeks. The field included the petrous portion of both temporal bones. Three years later, CT revealed residual tumor occupying the third ventricle and symmetric calcifications of the basal ganglia and both temporal lobes. Hearing loss was first suspected by the treating physicians 5 years after radiation, but the mother noted decreased hearing 2 years after initial surgery and irradiation. Pure-tone testing revealed a bilateral mixed hearing loss. The conductive component was approximately equal bilaterally, but the sensorineural component was greater on the left. Speech discrimination was moderately depressed on the left and normal on the right. There were no audiologic signs of retrocochlear pathology. The patient had temporal lobe and basal ganglia lesions compatible with radiation-induced cerebral necrosis, with more significant hearing loss lateralized to the side most affected. The pattern of delayed onset and gradual progression of hearing loss is consistent with reports in the literature.

Talmi et al. (72) reported on a 35-year-old woman evaluated for a neck mass. She had noted right blindness and right hearing loss since early childhood. She gave a history of radium treatment of a large, right-sided facial hemangioma with a total dose of 23,900 rads. Her audiogram showed SNHL on the right. The SRT was 25 dB with 90% discrimination.

Elwany (73) examined middle ear mucosa in six patients who exhibited conductive hearing loss 6–11 months after a total dose of 6500–8500 rads for treatment of head and neck malignancies not involving the temporal bone. Audiometry confirmed conductive hearing loss in all patients, two of whom also had a sensorineural component. Middle ear compliance and pressure were within normal limits and tympanometry confirmed eustachian tube patency. Exploratory tympanotomy was performed and a sample of mucosa overlying the promontory was excised. Ultrastructural changes were evaluated with electron microscopy. The author summarized:

> The epithelium showed marked reduction of cytoplasmic mass, variable degrees of ciliary loss and widening of inter-cellular spaces with disruption of the macula adherens. The connective tissue stroma showed increased production of collagenous fibrous tissue and increased numbers of synthetically active fibroblasts. New gland formation with reduced activity was observed. The endothelial cells of capillaries were swollen and the basal lamina duplicated. The lumina of other capillaries were completely obliterated and replaced by a fibrous cord.

In 1988, Evans et al. (74) reported on a series of 45 patients who had received postoperative radiotherapy for unilateral parotid tumors. The study comprised of 20 patients, but two were excluded for documented prior hearing loss. The remaining 18 subjects had received 5500–6000 rads, total dose, in 200–220 rads daily fractions over 5–6 weeks. The tumor and entire temporal bone received full treatment dose, and the contralateral auditory apparatus was calculated to receive <10% of the tumor dose as scatter radiation.

Pure-tone thresholds at 500, 1000, 2000, and 4000 Hz were evaluated, with the contralateral ear serving as the control. Thresholds were found to be within 10 dB of one another, and there were no statistically significant differences. The evaluation was conducted 2–16 years (mean 8 years) after irradiation. Age ranged from 27 to 75 years (mean 65 years). The authors concluded that permanent hearing loss is unlikely if the daily fraction is <220 rads and total dose is <6000 rads. The study included no preoperative audiometrics, and frequencies >4000 Hz, associated with greater incidence and severity of hearing loss in the literature, were not evaluated. This limits the ability to rely on the data in a general population.

Chowdhury et al. (75) conducted a prospective clinical trial, using Sheppard grommet ventilating tubes in patients enrolled for radiation therapy of nasopharyngeal carcinoma, stages I–IV. The radiation therapy schedule consisted of 6000 rads in 24 fractions >6 weeks. Fields included the temporal bones. All cases were assigned to the tube (T) group ($n = 58$) or no tube (NOT) group ($n = 57$). Pure-tone audiograms for air and bone conduction were performed at 2000, 4000, and 6000 Hz, and ventilation tubes were inserted under local anesthesia. Tinnitus was assessed using a 3-point self-ranking score for severity.

When the pretreatment air–bone gaps for the two groups were compared, there was no significant difference. The conductive hearing loss present before treatment in the NOT group increased by 3 dB during treatment and hearing improved by 7 dB in the T group. This was significant. A decrease of 3 dB in the averaged sensorineural threshold also occurred at 6 months in the NOT group but was not demonstrated in the T group. These findings were statistically significant. When individual data were evaluated, five patients in the NOT group had >10 dB loss when compared with three patients in the T group.

Tinnitus scores did not differ at the preradiation evaluation. At 6 months, the NOT group had significantly worse tinnitus, and the T group showed significantly better scores. Tinnitus may be related to negative pressure, fluid in the middle ear, or changes in the cochlea, nerve or brain. The mechanism of improvement is unknown. The 6-month follow-up period was insufficient to assess delayed SNHL. However, the authors recommend placement of ventilation tubes to prevent conductive loss secondary to "radiation otitis" and possible SNHL component related to chronic conductive loss in nasopharyngeal carcinoma.

Directed radiation to the eighth nerve has been utilized in the treatment of acoustic neuromas. Hirsch and Noren (76) report on 64 patients treated with stereotactic radiosurgery who had pure-tone thresholds better than 90 dB and were followed audiologically. Goals of radiosurgical treatment were eradication of the tumor, absence of complications, preservation of facial nerve function, and preservation of hearing. The size and shape of the tumor determined treatment decisions. Tumors of ≥18 mm could always be treated; larger tumors (up to 30 mm) had to be avoided to allow dose distribution; tumors >33 mm were not treated. Dose planning was performed and during treatment the head was positioned sequentially in the focus positions. Duration of treatment was 14–20 min and the dose was 1800–2500 rads at tumor periphery, 2200–5000 rads at tumor center.

Patients were followed with pre- and posttreatment audiometry, stapedius, reflex testing, and brain stem audiometry. Preoperatively, 51 ears had hearing thresholds (pure-tone average) <50 dB, and in 11 cases discrimination scores were better than 80%. Retrocochlear hearing loss was found by stapedius reflex testing. Brainstem evoked-response audiometry (BERA) was abnormal in all but one ear involved with tumor.

At 1 year postradiosurgery, 26% of patients showed preserved hearing with pure-tone averages (at 500, 1000, and 2000 Hz) no >5 dB worse than the preoperative value and discrimination scores equal to those obtained preoperatively. Severe hearing loss (pure-tone average >90 dB) occurred in 20%, and the remaining 54% showed pronounced deterioration of thresholds and/or discrimination scores.

Gradual hearing loss was noticed in the 49 months after radiation, as permanent or fluctuant loss or a change in sound quality. Two patients noticed a verified hearing improvement, and one patient with bilateral tumors caused by neurofibromatosis suffered sudden deafness within 24 h after treatment. Graphed data on pure-tone averages for long-term follow-up revealed a cluster between 35 and 65 dB at 2–8 years following irradiation.

Other audiologic symptoms included balance disturbance in the first year after treatment ($n = 8$). Tinnitus was unchanged. Facial weakness occurred in 15% of cases within 6–9 months after treatment and was transitory. Eighteen percent of patients experienced trigeminal dysfunction with a similar latency period. The radiation dose in these patients was 5000–10,000 rads. The authors recommended a minimum dose of 2000 rads for effective tumor shrinkage and a 3000 rads maximum for hearing preservation. However, it must be remembered that these data cannot be interpreted as reflecting the effects of radiation, because radiosurgery does not cure acoustic neuromas. At best, it slows or arrests growth. Consequently, all hearing changes may be attributable to the natural progression of the underlying disease.

As part of a study supported by a grant from the National Cancer Institute, Emami et al. (77) met a task force to update tolerance doses for irradiation in a variety of body tissues. The goal was to provide three-dimensional treatment planning and dose delivery, including partial volumes to tissue receiving variable dose levels.

Data were collected only on conventional dose fractions (180–200 rads/day) for one-third, two-thirds, and whole organ exposure. For each, TD 5/5 and TD 50/5 were determined: TD 5/5 reflects the probability of a 5% rate of the complication within 5 years after treatment, and TD 50/5 denotes a 50% rate of the complication probability >5 years.

The authors offered observations regarding brain tolerance to radiation. Radionecrosis of the brain typically occurs 3 months to several years after irradiation and presents as neurologic deficits not attributable to recurrent tumor. Histologically the authors noted "pallor of white matter secondary to diffuse cerebral edema and marked changes adjacent to the tumor. These include coagulation necrosis, vascular thickening, perivascular fibrosis, calcium deposition, fibrin deposition, and chronic inflammatory infiltrates." These were thought to be attributable to radiation effects on vasculature and oligodendroctye proliferation. The normal issue complication probability in this study cited an increase in TD 5/5 for the whole brain >4500 rads but noted "some patients developed significant necrosis from partial brain doses as low as 5000 rads. A sharp increase exists for doses >6000 rads."

According to these authors, there is no good evidence that the brain stem differs from the cerebrum, although traditionally the brain stem has been regarded as more radiosensitive. Because of the higher percentage of white matter in the brain stem, a 10% decrease from total brain tolerance dose was deemed advisable, and the recommendation given was a TD 5/5 of 6000 rads for one-third volume.

In the study, the sensitivity of the ear is divided into external/middle and inner portions. The TD 50/5 for acute radiation otitis was set at 4000 rads, and, for chronic otitis, at 6500–7000 rads. The author's review of "several studies" suggested a TD 5/5 of 6000 rads and a TD 50/5 of 7000 rads for SNHL or vestibular damage. It should be noted that the studies cited arc somewhat dated (see Gamble et al. (61), Leach (66), Dias (67), and Moretti (69).

In a British publication, Singh and Slevin (78) evaluated 28 patients irradiated for residual or recurrent parotid pleomorphic adenoma. A dosing schedule of 5000 rads in 15 daily fractions >20 days was used. The interval between radiotherapy and assessment was at least 5 years with a median of 14 years. Pure-tone audiometry at 500, 1000, 2000, 4000, 6000, and 8000 Hz was performed, and the contralateral ear was used as a control. Significance levels of >10 dB at 500, 1000, and 2000 Hz and >20 dB at 4000, 6000, and 8000 Hz were selected. Bithermal caloric testing (except in cases with otitis) was also conducted.

In the study, 15 of the 28 patients had significant hearing deficit: sensorineural in twelve and mixed in three. Seven of these had semicircular canal paresis by caloric response. The remaining 13 patients had no significant hearing loss. Losses occurred at both high and low frequencies, but losses in the 2000–8000-Hz range predominated. The mean difference between ears was between 52 and 63 dB at each of the four high frequencies tested. Three of the four patients who received a dose >5250 or 5500 rads had significant hearing loss.

In their discussion, the researchers hypothesize that the differences in dose per fraction utilized in their center (330 rads as opposed to 220 rads) may account for the greater incidence and severity of hearing loss than in the study by Evans et al. (74). They suggest that the biologic dose be adjusted based on the likelihood of tumor recurrence and that patients should be informed that hearing loss may occur as a late effect of radiotherapy.

Grau et al. (79) recommend that deleterious effects of irradiation on hearing should be kept in mind in both treatment planning and follow-up after radiotherapy. They base their conclusion on a prospective assessment of 22 patients evaluated prior to and 7–84 months following radiotherapy for nasopharyngeal carcinoma. Pure-tone audiometry, SRT, discrimination scores, impedance evaluation, and BERA were performed. Bone conduction scores at 500, 1000, 2000, and 4000 Hz were considered.

The treatment protocol used external beam on laterally opposed fields with a tumor radiation dose between 6000 and 6800 rads. Eight patients received split-course treatment. One patient's inner ear fell within the penumbra of the treatment field and dosimetric calculations reduced the inner ear dose to 75% of the central dose. Bone conduction pre- and posttreatment was compared for each frequency and ear. An absolute change in baseline was then calculated. There was no correlation between age and SNHL. Radiation dose did correlate significantly with SNHL in these patients ($p < 0.01$). A significantly higher incidence of SNHL was found for doses >5000 rad when compared with lower doses ($p < 0.05$). The absolute hearing loss was greatest at 4000 Hz, but age correction indicated that the differences between frequencies were related to age. Frequencies >4000 Hz were not evaluated. Latency of cochlear damage for doses >5000 rad was evidenced by a significantly higher incidence of SNHL at observation times >18 months. Auditory brainstem evoked response in four patients was "severely abnormal," although the specific changes were not described. Two of the four also had "clinical signs of brainstem dysfunction and severe sensorineural hearing effects."

20.7. Treatment of Brainstem Tumors

Two reports provide useful reviews of the management and prognosis of thalamic, brainstem, and spinal cord tumors. Grigsby's (80) group provides retrospective analysis of 83 adults treated with a combination of surgery and irradiation, or irradiation alone between 1950 and 1984.

All patients received external beam radiation therapy. Orthovoltage irradiation was used, with five patients receiving a three-field approach. The remainder were treated with right and left lateral opposed portals, using megavoltage equipment. Only ten patients received whole brain radiation with a boost to the tumor site. The treated volume encompassed the entire tumor bed and a generous margin (2–3 cm). Unless cerebellar invasion was evident, no attempt was made to irradiate the whole cerebellum.

The central tumor-axis dose was 5040 $\pm$ 207 rads (SD) with a range from 75 to 6300 rads. The median daily fraction was 180 rads, five fractions per week. The majority of patients received a mid-plane tumor dose of 5400 rads, five fractions per week. No morbidity data were provided on survivors with regard to auditory or vestibular function, except a 9.1% incidence of eighth cranial nerve paresis.

The only factor identified by univariate analysis to be critical for survival was primary location of the disease. Supratentorial tumors in patients were associated with a 10-year disease-free survival rate of 15.4% compared with 29.6% for those with infratentorial tumors. Grigsby et al. (81) reevaluated these data and additional pediatric data with regard to prognostic variables.

Wara et al. (82) described management of primary gliomas of the brain stem and cord. Only the recommendations for the brain stem are reviewed. Radiation has been the treatment of choice for gliomas in this site. Saggital MRI was used to demonstrate the lesion for treatment planning. The recommended field included a margin of 1–2 cm surrounding the tumor. In the cases reported, the superior border was the top of the third ventricle; inferiorly the bottom of C-2 was used. The anterior border was generally the anterior clinoid process, and the posterior extent, 2 cm beyond the tumor, by scan. Historically, treatment regimens utilized total doses $<$4500 rads, but improved survival rates have been seen with a total dose of 500 rads. Hyperfractionated irradiation allowed greater nervous system tissue tolerance. In patients with brainstem tumors with poor local control, increase in the total dose by using smaller fractions (100–120 rads) allowed doses as high as 7200 rads. This improved 2-year survival rates to 50%. According to the authors, most patients showed rapid neurologic improvement with conventional protocols but failed locally within 1–2 years. The overall 5-year survival rate cited was 30%. Adults and patients with focal lesions responded better than children and individuals with diffuse lesions.

20.8. Future Research Directions

Available literature suffers from flaws of research design, dated conclusions, and an excessive number of confounding variables. The pathophysiologic mechanism of delayed SNHL is unknown, although vascular changes are generally assumed to be involved. A properly controlled study with pretreatment assessment of all clinically relevant audiometric and vestibular parameters, long-term audiometric follow-up, and postmortem histologic examination of temporal bones is sorely needed. Such a study should include as many *in vivo* windows on brain and auditory function as possible, possibly incorporating not only audiometry and BERA, but also sequential MRI, magnetic resonance angiography (MRA), single photon emission computed tomography (SPECT), positron emission tomography (PET), brain electrical activity mapping (BEAM), and other studies. Until this information is available, responsible clinicians are forced to extrapolate from risk data that are inadequate in many ways for definitive guidance of treatment decisions.

20.9. Summary

Despite weaknesses in the available literature, it appears reasonable to assume that radiation may be associated with hearing losses in some patients. The mechanism of hearing loss remains unproved, but it is generally believed to be attributable to radiation-induced cochlear or vascular changes. Interestingly, most of the studies in the literature that show significant hearing change involve radiation to the nasopharynx or parotid, with direct cochlear radiation exposure. This suggests that radiation for posterior cranial fossa neoplasms may be less damaging, possibly because it is less likely to affect the inner ear directly. It is particularly important to note that none of the published studies provides confirmation of etiology in those patients believed to have radiation-related hearing loss beginning >18 months following radiation. None of the reports in the world literature has adequate, modern neurotologic assessment of patients with supposed radiation-induced hearing loss. Rather, they merely assume that if the patient was radiated prior to developing hearing loss and there is no evidence of recurrent tumor then the hearing loss must be related to the radiation. This post hoc, ergo propter hoc reasoning is classically flawed logic. There are innumerable causes of SNHL that are extremely common in modern society, and many of them are unilateral. Because all of the patients receiving radiation have been in hospital settings, they may be at increased risk for developing many of these conditions.

Available literature supports the existence of radiation-induced otitis and radiation-induced labyrinthitis, both of which occur during, or soon after treatment. However, proof is lacking for existence of radiation-induced delayed hearing loss. Statistics suggest that it is a possibility; but in some or all cases, the hearing loss may actually be related not to the radiation, but rather to the original tumor, underlying disease, the body's response to tumor, unrelated disease, or to other undetected causes. When hearing loss or other neurologic complications occur at >18–36 months following radiation therapy, there is actually no convincing evidence that they were caused by radiation. In fact, the literature reveals an appalling failure to work-up such patients to determine the true cause of their hearing deficits. In the future, interdisciplinary collaboration among neurotologists, radiation oncologists, and other specialists should address these issues, answer many remaining questions, and substantially improve the standard of care for hearing management in patients undergoing cranial irradiation.

20.10. Endolymphatic Sac Surgery

Surgery on the endolymphatic sac was first performed in January 1926, by George Portman (83) who exposed and incised the sac. In the discussion that followed his presentation of this procedure to the American Medical Association's section on Laryngology, Otology, and Rhinology, Eagleton mentioned good results in several cases of Meniere's disease after simply uncovering the posterior fossa dura. Since Portman's operation, numerous other techniques have been developed. In 1962, William House introduced the endolymphatic-subarachnoid shunt. Reviewing House's first 50 cases, Shambaugh (84) was impressed by certain cases in which House had been unable to identify the sac and had not incised the dura. The results of these and shunted cases were often equal. Sir Terrence Cawthorne reported by letter to Shambaugh a series in which he compared endolymphatic-subarachnoid shunt, incision of the saccus, and simple decompression without opening the sac. He noted that simple decompression worked in some cases. This was felt to be due to exposure of the sac to the atmospheric pressure of the middle ear, which becomes lower than intracranial pressure anytime the patient coughs, strains, or lies down. This intermittent relative negative pressure on one wall of the sac allows it to bulge slightly and probably results in hyperemia and increased fluid resorption. Encouraged by Cawthorne's findings, Shambaugh performed a series of simple endolymphatic sac decompressions and reported a 50% cure rate for vertigo and 45% cure rate for tinnitus in his initial primary typical cases.

Numerous other procedures are available for Meniere's disease. These include the tack operation (now performed rarely if ever) the endolymphatic-mastoid shunt, cryosurgery, cochleosacultomy, and others such as destructive procedures and nerve sections. Operations directed at the sac are favored because they preserve heating in most cases. Complication rates range from 0% to 6% and include complete loss of hearing in the operated ear, cerebrospinal fluid otorrhea, hydrocephalus, shunt tube occlusion, labyrinthitis, and meningitis (85). Most authors report improvement of vertigo in between 70% and 90%, with a median of ~80% reporting relief of vertigo and 25% reporting hearing improvement (all endolymphatic sac operations combined). Numerous endolymphatic sac series have now been reported. In 1977, Smyth et al. (86) reported results of endolymphatic sac decompression without incision, which he updated in 1981 (90). Using sac decompression, he reported improvement or stabilization of hearing in one-third and elimination of incapacitating vertigo in three-quarters of his patients. Like other authors, he has concluded that there is no difference between surgery in which the sac is exposed and surgery in which the sac is incised and entered. Ford (87) and Graham and Kemink (88) confirmed these findings. It is noteworthy that they found the Arenberg valve implant successful in a small number of revision cases. In most series except Smyth's, the incidence of worsened hearing and complications appears to be lower in the noninvasive or less-invasive procedures. Strikingly, Thomsen and associated reported a double-blind study in which no significant difference was found between mastoidectomy without exposure of the sac (Sham) and endolymphatic subarachnoid shunt, the active group showing 80% improvement and the placebo group showing 73% improvement (89).

In light of the clinical success rate in patients selected for surgery because of intractable, incapacitating vertigo, endolymphatic sac surgery seems justified. However, because of the apparent nonspecificity, we favor procedures with the lowest complication rate. Procedures that violate the sac show no advantage over procedures that simply expose it. Long-term follow-up is needed to determine whether, in fact, mastoidectomy without exposure of the sac provides equally satisfactory long-term results. Consequently, at the present time endolymphatic sac decompression is the recommended procedure.

Case 3 is a 56-year-old man with incapacitating vertigo and a 10-year history of bilateral Meniere's disease. He had been troubled by right-sided ear fullness, tinnitus, distortion, and hearing loss (Fig. 10.63). Three years following right endolymphatic sac decompression, he has remained free of vertigo. Moreover, although his pure-tone thresholds remain unchanged, his discrimination is better, his tinnitus is rarely present, and he is not troubled by fullness or distortion. This case was selected to emphasize that pure-tone threshold is not the only important parameter in assessing hearing. Improvement in discrimination and relief from distortion may be extremely helpful to the patient and may make the ear much more useful even if pure-tone thresholds do not improve.

20.11. Acoustic Neuroma

Since William House revolutionized otology by making translabyrinthine and middle-fossa surgery for acoustic neuroma practical and accepted, CPA tumors have been

NAME F.A.

DATE		Right ear air conduction 250	500	1000	2000	4000	8000	Left ear air conduction 250	500	1000	2000	4000	8000
2/81	PRE	60	65	65	65	65	55	60	65	65	50	50	55
1/82	POST	65	65	70	70	65	70						

	Right ear bone conduction 250	500	1000	2000	4000	8000	Left ear bone conduction 250	500	1000	2000	4000	8000
	NR	NR	65	70	65		NR	NR	70	NR	65	

SPEECH RECEPTION: Right XXX Left XXX

DISCRIMINATION: Right PRE 24% POST 52% Left 52% 52%

Figure 10.63 Preoperative and postoperative audiogram of Case 3, Meniere's disease.

discussed widely in the literature. Early diagnosis and treatment minimize morbidity. For this reason, most otolaryngologists look for them aggressively in selected instances. Nevertheless, misconceptions about the way in which CPA tumors present clinically may result in failure to suspect the tumor and initiate complete evaluation. The delay in diagnosis may have undesirable consequences. The following cases are selected from the author's series to illustrate particular diagnostic points.

Case 4 is a 57-year-old white man who worked at a dye-casting company for 39 years. He was exposed to high-intensity noise from various sources. Recently, he had been operating a screw conveyor (steel on steel) at an intensity of ~84 dBA with intermittent peaks of 91 dBA lasting for ~2 s. He was aware of gradually progressive high-frequency sensorineural hearing loss in both ears. In October 1979, he was exposed for between 30 and 90 s to an extremely loud metal noise due to a defective piece of machinery. Three days later he saw the plant nurse complaining of decreased hearing in the left ear and tinnitus. He denied vertigo. Previous audiograms confirmed that his hearing loss had been symmetrical. The audiogram following the incident showed a severe to profound sensorineural hearing loss in the left ear. He filed a legal claim for noise-induced hearing loss.

When he was referred for evaluation, his physical examination was within normal limits except for hearing acuity. There was no decreased sensation in his left ear canal. Audiogram confirmed the hearing loss (Fig. 10.64). His discrimination score was 12%. His serological test for syphilis (MHA-TP) was negative. He had a left reduced vestibular response of 23%. Brain stem evoked-response audiometry revealed low-amplitude responses with significantly prolonged wave I–V interval. Polytomograms of the internal auditory canals and CT scan were within normal limits. Pantopaque myelogram revealed a small filling defect within the left internal auditory canal (Fig. 10.65). The right internal auditory canal appeared normal. Translabyrinthine surgery revealed an 8-mm acoustic neuroma originating from the inferior vestibular nerve. This was excised without complication.

Neither normal internal auditory canal X-rays nor a normal CT scan rules out the presence of an acoustic neuroma. In many cases, MRI with contrast or air-contrast CT scanning is necessary, although CT is used very rarely now. Sudden deafness is a particularly important symptom of acoustic neuroma because it deprives us of the usual symptoms and signs that allow early diagnosis, particularly progressive hearing loss. Therefore, it the tumor is missed at the time it produces sudden deafness, it may not be diagnosed until it is large enough to cause serious neurological symptoms and signs.

AIR CONDUCTION (AC)

DATE AND TIME	RIGHT EAR: LEFT MASK	500	1000	2000	3000	4000	6000	8000	LEFT EAR: RIGHT MASK	500	1000	2000	3000	4000	6000	8000
7/2/81		20	20	20	60	60	60	55		70	70	75	70	75	70	75

BONE CONDUCTION (BC)

DATE AND TIME	RIGHT EAR: LEFT MASK	500	1000	2000	3000	4000	LEFT EAR: RIGHT MASK	500	1000	2000	3000	4000	Response Reliability: Right	Left
7/2/81		15	15	35	60			50	65	↓	60			

DISCRIMINATION

RIGHT: DATE	% SCORE	TEST LEVEL	LIST	LEFT MASK	LEFT: % SCORE	TEST LEVEL	LIST	RIGHT MASK
	96%	65	1AA		56%	MLL 80	1AB	*

SPEECH RECEPTION

DATE	RIGHT	LEFT MASK	LEFT	RIGHT MASK
7/2/81	20		65	*

Figure 10.64 Audiogram of Case 4 illustrating severe left-sided hearing loss. This pattern is unlikely to be due to free-field noise exposure.

Some authors feel that as many as 10% (90) to 15% (91) of all acoustic neuromas may present with sudden deafness. Reviews of this subject by Sataloff and associates (92) stress the need for greater awareness of this diagnostic problem.

Case 5 is a 78-year-old woman who had been followed for 7 years in an excellent teaching institution of bilateral hearing loss and severe left tinnitus. She had numerous cardiovascular problems and renal disease. Her workup for acoustic neuroma had included normal polytomograms of the internal auditory canals. However, a CT scan had not been performed, and MR was not available at the time.

When she was referred, her audiogram revealed profound deafness on the left and severe hearing loss on the right with 28% discrimination (Fig. 10.66). Her CT scan (Fig. 10.67) revealed a 5-cm acoustic neuroma. This was removed by a planned two-stage procedure. Translabyrinthine partial excision was used to debulk the tumor and preserve the facial nerve. Residual tumor was removed from the brain stem and lower cranial nerves ~2 weeks later. Often we perform such procedures in one stage.

It is important to remember that bilateral hearing loss even with "obvious" etiologies such as vascular disease and renal disease does not rule out acoustic neuroma, especially if the hearing loss is asymmetrical. Moreover, internal auditory canal bony architecture may be normal even in large tumors.

Case 6 is a 38-year-old attorney who presented with left-sided tinnitus and progressive hearing loss worsening >3 years (Fig. 10.68). CT scan 3 years earlier had been negative. A recent brain stem evoked-response audiogram was normal. CT scan revealed a left internal auditory canal neoplasm (Fig. 10.69). Because of bony erosion of the lateral portion of the internal auditory canal, no attempt was made to preserve hearing. Translabyrinthine surgery revealed a facial nerve neuroma that appeared to originate from the nervus intermedius.

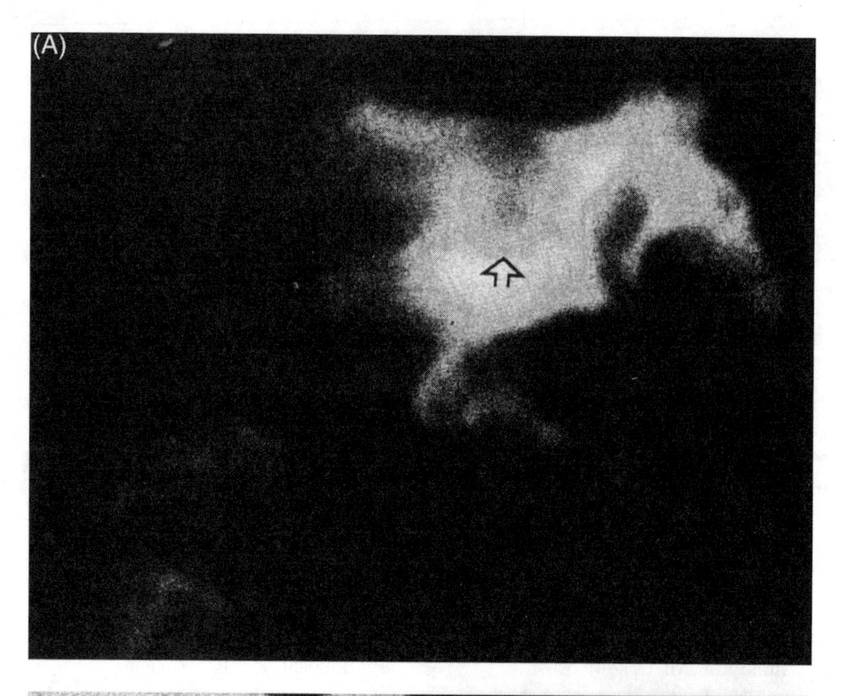

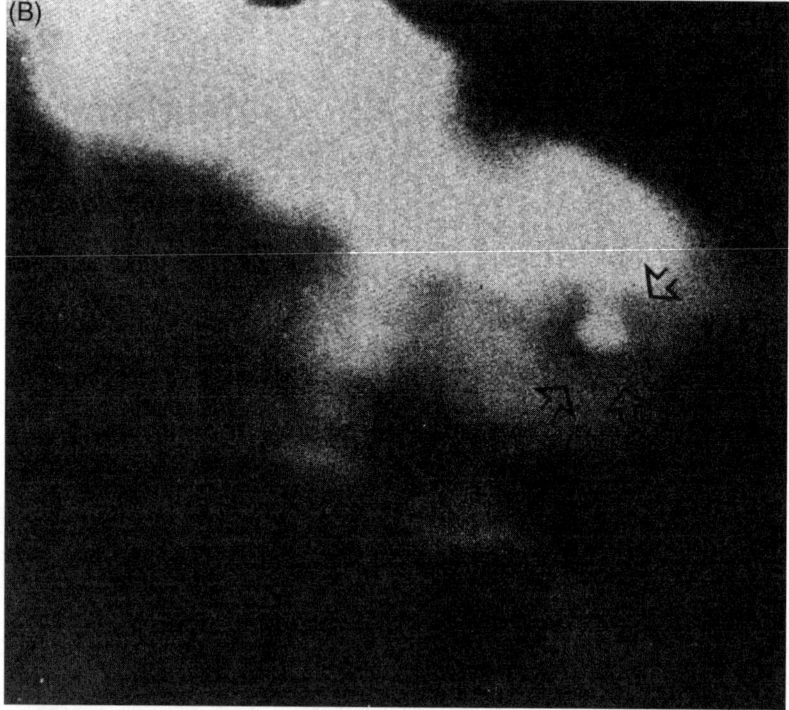

Figure 10.65 (A) Left internal auditory canal of Case 4. The size is normal, and the right internal auditory canal is the same size. (B) Myelogram of the left internal auditory canal of Case 4 illustrating Pantopaque entering, but no filling, the internal auditory canal. The medial wall of the vestibule is marked with a straight arrow. The superior semicircular canal is marked with a curved arrow. The contralateral side fulled completely. Myelograms have been replaced by MRI scans and air-contrast CT scans but still are the only study available in some places of the world.

NAME A.B.

DATE	RIGHT EAR AIR CONDUCTION							LEFT EAR AIR CONDUCTION					
	250	500	1000	2000	4000	8000		250	500	1000	2000	4000	8000
1/84	65	70	65	55	70	70		NR	NR	NR	NR	NR	NR

RIGHT EAR BONE CONDUCTION							LEFT EAR BONE CONDUCTION					
NR	65	65	55	55								

SPEECH RECEPTION: Right XXX Left XXX DISCRIMINATION: Right 28% Left ____

Figure 10.66 Preoperative audiogram of Case 5 showing profound hearing loss on the left and severe sensorineural hearing loss with poor discrimination on the right. The patient had a left acoustic neuroma.

Considering the findings of a basin-shaped audiogram, normal brain stem evoked-response audiogram, and previous normal CT scan, one could easily have missed this neoplasm. This case emphasizes the importance of constant suspicion and comprehensive evaluation until a definitive diagnosis is reached. Acoustic neuromas may be associated with any audiometric pattern.

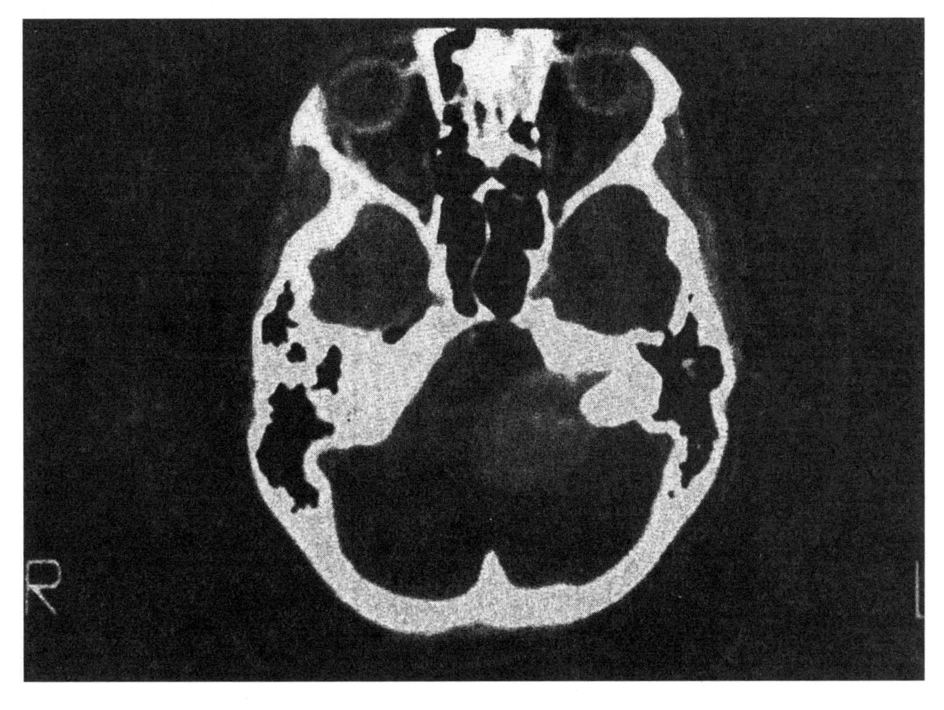

Figure 10.67 CT scan of Case 5 showing large left acoustic neuroma with normal internal auditory canal.

NAME M.W.M.

DATE	RIGHT EAR AIR CONDUCTION							LEFT EAR AIR CONDUCTION					
	250	500	1000	2000	4000	8000		250	500	1000	2000	4000	8000
10/81	10	20	20	15	15	25		20	30	75	65	30	55

RIGHT EAR BONE CONDUCTION							LEFT EAR BONE CONDUCTION					
3/83	10	10	10	0			10	20	20	0		

SPEECH RECEPTION: Right XXX Left XXX DISCRIMINATION: Right 100% Left 60%

Figure 10.68 Basin-shaped audiogram of Case 6, left facial nerve neuroma.

Case 7 is a 34-year-old woman whose only complaint was left-sided tinnitus. Her audiogram revealed left-sided hearing loss (Fig. 10.70). Her physical examination showed abnormalities of cranial nerves V, VII (sensory), VIII, and IX. Her ENG revealed left reduced vestibular response, and her brain stem evoked-response audiogram was

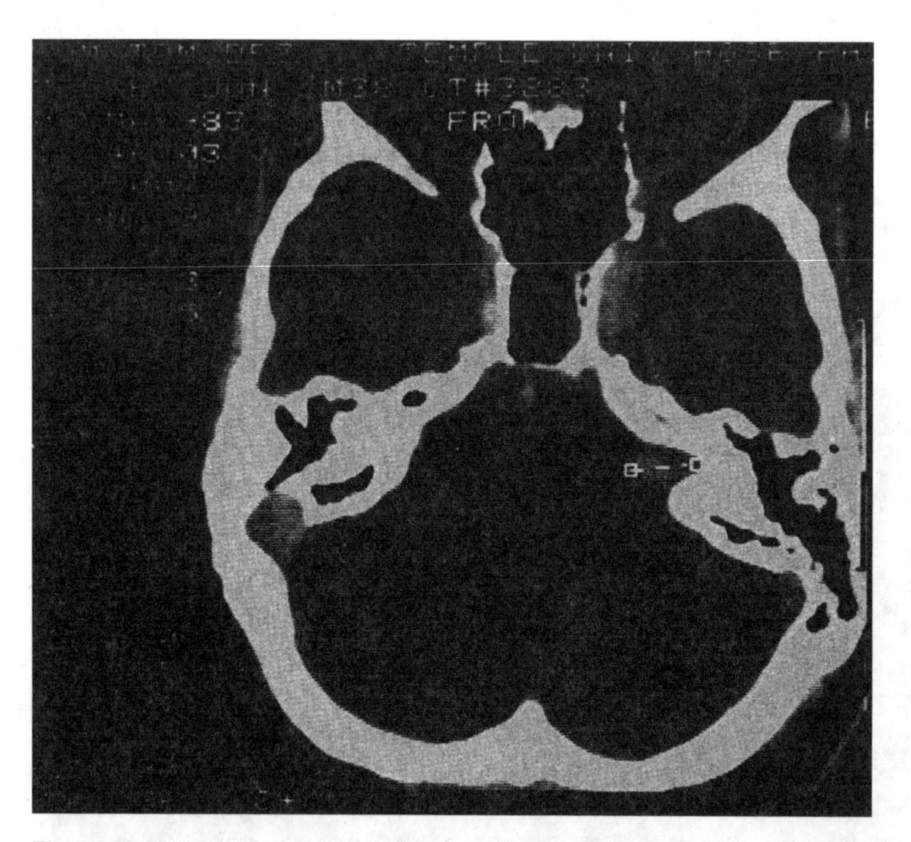

Figure 10.69 CT scan of Case 6 showing neoplasm of the left internal auditory canal, with erosion of the lateral aspect of the canal. This finding suggests a low probability of preserving hearing with total tumor removal. Patient underwent resection of a facial nerve neuroma.

NAME L.T.

DATE	RIGHT EAR AIR CONDUCTION 250	500	1000	2000	4000	8000		LEFT EAR AIR CONDUCTION 250	500	1000	2000	4000	8000
11/82	5	10	5	0	5	0		15	20	20	50	50	65

RIGHT EAR BONE CONDUCTION							LEFT EAR BONE CONDUCTION					

SPEECH RECEPTION: Right XXX Left XXX DISCRIMINATION: Right 100% Left 64%

Figure 10.70 Audiogram of Case 7.

abnormal. CT scan revealed a 6-cm acoustic neuroma (Fig. 10.71). The tumor was removed in two stages with preservation of the facial nerve. Because of the patient's young age, occasional viral infections and "cold sores," and her other cranial nerve abnormalities, it might have been tempting to diagnose postherpetic cranial polyneuropathy.

The case is presented to emphasize the occurrence of acoustic neuromas in young people, in whom this disease is relatively common (94). In the young, it is not unusual to see large tumors with minimal symptoms and signs.

Case 8 is a 71-year-old male with a long history of bilateral hearing loss. However, he had noted increased trouble hearing from his left ear for ~2 months. He also had had one episode of mild disequilibrium lasting 1–2 weeks, resolving completely. He had no tinnitus. His audiogram (Fig. 10.72) shows only mild asymmetry that could easily have been

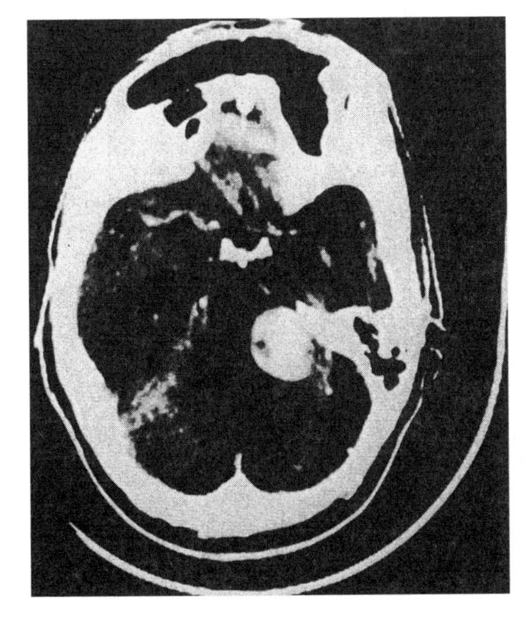

Figure 10.71 Large left acoustic neuroma of Case 7.

JOSEPH SATALOFF, M.D.
ROBERT THAYER SATALOFF, M.D.
1721 PINE STREET PHILADELPHIA, PA 19103

NAME N.K.

DATE	RIGHT EAR AIR CONDUCTION							LEFT EAR AIR CONDUCTION					
	250	500	1000	2000	4000	8000		250	500	1000	2000	4000	8000
	15	20	35	45	60	65		20	30	45	65	70	85

RIGHT EAR BONE CONDUCTION							LEFT EAR BONE CONDUCTION					
5	15	40	55	55	65		5	10	35	70	65	NR

SPEECH RECEPTION: Right — — Left ——— DISCRIMINATION: Right 88% Left 32%

Figure 10.72 Audiogram of 71-year-old male with a long history of bilateral hearing loss. Audiogram shows only mild asymmetry.

overlooked during routine testing. The patient's history of unilateral change, and his poor discrimination score on the left, resulted in additional testing. Brain stem evoked-response audiogram was abnormal, ENG revealed left reduced vestibular response, CT scan revealed mild enlargement of the left internal auditory canal (Fig. 10.73), and MRI with Gadolinium–DTPA contrast showed an acoustic neuroma filling the internal auditory canal and extending into the cerebellopontine angle (Fig. 10.74).

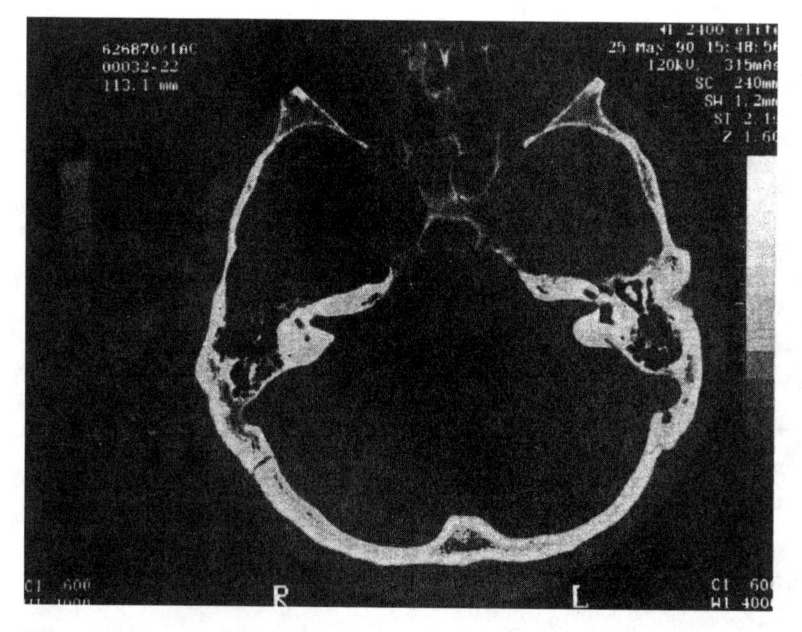

Figure 10.73 CT scan shows mild enlargement of the left internal auditory canal.

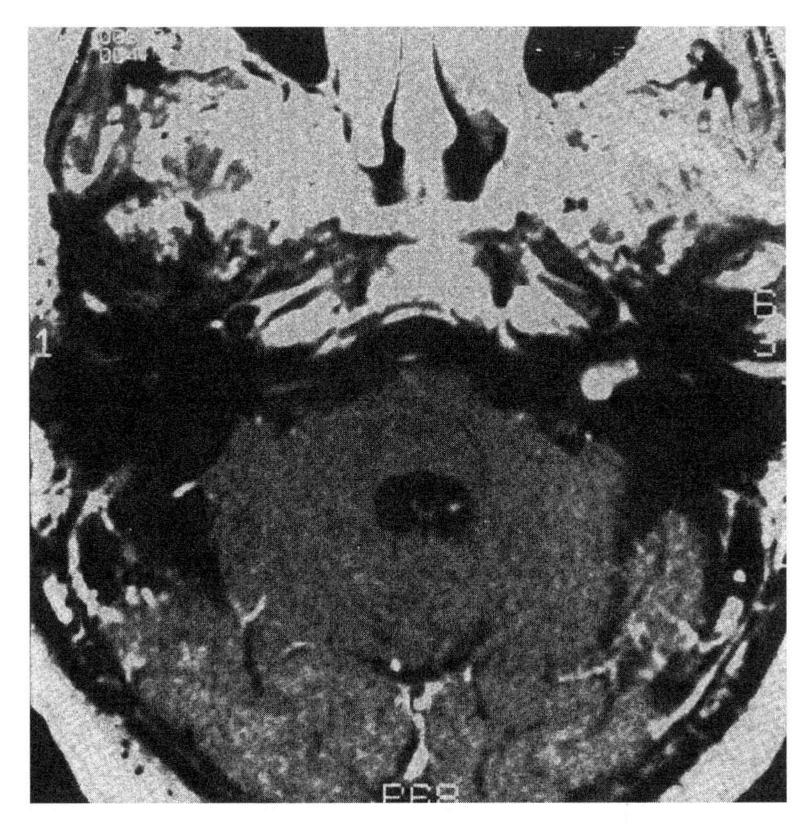

Figure 10.74 MRI with Gadolinium–DTPA contrast shows an acoustic neuroma filling the internal auditory canal and extending into the cerebellopontine angle.

Case 9 is a 50-year-old man. Fourteen years ago he had a 2-week episode of rotary vertigo without hearing loss or tinnitus. Thereafter, he was fine until 5 years ago when he developed sudden vertigo with no apparent etiology or antecedent precipitating event. He remained constantly off balance and had attacks of moderate to severe vertigo two to three times per week, thereafter. Progressive, fluctuating hearing loss began shortly after his attacks of vertigo 5 years ago. He also has right tinnitus. Comprehensive metabolic and autoimmune evaluation, CT, and MRI were normal. ENG showed right reduced vestibular response, and BERA revealed a cochlear pattern on the right. Middle-ear exploration for fistula revealed no perilymph leak and produced no improvement. Air-contrast CT showed a vascular loop [Fig. 10.75(B)] at the anterior inferior cerebellar artery (straight arrow) entering the internal auditory canal (IAC) and compressing the neurovascular bundle (curved arrow).

20.12. Cochlear Implant

In years since the first report of electrical stimulation of the auditory nerve (95), great progress has been made (96). Primarily because of the pioneering contributions of William House and colleagues in Los Angeles, cochlear implants now provide invaluable help to properly selected patients. Selection and rehabilitation are the essential elements

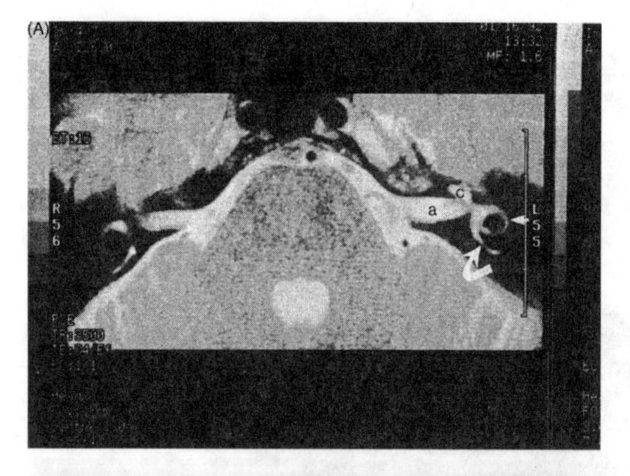

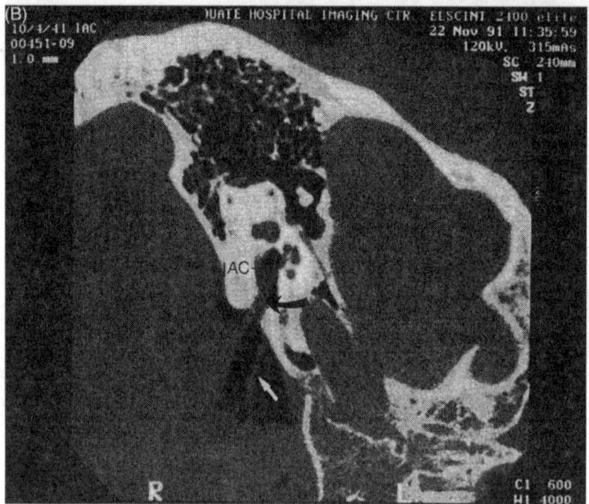

(C)

JOSEPH SATALOFF, M.D.
ROBERT THAYER SATALOFF, M.D.
1721 PINE STREET PHILADELPHIA, PA 19103

NAME ______________________________

DATE	RIGHT EAR AIR CONDUCTION							LEFT EAR AIR CONDUCTION					
	250	500	1000	2000	4000	8000		250	500	1000	2000	4000	8000
	40	35	65	60	55	65		10	5	0	-5	-5	0

RIGHT EAR BONE CONDUCTION							LEFT EAR BONE CONDUCTION					
40	35	55	55	55								

SPEECH RECEPTION: Right 45 Left 0 DISCRIMINATION: Right 76% Left 100%

(Each ear is tested separately with pure tones for air conduction and bone conduction, if necessary. The tones increase in pitch in octave steps from 250 to 8000 Hz. Normal hearing in each frequency lies between 0 and 25 db The larger the number > 25 db in each frequency the greater the hearing loss. When the two ears differ greatly in threshold, one ear is masked with noise to test the other ear. Speech reception is the patient's threshold for everyday speech, rather than pure tones. A speech reception threshold of > 30 db is handicapping in many situations. The discrimination score indicates the ability to understand selected test words at a comfortable level above the speech reception threshold.)

Figure 10.75 (Legend on facing page.)

(D)

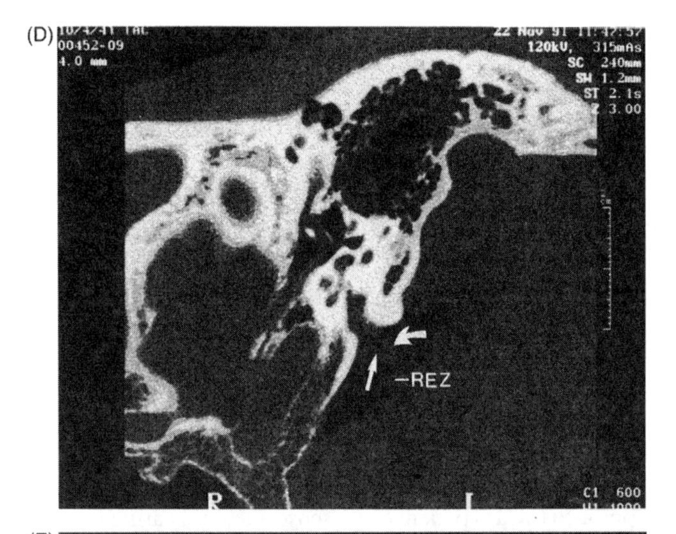

(E)

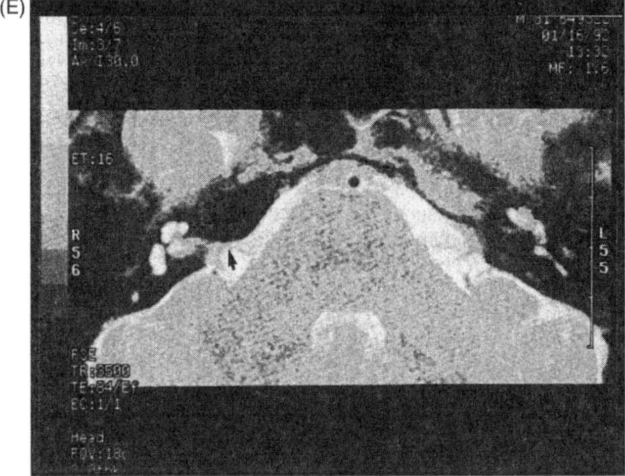

Figure 10.75 (*Continued*) (A) Normal, high-resolution MRI presented for comparison showing the eighth cranial nerve (a), cochlear (c), horizontal semicircular canal (straight arrow), and posterior semicircular canal (curved arrow). (B) Air CT reveals an anterior inferior cerebellar artery vascular loop (straight arrow) entering the internal auditory canal (IAC) and compressing the neurovascular bundle (curve arrow). (C) Audiogram of Case 9. (D) This air-contrast CT of another ear shows another anterior inferior cerebellar artery vascular loop (straight arrow) deflecting the eight nerve complex (curved arrow) and causing typical thickening at the root entry zone (REZ). (E) High-resolution MRI scan of another patient showing the position of the anterior inferior cerebellar artery (arrow). It is likely that improved MRI and MR angio will eventually replace air CT in the evaluation of vascular loop compression syndrome.

of a successful cochlear implant program. The surgery is technically easy for an experienced otologist. However, selecting appropriate patients requires extensive, highly specialized testing, and experience. Training a patient to use the implant postoperatively involves many days of highly skilled rehabilitation. In treating patients with profound sensorineural deafness who cannot be helped adequately with hearing aids, cochlear implants have proven worthwhile despite present technological limitations. Any patient with sensorineural hearing loss of profound severity deserves evaluation to determine his suitability for cochlear implant surgery.

In addition, criteria for cochlear implantation have changed as technology has improved. Now, even some patients with severe (as opposed to profound) sensorineural hearing loss are appropriate candidates for cochlear implantation. At present, if pure-tone thresholds are much worse than ~70 dB, and if discrimination is worse than ~30%, cochlear implantation may be appropriate. As technology advances further, criteria will continue to be modified.

21. CONCLUSION

Many causes of sensorineural heating loss lend themselves to medical or surgical therapy. In addition to those discussed, vascular compression and other entities may also be managed by appropriate operative intervention. It is important for physicians to maintain a positive attitude and diligent approach toward patients with sensorineural hearing loss. In this way, we will not only find more patients whom we can help now, but we will also find more ways to help other patients in the future.

REFERENCES

1. Rasmussen AT. Outlines of Neuroanatomy, Dubuque, Iowa: W.C. Brown, 1947.
2. Spoendlin H. The Organization of the Cochlear Receptor. Basel, Switzerland: S. Karger, 1966.
3. Lim DJ. Functional structure of the organ of Corti: a review. Hear Res 1986; 22:117–146.
4. Békésy G. Experiments in Hearing. New York, NY: McGraw-Hill, 1960.
5. Kim DO. Active and nonlinear cochlear biomechanics and the role of outer-hair-cell subsystem in the mammalian auditory system. Hear Res 1986; 22:105–114.
6. Dallos P. Cochlear Neurobiology: Revolutionary Developments, ASHA, Vol. June/July, 1988; 50–55.
7. Walters JG. The effect of salicylates on the inner ear. Ann Otol 1955; 64:617.
8. Harbert F, Young IM. Threshold auditory adaptation measured by tone decay test and Békésy audiometry. Ann Otol 1964; 73:48.
9. Schucknecht HF. Pathology of presbycusis in Geriatric Otolaryngology. In: Hamilton, Ont, eds. Canada: BC Decker Inc., 40–44.
10. Pauler M, Schucknecht HF, White JA. Atrophy of the stria vascularis as a cause of sensorineural hearing loss. Laryngoscope 1988; 98(7):754–759.
11. Moscicki EK, Elkins EF, Baum HM, McNamara PM. Hearing loss in the elderly: an epidemiologic study of the Framingham Heart Study Cohort. Ear Hear 1985; 6(4):184–190.
12. Susac JO, Hardiman JM, Selhost JB. Microangiopathy of the brain and retina. Neurology 1979; 29:313–316.
13. McCabe BF. Autoimmune sensorineural hearing loss. Ann Otol Rhinol Laryngol 1979; 88:585–589.
14. Boulassel MR, Deggouj N, Tomasi JP, Gersdorff M. Inner ear autoantibodies and their targets in patients with autoimmune inner ear diseases. Acta Otolaryngol (Stockh) 2001; 121:28–34.
15. Veldman JE. Immunology of hearing: experiments of nature. Am J Otol 1989; 10:183–187.
16. Hughes GB, Barna BP, Kinney SE, Calabrese LH, Nalepa NJ. Clinical diagnosis of immune inner-ear disease. Laryngoscope 1989; 98:251–253.
17. Veldman JE. Cochlear and retrocochlear immune-mediated inner ear disorders. Pathogenetic mechanisms and diagnostic tools. Ann Otol Rhinol Laryngol 1986; 95:535–540.

18. Sismanis A, Wise CM, Johnson GD. Methotrexate management of immune-mediated cochleovestibular disorders. Otolaryngol Head Neck Surg 1997; 116:146–152.
19. Hoistad DL, Schachern PA, Paparella MM. Autoimmune sensorineural hearing loss: a human temporal bone study. Am J Otolaryngol 1998; 19:33–39.
20. Veldman JE, Roord JJ, O'Connor AF, Shea JJ. Autoimmunity and inner ear disorders: an immune-complex mediated sensorineural hearing loss. Laryngoscope 1994; 94:501–507.
21. Arnold W. Systemic autoimmune diseases associated with hearing loss. Ann NY Acad Sci 1997; 830:187–202.
22. Harris JP. Autoimmunity of the inner ear. Am J Otol 1989; 10:193–195.
23. Ryan AF, Gloddek B. Harris JP. Lymphocyte trafficking to the inner ear. Ann NY Acad Sci 1997; 830:236–242.
24. Lasak JM, Sataloff RT, Hawkshaw M, Carey TE, Lyons KM, Spiegel JR. Autoimmune inner ear disease: steroid and cytotoxic drug therapy. ENT J 2001; 80:808–811, 815–816, 818.
25. Hirose K, Wener MH, Duckert LG. Utility of laboratory testing in autoimmune inner ear disease. Laryngoscope 1999; 109:1749–1754.
26. Carey TE, Nair TS, Cray JP et al. The search for the inner ear antigen of autoimmune sensorineural hearing loss. In: Veldman JE, Passali D, Lim DJ, eds. New Frontiers in Immunobiology. The Hague, the Netherlands: Kugler, 1999:67–74.
27. Harris JP, Ryan AF. Fundamental immune mechanisms of the brain and inner ear. Otolaryngol Head Neck Surg 1995; 112:639–653.
28. Clements PJ. Alkylating agents. In: Dixon J, Furst DE, eds. Second Time Agents in the Treatment of Rheumatoid Diseases. New York, NY: Marcel Dekker, 1991:336–361.
29. Songsiridej N, Furst DE. Methotrexate—the rapidly acting drug. Baillieres Clin Rheumatol 1990; 4:575–593.
30. Matteson EL, Tirzaman O, Facer GW et al. Use of methotrexate for autoimmune hearing loss. Ann Otol Rhinol Laryngol 2000; 109:710–714.
31. Salley LH Jr, Grimm M, Sismanis A, Spencer RF, Wise CM. Methotrexate in the management of immune mediated cochleovestibular disorders: clinical experience with 53 patients. J Rheumatoid 2001; 28:1037–1040.
32. Sismanis A, Thompson T, Willis HE. Methotrexate therapy for autoimmune hearing loss: a preliminary report. Laryngoscope 1994; 104:932–934.
33. Zavod MB, Sataloff RT, Rao VM. Frequency of cochlear enhancement on magnetic resonance imaging in patients with autoimmune sensorineural hearing loss. Arch Otoloryngol Head Neck Surg 2000; 126:969–971.
34. Gupta R, Sataloff RT. Noise-induced autoimmune sensorineural hearing loss. Otol Rhinol Laryngol 2003; 112(7):569–573.
35. Gates G, Schmid P, Kujawa SG, Nam B, D'Agostino R. Longitudinal threshold changes in older men with audiometric notches. Hear Res 2000; 141:220–228.
36. Yaremchuk KL, Dickson L, Burk K, Shivapuja BG. Noise level analysis of commercially available toys. JOHL 1999; 2(4):163–170.
37. Pearson BW, Barker HO. Head injury-same otoneurological sequelae. Arch Otol 1973; 97:81–84.
38. Rubin W. Whiplash and vestibular involvement. Arch Otol 1973; 97:85–87.
39. Tuohimaa P. Vestibular disturbances after acute mild head trauma. Acta Otol (Stockh) 1979; 87(suppl 359):1–67.
40. Schuknecht HF. A clinical study of auditory damage following blows to the head. Ann Otol 1950; 59:331–359.
41. House WF, Glorig A. Criteria for otosclerosis surgery and further experience with round window surgery. Laryngoscope 1960; 70:616–630.
42. Sheehy JL. Surgical correction of far advanced otosclerosis. Otolaryngol Clin North Am 1978; 11(1):121–123.
43. Frattali MA, Sataloff RT. Far-advanced otosclerosis. Ann Otol Rhinol Laryngol 1993; 102(6):433–437.

44. House WF. Oval window and round window surgery in extensive otosclerosis. Laryngoscope 1959; 69:693–701.
45. Wilis R. Severe otosclerosis. J Laryngol Otol 1963; 77:250–258.
46. Myers D, Wolfson RJ, Tibbels EW Jr, Winchester RA. Apparent total deafness due to advanced otosclerosis. Arch Otolaryngol 1963; 78:52–58.
47. Sheehy JL. Far-advanced otosclerosis. Diagnostic criteria and results of treatment; report of 67 cases. Arch Otolaryngol 1964; 80:244–248.
48. Sellars SL. Surgery of advanced otosclerosis. S Afr Med J 1972; 1:434–437.
49. Wiet RJ, Morganstein SA, Zwolan TA, Pitgon SM. Far advanced otosclerosis. Cochlear implantation vs. stapedectomy. Arch Otolaryngol Head Neck Surg 1987; 113:299–302.
50. Babighian G, Smadsi M, Galaverni G, de Min G. Extreme stapedectomy. An Otorrinolaringol Ibero Am 1991; 18:239–248.
51. Lindsay JR, Hemenway WG. Occlusion of the round window by otosclerosis. Laryngoscope 1954; 164:10–14.
52. Myers EN, Myers D. Stapedectomy in advanced otosclerosis: a temporal bone report. J Laryngol Otol 1968; 82:557–564.
53. Schuknecht HF, Barber W. Histologic variants in otosclerosis. Laryngoscope 1985; 95:1307–1317.
54. Von Glass W, Philipp A. Imaging of capsule otosclerosis using computerized tomography. HNO 1988; 36:373–376.
55. House WF, Glorig A. Criteria for otosclerosis surgery and further experiences with round window surgery. Laryngoscope 1960; 70:616–630.
56. Sataloff RT, Rosen DC. Effects of cranial irradiation on hearing acuity: a review of the literature. Amer J Otol 1994; 15(6):772–780.
57. Ewald CA. Die wirkung des radium auf das labyrinth. Zentralbl Physiol 1905; 10:298–299.
58. Girden E, Culler E. Auditory effects of roentgen rays in dogs. Am J Roentg 1933; 30:215–220.
59. Novotny O. Sull'azione dei raggi x sulla chiocciola della cavia. Arch Ital Otol Rhinol Laringol 1951; 62:15–19.
60. Kozlow MJ. Changes in the peripheric section of the auditory analyzer in acute radiation sickness. ORL J Otorhinolaryngeal Relat Spec 1959; 21:29–35.
61. Gamble JE, Peterson EA, Chandler JR. Radiation effects on the inner ear. Arch Otolaryngol 1968; 88:64–69.
62. Keleman G. Radiation and ear. Acta Otolaryngol Suppl (Stockh) 1963:184.
63. Bohne BA, Marks JE, Glasgow GP. Delayed effects of ionizing radiation on the ear. Laryngoscope 1985; 95:818–828.
64. Evans RA, Liu KC, Azhar T, Symonds RP. Assessment of permanent hearing impairment following radical megavoltage radiotherapy. J Larynol Otol 1988; 102:588–589.
65. Borsanyi S, Blanchard C. Ionizing radiation and the ear. JAMA 1962; 181(11):958–961.
66. Leach W. Irradiation of the ear. J Laryngol Otol 1965; 79:870–880.
67. Dias A. Effects on the hearing of patients treated by irradiation in the head and neck area. J Laryngol Otol 1966; 80:276–287.
68. Schuknecht HF, Karmody CS. Radionecrosis of the temporal bone. Laryngoscopy 1966; 76:1416–1428.
69. Moretti JA. Sensorineural hearing loss following radiotherapy to the nasopharynx. Laryngoscope 1976; 86:598–602.
70. Thibadoux CM, Pereira WX, Hodges JM, et al. Effects of cranial radiation on hearing in children with acute lymphocytic leukemia. J Pediatr 1980; 96:405–406.
71. Coplan J, Post E, Richman R, Grimes C. Hearing loss after therapy with radiation. Am J Dis Child 1981; 135:1066–1067.
72. Talmi Y, Kalmanowitch M, Zohar Y. Thyroid carcinomas, cataract and hearing loss in a patient after irradiation for facial hemangioma. J Laryngol Otol 1988; 102:91–92.

73. Elwany S. Delayed untrastructural radiation-induced changes in the human mesotympanic middle ear mucosa. J Laryngol Otol 1985; 99:343–353.
74. Evans RA, Liu KC, Azar T, et al. Assessment of permanent hearing impairment following radical megavoltage radiotherapy. J Larynol Otol 1988; 102:588–589.
75. Chowdhury CR, Ito JHC, Wright A, et al. Prospective study of the effects of ventilation tubes on hearing after radiotherapy for carcinomas of the nasopharynx. Ann Otol Rhinol Laryngol 1988; 97:142–145.
76. Hirsch A. Noren G. Audiological findings after stereotactic radiosurgery in acoustic neurinomas. Acta Otolaryngol (Stockh) 1988; 106:244–251.
77. Emami B, Lyman J, Brown A et al. Tolerance of normal tissue to theraputic irradiation. Int J Radiat Oncol Biol Phys 1991; 21:109–122.
78. Singh IP, Slevin NJ. Late audiovestibular consequences of radical radiotherapy to the parotid. Clin Oncol (R Coll Radio) 1991; 3:217–219.
79. Grau C, Moller K, Overgaard M, Overgaard J, Elbrond O. Sensorineural hearing loss in patients treated with irradiation for nasopharyngeal carcinoma. Int J Radiat Oncol Biol Phys 1991; 21:723–728.
80. Grigsby P, Garcia D, Simpson J, et al. Prognostic factors and results of therapy for adult thalamic and brainstem tumors. Cancer 1988; 11:2124–2129.
81. Grigsby P, Thomas P, Schwartz H, Fineberg B. Multivariate analysis of prognostic factors in pediatric and adult thalamic and brainstem tumors. Int J Radiat Oncol Biol Phys 1989; 16:649–655.
82. Wara W, Lindstadt D, Larson D. Management of primary brainstem gliomas and spinal gliomas. Semin Radiat Oncol 1991; 1(1):50–53.
83. Portman G. Vertigo: surgical treatment by opening of the saccus endolymphaticus, Arch Otol 1927; 6:309.
84. Shambaugh GE. Surgery of the endolymphatic sac. Arch Otol 1966; 83:29–39.
85. Snow B, Kimmelman CP. Assessment of surgical procedures for Meniere's disease. Laryngoscope 1979; 89:737–747.
86. Smyth GDL, Hassard TH, Kerr AG. The surgical treatment of vertigo. Am J Otol 1981; 2(3):179–187.
87. Ford CN. Results of endolymphatic sac surgery in advanced Meniere's disease. Am J Otol 1982; 3(4):339–342.
88. Graham MD, Kemink JL. Surgical management of Meniere's disease with endolymphatic sac decompression by wide bony decompression of the posterior fossa dura: technique and results. Laryngoscope 94(5):680–683.
89. Thomsen J, Bretlau P, Tos M, Johnsen NJ. Placebo effect in surgery for Meniere's disease. Arch Otolaryngol 1981; 107:271–277.
90. Summerfield MJ. Deafness in adults. NZ Med J 1978; 87:440–442.
91. Meyerhoff WL. When a person suddenly goes deaf. Med Times 1980; 108:25s–33s.
92. Sataloff RT, Davies B, Myers DL. Acoustic neuromas presenting as sudden deafness. Am J Otol 1985; 6(4):349–352.
93. Schmidt RJ, Sataloff RT, Newman J, Spiegel JR, Myers DL. The sensitivity of auditory brainstem response testing for the diagnosis of acoustic neuromas. Arch Oto HNS 2001; 127:19–22.
94. Graham MD, Sataloff RT. Acoustic tumors in the young adult. Arch Otolaryngol 1984; 110(6):405–497.
95. Djourno A, Eyries C. Prothese auditive per excitation electrique a distance due nerf sensorial a l'aide d'um bobinage inclus a demeure. Presse Med 1957; 35:14–17.
96. House WF, Berliner KI. Cochlear implants: progress and perspectives. Ann Otol Rhinol Laryngol 1982; 91(2, Part 3):7–24.

11
Sudden Sensorineural Hearing Loss

Jeffrey M. Zimmerman
Ear, Nose and Throat Specialists of Southern New Hampshire, Manchester, New Hampshire, USA

Robert T. Sataloff
Drexel University College of Medicine, American Institute for Voice and Ear Research, and Graduate Hospital, Philadelphia, Pennsylvania, USA

Heidi Mandel
Philadelphia College of Podiatric Medicine, Philadelphia, Pennsylvania, USA

Steven Mandel
Thomas Jefferson University, Philadelphia, Pennsylvania, USA

The term sudden sensorineural hearing loss (SSNHL) was introduced in 1944 by DeKleyn (1). A variety of definitions ensued and continue to exist in the literature today. However, most authorities would agree that SSNHL refers to a sensorineural hearing loss (SNHL) of ~30 dB over three contiguous audiometric frequencies occurring within a 3-day period or less. The condition affects ~4000 Americans yearly and accounts for ~1% of all cases of SNHL (2). All ages are affected; however, the incidence increases with age with a peak in the 50–60-year range (3). There is no sexual or geographic predominance and no laterality.

The proposed etiologies of sudden hearing loss can be categorized as follows: infectious, vascular, traumatic, neoplastic, immunologic, neurologic, metabolic, and toxic. In the overwhelming majority of patients there remains no identifiable etiology for SSNHL, despite the fact that over the years more than 100 possible causes have been suggested (3). In fact, a cause can be found in only 10–15% of patients with the remainder labeled "idiopathic" (4–6).

In addition to the conditions discussed in this chapter, additional causes, treatments and examples of SSNHL are discussed in Chapter 10.

1. INFECTIOUS

Bacteria, viruses, and fungi have all been implicated in SSNHL. Of these infectious agents, viruses are believed, by some, to be the most common culprit, causing viral cochleitis or viral neuritis. The most convincing evidence of virus-induced sudden hearing loss is the presence of viruses and/or viral particles in the cochlea. Westmore et al. (7) was able to isolate mumps virus from the inner ear of a patient with sudden hearing loss and Davis et al. (8) isolated cytomegalovirus (CMV) from the perilymph of an infant with congenital CMV. Davis and Johnson (9) were able to induce a mumps labyrinthitis in hamsters by inoculating the animals' subarachnoid space. Seroconversion and increased viral titers suggest a viral etiology in selected patients with SSNHL. Studies have demonstrated an increased incidence of seroconversion to various viruses including herpes simplex, herpes zoster, CMV, influenza, parainfluenza, mumps, measles, and adenovirus but fail to demonstrate a relationship between titer results, severity of hearing loss, and prognosis (10–12). Further evidence for a viral etiology is the finding of temporal bone histopathologic changes consistent with viral infection. Schuknecht et al. (13) demonstrated atrophy of the organ of Corti, tectorial membrane, stria vascularis, cochlear nerve, and vestibular organs in cases of known viral labyrinthitis, and multiple authors have demonstrated similar findings in patients with a history of sudden hearing loss (14–16).

Evaluation includes an extensive search for direct or indirect proof of an infection. Laboratory work-up includes a CBC with differential to evaluate for an increase in the white blood cell count in addition to an evaluation of viral titers. At the time of presentation titers are drawn for herpes, influenza A and B, coxsackie virus, toxoplasmosis and CMV. A repeat set is drawn ideally several weeks later, to evaluate during convalescence for a change in titre suggesting recent infection. Evaluation for Lyme disease can be performed, and syphilis should be ruled out with an fluorescent treponemal antibody absorption (FTA-ABS) or microhemagglutination assay-*Treponema palladium* (MHA-TP) which is to be run regardless of negative VDRL or RPR result, as luetic labyrinthitis is typically a form of late, tertiary syphilis.

Viral infection may be treated with acyclovir (400 mg po 5×/day × 10 days) in patients presenting with sudden hearing loss because of the high suspicion of a viral etiology. In addition, many physicians elect to prophylactically treat patients with doxycycline 100 mg po bid against Borrelia burgdorferi infection while awaiting lyme titres.

2. VASCULAR

Given the cochlea's dependence upon a single terminal branch of the labyrinthine artery, the idea of a vascular etiology for sudden hearing loss is very attractive. The cochlea has been shown to be highly intolerant of ischemic insult as demonstrated by the loss of cochlear action potentials after as little as 60 s of interrupted blood supply in animal models (17) and by the permanent loss of cochlear function following thirty minutes of labyrinthine artery occlusion (18). Histopathologic studies of the inner ear have been performed demonstrating characteristic changes following experimental interruption of the labyrinthine artery in animals (17,19). Complete occlusion of the artery results in severe degeneration of the membranous labyrinth whereas microembolization causes patchy areas of necrosis.

Multiple mechanisms of vascular occlusion exist including embolism, thrombosis, vasospasm, and hypercoaguable states. The association of sudden hearing loss with migraine, sickle cell disease, macroglobulinemia, thromboangitis obliterans (Buerger's disease), and cardiopulmonary bypass support the vascular theory (20,21). The high incidence of spontaneous recovery, losses frequently limited to few frequencies, paucity of vertigo in most cases, and a large number of patients without vascular risk factors, argues against ischemia as a *predominant* cause of sudden hearing loss. These phenomena may be explained in part because the cochlea is more vulnerable to ischemia than the vestibular labyrinth and sensory cells of the basal turn are more vulnerable than those at the cochlear apex. Nonetheless, in cases with no identifiable cause, the vascular theory influences management strongly.

Vascular evaluation includes CBC, PT/PTT, bleeding time, and lipid profile (22). Prophylactic therapy with vasodilators and/or volume expanders is advocated by some. Intravenous histamine can be administered to promote blood flow and eliminate vasospasm and is delivered intravenously at a rate adequate to cause facial flushing. Carbogen is a mix of 5% carbon dioxide and 95% oxygen which has been demonstrated to increase perilymphatic oxygen tension (23). Several short-term inhalations are performed over the course of a day with this mixture (usually 30-min inhalations every 3 h around the clock) and any improvement in subjective hearing is noted. Diatrizoate meglumine (Hypaque) was noted by Morimitsu to reverse a patient's sudden hearing loss (24) prompting further study of 60 patients in which a 37% recovery rate was reported (25). Hypaque continues to be used due to its volume expanding properties and its possible mechanical effect on injuries to the stria vascularis.

A number of different vascular-directed therapies have been described in the literature and continue to be employed including hyperbaric oxygen therapy, stellate ganglion block, papaverine, pentoxifylline, dextran, and nicotinic acid (26,27).

3. IMMUNOLOGIC

The immune system protects us from ourselves but perhaps at a cost. In fighting infection, the immune system initiates an inflammatory response. In the nonforgiving environment of the inner ear, this inflammatory response can be detrimental resulting in cochleovestibular symptoms. Harris et al. (28) demonstrated that, in an animal model infected with CMV, immunosuppression resulted in a lesser degree of hearing loss than in immunocompetent controls. Furthermore, the positive response often seen with corticosteroid administration lends support to the notion that the inflammatory response can be detrimental in sudden hearing loss.

Autoimmunity refers to the process by which the body recognizes its own antigens as "foreign" resulting in an immune response directed against itself. Autoimmune inner ear disease can occur both in isolation or in the context of systemic disease. History, clinical findings, response to immunosuppressive (e.g., steroids) medication, and immunologic evaluation of the patient's serum are all important in the diagnosis of autoimmune sensorineural hearing loss (AISNHL). Although autoimmune inner ear disease is characterized most often by a bilateral or unilateral progressive SNHL that occurs over a period of weeks to months, sudden SNHL may be the presenting complaint in as many as one-quarter of patients with AISNHL (29).

In addition to the history, laboratory findings may indicate an autoimmune etiology in the patient with sudden hearing loss. Nonspecific markers of immune system up-regulation include a complement panel and ESR in addition to rheumatoid factor, antinuclear antibody, antimicrosomal antibody, and anticardiolipin antibody. The presence or absence of four specific antigens has been associated with AISNHL: the *presence* of B35, CW4, and CW7 and the *absence* of DR4 (30). In addition to these markers, western blot analysis for antibodies against an inner-ear 68-kD protein has been demonstrated by Harris and Sharp (31). Disher et al. (32) found this protein to be present in ~50% of patients with possible AISNHL. The antibody in question may be the monoclonal antibody KHRI-3 which binds to an inner-ear-supporting cell antigen and has been demonstrated to precipitate a 68–70 kD antigen. The presence of antibodies to inner ear antigens supports the concept that an autoimmune process is responsible for inner ear disease in these patients.

One of the major criteria in determining hearing loss of autoimmune origin is response to steroid therapy. Therefore, patients should be started on high-dose prednisone (1 mg/kg per day) immediately upon presentation, but following initial lab draws, if possible. Close monitoring of hearing response, both subjective and objective, is necessary, and any drop in hearing during prednisone taper should alert the physician to increase the steroid dose. In 2001, Gianoli and Li (33) treated patients who had failed to respond or who could not tolerate systemic steroids. They reported a 44% hearing salvage in patients who underwent transtympanic steroid administration on four separate occasions over the course of 10–14 days and concluded that transtympanic steroid therapy may be an alternative treatment for those failing or unable to tolerate systemic therapy. For patients who are unable to tolerate steroid therapy or continue to relapse while on the taper, some recommend that cytotoxic therapy be added. In 2001, Lasak et al. (29) noted that 59% of patients with AISNHL demonstrated a positive response to this regimen with steroid-only responders demonstrating a greater improvement in pure-tone average (14.8 vs. 4.5 dB) whereas cytotoxic-agent responders demonstrated improvements in speech discrimination scores (26.2 vs. 6.9%). Methotrexate (MTX) 7.5–15 mg po weekly is often the first cytotoxic medication chosen due to its lower incidence of side-effects when compared with other agents. While on MTX, patients may be placed on folic acid and must have baseline laboratory tests before initiating therapy. Monitoring of their complete blood count as well as renal and hepatic function must be continued at regular intervals throughout the entire course of treatment. Cyclophosphamide is often chosen as a second-line agent, and patients should be kept well hydrated to help prevent hemorrhagic cystitis when receiving this drug. Traditionally, cyclophosphamide has been given orally on a daily basis. However, intravenous pulsed therapy has been used by the senior author (R.T.S) with excellent results and far fewer side effects.

4. TRAUMATIC

Traumatic disruption of the integrity of the otic capsule is a well-accepted etiology for sudden hearing loss. The mechanism of trauma may be obvious as in the event of temporal bone fracture with resulting disruption of the cochlea or its nerve. However, trauma may also result from relatively innocuous activities associated with straining or by exposure to a loud noise. Furthermore, endolymphatic hydrops with intracochlear membrane rupture may occur in the absence of a history of trauma.

Cochlear membrane breaks may occur in either an intracochlear fashion or via fistulization through the round or oval windows. For this reason, it is important that the otolaryngologist inquire about the events surrounding the onset of hearing loss. Certain patients are theoretically at increased risk for a traumatic etiology for their hearing loss including those with prior ear surgery or those with inner ear anomalies. Hughes et al. (34) asserts that in order to entertain the diagnosis of perilymph fistula (PLF), the hearing loss must be "closely associated with a well-defined event of trauma, exertion, or barotrauma." Although this assertion is not accepted universally, the diagnosis of perilymphatic fistula can be made with certainty only following the intraoperative observation of perilymph leakage.

A CT scan of the temporal bones should be obtained to rule out mastoid pathology or abnormalities. These abnormalities may be obvious, as in the event of temporal bone fracture or more subtle as in the event of a patent cochlear aqueduct. Treatment for endolymphatic hydrops with membrane rupture includes complete bedrest with a low sodium diet and diuretics. Management of perilymphatic fistula is controversial. However, with an adequate history, exploration with packing of the round and oval windows can be justified.

5. NEOPLASTIC

Cushing (35) first described a case of sudden hearing loss as the presenting complaint in a patient with vestibular schwannoma (VS) in 1917. The typical presentation of a VS is progressive, asymmetric hearing loss. Traditionally, the rate of sudden hearing loss in VS has been reported to be between 5% and 20%. However, one outlying study reported VS in 19 of 40 patients presenting with sudden deafness (36). Berenholz et al. (37) listed seven theories regarding the pathogenesis of SSNHL in VS. Of these, nerve and vascular compression by the tumor are the two theories that continue to receive the most attention.

Most otologists would agree that VS should be ruled out in patients presenting with sudden hearing loss. Gadolinium-enhanced MRI is the study of choice to detect VS owing to its sensitivity—detecting tumors as small as 3 mm—and a specificity which allows it to differentiate VS's from other cerebellopontine angle (CPA) lesions.

Reversal of SSNHL occurring both spontaneously and with steroid therapy has been described. Friedman et al. (38) presented a retrospective study examining the effect of a history of sudden hearing loss on hearing preservation in 45 patients undergoing surgery at the House Ear Clinic between 1990 and 1998. They concluded that patients with SSNHL who remain hearing preservation candidates have the same rate of hearing preservation when compared with those demonstrating a pattern of progressive hearing loss. Negative prognostic factors included preoperative tinnitus, advanced age, longer auditory brainstem response (ABR) latencies, and poorer speech discrimination scores.

Meiteles et al. (39) presented an interesting case of sudden hearing loss in an only hearing ear. The patient initially failed high-dose steroid therapy and underwent subsequent emergency decompression and resection of his vestibular schwanoma with a rapid improvement and restoration of hearing. The authors concluded that a therapeutic window may exist during which sudden hearing loss caused by intracanalicular tumors is reversible, and that in select patients emergency resection and nerve decompression should be considered.

6. NEUROLOGIC

Neurologic causes of sudden deafness such as multiple sclerosis, neurosarcoidosis, and cerebrovascular accident are rare, but have been reported. Ozunlu et al. (40) described a case of sudden hearing loss that was the presenting symptom in a patient with multiple sclerosis. The patient improved with steroid therapy but never reversed her initial ABR findings of absent waves II–V. Sarcoidosis is a chronic, idiopathic granulomatous disease with frequent pulmonary, ocular, and lymphatic system manifestations. Neurosarcoidosis affects 1–5% of all cases of sarcoidosis patients and rarely results in cranial neuropathies. Souliere et al. (41) presented two cases of steroid responsive neurosarcoidosis presenting with SSNHL. Neurosarcoidosis rarely occurs in the absence of systemic symptoms, and a review of the literature demonstrates that eighth nerve neurosarcoidosis invariably presents with ocular symptoms. The treatment for these patients involves high-dose steroids with an extended taper.

Gadolinium-enhanced MRI obtained at the onset of symptoms should include the brainstem and cerebral hemispheres, in addition to the internal auditory canals. In this way, abnormalities such as demyelinating plaques and focal areas of ischemia can be evaluated. Auditory brainstem response may be performed for site-of-lesion information, in addition to electroencephalography and single photon emission computed tomography (SPECT) scan to evaluate for any electrophysiological, metabolic, or vascular disturbances not picked up with other forms of imaging. In addition, a consultation with a neurologist is beneficial for comprehensive evaluation.

7. TOXIC

Ototoxicity as a result of exposure to certain medications is a well-accepted mechanism of hearing loss, both progressive and sudden. Aminoglycosides and loop diuretics are the most commonly cited medications and produce irreversible hearing loss whereas salicylates can cause reversible hearing loss.

Every patient presenting with hearing loss should be questioned about all medications and exposure to environmental toxins. Those agents known to cause hearing loss should be stopped immediately, and hearing change should be monitored closely.

8. METABOLIC

Electrolyte and hormonal imbalance may play a role in SSNHL as it has been reported in association with conditions such as hyperlipidemia, hypothyroidism, diabetes mellitus, and pregnancy (13,22,27,42). Because of the low incidence of SSNHL, it is difficult to link sudden hearing loss with such conditions definitively. Nonetheless, a complete metabolic panel should be ordered at the time of presentation, and many otologists recommend lipid profiling, thyroid function tests, and glucose testing.

9. MANAGEMENT

The management of SSNHL is highly controversial due to the condition's uncertain etiology in the vast majority of cases. Techniques vary from simple observation to a comprehensive "shotgun" approach in which multiple treatment modalities are implemented simultaneously and then progressively withdrawn (27).

Identification of a cause for the patient's hearing loss and subsequent directed therapy may improve the likelihood of hearing recovery. Therefore, every attempt should be made by the evaluating physician to find a causative factor. A thorough history is mandatory and should focus on the events surrounding the onset of hearing loss as well as any associated otologic symptoms including vertigo, tinnitus, or aural fullness. A history of trauma should be elucidated and should include questioning regarding episodes of increased straining associated with coughing, choking, or vomiting. Past medical history should be reviewed with particular attention paid to previous otologic disease and/or surgery as well as the presence of autoimmune disease in the patient or immediate family members. All medications should be listed and ototoxic agents should be discontinued immediately.

Physical examination in patients with sudden hearing loss is often normal except for hearing assessment but should be performed with care in hopes of detecting abnormalities suggestive of a cause for the patient's symptoms. A thorough evaluation of the external auditory canal and tympanic membrane can help rule out obvious conditions such as cerumen impaction, herpes zoster, and otitis media. Decreased sensation of the lateral, posterosuperior aspect of the ear canal (i.e., Hitselberger's sign) can suggest the presence of a VS and should be documented. Tuning-fork evaluation helps to confirm laterality and to determine the nature of hearing loss, and it should correlate with audiologic findings. A comprehensive vestibular evaluation should be performed to determine the involvement of the vestibular system. Neurologic evaluation follows with complete cranial and peripheral nerve examination, and consultation with a neurologist may be an important adjunct.

Diagnostic studies are an indispensible component of the workup of sudden hearing loss. An audiogram with tympanometry must be performed immediately upon presentation to quantify the degree of hearing loss and to establish a reference point for the patient. Subsequent frequent (sometimes daily) audiometry should be performed to monitor changes in hearing and response to therapy. Electronystagmography can detect vestibular impairment and is helpful in quantifying the degree of impairment in the vestibulocochlear apparatus and differentiating conditions that usually involve only hearing (such as mumps) from those that may affect hearing and balance (such as VS and AISNHL).

Imaging of the temporal bone with computed tomography evaluates for bony disease and integrity of the otic capsule, whereas MRI and MRA of the IAC's, brainstem and cerebral hemispheres evaluate for the presence of acoustic neuroma, skull base lesion, ischemia, demyelination, or other central pathology. A chest radiograph may also be ordered to rule out mediastinal lymphadenopathy associated with sarcoidosis.

A number of serologic tests are ordered by the senior author (R.T.S) in nearly all cases to complete the evaluation. A complete blood count with differential can detect early changes associated with hematologic diseases such as leukemia and multiple myeloma. Chemistries can demonstrate electrolyte imbalances, and a lipid profile will often detect those at increased risk for vascular disease. Immune system involvement is evaluated with antigen-nonspecific serologic tests, acute phase reactants, western blot immunoassay, and HLA typing for the markers for the presence of B35, CW4, CW7, and absence of DR4. Other tests for autoimmune inner ear disease also are ordered. The association of syphilis with hearing loss is well known and treponemal disease is evaluated with either FTA-ABS or MHA-TP tests. These tests should be run regardless of RPR or VDRL results. Lyme titers should be drawn according to geography and season. Acute and convalescent viral titers should be drawn for herpes simplex, herpes zoster, CMV, parainfluenza, mumps, measles and adenovirus, and other tests for thyroid function and collagen vascular diseases also may be appropriate, in addition to coagulation studies discussed previously (see Vascular Etiology).

The management of cause-specific SSNHL is relatively straightforward. Unfortunately, 85–90% of cases remain idiopathic. The probable multifactorial pathogenesis of idiopathic SSNHL makes the management of this condition extremely difficult. While few advocate simple observation, those who do cite improvement in approximately two-thirds of patients who receive no intervention (43). Others, including the senior author (R.T.S), advocate a multifaceted approach with therapy directed at all hypothesized etiologic factors in patients who fail to respond to an initial trial of steroid therapy and other oral medicines.

Most patients are directed to adhere to a low sodium (2 g) diet and are placed on diuretic therapy to address endolymphatic hydrops. Systemic steroid therapy is begun immediately at a dose =1 mg/kg per day of prednisone to address inflammatory and autoimmune factors. This dose is tapered slowly after ~1 week, any change in hearing is noted, and adjustments in dose are made accordingly while following audiograms at least weekly. Patients in whom systemic steroids are contraindicated, or in who are unable to tolerate such therapy, may benefit from transtympanic steroid administration as outlined by Gianoli (33). Because a high degree of suspicion surrounds viral involvement in SSNHL we advocate immediate implementation of acyclovir 1–2 g divided five times daily for at least 10 days (44), or Valacyclovir 1 g bid for 10 days. Doxycycline is also started as prophylaxis against *Borrelia burgdorferi* while awaiting acute and convalescent Lyme titers.

To address possible vascular causes, a combination of vasodilators and volume expanders is used to treat patients with SSNHL. During the initial outpatient trial, an oral vasodilator is used. Carbogen (5% carbon dioxide–95% oxygen) inhalations have been demonstrated to increase partial pressure of arterial oxygen and increase perilymphatic oxygen tension (23). Carbogen therapy is administered as 10 min inhalations, 8× daily for 3–7 days and improvements in symptoms during treatment are noted. Diatrizoate meglumine (Hypaque) acts as a volume expander, although the exact mechanism in SSNHL is unknown. This agent, in addition to a continuous histamine infusion—at a rate which causes facial flushing—is used by the senior author to promote cochlear blood flow and helps eliminate vasospasm.

Despite earlier literature indicating that a "shotgun" approach to treatment is no better than spontaneous recovery (27), the experience of the senior author and numerous other otologists of recovery following the addition of each of the aforementioned agents has led, multiagent therapy especially in patients who have failed an outpatient steroid trial or have SSNHL in a better-hearing-ear to be comprehensive. Each agent is removed sequentially based upon subjective and objective response until the patient remains on diuretics, antivirals, oral vasodilators, and systemic steroids. Adjustments in therapy are made with any improvements or worsening of symptoms as therapies are discontinued.

A study in 1984 outlined prognostic indicators for patients presenting with SSNHL (43). In 225 consecutive patients treated over 8 years, Byl reported 69% of patients had a meaningful improvement in hearing. Severity of initial hearing loss, severity of vertigo, and time from onset to initial visit were determined to be the most important factors in predicting hearing recovery. Also important, but more variable, were patient age, audiogram pattern, erythrocyte sedimentation rate (ESR) and the state of the opposite ear. Byl determined that extremes of age (>60 or <15) divided the patient's chances of recovery in half. Patients with up-sloping or mid-frequency losses were found to be more likely to recover hearing, as were those with an elevated ESR. Finally, a below-normal pure tone average in the contralateral ear was found to be a negative predictor for recovery.

10. CONCLUSION

SSNHL presents a diagnostic dilemma in the overwhelming majority of cases. The condition should be treated as an otologic emergency and every effort made by the evaluating physician to identify a cause for hearing loss. When no cause is demonstrated, the treatment remains controversial. Unfortunately, given the paucity of cases, an incomplete understanding of pathogenesis, and the serious nature of hearing loss, a randomized, controlled study is impossible. Therefore, until a more complete understanding of SSNHL comes to fruition, we recommend a therapeutic regimen which is comprehensive in nature and which addresses simultaneously all hypothesized etiologies of the disease.

REFERENCES

1. DeKleyn A. Sudden complete or partial loss of function of the octavus system in apparently normal persons. Acta Otolaryngol 1944; 32:407–429.
2. Jaffe BF. Clinical studies in sudden deafness. Adv Otorhinolaryngol 1973; 20:221–228.
3. Byl FM. 76 cases of presumed sudden hearing loss occurring in 1973: prognosis and incidence. Laryngoscope 1977; 87:817–825.
4. Jaffe BF, Maassab HF. Sudden deafness associated with adenovirus infection. New Engl J Med 1967; 276:1406–1409.
5. Mattox DE. Medical management of sudden sensorineural hearing loss. Otolaryngol Head Neck Surg 1980; 88:111–113.
6. Saeki N, Kitahara M. Assessment of prognosis in sudden deafness. Acta Otolaryngol Suppl (Stockh) 1994; 51:56–61.
7. Westmore GA, Pickard BH, Stern H. Isolation of mumps virus from the inner ear after sudden deafness. Br Med J 1979; 1:14–15.
8. Davis LE, James CG, Fiber F, McLauren LC. Cytomegalovirus infection from a human inner ear. Annals Otol Rhinol Laryngol 1979; 88:424–426.
9. Davis LE, Johnson RT. Experimental viral infections of the inner ear. I. Acute infections of the newborn hamster labyrinth. Lab Invest 1976; 34:349–356.
10. Koide J, Yanagita N, Hondo R, et al. Serological and clinical study of herpes simplex virus infection in patients with sudden deafness. Acta Otolaryngol Suppl (Stockh) 1988; 456:21–26.
11. Wilson WR. The relationship of the herpesvirus family to sudden hearing loss: A prospective clinical study and literature review. Laryngoscope 1986; 96:870–877.
12. Wilson WR, Veltri RW, Laird N, et al. Viral and epidemiologic studies of idiopathic sudden hearing loss. Otolaryngol Head Neck Surg 1983; 91:653–658.
13. Schuknecht HF, Donovan ED. The pathology of idiopathic sudden sensorineural hearing loss. Arch Otorhinolaryngol 1986; 243:1–15.
14. Khetarpal U, Nadol JB, Glynn RJ. Idiopathic sudden sensorineural hearing loss and postnatal viral labyrinthitis: a statistical comparison of temporal bone findings. Ann Otol Rhinol Laryngol 1990; 99:969–976.
15. Lindsay J, Davey P, Ward P. Inner ear pathology in deafness due to mumps. Ann Otol Rhinol Larygol 1960; 69:918.
16. Yoon TH, Paparella MM, Scharchern PA. Systemic vasculitis: temporal bone histopathologic study. Laryngoscope 1990; 99:600–609.
17. Suga F, Preston J, Snow JB. Experimental microembolization of cochlear vessels. Arch Otolaryng 1970; 92:213–220.
18. Kimura R, Perlman HB. Arterial obstruction of the labyrinth. Part I. Cochlear changes. Ann Otol Rhinol Laryngol 1958; 67:5–40.
19. Kim JS, Lopez I, DiPatre PL, et al. Internal auditory artery infarction: clinicopathologic correlation. Neurology 1999; 52:40–44.

20. Arenberg IK, Allen GN, Deboer A. Sudden hearing loss immediately following cardiopulmonary bypass. J Laryngol Otol 1972; 86:73.
21. O'Keeffe LJ, Maw AR. Sudden total deafness in sickle cell disease. J Laryngol Otol 1991; 105:653–655.
22. Ullrich D, Aurbach G, Drobik C. A prospective study of hyperlipidemia as a pathogenic factor in sudden hearing loss. Eur Arch Otorhinolaryngol 1992; 249:273–276.
23. Fisch U. Management of sudden deafness. Otolaryngol Head Neck Surg 1983; 91:3–8.
24. Morimitsu T. New theory and therapy of sudden deafness. In: Shambaugh GE, Shea JJ, eds. Proceeding of the Shambaugh Fifth International Workshop on Middle Ear Microsurgery and Fluctuant Hearing Loss. Huntsville, AL, USA: Strode Publishers, Inc., 1977:312–421.
25. Emmett JR, Shea JJ. Diatrizoate meglumine (hypaque) treatment for sudden hearing loss. Laryngoscope 1979; 89:1229–1238.
26. Probst R, Tschopp K, Kellerhals B, et al. A randomized, double-blind, placebo-controlled study of dextran/pentoxifylline medication in acute acoustic trauma and sudden hearing loss. Acta Otolaryngol (Stockh) 1992; 112:435–443.
27. Wilkins SA, Mattox DE, Lyles A. Evaluation of a "shotgun" regimen for sudden hearing loss. Otolaryngol Head Neck Surg 1987; 97:474–480.
28. Harris JP, Fan JT, Keithley EM. Immunologic responses in experimental cytomegalovirus labyrinthitis. Am J Otol 1990; 11:304–308.
29. Lasak JM, Sataloff RT, Hawkshaw M, et al. Autoimmune inner ear disease: steroid and cytotoxic drug therapy. Ear Nose Throat J 2001; 80:808–822.
30. Bowman CA, Nelson RA. Human leukocytic antigens in autoimmune sensorineural hearing loss. Laryngoscope 1987; 97:7–9.
31. Harris JP, Sharp PA. Inner ear autoantibodies in patients with rapidly progressive sensorineural hearing loss. Laryngoscope 1990; 100:516–524.
32. Disher MJ, Ramakrishnan, Nair TS, et al. Human autoantibodies and monoclonal antibody KHRI-3 bind to a phylogenetically conserved inner-ear-supporting cell antigen. Ann NY Acad Sci 1997; 830:253–265.
33. Gianoli GJ, Li JC. Transtympanic steroids for treatment of sudden hearing loss. Otolaryngol Head Neck Surg 2001; 125:142–146.
34. Hughes GB, Freedman MA, Haberkamp TJ, Guay ME. Sudden sensorineural hearing loss. Otolaryngol Clin North Am 1996; 29:393–405.
35. Cushing H. Tumors of the Nervus Acusticus and the Syndrome of the Cerebellopontine Angle. Philadelphia, PA: WB Saunders, 1917.
36. Chaimoff M, Nageris BI, Sulkes J, et al. Sudden hearing loss as a presenting symptom of acoustic neuroma. Am J Otolaryngol 1999; 20:157–160.
37. Berenholz LP, Eriksen C, Hirsh FA. Recovery from repeated sudden hearing loss with corticosteroid use in the presence of an acoustic neuroma. Ann Otol Rhinol Laryngol 1992; 101:827–831.
38. Friedman RA, Kesser BW, Slattery WH, et al. Hearing preservation in patients with vestibular schwannomas with sudden sensorineural hearing loss. Otolaryngol Head Neck Surg 2001; 125:544–551.
39. Meiteles LZ, Liu JK, Couldwell WT. Hearing restoration after resection of an intracanalicular vestibular schwannoma: a role for emergency surgery? Case report and review of the literature. J Neurosurg 2002; 96:796–800.
40. Ozunlu A, Mus N, Gulhan M. Multiple sclerosis: a cause of sudden hearing loss. Audiology 1998; 37:52–58.
41. Souliere CR, Kava CR, Barrs DM, Bell AF. Sudden hearing loss as the sole manifestation of neurosarcoidosis. Otolaryngol Head Neck Surg 1991; 105:376–381.
42. Wilson WR, Laird N, Young FM, et al. The relationship of sudden hearing loss to diabetes mellitus. Laryngoscope 1982; 92:155–160.
43. Byl FM. Sudden hearing loss: eight years' experience and suggested prognostic table. Laryngoscope 1984; 94:647–661.
44. Tucci DL, Farmer JC, Kitch RD, Witsell DL. Treatment of sudden sensorineural hearing loss with systemic steroids and valacyclovir. Otol Neurotol 2002; 23:301–308.

12

Mixed, Central, and Functional Hearing Loss

Robert T. Sataloff
Drexel University College of Medicine, American Institute for Voice and Ear Research, and Graduate Hospital, Philadelphia, Pennsylvania, USA

Joseph Sataloff
Thomas Jefferson University, Philadelphia, Pennsylvania, USA

1. MIXED HEARING LOSS

Whenever the hearing loss of a patient includes a mixture of both conductive and sensorineural characteristics, he/she is said to have a mixed hearing loss. The hearing deficiency may have started originally as a conductive failure, such as otosclerosis, and later developed a superimposed sensorineural component; or the difficulty may have been sensorineural in the beginning, such as presbycusis, and a conductive defect, perhaps resulting from middle-ear infection, may have developed subsequently. In some cases the conductive and the sensorineural elements may have started simultaneously, as in a severe head injury affecting both the inner ear and the middle ear.

In clinical practice, most cases with an original sensorineural etiology remain in that classification without an added conductive element. In contrast, most cases that start as conductive hearing impairment later develop some sensorineural involvement. Familiar examples are otosclerosis with presbycusis or cochlear otosclerosis and chronic otitis media with labyrinthitis.

1.1. Sensorineural Involvement

Otosclerosis was at one time thought to retain its purely conductive character for years; today it is recognized that this condition develops sensorineural impairment in same cases. Figures 9.5 and 12.1 show examples of sensorineural features in otosclerosis.

Similar evidence of sensorineural involvement often is seen in chronic otitis media. According to one hypothesis, some toxic inflammatory metabolite produces a cochleitis or labyrinthitis (Fig. 12.2).

Mixed hearing loss is becoming more common also in otosclerosis after stapedectomy. The sensorineural deficit may be caused by penetration of the oval window

JOSEPH SATALOFF, M.D.
ROBERT THAYER SATALOFF, M.D.
1721 PINE STREET PHILADELPHIA, PA 19103

NAME ____________________

DATE	RIGHT EAR AIR CONDUCTION							LEFT EAR AIR CONDUCTION					
	250	500	1000	2000	4000	8000		250	500	1000	2000	4000	8000
	70	80	70	75	90	NR	PRE-OP.						
	30	35	45	45	65	70	POST-OP.						

RIGHT EAR BONE CONDUCTION						LEFT EAR BONE CONDUCTION				
NR	50	50	60	NR	PRE-OP.					
30	30	35	40	NR	POST-OP.					

SPEECH RECEPTION: Right ______ Left ______ DISCRIMINATION: Right ______ Left ______

Figure 12.1 *History*: A 67-year-old man with insidious deafness for 25 years. No tinnitus or vertigo. Wears a hearing aid on the right ear. *Otologic*: Normal. Complete stapes fixation confirmed at surgery. *Audiologic*: Bilateral reduced air and bone conduction thresholds with some air-bone gap in the right ear. The left ear was not as severely involved as the right but did not exhibit any air-bone gap. Tuning fork tests showed bone conduction better than air in the right, with lateralization to the right ear. The tuning fork sounded louder on the teeth than on the mastoid. *Classification*: Mixed hearing loss. *Diagnosis*: Otosclerosis with secondary sensorineural involvement. *Aids to diagnosis*: The presence of an air–bone gap, good tuning fork responses, especially by teeth, and satisfactory hearing aid usage in conjunction with negative otoscopic findings are important in making the diagnosis. In this case the air–bone gap was closed. The apparent improvement in postoperative bone conduction is seen commonly, but it does not mean that the sensorineural hearing loss has been improved.

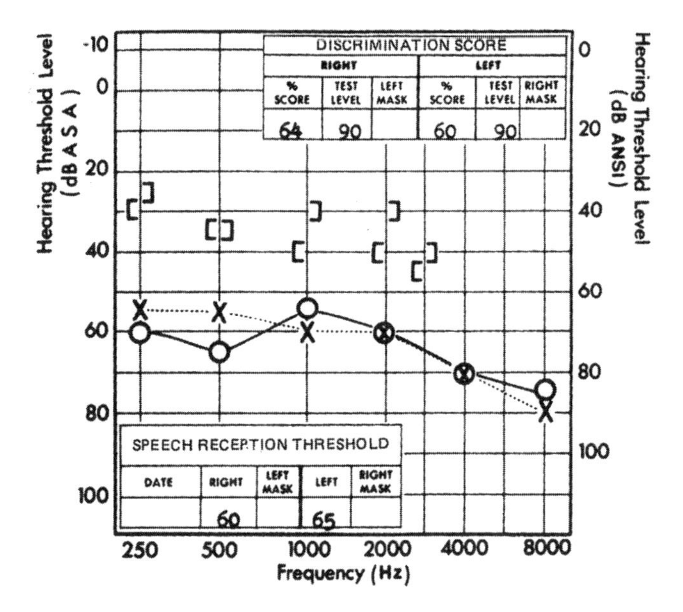

Figure 12.2 *History*: A 60-year-old man with bilateral chronic otorrhea for 40 years. Insidious hearing loss for many years, which is now stable. No tinnitus or vertigo. *Otologic*: Putrid discharge with evidence of cholesteatoma. *Audiologic*: Moderate to severe bilateral flat loss. Bone conduction is reduced at all frequencies but a 20- to 30-dB air-bone gap remains. Discrimination is reduced bilaterally. *Classification*: Mixed hearing loss. *Diagnosis*: Chronic otitis media with neural or cochlear involvement.

with exposure of the perilymph. Despite meticulous surgical care, the inner ear also can be traumatized readily and made more susceptible to infection by this procedure. In surgical trauma to the inner ear, high-frequency hearing loss often falls below the preoperative level, and the patient may complain of reduced discrimination though his pure-tone threshold is improved. Other symptoms suggestive of sensory damage, such as recruitment, distortion, and a lower threshold of discomfort, may be noted.

Figure 12.3 illustrates an important aspect of mixed hearing loss. The patient attributed his hearing loss to wax in his ears, but when the wax was removed, we noted an underlying sensorineural deafness of which the patient had not been aware. This example serves as a warning to physicians to avoid assuring any patient that his hearing loss can be corrected merely by removing cerumen, because mixed hearing loss may be found subsequently.

1.2. Evaluating Conductive and Sensorineural Components

On the other hand, pure conductive hearing loss may be misdiagnosed as mixed hearing loss because the high frequencies and the bone conduction are somewhat reduced. (See Figure 9.20.) In the case cited, hearing returned to normal when the fluid was removed from the middle ear; actually, there was no sensorineural damage. Whenever there is any possibility of fluid in the middle ear, a diagnostic myringotomy should be considered to avoid an erroneous diagnosis of sensorineural hearing loss. A high-frequency hearing loss with reduced bone conduction may create a mistaken impression on sensorineural damage.

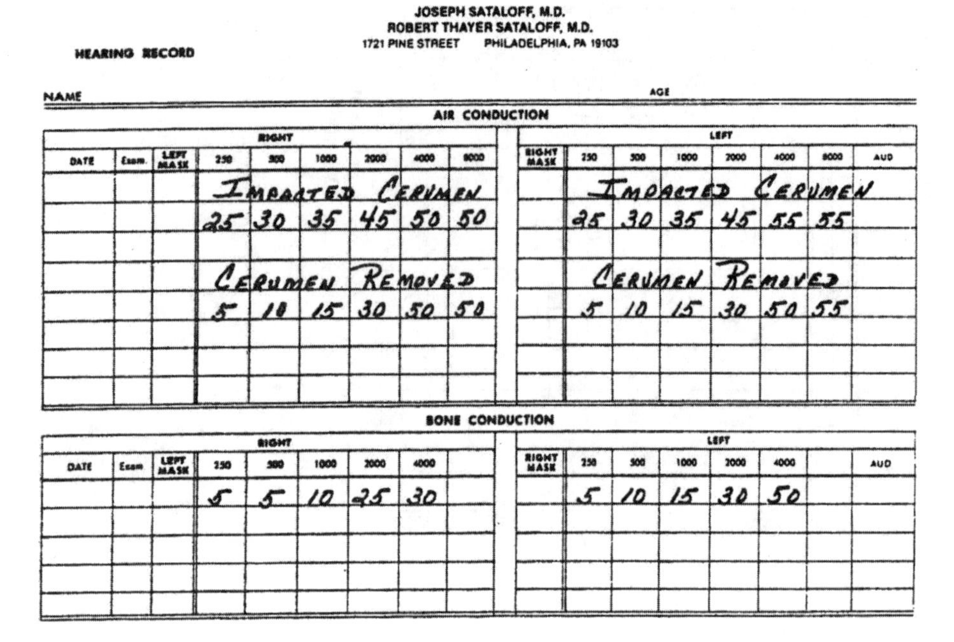
JOSEPH SATALOFF, M.D.
ROBERT THAYER SATALOFF, M.D.
1721 PINE STREET PHILADELPHIA, PA 19103

HEARING RECORD

NAME AGE

AIR CONDUCTION

DATE	Exam.	LEFT MASK	RIGHT 250	500	1000	2000	4000	8000	RIGHT MASK	LEFT 250	500	1000	2000	4000	8000	AUD
			Impacted Cerumen							Impacted Cerumen						
			25	30	35	45	50	50		25	30	35	45	55	55	
			Cerumen Removed							Cerumen Removed						
			5	10	15	30	50	50		5	10	15	30	50	55	

BONE CONDUCTION

DATE	Exam	LEFT MASK	RIGHT 250	500	1000	2000	4000		RIGHT MASK	LEFT 250	500	1000	2000	4000		AUD
			5	5	10	25	30			5	10	15	30	50		

Figure 12.3 *History*: A 45-year-old man with fullness in both ears and hearing loss for one week. No tinnitus or vertigo. No history of ear infections. Wanted the wax removed to restore his hearing. *Otologic*: Bilateral impacted cerumen. Removed, and eardrums normal. *Audiologic*: Bilateral reduced air conduction thresholds with greater loss in high frequencies. No bone thresholds obtained before removal of cerumen, but tuning-fork tests showed bone better than air, bilaterally. After cerumen was removed, air conduction thresholds improved, but a high-frequency loss remained. Bone conduction thresholds approximated the air conduction thresholds. Tuning-fork tests showed air better than bone after removal of wax. *Classification*: Mixed hearing loss before removal of cerumen. Sensorineural loss after removal of cerumen. *Diagnosis*: Impacted cerumen with progressive nerve deafness. *Aids to diagnosis*: It is always advisable to do audiometric studies before and after removing impacted cerumen and before reaching a diagnosis.

The bone conduction test as now performed is not a completely reliable measure of sensorineural efficiency. Excessive reliance on this test may mislead the physician. Figure 12.4 illustrates a case in which the bone conduction was almost undetectable audiometrically with a vibrating tuning fork on the patient's mastoid bone or forehead. Yet, when the instrument was placed directly on the patient's teeth, a good bone conduction response was obtained. This was a case of mixed hearing loss followed by a satisfactory surgical result.

In every case of mixed hearing loss one should determine how much of the deficit is conductive and how much is sensorineural. The prognosis depends largely on this estimate. For instance, a patient's chronic otitis media may be resolved and leave a 65-dB hearing loss. The tentative conclusion that a tympanoplasty is likely to restore the patient's hearing must be revised when bone conduction and discrimination studies show substantial sensorineural involvement, so that the chances of restoring hearing are poor. On the other hand, in some cases of otosclerosis with air conduction levels almost above the measurable limits of the audiometer, the prognosis for restoring hearing to the bone conduction hearing level by stapes surgery may be very good.

The best way to approximate the conductive and the sensorineural components of a hearing loss is to perform all possible tests for estimating the patient's sensorineural potential or "cochlear reserve." In addition to routine bone conduction, speech discrimination scores are essential. A good rule to follow is: If the patient hears and discriminates well when speech is made louder, then the conductive element probably is a major cause of the hearing difficulty, and there is a good chance that surgery will improve the hearing. If, on the other hand, the patient does not understand any better with a hearing aid or when the voice is raised, then the outlook for improved hearing is not as favorable even if the conductive portion of the mixed hearing loss is corrected. For example, the patient described in Fig. 12.4 heard reasonably well with a hearing aid in the right ear, though there seemed to be no useful residual hearing. Successful stapes surgery confirmed the preoperative evaluation. Because of the severe hearing loss in the right ear, it was not possible under office conditions to amplify speech sufficiently to test the patient's discrimination. However, the results with a hearing aid indicated fairly good discrimination. In another patient with mixed hearing loss and almost as much neural involvement, the discrimination was not good. Consequently, after a successful stapes mobilization, the hearing level was improved, but the discrimination was not helped, and so that patient was not nearly so pleased with the results.

1.2.1. *Range of Conditions*

Mixed hearing loss usually includes the following range of features: (a) visible pathology in the external ear canal or in the middle ear, associated with reduced bone conduction and other findings of sensorineural hearing loss; (b) normal otologic observations,

JOSEPH SATALOFF, M.D.
ROBERT THAYER SATALOFF, M.D.
1721 PINE STREET PHILADELPHIA, PA 19103

HEARING RECORD

NAME AGE

AIR CONDUCTION

			RIGHT							LEFT						
DATE	Exam.	LEFT MASK	250	500	1000	2000	4000	8000	RIGHT MASK	250	500	1000	2000	4000	8000	AUD
1st TEST			80	95	NR	NR	NR	NR		75	75	85	95	95	NR	
2 mos			40	45	65	70	75	NR								
1 YR.										45	55	65	65	75	NR	

Figure 12.4 *History*: A 62-year-old woman with severe deafness. Using a powerful aid for many years. Voice is normal. No tinnitus or vertigo. Several aunts also use hearing aids. *Otologic*: Normal. *Audiologic*: Bone conduction is better than air, but patient denied hearing the tuning fork on the mastoid or the forehead but heard it fairly well on the teeth. *Classification*: Mixed hearing loss. *Diagnosis*: Otosclerosis with sensorineural hearing loss. *Comment*: Both oval windows were overgrown with otosclerosis which required drilling. Note the good hearing improvement in spite of poor response to tuning fork. Surgery was done only because she did well with a hearing aid and had a good air–bone gap. A 1-year interval was allowed between operations on the two ears.

somewhat reduced bone conduction, but a significant air–bone gap; (c) reduced speech discrimination, though usually of a mild degree, with improved discrimination as the intensity of speech is raised; and (d) unilateral hearing loss with the conductive element predominating and the tuning fork lateralizing to the more severely impaired ear. In such cases, there is always an air–bone gap.

1.3. Prognosis

In mixed hearing loss the prognosis depends on the relative proportion of conductive and sensorineural pathology. If the sensorineural component is slight, the surgical prognosis is good, and under favorable circumstances the hearing may approximate the level of the bone conduction. However, the discrimination is not improved much, even after correction of the conductive defect.

2. CENTRAL HEARING LOSS

For the purpose of this book, a hearing loss is classified as "central" if it is caused by a lesion that affects primarily the central nervous system from the auditory nuclei to the cortex. The process of verbal communication is complex. The auditory pathway consists of a series of transducers that repeatedly change the speech stimulus so that it can be handled effectively by the cortex. The eardrum and the ossicular chain modify the amplitude of the sound waves, and the cochlea analyzes these waves into fundamentals that then are reflected as impulses to the cortex. The chief function of the auditory cortex is to interpret and to integrate these impulses and to provide the listener with the exact meaningful information intended by the speaker, or to permit the listener to react appropriately to the actual implication of the sound.

2.1. Reaction to Tests in Diagnosis

It appears, then, that interference with the neural impulse pattern traveling up the central pathways to the cortex would manifest itself not so much by a lowering of the hearing threshold for pure tones as by reduced ability to interpret information. On the strength of this reasoning, otologic authorities have suggested various techniques to diagnose central hearing impairment. Since damage to the central auditory pathway causes little or no change in the pure-tone threshold, the tests are designed to measure more complex function. For example, one test involves filtering out the higher frequencies of certain speech samples and then comparing the ability of persons with normal hearing to understand such speech sequences with that of patients with central hearing loss. It has been found that when this test is given to patients with unilateral central hearing loss (due to a brain tumor, for example), the *opposite* ear, which presumably hears normally, may have much poorer discrimination than the ear of a person who has normal hearing. This means that in unilateral central hearing loss the contralateral ear can be affected. A similar adverse effect is noted in the ear opposite the tumor when certain words are interrupted periodically or accelerated. Interestingly enough, the patient with central hearing impairment has no difficulty perceiving high-frequency sounds such as the letter *s* and *f* that are affected so characteristically in peripheral sensorineural lesions, but he/she has difficulty interpreting what is heard.

2.2. Characteristic Features

The prinicipal characteristics of central hearing loss are: (a) hearing tests do not indicate peripheral hearing impairment; (b) the pure-tone threshold is relatively good compared with the ability of the patient to discriminate, and especially to interpret, what he/she hears; (c) the patient has difficulty interpreting complex information; (d) there is usually an accompanying shortened attention span other processing on memory problems, and other neurological findings; and (e) apart from unusual cases with unilateral vascular lesions or neoplasms, hearing loss of this type resembles a bilateral perceptive disorder without any evidence of recruitment.

Some authors include sensory aphasia in the classification of central hearing loss, but most otologists consider this condition to be beyond the scope of their specialty.

The prognosis for central hearing impairment is poor, but reeducation seems to offer a useful approach. There is no characteristic audiometric pattern, except that the disparity between the hearing level and the speech interpretation is quite marked.

Present knowledge of central hearing loss is rather meager. There can be extensive brain damage without apparent hearing abnormality. When symptoms do occur, they usually are associated with some general disease such as encephalitis (Fig. 12.5), vascular accident, or neoplasm. Other causes include brain tumors and infections.

In certain cases, central hearing loss may mimic peripheral causes of deafness, including occupational deafness. This is particularly true of impairment associated with pathology in the cochlear nucleus. In particular, the spheroid cells of the superior ventral cochlear nucleus (SVCN) show an anatomical frequency gradient, low ventral to high dorsal (1,2). For example, erythroblastosis typically causes hearing loss that centers around 3000–4000 Hz. This may be caused by injury to SVCN spheroid cells, the second-order neurons of the ascending auditory pathway, even in the presence of normal hair cells in the organ of Corti (2). Central pathology must be included in the differential diagnosis of the hearing loss producing a 4000-cycle dip on the audiogram.

2.3. Central Auditory Processing Disorders (CAPD)

Central auditory processing disorders are common, relatively minor abnormalities usually characterized by decreased understanding of speech in the presence of background noise. Most of the research in this area has concentrated on school-aged children where these difficulties are now often recognized first. Children with severe CAPD may appear immature in the classroom, or may even seem to have a hearing handicap. The entity has become well recognized relatively recently. Consequently, many adults have this problem undiagnosed. Common complaints include inability to study or read in the presence of noise, slowing of reading speed caused by noise from vacuum cleaners or air conditioners, and suspicion by family members and friends that the patient has a hearing loss. If people talk to a person with CAPD while he or she is involved in auditory concentration such as listening to a television program, the person will often "not hear." However, if one gets the persons attention by calling his or her name before speaking, the first sentence will not be missed, and hearing is normal.

Various test batteries have been developed to assess specific areas of auditory behaviors. These include selective attention, auditory closure, rate of perception, and sequencing ability. Once a person has been diagnosed as having CAPD, treatment involves altering the listening environment to obtain the best possible signal-to-noise ratio, as

JOSEPH SATALOFF, M.D.
ROBERT THAYER SATALOFF, M.D.
1721 PINE STREET PHILADELPHIA, PA 19103

HEARING RECORD

NAME AGE

AIR CONDUCTION

	RIGHT								LEFT							
DATE	Exam.	LEFT MASK	250	500	1000	2000	4000	8000	RIGHT MASK	250	500	1000	2000	4000	8000	AUD
1ST TEST			55	55	50	35	25	25		45	45	40	30	20	20	
2 YRS.			60	45	50	20	20			20	25	0	15	15		
3 YRS.			50	40	45	15	20			35	30	10	10	15		
5 YRS.			35	25	45	15	15	15		15	15	0	10	10	15	

BONE CONDUCTION

	RIGHT									LEFT						
DATE	Exam	LEFT MASK	250	500	1000	2000	4000		RIGHT MASK	250	500	1000	2000	4000		AUD
1ST TEST			NR	50	40	30	30			NR	45	20	20	15		
2 YRS.			NR	45	55	25	20			25	40	15	30	30		

SPEECH RECEPTION

DATE	RIGHT	LEFT MASK	LEFT	RIGHT MASK	FREE FIELD	MIC.

DISCRIMINATION

	RIGHT				LEFT				
DATE	% SCORE	TEST LEVEL	LIST	LEFT MASK	% SCORE	TEST LEVEL	LIST	RIGHT MASK	EXAM.
1ST TEST	18	95		80	56	80		–	
3 YRS.	56	65			60	40		–	

HIGH FREQUENCY THRESHOLDS

	RIGHT						LEFT					
DATE	4000	8000	10000	12000	14000	LEFT MASK	RIGHT MASK	4000	8000	10000	12000	14000

RIGHT		WEBER	LEFT	
RINNE	SCHWABACH		RINNE	SCHWABACH

HEARING AID

DATE	MAKE	MODEL
RECEIVER	GAIN	EXAM.
EAR	DISCRIM	COUNC.

REMARKS

Figure 12.5 *History*: At age 8 this patient had to be helped to walk because of muscular incoordination and vestibular imbalance resulting from varicella encephalitis at age 18 months. Her eyesight and walking were affected, and hearing loss started 3 days after the encephalitis was diagnosed. The hearing gradually improved, and her speech is very good with only a slight voice defect. Prior to the initial visit she had another encephalitis attack, and hearing was depressed but gradually improved. *Otologic*: Normal, and caloric test was normal. *Audiologic*: The very poor discrimination score in the right ear was confirmed on several subsequent studies. The improvement in discrimination scores is not uncommon in such cases. There was no tone decay, and recruitment was not significant. *Classification*: Central hearing loss. *Diagnosis*: Encephalitis.

well as counseling the patient and family members. Good listening and visual behaviors are also stressed. Compensation for, and tolerance of central auditory processing disorders may diminish somewhat in the presence of advancing age, or deteriorating hearing. Evaluation and treatment by a neuropsychologist can be extremely helpful, especially if other cognitive problems are present or suspected.

3. FUNCTIONAL HEARING LOSS

Functional or psychogenic hearing loss is the customary diagnosis when there is no organic basis for the patient's apparent deafness. The inability to hear results entirely or mainly from psychological or emotional factors, and the peripheral hearing mechanism may be essentially normal. If there is some slight damage to the peripheral end-organ, the observed hearing loss is disproportionate to the organic lesion.

3.1. Cause and Characteristic Features

The basis for functional hearing loss in most patients is psychogenic, the product of emotional conflict. For example, anxiety is to the emotions what pain is in the physical realm. *Normal* anxiety is the natural reaction to an actual threat to one's welfare; it is recognized as such and recedes when the cause is removed or with the passage of time. In contrast, *neurotic* anxiety is an excessive reaction and may exist even in the absence of an external threat. Seldom is the true cause recognized consciously by the patient, and the anxiety persists beyond any recognizable need.

When anxiety is converted in part to a somatic symptom such as deafness, there generally are other evidences of emotional disturbance, such as insomnia. Tinnitus is a characteristic feature of "hysterical deafness," and patients often claim that the noise is unbearable. Hearing acuity usually varies, depending on the patient's emotional state at the time of testing. Patients may appear to be overly concerned about their auditory symptoms when in reality some or all of the tinnitus and the deafness is caused by anxiety, the origin of which lies elsewhere.

When all or nearly all of the anxiety is transferred to the ear, the case is one of true conversion or hysteria. The patient then usually is indifferent to the symptoms despite their apparent severity, and may delay seeking medical advice until persuaded to do so by his/her associates. The patient underreacts emotionally, the reason is that he/she has partially solved an emotional conflict by permitting it to assume a somatic form. This illusion is incomplete, and careful scrutiny will show residual emotional symptoms. Psychiatric evaluation is indicated.

3.1.1. A Product of Both Military and Civilian Life

Functional hearing loss, then, is an unconscious device by which the patient seeks to escape from an intolerable problem that he/she cannot face consciously. Hysterical blindness and paralysis are other examples of the same type of somatization or "conversion reaction." Often seen in military life during wartime, such situations occur also in civilian life. For example, the patient may go with his wife to consult the physician. The physician asks the patient about his problem, but before he can answer, his wife says, "He just doesn't hear me, doctor." When hearing tests indicate no hearing impairment, the physician talks to the patient alone. It then may become apparent (though the process of interrogation may take considerable time) that the patient subconsciously does not want to listen to his wife and therefore has developed a psychogenic hearing loss as a defense mechanism. Probably the classic example of psychogenic deafness is the young soldier in battle, too frightened to charge and yet ashamed of retreating while his buddies bravely go forward. In the absence of a rational way out, his unconscious mind conjures up the concept of deafness or blindness.

The chief statistics of functional hearing loss originate in the Armed Forces and the Veterans Administration hearing centers. Here some 25% of hearing-impaired patients are reported to have significant functional hearing losses. Strangely enough, a large percentage of such patients during World War II had little or no combat service. Disruption of family life and subjection to military discipline produced sufficient trauma to bring on psychogenic hearing loss.

The complexities of civilian society also have produced an abundance of emotional conflict and insecurity—sufficient to account for a complete spectrum of emotional disturbance, including psychogenic hearing loss. Too often such cases escape medical attention or diagnosis.

3.2. Functional Overlay

It is, of course, entirely possible for hearing loss of functional origin to be superimposed on true organic hearing impairment, in which case the term "functional overlay" is used. The problem then is to recognize the two components in the patient's hearing impairment.

The history and the otologic examination often provide important clues such as the unrealistic attempts of a patient to account for the difficulty. For example, the patient may claim that his/her hearing was excellent until a physician cleaned out his/her ears with such force that he/she suddenly went stone deaf. Another patient may carry on a perfectly normal conversation with the physician and hear everything said, while repeated hearing tests consistently suggest a severe deafness quite inconsistent with the patient's conversational accomplishments.

3.3. Diagnosis by Specific Features

The diagnosis of functional hearing loss should not be made solely by exclusion or merely because tests performed reveal no organic evidence of abnormality. There are specific features that characterize functional hearing impairment, and upon these a positive diagnosis must depend.

A critical appraisal of routine hearing tests usually will justify a diagnosis of functional hearing loss if this is present. In an organic lesion, all tests must not only give fairly consistent results when they are repeated, but they also must correlate with one another. It is a mistake to attribute well-marked discrepancies to individual variations. Several authors have suggested critical observations that should alert physicians to the possible presence of a functional hearing impairment. These leads have been found useful:

1. A medical history that experience indicates could not possibly explain the patient's condition, such as the sudden onset of profound deafness following instillation of drops into the ears. Care must be taken to exclude all organic causes and not simply to disprove the patient's explanation.
2. Too spectacular an improvement with a hearing aid, especially when the patient has set the controls at minimal amplification, or a sudden disproportionate improvement in hearing after a simple procedure such as drum massage or insufflating the eustachian tube. In such cases the power of suggestion rather than the mechanical procedure probably should receive the credit.
3. Decided fluctuations in hearing acuity as determined by any single test. The importance of repeated tests cannot be overemphasized. They are needed especially to establish basic hearing against which to evaluate the results of any treatment to be undertaken.
4. Inconsistency in the results of two or more tests. For example, the patient may hear everything when spoken to, and yet his audiogram may show a very severe

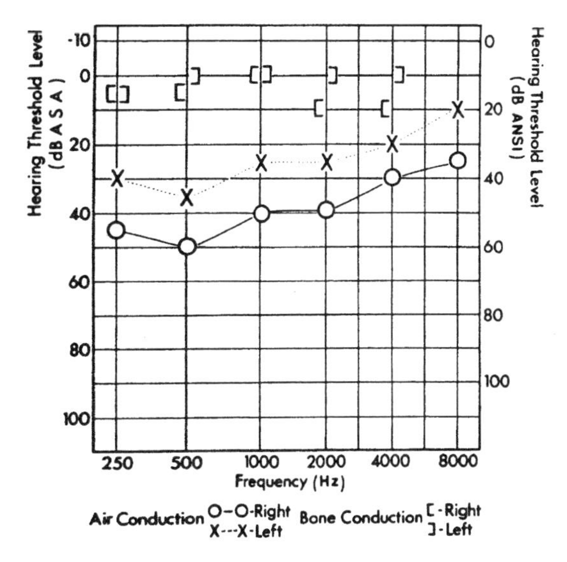

Figure 12.6 *History*: A 34-year-old woman with otosclerosis and hearing loss for over 10 years. Mother and aunt have same difficulty, but all refuse to use hearing aids. Patient reluctantly admits having some hearing loss. She often says, "What?" even when addressed loudly and habitually asks for repetition even though she evidently hears. She often repeats a question before answering and obviously has much better hearing than her responses indicate. The patient appears to be frustrated and emotionally disturbed and does not use her residual hearing effectively. Her associates have been led to believe that her hearing loss is worse than it actually is. This is a functional overlay on an organic otosclerosis. After positive suggestion this patient acquired a hearing aid and is doing much better. She refuses ear surgery.

JOSEPH SATALOFF, M.D.
ROBERT THAYER SATALOFF, M.D.
1721 PINE STREET PHILADELPHIA, PA 19103

HEARING RECORD

NAME AGE

AIR CONDUCTION

			RIGHT						LEFT							
DATE	Exam	LEFT MASK	250	500	1000	2000	4000	8000	RIGHT MASK	250	500	1000	2000	4000	8000	AUD
		–	25	15	15	20	30	35	-	45	50	55	55	60	75	
									80	65	70	75	70	70	NR	
									–	70	75	75	80	85	NR	
RUDMOSE AUDIOMETRY			15	15	25	15	15	15	–	40	40	40	40	50	60	
PGSR									–	10	15	15	21	25	35	

BONE CONDUCTION

			RIGHT						LEFT							
DATE	Exam	LEFT MASK	250	500	1000	2000	4000		RIGHT MASK	250	500	1000	2000	4000		AUD
		–	15	20	25	25	25		–	45	55	55	45	45		
									80	NR	55	NR	NR	NR		

Figure 12.7 *History*: A 37-year-old construction worker knocked to ground by a beam. No unconsciousness, but left ear required sutures. Noted some hearing loss in left ear after accident, which has progressively worsened. Denies ever having tinnitus or vertigo. This is a medicolegal problem, and he is suing for compensation of deafness. *Otologic*: Normal. Normal caloric findings. *Audiologic*: Note varying and inconsistent hearing levels during repeated audiograms. It is difficult to determine how much of the hearing loss is organic and how much functional. Psychogalvanic skin resistance testing confirmed a marked functional overlay, with only a 15-dB loss in all frequencies. *Classification*: Functional hearing loss. *Etiology*: Malingering.

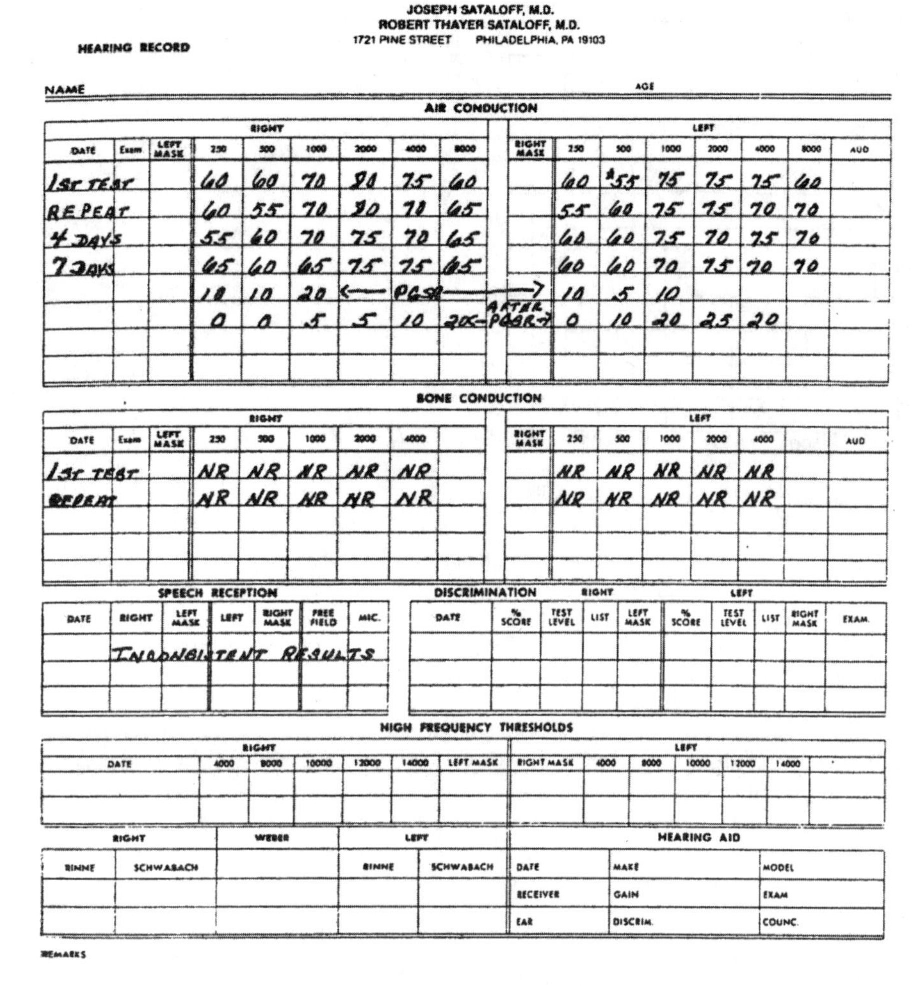

JOSEPH SATALOFF, M.D.
ROBERT THAYER SATALOFF, M.D.
1721 PINE STREET PHILADELPHIA, PA 19103

HEARING RECORD

NAME AGE

AIR CONDUCTION

		RIGHT								LEFT							
DATE	Exam	LEFT MASK	250	500	1000	2000	4000	8000	RIGHT MASK	250	500	1000	2000	4000	8000	AUD
1st TEST			60	60	70	80	75	60		60	*55	75	75	75	60	
REPEAT			60	55	70	80	70	65		55	60	75	75	70	70	
4 DAYS			55	60	70	75	70	65		60	60	75	70	75	70	
7 DAYS			65	60	65	75	75	65		60	60	70	75	70	70	
			10	10	20	←— PGSR			—→	10	5	10				
			0	0	5	5	10	20 ←	AFTER PGSR →	0	10	20	25	20		

BONE CONDUCTION

		RIGHT								LEFT						
DATE	Exam	LEFT MASK	250	500	1000	2000	4000		RIGHT MASK	250	500	1000	2000	4000		AUD
1st TEST			NR	NR	NR	NR	NR			NR	NR	NR	NR	NR		
REPEAT			NR	NR	NR	NR	NR			NR	NR	NR	NR	NR		

SPEECH RECEPTION

DATE	RIGHT	LEFT MASK	LEFT	RIGHT MASK	FREE FIELD	MIC.
	INCONSISTENT RESULTS					

DISCRIMINATION

	RIGHT				LEFT				
DATE	% SCORE	TEST LEVEL	LIST	LEFT MASK	% SCORE	TEST LEVEL	LIST	RIGHT MASK	EXAM.

HIGH FREQUENCY THRESHOLDS

	RIGHT						LEFT					
DATE	4000	8000	10000	12000	14000	LEFT MASK	RIGHT MASK	4000	8000	10000	12000	14000

RIGHT		WEBER	LEFT		HEARING AID		
RINNE	SCHWABACH		RINNE	SCHWABACH	DATE	MAKE	MODEL
					RECEIVER	GAIN	EXAM
					EAR	DISCRIM.	COUNC.

REMARKS

Figure 12.8 *History*: A 21-year-old woman with a series of emotional conflicts including breakup with her boyfriend, flunking out of college, and pending divorce of her parents. She now claims she cannot hear what goes on around her and for that reason flunked out of school. Her responses are generally delayed, and she seems to be "distant." *Otologic*: Normal. *Audiologic*: Even though she gave consistent pure-tone thresholds which showed a severe hearing loss, she often seemed able to hear soft voices behind her back. She denied hearing by bone conduction with a tuning fork. PGSR showed normal hearing. *Classification*: Functional. *Etiology*: Emotional disturbance. Her hearing returned to normal after psychotherapy.

hearing loss, such as 80 dB or even greater. In functional hearing loss of the psychogenic type, these inconsistencies usually are constant and repeatable, whereas in malingering they usually are inconstant, and the results of the tests vary considerably when they are repeated.

5. In alleged complete deafness the presence of cochlear nerve reflexes with loud noises indicates either malingering or hysteria. Psychogalvanic skin resistance tests, impedance audiometry, delayed anditory feedback tests, otoacoustic emissions testing, and evoked-response audiometry also are used to establish true hearing when subjective responses are unreliable or appear invalid.

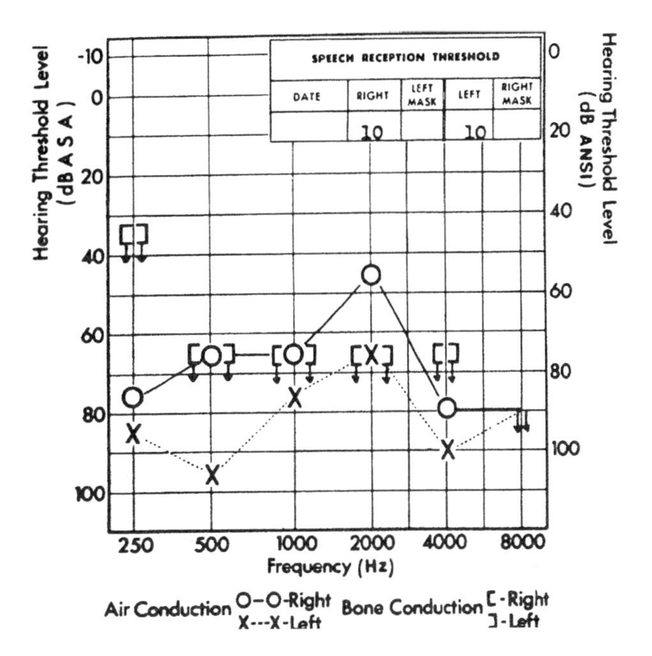

Figure 12.9 *History*: A 11-year-old girl who is doing poorly at school, and parents are concerned because she does not seem to hear them. No history of otologic disturbance, but an aunt uses a hearing aid. *Otologic*: Normal. *Audiologic*: In spite of an apparent severe bilateral hearing loss, the girl responds to speech at 10 dB. *Classification*: Functional. *Diagnosis*: Emotional conflict. This child was using a hearing loss (on the basis of her aunt's handicap) to solve her school and home difficulties. She was helped by several discussions in the office.

Psychogenically induced hearing loss usually is a uniform flat-tone loss in all frequencies, suggestive of a well-marked conductive impairment. However, in these patients, the bone conduction is practically absent. In a patient with unilateral functional deafness there may be complete absence of bone conduction on the side of the bad ear, though the good ear has normal acuity. Such a patient even may deny hearing shouts directed at the bad ear in spite of the good hearing in the opposite ear.

3.4. Audiometric Patterns in Functional Hearing Loss

There is no characteristic audiometric pattern in functional hearing loss, but the consistent inconsistencies serve to alert the physician. Usually, the hearing impairment is bilateral, and the bone conduction level is the same as the air conduction level. Figure 12.6 shows a hearing loss due to a functional overlay. Here the patient has some organic hearing loss caused by otosclerosis, but she does not use her residual hearing effectively and actually hears much less than she should. This is not rare in otosclerosis.

3.5. Should the Physician Undertake Psychotherapy?

The general practitioner or otologist must decide whether psychiatric involvement in patients with functional hearing loss is or is not too profound and complex to handle personally. The physician is more likely to assume the responsibility of limited psychotherapy

JOSEPH SATALOFF, M.D.
ROBERT THAYER SATALOFF, M.D.
1721 PINE STREET PHILADELPHIA, PA 19103

HEARING RECORD

NAME AGE

AIR CONDUCTION

RIGHT										LEFT							
DATE	Exam	LEFT MASK	250	500	1000	2000	4000	8000	RIGHT MASK	250	500	1000	2000	4000	8000	AUD
			50	45	45	35	30	35		5	5	5	0	0	5	

BONE CONDUCTION

RIGHT									LEFT						
DATE	Exam	LEFT MASK	250	500	1000	2000	4000	RIGHT MASK	250	500	1000	2000	4000	AUD
			50	45	55	45								

SPEECH RECEPTION

DATE	RIGHT	LEFT MASK	LEFT	RIGHT MASK	FREE FIELD	MIC.

DISCRIMINATION

	RIGHT				LEFT				
DATE	% SCORE	TEST LEVEL	LIST	LEFT MASK	% SCORE	TEST LEVEL	LIST	RIGHT MASK	EXAM.
	88	40			90	5			

Figure 12.10 Audiogram of a 13-year-old male with moderate sensorineural hearing loss on first evaluation. *History*: The boy was an in-patient on an adolescent medical service when he came to the attention of the author. He had been admitted with paralysis and anesthesia of his right leg. After full evaluation, he had been found to have conversion hysteria. During the process of his workup, he was found to have a right-sided hearing loss. Before his otologic evaluation, he had been evaluated by the neurology service and found to be normal except for a right-sided hearing loss. An ophthalmologist has found a mild refractive error and a peculiar, functional defect. Psychiatric consultation had confirmed the diagnosis of conversion reaction. He had had a normal lumbar puncture, electromyogram, electroencephalogram, skull series, internal auditory canal x-rays, CT scan, and multiple normal blood studies, including a full evaluation for collagen vascular disease. An otoscopic examination had been performed at the time of his admission and reportedly had been normal. Audiometry revealed a right-sided sensorineural hearing loss with a 40-dB speech reception threshold and 88% speech discrimination score. Tympanometry was normal. However, he showed Metz recruitment and reflex decay at 500 Hz and 1000 Hz in the right ear. A brain stem evoked-response audiogram was performed and revealed a normal pattern in the left ear and no normal waves including wave 1 in the right ear. His primary physicians were considering scheduling him for myelogram and arteriogram at the time he was sent for otologic consultation. At the time of examination, the patient admitted to having noticed a hearing loss approximately 2 weeks earlier. He denied tinnitus. He had some fainting feeling, but no spinning, vertigo, or sensation of motion. He admitted to mild right otalgia within the last few days. He denied any prior history of ear disease, trauma to his ear or any special concern about his ears. Except for his phychogenic paralysis and anesthesia, he appeared to be in good health. Physical examination was entirely normal except for a massive amount of tissue paper completely blocking the right ear canal and pressed against the eardrum. The child denied having put the tissue paper in his ear. After removal of the paper, his ear appeared to be normal except for a very mild external otitis. Repeat audiometric testing showed an inconsistent right sensorineural hearing loss. A Stenger test was performed and was positive. Hearing was estimated to be

JOSEPH SATALOFF, M.D.
ROBERT THAYER SATALOFF, M.D.
1721 PINE STREET PHILADELPHIA, PA 19103

HEARING RECORD

NAME AGE

AIR CONDUCTION

RIGHT									LEFT							
DATE	Exam.	LEFT MASK	250	500	1000	2000	4000	8000	RIGHT MASK	250	500	1000	2000	4000	8000	AUD
			20	15	10	20	15	15		5	5	10	5	10	5	

Figure 12.11 Audiogram of the 13-year-old male after tissue paper was removed from right ear.

if any of the following favorable factors are present: (a) the duration of the functional disturbance is short, and there is a history of previous stability, (b) the history shows that an important emotional crisis is now past, and (c) the physician is reasonably sure of his/her ability to win the patient's confidence and resolve the problem quickly. However, if the patient shows evidence of chronic or repeated emotional disturbances, psychiatric attention is essential, and it is advisable in nearly all patients with significant functional hearing loss.

3.6. Malingering

Malingering is the deliberate fabrication of symptoms that the patient knows do not exist. The patient is motivated by the desire to seek some advantage: financial compensation, escape from military service, or evasion of responsibility for failure. Malingering is becoming increasingly common in school children.

Characteristically, the malingerer abandons the symptoms when he/she thinks he/she is no longer being observed. By contrast, a psychogenic patient with a functional hearing loss believes in the symptoms, and they interfere with pleasures as well as work.

A minor form of malingering, pleading a headache to forego a dull social affair, generally is tolerated as a "white lie." However, to sham a disability as severe as deafness transcends normal behavior.

Figure 12.10 (*Continued*) within the normal range. A repeat brain stem evoked-response audiogram was scheduled. At the time this was performed, the right ear was again found to be filled with tissue paper which he had inserted since his recent otoscopic examination. Without the foreign-body occlusion, he was found to have a normal evoked-response audiogram. *Summary*: This 13-year-old patient managed to document his functional hearing loss with an initially normal otoscopic examination, a right-sided sensorineural hearing loss, stapedius reflex decay, and an abnormal brain stem evoked-response audiogram. Only through the good fortune of a repeated otoscopic examination and careful special testing was his normal hearing documented before he was subjected to myelography.

Unlike the individual who believes the symptoms are real, the malingerer who pretends deafness usually has no "pattern" in the alleged impairment. The hearing tests are replete with inconsistencies. When subjected to tests that the patient does not understand, for example, a malingering drop forge operator suspects he may be tripped up by the doctor and his/her testing machine. He wishes to preserve the fabrication that he is deaf, but when he is asked whether he can hear a signal of a given strength, he does not know when to say "yes," and when to say "no." He falters in his answers. Yesterday's audiogram may have shown a 70 dB hearing level for pure tones but a loss of only 10 dB for speech-reception. When the tests are repeated, his answers may vary by as much as 30–40 dB.

If the patient claims he has one "good" and one "bad" ear, and the examiner obstructs his "good ear" with a finger, then shouts into it loudly enough to be heard easily by bone conduction alone, the malingerer claims he hears nothing.

When a patient malingers by exaggerating a true organic hearing loss, the task of learning the truth becomes more difficult. Such a problem may assume considerable importance in medicolegal cases, particularly if they involve compensation claims for occupational deafness. Figures 12.7–12.11 show audiograms of several patients with false hearing impairments; the legends explain the motivating factors.

REFERENCES

1. Dublin WB. The cochlear nuclei revisited. Otolaryngol Head Neck Surg 1982; 90:744–760.
2. Dublin WB. The cochlear nuclei—Pathology. Otolaryngol Head Neck Surg 1985; 93:447–462.

13
Systemic Causes of Hearing Loss

Robert T. Sataloff
Drexel University College of Medicine, American Institute for Voice and Ear Research, and Graduate Hospital, Philadelphia, Pennsylvania, USA

Joseph Sataloff
Thomas Jefferson University, Philadelphia, Pennsylvania, USA

1. HEARING LOSS ASSOCIATED WITH NONHEREDITARY SYSTEMIC DISEASE

Although a great deal of attention has been paid to hereditary diseases associated with hearing loss, relatively little emphasis has been placed on recognizing the many nonhereditary diseases that are linked with deafness. Consequently, even some otolaryngologists may overlook important diagnostic information. Awareness of these illnesses is essential in distinguishing them from the many hereditary conditions they may mimic. Moreover, this knowledge often allows the physician to make important systemic diagnoses or to diagnose hearing loss early.

For example, the alert pediatrician always screens a child for hearing loss following an episode of meningitis. Similarly, internists and family practitioners should be alert to hearing problems in patients with syphilis, hypothyroidism, renal disease, and many other conditions. This chapter summarizes the more common and more serious systemic diseases associated with loss of hearing.

1.1. Rh Incompatibility

Differences in blood type between mother and child may produce deafness. If the father's blood type is Rh positive and the mother's is Rh negative, the fetus may carry the Rh factor on its blood cells. When fetal and maternal circulations mix, the mother will form antibodies directed at the Rh antigen on her child's red blood cells. The consequent immunologic attack on the infant's erythrocytes produces hemolysis and may occur in the inner ear to produce profound congenital deafness. The hearing loss is usually bilateral and most severe in the high frequencies (1). Genetically, the infant born of a woman's first pregnancy is not affected because the mixing of fetal and maternal blood that initiates the immunologic process usually occurs at the time of delivery of the first child. Subsequent Rh-positive children may then suffer severe hemolysis during fetal development, because of the persistence of maternal antibodies to the Rh factor which recognize antigenic fetal red blood cells.

Modern developments have made this type of hearing loss preventable in most patients. Rh screening should be done routinely before parenthood. When Rh incompatibility exists, the mother should be treated in the immediate postpartum period with a drug such as RhoGAM (Ortho Diagnostics, Inc.). This immunoglobulin effectively suppresses the formation of maternal antibodies to the Rh factor. The drug also should be given following miscarriage, abortion, or ectopic pregnancy, because these events may initiate the immunologic mechanism by introducing a small number of Rh-positive red blood cells into the maternal circulation.

1.2. Hypoxia

Oxygen deprivation in the neonatal period may produce sensorineural hearing loss. This connection may explain the increased incidence of deafness associated with traumatic, cyanotic, or premature birth, although the causal relationship has not been proven (2). Children who have suffered a complicated delivery should be screened.

1.3. Neonatal Jaundice

Kernicterus (encephalopathy associated with severe unconjugated hyperbilirubinemia) has long been known to cause sensorineural hearing loss (3). It is unclear whether the primary site of damage is peripheral, central, or both (2,4–6). In cochlear lesions (2), the audiogram usually reveals mild sensorineural hearing loss in the lower frequencies, gradually falling off to a severe hearing loss from 2000 Hz up (Fig. 13.1). Rh or ABO blood-group incompatibility between mother and child is one of the most common causes, although several others, such as hepatic and biliary dysfunction, can be responsible. Acute bilirubin encephalopathy is most likely to cause damage between 3 and 7 days of life, but damage may occur at older ages, even in adolescence (7). Regardless of the cause, hyperbilirubinemia should arouse suspicion of hearing loss. Hearing screening is recommended.

1.4. Rubella

German measles is the classic example of in utero disease producing severe sensorineural hearing loss, although other viruses may produce similar deafness. Rubella is caused by an RNA virus which is present in the throat secretions, blood, and stool of infected persons. It probably enters the body by penetrating the upper respiratory mucosa (8). Congenital rubella is caused by transplacental transmission of the virus to the fetus. The disease is

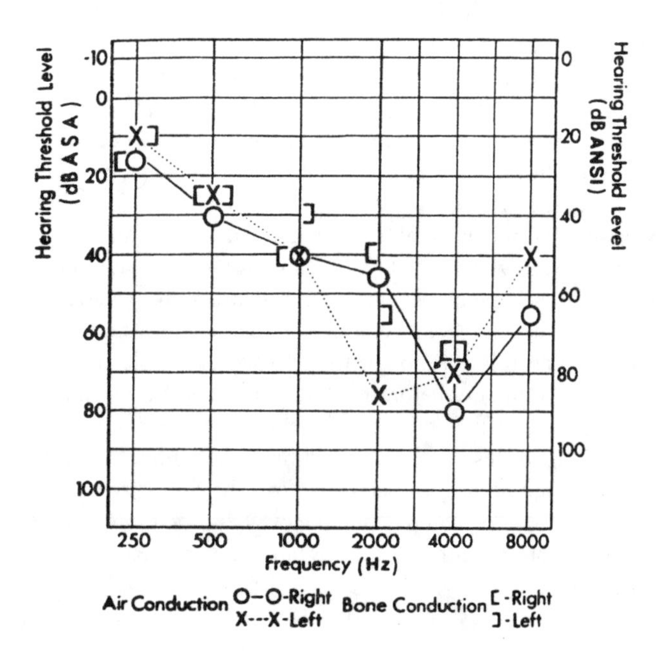

Figure 13.1 Bilateral sloping sensorineural hearing loss with some recovery at 8000 Hz in a child with a history of neonatal jaundice but no history of noise exposure. There was no recruitment of abnormal tone decay.

most common in children 5–9 years of age, but many cases occur in younger children, adolescents, and young adults.

The incubation time between exposure an appearance of the rash of rubella is from 14 to 21 days. Headache, fever, malaise, lymphadenopathy, and mild conjunctivitis may precede the rash by as much as a week, particularly in adults. The exanthem often is the first sign of the disease in children. Rubella may cause lymph node enlargement alone, without skin lesions, and it may be unrecognized until serologic study. Respiratory symptoms are not prominent. Forschheimer spots are small red lesions on the soft palate. These spots may be present but are not pathognomonic.

The rash of German measles is characterized by small, maculopapular, pink lesions, which are usually discrete. Sometimes they coalesce to form a diffuse erythematous exanthem. The rash starts on the forehead and face and spreads to the trunk and extremities. Usually, it is present for ~3 days and is preceded by tender lymphadenopathy which persists for several days after resolution of the rash. Postauricular and suboccipital nodes are most dramatically involved. Arthralgias and swelling of small joints may accompany the exanthematous period and may persist longer than other signs and symptoms. Purpura, hemorrhage, and encephalomyelitis also may occur.

Congenital rubella typically includes corneal clouding, chorioretinitis, cataracts, microphthalmia, microcephaly, mental retardation, patent ductus arteriosus, intraventricular septal defeat, pulmonic stenosis, and deafness. The *expanded rubella syndrome* described after the American epidemic of 1964 also includes thrombocytopenia, purpura, hepatosplenomegaly, interstitial pneumonia, metaphyseal bone lesions, and intrauterine growth retardation. In the 1964 epidemic, ~10% of women with rubella discovered during the first trimester delivered infants with the rubella syndrome. However, asymptomatic maternal rubella may produce the disease also.

Rubella deafness is characterized by sensorineural hearing loss with a flat audiometric pattern. The severity of hearing loss may differ substantially between the two ears (9). Severe to profound deafness has been found in 4–8% of children with histories of maternal rubella, and it has also been recognized following asymptomatic maternal infections (2). Histopathological studies reveal cochleosaccular aplasia and occasional middle-ear anomalies (Fig. 13.2).

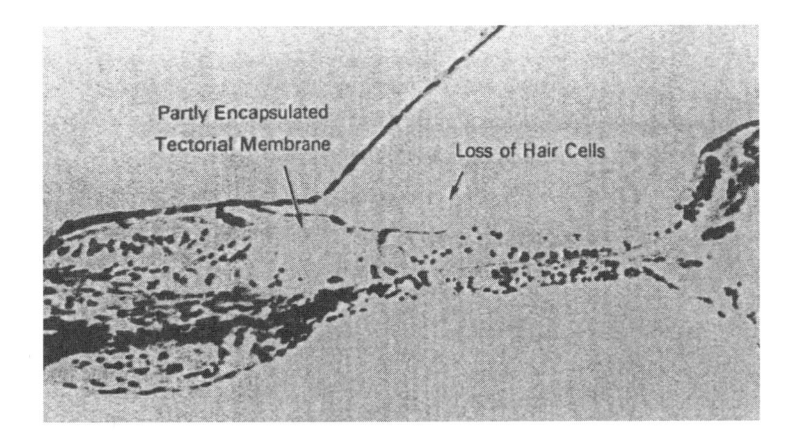

Figure 13.2 Maternal rubella producing temporal bone pathology. The organ of Corti is slightly flattened, and the hair cells are missing. The tectorial membrane is rounded, retracted into the inner sulcus, and partially encapsulated. [From Schuknecht (2), p. 180. Courtesy Harvard University Press.]

Rubella may be confused with infectious mononucleosis and viral disease, such as erythema infectiosum and enteroviral exanthems, which do not have the same teratogenic potential. The diagnosis of rubella may be confirmed by isolating viruses or documenting changes in antibody titer. The antibodies generally are present by the second day of the exanthem, and titers rise for 2 or 3 weeks. The initial determination of antibody titer should be performed as soon after exposure as possible to help distinguish rising titers of acute infection from persistent elevation secondary to prior immunity. Lymphocytosis and atypical lymphocytes may be present, but are nonspecific. In congenital rubella, serologic testing may revert to negative by the age of 3–4 years (8). Hence, a negative serologic test in an older child does not exclude a diagnosis of congenital rubella. Attenuated live viral vaccines have been given to young children in the United States since 1969. Although the attenuated virus can be detected for up to a month after immunization, transmission to other people is rare. The purpose of the vaccination program is to decrease the incidence of the disease, thereby decreasing the probability of pregnant women coming into contact with the infection. Adult women who are shown to be susceptible to rubella by serologic testing may be vaccinated also. Arthralgias and joint swelling occur in ~25% of immunized adult women, sometimes beginning as long as 2 months after vaccination (8). Subclinical rubella may develop following immunization, but viremia and fetal infection generally do not ensue. However, the attenuated virus vaccine itself can produce fetal damage (8). Therefore, it must never be given to pregnant women or those who may become pregnant within 2 months following immunization.

Gamma globulin may be given to patients following exposure to the disease and may prevent clinical rubella. However, serologic titers may rise, and fetal infection still may occur (8,10). Amniocentesis and culture of amniotic fluid may confirm fetal infection by recovery of the virus. However, negative cultures do not rule out infection. Because of the high incidence of birth defects, abortion should be considered seriously in any case of rubella found during the first 3–4 months of pregnancy, when infection is most likely to result in congenital anomalies.

1.5. Mumps

Mumps appears to be the most common cause for unilateral hearing loss in the United States (11). Interestingly, the vestibular system is affected by mumps rarely. Because the disease often occurs in childhood and because children are so adaptable, deafness may not be recognized for many years if hearing in the unaffected ear is normal (Fig. 10.51). The deafness is almost always total and unilateral. When these patients are tested, the ear that has normal hearing must be masked carefully to avoid crossover and the false impression that patients have residual hearing.

1.6. Rubeola and Other Infections

Measles (rubeola), cytomegalic inclusion disease, herpes, roseola, infectious mononucleosis, varicella, *mycoplasma* pneumonia, typhoid fever, scarlet fever, influenza, and other infections have also been associated with sensorineural hearing losses (12–14) (Fig. 10.60). The hearing loss may be severe or profound and may be sudden or gradually progressive. So far, only symptomatic and preventive therapy is available. These diseases may occur in adults, in children, and in utero. Particular effort should be made to protect pregnant mothers from exposure to these infectious agents. Measles and scarlet fever also are notorious for their destruction of the eardrum and middle ear.

1.7. Flu

The "common cold" often has an associated earache which may result from referred pain caused by pharyngeal inflammation or from otitis media. When the illness is bacterial, otalgia may be due to an otitis media secondary to pneumococcus, *Hemophilus* (especially, but not exclusively, in children), *Streptococcus*, or *Staphylococcus*. Anaerobic organisms also may be involved (15). In newborns, gram-negative organisms such as *Escherichia coli* are common pathogens. When the illness is viral, viruses often can be cultured from middle-ear fluid, although this is rarely necessary. Otitis media is especially common following epidemic influenza (Fig. 10.20). In addition to the conductive hearing loss caused by otitis media, both bacterial and viral infections can lead to labyrinthitis with sensorineural hearing loss, tinnitus, and vertigo. One of the key symptoms of the labyrinthine involvement is a feeling of fullness in the ear; consequently, this symptom should not always be attributed to middle-ear fluid.

1.8. Fungal Diseases

Fungal infections may invade the ear and produce conductive or even profound sensorineural hearing losses. This reaction is most common in immune-compromised or seriously ill patients, or in severe diabetics. Aspergillosis, candidiasis, blastomycosis, cryptococcosis, and other fungal infections occur. Mucormycosis is a particularly devastating infection. The ear involvement can occur with or without fungal meningitis.

1.9. Lassa Fever

Lassa fever is an acute febrile illness caused by infection with an arenavirus endemic in West Africa, but reported in the United States (16,17). Symptoms include malaise, weakness, arthralgia, and low back pain initially. In the ensuing several days, cough, sore throat, headache, epigastric pain, and chest discomfort are common. Vomiting, diarrhea, and fever usually occur by day 5. As the illness worsens, respiratory distress and bleeding, head and neck edema, pleural and pericardial effusions, shock, and death may be seen. Most patients begin recovering within ~10 days. Hearing loss occurs in ~18% of patients who are serologically positive for the Lassa virus. In an endemic area, 81% of local inhabitants with sudden deafness were found to have antibodies to Lassa virus (17).

1.10. Lyme Disease

Lyme disease is caused by *Borrelia burgdorferi*, a spirochete transmitted by the tick *Ixodes dammini*. A red papule appears 3–20 days following a tick bite and expands to a large annular red lesion associated with fever, backache, malaise, stiff neck, arthritis (particularly in the knees), lymphadenopathy, a complete heart block (in 8%), and neurological abnormalities. Neurologic disorders associated with Lyme disease may include encephalitis, radiculoneuritis, and neuropathies of any cranial nerve. Unilateral and bilateral facial paralysis has been reported. Hearing loss has not been recognized commonly, but it may occur (18,19). It has been identified in a patient with nonluetic interstitial keratitis, vestibuloauditory dysfunction, and bilateral recurrent facial paralysis previously thought to have been Cogan's syndrome, and in patients with sudden hearing loss, Meniere's syndrome, or hearing loss in combination with vertigo and/or facial paralysis. In rare cases, it may also cause retrocochlear hearing loss similar to that caused by acoustic neuroma, due to expansive granulomatous lesions in the posterior cranial fossa (20). Treatment includes antibiotics, and steroids in selected cases.

1.11. HIV/AIDS

The AIDS clinical syndrome is characterized by immunodeficiency, frequently complicated by opportunistic infection and neoplasia. AIDS-related diseases may affect every body system, including the head and neck. Patients with AIDS are particularly susceptible to infectious agents including viruses, bacteria, and fungi. Pneumocystic carinii infection has been found in the external and middle ear (21,22). These infections are associated with mixed conductive and sensorineural hearing loss.

Viral infections of the head and neck are common in AIDS patients, particularly those caused by cytomegalovirus (CMV), Epstein–Barr virus (EBV), human papilloma virus (HPV), and the herpes virus, both simplex and zoster. All cranial nerves may be affected. Herpes zoster is particularly likely to involve the eighth cranial nerve, causing hearing loss, vertigo, and often severe pain and facial paresis or paralysis. HIV-associated syphilis may also be responsible for sensorineural hearing loss.

Hearing loss may be caused not only by the great number of opportunistic otologic infections associated with AIDS, but also with drug-induced ototoxicity, central nervous system toxoplasmosis, and meningitis (especially that caused by tuberculosis or cryptococcus). The HIV virus itself is known to be neurotropic and may be itself capable of causing eighth nerve dysfunction including hearing loss.

1.12. Meningitis

Meningitis is still a relatively common severe infection in adults and children. When it causes hearing loss in an adult the patient usually reports it; however, it may go unrecognized in children. The incidence of associated deafness is impressively high: ~40% in fungal meningitis; estimated between 6% and 35% in bacterial meningitis; and uncommon in aseptic (viral) meningitis (23,24). The hearing loss is bilateral in ~80% of cases and partial in ~70%. Many patients suffering partial hearing loss recover to some extent. However, any patient with meningitis—especially a child—should have a hearing evaluation upon recovery.

1.13. Tuberculosis

Although tuberculosis affecting the ear is relatively uncommon today, it still is encountered, especially in the inner cities. Most commonly, it occurs as a chronic ear infection resistant or treatment. Multiple perforations of the tympanic membrane and watery otorrhea are typical. Complications, such as meningitis and facial paralysis, may ensue (25). Usually, because the diagnosis is not suspected, patients undergo numerous courses of various antibiotics, and occasionally even surgery, without improvement. Ear disease may be the first sign of tuberculosis. Once the diagnosis is made (by acid-fast staining of aural drainage or granulation tissue, or appropriate culture), a careful evaluation for systemic involvement is required, and antituberculosis therapy should be instituted.

1.14. Sarcoidosls

Sarcoidosis is seen most commonly in black patients and is generally diagnosed by its characteristic chest X-ray results (Fig. 13.3). It can occur in virtually any organ, however, including the ear (26). It is a granulomatous disease of unknown etiology and may cause sensorineural hearing loss which may be sudden or fluctuant or both. Sarcoidosis should be suspected, particularly in people who are known to have the systemic disease. It also must be in the differential in patients with chronic ear disease resistant to therapy, however, particularly when associated with uveitis, facial nerve paralysis, other neuropathy, diabetes insipidus, or meningitis.

1.15. Wegener's Granulomatosis

This disease is characterized by necrotizing vaculitis and granulomatous lesions of the nose, paranasal sinuses, lungs, and kidneys. It may occur as a granulomatous otitis resistant to conventional antibiotic therapy (27,28). Bleeding from the ear has even been the initial symptom in some cases. Conductive hearing loss is associated with serous otitis media or middle-ear granulomas, and sensorineural hearing loss may infrequently result from inner-ear granulomas or vascular causes. Other unusual granulomatous diseases must also be considered in any case of persistent middle-ear disease that does not respond as expected to routine therapy. The occurrence of *lethal midline granuloma* in the ear has also been reported but is quite unusual and may represent an incorrect diagnosis. Treatment is now available for both these previously fatal diseases: Wegener's granulomatosis is currently best treated with immunosuppressants (particularly, cyclophosphamide), and midline granuloma is responsive to orthovoltage irradiation.

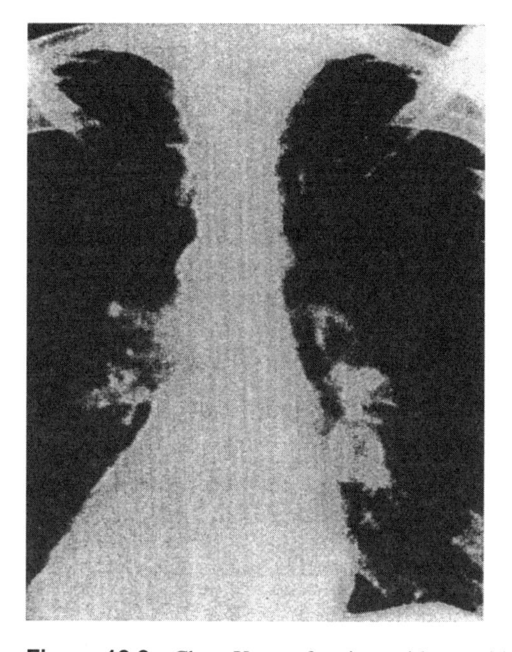

Figure 13.3 Chest X-ray of patient with sarcoidosis shows typical hilar adenopathy.

1.16. Vasculitis

Rheumatoid arthritis, giant-cell arteritis, polyarteritis nodosa, leukocytoclastic angitis (29), and various other vasculitides have been associated with hearing loss. Middle-ear fluid with conductive hearing loss is common, and sensorineural hearing loss also may occur. Occasionally, middle-ear disease may precede other manifestations of a vasculitis syndrome, or it may persist following otherwise successful therapy with steroids or other medications. In patients with known systemic vasculitis and conductive hearing loss that does not respond to conventional therapy, exploratory tympanotomy and middle-ear biopsy may be indicated. This combination may be performed as a diagnostic measure early in the course of the disease if primary vasculitis is suspected. Early detection and prompt treatment are the mainstays of therapy.

In fact, any systemic disease which affects blood vessels adversely may be associated with hearing loss. Such entities include diabetes, atherosclerosis and other vascular diseases, and syphilis which are discussed elsewhere in this chapter, as well as collagen vascular diseases including rheumatoid arthritis, lupus, Sjögren's disease and others.

1.17. Histiocytosis X

Letterer–Siwe disease, *Hand–Schuller–Christian disease*, and *eosinophilic granuloma* are similar to congenital lipid-storage diseases such as *Gaucher's disease* and *Neimann–Pick disease*, but there is no familial tendency, and the accumulation of lipid appears to be a secondary occurrence. *Letterer–Siwe disease* (Fig. 13.4) occurs with destructive skeletal lesions (particularly in the skull), anemia, purpura, hepatosplenomegaly, and adenopathy. Temporal bone lesions may be found (30) but are usually not isolated lesions. Death supervenes generally before the age of 2.

Hand–Schuller–Christian disease, a less severe form of histiocytosis, also occurs with destructive skull lesions that often involve the temporal bone (Fig. 13.5).

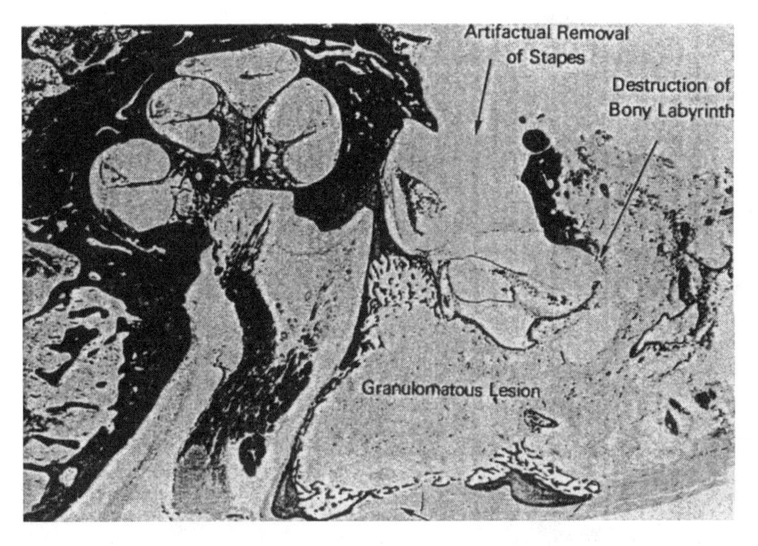

Figure 13.4 Letterer–Siwe disease with an extensive destructive lesion involving the posterior right temporal bone. The mastoid bone and the bony labyrinth surrounding the canal have been destroyed and replaced by viable vascular granulation tissue. The perilymphatic and endolymphatic spaces contain fibrinous precipitate. [From Schuknecht (2), p. 387. Courtesy Harvard University Press.]

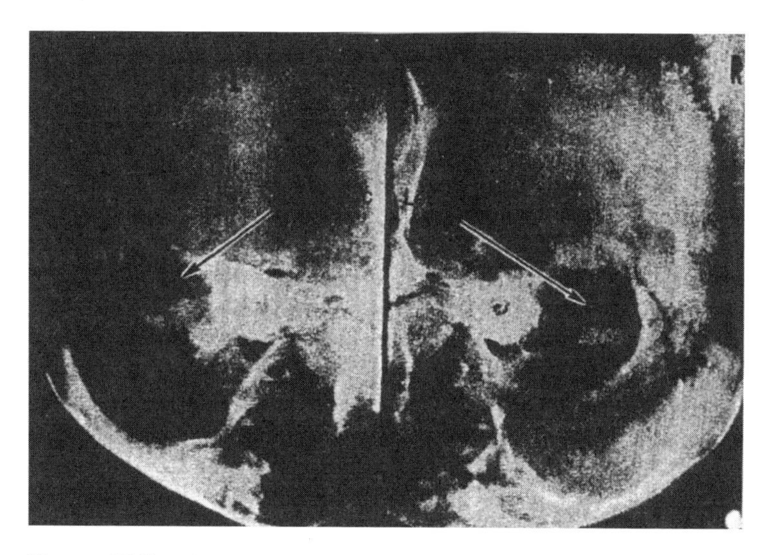

Figure 13.5 Bilateral destructive lesions of the mastoid in Hand–Schuller–Christian disease. [From Schuknecht (12), p. 387. Courtesy Harvard University Press.]

Classically, diabetes insipidus, exophthalmos, and defects in the calvarium are present. Apparent chronic ear infection with otorrhea is common, so that mastoidectomy often is performed before the correct diagnosis is made (2,31,32). However, preferred treatment is radiation therapy rather than surgery in most cases. Early childhood death is not as prominent as it is with *Letterer–Siwe* disease, and the disease may even arise as late as the second or third decade of life. Growth retardation, anemia, hypogenitalism, and pathological fractures may be other features. The mortality rate is ~30%.

Eosinophilic granuloma is the mildest form of histiocytosis. It occurs with one or two lytic lesions of the skull, without more serious systemic involvement. The disease usually becomes apparent in childhood or young adulthood, with 80% of patients >30 years of age. The long bones, ribs, and vertebrae may be involved, rather than the skull, and localized pain is the most common symptom. A draining ear also may be a feature.

1.18. Hypoparathyroidism

Hypoparathyroidism is one of the metabolic diseases related to calcium metabolism. The disease causes vitamin D deficiency and subsequent hypocalcemia. Long-standing untreated hypoparathyroidism appears to be associated with a high incidence of sensorineural hearing loss (33). Hearing loss is much less common in well treated hypoparathyroidism. It is unclear whether hearing loss associated with hypoparathyroidism is reversible (33).

1.19. Allergy

It has long been recognized that allergy can be associated with conductive hearing loss, especially serous otitis media in children. Eustachian-tube congestion and dysfunction may be the mechanism of action, but an allergic response within the middle-ear mucosa itself may also play a role. More recently, an association between allergy and sensorineural hearing loss, particularly Meniere's syndrome, has been suggested (34).

1.20. Hyperlipoproteinemia

A significant percentage of patients with inner-ear disease have been found to have hyperlipoproteinemia (35). Frequently, these patients have symptoms similar to those of Meniere's disease, particularly fluctuating sensorineural hearing loss in the low frequencies. However, the hearing loss may follow any sensorineural pattern. Hyperlipoproteinemia may be associated with dietary habits, pregnancy, and a number of diseases, including diabetes, hypothyroidism, myeloma, biliary obstruction, nephrotic syndrome, obesity, pancreatitis, and dysgammaglobulinemia. Oral contraceptives may cause a type 4 hyperlipoproteinemia pattern on lipoprotein electrophoresis. Vestibular abnormalities also may be prominent. High suspicion and early detection are essential, not only to help manage otologic disease, but also to prevent the development of atherosclerotic cardiovascular disease.

1.21. Hypertension

Hypertension is common, and associated with numerous disease. Several studies have suggested a correlation between noise-induced hearing loss and high blood pressure (36–54). Some studies have shown a high correlation between hypertension and noise exposure, others have shown no correlation. Some have suggested that noise exposure may increase stress and blood pressure (although it has not been shown that this response is abnormal, or damaging), and others have suggested that the tendency toward hypertension is associated with greater risk of hearing impairment. Talbott et al. (55) has even suggested that noise-induced hearing loss may be a marker of hypertension in older noise-exposed populations. It is certainly recognized that some conditions such as hyperlipoproteinemia are associated with increased risks of both hypertension and hearing loss. At present, it appears that people with hypertension have a higher incidence of hearing loss (with or without noise exposure). There is no compelling evidence to indicate that these persons are at greater risk of sustaining noise-induced hearing loss than others, nor that noise is capable of causing significant, prolonged hypertension in humans. Additional research in this area is needed.

1.22. Syphilis

At the present time, secondary syphilis is still a relatively rare cause of otologic problems. The classic forms of late (or tertiary) syphilitic disease, such as gumma formation, which may involve the ear and produce hearing loss, are also uncommon. However, *inner-ear syphilis*, may occur more frequently (56). It is a specialized form of tertiary syphilis. Diagnosis requires an FTA-absorption test MHA-TP or other sophisticated assay RPR and VDRL tests are generally negative. Even following treatment for congenital syphilis or in the presence of uninfected cerebrospinal fluid, live spirochetes can be sequestered in the fluid within the ear (57,58).

Spirochetes show unusual resistance to antibiotic therapy in the ear, in the anterior chamber of the eye, and in joint spaces. Of special importance in treatment rationale is their dividing time of 90 days (59), when compared with 33 h in early syphilis.

Symptoms may be Meniere's-like with fluctuating hearing loss, vertigo, and tinnitus, or there may be a sensorineural hearing loss of any pattern. As many as 6–7% of adults with Meniere's-like syndrome or with sensorineural hearing loss of unknown etiology may have this entity (34,56) although the percentage is much lower in most practices. Rapidly progressive sensorineural hearing loss, worse on one side than the other and with poor discrimination, is particularly characteristic (Fig. 13.6).

JOSEPH SATALOFF, M.D.
ROBERT THAYER SATALOFF, M.D.
1721 PINE STREET PHILADELPHIA, PA 19103

HEARING RECORD

NAME AGE

AIR CONDUCTION

RIGHT										LEFT							
DATE	Exam	LEFT MASK	250	500	1000	2000	4000	8000		RIGHT MASK	250	500	1000	2000	4000	8000	AUD
10/77			30	40	55	70	70	60			10	30	55	70	70	50	
11/77			15	20	35	55	55	35			15	20	45	65	55	50	
1/78			15	20	30	55	45	30			10	20	20	65	60	50	
3/79			15	20	30	55	45	30			10	15	20	55	55	50	

BONE CONDUCTION

RIGHT										LEFT							
DATE	Exam	LEFT MASK	250	500	1000	2000	4000			RIGHT MASK	250	500	1000	2000	4000		AUD
10/77			25	30	50	↓	↓				10	25	55	60	↓		
11/77			15	15	35	55	50				10	20	45	55	55		
3/79			15	20	30	55	40				10	10	20	55	50		

SPEECH RECEPTION

DATE	RIGHT	LEFT MASK	LEFT	RIGHT MASK	FREE FIELD	MIC.
10/77	45		30			
11/77	30		25			
3/79	20		20			

DISCRIMINATION

	RIGHT				LEFT				
DATE	% SCORE	TEST LEVEL	LIST	LEFT MASK	% SCORE	TEST LEVEL	LIST	RIGHT MASK	EXAM.
10/77	74				60				
11/77	84				70				
3/79	96				78				

Figure 13.6 *History:* A 50-year-old man with ringing tinnitus and rapidly progressive hearing loss within the last 3 months, worse on the right. Sounds are muffled and "garbled." No history of noise exposure, infection, ototoxic drugs, or trauma. No family history of deafness. Firmly denied exposure to gonorrhea or syphilis. *Otologic*: Normal. *Audiologic*: Bilateral asymmetrical sensorineural hearing loss with somewhat depressed discrimination. Diaplacusis present. Tone decay absent. *Laboratory*: VDRL negative, FTA-abs, and MHA-TP strongly positive. Thyroid-function tests, glucose-tolerance test, cholesterol, triglycerides, and internal auditory canal X-ray films normal. *Diagnosis*: Syphilitic hearing loss. *Course*: The patient was treated initially with steroids while waiting for the serological tests because of the rapidly progressive hearing loss and the strong suspicion of syphilis. Hearing improved within 3 weeks. Subsequently, antibiotics were started after the 11/77 audiogram. He received 2.4 million units of Bicillin i.m. weekly for 6 months, and steroids were tapered. His hearing remained good a year later on no medication.

Syphilis should be investigated in all cases of sudden deafness because it is one of the few causes that responds well to therapy, and if it is untreated, deafness in the remaining ear may follow. As such, some otologists advocate the use of high-dose steriods while the results of the FTA-absorption test are pending.

Syphilis otopathlogy shows endolymphatic hydrops, as in Meniere's disease, and osteitis of the otic capsule (Fig. 13.7) (2). Often, vestibular function is reduced bilaterally. The question of optimum treatment remains unanswered, although steroids and intensive antibiotic therapy are advocated most widely. However, it is clear that the usual antibiotic regimens for tertiary syphilis or neurosyphilis are not sufficient to eradicate infection of the inner ear. Prolonged treatment (a year or more) is recommended.

1.23. Hypothyroidism

Endemic cretinism (congenital hypothyroidism associated with inadequate iodine ingestion) may occur with abnormally long persistence of neonatal jaundice, poor feeding, hoarse crying, lethargy, delayed development, short stature, coarse features with protruding tongue, broad flat nose, sparse hair, dry skin, retarded bone age, and other well-known features. The auditory defect experienced in these children is generally a sensorineural hearing loss, although it may be mixed (60).

About 50% of adult patients with myxedema have hearing losses, which may be conductive, sensorineural, or mixed (61,62). The sensorineural hearing loss is typified by a low discrimination score, which may respond well to thyroid therapy. There may not be any change in pure-tone threshold, although threshold improvement may occur, as well. Hypothyroidism is found in up to 3% of patients with symptoms of Meniere's disease and is believed to produce endolympahatic hydrops (34). Thyroid replacement also appears to be effective in eliminating Meniere's symptoms in patients with this etiology. Pendred's syndrome, Hashimoto's thyroiditis, and other causes of thyroid dysfunction must be included in the differential diagnosis.

1.24. Hypoadrenalism and Hypopituitarism

Pituitary or adrenal hypofunction may be associated with symptoms of Meniere's disease (34). A flat 5-h glucose-tolerance test curve should alert the clinician to this possibility. The otologic symptoms are bilateral usually. Insulin stimulation test, or ACTH plasma cortisol stimulation test, can aid in the diagnosis. Hormone-replacement therapy should be instituted as soon as the diagnosis is made.

1.25. Autoimmune Sensorineural Hearing Loss

Autoimmune senorineural hearing loss was originally described as an entity of young adults characterized by bilateral, asymmetric, rapidly progressive sensorineural hearing loss with marked vestibular dysfunction (63). It is often associated with tissue destruction of the mastoid, middle ear or ear drum. There is more some indication that minor trauma such as a myringotomy to evacuate middle-ear fluid may rarely trigger this disease process.

More recently, it has become apparent that autoimmune sensorineural hearing loss may occur in any age group and may produce almost any audiometric pattern. Although it need not be accompanied by tissue destruction, bilateral involvement, and rapid progression are most common. This disease entity may cause bilateral total deafness, with or without substantial dizziness. Various laboratory tests may be helpful in establishing the diagnosis including tests of the humoral and cell-mediated immune systems, complement tests, haplotype, and others. Current test protocol includes quantitative serum immunoglobins by group, C-3, C-4, CH-50, C-1Q, T-cell subsets; HLA-A,B,C typing; HLA-DR typing, western blot for cochlear antibodies and other tests. This evaluation is ordered only in patients in whom there is a reasonably high clinical suspicion of autoimmune disease. Bowman and Nelson (64) has shown that increased Cw7 correlates positively with immune hearing loss, or possibly with steroid responsiveness. Absence of DR4 suggests an increased disease susceptibility; and the presence of Cw4 and B35 are associated somewhat more weakly with autoimmune sensorineural hearing loss. Treatment includes steriods and cytotopic drugs such as methotrexate or Cyclophosphamide. Rarely plasmaphersis may be necessary.

A special type of autoimmune sensorineural hearing loss called sympathetic cochleitis involves bilateral hearing loss following injury or surgery to one ear. It is believed due to exposure of inner-ear protein to the immune system. It is also treated with steroids and cytotopic agents.

1.26. Renal Failure

High-frequency sensorineural hearing loss is found not infrequently in patients with severe renal disease. Dialysis patients have been studied in particular. Their audiograms show a dip at 6000 Hz, with some depression at 4000 and 8000 Hz, as well. The loss usually does not go $<$2000 Hz, and wide fluctuations in threshold may occur during a single dialysis period (65). The pathogenesis of this otopathology is unclear. Hyperlipidemia has been evaluated and does not seem to be the cause. Prior exposure to ototoxic drugs frequently complicates evaluation, and its significance is unclear. Nevertheless, patients with renal failure require monitoring of auditory function.

1.27. Aging

The general aging process affects all parts of the body, and the ear is no exception (66). Presbycusis actually begins in childhood as a progressive loss of hair cells and nerve fibers within the inner ear. This process starts in the highest frequency regions and gradually progresses to the speech range. When bilateral, symmetrical, sloping, high-frequency

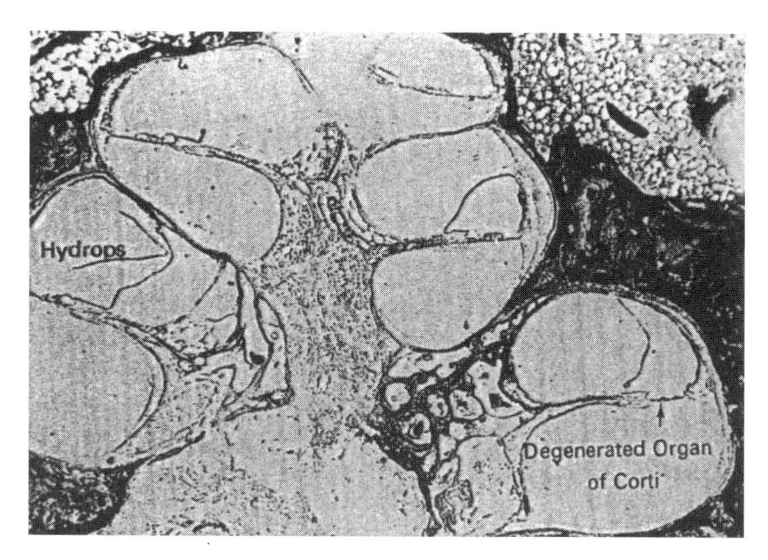

Figure 13.7 Congenital syphilitic labyrinthitis in a 70-year-old woman. Severe endolymphatic hydrops can be seen, as well as coalesced areas of bone destruction, severe degeneration of the organ of Corti, and absence of the organ of Corti in the basal 10 mm. Absence of cochlear neurons in the lower basal turn and ~50% degeneration of cochlear neurons elsewhere can be seen. [From Schuknecht (2), p. 265, Courtesy Harvard University Press.]

sensorineural hearing loss is identified in elderly persons, presbycusis is likely. Caution must be exercised because of the great tendency to assign all such patients to this category without proper evaluation. Actually, a number of such patients have hereditary forms of hearing loss. These forms are sometimes associated with other manifestations, which should be diagnosed properly. Others may have a variety of nonhereditary causes, such as syphilis or acoustic neuroma. The tendency to ascribe an "obvious" diagnosis to a given hearing loss must be restrained unless a thorough evaluation has been performed.

1.28. Psychosis

The presence of hearing loss has been studied in adult patients with paranoid and affective psychosis (67). Sensorineural hearing loss was found in ~60% of paranoid psychotics and >70% of the affective group. Conductive hearing loss was found in ~20% of the paranoid group and <2% of people with affective psychosis. It has also been suggested that the duration of hearing loss is longer in the paranoid group and frequently may precede the onset of psychosis. The effect of hearing loss on psychological development in early childhood is well recognized. The proclivity for relatively mild paranoid tendencies, neurosis, and other psychological disturbances in older hard-of-hearing patients has been well established also. However, the association with frank psychosis is relatively recent and is an interesting subject for further clarification.

1.29. Malignancy

Primary carcinomas and sarcomas of the ear occur (68–70) and are discussed in separate chapters, but tumors from distant sites also may metastasize to the temporal bone. This fact is frequently unrecognized but should be considered if otitis develops in a patient with a known cancer. It also emphasizes the need to look for a primary tumor elsewhere when a carcinoma of the ear is found. Metastatic carcinoma to the ear has been reported from breast, kidney, lung, stomach, larynx, prostate, thyroid, nasopharynx, uterus, meninges, scalp, rectum, the parotid gland, intestine, brain, carotid chemodectoma, spinal cord, and other sites (71). Tumors of the *skull base* also can produce hearing loss by direct involvement of the ear or by interference with eustachian tube function, leading to fluid in the middle ear and conductive hearing loss. Direct extension from adjacent basal-cell carcinomas, melanomas, meningiomas, benign, or malignant neural tumors of nearby cranial nerves, glomus tumors, hemangiomas (which may be multiple), and a variety of other neoplasms also may be implicated. Nasopharyngeal cancers classically occur with unilateral serous otitis media in an adult, secondary to eustachian tube occlusions. This malignancy is more common among Orientals but must be searched for in any patient with unexplained serous otitis.

Malignancies such as Hodgkin's disease, leukemia, lymphoma, and myeloma, which produce defects in the immunologic system, increase the incidence of ear infection and resultant hearing loss and serious otologic complications. The importance of this possibility may be overlooked in patients with dramatic systemic disease. Untreated otitis media can lead not only to a progressive hearing loss but also to meningitis and death, particularly in an immune-compromised patient.

Treatment for malignancy may involve *radiation therapy*, with its complications including hearing loss. Dryness and scaling of the skin of the external auditory canal may lead to buildup of debris and conductive hearing loss. *Osteoradionecrosis* of the temporal bone may produce chronic infection and may result in conductive or even severe sensorineural hearing deficit (2,72).

1.30. Coagulopathies

A few conditions such as cochlear artery occlusion may be associated with vascular dysfunction of the inner ear. Hypercoagulable states such as those associated with certain tumors, polycythemia, Buerger's disease, macroglobulinemia, and some viral infections have been implicated as etiological factors in certain cases of sudden hearing loss (73). Coagulation defects coincident with primary coagulopathies or secondary to diseases such as leukemia can cause inner-ear hemorrhage and consequent deafness (2). Such hearing losses are not reversible but may be prevented by appropriate management of the underlying disease. Therapy is especially critical if hearing has been lost in one ear already.

1.31. Glomus Jugulare and Glomus Tympanicum

Glomus jugulare tumors are rare, but when they occur, hearing loss and tinnitus are frequently the only symptoms. This peculiar neoplasm (paraganglioma) arises from cells around the jugular bulb and expands to involve neighboring structures (2). In doing so, the neoplasm most frequently extends to the floor of the middle ear, causing conductive hearing loss and pulsating tinnitus. As the disease progresses, it may appear as chronic otitis media and may even extend through the eardrum and appear to be granulation tissue in the ear canal. Unsuspecting biopsy of this apparent granulation tissue may cause profuse bleeding because of the marked vascularity of the tumor. As the disease extends, it may destroy portions of the temporal bone and jugular bulb and can extend intracranially.

Glomus tumors may also arise from cells along the medial wall of the middle ear. These are called glomus tympanicum tumors and are generally somewhat easier to manage surgically. It is essential to distinguish between glomus tympanicum and glomus jugulare before attempting surgical intervention.

As in any expanding neoplasm, early diagnosis of a glomus tumor facilitates surgical cure. Since conductive hearing loss may be the only symptom in many patients, the physician is obligated to establish a cause for every case of unilateral conductive hearing loss.

Physical examination may disclose a pinkish mass in the middle ear. Positive pressure on the eardrum may reveal blanching of the mass. Pulsating tinnitus may be audible to the examiner by using a Toynbee tube or a stethoscope placed over the ear. The finding of objective tinnitus may occur not only with glomus tumors, but also with carotid artery aneurysms, intracranial arteriovenous malformations, carotid artery stenosis, and other conditions. Glomus tumors must be distinguished from other masses, such as carotid artery aneurysms, high jugular bulbs, meningiomas, and adenomas, that may appear in the middle ear.

Radiological evaluation is now the mainstay of glomus tumor diagnosis. Biopsy is rarely indicated. CT scans of the temporal bone are used to assess bone erosion, and MRI, MR angiography, traditional arteriography and retrograde jugular venography are used to define the extent of the neoplasm (Fig. 13.8). Four vessel arteriograms are now being recommended by some otologists because of the high incidence of associated tumors. Up to 10% of patients with glomus tumors will have associated bilateral

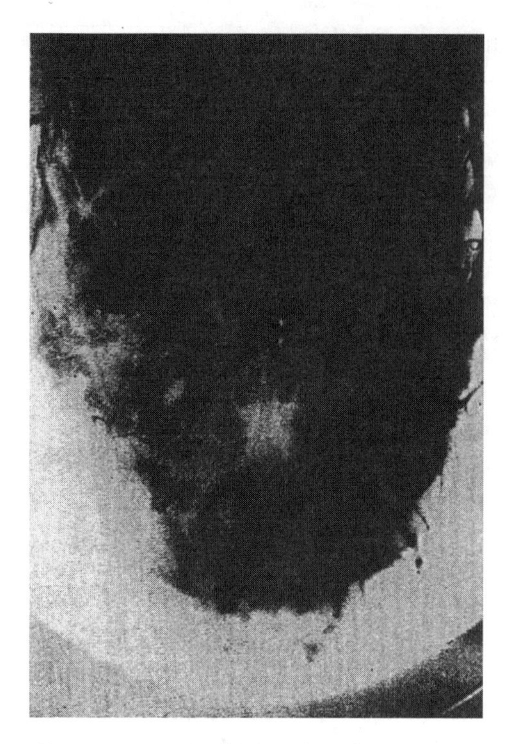

Figure 13.8 Angiogram shows the vascular blush of a glomus jugulare tumor.

glomus tumors, glomus vagale, carotid body tumor, or thyroid carcinoma (74). The vast majority of glomus patients are female, and the tumor is extremely rare in children. For this reason, biopsy is appropriate to rule out other lesions if the diagnosis is considered seriously in a child. Biopsy is also used in patients who are not surgical candidates prior to instituting palliative radiation therapy. However, such biopsies must be carefully performed in the operating room, and with blood available for replacement if necessary.

1.32. Internal Carotid Artery Aneurysm

Although aneurysmal presentations in the temporal bone are rare (75), they require accurate diagnosis and well-planned surgical management when they occur. Symptoms of hearing loss, vertigo, tinnitus, a feeling of fullness in the ear, and even facial nerve paralysis may be found. Central symptoms include headache, nausea, vomiting, and convulsions and may occur as the disease progresses. Early ear involvement may suggest *glomus tumors* symptomatically. Carotid artery aneurysm must be kept in the differential diagnosis whenever a middle-ear mass is found, especially if it is pulsatile or associated with pulsating tinnitus.

1.33. Vascular Disease

Patients with advanced atherosclerosis, particularly those who have suffered myocardial infarctions, have a higher incidence than the normal population of high-frequency sensorineural hearing loss. The pathogenesis is undetermined, but it is believed to be related to vascular changes within the inner ear (76). The *subclavian steal syndrome* involves collateral circulation from the vertebral artery in the presence of proximal left subclavian artery stenosis. Hearing loss occurs in ~10% of patients (77) and results from compromise of the vertebrobasilar system, which provides blood to the inner ear. Other otologic symptoms, such as vertigo and facial paralysis, also may occur with this syndrome because the vertebral artery supplies the pons, medulla, cerebellum, vestibular and cochlear labyrinth, and portions of the temporal bone, as well as the upper spinal cord, thalamus, and occipital cortex. Similar symptoms may occur with more limited dysfunction of the vertebral system, such as the *lateral medullary infarction syndrome*. Surgical treatment is available, but is beyond the scope of this chapter.

Sudden hearing loss is often ascribed to "vascular causes." Although this explanation is tempting, histological confirmation is scarce, although the phenomenon certainly exists—at least in association with larger cerebrovascular occlusive events. Many diseases which may cause anoxia of tissues may be responsible for sensorineural hearing loss, but more research is needed to prove the relationship. Such conditions include chronic hypotension, anemia, vasovagal abnormalities, and other similar maladies.

1.34. Stroke

Hemorrhage into the ear produces deafness, as noted in the discussion of coagulopathies. Similar findings occur following spontaneous *subarachnoid hemorrhage*, which produces blood in the internal auditory canal and cochlea. Major cerebrovascular occlusions may produce severe hearing deficits. *Occlusion of the vertebral* or *posterior-inferior cerebellar artery* produces lateral medullary syndrome, or *Wallenberg's syndrome*. This syndrome is characterized by ipsilateral ptosis and miosis; enophthalmos; facial hypesthesia; palatal, pharyngeal, and laryngeal paralysis; contralateral hypesthesia and decreased thermal sensation in the trunk and extremities; as well as occasional involvement of the sixth, seventh, and eighth cranial nerves (78,79). Sensorineural hearing loss occurs, and vestibular function is abnormal (80).

Occlusion of the anterior vestibular artery alone produces vestibular symptoms without hearing loss (2). *Occlusion of the anterior–inferior cerebellar artery* generally produces sudden vertigo with nausea and vomiting, hearing loss, facial paralysis, and cerebellar and sensory disturbances. Degeneration of the membranous labyrinth and brain stem auditory and vestibular nuclei occurs. Ipsilateral loss of pain and temperature sensation on the face are common, associated with decreased pain and temperature sensation on the opposite side of the body. Patients who survive usually improve slowly.

Vertebrobasilar ischemia may have similar, but transient symptoms, of which vertigo is the most prominent. Other associated findings may be hearing loss, diplopia, headaches, and speech difficulties. Although atherosclerotic vascular disease is the usual etiology, arthritis, syphilis, aneurysms, and subclavian steal syndrome also must be kept in mind.

Lateral venous sinus thrombosis and *thrombophlebitis of the jugular bulb* or the *internal jugular vein* may cause mastoid infection, brain abscess, or septicemia and meningitis due to septic emboli, and they may lead to hearing loss. In the past, these diseases generally have been seen as complications of ear surgery. Recently, however, jugular thrombophlebitis and its complications (including retrograde extensions) have been seen in heroin addicts who use the subclavian or internal jugular veins as access routes.

1.35. Multiple Sclerosis

Diffuse demyelinating disease, such as multiple sclerosis, may involve all parts of the nervous system. The first symptoms often are transient episodes of blurred vision, vertigo, clumsiness, or transient cranial nerve palsy. Later in the disease, intention tremor, scanning speech, and nystagmus may develop, as may diffuse neurological weakness. Sensorineural hearing loss and tinnitus are common. The disease usually occurs in the third or fourth decade of life; it is progressive and may be fatal. Most commonly, the hearing loss is high frequency, progressive, and bilateral (81), but it may be sudden, unilateral, and profound (2). Some recovery frequently occurs. The hearing loss often fluctuates and may even return to normal after severe depression. Often, abnormal tone decay may persist even when hearing is nearly normal (15). Brain stem evoked-response audiometry may be useful in establishing the diagnosis.

1.36. Infestations

A variety of tropical diseases can produce hearing loss, although these are rarely seen in the United States (82). *Halzoun*, caused by *Fasciola hepatica*, may be found anywhere where sheep and goats are raised, but it is especially common in South America, Latin America, and Poland. The disease has been found also in Africa, Asia, Europe, and North America. Infection usually occurs by eating raw aquatic plants, such as watercress, and causes hepatic distomiasis. Symptoms include dyspnea, dysphagia, deafness, and occasionally asphyxiation. *Myiasis*, caused by infestation of human tissue by fly larvae, may produce ear infection or more serious disease by penetration of the body through the ear.

Cochliomya hominivorax, or primary screwworm flies, are found in the American tropics and throughout the southern United States. They infect humans by laying eggs in open wounds or discharging orifices. They create deep malodorous wounds and may occur as an otitis or mastoiditis. Infestation is fatal in ~10% of cases. Other forms of myiasis affecting the ears may be found elsewhere in the world and should be suspected in known travelers. Parasitic infections of the external auditory canal are common in the tropics, in warmer portions of the United States, and may also be found sporadically throughout the country, especially in swimmers.

1.37. Conclusion

The great many diseases that may be associated with hearing loss highlight the need for a thorough history and physical examination in each patient with hearing impairment. Moreover, they remind us to search for unsuspected hearing loss early in patients with these maladies. Much more information is needed to clarify the nature of hearing loss associated with systemic diseases. Of particular importance is the need for temporal bone specimens for further research. Only through constant clinical attention and diligent investigation can we hope to diagnose, understand, and prevent hearing loss of all causes.

1.38. Summary

Hearing loss may accompany many systemic diseases. Familiarity with the otologic manifestations of these conditions facilitates early diagnosis and treatment of hearing impairment. Moreover, attention to these relationships often leads to the diagnosis of otherwise unsuspected, potentially serious systemic diseases in patients who complain of hearing loss.

2. HEARING LOSS ASSOCIATED WITH HEREDITARY DISEASES AND SYNDROMES

Most disease can be classified as hereditary or nonhereditary. Either type may be present at birth or may manifest later in life. Hereditary problems are frequently classified according to the mode for inheritance. An autosomal-dominant disease will be apparent in the person who carries the gene; the chance that it will be transmitted to any given offspring is 50%. In autosomal-recessive inheritance, neither parent may show a trait, but both are carriers. The chance that they will pass the trait to any given offspring is 25%.

Each person carries a great many genes for recessive traits, but these may be expressed only if mating occurs with someone with the same traits. The chances of this occurring are greatly increased in consanguinity. The chances are one in two that a carrier mother will pass an X-linked trait to any of her sons. An affected male transmits the carrier state to all of his daughters but none of his sons.

Understanding these patterns and the diseases with which they are associated allows us to predict the birth of affected children and helps us to counsel the parents of an afflicted child about the likelihood of having a second abnormal birth. Thus, this knowledge is critical not only in minimizing the occurrence of hereditary deafness but also in reassuring parents of children whose deafness or malformation resulted from other causes that the chance of having a second affected child is very small.

A nationwide census found that ~13,400,000 people in the United States admit to having a significant hearing loss (83). This total represents 66 people out of each 1000, or 6.6%, although undoubtedly, the actual number is higher. If all people with hearing loss are included regardless of the definition of "significant," the number is probably between 28 and 35 million people in the United States. Hereditary factors can be implicated in about one-third of hard-of-hearing patients. However, only a very small number of these people develop hearing loss before the age of 19; even fewer have it at birth. Nevertheless, it is important to consider genetic or hereditary hearing loss in the differential diagnosis of all cases of hearing impairment. These conditions may produce hearing loss of any audiometric pattern, including bilateral dips that are indistinguishable from audiograms from patients with noise-induced hearing loss. Such conditions may occur in young men or women with no history of noise exposure. When inheritance is recessive rather than dominant, there is no family history of hearing loss. So, care must be exercised in order to avoid missing the diagnosis.

2.1. Diabetes Mellitus

At least 6 million diabetics live in the United States, and the number appears to be growing by ~6% each year (84). The disease appears to follow autosomal recessive or polygenetic inheritance patterns, although this theory is still controversial. Up to 40% of diabetics (85) develop hearing loss, although estimates vary from study to study. The hearing loss generally is sensorineural, progressive, bilateral, most severe in the high frequencies (86), and tends to be worse in the elderly diabetic (87). A syndrome similar to Meniere's disease may be associated with diabetes and presents symptoms of fluctuating sensorineural hearing loss, episodic vertigo, tinnitus, and a feeling of fullness in the ear (88). Sudden, profound hearing loss may occur in diabetic patients also, but the causal relationship has not been proven (87,89).

Pathological examination has revealed small blood vessel changes in the inner ear similar to those found in the kidney and elsewhere (90). In addition, atherosclerosis of large blood vessels, possibly associated with elevated serum cholesterol and triglyceride levels, develops prematurely. Laboratory research also supports and association of diabetes and hearing loss (91).

Because of the significant incidence of diabetes, any person over age 18 with an unexplained sensorineural hearing loss or Meniere's-like symptom complex should have a fasting blood sugar and a 2-h postprandial blood sugar determination. If these tests are abnormal, a glucose-tolerance test should be carried out. Even if the tests are normal, it often is worthwhile to repeat them periodically. If this is done, it is not uncommon for the otologist to be the first clinician to diagnose diabetes. In addition, annual audiometric screening is recommended for all known diabetics.

2.2. Malignant Otitis Externa

This disease is life-threatening and invariably found in diabetic patients. Pain, rather than hearing loss, is usually the most prominent sign. It is a noncancerous infection, usually caused by *Pseudomonas aeruginosa*, extending to the temporal bone and the skull base (Fig. 13.9). Treatment requires hospitalization, high-dose intravenous antibiotics, and, sometimes, extensive surgical debridement of the temporal bone. The keystones of prevention are vigorous treatment of any diabetic with otitis externa and early recognition of referral for infections that do not resolve.

2.3. Hypoglycemia

Hearing losses may also be found in patients with hypoglycemia and can be seen with symptoms comparable to Meniere's disease (92,93). Glucose intolerance is another finding in a high percentage of patients with fluctuant hearing.

2.4. Familial Hyperlipoproteinemia

This condition exists in ~1% of the general population and is commonly associated with hearing loss (94,95). Generally, it is transmitted in an autosomal dominant pattern, although occasionally it appears to be recessive. It can be classified into one of three

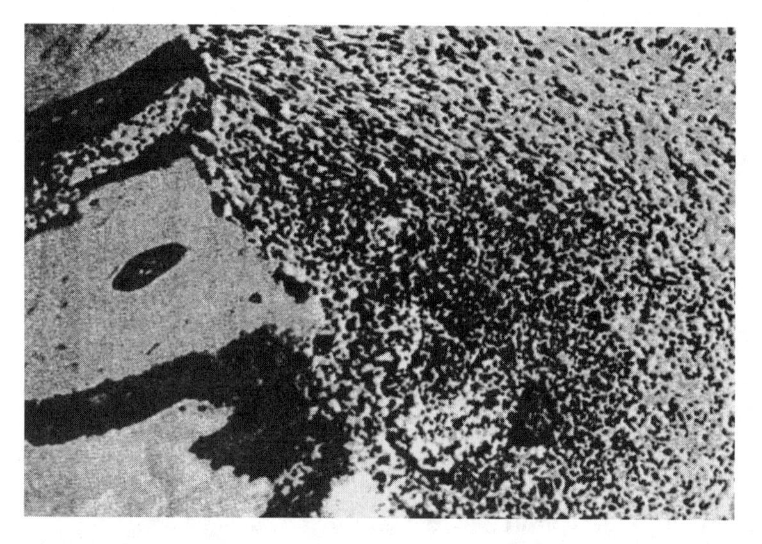

Figure 13.9 Malignant otitis externa with an abscess in the area of the cochlear aqueduct. The infection destroyed the bone beneath the cochlea and entered the posterior fossa. [From Linthicum and Schwartzman (126), p. 43. Courtesy W.B. Saunders.]

groups (familial hypercholesterolemia, familial combined hyperlipidemia, or familial hypertriglyceridemia) or subclassified as one of six types (types I, IIa, IIb, III, IV, and V).

It is important to distinguish between the hereditary hyperlipoproteinemias and acquired disease secondary to diet, nephrotic syndrome, hypothyroidism or other disease states. Types IIa, IIb, and IV are particularly important because of their association with sensorineural hearing loss. The pathology and implications can be similar to those seen in diabetic patients. Screening laboratory studies include fasting serum cholesterol and serum triglyceride determination and should be repeated at defined intervals. If either test is abnormal, a lipoprotein electrophoresis should be obtained after a 14-h fast. These tests should be done in children with sensorineural hearing loss also, and especially in children of hyperlipidemic parents.

Treatment is available for many patients with hyperlipoproteinemia and includes diet control and medications such as cholestryramine, nicotinic acid, and clofibrate.

2.5. Cleft Palate

Cleft lip, with or without cleft palate, occurs about once in every 1000 births in the Caucasian American population (96). Isolated cleft palate occurs once in ~2500 births (97). Both are somewhat less commonly found in black Americans. Although clefts have a familial tendency, they do not follow a classic pattern of inheritance. Moreover, there seems to be interaction between genetic and in utero hormonal or drug factors. Clefts also may be associated with numerous other congenital malformations (98). If parents of an affected child are normal, their chances of having another affected child are ~2%. If one of the parents is affected or if both a parent and child are affected, the incidence is considerably higher.

Nearly all cleft palate patients have abnormal eustachian tube function and a high incidence of serous otitis media (99). If untreated, this problem will produce conductive hearing loss and may lead to suppurative or adhesive otitis media. Similar problems may even occur with a submucous cleft palate, which an examiner may feel but not see. This disorder should be suspected and searched for in the presence of the bifid uvula. Treatment includes decongestants with or without antibiotics and the placement of tympanotomy tubes when middle-ear ventilation fails to improve.

2.6. Retinitis Pigmentosa

Usher's syndrome is a combination of retinitis pigmentosa and congenital sensorineural hearing loss and is usually recessive (100). The hearing loss is congenital, although it frequently is not discovered until the child is 1 or 2 years of age and sometimes even older. It is cochlear, bilateral, and more severe in the high frequencies (100–103). It is moderate in ~10% of cases and severe in 90%. Overall, ~20% or 25% of people with retinitis pigmentosa are found to have hearing loss, although only those with congenital sensorineural hearing loss have Usher's syndrome. The retinitis pigmentosa often is not diagnosed until about age 10, when progressive lose of eyesight becomes apparent, frequently beginning as night blindness. A decreased sense of smell is often present, and vestibular abnormalities may occur (104,105). Retinitis pigmentosa and hearing loss may be combined with a number of other maladies. In *Refsum's syndrome* (106), they are associated with progressive peripheral neuropathy, mental deterioration, and elevated serum phytanic acid. The lipid-storage defect may be treated with dietary restriction of phytanic acid (107). The *Bardet–Biedl syndrome* combines retinitis pigmentosa and hearing loss with hypogonadism polydactyly, obesity, and mental retardation (108). Patients with

Laurence–Moon syndrome have mental retardation, hypogenitalism, and spastic paraplegia (108). In *Alstrom's syndrome*, obesity and diabetes mellitus are combined with hearing loss and retinitis pigmentosa (109). In *Cockayne's syndrome*, they are combined with retinal atrophy, mental retardation, dwarfism, and prematurely senile ("birdlike") facies. These patients usually die in their twenties (110). *Kearn's syndrome* combines retinitis pigmentosa and mixed hearing loss with progressive external ophthalmoplegia and cardiac conduction defects (108).

In general, all patients with these syndromes are functionally blind by the time they reach age 50, and often earlier. They comprise ~10% of people with hereditary deafness. Some can be helped with hearing aids, but often the hearing loss is too severe, even though it generally is not progressive. As such, any deaf child should have a careful ophthamological evaluation. Conversely, any person with retinitis pigmentosa deserves a hearing test.

2.7. Glaucoma

A familial tendency is associated with glaucoma, especially of an open-angle type. Children and siblings of patients with glaucoma may have higher ocular pressures than normal and may have anatomical features that predispose them to glaucoma (111).

The relationship between glaucoma and hearing loss has been controversial. Data suggest a very high incidence of auditory vestibular dysfunction in glaucoma patients (112). Combined cochlear and vestibular hypofunction has been found in up to 60% of cases; pure cochlear hearing loss has been found in ~25%. Only ~25% of glaucoma patients with otologic dysfunction were symptomatic, the most common complaint being loss of hearing. The usual audiologic picture is bilateral cochlear hearing loss, although this may be found coincidentally in the older patient. In acute congestive glaucoma, nearly all patients tested have had bilateral hearing losses, and about one-third have had vestibular dysfunction as well.

Additionally, the majority of patients who have had glaucoma for >2 years have been found to have hearing losses; the severity of this loss appears to correlate with the severity of the glaucoma. Because of the frequently delayed recognition of hypoacusis, routine audiograms in patients with glaucoma may be useful until the relationship is better understood.

2.8. Other Ophthalmological Abnormalities

At least 20 other syndromes manifest combinations of eye and ear disease. They include: foveal dystrophy and sensorineural hearing loss (*Amalrio's syndrome*), which may involve as many as 5% of children with hereditary deafness (113); optic atrophy and deafness (*Leber's disease*) (114); mental retardation, retinal pseudotumor, and deafness (*Norrie's disease*) (115); nonsyphilitic keratitis and auditory vestibular abnormalities (*Cogan's syndrome*) (116); vestibular dysfunction, uveitis, alopecia, white eyelashes and hair, and elevated cerebrospinal fluid pressure in the early stage (*Vogt–Koyanagi syndrome*) (117); saddle nose, myopia, cataract, and hearing loss (*Marshall's syndrome*); vestibulocerebellar ataxia, retinitis pigmentosa, and nerve deafness (*Hallgren's syndrome*) (101); and others.

Hallgren's syndrome is particularly interesting because it accounts for ~5% of all hereditary deafness (113). Ninety percent of these patients are profoundly deaf, apparently at birth; 90% have ataxia and 10% have nystagmus. Twenty-five percent appear to be mentally deficient, usually schizophrenic.

2.9. Alport's Syndrome

This partially sex-linked, autosomally transmitted syndrome (84,117) may be detected during the first week of life by the presence of hematuria and albuminuria. Hypertension, renal failure, and death usually occur before the age of 30 in males. Females demonstrate a much less severe form of the disease. The hearing loss is progressive, bilateral, and cochlear. It generally is first detected when the patient is ~10 years old. This syndrome is estimated to account for ~1% of hereditary deafness (113). Treatment consists primarily of controlling urinary tract infections and renal failure. Frequently, the need for ototoxic drugs for urinary tract infection complicates the hearing loss.

Muckle–Wells syndromes, a variant of Alport's includes urticaria, amyloidosis, and relative infertility (118,119). This disease usually begins during adolescence. *Herrman's syndrome* combines hereditary nephritis and autosomal-dominant nerve deafness with mental retardation, epilepsy, and diabetes mellitus (120).

Approximately 10 other recognized syndromes combine hearing loss with renal disease. This frequent association warrants a routine urinalysis in the evaluation of sensorineural hearing loss, particularly in children and young adults. A number of these syndromes are associated with hypertension as well, so blood pressure also should be checked.

2.10. Waardenburg's Syndrome

This dominant syndrome (Fig. 13.10) includes partial albinism (classically seen as a white forelock of hair), laterally positioned medial canthi, different colored irises, and congenital nonprogressive sensorineural hearing loss (121). Vestibular abnormalities and temporal bone radiological abnormalities may occur (84). The deafness may be total, with only slight residual hearing in the low frequencies; moderate, with near-normal hearing in the higher frequencies and severe loss in the low frequencies; or unilateral, with near-normal hearing on one side (86). Only 20% of patients with Waardenburg's syndrome demonstrate hearing loss; however, Waardenburg's syndrome

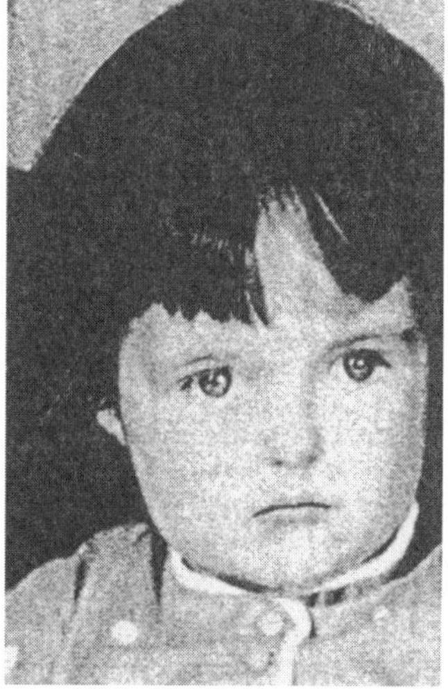

Figure 13.10 Mother and daughter with Waardenburg's syndrome showing white forelock and isochromic light iris. [From Smith (158), p. 143. Courtesy W.B. Saunders.]

accounts for ~1% of all hereditary deafness (113). At present, no treatment exists other than sound amplification when applicable. Genetic counseling is relevant in these cases.

2.11. Albinism and Nerve Deafness

Generalized albinism and nerve deafness (as opposed to the localized albinism of Waardenburg's syndrome) usually is recessive (94), although dominant forms (*Tietze's syndrome*) have been described (122). In contrast to Waardenburg's syndrome, with its localized areas of albinism, this syndrome is characterized by totally white skin, white hair, as well as absence of pigment in the iris, the sclera, and the fundus of the eye. Nystagmus and progressive high-tone, bilateral, sensorineural hearing loss also usually begin from ages 6–12 years. The disease is caused by the absence of the copper-containing enzyme tyrosinase. Thus far, no treatment is available other than symptomatic measures (inclusive of a hearing aid).

2.12. Leopard Syndrome

Leopard in an acronym for: *lentigines*, *electrocardiographic defect*, *ocular hypertelorism*, *pulmonary stenosis*, *abnormalities of genitalia*, *retardation of growth*, and *sensorineural deafness* (123). The syndrome is transmitted as an autosomal dominant. The lentigines usually are absent at birth but develop progressively (Fig. 13.11). Sensorineural hearing loss occurs in ~25% of cases and is usually mild. Treatment includes hearing aids where applicable, surgical correction of pulmonary stenosis, dermabrasion of the lentigines, and correction of other associated abnormalities as necessary.

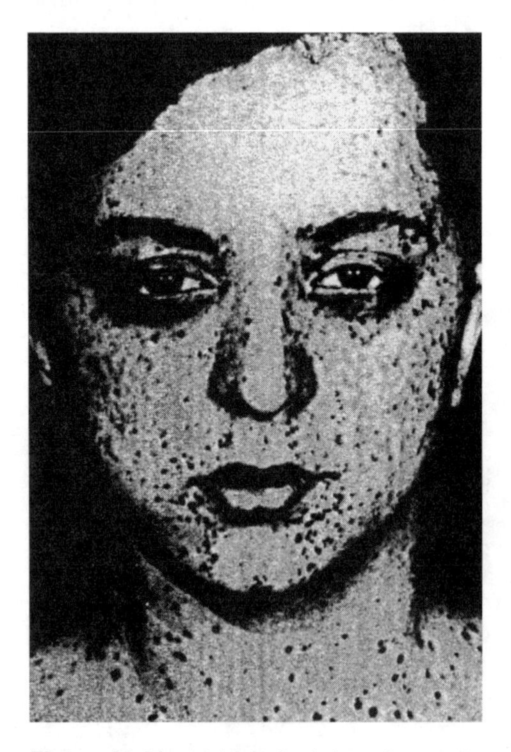

Figure 13.11 Multiple lentigines (Leopard) syndrome. [From Konigsmark and Gorlin (108), p. 239. Courtesy W.B. Saunders.]

2.13. Von Recklinghausen's Disease

Among the other syndromes that combine abnormalities of skin and hearing, generalized von Recklinghausen's disease (Fig. 13.12) has been well recognized since 1882. Multiple neurofibromas and café-au-lait spots are the most common features. Epilepsy frequently accompanies the syndrome, and mental retardation occurs in some cases. Neurofibromas may occur anywhere, including the eighth cranial nerve, sometimes bilaterally. Malignant degeneration of the neurofibromas has also been reported (124). Inheritance is as an autosomal dominant. A localized form of von Recklinghausen's disease exists and may arise as bilateral acoustic neurofibromas (125). These behave somewhat differently from acoustic neuromas unassociated with generalized von Recklinghausen's disease and may be quite large at the time of discovery (126). Vestibular testing may show decreased or absent caloric response. Treatment requires resection of the acoustic tumors if feasible.

2.14. Paget's Disease

Osteitis deformans is found at autopsy in ~3% of people >40 years of age and may occur in as much as 10% of the population >80 (108). In general, Paget's disease is felt to be an autosomal dominant syndrome. Bone pain is the most frequent symptom. The pathology involves a combination of abnormal deposition and resorption of bone in a mosaic pattern (Fig. 13.13) and manifests itself with bony deformation, particularly of the weight-bearing portions of the skeleton. Enlargement of the cranium is classic. Sarcomatous changes have been found in 1–2% of cases (127).

Various neurological changes have been reported. The hearing loss, which is not a constant feature of Paget's disease, may be conductive (caused by malformation of the ossicles or oval window area) or sensorineural (86). One mechanism for sensorineural hearing loss is narrowing of the internal auditory canal, producing pressure on the auditory nerve. However, this does not seem to occur as commonly as does the cochlear hearing loss. The diagnosis is made by X-ray films and detection of elevated serum alkaline

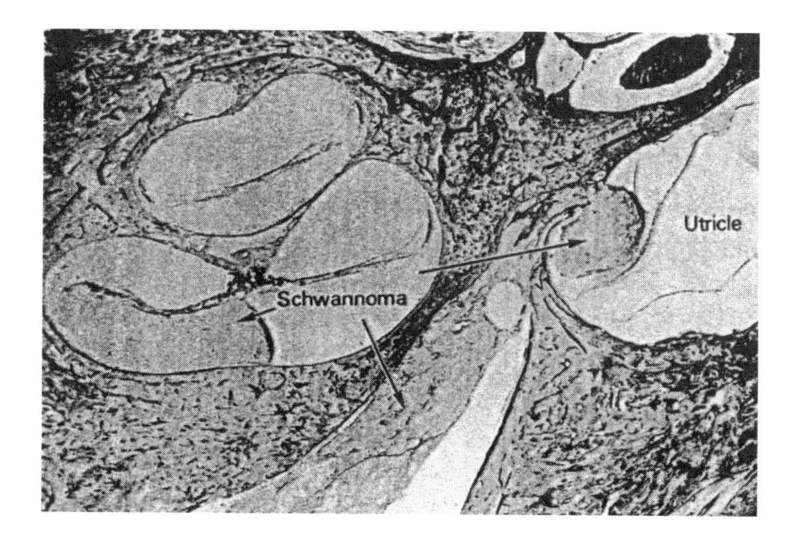

Figure 13.12 Von Recklinghausen's disease with bilateral vestibular schwannomas. In this photomicrograph, the neoplasm extends from the internal auditory canal into the basal turn of the cochlea and into the vestibule. [From Schuknecht (86), p. 376. Courtesy Harvard University Press.]

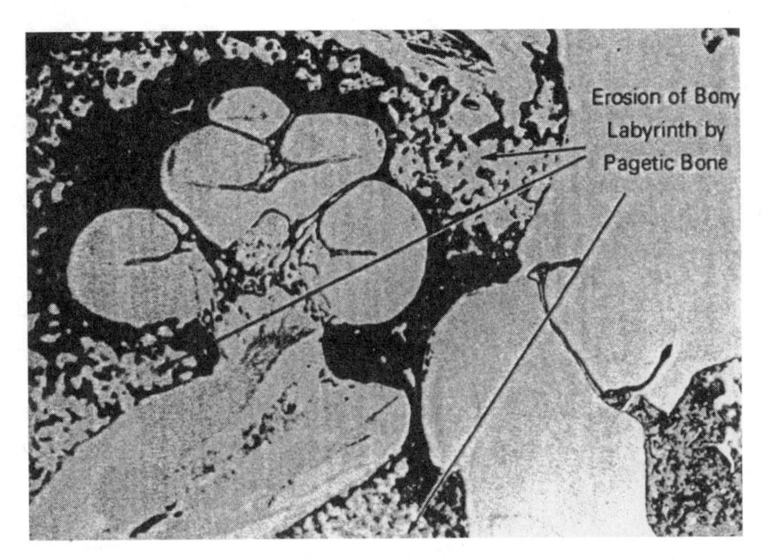

Figure 13.13 Paget's disease with extensive involvement of the bony labyrinth but with a normal-appearing membranous labyrinth. [From Schuknecht (86). Courtesy Harvard University Press.]

phosphatase and urinary hydroxyproline levels. Treatment may include reconstructive middle-ear surgery, although this procedure is often unrewarding because of the progression and unpredictability of the disease. Drugs that inhibit excessive bone resorption, such as mithramycin, sodium etidronate, and calcitonin, are being tried as agents to induce remission (128).

2.15. Fibrous Dysplasia

Fibrous dysplasia of the temporal bone is commonly associated with hearing loss (129). The process may be monostotic or polyostotic. Disseminated fibrous dysplasia is usually a feature of Albright's syndrome, in which fibrous dysplasia, skin pigmentation, and various endocrine disturbances occur, almost always in females. The monostotic and polyostotic forms occur in males and females, and both may be associated with hyperparathyroidism. Fibrous dysplasia may occur in the temporal bone and be associated with hearing loss (Fig. 13.14). Hearing may be improved by surgery in selected cases.

2.16. Osteogenesis Imperfecta

This autosomal dominant disease (Fig. 13.15) occurs in varying severity in two forms (86). *Osteogenesis imperfecta congenita* is present at birth, is more severe, and leads to multiple fractures and, frequently, early death, *Osteogenesis imperfecta tarda*, much less severe, becomes evident later in life and is more localized. Blue sclera, bone fragility, and hearing loss are the principal features (van der Hoeve–deKlein syndrome (130) or Lobstein's disease); but other systems, particularly the teeth, may be involved. Marked hearing impairment is found in 30–60% of patients with the tarda type (Fig. 10.17) (131,132). The hearing loss usually is conductive and bilateral, although mixed and purely sensorineural hearing losses have been reported. The tympanic membrane may be bluish and quite thin. At one time, osteogenesis imperfecta and otosclerosis were thought to be the same disease, since stapes fixation occurs in both,

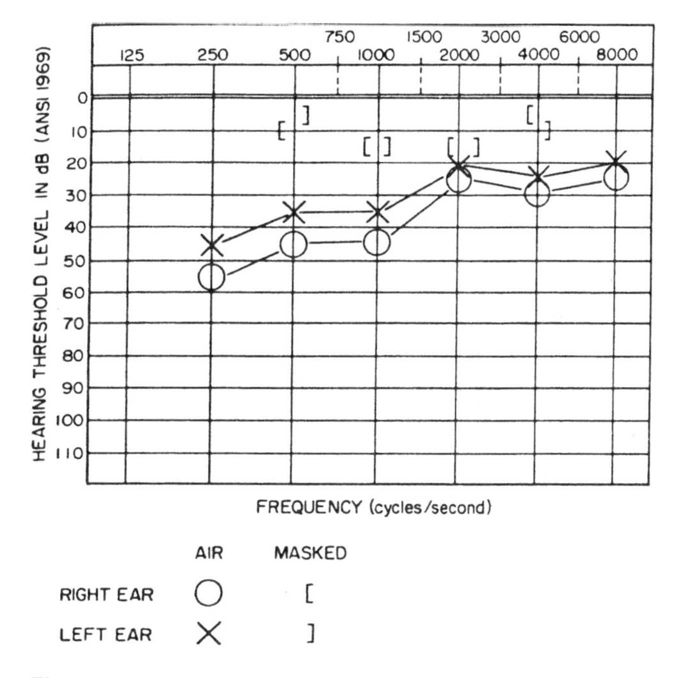

Figure 13.14 Fibrous dysplasia of the temporal bone in a 14-year-old girl. Her bilateral conductive hearing loss was due to involvement of the ear canal impinging on the malleus and incus.

but this does not appear to be the case. Their surgical and histopathological features probably differ, although controversy still exists concerning the pathology (86,126). The surgical treatment (stapedectomy) is the same, however, with possible technical modifications (133).

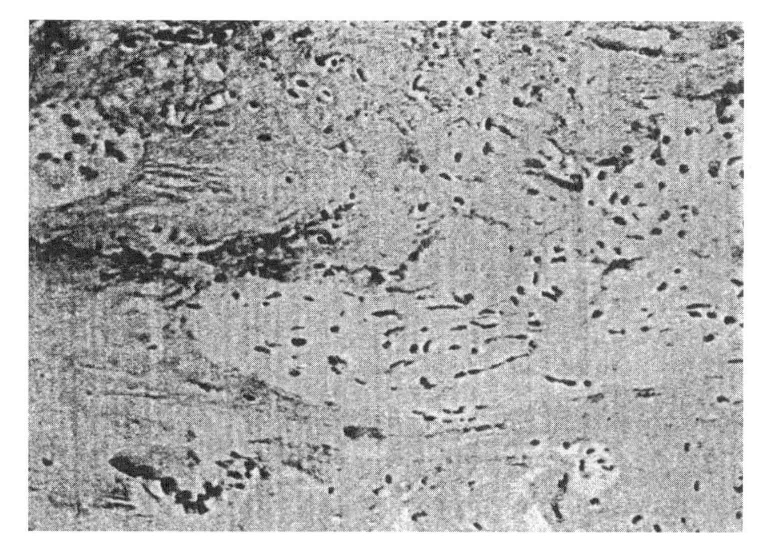

Figure 13.15 Osteogenesis imperfecta. The calcium and phosphorus content of this temporal bone is below normal and the bone matrix appears immature. [From Linthicum and Schwartzman (126). Courtesy W.B. Saunders.]

2.17. Crouzon's Disease

Premature synostosis of cranial sutures results in this distinctive autosomal dominant syndrome (Fig. 13.16) (134). The shape of the cranium varies, depending on the sutures involved. Ocular hypertelorism, exophthalmos, shallow orbits, beaked nose, and maxillary hypoplasia are common. About one-third of the patients with this disease have a hearing loss (generally conductive) associated with ossicular deformity or stapes fixation (135). Bilateral external auditory canal atresias have been noted in some cases, as have mixed hearing losses. The treatment is surgical in most cases. Craniotomy is performed in infancy to decrease cerebral compression, with cosmetic maxillofacial reconstruction becoming more and more satisfactory (136).

2.18. Cleidocranial Dysostosis

This dominant syndrome is occasionally associated with conductive or mixed progressive hearing loss and concentric narrowing of the external auditory canals (108). It occurs with hypoplasia or aplasia of one or both clavicles or with pseudoarthrosis. Abnormal eruption of teeth and incomplete ossification of the skull are further manifestations of this ossification defect of membranous bone. On physical examination, patients often can bring their shoulders together in front of their sternum. Facial bones are underdeveloped, the palate is high and arched, and the frontal sinuses do not develop.

2.19. Treacher Collins and Franceschetti–Klein Syndromes

Mandibulofacial dysostosis was first described in the 1840s, and the facial appearance is classic (Fig. 13.17) (137): hypoplastic zygomas produce downward-sloping palpebral

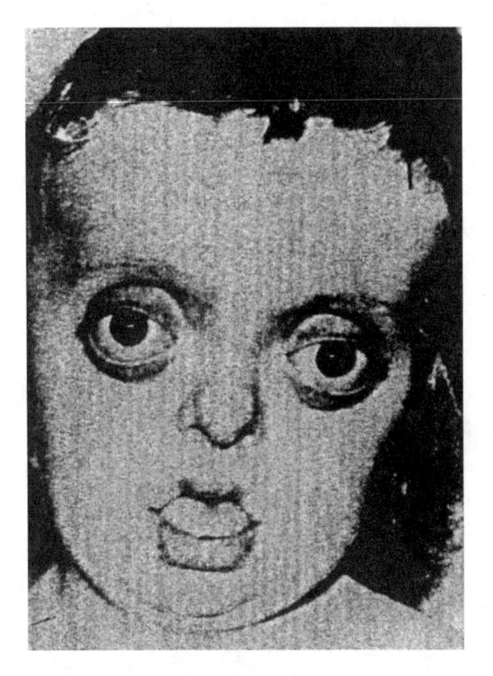

Figure 13.16 Crouzon's syndrome showing exophthalmos, hypertelorism, and under-developed maxillae. The patient had a 60-dB conductive hearing loss and an IQ of 110. [From Schuknecht (86), p. 176. Courtesy Harvard University Press.]

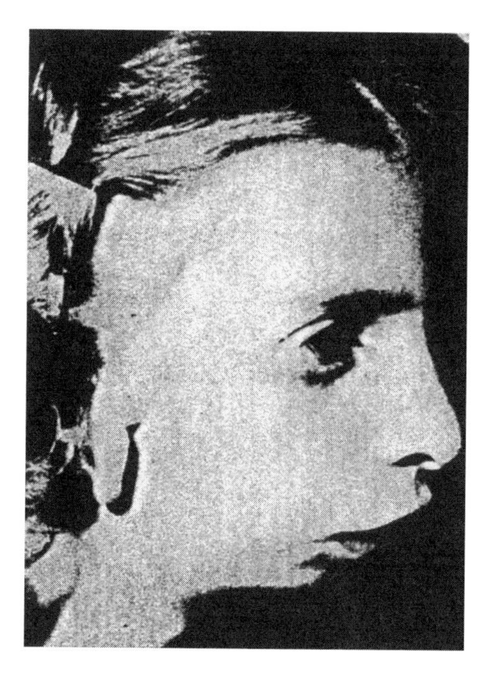

Figure 13.17 Treacher Collins syndrome with auricular dysplasia, malar hypoplasia, and defect of the lower eyelid. [From Smith (158), p. 111. Courtesy W.B. Saunders.]

fissures. Cheekbones are depressed, the chin recedes, the mouth has a large "fishlike" appearance, the mandible is hypoplastic, and coloboma of the lower eyelids with lack of cilia is common. Auricular malformations occur in ~85% of such patients (138). About one-third have external auditory canal atresia or an ossicular defect. Conductive hearing loss is most common, although sensorineural deafness has also been reported (139). Surgical treatment is rewarding in carefully selected patients, but early amplification should be used in patients with bilateral hearing losses. The inheritance pattern is autosomal dominant.

2.20. Pierre Robin Syndrome

The striking facies in this apparently autosomal dominant syndrome is characterized by an "Andy Gump" appearance (Fig. 13.18). A receding mandible and relatively large protruding tongue may produce respiratory obstruction. Cleft palate, cardiac, skeletal, and ophthalmological anomalies are also associated. About 20% have marked mental retardation. The patients have low-set ears with auricular malformations and conductive hearing loss (86,140). Genetic counseling is required. Management of the hearing loss may involve hearing aids or surgical intervention.

2.21. Albers–Schönberg Disease

This disease, known also as osteopetrosis and "marble bone" disease, may be transmitted as an autosomal dominant or an autosomal recessive disorder, but hearing loss has been associated primarily with the recessive form (108). The disease produces increased density of the entire bony skeleton (Fig. 13.19). Cranial enlargement may occur, and mild hypertelorism has been noted. In 35% of the cases, growth is retarded. Fractures

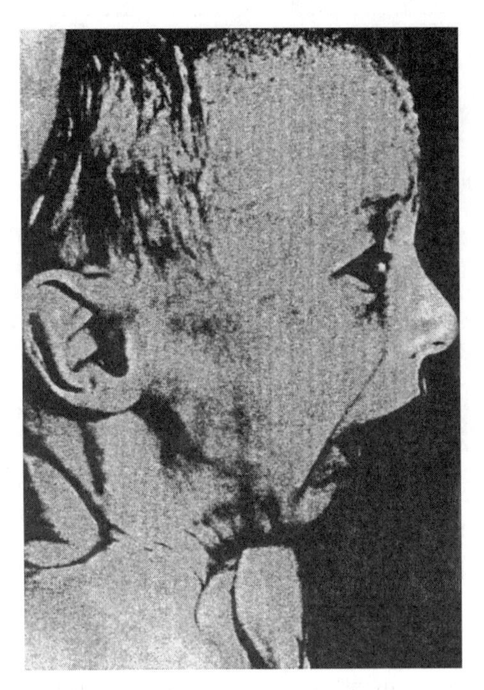

Figure 13.18 The classic "Andy Gump" appearance in the Pierre Robin syndrome. [From Smith (158), p. 8. Courtesy W.B. Saunders.]

are common. Foramina of the cranial nerves may be narrowed, intracranial pressure may be increased, and visual loss occurs in 80% of cases. Facial paralysis also is common.

Mental retardation is seen in ~20% of cases. Hemolytic anemia, thrombocytopenia, lymphadenopathy, hepatosplenomegaly, and osteomyelitis (frequently following dental extraction) occur. Between 25% and 50% of patients have a moderate mixed hearing

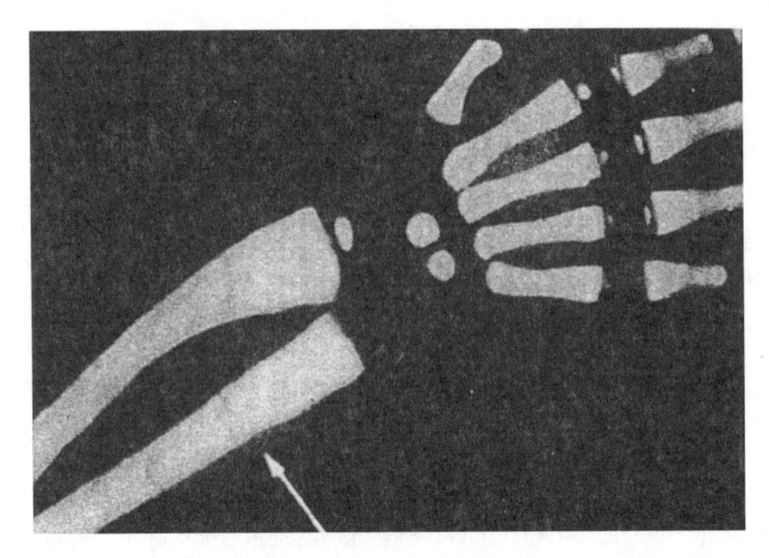

Figure 13.19 Markedly increase bony density with fractures of the radius and ulna in a patient with Albers–Schönberg disease. [From Kongismark and Gorlin (108), p. 157. Courtesy W.B. Saunders.]

loss beginning in childhood (141–143). The incidence of otitis media is increased in this disease. The diagnosis is made by X-ray films, and treatment is symptomatic. Frequent examination for otitis media should be performed in children with this disease. Hearing aids may prove helpful in some cases.

2.22. Klippel–Feil Syndrome

This well-known syndrome of cervical–vertebral fusion, often associated with spina bifida, cervical ribs, neurological abnormalities, strabismus, and other features, may be associated with hearing loss (86). When the Klippel–Feil anomalies are combined with abducens nerve paralysis, a retracted ocular bulb, and hearing loss, the condition is called *Wildervanck's syndrome* (108,144,145). A cleft palate or torticollis may be associated as well, and inheritance probably is multifactorial. Hearing loss may be unilateral or bilateral, moderate to severe. Both conductive and sensorineural deafness have been described, and vestibular studies are usually abnormal. Incomplete expression of the disease is common. The hearing loss is congenital and does not generally progress. Treatment may be surgical or may require a hearing aid.

2.23. Dwarfism

A great many musculoskeletal anomalies are associated with hearing loss. As with all congenital anomalies, the presence of short stature or other obvious physical malformations should alert the clinician to look early for a hearing defect.

2.24. Cornelia de Lange Syndrome

Cornelia de Lange syndrome (CDLS) is characterized by multiple congenital malformations and mental retardation (Fig. 13.20). Sataloff et al. (146) examined 45 patients with this rare disease and found numerous otolaryngologic abnormalities. Virtually all had hearing loss, and most had impaired language development. Importantly, there was

Figure 13.20 A 4-year-old girl with typical appearance of Cornelia de Lange syndrome: small nose, prominent filtrum, convergent eyebrows, and long, curly eyelashes. [Republished from Sataloff (146) with permission.]

a direct correlation between the severity of hearing loss and the severity of language and other problems. It appeared that unrecognized or untreated hearing loss may have exacerbated the difficulties experienced by these patients.

2.25. Huntington's Chorea

This autosomal dominant degenerative disease has its onset at ~35 years of age. Death generally ensues within 10–15 years. Emotional disturbance is followed by the onset of choreiform movements, seizures, dementia, and death. Cranial nerve deficits, including auditory nerve dysfunction, occur later in the disease (113).

2.26. Friedreich's Ataxia

An autosomal recessive degenerative disease, this entity usually appears in childhood and culminates in death in the midteen years. Early symptoms include ataxia, clumsiness, tremor, ataxic gait, and slurring of speech. Neurological deficits become more severe as the disease progresses. Optic atrophy and, rarely, retinitis pigmentosa occur. Hearing loss is sensorineural, mild, and progressive. These are several other closely related syndromes in which hearing loss is more prominent. Vestibular function testing may show central or peripheral abnormalities (147).

2.27. Bassen–Kornzweig Syndrome

A betalipoproteinemia may occur as progressive ataxia in childhood, much like Friedreich's ataxia (94,148). This syndrome is recessive and is associated with sensorineural hearing loss and progressive central nervous system demyelination secondary to the inability to transport triglycerides in the blood. Affected children have fatty stools, weakness, sensory losses, and atypical retinitis pigmentosa. Treatment is dietary but not very successful.

2.28. Unverricht's Epilepsy

Sensorineural hearing loss may be associated with this recessive syndrome (149). The disease usually develops in childhood and starts as a seizure disorder. Mental regression, cerebellar ataxia, choreoathetosis, and extrapyramidal signs develop along with massive myoclonic epilepsy. Treatment consists of anticonvulsant drug therapy.

2.29. Schilder's Disease

This syndrome probably is recessive (113). It involves massive disruption of nerve myelin, especially in the cerebral hemispheres. The onset generally happens in childhood and is always fatal. It manifests as a gait disorder, along with increased intracranial pressure, papilledema, abducens nerve paralysis, and optic neuritis (150). Deafness and cortical blindness may be initial symptoms. The deafness is progressive, and in some cases, it appears to be cortical (151). As yet, no treatment exists. The disease is also known as *encephalitis periaxialis diffusa*, and it is one of a group of diffuse cerebral scleroses.

2.30. Neurological Deficiency

In at least 20 additional syndromes, hearing loss is associated with other neurological deficits. The intimate relationship between the ear and the central nervous system makes this association natural. Despite this, it is tempting to omit a thorough neurological evaluation in patients with hearing loss, particularly in children, who are difficult to test. However, the abundance and gravity of these diseases should encourage diligent attention to a thorough, complete evaluation.

2.31. Pendred's Syndrome

Nearly 10% of hereditary sensorineural hearing loss is associated with this recessive syndrome of nonendemic goiter and deafness (113). The disease involves abnormal iodine metabolism and appears to be caused by absence of the enzyme iodine peroxidase. The thyroid enlargement usually appears ~8 years of age, but in some cases it may be present at birth. Later in life the goiter frequently becomes nodular.

Although patients generally are euthyroid, some may be mildly hypothyroid. Mental retardation is noted in some, but not all cases of Pendred's syndrome. The hearing loss usually is congenital, bilateral, sensorineural, and moderate to profound, and it usually is worse in the high frequencies. There may be slow progression during childhood. More than 50% of these patients have a severe hearing deficit (108). Vestibular function frequently is abnormal, although vertigo is uncommon. Inheritance is autosomal recessive. Polytomography or CT scan may reveal a Mondini malformation of the cochlea (152,153).

The disease must be differentiated from *endemic cretinism with deafness*, which may be found in areas where iodine is missing from the diet. However, cretinoid features are absent in Pendred's syndrome.

Several laboratory tests, including the perchloride test and the fluorescent thyroid-image test, aid in establishing the diagnosis of this syndrome. Therapy should include thyroid replacement, and the hearing loss must be treated symptomatically.

2.32. Hyperprolinemia

The mode of inheritance for this disorder of amino acid metabolism still is unclear. The disease is characterized by elevated plasma proline and hyperprolinuria (iminoglycinuria). There are two types (154). In *type 1*, activity of the enzyme proline oxydase is decreased. In *type 2*, proline-5-carboxylate dehydrogenase is deficient. Sensorineural hearing loss is seen in some patients who have *type 1* disease. Prolinemia with mental retardation and hearing loss constitutes *Schafer's syndrome*. However, much more work is needed to clarify the nature of this disease and to establish therapy.

2.33. Homocystinuria

This autosomal recessive disease is caused by deficiency of the enzyme cystathionine synthase (155). Nearly 60% of patients have mental retardation (156). The syndrome is characterized by ectopia lentis (occurring by the age of 10 years), tall stature, long extremities, medial degeneration of the aorta and elastic arteries, malar flush, often glaucoma, high-arched palate, cataracts, hepatomegaly, and other anomalies.

Sensorineural hearing loss occurs (94). Homocystinuria is now being treated with low-methionine diet, with supplemental cystine, and massive doses of vitamin B_6 (154,157).

2.34. Marfan's Syndrome

Arachnodactyly, ectopia lentis, and an appearance similar to that of the homocystinuria patient characterize this autosomal dominant disease. These patients have similar problems and generally die in their early thirties secondary to aortic and cardiac valvular disease. However, mental retardation is not characteristic, and no enzymatic defect has been found. A conductive hearing loss may be present, secondary to collapse of the external auditory canals and auricles, caused by cartilaginous abnormalities and inadequate support (86). A few reports of associated sensorineural hearing loss exist.

2.35. Mucopolysaccharidoses

Classically, six varities of inborn metabolic errors involving mucopolysaccharide metabolism can be described (94,108,158). *Hurler's syndrome* (gargoylism) is characterized by cloudy corneas, mental retardation, small stature, stiff joints, gargoylelike facies (Fig. 13.21), clawlike hand deformities, and death before the age of 10 years. The condition usually is recognized during the first year of life, and the inheritance is autosomal recessive. The nasopharynx is deformed and lymphoid tissue is markedly increased, leading to nasal obstruction and chronic nasal discharge. This discharge may cause eustachian tube obstruction, compounding middle-ear disease. Primary pathology may also be found within the middle ear, however, apparently resulting from the presence of the disease in utero. Hurler's syndrome can be diagnosed by amniocentesis. Sensorineural hearing loss may occur in this disease, but it is usually mild (86,108).

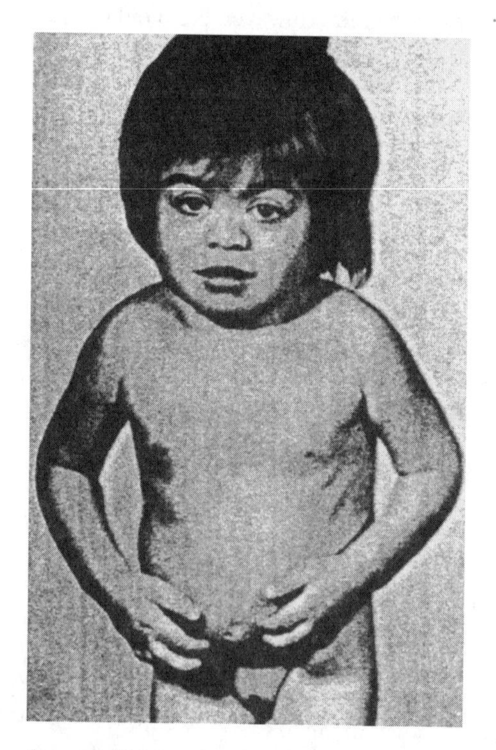

Figure 13.21 Patient with Hurler's syndrome showing scaphocephalic cranial enlargement, coarse facies with full lips, flared nostrils, low nasal bridge, and hypertelorism. [From Smith (158), p. 245. Courtesy W.B. Saunders.]

Hunter's syndrome has a similar habitus but does not show corneal clouding. It is X-linked, so the disease is expressed only in males. The onset of signs and symptoms usually occurs around age 2, and most patients die by the age of 20. However, some have lived into their sixties. About 50% of these cases are accompanied by progressive hearing loss. The loss usually is not severe (159) and is most commonly mixed or sensorineural.

In *Sanfilippo's syndrome*, an autosomal recessive anomaly, hearing loss is uncommon (108). When present, it appears around the age of 6 or 7, and then it progresses. These patients live a nearly normal life-span and develop their symptoms early in childhood. They show progressive mental deterioration, mild coarsening of the facial features, and stiffening of joints.

In *Morquio's syndrome*, also autosomal recessive, mixed hearing loss is common and usually begins in the teen years (160,161). Onset of symptoms is between ages 1 and 3, and the features are very similar to those of Hurler's syndrome, including the corneal clouding. The coarsening of facial features is milder, however. Severe kyphosis and knock knees are characteristic.

Scheie's syndrome is an allelic form of Hurler's syndrome. The broad mouth and full lips characteristic of gargoylism are present by the age of 8 years. Corneal clouding occurs, as do retinal pigmentation, hirsutism and aortic valvular defects, which also may be found in the other mucopolysaccharidoses. Psychoses and mental retardation may occur but are not as striking as in the related syndromes. The life-span is long, and the inheritance is autosomal recessive. Although documentation is inadequate, it is suspected that mixed hearing loss develops in 10–20% of these patients, usually in middle age (108).

Patients with *Maroteaux–Lamy syndrome* show features similar to those of Hurler's syndrome, but they do not have mental deterioration. They usually develop their symptoms and signs somewhat later than in Hurler's syndrome, and deformities are generally less severe. By about the age of 8 years, ~25% will exhibit hearing loss (probably conductive), apparently associated with recurrent otitis media (108).

Diagnosis of the mucopolysaccharidoses is confirmed by detecting stored or excreted specific mucopolysaccharides. As yet, treatment is not available for most of these diseases. However, in Hurler's syndrome, the enzyme, α-L-iduronidase is now being used with some favorable results (162).

Pseudo-Hurler's syndrome is an autosomal recessive generalized gangliosidosis, rather than a mucopolysaccharidosis (157). The defects are severe, similar to those of Hurler's syndrome, and they include a cherry-red spot in the macula of the eye in about half of the patients. Death generally occurs by 2 years of age. Information regarding hearing loss is lacking and will not be truly relevant until some effective therapy is found for the underlying disease.

2.36. Errors of Metabolism

Most of these diseases are autosomal recessive. *Tay–Sachs disease*, or amaurotic familial idiocy, is a sphingolipidosis caused by absence of hexosaminidase-A (163). This enzyme is now available for treatment of this disease. In addition, a blood test is available to detect carriers. In some areas, the test is performed routinely prior to marriage between Jewish people, among whom the disease is most common in this country (94). In the infantile form of the disease, flaccidity, motor regression, blindness with a macular cherry-red spot, and apathy begin at ~6 months of age. The disease progresses to spasticity and death by age 3 or 4. In the juvenile form, loss of vision usually is the symptom; the course is slower and death usually occurs in the twenties. High-frequency sensorineural hearing loss is felt to result from the basic metabolic anomaly (108,164). Otitis media also is frequent, resulting in a possible mixed hearing loss (165).

Wilson's disease, or hepatolenticular degeneration, affects the brain, liver, and kidney and may result in deafness (166). The Kayser–Fleischer ring, involving Descemet's membrane of the cornea, is pathognomonic. The disease is caused by deficiency of plasma ceruloplasmin, the primary copper-containing plasma protein. This deficiency produces excess serum copper. The inheritance pattern is autosomal recessive, and a test to detect carriers is available (167). Treatment with fair results relies on low-copper diets and attempts to remove serum copper with chelating agents such as penicillamine dimercaprol (BAL) and versene (113).

Fabry Anderson syndrome, a lipid-storage disease, produces mild sensorineural hearing loss in ~50% of patients (108). The disease is known also as *cardiovasorenal syndrome of Ruiter–Pompen and diffuse angiokeratitis.* It is characterized by elevated blood pressure, heart enlargement, angiokeratomata of the skin, pain in the extremities, abnormalities of sweat secretion, and albuminuria. Frontal bossing and a prominent mandible and lips are common. Corneal clouding may occur, and macular purplish spots on lips and near skin mucosal junctions are common. Death frequently is caused by myocardial infarction or renal failure (168).

Other inborn errors of metabolism, including mannosidosis, other mucolipidoses, and other diseases, may be associated with hearing loss. Some of these diseases may be relatively mild and are easily overlooked unless they are in the differential diagnosis.

2.37. Jervell and Lange–Nielsen Syndrome

This syndrome accounts for ~1% of hereditary deafness (113). It is autosomal recessive and is associated with sudden death (169,170,171). It is known also as *cardioauditory syndrome* and *surdocardic syndrome.* It is characterized by profound congenital hearing loss and electrocardiographic abnormalities, particularly a prolonged Q–T interval and large T waves (Fig. 13.22). No evidence of organic heart disease is found. Often, the disease is first suspected after an episode of fainting early in childhood. This episode is

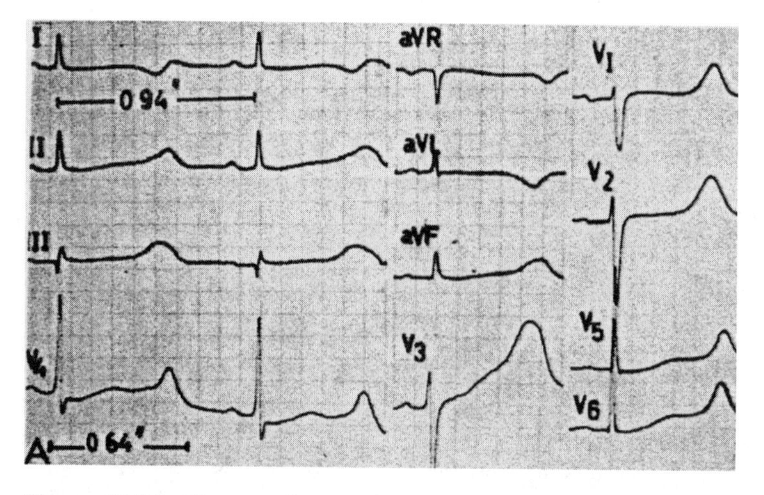

Figure 13.22 Electrocardiogram from a patient with Jervell and Lange–Nielsen syndrome showing prolonged QT intervals of 0.64 s, with 0.41 s begin the upper limit of normal. [From Kongismark and Gorlin (108), p. 360. Courtesy W.B. Saunders.]

probably caused by a Stokes–Adams attack, similar to that seen in adult cardiac patients. Once the diagnosis has been made by electrocardiography, appropriate treatment measures can be taken. In patients with this disease, a high index of suspicion may be lifesaving.

Leopard syndrome and *Kearn's* syndrome, already discussed, and a few other diseases, such as Refsum's syndrome, may also combine hearing loss with electrocardiographic or cardiac anomalies. This association also occurs in mucopolysaccharidoses. In these severe diseases, however, gross pathology is obvious.

2.38. Familial Streptomycin Ototoxicity

In several families even low doses of streptomycin have produced severe ototoxicity with impairment of vestibular functions (108). Inheritance appears to be autosomal dominant but may be multifactorial. Before prescribing any ototoxic drugs especially streptomycin, a careful history of prior exposure to ototoxic drugs, as well as familial sensitivity to such drugs, is required.

2.39. Sickle Cell Disease

In the United States, ~7–9% of African Americans carry sickle cell traits. About 1 in 400 has sickle cell disease, an autosomal recessive condition (108). Anemia, splenomegaly, and attacks of abdominal pain, jaundice, weakness, and anorexia develop. Sensorineural hearing loss occurs in ~20–25% of patients with the disease (172). Pathology in the inner ear is consistent with ischemic changes and is believed to be due to thromboembolic disease secondary to sickling (173). Sudden total deafness also has been reported and is believed to be caused by a vascular occlusion. In some cases, severe sensorineural hearing loss associated with sickle cell crisis has spontaneously returned to normalcy (174). Treatment is generally symptomatic.

2.40. Cystic Fibrosis

Cystic fibrosis is an autosomal recessive disease and is known also as *mucoviscidosis* or *fibrocystic disease of the pancreas.* Although a simple clinical test is available to detect the disease, the pathogenesis remains obscure. Pancreatic enzyme deficiency produces malabsorption and steatorrhea. Children with cystic fibrosis fail to gain weight despite increased appetite. Recurrent cough or wheezing during the first 6 months of life should prompt a diagnostic consideration of cystic fibrosis. Recurrent respiratory tract infection may be followed by respiratory failure, other organ-system involvement, and death. Recent improvements in therapy have led to a better prognosis.

Although recent data refute the belief that children with cystic fibrosis have a high incidence of hearing loss (175), such losses may result from pharyngeal infection or inflammation compromising eustachian tube function and middle-ear aeration. As such, a sweat chloride test should be performed on any child who manifests repeated upper respiratory tract infections, failure to thrive, and conductive hearing loss.

2.41. Kartagener's Syndrome

The association of situs inversus, bronchiectasis, and sinusitis has been recognized as an entity since 1933 (176), although earlier reports exist in the literature. Inheritance probably is multifactorial, although an autosomal recessive pattern predominates. In addition to the classic findings (Fig. 13.23), poor pneumatization of the mastoid air cells and bilateral conductive hearing loss in the 30- to 40-dB range are common (177).

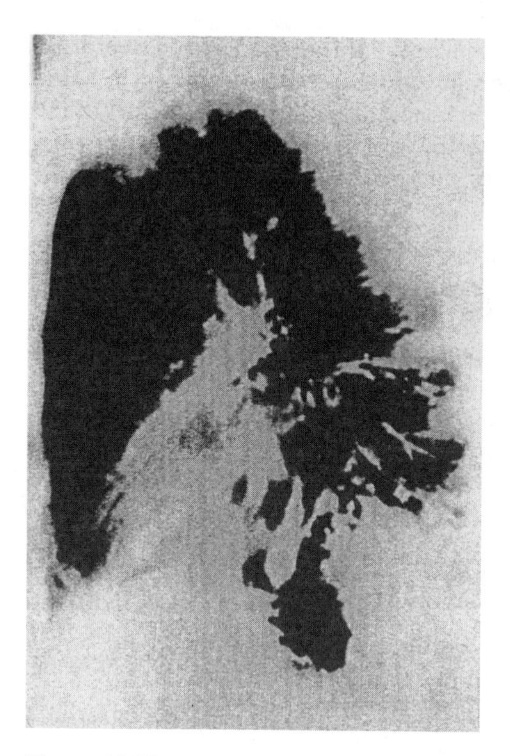

Figure 13.23 Typical bronchiectasis in a patient with Kartagener's syndrome.

The conductive hearing loss usually is due to middle-ear fluid, implicating eustachian tube obstruction. Middle-ear mucosal biopsy has shown chronic inflammatory changes. All patients with this disease need to be screened for hearing loss and to have their hearing restored medically, surgically, or, if needed, through use of a hearing aid.

2.42. Immune Deficiency Syndromes

The numerous syndromes that involve hypofunction of the immune system, such as *Wiskott–Aldrich syndrome*, *ataxia–telangiectasia*, *Bruton's agammaglobulinemia*, *hypogammaglobulinemias*, *thymic dysplasia*, and others, are associated with an increased incidence of infection. The ear is frequently involved, and the physician should check for otitis media. This examination is particularly important because in cases of severely depressed immune response, the signs of ear infection may be absent. Nevertheless, the presence of fluid may cause substantial hearing loss and may even lead to more serious otopathology.

2.43. Chromosome Anomalies

Turner's syndrome is characterized by sexual infantilism, streak gonads, short stature, webbed neck (Fig. 13.24), coarctation of the aorta in 70% of cases, and other stigmata. It is associated with an XO karyotype, and patients are phenotypic females. However, other chromosome anomalies with Turner-like syndromes have been found. About one- to two-thirds of patients exhibit a hearing loss (108). Sensorineural impairment with a bilateral symmetrical dip centered ~2000 Hz is common (178). Conductive hearing loss also is common, and bouts of otitis media may occur frequently. Severe deafness can be found in ~10% of cases.

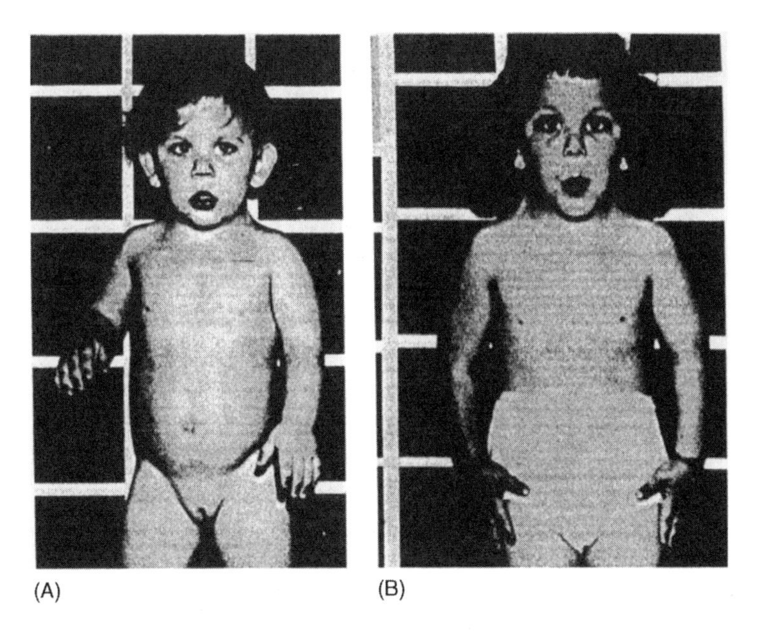

Figure 13.24 A patient with Turner's syndrome at age 2 years (A) and 4 years (B), with height ages of 17 months and 3 years, respectively. Note prominent ears, lateral neck web, and hyperconvex deep-set fingernails. [From Smith (158), p. 59. Courtesy W.B. Saunders.]

Similar otologic findings occur in *Klinefelter's syndrome* (XXY karyotype), which is characterized by male phenotypes, medullary gonadal dysgenesis, gynecomastia, and often mental retardation (179). Other genotypes have also been reported. *Noonan's syndrome* is known also as male Turner's syndrome without chromosome abnormality. The features are similar, except that in Noonan's syndrome (157), mental retardation is more likely. Congenital heart disease is also more common and usually consists of pulmonic stenosis. Sensorineural hearing loss has been reported in Noonan's syndrome. Patients with *XX gonadal dysgenesis* are tall females with sexual infantilism. As in Turner's syndrome, they have streak gonads. Deafness has been noted in some cases and may be of a severe congenital variety (108).

Down's syndrome, or *trisomy 21*, is responsible for ~10% of institutionalized mentally deficient patients. It occurs in 0.1–0.2% of the population. The more familiar features (Fig. 13.25) include the mongoloid slanted and widespread eyes, epicanthus, nystagmus, abnormal earlobes, mental retardation, short stature, gap between the great and second toe, broad, flat hand, and protuberant abdomen. Documentation of hearing loss in Down's syndrome is remarkably deficient in the literature. Estimates of sensorineural hearing loss range between 10% and 50%, and conductive or mixed hearing loss vary from 3% to 50% (180–182).

Trisomy 13, *trisomy 18*, *chromosome 18 long-arm deletion syndrome*, *chromosome 4 short-arm deletion syndrome*, and *cri-du-chat syndrome (chromosome 5 short-arm deletion syndrome)* are other major chromosome anomalies that may be associated with hearing loss.

2.44. No Associated Anomalies

Recessive sensorineural hearing loss without associated defects is the most common form of hereditary hearing loss (86). It usually is not progressive and generally is congenital, bilateral, and may vary from mild to severe. Certain audiometric patterns, such as

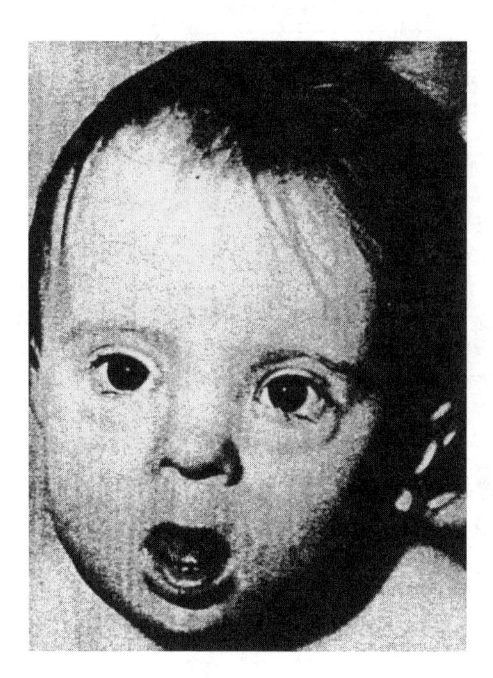

Figure 13.25 Infant with Down's syndrome showing flat facies, straight hair, protrusion of tongue, inner canthal folds, small auricles, and speckling of iris with lack of peripheral patterning. [From Smith (158), p. 35. Courtesy W.B. Saunders.]

the Menasse type (Fig. 13.26) and basin-shaped curve (Fig. 13.27), are particularly characteristic of hereditary hearing loss, but nearly any pattern may occur. Vestibular function is normal usually. X-linked inheritance may also occur (108,183) (Figs. 13.28 and 13.29). Dominant sensorineural hearing loss first occurs between the ages of 6 and

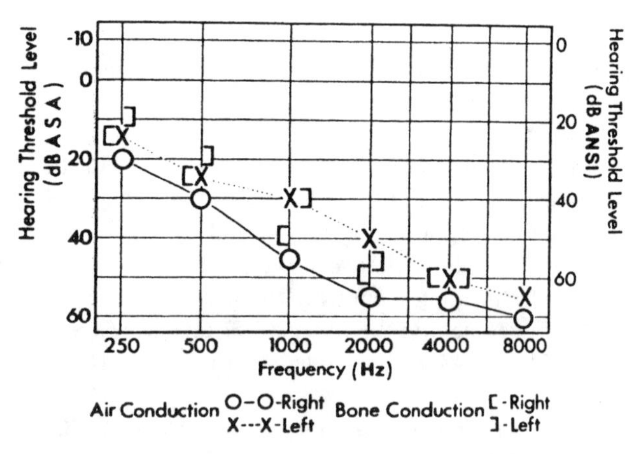

Figure 13.26 *History*: A 7-year-old boy whose mother noted poor hearing when the child was 2. Speech development was slow. Has difficulty pronouncing sibilants, and *s* is slurred. An older brother has a similar hearing loss. The hearing problem is nonprogressive in both children. *Otologic*: Ears normal. *Audiologic*: Pure-tone air and bone conduction thresholds indicate a bilateral, gradually sloping loss with no air–bone gap. *Speech reception threshold*: Right, 38 dB; left, 38 dB. *Discrimination score*: Right, 78%; left, 80%. There was no abnormal tone decay. *Classification*: Sensorineural hearing loss. *Diagnosis*: Menasse-type congenital hereditary hearing loss.

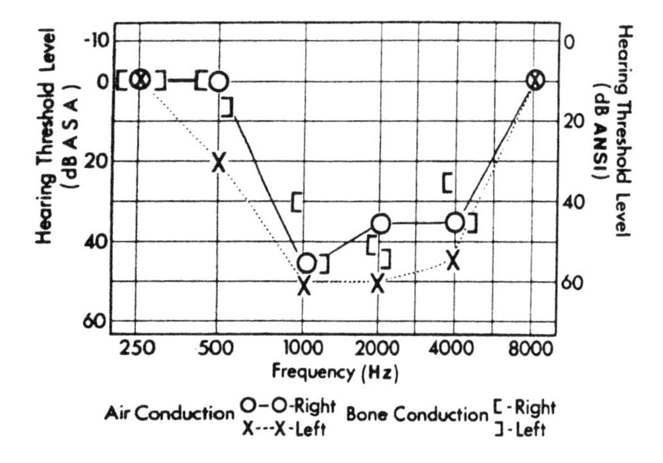

Figure 13.27 *History*: This child's father has a similarly shaped audiogram and a deep monotonous voice, which is apparent also in the child. Enunciation is good. *Otologic*: Ears normal. *Audiologic*: Air and bone conduction thresholds reveal a bilateral, basin-shaped curve with no air–bone gap. Right ear discrimination was 80% at 70 dB. *Classification*: Sensorineural. *Diagnosis*: Congenital and inherited.

12 years. The hearing loss is most commonly bilateral, progressive, and high frequency (84). As in recessive and X-linked hearing loss, however, any pattern may occur.

Hereditary progressive sensorineural hearing loss is also quite common and may be dominant or recessive. Colloquially, this condition is often referred to as *genetic deafness*.

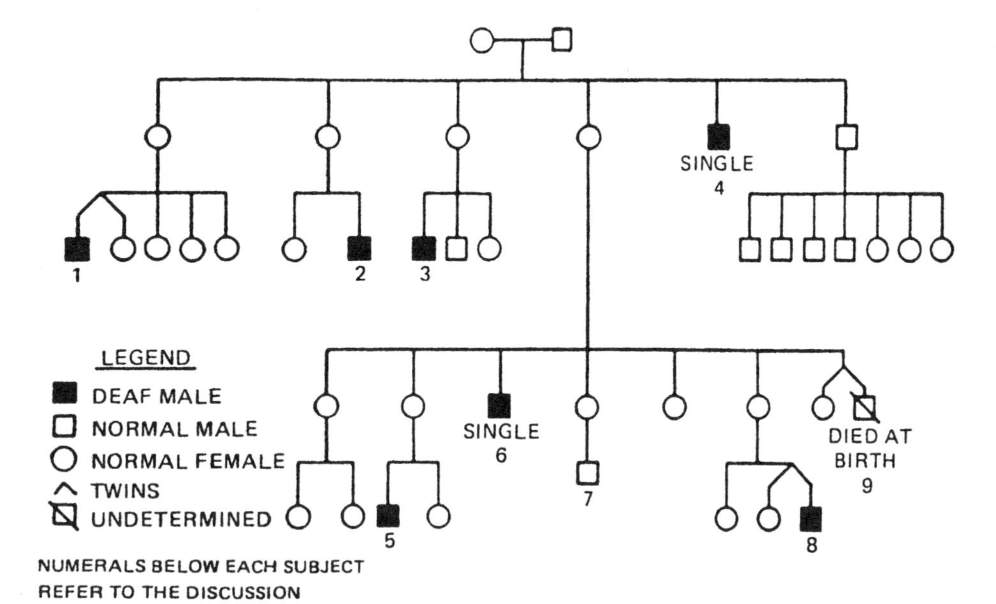

Figure 13.28 This family tree indicates some interesting genetic aspects of sex-linked hereditary deafness. Audiograms were available for this family on all of the living persons with profound congenital nerve deafness. Not all the members of this family have been given hearing tests, but those reported as not deaf have normal speech and clinically normal hearing as established from information supplied by close relatives. [From Sataloff et al. (183).]

This condition is characterized by sensorineural hearing loss that gets gradually worse over time. It is most commonly worse in the higher frequencies, but patterns resembling a 4000-Hz dip are not rare. The audiogram pattern frequently resembles that of presbycusis, but the condition usually becomes apparent in the second, third, fourth, or fifth decades. When similar hearing loss has occurred in previous generations of the same family, dominant hereditary progressive hearing loss may be diagnosed. However, absence of a positive family history does not rule out this condition. A recessive inheritance pattern is common.

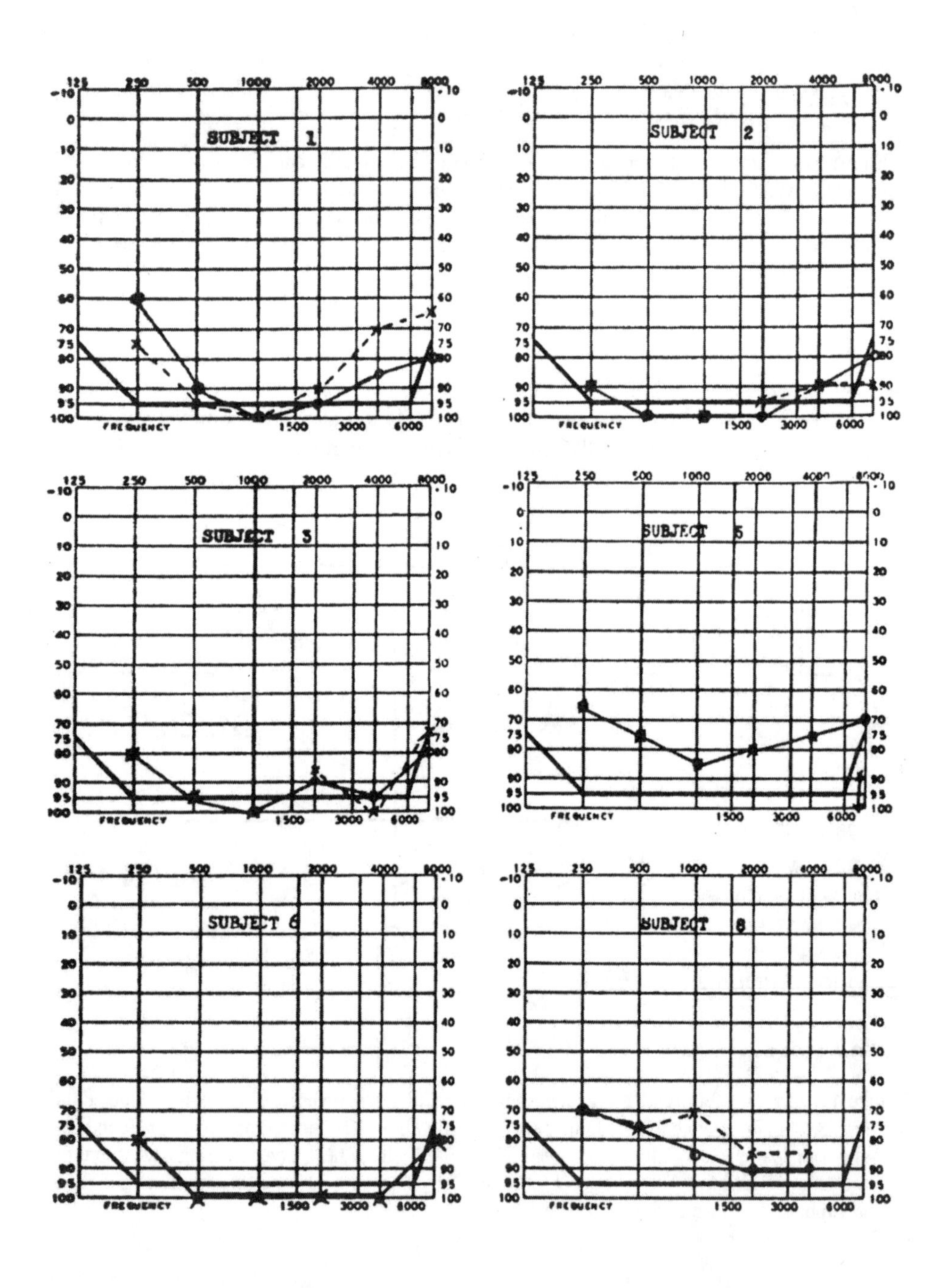

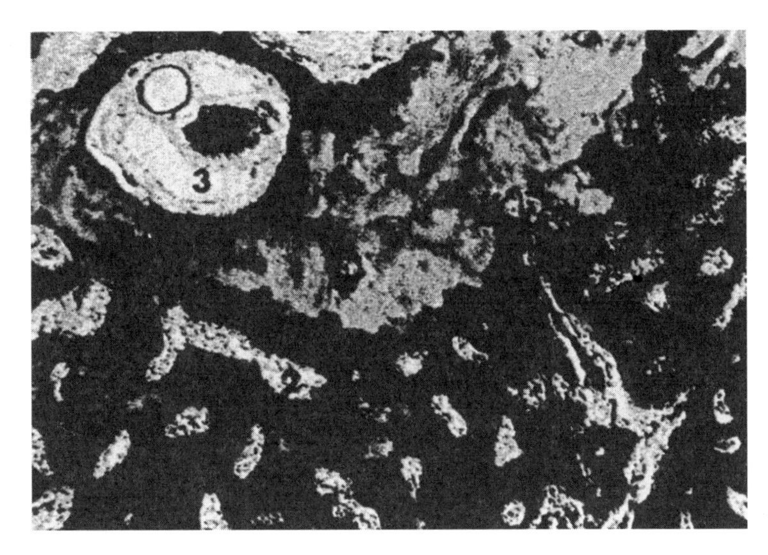

Figure 13.30 Otosclerotic bone showing a focus of otospongiosis (1), otosclerosis (2), and adjacent normal bone. Jacobson's nerve is seen on the promontory (3). [From Linthicum and Schwartzman (126), p. 49. Courtesy W.B. Saunders.]

Hereditary Meniere's syndrome is a dominant disorder but is rare (184). Like sporadic Meniere's disease, it is characterized by fluctuating hearing loss, episodic vertigo, and tinnitus.

Otosclerosis is found histologically at autopsy in $\sim$10% of white people and in $\sim$20% of patients $>$60 years (Fig. 13.30) (86,185). The incidence is much lower in

---

Figure 13.29 The audiogram obtained on six of the seven deaf individuals in the family referred in Fig. 13.28 (Subjects 4, 7, and 9 are not shown.) *Subject 1* (in Figs. 13.28 and 13.29) is 53 years of age and has little intelligible speech. He was the first-born and only male child in his family and has a twin sister. *Subject 2* is a 44-year-old deaf mute. He is the first-born and only male. *Subject 3* is a 37-year-old first-born man who received his education at a school for the deaf. *Subject 4* disappeared $>$40 years ago at the age of 7. At that time he was known to be deaf and unable to speak. *Subject 5* is a 16-year-old attending a school for the deaf. He is the first-born son and has a sister, whose hearing is normal. He uses a hearing aid with fair results and has fairly intelligible speech. The usefulness of early amplification and adequate training is evident in this subject. *Subject 6* is a 37-year-old bachelor classified as a deaf-mute with little intelligible speech. *Subject 7* is a 5-year-old first-born male child with normal hearing and speech. He was conceived after a 5-year period of apparent sterility requiring D & C. The mother denies any miscarriages or abortions prior to the subject's birth. *Subject 8* is a 3-year-old first-born male and has a twin sister. He has no speech and has profound nerve deafness. Both his twin sister and older sister have normal hearing and speech development. The hearing threshold on this youngster was obtained with repeatedly consistent psychogalvanic skin resistance tests. *Subject 9* was a male twin who died at birth. There are three sets of twins in the pedigree. In each set there is a male and female, and the male was the first-born son and demonstrated deaf-mustism in two of the three sets. It is apparent from this pedigree that deafness is manifest only in the male child and is transmitted through the maternal side. There are seven first-born males with profound congenital nerve deafness, a condition not present in subsequently born males or found in any females. There are insufficient subsequent males to establish statistically that deafness is restricted to the first-born males. Deafness does not exist in the children of the normal male member of the pedigree. The family pedigree exhibited demonstrates clearly that profound congenital nerve deafness can be hereditary and sex-linked. Early recognition and adequate educational measures are essential in such instances. [From Sataloff et al. (183).]

blacks. Inheritance follows an autosomal dominant pattern with variable penetrance and female predilection. It can occur as conductive or mixed hearing loss. Although controversial, there is reason to believe that otosclerosis can occur purely as sensorineural hearing loss with cochlear involvement alone.

In general, the hearing loss in otosclerosis is slowly progressive and usually occurs in early adult life. Vestibular function is usually normal. Frequently, otosclerosis is accelerated by pregnancy. The conductive component of the hearing loss can be treated by stapedectomy or with a hearing aid. Sodium fluoride may be helpful in selected cases (186).

Numerous other patterns of isolated hereditary hearing loss or hereditary hearing loss associated with malformations of the ears occurs.

2.45. Summary

Fortunately, many of the hereditary causes of hearing loss are preventable or treatable. Most of the syndromes involve known inheritance patterns, which makes genetic counseling useful in their management. Recessive syndromes can be minimized by avoiding consanguineous marriages. Screening programs, such as for sickle cell trait and Tay–Sachs disease, are also helpful.

Many of the inborn errors of metabolism and chromosome anomalies can be detected in utero by amniocentesis, allowing for possible elective abortion. The effects of maternal and paternal advanced age and other factors associated with increased appearance of congenital anomalies can be minimized as physician and patient populations becomes more familiar with these diseases and their causes.

REFERENCES

1. Asher P. A study of 63 cases of athetosis with special reference to hearing defects. Arch Dis Child 1952; 27:475–477.
2. Schuknecht H. Pathology of the Ear. Cambridge, MA: Harvard University Press, 1974: 262–266, 311–330, 383–388, 420–424, 444–446.
3. Coquette M. Les sequelles neurologiques tardives de l'ictere nucleaire. Ann Pediatr 1944; 163:83–104.
4. Goodhill V. The nerve-deaf child: significance of Rh, maternal rubella and other etiologic factors. Ann Otol Rhinol Laryngol 1950; 59:1123–1147.
5. Dublin W. Neurologic lesions of erythroblastosis fetalis in relation to nuclear deafness. Amt J Clin Pathol 1951; 21:935–939.
6. Gerrard J. Nuclear jaundice and deafness. J Laryngol Otol 1952; 66:39–46.
7. Barnett HL, Einhorn AH. Pediatrics. 15th ed. New York, NY: Appleton-Century-Crofts, 1972:1676–1682.
8. Wintrobe MM, Thorn GW, Adams RD et al. Harrison's Principles of Internal Medicine. 7th ed. New York, NY: McGraw-Hill, 1974:964–966.
9. Barr B, Lundstrom R. Deafness following maternal rubella. Acta Otolaryngol 1961; 53:413–423.
10. Vaughan VC, McKay RJ, Nelson WE. Textbook of Pediatrics. 10th ed. Philadelphia: Saunders, 1969:659–663.
11. Sataloff J. Hearing Loss. Philadelphia: Lippincott, 1966:142, 379–381.
12. Paparella MM. Otologic manifestations of viral disease. Adv Otorhinolaryngol 1973; 20:144–154.
13. Linday JR. Histopathology of deafness due to postnatal viral disease. Arch Otolaryngol 1973; 98:218–227.

14. Hardy JB. Fetal consequences of maternal viral infections in pregnancy. Arch Otolaryngol 1973; 98:218–227.
15. Neu HC, ed. Studies reveal the presence of coliforms and anaerobes in acute otitis media. Infect Dis 1977; 7:1–21.
16. Holmes GP, McCormick JB, Trock SC, Chase RA et al. Lassa fever in the United States. N Engl J Med 1990; 323(16):1120–1123.
17. Ryback LP. Deafness associated with Lassa fever. JAMA 1990; 264(16):2119.
18. Hanner P, Rosenhall U, Edstrom S, Kaijser B. Hearing impairment in patients with antibody production against Borrelia burgdoferi antigen. Lancet 1989; 1(8628):13–15.
19. Fox GM, Heilskov T, Smith JL. Cogan's syndrome and seroreactivity to Lyme borreliosis. J Clin Neuro Ophthalmol 1990; 10(2):83–87.
20. Mokry M, Flaschka G, Kleinert G, Kleinert R, Fazekas F, Kopp W. Chronic Lyme disease with an expansive granulomatous lesion in the cerebellopontine angle. Neurosurgery 1990; 27(3):446–451.
21. Coleman CU, Green I, Archibold RAK. Cutaneous pneumocystosis. Ann Inter Med 1987; 106:396–398.
22. Sandler ED, Sandler JM, Leboit PE, Wening BM, Mortensen N. Pneumocystosis carnii otitis media in AIDS: A case report and review of the literature regarding extrapulmonary pneumocystosis. Otolaryngol HNS 1990; 103(5)Part 1:817–821.
23. Nadol JB. Hearing loss as a sequela of meningitis. Laryngoscope 1978; 88:739–755.
24. Keane WM, Potsic WP, Rowe LD et al. Meningitis and hearing loss in children. Arch Otolaryngol 1979; 105:39–44.
25. Paparella MM, Shumrick PA. Otolaryngology. Vol. 2. Philadelphia: W.B. Saunders, 1973:161–167.
26. Hybels RL, Rice DH. Neuro-otologic manifestations of sarcoidosis. Laryngoscope 1976; 86:1873–1878.
27. Blarr IM, Lawrence M. Otologic manifestations of fatal granulomatosis of the respiratory tract. Arch Otolaryngol 1961; 73:639–643.
28. Karmody CS. Wegener's granulomatosis: presentation as an otologic problem. Otorhinolaryngol 1978; 86:574–584.
29. Hill JH, Graham MD, Gikas PW. Obliterative fibrotic middle ear disease in systemic vasculitis. Ann Otol Rhinol Laryngol 1980; 89:162–164.
30. Keleman G. Histiocytosis involving the temporal bone (Letterer-Siew, Hand-Schuller-Christian). Laryngoscope 1960; 70:1284–1304.
31. Schwartzman JA, Pulec JL, Linthicum FH. Uncommon granulomatous disease of the ear. Ann Otol Rhinol Laryngol 1972; 81:389–393.
32. Schuknecht NF, Perlman HB. Hand-Schuller-Christian disease of the skull. Ann Otol Rhinol Laryngol 1948; 57:643–676.
33. Ikeda K, Kobayashi T, Kusakari J, Takasaka T, Yumita S, Furukawa T. Sensorineural hearing loss associated with hypoparathyroidism. Laryngoscope 1987; 97:1075–1979.
34. Pulec J. Symposium on Meniere's disease. Laryngoscope 1972; 82:1703–1715.
35. Spencer JT Jr. Hyperlipoproteinemia in the etiology of inner ear disease. Laryngoscope 1973; 83:639–678.
36. Kent SJ, von Gierke HE, Tolan GD. Analysis of the potential association between noise-induced hearing loss and cardiovascular disease in USAF aircrew members. Aviation Space Environ Med April 1986; 348–361.
37. Takala J, Varke S, Vaheri E, Sievers K. Noise and blood pressure. Lancet November 1977; 974–975.
38. Hedstrand H, Drettner B, Klockhoff I, Svedberg A. Noise and blood-pressure. Lancet December 1977; 1291.
39. Manninen O, Aro S. Noise-induced hearing loss and blood pressure. Int Arch Occup Environ Health 1979; 42:251–256.
40. Manninen O. Cardiovascular changes and hearing threshold shifts in men under complex exposures to noise, whole body vibrations, temperatures, and competition-type psychic load. Int Arch Occup Environ Health 1985; 56:251–274.

41. Pillsbury HC. Hypertension, hyperlipoproteinemia, chronic noise exposure: Is there synergism in cochlear pathology? Laryngoscope 1986; 96:1112–1138.
42. Colletti V, Fiorino FG. Myocardial activity during noise exposure. Acta Otolaryngol (Stockh), 1987; 104:217–224.
43. Verbeek JHAM, van Dijk FJH, de Vries FF. Non-auditory effects of noise in industry. Int Arch Occup Environ Health 1987; 59:51–54.
44. Carter NL, Beh HC. The effect of intermittent noise on cardiovascular functioning during vigilance task performance. Psychophysiology 1989; 26(5):548–559.
45. Andriukin AA. Influence of sound stimulation on the development of hypertension. Cor VASA 1961; 3(4):285–293.
46. Carter NL. Heart-rate and blood-pressure response in medium-artillery gun crews. Med J Aust 1988; 149:185–189.
47. Cavatorta A, Falzoi M, Romanelli A, Cigala F, Ricco M, Bruschi G, Franchini I, Borghetti A. Adrenal response in the pathogenesis of arterial hypertension in workers exposed to high noise levels. J Hypertension 1987; 5(suppl. 5):S463–S466.
48. Wu TN, Ko YC, Chang PY. Study of noise exposure and high blood pressure in shipyard workers. Am J Industr Med 1987; 12:431–438.
49. Flynn AJ, Dengerink HA, Wright JW. Blood pressure in resting, anesthetized and noise exposed guinea pigs. Hear Res 1988; 34:201–206.
50. Gold S, Haran I, Attias J, Shapira I, Shahar A. Biochemical and cardiovascular measures in subjects with noise-induced hearing loss. J Occup Med 1989; 31(11):933–937.
51. Michalak R, Ising H, Rebentisch E. Acute circulatory effects of military low-altitude flight noise. Int Arch Occup Environ Health 1990; 62:365–372.
52. Theorell T. Family history of hypertension—an individual trait interacting with spontaneously occurring job stressors. Scand J Work Environ Health 1990; 16(suppl. 1):74–79.
53. Milkovic-Kraus S. Noise-induced hearing loss and blood pressure. Int Arch Occup Environ Health 1990; 62:259–260.
54. Tarter SK, Robins TG. Chronic noise exposure, high-frequency hearing loss, and hyptertension among automotive assembly workers. J Occup Med 1990; 32(8):685–689.
55. Talbott EO, Findlay RC, Kuller LH, Lenkner LA, Matthews KA, Day RD, Ishii EK. Noise-induced hearing loss: A possible marker for high blood pressure in older noise-exposed populations. J Occup Med 1990; 32(8):690–697.
56. Zoller M, Nadol JB, Girard KF. Detection of syphilitic hearing loss. Arch Otolaryngol 1978; 104:63–65.
57. Wiet RJ, Milko DA. Isolation of spirochetes in the perilymph despite prior antisyphilitic therapy. Arch Otolaryngol 1975; 101:104–106.
58. Mach LW, Smith JL, Walter EK et al. Temporal bone treponemes. Arch Otolaryngol 1969; 90:11–14.
59. Smith JL. Spirochetes in late seronegative syphilis, despite penicillin therapy. Med Times 1968; 96:621–623.
60. Meyerhoff WL. Hearing loss and thyroid disorders. Minn Med 1974; 57(11):987–998.
61. Trotter WR. The association of deafness with thyroid dysfunction. Br Med Bull 1960; 16:92–98.
62. Bataskis JG, Nishiyama RH. Deafness with sporadic goiter. Pendred's syndrome. Arch Otolaryngol 1962; 76:401–406.
63. McCabe B. Autoimmune sensorineural hearing loss. American Neurotological Society Meeting. Los Angeles 1979; 30–31.
64. Bowman CA, Nelson RA. Human leukocytic antigens in autoimmune sensorineural hearing loss. Laryngoscope 1987; 97:7–9.
65. Johnson DW, Mathog RH. Hearing function and chronic renal failure. Ann Otol Rhinol Laryngol 1976; 85:43–49.
66. Johnson LG, Hawkins JE. Sensory and neural degeneration with aging as seen in microdissections of the human ear. Ann Otol Rhinol Laryngol 1972; 81:179–183.

67. Cooper AF, Curry AR. The pathology of deafness in the affective and paranoid psychoses of later life. J Psychosom Res 1976; 20:97–105.
68. Naufal PM. Primary sarcomas of the temporal bone. Arch Otolaryngol 1973; 98:44–50.
69. Conley JJ. Cancer of the middle ear and temporal bone. NY State J Med 1974; 74:1575–1579.
70. Clairmont AA, Conley JJ. Primary carcinoma of the mastoid bone. Ann Otol Rhinol Laryngol 1977; 86:306–309.
71. Schuknecht HF, Allan AF, Murakami Y. Pathology of secondary malignant tumors of the temporal bone. Ann Otol Rhinol Laryngol 1968; 77:5–22.
72. Ramsden RT, Bulman CH, Lorigan BP. Osteoradionecrosis of the temporal bone. J Laryngol Otol 1975; 8(9):941–955.
73. Jaffe BF, Penner JA. Sudden deafness associated with hypercoagulation. Trans Am Acad Ophthalmol Otol 1968; 72:774–778.
74. Glasscock ME. Surgery of glomus tumors of the temporal bone. Middle Section Meeting, American Laryngological, Rhinological, and Otological Society, Inc., Indianapolis, IN, Jan 1979:19–21.
75. Conley J, Hildyard V. Aneurysm of the internal carotid artery presenting in the middle ear. Arch Otolaryngol 1969; 90:61–64.
76. Podoshin L, Fradis M, Rillar T et al. Senso-neural hearing loss as an expression of arteriosclerosis in young people. Eye Ear Noise Throat Monthly 1975; 54:18–23.
77. Shapiro SLP. Otologic aspects of the subclavian steal syndrome. Eye Ear Nose Throat Monthly 1971; 50:28–31.
78. Hiller F. The vascular syndromes of basilar and vertebral arteries and their branches. J Nerv Ment Dis 1952; 116:988–1016.
79. Vick NA. Grinker's Neurology, 7th ed. Charles C Thomas, Springfield, IL, 1976; 486–492.
80. Hallpike C. Clinical Otoneurology and its contributions to theory and practice. Proc R Soc Med 1965; 58:185–196.
81. von Leden H, Horton B. Auditory nerve in multiple sclerosis. Arch Otolaryngol 1948; 48:51–57.
82. Imperato PJ. Tropical diseases of the ear, nose and throat. In: English GM, ed. Otolaryngology. Vol. 5. Philadelphia: Harper & Row, 1979:32–41.
83. Schein JD, Delk MT Jr. The Deaf Population of the United States, National Association of the Deaf. Silver Spring, MD, 1974:1–34.
84. Proctor C. Hereditary Sensorineural Hearing Loss, American Academy of Ophthalmology and Otolaryngology. Rochester, MN, 1978:5–24.
85. Jorgensen M, Buch N. Studies on inner ear function and cranial nerves in diabetes. Acta Otolaryngol 1961; 53:350–364.
86. Schuknecht H. Pathology of the Ear, Harvard University Press, Cambridge, MA, 1974; 168–184, 262–266, 311–330, 374–379, 383–388, 420–424, 444–446, 488.
87. Axelsson A, Fagerberg SE. Auditory function in diabetes. Acta Otolaryngol 1988; 66:49–64.
88. Kitabchi AE, Shea JJ, Duckworth WC et al. High incidence of diabetes and glucose intolerance in fluctuant hearing loss. J Lab Clin Med 1971; 78(6):995–996.
89. Jorgensen MB. Sudden loss of inner ear function in the course of long standing diabetes mellitus. Acta Otolaryngol 1960; 51:579–584.
90. Rosen Z, Davis E. Microangiopathy in diabetics with hearing disorders. Eye Ear Nose Throat Monthly 1971; 50:31–35.
91. Triana RJ, Suits GW, Garrison S, Prazma J, Brechtelsbauer PB, Michaelis OE, Pillsbury HC. Inner ear damage secondary to diabetes mellitus. Arch Otolaryngol Head Neck Surg 1991; 117:635–640.
92. Weille F. Hypoglycemia in Meniere's disease. Arch Otolaryngol 1968; 87:555–557.
93. Parkin JL, Tice R. Hypoglycemia and fluctuating hearing loss. Ann Otol Rhinol Laryngol 1970; 79:992–997.
94. Proctor C. Diagnosis, prevention and treatment of hereditary sensorineural hearing loss. Laryngoscope 1977; (Oct. suppl.):87.

95. Spencer JT Jr. Hyperlipoproteinemia in the etiology of inner ear disease. Laryngoscope 1973; 83:639–678.
96. Grace LG. Frequency of occurrence of cleft palages and harelips. J Dent Res 1943; 22:495–497.
97. Fraser FC, Walker BE, Trasler DG. Experimental production of congenital cleft palate: Genetic and environmental factors. Pediatrics 1957; 19:782–787.
98. Gorlin RJ, Cervenka J, Pruzansky S et al. Facial clefting and its syndromes. Birth defects. Original Article Series 1971; 7:3–49.
99. Sataloff J, Fraser M. Hearing loss in children with cleft palates. AMA Arch Otolaryng 1952; 55(1):61–64.
100. Kloepfer H, Lagvaite J. The hereditary syndrome of congenital deafness and retinitis pigmentosa (Usher's syndrome). Laryngoscope 1966; 76:850–862.
101. Hallgren V. Retinitis pigmentosa combined with congenital deafness; with vestibulocerebellar ataxia and mental abnormality in a proportion of cases. A clinical and geneticostatistical study. Acta Psychiatr Scand 1959; 138(suppl.):1–101.
102. McLeod AC, McConnel FE, Sweeney A et al. Clinical variation in Usher's syndrome. Arch Otolaryngol 1971; 94:321–334.
103. Landau J, Feinmesser M. Audiometric and vestibular examination in retinitis pigmentosa. Br J Ophthalmol 1956; 40:40–44.
104. Russo C, Zibordi F, DeVita R et al. Cochlear-vestibular aspects of retinitis pigmentosa. Ann Laryngol 1968; 67:174–185.
105. Davenport SLH, Omenn GS. The heterogeneity of Usher syndrome [abstract 215]. Fifth International Conference on Birth Defects, Montreal, 1977:21–27.
106. Refsum S. Heredopathia atactia polyneuritiformis. Acta Psychiatr Scand 1946; 38(suppl.):1–303.
107. Steinberg D, Mize CE, Herndon JH Jr. et al. Phytanic acid in patients with Refsum's syndrome and response to treatment. Arch Intern Med 1970; 125:75–87.
108. Konigsmark BW, Gorlin RJ. Genetic and Metabolic Deafness. Philadelphia: W.B. Saunders, 1976:40–41, 76, 98–100, 156–159, 164–168, 188–191, 221–222, 311–312, 330–335, 345–351, 355, 364–370.
109. Alstrom CH, Hallgren B, Nilsson LBN et al. Retinal degeneration combined with obesity, diabetes mellitus and neurogenous deafness. Acta Psychiatr Scand 1959; 129(suppl.):1–35.
110. Paddison RM, Moossy J, Derbes VJ et al. Cockayne's syndrome. Derm Trop 1963; 2:195–203.
111. Newell PW, Ernest JT. Ophthalmology. 3rd ed. Mosby, St. Louis, MO 1974:332.
112. Seth RRS, Dayal D. Inner ear involvement in primary glaucoma. Ear Nose Throat J 1978; 57:69–75.
113. Proctor C. Hereditary deafness. American Academy of Ophthalmology and Otolaryngology, 1975 Instructional Section, Course 543; 2–4, 14.
114. Wilson J. Leber's hereditary optic atrophy: Some clinical and aetiological considerations. Brain 1963; 86:347–362.
115. Warburg M. Norries's disease (atrofia bulborum hereditaria). Acta Ophthalmol 1963; 41:134–146.
116. Cogan DG. Syndrome of non-syphilitic interstitial keratitis and vestibulo-auditory symptoms. Arch Ophthalmol 1945; 33:144–149.
117. Alport AC. Hereditary familial congenital hemorrhagic nephritis. Br Med J 1927; 1:504–506.
118. Muckle TJ, Well M. Urticaria, deafness and amyloidosis: A new heredo-familial syndrome. Q J Med 1962; 31:235–248.
119. Proctor C. Hereditary deafness, American Academy of Ophthalmology and Otolaryngology, 1971, Instructional Section, Course 516, 6.
120. Herrman C Jr, Aquilar MJ, Sacks OW. Hereditary photomyoclonus associated with diabetes mellitus, deafness, nephropathy, and cerebral dysfunction. Neurology 1964; 14:212–221.

121. Waardenburg PJ. A new syndrome combining developmental anomalies of the eyelids, eyebrows and nose root with pigmentary defects of the iris and head hair and with congenital deafness. Am J Hum Genet 1951; 3:195–253.
122. Tietz W. A syndrome of deaf-mutism associated with albinism showing dominant autosomal inheritance. Am J Hum Genet 1963; 15:259–264.
123. Gorlin RJ, Anderson RD, Moller JH. The LEOPARD (multiple lentigines) syndrome revisited. Birth Defects 1971; 7(4):110–115.
124. Batsakis JG. Tumors of the Head and Neck. Baltimore: Williams & Wilkins, 1974:231–235.
125. Kongismark BW. Hereditary deafness in man, part 2. N Engl J Med 1969; 281(14):776–777.
126. Linthicum FH, Schwartzman JA. An Atlas of Micropathology of the Temporal Bone. Philadelphia: W.B. Saunders, 1974:58–60, 70–72, 74–75.
127. Poretta CA, Dahlin DC, James JM. Sarcoma in Paget's disease of bone. J Bone Joint Surg 1957; 39A:1314–1329.
128. Khairi MRA, Johnson CC Jr, Altman RD et al. Treatment of Paget's disease of bone (osteitis deformans). JAMA 1974; 230:562–567.
129. Sataloff RT, Graham MD, Roberts BR. Middle ear surgery in fibrous dysplasia of the temporal bone. Am J Otol 1985; 6(2):153–156.
130. Van der Hoeve J, de Kleijn A. Blaue Sclirae, Knochenbruchigkeit und Schwerhorigkeit. Arch Ophthalmol 1918; 95:81–93.
131. Dessoff J. Blue sclerotics, fragile bones and deafness. Arch Ophthalmol 1934; 12:60–71.
132. Seedorff KS. Osteogenesis Imperfecta: A Study of Clinical Features and Heredity Based on 55 Danish Families Comprising 180 Affected Persons, Thesis, Copenhagen. Munksgaard Press, 1949:1–229, cited in Konigsmark and Gorlin.
133. Kosoy J, Maddox HE. Surgical findings in van der Hoeve's syndrome. Arch Otolaryngol 1971; 93:115–122.
134. Lake M, Kuppinger J. Craniofacial dysostosis (Crouzon's disease). Arch Ophthalmol 1950; 44:37–46.
135. Boedts D. La surdite dans la dysostose craniofaciale ou maladie de Crouzon. Acta Otorhinolaryngol Belg 1967; 21:143–155.
136. Tessier P. The definitive plastic surgical treatment of the severe facial deformities of craniofacial dysostosis. Crouzon's and Apert's diseases. Plast Reconstr Surg 1971; 48:419–442.
137. Franceschetti A, Klein D. Mandibulofacial dysostosis. New hereditary syndrome. Acta Ophthalmol 1949; 27:143–224.
138. Stovin JJ, Lyon JA, Clemmens RL. Mandibulofacial dysostosis. Radiology 1960; 74:225–231.
139. Hutchinson JC, Caldarelli DD, Valvassori GE et al. The otologic manifestations of mandibulofacial dysostosis. Tr Am Acad Ophthalmol Otolaryngol 1977; 84:520–528.
140. Igarashi M, Filippone MV, Alford BR. Temporal bone findings in Pierre Robin syndrome. Laryngoscope 1976; 86:1679–1687.
141. Johnston CC, Lavy N, Lord T et al. Osteopetrosis. A clinical genetic, metabolic and morphologic study of the dominantly inherited benign form. Medicine 1968; 47:149–167.
142. Myers EN, Stool S. The temporal bone in osteopetrosis. Arch Otolaryngol 1969; 89:460–469.
143. Wong ML, Balkany TJ, Reaves J et al. Head and neck manifestations of malignant osteopetrosis. Otolaryngology 1978; 86:585–594.
144. Wildervanck LS, Hoeksema PE, Penning L. Radiological examination of the inner ear of deaf mutes. Acta Otolaryngol 1966; 61:445–453.
145. Everberg G. Wildervanck's syndrome: Klippel-Feil's syndrome associated with deafness and retardation of the eyeball. Br J Radiol 1962; 36:562–567.
146. Sataloff RT, Spiegel JR, Hawkshaw MJ, Epstein JM, Jackson L. Cornelia de Lange syndrome. Arch Otolaryngol Head Neck Surg 1990; 116:1044–1046.
147. Monday LA, Lemieux B. Etude audiovestibulaire dans l'ataxie de Friedreich. J Otolaryngol 1978; 7:415–423.

148. Aggerbeck LP, McMahon JP, Scano AM. Hypobetalipoproteinemia: Clinical and biochemical description of new kindred with Friedreich ataxia. Neurology 1974; 24:1051–1063.
149. Latham AD, Munro TA. Familial myoclonus epilepsy associated with deaf mutism in a family showing other psychobiological abnormalities. Ann Engen 1938; 8:166–175.
150. Globus JH, Strauss L. Progressive degenerative subcortical encephalopathy (Schilder's disease). Arch Psychiatry 1928; 20:1190–1228.
151. Lichtenstein BW, Rosenbluth PR. Schilder's disease with melanoderma. J Neuropathol Exp Neurol 1959; 15:229–231.
152. Anderson PE. Radiology of Pendred's syndrome. Adv Otorhinolaryngol 1974; 21:9–18.
153. Lindsay J. Profound childhood deafness: inner ear pathology. Ann Otol Rhinol Laryngol 1973; 5(suppl.):1–21.
154. Selkae DJ. Familial hyperprolinemia and mental retardation. Neurology 1969; 19:494–502.
155. McKusick VA, Claiborne R, eds. Medical Genetics. New York, NY: HP Publising, 1973:63–78.
156. McKusick V. Heritable Disorders of Connective Tissue. 3rd ed. St. Louis: Mosby, 1966:150.
157. Kang ES, Byers RD, Gerald PS. Homocystinuria: response to pyriodoxine. Neurology 1970; 20:503–507.
158. Smith DW. Recognizable Patterns of Human Malformation. Philadelphia: W.B. Saunders, 1970:60–61, 242–255.
159. Leroy JG, Crocker AC. Clinical definition of Hunter-Hurler phenotypes. A review of 50 patients. Am J Dis Child 1966; 112:518–530.
160. Robbins MM, Stevens HF, Linker A. Morquio's disease: an abnormality of mucopolysaccharide metabolism. J Pediatr 1963; 62:881–889.
161. Van Noorden GK, Zellweger H, Parseti IV. Ocular findings in Morquio-Ullrich's disease. Arch Ophthalmol 1960; 64:585–591.
162. Medical News (editorial): Enzyme infusions to help Hurler's syndrome patients. JAMA 1973; 224:597–604.
163. Kaloduy EH, Kolodny EH, Brady RO et al. Demonstration of an alteration of ganglioside metabolism in Tay-Sachs disease. Biochem Biophys Res Commun 1969; 37:526–531.
164. Steinberg G. Erblinche Augenkrankheiten und Ohrenleiden. V Ohr Nas Kehlkopfheik 1937; 42:320–345.
165. Keleman G. Tay-Sachs-Krankeit und Gehororgan. Z Laryngol Rhinol Otol 1965; 44:728–738.
166. Danish JM, Tillson JK, Levitan M. Multiple anomalies in congenitally deaf children. Eugen Quart 1963; 10:12–21.
167. O'Reilly S, Weber PM, Pollycove M et al. Detection of carrier of Wilson's disease. Neurology 1970; 20:1133–1138.
168. Gorlin RJ, Pindborg JJ. Symptoms of the Head and Neck. New York, NY: McGraw-Hill, 1964:41–46.
169. Jervell A, Lange-Nielsen F. Congenital deaf-mutism functional heart disease with prolongation of the QT interval and sudden death. Am Heart J 1957; 54:59–68.
170. Fraser GR, Froggatt P, James TN. Congenital deafness associated with electrocardiographic abnormalities, fainting attacks, and sudden death: A recessive syndrome. Q J Med 1964; 33:361–385.
171. Levine SA, Woodworth CR. Congenital deaf-mutism, prolonged QT interval, syncopal attacks and sudden death. N Engl J Med 1958; 259:412–417.
172. Todd GB, Serjeant FR, Larson MR. Sensorineural hearing loss in Jamaicans with SS disease. Acta Otolaryngol 1973; 76:268–272.
173. Morgenstein KM, Manace ED. Temporal bone histopathology in sickle cell disease. Laryngoscope 1969; 79:2172–2180.
174. Urban GE. Reversible sensorineural hearing loss associated with sickle cell crisis. Laryngoscope 1973; 83:633–638.
175. Forman-Franco B, Abramson AL, Gorvoy JD et al. Cystic fibrosis and hearing loss. Arch Otolaryngol 1979; 105:338–342.

176. Kartagener M. Zur Pathogenese der bronkiektasien, bronkiektasien bei Situs viscerum inversus. Beitr Klin Tuberk 1933; 83:489–501.
177. Sethi BR. Kartagener's syndrome and its otological manifestations. J Laryngol Otol 1975; 89:183–188.
178. Anderson H, Filipsson R, Fluur E et al. Hearing impairment in Turner's syndrome. Acta Otolaryngol 1969; 247(suppl.):1–26.
179. Anderson H, Lindsten J, Wedenberg E. Hearing deficits in males with sex chromosome anomalies. Acta Otolaryngol 1971; 72:55–58.
180. Glovsky L. Audiological assessment of a mongoloid population. Train Sch Bull 1966; 63:27–36.
181. Fulton RT, Lloyd LL. Hearing impairment in a population of children with Down's syndrome. Am J Ment Defic 1968; 73:298–302.
182. Balkany TJ, Mischke RE, Downs MP et al. Ossicular abnormalities in Down's syndrome. Otolaryngol Head Neck Surg 1979; 87(3):372–384.
183. Sataloff J, Pastore PN, Bloom E. Sex-linked hereditary deafness. Am J Hum Genet 1955; 7:201–203.
184. Bernstein JM. Occurrence of episodic vertigo and hearing loss in families. Ann Otol Rhinol Laryngol 1965; 74:1011–1021.
185. Jorgensen MB, Kristensen HK. Frequency of histological otosclerosis. Ann Otol Rhinol Laryngol 1967; 76:83–88.
186. Linthicum FH, House HP, Althaus SR. The effect of sodium fluoride on otosclerotic activity as determined by strontium. Ann Otol Rhinol Laryngol 1972; 4:609–615.

14

A Brief History of Occupational Hearing Loss: A Personal Perspective

Joseph Sataloff
Thomas Jefferson University, Philadelphia, Pennsylvania, USA

For many centuries, even as far back as Biblical days, it has been known that long exposure to very loud noise can cause hearing loss. However, it was not until the Industrial Revolution in the 19th century, with its overwhelmingly noisy machines, that noise-induced hearing loss became of concern to society. As a result of their occupations thousands of workers were exposed for many months and years to very high noise levels. Hearing loss among employees performing certain jobs became so common that terminology connected the job to the symptoms. For instance, boiler maker's deafness and gun shooting deafness were common terms, as were many others. Although many workers recognized that hearing loss was being caused by their jobs, little effort was made to prevent its occurrence. In fact, during the Industrial Revolution little effort was made to prevent many diseases and injuries associated with occupational causes.

Hearing conservation really started during World War II. During the period of some 60 years following World War II, countless contributions have been made in many disciplines. Thousands of scientific publications and countless meetings and conventions have served to advance the practice of hearing conservation in the workplace. Important standards and guidelines have been set. New problems have arisen, particularly concerning the financial aspects of Workers' Compensation for occupational hearing loss. In a chapter of this size, it is hardly possible to discuss all of the influential individuals who have made significant contributions to the field of occupational hearing loss significance. However, it seems appropriate to highlight at least some of the most important features of this evolving field and to concentrate on the individuals whose positive contributions were particularly important in shaping the present state of hearing conservation. Unfortunately, many individuals had to be omitted from this historical review, but this in no way detracts from their importance or efforts. As in all written histories, the author's personal experience and activities influence the content of this chapter. Since I have been involved from the very beginning in many aspects of hearing conservation, my experience, observations, and personal contacts provide a broad base for this chapter. They are summarized to make available not only my personal perspective on the last half century, but more importantly to recognize the contributions of many key individuals, some of whose accomplishments have never been acknowledged in print.

The foundation for hearing conservation came from several different disciplines: *basic scientists* researched thc physiology of hearing, the physics of sound, acoustics, and equipment to measure hearing and hearing loss. These scientists include physiologists, acoustical engineers, physicists, psychologists, anatomists, medical researchers, and others. The *medical aspects* of hearing and occupational hearing loss were investigated by physicians, audiologists, industrial hygienists, safety engineers, and others. They concerned themselves chiefly with diagnosing noise-induced hearing loss, learning how to measure it clinically, developing standards, and the use of ear protectors, clinical testing equipment, rehabilitation measures, and many other subjects. A third group concerned itself chiefly with *implementing* hearing conservation in government and industry and included occupational physicians, otologists, occupational nurses, hygienists, safety engineers, audiologists, hearing conservationists, safety personnel, legislators, hearing protection company personnel, and others.

These groups were confronted early with serious questions that required urgent answers including the following, among many others: (*i*) What is normal hearing and how do you measure hearing and hearing loss accurately? (*ii*) How much noise should be considered safe, and how much should be considered harmful to hearing? (*iii*) How long an exposure to loud noises (at a sitting or during a day) is safe, and after how long (days, months, or years) is harmful? (*iv*) What equipment is necessary to measure noise in industry? (*v*) What are hearing impairment and handicap? (*vi*) Where does noise do damage in the auditory pathway? (*vii*) How do you test hearing levels and train personnel to do this testing in industry? (*viii*) In what kind of environment can hearing testing be done to produce valid and reliable results? (*ix*) Do you test a person's ability to hear and interpret speech or should testing be limited to pure tones? (*x*) How do you develop good ear protectors?

For many years prior to World War II, researchers in basic science investigated the physiology of hearing and established a clear understanding of how the auditory mechanism works. Scientists in this area included Leo Beranek, Hallowell Davis, Don Eldridge, Harvey Fletcher, Bob Galambos, Joe Hawkins, Don Henderson, Merle Lawrence, Rasmussen, S. Smith Stevens, Tasaki, Von Beckesy, Bray Weaver, Joe Zwislocki, and many others.

Without question, Dr Hallowell Davis was the most distinguished contributor to the field of hearing and hearing loss, and should be considered the "Father of Modern Science of Hearing and Hearing Loss." All of us should acquire some familiarity with Davis' achievements by reading his autobiography *The Professional Memoirs of Hallowell Davis* (1). This book includes information about the development of auditory neural physiology, the effects of intense noise on man, and Davis becoming the Director of the Psycho Acoustic Laboratory and the Director of the Central Institute of the Deaf in St. Louis, as well as the formation of CHABA (Committee on Hearing and Bioacoustics) and practically every organization devoted to hearing and hearing loss. Davis also was an extraordinarily kind human being and leader. In the Physical Acoustic Lab at Harvard, and the Central Institute for the Deaf (CID), he surrounded himself with competent and dedicated scientists and investigators to explore numerous problems and solutions, including Bill Clark, Don Eldridge, Ira Hirsh, Walter Rosenblith, Wayne Rudmose, and Dick Silverman.

In the medical and paramedical area, Dr Aram Glorig was an outstanding and distinguished leader in all aspects of hearing conservation. He was involved in hearing conservation for over 50 years until his very last day. He was a trained otolaryngologist but did not practice clinically or treat patients. Instead, he preferred to devote all of his time to hearing and hearing conservation. There was hardly an important committee on any aspect of hearing that Aram did not chair or participate in. He helped set up the Hearing Rehabilitation Center at Walter Reed Hospital in Bethesda, MD, during World War II, and directed many research projects including a National Survey to determine what is normal hearing. He was the force behind most of the hearing committees established by the American Academy of Otolaryngology, and many national committees. His input was always meaningful and farsighted. His commitment and enthusiasm were enormous, and sometimes slightly overzealous, such as the time he interrupted Professor Wayne Rudmose's speech with "What do you really know about the physics of sound?" Apparently, Aram was unaware that Wayne Rudmose was the professor of physics at Southern Methodist University, and one of the most knowledgeable men in the country. He was an important contributor to establishing standards and criteria, and he brought quality and validity to hearing issues. Rudmose often stated that Aram "was responsible for hearing conservation on the West Coast, and that I was responsible for it on the East Coast"; but Aram spent his efforts on both coasts and made great contributions. I had the good fortune to observe and appreciate the work of Hal Davis and especially to work collaboratively with Aram Glorig on hearing conservation during the past 50 years.

During and after World War II, concern for occupational hearing loss in the government and the general public increased rapidly especially with so many military personnel developing hearing loss due to exposure to gun fire. The military established two large medical centers in 1943 to rehabilitate personnel and veterans with hearing loss: the Walter Reed Army Hospital and the Philadelphia Naval Hospital. Dr Aram Glorig played a major role in developing and operating the military center at Walter Reed. Dr Frances Lededer and, later, Dr Fred Harbert directed the services at the Philadelphia Naval Hospital. I had the opportunity to serve as otologist for the Auditory Rehabilitation Center of the Philadelphia Naval Hospital during and after World War II.

It was during this period that the field of audiology developed, along with professional audiologists. Universities established hearing centers, performed clinical research, and contributed immeasurably to the knowledge of hearing testing and its many problems. The outstanding positive contributors in this area included Ray Carhart, Joe Chiarella, John Corbo, Steve Cofer, Bob Dobie, Leo Doerfler, A. Feldman, George Gates, Cyril Harris, Bill Hardy, Bud Inbus, Jim Jerger, H.H. Jones, Karl Kryter, Emily Letts, Dave Lipscomb, Roger Maas, Bill Melkin, Maurice Miller, Jeff Morill, Chuck Nelson, Miriam Pauls, Scott Reger, Julia and John Royster, Michael Seideman, Rick Stepkin, Alice Suter, Larry Vasallo, Henning Von Gierke, and countless others. Special credit should be given to Dr Dixon Ward, whom I believe was the most brilliant investigator in the field and who made outstanding contributions and helped elevate much of the research being performed to a more valid, reliable level.

Probably the earliest major step toward the development of hearing conservation in American industry occurred in 1952 when John Zapp and I presented a 4-day Institute on Hearing Conservation at Colby College in Waterville, ME, and later at the University of Maine in Orono. The course ran annually for 40 years and was attended by thousands of personnel from industry, medicine, law, government, science, audiology, and other related disciplines. Faculties for these courses provided intensive education on all aspects of occupational hearing loss and the implementation of hearing conservation programs in industry. The faculty consisted of persons already mentioned in this chapter, all of whom held outstanding leadership positions in their particular disciplines. Course graduates supplied industry and government with knowledgeable personnel to initiate and implement hearing conservation, as well as to manage claims for occupational hearing loss.

The National Safety Council, the Industrial Hygiene and Acoustical Societies, and many other societies assumed major roles in hearing conscrvation. They supported research, conducted meetings and seminars, and published many papers and pamphlets, as well as manuals to train personnel in hearing conservation. Some of the outstanding leaders in these areas included Joe Anticaglia, John Barry, Elliot Berger, K. Bohl, Jim Botsford, Sidney Busis, Fran Catlin, Bill Clark, Alex Cohen, John Corso, Alan Cudworth, Terry Dear, Bob Dobie, Bill English, Alan Feldman, Forest Fisher, H. Fletcher, Mike Fox, Floyd Fraser, D. Gassaway, George Gates, Charles Jackson, Dan Johnson, Bruce Karrh, Karl Kryter, Bob Ligo, Dave Lipscomb, Roger Maas, Bob McLean, Paul Michaels, Ken Morse, Jeff Morrill, Ann Murphy, Charles Nixon, Sid Pells, Joe Schwerha, Michael Seidman, Jim Spencer, Ray Strocko, Alice Sutter, Ed Toothman, Floyd Van Atta, H. Walworth, Dixon Ward, Chuck Williams, Ray Yerg, John Zapp, and many, many others, including my son otologist/neurotologist Dr Robert Sataloff, and myself.

Interest in hearing loss, especially the conservation of hearing, spread rapidly among numerous medical and related societies, especially the American Medical Association (AMA), American Academy of Otolaryngology, American Academy of Occupational Medicine, The American Audiological Association, The Acoustical Society, and the Industrial Hygiene Society. These societies established hearing conservation committees and sponsored research and meetings to find answers to the many questions confronting everyone involved in hearing conservation. The government also established a committee called CHABA to study hearing loss in the military. An important committee sponsored by the Otolaryngologic Society was chaired by Dr Stewart Nash, who provided a great deal of valid information on the subject of Workers' Compensation and how to determine hearing impairment and hearing handicap. Members of this committee included Dr Norton Canfield, Dr Ed Fowler, Dr Aram Glorig, Dr Fred Hill, Dr Gordon Hoopel, and Dr Stewart Nash.

The Bell Telephone Company and the Western Electric Company helped gather a great deal of information about the physics of sound, its transmission, and its measurement. The Western Electric Audiometer was a very good instrument that enabled physicians and audiologists to measure hearing levels. This was the first practical audiometer. Subsequently, this also resulted in developing the audiogram, used clinically to chart responses from 250 to 8000 cycles (hertz, Hz). Equipment to measure speech reception and discrimination ability was not available initially. However, as a result of research by audiologists at various institutions and the Central Institute for the Deaf (CID), equipment and tests were developed that enabled audiologists to provide stability and uniformity in hearing testing.

The Hearing Conservation Committee of the American Academy of Otolaryngology played an early and major role in developing hearing conservation programs and the dissemination of occupational hearing loss information. This group was led by such distinguished otologists as Fran Catlin, Ed Fowler, Mike Fox, Aram Glorig, Fred Harbert, Ted Hill, Gordon Hoople, Howard House, Stuart Nash, myself, and numerous others. This committee helped establish standards, define philosophy, and develop the diagnosis of hearing loss. The committee also sponsored numerous research projects and national meetings to discuss problems in American industry and how to solve them.

Mention should also be made of the companies and individuals who provided services to American industries, directly or through hearing testing vans, including such companies as CHSL, TK Group, Summar & Co., EPIC, Enviromed, Hearing Conservation Noise Control Inc., and Intel. A number of audiologists, including Joe Chiarello, Steve Cofer, Bob Gore, Gary Rau, Charles Russell, Bill Schnauffer, Dick Shackelford, Rick Stepkin, and many others, used vans to provide hearing tests in plants throughout the country. Many other companies and scientists contributed to standardizing hearing protectors, particularly American Optical, Elliot Berger, EAR, and Paul and Kevin Michaels, among many others.

When compensation for hearing loss became an important factor, a few interested attorneys and associates made outstanding contributions to stabilize and help provide fairness in the courts, including Jeff Adler, Steve Anthony, Howard Bunn, Mike Burns, Bob Collier, Stanley Frazee, Tem Fitch, David Hickton, Tom Jayne, Charles Katz, Jim Munro, Ed Palmer, Jim Savage, Jeff Sidebottom, Joella Szabo, Charles Weiner, and many others.

Countless other events helped establish hearing conversation, far too many to describe in a brief history such as this. Most have appeared in the literature, but many have received much less attention than they merit.

Professor Wayne Rudmose made a major contribution in 1954 when he published the results of Z 24 X2 Committee that examined thousands of audiograms being performed in industry (2). He explained that there was a great need for more accurate audiometry and reliable hearing testing. The vast majority of audiograms examined were inaccurate and could not be relied on to establish a standard or correlate hearing loss and noise exposure.

In 1964, the American Mutual Insurance Alliance published a brochure entitled "Background for Loss of Hearing Claims" by Floyd Fraser. The brochure pointed out that before 1948 hearing loss was not regarded as a significant factor in Workers' Compensation. From 1948 to 1959, hearing loss began to play a very prominent role in Workers' Compensation. The document describes the work of a subcommittee on noise of the American Academy of Otolaryngology and the development of claims in such states as Missouri, New York, and Wisconsin. This brochure in 1964 also listed 35 references, many of which reflected important contributions to hearing conservation.

The first book on industrial deafness was written by me with advice and collaboration from John Zapp in 1957 and published by the Blakeston division of the McGraw-Hill Company (3). In 1958, Aram Glorig published a pamphlet, "Noise and Your Ear" (4). In 1959, the Subcommittee on Noise of the American Academy of Otolaryngology published a guide for conservation of hearing (5) and pointed out the great need for better hearing testing in industry. In 1966, the American Industrial Hygiene Association published the *Industrial Noise Manual* (6).

In 1954, the status of compensation for occupational hearing loss was included in many reports, as in the "Workmen's Compensation Board Reports," prepared by New York's Workmen's Compensation Board. At about the same time, the United States Longshoremen and Harbor Workers' Compensation Act was instituted.

The Federal Government first showed its concern for occupational hearing loss by publishing the Walsh–Healy Act. This was the precursor to the Occupational Noise Regulation that appeared in the Federal Register in 1973 (7). This provided a major motivation for American industries to start taking measures to prevent hearing loss from developing in employees.

In 1952, Dr James Sterner, Medical Director of Eastman Kodak, conducted an opinion poll of experts in noise and hearing to determine the maximum intensity level of industrial noise that they considered safe for hearing preservation. There were wildly disparate estimates and lack of agreement even among the most experienced and knowledgeable experts. The futility of any attempt to establish meaningful guidelines by use of such polls was evident. It was not until 1973 that the Department of Labor appointed an OSHA (Occupational Safety and Health Administration) committee to establish noise levels considered safe and above which noise might be considered hazardous to hearing. The committee consisted of several representatives from labor, management, and government, and I was appointed Scientific Advisor of this committee. The committee spent over a year reviewing research reports and meeting with leading knowledgeable experts to try to establish the critical noise level. We gave most consideration to the three most important research reports available at the time, by Bill Baughn (8), Burns and Robinson (9), and W. Passchier-Vermeer (10). After a year of consideration, the OSHA Committee determined that these reports were not sufficiently reliable or valid and could not be used as a basis for a noise standard. An important critique of these reports published by Drs Glorig and Ward (11) discussed the shortcomings of each report. The critique should be carefully considered by everyone, since many reports still quote those studies as valid, and they are also unjustifiably used in Workers' Compensation cases and presented as accurate studies. Glorig and Ward advised that these practices should be discontinued. The final report of this OSHA Noise Committee was published in the Federal Register and stipulated 90 dB as the critical noise level.

Perhaps the least known or appreciated legacy to hearing conservation was made by Dr. Floyd Van Atta, who oversaw all the meetings of the Secretary of Labor's OSHA Noise Committee. His respect for science, as well as for workers and management, enabled the Noise Regulations to be practical and based on the best scientific evidence and good judgment. He resisted all outside efforts to make the critical noise level as low as 80 or 85 dBA because of lack of sufficient evidence to justify such low levels, and because of the practical implementation problems they would cause. A 90-dB level, and many other features in the final OSHA Noise Regulation, were accomplished largely because of Floyd Van Atta's firm efforts and guidance. The OSHA Noise Regulation, along with its 1983 Amendment, are now chiefly responsible for preventing occupational hearing loss from developing in millions of American employees. The Federal Railroad Administration (FRA), the Mining Industry (MSHA), and the Maritime Industry developed their own standards and criteria for occupational hearing loss. These differed somewhat from one another and the OSHA Noise Regulation, but they all required industries and their plants to come into compliance with requirements similar to those listed in the OSHA regulation.

Aram Glorig, Dixon Ward, and many others, including Ray Carhart, John Corso, Leo Doerfler, Ed Fowler, Dr Letts, and others, conducted numerous research programs to determine what should be considered normal hearing and at what level it becomes abnormal and impaired, as well as what noise level is capable of causing such a loss. Many hundreds of papers reporting results of such studies were produced. However, from the publication of these papers it became obvious that many were in disagreement with each other and valid and reliable information was difficult to distinguish. To overcome many of the shortcomings of existing research reports, industry and government established an advisory committee and a monitoring committee to sponsor research projects that would establish a critical noise level for American industry. A research team to do "an inter-industry noise study" was established. This research team and the guiding committee published a protocol for future researchers so that data produced would be valid, reliable, and comparable to other research projects.

The Inter-Industry Noise Study (12) extended for well over 3 years, and in 1978 the results of this major investigation were published. Following the guidelines of the protocol, some 250,000 employee records and noise measurements were reviewed in an effort to find some 348 experimental subjects who met the rigid requirements of the protocol. It was the purest study possible under the conditions existing in industry. It confirmed 90 dBA as the critical noise level for employees' exposure on a daily basis over a lifetime and confirmed that the noise regulation was a reasonable standard. The Inter Industry Noise Study was later evaluated critically by Dr T.R. Schori and Col. Dan Johnston at the University of

Dayton Research Institute, sponsored in part by the American National Standards Institute (13). They found the study valid and reliable and provided no serious criticism of the conclusions.

The need for expert otologic consultation in cases of workers' compensation claims also brought to light the fact that very few medical schools or otolaryngology residencies provided adequate training to physicians in occupational hearing loss. As a result, the American College of Occupational Medicine Hearing Conservation Committee developed criteria to help individual doctors diagnose occupational noise-induced hearing loss accurately. They specified eight characteristics of occupational hearing loss that remain a useful guide for clinicians (14). Although the criteria have been helpful in many cases, it is important to realize that some of the criteria are based on clinical experience, and definitive research confirmation is needed.

In addition to the military, medical, and scientific communities, American industries also became increasingly interested in and concerned with occupational hearing loss, not only with the prevention of hearing loss and the need to comply with the new OSHA regulation, but also with the potential cost and medical legal problems that the leaders of industry foresaw would develop. In fact, these problems did develop. The need became obvious for more trained individuals able to test hearing in industry and medical practice. This resulted in a new type of specialist called hearing testing technicians and hearing conservationists who could measure hearing loss and obtain valid and reliable results. To satisfy this need of industry and otology, Glorig and I convened a meeting in Maine with industrial audiologists and industrial nurses, including Betty Foster, Ann Murphy, Millie Sitner, Larry Vassallo, and several others. They developed criteria and courses designed to train high school graduates to supply the needs of industry and medicine for basic hearing testing. This effort evolved into the Council for Accreditation in Occupational Hearing Conservation (CAOHC). Programs were established in collaboration with schools and hospitals, and they resulted in thousands of trained personnel who could be hired by industry; and physicians could be reasonably certain that the audiograms performed were accurate. These trainees, along with audiologists also served as the staff for the numerous hearing testing vans that are now in use to make it easier for industry to do hearing testing without interfering with the employees' work, at a very affordable cost. Audiologists throughout the country are still playing a major role in training thousands of hearing conservationists.

In 1948, the U.S. Circuit Court of Appeals passed down a decision that industrial personnel with occupational hearing loss were entitled to compensation even though there was no loss of time and no loss of earnings (Slawinski *v* Williams & Company). This, of course, resulted in the development of OSHA Noise Regulations and other statutes; a huge increase in the number of claims; and the cost for Workers' Compensation for hearing loss reaching billions of dollars. Now practically every state in the United States has a statute that makes occupational hearing loss compensable. Most use the AMA formula for calculating hearing impairment, but a number of states deviate and have made their own methods of calculating impairment and payment for hearing loss. The Veterans Administration F, Federal Employee Liability Act 5 (FELA) and the Maritime and Mining Industries also have developed their own distinctive methods for quantifying occupational hearing loss and for paying compensation to employees with occupational noise-induced hearing loss.

As a result of numerous military and industrial personnel developing hearing loss, the hearing aid industry started to expand because of the commercial potential. Initially, hearing aids were large, cumbersome, and not very helpful. As time went on, a great deal of progress was made in producing hearing aids that could be worn comfortably and produced reasonably clear amplification. Dr Davis and his group at Psycho Acoustics Laboratory in Boston, MA, played an important role in helping manufacturers to develop practical and effective hearing aids.

It is difficult to measure the excellent effect of preventative measures, and we forget too quickly and too easily the many illnesses and injuries that compelled the development of preventative health care and legislation. One of the leading federal agencies to stress safety in the workplace is OSHA, which has established regulations to prevent accidents and exposure to disease and illness among our millions of industrial employees. Unfortunately, it is human nature to fail to appreciate fully what we have until we lose it. This is particularly true in matters of our health, such as good hearing, and the importance of hearing loss prevention, especially in the workplace. OSHA has produced remarkable achievements by preventing hearing loss in millions of American workers and by advancing society's appreciation of the physical, psychological, and social impairment and handicap experienced by an individual, as well as his/her family and friends, as a result of occupational noise-induced hearing loss.

Many industries complied promptly with the OSHA Noise Regulation and saved the hearing of millions of workers. Sadly, some industries still seem to have difficulty understanding the importance of OSHA's regulation and the need to implement it, and there are still workers in American industry developing hearing impairments that are completely (and cost effectively) preventable. Fortunately, many of the really noisy jobs have been eliminated or modified. For example, the hearing impact of some very noisy machines has been controlled by isolating the employees or the machines themselves in separate rooms. Where this is not feasible, hearing protection has proven effective. It is now scientifically and economically feasible to prevent occupational hearing loss from developing in practically all industrial employees by implementing an effective and inexpensive hearing conservation program as mandated by OSHA. These guidelines are appropriate not only for industries regulated by OSHA, but also for other industries where compliance is voluntary.

It is difficult to emphasize how much emotional, psychological, and familial strife has been prevented by the federal and state governments' emphasis on the prevention of occupational hearing loss. The results of OSHA's regulations and related efforts are monumental and stand as a great tribute to the scientific and political pioneers, and to the companies enforcing effective hearing conservation programs.

Many of us have helped accomplish all of this, but several pioneers warrant special remembrance for their leadership, commitment, persistence, and excellent results, especially Hallowell Davis, Terry Dear, Aram Glorig, Floyd Van Atta, and the all of the otologists, occupational physicians and nurses, audiologists, industrial hygienists and safety personnel, and hearing conservationists, who have devoted decades to preventing all cases of occupational hearing loss. Our mission is not yet accomplished, but great strides have been made. In retrospect, we should be proud of a job well done, and excited by the innumerable possibilities for even greater advances in the future.

REFERENCES

1. Davis H. The Professional Memoirs of Hallowell Davis. St. Louis, MO: Central Institute for the Deaf, 1991.
2. Rudmose W. Results of Z 24 X2 Committee. American National Standards Institute, 1957.
3. Sataloff J. Industrial Deafness. Blakeston division of the McGraw-Hill Company, 1957.
4. Glorig A. Noise and Your Ear. New York: Grune & Stratton, 1958.

5. American Academy of Ophthalmology and Otolaryngology, Subcommittee on Noise of the AAOO Committee on Conservation of Hearing. Guide for Conservation of Hearing. Trans Am Acad Ophthalmol Otolaryngol 1959; 63:236–238.
6. American Industrial Hygiene Association. Industrial Noise Manual. 2nd ed. Detroit, MI: AIHA, 1966.
7. Occupational Noise Regulation. Federal Register, 1973.
8. Baughn WH. Relation between daily noise exposure and hearing loss based on the evaluation of 6,835 nose exposure cases. Aerospace Medical Research Lab, Wright Patterson AFB, Ohio AMRL-TYR-73-53, June 1973.
9. Burns W, Robinson DS. Hearing and Noise Industry, London: Her Majesty's Stationery Office, 1970.
10. Passchier-Vermeer W. Hearing loss due to steady state broadband noise, Report No. 55. Leiden, the Netherlands: Institute for Public Health, Eng. 1968.
11. Ward WD, Glorig A. Protocol of Inter-Industry Noise Study. J Occup Med 1975; 17:760–770.
12. Yerg R, Sataloff J, Glorig A, Menduke H. Inter Industry Noise Study. J Occup Med 1978; 20(5):351–358.
13. Schori TR, Johnson DL. Inter-Industry Noise Study—Another Look. Sound Vibration. 1979; 161–162.
14. American College of Occupational Medicine Noise and Hearing Committee. ACOM Committee Report. J Occup Med 1989; 31:996.

15
Diagnosing Occupational Hearing Loss

Robert T. Sataloff
Drexel University College of Medicine, American Institute for Voice and Ear Research, and Graduate Hospital, Philadelphia, Pennsylvania, USA

Joseph Sataloff
Thomas Jefferson University, Philadelphia, Pennsylvania, USA

Tracy M. Virag
American Institute for Voice and Ear Research, Philadelphia, Pennsylvania, USA

Occupational hearing loss is a specific disease due to repetitive injury with established symptoms and objective findings. The diagnosis of occupational hearing loss cannot be reached reliably solely on the basis of an audiogram showing high-frequency sensorineural loss and patients' histories that they worked in a noisy plant. Accurate diagnosis requires a careful and complete history, physical examination, and laboratory and audiologic studies. Numerous entities such as acoustic neuroma, labyrinthitis, ototoxicity, viral infections, acoustic trauma (explosion), head trauma, hereditary hearing loss, diabetes, presbycusis, and genetic causes must be ruled out, as they are responsible for similar hearing loss in millions of people who were never employed in noisy industries.

1. DEVELOPMENT OF A NOISE STANDARD

Comprehensive understanding of the nature of occupational hearing loss has been hindered by the difficulties associated with scientific studies in an industrial setting. A brief review of the old literature and an in-depth discussion of the most comprehensive recent studies highlight the complexities of the problem and the clinical and scientific findings that form the basis for the guidelines set forth in this chapter.

In 1952, James H. Sterner, MD, conducted an opinion poll among a large number of individuals working with noise and hearing investigating the maximum intensity level of industrial noise that they considered safe to hearing (Fig. 15.1) (1). The wide range of estimates demonstrated clearly the lack of agreement even among knowledgeable individuals and the futility of any attempt to establish meaningful guidelines by means of such polls.

In 1954, the Z24-X-2 Subcommittee of American Standards Association (now ANSI) published its exploratory report on the relations of hearing loss to noise exposure (2). They concluded that on the basis of available data, they could not establish a "line" between safe and unsafe noise exposure. They presented questions that required answers before criteria could be formulated, such as: (a) What amount of hearing loss constitutes a sufficient handicap to be considered undesirable? (b) What percentage of workers should a standard be designed to protect? The report emphasized the need for considerably more research before "safe" intensity levels could be determined.

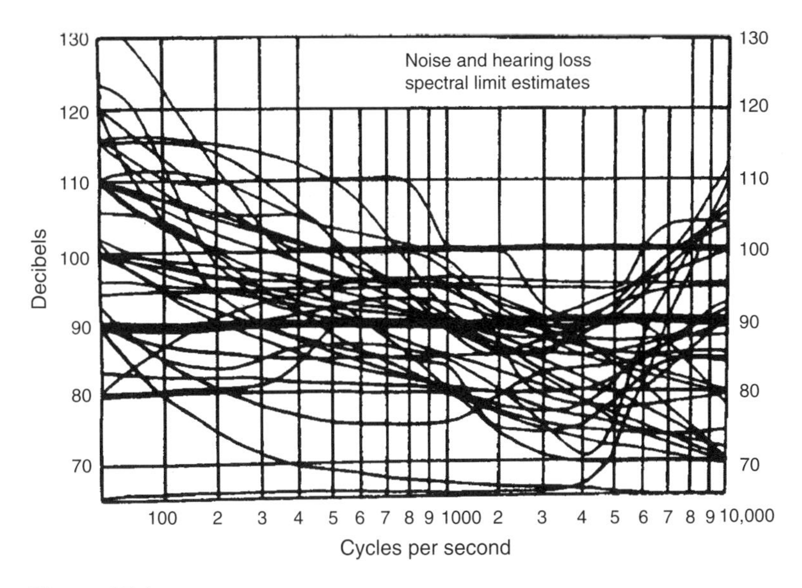

Figure 15.1

Many authors between 1950 and 1971 proposed damage risk criteria, only some of which were based on stated protection goals. Articles are referenced in Table IX of the National Institute of Occupational Safety and Health's (NIOSH's) Criteria for Occupational Exposure to Noise (3). All these reports had limitations that precluded the adoption of any one of them as a basis for the establishment of standards. In 1973, Baughn (4) published an analysis of 6835 audiograms from employees in an automobile stamping plant, with employees divided into three groups on the basis of estimated intensity of noise exposure. Its validity as the basis for a national noise standard was seriously questioned by Ward and Glorig (5) and others because of shortcomings of non-steady-state noise exposures, vague estimates of noise dosage, auditory fatigue, and test room masking. Baughn's raw data were never made available to the Secretary of Labor's Advisory Committee for Noise Standard despite a formal request from that group.

A study by Burns and Robinson (6) avoided many of the deficiencies of previous studies but was based on a very small number of subjects exposed to continuous steady-state noise, particularly in the 82–92 dBA range. Workers were included who "change position from time to time using noisy hand tools for fettling, chipping, burnishing, or welding"—hardly continuous or steady state. Their report admitted to the inclusion of workers exposed to non-steady-state levels below 90 dBA. In fact, some workers were included whose noise exposure range varied by 15 dBA. The oft-quoted Passchier-Vermeer report (7) was not based on an actual field investigation but was rather a review of published studies up to 1967. Some of these studies addressed the validity of measuring sound levels in dBA; none was really designed to be used as the basis for a noise standard.

As early as 1970, interested individuals from industry, labor, government, and scientific organizations discussed the concept of an Inter-Industry Noise Study. The project was started in 1974 for the stated purpose of gathering data on the effect of steady-state noise in the range of 82–92 dBA. While the results of such a study would obviously be of interest to those involved in noise regulation, the basic purpose of the study was scientific rather than regulatory. The detailed protocol has already been published (5) and will not be repeated here. Some of the important points are: (a) clear definitions of the temporal and spectral characteristics of the noise; (b) noise exposures must fall between 82 and 92 dBA, with no subject exceeding a 5-dBA range (later modified to a 6-dBA range); (c) noise environment must be steady state throughout a full shift, with few, if any, sharp peaks of impact noise; (d) subjects, both experimental and control, must include men and women; (e) no prior job exposure to noise greater than 92 dBA for experimentals and 75 dBA for controls; (f) minimum of 3 years on present job; (g) all audiometric testing, noise measurement, equipment calibration, otological examinations, histories, and data handling to be done in a standardized manner, as detailed in the protocol; and (h) the original raw data to be made available to all serious investigators upon request at the conclusion of the study. Hearing levels were measured in 155 men and 193 women exposed to noise levels ranging from 82 to 92 dBA for at least 3 years, with a median duration of approximately 15 years; they were also measured in 96 men and 132 women with job exposure that did not exceed 75 dBA. Noise exposure was considered steady state in that it did not fluctuate more than ± 3 dB from the midpoint as of the time of the first audiogram. As many subjects as possible were reexamined 1 year later and 2 years later.

Jobs involving some 250,000 employees were examined to find the 348 experimental subjects who met the criteria of the Inter-Industry Noise Study as of the time of entry. Within the range of 82–92 dBA, differences in noise intensity had no observable "effect" on hearing level. That is, the hearing levels of workers at the upper end of the noise intensity exposure were not observably different from the hearing levels of workers at the lower end of the noise exposure. Age was a more important factor than duration on the job in explaining differences in hearing level within any group. Comparisons between experimental and control subjects were made on an age-adjusted basis.

Differences between women exposed to 82–92 dBA and their controls were small and were not statistically significant. Differences between men exposed to 82–92 dBA and their controls were small and were not statistically significant at 500, 1000, and 2000 Hz. Levels in the noise-exposed group significantly exceeded those in the control group at 3000, 4000, and 6000 Hz by approximately 6–9 dB. At 8000 Hz, differences again became not significant.

There was no real evidence of a difference between noise-exposed workers and their controls with respect to the changes in hearing level during the course of their follow-up 1 and 2 years after initial audiograms. Changes were negligible for both groups.

It is important to note that the studies discussed and the regulations promulgated to date concern themselves with exposure to continuous noise. Recent reports demonstrate that intermittent exposure to noise results in different effects on hearing (8). Although it may produce marked, high-frequency, sensorineural hearing loss, it does not have the same propensity to spread to the speech frequencies even after many years of exposure, as is seen with continuous noise exposure.

2. AUDIOMETRIC FEATURES

The American College of Occupational Medicine Noise and Hearing Conservation Committee promulgated a position statement on the distinguishing features of occupational noise-induced hearing loss (9). This statement summarizes the currently accepted opinions of the medical community regarding diagnosis of occupational hearing loss. The American Occupational Medicine Association (AOMA) Committee defined occupational noise-induced hearing loss as a slowly developing hearing loss over a long time period (several years) as the result of exposure to continuous or intermittent loud noise. The committee stated that the diagnosis of noise-induced hearing loss is made clinically by a physician and should include a study of the noise exposure history. It also distinguished occupational hearing loss from occupational acoustic trauma, an immediate change in hearing resulting from a single exposure to a sudden burst of sound, such as an explosive blast. The committee recognized that the principal characteristics of occupational noise-induced hearing loss are as follows:

1. It is always sensorineural, affecting the hair cells in the inner ear.
2. It is almost always bilateral. Audiometric patterns are usually similar bilaterally.
3. It almost never produces a profound hearing loss. Usually, low-frequency limits are about 40 dB and high-frequency limits about 75 dB.
4. Once the exposure to noise is discontinued, there is no substantial further progression of hearing loss as a result of the noise exposure.
5. Previous noise-induced hearing loss does not make the ear more sensitive to future noise exposure. As the hearing threshold increases, the rate of loss decreases.
6. The earliest damage to the inner ears reflects a loss at 3000, 4000, and 6000 Hz. There is always far more loss at 3000, 4000, and 6000 Hz than at 500, 1000, and 2000 Hz. The greatest loss usually occurs at 4000 Hz. The higher and lower frequencies take longer to be affected than the 3000–6000 Hz range.
7. Given stable exposure conditions, losses at 3000, 4000, and 6000 Hz will usually reach a maximal level in about 10–15 years.
8. Continuous noise exposure over the years is more damaging than interrupted exposure to noise, which permits the ear to have a rest period.

On October 27, 2002, the American College of Occupational and Environmental Medicine (ACOEM) issued an evidenced-based statement regarding noise-induced hearing loss which can be viewed on the ACOEM website: http://www.acoem.com/guidelines/article.asp?ID=53. In this statement, minor revisions in the criteria were suggested. The current language of the ACOEM criteria about noise-induced hearing loss is as follows:

- It is always sensorineural, affecting hair cells in the inner ear.
- Since most noise exposures are symmetric, the hearing loss is typically bilateral.
- Typically, the first sign of hearing loss due to noise exposure is a "notching" of the audiogram at 3000, 4000, or 6000 Hz (Hertz), with recovery at 8000 Hz. The exact location of the notch depends on multiple factors including the frequency of the damaging noise and the length of the ear canal. Therefore, in early noise-induced hearing loss, the average hearing thresholds at 500, 1000, and 2000 Hz are better than the average at 3000, 4000, and 6000, and the hearing level at 8000 Hz is usually better than the deepest part of the "notch." This "notching" is in contrast to age-related hearing loss, which also produces high frequency hearing loss, but in a down-sloping pattern without recovery at 8000 Hz.

- Noise exposure alone usually does not produce a loss greater than 75 decibels (dB) in high frequencies, and 40 dB in lower frequencies. However, individuals with superimposed age-related losses may have hearing threshold levels in excess of these values.
- The rate of hearing loss due to chronic noise exposure is greatest during the first 10 to 15 years of exposure and decreases as the hearing threshold increases. This is in contrast to age-related loss, which accelerates over time.
- Most scientific evidence indicates that previously noise-exposed ears are not more sensitive to future noise exposure and that hearing loss due to noise does not progress (in excess of what would be expected from the addition of age-related threshold shifts) once the exposure to noise is discontinued.
- In obtaining a history of noise exposure, the clinician should keep in mind that the risk of noise-induced hearing loss is considered to increase significantly with chronic exposures above 85 dBA for an 8-hour time-weighted average (TWA). In general, continuous noise exposure over the years is more damaging than interrupted exposure to noise which permits the ear to have a rest period. However, short exposures to very high levels of noise in occupations such as construction or firefighting may produce significant loss, and measures to estimate the health effects of such intermittent noise are lacking. When the noise exposure history indicates the use of hearing protective devices, the clinician should also keep in mind that the real world attenuation provided by hearing protectors may vary widely between individuals.

The ACOEM document also discusses additional considerations in the evaluation of workers with suspected noise-induced hearing loss, addressing several issues such as exposure to noise sources that affect the ear asymmetrically; co-exposure to ototoxic agents; spreading of hearing loss to frequencies adjacent to those first affected; individual problems that may be associated with hearing loss; early recognition of noise-induced hearing loss; and age-correction. The document also mentions research priorities and evaluation of the effectiveness of a hearing conservation program.

2.1. Sensorineural Hearing Loss

Habitual exposure to occupational noise damages the hair cells in the cochlea, causing a sensory hearing loss. No damage to the outer or middle ear (conductive loss) can be caused by routine daily exposure to loud industrial noise. Ultimately, some of the nerve fibers supplying the damaged hair cells may also become damaged from many causes and result in a neural loss of hearing, as well.

2.2. Bilaterality of Occupational Hearing Loss

Both ears are equally sensitive to temporary threshold shift (TTS) and permanent threshold shift (PTS) hearing loss due to free-field occupational noise exposure and, therefore, damage is equal or almost equal in both ears. If an employee working in a very noisy environment develops substantial one-sided nerve deafness, it is essential to find the cause and to rule out an acoustic neuroma, which commonly presents as unilateral sensorineural deafness. In weapons and range fire, the ear nearest the stock (left ear in a right-handed shooter) sustains damage before and to a somewhat greater degree than the other ear; however, a loss will generally be present to some degree bilaterally.

2.3. The 4000-Hz Audiometric Dip

Noise-induced hearing loss has been shown to be sensorineural in nature and usually bilateral; it also contains a characteristic frequency response known as the 4000-Hz dip. Figure 15.2 shows a composite audiogram of the classic progress of many cases of occupational hearing loss. This pattern is actually more common in hearing loss caused by gunfire, but exposure to continuous noise, such as in weaving mills, some metal plants, etc., also produces this pattern in which the earliest damage occurs between 3000 and 6000 Hz. Some noise sources, such as papermaking machines, can damage the 2000 Hz frequency somewhat before the higher frequencies, while noise exposures such as chipping and jackhammers characteristically damage the higher frequencies severely before affecting the lower ones. However, in general, the frequencies below 3000 Hz are almost never damaged by occupational noise without earlier damage to the higher frequencies.

It has been known for many years that prolonged exposure to high-intensity noise results in sensorineural hearing loss that is greatest between 3000 and 6000 Hz. In such cases, the classic audiogram shows a 4000-Hz dip in which hearing is better at 2000 and 8000 Hz (Fig. 15.3). Unfortunately, the fact that noise produces this 4000-Hz dip has led some physicians to assume that any comparable dip is produced by noise. This error can lead to misdiagnosis and can result in undesirable medical and legal consequences.

Although there are numerous hypotheses that attempt to explain the 4000-Hz dip in noise-induced hearing loss (10–13), its pathogenesis remains uncertain. However, it is known that in most cases this loss initially affects hearing between 4000 and 6000 Hz and then spreads to other frequencies (14,15). Frequencies higher than those usually measured clinically may be tested on special audiometers and are helpful in diagnosing

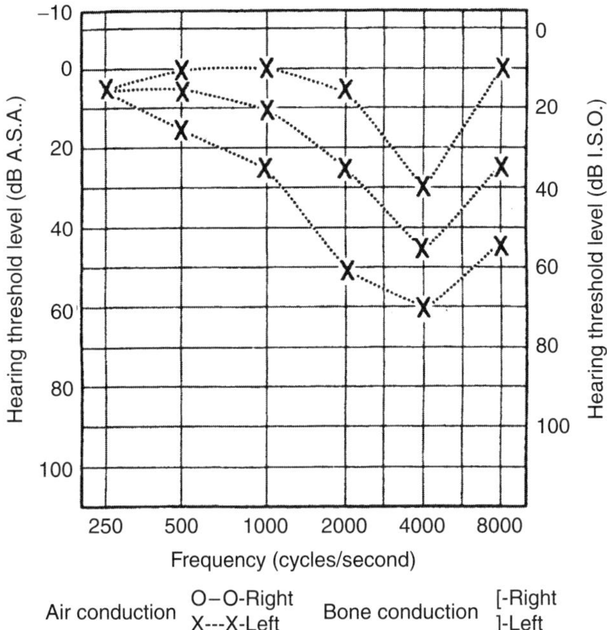

Figure 15.2 Series of audiometric curves showing a "classic" progressive loss that may be found in employees with excessive noise exposure.

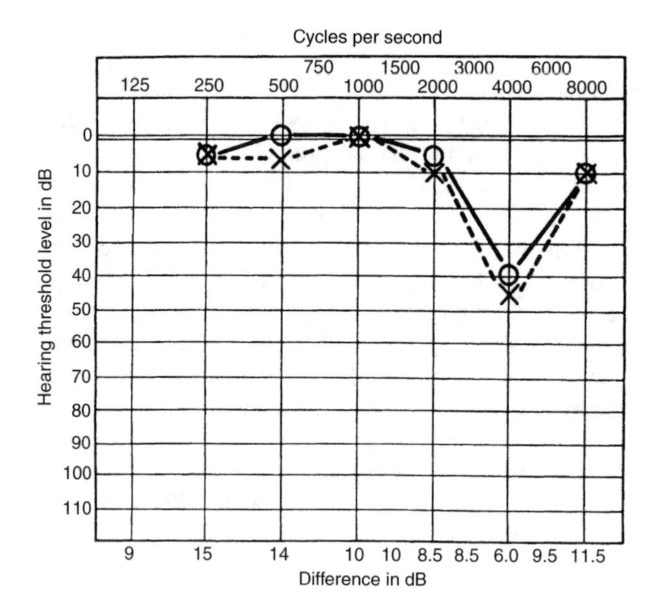

Figure 15.3 Typical 4000 Hz dip. In this audiogram and in all other audiograms in this chapter, bone conduction equals air conduction.

noise-induced hearing loss in selected cases (16). This hearing loss may result from steady-state or interrupted noise, although the intensities required to produce comparable hearing losses differ (17), and controversy exists as to the nature of the actual cochlear damage (17–20). Other types of acoustic trauma, such as that from blast injuries, may result in other audiometric patterns or in a 4000-Hz dip, but they will not be considered in this discussion.

2.4. Discrimination Scores

In almost all cases of occupational hearing loss where the high frequencies are affected (even severely), the discrimination scores are good (over 85%) in a quiet room. If patients have much lower discrimination scores, another cause in addition to occupational hearing loss should be suspected.

2.5. Gradual Hearing Loss with Early Onset

In addition to having the characteristics of a bilateral sensorineural hearing loss with a 4000-Hz dip, occupational hearing loss begins early with noise exposure and progresses gradually. Sudden deafness is not caused by noise that patients are exposed to regularly at their job: there are, of course, incidents of unilateral sudden deafness due to acoustic trauma from an explosion or similar circumstance. Other causes must be sought in sudden deafness in one or both ears regardless of occupational noise exposure.

Occupational hearing loss characteristically develops in the first few years of exposure and may worsen over the next 8–10 years of continued exposure, but the damage does not continue to progress rapidly or substantially with additional exposure beyond 10–15 years. Rarely will an employee working in consistent noise have good hearing for 4 or 5 years and then develop progressive hearing loss from occupational causes. Employees who retire after age 60 and develop additional hearing loss without continued noise exposure generally should not attribute this to their past jobs (21). The same pertains to employees who wear hearing protectors effectively and either develop hearing loss or have additional hearing loss.

2.6. Asymptotic Hearing Loss

Another characteristic of occupational hearing loss is that specific noisy jobs produce a maximum degree of hearing loss. This has been called asymptotic loss. For example, employees using jackhammers develop severe high-frequency, but minimal low-frequency, hearing losses. Employees working for years in about 92 dBA generally do not have over 20-dB losses in the low frequencies, and once they reach a certain degree of high-frequency hearing loss, little additional loss occurs. Many employees exposed to weaving looms experience a maximum of about a 40-dB loss in the speech frequencies, but they rarely have greater losses. If an employee shows a loss much greater than is typical for similar exposure, the otologist should suspect other causes.

3. LIMITATIONS OF THE AUDIOGRAM

An audiogram showing a 4000-Hz dip is not, by itself, sufficient evidence to make a diagnosis of noise-induced hearing loss. In order to do so, one must have at least a history of sufficient exposure to noise of adequate intensity to account for the hearing loss. In the absence of this history, or with a history and findings suggestive of another origin, a thorough investigation must be done to establish the true cause of the hearing loss. It must be understood that it is not always possible to ascribe a hearing loss to a noise or to completely rule out other causes. If, however, the patient's noise exposure has been sufficient, and if investigation fails to reveal other causes of hearing loss, a diagnosis of noise-induced hearing loss can be made with reasonable certainty in the presence of supportive audiometric findings.

3.1. Other Causes of the 4000-Hz Dip

3.1.1. Viral Infections

It is well known that viral upper-respiratory infections may be associated with hearing loss, tinnitus, and a sensation of fullness in the ears. This fullness is frequently due to inner ear involvement, rather than middle ear dysfunction. Viral cochleitis may also produce either temporary or permanent sensorineural hearing losses, which can have a variety of audiometric patterns, including a 4000-Hz dip (Fig. 15.4) (22). In addition to viral respiratory infections as causes of sensorineural hearing loss, rubella, measles, mumps, cytomegalic inclusion disease, herpes, and other viruses have been implicated (Fig. 15.5).

3.1.2. Skull Trauma

Severe head trauma that results in fracture of the cochlea produces profound or total deafness. However, lesser injuries to the inner ear may produce a concussion-type injury, which may be manifested audiometrically as a 4000-Hz dip. Human temporal

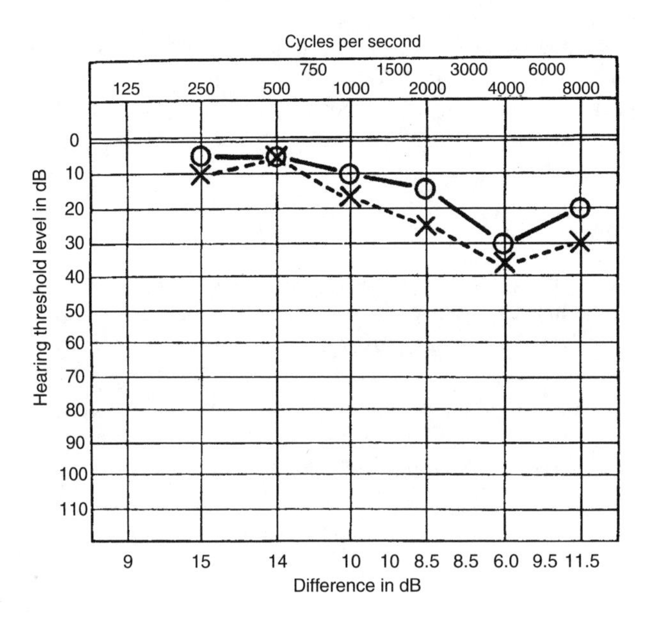

Figure 15.4

bone pathology in such cases is similar to that seen in noise-induced hearing loss (23). Similar findings also can be produced by experimental temporal-bone injury (13).

3.1.3. Hereditary (Genetic) Hearing Loss

Hereditary sensorineural hearing loss results commonly in an audiometric pattern similar to that of occupational hearing loss (24–26). This may be particularly difficult to diagnose, since hereditary deafness need not have appeared in a family member previously; in fact,

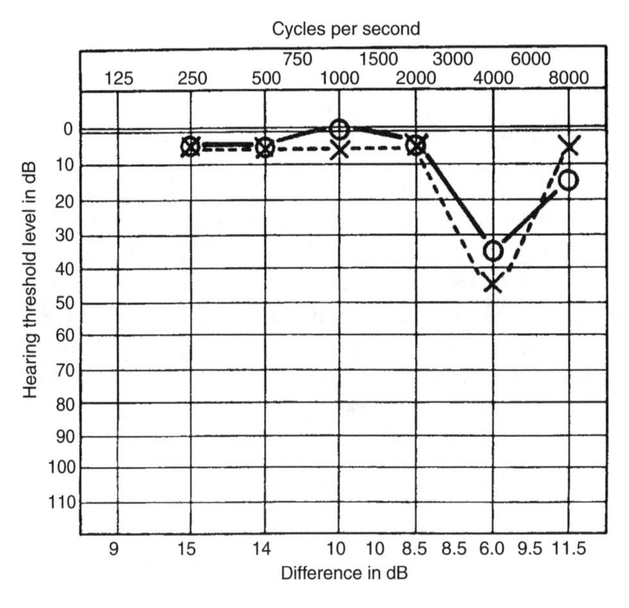

Figure 15.5

many cases of hereditary hearing loss follow an autosomal-recessive inheritance pattern. There have been new developments in identification of genes associated with hereditary hearing loss, making it possible currently to identify specifically some forms of genetic hearing loss.

3.1.4. Ototoxicity

The most commonly used ototoxic drugs at present are aminoglycoside antibiotics, diuretics, chemotherapeutic agents, and aspirin (in high doses) (see Table 10.2). When toxic effects are seen, high-frequency sensorineural hearing loss is most common and profound deafness may result, although a 4000-Hz dip pattern also may be seen (Fig. 15.6) (20). Unlike damage caused by the other ototoxic drugs listed above, aspirin-induced hearing loss usually is only temporary, recovering after cessation of the medication.

3.1.5. Acoustic Neuroma

Eighth-nerve tumors may produce any audiometric pattern, from that of normal hearing to profound deafness, and the 4000-Hz dip is not a rare manifestation of this lesion (Figs. 15.7 and 15.8) (27). In these lesions, low speech discrimination scores and pathological tone decay need not be present and cannot be relied on to rule out retrocochlear pathology. Nevertheless, asymmetry of hearing loss should arouse suspicion even when a history of noise exposure exists. There are several cases in which patients were exposed to loud noises producing hearing losses that recovered in one ear but not in the other because of underlying acoustic neuromas (M.D. Graham, personal communication).

3.1.6. Sudden Hearing Loss

Each year, clinicians see numerous cases of sudden sensorineural hearing loss of unknown origin. Although the hearing loss is usually unilateral, it may be bilateral and show a

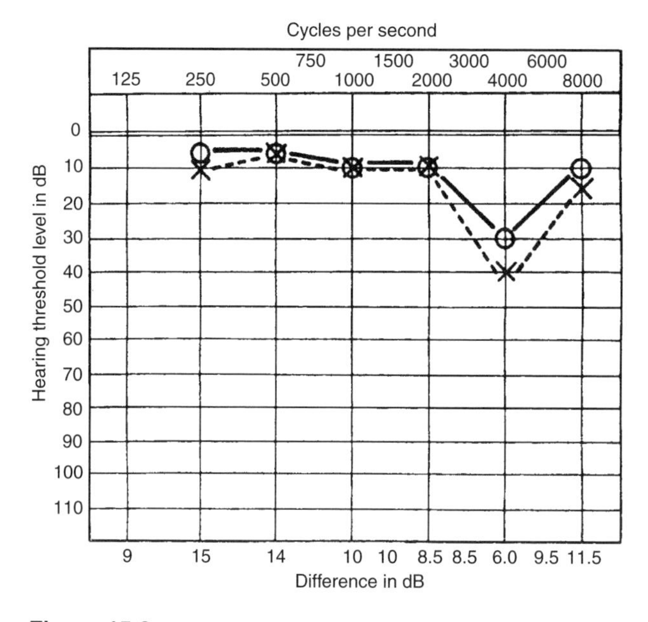

Figure 15.6

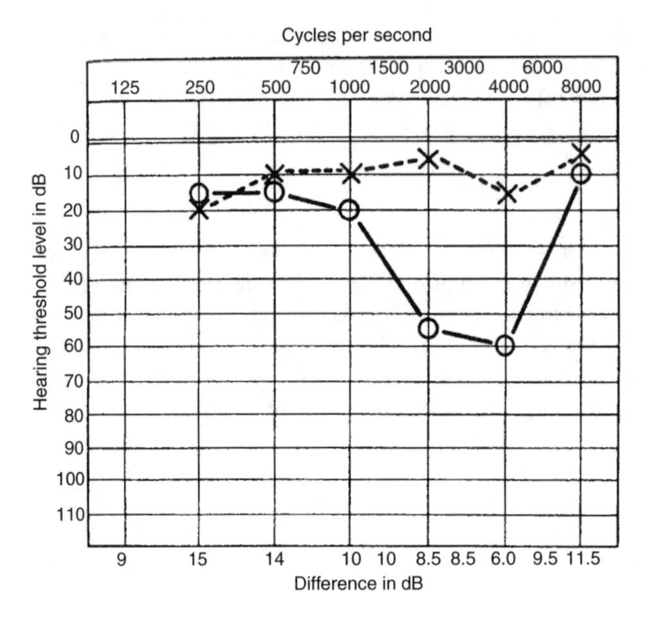

Figure 15.7

4000-Hz dip. This audiometric pattern may also be seen in patients with sudden hearing loss due to inner-ear-membrane breaks (28,29) and barotrauma (30,31).

3.1.7. Multiple Sclerosis

Multiple sclerosis also can produce sensorineural hearing loss that may show almost any pattern and may fluctuate from severe deafness to normal threshold levels. An example demonstrating a variable 4000-Hz dip is shown in Fig. 15.9.

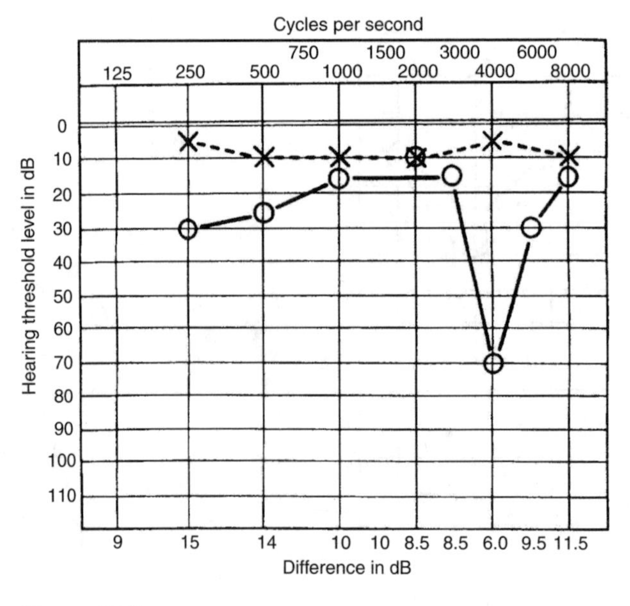

Figure 15.8

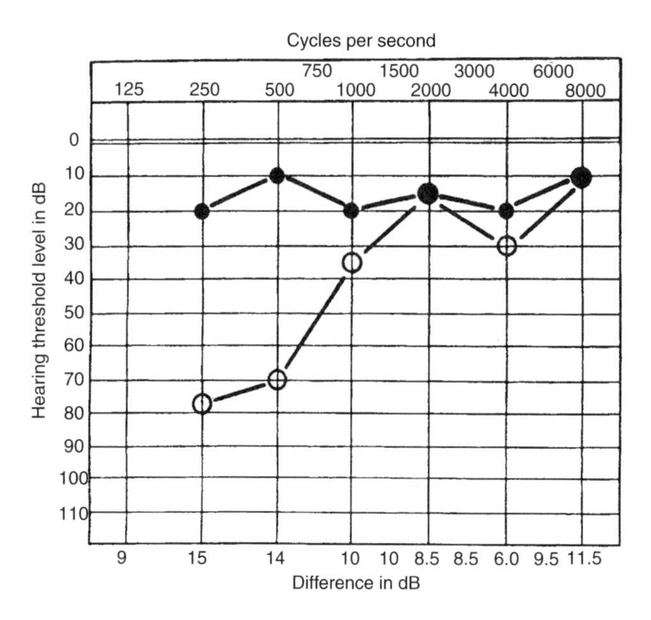

Figure 15.9

3.1.8. Other Causes

A variety of other causes may produce audiograms similar to those seen in noise-induced hearing loss. Such conditions include bacterial infections such as meningitis, a variety of systemic toxins (32), and neonatal hypoxia and jaundice. Figure 15.10 illustrates one such sensorineural hearing loss that resulted from kernicterus due to Rh incompatibility.

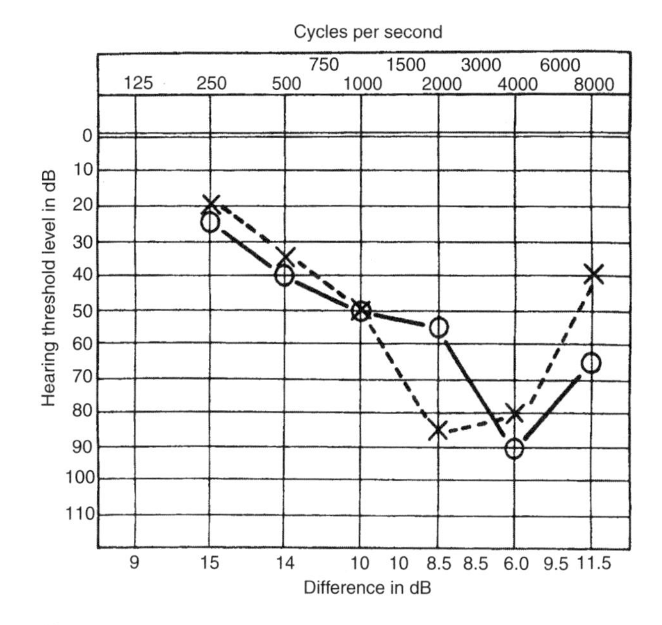

Figure 15.10

4. NOISE EXPOSURE HISTORY

It is important to recall that sound of a given frequency spectrum and intensity requires a certain amount of time to produce hearing loss in most subjects. While the necessary exposure varies from person to person, a diagnosis of noise-induced hearing loss requires a history of sufficient noise exposure. Guidelines for estimating how much noise is necessary to cause hearing loss in most people have been established by the scientific community and the federal government and are reviewed in this chapter. However, a reasonable assessment of patients' occupational noise exposure cannot be obtained solely from their histories, especially if compensation is a factor.

Patients who have worked for many years on weaving looms, papermaking machines, boilers, sheet metal, riveters, jackhammers, chippers, and the like nearly always have some degree of occupational hearing loss. However, many other patients have marked hearing losses that could not possibly have been caused by their minimal exposures to noise. Almost all patients working in industry can claim that they have been exposed to a great deal of noise. It is essential, especially in compensation cases, to get more accurate information by obtaining, if possible, a written work history and time-weighted average of noise exposure from the employer. If a physician does not have firsthand knowledge of the noise exposure in a patient's job, definitive diagnosis should be delayed until such information is made available.

Some publications (5,8) have perpetuated the idea that exposures below 90 dBA can produce handicapping hearing losses in the speech frequencies. A critical review of the most quoted publications (33,34) reveals that all these reports contain serious shortcomings casting considerable doubt on their conclusions. The Inter-Industry Noise Exposure studies are undoubtedly the best-conducted and monitored research projects relating hearing loss and noise exposure, but even these authors emphasize the need for additional valid and reliable research.

The 85- and 90-dBA noise exposure levels designated by the Occupational Safety and Health Administration (OSHA) are the levels at which initiation of a hearing conservation program and use of hearing protection is recommended. They are not necessarily the levels at which hearing becomes impaired in the speech frequencies even after years of exposure. Individuals who have handicapping hearing loss in their speech frequencies and are habitually exposed to less than 90 dBA probably have hearing losses from other causes. These losses have developed regardless of their jobs. It is important to find the specific causes for their hearing losses rather than make misleading, unjustified, and hasty diagnoses of occupational hearing loss.

The term biological hypersensitivity to noise is often misused and requires clarification. Many physicians and attorneys have attributed patients' substantial sensorineural hearing loss to hypersensitivity to noise, even though the exposure was 85 dBA or less. There is no basis for such an opinion. Prolonged exposure to this type of noise level will not cause handicapping hearing loss in the speech frequencies. Biological hypersensitivity to noise does not mean that individuals exposed to mild levels below 90 dBA can sustain substantial hearing losses, but rather that in a group of employees habitually exposed to very loud noise (over 95 dBA) without hearing protectors, a few will have little or no hearing loss (so-called "hard ears"), most will have a fair amount of loss, and a few will sustain substantially greater losses because they are hypersensitive.

Many years of otologic studies and clinical experience have demonstrated certain symptoms and findings that are characteristic of occupational hearing loss. For instance, we know that employees do not suffer total or very severe sensorineural deafness in the speech frequencies even if they work for years in the loudest industrial noise areas.

Several explanations have been proposed for this observation, for example: "the nerve-deafened ear acts as a hearing protector" and "what you don't hear doesn't hurt you." Even when noise exposure is very high and undoubtedly a contributing cause, all patients with severely handicapping losses in the speech frequencies should be studied carefully to find the underlying etiology.

Otologic history should include use, duration, effectiveness of hearing protection, type of noise exposure (including continuous or intermittent), dosage of exposure (daily hours and years), and presence of recreational noise exposure such as target practice, trap shooting, hunting, snow-mobile use, motorcycling, chain saw or power-tool use, etc. Recreational exposure may contribute to noise-induced hearing loss. Employees should be advised to use hearing protectors during recreational exposure to loud noise. Infrequent exposures and intermittent exposures are far less hazardous than continuous daily exposures. It seems that if the ear has sufficient rest periods, damage to the speech frequencies is minimized.

5. HISTOPATHOLOGY OF NOISE-INDUCED HEARING LOSS

Histologic studies of human inner ears damaged by noise reveal diffuse degeneration of hair cells and nerves in the second quadrant of the basal turn of the cochlea—the area sensitive to 3000–6000 Hz sounds (Figs. 15.11–15.13) (20). Similar findings have been demonstrated in cochlear hair cells and first-order neurons in experimental animals exposed to loud noises, as discussed below. Further experimental studies in rodents have shown noise-induced injury to the stria vascularis as well (36), but there is some question as to the applicability of this finding to clinical medicine.

Comprehensive discussion of the histopathology of noise-induced hearing loss is beyond the scope of this book. The subject is controversial and many questions remain unanswered. However, certain recent observations are of particular clinical interest and worthwhile reviewing, particularly those of Professor H. Spoendlin, of Innsbruck, Austria (36).

The psychophysical effects of sound stimulation at various intensities include the following:

1. Adaptation is an immediate and rapidly reversible threshold shift proportional to the sound intensity at the frequency of stimulation.

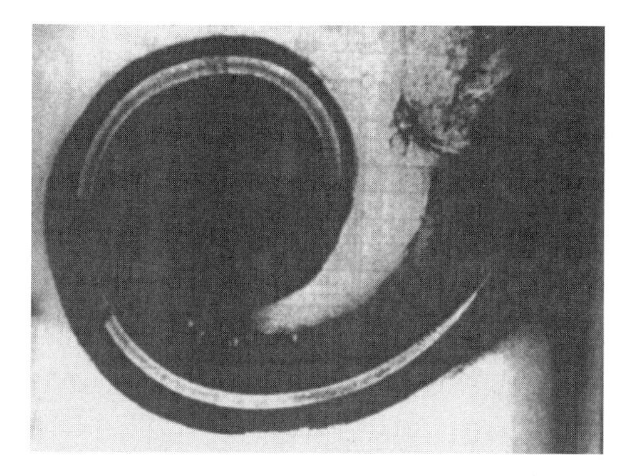

Figure 15.11

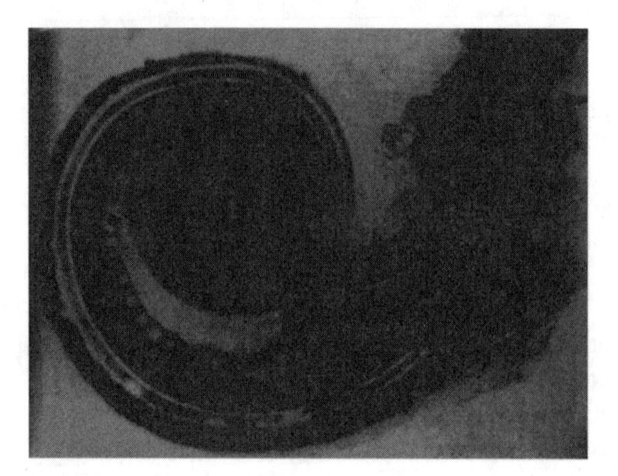

Figure 15.12

2. Temporary threshold shift (TTS) is pathological, metabolically induced fatigue. Its development and recovery are proportional to the logarithm of exposure time. It reverses slowly over a period of hours.
3. Permanent threshold shift (PTS), such as occupational hearing loss, develops by exposure to excessive noise for sufficiently long period of time.

The extended exposure time results in the destruction and eventual loss of the cells of the organ of Corti. Two mechanisms appear to be involved in this process:

1. In high-intensity noise exposure, there may be direct mechanical destruction.
2. In exposure to moderately intense noise, there is a metabolic decompensation with subsequent degeneration of sensory elements.

If a cochlea is examined shortly after even short-term exposure to extremely intense noise, it will show entire absence of the organ of Corti in the epicenter of the injured area. Moving laterally from the epicenter, the hair cells are swollen and severely distorted with cytoplasmic organelle displacement. More laterally, the cells reveal bending of sensory hairs. Further from the epicenter, but still in the area of damage, one finds only slight distortion of the outer hair cells. In addition to sensory damage, dislocation of the

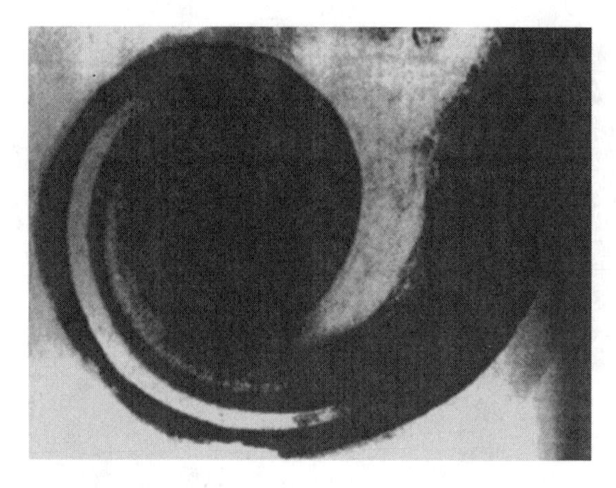

Figure 15.13

tympanic lamina cells, disruption of the heads of the pillar cells, holes in the basilar membrane, and other findings may be found.

After exposure to moderately intense acoustic stimuli, the nuclei of outer hair cells become extremely swollen. Swelling is also seen in the terminal unmyelinated portion of the afferent nerve fibers to the inner hair cells. This pathological condition of afferent nerve fibers is also seen in hypoxia. In both instances, it appears to result from metabolic derangement. Degeneration of mitochondria and alteration of the synaptic vesicles may be observed in the efferent nerve endings of the outer hair cells after long exposure (37).

Temporary threshold shift appears to cause only an increase in number and size of lysosomes, primarily in the outer hair cells, a finding probably also related to increased metabolic activity.

The localization of damage depends on the type of noise exposure. Exposure to white noise, multifrequency noise encountered under most industrial circumstances, usually produces damage in the upper basal turn of the cochlea, the 3000–6000 Hz frequency range in humans. Narrow-band noise causes damage in different areas depending on the frequency of the noise; the extent of the damage increases with increasing intensity of the center frequency. These observations are true for continuous noise; impulse noise results in much more variability in site of damage. The rise time of the impulse appears to be an important factor in the amount of hearing damage. Histologically, square-noise impulses appear to produce less damage than impulses with a gradual rising time of 25 ms (36). The practical implications of this finding remain unclear.

The histological progression of cochlear injury over time is also of interest. Some of the mechanically or metabolically induced structural changes are reversible, whereas others progress to degeneration. Metabolically induced changes such as swollen nuclei in the outer hair cells and swollen nerve endings usually reverse, although scattered degeneration occurs. Severe distortion of cells often leads to degeneration and membrane ruptures, which may result in the disappearance of the organ of Corti in the area of injury. In places where the organ of Corti is completely destroyed, the cochlear neurons undergo slow, progressive retrograde degeneration over months. Eventually, about 90% of the cochlear neurons associated with the injured area disappear, including their spiral ganglion cells of origin. This retrograde neural degeneration usually is not observed when only outer hair cells are missing (38). No significant neural degeneration has been noted in adjacent areas. Thus, although histopathological changes progress even after acoustic stimulation has stopped, significant degeneration occurs only in regions where substantial hair cell destruction (hence, substantial hearing loss) has already occurred. Consequently, progressive histological changes do not imply progressive clinical hearing deterioration.

Histologically, maximal damage to the cochlea secondary to noise exposure is never total. Even under laboratory circumstances, there are always some sensory elements preserved in the cochlea. Moreover, for each exposure intensity, "saturation damage" occurs after a certain time. The time interval is short for high-intensity noise exposure and long for low-intensity noise exposure. Additional exposure to noise of the same intensity does not produce additional observable damage beyond this "saturation damage" limit. This histological finding corresponds to asymptotic hearing loss. Moreover, even under laboratory conditions, there is a great variability of cochlear damage produced by controlled noise exposure, especially following stimulation by moderately intense noise or impulse noise. These are the conditions encountered most commonly in an industrial environment. This histological and experimental evaluation corresponds to the biological variability (hard ears and soft ears) observed in industrial populations.

6. OCCUPATIONAL NOISE EXPOSURE DURING PREGNANCY

During approximately the last decade, questions have been raised about whether significant noise exposure during pregnancy has any adverse effect upon the fetus. While definitive answers have not been established, some literature suggests that there may be an effect. Murato et al. looked at the effects of noise and cadmium in mice (39). They found (not surprisingly) that cadmium was teratogenic; and there was no synergistic effect between cadmium and noise. In the groups exposed to continuous noise for 6 hours, total percentages of malformed fetuses were significantly higher than those in the control group, but there were no significant increases in total percentages of fetal malformations in the combined treatment groups in comparison with the groups given the same dose of cadmium alone. The authors speculated that the magnitude of the teratogenicity due to noise might be so much weaker than that of cadmium that the effects were masked statistically. However, one must also consider the possibility that there was no effect. Nurminen suggested the possibility that noise-induced stress in the mother might cause reproductive disturbances (40). It has been noted that high noise levels and shift work (especially rotating schedules that interfere with circadian rhythm) may have negative effects on both weight and length of gestation, and that there have been suggestions that noise exposure and shift work correlated with menstrual disturbance and infertility. However, although Nurminen suggested that it would be prudent to consider high noise levels and shift work as reproductive risks, no evidence was presented to confirm or refute the hypothesis. Schell evaluated data from 115 births to determine whether noise exposure had an effect upon birth weight and gestation length (41). In females, there was a significant effect. In males, the effect did not quite reach the statistical significance. The author suggested that noise may reduce prenatal growth. Lelande et al. (42) examined the hearing of 131 children whose mothers had worked while pregnant with that child, in noise ranging from 65 to 95 dB over 8-h periods. They found a threefold increase in the risk of children having high-frequency hearing loss when their mothers had been exposed to noise in the range of 85–95 dBA, and a significant increase in the risk of hearing loss at 4000 Hz when maternal exposures had involved a "strong component" of low-frequency noise. While these papers certainly raise interesting questions, confirmatory data have been scarce. Further study will be necessary to determine with certainty whether there is any fetal risk associated with maternal noise exposure.

7. COFACTORS ASSOCIATED WITH OCCUPATIONAL HEARING LOSS

7.1. Non-Occupational Noise Exposure

Habitual exposure to loud rock-and-roll and amplified music can produce hearing damage between 2000 and 8000 Hz. Occasional exposure, however, can be annoying to unaccustomed listeners, but does not cause significant hearing damage in most cases. Sataloff et al. found potentially damaging noise levels at audience locations at a concert by the band Metallica™ (43). Peak levels reached 140 dBA and audiodosimetry measurements revealed loudness equivalent levels between 95 and 111 dBA. In a separate preliminary study, Sataloff et al. looked at noise levels in two movie theaters (44). They found maximum sound pressure levels of approximately 97 dBA, but exposures over time were safe. Household noises such as vacuum cleaners, fans, air conditioners, etc., generally do not damage hearing even though they may be disturbing. Despite some parental arguments to the contrary, it also appears that prolonged telephone use does not impair hearing. Mendes et al. (45) examined the hearing of 50 female telephone operators between 35 and 50 years of age who had been using ear phones at work for 6-h shifts for at least 10 years. They compared these subjects with an age-matched control group of 291 women who had not been exposed to ear phone noise and whose telephone use was typical of that practiced by the general public. The control group also had not been exposed to noise levels in excess of 85 dB. The authors found that even telephone operators using ear phones had no greater risk of hearing loss than the control group (45).

Exposure to ultrasonic noise, such as in certain commercial cleaners, does not affect hearing in the usually recorded frequencies (up to 8000 Hz). Community noises such as trolley cars, airplanes, noises from industrial plants, sirens, etc., also ordinarily do not cause measurable hearing damage. Yaremchuk et al. wrote a particularly interesting paper on noise hazards associated with commercially available toys (46). Guidelines for noise production of toys are regulated by the Voluntary Product Standards, PS 72-76, the Toy Safety Act of 1969. Yaremchuk and her coauthors evaluated 25 toys purchased at a national toy store chain and measured sound pressure levels at distances approximating ear level (2.5 cm) and a child's arm length (25 cm) from the surface of the toy. They found peak sound levels ranging from 81 to 126 dBA at 2.5 cm, and from 80 to 115 dBA at 25 cm. Some of their findings are summarized in Tables 15.1–15.3. They found that at ear level, seven of the 25 toys tested exceeded the 115 dBA OSHA-permissible noise level for

Table 15.1 Preliminary Measure Peak dBA of Three Trials with Portable SLM in Double-Wall Sound Booth

Toys	dBA 2.5 cm	dBA 25 cm
Bikehorn #1	120	108
Bikehorn #2	122	110
Headset portable radio (auto volume limiter)	81	NA
Guitar	106	92
Boom box music	112	106
Laser gun	114	96
Army gun	113	110
Whistle	114	106
Sonic shooter	117	101
X-ray phaser	108	88
Gatling gun	110	104
Toy drill	106	91
Telephone	95	89
Flute	126	102
Pop gun	92	90
Bicycle horn	120	90
Camcorder	94	NA
Keyboard #1	103	101
Siren/radio	110	96
Tape player #1	100	90
Tape player #2	114	101
Keyboard #2	104	86
Beeper/pager	122	93
Cassette recorder	126	115
Sound effects keyboard	98	80

Table 15.2 Leq Measures and Spectral Analysis of the 10 Toys Studied in Detail

Toys	Leq (dBA) at 2.5 cm	Leq (dBA) at 25 cm	Frequency spectrum (Hz)
Flute	127	118	1000–12000
Bicycle horn	126	111	1000–3000
Gatling gun	119	116	800–4000
Pop gun	122	116	300–5000
Army gun	119	116	500–3000
Telephone	117	115	1500–2000
Sonic shooter	119	115	
Cassette player (feedback)	119	116	700
Machine laser gun	117	116	700–8000
Guitar	117	115	500–5000

adult occupational noise exposure for any period of time, and six toys exceeded this level at the "arm's length" distance of 25 cm. Yaremchuk et al. concluded that these toys posed significant noise hazards and advocated a change in the Voluntary Product Standard PS 72-76, Toy Safety to measure sound pressure levels at ear level, as well as at 25 cm. We concur with their findings and recommendations.

In our automobile-dependent society, motor vehicle accidents and airbag deployment are common. All of us with busy otology practices encounter claims in which hearing loss, tinnitus, and/or dizziness have allegedly been caused by an airbag deployment. Various literature summarized this subject (47,48). In the report by Yaremchuk and Dobie (47), the authors noted that there had been an estimated 2.1 million airbag deployments in the preceding 10 years. Although the possibility of airbag-induced hearing impairment had been dismissed initially, later experience indicated that such hearing injury is possible. Airbags are deployed using either gas inflation, an explosive device, or hybrid devices. They deploy at speeds of approximately 200 miles per hour over 25–50 ms. Hence, they produce impulse noise (a high-amplitude, short-duration pressure wave). This noise is comparable to a single rifle shot and is generally in the range of 160–170 dB SPL. Just as with a single rifle shot, most people exposed to airbag deployment do not sustain hearing damage. However, such an event is capable

Table 15.3 Leq Measurements, Peak dBA, and Peak Latencies for the Pop Gun

	2.5 cm	25 cm
Sample rate	25 fs	25 fs
Leq	122.045 dBA	115.833 dBA
L90	122.265 dBA	115.902 dBA
L10	115.109 dBA	115.09 dBA
A duration	38.0711 ms	3.56916 ms
A peak[a]	123.217 dBA	124.222 dBA
B duration	96.1777 ms	21.415 ms
B peak[b]	123.105 dBA	123.709 dBA

[a]"A peak" is the first major amplitude peak of the sound and A "duration" is its latency.
[b]"B peak" is the second major peak and B duration is its latency.

of causing otologic symptoms in some people under specific circumstances. For example, in drivers whose heads were turned to the side when the front airbag deployed and whose ears were struck directly by the inflating airbag, otologic injury is common. Similar contact may occur from a side airbag, but the position of side airbags usually precludes such an occurrence and usually places the driver's or passenger's ear more than 1 ft from the airbag deployed. Interestingly, although it seems counterintuitive, a closed vehicle actually helps protect against hearing injury from airbag deployment by preserving protective effects from low-frequency components that are lost through open windows. Most airbag noise is low frequency. People who sustain cochlear injury from acoustic trauma such as airbag deployment ordinarily recognize the injury promptly (at the scene, or within approximately a day) unless there is some other major, related injury that distracts them. Readers who encounter this complex problem are encouraged to consult the articles by Yaremchuk and Dobie (47) and by Nixon (48) as well as the classic literature cited in those publications, and more recent studies.

7.2. Aspirin

Limited research exists that investigates the interaction of aspirin and noise exposure on hearing loss. A study by Carson et al. (49) explored the interaction of different aspirin doses and presence of noise on the hearing of rats. Permanent hearing losses and hair cell damage were noted in all animals exposed to the noise condition; however, a greater amount of hair cell loss was observed in the animals with the highest dose of aspirin. Although some animal studies, like the one discussed above, have shown a relationship between noise-induced hearing loss and aspirin, its interaction on human hair cells and hearing loss is unknown at this time.

7.3. Smoking

Research studies have implicated smoking as a cofactor that may affect the degree and risk of occupational hearing loss. Past experiments investigating the correlation of smoking and high-frequency noise-induced hearing loss illustrate a higher risk of hearing loss (50–54). One study revealed that current smokers had an increased rate of hearing loss by a ratio of 1.39 : 1 with an increase in the trend based on the number of packs smoked per day (50). However, the results of these investigations have been inconsistent regarding the exact role smoking plays in occupational hearing loss, in some part due to poor controls of noise exposure levels in participants. Recent studies have suggested that smoking, by itself, does not increase the risk of occupational hearing loss; however, when coupled with elevated blood pressure and other cardiovascular risk factors, those workers have a higher risk of hearing loss (55,56).

7.4. Industrial Solvents

Another group of cofactors has been the topic of recent experimental research in both humans and animals. Industrial solvents such as toluene, styrene, xylene, trichloroethylene, and carbon disulfide, by themselves or in mixtures, have been implicated in affecting the inner ear both with and without the presence of noise. A review of audiograms and electronystagmography (ENG) results for workers in a paint and varnish plant revealed that 42% of those exposed to industrial solvents had high-frequency hearing losses as compared to 5% in the age-matched control group; 47.5% of that same group demonstrated abnormal ENGs, vs. 5% in the control group (57). In investigating the effects of styrene and styrene plus noise exposure, Sliwinska-Kowalska et al. (58) found that exposure to the styrene increased the risk of hearing loss in humans; in addition, the styrene plus noise exposed group had two to three times higher odds of developing hearing loss. Both of these studies support the conclusion that exposure to some industrial solvents can increase the risk of high-frequency hearing loss. More research is needed to investigate other industrial solvents and their possible effects on workers and their hearing.

8. NON-ORGANIC HEARING LOSS

Due to the compensation and legal factors involved in occupational hearing loss, there are much larger numbers of individuals who present with non-organic hearing loss in supposed noise-induced loss cases. The diagnosing physician needs to be aware of possible non-organic components including functional hearing loss, functional overlay, and malingering.

Functional or psychogenic hearing loss results from the influence of psychological or emotional factors resulting in an inability to hear when the peripheral mechanism may be essentially normal. Patients with extreme anxiety or emotional conflicts may outwardly manifest their mental disturbances by converting them into hearing loss or other conversion disorders. Conversion disorders are an involuntary response, escaping from the extreme emotions that cannot be faced voluntarily. Functional hearing loss can also be noted in addition to actual organic, peripheral hearing loss. When this is evident, it is noted as functional overlay. Careful history taking and accurate audiometry will assist in the diagnosis of these non-organic components.

Unlike functional hearing loss, malingering is the intentional misrepresentation of one's hearing, usually for monetary or emotional gain. Individuals have been known to misrepresent their hearing for financial gain, to avoid work environments such as military deployment, to evade responsibility, or for purposes of gaining attention. Functional overlay can also exist in malingering hearing losses. Most malingerers are revealed by their inconsistent test results. In cases of unilateral malingering, the shadow curve that should be present due to interaural attenuation is often absent. For other cases, the pure tone average is not consistent with the speech reception threshold. To try to find the malingerer's true thresholds, a number of techniques are useful in audiology. An ascending method of test presentation decreases their ability to estimate loudness; this is also reduced if smaller 1- to 2-dB steps are used instead of the traditional 5-dB steps. Physiological measures such as otoacoustic emissions and auditory brainstem response (ABR) also can be helpful in determining normal hearing, as can other special tests.

9. CASE REPORTS

Chiefly because of OSHA requirements and greater emphasis on workers' compensation for occupational hearing loss, hundreds of thousands of employees will ultimately be referred to otologists for consultations. Otologists must provide expert advice on matters such as employing people with possible safety and communication problems, managing numerous otologic problems, and determining whether hearing loss is due to occupational noise or some other cause. The general characteristics of occupational hearing loss described can help guide physicians to a reasonably accurate diagnosis. In order to illustrate some of the numerous problems that have arisen in managing claims for occupational hearing loss, it may be helpful to review a series of actual cases that have come to workers' compensation court for adjudication. The histories are abstracted and the findings of plaintiff's and defense's experts are abbreviated, but included in each case are important features that illustrate both justified and unjustified contentions.

In all these cases, the physical aspects of the otologic examination revealed no abnormalities unless specifically stated in the case report. Appropriate complete physical examination, blood studies, and other examinations were performed with no abnormalities found unless specifically stated in the report.

9.1. Case Report 1

A 63-year-old pipefitter, employed by a shipbuilding company for 40 years, had 10 years' exposure to chippers and ship "scraping" noise. Hearing loss developed gradually over many years, most pronounced after chipping exposure. He denied tinnitus, vertigo, and gunfire exposure. Audiometry (Fig. 15.14) showed bilateral sensorineural hearing loss with fairly good residual hearing at 10,000 and 12,000 Hz, discrimination between 80

HEARING RECORD

JOSEPH SATALOFF, M.D.
ROBERT THAYER SATALOFF, M.D.
1721 PINE STREET PHILADELPHIA, PA 19103

Name Age

Air conduction

			Right							Left						
Date	Exam	Left mask	250	500	1000	2000	4000	8000	Right mask	250	500	1000	2000	4000	8000	AUO
			35	40	45	50	70	60		40	40	40	55	65	60	

Bone conduction

			Right							Left						
Date	Exam	Left mask	250	500	1000	2000	4000		Right mask	250	500	1000	2000	4000		AUO
			30	40	40	50	65			35	35	35	50	65		

Speech reception

Date	Right	Left mask	Left	Right mask	Free Field	Mic

Discrimination

	Right				Left				
Date	% Score	Test level	List	Left mask	% Score	Test level	List	Right mask	Exam
	82	60			88	60			

High frequency thresholds

	Right						Left					
Date	4000	8000	10000	12000	14000	Left mask	Right mask	4000	8000	10000	12000	14000
		60	40	40					60	45	45	

Right		Weber	Left	
Binne	Schwabach		Binne	Schwabach

Hearing aid		
Date	Make	Model
Receiver	Gain	Exam
Ear	Discrim	Counc

Remarks

Figure 15.14

and 88%, and good speech reception. These findings are characteristic of occupational hearing loss due to prolonged exposure to intense noise such as chipping, and the history is consistent with the diagnosis. If presbycusis or hereditary deafness were factors, the highest frequencies would be more seriously involved and the discrimination score might be worse.

9.2. Case Report 2

A 67-year-old railroad brakeman had worked for 35 years and retired 2 years ago. About 7 years prior to this otologic examination, he had been examined for hearing loss by an otologist, who concluded, on the basis of the patient's history of working with excessive noise and vibration, that the patient had "sensorineural hearing loss due to prolonged exposure to noise and vibration." Only two audiograms were available (Fig. 15.15), one taken in 1974 about 5 years prior to retirement and one in 1981, 2 years after retirement. Note the rather late onset of hearing loss and the progressive nature of the condition even over the past 4 or

JOSEPH SATALOFF, M.D.
ROBERT THAYER SATALOFF, M.D.
1721 PINE STREET PHILADELPHIA, PA 19103

HEARING RECORD

Name Age

Air conduction

			Right							Left						
Date	Exam	Left mask	250	500	1000	2000	4000	8000	Right mask	250	500	1000	2000	4000	8000	AUO
1974			20	25	40	45	50	50		15	25	45	45	50	55	
1982			30	30	45	60	65	70		25	25	50	55	65	65	

Bone conduction

			Right							Left						
Date	Exam	Left mask	250	500	1000	2000	4000		Right mask	250	500	1000	2000	4000		AUO
1974			20	20	35	50	45			10	25	40	45	55		
1982			25	25	45	60	70			25	30	45	55	70		

Speech reception

Date	Right	Left mask	Left	Right mask	Free field	Mic

Discrimination

	Right				Left				
Date	% Score	Test level	List	Left mask	% Score	Test level	List	Right mask	Exam

High frequency thresholds

	Right						Left					
Date	4000	8000	10000	12000	14000	Left mask	Right mask	4000	8000	10000	12000	14000

Right		Weber	Left	
Binne	Schwabach		Binne	Schwabach

Hearing aid		
Date	Make	Model
Receiver	Gain	Exam
Ear	Discrim	Counc

Remarks

Figure 15.15

5 years. This employee had been working in the same noise environment for many years, yet he did not notice the hearing loss until a few years prior to retirement. These factors help indicate that the diagnosis is presbycusis rather than occupational hearing loss. This is further substantiated by the fact that measurements revealed that this patient's exposure did not exceed 87 dBA, and he had not been exposed to the especially loud noises occasionally found in railroad employment.

9.3. Case Report 3

A 65-year-old railroad machinist worked around diesel engines for 25 years. He said he had been exposed to roaring diesel train engines for many years and that his hearing loss started many years ago and gradually got worse. He had occasional vertigo but no tinnitus. He denied any family history of hearing loss and was in good health. The audiologic studies showed a flat mixed hearing loss in both ears of about 60 dB, slightly worse in the higher frequencies. The bone conduction threshold was reduced but slightly better than the air conduction in the lower frequencies.

The otologist who first evaluated this patient diagnosed nerve deafness due to occupational exposure to diesel engines.

A later examination by another otologist showed the same audiologic findings but noted excellent bone conduction when a 500-Hz tuning fork was applied to the upper teeth. Discrimination was excellent. The diagnosis was otosclerosis with sensorineural as well as conductive hearing loss. In many older patients, bone conduction tests on the mastoid are not necessarily good indications of actual sensorineural function. Occupational noise did not cause this patient's hearing loss.

9.4. Case Report 4

A 36-year-old man began work in a paper mill in 1976. His occupation often involved exposure to noise levels in excess of 95 dBA throughout the workday. He also had a history of exposure to firearms, discharging a shotgun approximately 200–300 times annually. He is left-handed. In addition, he listened to loud music daily. Serial audiograms revealed development between 1976 and 1984 of an obvious dip at 4000 and 6000 Hz. Despite continued exposure to the same occupational and extracurricular noise sources, there is no evidence of significant deterioration after approximately his first 8 years on the job (Fig. 15.16). This is typical of occupational hearing loss.

9.5. Case Report 5

A 45-year-old man gave a complex history of having been evaluated by four otologists. The patient claimed that he had been hit in September 1981 on the right side of his face by a landloader. He was not dazed or unconscious, and there was no bleeding or visible trauma, but he could not hear with his right ear immediately after the incident and had been unable to hear since. He denied any hearing loss prior to the accident and also denied any familial hearing loss. In April 1982, an otologist performed a stapedectomy and the patient's hearing subjectively improved. He developed recurrent vertigo postoperatively. However, audiograms performed shortly after the surgery showed no evidence of hearing improvement compared with preoperative thresholds. The otologist stated that surgery was for otosclerosis and that, in his opinion, the otosclerosis had no relation to the accident. Another otologist who examined the patient because of his persistent vertigo agreed that the vertigo was postoperative and that the diagnosis was otosclerosis,

NAME:

	RIGHT EAR AIR CONDUCTION								LEFT EAR AIR CONDUCTION						
DATE	500	1000	2000	3000	4000	6000	8000		500	1000	2000	3000	4000	6000	8000
10/76	5	5	5	10	20	40			5	5	10	10	25	25	
8/81	5	0	5	10	50	35			5	0	0	30	65	75	
8/82	5	10	5	20	50	30	30		5	5	5	20	55	35	25
8/83	5	0	5	30	55	35	40		10	10	10	20	35	30	20
7/84	0	0	0	20	40	65	30		5	5	10	10	50	30	20
1/85	0	5	0	35	60	65	25		0	0	10	25	50	25	10
1/86	0	0	0	25	70	50	25		0	5	0	20	50	15	10
2/87	5	0	0	20	70	80	25		0	5	5	25	50	20	10
8/87	5	0	0	25	65	65	20		5	0	0	15	40	30	20
9/88	0	0	0	25	65	60	20		0	0	0	25	35	25	20
9/89	5	5	5	30	70	70	30		5	5	5	30	35	20	15
11/90	0	0	0	25	65	65	30		0	0	0	25	40	20	10
5/91	5	0	0	25	65	60	25		0	0	0	25	40	20	10

Figure 15.16

not related to the accident. The patient underwent unsuccessful right revised stapes surgery in an attempt to resolve his vertigo and restore his hearing. At the request of the plaintiff's attorney, another otologist examined this patient in March 1983. Remarks from his report are as follows: "It is apparent from reviewing the records that the patient had a dormant otosclerosis which has been activated and aggravated by the head injury suffered [in 1981]. It is my opinion that without this trauma the otosclerosis would have remained dormant for many years."

In April 1984, another otologic evaluation was performed and the audiologic findings revealed conductive hearing loss with no sensorineural involvement on the right side, excellent bone conduction and discrimination, and a good chance of improving this patient's hearing and clearing up his other symptoms. The difference of interpretation of the etiology of this otosclerotic process is an important issue for otologists. There is no question that all otologists have seen otosclerosis progress to this degree and in this manner without being aggravated by trauma. The contention that trauma could produce this sudden aggravation of otosclerosis has not been substantiated by any valid and reliable otologic study. An otologist may form an opinion that this could have happened, but in order for such an opinion to carry weight or be defensible, it is important that it be creditably based on scientific studies and accepted by the medical community.

9.6. Case Report 6

A 65-year-old diabetic man worked in a noisy cannery for over 40 years, using hearing protectors only in the last 10 years. His ears had drained intermittently over a 40-year period, most severely in the left ear for the last 7 years. Otologic examination revealed a large left perforation, with scarring and thickening of the residual tympanic membrane. Audiometry of the left ear revealed better bone conduction than air conduction. Bone conduction was excellent by Weber's test. Audiometry of the right ear revealed sensorineural hearing loss. Discrimination scores were good bilaterally. Serial audiograms from 1975 to 1981 showed progressive deterioration of hearing in the left ear. One otologist diagnosed right-sided occupational sensorineural hearing loss and left-sided mixed hearing loss due to occupational exposure with superimposed infection. Another otologist, basing his opinion on the most recent (1981) audiogram, claimed the entire hearing loss was due to occupational noise exposure. At the deposition, it became evident that the deterioration of hearing in the left ear between 1975 and 1981 was not a result of cannery noise, because the worker had worn hearing protection and had not been exposed to loud noise in that time period. Good discrimination scores helped indicate that a large portion of the hearing loss present before 1978 was attributed to occupational noise exposure, although superimposed chronic otitis, tympanosclerosis, and presbycusis were contributory. It was determined that the actual amount of noise exposure that warranted compensation should be determined by the 1975 audiogram.

9.7. Case Report 7

A 57-year-old dye caster and screw conveyor operator worked in a machine shop for many years. In 1979, while working around a screw conveyor, he was subjected to an extremely loud sound for about 30 s. He had no discomfort, but his wife noted marked hearing loss that evening. His physician diagnosed "vascular accident resulting from intense noise exposure" and prescribed medication.

The consulting otologist noted disparity in hearing tests taken before and after the incident and expressed the following opinion: "The cause and effect relationship between the loud noise exposure and sudden hearing loss is real, particularly in view of the fact that he has had normal hearing previously." He advised studies to rule out the possibility of other "disease processes."

The employee put in a claim for occupational hearing loss compensation because of deafness produced by loud noise, particularly in his left ear. Studies revealed the presence of a left acoustic neuroma that was confirmed operatively. Details of this case are reported in Chapter 9. It is important to note that all studies, including a computed tomography (CT) scan, had missed this small tumor. It was diagnosed only because of the otologist's insistence that the patient undergo a myelogram or air-contrast CT scan. Currently, a MR would be used. An estimated 11% of acoustic neuromas can present as "sudden deafness" (59).

9.8. Case Report 8

A 50-year-old employee had worked since 1969 in a plant making tires. His annual audiograms showed normal hearing until 1973, when annual testing of his hearing was discontinued. He gave a vague history of a "press explosion" in 1975 with some ringing in his right ear, but he did not complain of hearing loss. The explosion was not confirmed either at the plant or in any medical records. In 1976, he had fullness in his right ear that was diagnosed by his otologist as "eustachian tube blockage due to a temporomandibular joint problem." In 1976, he developed hearing loss and tinnitus in his right ear, causing the otologist to rule out an acoustic neuroma based on normal calorics, tomograms, and posterior fossa myelograms. In 1979, a myringotomy was done for right-ear blockage. In 1981, he had two vertigo attacks and his otologist diagnosed Meniere's disease. The employee retired in 1981 because of physical disabilities. In 1983, he applied for workers' compensation for hearing loss after being examined by his otologist and audiologist. His otologist's report included the following:

The patient shows a bilateral sensorineural hearing loss, which appears to be worse on the right side. The Weber test lateralized to the left ear as would be expected with this audiometric configuration. Speech discrimination is reduced in both ears. Based on the long time history of noise exposure in his occupation, it is very likely that a significant amount of the current hearing loss is probably related to noise.

The otologist representing the defense during litigation demonstrated clearly that:

1. The employee actually was not exposed to noise exceeding 88 dBA during his work and generally worked at much lower levels, chiefly in the loading department.
2. He worked for at least 4 years at the same job before developing any hearing loss.
3. The hearing loss started and became severe in his right ear long before the left ear became involved.
4. The hearing loss continued to get worse even after he retired in 1981.

The real cause for his hearing loss was probably related to his generalized arteriosclerosis, hypertension, peripheral vascular disease, and long-standing diabetes. In February 1981, he had transient cerebral ischemia attacks, arterial insufficiency of the left leg with occlusion of the left femoral artery, and stenosis of the common iliac artery, treated surgically. The employee's otologist and audiologist were apparently unaware of the patient's diabetes and peripheral vascular problems and surgery. Their impression of his job-related noise exposure, which they obtained from his history, was inaccurate. They were not aware that his actual noise exposure was not capable of producing his hearing loss. There is no question that this hearing loss was neither caused nor aggravated by the worker's job.

10. SUMMARY

A diagnosis of occupational hearing loss must be based on specific criteria. Otologists rendering medical diagnoses or legal opinions for patients alleging occupational hearing loss must be careful to base their opinions on facts. The potential medical, legal, and economic consequences of lesser diligence are likely to be serious.

REFERENCES

1. Fleming AJ, D'Alonzo CA, Zapp, JA. Modern Occupational Medicine. Philadelphia, PA: Lea & Febiger, 1954.
2. Exploratory Subcommittee Z24-X-2, American Standards Association. The relations of hearing loss to noise exposure, 1954.
3. National Institute for Occupational Safety and Health (NIOSH). Criteria document: Recommendation for an occupational exposure standard for noise, 1972.
4. Baughn WL. Relation between daily noise exposure and hearing loss based on the evaluation of 6,835 industrial noise exposure cases. Aerospace Medical Research Lab., Wright Patterson AFB, Ohio. AMRL-TR-73-53, June 1973.
5. Ward WD, Glorig A. Protocol of inter-industry noise study. J Occup Med 1975, 17:760–770.
6. Burns W, Robinson DS. Hearing and Noise in Industry. London: Her Majesty's Stationery Office, 1970.

7. Passchicr-Vcrmccr W. Hcaring loss duc to stcady statc broadband noisc. Rcport No. 55, Institute for Public Health, Eng., Leiden, The Netherlands, 1968.
8. Sataloff J, Sataloff RT, Yerg A, Menduke H, Gore RP. Intermittent exposure to noise: effects on hearing. Ann Otol Rhinol Laryngol 1983; 92(6):623–628.
9. Orgler GK, Brownson PJ, Brubaker WW, Crane DJ, Glorig A, Hatfield TR, Hanson R, Holthouser MG, Ligo RN, Markham T, Mote WR, Sataloff J, Yerg RA. American Occupational Medicine Association Noise and Hearing Conservation Committee Guidelines for the Conduct of an Occupational Hearing Conservation Program. J Occup Med 1987; 29:981–982.
10. Schuknecht HF, Tonndorf J. Acoustic trauma of the cochlea from ear surgery. Laryngoscope 1960; 70:479–505.
11. Lawrence M. Current concepts of the mechanism of occupational hearing loss. Am Ind Hyg Assoc J 1964; 25:269–273.
12. Kellerhals B. Pathogensis of inner ear lesions in acute acoustic trauma. Acta Otolaryngol 1972; 73:249–253.
13. Schuknecht HG. Pathology of the Ear. Cambridge, MA: Harvard University Press, 1974: 295–297, 300–308.
14. Gallo R, Glorig A. Permanent threshold shift changes produced by noise exposure and aging. Am Ind Hyg Assoc J 1964; 25:237–245.
15. Schneider EJ, et al. The progression of hearing loss from industrial noise exposure. Am Ind Hyg Assoc J 1970; 31:368–376.
16. Sataloff J, Vassallo L, Menduke H. Occupational hearing loss and high frequency thresholds. Arch Environ Health 1967; 14:832–836.
17. Sataloff J, Vassallo L, Menduke H. Hearing loss from exposure to interrupted noise. Arch Environ Health 1969; 17:972–981.
18. Salmivalli A. Acoustic trauma in regular Army personnel: clinical audiologic study. Acta Otolaryngol (Stockh) 1967; (Suppl. 22):1–85.
19. Ward WD, Fleer RE, Glorig A. Characteristics of hearing losses produced by gunfire and steady noise. J Audiol Res 1961; 1:325–356.
20. Johnsson L-G, Hawkins JE Jr. Degeneration patterns in human ears exposed to noise. Ann Otol Rhinol Laryngol 1976; 85:725–739.
21. Gates GA, Cooper JC Jr, Kannel WB, Miller NJ. Hearing in the elderly: the Framingham cohort, 1983–1985. Part I. Basic audiometric test results. Ear Hear 1990; 11(4):247–256.
22. Sataloff J, Vassallo L. Head colds and viral cochleitis. Arch Otolaryngol 1968; 19:56–59.
23. Igarashi M, Schuknecht HF, Myers E. Cochlear pathology in humans with stimulation deafness. J Laryngol Otol 1964; 78:115–123.
24. Anderson H, Wedenberg E. Genetic aspects of hearing impairment in children. Acta Otolaryngol (Stockh) 1970; 69:77–88.
25. Fisch L. The etiology of congenital deafness and audiometric patterns. J Laryngol Otol 1955; 69:479–493.
26. Huizing EH, van Bolhuis AH, Odenthal DW. Studies on progressive hereditary perceptive deafness in a family of 335 members. Acta Otolaryngol (Stockh) 1966; 61:35–41, 161–167.
27. Graham MD. Acoustic tumors: selected histories and patient reviews. In: House WF, Lutje CM, eds. Acoustic Tumors. Baltimore, MD: University Park Press, 1979, 192–193.
28. Facer GW, Farell KH, Cody DTR. Spontaneous perilymph fistula. Mayo Clin Proc 1973; 48:203–206.
29. Simmons FB. Theory of membrane breaks in sudden hearing loss. Arch Otolaryngol 1968; 88:67–74.
30. Soss SL. Sensorineural hearing loss with diving. Arch Otolaryngol 1971; 93:501–504.
31. Freeman P. Edwards C. Inner ear barotrauma. Arch Otolaryngol 1972; 95:556–563.
32. van Dishoeck HAE. Akustisches Trauma. In: Berendes J, Link R, Zollner F, eds. Hals-Nasen-Ohren-Heilkunde. Band III. Stuttgart: Georg Thieme, 1966; 1764–1799.
33. Yerg RA, Sataloff J, Glorig A, Menduke H. Inter-industry noise study. J Occup Med 1978; 20:351–358.

34. Cartwright LB, Thompson RW. The effects of broadband noise on the cardiovascular system in normal resting adults. Am Ind Hyg Assoc J 1978; 653–658.
35. Johnsson L-G, Hawkins JE Jr. Strial atrophy in clinical and experimental deafness. Laryngoscope 1972; 82:1105–1125.
36. Spoendlin H. Histopathology of nerve deafness. J Otolaryngol 1985; 14(5):282–286.
37. Spoendlin H, Brun JP. Relation of structural damage to exposure time and intensity in acoustic trauma. Acta Otolaryngol 1973; 75:220–226.
38. Spoendlin H. Primary structural changes in the organ of Corti after acoustic over-stimulation. Acta Otolaryngol 1971; 71:166–176.
39. Murato M, Takigawa H, Sakamoto H. Teratogenic effects of noise and cadmium in mice: does noise hanve teratogenic potential? J Toxicol Environ Health 1993; 339(2):237–245.
40. Nurminen T. Female noise exposure, shift work, and reproduction. J Occup Environ Med 1995; 37(8):945–950.
41. Schnell LM. Environmental noise and human prenatal growth. Am J Phys Anthropol 1981; 1:63–70.
42. Lalande NM, Hetu R, Lambert J. Is occupational noise exposure during pregnancy a risk factor of damage to the auditory system of the fetus? Am J Ind Med 1986; 10(4):427–435.
43. Sataloff RT, Hickey K, Robb J. Rock concert audience noise exposure: a preliminary study. J Occup Hear Loss 1998; 1(2):97–99.
44. Sataloff RT, Rau G, Preston L. Noise exposure in movie theaters. J Occup Hear Loss 1998; 1(4):281–282.
45. Mendes RC, Barrionuevo CE. Risk of occupational hearing loss in telephone operators. J Occup Hear Loss 1998; 1(3):179–184.
46. Yaremchuk KL, Dickson L, Burk K. Noise level analysis of commercially available toys. J Occup Hear Loss 1998; 2(4):163–170.
47. Yaremchuk KL, Dobie R. The otologic effects of airbag deployment. J Occup Hear Loss 1999; 2(2–3):67–74.
48. McKinley RL, Nixon CW. Human auditory response to an air bag inflation noise: has it been 30 years. J Acoust Soc Am 1998; 104(3), Pt2:1769.
49. Carson SS, Prazma J, Pulver SH, Anderson T. Combined effects of aspirin and noise in causing permanent hearing loss. Arch Otolaryngol Head Neck Surg 1989; 115:1070–1075.
50. Barone JA, Peters JM, Garabrant DH, Bernstein L, Krebsbach R. Smoking as a risk factor in noise-induced hearing loss. J Occup Med 1989; 29(9):741–746.
51. Thomas GB, Williams CE, Hoger NG. Some non-auditory correlates of the hearing threshold levels of an aviation noise exposed population. Aviat Space Environ Med 1981; 9:531–536.
52. Chung DY, Wilson GN, Gannon RP, et al. Individual susceptibility to noise. In: Hamernik RP, Henderson D, Salvi R, eds. New Perspectives in Noise-Induced Hearing Loss. New York: Raven Press, 1982, 511–519.
53. Drettner B, Hedstrand H, Klockhoff I, et al. Cardiovascular risk factors and hearing loss. Acta Otolaryngol 1975; 79:366–371.
54. Siegelaub AB, Friedman GD, Adour K, et al. Hearing loss in adults. Arch Environ Health 1984; 29:107–109.
55. Cocchiarella LA, Sharp DS, Persky VW. Hearing threshold shifts, white-cell count and smoking status in working men. Occup Med (London) 1995; 45:179–185.
56. Starck J, Toppila E, Pyykko I. Smoking as a risk factor in sensory neural hearing loss among workers exposed to occupational noise. Acta Otolaryngol (Stockh) 1999; 119:302–305.
57. Sulkowski WJ, Kowalska S, Matyja W, Guzek W, Wesolowski W, Szymczak W, Kostrzewski P. Effects of occupational exposure to a mixture of solvents on the inner ear: a field study. Int J Occup Med Environ Health 2002; 15:247–256.
58. Sliwinska-Kowalska M, Zamyslowska-Szmytke E, Szymczak W, Kotylo P, Fiszer M, Wesolowski W, Pawlaczyk-Lusczynska M. Ototoxic effects of occupational exposure to styrene and co-exposure to styrene and noise. J Occup Environ Med 2003; 45:15–24.
59. Sataloff RT, Davies B, Myers DL. Acoustic neuromas presenting as sudden deafness. Am J Otol 1985; 6(4):349–352.

16
Hearing Loss: Handicap and Rehabilitation

Robert T. Sataloff
Drexel University College of Medicine, American Institute for Voice and Ear Research, and Graduate Hospital, Philadelphia, Pennsylvania, USA

Joseph Sataloff
Thomas Jefferson University, Philadelphia, Pennsylvania, USA

Tracy M. Virag, Caren J. Sokolow, and John Luckhurst
American Institute for Voice and Ear Research, Philadelphia, Pennsylvania, USA

The most obvious effect of hearing loss is its impact on auditory communication; however, the damaging effect of hearing loss on an individual's ability to function effectively in social and business life is far more critical. In some people, hearing loss can undermine natural optimism and confidence in one's ability to interact successfully with other people. If hearing loss is not addressed through medical treatment and/or rehabilitation, it can lead to insecurity, social withdrawal, paranoia, and impaired function in professional settings.

1. EFFECT ON THE PERSONALITY

The impact of hearing loss varies by degree of loss, personality of the individual, and the individual's activity level. Interestingly, hearing loss can manifest as a great disability in people with comparatively mild hearing loss (e.g., otosclerosis or Meniere's disease), whereas some individuals with profound hearing losses may have severe disruption of communication without seriously affecting personality. The degree of impact depends on an individual's character—including mental, spiritual, societal, and economic resources—and other factors that determine one's reaction to hearing loss and the level of handicap it generates.

The effects of hearing loss cannot be restricted to physiology alone, as the mechanics of the ear cannot be divorced from the social aspects of hearing. Hearing is a phenomenon that utilizes peripheral pathways from the ears to the central pathways of the brain to form a compete understanding of acoustic signals; however, hearing is also our method of socialization. In any discussion of hearing, communication, deafness, and handicap, it is necessary to think of the person as a whole and not merely as a mechanism of hearing. For this reason, hearing loss concerns the otologist and the general practitioner, the pediatrician, the psychologist, the psychiatrist, and many others, including members of the legal profession.

1.1. The Relationship between Hearing and Speech

To understand the basis for personality changes and communication handicaps that hearing loss may produce, it is necessary to recall the relationship between hearing and speech. It is known that the ear is sensitive to a certain frequency range and that speech falls within that range. Speech can be divided into two types of sounds: vowels and consonants. Roughly speaking, vowels fall into the frequencies <1500 Hz, and consonants >1500 Hz. Vowels are relatively powerful sounds, whereas consonants are weak and sometimes dropped within everyday speech or not pronounced clearly. Vowels give power to speech; that is, they signify that someone is speaking, but they give very little information about what the speaker is saying. To give specific meaning to words, consonants are interspersed among the vowels. Thus, it can be said that vowels alert the listener to the presence of speech, and consonants help the listener to understand or discriminate what the speaker is saying.

For example, the hard-of-hearing individual (B), whose audiogram is shown in Fig. 16.1, would have difficulty hearing speech unless it was raised in volume. The difficulty lies in the low frequencies; causing the individual to have trouble hearing vowels and soft voices. However, if a voice were raised in volume, it would be heard and understood clearly. This illustrates a principal problem of loudness or amplification.

The person (A), whose hearing impairment is portrayed in Fig. 16.1, has a high-frequency loss with almost normal hearing in the low frequencies. In applying this hearing loss to speech inputs, person (A) would hear vowels almost normally, but would have difficulty in hearing and discriminating consonants. If a speaker raised his/her voice volume, the missing consonants would be emphasized only slightly and the loudness of the vowels may increase to a disturbing degree. The individual's primary problem is not *hearing*, but *distinguishing* what is heard. The vowels are heard and he/she is aware of someone speaking, but the patient cannot distinguish some of the consonants and misses the meaning of the conversation. This type of person would want the speaker to enunciate more clearly, pronouncing the consonants more distinctively rather than speaking louder. Hearing loss of this type, with its accompanying handicaps, is caused commonly by presbycusis, hereditary hearing loss, and certain types of congenital deafness.

1.1.1. High-Frequency Hearing Loss and Distortion

High-frequency sensorineural hearing loss often causes deterioration of a person's ability to understand speech. In addition, music, certain voices, and especially amplified sound may sound hollow, tinny, and muffled. The hard-of-hearing person affected in this manner may first ask his/her companion to speak louder, but in spite of the louder speech, he/she may understand even less. Loudness may actually reduce discrimination in such individuals with high-frequency hearing loss due to distortion of the signal. Distortion occurs most frequently in people with high-frequency losses because overall loudness also amplifies the low-frequency sounds, such as vowels, which they can hear at a normal level. Speaking in a louder voice creates overpowering vowels with relatively

JOSEPH SATALOFF, M.D.
ROBERT THAYER SATALOFF, M.D.
1721 PINE STREET PHILADELPHIA, PA 19103

NAME ____________________

DATE	AIR CONDUCTION							AIR CONDUCTION					
	250	500	1000	2000	4000	8000		250	500	1000	2000	4000	8000
(A)	15	20	40	60	65	NR							
(B)	50	50	40	30	30	35							

RIGHT EAR BONE CONDUCTION						LEFT EAR BONE CONDUCTION					

SPEECH RECEPTION: Right ______ Left ______ DISCRIMINATION: Right ______ Left ______

(Each ear is tested separately with pure tones for air conduction and bone conduction, if necessary. The tones increase in pitch in octave steps from 250 to 8000 Hz. Normal hearing in each frequency lies between 0 and 25 decibels. The larger the number above 25 decibels in each frequency the greater the hearing loss. When the two ears differ greatly in threshold, one ear is masked with noise to test the other ear. Speech reception is the patient's threshold for everyday speech, rather than pure tones. A speech reception threshold of over 30 decibels is handicapping in many situations. The discrimination score indicates the ability to understand selected test words at a comfortable level above the speech reception threshold.)

Figure 16.1 Patients A and B have an average pure-tone loss of 40 dB, but their clinical hearing is substantially different.

weaker consonants and does not improve the clarity of speech. It is easy to understand that people who have a high-frequency loss could easily become frustrated and confused in trying to follow a conversation.

1.1.2. Dynamic Aspects of Hearing-Loss Handicap

Although compensation for occupational hearing loss demands the establishment of standardized values for hearing loss, from a medical and social aspect, a hearing loss is a personal issue to each individual and cannot be measured on a universal scale. Furthermore, just as hearing loss usually varies over time for an individual, the handicapping effects are also dynamic. They are changing even as this book is being written. With the development of new media for sound communication, an individual's hearing comes to assume even greater importance. For example, a hearing loss today is much more handicapping than it was before television, radio, and telephones began to play such major roles in education, leisure, and the business world. Today, the inability to understand on a telephone is indeed a handicap for the majority of people. The loss of just high-frequency sounds, to a professional or an amateur musician, or even to a high-fidelity fan, is also a handicap. The hearing loss of tomorrow probably will have a different handicapping effect from the hearing loss of today.

1.2. Reactions to Hearing Loss

The manner in which people react to hearing loss varies considerably. Some may try to minimize or hide their impairment by making strenuous listening efforts, filling in missing information by guessing, and carefully concealing frustration by acting particularly pleasant and affable. His/her effort to "save face" may lead to numerous embarrassing situations through incorrect guessing. Toward evening, the person may be worn out from the efforts to hide his/her hearing handicap. Maintaining this façade often becomes fatiguing and can lead to nervousness, irritability, and instability.

More commonly, some people react to hearing loss by withdrawing and losing interest in their environment, particularly with impairments that have slow and insidious onsets. The personality change is reflected in an avoidance of social contacts and in a preoccupation with the individual's own misfortunes. These individuals may dismiss friends and make excuses to avoid social contacts that might cause the handicap to become more apparent to friends and to themselves.

1.3. Psychosocial Impact of Hearing Loss

1.3.1. Economic and Family Aspects

Hearing loss can affect an individual economically if the impairment affects job performance directly, or if a decrease in social activity negatively affects one's professional effectiveness. For example, withdrawing from societal interactions during meetings may negatively affect an individual's ability to perform at work. The handicapped person may realize that he/she cannot perform the duties of the job and, rather than tolerate criticism regarding his/her alertness and proper interest in business, he/she may resign and step down to a position of less worthy potential. In another scenario, he/she may be passed over for a promotion or asked to resign due to lack of participation and apparent lack of interest in employment.

When a salesperson becomes hard-of-hearing, business often suffers and ambitions frequently are suppressed or surrendered if the individual does not proactively seek assistance. On the contrary, a hearing impairment may be regarded as advantageous instead of a handicap to a chipper or a riveter while at work. The hearing loss may make the noise from work appear to be not as loud as it is to fellow workers with normal hearing. Because there is little or no verbal communication in most jobs that produce intense noise, a hearing loss will not be apparent by an inability to understand verbal directions. However, when the hard-of-hearing person returns to a family after work, the situation assumes a completely different perspective. The individual may have trouble understanding what is said by children or a spouse. Conversation becomes more difficult when in the presence of background noise, such as running water or a television or if attention is focused elsewhere. This kind of situation frequently leads at first to mild disputes and later to serious family tension.

Marital strain is often a serious social consequence of an unaided hearing loss. The hearing spouse may accuse the individual with a hearing loss of inattention. The accusation is then denied with the hearing-impaired spouse claiming that the other mumbles. Over time, the frustration and strain of listening with a hearing loss causes the person to become inattentive, weakening the marital bond. The same difficulties noted at home are evident at meetings, visits with friends, and at religious activities. A person with hearing loss often withdraws from the places where he/she feels pilloried by the handicap. This includes social activities such as the movies, theater, or concerts. Little by little, family life and social interactions can be undermined by an untreated hearing loss.

1.3.2. The Plight of the Elderly Hearing Impaired

Hearing loss in older persons, whether from causes associated with aging or owing to other sensorineural etiologies, is often quite severe. More severe to profound losses usually create a larger change in personality and lifestyle due to the challenges that listening now presents. Elderly individuals with hearing loss also find listening in the presence of multiple speakers or background noise especially difficult, as our ability to detect signals in noise diminishes with age. All too often, the unfortunate elderly person begins to believe that the inability to hear and understand a conversation is due to a deterioration of the brain. Family, friends, and stereotypes of the elderly population may reinforce this belief. People may ignore the hearing-impaired person in group conversations and assume that he/she does not know what is going on. Stereotypes of aging, such as physical and mental slowing, further undermine the elderly person's weakened self-confidence and hasten his/her withdrawal from society. The isolation caused by hearing loss can further delay elderly individuals from seeking medical attention to address their hearing handicap.

1.3.3. Effect of Profound Hearing Losses

In general, people with profound hearing losses are somewhat easier to help than borderline cases, as the former are under greater compulsion to admit that they have a handicap. People with borderline losses are more likely to hide their handicap and deny it, even to themselves. They conceal their hearing loss just as they try to conceal hearing aids if they can be induced, or are able, to use them. For people with mild losses, a hearing aid often becomes part of the handicap due to the stigma it contains. In contrast, people with severe losses perceive their major handicap as the communication difficulty that results from their hearing loss, in addition to some economic and social conflicts. Often individuals with profound hearing loss cannot hear warning signals such as a fire alarms or telephone rings, unless they use an assistive device. They may be unable to maintain their jobs if listening for sounds or reliance on verbal communication are an integral part of their duties. Another challenging handicap for people with severe hearing losses is the inability to localize sound. This difficulty is particularly prominent when the hearing loss exists unilaterally or in the case of an asymmetrical hearing loss, because two fairly symmetrical hearing ears are necessary to be able to localize the source and direction of sound.

Another interesting aspect of profound hearing loss is the change in the individual's clarity of speech over time. Speech deteriorates as he/she is unable to self-regulate tone, volume, and sound production due to the degree of hearing loss. Primary characteristics of speech become slurred s's and a rigid and somewhat monotonous vocal quality. To counteract the inability to hear one's own voice, the individual may raise vocal volume, often to the point of shouting. After a while, he/she may find this unsatisfactory and lose interest in monitoring his/her speech, not realizing the deterioration that is occurring. Without the ability to monitor one's voice, the individual is unable to control vocal volume, modulation, and pronunciation accuracy. With the loss of this important monitoring system in people with profound hearing loss, various speech and voice changes usually occur.

1.3.4. Effect on Social Contacts

Unlike other handicapped individuals, the hard-of-hearing have no outward signs of disability, and strangers can sometimes confuse imperfect hearing with imperfect intelligence. This misconception and similar attitudes can create a strained relationship between a speaker and a listener. The hard-of-hearing person may miss the depth of conversations, usually enriched by side remarks and innuendoes which are often unheard. This lack of a social connection eventually makes the person feel shut off from the normal-hearing world, making him/her prey to discouragement and hopelessness. Until a person loses some hearing, he/she hardly can realize how important it is to hear the small background sounds. These sounds help us feel alive and their absence makes life seem to be rather dull. Imagine missing the sounds of rustling leaves, footsteps, keys in doors, motors running, and the thousands of other little sounds that make human beings feel that they "belong."

2. AURAL REHABILITATION

Hearing loss may be medically or surgically treatable, but for more permanent hearing losses, like most sensorineural hearing loss, a cure does not exist. However, a great deal can be done to help the individual compensate for hearing handicap and lead as normal a life as possible, minimizing effects on personality as well as social and economic status. This is all done through a method emphasized during World War II and described as "aural rehabilitation". Thousands of servicemen with hearing impairments were rehabilitated successfully through the large hearing centers established by the Army, the Navy, and the Veteran's Administration. Although few such centers are available to civilians, many private otologists, audiologists, university centers, and hearing aid centers can provide rehabilitation measures for persons with handicapping hearing losses that cannot be corrected medically or surgically.

Almost everyone with a handicapping hearing loss can be helped by effective aural rehabilitation. The principal objective of such a program is to help the individual overcome hearing handicap through multiple components. The program includes the following:

1. *Giving the individual a clear understanding of the hearing problem* and explain why he/she has trouble hearing or understanding speech. This requires the otologist or audiologist to demonstrate to the patient how the hearing mechanism works and where the patient's pathology lies. The patient also should be given a clear understanding of the difference between hearing trouble and trouble in understanding what is heard. It also should be explained why he/she has more difficulty understanding speech in the presence of background noise or when several people are speaking simultaneously. The problems that might easily lead the patient to develop frustrations and personality changes should be explained clearly so that these problems can be met forthrightly and intelligently. The goal of aural rehabilitation is to prevent or mitigate psychosocial changes that may result from hearing loss.

2. *Psychological adjustment for each patient*, which involves giving the patient more penetrating insight into the personality problems that are already in evidence or likely to develop as a result of hearing loss. The individual also must be treated in relation to his/her job, family, friends, and way of life. Therapy does not use a generalized technique, but must be designed specifically to meet the needs of the hearing-impaired individual. Frequently, it is advisable to carry out this part of the program not only with the patient, but also with the patient's spouse or family, because it is impossible to separate a person's individual problems from the problems of his/her family. At this point in the program the patient must accept his/her hearing impairment as a permanent situation and not wait for a medical or surgical cure. Above all, confidence and self-assurance must be instilled in the patient. The patient must be encouraged to associate with friends and not become isolated because of difficulties in communication. It must be impressed on him/her that effectively using the hearing that remains allows him/her to

achieve ambitions and carry on as usual with only minor modifications. The assistance of a psychologist specializing in hearing impairment can be invaluable at this stage.

3. *The fitting of a hearing aid, when it is indicated.* This is a vital part of the program, but before a patient can be expected to accept a hearing aid, he/she must be realistically prepared for it. Many people are reluctant to use hearing aids and many who have purchased hearing aids never use them or use them ineffectively. Before a hearing aid is recommended, it is necessary to determine whether the patient will be helped by it enough to justify purchasing one. This is particularly important in a sensorineural impairment in which the problem is more one of discrimination than of amplification. Older hearing aids did very little to improve a person's ability to understand, but improved abilities to hear by making sounds louder. Recent hearing aid technology targets sensorineural losses with poorer discrimination, greatly improving the amount of benefit these individuals can receive from a hearing aid.

One of the most important things that a hearing aid does for people with hearing loss is it permits the individual to hear sounds with greater ease, removing the severe it strain of listening. Although the individual may not be able to understand more with an aid than without one, he/she nevertheless receives great benefit from the device because it relieves tension, fatigue, and some of the complications of a hearing impairment. It also calls other people's attention to the hearing loss and encourages them to speak more clearly.

The patient who seeks early medical attention for hearing loss is wise in many respects. If the condition can be benefited by medical or surgical means, the patient often has a better chance of being helped if a diagnosis is established early in the condition. If a hearing aid is necessary and the patient acquires it quickly, he/she will experience a less drastic adjustment period while wearing the hearing aid. An adjustment period is necessary because the individual will hear sounds that have not been heard for a long time when wearing the hearing aids, such as the refrigerator hum, barking dogs, and shoes clicking on the floor. These sounds can be distracting at first, but the brain will adapt to a noisy environment if given an appropriate adjustment period.

Thousands of hearing aids, bought and then given a too brief or half-hearted trial, can be relegated to a bureau drawer. Overlong postponement in acquiring the aid is sometimes a factor in disuse. In addition, a patient's misguided expectations can lead to a diminished trial period. The patient often expects to hear normally with a hearing aid, when the condition of his/her residual hearing makes such a result impossible. Both the physician and the audiologist should make it clear to the patient that a hearing aid cannot be a perfect substitute for a normal ear, especially in the case of sensorineural hearing loss. Other common causes of disappointment with a hearing aid are an incompetent hearing aid salesperson and the patient's preoccupation with an "invisible" or inconspicuous aid, when a hearing aid that will be most effective in improving the ability to understand conversation should be purchased.

4. *Auditory training to teach the patient how to most effectively use residual hearing with and without a hearing aid.* If the patient can be helped with a hearing aid, he/she should also be made to realize that merely wearing the aid will not solve all his/her hearing and psychosocial problems. The patient needs to learn to use the hearing aid with maximum efficiency in such situations as person-to-person conversation, listening to people in groups and at meetings, and on the telephone. Above all, the patient must recognize the limitations of an aid and when to reduce its use in situations in which it would be a detriment.

Use of gestural cues and environmental cues can be beneficial to both the non-hearing aid user and the hearing aid user. The patient can be taught to use hearing more effectively by looking purposefully at the speaker's face and to develop an intuitive grasp of conversational trends so that he/she can fill in the gaps better than the average person.

5. *Speech reading—a broader concept of lip reading.* This is particularly important in patients who have profound hearing losses. It teaches the patient to obtain information from the speaker's face that cannot be obtained by sound communication. Patients can be taught to read the emotional tone of the conversation. They can also discern if a question or a statement was spoken based on facial expressions. Individuals with hearing loss can learn to use consonant information read from the speaker's lips to fill in missing pieces not heard by the patient. All people do a large amount of speech reading naturally and a person can develop this faculty extremely well through excellent training, although some individuals have a greater aptitude for speech reading than others.

By cooperating with a carefully planned and competently presented rehabilitation program, almost all people with handicapping hearing loss can be aided not only to hear better, but, more importantly, they can be helped to overcome the many personal problems and psychosocial difficulties that may result from a hearing impairment.

One factor that often complicates helping hearing-impaired persons is the long delay in obtaining medical attention. It is difficult to get some people with hearing loss to admit that they have a hearing problem and it is even more difficult to convince them that they should consult a physician about it. This is one of the reasons that physicians often do not see patients until their hearing impairment has had years to create marked social and communicative problems, both for the patient and for family and friends.

In older people, this delay usually is the result of pride and denial (they think, "This could not happen to me!"). When hearing handicaps are neglected in children, it is more often due to a lack of adult recognition of the problem. The delay in diagnosis is more regrettable for children as many of the conditions that cause hearing loss in the pediatric population are treatable if they are detected early enough and other conditions can be prevented from progressing further. Too often, the symptom of inattentiveness is attributed to a child who fails to respond when spoken to, overlooking the possibility of a hearing loss.

The emphasis in this chapter on rehabilitative measures short of a medical or surgical intervention reflects the fact that a total cure—especially in adults—is not possible in many cases of hearing loss. The often dramatically successful middle-ear surgery is limited mainly to the treatment of otosclerotics who have a reasonably good spread between air- and bone-conduction levels (air–bone gap). For such a surgery to have a chance of success, the patient must have a cochlea in at least fair working order and have a functioning auditory nerve. Unfortunately, these requirements are not met by a majority of hard-of-hearing patients. Yet, for those for whom medical surgery is not an option, rehabilitation often can do a great deal in helping them cope.

The physician can play an important part in helping patients overcome their psychological hurdles after a hearing aid, speech reading, and similar measures have been provided. The patient may still need help in adjusting socially, economically, and emotionally to the continuing handicap. One of our patients wrote a book for the hard-of-hearing public on living with a hearing handicap. He theorized that "above the ears there is a brain," an organ of often inadequately explored possibilities that can be used to solve many problems if properly used. He points out that people who do not use their brain for all it is worth really suffer from a handicap far more severe than a mere hearing impairment, leading him to a series of case histories of men, women, and children who have lived successfully with their hearing losses. Sometimes, these individuals have had the help of understanding parents and marriage partners, but, in other cases, they had success in the face of misunderstanding and discouragement. Sometimes, we prescribe this book for our patients (1).

The informed physician, working in close collaboration with an audiologist, is the ideal person to share his/her knowledge of the causes of hearing handicaps with patients to help them overcome the many hurdles that loom so much larger in the patient's mind than the hearing loss itself.

3. HEARING AIDS

A hearing aid is a portable personal amplifying system used to compensate for a loss of hearing. Almost all hearing-impaired patients are candidates for a hearing aid, although some will receive greater benefits from their aid than others. Any patient who is motivated to use a hearing aid deserves a thorough evaluation and a trial with an appropriate instrument.

Patients with conductive hearing losses generally are the best candidates for hearing aids because they do not have a distortion problem and just need amplification. Many of these patients may also qualify for corrective surgery. However, all such patients should be advised that a hearing aid might be an effective, if somewhat bothersome, alternative to surgery. Because surgery for conductive hearing loss has been so satisfactory, most people wearing hearing aids are those who have nonmedically treatable losses, such as sensorineural hearing loss, or in whom corrective surgery has failed.

Patients with sensorineural hearing loss not only have difficulty perceiving loudness, but also have trouble discriminating speech because of distortion present in their auditory system. Distorted voices become more difficult to understand when they are in the presence of background noise. A sophisticated hearing aid can help compensate for these difficulties by improving loudness perception and boosting the signal-to-noise ratio in noisy environments.

Modern technology allows for specific modifications of hearing aids, which can make a suitable hearing aid useful to almost any patient with hearing loss in the speech frequencies. For example, the frequencies of greatest loss can be emphasized selectively, while others are left unamplified. This is particularly helpful for patients who can hear low frequencies (vowels) but have substantial hearing loss for higher frequencies (consonants). Patients with recruitment may benefit from hearing aids with limited output to protect them from uncomfortably loud sounds. In addition, wearing a hearing aid alerts other people to the patient's hearing loss and frequently prompts them to speak more clearly.

3.1. Hearing Aid Evaluation

Any patient with a hearing loss should have a thorough otologic and audiologic examinations before purchasing a hearing aid. Hearing loss may be merely a symptom of a more serious underlying disease. After being medically cleared, the patient should undergo a formal "hearing-aid evaluation."

It is necessary to know the patient's usable residual hearing for pure tones and speech, ability to hear speech in a noisy environment, and tolerance for loudness. With this information, the audiologist can choose the most audiometrically appropriate hearing aid. The final decision of the purchased hearing aid is based on the above audiometric considerations and the patient's lifestyle, activity level, preference, and income.

The ultimate criteria for a successful fitting are user acceptability and satisfaction. A hearing aid is traditionally given a trial period of 30 to 45 days, depending on individual state regulatory guidelines to determine successfulness of the fitting. A comprehensive follow-up education program should be arranged to help the user obtain maximum communication with amplification. Such a program might include auditory training (or retraining) and speech-reading lessons. If the patient's speech has deteriorated as a result of the hearing loss, speech-retraining therapy also is recommended.

3.2. Hearing Aid Components

The basic parts of a hearing aid include a *microphone*, an *amplifier* circuit, a *receiver*, and a *power supply*. A *microphone* transduces acoustic energy into electrical energy, activated by sound waves impinging on the microphone diaphragm. The electrical energy is then fed into an *amplifier* circuit, which increases and adjusts the power of the electrical signal. The amplified signal activates the *receiver*, which changes the signal back into amplified acoustical energy to be delivered to the ear canal with increased intensity. The *power supply* of a hearing aid is usually a zinc-air battery cell.

3.2.1. Ear Molds

Ear molds provide a connection for the delivery of the sound to the ear from an amplifier source. Unless it is specifically designed as an "open" mold, the tight-fitting seal it provides is important in preventing acoustic leakage, known as feedback, between the microphone and the receiver. When a patient complains that his hearing aid is not working because it whistles, the ear mold is usually not sealing properly or has not been inserted correctly. When a person complains that the aid is not amplifying well or not working at all, it may be because the canal portion is occluded with cerumen, the battery is dead, or for other mechanical reasons.

The ear mold may be made of hard material such as Lucite or of soft materials with various trade names. Nonallergenic materials are also available for individuals with sensitive skin. The acoustic properties of the ear mold and the length and the inside diameter of the connecting tube play an important part in the final acoustical characteristics of the amplified sounds that reach the ear. An improperly designed mold can produce discomfort and reduce markedly the acoustical response of the most carefully selected and adjusted hearing aid.

3.2.2. Special Considerations for Abnormal Ear Anatomy and Earmolds

Surgically altered external auditory canals and/or auricles may pose special challenges when ear molds are required. Normal anatomy can be quite variable; however, anatomy is even more unpredictable in patients with congenital anomalies or those who have undergone ear surgery. An exhaustive review of possible anatomic distortions is beyond the scope of this discussion, but familiarity with basic principles should be helpful for anyone trying to create ear molds in patients with anatomic abnormalities.

Congenital abnormalities of the auricle vary from total absence to minimal disfigurement; birth defects of the external auditory canal range from total absence to mild narrowing. The ear canal may be not only narrow, but also short in distance from the meatus toward the ear drum. In such cases, there may be a bridge of tissue at any point along the ear canal, walling off the ear drum; absence of the medial portion of the ear canal and ear drum, so the ear canal ends as a shallow pouch; or the ear drum may be laterally placed (uncommon except following surgery). In patients with atresia (absence of the ear canal), a bone-conduction aid or implantable hearing aid is required unless an ear canal is created surgically. Reconstructive surgery is possible in virtually all cases, nearly always without injury to the facial nerve (which often is also abnormal). The amount of hearing loss present depends on the abnormality as well as on inner ear function.

During surgical reconstruction for congenital atresia, the surgeon faces several limitations, which help explain the challenges these patients pose when they require ear molds. First, the surgeon must be concerned about both function and cosmetics. Often there are two surgeons involved in caring for congenital atresia, particularly if it involves both the auricle and the ear canal. One creates an external ear, attempting to make it appear as normal as possible. The ear is often made of rib cartilage with a thin layer of skin over it. Even slight irritation from an ear mold may cause erosion, infection, and loss of an entire reconstructed auricle. From a functional standpoint, the surgeon (often a different member of the surgical team) is interested in creating an ear canal, making an ear drum, and possibly creating or reconstructing ossicles. This processes is usually staged. The functional surgeon wants to make an ear canal that is as large as possible because they always re-stenose and end up smaller than they were when they left the operating room. The position of the ear canal is dictated by the position of the middle ear, which often is misplaced anteriorly and inferiorly and does not coincide with the cosmetically desirable ear canal position. The cosmetic and functional otological surgeons create compromises that permit acceptable appearance and function.

An ear mold can be used postoperatively as a surgical packing and/or as an assistive device used with a hearing aid. One of the authors (R.T.S.) often uses an earmold postoperatively as a stent to help maintain a large ear canal; this technique has not been reported previously. Often, the ear mold is created in the operating room. The ear block is cut down so that it is as small as possible and then placed against the grafted tympanic membrane. A mold is shaped to the contours of the entire surgical defect. A thin coating of antibiotic drops or ointment is placed in the ear prior to application of ear mold material, as the procedure is "clean" rather than completely sterile. When the mold is sent for processing, it is important to specify that it should not be shortened. This mold is intended to go all the way to the ear drum to help maintain ear canal size. Additionally, a large bore in the ear mold will enable passage of some air and low-frequency sound into the ear canal. We make a second mold immediately in the operating room and leave it in the ear to function as surgical packing and stenting. We use a medium viscosity silicone impression material to make this mold. This will prevent the material from distorting or stretching the ear canal after surgery, without generating too much pressure from the application syringe. It is removed 1 week following surgery and the permanent mold with the large bore is placed in the ear. It can be removed for cleaning once or twice daily, coated lightly with antibiotic ointment and reinserted. This technique usually results in final ear canal anatomy that is not only adequate in size, but also of a shape that will hold an ear mold. However, this technique is not used widely, yet.

Ear molds for surgical ears can also have retention problems due to the nature of reconstructive surgery. Most ear canals that have been created surgically are fairly straight and they may be somewhat flared at the meatus. This can sometimes create problems with mold retention. If they cannot be overcome by mold design, surgery is sometimes necessary to narrow the lateral aspect with the meatus slightly to create a small ridge to help hold the ear mold, or to widen the medial aspect of the ear canal, also creating an anatomic indention that will stabilize the mold. If the meatus is soft and "floppy" after surgery, and if there is plenty of soft tissue between the skin and cartilage or bone, a hard, acrylic material may be the most suitable for an ear mold, although a softer material can certainly be used. However, if the postsurgical anatomy presents a hard, inflexible meatus with thick skin adherent to bone, a softer, silicon material may be preferable, providing a better seal and greater comfort. In addition, softer material also may work better when the anterior wall of the external auditory canal has been removed during surgery, and the remaining anterior canal wall communicates with the region of the temporomandibular joint. In such cases, there is more ear canal motion than normal during chewing and talking; soft mold materials tend to work better than acrylic molds for excessive motion.

Chronic ear surgery in adults or children also may result in anatomic abnormalities. Simple mastoidectomy or "intact canal wall" mastoidectomy ordinarily leaves the ear canal undisturbed. However, both modified radical mastoidectomy and radical mastoidectomy involve removing the ear canal and creating a cavity. The difference between the two classifications is determined primarily by which structures are left in the patient. In a radical mastoidectomy, the ear drum, malleus, and incus are removed, and only the stapes (with or without its superstructure) is preserved. If a portion of the ear drum or additional ossicles can be saved, then the procedure becomes a modified radical mastoidectomy. In either case, the posterior wall of the external auditory canal should be removed to the level of the vertical portion of the facial nerve, and the inferior wall of the ear canal should be widened and beveled toward the mastoid tip. Because it is usually necessary to clean mastoid cavities, the meatus is also widened markedly (meatoplasty). It is often possible to place a mold and begin using a hearing aid within 8–12 weeks following the time of surgery; but healing will cause alterations in size and shape for 6–12 months. So, it may be necessary to remake ear molds periodically and patients should be warned of this potential inconvenience. Making molds for patients with mastoid cavities is not usually a major problem. Additional ear block material may be required, but there is usually enough irregularity in the shape of the canal, meatus and cavity to assure that there are curves and indentations that will help stabilize a mold.

Many patients who have mastoid cavities suffer repeated ear infections because of infections medial to the ear mold. Although this problem may be solved by venting when it is appropriate, the repeated infections are usually not the fault of the ear mold (assuming that there is no allergic reaction to the mold material). Rather often, they are related to pockets of residual disease, or moisture from the upper respiratory tract and middle ear that has not been sealed off from the rest of the mastoid cavity (a "tubal ear"). In the vast majority of cases, if hearing aid use is necessary in patients who are plagued with repeated otitis, the problem can be solved by revision surgery.

3.3. Types of Aids

There are four types of wearable hearing aids: the body aid, the eyeglass aid, the behind-the-ear (BTE) aid, and the in-the-ear (ITE) aid.

3.3.1. The Body Aid

Owing modern micro-technology, the body aid is now rarely dispensed. It is a large, high-powered instrument worn on the body and connected to the earmold. Body aids offer wide ranges of amplification and usually are used by patients who have severe hearing impairments. The microphone, amplifier, and battery are located in the case, which is worn on the body or carried in a pocket. The receiver is connected to the amplifier by a long wire and is attached directly to the ear mold. This separation of receiver and microphone helps eliminate acoustical feedback in high-amplification instruments. Body aids can be fit to losses of 40–110 dB HL.

3.3.2. *The Eyeglass Aid*

Another classic aid that is rarely dispensed is the eyeglass aid. The amplifying unit is housed in the arm of the eyeglasses and connects to the ear mold by a short length of tubing. Eyeglass hearing aids are appropriate for hearing losses of up to >70 dB HL. With special modifications, they can be used for even greater losses. Their use is limited due to poor cosmetics of the device and because of better technology available currently.

3.3.3. *The Behind-the-Ear Aid*

BTE hearing aids currently dominate the market for fitting of severe to profound losses, an application previously reserved for body aids and eyeglass aids. All of the necessary components of the amplifying system, including the battery, are held in a single case worn BTE. The amplified sound is then fed to the ear via a plastic hook attached to the tubing of an ear mold. This design provides adequate separation of microphone and receiver to reduce acoustical feedback, which is common in severe hearing losses. These aids can be used for losses in the 25–110 dB HL range, making them the most flexible instruments available on the market.

3.3.4. *The In-the-Ear Aid*

ITE instruments are the most widely dispensed hearing aids in North America today. Micro-digital technological developments allow for the entire hearing aid system to be housed inside the earmold shell. The aids can be used for losses in the 25–80 dB HL range. There are several different styles of ITE instruments available for dispensing: full shell, half shell, canal, and completely-in-the-canal (CIC), listed, respectively, from largest to smallest shell size. One drawback to the smaller ITE sizes is that they cannot provide as much amplification as the larger shells, making them inappropriate for more severe of hearing loss.

3.4. Performance Characteristics

All hearing aids have certain performance characteristics that must be taken into account when matching a patient to the best possible aid. The five most commonly considered characteristics are acoustic gain, acoustic output, basic frequency response, frequency range, and distortion. *Acoustic gain* is the decibel difference between the incoming signal reaching the hearing aid microphone and the amplified signal reaching the ear. *Acoustic output*, also called *saturation sound pressure level* (SSPL), is the highest sound–pressure level an aid is capable of producing. This parameter is important in assuring that the aid will not produce uncomfortably loud sounds, especially for patients with sensorineural hearing loss and recruitment, and that it will amplify the sound adequately for the patient. The *basic frequency response* is the curve found most commonly on the manufacturer's specification sheets. It describes the relative gain achieved at each frequency. Many conventional hearing instruments have external tone controls that can amplify or suppress certain frequencies, according to the needs of the user. For example, a hearing loss may involve only the frequencies of $\geq$2000 Hz. By appropriate selection or by manipulation of the tone controls, the hearing aid can be adjusted so that it will not amplify frequencies <2000 Hz. The *frequency range* is a calculated measure of the high- and low-frequency limits of usable amplification for an individual instrument. This is also important to determine if an aid is adequate for a user's hearing loss. *Distortion* of the acoustic signal is measured electronically and the output of the hearing aid must meet specifications established by the American National Standards Institute (ANSI).

3.5. Hearing Aid Circuitry

3.5.1. Analog Hearing Aids

Analog hearing aids were the first type of hearing aids available to consumers. They were best suited for amplification of signals in a generally quiet environment. Conventional instruments were adjusted manually (with a screwdriver) through a variety of potentiometers. External output and gain tone potentiometers extended the applicability of the hearing aid to a wider range and patterns of hearing loss. Once these controls were adjusted properly, the wearer should not change them.

3.5.2. Analog–Digital Hybrids

The next technology to become available in the hearing aid industry was analog–digital hybrid, also known as programmable hearing aids. These hearing aids allow for digital, computerized control of analog circuits and offer far more precision in the fitting process than is available in conventional analog hearing aids. Initial programming was more involved and time consuming, which may increase the price of the fitting. The improved sound quality, frequency shaping capabilities, and enhanced output controls make the analog–digital hybrid a more flexible option for hearing aid users.

3.5.3. Digital Hearing Aids

The most current state of the art in hearing aid circuitry is the digital hearing aid, the most dynamic and flexible instrument available to consumers. A digital instrument takes an acoustic input, changes it into binary data, performs any amplification and special modifications, and then transduces it back to an acoustic form to be delivered to the patient's ear. As the signal is in a binary form, an unlimited number of enhancements can be performed by a digital hearing aid. For example, digital hearing aids are designed to amplify a broader set of inputs—keeping soft sounds soft and audible, medium sounds are comfortably amplified, and loud inputs are perceived as loud but not uncomfortable. These instruments provide for an expanded dynamic range to sounds, especially useful for patients with recruitment. Digital instruments clearly offer the best sound quality and the most flexibility in fitting.

3.6. Controls

Digital technological advances have made signal processing capabilities far superior to those designs of the past. In particular, the features listed below have improved the quality of sound to make it seem natural and far more pleasing to the listener.

3.6.1. Proprietary Speech Enhancing Algorithms

With the digitalization of amplification, each manufacturer has developed a mathematical formula to improve speech sounds prior to the amplification stage. Utilizing the splitting of the speech band into smaller elements, these areas are controlled to package sound for better understanding and enjoyment.

3.6.2. Automatic Gain Control

Automatic gain control (AGC) circuits maintain the overall intensity of the output and prevent it from reaching an individual's uncomfortable listening level, effectively keeping sound within a maximum comfort range. The level at which this circuit is activated can be preset by an audiologist or a hearing-aid dispenser. The AGC is especially useful for those individuals with sensorineural hearing loss who do not tolerate loud sounds well. Digital technology now allows this control to occur inside specific frequency sections providing a superior means of adjustment than in the past.

3.6.3. Noise Suppression Features

Noise suppression circuits and dual microphone technology function to reduce unwanted background noise by either improving the signal-to-noise ratio using noise suppression, or using noise cancellation. This technology allows for improved speech understanding by modifying the speech areas. These features can be particularly useful in noisy environments such as a crowded restaurant or party. Noise suppression features should be considered especially for patients who are actively presented with speech-in-noise difficulties.

3.6.4. Feedback Reduction

Feedback is an acoustical output from a hearing aid, often characterized as a "whistle" or a "squeal." Feedback often occurs at high frequencies with a loose fitting ear mold or in the presence of an occluded ear canal. Feedback reduction circuits eliminate acoustic feedback caused by large peaks in the frequency response of hearing aids. These nonlinear peaks can cause harmonic and intermodulation distortion, generating user dissatisfaction with the instrument. There are other ways of reducing acoustic feedback, but they involve compromising overall acoustic gain. Some methods of feedback reduction are low-pass notch filters (hard rejects), which filter the peaks in the frequency response, frequency shifting circuits, and phase cancellation. These methods of feedback reduction generally increase user satisfaction with the instrument by reducing feedback during everyday activities such as chewing and smiling. Feedback reduction features are only available in digital and programmable instruments.

3.6.5. Volume

Most hearing aids have a volume control by which the user can vary the intensity of the signal reaching the ear. In some aids, the volume control also acts as an on–off switch. In the newer circuitry, the hearing aid may not have a volume control because the circuit modifies its output based on the loudness of the input, automatically adjusting the volume of the hearing aid. The digital circuits, depending on the size of the shell, have the option of adding a manual volume control, or activating the control allowing for some wearer control over the function.

3.6.6. Switches and Buttons

Some hearing aids have a separate switch for turning the aid on and off. Also incorporated into this switch may be a telephone (T) position for using the telephone. The telephone receiver is placed over the magnetic field of the hearing aid with the switch in this position.

The hearing aid is then set to receive only the magnetic signal of the telephone. After completion of the call, the switch is returned to the microphone (M) position to receive acoustical signals again. Some instruments have an MT position that allows the user to receive both magnetic and acoustical signals through the hearing aid at the same time. The telephone position also can be used with various assistive listening devices that use a magnetic field to transmit signals.

Recent technology has developed memory or program buttons on digital and programmable hearing aids. Multiple-memory hearing aids allow the user to scroll through preprogrammed settings in the hearing aid based on the person's listening environment. An audiologist can load acoustical programs, such as a quiet, noise/party, telephone, or music program, which modifies the output of the hearing aid. In one memory setting, the hearing aid may provide a generous low-frequency response; in another, the hearing aid changes directional characteristics to suppress background signals. Switching between memory positions allows the user to optimize hearing aid function in different listening environments.

3.7. The Hearing Aid User

Any patient with abnormal hearing may be a candidate for amplification. In general, when the thresholds in the speech frequencies are $\geq$25 dB, a hearing aid could be helpful, even if only one ear is involved. When both ears are involved and have usable residual hearing, binaural amplification is often recommended. Binaural amplification has been shown to be advantageous for three reasons—binaural summation, elimination of the head shadow, and binaural squelch. Binaural summation is the perception of a sound as louder ($\sim$3 dB) when it is presented binaurally vs. monaurally (2,3). When an individual is aided monaurally and speech is presented to an unaided ear, a head shadow effect is created where high-frequency speech signals are attenuated up to 16 dB because they cannot travel around objects as well as low-frequency sounds due to short wavelengths (4). Binaural hearing aids eliminate the head shadow effect by abolishing the need for sound to travel around the head to be detected. The third advantage to a binaural hearing aid fitting is called binaural squelch, which is the reduction in perceived noise or reverberation by the auditory system when the two ears are symmetrically stimulated, thus enhancing speech detection in noise (5,6). All three of these well-researched advantages should be presented to patients when discussing monaural vs. binaural hearing aid fittings.

Not all patients are receptive to using a hearing aid, as a stigma is often attached to hearing loss. It is not uncommon for a patient to refuse an aid or become upset over the recommendation. A few moments of calm, sensitive counseling may be invaluable. Associations of hearing aids with "getting old," mental incompetence and physical unattractiveness (especially in teenage patients) can be dispelled by a sensitive physician or audiologist.

3.7.1. Hearing Aids in Children

Studies have demonstrated that even a mild hearing loss in infancy and early childhood can have an effect on learning and development. A child with a hearing loss that cannot be corrected medically or surgically should be fitted with a hearing aid as soon as the hearing loss is diagnosed. This may be even as early as the age of $\geq$6 months. With the advent of newborn hearing screenings, more children with hearing loss are identified at an earlier age, decreasing the impact the hearing loss might have had on their language and development. A properly fitted hearing aid is usually well tolerated by the child because of the benefit he/she receives from it. In general, children should be fitted with binaural amplification in order to maximize auditory input for speech–language development and to provide sound localization capabilities.

3.8. Noncandidacy for a Hearing Aid

There are a few circumstances under which fitting a proper hearing aid may be difficult or impossible. In certain ears, the use of an ear mold may be medically contraindicated. This precludes the use of the majority of hearing aids, except those with open canal ear molds or bone-conduction receivers. Some ears, particularly in cochlear hearing loss, are extremely sensitive to loud sounds, despite the fact that they have substantially reduced hearing thresholds. Such a narrow *dynamic range* may make a hearing aid more bothersome than helpful. A similar situation is seen in ears with severely reduced discrimination ability. A hearing aid, in the latter scenario, would only be useful for sound awareness.

On August 25, 1977 the Food and Drug Administration established rules regulating professional and labeling requirements and conditions for the sale of hearing aids. In accordance with this regulation, the hearing aid dispenser or audiologist should advise a prospective user to consult a licensed physician (preferably an ear specialist) if any of the following conditions exist:

1. Visible congenital or traumatic deformity of the ear;
2. History of active drainage from the ear within the previous 90 days;
3. History of sudden or rapidly progressing hearing loss within the previous 90 days;
4. Acute or chronic dizziness;
5. Unilateral hearing loss of sudden or recent onset within the previous 90 days;
6. Audiometric air–bone gap $\geq$15 dB at 500, 1000, and 2000 Hz;
7. Visible evidence of significant cerumen accumulation or a foreign body in the ear canal;
8. Pain or discomfort in the ear.

The law "permits a fully informed adult to sign a waiver statement declining the medical evaluation for religious or personal beliefs." The dispenser or audiologist shall not sell a hearing aid until he/she receives a signed waiver or statement from a licensed physician indicating that the patient's "hearing loss has been medically evaluated and the patient may be considered a candidate for a hearing aid." This FDA law also requires a 30-day trial period during which the aid(s) can be returned.

3.9. Special Hearing Aid Systems

3.9.1. CROS Systems

When a patient has one ear that is not suitable for amplification, sound can be routed from the unaidable side to the better hearing side. This is especially helpful when trying to hear a conversation from the unaidable side. However, it does not restore the ability to localize the direction of the sound source.

CROS is an acronym for contralateral routing of signals. The microphone is placed on the unaidable side and the amplifier and receiver are located on the better hearing ear. The amplified sound is carried across the head via a wire, through a pair of eyeglasses, or by a radio signal, and fed to the better ear by a tube to an open ear mold. This arrangement is most useful in cases of unilateral hearing loss with up to a mild loss on the better hearing ear. A variety of modifications of the CROS principle have extended its applicability.

3.9.2. BICROS Systems

Bilateral contralateral routing of signals (BICROS), is used for patients with an asymmetric bilateral hearing loss. Microphones are placed on both sides of the head and the amplifier is placed on the more aidable ear. The amplified sound is fed only to the aidable ear via tubing and an occluding ear mold. In cases where the aidable ear has a high-frequency hearing loss, a similar arrangement with an open ear mold or an open BICROS is more appropriate. The BICROS can be thought of as a microphone on the unaidable ear and a hearing aid on the aidable side that also receives information from the contralateral side of the head.

3.9.3. Transcranial CROS

A *transcranial CROS* is similar to the CROS system, except that the signal is delivered to the better hearing ear by vibrating the bones of the skull. A deeply fit CIC hearing aid is manufactured and placed in the unaidable ear. The sound output of the CIC then travels to the better hearing ear via bone conduction. A transcranial CROS is indicated generally for individuals who have normal bone-conduction thresholds in their better hearing ear.

3.9.4. IROS System

In patients with mild hearing losses or in whom hearing is normal in the lower frequencies with a precipitous drop off in the higher frequencies, the ipsilateral routing of signals (IROS) system is used. This employs an open ear mold to allow the normally heard low-frequency signals to enter the ear naturally and vent off unwanted amplified frequencies. The hearing aid is then set to provide enhanced high-frequency emphasis.

3.9.5. High-Frequency Hearing Aids

Several companies are now producing an alternative to IROS hearing aids for users with precipitous, sloping hearing losses. Instead of using an open ear mold, these hearing instruments are designed with special tubing that is inserted into the ear canal, leaving the canal more open than a large IROS vent. The specially designed tubing allows for more low-frequency information to pass into the ear canal naturally where the patient's hearing is normal. The hearing aid is either manually or computer programmed to provide amplification in the high frequencies where the patient's hearing loss is located. In order to measure the response of this special hearing aid, the audiologist or dispenser must use the new ANSI average called the SPA or "Special Purpose Average" to determine the gain and SSPL output. These new high-frequency emphasis aids show promise of providing benefit to those who have previously not been able to wear amplification.

3.9.6. Bone-Conduction Hearing Aids

Bone-conduction hearing aids are useful for patients with atresia or stenosis of the ear canal, microtia of the pinna, or those with chronic drainage. They are a nonsurgical alternative to deliver sound to an affected ear with a conductive hearing loss. A bone-conduction vibrator attached to a headband is placed on the mastoid bone. The vibrator is connected to a body-worn processor that transmits the acoustic signal and transfers it to a mechanical one. Although the bone-conduction hearing aid is a nonsurgical method of delivering sound to an ear, its output is limited by the presence of distortion at high intensity levels and the inability to provide large amounts of amplification.

3.10. Implantable Devices

3.10.1. Bone-Anchored Hearing Aids

An improvement on bone-conduction hearing aids is the *bone-anchored hearing aid*, BAHA™ (Entific Medical Systems Inc., Sweden). Similar to the bone-conduction instruments, it is useful for patients with conductive hearing losses that are unaidable due to ear malformations or chronic drainage. The BAHA utilizes an external bone-conduction processor that is attached to a titanium screw implanted in the temporal bone. Because the skull is directly stimulated, more amplification can be delivered to the better hearing ear without as much distortion. Besides being used for conductive losses, the BAHA has been approved by the FDA for patients with unilateral sensorineural hearing loss to be applied as an alternative to a CROS hearing aid. Currently, the device is recommended for patients with bone-conduction thresholds in the speech frequencies no worse than 25 dB HL (with no speech frequency worse than 40 dB HL) and air-conduction thresholds not better than 40 dB HL. Its principle disadvantage is the presence of the bone anchor that protrudes permanently though the skin.

3.10.2. Middle Ear Implants

A *middle ear implant*, known as the Vibrant® Soundbridge®, is used to treat mild-to-severe sensorineural hearing losses. It involves a vibrating ossicular prosthesis device implanted through the mastoid bone in a surgical procedure under local or general anesthesia. The prosthesis is then controlled by a magnetically attached external processing device, similar to a BTE hearing aid. The processor transmits the amplified signal to the prosthesis and is adjusted based on the individual's hearing loss.

3.10.3. Cochlear Implant

A cochlear implant is analogous in many ways to a hearing aid. It is used in very carefully selected patients with severe profound bilateral sensorineural hearing loss, too severe to be aided effectively with hearing aids. A cochlear implant involves surgical placement of electrode bundles directly into the cochlea. The electrodes are controlled by a processor worn on the outside of the head, either BTE or on the body, which sends electrical impulses along the bundles to the acoustic nerve. The cochlear implant initially began with only a single electrode array device. This design has been supplanted by multichannel devices (up to 24 electrodes), which provide a more complex signal and generally better speech discrimination than the original single channel system. A newer single channel system is under investigation and appears to provide nearly comparable hearing with simpler postoperative management and much lower cost (7). Cochlear implants produce true sound; however, the outcome (hearing ability) with an implant varies by patient. Some individuals are very successful with cochlear implants, even being able to use the telephone with their implant; other patients are limited to only having sound awareness. Patients who undergo cochlear implant surgery require intense rehabilitation in order to learn how to use the device optimally. Cochlear implants are currently approved for use in adults and children.

3.10.4. Brainstem Implants

Auditory brainstem implants (ABI) are useful for patients who have undergone removal of acoustic tumors for Neurofibromatosis type II (NF2). The surgery for acoustic tumor removal in NF2 patients typically ablates hearing bilaterally. The ABI provides sound awareness for those patients by implanting electrodes in the brainstem. This implantation is typically done during the surgery to remove the acoustic tumors. The electrodes bypass the severed acoustic nerves and transmit sound directly to the auditory brainstem nuclei. Through extensive auditory re-training therapy, the implant recipients learn to associate the new acoustical information to assist in detection of environmental sounds and some speech information.

3.11. Assistive Devices

There are many inexpensive convenience items available to hearing-handicapped persons. If a patient's hearing aid is not equipped with a telephone switch, the patient may benefit from a volume control inset placed on his/her telephone or an amplified telephone. A battery-operated, pocket-sized telephone amplifier is also available for travel. For patients with profound hearing losses, an auxiliary receiver may be installed in home and office telephones. This allows the patient's secretary or spouse to hear the incoming message and repeat it so that the patient can lip-read and respond directly into the telephone. Specialized aids are available for patients with particular occupational needs, such as transistorized switchboard amplifier for telephone operators.

TDD or TTY devices are telephone devices for the deaf or profoundly hearing-impaired person. TDDs and TTYs use a typewriterlike keypad with a printout or LED display. They can communicate with one another with text messages or can be used to communicate with a hearing person by using a relay system. Most states have a telephone relay service through which the typed message is transmitted to an operator who then reads the message to the hearing contact and types the replies back to the hearing-impaired individual.

Detecting routine auditory signals such as the telephone or doorbell is a special problem for the hearing-impaired. The telephone company can provide a variety of bells and buzzers, either frequency adjustable or amplified, which may help solve the problem. Auxiliary devices such as flashing lights also are available. Waking devices that vibrate can replace an alarm clock. Lights, vibrators, and specialized sound signals can be connected to sound-sensing devices. These may be placed, for example, in a baby's room so that they can be activated by the sound of a cry.

Personal amplification systems are available for listening to television or at a social event. These systems typically include an amplifying headset and direct audio input from a remote microphone. They allow the speaker to talk directly into the microphone and effectively cut back on the background noise. The reduction in background noise improves the signal-to-noise ratio, thus improving comprehension.

Many other convenience devices are available as well. Earphones and loudspeakers for radio and television listening are useful and relatively inexpensive. An induction coil device or telephone switch in the hearing aid can be used with the radio or television to selectively amplify the desired signal without amplifying the surrounding noise. Even electronic stethoscopes are available for the hearing-impaired physician and nurse. It is worthwhile for all professionals in the health-care delivery system to be familiar with the complexities of both the problems and the solutions of hearing loss. Only in this way can we provide optimum care and maximize the quality of life of the many afflicted members of our society.

REFERENCES

1. Van Itallie PH. How to live with a hearing handicap. New York, NY: Paul S. Erikson, 1963.
2. Hirsh I. The influence of interaural phase on interaural summation and inhibition. J Acoustical Soc Am 1948; 20:544–557.
3. Licklider J. The influence of interaural phase relations upon the masking of speech by white noise. J Acoustical Soc Am 1948; 20:150–159.
4. Feston J, Plomp R. Speech reception threshold in noise with one or two hearing aids. J Acoustical Soc Am 1986; 79:465–471.
5. Byrne D. Binaural hearing aid fitting: research findings and clinical application. In: Libby E, ed. Binaural Hearing and Amplification. Chicago, IL: Zenetron, Inc., 1980.
6. Ross M. Binaural versus monaural hearing aid amplification for hearing-impaired individuals. In: Libby E, ed. Binaural Hearing and Amplification. Chicago, IL: Zenetron, Inc., 1980.
7. Rubinstein JT, Parkinson WS, Tyler RS, Gantz BJ. Residual speech recognition and cochlear implant performance: effects of implantation criteria. J Otol 1999; 20:445–452.

17
Hearing Protection Devices

Sandra Markowitz
University of Pennsylvania, Philadelphia, Pennsylvania, USA

Robert T. Sataloff
Drexel University College of Medicine, American Institute for Voice and Ear Research, and Graduate Hospital, Philadelphia, Pennsylvania, USA

Joseph Sataloff
Thomas Jefferson University, Philadelphia, Pennsylvania, USA

The most effective means of preventing noise-induced hearing loss is to remove the worker from hazardous noise. Employers must try to reduce noise levels in industrial work environments when it is not possible to move the employee. High-intensity noise sources may need to be treated by covering surfaces with acoustically absorbent materials. If engineered controls are insufficient, OSHA requires employers to provide employees with hearing protection devices. Hearing protectors should provide immediate, effective protection against occupational hearing loss.

An effective hearing protection device serves as a barrier between the noise and the hair cells of the cochlear, where noise-induced damage to hearing occurs. A wide variety of hearing protection devices is available, including ear inserts, earmuffs, and canal caps. Ear inserts or earplugs provide an acoustical seal in the outermost portion of the ear canal. Earmuffs are worn over the external ear and provide an acoustical seal against the head. Canal caps provide an acoustical seal at the entrance of the external ear canal.

1. PROTECTOR PERFORMANCE CHARACTERISTICS AND LIMITATIONS

The protection afforded by a hearing protector depends on its design and on several physiological and physical characteristics of the wearer (1). While wearing a hearing protection device, sound energy can reach the inner ears by four different pathways: bone conduction, hearing protection device vibration, hearing protection device leaks, and leaks around the hearing protection device (Fig. 17.1).

At sufficiently high-noise levels, noise will reach the inner ear through regular air conduction travel, as well as through bone conduction. The practical limits set by the bone- and tissue-conduction thresholds and the vibration of the protector vary considerably with the design of the protector and with the wearer's anatomy. Approximate limits for plugs and muffs are shown in Fig. 17.2 (2,3).

If hearing protectors are to provide optimum noise reduction, acoustical leaks through and around the protectors must be minimized. A well-fit, imperforate hearing protection device will provide a higher noise reduction rating (NRR) than a porous hearing protection device (Figs. 17.3 and 17.4).

The following rules should be followed to minimize loss of attenuation due to acoustical leaks:

1. Hearing protection devices should be made of imperforate materials. If it is possible for air to pass freely through the material, noise will be able to pass as well.

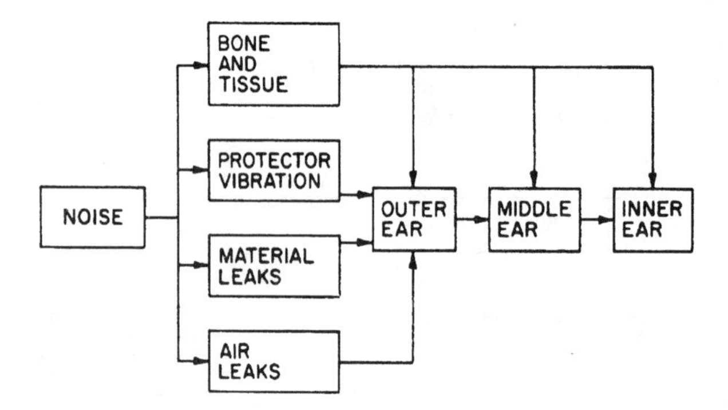

Figure 17.1 Noise pathways to the inner ear.

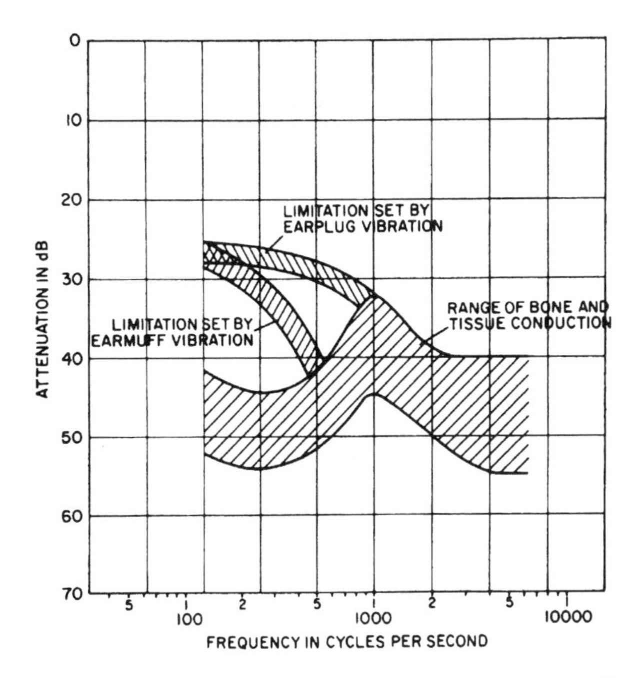

Figure 17.2 Practical protection limits for plugs and muffs.

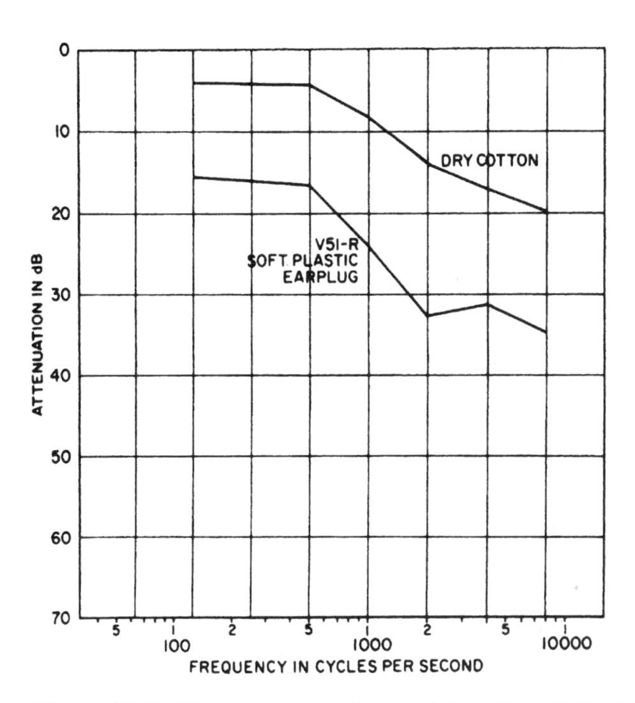

Figure 17.3 Mean attenuation characteristics of a well-fitted, imperforate earplug and an earplug made of dry cotton.

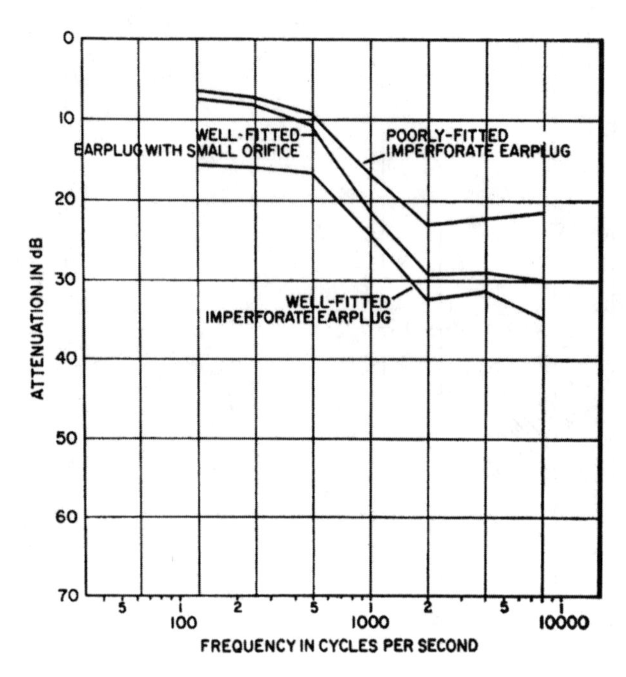

Figure 17.4 Mean attenuation characteristics of an earplug.

2. The hearing protection device should make a good seal in the ear canal or over the external ear. Reduction of a sound is dependent on preventing leakage.
3. Hearing protectors should be comfortable to wear. Uncomfortable devices are not used consistently.
4. Earmuffs should not be worn over long hair, poorly fit eyeglasses, or other obstacles.

2. PROTECTOR TYPES

2.1. Ear Canals

Ear canals differ widely in size, shape, and position among individuals, and between ears of the same individual. Earplugs should be chosen that are adaptable to a wide variety of ear canal configurations.

Ear canals vary in cross-section from ~3 to 14 mm, with a large majority falling in the range of 5–11 mm. Most ear canals are elliptically shaped, with a bend to direct sound toward the front of the head.

The space for an insert earplug is small. The entrance of the ear canal can be opened and straightened by pulling the external ear directly from the head (Fig. 17.5). For comfort and plug retention, the ear canal must return approximately to its normal configuration once the protector is seated.

Hearing protection devices generally are classified according to the manner in which they are worn. The three best-known types are ear inserts, earmuffs, and canal caps.

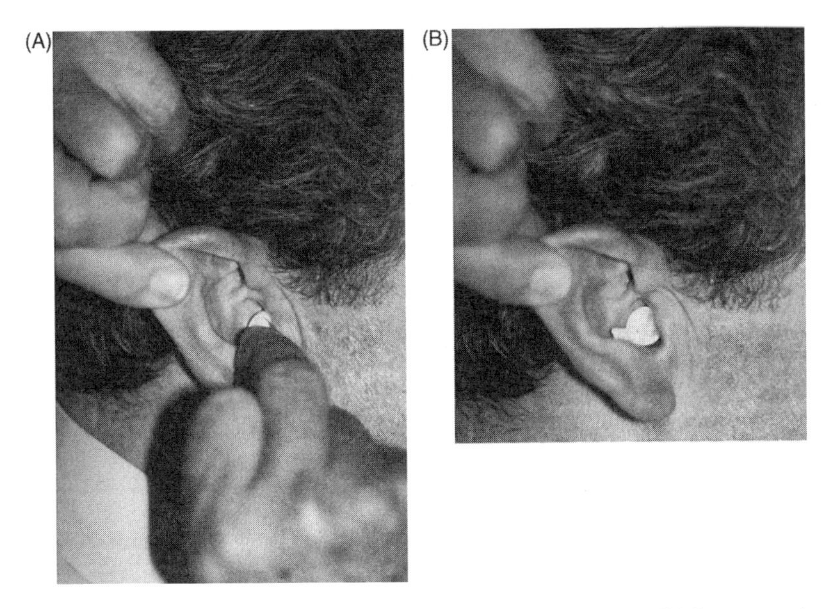

Figure 17.5 (A) Recommended method for inserting earplugs. (B) Proper seating of ear plugs.

2.2. Inserts

Ear inserts or earplugs fit directly inside the ear canal. Earplugs can be subcategorized into formable, premolded, and custom-molded. A properly fit earplug depends on a seal along the entire circumference of the ear canal walls. Ear inserts are supplied in three general configurations (Fig. 17.6). Cotton is the least effective protector, providing little attenuation, and it is not recommended.

Earplugs can be made from a variety of materials. Custom-made hearing protection devices generally are made from a thermoplastic or silicone material. Use of these materials allows earplugs to be made to fit the exact size and shape of the user's ear canal. Custom earplugs are able to provide a significant amount of attenuation. These earmolds tend to fit the most comfortably, making them the most likely to be worn as intended. Custom earmolds are reusable. However, the material does deteriorate after long-term usage.

Formable ear inserts are made from soft, compressible foam. These earplugs come in one size. The material is compressed to insert into the ear canal. Formable earplugs should be held in place while the foam decompresses to form an airtight seal. Once the foam seals the ear canal completely, formable earplugs provide a high attenuation value. Foam plugs tend to be difficult to insert properly into the ear. Annual employee education should instruct users to proper use of the earplugs. These earplugs may be washed and reused, but do need replacement often.

Premolded earplugs are available in specific sizes and shapes. The most common premolded earplug styles are the single flange and triple flange. These styles are available in various sizes to fit most people. Premolded earplugs are reusable, but do deteriorate and should be replaced periodically.

All earplugs have the advantage of being small and lightweight. In warm environments, ear inserts tend to be the most comfortable hearing protection device. It is easy to combine earplugs with other safety equipment, such as hard hats and safety goggles.

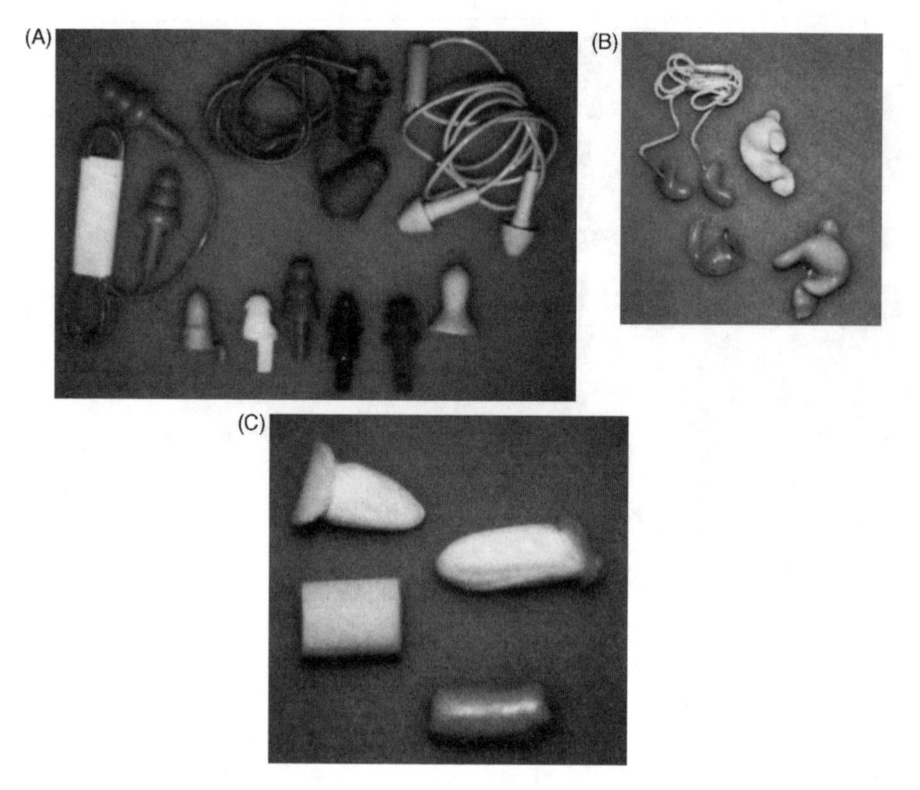

Figure 17.6 (A) Sized (B) custom molded, and (C) moldable, clockwise from top left: spun glass, spun glass, elastomer, expandable foam.

Earplugs require specific fitting instructions. Improperly seated earplugs will not provide the intended attenuation. Occasionally, ear inserts come loose. Excessive jaw movement may unseat earplugs. Often, earplugs are soiled. A custom-made earmold will need proper cleaning. Foam inserts are disposable and are thrown away after use. It is difficult to monitor earplug use as inserts sit deeply inside the worker's ear canal.

Some people are more prone to cerumen impaction than others. Earwax naturally migrates from inside the ear canal to the outside. Earplugs that temporarily occlude the ears rarely cause significant problems with wax impaction. If the ear becomes filled with earwax, it should be removed by a doctor.

2.3. Earmuffs

Earmuffs are designed to be worn over the ears to reduce the level of noise reaching the inner ear [Fig. 17.7 (A–D)]. The effectiveness of earmuffs depends on a hermetic seal between the ear cushion and the outer ear. The ear cups are made of plastic and lined with acoustic foam. The ear cups are held snugly in place with a headband. The headband can be worn over the top of the head, to the rear of the head, or under the chin. With proper hardware, earmuffs can be attached to hardhats. Sponge wedges are available to fill any openings between the earmuffs and safety glasses. Proper headband tension needs to be present to ensure a tight seal over the outer ears.

Most earmuffs have similar designs. Seal materials lining the ear cups that are placed against the skin should be made of a nontoxic material. Fit, comfort, and general performance of comparable models do not vary widely.

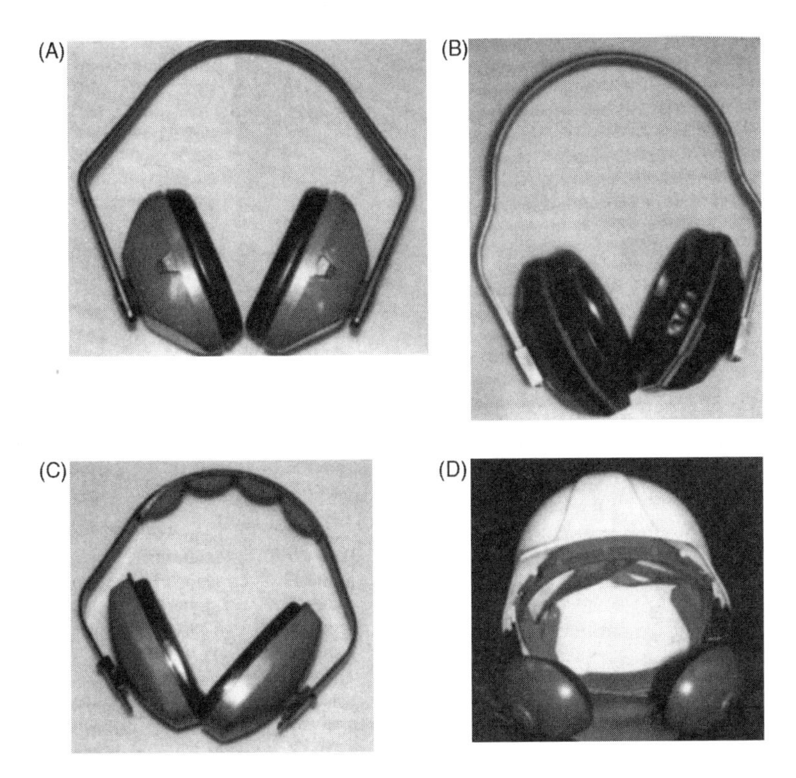

Figure 17.7 (A)–(D) Muffs.

If maximum protection is required, the protector ear cups must be formed of a rigid, dense, imperforate material. The size of the enclosed volume within the muff shell is directly related to the low-frequency attenuation. Ear seals should have a small circumference so that the acoustic seal takes place over the smallest possible irregularities in head contour. This minimizes air leaks caused by jaw and neck movements.

The ear cup is filled with an open-cell material to absorb high-frequency resonant noises. The material placed inside the cup should not contact the external ear. Contact may cause discomfort or soiling of the lining. Perspiration may cause the ear cup cushion material to become stiff or shrink. Fluid-filled cushions have the additional problem of leakage. Most earmuffs are equipped with easily replaceable seals to provide adequate protection.

Maintenance of the earmuffs is essential for maximum protection. The plastic of foam cushions can be cleaned with warm, soapy water. The inside of each earmuff should not get wet. When not in use, earmuffs should be placed in open air to allow moisture to evaporate.

Earmuffs have the advantage of having one size fits most users. Adjustable arms connected to the ear cups let the user adjust to a comfortable fit. The attenuation provided from earmuffs tends to be consistent among wearers. Use of earmuffs is easy to monitor because of their size. Employers can easily see whether employees are wearing earmuffs due to their visibility. Ear infection does not hinder earmuff use.

Less protection is expected when earmuffs are worn over hair or glasses. Obstructions should be minimized, otherwise the expected attenuation will not be provided. Earmuff use is difficult in warm environments and in close quarters. Warm environments make earmuff use uncomfortable. Perspiration tends to stick to the ear cup lining. Head maneuvering while wearing earmuffs in close quarters is restricted. Headband suspension forces can become reduced with usage. Reduced tension means that the ear cups are not being held to the ears tightly enough to ensure maximal attenuation.

2.4. Concha-Seated Ear Protectors

Protectors that cannot be strictly classified as earplugs or earmuffs include concha-seated ear protectors or canal caps. Sound attenuation with canal caps is achieved by sealing the external opening of the ear canal [Fig. 17.8 (A–C)]. The caps are made of two soft plastic conical caps and held in place by a springy headband. The adjustable headband allows canal caps to be worn in a variety of positions and can be combined to use evên with other protective equipment. Canal caps are ideal for employees who are constantly removing and replacing their hearing protection devices. They are not designed for continuous, long-term use.

Canal caps do not fit deeply into the ear canal, only close the ear opening. Therefore, they do not provide as much attenuation as earmuffs or earplugs. These hearing protection devices provide maximum protection at frequencies >1000 Hz. The attenuation <1000 Hz is considerably less by ~15–20 dB.

Although sizing is not a problem, good insertion instruction is essential with this type of hearing protection device. Concha-seated hearing protection devices are inserted similar to earplugs. The outer ear is pulled up and back. The tip of the cap is inserted into ear, pushing it into place. Canal caps are reuseable, but the caps do need replacing periodically.

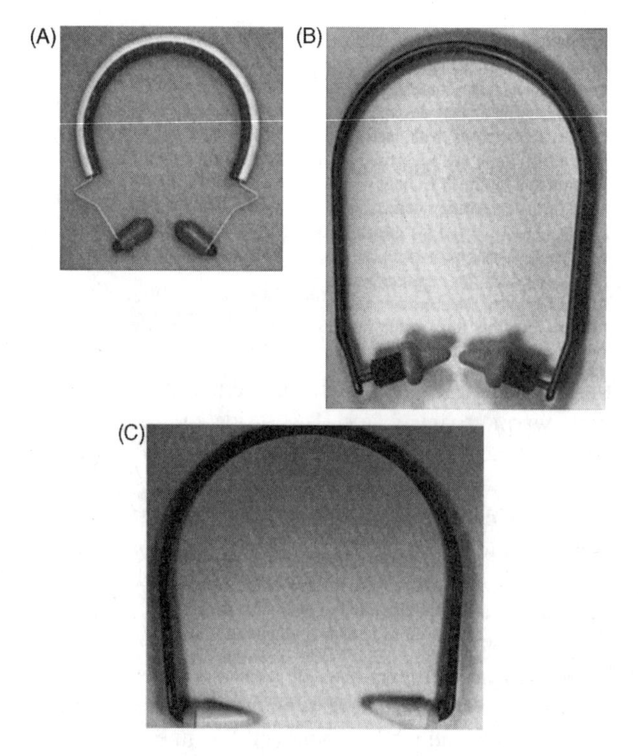

Figure 17.8 (A)–(C) Canal caps.

2.5. Combining Earplugs and Earmuffs

Both earplugs and earmuffs should be worn if the workers time-weighted noise exposure exceeds 100 dBA (Fig. 17.9). However, using double protection does not provide double attenuation. Use of earplugs and earmuffs provides an additional 5 dB of protection. The employee wearing both earmuffs with an NRR of 25 and earplugs with the NRR of 30 would be provided with ~35 dB of total attenuation. The National Institute for Occupational Safety and Health (NIOSH) cautions that even double protection may be insufficient if time-weighted average exposures exceed 105 dBA.

3. CONCERNS

Common reasons for workers to not wear their hearing protection devices include discomfort, interference in hearing speech and warning signals, and the belief that hearing loss is going to occur, regardless of hearing protection device use. Special care must be taken to ensure that the hearing protection device is fit to the user. A comfortable device that provides appropriate attenuation (neither less nor more) has the best chance of being used.

Wearing hearing protectors in very noisy areas makes it easier for normal hearing employees to understand conversations, instructions, and signals. Workers wearing protectors do not need to raise their voices as much as employees with no protection in the same noisy environment. Workers with hearing loss often state that use of a hearing protection device adds to existing hearing loss. Noise-induced hearing loss affects higher frequencies, between 3000 and 6000 Hz. Many consonant sounds are heard in at ~3000 Hz. A hearing loss in the high frequencies may cause speech discrimination problems. The attenuation curves provided for hearing protection devices show that

f(Hz)	Correction
25	− 44.7
32	− 39.4
40	− 34.6
50	− 30.2
63	− 26.2
80	− 22.5
100	− 19.1
125	− 16.1
160	− 13.4
200	− 10.9
250	− 8.6
315	− 6.6
400	− 4.8
500	− 3.2
630	− 1.9
800	− 0.8
1,000	0.0
1,250	+ 0.6
1,600	+ 1.0
2,000	+ 1.2
2,500	+ 1.3
3,150	+ 1.2
4,000	+ 1.0
5,000	+ 0.5
6,300	− 0.1
8,000	− 1.1
10,000	− 2.5
12,500	− 4.3
16,000	− 6.6
20,000	− 9.3

Figure 17.9 A-frequency weighting adjustments (4).

most of the attenuation is obtained in the higher frequencies. For this reason, the device may add to existing discrimination problems and the audibility of warning signals in selected cases, and hearing protectors should be individualized accordingly.

Education programs should stress that noise-induced hearing loss is avoidable. Ear protection can protect employees from future noise-induced hearing loss. Discomfort can be avoided with properly fit hearing protection devices.

4. OTHER HEARING PROTECTION DEVICES

4.1. Musician's Earplugs

Musicians are encountering hearing loss, though to a lesser extent than industrial workers. Whenever possible, the environment should be altered. Alterations include distancing the musician from the sound source and using acoustic baffles. When environmental modifications are not enough, a hearing protection device should be used. Musicians tend to use earplugs more frequently than earmuffs. Normal earplugs tend to have a high-frequency emphasis because these frequencies are damaged first by noise. The high-frequency emphasis compromises the musician's perception of sound quality. Musician's earplugs are custom molded earplugs with a passive filter to attenuate with a flatter frequency response. This gives sound a truer quality. By adding an acoustic filter to earmolds, high-frequency sounds are pre-emphasized, enabling uniform attenuation across the frequency range. A slightly different filter can be used to roll-off high-frequency information slightly. Another method of adjusting earplugs is the addition of a tuned vent. Tuned vents let low frequencies pass and high frequencies become attenuated. Musician's earplugs worn in high-level industrial noise may not be appropriate for hearing protection.

4.2. In-Ear Monitors

In-ear monitors are earpieces with high fidelity speaker systems built inside. In-ear monitors are used instead of conventional floor monitors. The earpiece itself can act as a hearing protection device. The device reduces unwanted noise so that the user can monitor himself or herself and the instruments at lower levels. This lower signal-to-noise ratio may provide some hearing protection, but may not be appropriate for industrial hearing protection.

4.3. Communication Headsets

Communication headsets are used primarily for broadcast, studio, film, and television applications. The headsets allow for increased communication, while providing sound attenuation through the use of dual headsets with a microphone. The headphone transducers feature extended frequency responses.

Noise cancellation from the microphone provides clear, audible speech. Noise cancellation can improve the signal-to-noise ratio at the microphone and at the receiver. This can be achieved with noise cancellation principles. A noise-canceling microphone picks up ambient noise through two apertures. This allows a portion of the low-frequency noise to be cancelled. One of the openings is oriented close to the mouth so that speech is not cancelled.

These electronic communication systems may improve speech clarity when used in high-level noise environments. The improved perceptions of the speaker's own voice allows better voice modulation. Also, some distortion that accompanies loud speech and shouting can be avoided. The listener can adjust the electronic gain control to obtain a level of undistorted speech that will provide the best reception. An exception may be found in very high levels, above ~130 dB overall, where speech perception is not always improved by the use of earplugs under communication headsets.

In noise fields >120 dB re 0.0002 μbar overall, noise may be attenuated at the microphone by noise shields encasing the sensing element. Efficient noise-attenuating shields tightly held around the mouth may be used with microphone systems to transmit intelligible communication in wide-band noise levels exceeding 140 dB.

4.4. Electronic Ear Protection

Electronic hearing protection can be divided into two categories, devices to improve short-range communications in noisy environments, and devices to provide entertainment during work or play. These devices are also known as noise canceling earmuffs and active ear muffs. The technology used in electronic ear protection devices is not noise cancellation. Various compression and filter techniques are used to improve communication in noise. Users can adjust the frequency range that is transmitted and the filtration level of high and low frequency noise. FM radio earmuffs are included in the electronic hearing protection category. These hearing protectors have an FM radio receiver built into the headphones. Other devices (such as a CD player) can be connected to these headphones.

5. HEARING PROTECTION, HEARING LOSS, AND HEARING AIDS

5.1. Use of Hearing Aids in Noise

A problem frequently faced by the hearing conservationist is the employee who uses a hearing aid, works in noise, and is required to use hearing protection. Workers with hearing loss may have difficulty hearing warning signals, posing a safety hazard to themselves and others.

Hearing aids are personal amplifiers that compensate for the deficit in hearing. Sounds reaching the hearing aid microphone are amplified. Use of hearing aids may make warning signals easier to hear. However, use of hearing aids in noisy industrial settings is not recommended. High-intensity sounds will be amplified by the hearing aid. The exposure may be traumatic to existing hearing. If workers are unable to perform their job because of the severity of their hearing loss, a transfer to a less noisy area is recommended. Efforts should be made to assist the employee before a transfer occurs.

5.2. Other Precautions

Earmolds worn with the hearing aid fail to provide adequate noise attenuation. Also, hearing aids advertised as having automatic noise suppression are not adequate for use in industrial noise. The circuit cannot work fast enough to reduce intense, transient noise.

6. NOISE REDUCTION RATING

Code of Federal Regulations, Title 40, Section 211, states that manufacturers are required to show the NRR on the label of hearing protection devices. The NRR is a simplified method of describing the amount of hearing protection a device provides. There are three standards for estimating the amount of attenuation provided. None of the methods specify minimum performance qualities or approve of a single type of hearing protection device.

The most common method of characterizing sound attenuation (ANSI Z24.22) is the real ear attenuation at threshold (REAT) method. This method finds the difference between the minimum levels of sounds that can be detected while ears are occluded and unoccluded. The change in hearing between the two conditions is called a threshold shift. This standard was revised with ANSI S3.19, and again with ANSI S12.6. The original standard data used puretone stimuli in an anechoic chamber. The revisions are based on 1/3 band octave noise presented at random incidence in a nondirectional sound field.

A problem with the NRR system is to ensure that the descriptor value does not sacrifice the estimated attenuation for simplicity. These methods often underestimate laboratory-derived methods for sound attenuation. The methods tend to give high weighting to the low-frequency noise, whereas high-frequency performance is not given the same consideration. Other methods have been developed to get around the shortcomings of the NRR.

The intention of the NRR was to calculate the exposure under the hearing protector by subtracting the NRR from the C-weighted unprotected noise level. However, noise levels are measured using the A-weighted scale. An additional 7 dB should be subtracted from the labeled NRR when working with the A-weighted noise levels to estimate the A-weighted noise level under the protector.

7. AMOUNT OF PROTECTION PROVIDED IN PRACTICE

A comparison of the noise analysis of a particular noise exposure and the levels specified by the chosen hearing conservation criterion should be used to determine the amount of bands noise reduction required. The hearing conservation criteria are often expressed in octave bands. The amount of noise reduction required can be determined by subtracting the sound–pressure levels specified by the criteria from the exposure levels measured in corresponding octave bands. Hearing protection should provide attenuation within each of these bands to meet the noise reduction requirements. Mandatory methods for estimating the adequacy of hearing protector attenuation are specified in Appendix B of the OSHA Hearing Conservation Amendment (see Chapter 14).

Hearing conservation criteria based on the A-frequency weighting may be used following the NIOSH method described subsequently:

Step 1: Take sound level measurements in octave bands at the point of exposure.

Step 2: Subtract from the octave-band levels (in decibels) obtained in Step 1 the center frequency adjustment values for the A-frequency weighting shown in Fig. 17.10.

Step 3: Subtract the A-weighted octave bands calculated in Step 2 the attenuation values provided by the protector for each corresponding octave band to obtain the A-weighted octave-band levels reaching the ear while wearing the hearing protector.

Step 4: Calculate the equivalent A-weighted noise level reaching the ear while wearing the hearing protector by adding the octave band levels

Sum (L_R) of dB Levels L_1 and L_2 Numerical Difference Between Levels L_1 and L_2	L_3: Amount to Be Added to the Higher of L_1 or L_2	
0.0 to 0.1	3.0	Step 1: Determine the difference between the two levels to be added (L_1 and L_2).
0.2 to 0.3	2.9	
0.4 to 0.5	2.8	
0.6 to 0.7	2.7	
0.8 to 0.9	2.6	
1.0 to 1.2	2.5	Step 2: Find the number (L_3) corresponding to this difference in the Table.
1.3 to 1.4	2.4	
1.5 to 1.6	2.3	
1.7 to 1.9	2.2	
2.0 to 2.1	2.1	
2.2 to 2.4	2.0	Step 3: Add the number (L_3) to the highest of L_1 and L_2 to obtain the resultant level ($L_R = L_1 + L_2$).
2.5 to 2.7	1.9	
2.8 to 3.0	1.8	
3.1 to 3.3	1.7	
3.4 to 3.6	1.6	
3.7 to 4.0	1.5	
4.1 to 4.3	1.4	
4.4 to 4.7	1.3	
4.8 to 5.1	1.2	
5.2 to 5.6	1.1	
5.7 to 6.1	1.0	
6.2 to 6.6	0.9	
6.7 to 7.2	0.8	
7.3 to 7.9	0.7	
8.0 to 8.6	0.6	
8.7 to 9.6	0.5	
9.7 to 10.7	0.4	
10.8 to 12.2	0.3	
12.3 to 14.5	0.2	
14.6 to 19.3	0.1	
19.4 to ∞	0.0	

Figure 17.10 Table for combining decibel levels of noise with random frequency characteristics.

as shown in Fig. 17.11. An alternative method for adding decibels is as follows:

$$\text{Equivalent dBA} = 10\log_{10}\left(\text{antilog}_{11}\frac{L_{125}}{10} + \text{antilog}\frac{L_{250}}{10} + \cdots + \text{antilog}\frac{L_{8000}}{10}\right)$$

where L_{125} is the sound–pressure level of the A-weighted octave band centered at 125 Hz; L_{250} the A-weighted octave-band level at 250 Hz etc.

Example: The first and second columns in Fig. 17.12 contain octave band sound pressure level data measured in a textile mill weaving room. The third column shows the same octave-band data, but with an A-frequency weighting (using Fig. 17.10). The mean and mean −1 SD attenuation values for a good earmuff hearing protection device (Fig. 17.13) are listed under 1 and 2 of the fourth column heading. By definition, 50% of the persons wearing this hearing protector can be expected to have less than the mean attenuation values, and ~14% can be expected to have less than the mean −1 SD. Octave-band levels reaching the ear canal while wearing the hearing protection device are listed for mean and mean −1 SD attenuation values under the last column heading. The octave-band exposure levels listed in each of the last two columns may be added (using Fig. 17.11) to determine the

Octave Band Center Frequency in Hz	Weaving Room Spectra in dB		Weaving Room Spectrum with A-Weighting in dB	Muff-Type Protector Attenuation in dB (1)	(2)	Resultant Exposure to Inner Ear in dB (1)	(2)
125	90	(less 16 =)	74	16	9	58	65
250	92	(less 9=)	83	21	15	62	68
500	94	(less 3=)	91	31	23	60	68
1,000	95	(less 0=)	95	42	30	53	65
2,000	97	(plus 1=)	98	43	32	55	66
4,000	95	(plus 1=)	96	45	35	51	61
8,000	91	(less 1=)	90	34	22	56	68
Overall	103 dB		102 dB(A)			66	75 dB(A)

Figure 17.11 Protection provided by ear protectors worn in a weaving room noise environment.

decibel exposure levels while wearing hearing protectors. In this example, the exposure level for those receiving the mean attenuation values from the protector would be 66 dBA, and for those receiving mean −1 SD values, the exposure level would be 71 dBA. Either exposure level while wearing hearing protectors is obviously well <90 dBA.

Two factors must be considered when selecting hearing protectors to meet the noise reduction requirements. First, the hearing protection device attenuation values determined in the laboratory are not always an accurate representation of the noise reduction capability of the protector. Second, the protection provided by hearing protectors varies

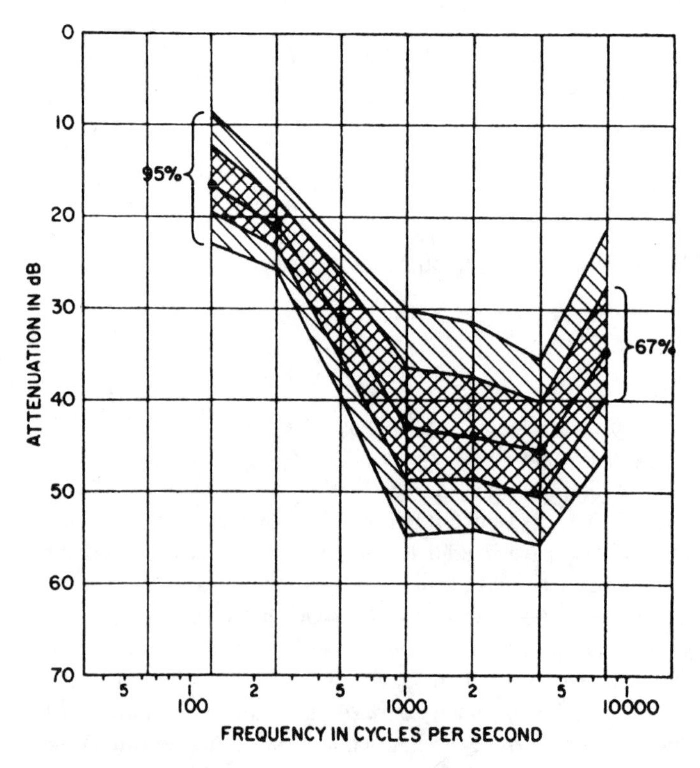

Figure 17.12 Mean attenuation characteristics of a good muff-type protector plotted with one and two standard deviation shaded areas (67% and 95% confidence levels). Attenuation values determined according to the ANSI specifications using pure-tone threshold shift technique (5).

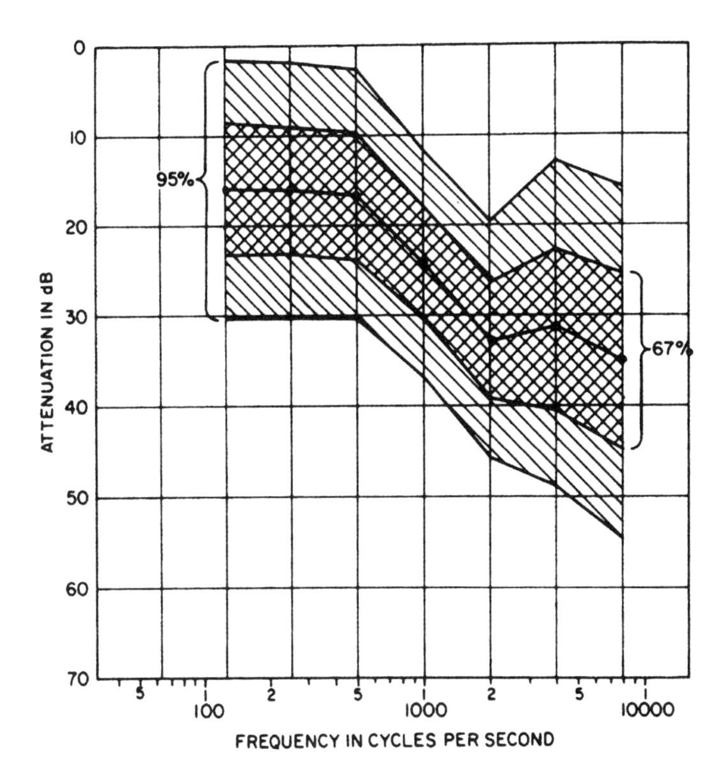

Figure 17.13 Mean attenuation characteristics plotted with one and two standard deviation shaded areas for a well-fitted imperforate earplug. Attenuation values determined according to the ANSI specifications using pure-tone threshold shift technique.

considerably among wearers and among the ways the protectors are worn. There are also significant differences in performance among some protectors of the same model. Standard deviations of 3–7 dB are commonly found in subjective measurements of protector attenuation at any test frequency. Therefore, attenuation values may have a range of ±14 dB for 95% confidence limits. The variability in the amounts of protection provided must be considered along with the mean attenuation values when selecting a hearing protector for a particular application.

Confidence limits of 100% of employees protected all the time are desirable. However, the spread of attenuation values is so great that very high-level confidence limits are not plausible. A practical choice is the mean attenuation minus one standard deviation, which would provide confidence limits ~86% level. This confidence limit would be very similar to the limits set by most of the present rules and regulations concerning noise exposure levels, which also have been limited by practical considerations.

In common practice, a much more simple method is used and provides fairly comparable results. To estimate the effective noise reduction, subtract 7 dB from the supplied NRR and divide that number by 2. For example, if an ear protector is reported to have a NRR of 27 dB, subtract 7 dB, then divide that value by 2, leaving 10 dB. So, the expected attenuation in the field would be 10 dB. If the environment exposed the person to noise levels of 95–99 dB, this would be an appropriate ear protector.

Additional methods of determining NRRs are presented in Appendix B of the OSHA Noise Amendment. An update to OSHA's Field Operations Manual on January 27, 1984, described administrative guidelines that only required protective measures in most situations. Citations will only be issued if engineering controls are feasible and either the "employee exposure levels are so high that the hearing protectors alone may not reliably reduce noise levels," or "the costs of controls are less than the cost of an effective hearing conservation program."

The point at which hearing protectors can reliably reduce a noise level is 100 dBA. Above this level, hearing protector NRR must be derated by 50% and must reduce exposures below 90 dB.

When calculating hearing protection effectiveness for comparison to engineering controls (50% derating), the 7-dB adjustment must be used along with the 50% derating. Therefore, to be effective for levels of 106 dBA, the NRR published by the manufacturers must be at least 39 dB, that is, $(39 - 7)/2 = 16$ dB. This will result in an exposure level of 90 dBA $(106 - 16 = 90)$.

It is possible for a particular hearing protector with a published NRR to have four different NRRs used by OSHA. A protector with a NRR of 30, would be under Appendix 2, maintain the 30 NRR of noise is measured on the C scale (dBC) and only 23 if measured on the A scale (dBA). For comparison to engineering controls, the NRR is 15 dBC and 11.5 NRR (dBA).

Because the effective protection provided by a given set of hearing protectors can only be approximated in most field applications, a hearing monitoring program is essential for all persons wearing hearing protectors. Fortunately, a very large majority of noise exposures are at relatively low levels, and the proper use of good hearing protection devices can provide adequate attenuation for a large majority of persons exposed to these noises (Figs. 17.13 and 17.14).

In summary, the use of a good ear protector can provide sufficient attenuation in the majority of work environments where engineering controls cannot be used successfully. For the few workers exposed to noise levels in excess of 115 dBA over long periods of

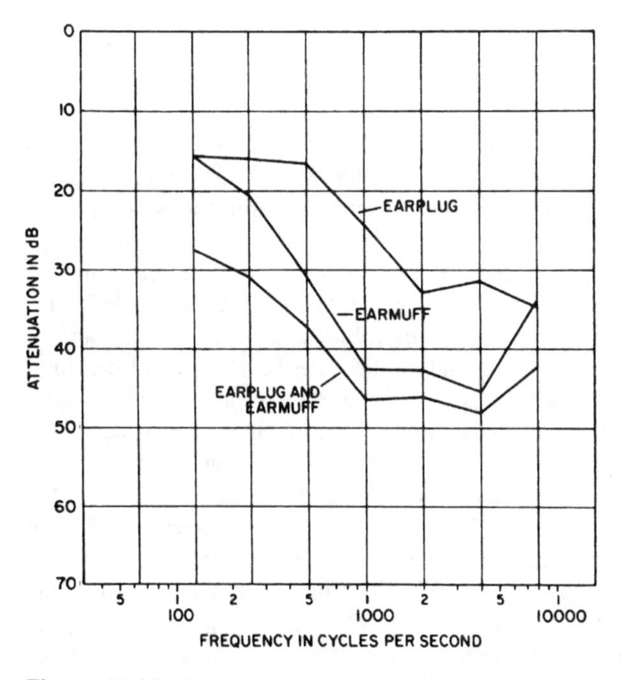

Figure 17.14 Mean attenuation characteristics of a muff and an earplug worn separately and together.

time, special care should be taken to ensure that the best hearing protection devices are used properly and hearing thresholds are monitored regularly. For higher exposure levels over extended time periods, it may be necessary to use a combination of ear inserts and earmuffs and/or to limit the time of exposure. Hearing protectors will provide adequate protection for only a small percentage of wearers when worn in levels >125 dBA for 8 h/day.

7.1. Noise Reduction and Communication

Workers should be able to communicate with each other and to hear warning signals in many different high-noise environments while wearing hearing protection devices. The effect of noise on communication depends on the spectrum of the noise. Communication difficulties tend to occur when the noise has high-level components in the speech frequency range, 400–3000 Hz. A review of speech interference studies (6,7) shows that conversational speech begins to be difficult when the speaker and listener are separated by ~2 ft in noise levels of 85 dBA. Many noisy areas exist in work environments, and the additional attenuation of hearing protection often makes communication difficult. However, some studies indicate that the use of hearing protection devices in noisy environment may improve speech perception.

7.2. Communication While Wearing Protectors

Wearing hearing protection devices obviously interferes with speech communication in quiet environments. Wearing a conventional set of earplugs or earmuffs in noise levels >90 dB in octave bands (or 97 dBA for flat spectra) may improve speech intelligibility for normal hearing ears (8,9). The improvement in speech perception occurs because speech-to-noise ratios are kept nearly constant. Additionally, the protected ear does not distort from overdrive caused by the high speech and noise levels. The efficiency in abnormal ears has not been demonstrated as conclusively.

The concept of an occluded ear hearing better in noise is difficult for the worker to accept. To gain acceptance, the hearing protection device should be tried in a quiet environment. Workers tend to be attracted to hearing protectors advertised as providing a *filter*, allowing the low frequencies in the speech range to pass but, blocks the noise. Some of these filter-type devices do provide better communication scores in quiet environments, and they may be a good choice for use in areas where there are intermittent exposures to moderately high noises. However, in steady-state noise levels greater than ~90 dB in octave bands, the conventional ear inserts or earmuffs provide communication scores that are at least as good as filter type (9) (Figs. 17.15 and 17.16), and provide better overall protection.

8. CHECKING HEARING PROTECTION ATTENUATION

When an employee is fitted with hearing protection, how can the technician determine that the hearing protection device is reducing noise and to what degree? Frequent audiometric testing can show whether any change in hearing thresholds can be attributed to noise exposure. Some companies conduct a hearing test after plugs have been fitted and compare these test results against those obtained without the earplugs in place. This *fit-test* gives the technician an idea of the attenuation provided by the hearing protection device.

One method of determining whether the hearing protection device is working properly is to perform a Rinne tuning-fork test. Workers should notice a temporary

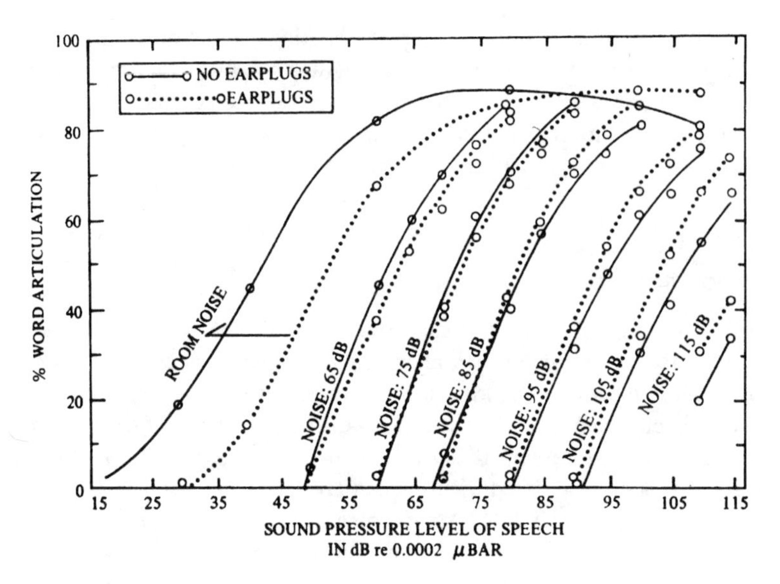

Figure 17.15 The relationship between articulation and speech level with noise level as the parameter.

conductive hearing loss while wearing their hearing protection devices. With the hearing protection device in place, a vibrating fork is placed on the mastoid and then moved to the entrance of the ear canal. If the protector is properly sized and fitted, the subject will indicate that the sound was louder at the mastoid than at the ear canal.

9. CHARACTERISTICS OF SUCCESSFUL EAR PROTECTOR USERS

There is a significant difference in the age, level of education, and years of noise exposure between workers identified as successful hearing protector users and those identified as unsuccessful users, according to Lt. Col. Kenneth Aspinall, Brooke Army Medical Center, Fort Sam Houston, TX.

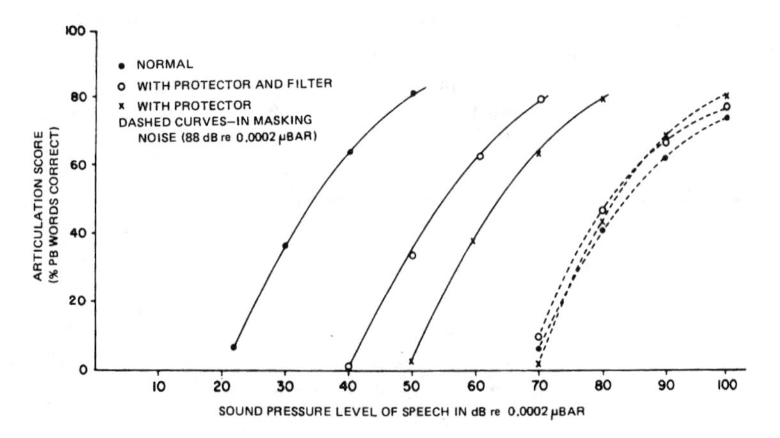

Figure 17.16 Average articulation curves with and without protector in quiet and with thermal masking noise.

Aspinall discussed the results of a questionnaire he administered to workers to determine which personality traits and attitudes are associated with the successful use of hearing protectors. Successful users were well informed about the dangers of noise exposure and the use of hearing protectors, concerned about their overall health, and have a higher incidence of hearing loss. They tended to be motivated to wear hearing protection devices, without reminder from supervisors. These workers perceived their supervisors as being highly interested in their compliance with the hearing protection plan.

Aspinall said the old adage *the older you are, the wiser you get* seems to apply to the use of hearing protectors. The average age of the 200 workers he identified as successful hearing protection users was 39 years, with an average education level of 12.7 years, and average noise exposure of 14.1 years.

The study found that successful users wore their hearing protection devices during unannounced inspections, were rated successful users by their supervisors, and rated themselves as successful users. Unsuccessful hearing protection device users were not observed wearing hearing protectors, were rated poorly by their supervisors, and rated themselves as unsuccessful users.

Some employees express concern over the possibility of the use of hearing protection causing ear infections. Regular wearing of hearing protection devices does not normally increase the likelihood of contracting ear infections. Some users may develop irritation from earplug use. Nondisposable earplugs should be washed before reuse. Earmuff cushions should be wiped periodically.

10. DOCUMENTS

The OSHA Noise Standard, 29 CFR 1910.95 and Regulation 7 of the Environmental Regulations for Workplace (1993) standard states that employees exposed to 8 h of $\geq$85 dB should be provided with hearing protection devices from their employers at no charge. To encourage the employees to wear the hearing protectors, employers should provide a variety of hearing protection options. The employers need to ensure that the hearing protection devices are worn properly. Educating employees of how to use and care for their hearing protection device is recommended.

11. EMPLOYEE EDUCATION

The Amendment to the Noise Standard is very specific concerning the need for employee education and the steps necessary to secure that education. Employers are required to institute training programs for employees exposed to noise levels at $\geq$85 dB for an 8 h time-weighted average. Training must be completed individually or in groups annually. The hearing conservation program should be updated to educate about current hearing protection devices and work processes. Each training session should address the effects of noise on hearing. Education programs should discuss the purpose of hearing protection devices, their advantages and disadvantages, attenuation of different types of devices, and instructions on fitting, care, and use of hearing protection devices. Additionally, the purpose and importance of audiometric monitoring should be explained.

If audiometric testing indicates the presence of a standard threshold shift (STS), the Amendment delineates follow-up procedures. Employees should be notified in writing within 21 days of the presence of STS. Employers must ensure hearing protection device use, refit and re-instruct regarding hearing protection device use, refer for an audiological and otological examination, and informed of medical evaluation results in the event of a threshold shift.

If further audiometric testing on employees exposed to noise levels <8-h time weighted average of 90 dBA indicates that STS is not persistent, the employee should be informed. The employee may also discontinue required use of hearing protection devices.

12. NEW EMPHASIS ON ENGINEERING CONTROLS

OSHA compliance officers will be taking a closer look to determine if engineering and administrative controls are technically and economically feasible for protecting workers from excessive noise. Citations will be issued if plants have been relying on hearing protection only rather than reducing noise at the source. Tests to determine feasibility are:

1. Hearing protectors alone cannot reliably reduce the noise levels, especially when noise levels exceed 100 dBA.
2. Engineering and administrative controls would cost less than that required to run the hearing conservation program.
3. Engineering and administrative controls do not necessarily have to reduce noise to or below the exposure limit. Controls and hearing device protection use are required to reduce the levels.

Engineering controls will not be required if the employer can show that the ongoing hearing conservation program is protecting employees. It is not clear what will be accepted as an effective program. Also, if the hearing conservation program is less costly than engineering controls, then those controls will not be required. If it would cost less to bring a plant in compliance with a hearing conservation program or if that program can be improved, engineering controls will not be required. However, if improvements to the program cannot be made and engineering controls are not available, the employer will be cited. The employer can be certain of a citation for a serious violation if hearing protection is required, but not used.

REFERENCES

1. von Gierke HE, Warren DR. Protection of the Ear from Noise: Limiting Factors, Benox Report, University of Chicago, pp. 47–60 (Dec. 1, 1953).
2. Nixon CW, von Gierke HE. Experiments on the bone-conduction threshold in a free sound field. J Acoust Soc Am 1959; 31:1121–1125.
3. Unpublished work by Paul L. Michael at The Pennsylvania State University.
4. American National Standard Specification for Sound Level Meters, S1.4-1971, American National Standards Institute, New York, 1971.
5. American Optical Company, Safety Products Division, Southbridge, MA.
6. Webster JC. Updating and interpreting the speech interference level (SIL). J Audio Engrg Soc 1970; 18:114–188.
7. Pollack I, Pickett JM. Making of speech by noise at high sound levels. J Acoust Soc Am 1958; 30:127–130.
8. Kryter KD. Effects of hearing protective devices on the intelligibility of speech in noise. J Acoust Soc Am 1946; 18:413–417.
9. Michael PL. Hearing protectors—Their usefulness and limitations. Arch Environ Health 1965; 10:612–618.

18
Tinnitus

Robert T. Sataloff
Drexel University College of Medicine, American Institute for Voice and Ear Research, and Graduate Hospital, Philadelphia, Pennsylvania, USA

Tinnitus, or noise in the ear, is one of the most challenging problems in otology and medicine. It has been speculated that tinnitus may be the result of a continuous stream of otoacoustic discharges along the auditory nerve to the brain caused by abnormal irritation in the sensorineural pathway. Although no sound is reaching the ear, the spontaneous nerve discharge may cause the patient to experience a false sensation of sound. Although this theory sounds logical, there is as yet no scientific proof of its validity.

Tinnitus is a term used to describe perceived sounds that originate within the person rather than in the outside world. Although nearly everyone experiences mild tinnitus momentarily and intermittently, continuous tinnitus is abnormal, but not unusual. The National Center for Health Statistics reported that $\sim$32% of all adults in the

United States report having had tinnitus at some time (1). Approximately 6.4% of people with tinnitus describe it as debilitating or severe. The prevalence of tinnitus increases with age up until ~70 years and declines thereafter (2). This symptom is more common in people with otologic problems, although tinnitus also can occur in patients with normal hearing. Nodar (3) reported that ~13% of school children with normal audiograms reported having tinnitus at least occasionally. Sataloff et al. (4) studied 267 normal elderly patients with no history of noise exposure or otologic disease and found 24% with tinnitus. As expected, the incidence is higher among patients who consult an otologist for any reason. Fower (5) questioned 2000 consecutive patients, 85% of whom reported tinnitus. Heller and Bergman (6) and Graham (7) found that 75% of patients complaining of hearing loss reported tinnitus. According to Glassgold and Altmann (8), ~80% of patients with otosclerosis have tinnitus and House and Brackmann (9) reported that 83% of 500 consecutive patients with acoustic neuromas had tinnitus.

One of the surprising features about tinnitus is that not everybody has it. After all, the cochlea is exquisitely sensitive to sounds, and relatively loud sounds are being produced inside each human head such as the rushing of blood through the cranial arteries, and the noises made by muscles in the head during chewing. The reason an individual rarely hears these body noises may be explained partially by the way that the temporal bone is situated in the skull and by the depth at which the cochlea is embedded in the temporal bone. The architecture and the acoustics of the head ordinarily prevent the transmission of these noises through the bones of the skull to the cochlea and thus to consciousness, even though the cochlea is built and situated in a way so that it can respond to very weak sounds normally transmitted by the air from outside the head. Only when there are certain changes in the vascular walls, perhaps caused by atherosclerosis, or in the temporal bone structure, does the ear pick up these internal noises. The patient may hear his/her own pulse as a result of a vascular disorder, and it may seem to be louder when the room is quiet or at night when trying to go to sleep. Pressing on various blood vessels in the neck occasionally stops this type of tinnitus.

Although it is usually just troublesome, tinnitus also may serve as an early sign of auditory injury. For example, high-pitched ringing or hissing may be the first indication of impending cochlear damage from ototoxic drugs, a clear signal that a drug should be stopped or its dosage reduced whenever possible. Generally, the tinnitus disappears, and no measurable hearing loss results, though in some instances the noises may persist for months or even years.

Among the common drugs capable of producing tinnitus are aspirin and quinine in large doses, and especially the diuretics, aminoglycoside antibiotics and chemotherapy drugs. These drugs should be used with extreme caution, especially when kidney function is deficient.

Among the common misconceptions about tinnitus is that it is idiopathic and incurable. Neither of these assumptions is always correct. Awareness of conditions that cause tinnitus, however, has not been as helpful to tinnitus research as might be expected. Recognizing a causal relationship has not shed much light on the actual mechanisms by which internal sounds are created.

Tinnitus is a difficult problem for the physician and patient in all cases. Tinnitus may be either subjective (audible only to the patient) or objective (audible to the examiner, as well). Objective tinnitus is comparatively easy to detect and localize because it can be heard by the examiner using a stethoscope or other listening device. It may be caused by glomus tumors, arteriovenous malformations, palatal myoclonus, and other conditions. Subjective tinnitus is much more common, unfortunately, it cannot be confirmed with current methods of tinnitus detection. Consequently, it is usually difficult to document its presence and quantify its severity, although a few tests are available to help with this problem. Although the character of tinnitus is rarely diagnostic, certain qualities are suggestive of specific problems. A seashell-like tinnitus is often associated with endolymphatic hydrops, swelling of the inner ear membranes associated with Meniere's syndrome, syphilitic labyrinthitis, trauma, and other conditions. Unilateral ringing tinnitus may be caused by trauma, but it is also suggestive of acoustic neuroma. Pulsatile tinnitus may be caused by arteriovenous malformations or glomus jugulare tumors, although more benign problems are more common.

When evaluating the history of a tinnitus (ear noise) problem, the following questions should be asked:

Yes/No

___ ___ 1. Are your noises localized?
___ Right ear ___ Head
___ Left ear ___ Cannot localize
___ Both ears ___ If both, which ear is worse? Right/Left/Neither

___ ___ 2. How long have you had ear noises? (Indicate which ear, and for how long)
Right/Left/Both
How long have they been bothersome?
___ Days
___ Weeks
___ Months
___ Years
Right/Left/Both

___ ___ 3. Was there a particular incident (head cold or other URI, explosion, etc.) that seems to have started your noises?
If yes, please describe:

___ ___ 4. Has the noise changed since it first appeared?
If yes, please describe:

___ ___ 5. Is it constantly present, or does it fluctuate (good days/bad days)?

___ ___ 6. Is it episodic (comes and goes)?
If yes, during what percentage of the time is it usually present during waking hours?_____%

___ ___ 7. If it is episodic, are you completely free of noises between attacks?

___ ___ 8. Recently, have attacks occurred *more* frequently?

___ ___ 9. Recently, have attacks occurred *less* frequently?

___ ___ 10. Are the noises more apt to occur at a particular time of the day?
If yes, when? ___ Morning ___ Day time ___ Evening ___ Night ___ None

___ ___ 11. Is there any activity that brings on the noises or makes them worse?
If yes, describe:

___ ___ 12. Are the noises worse when you are under stress?

___ ___ 13. Are the noises worse when you are tired?

___ ___ 14. Are there any foods or substances to which you are exposed that aggravate the noises?
If yes, please check the following that apply:
___ Alcohol ___ Coffee
___ Cigarettes ___ Chocolate
___ Excessive salt ___ Other, describe:

___ ___ 15. Are the noises worse during any season?
If yes, when: ___ Summer ___ Fall ___ Winter ___ Spring

___ ___ 16. Is there anything you can do to decrease the noises or make them go away?
If yes, describe:

Yes/No

___ ___ 17. Are there any activities or sounds that make the noises less disturbing? If yes, describe:

___ ___ 18. Does the noise sound the same in both ears? Please characterize the noise.

___ Ringing	___ Heart beat
___ Whistling	___ Bells
___ Buzzing	___ Hissing
___ Seashell-like (ocean roar)	___ Voices

___ ___ 19. What medications or treatments (medications, maskers, biofeedback, etc.) for the noises have you tried? Please list:

___ ___ 20. Have any of them helped? If Yes, please list:

___ ___ 21. To which of the following would you compare the loudness of your noise?

___ A soft whisper	___ A diesel truck motor
___ An electric fan	___ A jet taking off

___ ___ 22. Is the loudness fairly constant?

___ ___ 23. Does it vary *slightly* from day-to-day?

___ ___ 24. Does it vary *widely* from day-to-day?

___ ___ 25. Please rate the severity of the noises on a scale from 1 to 10. (Please check)

___ Mild (1, 2)	Aware of it when you think about it
___ Moderate (3–5)	Aware of it frequently, but able to ignore most of the time; occasionally interferes with falling asleep.
___ Severe (6–8)	Aware of it all the time, very disturbing; often interferes with activities, communication, etc.
___ Very Severe (9, 10)	Aware of it all the time, interferes with daily activities, communication and sleep; has changed your behavior.

___ ___ 26. Do you think your tinnitus interferes with your ability to hear?

___ ___ 27. Do you think other people should be able to hear the noises?

___ ___ 28. Do the noises sound as if they are coming from: ___ Inside your head ___Outside your head

___ ___ 29. Are your head noises ever voices? If yes, what do they say to you?

___ ___ 30. Do you have a feeling of fullness in your ears?

___ ___ 31. If yes, does it fluctuate with the noises?

___ ___ 32. Has anyone else in your family had tinnitus?

___ ___ 33. Do you have a hearing loss?

___ ___ 34. Do you have dizziness?

In some cases, the answers to these questions, combined with other information obtained through the history, physical examination and audiologic testing, permit identification of a specific cause of tinnitus. For example, ringing tinnitus associated with fluctuating ear fullness and hearing loss during straining can be caused by a perilymph fistula. This is a fairly common injury following trauma. Seashell-like tinnitus associated with ear fullness and fluctuating hearing loss unassociated with straining or forceful nose blowing suggests endolymphatic hydrops. Both types of tinnitus may be amenable to treatment in some cases.

Information also can be obtained about the effect of tinnitus on the individual. Recently, instruments have been developed to measure tinnitus handicap. One of the best divides is a 25-question assessment divided into functional, emotional, and catastrophic response categories (10,11). This Tinnitus Handicap Inventory (THI) is well-designed and psychometrically validated (12), but it still depends entirely upon voluntary, subjective responses. Objective criteria for detecting exaggerated responses (malingering) have not been developed, yet. Similar handicap indices have also been proposed for dizziness (13) and hearing loss (14,15).

1. DIAGNOSTIC SIGNIFICANCE OF THE DESCRIPTION OF TINNITUS

Because tinnitus, like pain, is subjective, it can be described by the patient only by comparing it with some familiar noise. The patient may say it sounds like the hissing of steam, the ringing of bells, the roar of the ocean, a running motor, buzzing, or a machine shop noises. Very often it is difficult for the patient to localize the noise in the ears; he/she may not be able to tell from which ear it is coming, or it may sound as though it were in the center of his/her head. Some people report that the noises that are not in their ears at all but inside the head. Quite frequently, a patient claims that the noise is in one ear and not in the other; yet, when by some surgical or medical intervention the noise is stopped in the symptomatic ear, the patient notices the noise in the contralateral ear. This means that the patient heard the tinnitus in the ear in which it was louder but did not realize that it was present also in the opposite ear.

How the patient describes the tinnitus is sometimes of diagnostic significance, although the description alone cannot be depended upon for etiological judgments. For example, a low-pitched type of tinnitus is more common in otosclerosis and other forms of conductive hearing loss. Sounds like ringing and hissing are more common in sensorineural hearing loss. The ocean-roaring type of noise or a noise like a hollow seashell held to the ear is reported most often in patients with Meniere's syndrome.

Patients sometimes say that the ear noises are so loud that they are unable to hear what is going on around them. They claim also that if only the ear or head noises would stop, they would be able to hear better. Unfortunately, this is not usually the case. It is possible to measure how loud these noises actually are. These measurements show that tinnitus rarely is louder than a very soft whisper (5–10 dB above hearing threshold), and it is actually the concomitant hearing loss or a psychological disturbance, rather than the masking effect of the tinnitus, that prevents patients from hearing.

2. PHYSICAL EXAMINATION AND TESTING

Patients with tinnitus deserve a comprehensive neurotologic assessment in order to determine whether there is any serious or underlying cause of the tinnitus. After a thorough history, physical examination includes a complete head and neck assessments, including examination of the cranial nerves. Other neurotological testing of the cerebellar and other functions should be performed, as indicated. In addition to a routine audiogram, testing commonly includes assessment of hearing at frequencies not tested routinely; brainstem evoked-response audiogram (BERA); balance tests such as electronystagmogram (ENG) and computerized dynamic posturography (CDP); otoacoustic emission (OAE) testing; imaging studies [magnetic resonance imaging (MRI), computerized tomography (CT), single photon emission computed tomography (SPECT), and positron emission tomography (PET)]. Blood tests for diabetes, hypoglycemia, hyperlipoproteinemia, Lyme disease, syphilitic labyrinthitis, thyroid dysfunction, collagen vascular disease, autoimmune inner-ear disease, and other conditions (many others). Such comprehensive evaluation can identify a treatable cause of tinnitus in a substantial number of patients, although not a majority.

3. AUDIOMETRY IN THE PRESENCE OF TINNITUS

Occasionally, when an audiogram is being performed on a patient with marked tinnitus, he/she complains of inability to detect the tone produced by the audiometer because of the tinnitus. That is a real problem, and it can be handled best by modifying the technique of audiometry. Instead of sounding a tone on the audiometer for a second or less, it is best to present quickly interrupted or warbled tones so that the patient can distinguish the discontinuous audiometer tone from the constant tinnitus. In this way the test will determine threshold more accurately than by using the routine method of tone presentation.

If the tinnitus is unilateral, the character of a patient's tinnitus can be determined by asking him/her to match it with sounds applied to the good ear. This is done using tinnitus matching audiometer circuitry. Both the frequency and intensity usually can be identified. This information is helpful for quantitative diagnostic and rehabilitation purposes. The presence of residual tinnitus inhibition after masking also has useful therapeutic implications, as discussed subsequently.

4. TINNITUS WITH A NORMAL AUDIOGRAM

It is standard procedure in otologic practice to perform an audiogram on every patient who complains of tinnitus. If the otoscopic findings are normal, and the audiogram shows normal hearing from the lowest frequencies to 8000 Hz, and yet the patient complains of tinnitus, several causes should be considered: (1) hearing defects >8000 Hz or at "in-between" frequencies not tested during routine audiogram; (2) vascular and neurologic disorders; (3) functional causes; (4) retrocochlear disease such as an acoustic neuroma; and (5) temporomandibular joint (TMJ) abnormality.

5. TINNITUS WITH HIDDEN HIGH-FREQUENCY HEARING LOSSES

Tinnitus and hearing loss are associated so frequently with each other that some physicians attribute tinnitus to a psychological disturbance if the audiogram shows normal hearing. Such a diagnosis is hardly justified. A diagnosis of psychological problems should be made on the basis of positive clinical findings rather than on the fact that the audiogram is normal. Studies show that many patients who complain of ringing or hissing tinnitus may have perfectly normal hearing in the frequencies that routine audiometers test, that is, 250, 500, 1000, 2000, 3000, 4000, 6000, and 8000 Hz. However, when the hearing is tested at higher frequencies up to 20,000 Hz, it is not uncommon to find a hearing deficit in the ear in which the patient claims to have tinnitus. In addition, the standard audiogram shows thresholds only at a few fixed octave frequencies. The hearing damage may well be at a point intermediate between these routinely tested frequencies. For example, a 40 dB loss at 3500 Hz would not be seen on a standard audiogram, because hearing is routinely measured at 2000 and 4000 Hz.

Left Ear - Air Conduction

250	500	1000	2000	4000	8000	10000	12000	14000
5	5	−5	−10	−10	−5	15	10	5

Right Ear - Air Conduction

250	500	1000	2000	4000	8000	10000	12000	14000
5	0	−5	−10	5	5	35	30	35

Figure 18.1 *History*: Audiogram of a 38-year-old woman with fullness in her right ear and high-pitched constant tinnitus for several months. No obvious cause. She had had eustachian tube inflations, allergic desensitization, bite correction, and nose treatments without help. *Otologic*: Normal. *Audiologic*: Normal hearing with high-tone hearing loss >8000 Hz. *Classification*: High-tone sensory hearing loss. *Diagnosis*: After further questioning the patient subsequently associated the onset of tinnitus with a bad cold. Hearing loss and subsequent tinnitus probably due to a viral cochleitis.

Continuous frequency audiometry is a tremendous help in cases of this type. Figures 18.1 and 18.2 relate cases of tinnitus that were diagnosed as functional because no hearing loss was found during routine audiometry. However, more detailed hearing tests revealed the presence of auditory damage which could well account for the patients' tinnitus.

6. TINNITUS AND OTOSCLEROSIS

Tinnitus usually is low-pitched and described as buzzing or, occasionally, a roaring sound in patients with otosclerosis. Sometimes the patient may say he/she hears a pulsing noise timed to his/her heartbeat. In some patients with otosclerosis, the tinnitus is even more disturbing than the hearing loss. However, not all patients with otosclerosis have tinnitus, and some with very severe otosclerosis deafness deny ever having had tinnitus. In most instances, tinnitus will disappear or diminish during the course of many years of hearing impairment. Tinnitus is uncommon in older patients who have had long-standing otosclerosis.

In view of all that has been learned about the pathology of otosclerosis, it might be thought that it would be possible to eliminate the tinnitus by correcting the fixation of the stapedial footplate. This is not necessarily so. Many patients who have had successful stapes surgery and a return to almost normal hearing have found that the tinnitus seems to persist, although it may not be as loud as it was prior to surgery. Yet in other patients tinnitus seems to subside completely when hearing is restored surgically. Rarely, it even may worsen. It would appear, then, that there are other factors besides stapes fixation in the etiology of tinnitus associated with otosclerosis. Otosclerosis eventually produces a sensorineural hearing loss in many cases, and tinnitus may be related to the same etiology.

Left Ear - Air Conduction

250	500	1000	2000	4000	8000	10000	12000	14000
0	−5	−5	−5	0	−10	−5	0	5

Right Ear - Air Conduction

250	500	1000	2000	4000	8000	10000	12000	14000
−10	−5	−5	−10	−5	45	30	35	60

Figure 18.2 *History*: An 18-year old woman with ringing tinnitus 8 weeks after fall on right side of head, with unconsciousness. Skull was fractured, and she had vertigo and bleeding from right ear. For 6 weeks she had light headedness. A persistent ringing tinnitus is present. *Otologic*: Normal. *Audiologic*: Note the loss in the very high frequencies in the right ear. *Classification*: Sensorineural hearing loss. *Diagnosis*: Direct trauma to the inner ear.

7. TINNITUS IN MENIERE'S DISEASE

One of the most disturbing and persistent types of tinnitus is associated with in Meniere's disease. It generally is described as an ocean roar or a hollow seashell sound. In the early stages of Meniere's disease, tinnitus often persists all the time and becomes the most disturbing symptom of the disease. Many patients would even sacrifice their hearing to get rid of the tinnitus, but unfortunately there is no guaranteed therapy, not even surgery, that will insure the resolution of the tinnitus. Fortunately, the tinnitus and the inner-ear involvement are restricted in most patients to one ear, but when both ears are affected, the psychological impact on the patient may become a serious challenge to both the patient and the physician.

8. TINNITUS AFTER HEAD TRAUMA, EXPOSURE TO NOISE

Ringing tinnitus is experienced commonly after a blow to the external ear or close exposure to a sudden very loud noise such as the explosion of a firecracker or the firing of a gun. In most instances the tinnitus is accompanied by a high-frequency hearing loss. If the hearing loss is temporary, the tinnitus usually subsides in a few hours or days. If permanent hearing loss has occurred from damage to the inner ear, ringing tinnitus may persist for many years or permanently.

The close relationship of head injury and tinnitus is clearly evident in the classic cartoon that portrays an individual knocked out by a punch on the jaw or a blow on the head, complete with "seeing" stars and bells ringing in his ears. The bells characterize the tinnitus that the patient hears after a severe blow. The noise probably is caused by concussion of the cochlea. The damage seems to be reversible most of the time, because the ringing gradually subsides in most patients.

9. "ACOUSTIC NEURITIS," ACOUSTIC NEUROMA, AND MISCELLANEOUS CAUSES

For want of a more specific etiology, tinnitus often is attributed to neuritis (inflammation) of the acoustic nerve inflammation, especially if a high-tone deafness accompanies it. It is not known precisely whether the irritation or inflammation actually injures the nerve itself, but at least some of the damage probably occurs in the cochlea. Diseases such as hepatitis, influenza, and other viral diseases frequently result in a high-pitched tinnitus, tentatively attributed to cochleitis or acoustic neuritis. Tinnitus may also be associated with presbycusis, impacted cerumen on the eardrum, and many other causes. It rarely is associated with infection of the middle ear.

Acoustic neuroma is a benign tumor of the cerebellopontine angle. It may cause tinnitus (often the first symptom) by damaging the acoustic nerve, or auditory pathways in the brainstem by compression. Tinnitus may be caused by other cerebellopontine angle neoplasms, as well, including meningioma, cholesteatoma, vascular malformation, and others. Similar compression and nerve damage/irritation can be caused by the anterior–inferior cerebellar artery a loop if this blood vessel is in contact with the auditory nerve or its root-entry zone. Tumors within the brainstem are also capable of causing tinnitus. These serious conditions highlight the need for thorough evaluation in patients who complain of tinnitus; and MRI should be considered especially in any patient with unilateral tinnitus.

10. TEMPOROMANDIBULAR JOINT PROBLEMS

Just how malocclusion or other TMJ disparity can cause tinnitus is not yet known, but there are some patients whose tinnitus has resolved with adequate TMJ treatment. The evidence conflicts with the assumption that all these cases are of emotional origin. Only a few patients with malocclusion ever complain of tinnitus, and even when malocclusion and tinnitus are present in the same patient, it does not follow necessarily that the dental abnormality causes the tinnitus. As dental correction is sometimes a formidable undertaking, one might consider a noninvasive treatment to indicate whether a complete corrective procedure would be likely to stop the tinnitus. Such a procedure may consist of placing a plastic prosthesis (bite block) over the lower molars at night. If there is a noticeable improvement in the tinnitus after a satisfactory trial period, then more permanent measures may be indicated. Indiscriminate or radical efforts to alleviate tinnitus by correcting malocclusion and subsequent TMJ symptomatology without a valid therapeutic test to establish the causal relationship should be avoided. Many otologists question the reality of any true relationship between dental malocclusion or TMJ dysfunction and tinnitus, although the association of TMJ problems with ear pain is common.

11. FUNCTIONAL CAUSES

For all reasons enumerated, a diagnosis of functional or psychological tinnitus should be made with great caution. Nevertheless, in some patients with normal hearing and tinnitus, a tentative diagnosis of functional tinnitus may have to be considered. The patients should be differentiated into two categories. Malingering occurs when the patient does not perceive an abnormal noise and is aware of the fact, but claims to have tinnitus, usually for secondary gain (such as a lawsuit). True functional or "hysterical" tinnitus is psychiatric in etiology and disappears following effective psychotherapy. It is also extremely rare. Excessive functional response to existing tinnitus is the most common category of the three. Most such patients become so preoccupied with their tinnitus that they continue to talk about it to their friends and will consult numerous doctors and nonmedical persons for relief that is not forthcoming. Unfortunately, the otologist can do little to cure such patients of their tinnitus, when the cause is not known. However, it may be possible to help the patient to adjust to the auditory disorder through psychological counseling, tinnitus maskers, habituators, and sometimes medications. Although it has not been proven, in the author's experience, patients with severe auditory processing disorders tend to be particularly disturbed by tinnitus that might be ignored by many patients. In these cases, neuropsychological testing and intervention may prove extremely valuable. It is helpful in such cases to bear in mind that tinnitus itself is really not very loud, but many patients overreact to their symptoms and allow their problem to assume distressing proportions.

12. MANAGING THE PATIENT WITH TINNITUS

Unless a correctable, structural, or metabolic cause is found, tinnitus is usually not curable. Most patients adjust well to their tinnitus, but some are extremely disturbed by it. An enormous number of medications have been used to treat tinnitus, most without success. Tinnitus maskers (devices similar to hearing aids) are recommended by some physicians, but their value is also limited for the vast majority of tinnitus sufferers. They introduce a noise into the ear that the patient is able to control. Some patients find this helpful, but most do not. External masking with a radio or fan is helpful to many people, especially at night if the tinnitus interferes with their ability to fall asleep. Many patients find tinnitus less disturbing if they wear a hearing aid. Consequently, in a patient with tinnitus and even a mild hearing loss, it may be worthwhile trying amplification sooner than one ordinarily would in a patient not troubled by tinnitus.

Some patients become so obsessed with the noises in their ears that they become emotionally disturbed and are unable to sleep at night, reporting that the noise keeps them awake. This happens mainly in highly "anxious" individuals, but the problem requires much patience and understanding on the part of the physician. Reassurance and encouragement are very helpful. Whatever symptomatic therapy may be available for the tinnitus should be provided. One practical suggestion is for the patient to put an automatic timer on his/her radio and play it when going to sleep. After an hour or two the automatic timer shuts off the radio, and by then the patient is fast asleep. Music from the radio may mask the patient's tinnitus by masking and distracting his/her attention from endogenous noises. A similar effect may be achieved by running a fan or a vaporizer in the bedroom, white noise, or by an electronic sea noise generator.

In managing the patient with tinnitus, it is advisable to have a forthright talk with the patient and to explain the most likely cause of the tinnitus and the fact that as yet there is no specific cure for it. Nevertheless, it is possible in most instances to mitigate the problem and particularly to alleviate the patient's excessive concern about the symptoms. If a hearing aid is indicated, its daily use usually makes the patient less aware of the tinnitus by focusing attention on other matters. Everyday noises definitely help to mask tinnitus in most patients. However, when the patient takes off the hearing aid at night, the tinnitus may become more noticeable. Electronic tinnitus maskers and biofeedback techniques have been developed and proven helpful for some patients. Periods of tension and stress also tend to make tinnitus more troublesome. Most people with tinnitus who keep themselves occupied, especially at work, are less bothered by their noises.

The use of tranquilizers, constructive suggestions, and specific cures for such conditions as otosclerosis and Meniere's disease are important therapeutic regimens in managing the patient with tinnitus. In many cases the tinnitus subsides gradually or even suddenly for no obvious reason. Medical attention to related vascular abnormalities such as hypertension, Buerger's disease, and atherosclerosis is recommended. Hypnosis is helpful in some instances. Psychotherapy and encouragement may be of tremendous value.

13. TINNITUS MASKERS, HABITUATORS AND RETRAINING

There are various methods of temporary relief from tinnitus such as hypnosis and biofeedback. One method may prove quite effective for one patient and provide no relief for another. The use of a tinnitus masker may alleviate symptoms for some people. A masker is almost always considered but usually prescribed only after other methods have failed.

A tinnitus masker is a hearing aid-like instrument that produces a narrow band noise centered around the pitch of the patient's tinnitus. These instruments are available as a masker alone or as a combination hearing aid/masker.

Before a masker can be fitted, a tinnitus evaluation must be performed. The evaluation involves taking a detailed history of the individual's tinnitus, as well as matching the pitch and intensity of the tinnitus as closely as possible. It should be noted also whether the tinnitus is unilateral or bilateral. After the tinnitus is matched, the minimal masking level is determined. This is the level at which the masking noise is perceived above the patient's tinnitus.

Finally, the patient is assessed to determine whether residual inhibition is present. This is a temporary cessation or lessening of the tinnitus for a few seconds to several minutes following the presentation of a masking noise to the affected ear. The presence of residual inhibition suggests a good prognosis for successful use of a masker.

An habituator is a device somewhat similar to a tinnitus masker. Habituators are used primarily to treat patients with hyperacusis (hypersensitivity to noise); but they may be helpful in some tinnitus patients, as well. An habituator supplies the ear with a constant, soft sound stimulus, so that the ear is never in silence. Theoretically, this activates or "primes" inhibitory neurons and improves their ability to respond to and suppress external stimuli. In some cases, this approach also seems to help suppress tinnitus. Tinnitus retraining programs have proven useful and are becoming progressively more sophisticated.

14. SURGERY

A few patients are truly distraught and disabled by tinnitus. In some patients, tinnitus persists even in a totally deaf ear. Such patients may be candidates for eighth nerve section, an intracranial procedure in which the nerve is cut. However, this procedure is only successful in relieving tinnitus 50–70% of the time. If the tinnitus is believed to be associated with a vascular loop (malpositioned blood vessel) compressing the eighth nerve, microvascular decompression can be performed through the posterior fossa. This procedure may be helpful, but the tinnitus must be extremely disturbing in order to justify an operation of this magnitude. Patients should also be informed that tinnitus may worsen following any ear surgery, as can one's hearing.

15. RESEARCH

Research into the causes of tinnitus poses unique problems. Paramount is the fact that tinnitus is a purely subjective phenomenon. Unlike hearing loss, which can be measured objectively, the only way to know that a patient hears noises is for the subject to report that fact. Naturally, experimental animals cannot make such a report even if researchers do something to them that is known to cause tinnitus. For example, large quantities of aspirin predictably produce a temporary tinnitus. In addition, exposure to extremely loud noise, sufficient to cause measurable temporary threshold shift generally produces temporary tinnitus. If an experimental animal could report the tinnitus, various techniques could be used to try to make it go away.

In hearing research, animals frequently are trained and conditioned to perform accurate audiograms. Unfortunately, similar techniques are not suitable for tinnitus. Conditioning for hearing loss works because the animals are trained to respond to an intermittent signal such as the tone from an audiometer. As experimentally induced tinnitus is present constantly, a trained animal will adapt to it and cease to respond. Although some of these technical difficulties may be solved, at present research depends primarily on either histologic studies or the use of living humans. Histologic, pharmacologic, and biochemical studies directed toward understanding the hair cells of the inner ear, nerves, blood flow, and neurotransmitters are underway and have produced interesting information. So far, however, very little of it has had proven clinical value.

Anyone who provides health care for severe tinnitus sufferers knows that finding volunteers for experimental treatments is easy. Some patients with tinnitus are miserable and anxious to try anything that may help. Investigators have used everything from garlic to electrical stimulation of the cochlea (first in the 19th century and more recently in the 1970s and 1980s), to masking to surgery. A long list of drugs has been tried, all of them without consistent success. These range from benign medications such as vitamin A to ototoxic drugs used therapeutically for medical labyrinthectomy (16).

The phenomenon of otoacoustic emissions (OAEs) is also interesting for tinnitus researchers and clinicians. OAEs are acoustic energies that can be detected by inserting a miniature microphone directly into the external auditory meatus and measuring the response emitted back through the eardrum. It has been established that many people do have OAEs that they cannot hear (not tinnitus). Therefore, these emissions cannot be considered a dependable guide to the presence or absence of tinnitus in experimental animals. It should be noted that OAEs are only present and recordable in people with hearing levels of 35–40 dB HL or better. With worse hearing loss, detectable OAEs should not be expected.

16. FUTURE RESEARCH

To better understand tinnitus, additional research is needed. Certain facts, however, have become clear. First, the sounds that encompass tinnitus are produced by many different causes. Their mechanisms may be located anywhere in the auditory pathway from the eardrum to the cortex. Mechanical causes of tinnitus, such as earwax impacted against an eardrum and palatal myoclonus, may be understood and treated. Other causes and mechanisms located in the cochlea or central nervous system are more mysterious. Future human and laboratory researchs are essential. Research projects using humans need to be carefully designed, involve minimal risks, and be approved by the Federal Food and Drug Administration if they are performed in the United States. This research should attempt to systematically localize and categorize tinnitus sources and to interrupt the tinnitus through pharmacologic (preferably), electrical or surgical means. Laboratory research probably will not be fruitful or practical for the professional investigator until a useful animal model has been developed. Until that time, the greatest benefit probably will come from further study of the neuropharmacology that effects the ascending and descending auditory pathways. Tinnitus remains among the most challenging problems for hearing healthcare professionals, researchers, and especially for patients.

REFERENCES

1. National Center for Health Statistics. Hearing status and ear examination: Findings among adults. United States 1960–1962. Vital and Health Statistics, Series 11, No. 32. Washington, DC: U.S. Department of HEW, 1968.
2. Reed GF. An audiometric study of two hundred cases of subjected tinnitus. Arch Otol 1960; 71:94–104.
3. Nodar RH. Tinnitus aurium in school-age children: survey. J Aud Res 1972; 12:133–135.
4. Sataloff J, Sataloff RT, Luenenburg W. Tinnitus and vertigo in health senior citizens with a history of noise exposure. Am J Otol 1987; 8(2):87–89.
5. Fower EF. Tinnitus aurium: its significance in certain diseases of the ear. NY State J Med 1912; 12:702–704.
6. Heller MR, Bergman M. Tinnitus aurium in normally hearing persons. Ann Otol Rhinol Laryngol 1953; 62:73–83.
7. Graham JM. Tinnitus in children with hearing loss. CIBA Foundation Symposium 85, Tinnitus, London, UK, 1981:172–181.
8. Glassgold A, Altmann F. The effect of stapes surgery on tinnitus in otosclerosis. Laryngoscope 1966; 76:1624–1632.
9. House JW, Brackmann DE. Tinnitus: Surgical treatment. CIBA Foundation Symposium, Tinnitus, London, UK, 1981:204–212.
10. Newman CW, Jacobson GP, Spitzer JB. Development of the Tinnitus Handicap Inventory. Arch Otol-HNS 1996; 122:143–148.
11. Newman CW, Wharton JA, Jacobson GP. Retest stability of the Tinnitus Handicap Questionnaire. Ann Otol Rhinol Laryngol 1995; 104:718–723.
12. Newman CW, Sindridge SA, Jacobson GP. Psychometrically adequacy of the Tinnitus Handicap Inventory (THI) for evaluating treatment outcome. J Am Acad Aud 1998; 9:153–160.
13. Jacobson GP, Newman CW. The development of the Dizziness Handicap Inventory. Arch Otol HNS 1990; 116:424–427.
14. Ventry I, Weinstein BE. The Hearing Handicap Inventory for the elderly: a new tool. Ear Hearing 1982; 3:128–134.
15. Newman CW, Weinstein BE, Jacobson GP, Hug GA. The Hearing Handicap Inventory for adults: psychometric adequacy and audiometric correlates. Ear Hear 1990; 11:176–180.
16. Graham MD, Sataloff RT, Kemink JL. Tinnitus in Meniere's disease: response to titration streptomycin therapy. J Laryngol Otol 1984; 98(12 suppl 9):281–286.

19
Dizziness

Robert T. Sataloff
Drexel University College of Medicine, American Institute for Voice and Ear Research, and Graduate Hospital, Philadelphia, Pennsylvania, USA

In addition to deafness and tinnitus, vertigo is an important symptom associated with disorders of the ear. The ultimate relationship of the vestibular portion of the labyrinth to the cochlea makes it easy to understand the reason why many diseases and lesions, such as Meniere's disease, affect both hearing and balance. Some diseases, like mumps, classically affect only the cochlea. Certain toxins and viruses affect only the vestibular portion without affecting the hearing. Intense noise affects only the cochlea.

1. VERTIGO

Vertigo, like deafness and tinnitus, is a subjective experience and is a symptom, not a disease. Its cause must be sought carefully in each case. The term "dizziness" or "vertigo" is used by patients to describe a variety of sensations, many of which are not related to the vestibular system. It is convenient to think of the balance system as a complex conglomerate of senses that each send the brain information about one's position in space. Components of the balance system include the vestibular labyrinth, the eyes, neck muscles, proprioceptive nerve endings, cerebellum, and other structures. The supranuclear oculomotor systems integrate information from various sites and are especially important for maintaining spatial orientation and permitting an individual to function during complex motion. The vestibular system and related structures are integral components of these systems. Within the inner ear, the cochlea performs audition, and the semicircular canals and other structures of the labyrinth perform vestibulation.

Vestibulation is the automation stabilization of the eyes relative to material space during normal patterns of movement, such as locomotion. It is a prerequisite for clear vision in these circumstances. The *vestibular system* includes structures responsible for several reflexes that are essential for normal balance function. These include the vestibulo-ocular reflex (VOR); vestibulospinal reflex (VSR) which includes the vestibulocolic reflex (VCR); and the cervico-ocular reflex (COR). The vestibular system works in concert with the saccadic, pursuit, optokinetic, fixation reflex, and vergence systems. If all sources provide information in agreement, one has no equilibrium problem. However, if most of the sources tell the brain that the body is standing still, for example, but one component says that the body is turning left, the brain becomes confused and we experience dizziness. It is the physician's responsibility to systematically analyze each component of the balance system to determine which component or components are providing incorrect information, and whether correct information is being provided and analyzed in an aberrant fashion by the brain. Typically, labyrinthine dysfunction is associated with an abnormal sense of motion. It may be true spinning, a sensation of being on a ship or of falling, or simply a vague sense of imbalance when moving. In many cases, it is episodic. Fainting, light headedness, body weakness, spots before the eyes, general lightheadedness, tightness in the head, and loss of consciousness are generally not of vestibular origin. However, such descriptions are of only limited diagnostic help. Even some severe peripheral (vestibular or eighth nerve) lesions may produce only mild unsteadiness or no dizziness at all, such as seen in many patients with acoustic neuroma. Similarly, lesions outside the vestibular system may produce true rotary vertigo, as seen with trauma or microvascular occlusion in the brain stem, and with cervical vertigo.

Dizziness is a relatively uncommon problem in healthy individuals. In a study of 267 normal senior citizens, Sataloff et al. (1) found only a 5% incidence of dizziness in contrast to a 24% incidence of tinnitus. However, most of the population is not as healthy as the highly selected sample in this study. Initial physicians' visits involve a complaint of dizziness or imbalance in 5–10%, accounting for >11 million physician visits annually (2).

Dizziness is the most common reasons for a visit to a physician in patients >65 years of age, and this symptom may cause patients to fall. Approximately one-third to one-half of people age ≥65 fall each year, and the consequences can be serious (3). Falls result in approximately 200,000 hip fractures per year, and this injury carries a 10% mortality rate.

Falls are the leading cause of death by injury in people >75 years of age. Dizziness is also a common consequence of head injury. Over 450,000 Americans suffer serious head injuries annually (4). A majority of these people complain of dizziness for up to 5 years following the injury and many are disabled by this symptom (5). Dizziness may also persist for long periods of time following minor head injury (6).

Causes of dizziness are almost as numerous as causes of hearing loss, and some of them are medically serious (multiple sclerosis, acoustic neuroma, diabetes, cardiac arrhythmia, etc.). Consequently, any patient with an equilibrium complaint needs a thorough examination. For example, although dizziness may be caused by head trauma, the fact that it is reported for the first time following an injury is insufficient to establish causation without investigating other possible causes. While recording the history from a patient with equilibrium complaints, at least the following questions should be asked:

1. When did you first develop dizziness?
2. What is it like (light headedness, blacking out, tendency to fall, objects spinning, you spinning, loss of balance, nausea, or vomiting)?
3. If you or your environment is spinning, is the direction of motion to the right or left?
4. Is your dizziness constant or episodic?
5. If episodic, how long do the attacks last?
6. How often do you have attacks?
7. Have they been more or less frequent recently?
8. Have they been more or less severe recently?
9. Under which circumstance did your dizziness first occur?
10. Exactly what were you doing at the time?
11. If you first noted dizziness after a head injury, how many hours, days, or weeks elapsed between the injury and your first imbalance symptoms?
12. Did you have any other symptoms at the same time, such as neck pain, shoulder pain, jaw pain, ear fullness, hearing loss, or ear noises?
13. Did you have a cold, the flu, or "cold sores" within the month or two prior to the onset of your dizziness?
14. Are you completely free of dizziness between attacks?
15. Do you get dizzy rolling over in bed?
16. If so, to the right, to the left, or both?
17. Do you get dizzy with position change?
18. If so, does your dizziness occur only in certain positions?
19. Do you get dizzy from bending, lifting, straining, or forceful nose blowing?

20. Do you have trouble walking in the dark?
21. Do you know of a causc for your dizzincss?
22. Is there anything that will stop the dizziness or make it better?
23. Is there anything that will bring on an attack or make your dizziness worse (fatigue, exertion, hunger, certain foods, menstruation, etc.)?
24. Do you have any warning that an attack is about to start?
25. Once an attack has begun, does head movement make it worse?
26. Do you have significant problems with motion sickness?
27. Do you get headaches in relation to attacks of dizziness?
28. Do you get migraine headaches?
29. Are there other members of your family with migraine headaches?
30. Does your hearing change when you are dizzy?
31. Do you have fullness or stuffiness in your ears?
32. If yes, does it change when you have an attack of dizziness?
33. Have you had previous head injuries?
34. Have you ever injured your neck?
35. Do you have spine disease like arthritis (especially in the neck)?
36. Have you had any injuries to either ear?
37. Have you ever had surgery on either ear?
38. What drugs have been used to treat your dizziness?
39. Have they helped?
40. Do you have hearing loss or tinnitus?
41. Do you have any other medical problems (diabetes, high or low blood pressure, history of syphilis, other)?
42. Does your occupation involve head or neck strain?
43. Do you do any recreational activities that could injure your head and neck including (soccer, wrestling, and others)?
44. Is there any association between eating and your dizziness?
45. Have you ever received psychotherapy?
46. Do you have any family members with dizziness?
47. Are you pregnant?

It is important to pursue a systematic inquiry in all cases of disequilibrium not only because the condition is caused by serious problems in some cases, but also because many patients with balance disorders can be helped. Many people believe incorrectly that sensorineural hearing loss, tinnitus, and dizziness are incurable; but many conditions that cause any or all of these may be treated successfully. It is especially important to separate peripheral causes (which are almost always treatable), from central causes such as brain stem contusion in which the prognosis is often worse.

1.1. Physical Examination

In patients with dizziness, physical examination should include a complete assessment of the head and neck, hearing evaluation, evaluation of extraocular motion and the presence of nystagmus (discussed subsequently) or papilledema in selected cases; testing for Hitselberger's sign (decreased sensation in the external auditory canal associated with pressure on the facial nerve from lesions such as acoustic neuromas) and Hennebert's sign (clinically, subjective dizziness caused by air pressure applied to the tympanic membrane during pneumotoscopy) and evaluation of the cranial nerves. These patients should also be tested for dysdiadochokinesis, dysmetria, and drift. Additional balance and gait testing should be performed, including at least Romberg and/or Tandem Romberg maneuvers. The patients should be evaluated for tremor, ticks, slurring of speech, and any other symptoms or signs of neurological dysfunction.

Every examination of the head and neck must include a careful search for spontaneous nystagmus, especially if vertigo is a complaint. A brief test can be done by first asking the patient to look straight ahead, holding his/her head straight, and then asking the patient to move his/her eyes from side to side. Normally, no nystagmus is present in an individual who is gazing straight ahead, but >50% of people have a slight unsustained nystagmus when they are looking >20–30 degrees to the right or the left. This is a normal reaction called end-point nystagmus and generally lasts only a few seconds. In pathological nystagmus of vestibular pathway origin there is generally a slow and a fast component. The nystagmus is aggravated by looking in the direction of the fast component, and the nystagmus is named (to the right or the left), depending on the direction of the fast component. The nystagmus may be prolonged when it is pathological. In central nervous system disorders it may persist for months. In acute attacks of Meniere's disease it may last for several days or as long as the attack continues. In positional vertigo the episode may be fleeting and last only as long as the head is held in a certain position. The nystagmus usually is weaker if the test position is repeated quickly. When marked nystagmus occurs spontaneously, and the patient does not complain of vertigo, it is suggestive of damage to the central nervous system or of congenital ocular nystagmus. Vertigo that accompanies spontaneous nystagmus also may suggest a central nervous system defect, but other conditions may produce the same symptoms, such as acute episodes of Meniere's disease, toxic labyrinthitis, and positional vertigo. Certain drugs, notably barbiturates and alcohol, also may produce nystagmus. Whenever pathological nystagmus is present, comprehensive otologic and neurological studies are indicated. Nystagmus can normally be induced by warm or cool irrigation. The caloric response causes nystagmus. Electronystagmography (ENG) will document the presence and direction of nystagmus, particularly if it is severe. ENG testing has the advantage of being able to study eye movements in the dark or behind closed eyes. This is important because visual fixation suppresses nystagmus or peripheral origin. However, the physician's eye is an order of magnitude more sensitive than the ENG machine in detecting fine nystagmus, and there is no substitute for good clinical examination.

1.2. Blood Tests

A high percentage of the numerous conditions that cause hearing loss may also cause dizziness. Therefore, in many cases evaluation of a dizzy patient includes blood tests such as a complete blood count, fasting glucose or 5-h glucose tolerance test (looking for hypoglycemia that provokes dizziness), cholesterol, triglycerides, FTA, Lyme titre, thyroid function tests, ANA, rheumatoid factor, sedimentation rate, and tests for autoimmune inner ear disease. Tests such as B12 and Folate levels, screening blood tests for lead or other toxic substances and viral screens also may be appropriate. Occasionally, dizziness also may be a presenting symptom of HIV/AIDS.

2. VESTIBULAR TESTING

The balance system is extremely complicated, and ideal tests have not been developed. Research is currently underway to develop tests that will assess accurately the entire, composite functioning of the balance system, and test each component in isolation. At present, the most commonly performed test is electronystagmography. Posturography is also used widely and vestibular evoked potential testing is under investigation.

A brief review of vestibular physiology is helpful in understanding balance tests. The semicircular canals are arranged in three planes at right angles to each other (*x*, *y*, and *z* axes), and work in pairs. The cupulae of the semicircular canals are stimulated by movement of endolymphatic fluid, and each canal causes the nystagmus in its own plane. That is, the horizontal canal produces horizontal nystagmus; the superior canal causes rotary nystagmus; and the posterior canal produces vertical nystagmus. Caloric testing stimulates primarily the horizontal semicircular canal and gives little or no information about function of the superior and posterior semicircular canals in most cases. The position of gaze and plane of the head are of great importance in vestibular testing, especially when the semicircular canals are stimulated by rotation, rather than water irrigation. For rotary testing, it is useful to remember that ampullipetal flow (toward the cupulae) produces greater stimulation than ampullifugal flow (away from the cupulae) in the horizontal semicircular canal, but the opposite is true for the superior and posterior canals. Therefore, the response to rotation represents the combined effect of stimulating the sensory system of one semicircular canal while suppressing that of its counterpart on the other side of the body. Using rotational excitation and selective head position, may be used to give information about the semicircular canals that is difficult or impossible to obtain from caloric stimulation. The clinical value of this additional information is controversial, and rotational testing is not common in the United States (although it is used more frequently in the UK).

3. ELECTRONYSTAGMOGRAPHY

ENG is a technique for recording eye movements and detects spontaneous and induced nystagmus. It actually is comprised of a battery of tests of the supranuclear oculomotor systems, excluding vergence, and with emphasis on the vestibular system. It is used for recording and detecting spontaneous and induced nystagmus. Electro-oculography (ECoG) is used to evaluate nonvestibular supranuclear oculomotor systems. These portions of the test battery include calibration, which assesses cerebellar function, gaze nystagmus, sinusoidal tracking, and optokinetic nystagmus. Vestibular tests include the detection and measurement of spontaneous nystagmus, Dix–Hallpike testing for positioning nystagmus, positional nystagmus testing, and caloric irrigations. Caloric irrigations usually are performed on one ear at a time, alternately using warm or cool stimuli. Simultaneous, bithermal stimulation and ice-cold stimulation are valuable in selected cases. The test may give useful information about peripheral and central abnormalities.

Caloric testing stimulates primarily the horizontal semicircular canal and gives little or no information about the function of the superior and posterior semicircular canals in most cases. The direction of gaze and plane of the head are of great importance investibular testing, especially when the semicircular canals are stimulated by rotation rather than thermal stimulus. For rotary testing, the response represents the combined effect of stimulating the sensory system of one semicircular canal while suppressing that of its counterpart on the other side of the body. Although this test does give some peripheral lateralizing information, it is especially informative about the integration of the vestibular signals in the brain stem (the central portion of the VOR). Rotational excitation and selective head position may be used to obtain information about the vertical semicircular canals which is difficult or impossible to deduce from caloric stimulation. Test interpretation relies on the knowledge that ampullofugal endolymphatic flow (away from the cupulae) produces nystagmus towards the opposite side in the horizontal semicircular canal, whereas the opposite is true for the vertical canals. Rotational testing is not widely available in the United States. It allows measurement of eye movements with eyes opened or closed, and permits quantification of the fast and slow phases, time of onset and duration, as well as other parameters. Although some centers use only horizontal leads, the use of both horizontal and vertical electrodes is preferable. ENG must be done under controlled conditions with proper preparation which includes avoidance of drugs (especially those active in the central nervous system). Even a small drug effect may cause alterations in the electronystagmographic tracing. The test is performed in several phases. These include calibration which assesses cerebellar function, gaze nystagmus, sinusoidal tracking, optokinetic nystagmus, spontaneous nystagmus, Dix–Hallpike testing, position testing, and caloric testing. The test may give useful information about peripheral and central abnormalities in the vestibular system. ENG measures the VOR whereas CDP measures not only the VOR, but also VSR and other functions, as well. Research suggests that CDP is also valuable in distinguishing organic from nonorganic dysequilibrium, an asset that is particularly valuable in some cases of alleged disability (7–11). Interpretation is complex. Table 19.1 summarizes ENG findings and their meaning. ENG is especially helpful when a unilateraly reduced vestibular response is identified in conjunction with other signs of dysfunction in the same ear. In such cases, it provides strong support for a peripheral (eighth nerve or end organ) cause of balance dysfunction.

4. DYNAMIC POSTUROGRAPHY

For approximately 30 years, platforms have been used to try to assess more complex integrated functioning of the balance system. Until recently, most were static posture platforms with pressure sensors used to measure body sway while patients tried to maintain various challenging positions such as Romberg and Tandem Romberg maneuvers. Movement was measured with eyes closed and opened. The tests had many drawbacks including inability to separate proprioceptive function and to eliminate visual distortion. In 1972, Nasher (12) introduced a system of dynamic posturography which has been developed into a test system which is now available commercially.

Dynamic posturography uses a computer controlled moveable platform with a sway referenced surrounding visual environment. In other words, both the platform and visual surround move, tracking the anterior–posterior sway of the patient. The visual surround and platform may operate together or independently. It is capable of creating visual distortion or totally eliminating visual cues. The platform can perform a variety of complex motions, and the patient's body sway is detected through pressure sensitized strain gauges in each quadrant of the platform. The typical test protocol assesses sensory organization through 6 test procedures and movement coordination through a variety of sudden platform movements. Balance strategies are assessed using both the sensory organization and movement coordination test batteries. Dynamic posturography provides a great deal of information about total balance function that cannot be obtained from tests such as ENG alone.

4.1. Imaging Studies

Brain and internal auditory canal imaging may be extremely valuable in documenting the nature and extent of physical abnormalities causing dizziness in selected cases, and in

(*text continued on p. 512*)

Table 19.1 Interpretation of ENG

Test	Finding	Description	Interpretation	Comments
Calibration	Ocular dysmetria (Calibration overshoot)	Greater than 50% of calibrations	Cerebellar of brainstem	Caution: Eyeblink artifact, alcohol intoxication
Gaze nystagmus		Use 20° and/or 30°		Test detects gaze nystagmus and paresis of ocular deviation
	Normal or end-point	Nystagmus at 40° or more, rarely at 30°	Normal	
	Vertical		May be normal	Upward more common than downward. (See "Spontaneous Nystagmus")
	Vertical without associated horizontal gaze nystagmus		Suggests midline or bilateral lesion in the upper pons or midbrain	Opiates and demerol may cause vertical nystagmus
	Vertical upbeating		Medulla, pons, or anterior vermis	May occur in metabolic derangements
	Vertical downbeating		Usually caudal brainstem	May be enhanced by lateral gaze and may beat obliquely
	Rotary nystagmus		CNS lesion	
	Bilateral, equal horizontal gaze nystagmus	Fast phase in direction of eye deviation. Intensity increased with increasing eye deviation (except congenital nystagmus)	Brainstem lesion or drug effect (especially barbiturates, diphenylhydantoin, and alcohol)	
	Bilateral, unequal horizontal gaze nystagmus		Organic CNS pathology, probably *not* drug effect	
	Unilateral horizontal gaze nystagmus		May be from intense spontaneous nystagmus (vestibular)	May elicit spontaneous nystagmus that has been suppressed. If there is spontaneous nystagmus in the same direction >8° with eyes closed, then it is not a central sign

Periodic alternating nystagmus			CNS lesion, usually caudal brainstem	
Rebound nystagmus			CNS lesion, usually cerebellar	
Pendular nystagmus			Congenital or severe visual impairment	
Square wave jerks			If present with eyes opened, CNS lesion usually in cerebellar system	May be normal behind closed eyes, especially in anxious patient
Ocular myoclonus			CNS lesion, usually dentatorubral	Usually associated with chronic movements of the larynx and palate
Internuclear ophthalmoplegia			CNS lesion, usually medial longitudinal fasciculus	If unilateral, is usually of vascular etiology. Bilateral is associated with multiple sclerosis
Sinusoidal tracking	Saccadic pursuit	Saccadic eye jerks in the direction of stimulus movement, break up of smooth pursuit	Central oculomotor lesion usually involving the brainstem. May be caused by barbiturates	Enhanced gaze nystagmus may appear toward the extremes of the sinusoidal pattern. Usually associated with gaze nystagmus and bilateral gaze optokinetic diminution
Spontaneous nystagmus		Recorded in vertical position with eyes closed and open		
	Normal vertical		May be normal	Occurs normally in ~8% with eyes closed, usually up beating may be more intense than 10°/s
	Normal horizontal with eyes closed		Normal	Occurs normally in 15–30%, speed <7°
	Normal horizontal eyes open		"Normal" or functional	Only voluntary nystagmus
	Vestibular	1. "Jerks," with slow and fast phases 2. Primarily horizontal	Usually is peripheral, but may be central in the area of the vestibular nuclear complex	

(continued)

Table 19.1 *Continued*

Test	Finding	Description	Interpretation	Comments
		3. Conjugate 4. Suppressed by visual fixation. Often enhanced by eye closure. Must be >7°/s		
	Congenital	1. May be pendular or spikelike 2. May change with gaze or fixation distance 3. Eye closure direction or abolishes it 4. There is no oscillopsia 5. Convergence usually suppresses the nystagmus 6. Almost always horizontal		On vertical gaze horizontal pendular component may disappear, but a vertical component rarely develops. Makes vestibular and optokinetic nystagmus very difficult to see
	Ocular or “fixation” noncongenital	May resemble congenital nystagmus	From chronic visual abuse, such as “miner’s nystagmus.” Pendular may also be caused by opiates	
	Central	Failure of fixation suppression May be horizontal, vertical, or rotary	Usually brainstem or cerebellar	
	Convergence nystagmus	Jerk-type, course, disconjugate	Usually a dorsal midbrain lesion	
Optokinetic nystagmus (OKN)		Slow phase in direction of movement, fast phase refixation in opposite direction	Normal	Larger stimulus pattern gives less variability. Abnormalities must be present at both stimulus speeds. Abnormality may be asymmetry (difference between the two directions) or bilateral diminition. Vertical asymmetry is often more severe. Test should be done at two speeds

Isolated or predominant vertical OKN abnormality		Indicates high midbrain lesion or upper pons, bilateral or midline Central oculomotor pathology Vergical OKN abnormality is rare from hemisphere lesions unless very diffuse	Slight OKN asymmetry may be normal (usually down-beating predominant). Beware of eye blinks
Horizontal OKN asymmetry	*Asymmetry* 1. Poorly formed—slow phase broken up by rapid jerks 2. Slow phase speed asymmetry. This may be converted in some cases to poorly formed by increasing stimulus speed		Greater than 8°/s nystagmus may appear to produce horizontal OKN asymmetry, as may strabisimus, longstanding unilateral blindness, extraocular muscle paresis and peripheral ocular pathology. Horizontal OKN abnormality with normal gaze test usually indicates cerebral hemisphere lesion. Lateral lesions of pons and midbrain below the oculomotor nucleus cause OKN predominance away from the lesion. (Clinical usefulness is low.) In cerebral hemisphere lesions (generally temporal parietal or occipital), OKN abnormalities beat toward the side of the lesion
Horizontal OKN abnormality with gaze nystagmus paresis		Usually indicates brainstem or cerebellar lesion	
Bilateral diminution		Bilateral or midline brainstem lesions or lack of cooperation	

(*continued*)

Table 19.1 *Continued*

Test	Finding	Description	Interpretation	Comments
Dix–Hallpike test		No nystagmus, or weak nystagmus with eyes closed	Normal	Rapid movements from sitting to head-hanging and turned position for 30 s or duration of nystagmus up to $1\frac{1}{2}$ min. If longer, is considered persistent. If positive, is repeated to test fatigability. Should be substantial decrease by third trial Nystagmus is often rotary, but measured on horizontal and vertical leads. Vertical is often greatest
	Classic response	*Classic response* 1. Latency of 0.5–8 s in most cases 2. Transient "paroxysmal" response, rapid crescendo in 2–10 s, then diminuendo 3. Dizziness often severe 4. Fatigability. Additional characteristics: a. Hearing and calorics often normal b. Usually unilateral c. Nystagmus usually downward pathologic ear	Classic response is a peripheral sign, although some cerebellar lesions may produce a similar response	Much more common in patients >55 years of age. Posttraumatic dizziness, chronic middle-ear infection, fistula and idiopathic (benign) paroxysmal positional vertigo are associated with a classical response that may occasionally be seen following stapes surgery
	Nonclassical response		Nonlocalizing, but is more likely to be central than is classical response	

Direction-changing nystagmus with eyes opened			CNS pathology	
Position test		Slow movement to various head positions (sitting, supine, head right, head left, right lateral, left lateral, head hanging) for 30 s each		
	Positional nystagmus	Slow phase of $<7^\circ/s$, in position other than sitting eyes forward. Like spontaneous nystagmus (present head upright with eyes centered), but present in at least one other position or changing direction or intensity of a spontaneous nystagmus	Idiopathic	
	Type IA	Nylen's classification (Aschan modification) I. Persistent (at least 1 min). (A) Type IA—direction changing (with different positions)	Greater than $7^\circ/s$ is pathologic but nonlocalizing, although IA tends toward central	Drugs may cause, type IA especially alcohol, barbiturates, and other sedatives, aspirin and quinine
	Type IB	(B) Type IB—direction fixed	IB tends toward peripheral	
	Type II	Transitory (gone within 1 min). It is also known as "positioning" nystagmus	Type II is usually a partially elicited "paroxysmal nystagmus"	

(*continued*)

Table 19.1 *Continued*

Test	Finding	Description	Interpretation	Comments
Caloric test				May be done with cold stimulus only, or bithermal. Bithermal widely practiced, with temperatures equally above and below body temperature (30° and 44°). Maximum speed of slow component is measured
	Unilateral weakness	Compare right ear vs. left ear. Unilateral weakness is >20% difference	Peripheral sign involving primary vestibular fibers or end-organ Could include a vestibular nucleus lesion	
	Directional preponderance	Compare right-beating vs. left-beating responses Directional preponderance is >30% difference	Nonlocalizing	
	Bilateral weakness	Bilateral weakness: <7°–10°/s	Either bilateral peripheral pathology (such as ototoxicity) or central pathology involving the vestibulo-ocular arc reflex area. If central, optokinetic test is usually abnormal	Antihistamines, tranquilizers, and barbiturates may also decrease response
	Failure of fixation suppression	Test for suppression by visual fixation. Abnormal if nystagmus with eyes open is equal to or greater than with eyes closed	Unusually central lesion, but may be caused by peripheral ocular pathology, contact lenses or sedation (especially barbiturates)	

Simultaneous bilateral bithermal calorics		Patient in 30° supine position Bilateral simultaneous 30° stimulation, then 44° stimulation 5 min later. Eye movement recorded for 30 s following irrigation. Position if there are three beats of nystagmus in the most active 10 s period (correcting for pre-existing nystagmus)	
	Type I	No nystagmus	Type I—Normal or symmetrically hyperactive or symmetrically hypoactive
	Type II	Nystagmus in one direction with cold and opposite direction with warm	Type II—Usually hypoactive labyrinth; rarely hyperactive
	Type III	Direction of nystagmus is constant	Type III—Usually vestibular abnormality, but nonlocalizing
	Type IV	Nystagmus with one temperature, but not the other	Type IV—Usually vestibular abnormality, but nonlocalizing

establishing the organic basis of the equilibrium complaint. When CT, MRI, or MRA shows a definitive lesion within the vestibular pathway consistent with the individual's history, this information can be extremely helpful. When these studies are negative, however, one cannot conclude that the dizziness is not organic in etiology. New dynamic imaging studies such as positron emission tomography (PET) and single photon emission computed tomography (SPECT) may be helpful in some such cases, as may other techniques or brain mapping (including EEG) (13).

Vascular ultrasounds also may be extremely helpful in selected cases, especially in patients complaining of positional vertigo. For example, vertebral ultrasound can be performed with the neck in neutral, flexed, and extended and turned positions. The position that the patient associates with dizziness should always be tested. If positioning provokes symptoms that occur simultaneously with cessation of vertebral artery blood flow, and if the dizziness resolves when the position is changed and vertebral artery flow is restored, it is reasonable to diagnose vertebrobasilar insufficiency. In some cases, this problem can be resolved surgically, although noninvasive measures may be adequate.

In addition to cervical vascular compromise, vertigo can be caused by abnormalities in the cervical spine, especially in the first three cervical vertebrae. MRI of the cervical spine, spine X-ray, and/or cervical spine CT may be appropriate in selected patients.

5. EVOKED VESTIBULAR RESPONSE

Evoked vestibular response testing is analogous to brain stem auditory-evoked testing. However, vestibular evoked potentials are still not being used clinically. Current research indicates that this test is likely to be valuable in the near future, and clinical trials to assess its efficacy are already underway.

For selected patients, vestibular therapy exercises (short of a comprehensive, extended physical therapy program) can be extremely helpful. They usually are supervised by an otologist or audiologist. Such exercises have proven particularly effective for benign position paroxysmal vertigo (BPPV). Classically, such patients develop vertigo in the supine position when their heads are turned to one side, especially when their necks are extended so that the involved ear is below the horizontal. Usually, the onset of the vertigo and torsional nystagmus is delayed from 1 to 40 s. It usually disappears within 1 min, but its intensity varies during the episode. Frequently, the vertigo is less intense if the patient assumes the position again immediately following resolution of symptoms. The etiology of BPPV usually is attributed to either cupulolithiasis (debris or otoconia from the utricle adherent to the cupula of the posterior semicircular canal, increasing gravity-induced sensitivity), or canalithiasis (debris or otoconia that are free floating in the endolymph of the posterior semicircular canal, causing endolymphatic fluid movements as they fall when the head is turned into the provoking position, thereby creating fluid movement that affects the cupula). The therapeutic approaches to BPPV involve either dislodging the debris embedded on the cupula and/or floating the debris out of the semicircular canal, or habituating the response of the central nervous system to the positional vertigo. There are several different approaches, only a few of which will be summarized briefly here. Brandt–Daroff habituation exercises must be performed several times a day. The patient starts in the sitting position and is moved rapidly into the position that causes vertigo. The patient remains in that position until the vertigo stop, after which he/she sits up again. Sitting up frequently results in rebound vertigo. After remaining sitting for 30 s, the patient moves rapidly to the mirror image position on the opposite side for 30 s and then sits up. The sequence is repeated again until the vertigo diminishes. The exercises are repeated every 3 h until the patient has two consecutive days without vertigo.

The Semont maneuver is based on a single treatment. The side of involvement is determined, and the patient is moved quickly into the position that causes dizziness, with the head turned into the plane of the posterior semicircular canal. This position is maintained for 2–3 min. The patient is then moved rapidly through the sitting position and then down into the opposite position, taking care to maintain alignment of the neck and head. If moving into the second position does not cause vertigo, the head is shaken abruptly a few times, theoretically to free the canal debris. This position is maintained for 5 min. The patient is then moved slowly to a seated position and must remain upright (including while asleep) for 48 h. The provoking position must be avoided for a week.

The Epley maneuver is probably the most popular approach at present. It also theorectically moves the debris by placing the patient in the provoking position for 3–4 min. Then, the head is extended further and turned to the opposite position (the patient may be turned onto his/her side so that the head is turned towards the floor). This position is also held for 3–4 min. After sitting up slowly, the patient must remain upright for 48 h and avoid the provoking position for at least 5–7 days. Epley has advocated using vibration over the mastoid during the procedure to help move the debris out of the region of the posterior semicircular canal. Most practitioners have achieved satisfactory results without the use of vibration.

Single treatment approaches like the Epley maneuver utilize Dix–Hallpike position. Habituation approaches generally do not depend on Dix–Hallpike positions. The Brandt–Daroff habituation exercises are summarized briefly. Methods advocated by Norre and DeWeerdt, and those of Norre and Becker, use specific movements to determine treatment positions and protocols that are individualized depending upon patients response (although the Dix–Hallpike positions may be included, as well, and are utilized routinely in the Brandt–Daroff method). The habituation exercises are similar in many ways to the Cawthorne–Cooksey exercises (and other protocols) that have proven useful for unilateral vestibular hypofunction.

For more specific information regarding these exercises and other approaches to vestibular rehabilitation, the interested reader is referred to the book *Vestibular Rehabilitation* by Susan Herdman which provides an excellent, practical and well-referenced review of the topic.

6. CT AND MRI STUDIES

In all patients with dizziness, imaging studies should be considered. In particular, MRI scan is invaluable to detect acoustic neuroma, other cerebellar pontine angle tumors, multiple sclerosis, and advanced ischemia as well as other conditions. Temporal bone CT may be helpful to detect not only middle-ear disease, but also cochleovestibular anomalies, and particularly, abnormalities in the cochlear aqueduct or vestibular aqueduct. Such anomalies have important diagnostic and therapeutic implications.

7. DYNAMIC IMAGING STUDIES

Although the value of new dynamic imaging studies such as PET and SPECT in patients with dizziness and tinnitus has not been fully clarified, it is important to be aware of these tests and their possible applications.

PET is a technique that permits imaging of the rates of biological processes, essentially allowing biochemical examination of the brains of patients *in vivo*. PET combines CT with a tracer kinetic assay method employing a radiolabeled biologically active compound (tracer) and a mathematical model describing the kinetics of the tracer as it is involved in a biological process. The tissue tracer concentration measurements needed by the model is provided by the PET scanner. A three-dimensional (3D) image of the anatomic distribution of the biological process is produced. The technique is feasible because four radio isotopes (^{11}C, ^{13}N, ^{15}O, and ^{18}F) emit radiation that will pass through the body for external detection. Natural substrates can be labeled with these radio isotopes, preserving their biochemical integrity. Positrons are emitted from an unstable radio isotope, forming a stable new element of anatomic weight one less than the original isotope. The emitted positron eventually combines with an electron, forming positronium. Because a positron is essentially an anti-electron, the two annihilate each other, and their masses are converted to electromagnetic energy. This process emits two photons 180° apart, which can escape from the body and can be recorded by external detectors.

The value of PET in patients with dizziness and tinnitus has not been investigated as thoroughly as the value of other imaging modalities, including CT, MRI, and SPECT (discussed subsequently). However, as Jamison et al. (14) point out, PET adds another dimension to our understanding of the brain by demonstrating metabolic alterations. PET quantitates local tissue distribution of radio nucleotides that are distributed through the brain according to function. Commonly measured functions include local cerebral metabolism, cerebral blood flow, cerebral oxygen utilization, and cerebral blood volume. Langfitt et al. (15) compared CT, MRI, and PET in the study of brain trauma, which often results in dizziness and tinnitus. Although they studied only a small number of patients, they found that PET showed metabolic disturbances that extended beyond the structural abnormalities demonstrated by CT and MRI and missed by xenon-133 measurement of cerebral blood flow. Although much additional research is needed to determine the appropriate uses of PET and to interpret meaningfully the abnormalities observed, PET appears sensitive to dysfunctions that are not detected by more commonly available studies.

SPECT is also a form of emission tomography. It utilizes a technique that detects photons one at a time, rather than in pairs like the photons detected with PET. SPECT also uses a radioactive tracer introduced intravenously. Because the photons are detected one at a time, in the presence of multiple photon emission, there is no simple spatial correlation. Therefore, the origin of photons in SPECT has to be traced using collimation, a process through which electromagnetic radiation is shaped into parallel beams. ECT provides images that are not nearly as sharp as transmission CT, and ECT images must be collected over a much longer period. Both static and dynamic SPECT studies may be performed. One particularly useful technique involves the use of technetium-99m HM-PAO (Tc-99m HM-PAO), a lipophilic chemical microsphere that crosses capillary walls freely. Within the brain, it is converted to a hydrophilic form that cannot leave the brain. Only a portion of the Tc-99m HM-PAO is converted, with the remainder diffusing back into the blood stream. The amount cleared by back diffusion depends on blood flow. As with PET, good studies on SPECT in dizziness and tinnitus are lacking. One study looked at the value of SPECT in assessing dizzy patients with suspected vascular etiology and found abnormalities in 15 of 18 patients. Significant alterations in cerebral blood flow were identified in the absence of any structural lesions identified by CT or MRI studies (16–18). Useful inferences can also be drawn from research on head injury. Abdel-Dayem et al. (19) compared Tc-99m HM-PAO SPECT with CT following acute head injury. SPECT has the following advantages: (1) it reflects profusion changes; (2) it was more sensitive than CT in demonstrating lesions; (3) it demonstrated lesions at an earlier stage than CT; (4) it was more helpful in separating lesions with favorable prognosis than those with unfavorable prognosis. Roper et al. (20) also compared Tc-99m HM-PAO SPECT with CT in 15 patients with acute closed head injury. They also found that SPECT can detect focal disturbances of cerebral blood flow that are not seen on CT. They observed that SPECT distinguished two types of contusions: those with decreased cerebral blood flow and those with cerebral blood flow equal to that of the surrounding brain. Bullock et al. (21) found SPECT useful in mapping blood brain barrier defects (including delayed blood brain barrier lesions) in 20 patients with acute cerebral contusions and four with acute subdural hematomas. Ducours et al. (22) found SPECT abnormalities on 9 out of 10 patients with normal transmission CT following craniofacial injury. Oder et al. (23) suggested that SPECT may help improve outcome prediction in patients with persistent vegetative state following severe head injury; and Morinaga et al. (24) used SPECT to demonstrate regional brain abnormalities in six patients with hyponatremia following head injury. The abnormalities observed on SPECT improved when the hyponatremia corrected.

Although extensive additional research and clinical experience are needed to clarify the roles of dynamic imaging studies, it appears likely that they may be useful to document abnormalities in patients with dizziness and/or tinnitus. The author (R.T.S.) has now had several patients in whom SPECT revealed the otherwise undetected organic basis for dizziness. A few cases involved posttraumatic dizziness, but others did not. For example, SPECT on a woman with episodic vertigo and normal ear function revealed a perfusion abnormality in the temporal lobe. Subsequent EEG confirmed an epileptic focus in this area, and her dizziness was controlled with antiseizure medication.

As Duffy observed, brain-imaging procedures, such as CT scan, PET, and MR (nuclear magnetic resonance), and BEAM (brain electrical activity mapping) represent different windows upon brain and function. They provide separate but complimentary information (25).

8. PERIPHERAL CAUSES

Vestibular disturbance may be suspected with a history of vertigo described as motion, particularly if associated with tinnitus, hearing loss or fullness in the ear. However, even severe peripheral disease of the eighth nerve or labyrinth may produce vague unsteadiness rather than vertigo, especially when caused by a slowly progressive condition such as acoustic neuroma, as mentioned earlier. Vascular loop compression (usually by the anterior inferior cerebellar artery) and internal auditory canal stenosis may cause the same symptoms. Similarly, central disorders such as brain stem vascular occlusion may produce true rotary vertigo typically associated with the ear. Therefore, clinical impressions must be substantiated by thorough evaluation and testing (26).

One of the most common causes of peripheral vertigo is Meniere's disease. The vertigo is classically of sudden onset, relatively brief in duration, and reoccurs in paroxysmal attacks. During the attack it generally is accompanied by an ocean-roaring tinnitus, fullness in the ear, and hearing loss. Occasionally, there may be some residual imbalance between attacks, but this does not happen often. Similar symptoms may accompany inner-ear syphilis and certain cases of diabetes mellitus, hyperlipoproteinemia, or hypothyroidism. The symptoms are caused by endolymphatic hydrops that distends Reissner's membrane in the cochlea.

Another type of vertigo associated with some abnormality in the vestibule is being recognized more frequently. In BPPV, the attack often occurs briefly with sudden movements of the head. Classically, there is a slight delay in onset, and if the maneuver is repeated immediately after the vertigo subsides, subsequent vertiginous responses are less severe. It generally is not associated with deafness or tinnitus. The posterior semicircular canal is most often the source of the difficulty.

Therapy for this condition is symptomatic. Vestibular exercises help more than medications. When the condition is disabling, it may be cured surgically by dividing all or part of the vestibular nerve. In most cases, hearing can be preserved. If the problem can be localized with certainty to one posterior semicircular canal, an alternate surgical procedure called singular neurectomy may be considered.

Certain viruses such as herpes classically produce vertigo by involving the peripheral end-organ or nerve. The attack usually is of sudden onset associated with tinnitus and hearing loss, and it subsides spontaneously. In the absence of tinnitus and hearing loss, the virus is assumed to have attacked the nerve itself; and this condition is called vestibular neuritis. Certain toxins readily produce vertigo.

Whenever a patient complains of vertigo and there is evidence of chronic otitis media, it is essential to determine whether there is presence of a cholesteatoma that is eroding into the semicircular canal and causing the vertigo. Vertigo can be present also with certain types of otosclerosis that involve the inner ear. In all such cases the specific cause of the vertigo should be determined, and whenever possible, proper therapy based on the specific cause should be instituted. Perilymph fistula as a cause of vertigo is particularly amenable to surgical treatment, as are autoimmune inner ear disease, hypoglycemia, hypothyroidism, allergy, and other causes.

9. NERVOUS SYSTEM INVOLVEMENT AND OTHER CAUSES

Vertigo involving the central nervous system is an urgent problem and must be ruled out in every case, especially in those cases in which it is associated with other symptoms outside the ear.

Certain symptoms strongly suggest that the cause of the vertigo must be sought in the central nervous system. If spontaneous nystagmus is present and persists, the physician should look for associated neurological signs such as falling and papilledema. This is especially true if the condition has persisted for a long time, and if the nystagmus is not horizontal and directional. If the rotary vertigo is associated with loss of consciousness, the physician should suspect also that the vertigo originates in the central nervous or cardiovascular system. The association of intense vertigo with localized headache makes it mandatory to rule out a central lesion.

Among the common central nervous system causes of vertigo are: (a) vascular crisis, (b) tumors of the posterior fossa, (c) multiple sclerosis, (d) epilepsy, (e) encephalitis, and (f) concussion.

The vertigo in a vascular crisis is of sudden onset and generally is accompanied by nausea and vomiting as well as tinnitus and deafness. In many instances there is involvement of other cranial nerves. In a special type of vascular crisis such as Wallenberg's syndrome there is thrombosis of the posterior–inferior cerebellar artery or vertebral artery, resulting in sudden overwhelming vertigo, dysphagia, dysphonia, and Horner's syndrome, as well as other neurological findings. The episode often is so severe that death may supervene.

It is also possible to have a much more discrete vascular defect in the vestibular labyrinth without an involvement of the cochlea. This usually results in an acute onset of severe vertigo. Recovery is rather slow, and the vertigo may persist as postural imbalance. The nystagmus may subside, but unsteadiness and difficultly in walking may continue much longer.

A still milder form of vascular problem related to vertigo occurs in hypotension and vasomotor instability. These people have recurrent brief episodes of imbalance and instability, particularly after a sudden change of position such as suddenly arising from tying their shoelaces or turning quickly. Histamine sensitivity may play an important role in this type of vertigo, especially when migraine attacks also are present.

In multiple sclerosis, vertigo is commonly a symptom, but it is rarely severe enough to confuse the picture with other involvement of the vestibular pathways. In epilepsy, vertigo occasionally may be a premonitory sensation before an epileptic attack, but unconsciousness usually accompanies the attack and helps to distinguish the condition.

Epilepsy and migraine may also be associated with dizziness. So can many toxins (ingested, inhaled, and transdermal). A complaint of vertigo in postconcussion syndrome is extremely common. The vertigo usually is associated with movement of the head or the body, severe headache, and marked hypersensitivity to noise and vibration. The patients usually are jittery, tense, and very touchy.

Of special interest to the otologist is the vertigo that sometimes is associated with a lesion of the posterior cranial fossa and particularly with an acoustic neuroma. It is a standing rule in otology that when a patient complains of true vertigo, an acoustic neuroma must be ruled out. This is especially true when the vertigo is accompanied by a hearing loss or tinnitus in one ear. Vertigo is not always a symptom in all posterior fossa lesions, but when it occurs, a number of simple tests help to establish the possibility that a tumor is present.

Other occasional causes of vertigo are encephalitis, meningitis, direct head injuries, and toxic reactions to alcohol, tobacco, and other drugs such as streptomycin, as well as many other conditions.

10. CERVICAL VERTIGO

Cervical vertigo may be due to muscle spasm compromising neck motion or interfering with the perception of orientation of the neck, vertebral artery compression by osteophytes, or direct compression of the spinal cord. Interestingly, one half of the proprioceptive receptors for the vestibular spinal tracts are located in the deep cervical musculature, with the rest being in the joint capsules of cervical vertebrae one to three (C1–C3). A majority of the proprioceptive information required for maintaining equilibrium comes from the muscle and joint capsule receptors. Whiplash and other spinal injury affects of the output from these receptors, giving the brain a false impression of the neck's orientation in space. This information conflicts with information from elsewhere in the balance system, and the resulting confusion is cervical vertigo.

Basilar artery compression syndrome is another condition associated with head motion, particularly neck extension. It is due to compression of the basilar artery with resultant interference with blood flow to the brain stem. This diagnosis cannot be made with certainty without objective confirmation through arteriography or other tests.

Carotid sinus syndrome and psychic disturbances frequently produce vertigo that may be confused with vertigo of vestibular origin. Taking a careful history and testing the carotid sinus make it possible to distinguish these conditions.

11. TREATMENT

Dizziness of central etiology will not be discussed in detail. In general, when dizziness is caused by cerebral, cerebellar or brain stem contusion, it is associated with other neurologic problems and is difficult to treat. The outlook for dizziness caused by vestibular injury is better.

Many medications are helpful in controlling peripheral vertigo. Meclizine is among the most common. It causes drowsiness in many people, but it is often effective in controlling vertigo. Scopolamine, administered through a transdermal patch, is also effective in controlling dizziness of labyrinthine etiology.

However, its side effects limit its use in many patients who are bothered especially by mouth dryness and dilation of the pupils causing blurred vision. Diazepam is also effective in suppressing vertigo, but long-term use of this potentially habit-forming drug should be avoided when possible. Prochlorperazine also helps many patients, sometimes producing good clinical improvement with low doses of 5–10 mg once in the evening. In hydrops, diuretic therapy with hydrochlorothiazide decreases vertigo, stabilizes hearing, and decreases fullness and fluctuation in many patients.

Patients with positional vertigo present special problems. It is helpful to distinguish BPPV from cervical vertigo. BPPV occurs when the patient's head is turned in a certain position. Typically, vertigo is induced when the patient rolls to one side in bed. Typically, ENG shows a short delay before onset of nystagmus, and the severity of nystagmus decreases with repeat testing. BPPV is generally not helped by medications such as Meclizine and Diazepam. Vestibular exercises which provoke vertigo and develop compensatory pathways and suppression are preferable (Table 19.2).

Cervical vertigo usually is accompanied by limitation of neck motion and tenderness. Turning the neck into certain positions causes dizziness, and symptoms can sometimes even be provoked by pressure over certain tender points in the neck or over the greater occipital nerve. Many treatments have been suggested for cervical vertigo. However, cervical manipulation and physical therapy generally produce the best results.

If dizziness is caused by a perilymph fistula, and if the patient seeks medical attention promptly, a short period of bed rest may be the first line of treatment. If the symptoms persist after 5 days of bed rest, or if they have been present for a long period of time before the diagnosis is made, surgical repair is warranted. This is accomplished with local anesthesia through the external auditory canal. The perilymph leak in the oval window or round window (or both) is repaired. Because of the hydrodynamics of the ear, the incidence of recurrent fistula formation is high, and the risks of permanent sensorineural hearing loss, tinnitus, and disequilibrium are substantial.

12. PHYSICAL THERAPY

Although the value of physical therapy in patients with balance disorders has become apparent only recently, it should not be underestimated. It is useful in patients who have failed conventional treatment, and in those who have persistent minor equilibrium problems after partially successful treatment. Physical therapy for balance disorders is quite specialized. The physical therapy team must have considerable interest and expertise in balance disorders, special equipment for balance rehabilitation, and willingness to devote substantial energy and resources to balance rehabilitation. Most physical therapy departments do not have experience and expertise in this problem, and their results with patients having balance disorders are not encouraging. However, when an appropriately trained team is involved, it can be exceedingly helpful. Such therapy is useful not only in patients with vestibular disorders. In patients with central disequilibrium, it is often the only help we have to offer.

Table 19.2 Vestibular Exercises (Cawthorne's Exercises)

Aims of exercise

(1) To loosen up the muscles of the neck and shoulders, to overcome the protective muscular spasm and tendency to move "in one piece."
(2) To train movement of the eyes, independent of the head.
(3) To practice balancing in everyday situations with special attention to developing the use of the eyes and the muscle senses.
(4) To practice head movements that cause dizziness, and thus gradually overcome the disability.
(5) To become accustomed to moving about naturally in daylight and in the dark.
(6) To encourage the restoration of self-confidence and easy spontaneous movement.

All exercises are started in exaggerated slow time and gradually progress to more rapid time. The rate of progression from the bed to sitting and then to standing exercises depends upon the dizziness in each individual case.

(A) Sitting position—without arm rests
  (1) Eye exercises-at first slow, then quick.
    (a) Up and down.
    (b) Side to side.
    (c) Repeat (a) and (b), focusing on finger at arms length.
  (2) Head exercises-head movements at first slow, then quick.
  (3) Shrug shoulders and rotate, 20 times.
  (4) Bend forward and pick up objects from the ground, 20 times.
  (5) Rotate head and shoulders slowly, then fast, 20 times.
  (6) Rotate head, shoulders, and trunk with eyes open, then closed, 20 times.
(B) Standing
  (7) Repeat 1.
  (8) Repeat 2.
  (9) Repeat 5.
  (10) Change from a sitting position to standing position, with eyes open, then shut.
  (11) Throw ball from hand to hand (above eye level).
  (12) Throw ball from hand to hand under knees.
  (13) Change from sitting to standing and turn around in between.
  (14) Repeat 6.
(C) Walking
  (15) Walk across room with eyes open, then closed, 10 times.
  (16) Walk up and down slope with eyes open, then closed, 10 times.
  (17) Do any games involving stopping, or stretching and aiming, such as bowling, shuffleboard, etc.
  (18) Stand on one foot with eyes open, then closed.
  (19) Walk with one foot in front of the other with eyes open, then closed.

12.1. Surgery

If medical treatments for peripheral vertigo fail, several surgical approaches are available. For endolymphatic hydrops, endolymphatic sac decompression provides relief of dizziness in ~70% of people. Despite the 30% failure rate, it is appropriate in selected cases. Vestibular nerve section is a more definitive procedure. This may be performed by entering the posterior fossa through the mastoid or just behind the sigmoid sinus. The eighth nerve is divided, preserving cochlear fibers and sectioning the vestibular division of the eighth nerve. The author's experience parallels that of other authors reporting success rates >90% with this procedure. If there is no usable hearing, the entire eighth nerve can be divided. Interestingly, this procedure does not appear to improve the success rate substantially. Failures are probably due to disequilibrium produced by a lesion more central in the vestibular pathway, or located elsewhere in the vestibular system altogether (the other ear). Despite all efforts, it is not possible to identify such conditions in all patients prior to surgery.

Occasionally, it is impossible to determine which ear is responsible for dizziness, especially if there are signs of abnormality bilaterally. In such cases, bilateral medical labyrinthectomy is possible. This procedure takes advantage of the ototoxicity of streptomycin. Patients must be selected carefully (27), and the procedure must be carefully controlled.When bilateral labyrinthectomy is complete, patients generally adapt well so long as they have visual cues. However, they are not able to function in total darkness.

13. SPECIAL CONSIDERATIONS IN INDUSTRY

Tinnitus, and especially vertigo, present special problems in industry. They are both usually subjective complaints that may be difficult to document objectively. Tinnitus is common but rarely disabling. It may be quite disturbing to some individuals. However, only in rare cases is it severe enough to interfere with the ability to work, or even to interfere very much with the quality of life on a daily basis. There are exceptions, of course.

However, vertigo and other conditions of disequilibrium may be disabling, particularly for people working in hazardous jobs. A person who loses his/her balance even momentarily may injure himself or other severity if he is working around sharp surfaces, rotating equipment, driving a forklift, or working on ladders or scaffolding. There are many other examples of occupations that are not possible for people with disequilibrium disorders. The problem is even worse in conditions that typically cause intermittent severe disequilibrium, such as Meniere's disease. Our inability to test the balance system thoroughly makes it impossible to objectively disprove a worker's contention that he has spells of dizziness. For example, if a worker is struck in the head while performing his job and claims that he has intermittent "dizzy spells" afterwards, even if he has a normal electronystagmogram, his assertion may be true. Even if malingering is suspected, in the absence of objective proof, if a physician contests this claim and declares him fit to return to work, both the physician and the industry may incur substantial liability if the worker suffers a period of disequilibrium and seriously injures himself or other workers. Considerable research is needed to develop more sophisticated techniques of assessing equilibrium, despite recent advances (8–12).

13.1. Proposed Classification

Dizziness and especially vertigo can be very disabling symptoms, and are often accompanied by nausea, vomiting, anxiety, headache, and other complaints. Symptoms may be constant, occur in episodic "attacks," frequently without warning, or may be constant with exacerbations. Some findings on clinical examination may be entirely objective, such as a third degree spontaneous nystagmus, whereas others may be entirely feigned, for example, an alleged posttraumatic ataxia of gait. Many others fall between these extremes. Clinical and laboratory assessments are used to establish data that determine impairment and disability classification. Organizing the history, physical and laboratory findings helps simplify and clarify this process.

13.2. Symptoms Elicited by History

1. No complaint of dizziness/unsteadiness.
2. Mild dizziness, does not interfere substantially with daily activities.

3. Mild-to-moderate dizziness, interferes with some activities but they can be resumed.
4. Moderate dizziness, interferes with many activities, but important activities can usually be accomplished with effort and with life-style adjustments.
5. Severe dizziness, interferes with most activities. Unable to drive or ambulate without assistance.
6. Profound dizziness, bed-ridden or wheelchair-dependent.

13.3. Signs on Physical Examination

1. None.
2. Mild unsteadiness with reduced visual or proprioceptive input.
3. Paroxysmal positioning nystagmus or moderate unsteadiness with reduced visual or proprioceptive input.
4. Staggering when walking with the eyes closed.
5. Inability to stand or walk without the assistance of another individual, even with eyes open.

13.4. Laboratory Test Results

Laboratory test results correlate well with, and indeed are the primary measure of, impairment of function, but they may not correlate well with disability. This is especially true if the brain and related structures are intact, and also if the onset of the impairment occurred months or years prior to the evaluation.

Nevertheless, laboratory test results are usually helpful in determining whether there is a physical basis for the individual's complaints.

In correlating symptoms with laboratory tests, it is useful to rate the overall results of the tests as follows:

1. Normal.
2. Mildly impaired test results.
3. Moderately impaired test results.
4. Markedly impaired test results.
5. Complete loss of function on testing.
6. Consistency between history, examination and test results that is not in keeping with organic etiology.
7. Inconsistency between history, examination and test results that is not in keeping with organic etiology.

The outcome of the symptoms evaluation, the physical signs and the laboratory test results should be correlated, with emphasis on the validated symptoms, that is, those confirmed by the history, physical findings and the results of laboratory tests, as best representing the subject's true state of disability.

13.5. Consistency

1. Consistency between history, symptoms, physical examination, and laboratory tests.
2. Inconsistency between history, symptoms, physical examination, and laboratory tests.

13.6. Criteria of Disability Resulting from Impairment of Spatial Orientation and Equilibrium

13.6.1. Class 1: Impairment of the Whole Person (0%)

A designation of Class 1 is assigned when (a) symptoms of dysequilibrium are present, without supporting clinical or objective evidence of organic disease and (b) the dizziness has no effect on the usual activities of daily living.

13.6.2. Class 2: Impairment of the Whole Person (1–10%)

A designation of Class 2 is assigned when (a) symptoms of dysequilibrium are present with supporting clinical and/or objective findings and (b) dizziness may interrupt activities for a brief period of time, but activities can be resumed as soon as the dizziness resolves. Dizziness has not caused the patient to discontinue working, driving, or engaging in any activity without restriction, and it has not required changes in plans or activities to accommodate dizziness.

13.6.3. Class 3: Impairment of the Whole Person (11–30%)

A designation of Class 3 is assigned when (a) symptoms of dysequilibrium are present with supporting clinical and/or objective findings and (b) dizziness had required the patient to change plans and make allowances for dizziness. The patient can continue to perform activities of daily living such as self-care and household duties, drive, engage in most recreational activities and work, but is unable to perform safely activities such as climbing or walking on girders and scaffolds.

13.6.4. Class 4: Impairment of the Whole Person (31–50%)

A designation of Class 4 is assigned when (a) symptoms of dysequilibrium are present with supporting clinical and/or objective findings and (b) the patient's dizziness requires constant adjustments in activities and budgeting of the patient's energies. With great effort, the patient is able to take care of family, engage in most essential activities, drive, work, and travel.

13.6.5. Class 5: Impairment of the Whole Person (51–75%)

A designation of Class 5 is assigned when (a) symptoms of dysequilibrium are present with supporting clinical and/or objective findings and (b) dizziness prevents the patient from doing most of the activities that he/she could do prior to the onset of dizziness, including drive, work, and take care of a family in the usual fashion. Even essential activities must be limited.

13.6.6. Class 6: Impairment of the Whole Person (76–95%)

A designation of Class 6 is assigned when (a) symptoms of dysequilibrium are present with supporting clinical and/or objective findings, (b) the usual activities of daily living cannot be performed without assistance, except for self-care not requiring ambulation, and (c) confinement to the home or other residential facility is necessary under most circumstances.

13.7. Disability

Disability rating is beyond the scope of this discussion. However, it is an important and complex subject that requires knowledge of an individual's professional activities. The same impairment rating may produce substantial differences in disability. For example, a computer data entry clerk with Class 3 impairment may have little or no disability, whereas a sky-scraper construction worker with the same impairment may be totally disabled for his usual occupation.

Comprehensive neurotological evaluation permits an accurate diagnosis in many cases of disequilibrium. If the patient has work-related issues, reasonable assignment of an impairment rating based upon scientific evidence should be possible in most cases. Further investigation of other tools (such as the Dizziness Handicap Inventory[28]) may provide better clinical and medical-legal approaches to the assessment of the impact of dizziness in the future.

14. CONCLUSION

Although dizziness is a common and vexing problem, it is not hopeless. The clinician should search diligently for the etiology in every patient with dysequilibrium. If we are enthusiastic about evaluating and treating dizzy patients and assume that we will be able to find a treatable cause (until proven otherwise), that assumption will prove right surprisingly often. In virtually all cases, patients feel better following comprehensive evaluation. Under the best of circumstances a treatable etiology is found, and their dizziness is eliminated. Under the worst circumstances, at least they understand that "no stone has been left unturned," and that they have nothing serious or life-threatening that is responsible for their problem. Physical therapy, medications, and sometimes surgery can usually be used to at least improve each patient's ability to function, if not to eliminate symptoms altogether. Dizziness is a common problem that is challenging to both patients and physicians; and it is important for all of us to remember that every patient can be helped at least to some extent if we try hard enough.

REFERENCES

1. Sataloff J, Sataloff RT, Lueneberg W. Tinnitus and vertigo in healthy senior citizens with a history of noise exposure. Am J Otol 1987; 8(2):87–89.
2. U.S. Department of Health and Human Services. Public Health Services. National Center for Health Statistics Series 13, No. 56, 1978.
3. Jenkins HA, Furman JM, Gulya AJ, Honrubia V, Linthicum FH, Mirko A. Dysequilibrium of Aging. Otolaryngol Head Neck Surg 1989; 100:272–282.
4. U.S. Department of Health and Human Services. Head Injury: Hope Through Research. National Institutes of Health, Publication No. 84-2478, 1989.
5. Gibson WPR. Vertigo associated with trauma. In: Dix MR, Hood JD, eds. Vertigo. Chichester, New York: John Wiley & Sons, 1984.
6. Mandel S, Sataloff RT, Schapiro S. Minor Head Trauma: Assessment, Management and Rehabilitation. New York, NY: Springer-Verlag, 1993.
7. Goebel JA, Sataloff RT, Hanson JM, Nashner LM, Hirshout DS, Sokolow CC. Posturographic evidence of non-organic sway patterns in normal subjects, patients and suspected malingers. Am Acad Otolaryngol Head Neck Surg 1997; 117(4):293–302.
8. Cevette MJ, Puetz B, Marion MS, Wertz MS, Wertz ML, Muenter MD. A physiologic performance on dynamic posturography. Otolaryngol Head Neck Surg 1995; 112:678–688.
9. Coogler CE. Using computerized dynamic posturography to accurately identify non-organic response patterns for posture control. Neurol Rep 1996; 20:12–21.
10. Hart CV, Rubin AG. Medico-legal aspects of neurotology. Otolaryngol Clin North Am 1996; 29(3):503–516.

11. Krempl GA, Dobie RA. Evaluation of posturography in the detection of malingering subjects. Am J Otol 1998; 19:619–627.
12. Nasher LN. A model describing vestibular detection of body sway motion. Acta Otolaryngolog Scand 1971; 72:429–436.
13. Herdman SJ. Vestibular Rehabilitation. Philadelphia, PA: F.A. Davis, 1994:1–377.
14. Jamison D, Alavi A, Jolles P, Chawluk J, Reivich M. Positron emission tomography in the investigation of central nervous systems disorders. Radiol Clin North Am 1988; 26(5):1075–1088.
15. Langfitt TW, Obrist WD, Alavi A. Langfitt TW, Obrist WD, Alavi A et al. Computerized tomography, magnetic resonance imaging, and positron emission tomography in the study of brain trauma. Preliminary Observations. J Neurosurg 1986; 64(5):760–767.
16. Boni F, Fattori B, Piragine F, Bianchi R. Use of SPECT in the diagnosis of vertigo syndromes of vascular nature. Acta Otorhinolaryngol Ital 1990; 10(6):539–548.
17. Laubert A, Luska G, Schober O, Hesch RD. Die digitale Subtraktionsangiographie (DSA) und die Jod-123-Amphetamin-Szintigraphie (IMP-SPECT) in der Diagnostik von akuten und chronischen Innenohrschwerhörigkeiten. Digital substracton angiography and iodine 123 amphetamine scintigraphy (IMP-SPECT) in the diagnosis of acute and chronic inner ear hearing loss. HNO 1987; 35(9):372–375.
18. Tuohimaa P, Aantaa E, Toukoniitty K, Makela P. Studies of vestibular cortical areas with short-living 1502 isotopes. ORL J Otorhinolaryngol Relat Spec 1987; 45(6):315–321.
19. Abdel-Dayem HM, Sadek SA, Kouris K et al. Changes in cerebral perfusion after acute head injury: comparison of CT with Tc-99m HM-PAO SPECT. Radiology 1987; 165(1):221–226 [published erratum appears in Radiology 1988; 167(2):582].
20. Roper SN, Mena I, King WA, Schweitzer J, Garrett K, Mehringer CM et al. An analysis of cerebral blood flow in acute closed-head injury using Tc-99m HM-PAO SPECT and computed tomography [published erratum appears in J Nucl Med 1991; 32(9):1684–1687].
21. Bullock R, Statham P, Patterson J et al. The time course of vasogenic oedema after focal human head injury—evidence from SPECT mapping of blood brain barrier defects. Acta Neurochir Suppl (Wien) 1990; 51:286–288.
22. Ducours JL, Role C, Guillet J et al. Cranio-facial trauma and cerebral SPECT studies using *N*-isopropyliodo-amphetamine (123I). Nucl Med Commun 1990; 11(5):361–367.
23. Oder W, Goldenberg G, Podreka I et al. HM-PAO-SPECT in persistent vegetative state after head injury: prognostic indicator of the likelihood of recovery? Intens Care Med 1991; 17(3):149–153.
24. Morinaga K, Hayaski S, Matsumoto Y et al. CT and 123I-IMP SPECT findings of head injuries with hyponatremia. No To Shinkei 1991; 43(9):891–894.
25. Duffy FH, Burchfiel JD, Lombroso CT. Brain electrical mapping (BEAM): A method of extending the clinical utility of EEG and evoked potential data. Ann Neurol 1979; 5:309–321.
26. Sataloff RT, Hughes M, Small A. Vestibular "Masking": a diagnostic technique. Laryngoscope 1987; 7:885–886.
27. Graham MD, Sataloff RT, Kemink JL. Titration streptomycin therapy for bilateral Meniere's disease: a preliminary report. Otol Head Neck Surg 1984; 92(4):440–447.
28. Jacobson GP, Newman CW. The development of the Dizziness Handicap Inventory Arch Otol Head Neck Surg 1990; 116: 424–427.

20
Facial Paralysis

Robert T. Sataloff
Drexel University College of Medicine, American Institute for Voice and Ear Research, and Graduate Hospital, Philadelphia, Pennsylvania, USA

Mary Hawkshaw
American Institute for Voice and Ear Research, Philadelphia, Pennsylvania, USA

The facial nerve courses with the auditory nerve, and in a complex pattern through the ear and mastoid before distributing its fibers to the facial muscles. It also gives off several branches for functions other than facial motion. Because of the close anatomic and embryologic association of the facial nerve with the ear (1), physicians should be aware that facial nerve abnormalities may accompany other ear problems. Such abnormalities may include paralysis, hemifacial spasm, ear canal pain, taste distortion, and other problems. Facial paralysis is the most common malady, and physicians concerned with ear problems should at least be acquainted with this condition.

1. ANATOMY

The motor fibers of the facial nerve run from the facial nucleus in the brain stem, bending around the abducens nucleus, courses from the brain stem across a short expanse of posterior fossa, enters the internal auditory canal in its anterior/superior compartment, becomes covered with bone in its labyrinthine segment as it leaves the internal auditory canal; courses anteriorly, bends at the geniculate ganglion, courses horizontally and vertically through the mastoid, exits through the stylomastoid foramen and innervates muscles of facial expression. It also carries special sensory fibers for taste, preganglionic parasympathetic fibers to the lacrimal and submandibular glands, and sensory fibers to the skin of the posterior/superior aspect of the external auditory canal. Decreased sensation in this region is called a positive Hitselberger's sign and is common in the presence of lesions that cause pressure on the facial nerve, such as an acoustic neuroma. Several other branches within the temporal bone also can be tested, including, for example, the greater superficial petrosal nerve which supplies preganglionic parasympathetic fibers to the lacrimal gland for tearing, the nerve to the stapedius muscle, and the chorda tympani nerve which supplies taste to the ipsilateral two-thirds of the tongue, and preganglionic parasympathetic fibers to the ipsilateral submandibular gland. Any portion of the nerve may be involved by disease or injury.

Facial paralysis is a relatively common affliction. Facial paralysis creates cosmetic and functional deficits that may be extremely troublesome. Usually the condition involves one side of the face. If the muscles cease working altogether, the condition is facial paralysis. If the muscles are merely weak, the condition is facial paresis. For the rest of this chapter, we will use the word "paralysis" to refer to complete paralysis or severe paresis. This condition results in drooping of one side of the face, inability to close the eye (which exposes it to dryness and injury) and incompetence of the corner of the mouth, which may cause drooling of liquids and difficulty articulating some sounds.

One of the principle problems in management of facial paralysis is a widespread tendency within many sectors of the medical community to under-evaluate the problem, and to misdiagnose it. In 1821, Sir Charles Bell studied the innervation of the facial musculature and named the motor nerve of the face the "facial nerve." Soon, all conditions of facial paralysis came to be known as "Bell's Palsy." As time passed, the true cause of many cases of facial paralysis was discovered. Bell's Palsy is now the name used to describe only those cases of facial paralysis in which the cause cannot be determined.

Unfortunately, many physicians still diagnose "Bell's Palsy" without going through the comprehensive evaluation necessary to diagnose and treat important causes of facial paralysis.

A proper evaluation includes a complete history; careful physical examination of the ears, nose, throat, neck, and parotid glands; at least a partial neurological examination; imaging studies (usually CT scan and MRI); hearing tests; electrical facial nerve tests; and blood tests. These tests are designed to detect the true cause of the problem. If none is found, the condition may properly be called "Bell's Palsy." Bell's Palsy always recovers, at least partially, although recovery may take as long as a year. If the face remains paralyzed without recovery of facial function, the diagnosis of Bell's Palsy is generally wrong, and the true etiology should be sought again.

When examining patients with facial paralysis, the severity of their paralysis should be recorded using a standardized grading system. The House–Brackman grading system is used most widely. The grading systems are actually easy to remember. Grade I is normal. Grade VI is completely paralyzed. Grade II is almost normal. Grade V is almost completely paralyzed. The difficult distinctions come between Grade III and Grade IV. In Grade III/VI facial paresis, the paralysis is obvious, but there is still some motion in the forehead and the patient can close his or her eye completely, with effort. In Grade IV out of Grade VI paresis, the patient can no longer obtain complete eye closure, and the forehead is paralyzed.

2. EVALUATION

The details of a comprehensive ear, nose, and throat examination, cranial nerve assessment, and audiometric testing will not be reviewed in this chapter. They are similar to those assessments performed in patients with hearing problems or suspected facial neuroma. Metabolic evaluation for diabetes, hypothyroidism, Lyme disease, and the many other conditions mentioned later in this chapter should also be performed. Radiological assessment is also similar and generally includes MRI with enhancement of the entire course of the facial nerve from the brain stem, through the temporal bone, and into the parotid gland and facial muscles. Temporal bone CT also may be helpful in showing areas of bone destruction or injury that may compromise the facial nerve. Special clinical examination of the facial nerve and facial nerve electrical testing warrants further discussion, however.

3. FACIAL NERVE TESTING

Facial nerve testing is done routinely when a motor abnormality is observed (see Case Report). However, tests may also be helpful in the absence of obvious motor abnormalities, especially if a pressure lesion is suspected, such as acoustic neuroma or vascular compression.

3.1. Case Report

A 32-year-old male who was injured at work. He was thrown against the wall striking the right side of his head. He was stunned for a few seconds, and had short-term retrograde amnesia. He also had headache and minor memory deficits following the injury. One day after the accident, he noted complete right peripheral facial paralysis. Neurotologic examination was normal expect for facial paralysis and a positive Hitselberger's sign on the right. Audiogram revealed a small 4000-Hz dip on the right (Fig. 20.1) with an even smaller dip on the left. CT and MRI were normal. Electroneuronography 3 months after injury still revealed 100% electrical degeneration on the right. The patient had declined facial nerve decompression. Partial recovery of voluntary facial motion occurred over the next several months.

PATIENT:

	RIGHT EAR AIR CONDUCTION							LEFT EAR AIR CONDUCTION					
DATE	250	500	1000	2000	4000	8000		250	500	1000	2000	4000	8000
	20	15	15	15	30	15		5	10	5	5	20	10

SPEECH RECEPTION:

R: 15 dB L: 5 dB

Figure 20.1 Audiogram of case report.

Facial paralysis is common after penetrating trauma, fractures, or surgery. It is seen less frequently after minor head trauma and is felt to be due to edema within the facial nerve canal, particularly in the labyrinthine segment where the nerve leaves the internal auditory canal to enter the middle ear, because this is the most narrow segment. The slight 4000-Hz audiometric dip is typical following inner-ear concussion.

4. SIMPLE TOPOGNOSTIC TESTS

The validity and reliability of topognostic facial nerve testing have not been proven. Nevertheless, they often provide helpful information regarding the site of a facial nerve lesion. A modified Schirmer tear tests helps determine whether the lesion is proximately distal to the greater superfacial petrosal nerve and auricular ganglion. The stapedius reflex decay test establishes this relationship to the stapedius muscle. Testing of taste on the anterior two-thirds of the tongue, or testing salivary flow from the submandibular glands, locates the lesion in relation to the chorda tympani nerve. Peripheral branches of the facial nerve can be tested with a facial nerve stimulator. Although minimal nerve excitability testing is performed most commonly, maximal nerve excitability may provide more information. It usually requires 1 or 2 mA of current above the nerve threshold. Assessment of all of these tests is subjective.

5. ELECTROMYOGRAPHY

Standard facial electromyography records muscle action potentials produced by voluntary efforts to move muscles. Responses are recorded using needle electrodes inserted into the muscle, and the signal is monitored on an oscilloscope visually and acoustically. Neuromuscular function and degeneration can be detected.

6. ELECTRONEURONOGRAPHY

Electroneuronography is similar to electromyography except that it uses surface electrodes and evoked rather than volitional stimuli. A stimulating electrode is placed over the main trunk of the facial nerve at the stylomastoid foramen, and a recording electrode is placed over the muscle to be tested. This technique allows measurement of latency and conduction velocity, although latency has proven a more useful measure. Recording of threshold is possible. However, supramaximal stimuli are generally used. Several parameters can be measured and have proven useful clinically.

Electroneuronography (ENoG) is a measure of the facial nerve action potential and is quantified in terms of amplitude as measured in millivolt. A comparison is made between the nerve response on the "poor" side vs. the "good" side and a percent weakness is calculated for the "poor" side.

The protocol for the test is as follows: The patient should first be instructed to relax as much as possible in order to minimize any muscle artifact from the response following electrode application. Electrode impedance is measured to ensure adequate recordings. The patient is then instructed to smile or contort his or her face to allow the tester to determine and set an appropriate sensitivity level for recording. An electrical current is then applied to the "good" side. Stimulator placement is critical and often a small amount of pressure at the stimulation sites is required in order to obtain clear and reliable recordings. Stimulus intensity as measured in milliamperes is gradually increased until the maximum tolerance level of the patient is reached and recordings become stable. Stimulation is then stopped, and maximum and minimum points are selected and plotted along the response curve. These values are subtracted from one another in order to obtain the amplitude of the response. The same protocol is then followed for the opposite or "poor" side. It is critical that maximum stimulation to this side of the face meet but not exceed that of the better side. This allows for a more exact comparison between the two responses. Once amplitude measures are obtained for both nerves, a percent weakness on the "poor" or involved side may be calculated as follows:

$$\frac{\text{Poor side amplitude}}{\text{Good side amplitude}} \times 100 = (x)$$

then $100 - (x) =$ percent weakness on poor side. A difference of $\geq 17\%$ between the two responses is considered significant.

7. CAUSES OF FACIAL PARALYSIS

What kind of problems may cause facial paralysis? There are many, and this chapter will provide a brief overview of only a few of them. Facial paralysis can be congenital. That is, some people are born with paralysis of one side or both sides of the face. This problem may result from various causes. Acquired facial paralysis is much more common.

Infection and inflammation are well established causes. The infection may be viral, particularly herpetic, and is often associated with a herpetic cold sore or "fever blister." Herpes Zoster ("Shingles") may also be etiologic. This problem may also involve the hearing nerve which runs in close proximity to the facial nerve. Viral or bacterial infections of the ear (otitis media) or brain (meningitis) also cause facial paralysis, as do Lyme disease (from ticks) and HIV infection. Facial paralysis has also been associated with toxic effects from exposure to heavy metals such as lead, and immunologic response following injections for tetanus, rabies, and polio. Metabolic conditions, including diabetes, hypothyroidism, and pregnancy, may precipitate facial paralysis as well. So may vascular problems associated with vasculitis some of which may accompany collagen vascular syndromes diabetes and other conditions. Trauma is also an important cause of facial paralysis, particularly temporal bone fracture associated with motor vehicle accidents or falls, and surgical trauma (facial paralysis is a potential complication of ear, brain, or parotid gland surgery).

In addition, facial paralysis may be the first presenting sign of a tumor. The tumor may be benign or malignant. It may occur anywhere along the course of the facial nerve, which starts in the brain stem, courses circuitously through the ear encased in a tight, bony canal, exits the mastoid bone at the skull base, and spreads out along the face. Tumors of the facial nerve itself (facial neuromas) commonly present with facial paralysis. So do cancerous tumors of the middle ear, parotid gland, and upper neck.

Facial paralysis occurs rarely with acoustic neuromas, despite the fact that compression of the facial nerve by these tumors is common. Other growths of the temporal bone that may cause facial paralysis include cholesteatoma, glomus tumor, and neuromas of other nerves. Facial paralysis may also be caused by neoplasms outside the temporal bone. When a parotid neoplasm presents with facial paralysis, the lesion is almost always malignant. A variety of other conditions may be associated with facial paralysis or paresis including hemifacial spasms which may be due to vascular loop compression of the facial nerve, Milkerson Rosenthal syndrome, barotrauma, sarcoidosis, and of course, Bell's Palsy. However, because facial paralysis may be due to so many serious conditions, physicians should not accept a cavalier diagnosis for "Bell's Palsy," but should pursue a credible diagnosis based on systematic comprehensive evaluation. Tumors and many of the other conditions we have mentioned may be treated effectively, particularly if they are diagnosed early.

8. TREATMENT

Treatment for facial paralysis depends upon the cause. If a specific disease or tumor is identified, it is treated appropriately; and in many cases facial nerve function can be restored. Regardless of the cause, symptomatic treatment is important, particularly eye protection. The eye should be treated with artificial tears and/or a more long-lasting artificial lubricant. At night, the eye should be taped shut with tape running horizontally along the upper lid toward the side of the face. The tape should not cross the lash line. When it does, it may invert the lashes and scratch the cornea. Corneal problems are of particular concern when there is a poor Bell's phenomenon (the eyeball does not rotate up when the eye is closed), or when there is simultaneous fifth cranial nerve pathology resulting in decreased corneal sensation. In these cases especially, it is important to protect the eye aggressively, usually surgically. This may involve implantation of a gold weight, surgical tarsorrhaphy, and/or medial or lateral canthoplasty. If no cause is found and the condition is truly Bell's Palsy, various treatments are available.

The medical specialists most involved in caring for Bell's Palsy are neurotologists, subspecialists of otolaryngology (ear, nose, and throat) who deal with disorders of the ear and related structures (including the facial nerve), and the ear–brain interface. Neurologists also commonly evaluate patients with facial paralysis but few specialize in this particular nerve. Most neurotologists specializing in facial nerve disorders base our treatment recommendations for Bell's Palsy on electrical testing of the facial nerve called ENoG. Treatment may involve observation, use of high-dose corticosteroids or surgery.

Surgery is always a last resort. However, it should be considered if electrical excitability decreases by 90% or more; if facial paralysis is sudden, total, and associated with severe pain; and in selected other situations. In general, if recovery from facial paralysis begins within 3 weeks, recovery is excellent and total. If recovery does not start until after 3 months, it is usually imperfect. In these cases, the nerve usually develops some degree of synkinesis (inappropriate neuromuscular function). This is caused by recovering nerve fibers growing down the wrong channels. This results, for example, in a slight smiling of the lips when the person closes his/her eyes. "Crocodile tears" and "gustatory sweating" result similarly from misdirected nerve fibers that should have gone to salivary glands. Hence, when the person eats, tears fall from the eyes, and the side of the face perspires. These problems can often be minimized with early, accurate diagnosis.

Because any disturbance of facial function is potentially so problematic, it is essential for any physician encountering this affliction to follow through with the comprehensive evaluation appropriate for this potentially serious disorder. Simply looking at a paralyzed face and proclaiming it "Bell's Palsy" is not in keeping with the current standards of care.

REFERENCE

1. Sataloff RT. Embryology and Anomalies of the Facial Nerve. New York, NY: Raven Press, 1990.

21

Squamous Cell Carcinoma of Temporal Bone

Mikhail Vaysberg
Philadelphia College of Osteopathic Medicine, Philadelphia, Pennsylvania, USA

Robert T. Sataloff
Drexel University College of Medicine, American Institute for Voice and Ear Research, and Graduate Hospital, Philadelphia, Pennsylvania, USA

The estimated incidence of temporal bone (TB) malignancy by the National Cancer Institute is 0.8–6 cases per million population per year (1,2). Squamous cell carcinoma (SCCA) represents the majority of these cases (3–6). SCCA, a rare neoplasm with an overall aggressive nature and poor prognosis, is estimated to comprise 90% of all primary malignant tumors originating in the TB (7–9). Kuhel et al. (10) described the sites of origin for TB SCCA: auricle (60–70%), the external auditory canal (EAC) (20–30%), and the middle ear (10%). There is a report of isolated SCCA *in situ* of the tympanic membrane (11). The advanced nature of lesions frequently makes it harder to assess the site of origin. Most reports found a peak age incidence of SCCA of the EAC to occur in the mid-50s (5,10,12). However, the population developing middle ear malignancy appears to be ~10 years younger (13). From conflicting data of multiple studies on the sex distribution of SCCA of TB, it appears that the distribution is equal between men and women (3,5,14–17).

Definitive information about the treatment and prognosis of SCCA of the TB is not easy to obtain due to a multitude of factors: the rarity of this malignancy, the lack of an accepted staging system (prior to 1990), and the wide variety of individualized treatments. Also, many authors include tumors of multiple histology in their reports of treatment outcomes, which further complicates the issue.

1. ETIOLOGY

Various etiologic factors have been associated with SCCA.

For SCCA of the auricle, sun exposure has been implicated as the most common cause. Bendl et al. (18) found strong correlation (82%) between the actinic or solar damage to the epidermis of the auricle and the development of SCCA. Shifman (19) has noted the infrequency of auricular SCCA in women with long hair and Chen and Dehner (16) found no association between SCCA of EAC with prior actinic changes of the epidermis. Some authors continue to believe that sun exposure may play some role. However, they emphasize that other factors such as genetic susceptibility contribute to SCCA malignancy after exposure to ultraviolet light (20).

The middle ear is sheltered from the carcinogenic effect of the sun. Thus, investigators have been concentrating on other etiologies. In 1908, Whitehead described long-standing chronic otitis media as a possible cause of SCCA. Wagenfeld et al. (21) found a history of previous otitis media in 64% of SCCA of TB patients; and, many studies list chronic otorrhea as the most common presenting symptom of SCCA of the TB (10,12). It is possible that the presence of continuous inflammation induces increased proliferation and increases the risk of malignant transformation in the surrounding tissue. For example, pseudoepitheliomatous hyperplasia (HP) is a benign skin lesion that is associated with chronic inflammations caused by various infectious agents, autoimmune skin diseases, burns, or even prolonged exposure to irritants (UV light, allergens, alcohol, chlorinated disinfectants, etc.). It bares striking resemblance to SCCA (22–24). There are several reported cases of over-treatment as a result of such histological similarity. Furthermore, Gacek et al. (22) have presented a case of HP conversion into an invasive SCCA after a 6-year symptom-free interval. HP could be an intermediate step in the continuum of the SCCA disease process. Yet, the rarity of the SCCA of the TB does not correlate with the abundant prevalence of chronic otits. Hence, it is most likely that additional factors have to be present in order to promote malignant transformation. Viral causes have been implicated. Jin et al. (25) reported high prevalence of HPV in middle ear carcinoma associated with chronic otitis media. Some believe that chronic otorrhea is more likely a result than a cause of SCCA of the ear. The growth of the tumor may result in secondary infection of the middle ear via Eustachian tube and EAC.

Cholesteatoma has been suggested as a causative agent of SCCA. Occasionally, it is present with carcinoma in the middle ear (12). Michaels and Wells (1980), however, have reviewed the TBs of 28 patients with SCCA of the middle ear; none of these patients had concomitant cholesteatomas. It seems more likely that cholesteatoma is present coincidentally or as a result of carcinoma, rather than a cause of SCCA.

In 1952, Aub (26) and later in 1965 Beal et al. (27) identified radiation as a possible etiology for SCCA of the middle ear and mastoid. They noted that a significant percentage of the employees of clock and watch factories developed TB malignancy. During the 1940s, the clock industry utilized radium-containing paint to give the numerals attractive nightglow. In order to apply this paint accurately, workers would moisten the brushes with their tongue and lick the radioactive substance through out the day. Tumor induction may have been caused by radon gas trapped in the mastoid air cell system or from the radon deposits in the cortex of the TB.

Lim et al. evaluated patients with malignant tumors of the TB. They found 7 out of 18 patients (39%) had previous history of radiation therapy for nasopharyngeal carcinoma. Notably, five out of seven patients had SCCA. The presentation from the time of exposure averaged 13 years (28). Other authors have reported similar findings associated with >5000 cGy of radiation. Unfortunately, radiation-induced malignancies appear to be more aggressive and therapy-resistant than sporadic tumors (28–30). It appears that regional postradiation changes may alter local immunity as chronic ear infections are common after radiotherapy to nasopharynx (28,31). The authors used the criteria for diagnosis of radiation-induced malignancy proposed by Cahan et al. (32). The patient must have a history of radiation, the second neoplasm must arise in the irradiated field, and a latent period of at least 5 years must have elapsed between the radiation exposure and the development of the second neoplasm. Further, there must be histologic and roentgenographic evidence of the preexisting condition in addition to microscopic proof of a tumor; the second tumor must be a different histologic type from that previously irradiated in order to eliminate the possibility of recurrence of the original tumor.

Local malignancies from the sites adjacent to TB may extend into its structures. These sites include the parotid gland, TMJ, EAC, preauricular skin, and nasopharynx. The TB may also be infiltrated by metastatic tumors arising from the breast, lung, kidney, stomach, larynx, prostate, thyroid, and other organs.

2. CLINICAL CHARACTERISTICS

Most patients present with the diagnosis of SCCA of the TB in the 5th or 6th decade with equal distribution among sexes.

Unfortunately, clinical presentation is nonspecific. Kuhel et al. (10) evaluated 442 patients and identified otorrhea (61%), otalgia (51%), and EAC mass (37%) as the most common presenting symptoms. In the report by Kenyon et al. (12), evaluation of 19 patients revealed three patients with persistent otorrhea present for at least 30 years. Many authors have found long-standing otorrhea, of ~20 years, to be common (12,33,34). They also noted changes in both the quality and quantity of discharge prior to patients' presentation with tumor. The amount of discharge increased, became malodorous, and occasionally bloody. Otalgia is also common and noted to become worse weeks prior to the diagnosis; it is usually described as dull and deep-seated, most likely as a result of bony invasion. Presence of trismus may indicate tympano-mandibular joint (TMJ) involvement.

The majority of patients have polyps or granulation tissue within the EAC. Even though the pathology specimen can be obtained easily on an outpatient basis, the pathology findings can be misleading. Sometimes these tumors may remain asymptomatic until they become quite large. Neoplasms involving the middle ear may cause a sensation of fullness and conductive hearing loss. According to several reports, up to 50% of patients may have hearing loss of various patterns (2,9,35). Invasion of the otic capsule is uncommon due to the resistance of enchondral layers; yet, once the neoplasm involves the inner ear, sensorineural hearing loss and vertigo are common.

The presence of facial nerve involvement indicates more advanced disease and significantly decreases survival (2,9,14,27,36,37). The degree of facial nerve involvement also plays a prognostic role. The treatment failure rate for patients with mild paresis has been reported as 34%, moderate paresis is 50%, and complete paralysis is 62% (RT Sataloff and JE Medina, personal communication). Other signs of advanced disease include cervical lymph node metastasis and other neuropathies (3,5,15,35,36,38).

2.1. Diagnostic Challenge

Extremely low incidence, high frequency of ear infections, and absence of specific symptom make the malignancy difficult to recognize. Other factors may contribute to the challenge.

One diagnostic dilemma is malignant otitis externa. The clinical presentation of the TB osteomyelitis closely resembles the presentation of TB carcinoma. Both occur in the elderly and, with regard to clinical and radiologic examination, SCCA can be indistinguishable from malignant otitis externa. Although SCCA of the TB is painful, malignant otitis externa is uniquely characterized by an acute onset of severe, recalcitrant, disabling pain. Several case reports described the simultaneous presentation of SCCA and osteomyelitis (39,40). It is possible that the presence of necrotic or neoplastic cells may be responsible for superinfection. This highlights the importance of histological evaluation.

Another dilemma arises when one encounters pseudoepitheliomatous HP. It is a benign lesion that mimics the clinical and histologic presentation of epithelial malignancy of the external auditory canal. It usually is associated with aural fullness but not otalgia or bloody otorrhea. If discharge is present, it is usually of recent onset. Once the diagnosis of HP is made, it is imperative that a patient be followed long-term for early detection of possible malignant transformation. Gacek et al. (22) reviewed TB specimens retrospectively diagnosed with SCCA. They found several cases of devastating misdiagnosis with resultant over-treatment. As a result, they proposed criteria for biopsy and close cooperation between the surgeon and the pathologist.

Rarity of the disease and its nonspecific presentation requires a high level of suspicion on the part of the otorhinolaryngologist. Most of the time, it is the prolonged symptomatology unresponsive to treatment that prompts further investigation; thus, the diagnosis often is delayed from 1 month to 4 years (16,21,41). The red flags should be history of persistent aural discharge associated with change in its quality and/or quantity, development or increase in otalgia, carnial nerve (CN) VII paralysis, aural or mastoid polyp, and vertigo. If, complete resolution does not occur after 2–3 weeks of vigorous medical therapy, it might be advisable to offer biopsy as surveillance for TB malignancy. Multiple authors have cautioned against reliance on a single biopsy results. They recommended strongly obtaining multiple deep biopsies including a cuff of normal tissue, even if general anesthesia or CT guidance is required. Inadequate biopsy may show only extensive inflammation and lead to delayed or missed diagnosis (16,22,42,43).

3. RADIOLOGY

From the earlier discussion, it is clear that the diagnosis of SCCA of TB is extremely difficult if attempted by clinical presentation alone. Chronic discharge and the frequent presence of the EAC mass preclude accurate evaluation of the canal and TM. The rarity of tumors and limited physical examination make it extremely difficult to by estimate the extent of the disease. In 1983, Olsen et al. (44) evaluated the validity of CT scan in comparison to mastoid X-rays and tomography and reported its accuracy in predicting the extent of tumor. In 1990, Arriaga et al. heavily based his EAC malignancy staging system on the findings of the CT scan. In 1991, they conducted a study comparing 12 anatomical sites correlating the histopathalogic findings of the SCCA with preoperative CT scans of the TBs of 13 patients. In 94 out of 96 instances, the pathology found on histopathalogic slides coincided with pathology found on the preoperative CT scan. This confirmed the findings of other investigators that high-resolution, thin-section CT scan is the best modality for evaluating the extent of bone involvement by tumor (45,46). As a result, they concluded that high-resolution CT (HRCT) scan provides "a consistent framework to evaluate efficacy of treatment techniques" (45). They also realized the limitations of the HRCT in distinguishing tumor from soft tissue inflammation or edema. General findings of poorly demarcated infiltrative mass, obliterated facial planes, or loss of normal fat density are highly suggestive of malignancy.

The assessment of tumor invasion into soft tissue and along fascial planes, perineural spread, and CNS infiltration is accomplished best with MRI (47,48). Noncontrast T1-weighted images can be used to evaluate tumor invasion of bone marrow by the replacement of normal hyperintense fatty signal. The administration of contrast media may enhance the signal of the tumor. MRI with contrast is now the study of choice for evaluation of the skull base, especially the common sites of invasion: the petroclinoid fissure and foramen lacerum. Loss of smooth contour of the cortical bone indicates bone destruction. MRA may accurately determine involvement of the major vessels. Although invasive angiogram is preferable in many cases.

Despite the great sensitivity of radiologic studies, intra-operative evaluation is of undisputed importance. Multiple authors recognized the limitations of current radiography. Arriaga et al., while utilizing HRCT for the design of a staging system, reported the limitations of HRCT in differentiation of tumor from soft tissue inflammation, tumor extension along fissures, and tumor progression without bone erosion (45). Spector has described perioperative findings of tumor extensions along various nerves, bones, vascular and fascial planes that were not apparent on clinical and radiologic evaluations (49). He found the facial nerve canal to be the most common conduit for intracranial tumor extension.

Leonetti et al. concentrated on the evaluation of advanced disease (stage T3 and T4 by Pittsburgh classification). They found a fairly good correlation between preoperative radiographic and intra-operative findings for anterior and inferior tumor extensions. However, according to them, radiographic preoperative staging underestimated a significant number of patients with tumor extension to the medial, posterior, superior, and petrous carotid canal. As a result, patients were under-staged and did poorly after surgery due to high local recurrence rate (46,49,50). Wagenfeld et al. (21) commented "It appears that assessing the resectability of these tumors involves a significant amount of guesswork." In conclusion, Leonetti et al. (46) proposed the use of a combination of CT scan and MRI to optimize radiologic tumor mapping. Moffat et al. (20) reported a poor correlation between the extent of intracranial involvement and CT and MRI studies.

In 1998, Moharir et al. presented a computer technique of three-dimensional reconstruction based on the fusion of images from two-dimensional CT and MR. Combining the two modalities permits accurate depiction of the bone, soft tissue, blood vessels, and cartilage in one image. Several manipulations were achieved by this technology: 360° rotation in any axis, size adjustment, edge detection and transparencies capabilities, control of color and light intensity. The ability to control the light intensity, color, and transparency of each structure allows the evaluation of the exact relationship of the tumor to the skull base. The segmented slices measure the tumor volume and planar dimensions. The three-dimensional modality offers preoperative planning, intra-operative guidance, tumor surveillance, reconstruction design, radiation therapy field formulation, and medical education. If it proves to be as accurate as it is suggested, multiple limitations of current diagnostic modalities for SCCA of the TB will be overcome (51).

4. CLASSIFICATION

For any type of cancer, it is imperative to have a system that evaluates the extent of the tumor, and, based on that, accurately predicts the amount of surgery needed, as well as the associated survival. The rarity of SCCA of the TB has for years hampered the formulation of an accurate classification system. Adequate analysis of the results has been difficult due to the lack of sufficient data from any single institution (52). This fact is reflected in the absence of a staging system accepted by the International Union Against Cancer or the American Joint Committee on Cancer. Many authors expressed frustration with the inability to interpret and compare the reports. Because survival is determined largely by the tumor's extent, an accurate and unanimously accepted classification system for SCCA of the TB might facilitate the exchange of clinical data in a comparable form.

One of the earliest mentions of classification was proposed <30 years ago (14). Since then, multiple authors have proposed various staging systems: Goodwin and Jesse, Stell and McCormic, Shih and Crabtree, Kinney, Spector, Pensak et al., Manolidis et al., etc. (53). Unfortunately, they were often based primarily on clinical examinations, included various histologic types of cancers, and lacked specifics for the site of tumor extension. The behavior of various types of tumors varies considerably and the limited clinical examination of the TB is well known. Kinney and Wood (2) found that in 60% of patients the tympanic membrane could not be observed. The presence of so many classifications systems highlights the absence of uniform agreement.

In 1990, Arriaga et al. proposed a Pittsburgh staging system specific for SCCA lesions of the external auditory meatus (45). This staging system relies heavily on preoperative findings of the CT scan of the TB. They offered treatment plans for each stage and found statistically different survival rates for each tumor stage. In their later work, they describe the close correlation between histopathology of surgical specimens and preoperative CT scans (54). Several authors have used the Pittsburgh staging system and found it reproducible for both SCCA and non-SCCA tumors (8,9,53,55). In 2000, Moody et al. (53) noted the significance of facial nerve paralysis as bony invasion and recommended the modification of the Pittsburgh system by upgrading a tumor stage in the presence of facial nerve paralysis. In 2002, Breau et al. (56) offered a modification for classification of early-stage lesions based on the site of the lesion within the EAC. From in-depth review, this modification appears more cumbersome than useful. Currently, the modified Pittsburgh classification system is most widely accepted due to its increased objectivity based on radiologic evaluation.

Modified Pittsburgh Staging System

T1	Tumor limited to the EAC w/o bony erosion *or* evidence of soft tissue involvement
T2	Tumor limited to the EAC with partial thickness bony erosion *or* evidence of limited (<0.5 cm) soft tissue involvement
T3	Tumor eroding the osseous EAC (full thickness) *and* limited (<0.5 cm) soft tissue involvement or tumor involving the middle ear or mastoid
T4	Tumor eroding the cochlea, petrous apex, medial wall of the middle ear, carotid canal, jugular foramen, or dura, *or* with extensive soft tissue involvement (>0.5 cm), such as involvement of TMJ or styloid process or evidence of facial paresis
N status	Involvement of lymph nodes is a poor prognostic finding and automatically places the patient in a higher category (i.e., stage III [T1, N1] or stage IV [T2, T3, and T4, N1] disease)
M status	Distant metastasis indicates a very poor prognosis and automatically places a patient in the stage IV category

Note: In the rest of the discussion, all tumors will be staged according to the modified Pittsburgh classification system, unless otherwise specified.

5. MANAGEMENT

Rarity of the disease poses challenges not only with development of a uniform classification system, but also with management strategies. Despite technological advances, skull base surgery continues to be very complex. It requires cooperative efforts between a patient and an extensive team of professionals. The goal of management is to be curative; yet, due to the nature of the TB, the complete excision of the tumor, whether piecemeal or en bloc, is often difficult once the tumor extends beyond the external auditory canal. Adjuvant radiation therapy appears to play a considerable role in increasing survival, but for advanced lesions it does not appear curative as a single modality. Chemotherapy use has been reported. It is usually offered as a palliative therapy but has not yet been shown to improve rates of survival. Nevertheless, all efforts must be made to control such devastating disease.

6. SURGICAL TREATMENT

Ideally, the tumor classification system should reflect the pattern of tumor infiltration and dictate the amount of resection necessary for the best treatment. For example, en bloc resection of the EAC (sleeve resection) is indicated for T1 tumors, lateral temporal bone resection (LTBR) is for tumors limited to the external auditory canal, and subtotal temporal bone resection (STBR) or total temporal bone resection (TTBR) is for advanced tumors (57).

6.1. Sleeve Resection

Sleeve resection is designed to treat selected T1 lesions of the external auditory canal. It involves the circumferential removal of all skin and the lateral cartilaginous portion of the canal of the EAC (Fig. 21.1). The intact tympanic membrane and intact bony canal are the limits of resection (Fig. 21.2). Many have cautioned against the use of these procedures due to high recurrence rate of the disease (58,59). The term "sleeve resection" has been applied to various procedures described in the literature, thus reader must be cautioned to evaluate the extent of the procedure upon encountering its description (5,60).

6.2. Lateral Temporal Bone Resection

This procedure was summarized by Crabtree et al. (5). It is the workhorse for surgical management of most of the lesions confined to the EAC without middle ear involvement (intact tympanic membrane). LTBR is *en bloc* removal of the entire EAC (bony and cartilaginous), tympanic membrane, malleus and incus. The middle ear cavity, stapes, and

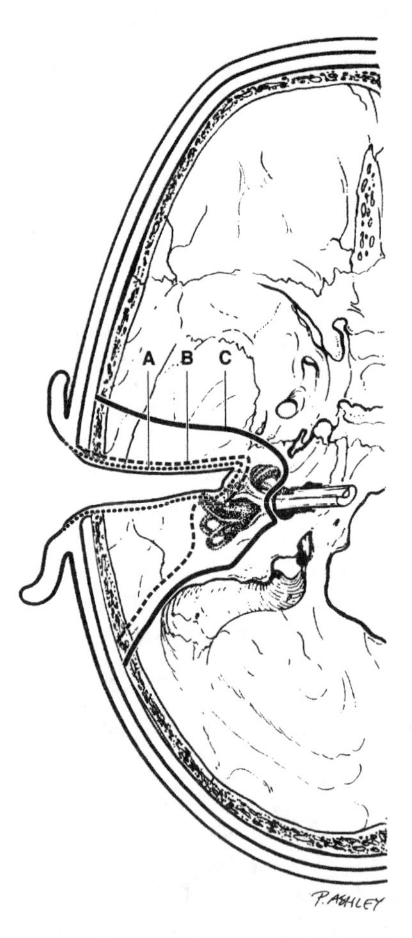

Figure 21.1 *Axial view*: (A) Dotted lines outline the extent of sleeve resection. It involves the circumferential removal of all skin and the lateral cartilaginous portion of the canal of the EAC. The intact tympanic membrane and intact bony canal are the limits of resection. (B) Dashed lines outline the extent of LTBR. It is en bloc removal of the entire EAC (bony and cartilaginous), tympanic membrane, malleus, and incus. The middle ear cavity, stapes, and facial nerve are the boundaries of the resection. (C) Solid lines outline the extent of partial or total TB resection. Unlike TTBR, STBR preserves the medial one-third of the petrous apex, carotid artery, and the contents of jugular foramen. The labyrinth, cochlea, facial nerve, zygoma, and mandibular condyle are sacrificed.

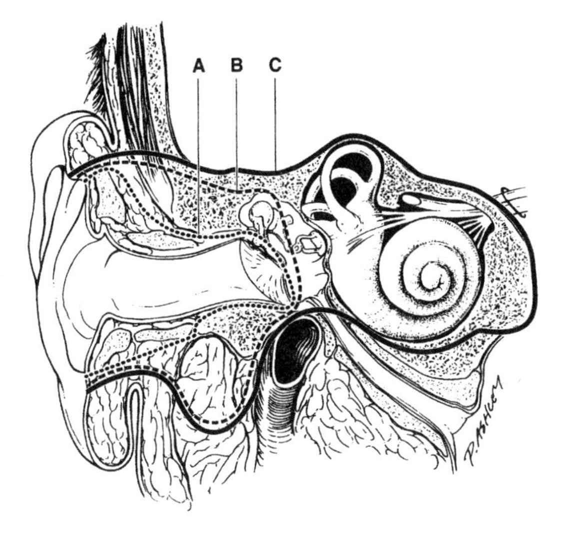

Figure 21.2 *Coronal view*: (A) Dotted lines outline the extent of sleeve resection. It involves the circumferential removal of all skin and the lateral cartilaginous portion of the canal of the EAC. The intact tympanic membrane and intact bony canal are the limits of resection. (B) Dashed lines outline the extent of LTBR. It is en bloc removal of the entire EAC (bony and cartilaginous), tympanic membrane, malleus, and incus. The middle ear cavity, stapes, and facial nerve are the boundaries of the resection. (C) Solid lines outline the extent of partial or total TB resection. Unlike TTBR, STBR preserves the medial one-third of the petrous apex, carotid artery, and the contents of jugular foramen. The labyrinth, cochlea, facial nerve, zygoma, and mandibular condyle are sacrificed.

facial nerve are the boundaries of the resection (Figs. 21.1 and 21.2). It is frequently accompanied by superficial parotidectomy and suprahyoid neck dissection. If a lesion extends anteriorly and is confirmed by frozen sections, resection of TMJ and condyle may be added.

A long modified preauricular/parotid incision extending to the anterior neck is used to gain access to the TB, parotid gland, and cervical structures. Postauricular connection is made to include posterior nodes. When a tumor lies deep in the EAC, the majority of the peripheral auricle is preserved. The cartilage of the tragus and a portion of concha are included with the specimen. If the tumor is present in the region of the concha, the entire auricle must be sacrificed. A canal wall-down radical mastoidectomy with facial recess exposure is performed. Incudo-stapedial joint is sectioned to prevent damage to stapes during removal of incus. The dissection continues anteriorly along the middle fossa tegmen into zygoma. The EAC is dissected of the TMJ without entering the posterior periosteum. The tympanic bone is cut below the tympanic ring, connecting TMJ with the facial recess. The bone in the hypotympanum is removed down to the area between the jugular bulb posteriorly and the carotid artery anteriorly. The styloid process is included with the specimen. The entire EAC is freed in circumferential fashion. The specimen may be left attached to the superficial lobe of the parotid gland, which is resected in continuity with the specimen. Gacek and Goodman recommend superficial parotidectomy for all en bloc procedures because a tumor arising in the EAC follows the path of least resistance to the preauricular soft tissue (6,7). Total parotidectomy is performed when gross disease involves the gland. The facial nerve is preserved if there is no connection to the tumor, and the deep lobe is removed as a separate specimen. Conventional or modified neck dissection may be performed at that stage depending on the stage of disease. A frozen section of lymph node may aid the decision. The soft tissue of the remaining tragus is used to cover posterior periosteum of the TMJ.

This procedure provides tumor margins, preserves facial nerve, has little blood loss, may not require craniotomy, and may preserve serviceable conductive hearing (5,10). Healing is completed usually by 5–6 weeks and results in a dry ear that requires no special care. Moody et al. (53) proposed modifying LTBR by leaving the uninvolved tympanic membrane intact.

6.2.1. *Procedures for Stage III and IV Diseases*

Tumors extending through the TM, arising in the middle ear or mastoid, or involving facial nerve are considered advanced because of their grave impact on the patient's prognosis. They are designated stages III or IV regardless of their size. As with other malignancies, as lesions become more advanced, the extent of surgical management becomes controversial. Various procedures have been employed to manage this disease: radical mastoidectomy, piecemeal removal of the tumor, subtotal resection of the TB, and total resection of the TB.

The general consensus is that TB cancer behaves as a local biologic process until very late in the disease. The high incidence of local recurrences and rare occurrence of distant metastasis illustrate this behavior. Because the cause of death in these patients is dominated by local recurrence, which is the highest if the tumor is not resected completely, author after author advocates an aggressive approach. In general, better survival is reported with more extensive surgery (9,16,52).

The extent of TB resection is the main determinant of postoperative morbidity (9,15). Thus, it is important to identify which lesions are treated best by which procedure.

6.3. Piecemeal Removal of Tumor

Many surgeons prefer piecemeal tumor removal. Several factors have made this procedure appear to be an attractive option. Many believe that the extent of a tumor cannot be sufficiently determined by preoperative evaluation. The amount of the excised healthy tissue is considerably less than with en bloc procedures; and, the close proximity of many vital structures to the TBs has made en bloc resection technically difficult and subject to significant morbidity and mortality. Thus, Kinney et al. (2) proposed piecemeal tumor resection after initial completion of LTBR. The gross tumor is removed until tumor-free margins are confirmed. Most structures, including the internal carotid artery and uninvolved facial nerve, that otherwise would be sacrificed are preserved. By their description, the entire parotid gland is removed and node picking is performed and submitted for frozen section to determine if there is metastatic spread. Neck dissection depends on the status of the neck nodes. If bony involvement is found, postoperative radiation therapy is recommended.

6.4. Subtotal Temporal Bone Resection

The very poor results obtained up to 1950 in the treatment of malignant tumors involving the TB prompted Parson and Lewis (61) to develop a more radical operation for their removal, in 1954. They reported en block resection of SCCA of TB in 100 patients via intracranial–extracranial approach. The overall survival rate continued to be 25%, but it signaled the return of TB resection into the therapy arsenal as a sound surgical option for advanced tumors. In the literature, one can find this procedure under different aliases, including TTBR. Unlike TTBR, STBR preserves the medial one-third of the petrous apex, carotid artery, and the contents of jugular foramen. The labyrinth, cochlea, facial nerve, zygoma, and mandibular condyle are sacrificed. Exposure is similar to LTBR. The medial resection of the petrosal ridge stops upon encountering the lateral portion of the petrosal internal carotid artery (Figs. 21.1 and 21.2). Both the middle and posterior fossa craniotomies are necessary to obtain access. The sigmoid sinus and posterior fossa dura are exposed, and a temporal craniotomy is performed to verify that there is no intracranial tumor extension. If dura involvement is noted, the dura can be resected and grafted. The zygoma and ascending ramus of mandibular are transected and removed. The posterior infratemporal space is inspected. Total parotidectomy, facial nerve transection, and radical neck dissection are performed. The facial nerve distal stump is tagged for future anastomosis procedures. Subtemporal craniotomy is performed to isolate the internal auditory canal. It is accessed and CN VII and CN VIII are transected. The resection of the TB can then be completed. CN IX–CN XII are preserved. Several complications are associated with the procedure: frequent CSF leak, postoperative vertigo, and significant blood loss (58). Some recommend modified STBR to preserve CN VII, cochlea and labyrinth unless they are involved with tumor.

6.5. Temporal Bone Resection

For extensive disease of the middle ear or involvement of the pneumatized spaces, TTBR may provide better oncologic control. In the original description of the procedure by Graham and Sataloff, the vascular and neural structures of the petrous apex are included with the resection. The goal of the procedure is en bloc removal of tumor without tumor transgression, providing tumor-free margins (Figs. 21.3 and 21.4). At the initial stages, tracheotomy, ventriculoperitoneal shun or lumbar drain, Greenfield filter, and possible intracranial bypass may be performed. The carotid artery is occluded preoperatively to assure cerebral blood flow will be adequate when the carotid artery is resected. Early in the procedure, a clip is placed on the internal carotid artery between the cavernous sinus and the origin of the opthalmic artery. This is a preventive measure to avoid an embolic event from cervical manipulation of the vasculature. Later, this maneuver permits dissection of the anterior petrous apex and carotid canal without needing to visualize the artery. Parotid gland excision, ascending ramus of the mandible resection, and neck dissections are than performed. The internal carotid artery, the jugular vein, and the cranial nerves IX–XII are resected. Pterygoid muscles are divided and left attached to the skull base. After identification, the vertebral artery is preserved. Deep cervical muscles are incised and the transverse process of C1 and/or C2 is divided if necessary for exposure. This allows medial dissection and exposure of the carotid entrance into the skull base. The mandibular division of the trigeminal nerve is identified, preserved, and traced to the foramen rotundum. A craniotomy is performed, raising a large bone flap that provides good suboccipital exposure and excellent exposure of the anterior middle fossa. The trunks and their divisions of the trigeminal nerve and carotid artery are resected, preserved. The dura of the middle fossa may be resected with the specimen. The transverse sinus is ligated. The superior petrosal sinus is occluded 0.5 cm before its entrance into the cavernous sinus. The brainstem is retracted gently to visualize the inferior petrosal sinus, jugular bulb, and foramen magnum. The anterior lateral bone cut is made through the superior petrosal sinus, the carotid canal just proximal to the cavernous sinus, and the skull base and lateral skull. This cut connects the anterior petrous apex with the neck and frees the anterior medial fossa. The eustachian tube lies inferolateral to the carotid canal. It is transected and sutured to prevent CSF leak. The final bone

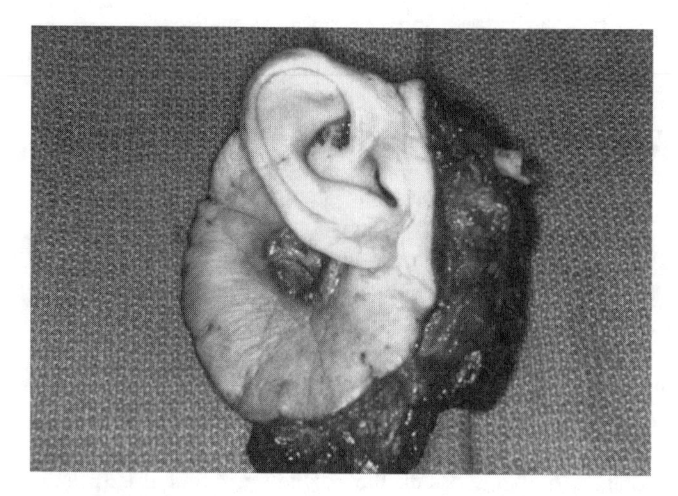

Figure 21.3 The entire TB, external auditory canal, adjacent neck tissue, and a portion of the mandible were removed.

cuts start posteriorly to connect lateral bone with the jugular foramen. Even with this procedure, it might be impossible not to violate a tumor margin. Despite such large resection, piecemeal removal of the small amounts of residual tumor may be necessary, especially in the occipital condyle, clivus, carotid canal, or cavernous sinus. Since the original report, further modifications have been made: clivus and the cavernous sinus were included, preoperative carotid artery screening with temporary balloon occlusion substituted for the Silverstem clamp, preoperative venous outflow evaluation as added,

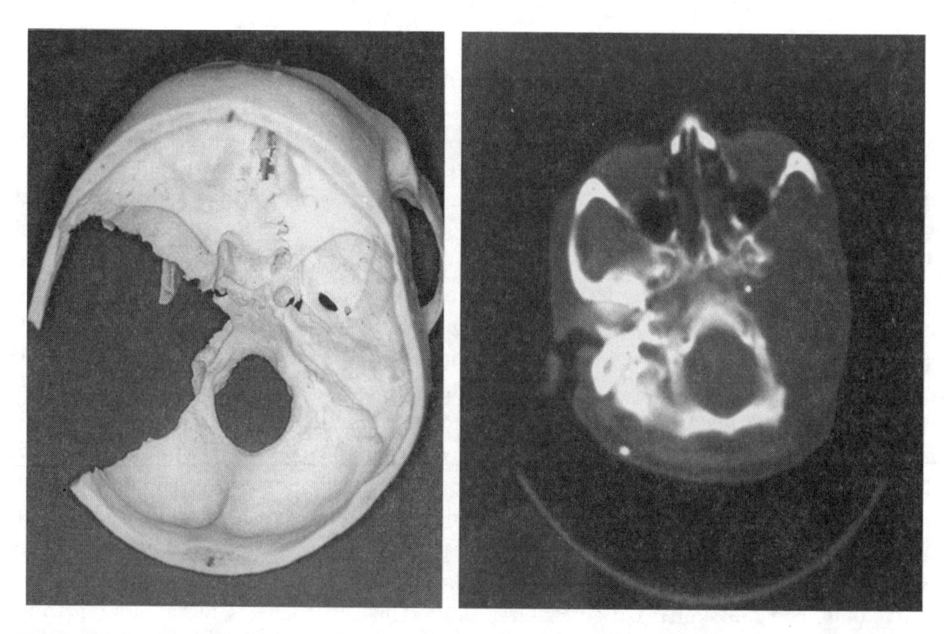

Figure 21.4 (Left) Skull shows the area of bone resected. (Right) A postoperative CT scan of the patient.

internal carotid artery and internal jugular vein sparing techniques have been advocated (14,34,61–64). But, are deemed inevitable in most cases by this author (R.T.S). Sacrifice of the internal carotid artery leads to significant morbidity and the risk of mortality (22,52,63,65). Unpublished data by Sobol indicates 0% 2-year survival after resection of internal carotid artery. We have not had the same experience. Moffat and Wagstaff (42) believe that even if the internal carotid artery appears to be involved, peeling the tumor off the adventitia of the arterial wall is appropriate to achieve adequate results without the neurologic sequela after carotid resection. We agree that this approach may be appropriate in the neck but do not believe it is ideal for extensive intratemporal tumor.

7. COMPLICATIONS OF TEMPORAL BONE RESECTION

Hemorrhage of major vessels are at risk of intra-operative injury. Some authors recommend reserving 6 units of autologous whole blood prior to the surgery. Intra-operative bleeding can be reduced by hypotensive anesthesia. Acute intracranial hemorrhage is one of the most feared complications. Rapidly progressing intracranial pressure may lead to significant brain injury and possible demise in a relatively short period of time. Patients' neurologic status must be monitored closely after surgery, and, if any suspicion arises, CT scan without contrast must be obtained immediately.

Wound infections are not common. Preoperative radiation therapy may increase the risk of wound infection. In this case, Pseudomonas aeruginosa is the most common causative organism, and may cause flap necrosis. In the absence of CSF leak, the risk of intracranial spread of infection is low. If meningitis is suspected, CSF culture must be obtained and IV antibiotics must be initiated until the microbiology report is available.

CSF leak. The main concern is risk of meningitis, thus watertight closure is of outmost importance. Particular attention must be paid to the occlusion of the eustachian tube and adequate closure of the incision. If CSF leak develops, lumbar drains should be placed and the patient should be monitored for signs of meningitis. In previous studies, ventriculoperitoneal shunts were required to control the leak. It is also important to avoid communication between tracheal secretions and the subcutaneous CSF leak, because retrograde contamination may result in intracranial infection.

Other complications include cerebral herniation, CSF fistula, hydrocephalus, carotid artery thrombosis, stroke, and others.

8. PATIENT SELECTION

TTBR is a formidable procedure, which commonly requires 18–24 h, and is associated often with substantial blood loss [although the author (R.T.S) have performed one case successfully without transfusion in a Jehovah's witness]. It should be performed only as a curative operation in patients who are physiologically and psychologically prepared to tolerate the procedure and a prolonged recovery. Paralysis of cranial nerves VI–XII is planned, and the patient and family must be thoroughly informed before this method is chosen. If the tumor encroaches upon the internal carotid artery, preoperative balloon test occlusion should be performed to determine the perfusion capacity of the contralateral side. This will help to determine whether the patient is able to tolerate the resection of the ICA or will require intracranial bypass surgery. On the basis of the ICA balloon occlusion test, Hirsch and Cheng (66) list three groups of patients based on probability of major cerebral ischemia from an acute ICA interruption. Venogram must be also performed to evaluate the venous anatomy. A unilateral dominance of the venous system is not uncommon. Significant cranial venous outflow problems may result in hydrocephalus and cerebral edema. The vein of Labbé provides the majority of venous outflow of the temporal lobe. If occluded at the entrance to the transverse sinus, severely debilitating consequences sometimes result (66). Radical neck dissection may have to be postponed or modified to prevent venous outflow problems. Sataloff et al. (64) described the transcranial sigmoid sinus balloon occlusion test in conjunction with arterial balloon occlusion to test for venous congestion.

9. RECONSTRUCTION

The destructive behavior and invasive character of SCCA in the TB dictate reconstructive options. A majority of tumors arising within the EAC have a tendency for deep spread into the TB. Thus, skin loss is not a significant issue. If the lesion involves the auricle, a large soft tissue deformity may result. The goals of reconstruction include correction of functional and cosmetic defect, protection of intracranial contents, and assurance of adequate healing.

Defects from mastoidectomy or sleeve resection can be corrected by canaloplasty with split-thickness skin graft. Defect after LTBR can be corrected with tympanoplasty. It provides easy tumor surveillance of the cavity and offers potentially serviceable hearing. Unfortunately, most current treatment protocols include postoperative radiation therapy. Once postoperative therapy is initiated, the skin or fascia graft may not provide adequate protection of bone or exposed dura, increasing the potential risk of osteoradionecrosis and fistula formation. To provide more substantial coverage, many authors recommend obliteration of the EAC with a vascularized flap. Various local and regional flaps have been used. One option is the superficial temporoparietal fascia flap. This flap is dependent on the superficial temporal artery. A temporalis muscle flap is also an excellent source of vascularized tissue in these defects (56). It is readily available in the operative field. The blood supply for this comes from anterior and posterior deep temporal arteries. Extensive dissection in the pre-auricular area, particularly during the resection of the ascending process of the mandible, or interruption of the external carotid artery, will compromise the viability of the flap. Gal et al., in their series, reviewed the results of seven patients with canal obliterations and two patients without, and found no cases of osteoradionecrosis. They found that by rotating a temporalis muscle flap into the defect, the loss of bulk of the parotid gland and neck contents is improved easily through adequate cutaneous coverage (67). Preoperative radiation therapy increases the risk for tissue loss and wound break down. In these patients, local rotation flaps are not recommended.

If the tumor has considerable skin involvement, the resection may result in a large soft tissue defect with exposed bone and dura. Cervical and facial rotational flaps can be used for small-to-medium-sized skin defects. They may be used also to suspend the face after resecting the facial nerve.

The defect created after STB and TTB resection is considerable (Fig. 21.4). Exposure of the internal carotid artery and the high likelihood of a CSF leak generally necessitate use of either pedicled or free myocutaneous flaps, although the value of galial advancement flaps should not be underestimated. With minimal donor site morbidity, all these flaps provide excellent soft tissue bulk for cosmesis and protect bone and dura from XRT-induced damages. The pectoralis major flap is a reliable option for reconstruction of head and neck defects (68). Its use for the lateral skull base may be limited by inadequate length and undue tension exerted by gravity (67). The lower island trapezius flap is a good alternative. It is based on the transverse cervical artery with a variable contribution from the dorsal scapular artery. It has several advantages: fairly good color match to the skin of the face, adequate thickness and length for surface coverage, and shorter operative time than free flaps (69,70). Several points must be considered. The history of ipsilateral neck dissection, especially in the posterior triangle, is a relative contraindication for the use of this flap due to the uncertain status of the transverse cervical artery. The harvest of this flap requires patient placement into the lateral decubitus position. If the blood supply to the lower island trapezius flap appears to be compromised, a latissimus dorsi flap may be harvested with the patient in the same position. It is based on the thoracodorsal artery and vein (71). The benefit of the latissimus flap is its versatility. It is excellent for broad skull defects, and may be used as a myofascial or myocutaneous flap. Also, if its length becomes an issue, it can be converted into a free flap (67). The rectus abdominis flap, like the latissimus, can be used as a myofascial or myocutaneous free flap. It provides excellent bulk and can cover any skull defect. Due to its remote location, a second team may harvest the flap in tandem with simultaneous TB resection. However, its use may result in ventral hernia, as it is a major support for the abdominal wall.

The radial forearm flap is a versatile flap with minimal donor site morbidity, but it lacks the bulk necessary for the obliteration of the dead space offered by other flaps. In certain instances it can be preferable because the use of bulky flaps interferes with tumor surveillance. Also, if complete tumor resection is not assured, it may be advisable to postpone the best reconstruction option until absence of recurrence is confirmed.

There are reports that pedicled and free flaps are associated with increased incidence of wound complications (72). Yet, if use of local tissue flaps is compromised by excessive tension, previous radiation, or other factors, one should not hesitate to use myocutaneous flaps.

Large cutaneous defects that include the resection of the auricle may compromise the social acceptability of the final result. With recent advances in reconstruction technology, repair of the defect should concentrate not only on the functional aspect, but also prepare the site for auricular replacement. Several options for auricular replacement exist: reconstruction of the ear from autologous tissue, glued 'tissue-borne' prostheses, and 'implant-borne' prostheses. Granstrom et al. (73) found low aesthetic acceptability of the reconstructed ear prosthesis. The glued tissue-borne prostheses may be unacceptable due to significant movement that may occur in the reconstructed region with mandibular movement. In contrast, implant-borne prostheses maintain their position and are much better tolerated by the patient. To enhance aesthetic results, Gliklich et al. (74) utilized free flaps and camouflage techniques with osseo-integrated implant technology. They describe anchoring the implant with a bar to the remaining skull. The prosthetic auricle and bone anchoring hearing aid may be secured on the bar at the level of the contralateral ear, recreating a natural appearance. In their study, all disease-free patients were offered prosthetic rehabilitation. They reported an 89% success rate of implants in previously radiated bone. Granstrom et al. have reported 35% implant loss rate when placed in irradiated bone, but they noted that improved vascularity improves implant success, as demonstrated with hyperbaric oxygen treatment (75). Unfortunately, prostheses age with sunlight and tobacco exposure, may have to be replaced every 2–3 years, prolong the rehabilitation process by several months, and require some dexterity to place. In their series, Gliklich et al. (74) found prosthetic acceptability by patients rose from 0% to 57%.

Vascular grafts and nerve grafts can be used to restore lost function. If the facial nerve is sacrificed, corneal protection and various reanimation techniques should be considered to improve both function and cosmesis.

10. PROSPECTIVE

To date, there have been no treatment–control studies reported. Most of the many treatment approaches are based on individual experience. In 1994, Prasad and Janecka (52) attempted to gain perspective on the role of surgery in the management of malignant tumors of the TB. They reviewed the available literature and selected 26 publications that had comparable data. From the data analysis of 144 patients, they made the following suggestions: (a) tumors limited to EAC have a 50% overall cure rate after mastoidectomy or LTBR or STBR; (b) addition of radiation therapy (XRT) after LTBR does not appear to be advantageous; (c) compared with mastoidectomy and LTBR, STBR improves survival once the tumor involves the middle ear. It may appear that conclusions made from such an in-depth study would hold true, and indeed, multiple authors continue to use their suggestions. However, upon closer examination, one finds that for each treatment modality the patients' sample size remained small. The staging system they used was limited: tumor confined to EAC, tumor extended into the middle ear, and tumor invading petrous apex. Moreover, the XRT protocols and techniques have changed. Furthermore, multiple studies (prospective and retrospective) published since 1994 have disputed their conclusions.

One of the suggestions made by Prasad and Janecka was that, for lesions limited to EAC, there was no significant difference in 5-year survival rates between radical mastoidectomy, LTBR and STBR (52). Other authors agree that, regardless of the treatment modality used, there is no statistical difference in the 5-year survival rates when the disease remains lateral to the tympanic membrane. Yet, much more promising cure rates ranging from 75% to 100% are reported (8,42,53,76).

Tumor extension in the middle ear or mastoid process mandates at least subtotal resection (77). Five-year survival rates among these patients vary from 50% to 100% (52,53,78). The morbidity from STBR is significant. Thus, some authors advocate piecemeal tumor removal after radical mastoidectomy, followed by high-dose postoperative radiation therapy (2,12,31,52,79). For example, Zhang et al. (80) reported the actuarial 5-year survival rate of 73% (8/11) for stage III disease and 12.5% (1/8) for stage IV disease. Nyrop and Grontved (76) applied piecemeal tumor removal for stage T3 and T4 followed by postoperative radiation therapy. They reported 0% cure rate due to 100% local recurrence most likely related to the incomplete resection. Austin et al. compared results of procedures other than en bloc excision for stages III and IV tumors with en block excision and noted significantly better survival with the more aggressive approach (9). They then compared en bloc resection with and without XRT and found better survival in the combination treatment group. Their conclusion was to proceed with en bloc excision followed by postoperative XRT. They concluded that a more radical approach with adequate margins will provide better tumor control for advanced stages. These findings were in agreement with the reports by Arena and Keen (36) and Spector (49). Moffat et al. evaluated 15 patients with extensive T3 and T4 tumors of the TB (42). All patients had been treated previously with mastoidectomy and radiation therapy. They criticized such treatment because of the difficulties associated with clinical and radiologic tumor surveillance of the operated site that subsequently delayed patients' presentation with recurrence. The general conclusion was that a more radical approach with adequate margins will provide better tumor control for advanced stages.

This correlates with a statement by Arriaga et al., "the basic principle of surgical oncology is complete tumor extirpation" (54).

Advanced tumors extending beyond the TB require TTBR. Reported 5-year survival ranges from 12.5% to 35% with combined surgery and XRT (9,42,57). Despite the technical advances in neuro-otological surgery, the morbidity and the mortality rates remain high. Arriaga et al. called results of TTBR disappointing; only one patient out of five had lived past 10 months (15). He questioned the significance of pathologic margins for these resections. Gacek et al. believed that if the only difference between STBR and TTBR is removal of petrous apex, then there is no benefit from a procedure with significantly increased postoperative morbidity (17). However, leaving gross tumor behind almost guarantees local recurrence and demise within 12 month. Moffat et al. recognized low 5-year survival rates, but advocated use of TTBR as a salvage surgery for patients with recurrence (97). They acknowledge the considerable negative impact on the patients' quality of life from recurrence of SSCA: foul otorrhea, severe intractable bone and meningeal pain from the tumor invasion. From their experience, TTBR sparing the internal carotid artery and postoperative radiation therapy provided palliative therapy with up to 47% 5-year survival rate.

Today, reported cure rates range from 75% to 100% for T1, T2 lesions with any single modality, 35% to 50% for T3, and 12.5% to 35% for T4 lesions with combined surgery and XRT modality (9,42,57).

11. RADIATION THERAPY

For the first half of the 20th century, poor results associated with resection of the TB diverted physicians' attention away from the surgical approach. Treatment was concentrated mainly on the use of radiation therapy (XRT) alone or in conjunction with radical mastoidectomy. Direct beam radiotherapy or radium implants in the mastoid cavity were used extensively. Radium implants were halted because of injury to the brainstem. Overall survival of patients with middle ear SCCA after treatment in 1930–1940s was 23% (81). Some authors continue to use XRT as a sole modality. Although survival rates up to 77% have been reported with use if this modality alone, poor bone penetration, concomitant chronic infections, and constantly changing protocols (depth, site, agents, dose, frequency, techniques, etc.) have left these claims subject to skepticism and criticism. It seems that once SCCA involves the middle ear, XRT alone is curative in ~23% of cases (6,22,49,81–84). The consensus is that radiation is an adjunctive therapy for SCCA of the TB.

The most common indications for XRT are large tumors, uncertain or positive margins, perineural invasion, regional adenopathy, parotid or soft-tissue extension, bony invasion, tumors involving the middle ear, mastoid, and more medial structures of the TB (9,85).

Hashi et al. retrospectively evaluated 20 patients with SCCA of the TB. After median follow-up for 43 months, they concluded that for T1 lesions in their eight patients, 100% disease control was achieved regardless of type of therapy used, radiation alone, surgery, or combination (86). Their finding correlated with the reports of other authors (2,9,37,53–55,59,87,88). Austin et al. (9) and Moody et al. (53) noted the benefit from adjunct radiation therapy in patients with T2 and T3 tumors. Spector (49) and Liu et al. (89) observed tumor extension beyond the operative field despite careful intraoperative monitoring, indicating the need for control of the residual disease. Most current therapeutic recommendations call for a combination of surgery and XRT for treatment of tumors stages T2–T4.

Treatment usually starts after the wound is healed, 2–4 weeks postoperative. A tumorcidal total dose of 54–60 Gy is commonly given without serious threat of brainstem and brain injury. Spector found higher doses with deeper penetration improve significantly survival after extensive resection. He recommended boost doses to 70 Gy if margins are positive (49). Pfreundner et al. stated that only a full course of external beam radiation target (EBRT) of >66 Gy should be administered to the tumor bed of lesions extending beyond surgical margins. In their series, the tumor recurred in regions, which received <66 Gy. They hypothesize that the tumor cells at the surgical margins are hypoxic and have decreased sensitivity to radiation (57). Liu et al. cautions that in order to avoid osteonecrosis, no more than 2000 ret should be given to any portion of the TB (87). The clinical target volume must take into account frequent perineural invasion (57).

There is ongoing debate whether confirmed metastatic lymph nodes should be included into the EBRT volume. Pfreundner et al. believe that the clinical target volume must take into account lymphatic drainage. They recommend incorporating an adjuvant dose of 54–60 Gy into the EBRT target volume (57). Arriaga et al. (54) point out that all of their patients who were dead from the disease died from local recurrence. Liu et al. (89) cautions against intentional coverage of distal node groups, as most of the patients fail at the primary site. It is important to note that the patient population is often old and can suffer significant morbidity from high-dose radiation treatments.

To control residual or recurrent lesions, brachytherapy with intracavitary implantation of radioactive material has been used as a form of long-term irradiation (49,57). Pfreundner et al. reported the addition of brachytherapy to external beam radiation in three patients with tumors beyond surgical margins and four patients with recurrent lesions. According to them, if the tumor is locally confined and small, brachytherapy can be used with the intent to cure (57).

Several authors emphasized that postoperative XRT is not an alternative to complete resection (9,49,54). Nyrop and Grontved (76) confirmed findings by Goodwin and Jesse (14); an inadequately excised tumor cannot be controlled by radiation therapy. In the experience of Wagenfeld et al. (21), treatment failure following radiotherapy was almost always within the tumor volume irradiated, suggesting that "geographic miss" was not a factor. The fact that there is a very low survival rate of patients with inoperative advanced disease despite the use of radiation therapy furthers strengthens these conclusions (8,9). Of course, isolated reports do exist of complete remission with postoperative external beam radiation therapy in patients with residual tumor, but, unfortunately, they are not generally the rule (57). The hypothesis behind radiation failure is that the invading tumor reduces already low oxygen tensions of the TB. The chronic infection associated frequently with SCCA of the TB, suggests the presence of hypoxic tumor cells. Thus, the reduced radiosensitivity of tumors shifts therapeutic responsibility for the patient's survival towards physical removal of the entire disease.

Complications of TB irradiation include CNS injury, osteoradionecrosis, osteomyelitis, neuropathies, delayed wound healing, and, in some instances, cochlear and vestibular damage.

12. PATTERNS OF SPREAD

Understanding patterns of invasion permits accurate radiologic assessment and adequate operative planning. One must also take into account the limitations of current radiologic tests: appreciating the areas of high probability of radiologic mistakes may reduce the number of "unexpected" intra-operative findings.

SCCA arises most commonly in the cartilaginous portion of the EAC. Natural pathways provide routes of spread for small tumors. The fissures of Santorini give flexibility to the cartilage and usually are filled with connective tissue. Together with a patent foramen of Huschke, they may permit tumor extension to the anterior soft tissues: preauricular skin, parotid gland, TMJ, or infratemporal fossa. Other points of weakness are the petrosquamous suture and the bony-cartilaginous junction. The density of the tympanic bone and the fibrous middle layer of tympanic membrane serve as barriers to tumor extension into the TB. As the tumor enlarges, it may spread in multiple directions through thin bone. Posteriorly, it can pass through the tympanomastoid suture and the retrofacial cells of Broca to involve the mastoid antrum and mastoid air cells. Medial spread through the TM permits penetration into the middle ear. Once in the middle ear, the tumor can extend into pneumotized spaces including the Eustachian tube and involve the bone overlying the carotid artery. In some areas, the bone separating the Eustachian tube from internal carotid artery is <1 mm thick. Involvement of the hypotympanum leads to extension into the jugular foramen. Inferior extension invades soft tissue of the neck, jugular foramen, foramen magnum, occiput and cervical vertebrae. Cephalad extension involves the tegmen tympani, or through the petrosquamous fissure, middle fossa dura, and the temporal lobe (45,46). Posterior spread can lead to invasion of the sigmoid sinus and the posterior fossa. The fallopian canal is the most common site of deep skull invasion (49).

Perineural pathways for tumor invasion beyond the middle ear space have been well described from evaluations of pathology specimens. The lesser petrosal nerve may serve as a route to the middle cranial fossa via the inferior tympanic canaliculus; and sympathetic fibers provide access through the roof of the tympanic cavity (90). This pattern of spread is impossible to appreciate perioperatively, making it most difficult to eradicate (91).

Rouviere (92) described the lymphatic network in EAC continuous with the network of the auricle and tympanic membrane. Three zones with different lymphatic drainage are found in the external auditory canal. The anterior zone drains into the intraparotid and preauricular nodes. The inferior portion of the canal drains inwardly. It terminates in deep intraparotid or infraauricular nodes. And, the posterior zone drainage takes a deep course below the sternoclaidomastoid muscle to the deep upper jugular nodes (57). The lymphatic network of the middle ear and mastoid empties into the retropharyngeal and deep upper jugular lymph nodes. Cervical lymph node metastases are more likely to occur once disease extends beyond the confines of the TB.

Nonlymphatic or hematogenous spread is usually a late event and results in deposits in the lung, bone, liver, and brain (80).

13. PROGNOSTIC FACTORS

Several factors are associated with decreased survival rates. These include the local extent of the tumor, facial paralysis, positive margins, dural involvement, and lymph node metastasis. Some studies have found that advanced age ($>$60–65-year-old) (8,89), multiple cranial nerve involvement, moderate-to-severe pain, and female sex may worsen prognosis (93,94).

Local extension is one of the most important issues if not the main prognostic factor (2,15,16,52,54,59,94). If SCCA is limited to the EAC without middle ear or CN VII involvement, it responds equally well to any kind of treatment, be a radical mastoidectomy with XRT or en bloc resection. The overall control rate for tumors limited to the EAC ranges from 80% to 100% (2,8,42,49,57,59,76,95). Limited extension beyond the EAC reduces the control rate to 60–70% (15). Gacek and Goodman (17) even proposed to separate and treat the lesions lateral to the tympanic membrane and the lesions extending to the middle ear as separate entities. For these patients, survival depends on aggressive tumor removal in conjunction with postoperative radiation therapy. Once the skull base is involved, radical TB resection with adequate margins and postoperative XRT is the best oncologically sound option. Unfortunately, even though it is technically feasible, survival rates drop dramatically to 25–38% (57). Pensak et al. (59) defined their contraindications for tumor resection. They include invasion of cavernous sinus, the internal carotid artery, the infratemporal fossa, or the paraspinal musculature. Others do find these factors limiting, but not contraindicating (42,64).

Facial nerve involvement is an ominous sign that has been long recognized (5). Moody et al. proposed to upgrade the SCCA lesions with evidence of facial nerve paralysis to stage IV of the Pittsburgh classification system. According to them, facial nerve paralysis can only be explained by two causes. First, the tumor invades the horizontal segment of the facial nerve within the fallopian canal. In order for this to occur, the tumor must erode through the medial wall of the middle ear. Second, the tumor affects the facial nerve at the level of the stylomastoid foramen. This is only possible when soft tissue extension is >0.5 cm (53). Both causes signal aggressive behavior of the tumor and high risk of perineural spread. Histological evaluations confirmed the role of the facial nerve as a direct pathway for the intracranial spread of various nonneurogenic neoplasms (96). Chee et al. (85) noted that tumors involving facial nerves were generally extensive and invade other vital structures of the skull base. Multiple other studies reported decreased survival rate associated with facial nerve involvement (7,8,66,89,93,94). The degree of facial nerve involvement has been reported to modify patients' prognosis. The treatment failure rate for patients with mild paresis is 34%, moderate paresis is 50%, and complete paralysis is 62% (64,78). These findings have not been reflected in any classification system and may need further investigation.

Positive tumor margins are one of the indicators of poor prognosis. In the series by Goodwin and Jesse (14), all 13 patients with incomplete resection of advanced SCCA had developed local recurrence despite postoperative radiotherapy. Arriaga et al. (54) noted an almost 50% decrease in survival in the patients with incomplete tumor removal. "Clear margins resulted in 9 with no disease and 2 failed treatment whereas pathologically involved margins accounted for 3 with no disease and 9 treatment failures" (15). Nyrop and Grontved quote 67% overall recurrence rate among patients with positive margins at surgery (25% in stages I and II and 100% in stages III and IV). None of the patients with T3 and T4 cancers had survived after incomplete tumor resection despite postoperative radiation therapy (76). Other authors confirmed these findings (54,55,57,86). Generally, once the tumor recurs, patients succumb to disease within an average of 12 months (42).

Other local sites of invasion such as the dura, temporal lobe, and carotid artery have been implicated as separate poor prognostic factors by several authors (15,42,53,54,57). On the basis of the individual experiences, authors decide whether these factors determine a tumor's resectability. For example, Pfreundner et al. (57) and Arriaga et al. (15) believe that extension of the disease beyond the dura renders it unresectable and incurable. Others, on the contrary, reported several patients surviving 5 years after resection of involved dura and/or brain (42,56,64,97). Similarly, Moffat et al. (42) believe that often peeling the tumor off the adventitia of the arterial wall is appropriate to achieve adequate results.

Several authors have attributed nodal involvement to have a negative impact on survival (7,49,54,80,93). The majority of reports describe the incidence of regional lymph node metastases to be 5–15% (14,35,42,59). It appears that cartilage invasion by the tumor is not correlated with metastasis (16). Yet, in the presence of metastatic lymphadenopathy, bony destruction is almost always present (3,15). The reported drop in survival ranges from 20% to 60% once the nodes are involved (15,49). This is reflected in the Pittsburgh tumor classification system. Once metastatic lymph nodes are identified, the tumor is immediately classified at least as a stage III. The question then becomes whether neck dissection is necessary. For head and neck tumors, prophylactic neck dissection in the absence of clinically and radiologically apparent metastatic lymph nodes is recommended once the risk of occult nodal involvement is at ~20%. For SCCA of the TB, some authors argue against neck dissection. Leonetti et al. (46) reported that no patients with metastatic lymph nodes die of regional lymph node disease, but instead all disease-related mortality is from the uncontrolled disease at the original site. Arriaga et al. also noted a lack of benefit from neck dissection in 2-year survival. All of their patients who were dead of the disease died from local recurrence (15,54). Furthermore, Moody et al. (53) reported neck dissection and/or parotidectomy to have no effect on survival for any stage of disease. In addition, Liu et al. (89) caution against intentional coverage of distal node groups during radiation therapy, as patients fail at the primary site. Other authors suggest sending the lymph nodes collected in the vicinity of the TB dissection for frozen section and proceeding with the neck dissection if they prove to have microinvasions. They believe that if the first echelon of the lymph nodes is negative for metastasis, the morbidity from the added neck dissection is unnecessary. There is a group of surgeons who perform routine selective neck dissection on every patient. Moffat et al. recommended supraomohyoid dissection for staging purposes, better access to the skull base, and incorporating soft tissue of the neck into the postoperative radiation therapy field (42,97). Moffat and Wagstaff (42) and Spector (49) reported the incidence of metastatic nodes to be 23–29%. This finding may reflect under-staging of the disease by authors who only treat TBs. In their latest discussion, Moffat and Wagstaff (42) attributed utmost prognostic importance to the presence of metastatic lymph nodes. They believe that lymph node involvement signifies more aggressive disease, which in turn elevates the risk of local recurrence and subsequent early demise (14,42,49,80). Once again, a lack of uniform information forces surgeons to use personal experiences and philosophies to design their treatment plans.

The degree of tumor differentiation has shown variable impact on survival. According to the observation by Kenyon et al. (12), patients with well-differentiated tumors developed CN VII paralysis and succumbed to the disease earlier than patients with poorly differentiated tumors. Moffat et al. (97) and Liu et al. (89) reported better survival with well-differentiated carcinoma. Arriaga et al. (15) and Chen et al. (16) found very little difference between the histologic differentiation and survival. As we can see, there is no consensus on this subject.

Tumor recurrence, whether it occurs after surgery or external beam radiation therapy, indicates a poor prognosis. Most treatment failures are due to local recurrence. Arriaga et al. (15) reported the incidence of local recurrences in 14 out of 17 patients with treatment failures. They also found tumor recurrence to be responsible for all the disease-related deaths (54). Because 90% of recurrence occurs locally, it must be the result of an incomplete resection. Indeed, Spector (49) notes that most of the local recurrences resulted from unexpected tumor extensions beyond the operative area. If for some reason residual tumor is left in the field and failed to respond to postoperative XRT, the majority of the patients will present within 24 months after initial surgery. The average time of presentation is 4–7 months (8,15,54). Once the tumor recurs, the preponderance of patients will succumb to SCCA within 18 months from the time of the diagnosis.

Moffat et al. reported an incredible 47% 5-year survival rate with salvage surgery and large doses of postoperative radiation therapy. Despite this, they continue to insist on the aggressive first treatment protocol because it offers the greatest chance for cure (42,97).

14. QUALITY OF LIFE ISSUES

Owing to contemporary advances in the field of microscopic otology, neurosurgery and pharmacology demonstrate the technical feasibility of TB resection, the results of such extensive procedures can be debilitating. They can cause disfigurement, social disturbances, hearing loss, facial nerve abnormality, communication difficulties, dysphagia, etc. We as physicians concentrate on technical aspects of the procedures and determine outcome of treatment by measuring tumor response rate, disease progression, and disease-free survival. However, none of these parameters reflect patients' own perception of successful outcome. Nowadays, there is increased awareness of the need for evaluating the impact of symptoms on the patient's quality of life. Kwok et al. (98) conducted a retrospective cross-sectional study of 23 patients with history of TB resection or parotidectomy with radiotherapy. They compared physicians' perceptions of success with patients' perceptions of the treatment outcome. According to their findings, with exception of hearing testing, clinical assessments did not generally correlate well with patients' ratings. Interestingly, facial disfigurement did not cause much patient distress. Several authors had reviewed life satisfaction among head and neck patients and noted that disfigurement is generally unrelated to the quality of life. Patients tend to adapt successfully and be grateful to be alive after confronting cancer (98–104). The difficulties with communication and swallowing were the most bothersome to patients and most under-estimated by physicians. As patient survival improves, the therapeutic decision must incorporate quality of daily life as experienced by the patients. The insight obtained from patients' responses will guide physicians and ancillary staff in the formulation of therapeutic and rehabilitation goals.

15. CHEMOTHERAPY

In the majority of cases, the increased morbidity associated with chemotherapy renders it as the last resort treatment. As a result, it is usually given in cases of aggressive, recurrent, or unresectable disease. The small number of patients receiving chemotherapy and the advanced stages of the disease for which chemotherapy is given, make it virtually impossible to define the impact of chemotherapy on survival. Arriaga et al. (15) did not find the addition of chemotherapy to enhance survival. Suzuki et al. reported five patients in whom intra-arterial (as opposed to intra-venous) chemotherapy was added to the combined surgery and XRT treatment. The superficial temporal artery or superior thyroid artery were used for vascular access. They found no tumor cells in three patients after completion of combined therapy (105–107). Knegt et al. conducted a prospective study utilizing meticulous tumor debulking followed by repeated applications of 5-fluoro-uracil cream and necrotectomy. They report 74% overall 5-year survival. The procedure demanded eight painful sessions of follow-up after initial surgery for dressing change, necrotectomy, and 5-fluoro-uracil reapplication. Patients were re-evaluated intraoperatively in 2–3 months and, if found to have residual tumor, had to repeat the entire protocol and undergo XRT. The data are hard to interpret because the authors used a limited classification system, did not describe the extent of the final debulking/necrotectomy procedure, and failed to report survival outcomes separately for each stage (108). Although none of the reports is adequate for evaluation, they reflect continued interest in the use of chemotherapy agents for treatment of SCCA of the TB. Also, in other areas of the head and neck, limited chemotherapy has been tried at the beginning stages of therapy in an attempt to decrease the tumor bulk and to predict radiosensitivity of the tumor. This has been successful in the soft tissues, but no reports have been published on such use of chemotherapy for the bones, in particular the TB.

16. CONCLUSION

SCCA of the TB remains a complex, challenging, and incompletely understood disease. However, it is not hopeless. Through aggressive treatment, even rare patients with advanced disease can be cured. Nevertheless, considerably more experience is needed before optimal treatment for all stages can be established with confidence.

REFERENCES

1. Horn J. Surveillance, Epidemiology and End Results. Rockvile, MD: National Cancer Institute, 1987.
2. Kinney SE, Wood BG. Malignancies of the external ear canal and temporal bone: surgical techniques and results. Laryngoscope 1987; 97:158–163.
3. Conley JJ, Schuller K. Malignancies of the ear. Laryngoscope 1976; 86:1147–1163.
4. Friedmann I. Pathology of the Ear. Oxford: Blackwell Scientific Publication, 1974:145–225.
5. Crabtree JA, Britton BH, Pierce MK. Carcinoma of the external auditory canal. Laryngoscope 1976; 86:405–415.
6. Lederman M. Malignant tumors of the ear. J Laryngol Otol 1965; 79:79–85.
7. Stell PM, McCormick MS. Carcinoma of the external auditory meatus and middle ear: prognostic factors and suggested staging system. J Laryngol Otol 1985; 99:847–850.
8. Testa JR, Fukuda Y, Kowalski LP. Prognostic factors of the external auditory canal. Arch Otolaryngol Head Neck Surg 1997; 123:720–724.
9. Austin JR, Stewart KL, Fawzi N. Squamous cell carcinoma of the external auditory canal: therapeutic prognosis based on a proposed staging system. Arch Otolaryngol Head Neck Surg 1994; 120:1228–1239.
10. Kuhel WI, Hume CR, Selesnick SH. Cancer of external auditory meatus and temporal bone. Otolaryngol Clin North Am 1996; 29:827–852.
11. Somers T, Vercruysee JP, Goovaerts G, et al. Isolated squamous cell carcinoma of the tympanic membrane. Otol Neurotol 2002; 23:808.
12. Kenyon GS, Marks PV, Scholtz CL, Dhillon R. Squamous cell carcinoma of the middle ear. Ann Otol Rhinol Laryngol 1985; 94:273–277.
13. Lewis JS. Temporal bone resection. Review of 100 cases. Arch Otolaryngol 1975; 101:23–25.
14. Goodwin WJ, Jesse RH. Malignant neoplasms of the external auditory canal and temporal bone. Arch Otolaryngol 1980; 106:675–679.
15. Arriaga M, Hirsch BE, Kamerer DB. Squamous cell carcinoma of external auditory meatus (canal). Otolaryngol Head Neck Surg 1989; 101:330–337.
16. Chen KTK, Dehner LP. Primary tumors of external (auricle and auditory canal) and middle ear. I: Introduction and clinicopathological study of squamous cell carcinoma. Arch Otolaryngol 1978; 104:253–259.
17. Gacek RR, Goodman M. Management of malignancy of the temporal bone. Laryngoscope 1977; 87:78–86.

18. Bendl BJ, Graham JH. New concepts on the origin of squamous cell carcinoma of the skin: Solar (senile) keratosis with squamous cell carcinoma: a clinicicopathologic and histochemichal study. Proceeding of the Sixth National Cancer Conference, Philadelphia, Lippincott, 1968:471–488.
19. Shifman NJ. Squamous cell carcinoma of the skin of the pinna. Can J Surg 1975; 18:279–283.
20. Moffat DA, Grey P, Ballagh RH, et al. Extended temporal bone resection for squamous cell carcinoma. Otolaryngol Head Neck Surg 1997; 116:617–623.
21. Wagenfeld DJH, Keane T, van Nostrand AWP, Bryce DP. Primary carcinoma involving the temporal bone: Analysis of twenty-five cases. Laryngoscope 1980; 90:912–919.
22. Gacek MR, Gacek RR, Gantz B, et al. Pseudoepitheliomatous hyperplasia versus squamous cell carcinoma of the external auditory canal. Laryngoscope 1998; 108:620–623.
23. Khan AS. Pseudoepitheliomatous hyperplasia of the external ear, middle ear and mastoid. Laryngoscope 1979; 89:984–987.
24. Monem SA, Moffat DA, Frampton MC. Carcinoma of the ear: a case report of a possible association with chlorinated disinfectants. J Laryngol Otol 1999; 113:1004–1007.
25. Jin YT, Tsai ST, Li C, et al. Prevalence of human papillomavirus in middle ear carcinoma associated with chronic otitis media. Am J Path 1997; 150:1327–1333.
26. Aub JC. Late effects of internally deposited radioactive materials in man. Medicine 1952; 31:221–229.
27. Beal D, Lindsay J, Ward PH. Radiation induced carcinoma of the mastoid. Arch Otolaryngol 1965; 81:9–16.
28. Lim LH, Goh YH, Chan YM et al. Malignancy of the temporal bone and EAC. Otolaryngol Head Neck Surg 2000; 122:882–886.
29. Goh YH, Chong VF, Low WK. Temporal bone tumors in patients irradiated for nasopharyngeal carcinoma. J Laryngol Otol 1999; 113:222–228.
30. Ruben RJ, Thaler SU, Holzer N. Radiation induced carcinoma of the temporal bone. Laryngoscope 1977; 87:1613–1621.
31. Lustig RL, Jackler RK, Lanser MJ. Radiation-induced tumors of the temporal bone. Am J Otol 1997; 18:230–235.
32. Cahan WG, Woodard HQ, Higinbotham NL. Sarcoma arising in irradiated bone: Report of eleven cases. Cancer 1984; 1:3–29.
33. Ward GE, Loch WE, Lawrence W. Radical operation for carcinoma of the external auditory canal and middle ear. Am J Surg 1951; 134:397–403.
34. Conley JJ. Cancer of the middle ear. Trans Am Otol Soc 1960; 53:189–207.
35. Hahn SS, Kim JA, Goodchild N, et al. Carcinoma of the middle ear and external auditory canal. Int J Radiat Oncol Biol Phys 1983; 9:1003–1007.
36. Arena S, Keen M. Carcinoma of the middle ear and temporal bone. Am J Otol 1988; 9:351–356.
37. Paaske PB, Witten J, Schwer S, et al. Results in treatment of carcinoma of the external auditory canal and the middle ear. Cancer 1987; 59:156–160.
38. Lewis JS. Cancer of the ear. CA Cancer J Clin 1987; 37:78–87.
39. Mattucci KF, Setzen M, Galantich P. Necrotizing otitis externa occurring concurrently with epidermoid carcinoma. Laryngoscope 1986; 96:264–266.
40. Grandis JR, Hirsch BE, Yu VL. Simultaneous presentation of malignant otitis externa and temporal bone cancer. Arch Otolaryngol Head Neck Surg 1993; 119:687–689.
41. Lewis JS. Surgical management of tumor of the middle ear and mastoid. J Laryngol Otol 1983; 97:299–311.
42. Moffat DA, Wagstaff SA. Squamous cell carcinoma of the temporal bone. Curr Opin Otolaryngol Head Neck Surg 2003; 11:107–111.
43. Suzuki K, Takahashi M, Ito Y, et al. Bilateral middle ear squamous cell carcinoma and clinical review of an additional 5 cases of middle ear carcinoma. Auris Nasus Larynx 1999; 26:33–38.
44. Olsen KD, DeSanto LW, Forbes GS. Radiographic assessment of squamous cell carcinoma of the temporal bone. Laryngoscope 1983; 93:1162–1167.

45. Arriaga M, Curtin H, Takahashi H, Hirsch BE, Kamerer DB. Staging proposal for external auditory meatus carcinoma based on preoperative clinical examination and computer tomography findings. Ann Otol Rhinol Laryngol 1990; 99:714–721.
46. Leonetti JP, Smith PG, Kletzker GR, et al. Invasion patterns of advanced temporal bone malignancies. Am J Otol 1996; 17:438–442.
47. Baker S, Latack JT. Magnetic resonance imaging of the head and neck. Otolaryngol Head Neck Surg 1986; 95:82–89.
48. Levine PA, Paling MR, Black WC, Cantrell RW. MRI vs. high resolution CT scanning: evaluation of the anterior skull base. Otolaryngol Head Neck Surg 1987; 96:200–267.
49. Spector JG. Management of temporal bone carcinoma: a therapeutic analysis of two groups of patients and long-term follow-up. Otolaryngol Head Neck Surg 1991; 104:58–66.
50. Kinney SE. Squamous cell carcinoma of the external auditory canal. Am J Otol 1989; 10:111–116.
51. Moharir VM, Fried MP, Vernick DM, et al. Computer-assisted three-dimensional reconstruction of head and neck tumors. Laryngoscope 1998; 108:1592–1598.
52. Prasad S, Janecka IP. Efficacy of surgical treatment for squamous cell carcinoma of temporal bone: a literature review. Otolaryngol Head Neck Surg 1994; 110:270–280.
53. Moody SA, Hirsch BE, Myers EN. Squamous cell carcinoma of the external auditory canal: an evaluation of a staging system. Am J Otol 2000; 21:582–588.
54. Arriaga M, Curtin H, Takahashi H, Kamerer DB. The role of preoperative CT scan in staging for external auditory meatus carcinoma: radiologic–pathologic correlation study. Otolaryngol Head Neck Surg 1991; 105:6–11.
55. Yeung P, Bridger A, Smee R, et al. Malignancies of the external auditory canal and temporal bone: Review. ANZ J Surg 2002; 72:114–120.
56. Breau RL, Gardner EK, Dorhoffer JL. Cancer of the external auditory canal and temporal bone. Curr Oncol Rep 2002; 4:76–80.
57. Pfreundner L, Scwager K, Willner J, et al. Carcinoma of EAC and middle ear. Int J Radiat Oncol Biol Phys 1999; 44:777–788.
58. Jackson CG, Marzo S, Ishiyama A, Canalis RF. Malignant tumors of the temporal bones. The Ear: Comprehensive Otology. Vol. 52. Philadelphia, PA: Lippincott, Williams, 2000:835–845.
59. Pensak ML, Gleich LL, Gluckman JL, Shumrick KA. Temporal bone carcinoma: Contemporary perspectives in the skull base surgical era. Laryngoscope 1996; 106:12324–12327.
60. Lewis JS. Cancer of the ear: a report of 150 cases. Laryngoscope 1960; 70:551–579.
61. Parson H, Lewis JS. Subtotal resection of the temporal bone for malignancy of the middle ear. Cancer 1954; 7:995–1001.
62. Campbell EH, Volk RM, Burkland CW. Total resection of the temporal bone for malignancy of the middle ear. Ann Surg 1951; 134:397–403.
63. Graham MD, Sataloff RT, Wolf GT, et al. Total en block resection of the temporal bone and carotid artery for malignant tumors of ear and temporal bone. Laryngoscope 1984; 94:528–533.
64. Sataloff RT, Myers DL, Lowry LD, Spiegel JR. Total temporal bone resection for squamous cell carcinoma. Otolaryngol Head Neck Surg 1987; 96:4–14.
65. Hilding DA, Selker R. Total resection of temporal bone carcinoma. Arch Otolaryngol 1969; 89:636–645.
66. Hirsch BE, Cheng CYJ. Carcinoma of the temporal bone. In: Myers EN, ed. Operative Otolaryngology Head and Neck Surgery. Philadelphia, PA: W.B. Saunders, 1997:1434–1458.
67. Gal TG, Kerschner JE, Futran ND, et al. Reconstruction after temporal bone resection. Laryngoscope 1998; 108:476–481.
68. Baek S, Biller HF, Krespi YP, et al. The pectoralis major myocutaneous island flap. J Head Neck Surg 1979; 1:293–300.
69. Urken ML, Naidu RK, Lawson W, Billr HF. The lower trapezius island musculocutaneous flap revised. Arch Otolaryngol Head Neck Surg 1991; 117:502–511.

70. Netterville JL, Panjie WR, Maves MM. The trapezius myocutaneous flap. Arch Otolaryngol Head Neck Surg 1987; 113:272–281.
71. Day TA, Davis BK. Skull base reconstruction and rehabilitation. Otolaryngol Clin North Am 2001; 34:1241–1257.
72. Nelligan PC, Mulholland S, Irish J, et al. Flap selection in cranial base surgery. Plast Reconstr Surg 1996; 98:1159–1166.
73. Granstrom G, Bergstrom K, Tjellstrom A. The bone-anchored hearing aid and bone-anchored epithesis for congenital ear malformation. Otolaryngol Head Neck Surg 1993; 109:46–53.
74. Gliklich RE, Rounde MF, Cheney ML, et al. Combining free flap reconstruction and craniofacial prosthetic technique for orbit, scalp and temporal defects. Laryngoscope 1998; 108: 482–487.
75. Granstrom G, Tjellstrom A, Albrektsson A. Postimplantation irradiation for head and neck cancer treatment. Int J Oral Maxillofac Implants 1993; 8:495–501.
76. Nyrop M, Grontved A. Cancer of the external auditory canal. Arch Otolaryngol Head Neck Surg 2002; 128:834–837.
77. Neely GJ, Forrester M. Anatomical considerations of medial cuts in the subtotal temporal bone resection. Otolaryngol Head Neck Surg 1982; 90:641–645.
78. Medina JE, Park AO, Neely GJ, et al. Lateral temporal bone resection. Am J Surg 1990; 160:427–433.
79. Chao CK, Sheen TS, Shau WY, et al. Treatments, outcomes, and prognostic factors of ear canal. J Formos Med Assoc 1999; 98:314–318.
80. Zhang B, Tu G, Xu G, et al. Squamous cell carcinoma of the temporal bone: reported on 33 patients. Head Neck 1999; 21:461–466.
81. Figi FA, Hempstead BE. Malignant tumors of the middle ear and mastoid process. Arch Otolaryngol 1943; 37:149–168.
82. Arthur K. Radiotherapy in carcinoma of the middle ear and auditory canal. J Laryngol Otol 1976; 90:753–762.
83. Million RR, Cassisi NJ. Management of Head and Neck Cancer. A Multidisciplinary Approach. Philadelphia, PA: Lippincott, 1984.
84. Boland J, Paterson R. Cancer of the middle ear and external auditory meatus. J Laryngol Otol 1955; 69:468–478.
85. Chee G, Mok P, Sim R. Squamous cell carcinoma of the temporal bone: diagnosis, treatment and prognosis. Singapore Med J 2000; 41:441–451.
86. Hashi N, Shirato H, Omatsu T, et al. The role of radiotherapy in treating squamous cell carcinoma of the external auditory canal, especially in early stages of disease. Radiother Onc 2000; 56:221–225.
87. Wang CC. Radiation therapy in management of carcinoma of external auditory canal, middle ear, or mastoid. Radiology 1975; 116:713–715.
88. Shih L, Crabtree JA. Carcinoma of the external auditory canal: an update. Laryngoscope 1990; 100:1215–1218.
89. Liu FF, Keane TJ, Davidson J. Primary carcinoma involving the petrous temporal bone. Head Neck 1993; 15:39–43.
90. Miller D. Cancer of external auditory meatus. Laryngoscope 1965; 65:448–461.
91. Soo K, Carter RL, O'Brian CJ, et al. Prognostic implications or perineural spread in squamous carcinoma of the head and neck. Laryngoscope 1986; 96:1145–1148.
92. Rouviere H. Anatomie des Lymphatiques de l'Homme. Paris: Masson et cie., 1932.
93. Birzgalis AR, Keith AO, Farrington WT. Radiotherapy in treatment of middle ear and mastoid carcinoma. Clin Otolaryngol 1992; 17:113–116.
94. Manolidis S, Pappas D, Von Doersten P, Jackson JJ, Glasscock ME. Temporal bone and lateral skull base malignancy: experience and results with 81 patients. Am J Otol 1998; 19:S1–S15.
95. Gillispe MB, Francis HW, Chee N, et al. Squamous cell carcinoma of the temporal bone: a radiographic–pathologic correlation. Arch Otolaryngol Head Neck Surg 2001; 27:803–807.

96. Selesnick SH, Burt BM. Regional spread of nonneurogenic tumors to the skull base via the facial nerve. Otol Neurotol 2003; 24:326–333.
97. Moffat DA, Chiossone-Kerdel JA, Da Cruz M. Squamous cell carcinoma. Tumors of the Ear and Temporal Bone. Philadelphia, PA: Lippincott, Williams, 2000:63–67.
98. Kwok HC, Morton RP, Chaplin JM, et al. Quality of life after parotid and temporal bone surgery for cancer. Laryngoscope 2002; 122:882–886.
99. Morton RP. Evolution of quality of life assessment in head and neck cancer. J Laryngol 1995; 109:1029–1035.
100. Morton RP, Witterick IJ. Rationale and development of a quality of life instrument for head-and-neck cancer patients. Am J Otol 1995; 16:284–293.
101. Dropkin MJ. Scaling of disfigurement and dysfunction in postoperative head and neck patients. Head Neck Surg 1983; 6:559–570.
102. Long SA, D'Antonio LL, Robinson EB, et al. Factors related to quality of life and functional status in 50 patients with head and neck cancer. Laryngoscope 1996; 106:1084–1088.
103. Kreitler S, Chaitchik S, Rapoport Y, et al. Life satisfaction and health in cancer patients, orthopedic patients and healthy individuals. Soc Sci Med 1993; 36:547–556.
104. Katz MR, Irish JC, Devine GM, et al. Rehabilitation and validity of an observer-rated disfigurement scale for head and neck cancer patients. Head Neck 2000; 22:132–141.
105. Harker GJ, Stephens FO. Comparison of intra-arterial versus intra-venous 5-fluorouracil administration on epidermal squamous cell carcinoma in sheep. Eur J Cancer 1992; 28:1437–1441.
106. Suzuki K, Tsuge I, Ito Y, et al. Histological efficacies and side effects of intra-arterial infusion of multiple drugs against head and neck malignant tumors. J Jpn Soc Cancer Ther 1990; 25:533.
107. Itoh Y, Kimura T, Motai H, et al. Evaluation on effectiveness of multi-drug combined intra-arterial chemotherapy as a neoadjuvant chemotherapy in the head and neck region. J Jpn Soc Cancer Ther 1994; 29:411.
108. Knegt PP, Ah-See KW, Meeuwis CA, et al. Squamous cell carcinoma of the external auditory canal: a different approach. Am J Otol 2002; 60:465–469.

22
Sarcomas of the Temporal Bone

Brian A. Neff
Department of Otorhinolaryngology, Mayo Clinic, Rochester, Minnesota, USA

Robert T. Sataloff
Drexel University College of Medicine, American Institute for Voice and Ear Research, and Graduate Hospital, Philadelphia, Pennsylvania, USA

A variety of benign and malignant lesions of the temporal bone have been described (Table 22.1) (1), but sarcomas that involve the temporal bone are rare. The literature contains scattered small series, case reports, and literature reviews of temporal bone sarcomas; however, most have too few cases to scientifically determine a superior treatment approach. However, most studies provide an opinion or theory on how to handle these rare and aggressive tumors. A few definitions are important to allow proper interpretation of these scientific reports. For the purposes of this chapter, radical surgery is defined as total tumor removal with sacrifice of a vital structure, such as a nerve or artery. *En bloc* resection refers to surgery, usually radical, designed to circumscribe and remove the entire tumor without violating the tumor margins. Complete or wide surgical excision refers to surgery that attempts to obtain clear margins, but does not necessarily imply radical surgery or avoidance of tumor transgression. In fact, tumor is often removed piecemeal during such procedures. Subtotal resection implies that gross residual disease remains after surgery. Examples of subtotal excision include surgical debulking and biopsy.

1. RHABDOMYOSARCOMA

1.1. Demographics

Rhabdomyosarcoma accounts for 8–20% of all sarcomas. It is also the most common sarcoma in children. The head and neck is the most common site, accounting for 30–50% of cases. Two-thirds of head and neck rhabdomyosarcomas are embryonal, with a mean age of presentation between 4 and 8 years of age. The most common head and neck subsite is the orbit. The temporal bone is a much rarer location. The Armed Forces Institute of Pathology (AFIP) found that only 6.9% of all rhabdomyosarcomas of the head and neck occur in the temporal bone. Yet, rhabdomyosarcoma is currently the most common sarcoma diagnosed in the temporal bone (2). Nuafel (3) found 64 cases of temporal bone rhabdomyosarcoma reported in the literature over a 100-year span ending in 1973. Numerous other cases have been described since that review (5,6,8,9,11,12,14,21). The temporal bone is considered a parameningeal subsite of the head and neck. Other parameningeal sites include the nasopharynx, nasal cavity, paranasal sinuses, infratemporal fossa, and pterygopalatine fossa (4). Rhabdomyosarcomas that affect these areas are frequently grouped together in order to obtain sufficient numbers to make any study statistically significant. Additionally, these parameningeal sites are

Table 22.1 Temporal Bone Neoplasms (1)

Primary temporal bone tumors	
Benign	Malignant
Meningioma	Epidermal
Paraganglioma	Squamous cell carcinoma
Neurofibroma	Verrucous carcinoma
Schwannoma	Basal cell carcinoma
Adenomas	Mucoepidermoid carcinoma
Eccrine cylindroma	Mesenchymal
Pleomorphic adenoma	Rhabdomyosarcoma
Mesenchymal	Chondrosarcoma
Chondroma	Osteosarcoma
Chondroblastoma	Liposarcoma
Chondromyxoid fibroma	Angiosarcoma
Hemangiomas	Malignant fibrous histiocytoma
Lipoma	Ewing's sarcoma
Myxoma	Neurogenic sarcoma
Fibroosseous tumors	Leiomyosarcoma
Fibrous dysplasia	Fibrosarcoma
Ossifying fibroma	Adenocarcinomas
Osteoblastoma	Adenoid cystic
Osteoma	Sebaceous cell
Giant cell granuloma	Papillary cystadenocarcinoma
Teratoma	Ceruminous adenocarcinoma
Dysontogenic	Melanoma
Choristoma	Plasmacytoma
Glioma	
Chordoma	
Metastatic tumors	
Prostate, GI, renal cell, lung, breast	
Lymphoma, leukemia (chloroma)	
Multiple myeloma	
Contiguous tumor invasion	
Meningioma	Pituitary neoplasms
Schwannoma	Craniopharyngioma
Neurofibroma	Chordoma
Gliomas	Nasopharyngeal carcinoma
Parotid neoplasms	

grouped since they are the most likely tumor areas to exhibit intracranial spread. However, such groupings make it difficult to glean information about temporal bone neoplasms specifically.

1.2. Pathology

There are four major histologic variants of rhabdomyosarcoma: embryonal, with its botryoid and spindle cell variant subtypes; alveolar; undifferentiated; and rhabdomyosarcoma not otherwise specified. Although the histologic appearance is distinct for the different rhabdomyosarcoma subtypes, all of them arise from a similar primitive myoblast or mesenchymal stem cell (5). The embryonal subtype has an overall histologic pattern that has been likened to that of normal embryologic muscle arrested at 3–10 weeks of

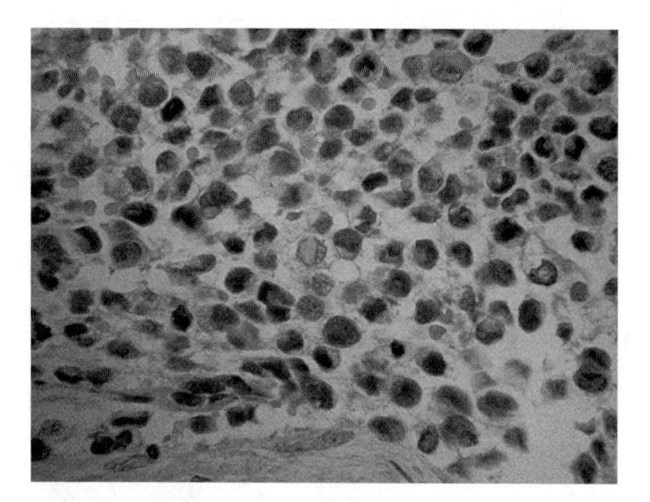

Figure 22.1 **(See color insert)** Rhabdomyosarcoma, embryonal 40× (400 magnification). Embryonal rhabdomyosarcomas are composed of small, oval to elongated cells arranged in loosely cohesive sheets. The cells can have varying amounts of eosinophilic cytoplasm that displace the nuclei.

gestation. It is composed of small spindle-shaped cells mixed with rhabdomyoblasts, which are small, oval, eosinophilic cells. These cells are arranged in a background consisting of a loose myxoid matrix and may have a botryoid appearance (Fig. 22.1). The alveolar histology consists of more rounded tumor cells arranged in ill-defined nests or alveolar spaces separated by fibrous strands (Fig. 22.2). The undifferentiated subtype contains more anaplastic appearing cells which are arranged in bizarre, tightly packed configurations (2) (Fig. 22.3). Rhabdomyosarcoma, not otherwise specified, is a category created as a catch-all for tumors not fulfilling histologic criteria for a specific subtype, but still believed to be a rhabdomyosarcoma. This relatively new classification has

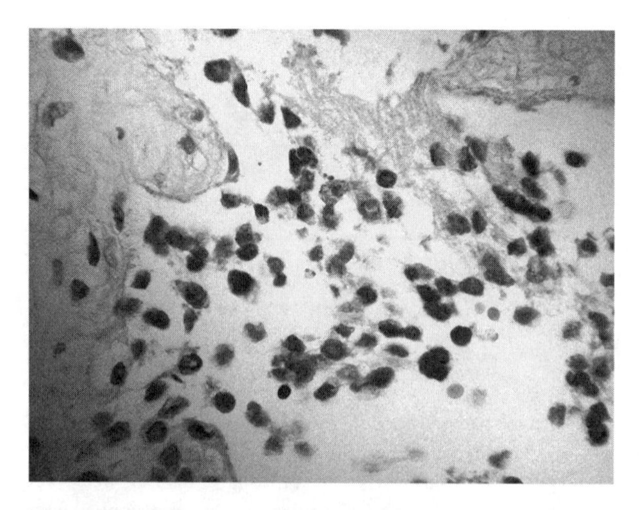

Figure 22.2 **(See color insert)** Rhabdomyosarcoma, alveolar 40× (400 magnification). Alveolar rhabdomyosarcomas are high-grade sarcomas composed of small, oval cells growing in nests or clusters separated by delicate fibrous septa. Discohesive cells are seen in the center of the nests while the periphery is characterized by the lepidic attachment of tumor cells simulating the morphology of pulmonary air spaces.

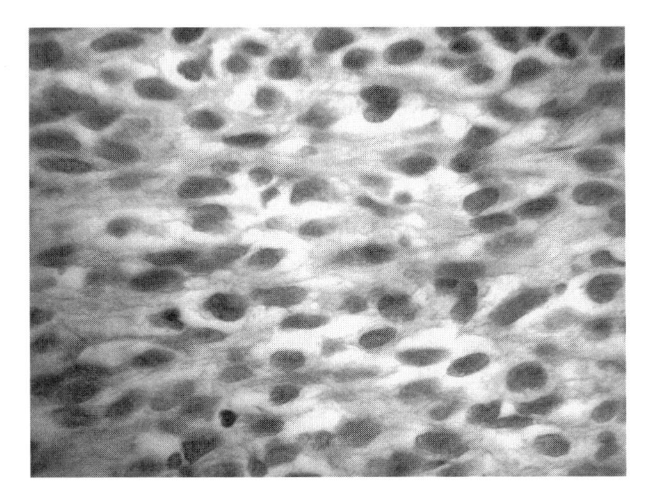

Figure 22.3 **(See color insert)** Rhabdomyosarcoma, undifferentiated 40× (400 magnification). The undifferentiated subtype shows no specific features. The growth pattern is closely packed and may be fascicular. Cells may have a moderate amount of cytoplasm and are typically larger than a mature lymphocyte. Nuclei tend to the large end of the spectrum and contain finely granular chromatin and prominent nucleoli.

deleted the previously reported pleomorphic subtype (5). Some pathologists have required the presence of skeletal muscle cross-striations to make the light microscopic diagnosis of rhabdomyosarcoma; however, insistence on them will lead to underdiagnosing these lesions (6). Immunohistochemical staining has shown that a large number of rhabdomyosarcomas do not contain these striations. Furthermore, immunohistochemical analysis aids in diagnosis in that rhabdomyosarcomas are usually positive for desmin, myoglobulin, muscle-specific actin, and vimentin (VIM) found in skeletal muscle (7).

1.3. Clinical Presentation/Diagnostic Work-Up

One of the most common early presentations of rhabdomyosarcoma can mimic otitis media or acute mastoiditis. This is true for all age groups, but it is especially true in children. Many series demonstrate an average delay of up to 7 weeks from the time of initial patient presentation until definitive diagnosis (8). The presence of acute onset otalgia, acute facial nerve palsy, and an opaque tympanic membrane has often led to the early false diagnosis of acute otitis media. Patients have been treated with prolonged courses of antibiotic therapy which delayed diagnosis. Additionally, there have also been reports of incidental aural polyps being found and biopsied at the time of tympanostomy tube placement, with biopsy results revealing rhabdomyosarcoma. In this setting, otorrhea, aural granulation, or aural polyps are frequently overlooked as harbingers of a growing rhabdomyosarcoma (9–11). When reviewing the presentation of all patients with rhabdomyosarcoma of the middle ear and mastoid, they presented with the following findings: 54% had a polyp in the external auditory canal, 40% developed otorrhea, 30% of patients exhibited bloody otorrhea or frank bleeding, and conductive hearing loss and facial nerve paresis occurred 14% of the time (11). Histologic studies have shown that rhabdomyosarcomas of the middle ear usually spread along the path of least resistance; additionally, they can demonstrate early invasion and destruction of the facial nerve in the fallopian canal, thus causing facial paresis. From that point, they can extend to the internal auditory canal and meninges (6). Virtually any cranial nerve may become involved with the tumor once it has spread to an intracranial location. Around 35% of patients with middle ear rhabdomyosarcoma develop meningeal spread which is almost uniformly fatal (12). Additionally, tumor spread can occur through the jugular fossa to cause a jugular foramen syndrome, also called Vernet's syndrome. This syndrome is caused by compression and subsequent paralysis of cranial nerves IX, X, and XI. In turn, the patient will usually manifest symptoms of hoarseness and dysphagia (due to paralysis of the palate, vocal fold, and superior pharyngeal constrictor).

Around 25% of rhabdomyosarcomas can arise in the area of the eustachian tube with extension to the petrous apex. These masses present differently from the more lateral lesions involving the mastoid and middle ear. Frontal headaches, retro-orbital pain, and diplopia from abducens nerve palsy are common. This clinical picture mimics petrous apicitis. Aural symptoms and facial nerve paresis are usually absent until late in the disease process. These lesions usually exhibit signs of intracranial spread earlier in the disease course. Common areas of spread include Meckel's cave and the carotid canal which can cause facial parasthesias and Horner's syndrome, respectively (6).

A history of otitis media in a child unresponsive to antibiotics, with a polyp in the external auditory canal, should raise the suspicion of tumor until biopsy proves otherwise. If cranial neuropathies are present, imaging is indicated as well as biopsy of any mass present on physical exam or radiograph. If the diagnosis of rhabdomyosarcoma is made at biopsy, work-up should include a computed tomography (CT) of the temporal bones if it has not already been obtained. If there is evidence of extensive soft tissue extension to the infratemporal fossa, other extracranial areas, or intracranially, a magnetic resonance imaging (MRI) may be useful. If meningeal spread is questioned, most oncologists will perform a lumbar puncture for cytology as well as a bone marrow biopsy. Around 14% of parameningeal rhabdomyosarcomas in the early Intergroup Rhabdomyosarcoma Study (IRS) had evidence of distant metastasis at the time of presentation (11). The most common site was the lungs, with other frequently involved areas including skeletal bone and liver (11). Consequently, the metastatic work-up should include a CT of the chest and abdomen, liver function tests, and a bone scan.

1.4. Historical Treatment

Early case reports and series prior to current multimodality therapy used surgical excision of varying degrees with radiation therapy or chemotherapy in some cases. Most surgical resections were not complete. Jaffe et al. (13) reported no 2-year survivors among 40 patients who underwent resection of their temporal bone rhabdomyosarcoma. Deutsch et al. reported similar, poor survival statistics with only five survivors out of 78 patients at 31–54 months follow-up. These patients had their temporal bone rhabdomyosarcomas treated by combinations of surgical excision, radiation, and chemotherapy (14). It was because of these dismal statistics that in 1972, the Intergroup Rhabdomyosarcoma Study Group (IRSG) was formed with the primary goal of performing large, randomized trials designed to try to improve survival rates with initial treatments. The study group evaluated treatment protocols for all body sites for patients <21 years of age. IRSG has completed and published four studies since 1972, and a fifth one has been started to evaluate the efficacy of newer chemotherapy agents in the treatment of rhabdomyosarcoma. These studies have developed treatment protocols using the multimodality approach of surgery, radiation therapy, and chemotherapy to almost triple the cure rate and survival in the last 30 years (15). Although often criticized, the IRS and European protocols are probably the most widely accepted treatment approaches for eradicating rhabdomyosarcomas of all sites, including the temporal bone. They are, however, not without flaws that make surgical decision-making more difficult.

Table 22.2 IRSG Staging (15)

Stage	Site	Size (cm)	Lymph node	Metastasis
I	Head and neck (excluding parameningeal sites), orbit or eyelid, GU (excluding prostate/bladder)	Any	N_0, N_1, or N_x	−
II	Parameningeal, prostate/bladder, other (retroperitoneum, trunk, etc.)	≤5	N_0, N_1	−
III	Parameningeal, prostate/bladder, other (retroperitoneum, trunk, etc.)	≤5	N_1	−
		>5	N_0, N_1, or N_x	−
IV	Any site	Any	N_0, N_1, or N_x	+

Note: N_0—no regional nodes, N_1—positive regional nodes, N_x—node status unknown.

1.5. Treatment

The IRSG IV system categorizes patients according to their tumor stage, completeness of resection following surgery (postsurgical pathologic grouping), and risk categories. The IRSG IV staging classification (Table 22.2) stages rhabdomyosarcomas by the location of the primary site, tumor size, the presence or absence of lymph node metastasis, and the presence or absence of distant metastatic disease. Rhabdomyosarcoma of the orbit and non-parameningeal head and neck sites are stage I independent of their size or lymph node status. Rhabdomyosarcoma of the parameningeal head and neck sites are stage II if they are ≤5 cm and have no regional or distant metastatic disease. Parameningeal sites include tumors of the nasopharynx, nasal cavity, middle ear, mastoid, temporal bone, paranasal sinuses, pterygopalatine fossa, and infratemporal fossa. These same sites are also termed unfavorable primary sites by the IRSG system. These same parameningeal tumors are upstaged to stage III if neck lymph nodes are present or if the tumor is >5 cm. All head and neck lesions with metastatic disease are stage IV.

IRSG postsurgical pathologic grouping (Table 22.3) classifies patients according to the completeness of their surgical resection, and thus, the amount of residual disease present. Group 1 contains patients whose lesions were completely resected with microscopically negative margins. Group 2 contains patients with residual microscopic disease. Group 3 includes patients with gross residual disease after subtotal surgical resection or biopsy. Lastly, group 4 encompasses patients who had distant metastatic disease at the time of their presentation.

The IRSG IV system finally assigns a risk category to each patient depending on their stage and postsurgical pathology group. Patients were separated into low-, intermediate-, and high-risk categories which were then used to determine disease-free survival (Table 22.4). The higher the risk category, the poorer the disease-free survival for the

Table 22.3 IRSG Postsurgical Pathology Group (15)

Group 1	Local disease, complete resection
Group 2	Persistant microscopic disease
Group 3	Incomplete resection or biopsy-persistant gross residual disease
Group 4	Distant metastasis present at initial patient presentation

patient. Disease- or failure-free survival at 3 years for all sites were 88% in low-risk patients and 70% for intermediate-risk patients (15). The IRSG IV study did not include analysis of high-risk patients. This risk schema can be applied to temporal bone rhabdomyosarcomas. Low-risk temporal bone lesions are stage II tumors of <5 cm, without neck lymphadenopathy. Additionally, low-risk patients need to be in the group 1 postsurgical pathology grouping, and consequently, the temporal bone rhabdomyosarcomas must be completely excised with negative margins. All remaining lesions are intermediate-risk tumors except for those presenting as stage IV with metastatic disease. Patients with stage IV rhabdomyosarcomas are placed in the high-risk category.

It can be seen from the IRSG system that the patient risk category and survival are dependent not only on the location and extent of disease, but also on the completeness of surgical resection. Therefore, theoretically, the patient's prognostic group and future treatment depends on the aggressiveness of the surgeon. As previously mentioned, the only way that a temporal bone lesion can be low risk is if it demonstrated localized disease of <5 cm, and was resected completely at the time of initial surgery (Table 22.4) (15). This scenario was extremely uncommon in any of the IRSG trials. With the evolving field of cranial base surgery, removal of previously unresectable lesions is now possible. How then should one decide how aggressive to be when resecting a temporal bone rhabdomyosarcoma? There are no definitive answers since there is a paucity of these lesions, making it nearly impossible to design a significant, randomized, prospective study to address this question. Future studies need to either redefine this staging system or attempt to evaluate the temporal bone and other parameningeal sites as a distinct subset of patients in order to assess whether modern techniques of skull base tumor

Table 22.4 Risk Categories (15)

Risk	Stage	Group	Residual tumor	Site	Size (cm)	Node	Comment
Low	I	1	None	Favorable	All	N_0, N_1, N_x	
	I	2	Microscopic	Favorable	All	N_0	
	I	3	Gross	Favorable	All	N_0, N_1, N_x	Orbit/eyelid only
	II	1	None	Unfavorable	≤ 5	N_0	
Intermediate	I	2	Microscopic	Favorable	All	N_1	
	I	3	Gross	Favorable	All	N_0, N_1, N_x	Not orbit/eyelid
	II	2, 3	Microscopic or gross	Unfavorable	≤ 5	N_0	
	III	1	None	Unfavorable	>5	N_0	
	III	2	Microscopic	Unfavorable	≤ 5	N_1	
					>5	N_0, N_1, N_x	
	III	3	Gross	Unfavorable	All	N_0, N_1, N_x	
High	IV	Any	Microscopic or gross	Any	All	Any	+Metastasis

Note: Favorable sites—head and neck (excluding parameningeal), orbit/eyelid, GU (excluding prostate/bladder); unfavorable sites—parameningeal, prostate/bladder, extremity, other (trunk, retroperitoneum, etc.).

resection will improve survival over patients undergoing subtotal removal or biopsy of the tumor mass.

Baker et al. have reported the latest IRS IV trial in which the authors evaluated 894 patients with rhabdomyosarcomas. Patients were treated with multiagent chemotherapy in 3-week cycles for a total of 46 weeks. All temporal bone rhabdomyosarcomas were treated with radiation except for the very rare instances in which they met the criteria for a low-risk lesion in which complete tumor excision was accomplished. Radiation therapy was started 3 weeks after the initiation of chemotherapy, and consisted of either hyperfractionated treatments twice a day to a total dose of 60 Gy or conventional radiation to a total dose of 50 Gy. Their 3-year disease-free survival in patients who had a biopsy or surgery that left gross residual disease was 75% for "unfavorable sites." This was in contrast to the 86% 3-year survival for these same unfavorable sites after complete resection or nearly complete resection with only microscopically positive margins. Excitement over these successful results should be tempered due to the fact that they reported survival for all unfavorable sites which included not only parameningeal sites such as the sinuses, temporal bone, and nasopharynx, but also the bladder, prostate, long bones, and retroperitoneum (15). Therefore, very little information can be drawn from their 3-year survival statistics when trying to predict survival for temporal bone lesions. In fact, these rates probably grossly overestimate the true survival rates for most skull base tumors. Additionally, from a closer look at the IRS IV 3-year survival statistics, it would appear that the aggressiveness of the resection might improve survival by up to 11%. This highlights the previously posed question of whether or not skull base procedures to completely resect these temporal bone lesions may help achieve a survival improvement.

Some physicians have tried to look at this controversial question. Prasad et al. (16) treated 14 rhabdomyosarcoma patients with complete skull base resection followed by postoperative chemotherapy. Radiation therapy was utilized for close or microscopically positive margins. They noted a 2-year survival of 42%. Healy et al. (17) reported nine patients with infratemporal fossa rhabdomyosarcoma treated with chemotherapy followed by complete radical resection using microsurgical techniques. With this radical approach they reported a 44% long-term survival. Both studies failed to include a less radical surgical or nonsurgical chemoradiation arm. The poor 2-year survivals could easily be explained by the fact that these radically resected lesions were, on average, more advanced than those who were treated in the IRSG studies. Additionally, the tumors were in locations that carry a worse prognosis than many of the cases in the IRSG studies. Consequently, the authors were unable to delineate any positive or negative effects that aggressive surgical resection may have.

The IRS trial physicians as well as many other researchers have discouraged aggressive surgery of the skull base in children. They make this statement based on the excellent response that they have seen with chemotherapy and radiation therapy in the IRS trials, and on their desire to avoid the morbidity associated with resection of cranial nerves. Raney et al. (18) reported the IRS I trial results in which there were 24 temporal bone rhabdomyosarcomas. All were treated with surgical biopsy followed by chemotherapy and radiation therapy. They reported an improved 5-year survival of 46% when compared to previous retrospective studies that used a variety of surgical approaches as the main treatment option. Consequently, it was concluded that surgical resection should be done only if it does not require extensive reconstruction or sacrifice of neurovascular and other functional structures; therefore, subtotal resection or biopsy of temporal bone lesions was deemed the appropriate initial surgical management (18). Again, the fault of this study is that they failed to ask whether their improved survival rate could have been even better if complete resection had been combined with their chemoradiation protocols. They also do not clearly defend their position that functional concerns seem to supercede the oncologic principle that any malignancy should be removed entirely en bloc. Surgical debulking is not appropriate in squamous cell neoplasms of the head and neck, and probably is detrimental to survival in the treatment of sarcomas. Unfortunately, the subsequent IRS II–IV studies have not examined temporal bone tumors as an isolated group. In addition to the above controversies, the IRS protocols created significant morbidity in the children treated with high-dose radiation therapy. Frequent episodes of xerostomia, osteoradionecrosis, and asymmetric growth disorders of the craniofacial skeleton were reported (15).

Not all proponents of limited surgery followed by chemoradiation had as much success as the IRS trials. Anderson et al. (19) treated six patients with rhabdomyosarcoma of the temporal bone with various surgical procedures; however, 86% received either debulking or biopsy. They were then treated with vincristine, adriamycin, and cytoxan (VAC) chemotherapy for 1 year along with 50–60 Gy of radiation to the temporal bone. This treatment protocol closely mirrored the IRS I protocol. They reported a 0% 5-year survival for their patients with rhabdomyosarcoma of the temporal bone. Even with these poor results and this small number of patients, they reported that there was no indication for initial radical surgical excision if vital structures need to be sacrificed (19).

Flamant et al. (20) reported on the European Second International Society of Paediatric Oncology (SIOP) MMT84 protocol for the treatment of rhabdomyosarcoma. After biopsy confirmed diagnosis, patients were given 6–10 courses of induction chemotherapy with ifosfamide, vincristine, and dactinomycin. Local therapy, which consisted of surgery or radiation, was undertaken after six courses of chemotherapy if residual tumor was still present. However, complete surgery that sacrificed cranial nerves was not performed at this time in favor of radiation therapy to produce local control. Rarely, radical surgery was used to salvage chemoradiation treatment failures. Unlike the IRS trials, the European MMT84 protocol group placed parameningeal lesions in their own specific treatment group regardless of initial surgical treatment; however, the authors also failed to compare total vs. subtotal excision within this group. They reported a 58% overall 5-year survival and a 42% 5-year disease-free survival in a subset of patients that included temporal bone lesions, but was not limited to these exclusively. Similar to the IRS trials, a criticism of the European trials was that significant postradiation complications were reported (20).

Schouwenburg et al. (21) looked to decrease the morbidity seen with treating rhabdomyosarcomas with high-dose radiotherapy, and they also hoped to reduce the morbidity associated with radical salvage surgery after chemoradiation treatment failures. They developed the AMORE protocol (Ablative surgery, Moulage technique with afterloading brachytherapy, and Reconstructive surgery) to treat initial chemotherapy nonresponders. It utilized complete macroscopic debulking surgery preserving vital neurovascular structures. At the same surgical setting, a thermoplastic rubber mold of the surgical defect is made and placed in the wound bed as well as brachytherapy catheters. Postoperatively, after-loading brachytherapy with iridium was given to a dose of 56 Gy. Finally, a staged reconstructive surgery was done 1–2 weeks after brachytherapy. A significant number of these reconstructions utilized a free flap to fill the surgical defect. The 3-year survival of 15 patients was reported at 70%. This is an impressive figure considering that they were treating initial chemotherapy failures. They did not report any treatment to parameningeal or temporal bone lesions, but did not state that these sites were a contraindication to the protocol (21). Additionally, despite stating that the radiation complications were greatly reduced with their brachytherapy regimen, they failed to prove this fact objectively. They also did not show objectively that their protocol preserved function without sacrificing survival compared to similar patients treated with complete excision and immediate reconstruction followed by radiation and chemotherapy.

2. CHONDROSARCOMA

2.1. Embryology

There are a few different theories proposed as to how chondrosarcomas develop in the temporal bone. The bones of the skull base mature predominantly by endochondral ossification while the bones of the skull vault develop primarily by intramembranous ossification (22). The areas of the petro-occipital, spheno-occipital, and spheno-petrosal synchondroses, as well as a large part of the petrous portion of the temporal bone, are sites in the mature skull that underwent endochondral development (23). It is hypothesized that islands of residual endochondral cartilage may be present in these areas and that chondrosarcoma develops from these chondrocytes (24,25). Alternatively, it is suggested that chondrosarcomas may arise from pluripotent mesenchymal cells involved in the embryogenesis of the skull base and temporal bone. Lastly, metaplasia of mature fibroblasts has been implicated as an inciting mechanism in the development of chondrosarcomas (24,25).

2.2. Demographics

Chondrosarcomas account for 0.1% of all head and neck tumors and ~6% of all skull base lesions (26). The most common skull base site varies between differing studies, with common sites of involvement including the tempero-occipital junction, middle cranial fossa, spheno-ethmoid complex, anterior cranial fossa, and clivus (25,27,28). The mean age at presentation is the fourth and fifth decades of life, with most studies not showing a sex predilection (24,28,29). Primary chondrosarcoma develops *de novo* in normal bone. Most temporal bone chondrosarcomas arise in this manner. Rarely, there are secondary chondrosarcomas which arise from preexisting cartilaginous tumors or abnormalities. Chondrosarcoma has been reported in association with Paget's disease, Mafucci syndrome, osteocartilagenous exostoses, Olliers's disease, and osteochondromas (30,31).

2.3. Pathology

There have been several subtypes of chondrosarcomas proposed. The conventional subtype consists of hyaline, myxoid, or a combination of these components. The hyaline component has been described as neoplastic chondrocytes residing within lacunar spaces surrounded by a hyaline matrix. The myxoid component features areas of chondrocytes that are surrounded by a frothy mucinous matrix (24,28) (Fig. 22.4). Mesenchymal and undifferentiated subtypes are rare histologies that reveal a more anaplastic appearance, and these lesions usually present with aggressive, advanced disease. The clear cell subtype (malignant chondroblastoma) is extremely rare with only a few cases reported in the head and neck (26,32).

Chondrosarcomas have been placed into three categories based on cellularity and nuclear atypia. Rosenberg reviewed 200 cases of skull base chondrosarcoma and found that 50.5% of tumors were grade 1 (well differentiated), 21% were grade 2 (moderately differentiated), 28.5% were a mixture of grades 1 and 2, and 0% were grade 3 (poorly differentiated) tumors (28). This grading system is important since it corresponds to prognosis based on tumor biology, distinct from its location or stage of presentation. Evans et al. (33) reported on chondrosarcomas from all body sites and found a correlation between survival and histologic grade. Five-year survival for chondrosarcomas of grades I, II, and III is 90, 81, and 43%, respectively. Koch et al. (26) reported additional factors that bode a poorer prognosis which included advanced stage, metastatic disease, and mesenchymal

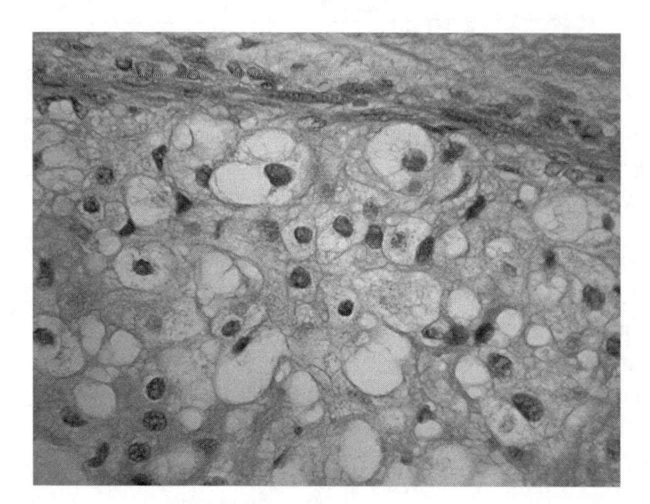

Figure 22.4 (**See color insert**) Chondrosarcoma (400 magnification). This conventional chondrosarcoma consists of neoplastic cells within lacunar spaces embedded in a hyaline matrix. Nuclei are large, sometimes binucleate, and have prominent nucleoli. Mitotic figures are rare in low-grade tumors and are readily identified in high-grade lesions.

or pure myxoid tumors. Others have stated that location (unresectable location), and dedifferentiated subtypes also bear a poorer prognosis (27,32,34). The differential diagnosis for chondrosarcomas of the temporal bone and skull base includes chordoma, chondroid chordoma, osteosarcoma, meningioma, enchondroma, and glomus jugulare (24,25). Most of these can be distinguished with a thorough history and physical as well as imaging studies. However, most radiologists agree that chordoma, chondroid chordoma, and chondrosarcoma are difficult to distinguish reliably by MRI and CT scan (30). Histologic differentiation is often difficult, as well. Chordomas differ from chondrosarcomas in that they contain an abundance of cells with bubbly eosinophilic cytoplasm called physaliphorous cells. Chondroid chordomas contain features of both chondrosarcoma and chordoma. They contain neoplastic chondrocytes along with physalipharous cells which are present but much rarer than in chordomas (22,25). Frequently, analysis by an immunohistochemical panel of antibodies is needed for definitive diagnosis. Typically chordomas stain positive for cytokeratin, epithelial membrane antigen, S-100 protein, and VIM. Conversely, chondrosarcomas stain positive for VIM and S-100 and negative for cytokeratin and epithelial membrane antigen (24,28,35). Wojno et al. (35) were able to demonstrate both staining patterns within a single tumor, and hence, they helped distinguish these masses as chondroid chordomas.

2.4. Clinical Presentation

Several clinical presentations have been reported for chondrosarcoma of the skull base. The symptoms correlate with the anatomic site of destruction or compression. Initial complaints may include hearing loss, pulsatile tinnitus, vertigo/unsteadiness, aural fullness, and headache. Multiple cranial neuropathies are common and present as diplopia, facial pain, parasthesias, hemifacial spasm, facial paresis, dysphagia, hoarseness, shoulder weakness, and hemi-tongue weakness and atrophy. A comprehensive neurotologic work-up is indicated when temporal bone lesions are suspected. The work-up includes pure tone and speech audiometry, CT of the temporal bones and brain, MRI/MRA, as well as any other tests that are indicated clinically (23,27).

2.5. Treatment

Treatment approaches have evolved throughout the years and include combinations of surgical debulking, complete surgical excision, radiation, and chemotherapy. Although most studies report a treatment bias, the paucity of patients with chondrosarcoma of the temporal bone makes it impossible to perform prospective trials that could lead to definitive treatment conclusions. Anecdotal protocols are common and will be reviewed here. Due to a concern that surgical debulking violates tumor boundaries and oncologic principles, the concept of total en bloc resection with total gross removal of disease has been suggested as the preferred surgical procedure when removing chondrosarcoma of the skull base and temporal bone. A few authors have reported small surgical series with successful outcomes. Kveton et al. (27) reported the House clinic's experience of five cases of skull base chondrosarcoma involving the temporal bone. They conveyed their preference for total en bloc resection and stated that they were successful in removing petrous apex lesions via an infratemporal fossa approach in most cases. Similarly, Charabi et al. (36) completely resected a temporal bone chondrosarcoma with jugular foramen extension through an infratemporal fossa approach. Al-Mefty et al. (37) described a combined infratemporal and posterior fossa approach for total removal of two temporal bone chondrosarcomas with posterior fossa extension. Additionally, Watters and Brookes (23) reported eight cases of petrous temporal bone chondrosarcoma and concurred that total en bloc resection was ideal. Siedman et al. (25) reviewed thirty-one cases of chondrosarcomas of the temporal bone, but they did not report a clear treatment protocol. However, the authors did suggest that optimal treatment would include complete surgical excision followed by radiation (25). Lastly, Burkey et al. (31) reported three cases of subtotal temporal bone resection followed by radiation therapy. They favored complete tumor resection, but pointed out that two of their patients who had undergone partial resections were alive with stable disease at 2 years. However, this follow-up does not address the fact that well-differentiated lesions are frequently slow growing and can recur up to 10 years after treatment; consequently, their reported follow-up was not long enough to evaluate their treatment protocol.

Radiation treatment for chondrosarcomas used to be considered futile. Harwood et al. reported on chondrosarcomas that occurred at multiple body sites and found that conventional external beam radiation of 5000 rads (cGy) over 4–5 weeks locally controlled 50% of 31 cases. Complete remission was seen in three of these cases after 15 years of follow-up. They stated that earlier claims that chondrosarcomas were radio-resistant failed to consider that clinical regression of tumor after radiation treatment was slow. Additionally, the radiographic appearance of bone involved with the lesion never completely returns to normal after radiation therapy despite prolonged evidence of clinical remission. Both these factors may have falsely led earlier investigators to claim treatment failures in patients who may have been treated successfully (34). Several recent series have reported on the advantage of proton beam radiation in the treatment of skull base chondrosarcomas. With this modality, very large radiation doses are deliverable to the mass while achieving a rapid reduction in dose outside the defined target volume (38). Rosenberg et al. (28) reported the largest series to date of skull base chondrosarcomas which included temporal bone lesions. They reported treating 200 patients with proton beam therapy of 64.2–79.6 Gy equivalents in 38 fractions. All patients had been treated previously with surgical procedures ranging from biopsy to total excision. They report the most successful 5- and 10-year local control rates of 99%

and 98%, respectively, as well as 5- and 10-year survival rates of 99% and 98%, respectively. Their mean follow-up was reported as 65 months. Austin-Seymour et al. (39) reported on 28 patients with low-grade chondrosarcoma of the skull base treated with 69 gray equivalents of proton beam therapy in 31–41 fractions. Their results were not as favorable as Rosenburg's results and demonstrated 5-year local control and survival rates of 82% and 76%, respectively. However, they failed to separate chondrosarcomas from chordomas when reporting these group figures, and this may have impacted negatively upon their 5-year statistics. They concluded that surgery followed by proton beam radiation was the treatment of choice for chondrosarcoma and chordoma of the skull base. Coltrera et al. (24) reported 13 cases of temporal bone chondrosarcoma. They generalized that proton beam therapy was the preferred radiation modality and that conventional external beam radiation has yielded poor results. However, they did not comment on radiation dosage, success of treatment in their series, or the specific role of surgery.

Stereotactic radiosurgery has also been described in the treatment of chondrosarcoma of the skull base (24). Radiosurgery combines the accuracy of stereotactic guidance with high-dose, single-session irradiation. There is steep radiation fall-off outside the image-defined tumor margin. Hence, the tumor gets a tumoricidal dosage while the surrounding brain and neural structure receive tolerable levels of radiation, usually without serious side effect. Muthukumar et al. (40) treated six cases of chondrosarcoma of the skull base with 24–40 Gy to the tumor while the tumor margins received 12–20 Gy. Tumor volume was the limiting factor for this treatment modality. They limited treatment to those with masses <3 cm diameter and 10.3 cm^3 of volume. They showed improvement in over half of their patients, but did not separately group their chondrosarcoma lesions from their patients with chordomas when reporting results (40). Debus et al. (29) used fractionated stereotactic radiotherapy to treat eight chondrosarcomas of the skull base. A linear accelerator was used with the head immobilized in a stereotactic frame to deliver 64.9 Gy to the tumor center. The advantage of fractionated delivery is that it allows treatment of larger tumor volumes than previous single-session radiosurgery approaches. All patients had undergone previous resection of their lesions, and when combined with stereotactic radiation, 100% 5-year local control and survival rates were achieved. They did acknowledge that they had a small number of patients, and consequently, they cautioned that fractionated proton beam therapy was still the standard for postoperative radiation therapy.

There is no agreement as to when and if chemotherapy should be utilized in the treatment of skull base chondrosarcoma. Several reports state that chemotherapy is not effective in head and neck and skull base chondrosarcomas, and therefore, it should not be utilized (31). Conversely, Finn et al. (32) stated that multi-agent chemotherapy protocols used to treat other soft tissue sarcomas may be useful in treating chondrosarcomas with a high probability of aggressive local spread or metastasis. Included were grade II or III conventional chondrosarcoma, mesenchymal chondrosarcomas, and undifferentiated chondrosarcomas. Chemotherapy may also offer palliation for chondrosarcomas that are unresectable. Koch et al. (26) concurred that advanced, high-grade lesions and mesenchymal subtypes may benefit from multi-agent chemotherapy.

3. OSTEOSARCOMA

3.1. Demographics

Osteogenic sarcoma or osteosarcoma is a highly aggressive malignant tumor that usually presents in the metaphysis of long bones. The majority of cases occur between the ages of

10 and 30 years with the median age of 28. There are ~7400 new cases and 4200 deaths from osteosarcoma occurring annually in the United States. Approximately 10% of cases occur in the head and neck, and this accounts for ~900 new cases per year (41,42). In a review of the world literature, Sataloff et al. (43) found 19 reported cases of osteosarcoma involving the temporal bone, with the largest series reporting three cases. An additional three have been reported since then for a total of 22 cases (42–45). Almost every bone of the skull has been involved as a primary site; however, the mandible was clearly the most common primary location in the head and neck (41,46,47).

3.2. Pathology

Osteosarcoma is thought to arise from immature bone-forming cells or through neoplastic differentiation of other immature mesenchymal cells into osteoblasts. Histologically, a tumor is considered to be an osteosarcoma if it demonstrates malignant spindle cells producing osteoid in various stromal backgrounds (Fig. 22.5). Subtypes are based on the predominant characteristic of the cells and stroma and include osteoblastic, chondroblastic, fibroblastic, small cell, and telangiectatic types (46,48).

Additionally, tumors are graded with regard to the degree of histologic differentiation with grade 1 being well differentiated and grade 4 being poorly differentiated (46). Ninety-five per cent of extragnathic craniofacial osteosarcomas present with an aggressive histology of grade 3 or 4; however, many authors are not able to clearly correlate a high-grade histology with poorer survival (41,45,48). There are a few commonly named factors that have been reported as significant in the development of osteosarcoma: previous radiation therapy, a history of hereditary retinoblastoma, fibrous dysplasia, and Paget's disease (41,48). Three cases of osteosarcoma of the temporal bone have been reported to arise in areas of previous radiation treatment. One of these occurred in a patient previously treated with 66 Gy for nasopharyngeal carcinoma (43,44). Patients with retinoblastoma have a 500 times increased risk of developing an osteosarcoma during their lifetime, independent of previous radiation treatment to their primary site.

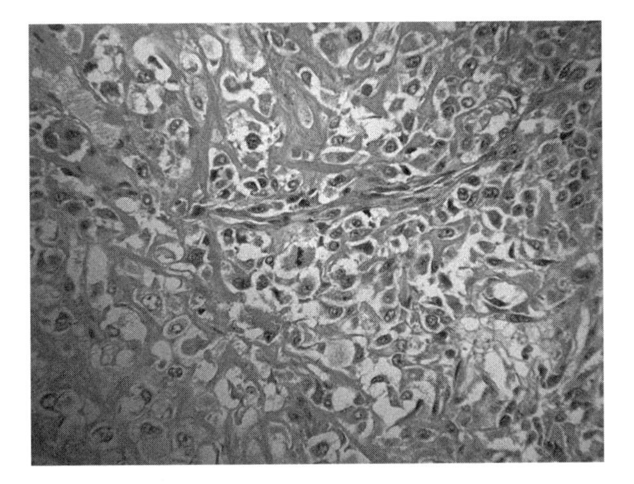

Figure 22.5 **(See color insert)** Osteosarcoma (200 magnification). The production of osteoid defines an intramedullary tumor as osteosarcoma. This definition holds regardless of histologic pattern. Cell size and shape can vary widely. Nuclei are typical large, pleomorphic and contain one or more prominent nucleoli. The chromatin pattern is usually coarse and clumped.

Both retinoblastoma and osteosarcoma patients have shown specific deletions in the long arm of chromosome 13, and a retinoblastoma gene that has been implicated in the development of both retinoblastoma and osteosarcoma has been reported (41). Approximately, 1% of patients with head and neck fibrous dysplasia will eventually develop osteosarcoma; however, these patients do not appear to have a poorer prognosis than patients with *de novo* osteosarcoma (41). On the other hand, patients who develop osteosarcoma in an area of Paget's disease seem to develop an aggressive form of the lesion with a reported 5-year survival of 2% (7).

3.3. Clinical Presentation/Diagnostic Work-Up

In most studies of head and neck osteosarcoma, the most common presentation is pain and swelling over the area of bone containing the lesion (42). In their review of 19 patients with temporal bone osteosarcoma, Sataloff et al. (43) demonstrate that the most common presenting symptoms or signs were a mass in the temporal fossa, mastoid, or external ear canal in 84%, facial paralysis in 47% (9/19), conductive hearing loss in 37% (7/19), otalgia in 32% (6/19), bloody or purulent otorrhea in 16% (3/19), and other cranial nerve deficits in 16% (3/19) (43). As with other skull base and temporal bone malignancies, symptoms caused by cranial nerve deficits are determined by the site of tumor involvement.

Work-up for temporal bone and skull base osteosarcoma again starts with a CT of temporal bones and brain to delineate areas of bone erosion. CT has supplanted plain films even though the latter are fairly good at detecting most lesions. The appearance of osteosarcomas on plain film is either lytic or sclerotic with ossific laminae radiating in a sunburst pattern around the periphery of the lesion. The only significant associated clinical laboratory finding is an elevated alkaline phosphatase, which is nonspecific and not always present (42,48). Regional metastasis to lymph nodes are very rare in osteosarcoma, but distant metastatic disease has been reported in 17–51% of head and neck lesions. The overwhelming majority of metastases are to the lungs, with less frequent sites including the liver, bone, and brain (41,48). Routine metastatic work-up should include a chest CT, as well as liver function tests, and a bone scan. Poor prognostic factors include osteosarcoma in association with Paget's disease, as well as site of lesion, positive surgical margins, and metastatic disease at presentation (41,46). Nora et al. (45) showed that head and neck osteosarcomas of the extragnathic bones had a much poorer 5-year survival of 10% than those that involved the mandible and maxilla whose 5-year survival approached 40%.

3.4. Treatment

Again, only 22 cases of osteosarcoma of the temporal bone have been reported in the English literature since 1910 (42–45). Over that time, treatment has widely varied and follow-up was inconsistent or not reported in nearly one-third of the cases. Consequently, treatment protocols for temporal bone osteosarcoma have to be extrapolated from series reporting osteosarcomas located in the head, neck, and other body sites. The two largest head and neck series were reported by Garrington et al. at the AFIP, and Caron et al. at Memorial Sloan Kettering Cancer Center. They looked at 56 and 43 patients, respectively, and concluded that radical resection was the treatment of choice, with 5-year survivals of 35% and 23.3%, respectively. However, both these studies failed to distribute patients equally into alternative therapy treatment arms with very small numbers receiving radiation, chemotherapy, or combinations of the two (48,49). Kassir et al. (46) performed a meta-analysis of 173 patients in 23 different reports and found an overall 5-year survival for head and neck osteosarcoma of 37%. They found that patients with negative surgical margins did the best and stated that radiation and chemotherapy still had an unproven role in the adjuvant treatment of osteosarcoma. However, they did admit that their retrospective analysis lent significant bias to their results since chemotherapy and radiation were used mainly in more advanced or unresectable lesions (46).

Several studies have shown that adjuvant radiation improves local control and survival. Chambers and Mahoney (50) reported a higher 5-year survival of 73% when they used high-dose preoperative radiation followed by wide surgical resection for osteosarcomas of all body sites. Additionally, Mark et al. (41) obtained better results with wide surgical excision of head and neck osteosarcomas followed by postoperative radiation and chemotherapy, but they did not compare results to surgery alone. There have been no prospective trials to evaluate chemotherapy efficacy in head and neck or temporal bone osteosarcomas. There have been a few prospective, randomized studies showing a survival benefit when utilizing chemotherapy to treat osteosarcoma of the extremities. Link et al. (51) enrolled 114 patients with extremity osteosarcomas and showed that 2-year disease-free survival was 56% in the surgery plus chemotherapy arm vs. 18% in the surgery alone group. Eibler et al. (52) randomized 59 patients with a surgery arm alone vs. surgery with adjuvant multi-agent chemotherapy. Patients received four, six-week cycles of chemotherapy postoperatively. Two-year disease-free and overall survivals were 55% and 80% in the adjuvant chemotherapy arm vs. 20% and 48% in the surgery alone group, respectively. For temporal bone lesions, ideally the patient should receive en bloc excision followed by 55–69 Gy of postoperative radiation therapy. However, Castro et al. (53) have shown some benefit in treating with heavy charged particle irradiation to deliver 66 gray equivalents to a more defined tumor target. Early success with chondrosarcomas and osteosarcomas may make this the radiation modality of choice for most skull base and temporal bone sarcomas. Chemotherapy can also be given concurrently with the postoperative radiation. Most recent reports including Sataloff et al. (42) and Sharma et al. (43) support the radical resection of temporal bone osteosarcoma with the use of adjuvant radiation and chemotherapy.

4. MALIGNANT FIBROUS HISTIOCYTOMA

4.1. Demographics

Malignant fibrous histiocytoma or MFH was introduced as a tumor category ~30 years ago. The category is a pathological redefinition that has created a new sarcoma type. Initially, this category drew from a pool of lesions that had been misdiagnosed as high-grade fibrosarcomas, liposarcomas, pleomorphic rhabdomyosarcomas, and undifferentiated sarcomas. The tumor is a fairly common sarcoma in adults and arises most frequently in the extremities, pelvis, and retro-peritoneum; however, it is a rare tumor of the head and neck. Rarely, MFH also can affect bone, but it is extremely rare in skull bones with only 17 reported cases in the English literature (54). There are three reported cases of MFH that involve the temporal bone (54–56). It is also possible that the number of MFH cases has been underestimated. Many of Nuafel's (3) 89 cases of undifferentiated sarcomas and 17 cases of fibrosarcomas might be reclassified as MFH if examined by today's pathologic criteria. None of the studies clearly summarized the clinical presentation of these temporal bone MFH, but there is no reason to believe that they would differ greatly from other sarcoma histologies. Work-up includes a thorough head and neck exam, cranial nerve exam, and CT and MRI imaging as a starting point.

4.2. Pathology

The microscopic appearance of MFH includes rounded histiocytic cells mixed with spindle fibroblastic cells in a matted pattern (Fig. 22.6). There are various subtypes based on microscopic characteristics which include the angiomatoid, inflammatory, and pleomorphic subtypes. It is thought that MFH arises from a primitive mesenchymal cell which has the ability to form both fibroblastic and histiocytic elements. Immunohistochemical staining is positive for CD 68 which is a histiocyte stain. It is negative for S-100 and EMA (54,56).

4.3. Prognosis

Head and neck MFH has an average age at presentation of ~43 years, with a slight predilection for males (56%). Male sex and increasing age were independently found to be associated with a poorer prognosis. Additionally, tumors arising from or extending into bone showed a more aggressive clinical course and a poorer prognosis than those restricted to soft tissues. Tumors >3 cm were also shown to do more poorly than smaller primary tumors. The metastatic rate was 32% for head and neck MFH, and these patients had a poorer survival (54,55). The lung was by far the most common metastatic site with brain, bone, liver, and regional neck lymph node metastases also reported. Regional neck metastasis was seen in 15% of patients; however, neck dissection is not suggested in the absence of clinically positive nodes (54,55).

4.4. Treatment

Raney et al. treated seven children with MFH of which three had head and neck lesions. The treatment protocol closely mirrored the IRS protocols used for treating rhabdomyosarcoma in children. They stated that every child should have as complete a resection as possible; however, all of the head and neck lesions were simply biopsied with no attempt at complete surgical resection. Biopsy was then followed by multi-agent

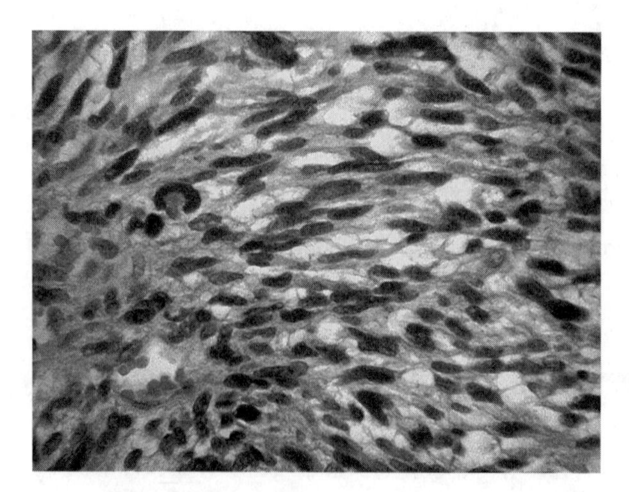

Figure 22.6 **(See color insert)** Malignant fibrous histiocytoma (400 magnification). Malignant fibrous histiocytomas can show any number of different morphologic patterns but all consist of varying proportions of myofibroblasts, histiocytoid cells, and fibroblasts. Typically the cells are arranged in random bundles with an intersecting fascicular pattern.

chemotherapy for 2 years. If initial surgical therapy left gross or microscopic disease, 40–50 Gy was delivered to the tumor over 5–6 weeks. Even with this aggressive treatment, three of seven children died of distant metastatic disease (56). Nakayama et al. (54) suggested that radical resection of head and neck MFH which involved bone was appropriate. They reported a total temporal bone resection with facial nerve sacrifice as an example of their treatment philosophy. They suggested that postoperative chemoradiation would also be beneficial. Their patient was alive without disease at 18 months. The same authors also reviewed several small series containing 15 other cases of MFH involving bones of the skull and reported a 34–50% 5-year survival. In their review of 87 cases of head and neck MFH, Blitzer et al. stated that wide surgical resection was the treatment of choice. They also advocated chemotherapy of some type, using Adriamycin since this agent seemed to provide the best partial tumor regression. Radiotherapy was not effective in their series, however they did support its use with chemotherapy in the inoperable patient or as palliation in those patients with distant metastatic disease. They reported a 45% mortality rate in their own 11 patients, but were unable to give 5-year survival statistics for the 87 reviewed cases due to incomplete and variable follow-up (55).

5. FIBROSARCOMA

5.1. Demographics

Fibrosarcoma used to be the most commonly diagnosed head and neck sarcoma, but the frequency has dropped with more rigid histologic criteria and the development of new pathologic categories. In reviews of previous fibrosarcomas, up to one-half would be regrouped into other diagnoses (2,57). There are numerous lesions to be considered in the differential diagnosis of fibrosarcoma-like or spindle cell tumors including such benign processes as fibromatosis, fibromas, nodular fascitis, and malignant sarcomas such as malignant fibrous histiocytoma, rhabdomyosarcoma, liposarcoma, monophasic synovial sarcoma, and neurogenic sarcoma. Some authors have reported that fibrosarcomas account for 5–10% of sarcomas, with 10–20% of these occurring in the head and neck (2). Fibrosarcomas commonly affect the 40–70-year age range with the average age being 44 years old. Like osteosarcoma, fibrosarcoma is common in areas of previous tissue injury. Up to 10% of all fibrosarcoma patients have a history of distant radiation therapy, 4% have a history of trauma to the site, and 0.8% have a history of thermal injury to the site. The soft tissue of the neck was the most common site involved with fibrosarcoma, and the soft tissue of the face and scalp followed closely in frequency (57). In a review of the literature, 17 cases of fibrosarcoma were found to involve the temporal bone. Temporal bone fibrosarcoma will also likely become a much rarer diagnosis due to the more current, stringent histiologic criteria (3,58,59). History, physical, and radiologic evaluations should mirror other skull base sarcomas.

5.2. Pathology/Prognosis

Fibrosarcoma is composed of spindle-shaped fibroblasts arranged in intersecting fascicles that give rise to a "herringbone pattern" (Fig. 22.7). This is very similar to neurogenic sarcoma which will be discussed briefly later. Immunohistochemistry is not diagnostic for fibrosarcoma, but it is helpful in excluding lesions in its pathologic differential. Pathologists designate these tumors as grade 1 (well differentiated)

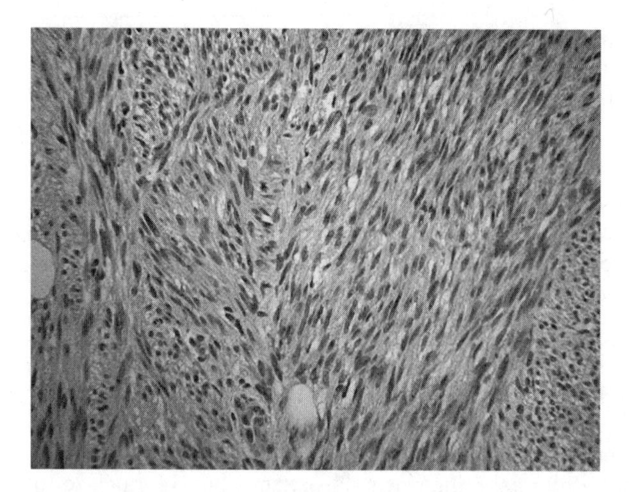

Figure 22.7 **(See color insert)** Fibrosarcoma (100 magnification). Fibrosarcomas are composed of slender spindle-shaped cells oriented in interweaving bundles. This interlacing arrangement gives rise to the term herring-bone pattern and is associated with lower grade tumors. Nuclear pleormorphism is minimal.

to 4 (poorly differentiated). Most tumors usually present with a grade 2 or 3 histology (2,57).

Unlike some of the other sarcomas, higher grades appear to correlate with increased local recurrences, increased metastatic rate, and decreased overall survival. Additional poor prognostic factors include tumors >5 cm, presence of metastatic disease, and incomplete surgical resection. Metastatic disease from head and neck fibrosarcoma occurs in 19% of cases (57,60).

5.3. Treatment

Frankenthaler et al. (57) reviewed 118 cases of head and neck fibrosarcoma at MD Anderson. They supported wide local excision, but were unable to obtain negative margins in 47% of their cases. If the surgical margins were microscopically positive or questionable, then 60 Gy of radiation therapy was given postoperatively over 6 weeks. Up to nine courses of multi-agent chemotherapy, consisting of doxorubicin-based regimens, were given preoperatively when the biopsy showed either grade 3 or 4 lesions, or for an unresectable tumor. Unresectable sarcomas were debulked surgically after induction chemotherapy, and then they were treated with radiation therapy. They reported a 63% 5-year overall survival with this approach. Freedman et al. (61) employed a bi-treatment approach. Their primary goal was obtaining wide surgical margins followed by postoperative radiation in cases when there were high-grade lesions, close margins, or microscopically positive margins. They reported a 5-year survival of 68%. Conley et al. (60) reported using surgery and postoperative radiation in his treatment approach, but he stated that radical *en bloc* resection should definitely be used for more aggressive, poorly differentiated lesions since radical resection was their only chance for survival. Extrapolating these treatment philosophies to temporal bone lesions, total "radical" resection of skull base lesions seems appropriate with postoperative radiation and possibly induction chemotherapy in high-grade lesions.

6. EWING'S SARCOMA

6.1. Demographics

Ewing's sarcoma represents 6–9% of primary malignant bone tumors and is encountered most commonly in the long bones, pelvis, ribs, or vertebrae (62,63). During the early IRS trials, pathologists discovered that there was a distinct subgroup of extra-osseous tumors that could be distinguished from undifferentiated rhabdomyosarcoma. They termed this group as extra-osseous Ewing's sarcoma. Consequently, future IRS trials excluded these from their study treatment groups (5,18). This extra-skeletal form is extremely rare in the head and neck. Ewing's sarcoma that involves the facial bones and skull is also extremely rare, accounting for <1% of all reported Ewing's sarcomas. The most common head and neck site affected is the mandible. Approximately 50 cases of Ewing's sarcoma of the skull or skull base have been reported in the literature. Most Ewing's sarcomas of the skull arise in the convexities of the calverium; however, a handful have been reported in the temporal bone (64,65). The literature reports nine lesions involving the temporal bone, of which most arise from the squamous portion of the temporal bone. Additionally, there are reports of five cases of petrous apex Ewing's sarcoma and three cases involving the mastoid portion of the temporal bone. The average patient age of those affected with head and neck Ewing's sarcoma was 14.5 years old, and the age range for those with temporal bone tumors was from 3 to 19 years of age (3,63,65–67).

6.2. Pathology

Grossly, Ewing's sarcomas are firm, rubbery vascular masses arising from bone. Microscopically, these lesions consist of small, blue, round cells arranged in sheets with a few pseudorosettes present (Fig. 22.8). Additionally, there are glycogen granules in the cytoplasm of these cells after periodic acid-Schiff staining; however, this is also seen in

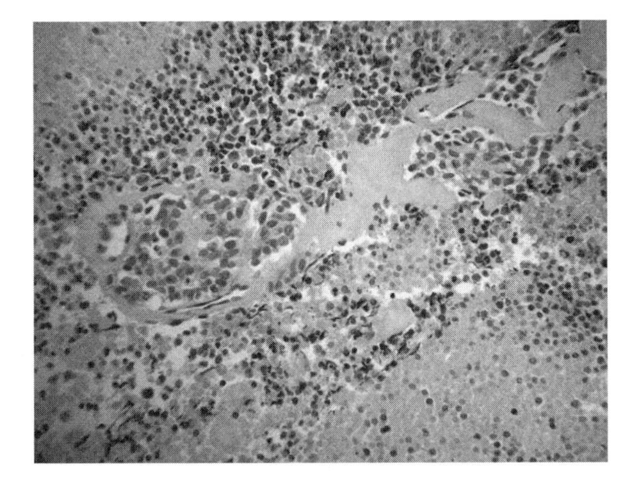

Figure 22.8 **(See color insert)** Ewing's sarcoma (200 magnification). Tumors cells are arranged in sheets and lobules with broad areas of necrosis. Individual cells have a small amount of amphophilic cytoplasm and contain well-defined, oval nuclei with finely dispersed chromatin and small nucleoli. Intracellular glycogen can be demonstrated by PAS stain.

embryonal rhabdomyosarcoma. The histologic differential for a cranial Ewing's sarcoma includes neuroectodermal tumor, rhabdomyosarcoma, metastatic neuroblastoma, and lymphoma. Immunohistochemical staining is important in helping to distinguish among these lesions. Ewing's sarcoma is positive for CD 99, VIM, and MIC-2 which is a specific marker for Ewing's sarcoma. These lesions are negative for the skeletal marker, desmin, as well as the neural markers S100 and chromogranin, and the lymphoma markers such as leukocyte common antigen, CD20, and CD3 (65–67).

6.3. Clinical Presentation and Work-Up

Patients presenting complaints differ depending upon the anatomic site within the temporal bone affected by the tumor. A mass arising from the squamous portion of the temporal bone will present as temporal scalp swelling and pain. Additionally, headache and signs of increased intracranial pressure such as vomiting and papilledema can be present. These latter symptoms will be more prominent with large intracranial extensions of tumor causing midline shift of the cerebrum. In the middle ear, Ewing's sarcoma will present with otalgia, facial nerve paresis, conductive hearing loss, and multiple cranial neuropathies (62,68). Work-up includes CT and MRI of the temporal bones and brain, and other neurotological studies as indicated based on symptoms and signs. Ewing's sarcoma will appear isodense to bone with enhancement after IV dye administration, and extensive bone erosion and destruction are usually present (64). A technetium-99 bone scan will show increased uptake at the primary site as well as any metastatic sites; however, metastatic disease is extremely uncommon in Ewing's sarcoma of the skull (62). A skeletal survey, in addition to the bone scan, is obtained to further evaluate for a non-head and neck Ewing's sarcoma primary that may have metastasized to the skull. Hadfield et al. (65) reported that metastatic Ewing's sarcoma to the skull and skull base is much more common than a primary Ewing's sarcoma of the skull base. A 24-h urine should be collected for VMA in order to exclude a metastatic neuroblastoma from the differential diagnosis. Osteosarcoma may also be in the radiologic differential of Ewing's sarcoma; however, biopsy clearly differentiates these two lesions. MRI has a limited role in the evaluation of these tumors due to poor delineation of bony detail (66).

6.4. Treatment

Desai et al. (66) reported 14 patients with primary Ewing's sarcoma of the cranium. They supported en bloc excision of each tumor; however, they did not report any cases involving the temporal bone. Surgical treatment was then followed by the round cell chemotherapy protocol (RCT II). It consisted of two alternating cycles of iodophamide and etoposide with vincristine, doxorubicin, and cyclophosphamide. At 8–9 weeks after starting chemotherapy, 40–50 Gy of radiation was delivered to the primary site. Six courses of maintenance chemotherapy were given after the radiation was complete. They were able to achieve a 57% 5-year survival with this protocol. Multiple other small series report that optimal treatment would include complete excision followed by chemoradiation. There were no reported cases in any series where radical skull base surgery was utilized to eradicate tumor from the middle ear cleft, mastoid, or petrous apex. The authors advocating radical removal of lesions were mainly basing their opinions on wide local excision of calverium lesions which are easier to remove without functional deficits (66–68). Horowitz et al. (69) supported a different treatment approach. They suggested that multi-agent chemotherapy, such as RCT II, should be the initial treatment after biopsy-proven diagnosis. This would be followed by surgical debulking of as much disease as possible without sacrificing vital structures. Lastly, postoperative chemoradiation would be given to treat the residual disease. The goal of this treatment approach was to reduce surgical morbidity while achieving the same survival as those series that utilized wide local excision as a primary treatment.

7. NEUROGENIC SARCOMA

7.1. Demographics

A brief review of definitions will hopefully help clarify this rare head and neck sarcoma. A neurofibroma is a benign peripheral nerve mass composed of perineural cells, fibroblasts, neurites, and schwann cells. The nerve fibers are essentially engulfed in this un-encapsulated mass and, therefore, the nerve is not separable from the tumor during surgery. Neurofibromas are usually multiple with 22–40% occurring in the head and neck. Additionally, neurofibromas are the tumor associated with Von Recklinghausen's disease or neurofibromatosis type I (NF1), which originates from a genetic abnormality on chromosome 17. This is not to be confused with neurofibromatosis type II (NF2) or acoustic neurofibromatosis which is manifested as multiple or bilateral acoustic schwannomas which are a different entity from a neurofibroma. Schwannomas arise from schwann cells and are separable from their nerve of origin at the time of surgery. The chromosomal lesion in NF2 is found on chromosome 22 instead of 17 (70).

Neurogenic sarcoma is the malignant counterpart of neurofibroma. Neurogenic sarcoma has been called by multiple names including malignant peripheral nerve sheath tumor, neurofibrosarcoma, and malignant schwannoma. The last two are currently discouraged because they can lead to incorrect inferences as to the origin of this tumor (70). In the early years of light microscopy, many pathologists found it very difficult to distinguish, histologically, between fibrosarcoma and neurogenic sarcoma. The only reliable way to clearly identify neurogenic sarcoma was its gross pathologic origin from a nerve or by a surgeon's report that the mass arose from a nerve. They then called the variant neurofibrosarcoma (71). Recently, however, immunohistochemical stains have made the distinction more clearly, and consequently, it has been determined that neurofibrosarcomas are not true fibrosarcomas. Malignant schwannoma is also discouraged because Enzinger (1988) stated that schwannomas have almost no malignant potential, except in a few cases; however, this point is debatable (70). Neurogenic sarcomas arise almost exclusively from preexisting neurofibromas, but very rarely they may arise *de novo*. A few studies disagree about which is more common; additionally, the frequency of these malignancies arising in the setting of NF1 remains controversial. Literature reports of neurogenic sarcomas arising in patients with NF1 range from 5% to 50%. The most common areas affected by neurogenic sarcomas are the trunk and extremities. Head and neck neurogenic sarcomas account for 6–17% of cases; and the neck is the most common area, with the brachial plexus and vagus nerve as the most common sites (71,72).

7.2. Histology

Neurogenic sarcoma is very similar to fibrosarcoma. Neurogenic sarcomas contain microscopic fields composed of spindle-shaped cells arranged in interlacing fascicles in a herringbone pattern (Fig. 22.9). Occasionally there may be a clue that some neurogenic features are present, but diagnosis really relies upon immunohistochemistry. Neurogenic sarcoma stains heavily for S-100 antigen (a neural marker which is not seen in fibrosarcoma.) It also stains for myelin basic protein and leu 7 which are not as specific.

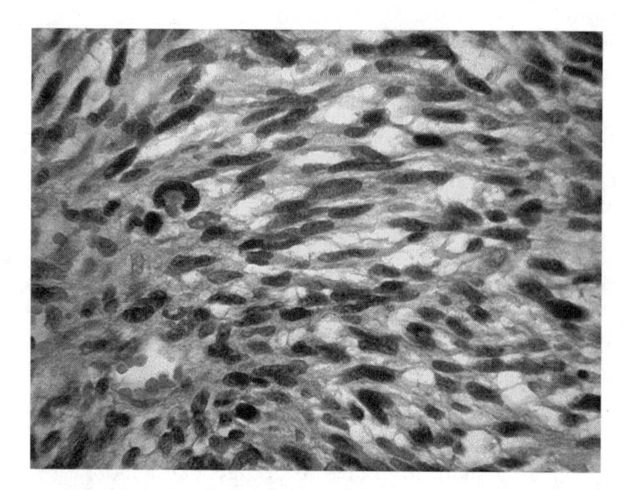

Figure 22.9 **(See color insert)** Neurogenic sarcoma (400 magnification). Neurogenic sarcomas are composed of tightly packed spindle cells arranged in interlacing bundles. Cell nuclei are large and can be pleomorphic. Often, there is a suggestion of nuclear palisading. Higher grade tumors are associated with increasing cellularity, nuclear atypia, and mitoses.

Pathologically, neurogenic sarcomas also appear quite similar to amelanotic melanoma, but melanoma stains positive for HMB-45, whereas neurogenic sarcoma does not (70,71).

7.3. Clinical Presentation

Neurogenic sarcoma of the temporal bone is an extremely rare lesion. Naufel in 1973 reviewed the world literature and found three neurogenic sarcomas in case reports that arose from cranial nerve VII (3,73,74). The masses were found on the mastoid segment of the facial nerve and none arose from an existing lesion per the authors report. Clinical presentation was similar to other middle ear sarcomas with symptoms which included preauricular or mastoid swelling or mass, polyps in the external auditory canal, otorrhea, conductive hearing loss, and facial nerve paralysis (73,74). Greager et al. (72) also reported a case of neurogenic sarcoma of the "ear," but did not document further details of anatomy or treatment.

7.4. Treatment

Treatment strategies need to be adapted from larger series of neurogenic sarcomas arising from other body sites. Hutcherson et al. (71) reviewed seven patients with head and neck neurogenic sarcoma. The patients varied widely in surgical treatment, and they also received various treatments of either radiation therapy, chemotherapy, or both. However, they concluded that wide surgical excision is the mainstay of treatment with postoperative radiation being helpful in local control and prevention of local recurrences. They had a 43% 5-year survival in this series. D'Agostino et al. (75) studied 24 cases of which six were located in the head and neck region. Again, no specific protocol was evaluated; however, they felt that wide surgical excision was the only effective treatment. They did not believe radiation conferred any survival advantage. Greager et al. (72) reported 17 patients with head and neck neurogenic sarcoma. They found that wide local excision with resection of the involved neurovascular structures were the mainstay of treatment. They did not find adjuvant radiation or chemotherapy to be helpful, but admitted that their experience with these techniques was limited and their role needed further study. Hellquist and Lundgren (70), and Goepfert et al. (76) both retrospectively reviewed small series of patients with head and neck neurogenic sarcoma and found radical surgery plus 60 Gy radiation afforded the best survival. Greager et al. (72) reported 2-, 5-, and 10-year overall survivals of 65%, 47% and 29%, respectively. A poor prognosis was also reported in patients with associated Von Recklinghausen's disease (NF1) which may be due to a more aggressive lesion. Other negative prognostic indicators included presence of metastatic disease which occurs in up to 47% of head and neck neurogenic sarcomas, tumors >5 cm in size, and possibly tumor grade; however, histologic differentiation did not always correlate well with survival (71,72).

8. ANGIOSARCOMA

Angiosarcoma is a malignant vascular tumor most commonly arising in the skin and soft tissue of the scalp in elderly men (Fig. 22.10). They are aggressive tumors that tend to recur locally and metastasize early despite early treatment. Most give a history of a several-month period of an enlarging purple skin lesion with pain, edema, and intermittent bleeding (77). Kinkade (78) reported the only clearly documented case of angiosarcoma involving the petrous portion of the temporal bone. Naufel and Mark et al. have reported nine patients with "ear" involvement by an angiosarcoma, but neither author delineated whether these lesions arose from the middle ear and mastoid or whether they primarily involved the skin of the pinna or external auditory canal. The latter is more likely the case since nearly all of these tumors arise in the subcutaneous tissues (3,77). However, Linthicum and Schwartzman (80) provided a pathology slide demonstrating an angiosarcoma adjacent to the stapes footplate; however, clinical presentation and treatment were not reported. Treatment is aimed at wide surgical excision followed by 50–60 Gy of postoperative radiation. Most head and neck series reported 5-year survivals of 10–41% (77,79).

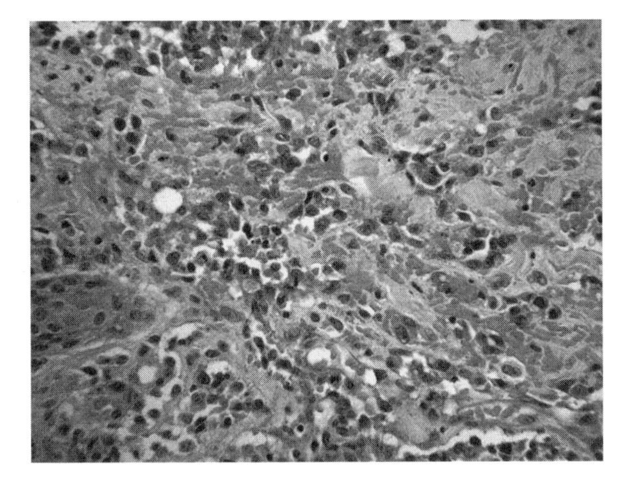

Figure 22.10 **(See color insert)** Angiosarcoma (400 magnification). These vasoformative tumors are composed of malignant endothelial cells that arrange themselves in complex anastomosing channels. The cellular nuclei are plump and tend to heap-up and project into the lumens. Mitoses are common and extravasated red cells are seen in the stroma.

9. LIPOSARCOMAS

Liposarcomas are composed of malignant lipoblasts situated in a loose stroma (Fig. 22.11). The head and neck is involved in only 2–6% of cases. The myxoid subtype was the most common subtype reported in the head and neck. The tumor arises *de novo*, not from benign lipomas. Statistics of 5-year survival depend on the histologic subtype. Well-differentiated myxiod liposarcomas have a 5-year survival of 75–100%. The more poorly differentiated tumors such as the round and pleomorphic subtypes have 5-year survivals of 6–27% (2). In Naufel's (3) review he reported one case involving the temporal bone which was reported in the German literature. They did not suggest any general treatment approaches; however, combination therapy consisting of surgery, radiation, and chemotherapy would probably be most effective.

10. LEIOMYOSARCOMA

Leiomyosarcoma is a malignant tumor of smooth muscle origin (Fig. 22.12). In the head and neck, smooth muscle is scarce and is located in the walls of blood vessels and errector pili musculature of skin. Immunohistochemical staining is positive for smooth muscle specific actin. The most common head and neck site is the sinonasal tract. Zbaren and Ruchti (81) reported one case involving the middle ear which was deemed unresectable and was treated unsuccessfully with chemotherapy. They recommended surgical excision, followed by postoperative radiation and chemotherapy. Prognosis is poor, with 1-year survival reported at 50% for tumors of the sinonasal tract.

11. CONCLUSION

In conclusion, sarcomas of the temporal bone are rare. With the exception of rhabdomyosarcoma, there is very little literature that gives a clear picture of how to optimally treat

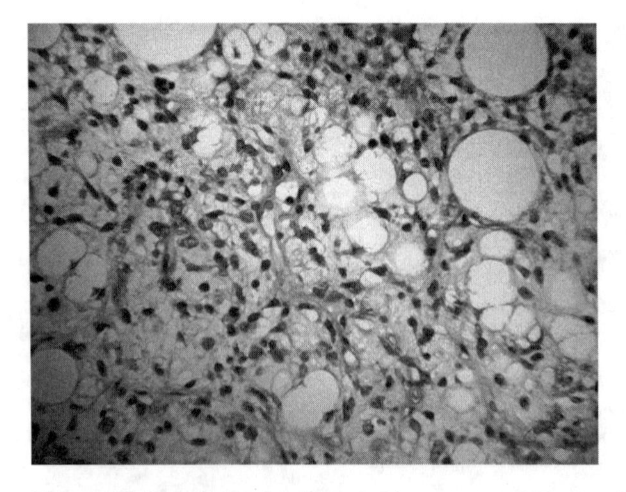

Figure 22.11 **(See color insert)** Liposarcoma (200 magnification). Malignant lipoblasts are the hallmark of this lesion. They appear as vacuolated cells with eccentrically displaced nuclei, resembling altered adipose tissue. Cell nuclei are large, variably pleomorphic, and may appear in the form of rings or florets.

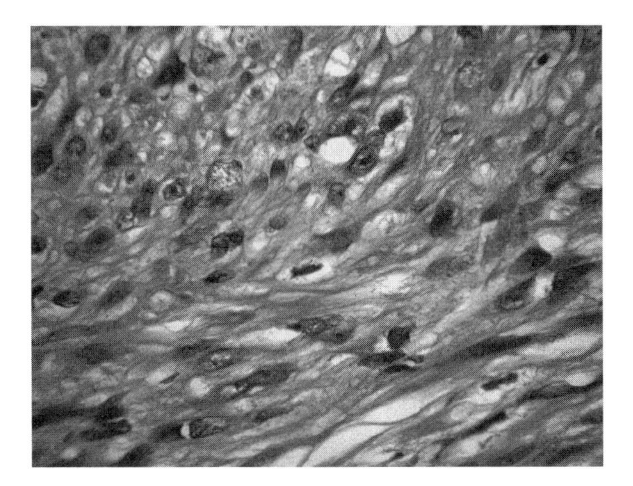

Figure 22.12 **(See color insert)** Leiomyosarcoma (200 magnification). Leiomyosarcomas consist of intersecting bundles of spindle-shaped smooth cells. Nuclei are cigar-shaped and have blunt ends. The neoplastic cells contain a moderate degree of eosinophilic cytoplasm that appears fibrillar in the longitudinal axis. When cut on end, tumor cells appear to have perinuclear halos.

these lesions. Using broad generalizations, rhabdomyosarcoma is the only temporal bone sarcoma where combined chemotherapy and radiation therapy is commonly accepted as the primary treatment modality, with nonradical surgery or biopsy being the only role for the surgeon. However, this has not been proven definitively to be the best approach. Other nonrhabdomyosarcoma sarcomas are usually treated with wide and/or radical surgical resection with varying protocols utilizing pre- and postoperative radiation and chemotherapy. Proton beam radiation has been used increasingly to treat chondrosarcomas and osteosarcomas of the temporal bone and skull base, with encouraging results. Further research and wider application may confirm this as a good radiation modality for all skull base sarcomas. As with most head and neck malignancies, true leaps in survivorship await treatments aimed at correcting the underlying molecular and genetic origins of these tumors.

REFERENCES

1. Marsh M, Coker N, Jenkins H. Temporal bone neoplasms and lateral cranial base surgery. In: Cummings C, ed. Otolaryngology Head and Neck Surgery. 3rd ed. St. Louis, MO: Mosby, 1996:3245–3281.
2. Kyriakos K, Batsakis JG, eds. Tumors of the Head and Neck. 3rd ed. Baltimore, MD: Williams & Wilkins, 1987.
3. Nuafel PM. Primary sarcomas of the temporal bone. Arch Otolaryngol 1973; 98:44–50.
4. Daya H, Chan HSL, Sirkin W et al. Pediatric rhabdomyosarcoma of the head and neck. Is there a place for surgical management? Arch Otolaryngol Head Neck Surg 2000; 126:468–472.
5. Newton WA, Gehan EA, Webber BL et al. Classification of rhabdomyosarcomas and related sarcomas. Cancer 1995; 76(6):1073–1085.
6. Canalis RF, Gussen R. Temporal bone findings in rhabdomyosarcoma with predominantly petrous involvement. Arch Otolaryngol 1980; 106:290–293.
7. Odell PF. Head and neck sarcomas: a review. J Otolaryngol 1996; 25(1):7–13.
8. Anderson GJ, Tom LWC, Womer RB et al. Rhabdomyosarcoma of the head and neck in children. Arch Otolaryngol Head Neck Surg 1990; 116:428–431.

9. Goepfert H, Cangir A, Lindberg R et al. Rhabdomyosarcoma of the temporal bone. Is surgical resection necessary? Arch Otolaryngol 1979; 105:310–313.
10. El-Gothamy B, Fujita S, Hayden RC. Rhabdomyosarcoma—a temporal bone report. Arch Otolaryngol 1973; 98:106–110.
11. Wiatrak BJ, Pensak ML. Rhabdomyosarcoma of the ear and temporal bone. Laryngoscope 1989; 99:1188–1192.
12. Tefft M, Fernandez C, Donaldson M et al. Incidence of meningeal involvement by rhabdomyosarcoma of the head and neck in children. A report of the intergroup rhabdomyosarcoma study. Cancer 1978; 42:253–258.
13. Jaffe BF, Fox JF, Batsakis JG. Rhabdomyosarcoma of the middle ear and mastoid. Cancer 1971; 27(1):29–37.
14. Deutsch M, Felder H. Rhabdomyosarcoma of the ear-mastoid. Laryngoscope 1974; 84:586–592.
15. Baker KS, Anderson JR, Link MP et al. Benefit of intensified therapy for patients with local or regional embryonal rhabdomyosarcoma: results from the Intergroup Rhabdomyosarcoma Study IV. J Clin Oncol 2000; 18(12):2427–2434.
16. Prasad S, Snyderman C, Janecka IP et al. Role for surgical resection of parameningeal rhabdomyosarcoma in childhood. In: Proceedings of the 1994 Joint Annual Meeting of the German and North American Skull Base Society. New York: Thieme; 1994:8.
17. Healy GB, McGill TJI, Janecka I. The Surgical Management of Sarcomas of the Head and Neck in Children. Proceedings of the 4th International Conference on Head and Neck Cancer. Madison, WI: Omni Press, 1996:379–384.
18. Raney RB, Lawrence W, Maurer HM et al. Rhabdomyosarcoma of the ear in childhood. A report from the Intergroup Rhabdomyosarcoma Study I. Cancer 1983; 51(12):2356–2361.
19. Anderson GJ, Tom LWC, Womer RB et al. Rhabdomyosarcoma of the head and neck in children. Arch Otolaryngol Head Neck Surg 1990; 116:428–431.
20. Flamant F, Rodary C, Rey A et al. Treatment of non-metastatic rhabdomyosarcomas in childhood and adolescence. Results of the Second Study of the International Society of Paediatric Oncology: MMT84. Eur J Cancer 1998; 34(7):1050–1061.
21. Schouwenburg PF, Kupperman D, Bakker FP et al. New combined treatment of surgery, radiotherapy, and reconstruction in head and neck rhabdomyosarcoma in children: The AMORE Protocol. Head Neck 1998; 20:282–283.
22. Lau DPC, Wharton SB, Antoun NM et al. Chondrosarcoma of the petrous apex. Dilemmas in diagnosis and treatment. J Laryngol Otol 1997; 111:368–371.
23. Watters GWR, Brookes GB. Chondrosarcoma of the temporal bone. Clin Otolaryngol 1995; 20:53–58.
24. Coltrera MD, Googe PB, Harrist TJ et al. Chondrosarcoma of the temporal bone. Diagnosis and treatment of 13 cases and review of the literature. Cancer 1986; 58:2689–2696.
25. Siedman MD, Nichols RD, Raju UB et al. Extracranial skull base chondrosarcoma. Ear Nose Throat J 1989; 68:626–632.
26. Koch BB, Karnell LH, Hoffman HT et al. National Cancer Database Report on chondrosarcoma of the head and neck. Head Neck 2000; 22:408–425.
27. Kveton JF, Brackmann DE, Glasscock ME et al. Chondrosarcoma of the skull base. Otolaryngol Head Neck Surg 1986; 94:23–31.
28. Rosenberg AE, Nielson GP, Keel SB et al. Chondrosarcoma of the base of the skull: a clinicopathologic study of 200 Cases with emphasis on its distinction from chordoma. Am J Surg Pathol 1999; 23:1370–1378.
29. Debus J, Schulz-Ertner D, Schad L et al. Stereotactic fractionated radiotherapy for chordomas and chondrosarcomas of the skull base. Int J Radiat Oncol Biol Phys 2000; 47:591–596.
30. Brown E, Hug EB, Weber AL. Chondrosarcoma of the skull base. Neuroimaging Clin North Am 1994; 4:529–541.
31. Burkey BB, Hoffman HT, Baker SR et al. Chondrosarcoma of the Head and Neck. Laryngoscope 1990; 100:1301–1305.
32. Finn DG, Goepfert H, Batsakis JG. Chondrosarcoma of the head and neck. Laryngoscope 1984; 94:1539–1544.

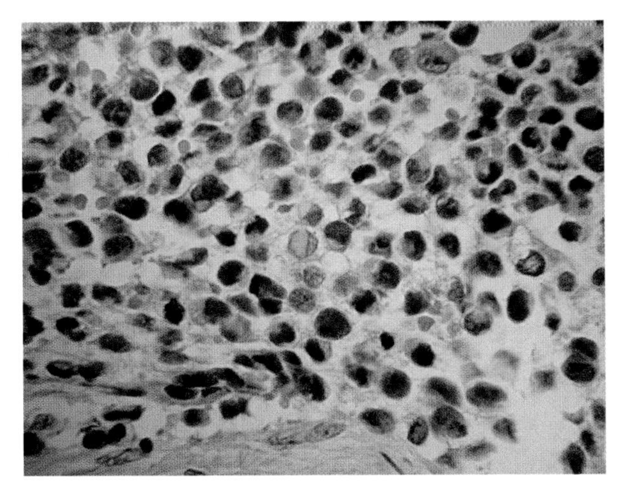

Figure 22.1 Rhabdomyosarcoma, embryonal 40× (400 magnification). Embryonal rhabdomyosarcomas are composed of small, oval to elongated cells arranged in loosely cohesive sheets. The cells can have varying amounts of eosinophilic cytoplasm that displace the nuclei.

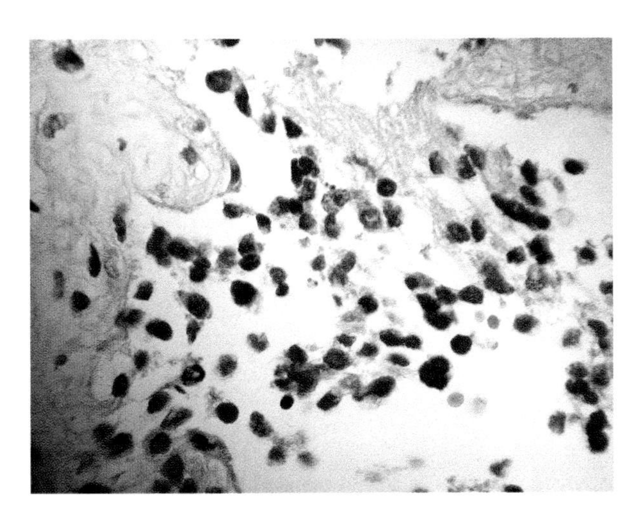

Figure 22.2 Rhabdomyosarcoma, alveolar 40× (400 magnification). Alveolar rhabdomyosarcomas are high-grade sarcomas composed of small, oval cells growing in nests or clusters separated by delicate fibrous septa. Discohesive cells are seen in the center of the nests while the periphery is characterized by the lepidic attachment of tumor cells simulating the morphology of pulmonary air spaces.

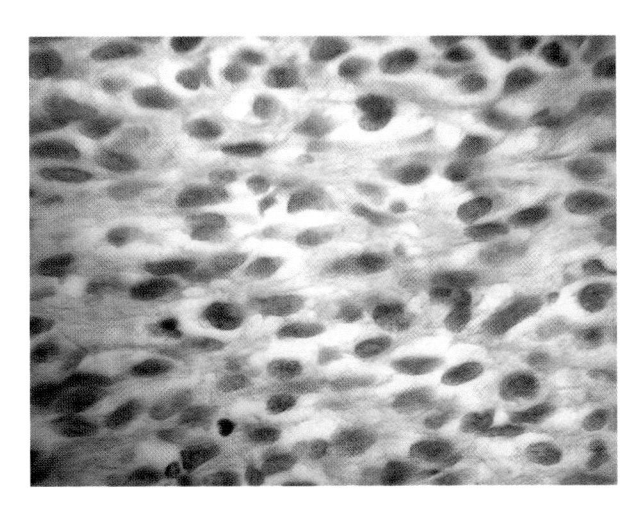

Figure 22.3 Rhabdomyosarcoma, undifferentiated 40× (400 magnification). The undifferentiated subtype shows no specific features. The growth pattern is closely packed and may be fascicular. Cells may have a moderate amount of cytoplasm and are typically larger than a mature lymphocyte. Nuclei tend to the large end of the spectrum and contain finely granular chromatin and prominent nucleoli.

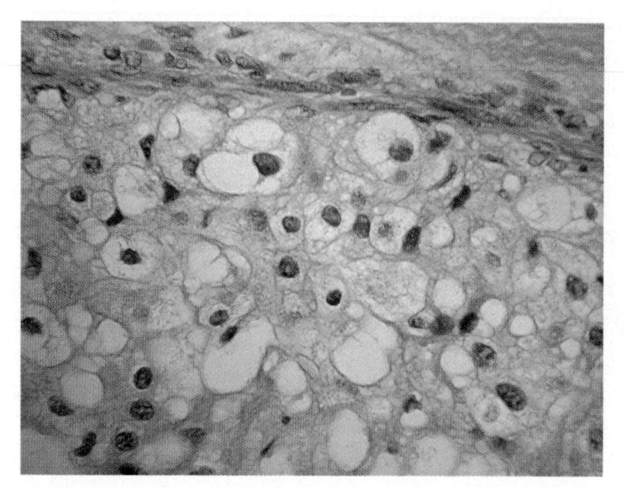

Figure 22.4 Chondrosarcoma 40× (400 magnification). This conventional chondrosarcoma consists of neoplastic cells within lacunar spaces embedded in a hyaline matrix. Nuclei are large, sometimes binucleate, and have prominent nucleoli. Mitotic figures are rare in low-grade tumors and are readily identified in high-grade lesions.

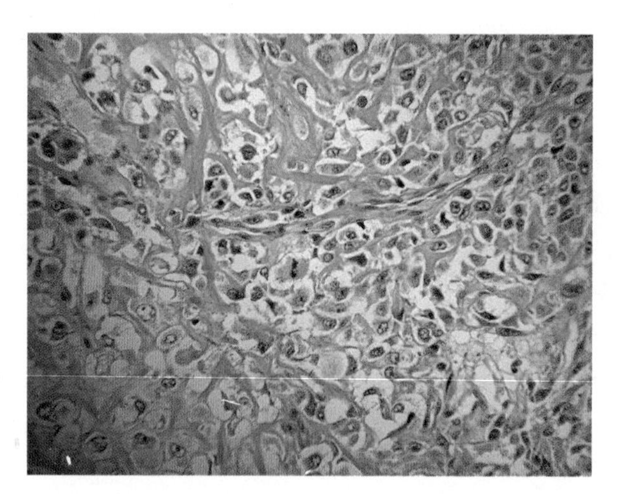

Figure 22.5 Osteosarcoma 20× (200 magnification). The production of osteoid defines an intramedullary tumor as osteosarcoma. This definition holds regardless of histologic pattern. Cell size and shape can vary widely. Nuclei are typical large, pleomorphic and contain one or more prominent nucleoli. The chromatin pattern is usually coarse and clumped.

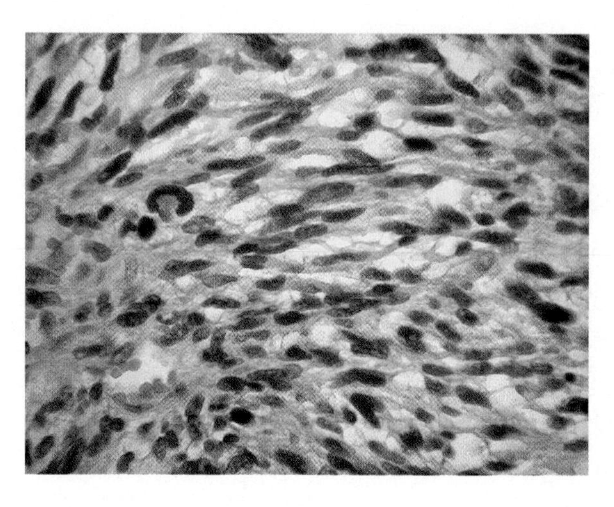

Figure 22.6 Malignant fibrous histiocytoma 40× (400 magnification). Malignant fibrous histiocytomas can show any number of different morphologic patterns but all consist of varying proportions of myofibroblasts, histiocytoid cells, and fibroblasts. Typically the cells are arranged in random bundles with an intersecting fascicular pattern.

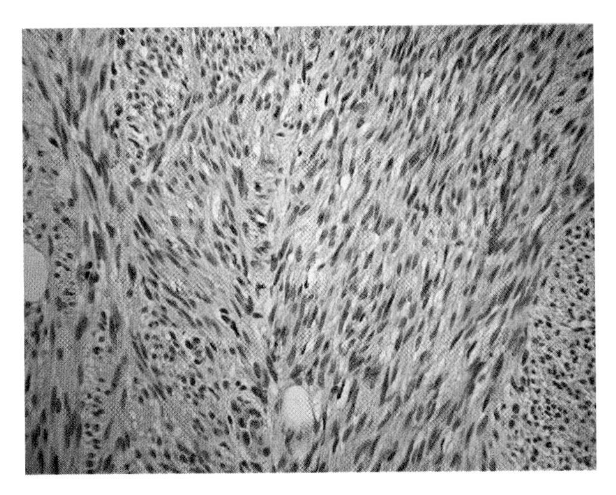

Figure 22.7 Fibrosarcoma 10× (100 magnification). Fibrosarcomas are composed of slender spindle-shaped cells oriented in interweaving bundles. This interlacing arrangement gives rise to the term herring-bone pattern and is associated with lower grade tumors. Nuclear pleormorphism is minimal.

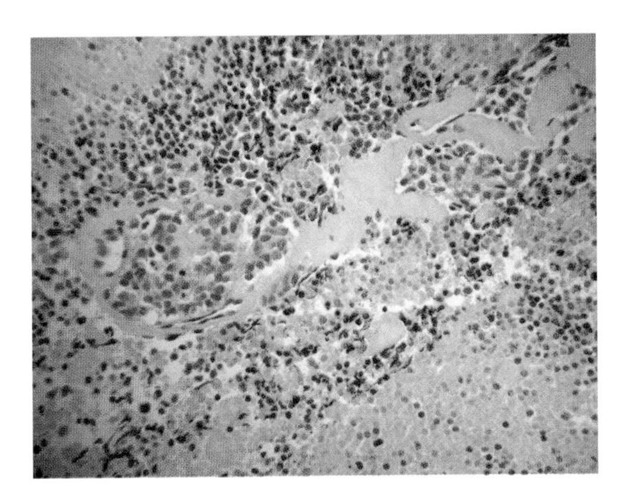

Figure 22.8 Ewing' sarcoma 20× (200 magnification). Tumors cells are arranged in sheets and lobules with broad areas of necrosis. Individual cells have a small amount of amphophilic cytoplasm and contain well-defined, oval nuclei with finely dispersed chromatin and small nucleoli. Intracellular glycogen can be demonstrated by PAS stain.

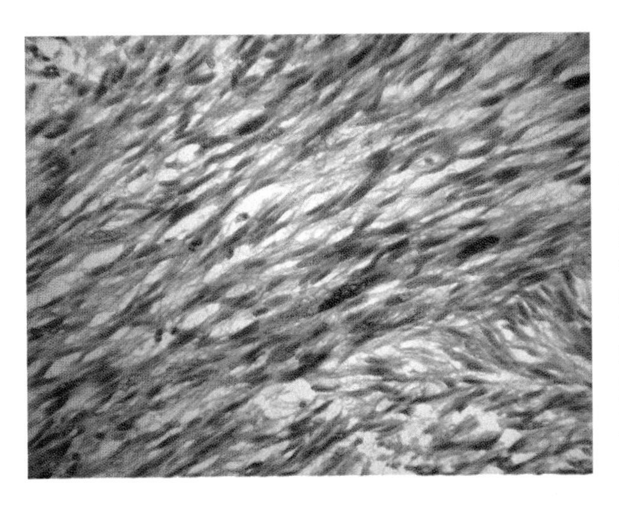

Figure 22.9 Neurogenic Sarcoma 40× (400 magnification). Neurogenic sarcomas are composed of tightly packed spindle cells arranged in interlacing bundles. Cell nuclei are large and can be pleomorphic. Often, there is a suggestion of nuclear palisading. Higher grade tumors are associated with increasing cellularity, nuclear atypia, and mitoses.

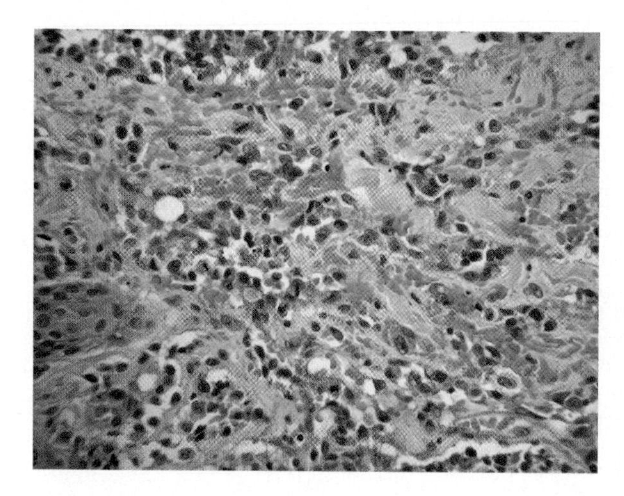

Figure 22.10 Angiosarcoma 40× (400 magnification). These vasoformative tumors are composed of malignant endothelial cells that arrange themselves in complex anastomosing channels. The cellular nuclei are plump and tend to heap-up and project into the lumens. Mitoses are common and extravasated red cells are seen in the stroma.

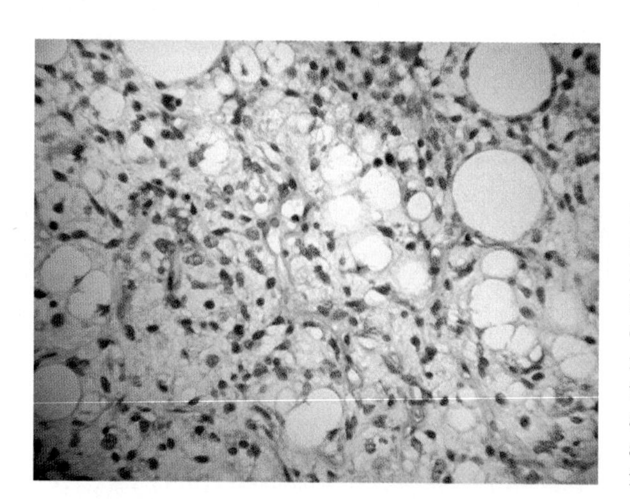

Figure 22.11 Liposarcoma 20× (200 magnification). Malignant lipoblasts are the hallmark of this lesion. They appear as vacuolated cells with eccentrically displaced nuclei, resembling altered adipose tissue. Cell nuclei are large, variably pleomorphic, and may appear in the form of rings or florets.

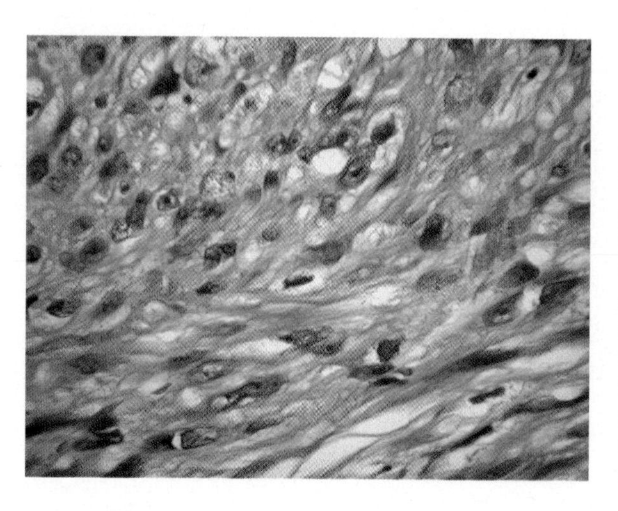

Figure 22.12 Leiomyosarcoma 20× (200 magnification). Leiomyosarcomas consist of intersecting bundles of spindle-shaped smooth cells. Nuclei are cigar-shaped and have blunt ends. The neoplastic cells contain a moderate degree of eosinophilic cytoplasm that appears fibrillar in the longitudinal axis. When cut on end, tumor cells appear to have perinuclear halos.

33. Evans HL, Ayala AG, Romsdahl MM. Prognostic factors in chondrosarcoma of bone: a clinicopathologic analysis with emphasis on histologic grading. Cancer 1977; 40:818–831.
34. Harwood AR, Krajbich JI, Fornasier VL. Radiotherapy of chondrosarcoma of bone. Cancer 1980; 45:2769–2777.
35. Wojno KJ, Hruban RH, Garin-Chesa P et al. Chondroid chordomas and low-grade chondrosarcomas of the craniospinal axis: an immunohistochemical analysis of 17 cases. Am J Surg Path 1992; 16:1144–1152.
36. Charabi S, Engel P, Bonding P. Myxoid tumours in the temporal bone. J Laryngol Otol 1989; 103:1206–1209.
37. Al-Mefty O, Fox JL, Rifai A et al. A combined infratemporal and posterior fossa approach for the removal of giant glomus tumors and chondrosarcomas. Surg Neurol 1987; 28:423–431.
38. Suit HD, Goitein M, Munzenrider J et al. Definitive radiation therapy for chordoma and chondrosarcoma of base of skull and cervical spine. J Neurosurg 1982; 56:377–385.
39. Austin-Seymour M, Munzenrider J, Goitein M et al. Fractionated proton radiation therapy of chordoma and low-grade chondrosarcoma of the base of the skull. J Neurosurg 1989; 70:13–17.
40. Muthukumar N, Kondziolka D, Lunsford LD et al. Stereotactic radiosurgery for chordoma and chondrosarcoma: further experiences. Int J Radiat Oncol Biol Phys 1998; 41:387–392.
41. Mark RJ, Sercarz JA, Tran L et al. Osteogenic sarcoma of the head and neck. The UCLA experience. Arch Otolaryngol Head Neck Surg 1991; 117:761–766.
42. Sharma SC, Handa KK, Panda N et al. Osteogenic sarcoma of the temporal bone. Am J Otolaryngol 1997; 18(3):220–223.
43. Sataloff RT, Myers DL, Spiegal JR et al. Total temporal bone resection for osteogenic sarcoma. ENT J 1998; 67:626–627, 630–632, 634.
44. Goh YH, Chong VFH, Low WK. Temporal bone tumours in patients irradiated for nasopharyngeal neoplasm. J Laryngol Otol 1999; 113:222–228.
45. Nora FE, Unni KK, Pritchard DJ et al. Osteosarcoma of extragnathic craniofacial bones. Mayo Clin Proc 1983; 58:268–272.
46. Kassir RR, Rassekh H, Kinsella JB et al. Osteosarcoma of the head and neck: meta-analysis of nonrandomized studies. Laryngoscope 1997; 107:56–61.
47. Hazarika P, Nayak DR, Sahota JS et al. Osteogenic sarcoma of sphenoid bone: an extended lateral skull base approach. J Laryngol Otol 1995; 109:1101–1104.
48. Garrington GE, Scofield HH, Cornyn J et al. Osteosarcoma of the Jaws. Analysis of 56 cases. Cancer 1967; 20(3):377–391.
49. Caron AS, Hadju SI, Strong EW. Osteogenic sarcoma of the facial and cranial bones: a review of forty-three cases. Am J Surg 1971; 122:719–725.
50. Chambers RG, Mahoney WD. Osteogenic sarcoma of the mandible: current management. Am Surg 1970; 36:463–471.
51. Link M, Goorin A, Miser A et al. The role of adjuvant chemotherapy in the treatment of osteosarcoma of the extremities: preliminary results of the multi-institutional osteosarcoma study. Proc Am Soc Clin Oncol 1985; 4:237.
52. Eibler F, Giuliano J, Eckardt J et al. Adjuvant chemotherapy for osteosarcoma: a randomized prospective trial. J Clin Oncol 1987; 5(1):216.
53. Castro JR, Linstadt DE, Bahary JP et al. Experience in charged particle irradiation of tumors of the skull base: 1977–1992. Int J Radiat Oncol Biol Phys 1994; 29(4):647–655.
54. Nakayama K, Nemoto Y, Inoue Y et al. Malignant fibrous histiocytoma of the temporal bone with endocranial extension. AJNR Am J Neuroradiol 1997; 18:331–334.
55. Blitzer A, Lawson W, Zak FG et al. Clinical–Pathological determinants in prognosis of fibrous histiocytomas of head and neck. Laryngoscope 1981; 91:2053–2069.
56. Raney RB, Allen A, O'Neill J et al. Malignant fibrous histiocytoma of soft tissue in childhood. Cancer 1986; 57:2198–2201.
57. Frankenthaler R, Ayala AG, Hartwick RW et al. Fibrosarcoma of the head and neck. Laryngoscope 1990; 100(8):799–802.
58. Pope EM. Fibrosarcoma of dura, middle fossa, radical mastoid operation. Case report. Laryngoscope 1931; 41:555–556.

59. Apte NK. Fibrosarcoma of the temporal bone. J Laryngol Otol 1965; 79(12):1101–1103.
60. Conley J, Stout AP, Healey WV. Clinicopathologic analysis of eighty-four patients with an original diagnosis of fibrosarcoma of the head and neck. Am J Surg 1967; 114:564–569.
61. Freedman AM, Reiman HM, Woods JE. Soft-tissue sarcomas of the head and neck. Am J Surg 1989; 158(4):367–372.
62. Fitzer PM, Steffey WR. Brain and bone scans in primary Ewing's sarcoma of the petrous bone. J Neurosurg 1976; 44:608–612.
63. Davidson MJC. Ewing's sarcoma of the temporal bone. A case report. Oral Surg Oral Med Oral Pathol 1991; 72:534–536.
64. Desai K, Goel A, Nadkarni TD. Primary petrous bone Ewing's sarcoma. Br J Neurosurg 2000; 14(2):143–145.
65. Hadfield MG, Luo VY, Williams RL et al. Ewing's sarcoma of the skull in an infant. A case report and review. Pediatr Neurosurg 1996; 25:100–104.
66. Desai KI, Nadkarni TD, Goel A et al. Primary Ewing's sarcoma of the cranium. Neurosurgery 2000; 46(1):62–69.
67. Wantanabe H, Tsubokawa T, Katayama Y et al. Primary Ewing's sarcoma of the temporal bone. Surg Neurol 1992; 37:54–58.
68. Sharma RR, Netalkar A, Lad SD. Primary Ewing's sarcoma of the greater wing of the sphenoid bone. Br J Neurosurg 2000; 14(1):53–57.
69. Horowitz ME, Malawer MM, Woo SY et al. Ewing's sarcoma family of tumors: Ewing's sarcoma of bone and soft tissue and the peripheral neuroectodermal tumors. In: Pizzo PA, Poplock DG, eds. Principles and Practice of Pediatric Oncology. 3rd ed. Philadelphia, PA: Lippincott-Raven Publishers, 1997:831–863.
70. Hellquist HB, Lundgren J. Nuerogenic sarcoma of the sinonasal tract. J Laryngol Otol 1991; 105:186–190.
71. Hutcherson RW, Jenkins HA, Canalis RF et al. Nuerogenic sarcoma of the head and neck. Arch Otolaryngol 1979; 105:267–270.
72. Greager JA, Reichard KW, Campana JP et al. Malignant schwannoma of the head and neck. Am J Surg 1992; 163:440–442.
73. Kettel K. Intratemporal sarcoma of the facial nerve. Arch Otolaryngol 1950; 52:778–781.
74. Guttman MR, Simon MU. Neurofibrosarcoma of the facial nerve involving the tympanomastoid. Arch Otolaryngol 1951; 54:162–166.
75. D'Agostino AN, Soule EH, Miller RH. Primary malignant neoplasms of nerve (malignant neurolemmomas) in patients without manifestations of multiple neurofibromatosis (von Recklinghausen's disease). Cancer 1963; 16:1003–1014.
76. Goepfert H, Lindberg RD, Sinkovics JG. Soft-tissue sarcoma of the head and neck after puberty. Arch Otolaryngol 1977; 103(6):365–368.
77. Mark RJ, Tran LM, Sercarz J et al. Angiosarcoma of the head and neck. The UCLA experience 1955 through 1990. Arch Otolaryngol Head Neck Surg 1993; 119:973–978.
78. Kinkade JM. Angiosarcoma of the petrous portion of the temporal bone. Ann Otol Rhinol Laryngol 1948; 57:235–240.
79. Aust MR, Olsen KD, Lewis JE et al. Angiosarcomas of the head and neck: clinical and pathological characteristics. Ann Otol Rhinol Laryngol 1997; 106:943–951.
80. Linthicum FH, Schwartzman JA. An Atlas of Micropathology of the Temporal Bone. San Diego, CA: Singular Publishing Group, Inc., 1994:78–79.
81. Zbaren P, Ruchti C. Leiomyosarcoma of the middle ear and temporal bone. Ann Otol Rhinol Laryngol 1994; 103:537–541.

23
Hearing in Dogs

D. Caroline Coile
Georgia Pines Station, Ochlocknee, Georgia, USA

The domestic dog serves in a variety of roles, including sentry, patrol, herding, hunting, and guide dog functions, all of which require acute hearing for optimal performance. A popular subject of early research in comparative hearing, the dog eventually fell from favor as a research subject, in part because of its lack of uniformity. Yet, it is exactly this diversity that makes the dog a particularly intriguing subject in which to examine intraspecies relationships in head size and pinna configuration. In addition, the presence of hereditary (particularly pigment related) deafness in dogs has both basic and applied implications. Despite such a wealth of incentives, a paucity of information about canine hearing exists; this report is a synopsis of that information.

1. INTERAURAL DISTANCE

Dogs and humans have similar low-frequency hearing thresholds, but the frequencies to which dogs are maximally sensitive are a little higher (~8000 Hz) than the maximal sensitivity in humans (~3000 Hz). The dog's high-frequency threshold is ~47,000 Hz (1).

As a rule of thumb, smaller species of land mammals can hear higher frequencies compared with the frequencies larger species can hear; this difference has been conjectured to be an adaptation for sound localization in smaller animals. The wider the ears are separated, the easier it is to use differences in arrival time of sounds to either ear as a localization cue. Sound shadowing from the animal's own head provides another important localization cue; the smaller the sound waves are in comparison to the head width, the greater this effect. This means that an animal with a narrower head must be able to perceive higher frequency sound in order to use this cue. The trend toward higher frequency hearing in smaller animals could be an adaptation enabling the animal to use high-frequency sound for localization cues despite its small head size. These interspecies observations might suggest that dogs with small or narrow heads should have either poorer sound localization or better high-frequency hearing than dogs with larger, wider heads.

Although there are no comparison studies between breeds in sound localization ability, in a comparison of the hearing thresholds of four dogs representing four breeds (Chihuahua, Dachshund, Poodle, and Saint Bernard) of different sizes, no differences in high- or low-frequency hearing were found. All had maximal sensitivity at ~8000 Hz, with a high frequency cut-off ~41,000–47,000 Hz—a far smaller difference than would be predicted by the variation in head size. When compared with other species of comparable sizes, small dogs have a slightly better ability to hear high frequencies, and large dogs have a much better ability to hear high frequencies, than would be expected on the basis of their size (1).

Silent dog whistles, ultrasonic pest deterrent devices, and ultrasonic dog training aids are all based upon the observation that different species have different high-frequency reception. Rodent and insect ultrasonic deterrents emit frequencies above the hearing range of humans, typically ranging from 25,000 to 65,000 kHz. Manufacturers claim that dogs and cats also cannot hear the high frequencies emitted by these devices—but they can. Dogs can hear frequencies as high as ~47,000 Hz, and cats can hear even higher frequencies, up to ~85,000 Hz (2). They are thus subjected to a barrage of high-frequency sound emitted from most of these devices. Ironically, objective testing of ultrasonic pest devices demonstrate they are ineffective on fleas (3).

2. OUTER EAR

Dog pinnae vary greatly in design. As if the natural variation in pinna conformation was not enough, humans have cropped dogs' ears for centuries, with various justifications. For example, breeds originated for fighting had cropped ears in order to reduce the opportunity for an opponent to bite their ears. Besides cropping simply to conform with style and breed standards, the most common modern reason is the belief that dogs with erect ears are less prone to ear infections. Dogs with pendulous ears that also have heavy hair growth in the ear canal do have significantly higher rates of otits externa than dogs with other ear types, but several other factors are also involved (4,5). Of all the reasons put forth for cropping, the most intriguing is the idea that it increases hearing ability in dogs that have naturally pendulous ears.

It is true that a hanging ear flap cannot function in the same manner as an erect ear. The erect ear can swivel to direct its opening toward a sound source. Its configuration directs the sound waves into the ear canal, and additionally, amplifies them slightly. In cats, the pinna produces up to 28 dB amplification of sounds (especially high-frequency sounds) emitted within the pinna's optimal axis of reception directly in front of its cup (6). The naturally erect dog ear probably has a comparable amplification ability that is not available to dogs with other ear configurations.

The hanging pinna also imposes a barrier between the sound waves and the ear canal, which could conceivably dampen sounds. Despite the rich array of dog ears waiting to be tested, only one study has addressed the effects of pinna configuration on hearing thresholds. In a comparison of four dogs, one of which had erect pinnae, and three of which had hanging pinnae, no significant differences in high- or low-frequency hearing were detected. When one dog with hanging ears was retested with its pinnae taped over its head, no differences in its low-frequency thresholds and only minor differences (>3 dB) in its high-frequency threshold were found (1). Taping the dog's ear back does not give the pinna the same acoustic powers that a normally erect pinna has; however, neither does ear cropping. A cropped ear is missing a good portion of the natural bell shape that a naturally erect ear has; it is thus questionable that cropping ears affords any worthwhile gain in auditory sensitivity.

The pinna does play a large role in sound localization, however, especially in discriminating front versus back and high vs. low sources (7). The asymmetry of the entire pinna, coupled with the various folds within it, transforms sound waves differently depending upon the angle at which they impinge upon the pinna, providing further cues for sound localization. Dogs with hanging ears may thus be compromised compared to erect-eared dogs at locating the source of a sound in the high vs. low dimension or front vs. back dimension, but should not be compromised at left vs. right localization. A mobile pinna would seem to be an asset in localization, but the movement of the pinna is too slow to aid in localization of brief sounds.

The pinna funnels the sound waves into the auditory meatus, which differs from the human canal because it initially travels downward and then turns abruptly inward. The horizontal section can be difficult to treat in cases of otitis externa. Chronic otitis externa can sometimes result in impaction or occlusion of the ear canal, and ultimately, ossification of the canal, all circumstances that can cause conductive deafness. Surgical procedures can often re-establish hearing in such dogs (8).

3. MIDDLE EAR

Chronic ear disease or trauma can also cause rupture of the tympanic membrane. Although such rupture undoubtedly decreases hearing sensitivity, dogs with ruptured eardrums can still hear. The intact tympanic membrane vibrates in a complex pattern in response to sounds. At low frequencies the entire membrane vibrates, but at higher frequencies different parts of the membrane vibrate somewhat independently of one another. Differences in the size of the tympanic membrane between species have long been thought to affect auditory response properties. The area of the tympanic membrane varies greatly in dogs, from 30 mm^2 in the Chihuahua to 55 mm^2 in the Saint Bernard (1). Despite this large difference, no differences in hearing ability between these breeds have been documented.

Vibration of the tympanic membrane causes a leverlike action of the ossicles of the middle ear; this, combined with the concentration of energy onto the tiny oval window, results in an increase in sound pressure level that approximately counteracts the loss of energy that would otherwise occur when going from the air-filled middle ear to the fluid-filled cochlea. Decreased articulation of the ossicles may contribute to hearing loss in older dogs (presbycusis), but is probably a comparatively minor factor.

The stapedius and tensor tympanus muscles are attached to the ossicles and serve to stiffen the tympanic membrane and retract the stapes from its normal position in the oval window in response to loud noise. These actions decrease the sound amplitude by as much as 30 dB in humans, and probably at least that much in dogs. This acoustic reflex, which has a threshold of 70–110 dB in dogs (9), protects the ear from damage from continued loud noise, though it cannot protect it from sudden loud noise or from continued extreme loud noise.

There is no published research of noise-induced hearing loss for dogs, although there are anecdotal reports of hunting dogs in close proximity to gunfire eventually incurring hearing loss. Dogs housed in kennels are often subjected to almost continuous high levels of noise from barking, a situation compounded by the lack of acoustic insulation in most kennel buildings. Dogs shipped by air are often subjected to high noise levels by airline personnel who place dogs along with baggage on the tarmac next to jet engines. Dogs employed as airline bomb or contraband detectors may have to work around jet engines but can be trained to wear ear protection.

4. INNER EAR

Movement of the ossicles produces movement of fluid in the cochlea via the oval window. Within the cochlea, the organ of Corti contains receptor cells called hair cells, each of which has many stereocilia that extend to and are embedded in another membrane, the tectorial membrane. Movement of cochlear fluid produces movement of the basilar membrane upon which the hair cell stereocilia of the organ of Corti are embedded; because the tectorial membrane remains stationary, a shearing force upon the hair cells is created. It is this shearing force that results in a nerve impulse from the hair cell. With repeated or sustained exposure to loud noise, the hair cells become damaged or disarranged, resulting in hearing loss.

Hair cells can also be damaged by certain drugs, many of which are routinely prescribed in veterinary practice. The most common of these ototoxic drugs are the aminoglycoside antibiotics, which include the frequently prescribed gentamicin and neomycin (10). Routine administration of these drugs seldom causes loss of hearing, but prolonged systemic administration, or topical administration in dogs with ruptured tympanic membranes, can result in irreversible hearing loss.

Hereditary deafness is known in many breeds of dogs. The Dalmatian is the best known example, with from 20% to 30% deafness in either one or both ears (11,12). Affected Dalmatians are born with normal auditory structure, but soon after birth the stria vascularis (the vascular bed of the cochlear duct) degenerates. The cochlear duct in turn collapses and the hair cells degenerate, so that these dogs are deaf by ~1 month of age (13). In some other breeds, deafness results from degeneration of hair cells without any circulatory involvement (14).

In dogs with normal hearing and hair cells, the hair cells send their signals to the brain by way of the auditory nerve fibers that leave the organ of Corti via the spiral ganglion. Acquired hearing loss in old dogs is often associated with loss of cells in the spiral ganglion (15). This structure also degenerates in the deaf Dalmatian, possibly secondarily to the loss of hair cells. Normal auditory nerve fibers leave the ear and travel to the brain, with most entering the brainstem, where they make contact with cells in the cochlear nucleus.

Because the auditory signals must travel by way of the brainstem, recordings of the electrical activity of the brainstem in response to auditory stimuli can indicate whether a dog's ear is functioning normally. The most widespread and useful test of hearing in dogs is the brainstem auditory-evoked response (BAER). This test is conducted in awake dogs by applying electrodes just under the skin of the head, and directing a series of click sounds into each ear separately by means of an earphone or head phone. The normal BAER response consists of a series of four waves, the first of which represents activity of the cochlea and cochlear nerve, whereas the later peaks represent activity within the brain. In the deaf Dalmatian (and other breeds with deafness originating in the ear) all of these peaks are absent. The BAER can be used to distinguish between cochlear deafness and conduction deafness by introducing the sound stimuli as a vibration held to the skull (16). This method bypasses the ear canal, tympanic membrane, and ossicles; if the BAER is still abnormal, the location of the problem is pinpointed to the cochlea or auditory nerve.

In dogs with normal hearing, the signals travel from the brainstem to several other structures of the brain, including the superior olivary complex, lateral lemniscus, inferior colliculus, and medial geniculate nucleus, all of which process the auditory information before sending it on, until it ultimately arrives in the auditory cortex. In the absence of auditory stimulation, the auditory cortex will deteriorate, an observation that led some early researchers to ascribe the hereditary deafness in the Dalmatian to brain abnormalities. These abnormalities are secondary to the initial cochlear losses, however, and do not suggest that deaf Dalmatians have abnormal brains. In fact, deafness in dogs due to brain abnormalities is rare, and in such cases usually the least of the dog's problems.

5. PIGMENT-RELATED DEAFNESS

The most common cause of deafness in young dogs is hereditary pigment-related deafness, such as that associated with genes responsible for lack of pigmentation in the coat, iris, and other body parts. The most prominent of these genes are the merle (M) and extreme piebald (s^w) alleles.

In merle dogs, the presence of one M allele results in the merle coat coloration (a patchwork of dark and light coat, such as that often seen in the Catahoula Leopard Dog and Australian Shepherd). The incidence of deafness in merle dogs is higher than that in nonmerle dogs of the same breed. The presence of two M alleles results in an almost all white dog, often with blue eyes. These dogs are very often deaf and often have visual anomalies.

In extreme piebald dogs, the presence of two s^w alleles results in a predominantly white dog with spots (such as the pattern seen in Dalmatians and English Setters). Dalmatians with a patch (which is a large pigmented area present at birth, rather than spots, that are not present at birth and are caused by genes at another locus) are less likely than those without a patch to be deaf (17). Unfortunately, from a dog show competition point of view patches on Dalmatians are undesirable. Dalmatians with one or two blue eyes, or with pigment missing from the tapetum of either eye, are also more likely to be deaf than those with brown eyes or fully pigmented tapeta. These observations point to an association between expressivity of the s^w allele and deafness, but the genetics of deafness in piebald breeds is far from clear. Dogs can be either unilaterally or bilaterally deaf; partial deafness within one ear is not seen. Offspring of normally hearing parents can be deaf, and offspring of bilaterally deaf parents can have normal hearing. In general, however, dogs with either parent either unilaterally or bilaterally deaf are more likely to be deaf than dogs with parents having a total of four good ears.

This association with coat color is more than just coincidental; animals with alleles that result not only in absence of coat pigmentation but in missing pigment on other body structures may also be lacking normal pigment on internal body structures. For example, normal blood vessels have pigment producing cells (melanocytes) within them; it is likely that blood vessels missing melanocytes do not develop normally. Melanocytes in the stria vascularis seem to be suppressed by the action of the merle or extreme piebald genes. The melanocytes may be responsible for maintaining potassium levels in endocochlear fluid, as well as for the maintenance of the cochlear duct (18,19). In addition, pigment molecules are thought to play a role in guiding some neural connections during development; abnormal connections are seen in association with some other alleles resulting in reduced pigment, such as those in the albino series (20). Not all white dogs are the result of such alleles; those with the greatest likelihood of having deafness associated with white or merle coloration seem to be those in whom coat color is often associated with absence of normal ocular pigment.

Nor is all hereditary deafness in dogs pigment related. Deafness in Doberman Pinschers, for instance, results from primary degeneration of the cochlear hair cells (without vascular involvement) and is not associated with merle or piebald genes. This type of deafness, which also affects the sense of balance, is caused by a single recessive allele (14).

Dogs suspected of or at risk of being deaf can have a BAER test performed at most veterinary teaching hospitals and several specialty clinics (21). Preliminary screening tests can be conducted in the home or veterinary clinic, but the results may not be very accurate. Saying a word to which the dog normally responds when the dog's attention is elsewhere and when the dog cannot see the person speaking is a simple preliminary test. Ringing a bell or sounding a novel stimulus from behind the dog should result in either the dog's attention (as evidenced by head turning and ear twitching) or a startle reaction. Dogs that are panting will characteristically stop panting momentarily in order to listen, so this can be used as a tentative sign of hearing. Testers must be careful that no wind currents from clapping hands or vibrations through the floor reach the dog, and that the dog cannot see the tester out of the corner of its eye (bearing in mind that dogs have a greater peripheral field of vision than humans), or cannot see the tester's shadow or reflection. Other dogs that could provide cues by their response should be removed from the area. Dogs habituate to auditory stimuli rapidly; repeating the test will often result in lack of responsiveness. These tests can generally detect a dog that is bilaterally deaf. A dog with unilateral deafness will still respond to words and sounds in a manner almost indistinguishable from a dog with normal hearing. It will swivel both pinnae in response to a sound, even though one pinna may be leading to a totally deaf ear. A dog with unilateral hearing loss may have some difficulty in localizing sounds in the left vs. right dimension, but this can be difficult to test without more sophisticated techniques.

Unilaterally deaf dogs make excellent companions and their owners often are never aware that the dog has any loss of hearing. Bilaterally deaf dogs present a more challenging, and controversial, situation. These dogs are at greater risk of being killed or injured in an accident because they cannot hear their owner's calls or cannot hear approaching danger. They may be more likely to snap because they can be easily startled by the sudden appearance of a dog or person. They can be difficult to train because special ways of first having the dog watch the owner must be taught, and then visual or tactile signals must be used. Owners who do not realize their dog is deaf too often conclude it is stupid or stubborn and may abuse or abandon it. Many people have raised deaf dogs successfully, however. They use hand signals, flashlights, vibrating collars, and even other dogs to help them communicate with their deaf pets. Nonetheless, living with a totally deaf dog presents hardships not encountered when living with a dog that can hear. Owners of deaf dogs can find advice through several sources (21,22).

REFERENCES

1. Heffner HE. Hearing in large and small dogs: absolute thresholds and size of the tympanic membrane. Behav Neurosci 1983; 97:310–318.
2. Heffner RS, Heffner HE. Hearing range of the domestic cat. Hear Res 1985; 19:85–88.

3. Hinkle NC, Koehler PG, Patterson RS. Egg production, larval development, and adult longevity of cat fleas (Siphonaptera: Pulicidae) exposed to ultrasound. J Econ Entomol 1990; 83:2306–2309.
4. Hayes HM, Pickle LW, Wilson GP. Effects of ear type and weather on the hospital prevalence of canine otitis externa. Res Vet Sci 1987; 52:294–298.
5. August JR. Otitis externa: a disease of multifactorial etiology. Vet Clin North Am Small Anim Pract 1988; 18:731–742.
6. Phillips DP, Calford B, Pettigrew JD, Aitkin LM, Semple MN. Directionality of sound pressure transformation at the cat's pinna. Hear Res 1982; 8:13–28.
7. Heffner RS, Koay G, Heffner HE. Sound localization in chinchillas III: effect of pinna removal. Hear Res 1996; 99:13–21.
8. Krahwinkel DJ, Pardo AD, Sims MH, Bubb WJ. Effect of total ablation of the external auditory meatus and bulla osteotomy on auditory function in dogs. J Am Vet Med Assoc 1993; 202:949–952.
9. Sims MH, Weigel JP, Moore RE. Effect of tenotomy of the tensor tympani muscle on the acoustic reflex in dogs. Am J Vet Res 1986; 47:1022–1031.
10. Merchant SR. Ototoxicity. Vet Clin North Am Small Anim Pract 1994; 24:971–980.
11. Wood JLN, Lakhani KH. Prevalence and prevention of deafness in the Dalmatian: assessing the effect of parental hearing status and gender using ordinary logistic and generalized random litter effect models. Br Vet J 1997; 154:121–133.
12. Strain G. Congenital deafness and its recognition. Vet Clin North Am Small Anim Pract 1999; 29:895–901.
13. Strain GM. Aetiology, prevalence and diagnosis of deafness in dogs and cats. Br Vet J 1996; 152:17–36.
14. Wilkes MK, Palmer AC. Congenital deafness and vestibular deficit in the doberman. J Small Anim Pract 1992; 33:218–224.
15. Knowles K, Blauch B, Leipold H, Cash W, Hewitt J. Reduction of spiral ganglion neurons in the aging canine with hearing loss. J Vet Med 1989; 36:188–199.
16. Strain GM, Green KD, Twedt AC, Tedford BL. Brain stem auditory evoked potentials from bone stimulation in dogs. Am J Vet Res 1993; 54:1817–1821.
17. Strain GM, Kearney MT, Gignac IJ, Levesque DC, Nelson HJ, Tedford BL, Remsen LG. Brainstem auditory evoked potential assessment of congenital deafness in Dalmatians: associations with phenotypic markers. J Vet Intern Med 1993; 6:175–182.
18. Steel KP, Barkway C. Another role for melanocytes: their importance for normal stria vascularis development in the mammalian inner ear. Development 1989; 107:453–463.
19. Cable J, Huszar D, Jaenisch R, Steel KP. Effects of mutations at the W locus (c-kit) on inner ear pigmentation and function in the mouse. Pigment Cell Res 1994; 7:17–32.
20. Witcop CJ, Jay B, Creel D, Guillery RW. Optic and otic neurological abnormalities in oculocutaneous and ocular albinism. In: Cotlier, Maumenee, Berman, eds. Genetic Eye Diseases: Retinitis Pigmentosa and Other Inherited Eye Disorders. Proceedings of the International Symposium on Genetics and Ophthalmology held in Jerusalem, Israel. New York, NY: Alan R. Liss, 1981.
21. Strain G. Deafness in dogs and cats. BAER Test Sites In: Deafness in Dogs and Cats. Available from http://www.lsu.edu/guests/senate/public_html/baersite.htm.
22. Cope-Becker S. Living With a Deaf Dog: A Book of Advice, Facts and Experiences About Canine Deafness. Cincinnati: Cope-Becker S, 1997.

24
Tables Summarizing Differential Diagnosis

Robert T. Sataloff
Drexel University College of Medicine, American Institute for Voice and Ear Research, and Graduate Hospital, Philadelphia, Pennsylvania, USA

Joseph Sataloff
Thomas Jefferson University, Philadelphia, Pennsylvania, USA

The preceding 23 chapters discuss a great many causes of hearing loss, associated symptoms, and diagnostic distinctions. The tables in this chapter help summarize the information so that similarities and differences are apparent at a glance. While they certainly do not cover all the information presented, the tables include most of the common and important conditions discussed so far, highlighting their most important distinguishing features.

Table 24.1 Audiologic Criteria for Classifying Hearing Loss[a]

	Air conduction pattern	Bone conduction pattern[b]	Air-bone gap	Lateralization of 500-Hz fork	Recruitment	Abnormal tone decay	Discrimination
Conductive	Greater low-tone loss, except when fluid is in the ear. Maximum loss is 60–70 dB ANSI	Normal or almost	At least 15 dB	To worse ear	Absent	Absent	Good
Sensory	Greater low-tone loss or high-tone dip	BC = AC	No gap	To better ear with low intensity To worse ear with high intensity	May be marked and continuous	Absent	Poor
Neural	Greater high-tone loss	BC = AC, or BC worse than AC	No gap	To better ear	Absent	Marked tone decay in acoustic neuroma and nerve injuries	Reduced
Sensori-neural	Greater high-tone loss or flat loss	BC = AC	No gap	To better ear	Absent or slight	Absent	Reduced
Functional	Flat	Usually no BC	No gap	Vague	Absent	Variable	Usually good or no response
Central	Variable or even normal threshold	BC = AC, or absent BC	No gap	None	None	Undetermined	Reduced

	Audiometric responses[c]	Tinnitus	Békésy tracings	Impedance audiometry	Patient's voice[c]	Other findings
Conductive	Vague and slow	Absent or low	Overlap of pulsed and continuous tracings	Often abnormal and diagnostic tympanogram	Soft or normal	No diplacusis Hears better in noisy environment
Sensory	Sharp	Low roar or seashell	Pulsed tracings slightly wider at higher frequencies Little or no separation	Normal typanogram Metz recruitment (stapedius reflex at low intensity)	Normal	Diplacusis Hears worse in noisy environment Lowered threshold of discomfort
Neural	Sharp	Hissing or ringing	Separation of tracings in acoustic neuroma	Normal tympanogram Stapedius reflex absent or decayed	Louder	No diplacusis Hears worse in noisy environment
Sensori-neural	Sharp	Hissing or ringing	Slight separation of tracing	Normal tympanogram Other test variable	Louder	No diplacusis Hears worse in noisy environment
Functional	Inconsistent	Absent	Separation of tracing with poorer thres-hold for pulsed tone	Normal	Normal	No diplacusis Hears worse in noisy environment
Central	Slow	None	Undetermined	Normal tympanogram Difference between ipsilateral and contralateral stapedius reflex responses	Normal	No diplacusis Hears poorly in noise Poor integration of complex stimulus

[a]These criteria are the usual ones, but many variations and exceptions are encountered.
[b]BC, bone conduction; AC, air conduction.
[c]These are common findings, but there are many exceptions.

Table 24.2 Causes of Conductive Hearing Loss with Abnormal Findings Originating in External Canal

Diagnosis	History	Onset of hearing loss	Otoscopic findings	Tinnitus	Audiologic and impedance	Special findings
Congenital aplasia	Ear deformed at birth with hearing impairment Unilateral or bilateral	Congenital	Auricular deformity and canal closed	None	Flat hearing loss about 60–70 dB, worse if inner ear is involved	If deformity is bilateral, speech development is impaired
Treacher Collins syndrome	Abnormal findings present at birth bilaterally	Congenital	Auricular deformity and canal closed	None	Flat hearing loss about 60–70 dB	Bilateral deformity: slanted eyes, receding jaw and malar bones
Stenosis	Ear blocked either since birth or following infection, trauma, or surgery to the ear	Congenital or slowly developing	Eardrum not visible due to closure of canal	None	Flat hearing loss about 60 dB Type B	Auricle is normal, but canal is closed uniformly Usually no inflammation is present
Cerumen	Ear blocked after attempting to clean canal or after chewing	Slow or sudden after attempting to clean ear canal	Wax blocking canal and drum not visible	Rarely	Flat loss, about 45 dB Type B	Wax is visible, and hearing returns after wax is removed
Fluid in canal	After swimming or bathing or applying medication in ear	Sudden	Fluid in canal	Occasionally	Mild loss with loss in higher frequencies Bone conduction normal Type B	Eardrum normal after fluid is removed

External otitis	Pain and itching in canal, aggravated by chewing or moving auricle Tenderness No pain on noseblowing	Insidious	Canal wall inflamed and debris present Eardrum intact	None	Flat loss, about 45 dB Type B	Tender canal walls and surrounding areas Hearing improves with removal of debris
Exostosis or osteoma	Either constant blockage or intermittent if small opening opens and closes with wax or debris	Sudden or intermittent	Hillocks of bony projections from canal wall or large bony occlusion	None	Flat loss, about 45 dB Type B	Canal closed by hillocks or mounds of bone from canal wall X-ray films: normal middle ears and external bony projections
Granuloma	Fullness in ears-often painless	Slow	Firm granulation with or without excessive bleeding Eardrum normal if visible Often no inflammation	None	Flat loss, about 45 dB Type B	Often no pain or inflammation Middle ear normal Occasional palpable bony defect in canal Positive biopsy
Cysts	Little or no discomfort but fullness	Slow	Soft mass in canal covered with skin	None	Flat loss, about 45 dB	Drum normal if visualized
Collapse of canal	Hearing loss only during testing	Only with earphones	Relaxed opening to external canal	None	Mild low-tone loss, sometimes apparent high-tone loss	Patient says he hears worse with earphones
Foreign body	Foreign body in ear In children, no clear history	Sudden	Foreign body in ear	None, except live insect in ear	Mild flat loss	Mass in ear not attached to canal wall and not covered with skin

Table 24.3 Conductive Hearing Loss with Abnormal Findings Visible in Tympanic Membrane and Middle Ear

Etiology and principal findings	History	Onset of hearing loss	Tinnitus	Audiologic and impedance	Otosclerotic	Special
Myringitis bullosa (blebs on drum)	Discomfort, fullness in ear, not aggravated by swallowing or chewing and not associated with general malaise	Slow	Slight	Very mild hearing loss of about 25 or 30 dB	Clear or hemorrhagic blebs on drum involving only outer layer	Drum is intact and moves with air pressure through canal or nose
Ruptured eardrum	Severe explosive, slap on ear, or foreign body poked into ear, with sudden pain, hearing loss, and possible bleeding and fullness	Sudden	Sometimes	Flat loss from 40 to 60 dB, sometimes with sensorineural component Type B	Jagged central perforation of drum with no inflammation if seen early	Perforation in drum is jagged and not associated with infection
Perforated drum caused by burns	Spark in ear from welding or exposure to fire	Sudden	None	Flat loss of about 60 dB	Usually complete destruction of drum with little infection	Marked destruction of drum with history of severe pain caused by burn
Dry perforation in drum Anterior or central	Previous otitis media due to adenoid hypertrophy, allergy, or eustachian tube pathology, or following secretory otitis media	Slow	None	Usually less than 40 dB and worse in lower frequencies	Central or anterior perforation No infection and normal middle-ear mucosa Edge of perforation usually smooth and regular	Discharge usually intermittently with colds or water in ears

Superior (Shrapnell's area) or posterior perforation on large portion of drum	Previous otitis media with chronic otorrhea or mastoid infection	Slow	None	Variable from 15 to 60 dB Type B	Posterior or superior perforation	The amount of hearing loss depends on damage to ossicular chain and other pathology in the middle ear
Healed perforation	Previous ear infections	Gradual	None	From minimal to 70-dB loss Type A_S or A_D or normal	Thick scars or transparent closure in drum that looks like perforations	Drum moves with gentle air pressure in canal or through nose; hearing loss depends on damage in middle ear
Hypertrophied adenoids	Intermittent ear blocking and fullness Some mouth breathing	Gradual	None	Maximum loss usually about 40 dB and often worse at higher frequencies Type B or C	Clear fluid level in middle ear or thickened or retracted drum	Large lymph tissue masses in lateral recesses of nasopharynx
Cleft palate	Recurrent otitis in childhood	Gradual	None	As above	Type B or C Opaque, thickened, or retracted drum	Congenital malformation leads to abnormal eustachian tube function
Retracted drum	Stuffiness in ears	Gradual	None	As above	Abnormalities in nasopharynx or eustachian tube, such as allergy, adenoids or neoplasm	Pressure disparity in middle ears and poor eustachian-tube function

(Continued)

Table 24.3 *Continued*

Etiology and principal findings	History	Onset of hearing loss	Tinnitus	Audiologic and impedance	Otosclerotic	Special
Serous otitis media	Stuffiness in ears, feeling of fluid	Gradual	None	As above Type C	Fluid in middle ear, sometimes with evidence of inflammation and associated with upper respiratory infection	Blocked eustachian tube and abnormal findings in nasopharynx or tubes
Acute otitis media	Stuffiness, pain, and fullness in ear, sometimes fever	In several hours	None	Maximum loss usually about 40 dB and often worse at higher frequencies	Inflamed or bulging drum with prominent vessels; absent landmarks	Associated with inflammation in nasopharynx and upper respiratory infection
Secretory otitis media	Generally without upper respiratory infection but may follow it; recurrent fullness; no pain or systemic symptoms	Slowly	None	As above; bone conduction may be slightly reduced Type B	Fluid level or bubbles and straw-colored fluid or even gellike mass	Eustachian tube is patent, and condition recurs Nasal mucosa also secretory
Aerotitis media	Sudden pain and fullness on descending in airplane or elevator	Sudden	None	Mild and usually mostly in higher frequencies Type B or C	Retracted drum with possible fluid level Hearing returns with politzerization or myringotomy	Resolution often is spontaneous, but early myringotomy resolves hearing loss

Chronic otitis media Dry with ossicular damage	Previous otitis media with prolonged otorrhea for many months before cessation; no pain	Gradual	None	Flat loss, 60–70 dB	Large marginal perforation in drum and disruption of ossicular chain by erosion	Usually, end of incus or crura of stapes are eroded X-ray films show sclerosis but no active infection in mastoid
Mucoid discharge	Intermittent otorrhea especially following upper respiratory infection, but dry in between; no pain	Gradual	None	Mild with maximum of 40 dB and mostly in lower frequency Type B	Usually anterior perforation	Associated with eustachian-tube and nasopharyngeal infections Ossicular chain intact X-ray films, no mastoid involvement
Putrid and purulent discharge	Persistent otorrhea with evidence of mastoid bone destruction; occasional discomfort	Gradual	None	Flat loss up to about 60 dB	Marginal perforation or no drum	Degree of hearing loss depends on damage to ossicles X-ray films show chronic mastoiditis
Putrid and purulent discharge with cholesteatoma	As above Usually discharge	Gradual	None	Flat loss up to about 60 dB	Marginal perforation or no drum; white cholesteatoma; debris in canal	As above, and cholesteatoma
Putrid and purulent discharge with cholesteatoma and erosion into semicircular canal	As above Vertigo	Gradual	Occasional tinnitus	Flat loss up to about 60 dB	As above; vertigo and eye deviation with pressure of air in ear canal	As above; positive fistula test

(*Continued*)

Table 24.3 *Continued*

Etiology and principal findings	History	Onset of hearing loss	Tinnitus	Audiologic and impedance	Otosclerotic	Special
Glomus jugular tumor	Gradual stuffiness in one ear or persistent discharge	Gradual	Often hears own heartbeat	Very mild hearing loss at first and later up to about 60 dB	Red appearance of drum and middle ear or hemorrhage; tissue appears granulomatous	Much bleeding in ear on manipulation, x-ray films show erosion
Tuberculosis	Mild hearing loss with chronic ear infection; may be associated with tuberculosis elsewhere	Gradual	None	Minimal to 60-dB flat loss	Granulation tissue that resists treatment; later, cervical adenopathy Multiple perforation of eardrum	Biopsy and culture show tuberculosis (Acid fact bacilli)
Granuloma	Chronic otorrhea with fullness in ear and little pain	Gradual	None	Minimal to 60-dB flat loss	Firm granulations that regrow after removal	Biopsy shows specific etiology
Carcinoma	As above, and sometimes some pain and nodes	Gradual	None	As above	As above	As above
Letterer-Siwe's disease	Generalized skin rash; chronic otorrhea	Gradual	None	As above	Bleeding and erosive granulations causing stenosis of canal	X-ray films show bone erosion and punched-out areas in skull

Tympanosclerosis	Chronic otitis media in the past	Gradual	None	Using flat and about 60–70 dB	Eardrum eroded or scarred and thick; deformed appearance in middle ear	Sclerosis in mastoid bone
Hemotympanum	Blow to head or ear, with pain and fullness	Sudden	Roaring	40–70 dB with greater loss at higher frequencies Type B	Bloody fluid in middle ear	No infection is present and eardrum does not move with pressure
Systemic diseases: measles, scarlet fever	Acute otitis media, often followed by chronic otitis and hearing loss	Sudden	None	30–70 dB flat loss Type B	Perforated drum with or without chronic otitis	Chronic ear infection since childhood disease
Adhesion in middle ear	Slight hearing loss and fullness in ear with colds	Gradual and fluctuating	None	Up to about 35 dB, or worse in lower frequencies	Drum retracted or scarred, reflecting previous otitis media	Hearing loss not corrected by inflation
Flaccid tympanic membrane	Feeling of flutter in eardrum corrected by self-politzerization and nose blowing	Gradual and fluctuating	None	As above Type A_D	Wrinkled and loose eardrum	Drum is easily blown out and seems to be loose and redundant
Blue eardrum	Fullness in ear	Fluctuates	None	Up to about 45 dB and worse in higher frequencies	Drum seems to be blue or purple and does not move	Drum is dark blue and does not politzerize easily Normal x-ray findings in mastoid No infection

(Continued)

Table 24.3 *Continued*

Etiology and principal findings	History	Onset of hearing loss	Tinnitus	Audiologic and impedance	Otosclerotic	Special
Simple mastoidectomy	Ear infection followed by surgery—usually postauricularly; no subsequent ear discharge	After infection	None	Often little or no hearing loss, but sometimes up to 70 dB if ossicular chain is disrupted	Eardrum often almost normal, or only small perforation and normal auditory canal	Usually postauricular scar and no progressive hearing loss
Modified radical mastoidectomy	Ear infection and surgery without removal of ossicles	After infection	None	About 40-dB flat loss	Eardrum slightly deformed and posterior canal wall taken down	Postauricular or endaural scar and malleus are visible
Radical mastoidectomy	Ear infection and surgery with removal of eardrum remnants and ossicles	After infection	None	70-dB flat loss	No eardrum or ossicles visible and mastoid cavity seen through canal wall	X-ray films shows surgical defect in mastoid
Fenestration	Hearing loss and surgical correction	Insidious hearing loss over years	May be low-pitched	Variable hearing loss from 30 to 70 dB, depending on success of surgery	Partially exenterated mastoid cavity with displaced eardrum	Positive fistula test

Myringoplasty	Hearing loss and hole in drum with surgical repair; often vein or skin was used	Mild and gradual	None	Flat loss up to 70 dB	Healed perforation or large, thick drum repair, landmarks may be missing	Drum is in good position, and patient has scar at donor site
Tympanoplasty	Chronic otitis and hearing loss followed by surgery for hearing and infection	Gradual	None	About 50–70 dB	Large middle ear may be covered with skin; ossicles may be absent, and cavity resembles radical mastoidectomy	Varying findings, depending on type of surgery done
Artificial prosthesis	Ear infection with large defect in eardrum	Gradual	None	50–70 dB flat loss	Much of drum is absent, and patient uses artificial prosthesis that is inserted into canal to middle ear	Patient improves hearing with prosthesis
Radiation therapy to nasopharynx or thyroid	Clear fluid in ear following irradiation	Gradual	None	30–40 dB, with greater loss in higher frequencies	Fluid level and pressure disparity	X-ray treatment

Table 24.4 Sensory Hearing Loss

History	Onset of hearing loss	Otoscopic	Tinnitus	Audiologic and/or special		Diagnosis
Recurrent intermittent vertigo, nausea Ear feels full Voices sound tinny and hollow Difficulty understanding speech	Intermittent and then permanent	Normal	Ocean roar or hollow seashell	Recruitment complete, continuous, and hyperrecruitment Diplacusis Poor discrimination compared with hearing loss Discrimination worse with intensity Small-amplitude (type II) Békésy tracings	Recruitment may be complete and often hyperrecruitment Diplacusis Patient distraught Usually unilateral	Meniere's disease with vertigo
Occasional mild imbalance Some ear fullness No noise exposure	Insidious or sudden	Normal	Absent or high-pitched	Marked recruitment in some patients Loss only in high frequencies in most cases	Complete recruitment Usually unilateral No noise exposure	Viral disease
Exposure to sudden noise such as gunfire or explosion	Sudden hearing loss with tinnitus and then improves	Normal	Ringing	Starts with 4000-Hz dip and widens if severe Sometimes permanent	Highest frequency normal unless advanced age or severe damage; nonprogressive	Acoustic trauma
Direct blow to head	Sudden with ringing tinnitus; may be some improvement	Normal	Ringing	4000-Hz dip, loss in high frequencies or "dead" ear Labyrinth also may be affected	X-ray film may reveal fracture of temporal bone Vertigo may be present	Head trauma

Surgery for otosclerosis Sound distortion and some imbalance	Following surgery for stapes mobilization or stapedectomy	Normal	Usually present with buzz or roar	High-tone hearing loss–worse postoperative; discrimination reduced	Discrimination worse postoperatively than preoperatively, even if hearing loss is impaired	Poststapedectomy
Daily exposure to intense noise for many months No vertigo	Insidious	Normal	Uncommon	Early 4000-Hz dip or slightly broader dip	Marked and continuous recruitment Only a little discrimination change because only high tones involved	Occupational deafness (early)
Slight difficulty in understanding speech	Insidious	Normal	Occasional hissing, sometimes ringing	High-tone drop at 8000, 6000, and 4000 Hz Mild recruitment but continuous Discrimination reduced Bilateral usually	Age group about 50–60 Bilateral hearing loss	Presbycusis
Taking ototoxic drugs, especially in presence of kidney infection	Insidious or sudden	Normal	Generally high pitched	High-tone loss bilaterally, but may progress to all frequencies	Recruitment present Hearing loss usually associated with prolonged administration of drug	Drug ototoxicity
Hearing loss, with or without imbalanced	Insidious or sudden, unilateral or bilateral	Normal	High pitched or sea shell	Asymmetric, progressive; but may be sudden and total, unilateral or bilateral	Recruitment common Decreased discrimination	Autoimmune, syphilis lyme disease

Table 24.5 Neural Hearing Loss

History	Onset of hearing loss	Otoscopic	Tinnitus	Audiologic and/or special		Diagnosis
Early unilateral high-tone hearing loss Sometimes persistent vertigo	Insidious	Normal	Variable	Abnormal tone-decay; wide separation on continuous and interrupted Békésy tracings; poor discrimination compared with hearing loss	Unilateral; marked tone decay decreased; no response to caloric test Spontaneous nystagmus; erosion visible by x-ray film late in disease Stapedius reflex decay; abnormal BERA; other neurological deficits	Acoustic neuroma
Trouble understanding some people with soft voices	Insidious	Early atrophic eardrum; more white than normal	Occasional hissing	Gradual high-tone hearing loss; reduced discrimination; no abnormal tone decay; bone conduction often worse than air conduction	Age range roughly 60–75 Bilateral progressive deterioration of hearing; no abnormal tone decay	Presbycusis
Severe head injury with loss of consciousness	Sudden	Normal, or blood in canal if fracture involves middle ear	Ringing or none	Total, usually unilateral loss of hearing from injury to auditory nerve	X-ray film may show fracture around internal auditory meatus	Skull fracture
Sudden hearing loss, occasionally with vertigo or pain in ear	Sudden	Normal	High-pitched	Severe unilateral hearing loss	Other symptoms of herpes may be present	Viral

Table 24.6 Sensorineural Hearing Loss

History	Onset of hearing loss	Otoscopic	Tinnitus	Audiologic	Special	Diagnosis
Difficulty in hearing and understanding speech	Insidious	Normal	Occasional	Usually reduced hearing in all frequencies, especially in higher range; reduced discrimination	Usually over 50 Bilateral progressive hearing loss No abnormal tone decay; starts at 3000–6000 Hz Presbycusis starts at highest frequencies	Presbycusis
Exposure to intense noise over many months or years	Insidious	Normal	Uncommon	Early high-tone loss (C-5 dip), later involving lower frequencies; reduced discrimination; Békésy tracings depend on stage	In working age group; bilateral Starts usually at 3000–6000 Hz and spreads: no abnormal tone decay	Noise-induced hearing loss
Severe head injury often with unconsciousness, subjective vertigo	Sudden	Normal or some middle-ear pathology due to fracture	Hissing	Hearing loss usually is severe but may be only high-tone dip or high-tone loss	Fractured temporal bone with absent caloric responses Eardrum often appears to be normal; no spontaneous nystgmus, except early	Head trauma

(*Continued*)

Table 24.6 *Continued*

History	Onset of hearing loss	Otoscopic	Tinnitus	Audiologic	Special	Diagnosis
Ototoxic drug, usually in large doses, or a small dose in presence of kidney disease	Sudden	Normal	High-pitched	Rapid, sometimes severe bilateral hearing loss	Rapid and severe bilateral hearing loss, worse in high frequencies	Ototoxicity (neomycin, streptomycin kanamycin, or other ototoxic drugs)
Exposure to intense noise	Sudden hearing loss and tinnitus, followed by gradual recovery	Normal	Temporary ringing	High-tone dip or more severe losses, mostly in high frequencies	Except in unusual cases, recovery occurs within several days	Auditory fatigue Temporary threshold shift
Rh incompatibility in parents; kernicterus and speech defect in child	Congenital	Normal	None	Descending audiogram; nonprogressive hearing loss	Speech defect; sometimes other neurological deficits	Rh factor incompatibility
Sudden unilateral hearing loss during or following mumps	Sudden	Normal	None	Total unilateral hearing loss	Normal vestibular reaction	Mumps

Severe hearing loss after meningitis with high fever	Sudden	Normal	None	Generally subtotal bilateral hearing loss	Labyrinth also is involved in many cases	Meningitis
Sudden unilateral hearing loss with or without dizziness	Sudden	Normal	High-pitched or motorlike	Subtotal high-tone loss, usually unilateral	Occasionally associated with hypotension but generally no specific vascular disease	Vascular disorders or membrane ruptures May follow barotrauma
Child has retarded or defective speech	Congenital	Normal	None	High-tone loss bilaterally	Often follows maternal rubella in first trimester of pregnancy, or anoxia, trauma, or jaundice at birth	In utero and birth lesions
Disturbing unilateral tinnitus and hearing loss	Sudden and sometimes progressive	Normal	May be hissing	High-tone loss or subtotal loss of hearing, usually unilateral; test for acoustic neuroma negative	Often follows a viral infection	Auditory neuritis

Table 24.7 Distinguishing External Otitis from Otitis Media

	External otitis	Otitis media
History	Onset after getting water in ear or irritating auditory canal	After rhinitis or blowing nose hard or sneezing
Pain	In auditory canal and around meatus; aggravated by moving auricle and chewing	Deep in ear; aggravated by sneezing and blowing nose
Tenderness	Around auricle	No tenderness
Otoscopic	Skin of canal infected and absence of normal cerumen; eardrum not inflamed	Skin of canal not infected, but eardrum injected or bulging
Discharge	Debris from skin of canal	Often mucoid or mucopurulent through tympanic perforation
With air pressure in canal	Eardrum moves with positive pressure	Eardrum does not move well, especially if perforation is present
Hearing	Hearing loss disappears when canal is cleared	Hearing loss is present even with clear canal
With politzerization	Eardrum moves	Eardrum does not move well
	If the external canal is infected and there is mucoid discharge, it is a combination of external otitis and otitis media, since the mucus comes from the middle ear through a perforated eardrum	
Fever	Comparatively little	Often general malaise and fever
Imaging	Normal mastoids and middle ears	Congestion in mastoid and middle ears

25
Noise Measurement

Paul L. Michael
Paul L. Michael and Associates, Inc., and The Pennsylvania State University, State College, Pennsylvania, USA

Kevin L. Michael
Michael and Associates, Inc., State College, Pennsylvania, USA

A wide variety of new sound measuring equipment has been made available over the past few years. There are sound-level meters that provide only the basic overall weighted measurements required by OSHA and others that provide a very wide range of functions including integration for dose and impulse noise measurements. The sophisticated measuring equipment can be used to obtain an enormous amount of data in a relatively short period of time. The weak link in sound measuring equipment, the microphone, has also been improved in both performance and durability in recent years.

Even with this much improved equipment, careful consideration of the objectives of the measurement must be made before the equipment is selected. The microphone must function reliably over the range of sound frequencies and levels that are to be measured, and it must be readily adaptable for use with any instruments selected (1–9). The microphone must also be capable of operating in the temperature and humidity conditions that may be encountered, and its operation must not be adversely affected by other environmental factors such as electromagnetic energy or vibration. These specifications should be included in the literature supplied with the microphone; if not, inquiries should be made directly to the manufacturer before the equipment is purchased.

If sound levels change significantly in a short time, conventional sound-level meters may not respond fast enough to afford accurate readings (1) (therefore special instrumentation—such as sound-level meters with integrating circuits, impulse meters, or oscilloscopes—is required (4)). As a rule of thumb, if there are more than ten impulses per second *and* if the differences between the peaks and troughs of the signal are less than 6 dB, a conventional sound-level meter will provide a reasonable measurement. In all cases, accurate sound-level measurements require a well-trained operator and calibration equipment. A calibrator, specifically designed for the microphone, must be used before and after the measurements to ensure the required accuracy (8).

1. THE SOUND-LEVEL METER

Basically, the sound-level meter consists of a microphone, an amplifier-attenuator circuit, and an indicating meter. The microphone transforms airborne acoustic pressure variations into electrical signals with the same frequency and amplitude characteristics and feeds the electrical signals to a carefully calibrated amplifier-attenuator circuit. The electrical signals are then directed through a logarithmic weighting network to an indicating meter where the sound pressure is displayed in the form of levels about 0.0002 μbar.

Most sound-measuring instruments present the sound-pressure levels in terms of its root-mean square (rms) value, which is defined as the square root of the mean squared displacements during one period. The rms value is useful for hearing-conservation purposes because it is related to acoustic power and its correlates with human response. Also, the rms value of a random noise is directly proportional to the bandwidth; hence, the rms value of any bandwidth is the logarithmic sum of the rms values of its component narrow bands. For example, octave-band levels may be added logarithmically to find the overall level for the frequency range covered. The rms value of pure tones or sine waves is equal to 0.707 times the maximum value (see Fig. 25.1).

Rms values cannot be used to describe prominent peak pressures of noise which extend several dB above a relatively constant background noise. Maximum, or peak, values are used for this purpose. On the other hand, peak readings are of relatively little value for measuring sustained noises unless the waveform is known to be sinusoidal because of the peak reading's relationship to acoustic power changes with the complexity of the wave. As the waveform becomes more complex, the peak value can be as much as 25 dB above the measured rms value.

In addition to the rms and peak values, a rectified average value of the acoustic pressure is sometimes used for noise measurements. A rectified average value is an average taken over a period of time without regard to whether the instantaneous signal values are positive or negative. The rectified average value of a sine wave is equal to 0.636 times the peak value. For complex waveforms, the rectified average value may fall as much as 2 dB below the rms value. In some cases, rectified average characteristics have been used in sound-level meters by adjusting the output to read 1 dB above the rms

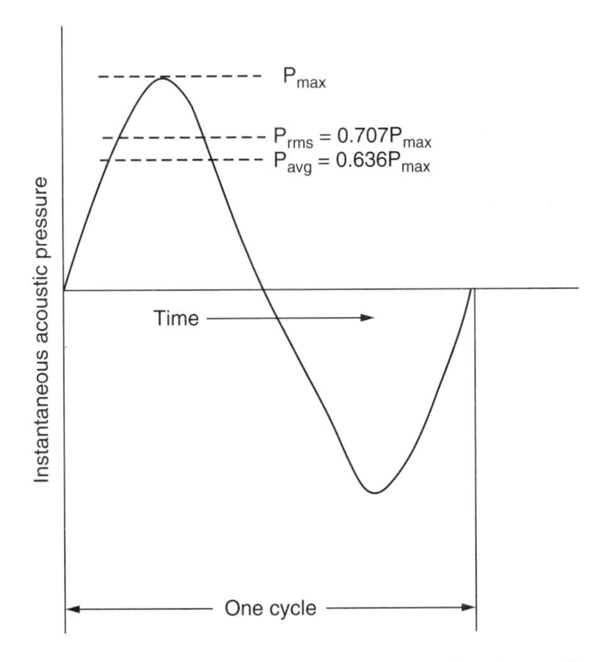

Figure 25.1 A comparison of maximum (P_{max}) or peak, root-mean square (P_{rms}), and rectified-average (P_{avg}) values of acoustic pressure.

level for sine wave signals, so that the average reading will always be within 1 dB of the true rms value. Fig. 25.1 shows a comparison of rms, maximum, and rectified average of a sinusoidal wave (10–12).

1.1. Meter Indication and Response Speed

The indicating meter of a sound-level meter may have ballistic characteristics that are not constant over its entire dynamic range, or scale, which will result in different readings depending on the attenuator setting and the portion of the meter scale used. When a difference in readings is noted, the reading using the higher part of the meter scale (the lowest attenuator setting) should be used, since the ballistics are generally more carefully controlled in this portion of the scale.

Most general-purpose sound-level meters have fast and slow meter response characteristics that may be used for measuring sustained noise (13). The fast response enables the meter to reach within 4 dB of its calibrated reading for a 0.2-sec pulse of 1000 cps; thus, it can be used to measure with reasonable accuracy noises whose levels do not change substantially in periods less than 0.2 sec. The slow response is intended to provide an averaging effect that will make widely fluctuating sound levels easier to read; however, this setting will not provide accurate readings if the sound levels change significantly in less than 0.5 sec.

1.2. Frequency-Weighting Networks

General-purpose sound-level meters are normally equipped with three frequency-weighting networks, A, B, and C, that can be used to approximate the frequency distribution of noise over the audible spectrum (13). These three frequency weightings, shown in Fig. 25.2, were chosen because (a) they approximate the ear's response characteristics at different sound levels (see Fig. 2.13) and (b) they can be easily produced with a few common electronic components. Also shown in Fig. 25.2 is a linear or flat response that weights all frequencies equally.

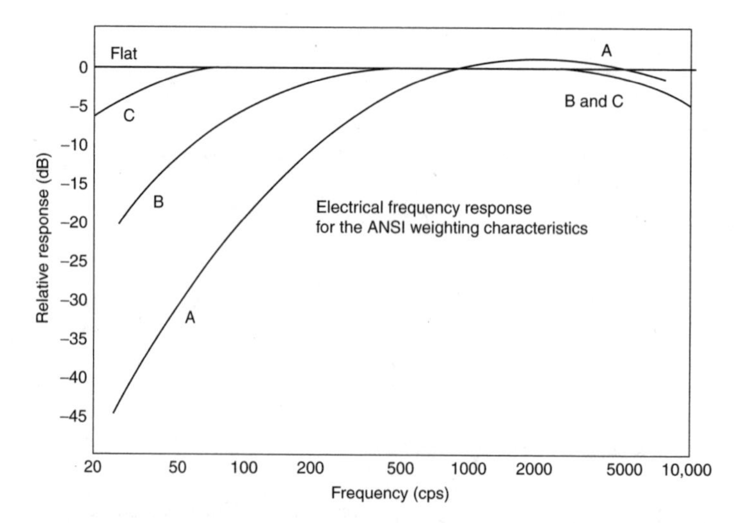

Figure 25.2 Frequency-response characteristics for sound-level meters [13].

The A-frequency weighting approximates the ear's response characteristics for low-level sound, below about 53 dB re 0.0002 μbar. The B-frequency weighting is intended to approximate the ear's response for levels between 55 and 85 dB, and the C weighting corresponds to the ear's response for levels above 85 dB.

In use, the frequency distribution of noise energy can be approximated by comparing the levels measured with each of the frequency weightings. For example, if the noise levels measured using the A and C networks are approximately equal, it can be reasoned that most of the noise energy is above 1000 Hz because this is the only portion of the spectrum where the weightings are similar. On the other hand, if there is a large difference between these readings, most of the energy will be found below 1000 Hz.

Many specific uses have been made of the individual weightings besides the frequency distribution of noise. In particular, the A network has been given prominence in recent years as a means of estimating annoyance caused by noise and of estimating the risk of noise-induced hearing damage.

2. MICROPHONES

The three basic types of microphones used for noise measurements are piezoelectric, dynamic, and condenser. Each type has advantages and disadvantages that depend on the specific measurement situation; all three types can be made to meet the American National Standard Specification for Sound Meters (S1.4.1971) (14,15). The dynamic microphone has not been used as much as the other two types in recent years, but it is still suitable for some applications.

New piezoelectric and condenser microphones (a) are much less expensive than earlier ones, (b) have excellent frequency responses from a few hertz to 10,000 Hz (and up to 1000 kHz if needed), and (c) have good reliability records. Piezoelectric and condenser microphones are now provided as original equipment with most sound-measuring instruments with tighter tolerance limits as included in the ANSI Specification for Sound Level Meters (1).

Most new sound measuring instruments cover a dynamic range of at least 40 to 150 dB re 0.0002 μbar. More expensive equipment will permit measurements in octave bands from about 0 dB to well above 160 dB, depending on the microphones selected. Most modern noise measuring instruments are not damaged by exposures to normal ranges of temperature and humidity; however, temporary erroneous readings may result from condensation when they are moved from cold locations to warm, humid areas. Also, special care should be taken when using dynamic microphones in areas with high levels of electromagnetic energy.

2.1. Temperature and Humidity Effects

The Rochelle-salt microphones that were supplied with older sound-measuring equipment are easily damaged by heat or humidity extremes. These piezoelectric microphones can be permanently damaged by heat such as that produced in a closed car on a hot day; therefore, extreme care should be taken in their use and storage. The piezoelectric microphones made with barium titanate or lead zirconate titanate that are furnished with most sound-measuring equipment today are not damaged by exposure to normal ranges of temperature and humidity. However, temporary erroneous readings may result from condensation if they are moved from very cold to warm, humid areas.

The dynamic microphone's response characteristics are somewhat dependent on the ambient temperature, but over most of the audible frequency range, the variation is less than about 1 dB for 50°F change in temperature. The dynamic microphone is affected relatively little by humidity extremes, except for temporary erroneous readings that may result from condensation. Specific temperature correction information for each microphone should be available from the manufacturer.

The condenser microphone is not permanently damaged by exposure to humidity extremes, but high humidity may cause temporary erroneous readings. The variation of sensitivity of condenser microphones with temperature is approximately −0.04 dB per °F. Here again, correction information for a specific microphone should be obtained from the manufacturer.

2.2. Microphone Directional Characteristics

Most noises encountered in industry are produced from many different noise sources and from their reflected energies. Therefore, at any given position in these areas, noise will be coming from many different directions and often may be considered to be randomly incident on any plane where a microphone diaphragm might be placed. For this reason, microphones are sometimes calibrated for randomly incident sound; however, depending on the design and purpose of the microphones, they may be calibrated for grazing incidence, for perpendicular incidence, or for use in couplers (pressure calibration). Thus, care must be taken to use microphones in the manner specified by the manufacturer in order to obtain the highest accuracy.

Microphones commonly used with sound-measuring equipment are nearly omnidirectional for frequencies below 1000 Hz; however, directional characteristics become important for frequencies above 1000 Hz (see Fig. 25.3). Therefore, when measurements are to be made of high-frequency noise produced by a directional noise source, i.e., where a high percentage of the noise energy is coming from one direction, the orientation of the microphone becomes very important even though this microphone may be described as omnidirectional. For a microphone calibrated with randomly incident sound, the microphone should be pointed at an angle to the major noise source that is specified by the

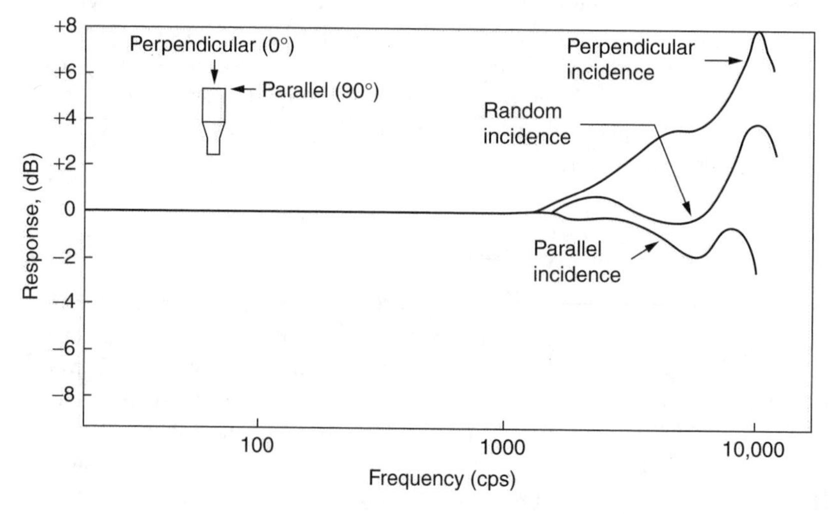

Figure 25.3 Directional characteristics of a piezoelectric microphone.

manufacturer. An angle of about 70° from the axis of the microphone is often used to produce similar characteristics to randomly incident waves, but the angle for each microphone should be supplied by the manufacturer. A free-field microphone is calibrated to measure sounds perpendicularly incident to the microphone diaphragm; thus, it should be pointed directly at the source to be measured. A pressure-type microphone is designed for use in a coupler such as those used for calibrating audiometers; however, this microphone can be used to measure noise over most of the audible spectrum if the noise propagation is at grazing incidence to the diaphragm and the microphone calibration curve is used.

Directional characteristics of microphones may be used to advantage at times. For example, an improved signal-to-noise ratio may be obtained for sound pressure-level measurements of a given source by using 0° incidence when high background levels are being produced by sources at other locations. Erroneous readings caused by reflected high-frequency sound emitted by other sources, but coming from the same direction, may be checked with directional microphones by rotating the microphone about an axis coinciding with the direction of incident sound. Reflected energy will be evidenced by a variation in level as the microphone is being rotated. The microphone orientation corresponding to the lowest reading should be chosen since the reflection error would be minimal at this position.

Special-purpose microphones with sharp directional characteristics may be used to advantage in some locations. These microphones are particularly useful for locating specific high-frequency noise sources in the presence of other noise sources.

2.3. Microphone Cables

Standard microphone cable with shielded and twisted wires should be used with a dynamic microphone to minimize electrical noise pickup. Usually, no correction is needed when this cable is used between a dynamic microphone and its matching transformer unless the cable is longer than 100 ft.

Cable corrections may, or may not, be required for condenser microphones, depending on the preamplifier design and the overall calibration. Some condenser microphones are calibrated when mounted directly on the sound-measuring equipment; others are calibrated with cables attached. Instructions with the microphones should provide this information.

A correction is normally required when a titanate-type piezoelectric microphone is used with a cable unless the microphone has a built-in circuit to lower its output impedance. For the case where there is no built-in impedance-reducing circuit, a correction of about +7 dB must be added to the titanate microphone output when used with a 25-ft cable. The exact correction factor should be supplied with the microphone.

A correction factor that is a function of temperature must be applied when a Rochelle-salt microphone is used with a cable. These correction factors are found in instruction manuals supplied by the instrument manufacturers.

3. FREQUENCY ANALYZERS

In many instances, the rough estimate of frequency-response characteristics provided by the sound-level-meter weighting networks does not give enough information. In these cases, the output of the sound-level meter can be fed into a suitable analyzer which will provide more specific frequency distribution characteristics of the sound pressure (16).

The linear network of the sound-level meter should be used when the output is to be fed to an analyzer. If the sound-level meter does not have a linear network, the C network may be used for analyses over the major portion of the audible spectrum (see Fig. 25.2).

3.1. Octave-Band Analyzers

The octave-band analyzer is the most common type of filter used for noise measurements related to hearing conservation. Octave bands are the widest of the common bandwidths used for analyses; thus, they provide information of spectral distribution of pressure with a minimum number of measurements.

An octave band is defined as any bandwidth having an upper band-edge frequency, f_1. The center frequency (geometric mean) of an octave band, or other bandwidths, is found from the square root of the product of the upper and lower band-edge frequencies. The specific band-edge frequencies for octave bands are arbitrarily chosen. The older instruments usually have a series of octave bands extending from 37.5 to 9600 Hz (37.5 to 75, 75 to 150, 150, to 300, . . . , 4800 to 9600). Newer octave-band analyzers may be designed for octave bands centered at 31.5, 63, 125, 250, 500, . . . , 8000 Hz, according to American Standard Preferred Frequencies for Acoustical Measurements (S1.6-1960) (17). Octave band-edge frequencies corresponding to the preferred center frequencies can be calculated using two equations with two unknowns. The first equation comes from the definition of an octave band; the upper band-edge frequency is equal to twice the lower band-edge frequency ($f_2 = 2f_1$). The second equation describes the center frequency, f_c, in terms of the band-edge frequencies ($f_c = \sqrt{f_1 f_2}$). For example, the band-edge frequencies corresponding to a center frequency of 1000 cps can be calculated as follows:

$$f_c = \sqrt{f_1 f_2} \tag{25.1}$$

and

$$f_2 = 2f_1 \tag{25.2}$$

From Eq. (25.1), $f_c = 1000 = \sqrt{f_1 f_2}$, and substituting Eq. (25.2) into Eq. (25.1), $1000 = \sqrt{f_1} \times 2f_1 = f_1\sqrt{2}$. Therefore $f_1 = 1000/1.414 = 707$ cps, and from Eq. (25.2), $f_2 = 2 \times 707 = 1414$ cps.

Most combinations of sound-level meter and octave-band analyzer have separate attenuators on each instrument. In these cases, it is always important to take a measurement of the overall noise on the sound-level meter first and leave its attenuator at this position for all analyzer measurements. This procedure prevents overloading of the sound-level meter and resulting erroneous readings. If the overall level changes during a series of measurements, the entire procedure must be repeated.

3.2. Half-Octave and Third-Octave Analyzers

When even more specific information of the spectral pressure distribution is desired than that provided by octave bands, narrower-band analyzers must be used. The number of measurements necessary to cover the overall frequency range will be directly related to the bandwidth of the analyzer; thus, a compromise must be reached between the resolution required and the time necessary for the measurements.

Half-octave and third-octave filters are the next steps in resolution above octave-band analyzers. A half-octave is a bandwidth with an upper-edge frequency equal to

$\sqrt{2}$ times its lower-edge frequency. A third-octave has an upper-edge frequency that is $3\sqrt{2}$ times its lower-edge frequency.

3.3. Adjustable-Bandwidth Broad-Band Analyzers

Some analyzers are designed with independently adjustable upper and lower band-edge frequencies. This design permits a selection of bandwidths in octaves, multiples of octaves, or fractions of an octave. The smallest fraction of an octave usually available on these adjustable-bandwidth analyzers is about one-tenth, and the largest extends up to the overall reading.

In addition to the obvious advantage of being able to select the proper bandwidth for a particular job, these analyzers permit the selection of any octave band, rather than a preselected series of octaves. For example, they can be adjusted to the older series of octaves (75–150, 150–300, etc.) or to bands with preferred center frequencies described in recent American Standards (125, 250, 500, etc.) (17). The disadvantage of these instruments is their relatively large size.

3.4. Narrow-Band Analyzers

Analyzers with bandwidths narrower than tenth-octaves are normally referred to as narrow-band analyzers. Narrow-band analyzers are usually continuously adjustable, and they are classified either as constant-percentage-bandwidth or as constant-bandwidth types.

The constant-percentage narrow-band analyzer is similar to the broad-band fractional-octave analyzer in that its bandwidth varies with frequency. As its name indicates, the bandwidth of the constant-percentage analyzers is a constant percentage of the center frequency to which it is tuned. Typically, a bandwidth of about 1/3-octave might be selected with these analyzers.

The bandwidth of a constant-bandwidth analyzer remains constant for all center frequencies over the spectrum. Provision may be made on some instruments to vary the bandwidth, but typically, the bandwidth of a constant-bandwidth analyzer remains constant at a few hertz.

The constant-bandwidth analyzer normally provides a narrower bandwidth and better discrimination outside the passband than the constant-percentage analyzer; therefore, it is often the best choice when discrete frequency components are to be measured. Also, it usually covers the entire spectrum with a single dial sweep, thus facilitating coupling to recorders for automatic analysis. Most constant-percentage analyzers require band switching to sweep the audible spectrum. On the other hand, caution must be used when constant-bandwidth analyzers are used to analyze noises that have frequency modulation, or warbling, of components, for serious errors may result (18). Frequency-modulated noises are commonly produced by reciprocating-type noise sources in some machinery. Frequency-modulated noise is not a major problem if constant-percentage analyzers are used.

4. DOSIMETERS*

A noise dosimeter is a sound-level meter that integrates noise samples over time. In other words, the dosimeter averages the noise observed in discrete periods of time and sums up

---

*The authors are indebted to Andrew G. Schauder, CIH, United Merchants and Manufacturers, for assistance in preparation of this section.

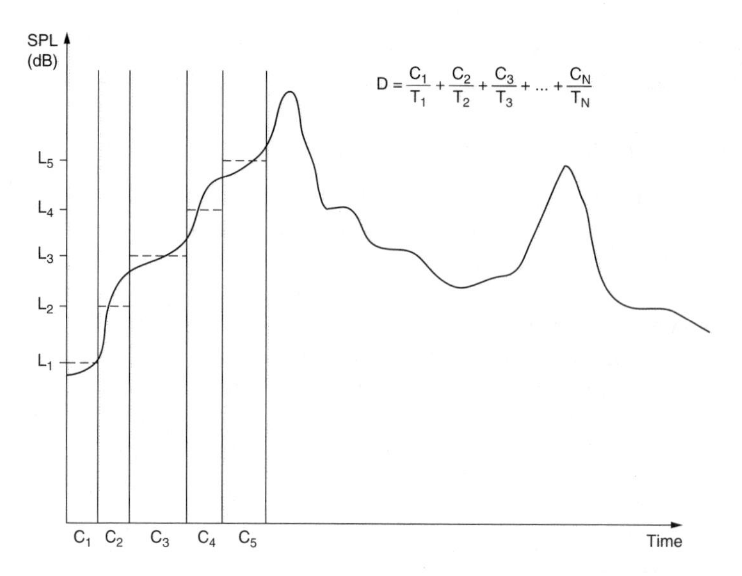

Figure 25.4 Method of dose computation used in dosimeters. SPL = sound-pressure level.

all of those averages to give a total (Fig. 25.4). Dosimeters take several noise samples per second. For example, the Quest Dosimeter takes 16 samples per second. Most dosimeters are of type II accuracy as defined by the American National Standards Institute S1.4-1983. Current dosimeters can be set at an array of parameters, but for OSHA compliance purposes, the unit should be set for the A-scale, 5-dB exchange rate, 80-dB criterion/threshold, slow response.

Figure 25.5 illustrates the components of a dosimeter. The noise signal enters a microphone and is fed into the amplifier. It then goes into a waiting network to conform the noise measures to the A scale. The signal then passes to an rms detector where the AC noise signal is converted to a DC signal representing the rms value of the original input. The DC signal is then fed into the microcomputer chip, which converts the signal to a number that represents the decibel level. This process is repeated several times per second. These data are stored and summed up over time.

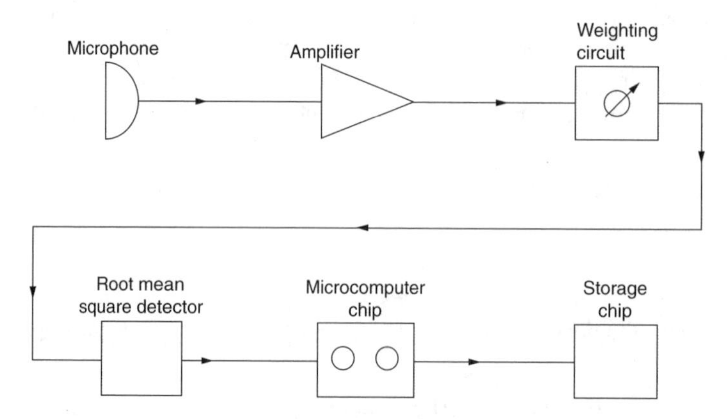

Figure 25.5 Audiodosimeter components.

If an employee works in a noisy environment that changes frequently, it is easier to use a noise dosimeter than to use a sound-level meter coordinated with time and motion studies. This is also true for employees who work in relatively constant noise environments but who are highly mobile. Either dosimeters or sound-level meters may be used for determining compliance with OSHA's noise exposure standard or Hearing Conservation Amendment. On the one hand, dosimeters have advantages over other noise-measuring equipment:

1. The procedure is simple and produces a single unit of exposure that is either less than, equal to, or greater than permissible exposure levels.
2. The device works well in varying noise levels.
3. The person making the determination of noise level need not be present during the entire measurement period.
4. The unit handles multiple determinations concurrently.
5. Dosimeters are cost effective relative to more expensive analytical equipment needed to provide comparable measures.

On the other hand, they have potential disadvantages:

1. Accuracy is limited (type I/2 dependent).
2. Dosimeters are susceptible to nonintentional or intentional errors that may influence readings (such as employees tapping on or singing into dosimeter microphones).
3. Measurements may be affected by body shielding.
4. Calibration may drift out of adjustment over long measurement periods.
5. With the exception of certain more expensive pieces of equipment, no information is provided on noise-level history.
6. Dosimeters are inaccurate if impulse or impact noise is present.
7. Dosimeters *may* be inherently inaccurate owing to the 5-dB doubling rate employed in the electronic circuits.

5. IMPULSE OR IMPACT NOISE MEASUREMENT

The inertia of the indicating meters of general-purpose sound-level meters prevents accurate, direct measurements of single-impulse noises that have significant level changes in less than 0.2 sec. Typical noises with short time constants are those produced by drop hammers, explosives, and other objects with short, sharp, clanging characteristics. A low-inertia device such as an oscilloscope must be used to measure these impulse-type noises if detailed information is required.

Measurement of impulse noise characteristics may be taken directly from a calibrated oscilloscope with a long persistence screen, or photographic accessories may be used to obtain permanent records. The oscilloscope is usually connected to the output of a sound-level meter having a wide frequency response and calibrated with a known sound level of sinusoidal characteristics. The screen of the oscilloscope is calibrated directly in decibels (rms) by comparing the oscilloscope deflection produced by a sinusoidal signal with the sound-level-meter reading. Several calibration points may be fixed on the oscilloscope screen by providing various signal levels into the sound-level meter, or the scale may be determined from a single calibration level by using linear equivalents to decibels. For example, for a sine wave signal, half of a given deflection on the oscilloscope will be equivalent to a 3-dB drop in level, and 0.316 times the deflection will be equivalent to a 10-dB drop in level. These equivalent values may be calculated from the equation

$$\text{Drop in level (dB)} = 20\log_{10}\frac{d_1}{d_2} \tag{25.3}$$

where d_1 and d_2 are the small and large linear screen deflectors being compared. It should be noted that this calibration using a sine wave is for convenience, and that there is a constant factor of 3 dB that must be added to the rms calibration to obtain the true instantaneous peak values for sine waves. The relationship of rms to peak values is more complex for nonsinusoidal waves (10–12).

Care should be taken while using an oscilloscope driven by a sound-level meter to prevent errors resulting from overloading. If the oscilloscope deflections show a sharp clipping action at a given amplitude, the attenuator settings on one or both instruments may require adjustment upward. Also, a check should be made to determine whether the indicating meter of the sound level may be done by switching the meter out of the circuit to a battery check position, and observing the waveform. If the oscilloscope waveform is changed in any way by the indicating meter, it should be removed from the circuit each time a deflection is measured on the oscilloscope.

The oscilloscope is inconvenient to use in many field applications because it is relatively large and complex. Also, most oscilloscopes requires AC power and, in the field, the supply may vary and cause changes in calibration. For these applications, it is often convenient to use peak-reading impact-noise analyzers which may be connected to the output of a sound-level meter. These battery-driven instruments do not provide as much information as the oscilloscope trace, but they are often adequate. The electrical energy produced by an impulse noise is stored for a short time by these instruments in capacitor-type circuits so that information may be gained on the maximum peak level, on the average level over a period of time, and on the duration of the impact noise. As with the oscilloscope, care must be taken not to overload the sound-level meter driving these impact-noise meters.

6. MAGNETIC FIELD AND VIBRATION EFFECTS

The response of sound-level meters and analyzers may be affected by the strong alternating magnetic fields found around some electrical equipment. Dynamic microphones, coils, and transformers are particularly susceptible to hum pickup from these fields. Some of the newer dynamic microphones have humbucking circuits that minimize this pickup, but caution should be used in all cases. To test for hum pickup, disconnect the suspected component and check for a drop in level on the indicating meter. It is good practice to follow the equipment manufacturer's procedure for this check.

The magnetic fields produced by dynamic microphones may attract metal filings that will change the frequency response characteristics; therefore, dynamic microphones are not a good choice for measurements in metal-shop areas.

Vibration of the microphone, or measuring instrument, may cause erroneous readings, and in some cases, strong vibrations may permanently damage the equipment. It is always good practice to mechanically isolate sound-measuring equipment from any vibrating surface. Holding the equipment in your hands or placing it on a foam rubber pad is satisfactory in most cases. Possible effects of vibration can be checked by observing the meter reading while the noise is shut off if this can be done without changing the vibration. If the meter reading drops by more than 10 dB, the effects of vibration are not significant. If the noise cannot be shut off without changing the vibration, the same result can be obtained by replacing the microphone cartridge with a dummy microphone. The equipment manufacturer can supply information necessary to build a dummy microphone.

7. TAPE RECORDING OF NOISE

It is sometimes convenient to record a noise so that an analysis may be made at a later date. This is particularly helpful when lengthy narrowband analyses are to be made, or when very short, transient-type noises are to be analyzed. However, extreme care must be taken in the calibration and use of recorders to avoid errors. Also, direct sound-pressure measurement and analysis should be made during the recording procedure so that the operator will be aware when additional measurements or data are necessary.

Many of the professional, or broadcast-quality, tape recorders are satisfactory for noise-recording application; however, care must be taken that the microphone has the proper characteristics. Many times, the specifications given for a tape recorder do not include the microphone characteristics and the microphone may be of very poor quality. When the tape recorder is not specifically built for measuring noise, it is usually good practice to connect the recorder to the output of a properly calibrated sound-level meter. As is the case when attaching any accessory equipment to sound-level meters, it is important that the impedances are properly matched. The bridging input of a tape recorder is satisfactory for the output circuits of most sound-level meters.

When a tape recorder is used to record noises that have no high prominent peaks, the recording level usually should be set so that the recorder meter (VU meter) reads between −6 and 0 dB. This setting assumes that a sinusoidal signal reading of +10 dB on the VU meter will correspond to about 2 or 3% distortion according to standard recording practice.

If the recorded noise has prominent peaks, it is good practice to make at least two additional recordings with the input attenuator set so that the recording levels are between −6 and 0, −16 and 0, and −26 and −20. If there is less than 10 dB between any two of these adjacent 10-dB steps, overloading has occurred at the higher recorded level and the lower of the two should be used.

It is important to calibrate the combination of tape recorder and sound-level meter at known level and tone control settings throughout the frequency range before the recordings are made. Prior to each series of measurements, a pressure-level calibration should be made by noting the overall sound-pressure-level reading corresponding to the recording along with a notation of the tape recorder dial settings. Also, it is good practice to note the types and serial numbers of the microphone and sound-level meter, the location and orientation of the microphone, the description of the noise source and surroundings, and other pertinent information for each recording. It is often convenient to record this information orally on the tape to be sure that the information will not be lost or confused with other tapes.

8. GRAPHIC-LEVEL RECORDING

A graphic-level recorder may be coupled to the output of a sound-level meter, or analyzer, to provide a continuous written record of the output level. Graphic-level recorders provide records in the conventional rms logarithmic form used by sound-level meters; thus, the data may be read directly in decibels. Some older recorders use rectified average response characteristics, so corrections must be made to convert these recordings to true rms values. As with sound-level meters, these recorders are intended primarily for the recording of sustained noises without short or prominent impact-type peak levels. The equipment manufacturer or instruction manual should be consulted to determine the limitations of each graphic-level recorder. Graphic-level records provide a valuable permanent record but necessitate expensive data reduction.

9. INSTRUMENT CALIBRATION

If valid data are to be obtained, it is essential that all sound-measuring and analyzing equipment be in calibration. When this equipment is purchased from the manufacturer, it should have been calibrated to the pertinent American Standards Specifications (13,16). However, it is the responsibility of the equipment user to keep the instrument in calibration by periodic checks.

Most general-purpose sound-measuring instruments have built-in calibration circuits that may be used for checking electrical gain. Most sound-level meters have built-in or accessory acoustical calibrators that may be used to check the overall acoustical and electrical performance at one or more frequencies. These electrical and acoustical calibrations should be made according to the manufacturer's instructions at the beginning and at the end of each day's measurements. A battery check should also be made at these times. These calibration procedures cannot be considered to be of high absolute accuracy, nor will they detect changes in performance at frequencies other than that used for calibration; however, they will serve to warn of the most common instrument failures, thus preventing a long series of invalid measurements.

Periodically, sound-measuring instruments should be sent back to the manufacturer, or to a competent acoustical laboratory, for a complete overall calibration at several frequencies throughout the instrument range. These calibrations require technical competence and the use of expensive chambers and equipment, which cannot be justified by the normal user of sound-measuring equipment. The frequency of these more complete calibrations depends on the purpose of the measurements and how roughly the instruments have been used. In most cases, it is good practice to have a complete calibration performed every 6 months, or at least once each year. In any case, a complete calibration should be made if any unusual change (more than 2 dB) is seen in the daily calibration.

10. NOISE SURVEYS

The importance of accurate and stable instrumentation for noise measurement is obvious. Not so obvious, but just as important, is the need for careful planning of the survey to make sure that all objectives will be satisfied from the measurement data collected.

The need for a comprehensive definition of the purpose and scope of a noise survey cannot be overemphasized. The choice of instruments, measurement techniques and locations, and data-recording procedures will be determined from this survey design.

10.1. Purpose and Scope

The purpose and scope of noise surveys may vary considerably; however, four general types may be considered:

1. A survey to determine hearing-damage risk
2. A survey to determine speech interference levels

3. A survey to determine disturbance levels
4. A survey for noise control purposes

The first three survey types require somewhat similar measurement procedures for these measures of man's reaction to noise. The fourth survey type usually requires a more detailed analysis of the noise, and measurement procedures are somewhat different than for the first three types.

10.1.1. Hearing-Damage-Risk Surveys

Current rules, regulations, and guidelines concerned with noise-induced hearing loss specify that the A-frequency weighting and slow meter response on sound-level meters be used to measure noise exposure levels (15). Most of these specifications are also concerned with time and level patterns of exposures. Therefore, the purpose and scope of a survey is often defined broadly by the pertinent safety regulation specification (19–22).

Sound-level meter measurements are generally required only at the position that will be occupied by persons exposed to noise in hearing-damage-risk surveys. The time necessary for the survey is determined by the time necessary to establish meaningful time and level patterns of noise exposure. If a man is exposed to a noise having nearly the same level continuously, or if there is a predictable on-off time pattern, a very short noise measurement sample should be adequate. If the on-off time, or levels, are not predictable, many days may be required to determine a meaningful exposure pattern. Dosimeters may be helpful in such situations.

Data-recording methods will vary depending on the time-level exposure patterns. For simple time-level patterns, direct manual readings and recordings are adequate. These measurements need not be repeated until some change in noise sources, job locations, or other changes in time-level patterns are indicated. Unexpected shifts in monitored hearing thresholds would, of course, be another indication for the need of additional noise measurements. Complicated time-level noise-exposure patterns may require an unreasonable amount of manual recording time, so that automatic recording of sound levels will be desired. Noise survey information should be recorded on a data sheet (Fig. 25.6). Damage-risk determination is discussed in detail in Chapter 20.

10.1.2. Speech-Interference-Level Surveys

Speech-interference-level surveys maybe made with narrow-band, octave-band, A-frequency weighting, or other frequency weightings (23). In many instances, the relatively simple A-frequency weighting is adequate. The purpose and scope of the survey and the physical characteristics of the noise will determine the degree of detail and, hence, the means of measuring the noise.

The positions of measurement will be fixed at the locations where speech must be understood. As with the measurements made for damage-risk assessment, the number of measurements will be determined by the noise-exposure pattern. Data recording will also be similar to the methods used in recording damage-risk data.

10.1.3. Disturbance-Level Surveys

Considerable flexibility is required in rules for estimating disturbances caused by noise because of the many psychological, physiological, and physical variables involved in most situations. Disturbances may vary from minor annoyances caused by very low noise levels to physical alterations in vision or tactile abilities caused by high levels of noise (24,25).

Date ____________

Location →							Sound measuring equipment Type ______ Model # ______ Serial # ______ Type # ______ Model # ______ Serial # ______
Octave band (center freq.)	Decibels	Decibels	Decibels	Decibels	Decibels	Decibels	Octave band (center freq.)
Overall-linear							Overall-linear
A - frequency weighting							A - frequency weighting
31 Hz							31 Hz
62 Hz							62 Hz
125 Hz							125 Hz
250 Hz							250 Hz
500 Hz							500 Hz
1000 Hz							1000 Hz
2000 Hz							2000 Hz
4000 Hz							4000 Hz
8000 Hz							8000 Hz
Taken by: ______	Time: ______ Remarks:	Time: ______ Remarks:	Time: ______ Remarks:	Time: ______ Remarks:	Time: ______ Remarks:	Time: ______ Remarks:	Field calibration Cal. type ______ Correct cal. Level ___ Freq. Mark Cal Lev (Hz) dB 125 250 500 1000 2000 Time

Figure 25.6 Noise survey data sheet.

In specific cases, a particular measuring means may be indicated for the best correlation of noise levels with man's response to noise; however, in many instances, one of several different measuring means will provide satisfactory data. If a particular noise source has a characteristic spectrum with a high percentage of energy in narrow-frequency bands, a third-octave or narrow-band analysis may provide much better correlation with responses to this noise than would measurements with wider-frequency bands. Narrow-frequency bands may also be required to measure or pinpoint the contribution of a particular source to a high background noise if the relationship of the listener and the source are bad. For example, if a man lives near a factory that has recently fired a member of his family, the noise made by this factory may be much more disturbing to him than higher levels produced by other sources such as traffic. Thus, measurements must be made to differentiate between the contributions of the various sources. However, in a very large number of cases where common broad-band noises are involved, measurements using A-frequency weighting, octave-bands, or other frequency weightings are just as effective as narrow-band analyses for measuring disturbance levels.

The positions and number of measurements will be determined by the purpose of the survey and the variability of the noise characteristics. A general survey will normally include measurements at the boundaries of all properties near the noise source and at other locations where complaints might be expected. A survey to investigate a specific complaint might be restricted to a single location. An adequate number of measurements should be made in each area to determine the level and time patterns produced by the particular source under investigation and the relationship of this source noise to background noises. This information is generally required for daytime, evening, and night for all conditions of source operation.

Data-recording requirements are much the same as for damage-risk and speech intelligibility. For simple noise exposure patterns, manual recording is satisfactory, but automatic recording is necessary to assess complex exposure patterns.

10.1.4. Noise Control Surveys

Noise control surveys normally require octave or narrow-band analyses to pinpoint and describe individual source contributions. Narrow-band analyses are required to differentiate between two or more major contributors to the overall noise levels when these contributors are closely spaced, or when noise control work is to be done on a particularly source located in high background noise.

The locations for measurements in a noise control survey will differ depending on the purpose. Two general purposes of noise surveys are: (a) to pinpoint and describe a particular noise source so that effective noise control measures can be selected, and (b) to determine the acoustical power output of a noise so that noise levels can be predicted in other locations. Details of these measurement procedures are given later in this chapter.

The on-off time pattern of a given noise source is usually well known or easily controlled during noise control procedures so that relatively short samples of noise are generally meaningful. When on-off time patterns are predictable, manual recording of data is normally satisfactory; however, automatic recording means may be helpful when narrow-band analyses are used or when multiple-point measurements are made in anechoic or reverberation chambers (26).

10.2. Measurement Techniques

10.2.1. Selection of Instrument and Measurement Locations

The kind of instruments needed for a particular survey should be determined from the purpose and scope, as described earlier. Measurement positions are also described generally by the purpose and scope, although each individual situation should be considered carefully to be sure that objectives of the survey will be met. Additional measurement locations, or measuring means, may be indicated by unexpected results during the survey. For example, if a reading is unexpectedly high when the microphone is pointed in one direction, it may be desirable to make additional measurements around that point. A survey must be flexible so that full advantage can be taken from leads provided by measurement data as the survey progresses.

Adjustments, modifications, battery replacements or recharging, etc., are much easier to do in the laboratory than in the field. Thus, it is very important that the following checks be made in the office or laboratory just prior to making a field trip.

1. Connect all equipment as it will be used in the survey, turn the power on, and allow sufficient time for stabilization (see equipment instruction manuals).
2. Check battery condition and replace or recharge if necessary.
3. Calibrate the equipment electrically and acoustically (see instruction manuals). Check each instrument separately and check the combination of instruments to be used.
4. Measure some familiar wide-band noise for a gross check on analyzer band performances.
5. Replace the microphone with a suitable dummy load (see instruction manual) and measure electrical background noise.

The first three steps listed above provide a satisfactory check of the sound-level-meter portion of the equipment. Step 4 is needed only when octave- or narrow-band analyzers are to be used. Step 5 should be done periodically and, in particular, whenever low sound-pressure levels are to be measured.

Special preparation procedures may be required in some instances. For example, if measurements are to be made outdoors where wind may cause erroneous readings, a wind screen should be provided for the microphone, and proper corrections must be applied to any data taken while using the wind screen. Data correction must also be applied if microphone cables, microphone accessories, or combinations of equipment are to be used. In most cases, these corrections may be found in instruction books provided for the equipment; however, the safest procedure is to recheck and recalibrate all equipment in the exact manner it is to be used immediately prior to leaving for a noise survey.

10.2.2. Travel and On-Site Preparation Procedures

All noise measurement instrumentation should be hand-carried throughout transportation from the laboratory to the survey site. Many pieces of this equipment may be damaged or its operational characteristics may be changed by excessive vibration, mechanical shock, humidity, and temperature cycling that might be encountered during normal shipping procedures. When traveling by public conveyance, the equipment should be carried in the passenger space. When traveling by automobile, the equipment should be placed on the seat or on resilient pads to reduce vibrations and shock.

Extremes of temperature and humidity should be avoided at all times, but, in particular, just prior to measurement. Microphones supplied with most modern noise measurement equipment will not be permanently damaged by exposure to normal temperature and humidity extremes; however, condensation that may result from bringing a cold instrument into a warm room may change the microphone response characteristics temporarily. For example, this condition may result when an instrument is taken from a car, where it had been stored overnight in cool temperatures, into a warm and humid area where the survey is to be made.

10.2.3. On-Site Checks

All measurement equipment should be rechecked and recalibrated on location before beginning measurements and at 2-hr intervals during the survey. Care must be taken during field calibrations to be sure that the acoustical calibration signal is at least 20 dB higher than the background noise measured with the calibrator (not operating) mounted on the microphone. It should be remembered that both electrical and acoustical calibrations are needed and that acoustical calibrations should be made with the microphone mounted on a cable, extension, or directly as it is to be used during the survey.

Field calibrations should always be made with calibrators recommended for the particular microphone in question. Other calibrators may be the right physical size; however, a physical fit does not ensure accuracy. Also, in some cases, a microphone cartridge may be permanently damaged from the use of calibrators intended for other microphones.

To serve as a reminder to perform the calibrations, and for medico-legal purposes, the field calibration data should be recorded along with noise measurement data. Unless the field calibration is recorded is recorded serially as an integral part of a noise survey, many operators will neglect this calibration and much measurement time may be wasted.

11. INDOOR SPACES: NOISE SPECIFICATION

Any indoor-noise-limit specification obviously must limit levels so that there is no danger of noise-induced hearing impairment (27–29). In addition, these specifications should cover noise disturbances that may take many forms, including interference with communication, annoyance, distraction, and interference with work or relaxation.

11.1. Speech Interference Guidelines

The frequency range from 200 to 6000 Hz, which contains most of the information in speech, may be divided into a large number of frequency bands, each having equal importance to speech intelligibility. If a dynamic range of about 30 dB is maintained above threshold in each of these bands, intelligibility scores approaching 100% should be possible for normal-hearing persons (30). A restriction of this speech range in any band will limit intelligibility scores. For example, a dynamic range of 15 dB will limit a specific speech contribution to about 50% of its potential value. The overall contribution of all bands in this range may be expressed in terms of the average of the contributions in each band. This single number percentage of the total possible contributions to speech is called the articulation index (30,31).

The masking of speech by noise has the effect of increasing a person's threshold of hearing with varying degrees at different frequencies depending on the spectra of the masking noise. Thus, speech must be made louder in some noise backgrounds if a high level of intelligibility is to be maintained. If it is impossible to maintain the required dynamic range of speech pressure levels because of distortion or potential danger from overloading the ear, or because of inadequate speech power, the overall speech intelligibility will be reduced.

11.1.1. Practical Speech Interference Calculations

One of the most widely accepted simplified procedures for determining the effect of noise on speech intelligibility makes use of the arithmetical average of three octave-band sound-pressure levels measured from the background noise. The average sound-pressure level in the original three octave bands (600–1200, 1200–2400, and 2400–4800 Hz) was proposed by Beranek as speech interference levels (SIL) (32–35) that could be used to determine when speech communication is easy, difficult, or impossible under specified conditions.

Recently, Webster has proposed that the SIL octave bands be shifted slightly to conform with ANSI preferred frequencies that are commonly used in modern instrumentation design.

The arithmetical average of the sound-pressure levels in the new octave bands centered at 500, 1000, and 2000 Hz is called preferred octave speech interference levels (PSIL) (36).

Other measures of noise, such as the A-frequency weighting, have also been shown to provide reasonably good estimates of speech interference levels for many common background noise spectra (36–38). The A-frequency weighting measure (dBA) is particularly appealing because it is an easily obtained single number and because of its widespread use in various rules, regulations, and standards pertaining to hearing conservation, noise control, and community noise control.

Figures 25.7 and 25.8 show some guidelines in terms of PSIL and dBA for maximum noise levels that can be tolerated if everyday speech is to be intelligible to normal-hearing persons when face to face and when using the telephone.

Caution must be used when applying the sound-pressure-level limits shown in Figs. 25.7 and 25.8. These data are intended only for common broadband background noises

Speaker to Listener Distance in Feel	Speech Interference Level Ratings in dB re 0.00002 n/m² Normal Speech		Raised Voice		Very Loud Speech		Shouting	
	PSIL	dBA	PSIL	dBA	PSIL	dBA	PSIL	dBA
1	70	75	76	81	82	87	88	93
3	60	65	66	71	72	77	78	83
6	54	59	60	65	66	71	72	77
12	48	53	54	59	60	65	66	71

Figure 25.7 Background noise levels that cannot be exceeded if face-to-face speech is to be intelligible at the distances and speech levels specified. These values are intended for normal-hearing persons located in common broadband background noises that do not have a high percentage of energy in narrow-frequency bands.

that do not have a high percentage of energy in narrow-frequency bands. Also, the data are based on male voices and for normal-hearing listeners.

11.2. Other Communication Interference Factors

Unfortunately, that portion of the speech frequencies containing most of the consonant power (above 2000 Hz), which provide much of the information in speech, is relatively easily masked with background noise because of the low speech-power levels in these frequencies. Thus, background noise may be particularly bothersome to those persons with high-frequency, sensorineural hearing losses in this same frequency range.

Background noise can also mask warning signals, thereby creating a potential injury hazard. It is impossible to set reasonable guidelines on the masking of warning signals unless their spectra and the acoustical characteristics of the space are defined. Some guidance may be provided by critical band relationship (39).

11.3. Indoor Noise Limits for Purposes Other Than Speech Communication

The many psychological, physiological, and physical variables involved in defining annoyance, distraction, and interference with work or relaxation that are found for different individuals in different noise exposure situations make it impossible to establish a

Quality of Telephone Speech Intelligibility	Speech Interference Level Ratings in dB re 0.00002 n/m² Calls Within a Single Exchange		Multiple Exchange Calls	
	PSIL	dBA	PSIL	dBA
Satisfactory	68	73	63	68
Difficult	68–83	73–88	63–78	68–83
Unsatisfactory	<83	<88	<78	<83

Figure 25.8 Background noise levels that cannot be exceeded for acceptable telephone conversation. These values are intended for normal-hearing persons located in common background noises that do not have a high percentage of energy and narrow-frequency bands.

single set of applicable rules or guidelines based on sound-pressure levels alone. For example, a dripping faucet, a hushed conversation, a child crying, or a piece of hard chalk scraped along a blackboard may cause a considerable amount of annoyance or distraction with very low sound levels, while much higher levels are not considered annoying under normal circumstances when people are attending ballgames, listening to music, etc.

11.3.1. Annoyance and Distraction Factors

The degree of annoyance provided by a given noise is significantly influenced by many factors, including:

1. The personal relationship of the individual with the noise-producing source. If the noise source is caused by or in some way for himself, the noise is less likely to be annoying than if it were created by a neighbor. If the noise is created by a neighbor, the amount of annoyance is often affected by the individual's personal relationship with the neighbor (39–41).
2. Annoyance caused by a given noise is usually greater indoors than outdoors (39–41).
3. Noises produced at night are usually more annoying than the same noise produced in the daytime (39–42).
4. Past exposure patterns influence reactions to specific noise exposures. An individual living in a highly industrialized area is less likely to be disturbed by noise than suburban area residents (39).
5. Annoyance generally increases as either the level or frequency of the noise increases.
6. Noises that are intermittent and occur randomly in time are normally judged more annoying than those that are continuous or unchanging.
7. A noise source that moves is usually judged to be more annoying than a stationary source (43).

It is obvious that all noise exposure circumstances must be fully described in each case before meaningful noise specifications can be established. The many variables involved prevent the establishment of a generally applicable guideline; however, several indoor noise limit guidelines have been proposed for a number of specific locations having average conditions. Although these guidelines do not always hold accurately, they do provide useful and necessary guidelines for architects, engineers, and others in many cases (see Fig. 25.9)

11.3.2. Measurement Procedures

Most of the procedures proposed for describing annoyance or distraction effects of noise make use of tabulations of noise measures that are correlated with different levels of human response in the specific activity or space considered. A large number of noise measures have been proposed for these subjective responses, some being better for specific purposes than others (30–38,44–47).

Two of the most widely used noise measures in indoor noise criteria are SIL (or PSIL) and A-frequency weight (dBA) measures (31–36,48,49). Figure 25.9 shows design goals that will provide acceptable indoor space environments in most instances. Maximum permissible levels may be 5–10 dB higher than the design goal values given under some circumstances where noises are continuous and broadband, and where communication distances are relatively small.

Location	A-Weighted Sound Pressure Level in (dBA)	PSIL in (dB)
Residences		
Rural and suburban	25–30	20–25
Urban	25–35	20–30
Industrial	30–40	25–35
Offices		
Conference rooms	25–35	20–20
Large	25–30	20–25
Small	20–35	25–30
Executive offices	30–40	25–35
Closed office (wall to ceiling)	30–45	25–40
Open office (half-walls)	35–50	30–45
Halls and corridors	35–55	30–50
Churches and schools		
Libraries	30–40	25–35
Classrooms	30–40	25–35
Laboratories	35–45	30–40
Halls and corridors	35–55	30–50
Kitchens	45–55	40–50
Auditoriums		
Lecture halls	35–40	30–35
Concert halls	25–30	20–25
Movie theaters	35–45	30–40
Lobbies	40–50	35–45
Restaurants	40–50	35–45
Cafeterias	45–55	40–50
Stores	40–50	35–45
Hospitals		
Private rooms	30–40	25–35
Operating rooms	35–45	30–40
Laboratories	40–50	35–45
Lobbies	40–50	35–45
Halls and corridors	40–50	35–45
Manufacturing areas	As low as practical but in all cases less than:	
	85	80

Figure 25.9 Table of noise specifications for indoor space.

11.3.3. Interference with Work

Noise can influence work in many ways, both directly and indirectly. The amount of interference with work may vary from small distractions caused by very low noise levels to alterations in visual and tactile perception that result from higher levels (40,50–53). Interference with communication is obviously another important factor to be considered when measuring work output.

As is the case with annoyance caused by noise, work interference cannot be described by noise levels alone, and generally applicable guidelines are not available because of the many variables involved. Experiments have shown that the effect of noise is more likely to result in an increased rate of errors or accidents rather than decreased total work output (50).

Generally, these effects are increased as the noise level is increased, particularly if the level rises above about 90 dB in the central octave bands.

Some studies have shown that broadband masking noise or instrumental music can be used effectively to mask interrupted noises, or hushed conversations, and thus reduce distractions. However, there is evidence of significant intersubject variability and other factors that preclude the general applicability of these data.

REFERENCES

1. *American National Standard Specification for Sound Level Meters, 1985*, American National Standards Institute, New York (ANSI), 1983–1985. ANSI S1.4A-1985 (ASA 47) Amendment to S1.4-1983.
2. *American Standard Specification for Octave-Band and Fractional-Octave-Band Analog and Digital Filters, 1986*, American National Standards Institute, New York (ANSI), ANSI S1.11-1986 (ASA 65).
3. *American National Standard Design Response of Weighting Networks for Acoustical Measurements, 1986*, American National Standards Institute, New York (ANSI), ANSI S1.42-1986 (ASA 64).
4. *American Standard Methods for Measurement of Impulse Noise, 1968*, American National Standards Institute, New York (ANSI), ANSI S12.7-1986 (ASA 62).
5. *Draft American National Standard Methods for the Evaluation of the Potential Effect on Human Hearing of Sounds with Peak A-Weighted Sound Pressure Levels and Peak C-Weighted Sound Pressure Levels Below 140 Decibels, 1986*, American National Standards Institute, New York (ANSI), ANSI S3.28-1986.
6. *American National Standard Specification for Personal Noise Dosimeters, 1978*, American National Standards Institute, New York (ANSI), ANSI S1.25-1978 (ASA 25).
7. *American Standard Methods for Measurement of Sound Pressure Levels, 1983*, American National Standards Institute, New York (ANSI), ANSI S1.13-1971 (R 1983).
8. *American National Specification for Acoustical Calibrators* (ANSI), S1.40-1984. Standards Secretariat, Acoustical Society of America, New York, 1984.
9. *Guidelines for Developing a Training Program in Noise Survey Techniques*, Committee on Hearing and Bioacoustics (CHABA), ONRC Contract No. N00015-67-A-0244-0021, U.S. EPA ONAC, 1975.
10. Snow, WE. Significance of reading acoustical instrumentation, *Noise Control, 5*:40, 1959.
11. Beranek, LL. *Acoustic Measurements*, Wiley, New York, 1949.
12. Harris, CM. *Handbook of Noise Control*, McGraw-Hill, New York, 1957.
13. *American Standard Specification for General-Purpose Sound Level Meters, S1.4-1961*, American National Standards Institute, New York.
14. Michael, PL. *Noise at the Source*, National Safety Congress Transactions, National Safety Council, 1971.
15. *American National Standard Specification for Sound Level Meters, 1971*, American National Standards Institute, New York (ANSI), ANSI S1.4-1971.
16. *American Standard Specifications for Octave, Half-Octave, and Third-Octave Filter Sets, S.11-1966*, American National Standards Institute, New York.
17. *American Standard for Preferred Frequencies for Acoustical Measurements, S1.6-1960*, American National Standards Institute, New York.
18. Scott, HH. The degenerative sound analyzer, *J. Acoust. Soc. A, 11*:225, 1939.
19. American Conference of Governmental Industrial Hygienists, *Threshold Limit Values of Physical Agents Adopted by ACGIH for 1970.*
20. Safety and Health Standards for Federal Supply Contracts. U.S. Department of Labor, *Fed. Register, 34*:7498–7949, 1969.
21. Guidelines for noise exposure control, *Am. Indust. Hyg. Assoc. J., 28*:418–424 (1967). Ibid., *Am. Assoc. Indust. Nurses J., 16*:17–21, 1968.

22. Occupational Safety and Health Standards (Williams-Steiger Occupational Safety and Health Act of 1970). U.S. Department of Labor, *Fed. Register, 36*:10518, 1971.
23. Webster, JC. SIL — Past, present and future, *Sound Vibration*, 22–26, August 1969.
24. Anticaglia JR, and Cohen, A. Extra-auditory effects of noise as a health hazard, *Am. Indust. Hyg. Assoc. J., 31*:277–281, 1970.
25. Glorig, A. *Noise and Your Ear*, Grune & Stratton, New York, 1958.
26. *ISO R.140 Field and Laboratory Measurements of Airborne and Impact Sound Transmission*, International Organization for Standardization, Geneva, Switzerland.
27. American Conference of Governmental Industrial Hygienists, *Threshold Limit Values of Physical Agents Adopted by ACIH for 1970.*
28. Safety and Health Standards for Federal Supply Contracts (Walsh-Healey Public Contracts Act). U.S. Department of Labor, *Fed. Register, 34*:7948–7949, 1969.
29. Guidelines for Noise Exposure Control, *Am. Indust. Hyg. Assoc. J., 28*:418–424, 1967. Ibid., *Arch. Environ. Health, 15*:674–678 (1967). Ibid., *J. Occup. Med., 9*:571–575, 1967. Ibid., *Am. Assoc. Indust. Nurses J., 16*:17–21 (1968).
30. French, NR and Steinberg, JC. Sound control in airplanes, *J. Acoust. Soc. Am., 19*:90–119, 1947.
31. Beranek, LL. *Acoustics*, McGraw-Hill, New York, 1954.
32. Beranek, LL. Airplane quieting II — Specification of acceptable noise levels, *Trans. ASME, 69*:96–100, 1947.
33. Beranek, LL. Criteria for office quieting based on questionnaire rating studies, *J. Acoust. Soc. Am., 28*:833–852, 1956.
34. Beranek, LL. Revised criteria for noise in buildings, *Noise Control, 3*(1):19–27, 1957.
35. Beranek, LL, Reynolds, JL, and Wilson, KE. Apparatus and procedures for predicting ventilation system noise, *J. Acoust. Soc. Am. 25*:313–321, 1953.
36. Webster, JC. SIL — Past, present and future, *Sound Vibration, 22*:26, 1969.
37. Young, RW. Don't forget the simple sound-level-meter, *Noise Control, 4*(3):42–43, 1958.
38. Kyter, KD. Concepts of perceived noisiness, their implementation and application, *J. Acoust. Soc. Am., 43*:344–361, 1968.
39. Fletcher, H. Auditory patterns, *Rev. Mod. Phys., 12*:47–65, 1940.
40. Parrack, H. Community reaction noise, Chapter 36 in *Handbook of Noise Control* (C. M. Harris, ed.), McGraw-Hill, New York, 1957.
41. Cohen, A. Noise effects on health, productivity and well-being, *Trans. N.Y. Acad. Sci., 30*:910–918, 1968.
42. Wilson, A. *Noise*, Chapter IV. Her Majesty's Stationers, London, 1963.
43. Kryter, KD. Psychological reactions to aircraft noise, *Science, 151:*1346–1355, 1966.
44. Kryter, KD and Williams, CE. Masking of speech by aircraft noise, *J. Acoust. Soc. Am., 39*:138–150, 1966.
45. Williams, CE, Stevens, KN, Hecker, MHL, and Pearson, KS. *The Speech Interference Effects of Aircraft Noise*, Federal Aviation Administration Report DS-67-19, 1967.
46. Klumpp, RG and Webster, JC. Physical measurements of equally speech-interfering navy noises, *J. Acoust. Soc. Am., 35*:1328–1338, 1963.
47. Kryter, KD. The meaning and measurement of perceived noise level, *Noise Control, 6*(5):12, 1960.
48. Peterson, APG and Gross, EE. *Handbook of Noise Measurements*, General Radio Company, West Concord, MA, 1963.
49. *ASHRAE Guide and Data Book — Fundamentals and Equipment*, American Society of Heating and Refrigeration and Air-Conditioning Engineers, New York, 1970.
50. Broadbent, DE. Effects of noise on behavior, Chapter 10 in *Handbook of Noise Control* (C. M. Harris, ed.), McGraw-Hill, New York, 1957.
51. Carpenter, A. How does noise affect the individual? *Impulse, 24*, 1964.
52. Broussard, IG et al., *The Influence of Noise on Visual Contrast Threshold*, U.S. Army Medical Research Laboratory Rept. No. 101, Fort Knox, KY, 1952.
53. Loeb, M. *The Influence of Intense Noise on Performance of a Precise Fatiguing Task*, Army Medical Research Laboratory Rept. No. 268, Fort Knox, KY, 1957.

26
Noise Control

Paul L. Michael
Paul L. Michael and Associates, Inc. and The Pennsylvania State University, State College, Pennsylvania, USA

Kevin L. Michael
Michael and Associates, Inc., State College, Pennsylvania, USA

Vibrations of any solid object will produce pressure variations in air that may be perceived as sound or noise when the vibration amplitudes are sufficiently high and the vibration frequencies are in the audible range. Vibration amplitudes are directly related to the noise levels produced; thus, reduction of mechanical vibration amplitudes may be a very effective noise and control measure.

If vibration of component parts of a machine cannot be reduced sufficiently to prevent noise problems from developing in surrounding areas, the noise levels must be controlled by enclosures, barriers, isolation procedures, or by the use of noise-absorbing materials. Often it is necessary to make use of combinations of these noise control procedures in order to obtain the required noise reduction.

1. REDUCTION OF RADIATED NOISE

The overall noise level radiated from a machine may be the product of a number of different individual noise sources within the machine, and in many cases, each individual noise source must be considered separately for the most efficient noise control measures. Because of the logarithmic nature of the individual noise source contribution, it is essential that any noise control measure be directed toward the individual sources in the order of their contributions to the overall noise level. Otherwise, much time, effort, and money can be wasted.

For example: Consider a machine that produces an overall sound-pressure level of 98 dB at a given location near the machine. If this machine has three individual noise source components capable of producing levels of 86, 91, and 96 dB, respectively, at the same location if operated singly, the 96 dB source should be treated first with attention being given to the 91 dB and 86 dB sources later, in that order. If noise control measures were applied to the 86 dB source first, this source could be completely removed and the overall level would be reduced by only about 0.3 dB. However, if the 96 dB and 91 dB sources were reduced by 12 and 10 dB, respectively, the overall level would be reduced by more than 9 dB.

Often it is impractical or impossible to determine the contribution of an individual noise because other noise-making parts are running at the same time. In these cases, the contributions of the individual sources can usually be determined from frequency analyses of the overall noise levels produced. Correlations can be made between the frequency bands having the highest levels and the running speeds of various machine components which might produce these noises. Another practical guide for pinpointing a noise source among other noise sources is to correlate dimensions of the radiating source with the frequency spectra because only those vibrating parts having dimensions similar to, or larger than, a quarter wavelength (in air) are capable of radiating noise efficiently.

Once the principal noise-making components are located and action priorities established, the most effective noise control procedures must be selected. More than one control procedure may be required to reduce the levels radiated by some sources. Where two or more control procedures may be ineffective when used singly, together they may produce significant results.

1.1. Component Size, Shape, and Material

When possible, machine component sizes should be held to dimensions that are small in comparison to the quarter wavelength (in air) of any vibrational energy that might be connected to the component so that noise will not be radiated efficiently. If it is not possible to hold the overall dimension of a machine part to a small size it may be possible to use two pieces instead of one in such a way that the second piece is isolated from the vibration source.

Any machine component should be made in a shape and size to resist vibration and, in particular, to avoid resonances* that may cause high-level noise radiation. Heavy, rigid parts are usually preferred in the design of quiet equipment; however, lightweight material can be used in many cases if properly supported and damped.

1.2. Vibration Damping

Materials used in the construction of machines have varying degrees of internal damping; however, the effect of internal damping is small in comparison with the damping effectiveness of specially developed damping materials that can be added. For example, lead and other materials have high internal damping characteristics, but specially compounded damping materials with the same weight are generally much more effective whenever damping is required.

Vibration-damping materials normally are effective in just two cases: (a) when forced vibration frequencies correspond with the resonant frequencies in component parts of the attached equipment, or (b) when impact-type shocks are applied to relatively thin surfaces.

Resonant conditions can be detected by slowly increasing the operating speed of a machine from below to above its normal operating range. If a significant increase in loudness is heard, or if a tone suddenly becomes clearly audible at certain speeds, a resonant condition is indicated and a vibration damping treatment probably will be effective. If there is not a marked change in character or loudness of the noise radiated as the machine speed is increased, it is doubtful that a vibration-damping treatment will be worthwhile. A gradual increase of loudness or pitch during an increase of operating speed does not indicate a resonant condition.

Metal panels often have many resonant frequencies that can be excited by either continued forced vibrations or a single blow. Examples of unwanted resonant panel noise may be found in many commonly used products, ranging from automobile bodies to porch furniture. Damping treatments may often be used to reduce the overall levels produced by resonant panels, and in addition, the damping treatment normally results in a shift of noise energy to lower frequencies where the ear is less sensitive.

In cases where damping treatment is indicated, the area of coverage is often critical. Complete coverage of a vibrating panel ensures good damping; however, it is not the most economical use of the damping material, and it sometimes adds significantly to the weight of the product. A single-point application of a damping device or material is seldom effective; however, spot-damping treatments (small-area coverage) can be effective if care is taken to place the damping material precisely on the areas having maximum vibration amplitude. Unfortunately, the spot treatment procedure has the obvious disadvantage of requiring tedious vibration measurements to determine the areas where spot damping must be applied.

The thickness (or weight per unit area) of a vibration-damping material is also related to its effectiveness; however, the many parameters involved in different

---

*Resonance of a component part exists when any change in the excitation frequency of forced vibration causes a decrease in the vibration amplitude of the component part.

applications make it impossible to establish definite rules on the thickness required for a specific problem.

A few general guidelines in the application of damping treatment are as follows:

1. A given damping treatment is usually more effective for high than for low vibration frequencies.
2. A heavy piece of equipment will require a heavier damping treatment than a light piece.
3. The stamping of ribs in flat sheet metal panels will raise the natural resonant frequencies of a panel without increasing its mass, thereby making possible the advantages noted in Items 1 and 2.

1.3. Tolerances

Vibrational amplitudes of considerably less than one-thousandth of an inch can produce high-level noise. Thus, excessive tolerances or worn parts in moving systems often are primary noise sources.

A first step in a tolerance-noise control procedure is to replace all worn parts and to properly align all moving pieces. In many cases, this maintenance of equipment provides the added advantage that a machine will run more efficiently and it will last longer. Noise is indeed wasted energy in many instances.

If proper maintenance of equipment does not provide enough noise reduction, it may be necessary to decrease the original tolerances between some of the moving parts or to select new materials for component parts. Reduced tolerances prevent excessive levels; however, care must be taken not to reduce tolerances too much because the machine's operation may be impaired or the higher friction and the resulting heat may shorten the life of component parts. Similarly, materials with high internal damping or relatively soft surfaces will usually reduce the noise produced from shock or impact, but the operational life of these materials may be very short. Compromise between noise radiation, cost, and wearability will often determine whether the radiated noise can be controlled at the source or if noise control measures must be taken external to the source.

2. VIBRATION ISOLATION

Vibration of a machine may be transmitted to its supporting surface if rigidly mounted, and these vibrations may in turn cause noise to be radiated from the floor, ceiling, walls, or other structures attached to the supporting surface. In these cases, one of the most effective noise control measures may be to mount the machine on vibration isolators.

Vibration isolators are made from a number of different materials, and their designs take many different forms. Coil or leaf springs, gas- or liquid-filled devices, and pads made from rubber, cork, felt, or fiberglass are common forms of vibration isolators. The choice of the most effective vibration isolator depends on such factors as the machine weight and size, the vibration frequency spectra, and the environmental conditions to which the isolator will be exposed. The proper isolator characteristics must be chosen for a particular job if it is to be effective and long lasting. In fact, the choice of the wrong mechanical characteristics of vibration isolators may not only be ineffective, but may amplify the force vibrations transmitted to the supporting surface of the machine.

All connections to a vibrating machine other than mounting points must also be properly designed for vibration isolation; otherwise, benefits from the vibration isolating mounts may be lost. Electrical, fuel, control, ventilation, and other connections may all require different kinds of flexible connections in order to obtain maximum effectiveness. Here again, definite rules for the design of flexible connectors cannot be drawn because of the many parameters involved. Generally, the following guides can be used for the design of flexible couplers:

1. If possible, the connection to the machine should be made where the vibration amplitudes are at a minimum, and the other end of the connection should be made on the most solid and massive support available.
2. A relatively rigid machine coupling can be tolerated if it is terminated on a solid and massive surface. However, a very flexible coupling may be necessary if the machine is coupled to a flexible and lightweight surface.
3. If connection are to be made with metal or nonmetal tubing, long loops or coils should be used that are flexible in all directions.
4. When nonmetals cannot be used because of high temperature or solvent problems, special flexible metal tubes made of stainless steel, brass, copper, or Monel are available. Special flexible metal tubing normally cannot withstand high pressures and should be carefully chosen for a particular application.

A complete treatment of the problem of vibration isolators is beyond the scope of this book. More details on the theory and application of vibration isolation may be found elsewhere (1–6).

3. REDUCTION OF NOISE AWAY FROM THE SOURCE

If sufficient noise reduction is not possible by direct treatment of the noise source or by mechanically isolating the source from surrounding structures, the next step is to use noise control measures in the surrounding areas. Two common noise control procedures that can be used singly or together to reduce radiated noise levels are (a) to partially or completely enclose the noise source with materials having high sound transmission loss characteristics, or (b) to absorb the noise with sound-absorbing materials placed in selected locations.

4. NOISE BARRIERS AND ENCLOSURES

The effectiveness of a material for use as a barrier to noise (transmission loss characteristics) is highly dependent on its weight per unit area. Normally, the kind of material used for a noise barrier is relatively unimportant if the material is not porous and is constructed so that the necessary weight per unit area is provided.

The transmission loss (TL) provided by a barrier normally is highly dependent on the frequency of sound, with the loss at high frequencies being considerably greater than at low frequencies. The TL of most single-wall type barriers for randomly incident noise increases about 5 dB for each doubling of frequency.

The relationship of transmission loss provided by a single-wall-type barrier to its weight per unit area is commonly expressed in terms of an average TL for frequencies between 125 and 2000 Hz. This single number value is sufficiently accurate for most practical purposes since the 5 dB/doubling of frequency holds reasonably well for most single-wall structures. Transmission losses for a large variety of single-wall barriers are available in the literature (4–10).

Multiple-wall construction with enclosed air spaces provides considerably more attenuation than the single-wall-mass law would predict (7–9). However, considerable care must be taken to avoid rigid connections between the multiple wall surfaces when they are constructed or any advantages in attenuation will be lost (9,10).

Noise leaks that may result from cracks, holes, windows, or doors in a noise barrier can severely limit noise reduction characteristics. In particular, care must be taken throughout construction to prevent acoustical leaks that may be caused by electrical outlets, plumbing connections, telephone lines, etc., in otherwise effective barriers. For example, a hole 1.5 in. square in a wall will transmit about the same amount of acoustical energy as 100 ft^2 of a wall area that has a TL of 40 dB.

The choice of a simple barrier, a partial enclosure, or a complete enclosure depends on several factors, including:

1. Position of the noise source with respect to the exposure area
2. Acoustical characteristics of the surrounding area
3. Frequency spectrum of the noise
4. Amount of noise reduction required

A simple barrier may be effective if the positioning of the noise source or sources and the acoustical characteristics of the surrounding area are such that the major noise contribution is coming from one general direction. Also, it must be feasible to build a barrier whose smallest cross-sectional dimension is large compared to the wavelengths of the major noise spectrum components (see Chapter 2, "The Physics of Sound") in a location between the source and the exposure area.

If a single barrier does not provide adequate noise reduction because of multiple angles of incidence, or because of too many low-frequency (large-wavelength) energy bends around the barrier, then additional wall or ceiling barriers are often effective. If the multiple-barrier, or partial-enclosure, structure is not effective, a complete enclosure may be necessary.

Any barrier or enclosure must be carefully isolated mechanically from the noise source, otherwise forced vibrations may cause noise to be radiated from the barrier or enclosure and its quieting effects will be nullified. Connections through the enclosure to the machine are particularly important. Tubing, wiring, and other small connections should be passed through rubber grommets that are placed near corners or other stiffening members of the enclosure. If needed, ventilating air should be supplied through ducts lined with sound-absorbing material. In addition, all portions of the enclosure should be carefully designed to avoid dimensions having resonant frequencies corresponding to the spectra of principal noise components of the source.

When barriers or enclosures confine the radiated noise in a relatively small volume that has hard acoustically reflecting surfaces, the radiated noise will combine with the reflected energy so that the overall levels around the source are substantially increased. For this reason, an enclosure constructed with panels having a TL of 25 dB may provide only 5 dB of noise reduction if the noise levels inside the enclosure are increased by 20 dB from the reflected energy buildup. In these cases, noise absorption materials must be used within the enclosure to minimize the reflected energy buildup.

4.1. Noise Absorption

Good sound-absorbing materials are normally light in weight and porous in contrast to the massive and nonporous requirements for a good noise barrier. Thus, a good sound absorber is usually a poor sound barrier, and vice versa.

To be a good sound absorber, the sound waves must penetrate into the absorbing material where the sound is dissipated in the form of friction and heat. The amount of sound entering a porous material (and the amount of sound energy absorbed) is dependent on the wavelength of the sound and its angle of incidence on the absorbing material.

The ability of a material to absorb sound of a particular frequency is often described by a sound-absorption coefficient (a) which is the ratio of the sound energy absorbed by the material to the amount of energy incident upon it. A surface that absorbs all energy incident upon its surface is said to have an absorption coefficient of one, while a surface that reflects all energy has an absorption coefficient of zero. An average sound absorption coefficient (a) in a room having several different surface materials is found for a given frequency by

$$a = \frac{a_1S_1 + a_2S_2 + a_3S_3 + \cdots + a_nS_N}{S_1 + S_2 + S_3 + \cdots + S_N} \tag{26.1}$$

where $a_1, a_2, a_3 \ldots a_n$ are the coefficients of absorption of the various surfaces of the room having corresponding areas S_1, S_2, $S_3 \ldots S_n$. The coefficients of absorption for most surface materials are readily available in the literature (6,11,12).

The relationship between average sound-pressure level, L_p, power level, L_P, and single-frequency absorption coefficients for a given semi-reverberant room may be written as

$$L_p = L_p + 10\log\left(\frac{Q}{4\pi r^2}\right) + 10.5 \text{ dB} \quad \text{re } 0.00002 \text{ n/m}^2 \tag{26.2}$$

where r is the distance in feet from the measurement point to the source, Q is the directivity factor, and $R = aS/1-a$ is the room constant in square feet. In a high-reverberant field, the average sound-pressure level can be written as

$$L_p = L_p - 10\log R + 16.6 \text{ dB} \quad \text{re } 0.00002 \text{ n/m}^2 \tag{26.3}$$

A rule of thumb that may be used to determine the amount of noise reduction possible from the application of acoustically absorbent materials on room surfaces is

$$\text{dB reduction} = 10\log\frac{\text{absorption units after treatment}}{\text{absorption units before treatment}} \tag{26.4}$$

where the absorption units are the sum of the products of surface areas and their respective noise absorption coefficients.

The overall noise-absorbing efficiency of acoustic materials is sometimes expressed by a single number known as the noise reduction coefficient (NRC). The NRC is found arithmetically by averaging four absorption coefficients between 250 and 2000 cps (usually 250, 500, 1000, and 2000 Hz). Small differences in NRC may not be detectable, so it is common practice to round off coefficients to the nearest 0.05.

5. ACTIVE NOISE CONTROL

Active noise control is a sound minimization technique that utilizes the generation of an out-of-phase signal (called the active source) to "cancel out" unwanted noise. Research in this area has been ongoing for several decades. Recent advances in microcomputers, dedicated signal processing circuitry, and adaptive filtering algorithms have enabled the practical development of real-time active noise control systems.

Active noise control is most effective on the low-frequency components of a noise source, usually below 1000 Hz (13,14). Effective *global* cancellation can be achieved if the active source is located near the noise source. In this case, the upper frequency limit of active attenuation in a free field will be about 70 divided by the separation of the sources. That is, if the sources are separated by 0.1 meter, the upper frequency limit would be about 700 Hz. Global minimization can also occur if the active source is acoustically coupled to the source. For example, active control is currently used to minimize duct noise. The active source and the noise source are coupled by the duct, which acts as a waveguide. Effective low-frequency *local* cancellation can be achieved at any point in space, but the benefits are limited to a small area, such as inside the ear cup of a hearing protector (15). Because active control is frequency-limited, it is usually used in conjunction with traditional passive noise control measures.

The noise control problem becomes much more complicated in reverberant fields and when there are multiple noise sources (13,14). Current research on global minimization of noise sources in typical work areas is utilizing arrays of sensors and active sources to achieve noise cancellation at many locations within a sound field. The result can be an overall reduction of sound pressure level, a reduction at a specific point in space, or a reduction in a plane in space. This is useful in situations such as minimizing machine or vehicle noise at the location of the operator's head or minimizing airplane noise at ear level in the passenger area.

A closely related field is the active reduction of vibration. As an example, accelerometers and shakers can be used as active sensors and sources to minimize unwanted vibration. The vibration generated by machinery is sampled and an inverse signal is introduced into the mounting system of the machine or to the surrounding structures. In an airplane, for example, noise is generated in the cabin from the vibration of the outer structure. Accelerometers and shakers can be used on the shell of the aircraft to minimize the vibration and noise transmitted to the pilot and passenger areas. Advanced sensors and actuators are currently being developed specifically for active vibration control.

6. SPECIFICATIONS FOR PURCHASING EQUIPMENT

General principles of noise control are well known, but unfortunately little effort has been expended toward applying this knowledge. The lack of progress in noise reduction can be attributed for the most part to the absence of a demand for quiet equipment. Quieting procedures normally will increase the cost of equipment, and manufacturers will not jeopardize sales by increased cost in a competitive market unless the demand for engineering control of noise warrants the modifications on new equipment throughout an industry.

Another very strong reason for noise limits in purchasing specifications is that noise control procedures are usually much more effective and less expensive when taken during the design and development stages rather than after the equipment is in use. Many prefer quiet machinery and will welcome reasonable noise limits in purchasing specifications.

6.1. Standards

Standard methods for measuring and reporting noise levels have been developed by several manufacturing groups or associations for use with their products. Organizations that have noise measurement and reporting specifications include the American Gear Manufacturers Association (16), the American Society of Heating, Refrigeration, and Air-Conditioning Engineers (17), the Air Moving and Conditioning Association (18), the Institute of Electrical and Electronic Engineers (19), the American Iron and Steel Institute (20), and the National Electrical Manufacturers Association (21).

Unfortunately, many manufacturers do not belong to organizations that have standard methods of measuring and reporting noise characteristics, and the specifications set forth by other groups may not be applicable (22–30). A comparison of noise characteristics from competitive machines from different manufacturers may be meaningless unless the measurements are clearly defined and applicable for describing the noise made by the equipment.

6.2. Noise Control Specifications

Standard procedures that are best for specifying noise characteristics of one piece of equipment may not be best for another because of size, use, levels, etc. Therefore, only general guidelines can be given in setting up overall engineering specifications.

A major objective of any engineering specifications is to make the equipment manufacturer aware of his responsibility for the noise produced by his equipment. If the manufacturer does not belong to a group that has its own noise measurement specifications which are acceptable, he should be guided by a reference in the purchasing specification to a pertinent standard procedure such as one of those referenced (16–20). Otherwise, a specific set of measurement and reporting instructions should be provided in the specification.

In any engineering specification for noise, the acceptable noise levels obviously must be listed but, in addition, the test signals, the instrumentation, the test procedures, and the test environment must also be carefully specified. The characteristics of acceptable noise levels should be specified in detail whenever possible; however, in some instances, it may not be possible to specify the levels precisely because of equipment size or unusual conditions of use. In these cases, a selection from available equipment can be made on the basis of the lowest noise levels produced.

6.3. Test Signals

Octave-band sound-pressure-level measurements are usually adequate to describe the noise characteristics of a machine; however, more specific information in the form of narrow-band analyses may be desired when a large portion of the energy is contained in narrow-frequency bands. In all cases linear and A-weighted overall measurements should be made before and after each series of octave- or narrow-band analyses.

6.4. Test Instrumentation

All instruments used for noise measurements should meet the latest standards of the American National Standards Institute. Also, these instruments should be calibrated electrically and acoustically immediately before and after the measurement on each piece of equipment to be tested.

6.5. Test Procedures

Noise measurements are the vendor's responsibility; however, to be sure that measurements are performed properly, the purchaser should reserve the right to send qualified representatives to the vendor's plant to observe or to conduct noise tests if necessary. General test requirements may include the following steps:

1. Noise measurements should be made when the equipment is operating at both normal and maximum running speeds. In all cases, the equipment should be mounted in the same manner as intended for permanent operation.
2. Noise measurement equipment should be located so that electric or magnet fields, mechanical vibrations, wind, or other extraneous factors will not affect the accuracy of the data.
3. Measurements should be made at locations corresponding to positions where human ears may be located when the equipment is in its proposed permanent location. In addition, measurements should be made around the equipment at 30° intervals, 5 ft above the floor level at a horizontal distance of from 3 to 6 ft from the equipment. All measurement positions should be accurately recorded. A sample noise survey data sheet is shown in Fig. 25.6.
4. Whenever mean noise levels vary by more than 6 dB during normal operation, measurements should be repeated to describe each operational phase that produces different noise levels.
5. All sound-pressure-level measurements should be taken with the slow-meter damping characteristics and average meter indications recorded when the range of meter deflections is less than 4 dB. When the meter deflections equal or exceed 4 dB, the range of meter deflections, and any other prominent level variation characteristics, should also be recorded.
6. Octave-band, A-frequency weighted, and flat (9,10) sound-pressure-level measurements should be made at each measurement location. When the noises produced contain pure-tone noise components, a narrow-band analysis may be necessary.
7. Other noise measurement data that should be recorded include: (a) the type, model, and serial numbers of all instruments used; (b) the microphone type and serial number; (c) the microphone mounting or cable length; (d) the microphone orientation; (e) calibration information; (f) the response speed of the indicating meter; and (g) any remarks covering any significant phase of the test procedure not covered elsewhere.

6.6. Test Location

The ideal test location for one noisemaker may not be ideal for another because of differences in size, frequency, or directional characteristics and levels produced. For most relatively small noise sources, a free or reverberant field for testing is best, and this requires specially treated anechoic or reverberation chambers.

Size of noisemaking devices may prevent testing in an anechoic room, reverberation chamber, or other carefully controlled or predictable test environment. However, meaningful measurements can often be made in other locations if the test environment can be described carefully in a simple manner, i.e., if there are no significant noise-reflecting surfaces other than the inner surfaces of the test room, or if other reflecting surfaces can be described acoustically in simple terms. A test room description should include:

1. Complete elevation and floor plan sketches of the equipment location in the test room along with room dimensions. Positions and descriptions of other equipment in the test room should be included.
2. Materials used on the floor, walls, and ceiling of the test room.
3. Floor supports used for the equipment under test, i.e., is it bolted down on concrete, or vibration mounts, etc.? The mounting should be the same as planned for the permanent operating position.
4. Ambient noise levels at the time of the test should be recorded for each frequency weighting, octave band, or narrow band used in the measurements procedure. Ambient levels should be more than 10 dB below any level recorded.

6.7. Requirements

The *Guidelines for Noise Exposure Control* (28) may be used as a guide for establishing noise exposure limits. The total noise contributed by all noise sources should be less than the established limits.

It must be possible to use the noise measurement data supplied by the equipment manufacturer to determine the levels that will be produced in the work area. Calculations using sound power or sound-pressure level can be made as described in Chapter 2 if adequate information is provided.

An alternate way of approaching this problem is to place the burden of performance on the manufacturer of the equipment being purchased. The purchasing specification can require that the purchased equipment will not produce sound-pressure levels greater than the specified level in the area where the equipment is to be used. This kind of specification has had only limited success because manufacturers are often unable or unwilling to estimate levels in a complex acoustical environment.

7. EXAMPLES OF NOISE CONTROL

Engineering procedures for the control of noise may take many forms. The most effective and economical means for achieving a reasonably quiet work environment is to use machines and equipment designed to produce a minimum amount of noise. Unfortunately, many long-lived and expensive machines now in use produce very high noise levels that must be controlled by engineering means. Some examples of machines and equipment that have been quieted by engineering means are listed in Fig. 26.1.

7.1. Examples of Noise Absorption

Machines that use moving parts such as cams, gears, reciprocating pieces, and metal stops are often located in large, acoustically reverberant areas that reflect and build up noise levels in the room. A significant reduction of noise levels can be accomplished at times, in locations away from the noise sources, by use of absorption materials. The type, amount, configuration, and placement of absorption materials must be considered

Materials	125	250	500	1,000	2,000	4,000
	Coefficients (Hz)					
Brick—glazed	.01	.01	.01	.01	.02	.02
Brick—unglazed	.03	.03	.03	.04	.05	.07
Brick—unglazed, painted	.01	.01	.02	.02	.02	.03
Carpet—heavy, on concrete	.02	.06	.14	.37	.60	.65
Same—on 40 oz. hairfelt or foam rubber (carpet has coarse backing)	.08	.24	.57	.69	.71	.73
Same—with impermeable latex backing on 40 oz. hairfelt or foam rubber	.08	.27	.39	.34	.48	.63
Concrete block—coarse	.36	.44	.31	.29	.39	.25
Concrete block—painted	.10	.05	.06	.07	.09	.08
Concrete block—poured	.01	.01	.02	.02	.02	.03
Fabrics						
Light velour—10 oz. per sq. yd. hung straight, in contact with wall	.03	.04	.11	.17	.24	.35
Medium velour—14 oz. per sq. yd. draped to half area	.07	.31	.49	.75	.70	.60
Heavy velour—18 oz. per sq. yd. draped, to half area	.14	.35	.55	.72	.70	.65
Floors						
Concrete or terrazzo	.01	.01	.015	.02	.02	.02
Linoleum, asphalt, rubber or cork tile on concrete	.02	.03	.03	.03	.03	.02
Wood	.15	.11	.10	.07	.06	.07
Wood parquet in asphalt on concrete	.04	.04	.07	.06	.06	.07
Glass						
Large panes of heavy plate glass	.18	.06	.04	.03	.02	.02
Ordinary window glass	.35	.25	.18	.12	.07	.04
Glass Fiber—mounted with impervious backing—3 lb /cu ft, 1" thick	.14	.55	.67	.97	.90	.85
Glass Fiber—mounted with impervious backing—3 lb /cu ft, 2" thick	.39	.78	.94	.96	.85	.84
Glass Fiber—mounted with impervious backing—3 lb / cu ft, 3" thick	.43	.91	.99	.98	.95	.93
Gypsum Board—½" nailed to 2 x 4's. 16" o.c.	.29	.10	.05	.04	.07	.09
Marble	.01	.01	.01	.01	.02	.02
Openings						
Stage, depending on furnishings			.25- .75			
Deep balcony, upholstered seats			.50-1.00			
Grills, ventilating			.15- .50			
Grills, ventilating to outside			1.00			
Plaster—gypsum or lime, smooth finish on tile or brick	.013	.015	.02	.03	.04	.05
Plaster—gypsum or lime, rough finish on lath	.14	.10	.06	.05	.04	.03
Same, with smooth finish	.14	.10	.06	.04	.04	.03
Plywood paneling—3/4" thick	.28	.22	.17	.09	.10	.11
Sand						
Dry—4" thick	.15	.35	.40	.50	.55	.80
Dry—12" thick	.20	.30	.40	.50	.60	.75
Wet—14 lb. water per cu. ft., 4" thick	.05	.05	.05	.05	.05	.15
Steel						
Water	.01	.01	.01	.01	.02	.02
As in a swimming pool	.008	.008	.013	.015	.020	.025

Figure 26.1 Sound absorption coefficients of materials. The absorption coefficient *a* of a surface that is exposed to a sound field is the ratio of the sound energy absorbed by the surface to the sound energy incident upon the surface. For instance, if 55% of the incident sound energy is absorbed when it strikes the surface of a material, the *a* of that material would be 0.55. Since the *a* of a material varies according to many factors, such as frequency of the noise, density, type of mounting, surface conditions, etc., be sure to use the *a* for the exact conditions to be used and from performance data listings such as acoustical materials (refer to the bulletin published yearly by the Acoustical Materials Association, 335 East 45th Street, New York, New York 10017).

specifically for each application; however, the choice of absorbing materials can be guided by the absorption coefficients shown in Fig. 26.1.

Example 1.1 The noise produced by 10 wire-cutting machines around the periphery of a 20 ft × 60 ft × 75 ft reverberant room was reduced as shown below by the installation of absorption material above the machines.

Octave-band center frequency (O.B.) (Hz) (29)	31.5	63	125	250	500	1000	2000	4000	8000
Noise reduction (N.R.) (dB)	—	—	—	2	5	5	10	12	10

Example 1.2 Several motor generator sets were producing excessive noise levels in a large, reverberant room. Noise levels at significant distances away from the generators were reduced as shown below by hanging 6 lb/ft (27). Fiberglass baffles in rows just above the level of lights on 3-ft centers. These baffles may be completely encased in a thin film of materials such a polyethylene or mylar without significantly reducing their effectiveness in many applications.

O.B. (Hz)	31.5	63	125	250	500	1000	2000	4000	8000
N.R. (dB)	—	4	7	9	10	7	8	8	3

Example 1.3 The noise levels produced in a large, reverberant textile mill weave room was reduced with Eloff Hanson Sonosorbers suspended above the lights as shown below.

O.B. (Hz)	31.5	63	125	250	500	1000	2000	4000	8000
N.R. (dB)	—	6	9	6	6	6	11	11	12

7.2. Examples Using Noise Barriers and Enclosures

The noise reduction that can be attained with barriers depends on the characteristics of the noise source, the barrier configuration and materials used, and the acoustical environment on either side of the barrier. The material used for noise barriers may be described generally in terms of its transmission loss (see Fig. 26.2), but all other factors must be considered for specific problems.

The noise reduction achieved by various configurations of specific barrier or enclosure materials may vary significantly. Generally, a single-wall barrier with no openings placed between the source and the person exposed might expect 2–5 dB reduction in the low frequencies and 10–15 dB in the high frequencies. Distance of the source and observer from the barrier is also a significant factor (Fig. 26.3). If both the source and observer are close to the barrier, higher noise reduction values are possible. The effects of two- or three-sided barriers are difficult to predict on a general basis; however, well-designed partial enclosures may provide about 5–10-dB noise reduction in the low frequencies and about 20–25 dB in the high frequencies. Complete enclosures of practical designs may provide in excess of 10–15-dB noise reduction in the low frequencies and in excess of 30 dB in the high frequencies. Caution must be taken with any barrier or enclosure to be sure there are no unnecessary openings. Figure 26.4 shows the average transmission losses of a single barrier as a function of barrier mass and percentage of open area.

Material or Structure	125	175	250	350	500	700	1,000	2,000	4,000
A. Doors									
1. Heavy wooden door—special hardware; rubber gasket at top, sides and bottom: 2.5" thick; 12.5 lb/sq ft	30	30	30	29	24	25	26	37	36
2. Steel clad door—well-sealed at door casing and threshold	42	47	51	48	48	45	46	48	45
3. Flush—hollow core; well-sealed at door casing and threshold	14	21	27	24	25	25	26	29	31
4. Solid oak—with cracks as ordinarily hung: 1.75" thick	12		15		20		22	16	
5. Wooden door (30" × 84"), special soundproof construction—well-sealed at door casing and threshold: 3" thick; 7 lb/sq ft	31	27	32	30	33	31	29	37	41
B. Class									
1. 0.125" thick; 1.5 lb/sq ft	27	29	30	31	33	34	34	34	42
2. 0.25" thick; 3 lb/sq ft	27	29	31	32	33	34	34	34	42
3. 0.5" thick; 6 lb/sq ft	17	20	22	23	24	27	29	34	24
4. 1" thick; 12 lb/sq ft	27	31	32	33	35	36	32	37	44
C. Walls—Homogeneous									
1. Steel sheet—fluted; 18 gage stiffened at edges by 2 × 4 wood strips; joints sealed; 4.4 lb/sq ft	30	20	20	21	22	17	30	29	31
2. Asbestos board—corrugated, stiffened horizontally by 2 × 8 in. wood beam; joints sealed; 7.0 lb/sq ft	33	29	31	34	33	33	33	42	39
3. Sheet steel—30 gage; 0.012" thick; 0.5 lb/sq ft	3	6		11		16		21	26
4. Sheet steel—16 gage; 0.598" thick; 2.5 lb/sq ft	13	18		23		28		33	38
5. Sheet steel—10 gage; 0.1345" thick 5.625 lb/sq ft	18	23		28		33		38	48
6. Sheet steel—0.25" thick; 10 lb/sq ft	23	28	38	33	41	38	40	43	48
7. Sheet steel—0.375" thick; 15 lb/sq ft	26	31	39	36	42	41	47	41	51
8. Sheet steel—0.5" thick; 20 lb/sq ft	28	33		38		43		48	53
9. Sheet aluminum—16 gage; 0.051" thick; 0.734 lb/sq ft	5	8		13		18		23	28
10. Sheet aluminum—10 gage; 0.102" thick; 1.47 lb/sq ft	8	14		19		24		29	34
11. Plywood— 0.25" thick; 0.73 lb/sq ft	20		19		24		27	22	
12. Plywood—0.5" thick; 1.5 lb/sq ft	8	14		19		24		29	34
13. Plywood—0.75" thick; 2.25 lb/sq ft	12	17		22		27		32	37
14. Sheet lead—0.0625" thick; 3.9 lb/sq ft			32		33		32	32	32
15. Sheet lead—0.125" thick; 8.2 lb/sq ft			31		27		37	44	.33
16. Glass fiber board—6 lb/cu ft; 1" thick; 0.5 lb/sq ft	5	5	5	5	5	4	4	4	3
17. Laminated glass fiber (FRP): 0.375" thick			26		31		38	37	38
D. Walls—nonhomogeneous									
1. Gypsum wallboard—two ½" sheets cemented together, joints wood battened; 1" thick; 4.5 lb/sq ft	24	25	29	32	31	33	32	30	34
2. Gypsum wallboard—four ½" sheets cemented together; fastened together with sheet metal screws; dovetail-type joints paper taped; 2" thick; 8/9 lb/sq ft	28	35	32	37	34	36	40	38	49

Figure 26.2 The second attenuation provided by a barrier to airborne diffuse sound energy may be described in terms of its sound transmission loss TL. TL is defined (in dB) as 10 times the logarithm to the base of 10 of the ratio of the acoustic energy transmitted through a barrier to the acoustic energy incident upon its opposite side. It is a physical property of the barrier material and not of the construction techniques used.

Material or Structure		125	175	250	350	500	700	1,000	2,000	4,000
3. ¼" plywood glued to both sides of 1 × 3 studs, 16 in. o.c.; 3" thick; 2.5 lb/sq ft		16	16	18	20	26	27	28	37	33
4. Same as (3) above, but ½" gypsum wallboard nailed to each face: 4" thick; 6.6 lb/sq ft		26	34	33	40	39	44	46	50	50
5. ¼" dense fiberboard on both sides of 2 × 4 woods studs, 16" o.c.; fiberboard joints at studs; ⅘" thick: 3.8 lb/sq ft		16	19	22	32	28	33	38	50	32
6. Soft-type fiberboard (¾") on both sides of 2 × 4 wood studs, 16" o.c.; fiberboard joints at studs; 5" thick; 4.3 lb/sq ft.		21	18	21	27	31	32	38	49	53
7. ½" gypsum wallboard on both sides of 2 x 4 wood studs, 16" o.c.; 4.5" thick; 5.9 lb/sq ft		20	22	27	35	37	39	43	48	43
8. Two ⅜" gypsum wallboard sheets glued together and applied to each side of 2 × 4 wood studs, 16" o.c.; 5" thick; 8.2 lb/sq ft		27	24	31	35	40	42	46	53	48
9. 2" glass fiber (3 lb/sq ft) + lead vinyl composite; 0.87 lb/sq ft				4		4		13	26	31
10. ⅜" steel + 2.375" polyurethane foam (2 lb/cu ft) + 1/10" steel				38		52		55	64	77
11. Same as (10) above, but 2.5" glass fiber (3 lb/cu ft) instead of foam				37		51		56	65	76
12. ¼" steel + 1" polyurethane foam (2 lb/cu ft) + 0.055" lead vinyl composite; 1.0 lb/sq ft				38		45		57	56	67
E. Masonry										
1. Reinforced concrete: 4" thick; 5.9 lb/sq ft		37	33	36	44	45	50	52	60	67
2. Brick—common; 12" thick; 121 lb/sq ft		45	49	44	52	53	54	59	60	61
3. 3-¾ × 4-⁷⁄₁₁ × 8 glass brick; 3.75" th		30	36	35	39	40	45	49	49	43
4. Concrete block—4" hollow, no surface treatment	Cinder Aggregate	27	29	32	35	37	42	45	46	48
5. Concrete block—4" hollow, one coat resin—emulsion paint	Cinder Aggregate	30	33	34	36	41	45	50	55	53
6. Concrete block—4" hollow, one coat cement base paint	Cinder Aggregate	37	40	43	45	46	49	54	56	55
7. Concrete block—6" hollow, no surface treatment	Cinder Aggregate	28	34	36	41	45	48	51	52	47
8. Concrete block—8" hollow, no surface treatment	Cinder Aggregate	18	24	28	34	37	39	40	42	40
9. Concrete block—8" hollow, one coat cement base paint	Cinder Aggregate	30	36	40	44	46	48	51	50	41
10. Concrete block—8" hollow, filled with vermiculite insulators	Cinder Aggregate	20	29	33	36	38	38	40	45	47
11. Concrete block—4" hollow, no surface treatment	Expanded Shale Aggregate	21	26	28	31	35	38	41	44	43
12. Concrete block—4" hollow, one coat rcsin-emulsion paint	Expanded Shale Aggregate	26	30	32	34	37	42	43	46	44
13. Concrete block—4" hollow, two coats rcsin-emulsion paint	Expanded Shale Aggregate	24	31	33	35	38	42	44	47	44
14. Concrete block—4" hollow, one coat cement-base paint	Expanded Shale Aggregate	23	30	35	38	42	43	44	48	43
15. Concrete block—4" hollow, two coats cement-base paint	Expanded Shale Aggregate	34	38	40	42	45	47	49	51	46
16. Concrete block—6" hollow, no surface treatment	Expanded Shale Aggregate	22	27	32	36	40	43	46	45	43
17. Concrete block—4" hollow, no surface treatment	Dense Aggregate	30	36	39	41	43	44	47	54	50
18. Concrete block—4" hollow, one coat cement base paint on face	Dense Aggregate	30	36	39	41	43	44	47	54	49
19. Concrete block—6" hollow, no surface treatment	Dense Aggregate	37	46	50	50	50	53	56	56	46
20. Concrete block—6" hollow, one coat resin-emulsion paint each face	Dense Aggregate	37	50	54	52	53	55	57	56	46
21. Concrete block—8" hollow, no surface treatment	Dense Aggregate	40	47	53	54	54	56	58	58	50
22. Concrete block—8" hollow, two coats resin-emulsion paint each face	Dense Aggregate	38	50	54	54	55	58	60	58	49

Figure 26.2 Continued.

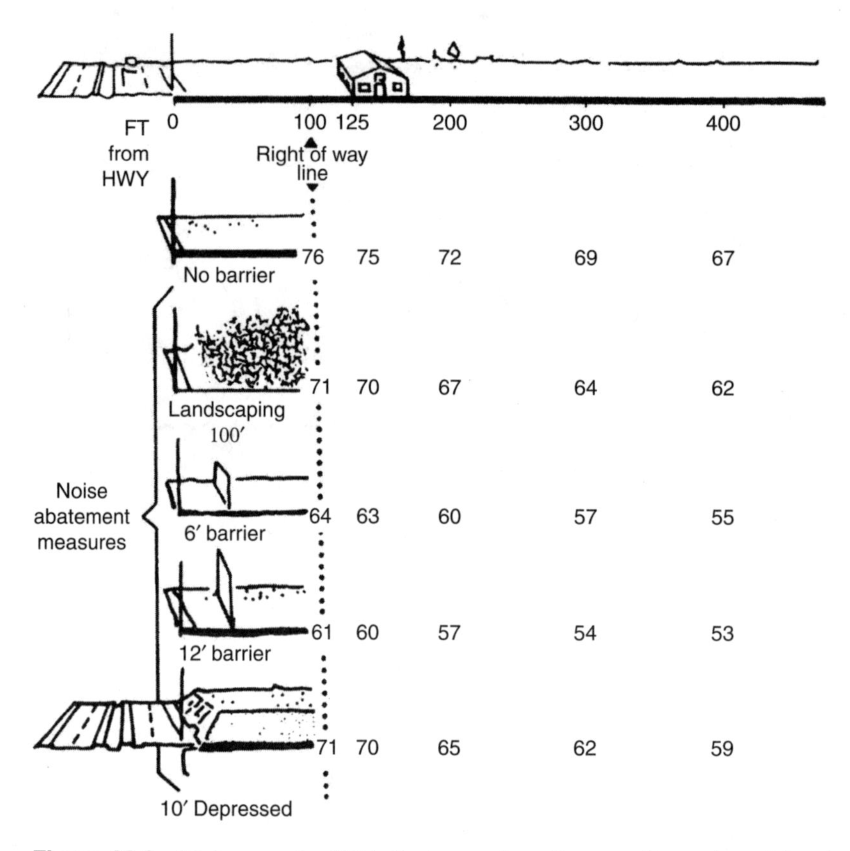

Figure 26.3 Highway noise (dBA, L_{10}) at various distances from edge of four-lane highway. Traffic: 5000 vehicles per hour, 5% trucks, 53 mph.

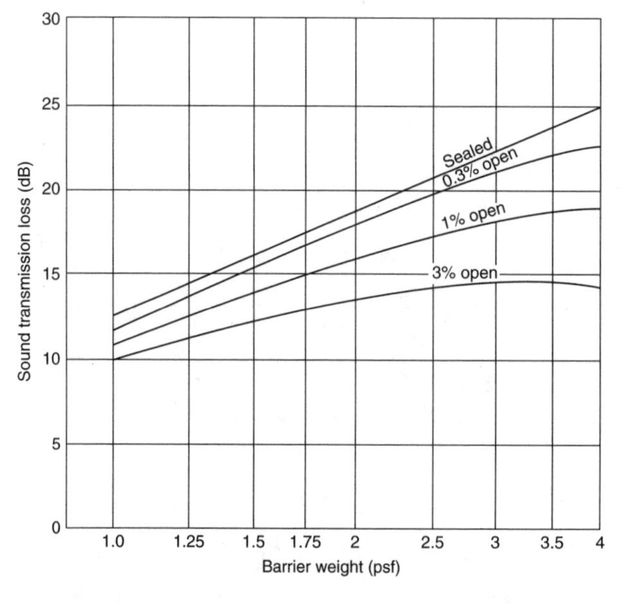

Figure 26.4 Average sound transmission loss of a single sound barrier as a function of barrier mass and percentage of open area.

Example 2.1 An operator positioned close to a punch press that used compressed air jets to blow foreign particles from the die was exposed to excessive noise levels. A 1/4-in.-thick safety glass provided good visibility and access to the work position and gave the following noise reduction at the operator's head position.

O.B. (Hz)	31.5	63	125	250	500	1000	2000	4000	8000
N.R. (dB)	—	—	1	2	3	9	14	20	22

Example 2.2 (Fig. 26.3.) Highway noise at various distances from the edge of a four-lane highway are plotted with no barrier and with different barrier configurations (30). The traffic density during these measurements averaged 5000 vehicles per hour with 5% trucks. The average vehicle speed was about 53 mph.

Example 2.3 A sheet metal belt guard was installed around a high-speed rubber-tooth belt. The noise reduction achieved is shown below.

O.B. (Hz)	31.5	63	125	250	500	1000	2000	4000	8000
N.R. (dB)	—	—	—	—	—	—	7	9	19

Example 2.4 An electric motor-gear drive assembly was enclosed in 1/8-in. steel with welded joints which was lined with 1-in. Fiberglass (No. 615) PF board spaced 1 in. apart. The noise reduction achieved is shown below.

O.B. (Hz)	31.5	63	125	250	500	1000	2000	4000	8000
N.R. (dB)	—	5	6	12	14	25	35	24	23

Example 2.5 A complete enclosure was constructed to enclose large sirens for production testing. The enclosure was made with sheet steel lined with Fiberglass, the inner side of which in turn was covered with an open-mesh protective surface. The noise reduction is shown below.

O.B. (Hz)	31.5	63	125	250	500	1000	2000	4000	8000
N.R. (dB)	—	—	—	15	13	27	33	38	43

Example 2.6 An operator of a pneumatic system that included compressors and ducts for conveying pellets spent a large portion of his time at a central location. An enclosure for the operator was designed using wood framing with 5/8-in. gypsum board inside and out. The open spaces between the boards were filled with Fiberglass, and all joints in the gypsum board were sealed. Double-glazed windows were provided for observation of equipment on all sides. The noise reduction is shown below.

O.B. (Hz)	31.5	63	125	250	500	1000	2000	4000	8000
N.R. (dB)	8	6	8	15	15	10	14	18	19

7.3. Examples Using Impact, Radiation, and Vibration Reduction

Example 3.1 A high-speed film rewind machine (15 HP) produced excessive noise from the metal-to-metal impacts between gear teeth. Fiber gears were substituted for the metal ones and the gears were flooded in oil. The noise reduction is shown below.

O.B. (Hz)	31.5	63	125	250	500	1000	2000	4000	8000
N.R. (dB)	—	10	6	5	5	8	20	16	14

Example 3.2 An 8-ft-diameter hopper with an electric solenoid-type vibrator coupled solidly to a bottom bin was causing excessive noise. A live bottom bin by Vibra Screw was installed that required less vibratory power since only the cone is vibrated. Also, the new system had less radiation area and there were no metal-to-metal impacts. The noise reduction achieved is shown below.

O.B. (Hz)	31.5	63	125	250	500	1000	2000	4000	8000
N.R. (dB)	—	7	6	20	22	16	12	12	19

Example 3.3 Screw machine stock tubes constructed of solid steel usually make excessive noise because there is nearly continuous impact between the tube and the screw stock. New tubes such as the Corlett Turner Silent Stock tube, constructed as a sandwich with an absorbent material between the outer steel tube and an inner helically wound liner, provide significantly lower noise levels. The noise reduction achieved with the new tube design operated at 4000 rpm with 1/2-in. hexagonal stock is shown below.

O.B. (Hz)	31.5	63	125	250	500	1000	2000	4000	8000
N.R. (dB)	—	12	15	15	14	20	29	34	30

7.4. Examples Using Acoustical Damping

Example 4.1 A metal enclosure around a rubber compounding mill vibrated freely, thus amplifying the motor, gear, and roll noises of the mill. An application of vibration damping material (1/4-in. Aquaplas F 102A) to the inner surface of the metal enclosure reduced the noise as shown below.

O.B. (Hz)	31.5	63	125	250	500	1000	2000	4000	8000
N.R. (dB)	10	9	9	13	9	7	8	10	11

Example 4.2 The guards and exhaust hoods of a 10-blade gang ripsaw were coated with MMM Underseal (EC-244). The noise reduction attained while the saw was idling is shown below.

O.B. (Hz)	31.5	63	125	250	500	1000	2000	4000	8000
N.R. (dB)	6	7	10	7	5	3	3	5	6

Example 4.3 A 3/8-in. steel casing of a 2000-HP extruder gear and its base were vibrating excessively, causing unwanted noise. Accelerometer measurements showed the casing and the 1-in. steel base were vibrating at about the same level. The casing was damped with a 1/4-in. felt (No. 11 Anchor Packing Co.) plus an outer covering of 1/4-in. steel. The felt-steel sandwich was bolted together on 8-in. centers. The steel base had 9-in.-deep ribs that made the felt-steel damping impractical, so the base was damped by a thick cover of sand. The noise reduction attained is shown below.

O.B. (Hz)	31.5	63	125	250	500	1000	2000	4000	8000
N.R. (dB)	—	—	—	—	4	17	26	24	18

7.5. Examples Using Reduced Driving Force

Any noise produced by a repetitive force that is caused by an eccentricity or imbalance of a rotating member will increase with rotational speed. Obviously, one very important noise control procedure is to dynamically balance all rotating pieces. Also, these pieces should rotate concentrically. Proper maintenance of all bearing and other rotating contact surfaces is essential to keep equipment running quietly.

No machine should be operated at an unnecessarily high speed. In many instances, a significant reduction in noise can be achieved by using a larger machine that can do the same job while operating at lower speeds.

Reduction of driving force in almost any form is an effective noise control procedure. In many instances, a reduction of driving force will provide the additional advantage of reduced radiation area.

Example 5.1 A blower exhaust system running at 705 rpm, 6 in. static pressure, and 13,800 cfm was badly out of balance and bearings needed replacing. After new bearings were installed and the system was balanced, the following improvement was found.

O.B. (Hz)	31.5	63	125	250	500	1000	2000	4000	8000
N.R. (dB)	—	3	3	11	12	11	10	8	10

Example 5.2 An oversized propeller-type fan (36 in.) mounted in the wall of a large, reverberant room produced excessive noise when operated at 870 rpm. It was possible to get the significant noise reduction shown below, while at the same time providing sufficient ventilation, by reducing the fan speed from 870 to 690 rpm.

O.B. (Hz)	31.5	63	125	250	500	1000	2000	4000	8000
N.R. (dB)	—	3	7	8	12	9	8	6	4

Example 5.3 Small metal parts were dropped several inches into a metal chute where they were moved by gravity onto another operation. The dropping distance and weight of the pieces should not be changed, so the chute surface was covered with a layer of

1/16-in. paperboard, and this layer was in turn covered by 18-gauge steel. The noise reduction of this sandwich covering is shown below.

O.B. (Hz)	31.5	63	125	250	500	1000	2000	4000	8000
N.R. (dB)	4	4	4	2	7	9	12	14	16

Example 5.4 Steel balls tumbling against the steel shell of a ball mill were producing excessive noise. The steel shell was lined with resilient material (rubber) to achieve the noise reduction shown below.

O.B. (Hz)	31.5	63	125	250	500	1000	2000	4000	8000
N.R. (dB)	—	3	4	6	7	11	12	15	19

7.6. Examples Using Mufflers and Air Noise Generation Control Means

Example 6.1 An air-driven impact gun usually makes excessive noise. A simple means of reducing this noise is to pipe the exhausted air to a remote location by means of a rubber hose. Another noise reduction means is to use an internal muffler. The following noise reduction figures were achieved with an air gun running free.

O.B. (Hz)	31.5	63	125	250	500	1000	2000	4000	8000
N.R. (dB) muffler	—	—	2	2	4	15	9	6	7
N.R. (dB) rubber hose	—	—	19	17	30	42	29	28	28

Example 6.2 The air intakes of reciprocating air compressors often create objectionable low-frequency noise. An intake filter muffler, such as the Burgess Manning Model Delta P-SDF, can reduce the noise in the 63-cps octave band by as much as 23 dB.

Example 6.3 The discharge of a Gast Air Motor (Model 4 AM and 6 AM) created excessive noise. A Burgess Manning Delta P CA type muffler installed on the discharge outlet produced the following noise reduction.

O.B. (Hz)	31.5	63	125	250	500	1000	2000	4000	8000
N.R. (dB)	—	2	7	7	9	10	23	29	23

Example 6.4 The blower noise from the discharge of a pneumatic conveying system handling synthetic fiber fluff was excessive. An absorbing-type muffler was not desired because of the possibility of snagging and plugging. A resonant-type muffler supplied by Universal Silencer Corporation provided the noise reduction shown below.

O.B. (Hz)	31.5	63	125	250	500	1000	2000	4000	8000
N.R. (dB)	—	12	23	13	11	10	—	—	—

Example 6.5 An air intake of a 7000-HP gas turbine operating at 5800 rpm and 62,000 HP created excessive noise. A parallel baffler muffler consisting of six plates, each 3.5 in.

wide, filled with Fibcrglass and faced with 18-gage perforated sheet steel, was attached to the intake, and the baffle was in turn fed by an unlined 0.25-in. duct made of steel plate. The cross-section of the duct was 7 ft × 8 ft. The noise reduction achieved is shown below.

O.B. (Hz)	31.5	63	125	250	500	1000	2000	4000	8000
N.R. (dB)	—	—	10	16	22	33	35	27	26

Example 6.6 The noise produced by a tube reamer was reduced by the following values by mounting a Wilson 8500 muffler on the exhaust.

O.B. (Hz)	31.5	63	125	250	500	1000	2000	4000	8000
N.R. (dB)	2	2	3	10	23	26	28	16	18

7.7. Examples Using Drive System Modifications

Example 7.1 A rubber-toothed belt used to drive a pump was replaced by a V-belt drive. The noise reduction achieved is shown below.

O.B. (Hz)	31.5	63	125	250	500	1000	2000	4000	8000
N.R. (dB)	—	5	4	4	2	—	8	17	18

Example 7.2 An edger-laner for trimming foamed plastic created noise levels as high as 102 dB in the 250-Hz octave band. The noise was caused primarily by the cutter blades chopping the conveying airstream. The clearance between the cutter blades and the casing was increased from 3/32 in. to 1 in., thereby lowering the air velocity and reducing the noise level to 84 dB in the 250-Hz band.

A single noise control procedure often may be ineffective by itself, but when coupled with one or more other procedures, it may produce significant results. As an example, a typical noise source having a frequency spectrum in which all octave-band pressure levels are essentially the same may have the following noise reduction values for the various noise control procedures shown below.

Noise Reduction (in dB) as a Function of Frequency (in Octave Bands)

N.R. Procedure	31.5	63	125	250	500	1000	2000	4000	8000
1. Mounted on vibration isolators	11	7	3	—	—	—	—	—	—
2. Single-wall barrier	—	—	3	5	6	6	6	6	7
3. Complete enclosure of absorbing material	—	—	—	4	5	5	6	7	7
4. Complete enclosure of solid material with no absorption inside	—	2	5	14	18	26	26	27	29
5. Complete enclosure of solid material with no absorption inside mounted on vibration isolators	11	8	7	16	21	29	34	35	40

(continued)

Continued

N.R. Procedure	31.5	63	125	250	500	1000	2000	4000	8000
6. Complete enclosure of solid material with absorption inside mounted on vibration isolators	11	11	13	25	32	38	40	42	45
7. Complete No. 6 procedure mounted on vibration isolators and enclosed in solid materials with absorption inside	20	17	22	44	50	57	57	59	64

Many of the examples of noise control in this chapter were taken from material prepared for the American Industrial Hygiene Association (AIHA) *Industrial Noise Manual* (22). A more extensive listing of examples and a more complete general discussion of engineering noise control can be found in the AIHA Manual.

REFERENCES

1. Jorgensen R, ed., *Fan Engineering*, 6th ed., Buffalo Forge Company, Buffalo, NY, 1961.
2. Den Hartog JP, *Mechanical Vibrations*, McGraw-Hill, New York, 1947.
3. Crede CE, *Vibration and Shock Isolation*, Wiley, New York, 1951.
4. Geiger PH, *Noise-Reduction Manual*, Engineering Research Institute, University of Michigan, 1953.
5. Harris CM, *Handbook of Noise Control*, McGraw-Hill, New York, 1957.
6. Beranek LL, *Noise Reduction*, McGraw-Hill, New York, 1960.
7. *Sound Insulation of Walls and Floor Construction*, National Bureau of Standards, U.S. Department of Commerce, Building Materials and Structure Report BMS 17 with 2 supplements.
8. Buckingham E, Theory and interpretation of experiments on transmission of sound through partition walls, *Sci. Papers Bur. Standards*, *220*:193–219, 1925.
9. *Recommended Practice for Laboratory Measurement of Airborne Sound Transmission Loss of Building Floors and Walls*, ASTME90–55, Philadelphia.
10. Bonavallet GL, Retaining high sound transmission loss in industrial plants, *Noise Control*, *3*(2):61–64, 1957.
11. *Sound Absorption Coefficients of the More Common Acoustic Materials*, National Bureau of Standards, U.S. Department of Commerce, Letter Circular LC870.
12. *Sound Absorption Coefficients for Architectural Acoustical Materials*, Acoustical Materials Association, New York.
13. Warnaka G, Active attenuation of noise—the state of the art, *J. Noise Control Eng.*, *18*(3):100–110, 1982.
14. Swanson DC, Active attenuation of acoustic noise: Past, present, and future, *ASHRAE Trans.*, *95* (Part II):63–76, 1989.
15. Dancer A, *Noise Induced Hearing Loss*, B. C. Decker, Philadelphia, Chap. 34, 1989.
16. *AGMA Standard 295.02-65, Specifications for Measurement of Sound on High Speed Helical and Herringbone Gear Units*, American Gear Manufacturers Association, Washington, D.C.
17. *ASHRA Standard 36–62, Measurement of Sound Power Radiated from Heating, Refrigerating and Air Conditioning Equipment*, American Society of Heating, Refrigerating and Air Conditioning Engineers, Inc., New York.
18. *AMCA Standard 300-67, Test Code for Sound Rating*, Air Moving and Conditioning Association, Inc., Park Ridge, IL.
19. *Test Procedure for Air-Borne Noise Measurements on Rotating Electrical Machinery*, Institute of Electrical and Electronic Engineers, Inc., New York, 1965.
20. Guidelines for noise control specifications for purchasing equipment, *Iron and Steel Engineer*, May 1970.

21. *Standards Publication, Gas Turbine Sound and Its Reduction*, Publication No. 33-1964, National Electrical Manufacturers Association, New York.
22. *Industrial Noise Manual*, 2nd ed., American Industrial Hygiene Association, Detroit, MI, 1966.
23. *ISO Recommendation R-495, General Requirements for the Preparation of Test Codes for Measuring the Noise Emitted by Machines*, International Organization for Standardization, American National Standards Institute, New York.
24. *American National Standards Specification for Sound Level Meters, ANSI S1.4-1971*, American National Standards Associations, New York, 1971.
25. *American National Standards Specification for Octave, Half-Octave, and Third Octave Band Filter Sets, ANSI S1.11-1966*, American National Standards Association, New York, 1966.
26. *American National Standard Method for Physical Measurement of Sound, ANSI S1.2-1962*, American National Standards Institute, New York, 1962.
27. *Proposed American National Standard Method for Rating the Sound Power Spectra of Small Stationary Noise Sources, ANSI S3.17-197X*, American National Standards Institute, New York (Third Draft Dec. 1971).
28. *Guidelines for noise exposure control*, Intersociety Committee on Guidelines for Noise Exposure Control, *Am. Indust. Hygiene Assoc. J.*
29. *Preferred Frequencies and Band Numbers for Acoustical Measurements (ANSI) S1.6-1967*, American National Standards Institute, New York.
30. Department of Transportation noise standards, *Fed. Register*, *37*(114):94–95, 1972.

27

Noise Criteria Regarding Risk and Prevention of Hearing Injury in Industry

Terrence A. Dear
E. I. du Pont de Nemours & Company, Wilmington, Delaware, USA

Prevention of hearing injury seems the obvious goal of occupational noise standards. Damage Risk Criteria (DRC) establish the degree of risk built into regulation according to the criteria numbers selected (1). Experience shows that a hearing conservation program can be extremely effective in the management of risk to protect most employees, particularly when participation is encouraged beyond criteria numbers (2). OSHA's amendment of March 8, 1983 (3) established a hearing conservation program that has many good points.

Workplace noise regulations typically establish criteria selected to limit the percentage of workers at risk of acquiring "beginning" hearing impairment over a working lifetime. Generally, this concept of impairment is defined in terms of one's reduced ability to communicate speech, owing to prolonged exposure to high noise levels. Correspondingly, specific requirements for measuring and computing noise levels and exposures (at frequencies between 63 and 8000 Hz, under OSHA regulation) are established. The results of these procedures are usually to be compared with criteria limits to determine when specific courses of action are to be taken according to regulatory requirements.

In principle, this approach seems to work reasonably well so long as the "numbers" resulting from measurement and computation have been determined in the same manner as those used in the original alignment of risk and criteria (4). However, noise measurement procedures and instrumentation have been modified over the years to the point where the measured "numbers" are now different from those used to establish the original relationships between risk percentages and numerical criteria (5–16). This process of fundamental change in measurement techniques has occurred without any compensating adjustment in the criteria numbers. DRC specifies the maximum allowable noise quantity to which persons may be exposed if risk of hearing impairment is to be avoided. Percentage Risk (PR) is defined as the difference between the percent of persons exposed to noise who reach impairment and the percent of those not exposed to any industrial noise who reach impairment. Both concepts are based on correlations of workplace measurements and audiometric test data for noise-exposed populations in U.S. industry before integrating instrumentation was available.

There are two key factors concerning the background for DRC and PR concepts, as follows:

1. They are based solely on workplace noise measurements traditionally made from point to point with a hand-held sound-level mater (ANSI S1.4-type 2) (21). Typically, the instrument microphone was at 3 ft (or 1 m) from the noise source, preferably without employees present.
   This type of noise measurement is described in Table 27.1 in comparison with the characteristics of individual exposure measurements as preferred by the OSHA amendment of 1983.
2. Selection of a DRC implies that there will be some percentage of the exposed population remaining at risk at that level.
   For example, by OSHA regulation, achievement of a 90-dBA criterion level through engineering controls can result in some of those exposed being at higher risk upon removal of their personal hearing protection. This assumes that employees are usually overprotected when using hearing protection devices, as observed in Pell's study (2).

To summarize, OSHA permits greater risk for some employees based on a shift from personal hearing protection to a higher-ranking compliance requirement.

Looking at these facts in light of current and proposed noise regulations raises some important questions.

In view of the history of DRC and the comparison of two different survey and monitoring methods for occupational noise measurements presented in Table 27.1, is it appropriate to assume that the criterion numbers should remain identical for each method as implied in OSHA's noise regulation and other current regulatory proposals?

Is it advisable to require compliance by methods that would take employees at negligible or lower risk due to use of personal hearing protection and place them at higher risk because some goal of engineering controls has been achieved?

Is it likely that historical alignment of DRC and percentages of employees at risk (PR) would be unchanged by OSHA-specified requirements for a hearing conservation program?

These questions all seem to require a negative answer. This conclusion suggests that OSHA's noise regulation and current regulatory proposals, such as those for Ontario and the EEC, need to be revised on the basis of these considerations alone.

Table 27.1 Occupational Noise Measurement Comparisons Chart (Reference ANSI S1.13 1971)[a]

	Application		
	Survey and Monitoring		
Measurement Item	Individual Exposure (Direct)	Workplace	Noise Spectrum analysis[b]
Measurement Objective	Measure time integrated noise exposure. Measurement can be converted to an equivalent dBA level (sometimes referred to as TWA, time-weighed average).	Measure sound levels at specific locations for computational evaluation of exposure and noise control needs. Special features required to measure impulse noise, (e.g., peak hold).	Measure noise vs. frequency. Octave band filter sets divide noise spectrum into several frequency (octave) bands. Other filter sets can further subdivide frequency bands.
Instrument Type	Dosimeter (worn by employee)	Survey or Sound Level Meter Type 25 (usually hand-held)	Sound-Level Meter with Filter Set or Analyzer Type 1 or Type 2. Real time analyzer/Fast Fourier transform system.
Instrument Specification	ANSI S1.25[a].	ANSI S1.4[a] with A-Scale, slow response, Type 1 or Type 2 (Preferred)	ANSI S1.4[a] Meter and ANSI S1.11 Filter Sets (Octave, 1/3 octave)
Example Measurement Applications	Measurement of dose. Evaluate administrative or dose control aspect of worker exposure. Statistically valid sampling procedures required.	OSHA noise survey. (Use A-weighing network, slow response.) Hearing conservation program survey.	Noise control analysis. Neighborhood noise survey. Noise source identification. Laboratory measurements.
Microphones	Designed specifically for dosimeter usage. Should have an 8000-Hz upper frequency limit to be in strict adherence to OSHA noise regulation.	Per ANSI S1.4[a] and ANSI S1.10[a]. In general, 1-in. microphones are supplied. For measurement of frequencies above 10,000 Hz, 1/2-in. or 1/4-in. may be required—check manufacturer. Impulse/impact noise may require special microphones.	
Microphone Placement	Near employee's ear (e.g., worn on shoulder).	Approximate location of employee's ears and/or the traditional 3 ft or 1 m from the source. Prefer that employee is not present.	As required for source identification. Direction is an important factor in some sound fields.

[a]ANSI—American National Standards Institute, Inc., 1430 Broadway, New York, NY 10018.

(continued)

Table 27.1 *Continued*

	Application		
	Survey and Monitoring		
Measurement Item	Individual Exposure (Direct)	Workplace	Noise Spectrum analysis[b]
Instrument Calibration	Acoustic calibration as recommended by instrument manufacturer (a separate instrument) Per manufacturer's instructions where manufacturer and unit are certified.	Use acoustical calibrator to check instrument calibration before each day's measurement at a minimum. Calibration to be traceable to National Institute for Standards and Technology.	
Surveyer	Measurements shall be made by, or under the supervision of, a person trained and skilled in the use of acoustic measurement instruments and documentation.		
Environmental Factor	Temperature, humidity, and vibration limited. Consult manufacturer. Integrates all sound. Air currents across the microphone on produce irregular noise measurements. Commercial wind screens are available from the instrument manufacturers and should be used.	Temperature, humidity and vibration limited. Consult manufacturer. Noise not due to pertinent sources can be excluded.	High background noise can interfere with noise from sources under investigation. Temperature, humidity, and vibration limited.

[b]May also be useful for evaluation of personal hearing protection requirements (ASA STD1–1975). ASA-Acoustical Society of America, 335 East 45th St., New York, NY 10017. © 1986, E. I. du Pont de Nemours & Company.

The occupational physician should be aware of the limitations of DRC in light of these facts because, once a specified exposure limit is reached in the workplace, it is still necessary to protect employees whose hearing remains at risk at or below the regulated level.

In summary, studies of worker exposure to noise (22) have historically relied on voluntary subjects, questionnaires, and estimates of time-dependent exposure or dose in the workplace and almost never on instrumentation capable of directly recording and analyzing statistically valid measurements of individual exposure. Furthermore, direct exposure assessment methods (see Table 27.1) have not been correlated to audiometric test results for test subjects as they were for historical workplace measurements used to develop DRC and PR.

Dr. F. A. Van Atta (23), who is generally considered the author of the initial OSHA noise regulation (1971), stated, "It is not scientifically possible to set a realistic standard for exposure to materials or to energy that will protect the whole of any population that is exposed."

As Dr. Sidney Pell (2) has shown, in one of the very few longitudinal noise exposure studies suitable as a basis for regulation, protection of almost all employees can indeed be accomplished through an effective hearing conservation program. Emphasis on effective programs must continue to grow in regulation and throughout industry to achieve maximum prevention of hearing impairment for employees. OSHA's hearing conservation amendment of March 8, 1983, though overly complex, was a step in the right direction.

Shaw's (24) attempt at providing evidence supporting a 3 dBA exchange rate relies heavily on questionable treatment of the Passchier-Vermeer data. Passchier-Vermeer (25) selected 20 of about 100 studies to pool data, and there was significant exposure to impact noise for subjects in the selected cohort. Therefore, Shaw's treatment should not be accepted as a basis for adopting 3 dBA and firmly rejecting the current 5 dBA exchange rate for steady-state noise exposure.

1. FACTORS THAT CONTRIBUTE TO PYRAMIDING OF NOISE MEASUREMENT "NUMBERS"

Several key factors will cause measured "numbers" to increase or result in equivalent penalties when comparisons of measurements with the criteria limits of regulation are made.

Factors that will increase numerical values of measurements made with instrumentation currently on the market and/or as proposed in the Shaw report (24) are as follows:

Change from present 5 dBA exchange rate to a 3 dBA exchange rate as proposed by EPA (26), Shaw, and ISO 1999 (1984) (27), for example.

Extended dynamic range of dosimeters as specified in the OSHA regulation of March 8, 1983 (3). For example, Table G-16a of that report specifies computation of dose over the range of 80–130 dBA as compared to the 90–115 dBA range of criteria per Table G-16 (see Chapter 34, this volume).

Increased crest factor, for example the crest factor of "30 dB" originally specified in the OSHA amendment of January 16, 1981 (3), and as prescribed for simultaneous integration of steady-state and impulse/impact noise in a new way (24).

Extension of the high-frequency response well beyond the speech range, particularly with type I instrumentation. This problem is inherent in all instrumentation that does not have a built-in cutoff at the 8000-Hz limit of OSHA regulation (20) (see Figure G-9 of Paragraph 1910.95, FR Vol. 36, No. 105, 5/29/71) or below. This problem is most notable for type 1 and type 2 instrumentation with microphones that measure above

8 kHz. It should be noted that the upper-frequency limit for audiometry and personal hearing protection evaluation methods of OSHA's current standard is also 8000 Hz.

Permitting unrestricted microphone locations and movements in sound fields for dosimeter measurements opposite restrictions inherent in measurements used as the basis for DRC.

There are also factors that will affect the comparisons of measured "numbers" with criteria limits.

Derating of personal hearing protection (PHP) values, for example by 50%, as proposed by OSHA (28) and the Shaw report (24).

Extrapolation of exposure data integration beyond 115 dBA, for example to 130 dBA, as per Table G16a of OSHA's amendment of March 8, 1983 (3). To our knowledge, there are no valid scientific data or evidence to support such extended integration.

There are additional sources of higher-dosimetry numbers, including the noise from coughing, speech, physical impacts on the microphone, whistles, and myriad other extraneous sources. These contributions, including malingering, cannot be excluded from directly integrated dose measurements as they were generally eliminated from historical workplace measurements by common-sense observations.

The range of potential for increases in measured numbers that could result from various combinations of these factors is shown incrementally and cumulatively in Fig. 27.1.

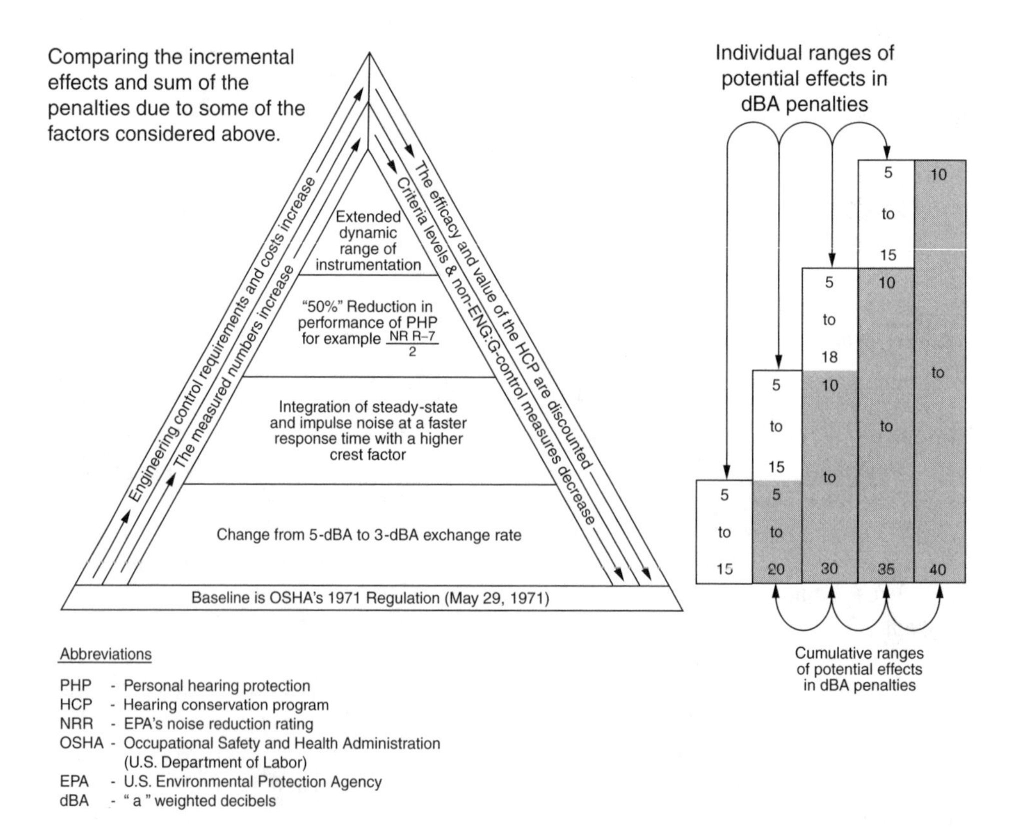

Figure 27.1 Pyramiding effects of factors increasing measurement "numbers." (Copyright 1986, E. I. du Pont de Nemours & Company.)

How these factors actually add up and their total impact will depend on the noise level involved, employee exposure, the degree of impulsive noise present, the specific instrumentation and measurement methodology used, and the type of personal hearing protection selected. Therefore, increases are to be expected in many workplace measurements and exposure determinations made without these factors involved.

The differences to be expected and the sources of these variations are important, particularly in evaluations of compliance with regulation. The estimated ranges and cumulative effects are presented in Fig. 27.1, but all factors may not be present in each case. Most of the increases should be in the lower ends of the ranges. On the other hand, even the upper ends of the ranges could be exceeded in some cases (e.g., highly impulsive noise).

Increases (and in some cases decreases) in the measured numbers have been commonly observed where the same industrial noise has been monitored with both a type 25 sound-level meter and dosimeter, in the absence of any of the factors mentioned above (7,9,12,19). These variations are mostly due to the differences in instrumentation shown in Table 27.1.

Except for the high-frequency cutoff at 8000 Hz for instrumentation, there is no compelling scientific evidence to suggest that any of the factors considered needs to be incorporated in regulation or instrumentation (22). All factors included in Fig. 27.1 represent requirements that are largely based on hypotheses. Therefore, it is highly questionable as to any benefits that might be derived from incorporating such requirements in occupational noise regulation. Certainly, none of the factors considered would add significantly to the benefits derived from an effective hearing conservation program alone. This fact offsets any claims that the factors provide desirable degrees of conservatism.

Therefore, it is recommended that proposals (24) and/or guidelines for changing the exchange rate from 5 dBA to 3 dBA, imposing a "30 dB" crest factor for simultaneous integration of impulse/impact noise, and reducing personal hearing protection effectiveness by 50% be avoided in setting workplace noise regulations.

To support this recommendation, each factor will now be considered in more detail.

2. REASONS TO AVOID PROPOSED CHANGES FROM 5-DBA TO 3-DBA EXCHANGE RATE

The exchange rate is defined as the increase or decrease in the permissible noise level criteria as the time of permissible employee exposure at that level is halved or doubled, respectively. For example, the current exchange rate of 5 dBA in the OSHA regulation causes the permissible exposure time to be reduced from 8 to 4 h when the exposure level increases from 90 to 95 dBA.

Proposals to change from 5-dBA to 3-dBA exchange rate rely on highly questionable studies and interpretations of studies (22,30). For example, such studies form the basis for a proposal in Ontario that the 5-dBA exchange rate of regulation be changed to 3 dBA (24). Unfortunately, the latter report (24) did not consider 10 years and more of available data concerning actual performance in industry under current Ontario and/or U.S. regulatory criteria to check predictions of risk based upon the earlier cross-sectional studies selected for analysis.

The Ontario proposal relies heavily on the flawed studies of Burns and Robinson (1970) (3) and Passchier-Vermeer (1973) (25), while ignoring the work of Sulkowski (22), Gosztonyi (32), Scheiblechner (33), Schneider (34), and Pell (2), among others. Furthermore, the findings of Passchier-Vermeer do not agree with those of Burns and Robinson.

As Ward advised the Ontario Ministry of Labour, it is also important to point out that there is a fundamental error in the procedure used by Passchier-Vermeer and presumably also by Shaw (24) to derive industrial noise-induced permanent threshold shift (NIPTS). Their values were obtained by subtracting from a worker's actual hearing threshold level (HTL) an appropriate "age correction" taken from A. Spoor. The Spoor curves, however, are empirically determined differences, in a nonindustrial-noise-exposed population, between people of a given age and those aged 25 years. In order for the worker's actual HTL minus the Spoor correction to be the NIPTS attributable to his noise exposure, it would have to be the case that the workers had average HTLs of 0 at age 25 (or at least that they would have had 0-dB HTLs at age 25 if they had not entered the industry concerned). Both recent and non-so-recent studies of nonindustrial-noise-exposed populations have shown this to be incorrect; instead, the median HTLs of 25-year-old persons who have never been exposed to workplace noise invariably are 3–4 dB at the lower audiometric frequencies (500–2000 Hz), 6–7 dB at 3000 and 4000 Hz, and 10–15 dB at 6000 Hz. Therefore, the inferred NIPTS values of Shaw's and Passchier-Vermeer's analyses are inflated by these amounts. It is an effect that cannot be eliminated from human data because of the action of socioacusis and nosoacusis as well as presbycusis on the hearing process. Since Shaw's (24) conclusion is largely founded on analysis incorporating this error, his recommendation to change from 5 dBA to 3 dBA should be rejected.

The question of intermittency of exposure is a main item of controversy in the exchange rate issue (35–37). It is clear from all studies to date that a 3-dBA exchange rate is only valid for zero intermittency, a workplace condition that does not generally exist in North America industry. Ward (38) and Bies (39) have more recently confirmed this and the validity of a 5-dBA exchange rate for intermittent exposures.

The fact that the Burns and Robinson and Passchier-Vermeer studies are seriously flawed is underscored in Sulkowski's *Industrial Noise Pollution and Hearing Impairment* (22) and S. F. Galeano's chart from "How O-I's Hearing Conservation Program Gives Useful Statistics for Future Analysis" (30) (see Table 27.2).

To quote Sulkowski directly (22, p. 98):

> It is stressed that despite the millions of audiometric records gathered from exposed workers, the relations between noise exposure and the resultant noise-induced permanent threshold shift are still imprecise and the data from field and laboratory studies, on which the present damage risk criteria and standards are based, are imperfect and controversial.

Table 27.2 Quality Checklist for Studies Intended to Establish a Noise Exposure versus Hearing Loss Relationship (30)

	Quality Parameters of the Study[a]				
Study	Otological screening	Elimination of STS	Sample size	Protocol	Steady-state condition
Passchier-Vermeer (61)	?	?	?	?	?
Robinson (62)	F	X	F	?	F
Baughn (63)	F	X	F	?	F
A. U. (Austria) (64)	X	F	X	?	?
IINS (63)	X	X	?	X	?

[a]F = failed; X = proper; ? = attempted.

In addition, there is ample evidence in the form of scientific critiques contained in the OSHA record (1975, 1976) to demonstrate conclusively that many of the studies cited here cannot be used to distinguish among exchange rates such as 3, 4, 5, and 6 dBA.

The following shortcomings and weaknesses of thc data [including such studies as those by Passchier-Vermeer (25), Burns and Robinson (31), and Baughn (40)] are considered to be the source of controversy and reservations with regard to their reliability and adequacy: not sufficiently accurate noise measurements; inadequate histories of exposure duration; otological examinations and histories performed by inexperienced persons; audiometry conducted in rooms with high background noise; incorrect calibration of instruments and audiometric technique; nonexclusion of temporary threshold shift due to recent noise exposure; nonexclusion of subjects with nosocusis; lack of use of proper controls; nontypical continuous or steady-state noise exposure; questionable statistical and interpretations of results.

In consideration of the need for reliable scientific data on noise-induced hearing loss as the basis for the establishment of valid damage risk criteria, the long-time national study in the U.S., the so-called The Inter-Industry Noise Study, was performed [Yerg et al. (41)].

The data from this and subsequent studies have provided accurate and reliable information and have been helpful in establishing a scientific basis of a safe occupational noise standard.

Sulkowski's final conclusion is as follows (22, p. 206): "[There is] no general agreement about trading relation between level and exposure time, but it seems that the 5 dB doubling rate is more appropriate than 3 dB time/intensity trade-off value."

The Burns and Robinson study (3) is critically flawed in this specific aspect since, for example, the authors could not "assign" exposures with any better accuracy than ±5 dBA "and more" for test subjects. Numerous other shortcomings in their study (Table 27.2) underscore this key flaw.

The fact that many subjects in these studies were clearly exposed to impulsive and impact noise further detracts from attempts to apply these studies to the development of steady-state noise criteria. In addition, a more fundamental technical flaw is the historical development of the exchange rate.

Shaw (24) states that the first regulation based on the total-energy theory "seems to have been that promulgated by the U.S. Air Force in 1956." The details concerning regulation AFR160-3, as presented by Jones (4), indicate a fundamental weakness in formation of the equal-energy hypothesis. Jones makes the following statement about the underlying reason for the 3-dBA exchange rate in AFR160-3: "Because, on the decibel scale, a doubling of the sound intensity or energy results in an increase of 3 dB, the allowable sound pressure could be increased by 3 dB, if the duration of exposure were cut in half."

Since it is also true that the doubling of sound intensity or energy can produce, depending on phase relationships, an increase of 6 dB, following the logic upon which AFR160-3 is said to be based would lead to a 6-dB exchange rate as a worst case. Since a 6-dB exchange rate is obviously less conservative, it must be concluded that the original basis for selecting 3 dB, according to AFR160-3, is not appropriate. Shaw did not consider this background and did not point out that AFR160-3 recommended auditory protection at a criterion level equivalent to about 90 dBA, with mandatory protection at a level equivalent to about 100 dBA (i.e., 95 dB at the four octave bands between 300 and 4800 Hz). A key fact is that the Air Force has evolved a 4-dBa exchange rate, which it presently uses.

In summary, the best evidence suggests that the Burns and Robinson (31) and Passchier-Vermeer (25) studies are not to be relied upon as scientific bases for proposing that the criterion exchange rate be changed from 5 dBA to 3 dBA in regulations pertaining to occupational noise exposure. On the other hand we know that the 5 dBA exchange rate with a 90 dBA criterion level works very well. Owing to the effectiveness of OSHA's Hearing Conservation Program, there is far less risk to employees than indicated in earlier DRC/PR predictions.

3. SIMULTANEOUS ASSESSMENT OF EXPOSURE TO IMPULSE/IMPACT AND STEADY-STATE NOISE AND HIGH CREST FACTOR INSTRUMENTATION

Impulse and impact noise exposures were included in data used as bases for developing damage risk and percentage risk criteria (40). Close examination of the Burns and Robinson (31) and Passchier-Vermeer (25) studies shows that the selected populations were exposed to noise from many types of impulse and impact noise sources and operations. Therefore, the effects analyzed in those studies incorporated exposures with significant impulsive noise content.

In its January 16, 1981, amendment proposal (42), OSHA specified a crest factor of "30 dB" designed to emphasize impulse/impact noise components in a new way relative to the measurement of steady-state noise.

As Shaw (24), Erdreich (43,44), and others have noted, large numerical penalties should be expected when instrumentation incorporating high crest factors to integrate impulse/impact noise with steady-state noise is required. Penalties of 10, 15 dBA, and more are the numbers generally discussed in this context. It is important to realize that a 10-dBA penalty represents 10 times the intensity of the noise measured by conventional methods.

Pfander et al. (45) show that "CHABA's upper limits for tolerable exposure to impulse noise are set below those established by the Federal Republic of Germany (FRG)." That is, the CHABA limits are too stringent by a significant degree. As Pfander points out. "Since the (FRG) values are based upon 10,000 such tests, we feel their validity has been amply demonstrated." This disputes Shaw's claims (24) that the CHABA limits are too lenient.

It should be emphasized that Pfander's work underscores the importance of measuring peak sound-pressure level in evaluating exposure to impulse noise according to DRC. Incidentally, Pfander's rating of personal hearing protection for impulse noise is in the 30-dB range, according to that reference.

Voigt and Ostlund (46) also point out that a 3-dBA exchange rate standard based on the equal-energy hypothesis is not appropriate to the measurement of impulse/impact noise. The results of other research agree with this viewpoint (47–49).

The stringency of measures incorporated in OSHA's original noise regulation seems to be ignored by those supporting untried and unproven changes in measurement instrumentation and procedures (24,50). It is important to appreciate fully the degree of protection afforded by procedures established in OSHA's noise regulation.

First, it is recognized that there is a component of impulse/impact noise that can be considered in the same manner as steady-state noise from the standpoint of auditory response and its relationship to progressive development of permanent threshold shift. This component is the root mean square or rms value. For this component of impulse/impact noise, the stringent 90-dBA criterion applies to 8-h exposure.

There is a difference in the loss mechanism and spectral aspects that must be taken into account (51), when impulse/impact noise is present in any exposure. It is also necessary to consider those components of impulse/impact noise that provoke a different auditory response. For example, there are those bursts of noise energy that occur in time frames for which consideration of the peak noise level in the time domain is essential.

In this context the noise level versus time is the most important aspect of the noise spectrum. Factors such as threshold level, rise time to peak sound pressure level, decay time, and peak-to-peak time intervals are also relevant to these considerations. According to the DRC basis for impulse/impact noise, a criterion of 140-dB peak sound-pressure level applies to measurements made on a "peak hold" circuit. The number of permissible impulse/impact events can be permitted to increase as the measures/observed peak level decreases from the 140-dB criterion according to McRobert and Ward (52).

Proposals by Shaw (24) and others (50) would virtually eliminate such key assessments and replace them with unproven integration methods. It is essential to recognize that OSHA requires that the steady-state component (rms) and the impulse/impact part (peak) of the same noise spectrum be met (i.e., complied with) *simultaneously*.

A simultaneously applied criterion of 90-dBA (rms) and 140-dB peak is considered the practical and proven method for assessing combined effects of steady-state and impulse/impact noise with respect to potential for injury to hearing. Our data show (2) that these criteria have resulted in reducing risk below historical DRC projections. Effectiveness is enhanced by application of the methodology in context of a hearing conservation program. This straightforward, practical, and protective approach is recommended as the best available at this time, pending further research, as suggested by Erdreich (44).

The industrial workplace should not be considered a laboratory where proven criteria and instrumentation can be suddenly replaced with experimental counterparts by regulatory fiat. Proposals to increase the crest factor handling capability of integrating instrumentation as a means of accounting for the effects of impulse/impact noise are not based on hard scientific evidence. Therefore, they should be rejected because of the added costs introduced without benefits, as well as the possibility of increasing risk.

4. REDUCTION IN PERFORMANCE VALUES FOR PERSONAL HEARING PROTECTION DEVICES

Some studies conclude that performance of personal hearing protection devices in selected industrial settings does not approach laboratory ratings for those devices. Since these ratings, the NRR for example, are primarily designed to provide a basis for selection of devices on a relative scale, like EPA's mileage ratings for automobiles, such findings are not surprising.

Some investigators [e.g., Shaw (24)] and regulatory bodies, such as OSHA, have concluded that the only immediate remedy is to cut the performance of personal hearing protection devices in half (28). The 50% concept is technically defective and misleading to users of such devices, as will be shown later.

Some specific facts to consider concerning these studies are as follows:

1. NIOSH studies considered only groups of workers who did not necessarily participate in a hearing conservation program as required by OSHA's amendment (March 8, 1983) to the noise regulation.
2. The studies do not contain well-defined measures of noise exposure of the subjects.
3. There is no evidence that appropriate fitting and training procedures were common to the subjects included in the field studies.

In the near future, we shall be able to examine longitudinally the results of OSHA's Hearing Conservation Program across industry and learn the facts. Early indications are that percentages of impairment will fortunately be much lower than the early DRC predictions would indicate. The question to be asked in the context of OSHA's amended regulation is not whether employees will achieve laboratory NRR values for their personal hearing protection devices. The correct question is whether the devices used in the context of a hearing conservation program provide adequate protection for the individual.

One of the primary benefits derived from an employee's participation in a hearing conservation program is the annual monitoring and evaluation of the performance of personal hearing protection devices for that individual. The instant 50% reduction concept not only ignores this benefit, but precludes its proper longitudinal evaluations.

It is inappropriate to require a reduction in performance of personal hearing protection devices before they have been properly evaluated for an individual user in a hearing conservation program over a period of several years. This does not imply that where the devices prove inadequate, nothing could be done in sufficient time to rectify the situation. In fact, the matter would be resolved annually in medical follow-up to any employee's audiometric test where the results indicate that personal hearing protection may be a problem. OSHA's standard threshold shift (STS) is designed to flag such problems. Furthermore, there is much evidence to support the efficacy of hearing protection (53–55).

Monitoring personal hearing protection on an individual basis in the context of a hearing conservation program is the best approach, according to our experience (2). Arbitrary reduction of performance values for these devices across the board is an unnecessary and costly action. Added costs will come from engineering controls required to make up the difference between the measured numbers and arbitrarily reduced effectiveness of protection (see Fig. 27.1).

In using noise reduction factors when evaluating personal hearing protection devices (PHPDs) and comparing these to exposure noise levels, one deals with logarithmic quantities. For example, if a personal hearing protector with a performance equivalent to 20 dBA is suddenly reduced to 10 dBA, this is much more than a 50% reduction in performance relative to the noise level in question. In fact, a derating from 20 dBA to 10 dBA is an order of magnitude (10 to 1) reduction in performance.

Such reductions can be extremely costly, particularly when the performance is close to providing the difference between the exposure and criteria levels. In any case, derating performance to one-tenth, or less, is unnecessary according to the results of our hearing conservation program (2). Certainly, the added costs of arbitrary application of the "halving" concept will not improve on those benefits available from the hearing conservation program alone.

4.1. Engineering Control Requirements

The focus of criteria numbers in occupational noise regulation to date has been on engineering controls. This appears to be the primary reason for historical controversy about the numbers and the cost to achieve them. Since any incorporation in regulation of the number inflating factors discussed earlier will create additional engineering control requirements, this aspect of compliance is discussed here.

Regarding regulatory emphasis on engineering controls, good judgment is essential for success over the long term. For example, the current 90-dBA-based code in the United Kingdom is being enforced according to Health and Safety Executive (HSE) guidelines (56) that appear practically states as follows:

> This standard will be enforced by all the HSE's Inspectorates taking common sense account of the different problems encountered in different workplaces.
> Stress was laid on engineering controls wherever they can be applied. There is however a realistic appreciation by the enforcement authorities that in the same way that Rome was not built in a day, industry's noise problem will not be solved in a day either. Where plant and machinery cannot be treated effectively to suppress noise levels, the provision and wearing a ear defenders (of one sort of another as appropriate in the circumstances) will be required of employees. Here managements are being

asked to consult with safety representatives so that a "constructive effort" can be made to tackle the problem jointly, it may also be noted that in countries that have adopted more stringent noise criteria on paper, there is often a corresponding lack of enforcement.

Comparing the different approaches to engineering controls currently used by OSHA and MSHA (57), it appears the MSHA's approach to common machinery problems will be most practical for the long term.

In contrast, OSHA's enforcement history indicates that each machinery user should be required to invent (or reinvent) the engineering control "wheel" in each specific instance at the enforcement/implementation stage.

Certainly, some portion, if not all, of the responsibility for engineering noise control of industrial machinery belongs to the supplier and not the user. It is, however, recognized that OSHA enforcement efforts are directed only at the user. This usually results in relatively short-term retrofit and fixes (e.g., enclosures) in most cases reaching the implementation phase. Retrofit, thus, represents a diversion of progress and funds from solving noise problems at source. It is perceived that OSHA's enforcement of engineering controls continues to perpetuate substantial capital investment for retrofit as its predominant objective, an unhappy prospect for industry and the economy.

It is suggested that the current approach be modified to focus on meaningful control at source for machinery common to industrial sites. The case for electrical motors exemplifies how this can be done (58). This basic approach to engineering controls at source would complement the hearing conservation program with prospects for significant long-term achievements. Implementation of front-end loading and concurrent engineering methods for designing new plants and facilities will enhance these opportunities.

Finally, incorporating noise control in the design of new plants and facilities to meet dosimeter-evaluated exposure criteria is much more difficult than designing to meet workplace levels. Workplace levels are much easier to predict at the design stage than individual employee exposures, particularly when a high degree of employee mobility is projected.

4.2. Developing Regulations

Currently, the European Economic Communities (EEC, now called the European Union) and the Canadian Province of Ontario are working on occupational noise regulations.

The European Parliament has recently (59) decided that the EEC noise directive will be based on a criterion level of 90 dBA for engineering controls versus the previously proposed 85-dBA level. This decision was based on economic practicality, recognizing that the hearing conservation program would be most important in protecting affected employees beginning at 85 dBA.

On the other hand, a well-advised shift from any onerous combination of Shaw's recommendations (24) toward increased reliance on the hearing conservation program would certainly provide a more practical approach to the protection of employees in Ontario, according to our experience.

5. OSHA'S AMENDMENT: A SHIFT FROM SOLE RELIANCE ON DAMAGE RISK CRITERIA

OSHA's amended regulation of March 3, 1983 (3,20) was the first positive step toward reliance on a hearing-conservation-program-based regulation for occupational noise exposure. OSHA should also be credited with having resisted EPA's inadequately supported demands to adopt an 85-dBA standard for engineering and administrative controls, with a 3-dBA exchange rate (26). Inappropriate extrapolation of the DRC affects OSHA's regulation in a complexity of dual criteria (see Tables G-16 and G-16a of Ref. 3; see also Ref. 20 and Chapter 34, this volume). On the other hand, OSHA requirements for a hearing conservation program could be viewed almost independently of complex monitoring methods specified for determining which employees should participate in the program.

For example, it is recognized that monitoring, noise measurements, and related computations are in the OSHA regulation to provide for selection and evaluation of personal hearing protection and notification of employees regarding exposure levels according to OSHA regulation. However, it appears appropriate and conservative simply to place all employees exposed to workplace noise levels at or above 85 dBA in the hearing conservation program. Based on this primary method of determination, procedures permitted by the OSHA regulation could then be more practically applied to remove employees from the program when (and if) advisable. This could eliminate much of the cost and difficulty associated with OSHA's initial determination procedures.

It costs between $20 (Du Pont's estimate) and OSHA's estimated $41 (3) per employee per year to participate in a hearing conservation program. Comparing these with costs of developing a basis for eliminating employees from the program by noise measurement and dose determination, removal will not generally prove economical by our experience. Certainly, 100% participation has some administrative appeal, particularly in large programs.

Fortunately, OSHA's first step toward the overriding efficacy of a hearing conservation program appears to permit more protective and cost-effective options.

6. MAJOR BENEFITS OF THE HEARING CONSERVATION PROGRAM FOR THE INDIVIDUAL EMPLOYEE

The history and application of DRC-based occupational noise regulations have been examined in the context of proposed changes in requirements, instrumentation, measurement methodology, and their anticipated effects on the measured numbers. At this point, a hearing conservation program is the best approach to avoiding most, if not all, of the problems considered.

A hearing conservation program could be implemented without criteria numbers and exchange rates. For example, the frequency of audiometric testing and follow-up could be varied according to job description and/or age, with all employees participating in the program. This would permit noise measurements to be made primarily for hearing preservation and long-term engineering control purposes.

Recognizing that this simplistic approach to worker protection and occupational noise regulation is not likely to be adopted in the near future, let's look at the benefits of a hearing conservation program.

A hearing conservation program, with medical surveillance, audiometric testing, engineering controls, and hearing protection when necessary and appropriate, is a tested and practical procedure at 90 dBA (2). It is possible to protect each worker on an individual basis through implementation of such a hearing conservation program. This degree of worker protection and risk avoidance is not necessarily achieved through compliance with present and proposed governmental regulations that prefer engineering and administrative or dose controls.

Hearing conservation is effective in preventing noise-induced hearing impairment due to exposure in the workplace (2,18,32).

A comprehensive hearing conservation program calls for:

1. Initial and follow-up audiometric testing.
2. Supplying and fitting personal protective devices such as earplugs or muffs on an individual basis.
3. Education, training, and monitoring of personnel wearing the devices, with supervision.
4. Medical surveillance based on evolution of serial audiometric testing.

A chart that may help in understanding how an effective hearing conservation program should work is presented in Fig. 27.2. Figure 27.3 presents a physician's hearing conservation flow chart that facilitates coordination with the elements of Fig. 27.2.

Our experience has shown that maximum effectiveness of hearing conservation results when employees understand the reasons for the program, cooperate and take initiative in its implementation, and conduct it on a daily basis for their own protection.

During the period 1966–1971, Dr. Sidney Pell, corporate biostatistician with the Medical Division, conducted an investigation of the Du Pont hearing conservation program (2). Dr. Pell concluded from his study "that a hearing conservation program whose components include periodic audiometric testing and ear protection, and which utilizes a hearing conservation criterion of approximately 90 dBA, is capable of protecting the hearing of noise-exposed workers."

Dr. Pell's findings are corroborated in a similar study reported in 1974 by Dr. R. E. Gosztonyi, Jr. (32). These results are as valid today as they were then.

Since it has been proven epidemiologically (2) that a hearing conservation program will provide physician-monitored protection of the individual employee's hearing, criteria numbers should be based on practical action levels that ensure proper functioning of the program. Beyond that, we should consider sensible goals for technologically and economically feasible engineering controls that will promote long-term progress in solving class noise problems at source.

5.1. Audiometric Testing

An essential part of the hearing conservation program is audiometric testing. Applicants are given preplacement physical examinations which include an audiometric test. Employees receive periodic medical examinations, which include an audiogram. After being hired, if assigned to a designated high-noise area, employees are then included in the hearing conservation program. This means that they are given an audiometric test immediately prior to assignment to the designated noise area, 3–6 months after placement, 1 year after placement, and annually thereafter. This initial frequency of testing helps to identify susceptibility as well as physiological problems at an early stage.

To avoid any misinterpretation of our program and the concerns on which it is based, it should be understood that, by definition, present noise exposure regulations are not intended to protect all workers. That approach is unacceptable in the context of our long-standing safety programs, particularly since our hearing conservation program has proven successful (2,18) beyond the regulatory goals.

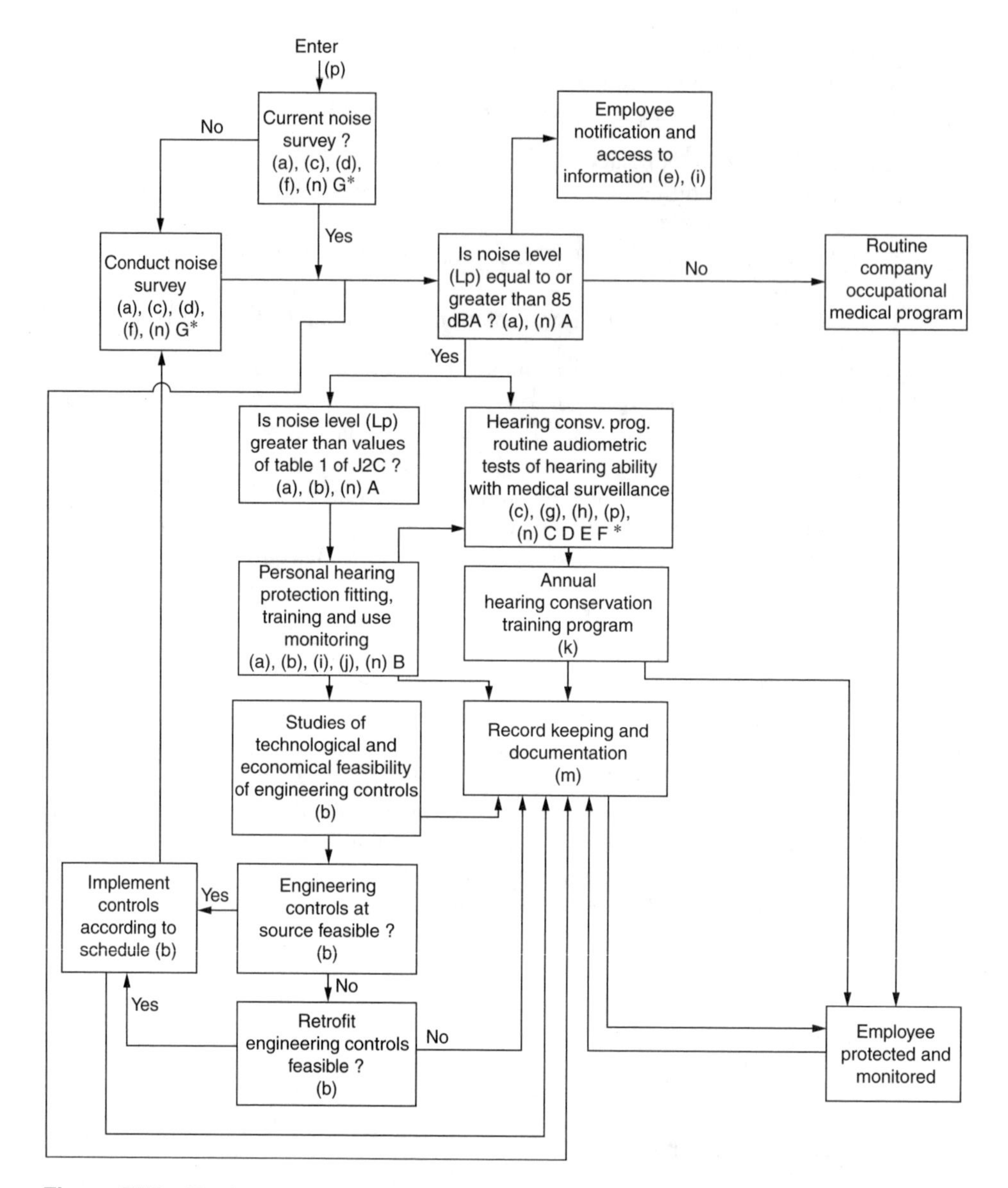

Figure 27.2 Hearing conservation program administration chart. Lower-case letters in parentheses refer to applicable paragraph designations of current amended (3/8/83) OSHA Noise Regulation at CFR 1910.95. Uppercase letters refer to the Appendices of the 3/8/83 Amendment. Those further designated * are nonmandatory appendices (F*,G*). (Copyright 1986, 9/14/83.)

5.2. Personal Hearing Protection

The use of hearing protectors as a means of compliance when reasonable and appropriate is often equal or superior to engineering and administrative controls in protecting the hearing of noise-exposed workers by providing noise attenuation to levels below 90 dBA. Personal protection used in context of our hearing conservation program to date has demonstrated maximum benefit-to-cost ratios in comparison with other available compliance methods.

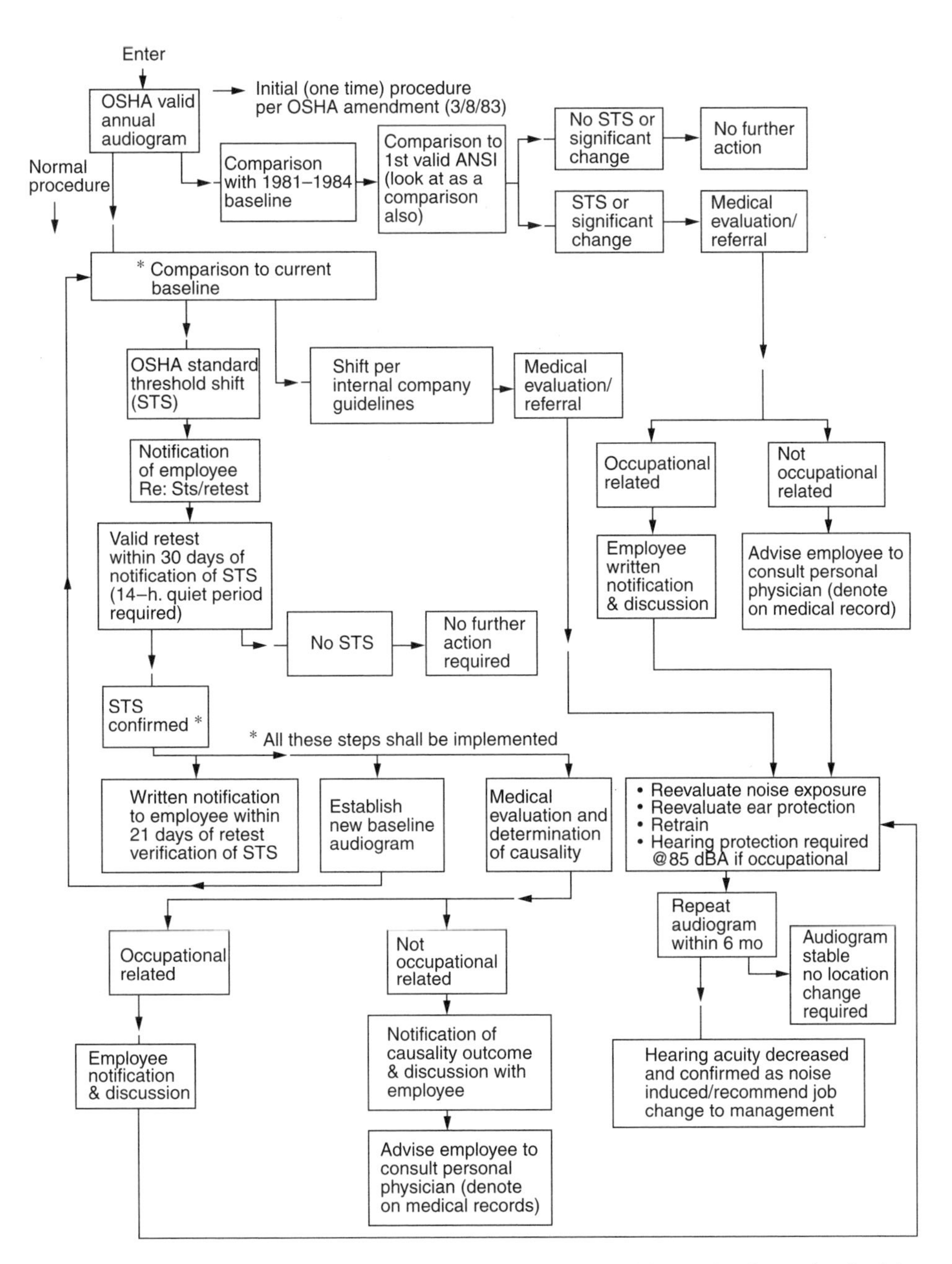

Figure 27.3 Physician's hearing conservation flow chart. In addition to the above, the physician should always keep track of individual hearing threshold levels relative to any indicated progress toward hearing impairment as defined by American Academy of Otolaryngology and American Medical Association criteria using complete audiometric test histories and should take appropriate action to prevent hearing impairment for employees. (Copyright 1986, E. I. du Pont de Nemours & Company.)

In our program, employees are fitted by registered nurses who have been adequately trained in this procedure and are under the supervision of a physician. Training in use and care of the devices is emphasized.

Supervisors are responsible for monitoring employee compliance with all safety rules and personal protective equipment requirements. This includes such items as safety glasses, hard hats, and hearing protective devices. If an employee is not using any designated protective equipment or not using it properly, the supervisor discusses this with the employee and explains again the reasons for the equipment and how it should be used. Plant medical personnel will assist the supervisor if requested. If necessary to gain compliance with the requirements for personal protective equipment, regardless of the type, appropriate disciplinary action will be taken.

To aid our physicians in administration of the hearing conservation program, a flow chart similar to that shown in Fig. 27.3 is included in their program guidelines.

Given that equal protective of hearing health is available by various methods, relative cost-effectiveness should be heavily weighted in selection of the best alternative. The cost of our hearing conservation program, on a per employee basis, is 1/400 of the cost estimated by Bolt, Beranek, and Newman (BBN) for industry to meet a 90-dBA standard. It is therefore urged that hearing protectors, as used in a hearing conservation program, be recognized as a means of compliance at least equal to engineering and administrative-dose controls in protecting the hearing of employees in industry.

6. CONCLUSION

Employees can be protected from risk of injury to hearing due to industrial noise exposure through participation in a medically administered hearing conservation program of the type required by OSHA's noise regulation.

Monitoring of each participating employee also provides annual evaluation of the effectiveness of personal hearing protection as worn by the individual.

Many opportunities for prevention of the long-term effects of noise on hearing are available to the physician and supporting health professionals over a working lifetime. This provides assurance to the physician that the risk of impairment can be identified and controlled for each employee under his or her care.

This approach represents a great improvement in terms of hearing health protection over sole reliance on engineering control implementation, particularly retrofit controls, at a selected criterion level characterized by reliance on DRC and/or PR.

Of major concern are the criterion numbers specified in standards and how these numbers are to be measured. Years ago (60), R. H. Bolt, of Massachusetts Institute of Technology, warned attendees of the First West Coast Noise Symposium to "Beware of single numbers" and more particularly how they are used. Today, we also need to be concerned about the numbers in terms of how they are measured and, more specifically how current measurements compare to those originally used to establish the DRC numbers.

As pointed out here, instrumentation and methodology have changed dramatically since the DRC and PR were established, but the OSHA criteria numbers, for example, have not changed.

There is every indication in comparing data from many studies (see Ref. 5–16, 61–64) that criteria "numbers" need to be revised upward to realign the DRC and PR with these numbers. This is particularly true if we are to enjoy the advantages of technological advances in instrumentation and measurement methodology, such as dosimetry.

It has been noted that some proposed regulatory changes, such as replacing the 5-dBA with a 3-dBA exchange rate, are not warranted by the scientific evidence on hand. Current suggestions that impulse and impact noise be combined with steady-state noise using high-crest-factor-based integrating instrumentation are highly questionable and oversimplified. These and other factors that would increase the measured numbers have been considered in some detail. Such increases and the changes that would bring them about must be avoided. Otherwise, the effect could be the same as lowering the original regulatory criterion limit by 5 or 10 dBA or more.

It is recognized that engineering controls have been the focus of, and to some extent the problem, with the numbers, at least in OSHA's regulation. Engineering controls are important and much can be accomplished, but not by the user-retrofit-emphasizing approach OSHA has taken to date in enforcement.

A much broader view of long-range engineering controls is essential to achieving progress opposite costly, short-term retrofit as the focus of regulatory enforcement.

Original equipment manufacturers have a lot to say about how much noise will be made in the industrial workplace. Until we all accept that fact, true progress in engineering controls will receive a setback every time a costly "fix" is implemented for enforcement purposes in the range of 90–115 dBA. This is particularly true where any retrofit implementation falls short of achieving the criterion limits (e.g., 90 dBA).

Finally, some concluding remarks about personal hearing protection. There has been much written about "real-world" performance of personal hearing protection. Most of those articles are concerned with what happens when attempts are made to use laboratory data as the predictors of field performances on a short-term basis. This type of study is the basis for derating (e.g., 50%) of the estimated performance of personal hearing protection devices.

Not considered in most articles is the actual real-world effectiveness measured annually on an individual user basis through serial audiometric testing in the context of a hearing conservation program. This methodology goes beyond estimates of general performance to continuous monitoring of specific effectiveness for the individual.

The fact is that personal hearing protection is best evaluated according to how it protects the hearing of the employee working in a noise environment, given the specific needs of the individual. Long-term evaluation based on medical review and surveillance is the key to the success of this proven approach (2). This straightforward approach is superior to assessment of prediction schemes as a method of evaluating real-world performance and effectiveness of personal hearing protection. In fact, this hearing-conservation-program-based approach restores confidence in performance and effectiveness that many of the real-world studies have questioned.

AFTERWORD

Many changes have taken place in the composition, character, and demographics of the U.S. industry since the first edition of this book was published.

The basic concepts discussed in this chapter and the best evidence supporting them have not been superseded by any new scientifically valid information that would require their modification and/or replacement. Rather, the intervening years of performance have served to reinforce their efficacy and essentiality to the protection of the hearing health of workers, where applied.

Many potential noise exposure hazards have been eliminated by various means including engineering controls, productivity enhancements, changes in workplace schedules and "off-shoring" of noise exposure related to employment.

Competent, biostatistically valid longitudinal studies of worker populations exposed to industrial noise, with or without personal hearing protection in context of medically supervised hearing conservation programs, retain their paucity in the historical literature (2).

In the late 1990s, the Mine Safety and Health Administration (MSHA) conducted a rule-making process that treated the relevant historical literature in a fallacious manner. This observation can be confirmed by comparing the list of references that were published in the *Federal Register* by MSHA for the proposed regulation and the promulgated rules, respectively. Many of the references in the proposed regulation were eliminated from the final document, including some that support the Agency's eventual retention of OSHA's criteria (e.g., Reference 2 of this chapter). Deletion of key references without apparent cause or explanation is a bibliographical blunder that raises serious questions concerning MSHA's approach to rulemaking, particularly where the scientific literature is most important.

In conclusion, the subject MSHA rulemaking process clearly demonstrated that there is no need to continue the dual bureaucracies OSHA and MSHA. The rationale, practicality, and cost savings of incorporating MSHA's activities under the OSHA umbrella strongly suggest that such a merger is long overdue. This opportunity is made even more evident and attractive by current trends in United States industry and related demographics of the labor force.

REFERENCES

1. Dar TA, and Karrh BW, An effective hearing conservation program—Federal regulation or practical achievement, *Sound Vibration 13*(9):12–19, 1979.
2. Pell S, An evaluation of a hearing conservation program—A five year longitudinal study, *Am. Ind. Hyg. Assoc. J., 34*:82–91, 1973.
3. *Federal Register*, Vol. 48, No. 46, pp. 9738–9785, March 8, 1983.
4. Jones HH, Standards and threshold limit values for noise, *Industrial Noise and Hearing Conservation, CH11* (J. B. Olishifski, ed.), National Safety Council (151.17), 1975.
5. Trethewey MW, Evensen HA, and Lord HW, *Measurement Discrepancies Associated with Evaluating Impulsive Noise*, Michigan Technological University, Houghton, MI, 1976.
6. Dotti NR, *Noise Dosimeters in the Industrial Environment*, Ostergaard Associates, Caldwell, NJ.
7. Rockwell TH, Noise compliance and the job shop, *Sound Vibration*, Sept. 1979.
8. Barry JP, Problems in enforcement of the occupational noise standard, USDOL/OSHA paper presented at NOISE-CON 79.
9. Rockwell TH, OSHA noise violations: Real or Imaginary? *Metal Stamping*, Aug. 1980.
10. Earshen JJ, On overestimating of noise dose in the presence of impulsive noise, paper presented at INTER-NOISE 80, Dec. 1980.
11. Kamperman GW, Dosimeter response to impulsive noise, paper presented at INTER-NOISE 80, Dec. 1980.
12. Rockwell TH, Real and imaginary OSHA noise violations, *Sound Vibration*, March 1981.
13. Stevin GO, Integrating sound level meter for analysis of impulse noise and continuous noise, *Acustica, 51*:55–57, 1982.
14. Fairman TM, *Noise Dosimetry and Typical Noise Doses*, U.S. Air Force Aerospace Medical Research Laboratory, AF AMRL-TR-82-21, 1982.
15. Rockwell TH, Rubber yardsticks, *INTER-NOISE 82 Proceedings*, pp. 861–864.

16. Earshen JJ, On overestimation of noise in the presence of impulse noise, *INTER-NOISE 80, II*:1007.
17. Dear TA, Calculating OSHA noise compliance, *Pollution Engineering, 5*(1):33–43, 1973.
18. Alberti PW, *Personal Hearing Protection in Industry*, Raven Press, New York, Chap. 32, 1982.
19. Dear TA, A common sense approach to workplace noise regulation, Proc. IV Int. Congr. on Noise as Publ. Health Problem, Turin, Vol. I, pp. 349–352, 1983.
20. Occupational noise exposure, *Federal Register, 36*(105), 1910.95, May 1971, Rules and Regulations.
21. *Specifications for Sound Level Meters*, ANSI S1.4 1974, American National Standards Institute, New York.
22. Sulkowski WJ, *Industrial Noise Pollution and Hearing Impairment—Problems of Prevention, Diagnosis and Certification Criteria*, NTIS USA, pp. 86–98, 1980.
23. Van Atta FA, Occupational noise exposure regulation, *National Safety News*, pp. 79–84, March 1973.
24. Shaw EAG, *Draft Report of the Scientific Advisor to the Special Advisory Committee on the Ontario Noise Regulation*, Ontario Ministry of Labour, Toronto, March 1984.
25. Passchier-Vermeer W, Hearing loss due to continuous exposure to steady-state broad-band noise, *J. Acoust. Soc. Am., 56*(5):1585–1593, Nov. 1974.
26. U.S. Environmental Protection Agency. Request for review and report on occupational noise-exposure regulation, *Federal Register, 39*, No. 244, Part II, Dec. 1974.
27. International Standards Organization (ISO), *Assessment of Occupational Noise Exposure for Hearing Conservation*, ISO Standard 1999–1984.
28. *Guidelines for Noise Enforcement*, OSHA, Dept. of Labor, CPL2-2.35A, 29CFR 1910.95 (b)[1], Dec. 19, 1983.
29. Hess PW, The GENRAD 1988–9001 Precision Integrating Sound Level Meter and Analyzers as a workplace sound monitor, Am. Ind. Hyg. Conf., Detroit, May 1984.
30. Galeano SF, How O-I's hearing conservation program gives useful statistics for future analysis, *TAPPI, 64*(4), April 1981.
31. Burns W and Robinson DW, *Hearing and Noise in Industry*, HMSO, London (1970).
32. Gosztonyi RE, Jr., The effectiveness of hearing protective devices, *J. Occ. Med., 17*(9):569–580, Sept. 1975.
33. Scheiblechner H, The validity of the "energy principle" for noise induced hearing loss, *Audiology, 13*:93–111, 1974.
34. Schneider EJ et al., The progression of hearing loss from industrial noise exposures, *Am. Ind. Hyg. Assoc. J.*, May/June 1970.
35. Sataloff J, Menduke H, Sataloff RT, et al., Effects of intermittent exposure to noise: Effects on hearing, *Ann. Otol. Rhinol. Laryngol., 92*:623–628, 1983.
36. Nielsen DW, Franseen L, and Fowler D, The effects of interruption on squirrel monkey temporary threshold shifts to a 96-hour noise exposure, *Audiology, 23*:297–308, 1984.
37. Benwell DA, Regulation occupational exposure to noise—A review, *Can. Acoust., 11*(3):25–44, 1983.
38. Ward WD, *Noise as a Public Health Problem*, Stockholm, Sweden, Vol. 4, pp. 167–177, 1990.
39. Bies DA and Hansen CH, *Engineering Noise Control*, Unwin Hyman Ltd., pp. 73–75, 1988.
40. Baughn WL, Noise control—Percent of population potential, *Int. Aud., 5*:331–338, 1966.
41. Yerg R, Sataloff J et al., Inter industry noise study: The effects upon hearing of steady-state noise between 82 and 92 dBA, *J. Occup. Med., 20*:351–356, 1978.
42. Occupational noise exposure: hearing conservation amendment, *Federal Register, 46*, No. 11, January 16, 1981, Rules and Regulations.
43. Erdreich J, Problems and solutions in impulse noise dosimetry, *Sound Vibration*, 28–32, March 1984.
44. Erdreich J, Impulse noise: Comparison of dose calculated by 5 dB rule and 3 dB rule, Abstract, 107th meeting ASA, Norfolk, 1984, *JASA, 75*(Suppl 1), 1984.

45. Pfander F, Bongartz H et al., Danger of auditory impairment from impulse noise: A comparative study of the CHABA damage-risk criteria and those of the Federal Republic of Germany, *J. Acoust. Soc. Am., 67*(2):628–633, 1980.
46. Voigt P and Ostlund E, *The Influence of Impulse Sound in Noise Measurement and The Risk Criteria for Occupational Hearing Loss*, Research Foundation for Occupational Safety and Health in the Swedish Construction Industry, Stockholm.
47. Hendersen D, Salvi RJ, and Hamernik RP, Is the equal energy rule applicable to impact noise? *Hearing and Prophylaxis* (H. M. Borchgrevink, ed.), *Scand. Audiol.*, Suppl. 16, 1982.
48. Rossi G, Acoustic reflex amplitude in response to continuous noise and impulse noise with the same energy content, Proc. IV Int. Congr. on Noise as Publ. Health Problem, Turin, Vol. I, pp. 185–191, 1983.
49. Price GR, Rating the hazard from intense sounds: Putting theory into practice, *Hearing and Prophylaxis* (H. M. Borchgrevink, ed.), *Scand. Audiol.*, Suppl. 16, 1982.
50. Von Gierke HE, Robinson DW, and Karmy SJ, Results of the workship on impulse noise and auditory hazard, 1981 ISVR, Memo 618.
51. Hodge DC and Price GR, Hearing damage risk criteria, *Noise and Audiology* (D. Lipscomb, ed.), University Park Press, Baltimore, Chap. 6, 1978.
52. McRobert H and Ward WD, Damage-risk criteria, the trading relationship between intensity and the number of non-reverberant impulses, *J. Acoust. Soc. Am., 53*:1297–1300, 1973.
53. Waugh R, dB(A) Attenuation of ear protectors, *J. Acoust. Soc. Am., 53*:440, 1973.
54. Waugh R, Simplified hearing protection ratings—An international comparison, *Sound Vibration, 93*(2):289–305, 1984.
55. Forshaw SE, Hearing protection practice in the Canadian forces, *Hearing and Prophylaxis* (H. M. Borchgrevink, ed.), *Scand. Audiol.*, Suppl. 16, 1982.
56. Noise—The future, *Health and Safety Monitor*, UK, Vol. F, Issue No. 6, Feb. 1984.
57. Sec.: 206, *Noise Standard, Federal Noise Safety and Health, Act of 1977* (MSHA), PL 95–164, 95th Congress S717, Nov. 9, 1977.
58. Dear TA, Tackling the problem of reducing motor noise, *Electrical Construction and Machinery*, McGraw-Hill, New York, 1981.
59. European Economic Communities (EEC), Amended proposal for a Council Directive on the protection of workers from the risks related to exposure to chemical, physical, and background agents at work, *Official J. Eur. Communities*, No. C214/11–16.
60. Bolt RH, Summary of symposium, *Noise Control, 64*, Sept. 1955.
61. Passchier-Vermeer W, Noise-induced hearing loss from exposure to intermittent and varying noise, International Conference on Noise as a Public Health Problem, Dubrovnik, Yugoslavia, May 13–18, 1973.
62. Robinson DW, *Occupation Hearing Loss*, Academic Press, New York, 1971.
63. Baughn WL, *Relation Between Daily Noise Exposure and Hearing Loss Based on the Evaluation of 6835 Industrial Noise Exposure Cases*, Aerospace Medical Research Laboratory (AMRL-TR-73-53).
64. Algemeine Unfallversichungsanstalt (A.U.,), A. Raber, The incidence of impaired hearing in relation to years of exposure and continuous sound level, Austria, unpublished report, 1973.

28

Hearing Loss and Other Otolaryngologic Complications of Scuba Diving

Michael C. Neuenschwander
Director, Southeastern Voice Disorders Center, Riverdale, Georgia, USA

Robert T. Sataloff
Drexel University College of Medicine, American Institute for Voice and Ear Research, and Graduate Hospital, Philadelphia, Pennsylvania, USA

Hearing loss, dizziness, and bleeding from the ears have been recognized as hazards associated with industrial exposure to high noise fields underwater (1). Vocational and avocational scuba diving may damage not only the ear, but also other areas of the head and neck. Otolaryngologists, occupational physicians, nurses, and other concerned with occupational hearing disorders should be familiar with scuba diving and its otolaryngologic complications.

1. CASE PRESENTATION

The patient is a 43-year-old white male in good health who went on a week-long recreational dive trip in the Caribbean. He had been suffering from an acute upper respiratory tract infection, but he decided not to cancel his dive trip. On the first day of diving in 30 ft of water, he experienced difficulty equalizing ear pressure on descent. He continued to descend despite increasing fullness and pain in his right ear. He attempted several times, unsuccessfully, to clear his ears using a Valsalva maneuver. At this point, he experienced overwhelming pain, a sensation of fullness, decreased hearing he described as "deafness," and tinnitus in his right ear. He did not have vertigo. He ascended to the boat and returned to dry land. He was treated by a local island doctor with decongestants and antibiotics. His symptoms of fullness in his right ear began to resolve by the third day, but he continued to have substantial hearing loss and tinnitus in his right ear. He noticed an improved ability to clear his ears and continued to dive on his trip for the next couple of days without difficulty. At home, he was evaluated by our otolaryngologist, who found hemotympanum with an intact but bulging tympanic membrane.

His audiogram revealed a moderate sensorineural hearing loss and a small air-borne gap in the right ear, consistent with a mixed hearing loss (Fig. 28.1A). He had a type B tympanogram bilaterally. Discrimination scores were good. This evaluation led to the diagnosis of middle and inner ear barotrauma.

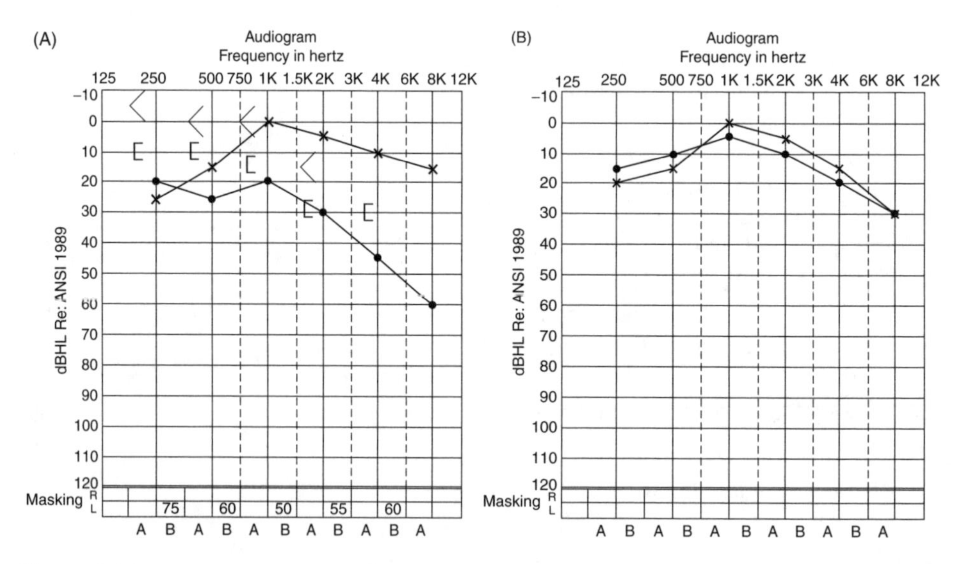

Figure 28.1 (**A**) Audiogram demonstrating mixed hearing loss in a patient with middle and inner ear barotrauma. (**B**) Post-treatment audiogram. Note resolution of conductive hearing loss and persistent high-frequency sensorineural hearing loss.

On bedrest and antibiotics, his hearing and tinnitus gradually improved. However, he had a persistent conductive loss due to middle ear fluid and underwent myringotomy. His hearing subjectively returned to baseline. On follow-up audiometry at 1 month, he showed recovery of his conductive hearing loss, although he had a persistent high-frequency sensorineural hearing loss (Fig. 28.1B). He did not return for further hearing evaluation, but reports normal hearing. This represents a typical case of otologic injury suffered during diving. However, not all patients enjoy such satisfactory outcomes.

2. HISTORY OF DIVING

The earliest evidence of breath-hold diving dates back to 4500 BC in Mesopotamia. The earliest device employing trapped air, similar to the modern diving bell, was an inverted container used by Alexander the Great in 330 BC. Over the following 1650 years, humans have been diving to greater depths for longer periods using various forms of air-containing tanks. The major breakthrough came in 1943 when Jacques Cousteau developed the Aqua Lung, which employed a regulator to release gas on demand at ambient pressures. This technology, still in use today, has become more modern and durable. It is not just limited to commercial and professional divers. Anyone today can travel to almost any destination, learn the basics of diving, and begin diving that same day. There are over 3.5 million divers with a growth rate of 2–12% per year (2,3). These people do not just live in tropical climates. Many dive avocationally, but a substantial number dive professionally. And as more and more people dive, otolaryngologists will have a growing number of patients who present with signs and symptoms of barotrauma, noise-induced hearing loss, and other problems.

3. DIVING BASICS

3.1. Open vs. Closed Systems

Scuba diving is based on one simple principle. It uses a self-contained, closed system of tanks, hoses, and valves that deliver air on demand at ambient pressure. Underwater air systems may be of the open or the closed type. In an open system, a given volume of gas in a container at the surface will be reduced in volume on descent. If a given volume of air is placed in an open system at depth, the volume of air expands as the container starts to ascend. The excess gas from gas expansion will escape and the total volume will remain the same. In a closed system, a volume of gas at the surface will decrease on descent just as in an open system. The difference between the two systems is on ascent. A given volume of air infused into a container at depth will expand in volume as the container ascends. The expanding gas will have no escape and overexpansion will be limited by the compliance of the system. If we take scuba diving as an example of a closed system, expanding gases are only allowed to escape when a diver breathes or equalizes. Thus, current scuba systems are closed systems and have attendant limitations, knowledge of which is important to understand the mechanisms of otolaryngologic injury (4).

3.2. Equipment

Typical scuba gear includes a tank attached to a vest, which is usually a buoyancy compensation device that acts to aid a diver in underwater control of buoyancy. The tank holds

50–80 ft^3 of air under 2250–3000 pounds of pressure. The air mixture is released via the regulator. The first stage of the regulator is attached to the tank and represents one stage of pressure reduction. A second stage, with a mouthpiece, further reduces pressure and is triggered by negative inspiratory pressure. The rest are extras such as extra regulators (or octopus), back-up tanks, masks, fins, weights, wet-suits, and dive computers.

Divers must know their equipment and some basic techniques. More importantly, though, divers must have an understanding of underwater physics as well as the limits of their dive plans in order to dive safely and minimize the risk of barotrauma.

3.3. Physics

3.3.1. Gases

Usually, divers breathe air, made primarily of nitrogen, oxygen, and a small portion of carbon dioxide. Helium (He) may be substituted for nitrogen in complex mixtures. Nitrogen (N) is inert and is not metabolized and, therefore, tends to accumulate in tissues. Nitrogen diffusion depends strictly on the partial pressure of the gas and its diffusion gradient. Nitrogen accumulation in the body depends on the depth of the dive, the length of the dive, and the time above the surface after the dive. These parameters have been calculated into dive tables currently in use by various organizations, the most common of which are the U.S. Navy Dive Tables. Based on these tables, the depth limit for recreational diving is 130 ft of salt water (fsw). At this depth a diver only has 8 min underwater without making a decompression stop. Deeper or longer dives require various mixtures of O_2, N, and He or decompression stops to allow time for nitrogen or helium to diffuse out of the lungs (4).

3.3.2. Decompression Sickness

If a diver ascends too rapidly after not observing depth and time limits, nitrogen bubbles form as nitrogen becomes insoluble too rapidly. These bubbles may cause mechanical obstruction of arteries and veins; damage nerves, arteries, and other tissues; and may activate the coagulation and complement systems. Symptoms range from a mild rash, fatigue, and painful joints to convulsions, stroke, paralysis, and even death. The symptoms observed are described as decompression sickness. It is also known as Caisson's disease or "the bends" and was first described in laborers at the St. Louis bridge in the early 1870s (5). More than 80% of individuals will have central nervous system symptoms and 95% of these individuals will have symptoms within 4 h of surfacing. Decompression sickness occurs on average once in every 5–10,000 dives in recreational divers and is 10 times more common in professional divers. The dive tables are based on probability; nevertheless, as many as half of all cases of the bends are suffered by divers who claim to have been diving within the "no decompression limits" (5).

3.3.3. Boyle's Law

Another fundamental principle in the pathophysiology of barotrauma is Boyle's law. As pressure of a gas increases, there is a proportional decrease in volume. This pressure is measured in atmospheres (atm). One atmosphere is the pressure of a 1 $in.^2$ column of air rising to the height of the Earth's atmosphere. Underwater, this pressure is equivalent to only 33 fsw or 10 m of salt water (msw). Thus, at 33 ft, the absolute pressure equals 2 atm, at 66 ft it equals 3 atm, and so on. Air-filled spaces in the body, such as the middle ear, sinuses, and lungs, are subject to pressure and volume changes. A diver equalizes the pressures in these spaces with ambient pressure either passively or by using the Valsalva and various other techniques (4). The inability to safely equalize predisposes a diver to most of the otolaryngologic problems seen.

3.4. Dive Profile

As mentioned above, divers use a dive profile to plan their dives and otolaryngologists should be familiar with some of its concepts. A dive profile is a map of a dive that allows a diver to calculate the length of time he can spend at any given depth. The time spent at the surface after each dive determines how much nitrogen is left in the body. The nitrogen left in the body, or the residual nitrogen time, is added into the calculations for depth and time of the next dive. This is useful when making repetitive dives (4).

Otolaryngologists should be familiar with these concepts as well as what a dive profile looks like. Dive profiles provide the physician with a common language with which a diver can communicate his problem. They can also aid the physician in determining the mechanism of injury, i.e., differentiating inner ear barotrauma from inner ear decompression sickness (6).

4. BAROTRAUMA: SYMPTOMS AND EVALUATION

In general, symptoms of otologic dysfunction include ear pain or fullness, tinnitus, hearing loss, and vertigo. Ear fullness may occur from occlusion of the external canal or from high or low middle-ear pressure. Pain is felt with increasing pressure and is increased further by rupture of the tympanic membrane. Hearing loss may be conductive, sensorineural, or mixed. Tinnitus usually occurs with inner ear or central pathway disease. Vertigo is usually accompanied by nystagmus unless the dysfunction is nonacute and suppressed by visual fixation. Vertigo may subside after an acute unilateral injury over a varying time period between several days to 4–6 weeks. This is a result of central nervous system compensation and does not mean that the injured part of the vestibular system has been restored. Vertiginous symptoms include nausea, vomiting, and loss of spatial orientation with certain positions and may intensify with loss of proprioception and visual orientation as is common underwater. Vertigo is one of the most hazardous symptoms in diving and it may occur at any phase of the dive (7). These symptoms may be found alone or in combination in either external, middle, or inner ear injuries.

Evaluation of a dive injury begins with an adequate history and examination. The history should focus on the timing of symptoms discussed, previous otologic problems, or previous otologic surgery. In the history, we should remember that dive injuries may be very subtle and can occur with free diving (8). The physician's knowledge of diving is very important in the history. The patient should have a complete otolaryngologic examination, including tuning fork tests using a 512-Hz fork (9). Lateralization to the affected ear may indicate a conductive loss and lateralization to the opposite ear may indicate a sensorineural loss. This can be done on initial evaluation to decide on the course of treatment, but an audiogram and tympanogram should be done as soon as possible (10). Finally, a patient complaining of dizziness after a dive injury should have electronystagmography (ENG) testing when feasible (11,12).

4.1. External Ear Barotrauma

External ear or barotrauma is the result of obstruction of the external ear and may occur on descent or ascent. It does not occur if water is allowed to gain entrance to the canal.

Obstruction of the external ear from cerumen, foreign bodies, otitis externa, exostosis of the canal, ear plugs, or tight-fitting diving hoods can lead to barotrauma of the canal and tympanic membrane (12–14). With obstruction of the external ear, there is a relative pressure difference between the middle ear and external ear canal. The difference in pressures result in tissue damage, including congestion, hemorrhage, inward or outward bulging of the tympanic membrane (TM), and possible rupture of the TM. Rupture of the TM may lead to a caloric response heralded by intense vertigo. Furthermore, unilateral obstruction of the external ear canal can also lead to vertigo because of the asymmetric stimulation of the vestibular system (12).

The best treatment is prevention. The external ear must remain patent. Cerumen and foreign bodies should be removed before a dive and the use of ear plugs or tight-fitting hoods should be avoided. Any free diver descending to more than 5 ft should avoid ear plugs since the pressure gradient between water and air leading to TM rupture may be as little as 100 mmHg or 4.3 fsw (2,12). Otic antibiotic and steroid drugs may be used if there is a tympanic membrane perforation contaminated by water. In any case, diving should be avoided until the external ear and/or tympanic membrane heal.

4.2. Otitis Externa

Otitis externa is the second most common diving complication after middle ear barotrauma (12). Prolonged exposure to water or humid conditions alters the normal skin and its flora. Thus, a greater number of Gram-negative species including Pseudomonas and Proteus in addition to Staph species are seen and may result in symptomatic infection. Fungi are seen primarily after prolonged antibacterial treatment or in immunosuppressed patients.

Again, the best treatment is prevention. Ear canals should be debrided and trauma should be avoided. Treatment is the same as other forms of otitis externa. All swimming or diving should be avoided until the infection has resolved.

4.3. Middle Ear Barotrauma

Once under water, the diver begins to feel the effects of atmospheric and water pressure. Soon, fullness and pain in the ears is a cue to the diver that he or she needs to equalize middle ear pressure. Successful divers begin to equalize before symptoms begin. Continued descent and failure to equalize may result in various degrees of middle ear barotrauma.

The ability to equalize pressure in the middle ear space with ambient pressure is controlled by the Eustachian tube. This has been reported by several authors (12–16) and prospectively studied by Miyazawa, who showed that patients with poor Eustachian tube function had a higher incidence of middle ear barotrauma (17). The Eustachian tube, which is normally closed, acts as a flutter valve between the middle ear and external environment. The diver must actively open the tube during compression.

Farmer depicted what happens to the middle ear during descent (7). If a diver descends to 2.6 ft and fails to equalize, a pressure differential of 60 mmHg develops in the middle ear. This may lead to mucosal congestion and edema. The Eustachian tube lumen is narrowed and middle ear pressure becomes more negative, which makes further attempts at equalizing more difficult. At 3.9 ft, a pressure differential of 90 mmHg occurs. At this pressure the Eustachian tube is "locked." Fullness and pain begin to appear at around pressure differentials of 60 mmHg. If the diver continues to descend, the tympanic membrane will rupture. The TM has been shown to rupture at pressure differentials of 100–500 mmHg. Furthermore, the inexperienced diver may think that a more forceful Valsalva may clear his ears. This maneuver may create a pressure differential between the inner ear and middle ear across the oval and round windows which is so great that rupture of these windows may occur with resultant perilymphatic fistula, vertigo, inner ear hemorrhage, and inner ear membrane breaks (6,8,18). These concepts are discussed below.

The symptoms of middle ear barotrauma range from a sensation of fullness in the ear to severe pain. A conductive hearing loss is usually present. Mild tinnitus and vertigo may occur, especially after rupture of the TM. If hearing loss, tinnitus, and vertigo occur, inner ear barotrauma should be assumed (12). The otoscopic signs of middle ear barotrauma include edema, erythema, serous otitis media, hemotympanum, and TM perforation. Edmonds classically described these findings as six grades, in a modification of Teed's original classification (12,19):

Grade 0: Symptoms without otoscopic signs
Grade 1: Erythema and retraction of the TM
Grade 2: Grade 1 plus slight hemorrhage in the TM
Grade 3: Grade 1 plus gross hemorrhage in the TM
Grade 4: Bulging TM with hemotympanum
Grade 5: Free hemorrhage and TM perforation

Poor nasal function has also been shown to affect Eustachian tube function. Becker and Parell state that conditions such as allergic rhinitis, sinusitis, or a deviated septum can hinder Eustachian tube function. It is possible that correction of a deviated septum can restore the ability to autoinflate the ears in most people, although this assertion remains controversial (13). Thus, a complete head and neck exam is necessary in the diagnosis and prevention of middle ear barotrauma.

Treatment of middle ear barotrauma has not been studied well in prospective trials. Treatment varies among authors and depends on the type of middle ear barotrauma. The disease is generally self-limited and most middle ear barotrauma may not require specific treatment (15). For Grades 0 and 1, Farmer (12) recommends avoiding further diving until all symptoms and existing nasal symptoms have cleared and the diver can easily autoinflate both ears. Systemic decongestants such as Sudafed may be given. Antihistamines and nasal steroids are recommended by some authors, although their usefulness has been questioned. Treatment for Grades 2–4 is similar to that for Grades 0–1 with some additions. Systemic steroids have been used with anecdotal success. Furthermore, the patients is at increased risk for otitis media, and since Eustachian tube function is poor, prophylactic antibiotics are recommended by most authors, although their use is controversial. Topical antibiotic and steroid ear drops are of no benefit because these substances do not readily cross an intact TM. Treatment of Grade 5 middle ear barotrauma includes avoiding diving until the middle ear process has resolved. Most perforations will heal spontaneously. The external canal should be cleared of any debris. Systemic decongestants as well as prophylactic antibiotics are recommended. If the TM perforation does not heal spontaneously, surgical repair may be necessary.

In all grades, prevention is the best treatment. Frequent otitis media, middle ear surgery, previous TM perforation, and the existence of cholesteatoma all suggest poor Eustachian tube function during nondiving conditions, and such individuals are likely to have greater difficulty under typical diving pressures. Divers should be taught various techniques of equalizing middle ear pressures (4,12). Divers should also be taught to equalize beginning immediately and every 2 ft or so to prevent a "locked" Eustachian tube. If ear pain is felt and various equalizing techniques have been used, the diver should ascend until proper equalization occurs (14). A forceful Valsalva maneuver should be avoided at depth, as this technique may lead to inner ear barotrauma as well. Divers may return to diving once the middle ear process has resolved but should be warned that they are at risk for continued diving problems.

Middle ear barotrauma usually occurs on descent, but it may also occur on ascent, in which case it is termed "reverse middle ear squeeze" (2). It is caused by the inability to equalize an expanding gas in the middle ear. It may be bilateral but is most commonly unilateral. This may occur, for example, when a diver has had difficulty equalizing on descent, producing Eustachian tube edema. The diver may have finally cleared, but the now-obstructed Eustachian tube will not allow the expanding gas to escape during ascent. Lateral bulging of the TM occurs and the signs and symptoms are otherwise the same as in descent-induced middle ear barotrauma.

4.4. Alternobaric Vertigo

Middle ear equilibration may be asymmetric. Lundgren first described the condition of alternobaric vertigo from asymmetric middle ear equilibration (20). It is a phenomenon described during ascent. A unilateral increase in middle ear pressure on ascent after a Valsalva maneuver can result in unequal stimulation of the vestibular end-organ. The vertigo disappears with stopping the ascent and descending further (2,7,13). The vertigo lasts from a few seconds to 10 min. In Lundgren's studies, experienced divers and those who reported middle ear equilibration difficulties were at greater risk. People who equilibrate equally but have a damaged vestibular end-organ may also be at risk. Vertigo at any depth may be hazardous and life-threatening to any diver. Any diver who experiences this condition should avoid diving, at least until the cause has been clarified and the condition has resolved.

4.5. Transient Vertigo

As mentioned earlier, unequal stimulation of the vestibular system may lead to transient vertigo. During most diving conditions, the vestibular system is stimulated symmetrically. Vertigo and nystagmus may occur when cold water unequally enters one ear, especially when a diver is in a position such that the horizontal semicircular canal is in a vertical position (7). This may occur with obstruction of the external ear canal or with a tympanic membrane perforation.

4.6. Alternobaric Facial Paralysis

Becker described facial paralysis in divers who experienced alternobaric vertigo (21). The symptoms were transient and occurred on the same side as the ear that was overpressurized. Two theories were given. The first is that a dehiscent facial nerve in the fallopian canal was compressed by excessive middle ear pressure. The other is that gas bubbles enter the nondehiscent fallopian canal through the fenestrum of the chorda tympani. This problem is infrequent but may be underreported because of its transient nature (12). Treatment consists of prevention and should focus on adequate ear clearing.

4.7. Inner Ear Injuries

Inner ear diving injuries were not recognized until the late 19th and early 20th centuries. As diving gained popularity and safety issues improved, more and more inner ear injuries were identified in the literature. Because they often accompanied decompression sickness, inner ear injuries were thought to be central in origin and relegated to secondary importance. Symptoms of hearing loss, vertigo, and tinnitus were not treated. In addition, vestibular symptoms would subside in a few weeks in otherwise healthy individuals because of central nervous system compensation, even though end-organ injury was permanent and no recovery of inner ear function occurred. These injuries were not life-threatening, were isolated, and were relatively rare. Therefore, patients often went unrecognized and without treatment (12).

The first reports of auditory acuity changes in divers in the literature date back to the 1800s. The reasons were not known, but in 1929 Vail suggested they were from "nitrogen bubbles" and in 1959, Taylor suggested the changes were from "various exposures of acoustic trauma" (15,22). Later, in 1968, Simmons postulated auditory changes in sudden hearing loss were due to cochlear membrane tears from raised intracranial pressure (23). In the same year, Fee was the first to document oval window fistulas surgically (24). More concrete evidence and theories for injuries to the inner ear were documented by Goodhill in 1971 and Freeman and Edmonds in 1972 (25,26).

In 1977, Farmer divided inner ear injuries into four categories (7):

1. Inner ear barotrauma, or injuries occurring during descent
2. Isobaric otologic barotrauma, or injuries occurring at stable depth
3. Inner ear decompression sickness, or injuries occurring during ascent
4. Sensorineural deafness related to high background noise

4.7.1. Inner Ear Barotrauma

Symptoms of inner ear barotrauma may include tinnitus, vertigo, and unilateral or bilateral hearing loss. Hearing loss is sensorineural, but if middle ear barotrauma coexists, then it may be mixed. Inner ear barotrauma occurs on descent and is usually accompanied by difficulty equalizing, and a history of middle ear barotrauma may be given. Deep diving is not necessary for inner ear barotrauma to occur (27). It has been reported in water as shallow as 7 ft (18).

On examination, there may be evidence of middle ear barotrauma. One of the most important findings may be a Weber test that lateralizes to the opposite ear. Thus, the history, timing of symptoms, tuning fork tests, and associated middle ear barotrauma can lead to a diagnosis of inner ear barotrauma.

Based on their experience and other reported cases, Parell and Becker divided the causes of inner ear barotrauma into three categories (8):

1. Inner ear hemorrhage
2. Cochlear membrane tears with or without oval or round window rupture
3. Perilymphatic fistula

Hemorrhage. When inner ear barotrauma is accompanied by middle ear barotrauma, inner ear barotrauma may occur with or without a perilymphatic fistula. Vail first reported hemorrhage in the basal turn of the cochlea in animal studies in 1929 (22). In 1983, Kelemen reported hemorrhage in the temporal bones of three drowning victims (28). Then, in a study of 18 temporal bones isolated from drowning and rapid ascent victims, Antonelli et al. in 1993 found middle ear hemorrhage to be nearly universal (29).

Hemorrhage was also found in the cochlea adjacent to the round window, in the cochlear aqueduct, and on both sides of Reissner's membrane, which they suggested may have been due to a membrane break. Landolt thought hemorrhage stemmed from the stria vascularis in his studies in monkeys (30). These reports do not support conclusions, but they suggest a possible mechanism for inner ear injury. Inner ear hemorrhage is thought to be one mechanism involved in diving accident cases that present with absent or transient vestibular symptoms and moderate sensorineural hearing loss with excellent spontaneous recovery.

Membrane Tears. The second theory behind the pathophysiology of inner ear barotrauma is that of cochlear membrane tears. There is no conclusive evidence that membrane breaks occur in the inner ear. In fact, there are only case reports and suggestions of this event occurring in the literature. In normal-hearing individuals who develop sudden sensorineural hearing loss without vestibular symptoms, Simmons stated it was reasonable to suppose that they could be caused by a mechanical defect by rupture, break, or dislocations of intracochlear membranes (23,31). These tears may occur from increased cerebro spinal fluid (CSF) pressure transmitted via the cochlear aqueduct or internal auditory canal.

Simmons stated that widespread mixing of endolymph and perilymph may have caused the initial hearing loss observed; but, as healing occurred, ionic gradients were restored and hearing was regained in frequencies that were not anatomically damaged. Audiograms of patients with membrane tears had hearing loss throughout most frequencies, which then removed spontaneously except for a small frequency range. These audiograms are the only evidence of membrane breaks occurring. In his study of 88 patients, the likelihood for spontaneous recovery for the group was 65% and was more common in patients who showed upward-sloping audiograms (31).

Perilymphatic Fistula. The previous theories of inner ear hemorrhage and intracochlear membrane tears are certainly difficult to prove surgically. Perilymphatic fistulas, however, have been studied more extensively. Freeman and Edmonds were the first to relate labyrinthine window rupture to inner ear barotrauma in divers in 1974 (32). Their cases were in relatively shallow water and were associated with evidence of middle ear barotrauma.

In a classic paper, Goodhill described two possible mechanisms whereby perilymphatic fistulas may form (25). These were the "explosive" and "implosive" theories. In the explosive theory, inadequate middle ear pressure equilibration on descent causes middle ear pressure to be negative relative to intralabyrinthine pressure. Straining in an attempt to equalize, as in the Valsalva maneuver, increases CSF pressure. This pressure increase, in theory, is transmitted via the cochlear aqueduct or the internal acoustic meatus to the oval and round windows (25). If a cochlear aqueduct is abnormally patent or wide, CSF pressures may be transmitted more readily, rupturing either the oval or round window.

In the implosive theory, Goodhill proposed that perilymphatic fistulas are created by forces acting in the opposite direction. In this theory, sudden Valsalva maneuvers may result in middle ear ventilation with a rapid increase in middle ear pressure. This rapid pressure increase can rupture the round window or sublux the stapes footplate into the vestibule, creating an oval window fistula.

Perilymphatic fistulas may lead to varying degrees of sensorineural hearing loss with or without varying degrees of vestibular dysfunction (8,25,33–35). These symptoms, in fact, are similar to those seen in inner ear decompression sickness. Any diver who presents with sensorineural hearing loss, vertigo, and tinnitus from a dive in which decompression sickness is unlikely should be treated as having inner ear barotrauma and a possibly oval or round window fistula (36). There are usually no other neurologic signs, dives are within the "no decompression limits," and injury occurs on descent. In clinical practice, patients presenting with hearing loss after a dive should be treated as if they have a fistula until proven otherwise, although other mechanisms may be involved.

Treatment of inner ear barotrauma consists of placing the diver on bedrest with head elevation, and care should be taken to avoid actions that may increase CSF pressure, such as coughing, nose blowing, and straining. Anticoagulants should not be used because of the possible danger of inner ear hemorrhage and ototoxic drugs should be avoided. Serial audiograms should be obtained and the patient's hearing should be followed closely.

The question of the timing of surgical exploration for cases of inner ear barotrauma and possible perilymphatic fistula is controversial (11,33,35). Most authors feel a trial of observation is warranted as inner ear barotrauma may be due to something other than perilymphatic fistula. If a fistula is present, some authors reason that it is common for the fistula to heal spontaneously (11). Certainly, exploratory tympanotomy with repair of the fistula is advised for cases in which symptoms persist over a few days of conservative management, worsen or fluctuate in the interim, where there is stable hearing loss but persistent vertigo, or where deafness is sudden and profound.

4.7.2. *Inner Ear Injuries at Depth*

Inner ear injuries occurring at depth are the second cause of otologic injury that may result in permanent inner ear dysfunction. They are rare and have been described in divers at deep depth in whom an exchange of inert gases has taken place. Specifically, oxyhelium mixtures replaced by nitrogen or neon at a depth of 600 ft caused symptoms of vertigo, nausea, and nystagmus. These symptoms were permanent in two of three divers (12). There were no changes in auditory function.

The explanation is not clear, but this condition may be related to the different solubilities of the inert gases between inner ear fluid compartments. There may be an actual displacement of inner ear structures or an osmotic flux of water due to a more rapid accumulation of the inert gas in the endolymph as described by Farmer (12). The mechanism is not known. Until this has been established, the exchange of inert gases at stable deep depths should be done with caution.

4.7.3. *Inner Ear Decompression Sickness*

As noted earlier, isolated decompression sickness was largely unrecognized until the 1960s and 1970s. The symptoms of inner ear decompression sickness are vertigo, nausea, tinnitus, and hearing loss. It is a phenomenon that has been described on ascent from a dive. Most patients will have symptoms once they reach the surface or shortly thereafter (2). Other neurologic symptoms may be present, indicating the presence of general decompression sickness or "the bends" (6,37).

Farmer described a series of cases of isolated inner ear decompression sickness in which none of the dives was associated with middle ear barotrauma, otologic symptoms at maximum depth, or with other central nervous system signs (36). After these divers were recompressed at varying times, Farmer found that those who were recompressed early had the best chance for complete recovery. In divers whose recompression treatment was delayed longer than 68 min, residual, permanent inner ear dysfunction was observed (36).

Theories as to the cause of inner ear decompression sickness involve the formation of gas bubbles in the inner ear (6,13,14,28,29,36). In Farmer's studies, a majority of divers exchanged inert gases in their breathing mixture in the latter stages of decompression. He speculated that the mechanism of bubble formation was similar to that seen at stable depths, where sudden decreases in partial pressure of inert gases or counterdiffusion of gases between fluid compartments led to bubble formation.

Landolt provided the largest series of evidence for the pathophysiology of inner ear decompression sickness (30). In a series of 90 squirrel monkeys rapidly decompressed in a hyperbaric chamber, 95% showed evidence of vestibular injury. Landolt thought it very likely that the stria vascularis may be particularly susceptible to bubble formation based on studies of ultrasound detection of bubbles in CSF and other tissues after decompression sickness and from other studies which indicated that bubble formation may be more frequent in areas having slow circulation such as the white matter of the cervical C-spine. Based on temporal bone analyses, he theorized that bubble formation may lead to vascular obstruction with subsequent protein leakage and resultant hemorrhage.

Later, appearance of ectopic bone growth was seen and termed fibro-osseus labyrinthitis (12,30). Thus, one current theory behind inner ear decompression sickness pathophysiology includes nitrogen bubble formation resulting in tissue damage seen as a protein exudate, hemorrhage, and ectopic bone growth in the inner ear.

Understanding this mechanism of inner ear decompression sickness is important to physicians attempting to differentiate between inner ear barotrauma (IEBT) and inner ear decompression sickness (IEDCS). The treatment of each is different and possibly permanently harmful if misdiagnosed (6,12). The important factors are summarized in Table 28.1.

Differentiating the two clinically may be difficult. Divers may not be able to recall the timing of the onset of symptoms. Decompression sickness may occur even when dive tables are followed strictly. Finally, spontaneous recovery may occur and obfuscate a final diagnosis.

Treatment of IEDCS involves early diagnosis, attention to ABCs of life support and prompt recompression (3,6,12). Recompression should be avoided in cases of suspected inner ear barotrauma because further damage to the inner ear may occur. The exact depth of recompression is not known, but authors such as Farmer theorize that an arbitrary depth of 3 atm deeper than the depth at which symptoms occurred is reasonable. When diving, returning to the depth at which symptoms occur may not be possible and attempts to be taken to the nearest hyperbaric chamber should be made. In reality, divers may not be able to get medical assistance for hours after an injury. In such cases, pure oxygen and fluid replacement should be administered early, especially if decompression sickness is

Table 28.1 IEBT vs. IEDCS

Factor	IEBT	IEDCS
Dive profile	Any depth, but more common in relatively shallow water, rapid descent, lack of middle ear pressure equilization	Deep and long dives, rapid ascent, decompression stops
Gas mixture	Air	Mixed gases
Onset of symptoms	Descent	Ascent or after decompression
Associated symptoms	Middle ear barotrauma, conductive hearing loss	Symptoms of decompression sickness

Abbreviations: IEBT, Inner Ear Barotrauma, IEDCS, Inner Ear Decompression Sickness.

immediately evident. Valium is useful for relief of vertigo, nausea, and vomiting. When possible, a complete otoneurological examination should be done, including audiometry and ENG. The use of anticoagulants, salicylates, and steroids has not been adequately studied. Anticoagulants are not recommended (36). Finally, any diver who suffers permanent inner ear dysfunction as a result of diving should not return to diving. Patients who have acoustic or vestibular dysfunction should also be cautioned against diving.

8. HIGH PRESSURE NERVOUS SYNDROME

The high pressure nervous syndrome (HPNS) is a syndrome in which symptoms of dizziness and nausea with intention tremor and decreased psychomotor performance were observed in divers who underwent rapid compression with helium–oxygen mixtures at deep depths (greater than 15 atm) (1,12). Symptoms could be reduced by adding nitrogen to the mixture of gases and slowing the rate of descent. Braithwaite and Farmer hypothesized that the problem was centrally located and occurred from a decrease in the normal cerebellar inhibitory modulation of the vestibular nuclei (12). Treatment includes avoidance of rapid descent and improper gas mixtures.

9. OCCUPATIONAL HEARING LOSS

The impact of noise exposure underwater may have implications for the scuba diver similar to those faced by workers exposed to airborne noise-induced hearing loss. Hollien and others (1) have demonstrated that the hearing mechanism underwater is via bone conduction and that there is a 60 dB SPL threshold shift and reduced dynamic hearing range. Increases in air conduction thresholds may be associated with increasing depth. Furthermore, central effects caused by high pressure nervous syndrome or varying gas mixtures may limit fine auditory discrimination. Sources for underwater noise include the dive chamber, dive helmets, underwater power tools, and underwater explosions. These may further be affected by dive hoods. Hollien states that divers may be less resistant to noise-induced hearing loss underwater. The short- and long-term effects of underwater noise exposure are not completely understood and hearing conservation programs (HCPs) have not been developed for this type of noise exposure (1,12). Until HCPs are developed, divers should be aware of the risks to hearing, follow safe diving procedures, and attempt to prevent and seek treatment for otologic injuries suffered while diving.

10. PARANASAL SINUS BAROTRAUMA

Paranasal sinus barotrauma is a relatively common condition sometimes referred to as "sinus squeeze" (14). The mechanism involves inadequate pressure equalization between the air-filled sinus cavities and the nasal airway. Ventilation of the sinuses in such cases is largely dependent on nasal function. Inflammation and congestion due to disorders of the nasal and sinus mucosa, structural deformities, or mass lesions may result in complete or partial obstruction of the sinus ostia (12). Paranasal sinus barotrauma may occur on descent or ascent.

When such obstruction occurs on descent, there is absorption of preexisting air and decreased intrasinus pressure. Inflammation and vascular congestion of the sinus mucosa occurs, which may lead to a transudate, mucosal hemorrhage, or a hemorrhage into the obstructed sinus. These changes in the mucosa may be responsible for the pain felt.

On ascent, the primary mechanism is thought to be related to a ball-valve type of effect. There is a one-way valve created by inflamed mucosa, a cyst, or polyps within the sinus. Pressure equalization may occur on descent but may be impaired on ascent.

Fagan reported 50 cases of paranasal sinus barotrauma (38). In about 70% of divers, symptoms occurred on descent. The most common sinus involved was the frontal, followed by the maxillary, ethmoid, and sphenoid. Pain and epistaxis were very common. Other symptoms included pain in the upper teeth and paresthesias over the distribution of the infraorbital nerve. Many patients also had a history of previous paranasal sinus barotrauma, recent upper respiratory tract infection, chronic nasal and sinus problems, and associated middle ear barotrauma.

When treating these patients, it is important to emphasize to divers that they should avoid diving with active conditions such as mechanical obstruction from intranasal deformities, polyps, or masses (12,38). Underlying allergy, sinusitis, or upper respiratory tract infection should be treated. It is also important to avoid smoking, as smoking may cause nasal and sinus mucosal inflammation. Topical and systemic vasoconstrictors can be used in stable sinonasal disease. These drugs may have side effects and may cause excessive drying of the nasal mucosa. Antihistamines are not recommended as their effect is unpredictable and they may cause drowsiness, which can be dangerous for divers.

Patients who have to use medications before diving should do so about half an hour before a dive. They should also be warned that they are not ideal diving candidates. Finally, patients with any history of the previously mentioned underlying factors occurring in the absence of altered atmospheric pressure may be less likely to equilibrate middle ear or sinus pressures in constantly changing pressure environments encountered in diving. These patients should be cautioned and counseled prior to diving, and diving safety issues should be stressed (Table 28.2).

Table 28.2 Diving Safety Issues

1. Know your equipment.
2. Have a dive plan and stick to it.
3. Always dive with a buddy.
4. Take your time—especially on ascent and descent.
5. Avoid foreign bodies in the external ear canal.
6. Equalize your ears early and often—every couple of feet.
7. Learn modified techniques to equalize ear pressure. These include the swallow, yawn, raising the soft palate, Frenzel maneuver (use the tongue to piston air into the nasopharynx with the mouth, nose, and glottis closed), and Toynbee maneuver (swallowing with the mouth and nose closed). These maneuvers are helpful, more physiologic, and less traumatic than the Valsalva maneuver.
8. Never hold your breath—even swells and troughs overhead may be enough to change ambient pressure and lead to injury.
9. Limit noise exposure while underwater.
10. Sport diving should be safe, fun, and exciting.

11. CONCLUSION

Industrial, professional, and recreational scuba diving is common and poses substantial risks for otolaryngologic injury. Professionals involved with the hearing health of patients and workers should have fundamental knowledge of the terminology and basic principles of diving, as well as its potential medical consequences. Every effort should be made to differentiate accurately between external, middle ear, and various forms of inner ear injury to provide prompt and appropriate treatment. It is especially important to differentiate between inner ear barotrauma and inner ear decompression sickness. Related conditions such as barotraumatic sinusitis must also be understood and treated expertly. Prospective divers with health conditions that increase their diving-related risks should be identified and counseled. These include patients with poor Eustachian tube function, tympanic membrane perforations, patients with a history of previous ear surgery such as stapedectomy or mastoidectomy, patients with endolymphatic hydrops (Meniere's syndrome), and others. It is also particularly important to be familiar with the special hazards associated with underwater exposure to loud noises, which pose risks not only of loss of hearing but also loss of life. With appropriate precautions and enlightened medical care, scuba diving can be a pleasant recreation and a safe and productive vocation.

REFERENCES

1. Hollien H, Underwater hearing conservation. In: Sataloff RT, Sataloff J., eds. Occupational Hearing Loss. 2nd ed. New York, NY: Marcel Dekker, 1993; 567–581.
2. Neblett LM, Otolaryngology and sport scuba diving: update and guidelines. Ann Otol Rhinol Laryngol 1985; 94(Suppl 115):2–12.
3. Neuman TS, Diving medicine. Clin Sports Med 1976; 6:647–661.
4. NAUI Textbook. 5th ed. Montclair, CA, 1987; 14–162.
5. Moon RE, Vann RD, Bennet PB, The physiology of decompression illness. Sci Am 1995; 273:70–77.
6. Shupak A, Doweck I, Greenberg E, et al. Diving related inner ear injuries. Laryngoscope 1991; 101:173–179.
7. Farmer JC, Diving injuries to the inner ear. Ann Otol Rhinol Laryngol 1977; 86(Suppl 36): 1–20.
8. Parell GJ, Becker GD, Inner ear barotrauma in scuba divers: a long-term follow-up after continued diving. Arch Otolaryngol Head Neck Surg 1993; 119:455–457.
9. Burgess LP, Frankel SF, Lepore ML, Yim DW, Tuning fork screening for sudden hearing loss. Military Med 1988; 153:456–458.
10. Green SM, Rothrock SG, Green EA, Tympanometric evaluation of middle ear barotrauma during recreational scuba diving. Int J Sports Med 1993; 14:411–415.
11. Caruso VG, Winkelmann PE, Correia MJ, Miltenberger GE, Love JT, Otologic and otoneurologic injuries in divers: clinical studies on nine commercial and two sport divers. Laryngoscope 1977; 87:508–521.
12. Farmer JC, Ear and sinus problems in diving. In: Bove AA, Davis JC, eds. Diving Medicine, 2nd ed. Philadelphia, PA: W.B. Saunders, 1990:200–222.
13. Becker GD, Parell GJ, Otolaryngologic aspects of scuba diving. Otolaryngol Head Neck Surg 1979; 87:569–572.
14. Jerrard DA, Diving medicine. Emerg Med Clin North Am 1992; 10:329–338.
15. Taylor GD, The otolaryngologic aspects of skin and scuba diving. Laryngoscope 1959; 69:809–858.
16. Shupak A, Ostfield E, Sharoni Z, Doweck I, Pressure chamber tympanometry in diving candidates. Ann Otol Rhinol Laryngol 1991; 100:658–660.

17. Miyazawa T, Ueda H, Yanagita N, Eustachian tube function and middle ear barotrauma associated with extremes in atmospheric pressure. Ann Otol Rhinol Laryngol 1996; 105:887–892.
18. Parell GJ, Becker GD, Conservative management of inner ear barotrauma resulting from scuba diving. Otolaryngol Head Neck Surg 1985; 93:393–397.
19. Edmonds C, Freeman P, Thomas R, et al. Otological Aspects of Diving. Sidney, Australia: Australian Medical Publishing Company, 1973.
20. Lundgren CE, Alternobaric vertigo—a diver's hazard. Br Med J 1965; 2:511–513.
21. Becker GD, Recurrent alternobaric facial paralysis resulting from scuba diving. Laryngoscope 1983; 93:596–598.
22. Vail HH, Traumatic conditions of the ear in workers in an atmosphere of compressed air. Arch Otolaryngol 1929; 10:113–126.
23. Simmons FB, Theory of membrane breaks in sudden hearing loss. Arch Otolaryngol 1968; 88:67–74.
24. Fee GA, Traumatic perilymphatic fistula. Arch Otolaryngol 1968; 88:471–480.
25. Goodhill V, Sudden deafness and round window rupture. Laryngoscope 1971; 81:1462–1474.
26. Freeman P, Edmonds C, Inner ear barotrauma. Arch Otolaryngol 1972; 95:556–563.
27. Soss SL, Sensorineural hearing loss with diving. Arch Otolaryngol 1971; 93:501–504.
28. Keleman G, Temporal bone findings in cases of salt water drowning. Ann Otol Rhinol Laryngol 1983; 92:134–136.
29. Antonelli PJ, Parell GJ, Becker GD, Paparella MM, Temporal bone pathology in scuba diving deaths. Otolaryngol Head Neck Surg 1993; 109:514–521.
30. Landolt JP, Money KE, Topliff EDL, Nicholas AD, Laufer J. Johnson WH, Pathophysiology of inner ear dysfunction in the squirrel monkey in rapid decompression. J Appl Physiol 1980; 49:1070–1082.
31. Simmons FB, The double membrane break syndrome in sudden hearing loss. Laryngoscope 1979; 89:59–66.
32. Freeman P, Tanker J, Edmonds C, Rupture of the round window membrane in inner ear barotrauma. Arch Otolaryngol 1974; 99:437–442.
33. Freeman P, Rupture of the round window membrane. Otolaryngol Clin North Am 1978; 11:81–98.
34. Pullen FW, Rosenberg GJ, Cabeza CH, Sudden hearing loss in divers and fliers. Laryngoscope 1979; 89:1373–1377.
35. Taylor PH, Bicknell PG, Rupture of the round window membrane. Ann Otol Rhinol Laryngol 1976; 85:105–110.
36. Farmer JC, Thomas WG, Youngblood DG, Bennett PB, Inner ear decompression sickness. Laryngoscope 1976; 86:1315–1326.
37. Francis TJR, Dutka AJ, Hallenbeck JM, Pathophysiology of decompression sickness. In: Bove AA, Davis JC, eds. Diving Medicine. Philadelphia, PA: W.B. Saunders, 1990:170–187.
38. Fagan P, McKenzie B, Edmonds C, Sinus barotrauma in divers. Ann Otol Rhinol Laryngol 1976; 85:61–64.

29
Hearing Conservation Underwater

Harry Hollien
University of Florida, Gainesville, Florida, USA

1. INTRODUCTION

It has been observed that some divers experience auditory problems as a result of exposure to high noise fields underwater. These difficulties include conditions such as hearing loss, vertigo, and bleeding from the ears. The noise sources that appear to be at the root of these problems are operation of underwater power tools, impact noise (such as explosions), engine noise, and air turbulence in diving helmets. While the actual correlation of these detrimental auditory effects with the cited sources is not known, it is believed that a causal link exists. Accordingly, the focus of this chapter will be on the potential threat of underwater noise to good hearing—and on theoretical explanations basic to these problems.

Before proceeding, however, it should be stressed that, while noise-induced hearing loss is recognized as a major occupational-based disease, only the effects of airborne sound have been assessed with respect to damage risk criteria (DRC) and hearing conservation. Admittedly, the subpopulation of the working diver is not a very large one, but it is of importance nonetheless. Moreover, the noxious noise hazard associated with this group appears to be a serious one and its negative impact is increasing. Unfortunately, only a few responsible/relevant agencies, organizations, or individuals seem to be aware of this threat to those underwater operations dependent on diving personnel for the well-being of the individual diver. Worse yet, very little pertinent research has been carried out and, of that, the relationships that have been generated have not always been helpful. For example, little information is available except for some reasonable data on underwater auditory thresholds plus a little on suprathreshold difference limens (DLs) and sound localization. Hence, practically all of the concepts about divers' hearing are, at best, theoretical or, at worst, based on questions and guesses. Indeed, very little is understood about the dynamics of underwater suprathreshold auditory function, the sounds that can or cannot be tolerated, and/or those factors that interact with the detrimental effects of noise on hearing. Accordingly, some of the information and models to follow are extrapolated and/or speculative.

2. AUDITORY SENSITIVITY IN AIR

Obviously, there is no need to review the basics of the hearing dynamics of airborne sound as information of this type has been quite adequately provided in other chapters. However, it would be useful to reiterate just a few of these relationships because they provide the substrate for a good understanding of the differences between hearing in air and hearing underwater. For example, among those dimensions that can be used to describe auditory sensitivity, two of the most relevant would appear to be absolute and differential thresholds. As should be obvious by now, an absolute threshold, is a "curve relating the smallest intensity required for detection to the frequency of a tone . . . for some arbitrary level of performance" (1). Since absolute threshold is one measure of the lowest intensity detectable by an individual, it is reasonable to infer that (a) it is a measure of signal registration by the sensory system and (b) it specifies the normal threshold sensitivity of the intact auditory system. Note also that this function can be applied to maximum as well as minimal sensitivity. Not included in this definition are the procedures by which the presence of sound—at absolute threshold—is detected. There are a number of methods that are utilized in these assessments. The relationships among them will have a bearing on the comparisons of hearing measurements obtained in air to those from the submerged ear.

The differential threshold, or DL, is the amount of *change* in a stimulus that is required for the observer to detect a particular shift "for some arbitrary level of performance." Consequently, the ability to discriminate between two stimuli that are similar in frequency or intensity is an indication that the sensory registration of the two signals is different and that it is independent of the amplification provided by the mechanical systems of the outer and middle ear. As will be seen, these relationships are basic to the discussions that follow.

3. HEARING UNDERWATER

3.1. Sensitivity

Information about underwater hearing sensitivity tends to be both sparse and (somewhat) primitive. One of the reasons for this situation is that it is difficult to impossible to duplicate ordinary psychophysical research techniques in this milieu. As Goerters (2) points out, a human simply cannot be immersed and tested. Rather, a great variety of life support gear must necessarily be attached to the diver/subject with all of the concomitant—and perhaps shifting—effects they will have on response to a heard stimuli. Moreover, heightened stress and demands on the divers' attention often will result in reduced or impaired underwater performance (3). On the other hand, it is now well established that immersion of the head in water results in a (frequency-independent) detection threshold at about 60-dB SPL (at 0.0002 dynes/cm^2). The basis for this statement is as follows.

In 1947, Sivian (4) published a theoretical paper in which he discussed the effects submersion of the head would have on human hearing. He speculated that water "plugging the ear canal would enhance hearing by bone conduction." He tested his predictions to some extent and estimated the hearing loss in water (relative to air) to be between 44–49 dB. Later, four studies were published in which investigators attempted to empirically establish human hearing thresholds underwater. First Hamilton (5) reported both upward threshold shifts of 35–45 dB in divers and no change in the loudness for occluded ears; these data provided evidence that supported a bone conduction theory and confirmed (to some extent) Sivian's predictions. The second investigation was by Wainwright (6) who found an upward threshold shift of 43–75 dB and that the occlusion of the era canal had no effect on perceived loudness; the third author, Reysenback de Haan (7), tended to concur with Wainwright. Finally, Montague and Strickland (8) tested divers' hearing with and without hoods. They reported an upward threshold shift of 40–70 dB relative to air, depending on the frequency. Furthermore, they noted an additional upward threshold shift of 20 dB at frequencies above 1000 Hz when the divers wore hoods.

Taken as a whole, these data are rather confusing. In all fairness, however, the authors involved were attempting to break new ground scientifically; hence, they had to devise novel techniques and conduct research in an area where there were many unknowns. They certainly employed a broad spectrum of experimental methods—some of which were not quite appropriate.

A little later, an effort was made by the present author and his colleagues to develop and carry out a coordinated research program on underwater hearing in order to resolve the cited conflicts. The apparatus used to create an appropriate underwater hearing laboratory may be seen in Fig. 29.1; the environment proved anechoic when no surface was within 40 ft of 1 DICORS. Calibration of all equipment was carried out, of course—underwater and with the diver in place. The first of these studies (9) established stable mean underwater thresholds at about +60-dB SPL. These data were well within the range of values reported by previous investigators, clarifying the earlier data and supporting those authors who argued that some sort of bone conductive mechanisms best explained the underwater hearing processes. It should be noted, however, that the diver's hearing was tested minimal audible pressure (MAP) in air but minimal audible frequency (MAF) underwater; hence, slight differences can be expected. On the other hand, it is the underwater thresholds themselves which are important and not the *shift* from waterborne to airborne detection.

Additional experiments were carried out to examine the validity of these findings and to further explore the possibility that underwater hearing is accomplished by bone conduction. First, Hollien and Brandt (10) obtained thresholds for divers with and without air bubbles in their external auditory canals. They reasoned (as Sivian had) that (a) if an air bubble rested against the tympanic membrane, the impedance characteristics of the canal would be altered and (b) if these structures were crucial to underwater hearing, a difference in threshold level would be found. No significant differences were observed between the two conditions. Thus, it appeared that (underwater) sound traveled directly to the cochlea by means of bone conduction. Later, Bauer (11) argued that bubble occlusion of the

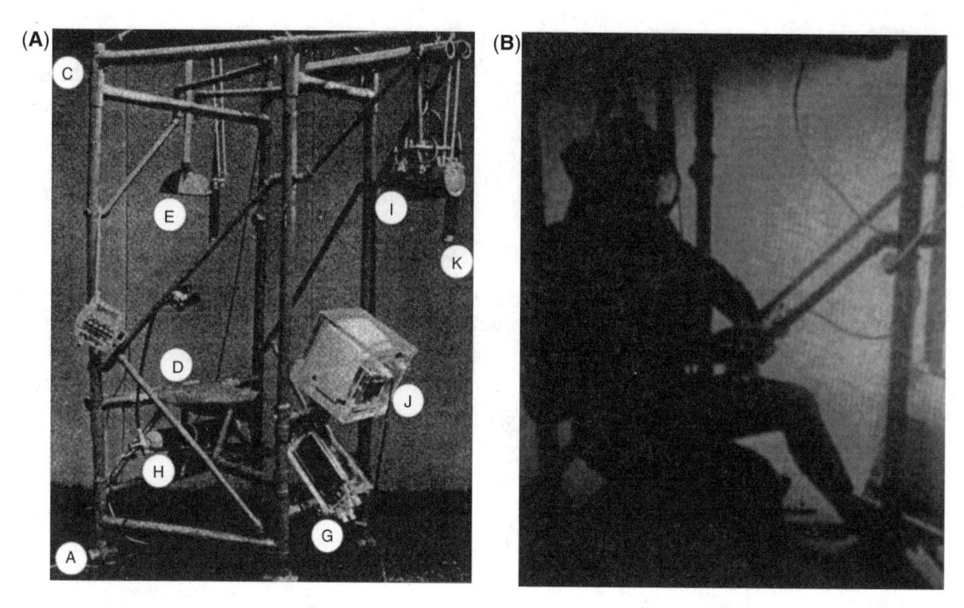

Figure 29.1 A pair of photographs showing the DICORS system; (**A**) the surface with all equipment attached and (**B**) in use but at a depth of only 20 ft. The support cable A, the stabilization weights B, and the diver/subject F are not shown in (A). However, C is the PVC framework, D the diver's seat, E the head positioner and calibration hydrophone, G/H are safety gear (the reserve tank and regulator plus a television camera for observing the subject), I is the projector, and J/K identify equipment (a hydrophone and TV monitor) used primarily in speech research.

external auditory canal does not constitute a valid test of middle-ear function. Accordingly, Hollien and Feinstein (12) further examined the question by obtaining thresholds for the divers who wore (a) no neoprene hood, (b) an intact hood, and (c) a hood wherein earholes and rubber tubes coupled the outside milieu to the external auditory meatus. When the divers wore either the hood or the hood with the earholes/tubes, the threshold values were greater than those previously reported (9,10). They also were greater than the values obtained for the no-hood condition (which, in turn, was consistent with prior curves). In other words, a difference was found between the hood (all types) and no-hood (present and prior) conditions. These findings supported the hypothesis that, when submerged, the human hearing mechanism is primarily bone conductive rather than tympanum, or "water," conductive.

Finally, the approach taken by Smith (13,14) in his assessment of the bone conduction question was somewhat different from those previously cited. He suggested that it would be necessary to determine if bone conduction thresholds in air were the same as those in water before it would be possible to decide that diver hearing is accomplished by this mechanism. Smith was able to demonstrate that bone conduction thresholds, obtained in either environment, were functionally the same.

3.2. The General Nature of Underwater Hearing

As data from these experiments accumulated, this author and his associates began to realize that there was a basic flaw in the general approach to the establishment of underwater hearing thresholds (UHT). That is, the relationships in this regard were specified as *shifts* from thresholds in air. At this juncture, it became apparent that, conversely, it would be necessary to consider them as *absolute* thresholds, and for two reasons. First, the underwater environment is an entity in and of itself—with its own physical properties—rather than a variation of space defined on the basis of the elastic, compressible nature of air. It held, then, that UHT also were valid entities, rather than simple variations of thresholds in air. Second, the approach that had to be used with the submerged head involved MAF of "free field" measurements, whereas most assessments in air are MAP. Hence, air-water relationships could be estimated only if this "error" was taken into account. It also was reasoned that, if underwater hearing "loss" was peripheral in nature, there should be no evidence of any neurological deficit—i.e., neural (auditory) components underwater should parallel those in air. In response, research was carried out on underwater thresholds for speech (15). It was found that mean (underwater) speech reception thresholds (SRTs) were 13–15 dB above the mean thresholds for the 0.5–2.0-kHz frequency range. This relationship roughly parallels the SRT/sinusoidal ratio found in air. Moreover, when all of the data reported by the University of Florida group were plotted, they confirmed the original relationship that underwater thresholds for sound detection were relatively flat and located at about 60-dB SPL (see Fig. 29.2).

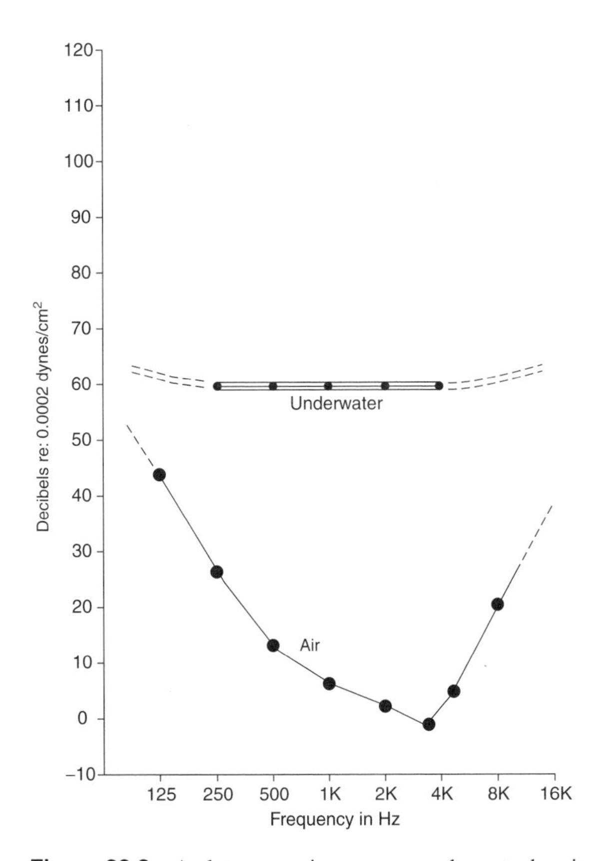

Figure 29.2 A plot contrasting average underwater hearing thresholds with those in air. While it is difficult to avoid viewing the underwater thresholds as deviations from those for airborne sound, the former actually should be considered as independent of the latter. That is, the pathway for auditory events is a different one underwater and, hence, should be dealt with as a unique but intact mechanism, rather than some sort of modification to the hearing modality in air.

Other evidence is available to demonstrate that neurological deficits are not associated with underwater hearing. For example, the ability of divers to localize sound underwater suggests that reductions here are due to mechanical rather than neurological constraints—especially since the speed of sound is increased, and the human head is acoustically transparent, in this environment. To be specified, as early as 1944, Ide (16) reported near-normal underwater sound localization for divers. His work has been confirmed and greatly expanded by a number of other investigators (17–23) who utilized minimum audible angle and direct angle estimations to compare the precision of divers' underwater localization judgments with those made in air. Their results demonstrated that humans have the capacity to localize sound sources underwater, and to do so with some degree of accuracy. The fact that Hollien et al. (24) have been able to establish a diver navigation procedure on the basis of sound "beacons" which appear to move (UAPP) is yet a further case in point (see also Refs. 25–28).

It should be noted that the few studies cited in this section are virtually all that have been reported on underwater hearing dynamics (see below for exceptions). To date, there are no available data on equal loudness contours, tolerance for loud sounds, or other issues germane to hearing dynamics. As would be expected, information about human tolerance for underwater sound is of critical importance to DRC and hearing conservation.

3.3. The Effects on Hearing of High Ambient Pressures

A number of investigators have examined the effects of increasing ambient pressure on auditory sensitivity. For example, Farmer et al. (29) carried out air and bone conduction threshold measurements at several simulated depths (in HeO_2) to 600 ft. As had Fluur and Adolfson (30), they found reversible increases in air conduction thresholds (up to 26 dB in the lower-frequency ranges) with increases in pressure. Finally, the effects of deep diving (and the concomitant pressures involved) may limit a diver's ability to make fine auditory discriminations. Hollien and Hicks (31) have theorized that High Pressure Nervous Syndrome (HPNS) may be partially responsible for breakdowns in communication task performance on deep dives. Furthermore, while there currently is no direct evidence of an HPNS influence on auditory sensitivity, it is plausible to assume that hyperbaric pressures, and perhaps the use of varying breathing gas mixtures, could temporarily affect at least some of a diver's sensorineural and cognitive processing functions (21,32–37).

3.4. Underwater Dynamic Hearing Range

While the dynamic hearing range for humans in air is well established, parallel data are not available for the submerged ear. For example, as early as 1947, Silverman (38) published a classic paper in which he established thresholds of tolerance in a variety of subjects for sinusoids and speech. His data also provide "normative" values for our model of underwater hearing (see below). As would be expected, many other aspects of the dynamic hearing range (in air) have been studied during the past four decades but the situation is substantially different when hearing underwater is considered. With the exception of the cited sound localization studies and the work of Thompson and Herman (39) and Klepper (40) on DLs, virtually no data at all are available in this area. Indeed, even the thresholds of tolerance for sound underwater remain undiscovered.

4. SOURCES OF UNDERWATER/HYPERBARIC NOISE

It is now well recognized that many divers have developed noise-induced hearing problems, especially those of a sensorineural nature (41–45). Of course, sources other than noise can also cause hearing loss (barotraumas of all types, decompression effects, and so on); however, they are not directly relevant to this discussion and will not be reviewed. Rather, it is important to understand the role noise plays as a cause of those induced hearing deficits that occur in divers. The sources of this hazard include power tools used underwater, explosions (including diver recall devices), the noise (especially airflow) in diving helmets, chambers and personnel transfer capsules, and engine noise (ships). These noises can be airborne (in chambers, helmets) or waterborne (power tools, ship engines, explosives). While both will be discussed, it must be remembered that airborne noise may not be very different from that at the surface as the only difference here would be that the diver is experiencing very high ambient pressures. Moreover, and as will be seen, most problems probably are associated with the waterborne noise. It also should be noted that the noise sources are of two types: steady state and impulse. Steady-state noise levels are more easily measured (both peak intensity and intensity by frequency band) and related to human hearing. However, as Bromer (46) points out, there is now evidence that the high peaking of impulse/impact sounds can cause hearing damage from virtually the first exposure. Hence, both the effects of high ambient noise *and* impact sounds must be taken into account.

4.1. Chamber Noise

The sources of chamber noise include the operation of pumps, compression systems, and rapid gas flow (venting, for example). Not a great deal is known about the energy levels of these sources and how they operate to create auditory traumas; however, some measurements have been carried out (45,47,48). For example, Brown et al. (47) report that overall ambient noise levels in chambers range from 78–84 dB SPL (at 0.0002 dynes/cm^2), but that these levels were higher below 1 kHz (and lower to much lower above that reference). Murry's (48) data were consistent with these findings but the levels reported by Summitt and Reimers (45) were somewhat higher. In any case, energy of this extent is probably hazardous especially since it extends over long periods of time. Furthermore, rather intense peak chamber noise levels were reported by Brown et al.; indeed, they observed magnitudes over 90 dB for the compression phase of "travel" and even higher ones for venting. Summitt and Reimers also measured chamber noise during "travel" (decompression) and when the chamber was being ventilated. Their data varied from 107–121 dB. As can be seen, ambient noise levels this high would be detrimental to hearing even if they lasted only for relatively short periods of time.

4.2. Noise in Diving Helmets

It has been long recognized that relatively high noise levels can exist in diving helmets; however, about the only data reported on this issue are those published by Summitt and Reimers (45). They measured ambient noise levels in six different types of diving helmets, with the intake values partially and fully open—and as a function of depth in sea water. All values were found to be unacceptable high—varying from 93–99 dB in the "best" helmet to 109–113 dB in the "worst." Of even greater concern was the fact that these values were found for continuous noise, a situation that may increase the hazard to hearing.

4.3. Underwater Power Tools

Most of the available information in this area has been gathered by Mittleman (49) and Molvaer and Gjestland (50). Many (if not all) of their measurements were made at ear level during simulated dives. Tools such as impact wrenches, high-pressure cleaning tools, rock drills, and so on were evaluated. They report energy levels ranging from 90–105 dB depending on the device employed. They demonstrate that even older power tools (much less those now in the development stage) create sound fields that clearly are detrimental to diver hearing (51).

4.4. Explosions

The tools described above produce (in most cases) both steady-state and impulse noises. Of course, it is difficult to measure the peak of an impact sound due to its very brief duration. Nevertheless, noise of this type clearly has a detrimental (if not devastating) effect upon diver auditory function. Additionally, the reports by Mittleman (49), combined with those of Hicks and Hollien (25), reveal that many explosives result in rather high (peak) energy levels. For example, explosives as small as an M-80 firecracker, detonated at a depth of 3 meters, can exhibit a peak energy level of as high as 144-dB SPL and this level drops only to 120 dB even at distances of over 1 km. Peak energy levels of a 38 caliber revolver (fired underwater) and stud guns are somewhat lower but not by very much (i.e., 137 dB at 75 m to 106 dB at over a kilometer for the 38 revolver). Admittedly, the durations of these explosions are fairly short (200–350 msec). Nevertheless, they do have high energy wave-fronts that impact the diver's head; they certainly exhibit sufficient energy to potentially damage his or her hearing. Worse yet, even at great distances, large explosions produce energy fields that certainly are dangerous to diver hearing (see Christian (52) and Goertner (53), among others). In summary, there are a rather substantial number of impulse-type noise sources where the signal is of sufficient strength and character to damage diver hearing and apparently this is what is happening. While adequate data on peak levels and frequency bands are available for only a few of these hazards, there are sufficient data to suggest they occur, and at dangerous levels.

5. ISSUES RELATED TO HEARING CONSERVATION AND DRC

As should be obvious by now, occupational noise standards were established to assist specialists in preventing injury to hearing. For example, the 1983 OSHA criteria [54] provide the standards necessary for measuring and computing noise levels and exposures; workplaces are required to have their noise emissions adjusted to meet these standards. In this regard, Sataloff and Michael (55) as well as Dear (56) define key constructs in noise regulation. These concepts include damage risk criteria and percentage risk. Damage risk criteria (DRC) specify the maximum allowable noise quantity to which persons may be exposed if risk of hearing impairment is to be avoided. Percentage risk (PR) is defined as the difference between the percent of persons exposed to industrial noise who reach impairment and the percent of those not exposed to any such noise who exhibit similar losses. Selection of a damage risk criterion implies that there will be some percentage of the exposed population remaining at risk at that level. While DRC and PR have long since been established for the normal work place (i.e., with respect to airborne noise), these metrics are not yet available for the submerged worker.

5.1. Effects of Noise on Hearing

A rather substantial literature is available in this general area—see, for example, the references cited plus Fausti et al. (57), Staiano (58), and Weltman and Fricke (59). Yet another dimension to the issue was reported by Sataloff et al. [60], who contrasted the effects of intermittent noise and steady-state noise on hearing thresholds. While previous investigators had suggested that intermittent noise would be less detrimental than continuous exposure noise of the same intensity, these authors found that "intermittent exposure to intense noise resulted in severe losses in the high frequencies and little to no loss in lower frequencies, even after many years of exposure." These investigators encouraged further research designed to determine if the cited effects were specific to jackhammer noise or were generalizable to other intermittent noises. In any case, these results are most relevant to the problems of waterborne sound; they also are particularly important when the model to follow is considered.

5.2. Hearing Conservation Programs

Effective programs of this type already have been developed for air but not for underwater noise control. To be specific, Pell (61) has shown that protection of almost all employees can be accomplished through an effective hearing conservation program. Programs of this type should be established for underwater work if reasonable hearing conservation is to be achieved for divers. However, it should be noted that, while several types of noise evaluation measures/procedures are available (i.e., routine noise exposure rates, noise exposure accumulation rates, and impulsive/continuous noise environments) they are for air. Further, OSHA standards only require measurement of continuous noise (an arguable concept) in order to provide a reasonable basis for DRC. In any event, even limited information of this type should assist in the development of underwater hearing conservation programs.

To summarize, when the available data are reviewed, it is obvious that they still are insufficient for development of even minimal underwater DRC for hearing. Of course, some existing data/concepts can be used: nevertheless, there is little question but specialized underwater DRC and hearing conservation techniques will have to be established if diver hearing is to be protected. However, before such work can be initiated, it would appear necessary to structure an appropriate theoretical model on which to base further inquiry.

6. A THEORETICAL FRAMEWORK

6.1. Theories of Hearing Underwater

Without doubt, some sort of an integrated theory is required as a basis for advances in underwater hearing conservation. In order to provide perspective, several of the older theories of underwater hearing should be considered first. One of these approaches was called the "tympanic" theory and was developed by Bauer (11). He states that underwater hearing is accomplished in essentially the same manner as hearing in air. Sound enters the submerged ear canal and vibrates the tympanic membrane; it consequently is transmitted to the cochlea through the ossicular chain. However, because the human ear is adapted (impedance match) to function in air, and because the characteristic acoustic impedance of water is much greater than that of air, a substantial mismatch exists between water and the immersed ear. It is for these reasons that Bauer contends the human ear is not as sensitive to waterborne sound as it is to sounds that are airborne. His model is employed to predict that this sensitivity loss is frequency dependent—i.e., that there will be no loss of sensitivity at 100 Hz but, rather, an almost linear drop in sensitivity (of about 12 dB per octave) as frequency increases from 100 Hz to 5000 Hz. Unfortunately, this model does not appear to be supported by available data on underwater hearing.

A second theory involves Sivian's "dual-path" approach [4]; he theorized that underwater hearing is mediated by both the tympanic and bone conduction mechanisms and that they are of approximately equal sensitivity at 1000 Hz. At other frequencies, one or the other of the two pathways may predominate. One implication of the dual-path theory is that, given two equally efficient routes by which underwater acoustic energy reaches the cochlea, a deficiency in only one route should not result in degraded underwater hearing. It is also stated that, in some circumstances, these two mechanisms may interact. Unfortunately, when underwater hearing sensitivity is compared to that in air, the two are *not* found to be equal.

A third theory is the "bone conduction" model; as such, it was first suggested by Reysenback de Haan (7). Basically, it can be said that, because the impedance of the human skull is very close to that of water, sound is readily transmitted from water to the cochlea through these tissues, and that it bypasses the (now) acoustically inefficient route of the external and middle ear. Further, this theory permits the suggestion that the ear canal is acoustically transparent in water and that the middle ear also is ineffective primarily because the ossicles lack appropriate mass. Finally, it is postulated that the two cochlea are not as independently stimulated underwater as they are in air due to the cross-conduction of sound through the skull—a condition that should somewhat impair the effectiveness of underwater sound localization in humans.

It is obvious that the data cited in the earlier sections of this paper can serve as the basis for easy rejection of the first two theories. However, the third appears attractive although somewhat incomplete. That is, even when it is generally modified by the early data reported by the University of Florida group, it cannot fully predict the nature of auditory function underwater. On the other hand, when Hollien's (1973) descriptions of underwater hearing mechanisms (19) are added to this third hypothesis, reasonably accurate predictions appear possible. Specifically, it is suggested that the so-called underwater hearing "loss" is not a loss at all but rather that the thresholds of sensitivity here should be based on their own set of principles—i.e., on the *mechanical* relationship between sound transmission in water and the anatomy of the human head. To be specific, the observed thresholds are the consequence of the different (mechanical) force/amplitude (FA) arrangements—i.e., those which exist underwater. As is well known, sound travels through air in a high amplitude, low force mode (Af), yet through a fluid such as water as high force, low amplitude (aF). The external and middle sections of the ear function to increase force from its airborne level to one that will interface properly with the viscous fluid of the vestibular system. Hence, hearing in air is Af to aF whereas hearing underwater involves a third sound transmission component (aF to Af to aF) with all the reductions in efficiency that this multiple change implies. In short, the external- and middle-ear mechanisms are not needed for either transduction or energy transformation and, underwater, sound waves enter the cochlea directly through the skull. The hearing mechanism simply functions differently when the human head is submerged.

6.2. A Model for Underwater Hearing

The postulated model for underwater hearing can best be understood by consideration of Fig. 29.3. A stylized representation of the threshold of human hearing (in air) can be seen in the lower portion of the figure; the thresholds of discomfort, tickle (feeling), and pain at the top. The bar that can be found at approximately the 60- to 65-dB level (SPL at

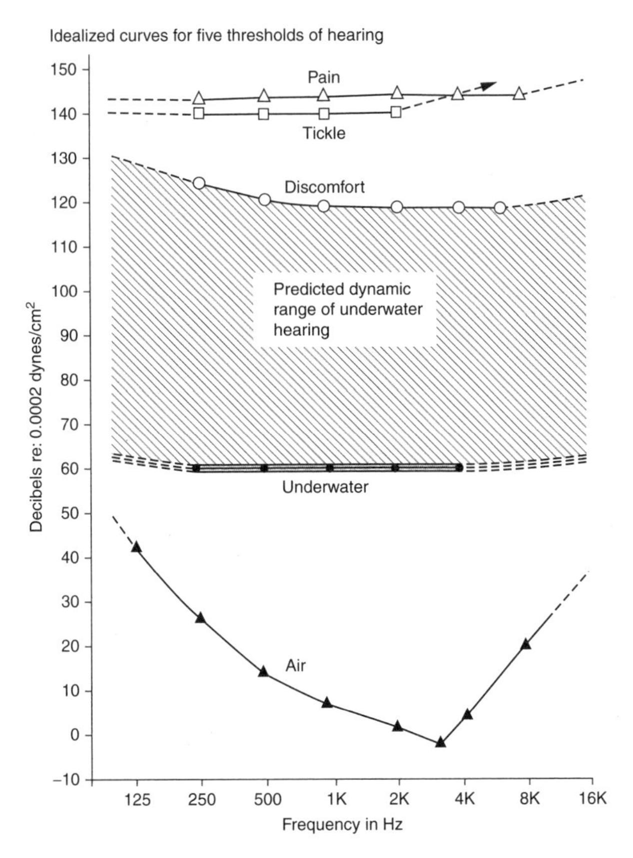

Figure 29.3 A model that can be employed to predict dynamic hearing range underwater (i.e., roughly the area between 60- and 120-dB SPL). The three thresholds of discomfort, tickle, and pain are based on extrapolations of data obtained in air.

0.0002 dynes/cm^2) is a compilation of available UHT data. Note that most of the frequency nonlinearities induced by the external and middle ears are not reflected in UHT. A 5-dB "bar" is included because it is hypothesized that variation in differential middle-ear pressure can reflect on cochlea efficiency by increasing or decreasing system impedance at the round and oval windows. To illustrate: divers sometimes experience a slight increase in hearing sensitivity when they "clear their ears" (i.e., equalize the pressure in the middle ear with that of the outside water).

In short, it is theorized that the underwater "shift" in the threshold of hearing is due to differing mechanics and the dynamic range of the submerged ear is sharply reduced; see, for example, the difference between UHT and the thresholds of tolerance. It is further hypothesized that, while hearing function within this frequency/intensity window is of the same general nature as that found in air, it is sharply restricted underwater. This postulate is based on the (yet untested) supposition that thresholds for tolerance do not shift upward underwater (as does UHT) as no neural or mechanical compensation is provided. Simply stated, this model specifies that the human dynamic hearing range is reduced from one in air which can exceed 130 dB (for some frequencies anyway) to one in water that ordinarily is reduced to 55–60 dB. In turn, this reduction causes the diver to be materially less resistant to the detrimental effects of waterborne noise because (a) underwater sound does not decay as rapidly as it does in air and

(b) because, due to the elevated thresholds of detectability, divers tend not to be aware of the actual intensity of some of the high energy sounds they experience. If these hypotheses are supported by experimental data, establishment of hearing DRC for underwater noise can be accomplished—primarily because compensatory elements can be included in the procedures adopted. It is recognized, of course, that the development of hearing conservation programs for underwater noise will be somewhat more complex than were those for airborne sounds. Nevertheless, it is argued that they can be established when appropriate data about dynamics of underwater hearing become available.

REFERENCES

1. Yost WA and Nielsen DW, *Fundamentals of Hearing*, Holt, Rinehart and Winston, New York, 1977.
2. Goerters KM, Hören unterwasser: Absolute Reizschwellen und Richungswahrnehmung, *Sonderdruck aus Meerestechnik*, *5*:191–197, 1972.
3. Weltman, G and Egstrom GH, Perceptual narrowing in novice divers, *Hum. Factors*, *8*:499–506, 1966.
4. Sivian LJ, On hearing in water versus hearing in air, *J. Acoust. Soc. Am.*, *19*:461–463, 1947.
5. Hamilton PM, Underwater hearing thresholds, *J. Acoust. Soc. Am.*, *29*:792–794, 1957.
6. Wainwright WN, Comparison of hearing thresholds in air and water, *J. Acoust. Soc. Am.*, *30*:1025–1029, 1958.
7. Reysenback de Haan FW, Hearing in whales, *Acta Oto-Laryngol. Suppl.*, *134*:1–114, 1957.
8. Montague WE and Strickland JF, Sensitivity of the water-immersed ear to high-and-low-level tones, *J. Acoust. Soc. Am.*, *31*:1121–1125, 1961.
9. Brandt JF and Hollien H, Underwater hearing thresholds in man, *J. Acoust. Soc. Am.*, *42*:966–971, 1967.
10. Hollien H and Brandt JF, The effect of air bubbles in the external auditory meatus on underwater hearing thresholds, *J. Acoust. Soc. Am.*, *46*:384–387, 1969.
11. Bauer BB, Comments on effect of air bubbles in the external auditory meatus on underwater hearing thresholds, *J. Acoust. Soc. Am.*, *47*:1465–1467, 1970.
12. Hollien H and Feinstein SH, Contribution of the external auditory meatus to auditory sensitivity underwater, *J. Acoust. Soc. Am.*, *57*:1488–1492, 1975.
13. Smith PF, Bone conduction, air conduction, and underwater hearing, U.S. Naval Submarine Medical Center, Groton, CT, *Memorandum Report*, *12*:1–7, 1965.
14. Smith PF, Underwater hearing in man: 1. Sensitivity, *NSMRL Report, 569*, Naval Submarine Base, New London, CT, 1969.
15. Brandt JF and Hollien H, Underwater speech reception thresholds and discrimination, *J. Aud. Res.*, *8*:71–80, 1968.
16. Ide JM, Signaling and homing by underwater sound for small craft and commando swimmers, *Sound Report No. 19*, Naval Research Laboratories, Washington, D.C., 1944.
17. Feinstein SH, Acuity of the human sound localization response underwater, *J. Acoust. Soc. Am.*, *53*:393–399, 1973.
18. Feinstein SH, Minimum audible angle underwater: A replication under different acoustic and environmental conditions, *J. Acoust. Soc. Am.*, *54*:879–881, 1973.
19. Hollien H, Underwater sound localization in humans, *J. Acoust. Soc. Am.*, *53*:1288–1295, 1973.
20. Hollien H, Feinstein SH, Rothman HB, and Hollien PA, Underwater sonar systems (passive) in humans, In *Proc. Speech Communication Seminar (SCS-74)*, Stockholm, Sweden, pp. 149–157, 1974.
21. Hollien H, and Feinstein SH, Hearing in divers, *Underwater Research* (E. Drew, J. Lythgoe, and J. Woods, eds.) Academic Press, London, pp. 81–138, 1976.

22. Smith PF, Yonowitz A and Dering G, Underwater hearing in man III: An investigation of underwater sound localization in shallow and noisy water, *Report No. 779*, U.S. Naval Submarine Med. Res. Lab, Groton, CT, pp. 1–3, 1974.
23. Stoufler JL, Doherty ET, and Hollien, H, Effects of training on human underwater sound localization ability, *J. Acoust. Soc. Am.*, *57*:1212–1213, 1975.
24. Hollien H, Hicks JW, Jr., and Klepper B, An acoustic approach to diver retrieval, *Undersea Biomed. Res.*, *13*:111–128, 1986.
25. Hicks JW, Jr., and Hollien H, A research program in diver navigation, Washington, D.C., *Proc. IEEE Acoust. Communications Workshop, D-5*:1–10, 1982.
26. Hollien H, New data on acoustic navigation by divers, in *Diving for Science-86* (C. T. Mitchel, ed.), pp. 145–152, 1987.
27. Hollien H. and Hicks JW, Jr, Diver navigation by sound beacon, *Sea Grant Today*, *13*:10–11, 1983.
28. Hollien H, Diver navigation by means of acoustic beacons, *J. S. Pacific Underwater Med. Soc.*, *17*:127–138, 1983.
29. Farmer JC, Jr., Thomas WG, and Preslar M, Human auditory responses during hyperbaric helium oxygen exposures, *Surg. Forum*, *22*:456–458, 1971.
30. Fluur E and Adolfson J, Hearing in hyperbaric air, *Aerosp. Med.*, *37*:783–785, 1966.
31. Hollien H and Hicks JW, Jr. Helium/pressure effects on speech: Updated initiative for research, Washington, D.C. *Proc. IEEE Acoust. Communications Workshop, D-4*:1–26.
32. Bennett PB, Performance impairment in deep diving due to nitrogen, helium, neon and oxygen, *Proc. Third Symp. Underwater Physiology* (C. L. Lambertsen, ed.) pp. 327–340, 1967.
33. Brauer RW, Johnson DO, Pessotti RL, and Redding R, Effects of hydrogen and helium at pressures to 67 atmospheres, *Fed. Proc. Am. Soc. Exp. Biol.*, *25*:202, 1966.
34. Hollien H, Rothman HB, Feinstein SH, and Hollien PA, Auditory sensitivity of divers at high pressures. In *Underwater Physiology* (C. Lambertson, ed.), FASEB, Bethesda, MD, pp. 665–674, 1976.
35. Miller KW, Paton WDM, Street WB, and Smith EG, Animals at very high pressure in helium and neon, *Science*, *157*:97–98, 1967.
36. Rostain JC, Gardetle-Chauffour MC, and Naquet R, HPNS during rapid compression of neon breathing HeO_2 and HeN_2O_2 at 300 m and 180 m, *Undersea Biomed. Res.*, *7*:77–94, 1980.
37. Zaltsman GL, Giperbariches kiye eiplepsiya I narkoz, *Rep. USSR Acad. Sci.*, Leningrad, USSR, 1968.
38. Silverman SR, Tolerance for pure tones and speech in normal and defective hearing, *Ann. Otol. Rhinol. Laryngol.*, *56*:658–677, 1947.
39. Thompson RKR, and Herman LM, Underwater frequency discrimination in the bottlenosed dolphin (1–140 kHz) and the human (1–8 kHz), *J. Acoust. Soc. Am.*, *57*:943–947, 1975.
40. Klepper B, Intensity difference limits in divers, unpublished MA thesis, University of Florida, 1981.
41. Edmonds C, Hearing loss with frequent diving (deaf divers), *Undersea Biomed. Res.*, *12*: 315–319, 1986.
42. Molvaer OI and Lehman EH, Hearing acuity in professional divers, *Undersea Biomed. Res.*, *12*:333–349, 1985.
43. Smith PF, Development of hearing conservation standards for hazardous noise associated with diving operations. *NSMRL Report 1020*, Naval Submarine Base, New London, CT, 1984.
44. Soss SL, Sensorineural hearing loss in divers, *Arch. Otolaryngol.*, *93*:501–504, 1971.
45. Summit JK and Reimers SD, Noise: A hazard to divers and hyperbaric chamber personnel, *Aerosp. Med.*, *42*:1173–1177, 1971.
46. Bromer N, Peak pressure of impact sounds and their potential effect on human hearing, *Progr. Rep.*, Vipoc, Ltd., Melborne, Australia, 1985.
47. Brown DD, Giordano F, and Hollien H, Noise levels in a hyperbaric chamber, *Sound and Vibration*, *11*:28–31, 1977.
48. Murry T, Hyperbaric chamber noise during a dive to 100 feet, *J. Acoust. Soc. Am.*, *51*:1362–1365, 1972.

49. Mittleman J, Stud gun sound pressure level study: Experimental and theoretical work, *NCSL Report 297–76*, Naval Coastal Systems Laboratory, Panama City, FL, 1976.
50. Molvaer OI and Gjestland T, Hearing damage risk in divers operating noisy tools underwater, *Scand. J. Work Environ. Health*, *7*:263–270, 1981.
51. Hollien H and Hollien PA, Effects of diver tool use on diver hearing, In *Diving for Science-90* (W. C. Jaap, ed.), pp. 163–178, 1990.
52. Christian EA, Source levels for deep underwater explosions, *J. Acoust. Soc. Am.*, *42*:905–907, 1967.
53. Goertner JF, Prediction of underwater explosion safe ranges for sea mammals, *Ad-Al39823/9*, Naval Surg, Weapons Center, Silver Spring, MD, 1984.
54. U.S. Department of Labor, *Occupational Noise Exposure Standard*, 29 CF 1910.95, Washington, D.C., Occupational Safety and Health Administration, 1983.
55. Sataloff J and Michael P, *Hearing Conservation*, Charles C Thomas, Springfield, IL, 1973.
56. Dear T, Noise criteria regarding risk and prevention of hearing injury in industry, in *Occupational Hearing Loss* (R. Sataloff, and J. Sataloff, eds.) Marcel Dekker, Inc., New York, 1986.
57. Fausti S, Erickson D, Frey R, Rapport B, and Schechter M, The effects of noise upon human hearing sensitivity from 8000 to 20,000 Hz, *J. Acoust. Soc. Am.*, *69*(5):1343–1349, 1971.
58. Staiano M, OSHA noise exposure due to intermittent noise sources, *Sound and Vibration*, *5*:1821, 1986.
59. Weltman WD and Fricke JE (eds.), Noise as a public health hazard, *ASHA Reports 4*, American Speech and Hearing Association, Washington, D.C., 1969.
60. Sataloff J, Sataloff R, Yerg R, Menduke H, and Gore R, Intermittent exposure to noise: Effects on hearing. *Ann. Otol. Rhinol. Laryngol.*, *92*:623–628, 1983.
61. Pell S, An evaluation of a hearing conservation program—A five-year longitudinal study, *Am. Ind. Hyg. Assoc. J.*, *34*:82–91, 1973.

30
Hearing Loss in Musicians

Robert T. Sataloff
Drexel University College of Medicine, American Institute for Voice and Ear Research, and Graduate Hospital, Philadelphia, Pennsylvania, USA

Joseph Sataloff
Thomas Jefferson University, Philadelphia, Pennsylvania, USA

Caren J. Sokolow
American Institute for Voice and Ear Research, Philadelphia, Pennsylvania, USA

Performing artists are frequently sensitive, somewhat "high-strung" people who depend upon physical perfection in order to practice their crafts and earn their livelihoods. Any physical impairment that threatens their ability to continue as musicians may be greeted with dread, denial, panic, depression, or similar responses that may be perceived as exaggerated, especially by physicians who do not specialize in caring for performers. In the case of hearing loss, such reactions are common even in the general public. Consequently, it is not surprising that psychological concomitants of hearing loss in musicians are seen in nearly all cases.

Many successful performers are communicative and gregarious and anything that impairs their ability to interact in their usual manner can be problematic. Their vocational hearing demands are much greater than those required in most professions and often musicians' normal reactions to hearing loss are amplified by legitimate fears about interruption of their artistic and professional futures. The problems involved in accurately assessing the disability associated with such impairments are addressed below in the discussion of occupational hearing loss in musicians.

1. HEARING LOSS IN MUSICIANS

Performing artists are required to accurately match frequencies over a broad range, including frequencies above those required for speech comprehension. Even mild pitch distortion (diplacusis) may make it difficult or impossible for musicians to play or sing in tune. Elevated high-frequency thresholds may lead to excessively loud playing at higher pitches, and to artistically unacceptable performance, which may end the career of a violinist or conductor, for example. It is extremely important for musicians to be protected from hearing loss. However, the musical performance environment poses not only critical hearing demands, but also noise hazards. Review of the literature reveals convincing evidence that music-induced hearing loss occurs, but there is a clear need for ongoing research to clarify incidence, predisposing factors, and methods of prevention. It is interesting to note that, in direct contrast to many other publications, Johnson and Sherman evaluated 60 orchestra members and 30 non-musicians from 250 to 20,000 Hz and found no substantive differences (1). This study suggested that there is no additional risk to hearing as a result of exposure to orchestral music. Similarly, Schmidt and Verschuure showed that the students of Rotterdam Conservatory did not show any decreased hearing loss when compared to a group of medical students of the same age, despite the music students' exposure to music (2).

As mentioned previously, noise exposure can cause both temporary and/or permanent hearing loss. In a study to evaluate temporary threshold shift in performers and listeners, Axelson and Lindgren determined that the performers showed less of a shift than the audience did (3). It was surmised by the authors that this may be explained by pre-exposure hearing levels. The performers had poorer hearing levels than listeners did before being exposed to the study noise. Another interesting finding was that the male listeners showed more of a temporary shift than the females. The authors suggested that exposure to live pop music should be limited to 100 dBA or less for no more than 2 h. When symphony orchestra musicians from the Royal Danish Threatre were studied, Ostri and Eller found that 58% of the 95 subjects demonstrated a hearing loss when using 20 dB HL as the "normal" cut-off value. The male subjects were more affected by noise exposure than the female subjects (4). The authors concluded that symphonic music does indeed cause hearing loss. In 1992, McBride et al. set out to determine whether noise exposure affected the classical musician (5). Contrary to other studies, audiograms showed no significant differences between participants of the same sex and age. They did prove that the musicians were exposed to high doses of noise that do pose an occupational hazard. Drake-Lee studied four heavy-metal musicians before and after performance (6). It was determined that exposure does cause a temporary threshold shift with a maximum effect at lower frequencies. An article by Bu in 1992 examined hearing loss in Chinese opera orchestra musicians (7). Bu discovered that the incidence of hearing loss in this group was exceedingly high and apparently associated with the type of instruments used (7). Bu has suggested measures to combat the noise exposure other than the use of ear protectors, such as seating percussion musicians 1 m lower than the other musicians in order to better preserve hearing of the instrumentalists around them (7). However, such suggestions have a variety of drawbacks and practical limitations.

Occupational hearing loss is usually bilateral, fairly symmetrical sensorineural hearing impairment caused by exposure to high-intensity work place noise or music. This subject has been discussed and specifically with regard to musicians in a previous review (8). Music is one of the professions that can produce a somewhat asymmetrical hearing loss in selected cases.

It has been well established that selected symphony orchestra instruments, popular orchestras, rock bands, and personal stereo headphones produce sound pressure levels (SPLs) intense enough to cause permanent hearing loss. Such hearing loss may also be accompanied by tinnitus and may be severe enough to interfere with performance, especially in violinists. The violin is the highest pitched string instrument in routine use. The amount of hearing loss is related to the intensity of the noise, duration and intermittency of exposure, total exposure time over months and years, and other factors. Rosanowski and Eysholdt published a case study on a violinist with bilateral tinnitus (9). They recorded peak sound pressure levels of over 90 dB. The violinist showed a 20-dB drop in hearing between 2 and 8 kHz on the side where the violin rests (left). This phenomenon (asymmetry) is produced by the head shadow, the same mechanism that causes asymmetrical hearing loss in rifle shooters. The authors point out the potential hazard of other auditory symptoms (i.e., tinnitus) as a result of noise exposure. In their 1999 study of hearing, tinnitus, and exposure to amplified recreational noise, Metternich and Brusis found a very high risk of tinnitus even when subjects were exposed to short durations of amplified music (10). This risk appears to be greater than the risk of permanent hearing loss from the same exposure. Various methods have been devised to help protect the hearing of performers. For example, many singers and other musicians (especially in rock bands) wear ear protectors. They may not feel comfortable wearing ear protection during a performance but may take precautionary measures during practice. Ear protectors have changed tremendously over the years, and there are more sophisticated and suitable models available now that cater to the musician. The importance of using new, more appropriate ear protectors for professional musicians should be stressed, especially since the previously unappreciated relationship between orchestral music exposure and noise-induced hearing loss has become clear. In their 1983 publication, "The Hearing of Symphony Orchestra Musicians," Karlsson et al. determined that the critera used to evaluate noise exposure in industry must be different from the criteria used to assess acoustic instrument levels in symphonic music (11). This complex issue is discussed later in this chapter. Singers need to be made more aware of the hazards of noise exposure and find ways to avoid or reduce its effects whenever possible. They should also be careful to avoid exposure to potentially damaging avocational noise such as loud music through headphones, chainsaws, snowmobiles, gunfire, motorcycles, and power tools. Hoppmann has reviewed the hazards of being an instrumental musician including hearing loss, and he emphasizes the need for a team approach to comprehensive arts medicine diagnosis and care (12). Noise exposure has a cumulative effect, and exposure to these other types of noise just compounds the damage to a performer. In his very interesting article entitled "Binaural Hearing in Music Performance," Woolford evaluated the effects of hearing impairment on performance and found that there is no direct correlation between degree of hearing impairment and level of performance (13).

Clinical observations in the authors' practice suggest that the rock performance environment may be another source of asymmetrical noise-induced hearing loss, a relatively unusual situation since most occupational hearing loss is asymmetrical. Rock singers and instrumentalists tend to have slightly greater hearing loss in the ear adjacent to the drum and cymbal, or the side immediately next to a speaker, if it is placed slightly behind the musician. Various methods have been devised to help protect the hearing of rock musicians. For example, most of them stand beside or behind their speakers, rather than in front of them. In this way, they are not subjected to peak intensities, as are the patrons in the first rows.

The problem of occupational hearing loss among classical singers and musicians is less obvious, but equally important. In fact, in the United States, it has become a matter of great concern among unions and management. Various reports have found an increased incidence of high-frequency sensorineural hearing loss among professional orchestra musicians as compared to the general public; and sound levels within orchestras have been measured between 83 and 112 dBA, as discussed below. The size of the orchestra and the rehearsal hall are important factors, as is the position of the individual instrumentalist within the orchestra. Players seated immediately in front of the brass section appear to have particular problems, for example. Individual classical instruments may produce more noise exposure for their players than assumed. In their study entitled "Hearing Assessment of Classical Orchestral Musicians," Kahari et al. reported that male musicians have a more pronounced high-frequency hearing loss than females exposed to the same musical noise (14). They also noted that percussion and woodwind players demonstrated a slightly more pronounced hearing loss than musicians of large string instruments.

Because many singers and instrumentalists practice or perform 4–8 h a day (sometimes more), such exposure levels may be significant. An interesting review of the literature may be found in the report of a clinical research project on hearing in classical musicians by Axelsson and Lindgren (15). They also found asymmetrical hearing loss in classical musicians, greater in the left ear. This is a common finding, especially among violinists. A brief summary of most of the published works on hearing loss in musicians is presented below.

In the United States, various attempts have been made to solve some of the problems of the orchestra musician, including placement of plexiglass barriers in front of some of the louder brass instruments, alteration in the orchestra formation, such as elevation of sections or rotational seating, changes in spacing and height between players, use of specialized musicians' ear protectors, and other measures. These solutions have not been proven effective, and some of them appear impractical, or damaging to the performance. The effects of the acoustic environment (concert hall, auditorium, outdoor stage, etc.) on the ability of music to damage hearing have not been studied systematically. Recently, popular musicians have begun to recognize the importance of this problem and to protect themselves and educate their fans. Some performers are wearing ear protectors regularly in rehearsal, and even during performance as noted in the press in 1989 (16). Considerable additional study is needed to provide proper answers and clinical guidance for this very important occupational problem. In fact, a review of the literature on occupational hearing loss reveals that surprisingly little information is available on the entire subject. Moreover, all of it is concerned with instrumentalists, and no similar studies in singers were found.

Study of the existing reports reveals a variety of approaches. Unfortunately, neither the results nor the quality of the studies is consistent. Nevertheless, familiarity with the research already performed provides useful insights into the problem. In 1960, Arnold and Miskolczy-Fodor studied the hearing of 30 pianists. Sound pressure level measurements showed that average levels were approximately 85 dB, although periods of 92–96 dB were recorded (17). The A-weighting network was not used for sound level measurements in this study. No noise-induced hearing loss was identified. The pianists in this study were 60–80 years of age, and, in fact, their hearing was better than normal for their age. Flach and Aschoff (18) and later Flach (19) found sensorineural hearing loss in 16% of 506 music students and professional musicians, a higher percentage than could be accounted for by age alone, although none of the cases of hearing loss occurred in students. Hearing loss was most common in musicians playing string instruments. Flach and Aschoff also noticed asymmetrical sensorineural hearing loss worse on the left in 10 of 11 cases of bilateral sensorineural hearing loss in instrumentalists (18). In one case (a flautist), the hearing was worse on the right. In 4% of the professional musicians tested, hearing loss was felt to be causally related to musical noise exposure. Histories and physical examinations were performed on the musicians and tests were performed in a controlled environment. This study also included interesting measurements of sound levels in a professional orchestra. Unfortunately, they are reported in DIN-PHONS, rather than dBA.

In 1968, Berghoff (20) reported on the hearing of 35 big band musicians and 30 broadcasting (studio) musicians. Most had performed for 15–25 years, although the string players were older as a group and had performed for as much as 35 years. In general, they played approximately 5 h per day. Hearing loss was found in 40- to 60-year-old musicians at 8000 and 10,000 Hz. Eight musicians had substantial hearing loss, especially at 4000 Hz. Five out of 64 (8%) cases were felt to be causally related to noise exposure. No difference was found between left and right ears, but hearing loss was most common in musicians who were sitting immediately beside drums, trumpets, or bassoons. Sound level measurements for wind instruments revealed that intensities were greater 1 m away from the instrument than they were at the ear canal. Unfortunately, sound levels were again measured in PHONS. Lebo and Oliphant studied the sound levels of a symphony orchestra and two rock and roll bands (21). They reported that sound energy for symphony orchestras is fairly evenly distributed from 500 through 4000 Hz, but most of the energy in rock and roll music was found between 250 and 500 Hz. The SPL for the symphony orchestra during loud passages was approximately 90 dBA. For rock and roll bands, it reached levels in excess of 110 dBA. Most of the time, during rock music performance, sound energy was louder than 95 dB in the lower frequencies, while symphony orchestras rarely achieved such levels. However, Lebo and Oliphant made their measurements from the auditorium, rather than in immediate proximity to the performers (21). Consequently, their measurements are more indicative of distant audience noise exposure than that of the musicians or audience members in the first row. Rintelmann and Borus also studied noise-induced hearing loss in rock and roll musicians, measuring sound pressure level at various distances from 5 to 60 ft from centerstage (22). They studied six different rock and roll groups in four locations and measured a mean sound pressure level of 105 dB. Their analysis revealed that the acoustic spectrum was fairly flat in the low- and mid-frequency regions and showed gradual reduction above 2000 Hz. They also detected hearing loss in only 5% of the 42 high school and college student rock and roll musicians they studied. The authors estimated that their experimental group had been exposed to approximately 105 dB (SPL) for an average of 11.4 h a week for 2.9 years.

In 1970, Jerger and Jerger studied temporary threshold shifts (TTSs) in rock and roll musicians (23). They identified TTSs greater than 15 dB in at least one frequency between 2000 and 8000 Hz in eight of nine musicians studied prior to performance and within 1 h after the performance. Speaks and coworkers (24) examined 25 rock musicians for threshold shifts, obtaining measures between 20 and 40 min following performance. In this study, shifts of only 7–8 dB at 4000 and 6000 Hz were identified. Temporary threshold shifts occurred in about half of the musicians studied. Six of the 25 musicians had permanent threshold shifts. Noise measurements were also made in 10 rock bands. Speaks et al. found noise levels from 90 to 110 dBA. Most sessions were less than 4 h, and actual music time was generally 120–150 min. The investigators recognized the hazard to hearing posed by this noise exposure. In 1972, Rintelmann et al. studied the effects of rock and roll music on humans under laboratory conditions (25). They exposed normal-hearing females to rock and roll music at 110 dB SPL in a sound field. They also compared subjects exposed to music played continuously for 60 min with others in which the same music was interrupted by 1 min of ambient noise between each 3-min musical selection. At 4000 Hz, they detected mean TTs of 26 dB in the subjects exposed to continuous noise, and 22.5 dB in those exposed intermittently. Both groups required approximately the same amount of time for recovery. Temporary threshold shifts sufficient to be considered potentially hazardous for hearing occurred in slightly over 50% of the subjects exposed to intermittent noise, and 80% of subjects subjected to continuous noise.

In 1972, Jahto and Hellmann (26) studied 63 orchestra musicians playing in contemporary dance bands. Approximately one-third of their subjects had measurable hearing loss, and 13% had bilateral high frequency loss suggestive of noise-induced hearing damage. They also measured peak sound pressure levels of 110 dB (the A scale was not used). They detected potentially damaging levels produced by trumpets, bassoons, saxophones, and percussion instruments. In contrast, in 1974 Buhlert and Kuhl (27) found no noise-induced hearing loss among 17 performers in a radio broadcasting orchestra. The musicians had played for an average of 20 years and were an average of 30 years of age. In a later study, Kuhl (28) studied members of a radio broadcasting dance orchestra over a period of 12 days. The average noise exposure was 82 dBA. He concluded that such orchestras were exposed to safe noise levels, in disagreement with Jahto and Hellmann (26). Zeleny et al. (29) studied members (ages were 19–59 years) of a large string orchestra with intensities reaching 104–112 dB SPL. Hearing loss greater than 20 dB in at least one frequency occurred in 85 of 118 subjects (72%), usually in the higher frequencies. Speech frequencies were affected in six people (5%).

In 1976, Siroky et al. reported noise levels within a symphony orchestra ranging between 87 and 98 dBA, with a mean value of 92 dBA (30). Audiometric evaluation of 76 members of the orchestra revealed 16 musicians with hearing loss, 13 of them sensorineural. Hearing loss was found in 7.3% of string players, 20% of wind players, and 28% of brass players. All percussionists had some degree of hearing loss. Hearing loss was not found in players who had performed for fewer than 10 years but was present in 42% of players who had performed for more than 20 years. This study needs to be reevaluated in consideration of age-matched controls. At least some of the individuals reported have hearing loss not causally related to noise (such as those with hearing levels of 100 dB HL in the higher frequencies). In a companion report, Folprechtova and Miksovska also found mean sound levels of 92 dBA in a symphony orchestra with a range of 87–98 dBA (31). They reported that most of the musicians performed between 4 and 8 h daily. They reported the sound levels of various instruments as given in Table 30.1.

Table 30.1 Sound Levels of Various Instruments

Instrument	Sound level (dBA)
Violin	84–103
Cello	84–92
Bass	75–83
Piccolo	95–112
Flute	85–111
Clarinet	92–103
French horn	90–106
Oboe	80–94
Trombone	85–114
Xylophone	90–92

Source: Folprechtova and Miksovska (31).

A study by Balazs and Gotze, also in 1976, agreed that classical musicians are exposed to potentially damaging noise levels (32). The findings of Gryczynska and Czyzewski support the concerns raised by other authors (33). In 1977, they found bilateral normal hearing in only 16 of 51 symphony orchestra musicians who worked daily at sound levels between 85 and 108 dBA. Five of the musicians had unilateral normal hearing, the rest had bilateral hearing loss.

In 1977, Axelsson and Lindgren studied factors increasing the risk for hearing loss in popular musicians (34). They reported that aging, length of exposure per musical session, long exposure time in years, military service and listening to popular music with head phones all had a statistically significant influence on hearing. They noted that the risk and severity of hearing loss increase with increasing duration of noise exposure, and increasing sound levels. In pop music, the exposure to high sound levels was felt to be limited in time, and less damaging low frequencies predominated.

Also in 1977, Axelsson and Lindgren published an interesting study of 83 popular musicians and noted a significant incidence of hearing loss (35). They reanalyzed previous reports investigating a total of 160 popular musicians, which identified an incidence of only 5% hearing loss. In their 1978 study, Axelsson and Lindgren tested 69 musicians, four disk jockeys, four managers, and six sound engineers (36). To have hearing loss, a subject had to have at least one pure-tone threshold exceeding 20 dB at any frequency between 3000 and 8000 Hz. Thirty-eight musicians were found to have sensorineural hearing loss. In 11, only the right ear was affected; in 5, only the left ear was affected. Thirteen cases were excluded because their hearing loss could be explained by causes other than noise. Thus, 25% of the pop musicians had sensorineural hearing loss probably attributable to noise. The most commonly impaired frequency was 6000 Hz, and very few ears showed hearing levels worse than 35 dB loss. After correction for age and other factors, 25 (30%) had hearing loss as defined above. Eleven (13%) had hearing loss defined as a pure-tone audiometric average greater than 20 dB HL at 3000, 4000, 6000, and 8000 Hz in at least one ear. Of these 11, 7 had unilateral hearing loss (8%). The authors concluded that it seemed unlikely that sensorineural hearing loss would result from popular music presented at 95 dBA with interruptions, and with relatively short exposure durations and low frequency emphasis. Axelsson and Lindgren published further articles on the same study (37,38). They also noted that temporary threshold shift measurements in popular music environments showed less shift in musicians than in the audience. They also found that female listeners were more resistant to temporary threshold shift than males. In a follow-up study to their 1975 work, Axelson and Eliasson reevaluated 53 out of the original 83 pop/rock musicians they had studied (39). Their findings indicate a rather slow progression of hearing loss. The authors were surprised to find that hearing remained so stable and fell within 20 dB of original thresholds. They surmised that this presentation may have something to do with the stapedius reflex, but this theory has yet to be proven. In our opinion, this finding is a manifestation of the asymptotic pattern seen in noise-induced hearing loss (NIHL) from other sources and is not surprising. In a 2001 follow-up study to a 1979 assessment done by Axelsson and Lindgren, Kahari and Axelsson reevaluated 56 out of the original 140 orchestral musicians in Sweden (40). Interestingly, 16 years later, those studies showed no significant pure-tone threshold decreases. The male participants continued to demonstrate a greater hearing loss than the females in the high frequency range.

In 1981, Westmore and Eversden studied a symphony orchestra and 34 of its musicians (41). They recorded SPL for 14.4 h. Sound levels exceeded 90 dBA for 3.51 h and equaled or exceeded 110 dBA for 0.02 h. In addition, there were brief peaks exceeding 120 dBA. They interpreted their audiometric testing as showing NIHL in 23 of 68 ears.

Only 4 of the 23 ears had a hearing loss greater than 20 dB HL at 4000 Hz. There was a "clear indication" that orchestral musicians may be exposed to damaging noise. However, because of the relatively mild severity, they speculated that "it is unlikely that any musician is going to be prevented from continuing his artistic career." In Axelsson and Lindgren's 1981 study, sound level measurements were performed in two theaters, and 139 musicians underwent hearing tests (15). Sound levels for performances ranged from 83 to 92 dBA.

Sound levels were slightly higher in an orchestra pit, although this is contrary to the findings of Westmore (41). Fifty-nine musicians (43%) had pure-tone thresholds worse than expected for their ages. French hornists, trumpeters, trombonists, and bassoonists were found to be at increased risk for sensorineural hearing loss. Asymmetric pure-tone thresholds were common in musicians with hearing loss, and in those still classified as having "normal hearing." The left ear demonstrated greater hearing loss than the right, especially among violinists. Axelsson and Lindgren also found that the loudness comfort level was unusually high among musicians. Acoustic reflexes also were elicited at comparatively high levels, being pathologically increased in approximately 30%. Temporary threshold shifts were also identified, supporting the assertion of noise-related etiology.

Also, in 1983 Lindgren and Axelsson attempted to determine whether individual differences of TTS existed after repeated controlled exposure to non-informative noise, and to music having equal frequency, time, and sound level characteristics (42). They studied 10 subjects who were voluntarily exposed to 10 min of recorded pop music on five occasions. On five other occasions they were exposed to equivalent noise. Four subjects showed almost equal sensitivity in measurements of TTS, and six subjects showed marked differences, specifically greater TTS after exposure to the non-music stimulus. This research suggests that factors other than the physical characteristics of the fatiguing sound contributed to the degree of TTS. The authors hypothesized that these factors might include the degree of physical fitness, stress, and emotional attitudes toward the sounds perceived. The authors concluded that high sound levels perceived as noxious cause greater TTS than high sound levels that the listener perceived as enjoyable.

In 1983, Karlsson and coworkers published a report with findings and conclusions substantially different from those of Axelsson and others (43). Karlsson investigated 417 musicians, 123 of whom were investigated twice at an interval of 6 years. After excluding 26 musicians who had hearing loss for reasons other than noise, he based his conclusions on the remaining 392 cases. Karlsson et al. concluded that there was no statistical difference between the hearing of symphony orchestra musicians and that of a normal population of similar age and sex (43). Those data revealed a symmetric dip of 20 dB at 6000 Hz HL in flautists, and a 30 dB HL left high-frequency sloping hearing loss in bass players. Overall, a 5 dB HL difference between ears was also found at 6000 and 8000 Hz, with the left side being worse. Although Karlsson and coworkers concluded that performing in a symphonic orchestra does not involve an increased risk of hearing damage, and that standard criteria for industrial noise exposure are not applicable to symphonic music, their data are similar to previous studies. Only their interpretation varies substantially.

In 1984, Woolford studied sound pressure levels in symphony orchestras and hearing (44). Woolford studied 38 Australian orchestral musicians, and measured sound pressure levels using appropriate equipment and technique. He found potentially damaging sound levels, consistent with previous studies. Eighteen of the 38 musicians had hearing losses. Fourteen of those had threshold shifts in the area of 4000 Hz, and four had slight losses at low frequencies only.

Johnson et al. studied the effects of instrument type and orchestral position on the hearing of orchestra musicians (45). They studied 60 orchestra musicians from 24 to 64 years in age, none of whom had symptomatic hearing problems. The musicians underwent otologic histories and examinations, and pure-tone audiometry from 250 through 20,000 Hz. Unfortunately, this study used previous data from other authors as control data. In addition to the inherent weakness in this design, the comparison data did not include thresholds at 6000 Hz. There appeared to be a 6000-Hz dip in the population studied by Johnson et al., but no definitive statement could be made. The authors concluded that the type of instrument played and the position on the orchestra stage had no significant correlation with hearing loss, disagreeing with findings of other investigators. In another paper produced from the same study (46), Johnson reported no difference in the high frequency thresholds (9000–20,000 Hz) between musicians and non-musicians. Again, because he examined 60 instrumentalists, but used previously published reports for comparison, this study is marred. This shortcoming in experimental design is particularly important in high-frequency testing during which calibration is particularly difficult and establishment of norms on each individual piece of equipment is advisable.

In their 1986 article entitled "The Level of the Musical Loud Sound and Noise Induced Hearing Impairment," Ono and Deguchi determined the importance of measuring the cumulative effects of noise over long periods of time on a single individual. They developed a compact noise dosimeter designed to evaluate long-term and varied exposure called "Noise Badge" (47).

In 1987, Swanson et al. studied the influence of subjective factors on temporary threshold shift after exposure to music and noise of equal energy (48), attempting to replicate Lindgren and Axelsson's 1983 study. Swanson's study used two groups of subjects, 10 who disliked popular music, and 10 who liked popular music. Each subject was tested twice at 48-h intervals. One session involved exposure to music for 10 min. The other session involved exposure to equivalent noise for 10 min. Their results showed that individuals who liked pop music experienced less TTS after music than after noise. Those who disliked the music showed greater TTS in music than in noise. Moreover, the group that liked popular music exhibited less TTS than the group that disliked the music. These findings support the notion that sounds perceived as offensive produce greater TTS than sounds perceived as enjoyable.

A particularly interesting review of hearing impairment among orchestra musicians was published by Woolford et al. in 1988 (49). Although this report presents only preliminary data, the authors have put forward a penetrating review of the problem and interesting proposals regarding solutions, including an international comparative study. They concluded that among classical musicians the presence of hearing loss from various etiologies including noise has been established, that some noise-induced hearing impairments in musicians are permanent (although usually slight), and that successful efforts to reduce the intensity of noise exposure are possible.

In addition to concern about hearing loss among performers, in recent years there has been growing concern about NIHL among audiences. Those at risk include not only people at rock concerts, but also people who enjoy music through stereo systems, especially modern personal headphones. Concern about hearing loss from this source in high school students has appeared in the lay press and elsewhere (50,51). Because young music lovers are potentially performers, in addition to other reasons, this hazard should be taken seriously and investigated further.

In 1990, West and Evans studied 60 people aged 15–23 years at the University of Keele, looking for hearing loss caused by listening to amplified music (52). They found widening of auditory band widths to be a sensitive, early indicator of noise-induced hearing loss that was detectable before threshold shift at 4000 or 6000 Hz occurred. They advocated the use of frequency resolution testing and high resolution Békésy audiometry for early detection of hearing impairment. Interestingly, West and Evans found that subjects extensively exposed to loud music were significantly less able to differentiate between a tone and its close neighbors. Reduced pitch discrimination was particularly common in subjects who had experienced TTS or tinnitus following exposure to amplified music.

In 1991, van Hees published an extensive thesis on noise-induced hearing impairment in orchestral musicians (53). He agreed that noise levels were potentially damaging in classical and wind orchestras. Unlike other researchers, van Hees found it more useful to classify the instruments by orchestral zone, rather than by instrument or instrument group. However, he found a much greater incidence of hearing loss among both symphony and wind orchestra musicians than reported in the previous literature. He also did not find evidence of asymmetric hearing loss in violinists and cello players, in contrast to previous investigators.

In 1991, the musicians of the Chicago Symphony Orchestra were evaluated by Royster et al. (54). Subjects were given individual dosimeters and 68 measurements were made during various rehearsals and performances. More than half of the musicians had audiograms consistent with NIHL and showed a high prevalence of "notch" patterns (54). The authors measured Leq values and hearing threshold levels on 32 musicians and found that pure tone thresholds between 3 and 6 kHz correlated to the measured Leq, which was determined to range from 79 to 99 dBA SPL (53).

In 1992, Camp and Horstman investigated sound exposure among instrumentalists during a performance of Wagner's *Ring Cycle*. Peak exposures were measured in the orchestra pit ranging as high as 100 dBA to 104 dBA, although measured dose for selected musicians ranged from only 20.2% to 46.1%. They also investigated the efficacy of free-standing clear plastic shields. Sound attenuation was found to be dependent upon the position and angle of the ear and measuring equipment microphone relative to the shield. Attenuation up to 17 dB at high frequencies was noted in some cases. However, the free-standing plastic shields provided little or no attenuation unless they were placed within 7 in. of the ear (55).

In 1995, Griffiths and Samaroo studied hearing sensitivity among professional pianists. The hearing of members of two steel pan orchestras was compared with a group of non-playing control subjects in Trinidad. Four-hour noise dosimetry revealed time-weighted averages of 98 dBA in a small group indoor setting and at 113 dBA in a large group outdoor rehearsal. Compared with controls, the pianists had significantly poorer hearing at 2000, 3000, 4000, and 6000 Hz. Audiometric dips were found in 72% of the pianists studied (56).

In their 1998 evaluation of choir singers and hearing loss, Steurer and Simak made some unusual discoveries that warrant additional research. The authors discovered that low-frequency hearing is most affected in these subjects (57), in particular, 250 to 1 kHz. They are not able to explain the demonstrated hearing losses below 100 Hz but have speculated that there may be increased endolymphatic pressure when singing that could account for the loss in this range.

Review of these somewhat confusing and contradictory studies reveals that a great deal of important work remains to be done in order to establish the risk of hearing loss among various types of musicians, the level and pattern of hearing loss that may be sustained, practical methods of preventing hearing loss, and advisable programs for monitoring and early diagnosis. However, a few preliminary conclusions can be drawn. First, the preponderance of evidence indicates that NIHL occurs among both popular and classical musicians and is causally related to exposure to loud music. Second, in most instances, especially among classical musicians, the hearing loss is not severe enough to interfere with speech perception. Third, the effects of mild high-frequency hearing loss on musical performance have not been established. Fourth, it should be possible to devise methods to conserve hearing among performing artists without interfering with performance. In 1991, Chasin and Chong reported on an ear protection program for musicians (58). They provide an interesting discussion of the use of ear protectors in musicians, although several aspects of their paper are open to challenge. In particular, their assertion that some vocalists (particularly sopranos) have self-induced hearing loss caused by singing has not been substantiated. In a subsequent paper, Chasin and Chong proposed four environmental techniques to reduce the effect of music exposure on hearing (59). They recommended that speakers and amplifiers should be elevated from the floor; treble brass instruments should be on risers; there should be 2 m of unobstructed floor space in front of the orchestra; and small stringed instruments should have at least 2 m of unobstructed space above them.

2. LEGAL ASPECTS OF HEARING LOSS IN SINGERS AND MUSICIANS

The problem of hearing loss in musicians raises numerous legal issues, especially the implications of occupational hearing loss; and hearing has become an issue in some orchestra contracts. Traditionally, workers' compensation legislation has been based on the theory that workers should be compensated when a work-related injury impairs their ability to earn a living. Ordinarily, occupational hearing loss does not impair earning power (except possibly in the case of musicians and a few others). Consequently, current occupational hearing loss legislation broke new legal ground by providing compensation for interference with quality of life. Therefore, all current standards for defining and compensating occupational hearing loss are based on the communication needs of the average speaker, and are usually compensated in accordance with the recommendations of the American Academy of Otolaryngology (8). Since music-induced hearing loss appears to rarely affect the speech frequencies, it is not compensable under most laws. However, although a hearing loss at 3000, 4000, or 6000 Hz with preservation of lower frequencies may not pose a problem for a boiler maker, it may be a serious problem for a violinist. Under certain circumstances, such a hearing loss may even be disabling. Because professional instrumentalists require considerably greater hearing acuity throughout a larger frequency range, we must investigate whether the kinds of hearing loss caused by music are severe enough to impair performance. If so, new criteria must be established for compensation for disabling hearing impairment in musicians, in keeping with the original intent of workers' compensation law.

There may also be unresolved legal issues regarding hearing loss not caused by noise in professional musicians. Like people with other handicaps, there are numerous federal laws protecting the rights of the hearing impaired. In the unhappy situation in which an orchestra must release a hearing-impaired violinist who can no longer play in tune, for example, legal challenges may arise. In such instances, and in many other circumstances, an objective assessment process is in the best interest of performers and management. Objective measures of performance are already being used in selected areas for singers, and they have proven very beneficial in helping the performer assess certain aspects of performance quality and skill development dispassionately. Such technologic advances will probably be used in the future more frequently to supplement traditional subjective assessment of performing artists for musical, scientific, and legal reasons.

3. EAR PROTECTORS FOR THE MUSICIAN

Current ear protectors offer a much more suitable solution to noise protection than their predecessors. There are several models available that cater to musicians and their specific requirements. The design strategy has improved to allow more accurate music and speech perception at lower intensity levels. There are various models that provide differing attenuation levels ranging from 9 to 25 dB. These protectors are custom fitted to the individual and thus provide better performance than pre-formed, in-the-ear hearing protectors. They can be purchased through an audiologist or hearing aid dispenser.

4. NOISE EXPOSURE AND THE AUDIENCE

The focus of this chapter thus far has been on the musician and noise exposure. A few more recent studies have looked at the effects of noise on an audience. In particularly, Gunderson and Moline, in a 1997 study (60), evaluated the effect of noise exposure on employees of urban music clubs. Average sound levels ranged from 94.9 to 106.7 dBA. Only 16% of the employees used ear protection. The authors recommended the development of a hearing conservation program for this often overlooked population. Similarly, in 1997, Sataloff et al. evaluated noise levels at an outdoor rock concert. Results showed audience exposure to levels of 119 dBA. These studies indicate that future effects need to focus on hearing conservation not only for the performer, but for audiences as well (61).

5. TREATMENT OF OCCUPATIONAL HEARING LOSS IN SINGERS AND MUSICIANS

For a complete discussion of hearing loss, the reader is referred to other chapters in this book. Most cases of sensorineural hearing loss produced by aging, hereditary factors, and noise cannot be cured. When they involve the speech frequencies, modern, properly adjusted hearing aids are usually extremely helpful. However, these devices are rarely satisfactory for musicians during performance. More often, appropriate counseling is sufficient. Musicians should be provided with a copy of their audiogram and an explanation of its correspondence with the piano keyboard. Unless a hearing loss becomes severe, this information usually permits musicians to make appropriate adjustments. For example, conductors with an unknown high-frequency hearing loss will call for violins and triangles to be excessively loud. If they know the pattern of hearing loss, this error may be reduced. Musicians with or without hearing loss should routinely be cautioned against avocational loud noise exposure without ear protection (hunting, power tools, motorcycles, etc.) and ototoxic drugs. In addition, they should be educated about the importance of immediate evaluation if a sudden hearing change occurs. When diplacusis (pitch distortion) is present, compensation is especially difficult, especially for singers and string players. Auditory retraining may be helpful in some cases. Hopefully, electronic devices will be available in the future to help this problem, as well.

6. SUMMARY

Good hearing is of great importance to musicians, but the effects on performance of mild high-frequency hearing loss remain uncertain. It is most important to be alert to hearing loss from all causes in performers, to recognize it early, and to treat it or prevent its progression whenever possible. Musical instruments and performance environments are capable of producing damaging noise. Strenuous efforts must be made to define the risks and nature of music-induced hearing loss among musicians, to establish damage–risk criteria, and to implement practical means of noise reduction and hearing conservation.

Musicians depend on their hearing almost as much as they do on their voices and fingers. It is important not to take such valuable and delicate structures as our ears for granted. Like the voice and the hand, the ear must be understood and protected in order for a musician to enjoy a long, happy, and successful career.

REFERENCES

1. Johnson DW, Sherman RE, Aldredge J, Lorraine A. Extended high frequency hearing sensitivity. A normative threshold study in musicians. Ann Otol Rhinol Laryngol 1986; 95(2 pt 1):196–202.
2. Schmidt JM, Verschuure J, Brocaar MP. Hearing loss in students at a conservatory. Audiology 1994; 33(4):185–94.
3. Axelsson A, Lindgren F. Temporary threshold shift after exposure to pop music. Scand Audiol 1978; 7(3):127–135.
4. Ostri B, Eller N, Dahlin E, Skylv G. Hearing impairment in orchestral musicians. Scand Audiol 1989; 18(4):243–249.
5. McBride D, Gill F, Proops D, Harrington M, Gardiner K, Attwell C. Noise and the classical musician. Br Med J 1992; 305(6868):1561–1563.
6. Drake-Lee AB. Beyond music: auditory temporary threshold shift in rock musicians after a heavy metal concert. J R Soc Med 1992; 85(10):617–619.
7. Bu G. Prevention of noise induced hearing loss in musicians of Chinese opera. Zhonghua Er Bi Yan Hou Ke Za Zhi 1992; 27(2):73–75, 124.
8. Cocchiarella L, Andersson GBJ. Guides to the Evaluation of Permanent Impairment. 5th ed. Chicago, IL: AMA Press, 2001:245–275.
9. Rosanowski F, Eysholdt U. In situ sound pressure measurement in a professional violinist with bilateral tinnitus. Laryngorhinootologie 1996; 75(9):514–516.
10. Metternich FU, Brusis T. Acute hearing loss and tinnitus caused by amplified recreational music. Laryngorhinootologie 1999; 78(11):614–619.
11. Karlsson K, Lundquist PG, Olaussen T. The hearing of symphony orchestra musicians. Scand Audiol 1983; 12(4):257–264.
12. Hoppmann RA. Instrumental musicians' hazards. Occup Med 2001; 16(4):619–631.
13. Woolford D. Binaural hearing in music performance. Aust Acoust Soc 1998; 26(2):68.
14. Kahari KR, Axelsson A, Hellstrom PA, Zachau G. Hearing assessment of classical orchestral musicians. Scand Audiol 2001; 30(1):13–23.
15. Axelsson A, Lindgren F. Hearing in classical musicians. Acta Otolaryngol, 1981; (Suppl 377):3–74.
16. Toufexis A: A firehose down the ear canal. Time 1989: 78.
17. Arnold GE, Miskolczy-Fodor F. Pure-tone thresholds of professional pianists. Arch Otolaryngol 1960; 71:938–947.
18. Flach M, Aschoff E. Zur Frage berufsbedingter Schwerhöbrigkeit beim Musiker. Z Laryngol 1966; 45:595–605.

19. Flach M. Das Gehör des Musikers aus ohrenarztlicher Sicht. Msch Ohrhk 1972; 9:424–432.
20. Berghoff F. Horleistung und berufsbedingte Horschadigung des Orchestermusikers mit einem Beitrag zur Pathophysiologie des larmtraumatischen Horschadens. 1968, Dissertation. Cited in Axelsson A, Lindgren F. Hearing in classical musicians. Acta Otolaryngol 1981; (Suppl 377):3–74.
21. Lebo CP, Oliphant KP. Music as a source of acoustic trauma. Laryngoscope 1968; 72(2):1211–1218.
22. Rintelmann WF, Borus JF. Noise-induced hearing loss in rock and roll musicians. Arch Otolaryngol 1968; 88:377–385.
23. Jerger J, Jerger S. Temporary threshold shift in rock-and-roll musicians. J Speech Hear Res 1970; 13:221–224.
24. Speaks C, Nelson D, Ward WD. Hearing loss in rock-and-roll musicians. J Occup Med 1970; 13:221–224.
25. Rintelmann WF, Lindberg RF, Smitley EK. Temporary threshold shift and recovery patterns from two types of rock and roll presentations. J Acoust Soc Am 1972; 51:1249–1255.
26. Jahto K, Hellmann H. Zur Frage des Larm- und Klangtraumas des Orchestermusikers. Audiologie und Phoniatrie HNO, 1972; 20(1):21–29.
27. Buhlert P, Kuhl W. Hobruntersuchungen im freien Schallfeld zum Altershorverlust. Acustica 1974; 31:168–177.
28. Kuhl W. Keine Gehorschabigung durch Tanzmusik, simfonische Musik und Maschinengerausche beim Rundfunk. Kampf dem Larm 1976; 23(4):105–107.
29. Zeleny M, Navratilova Z, Kamycek Z, et al. Relation of hearing disorders to the acoustic composition of working environment of musicians in a wind orchestra. Cesk Otolaryngol 1975; 24(5):295–299.
30. Siroky J, Sevcikova L, Folprechtova A, et al. Audiological examination of musicians of a symphonic orchestra in relation to acoustic conditions. Cesk Otolaryngol 1976; 25(5):288–294.
31. Folprechtova A, Miksovska O. The acoustic conditions in a symphony orchestra. Pracov Lek 1978; 28:1–2.
32. Balazs B, Gotze A. Comparative examinations between the hearing of musicians playing on traditional instruments and on those with electrical amplifications. Ful-orr-gegegyogyaszat 1976; 22:116–118.
33. Gryczynska D, Czyzewski I. Damaging effect of music on the hearing organ in musicians. Otolaryngol Pol 1977; 31(5):527–532.
34. Axelsson A, Lindgren F. Factors increasing the risk for hearing loss in "pop" musicians. Scand Audiol 1977; 6:127–131.
35. Axelsson A, Lindgren F. Does pop music cause hearing damage? Audiology 1977; 16:432–437.
36. Axelsson A, Lindgren F. Hearing in pop musicians. Acta Otolaryngol 1978; 85:225–231.
37. Axelsson A, Lindgren F. Horseln hos popmusiker. Lakartidningen 1978; 75(13):1286–1288.
38. Axelsson A, Lindgren F. Pop music and hearing. Ear Hear 1981; 2(2):64–69.
39. Axelsson A, Eliasson A, Israelsson B. Hearing in pop/rock musicians: a follow-up study. Ear Hear 1995; 16(3):245–253.
40. Kahari KR, Axelsson A, Hellstrom PA, Zachau G. Hearing development in classical orchestral musicians. A follow-up study. Scand Audiol 2001; 30(3):141–149.
41. Westmore GA, Eversden ID. Noise-induced hearing loss and orchestral musicians. Arch Otolaryngol 1982; 107:761–764.
42. Lindgren F, Axelsson A. Temporary threshold shift after exposure to noise and music of equal energy. Ear Hear 1983; 4(4):197–201.
43. Karlsson K, Lundquist PG, Olaussen T. The hearing of symphony orchestra musicians. Scand Audiol 1983; 12:257–264.
44. Woolford DH. Sound pressure levels in symphony orchestras and hearing. Preprint 2104 (B-1), Australian Regional Convention of the Audio Engineering Society, Melbourne, September 25–27, 1984.

45. Johnson DW, Sherman RE, Aldridge J, et al. Effects of instrument type and orchestral position on hearing sensitivity for 0.25 to 20 kHz in the orchestral musician. Scand Audiol 1985; 14:215–221.
46. Johnson DW, Sherman RE, Aldridge J, et al. Extended high frequency hearing sensitivity: a normative threshold study in musicians. Ann Otol Rhinol Laryngol 1986; 95:196–201.
47. Ono H, Deguchi T, Ino T, Okamoto K, Takyu H. The level of the musical loud sound and noise induced hearing impairment. J UOEH. 1986; 8(Suppl):151–161.
48. Swanson SJ, Dengerink HA, Kondrick P, Miller CL. The influence of subjective factors on temporary threshold shifts after exposure to music and noise of equal energy. Ear Hear 1987; 8(5):288–291.
49. Woolford DH, Carterette EC, Morgan DE. Hearing impairment among orchestral musicians. Music Percept 1988; 5(3):261–284.
50. Gallagher G. Hot music, high noise, and hurt ears. Hear J 1989; 42(3):7–11.
51. Lewis DA. A hearing conservation program for high-school level students. Hear J 1989; 42(3):19–24.
52. West DB, Evans EF. Early detection of hearing damage in young listeners resulting from exposure to amplified music. Br J Audiol 1990; 28:89–103.
53. van Hees OS. Noise Induced Hearing Impairment in Orchestral Musicians. Amsterdam: University of Amsterdam Press, 1991.
54. Royster JD, Royster LH, Killion MC. Sound exposures and hearing thresholds of symphony orchestra musicians. J Acoust Soc Am 1991; 89(6):2793–2803.
55. Camp JE, Horstman SW. Musician sound exposure during performance of Wagner's Ring Cycle. Med Prob Perform Arts 1992; 7:37–39.
56. Griffiths SK, Samaroo AL. Hearing sensitivity among professional pianists. Med Prob Perform Arts 1995; 10:11–17.
57. Steurer M, Simak S, Denk DM, Kjautzky M. Does choir singing cause noise-induced hearing loss? Audiology 1998; 37(1):38–51.
58. Chasin M, Chong J. An in situ ear protection program for musicians. Hearing Instrum 1991; 42(12):26–28.
59. Chasin M, Chong J. Four environmental techniques to reduce the effect of music exposure on hearing. Med Prob Perform Arts 1995; 10:66–69.
60. Gunderson E, Moline J, Catalano P. Risks of developing noise-induced hearing loss in employees of urban music clubs. Am J Ind Med. 1997; 31(1):75–79.
61. Sataloff RT, Hickey K, Robb J. Rock concert audience noise exposure: a preliminary study. J Occup Hear Loss 1998; 1(2):97–99.

31
Hearing Conservation in Industry

Robert T. Sataloff
Drexel University College of Medicine, American Institute for Voice and Ear Research, and Graduate Hospital, Philadelphia, Pennsylvania, USA

Joseph Sataloff
Thomas Jefferson University, Philadelphia, Pennsylvania, USA

Every plant in which employees are exposed to noise exceeding the OSHA guidelines must have a hearing conservation program. OSHA regulations are not the only, or even the most important reasons for having a hearing conservation program. The threat of creating hearing loss in millions of workers and the potential cost of claims for compensation for those losses carry greater urgency.

Physicians and allied health personnel are being asked to advise industry about if, when, and how to conduct a hearing conservation program. To provide such advice, these personnel must have special training and certification in hearing conservation. The minimum 20-h training program recommended by the Council for Accreditation in Occupational Hearing Conservation (CAOHC) is designed for industrial nurses and technicians. Physicians, audiologist, hygienists, and engineers, who provide the technical advice and considered judgment required by the industry, must have more extensive training.

The topics covered by such training courses should include the following:

The anatomy of the ear and how we hear

Bone conduction—how to measure it and avoid pitfalls in interpretation of audiogram and pitfalls in diagnosing solely on the basis of an audiogram
What a complete otologic evaluation is
How to interpret an otologist's report and what it should contain
Various types of hearing disorders and how to differentiate occupational hearing loss from other causes
Noise-induced hearing loss
Advanced hearing tests, including speech and site of tests
How to educate management and labor in hearing conservation
Advantages of an in-plant program
Disadvantages of using a van for annual testing
Aural rehabilitation and use of hearing aids
When a plant needs a hearing conservation program
Noise surveys and how to interpret them
Audiometric tests on employees: technic and pitfalls
Responsibility for supervising the program
Certification for hearing conservation
The cost of a hearing conservation program
Industrial relations with employees, labor, and management
The physics of sound, with special emphasis on how to interpret dBA and decibels
Examples of common noisy occupations
Limitations of threshold testing
Extra-auditory effects of noise
TTS and PTS
Status and susceptibility tests
Ambient noises and effects of masking
Scheduling audiometry
Self-recording audiometry
Keeping records and storing data
Computerized audiometry, advantages, and disadvantages
Preemployment audiograms
Baseline audiograms
Monitoring audiometry
Exposure histories
What a tester should tell the subjects
Practical experience in noise measurements and hearing testing
Calibration of equipment
Hearing protectors, their use and monitoring
Medicolegal aspects
Compensation regulations
OSHA
Longshore and Harbor Workers' Act
State laws
Common law
Federal Employment Compensation Act
How to measure impairment
How the otologist should testify in workers' compensation cases
The role of the physician, nurse, safety engineer, attorney, personnel director, and hygienist
Hiring personnel with hearing loss

How frequently to repeat audiograms
When to refer an employee or his records for evaluation
What to do with audiograms

Every course must include faculty members with academic training and practical experience inside industry.

1. WHAT IS A HEARING CONSERVATION PROGRAM?

Too frequently, those who operate industrial plants and the physicians advising them believe that they have an effective hearing conservation program because they have performed thousands of audiograms and numerous noise measurements. When asked what the audiograms show and whether all exposed employees are using hearing protectors effectively, the replies frequently are unsatisfactory or even embarrassing. Often, the industrial representative will point to a stack of audiograms, saying that he hasn't the slightest idea what the reports contain or mean and that he doesn't really know if the employees are using hearing protectors properly. Every effective hearing conservation program must include the following elements:

1. A *responsible and trained member of the plant must supervise the program*, especially the proper use of hearing protectors. He should solve problems as they arise, know where to seek expert help, explain the program to old and new employees, control the effectiveness and the cost of the program, and handle referrals to local physicians. In a small plant, the same individual also may be required to perform audiometry and take noise measurements.
2. A *complete noise survey of all areas* should be conducted. Such a survey is done with a sound-level meter of octave-band analyzer and can be performed by a well-trained person inside the plant or by a consultant. It is an inexpensive assessment and should be repeated annually or when new machinery is acquired or manufacturing processes are changed. On the basis of these noise measurements, management has to decide whether it requires a hearing conservation program and whether noise control via engineering is technically and economically feasible.
3. A *noise control program* should be implemented. An acoustical consultant usually is needed to decide whether present machinery can be quieted at a reasonable cost. When hiring an outside consultant, it is advisable to get a fixed-price estimate and written guarantees on the results. Noise specifications must be included for all new machinery.
4. An *educational program about hearing conservation is essential for management and employees*. Before starting the hearing testing program or providing hearing protectors, management and labor should be informed that a good hearing conservation program is oriented medically, and its objectives are to prevent all causes of hearing loss. The otologist must play a leading role in the education of employees. Special movies and audiovisual aids are extremely helpful. The educational program must be on a continuing basis and include all new employees.
5. *The hearing testing program must be run efficiently* without time delays or keeping employees away from their jobs. The testing must be performed by certified personnel using equipment and test rooms that meet ANSI standards. A hearing test must be performed on all employees before they are hired,

terminated, or moved into noisy jobs. Hearing tests on all employees must also be performed routinely, with the interval between tests depending on their noise exposure and audiometric evaluation.

6. *The interpretation of all audiograms should be done by an otologist experienced in hearing conservation.* The otologist must advise as to the reliability and validity of the hearing test, whether repeat or additional tests or referral are indicated, and whether the industry should hire certain applicants to work in a particular job. Referrals of individuals employees for outside consultation must be done with minimal disruption of work. Employees with hearing losses that can be corrected should be urged to visit their own otologist.
7. A *hearing protection program* is of utmost importance. Protectors should be distributed only after noise measurements have been taken and employees and management have been advised of the purpose of the protectors and how to use, care for, and replace them. The hearing protection program must be supervised continuously by management and labor. Protectors must be made available to all employees. Audiograms should be evaluated, at least annually, to monitor the effectiveness of the hearing protection program.

2. TEAMWORK

Every hearing conservation program, even in a small plant, requires teamwork. The team consists of the medical, hygiene, and safety departments, personnel director, supervisors, and labor representatives. Usually the hygienist or safety engineer is in charge of the noise measurement and noise control functions. The medical department is responsible for hearing testing, diagnosis, and referrals to outside consultants. The personnel director's job includes hiring personnel who may have hearing loss. Most of the time his decision is based on the information supplied by the medical department and hygienist or safety engineer. Although members of the medical department usually fit the hearing protectors, everyone else on the team, including the department supervisors, should help them monitor the proper use by employees at work. The education program for overall hearing conservation is generally done by the safety engineer in conjunction with the medical department.

3. IMPORTANT FEATURES

After more than 40 years of experience with hearing conservation program in hundreds of large and small plants, we recommend the following basics in conducting the program:

1. *Hearing tests should be performed routinely* on all employees, including management, and should not be restricted only to personnel exposed to high noise levels. The hearing conservation program is a medical one designed to prevent and detect all causes of hearing losses, not only those due to noise. Hearing testing is in the employee's best interest, just like an eye or heart examination, and should be part of a general health evaluation. Preemployment audiograms and otologic histories should be taken and interpreted on the day an applicant is to be hired, not weeks later. This is one advantage of having an in-plant hearing testing program rather than depending on outside services that may not always be available when needed.
2. *Monitoring audiometry must be kept up-to-date* and the results used for constant evaluation of the hearing conservation program.

3. *Every abnormal audiogram should be evaluated* and, if possible, a diagnosis should be established by an otologist. This should be done for all applicants as well as all employees, including management.
4. If possible, all *employees who are terminated should have an audiogram* just prior to leaving the company.
5. *Audiograms must be done at a time when the employee is free of temporary hearing loss.* In almost every industry this can be accomplished by using hearing protectors effectively. Only in rare instances do audiograms have to be performed just prior to the employee's starting work or on a Monday morning.
6. *Personnel with hearing handicaps have to be hired* not only because of government regulations, but more practically because of experience. Industry is hardly in a position to set very conservative or rigid standards for hiring employees with hearing loss. The tight labor market militates against this. Furthermore, it would screen out our most experienced and eligible applicants from noisy jobs. For example, most applicants with experience as chippers, weavers, or paper machine operators have hearing losses. Such workers can be hired safely. However, recording accurate histories or previous noise exposure, including off-the-job exposures, is essential. Audiograms should be performed frequently on such personnel to be certain that no additional hearing loss is occurring.

   There are applicants with hearing loss who should not be accepted for employment by some industries. For instance, an individual with a high-frequency hearing loss and a severe discrimination problem should not be hired where verbal communication is essential. Applicants with a hearing impairment should not be employed in a noisy job if it is likely to cause a safely hazard for himself or others. For their own benefit, those with a diagnosis of progressive hereditary sensorineural hearing loss should not be employed in a noisy area. To encourage management to hire hearing-handicapped workers, industry must be protected from being liable for the hearing loss that existed prior to employment, and compensation laws should be written accordingly. The final decision on hiring or not hiring rests with the personnel director.
7. *There is no valid evidence to show that an employee with sensorineural deafness is more susceptible to noise* than one with normal hearing. It is even more likely that the ear with sensorineural loss is less susceptible to certain types of noise. Industry should be free to hire such workers, provided they can be assured that issued hearing protection is used and that their hearing loss will not be aggravated by the job's noise exposure.
8. *In noisy areas that are marked as being hazardous, all employees must use hearing protectors.* To maintain the integrity of the program, management and visitors just walking through a noisy area also should be provided with hearing protectors. Without a *mandatory and continuing education policy*, many hearing protection programs have failed.
9. Employees should be permitted to choose from *several types of acceptable hearing protectors.* If the protectors are lost or damaged, *new ones should be readily available.* Employees who develop ear infections may be excused from using protectors only for brief periods of time under a doctor's advice until the infection is cleared. Sometimes a different type of protector can be utilized.
10. *Referrals to an otologist must be appropriate.* It is financially impractical for a plant to refer large numbers of employees for otologic examination outside of the plant. The cost of lost time from work would be prohibitive. *All employees*

who have reactions from hearing protectors or whose hearing is getting progressively worse, even though they are using protectors properly, should be referred to their otologist. *Employees who have external otitis and chronic middle-ear infections* generally should be referred to their own otologist for definitive treatment. Certainly, *all potential medicolegal problems* as well as other special problems, such as accident cases and employees with unusual symptoms, should be evaluated by an otologist.

11. *Full cooperation of management and labor leaders* is essential in maintaining a successful hearing conservation program.

4. HEARING PROTECTORS

Hearing protectors generally are classified according to the manner in which they are worn. The three best-known types are inserts, muffs, and canal caps.

Personal protection devices are the keystone of a hearing conservation program. They are immediate, effective protection against occupational hearing loss. A variety of types of hearing protectors must be available to satisfy the needs of all employees, including insert type protectors, muffs, and canal caps. Various types of ear protectors as well as their relative values and merits are discussed in Chapter 17.

4.1. Education

Management and supervisory personnel first should be instructed in the importance of wearing hearing protectors and in the responsibility of enforcing their proper use. In fact, enforcement policies should be prepared in joint meetings with management and labor representatives. As part of the education program, all management and supervisory personnel are fitted with hearing protectors. These should be worn whenever they enter a noisy area, even if only passing through. Muffs are especially effective because of their high visibility.

Hourly employees should attend educational sessions to learn about hearing and the wearing of protection. These can be run by the medical and safety personnel. Excellent films, booklets, and posters are available. These sessions also should be used for explaining how the hearing protection program will function, discussing its enforcement and replacement policies, describing means of caring for and replacing protectors, and answering questions (there will be many) concerning the wearing of hearing protection. The hearing testing program also should be explained at this time.

Quick collapse of a well-planned program can be expected if there is no follow-up to initial efforts. This entails continuing educational programs; strict enforcement policies; frequent inspection by the medical and safety personnel; training complaints seriously and working individually with employees who have problems; following up on medical excuse; maintaining accurate records, and using audiometric results as a guide to the effectiveness of the program.

5. PHYSICIAN'S RESPONSIBILITY

In many instances the industrial physician is the individual responsible for supervising the hearing conservation program in a large plant. Usually, the industrial physician will require a consultant otologist trained in hearing conservation to express an opinion as to the cause of deafness and the desirability of hiring an applicant. The consulting otologist should also be able to testify as an expert witness in medicolegal situations.

Because most industries eventually will be doing hearing tests, otologists will be asked more often to evaluate hearing losses uncovered by industrial examinations. The otologist's evaluation must be comprehensive and the report to the referring industry complete. It should include at least the following information:

1. An otologic history of the hearing loss and associated symptoms
2. A complete otolargyngological examination
3. Air and bone conduction thresholds, speech reception thresholds, discrimination scores, recruitment and tone decay studies when indicated, and other special tests when necessary
4. A definitive comment on the validity and reliability of the hearing test results
5. A definitive diagnosis and prognosis if possible
6. Advice as to available therapy
7. Restriction in the employment of the individual

Naturally, the employee and management are interested in learning whether the hearing loss is curable, whether it is attributable to excessive noise exposure, or whether the diagnosis is more serious. A diagnosis of occupational deafness must be made on positive findings, not merely by exclusion. The history and audiologic findings are important, but so is an accurate knowledge of the noise dosage. The latter can be obtained only by direct measurements taken at the employee's job. Furthermore, the fact that an employee works at a so-called noisy job does not necessarily mean that the hearing loss is caused by that exposure. The impairment may be due to ototoxic drugs, viral cochleitis, or numerous other causes that produce audiometric patterns similar to those in noise-induced hearing loss.

The characteristic features in differential diagnosis of occupational hearing loss have been described in Chapter 15. Physicians must be very careful to base their diagnoses on accurate historical information, thorough evaluation, and specific information. Casual diagnosis of occupational hearing loss may result in delay of diagnosis of a more serious etiology. In addition, it may result in unjustified legal and economic hardship.

5.1. Some Basic Questions and Criteria in Considering Applicants with a Hearing Handicap

One of the most important decisions a physician may have to make in examining a worker on behalf of a prospective industrial employer is whether or not to recommend his employment if he has a hearing handicap. In general, it is inadvisable to adopt a blanket policy against hiring individuals with mild high-tone hearing loss. To do so would deprive industry of many skilled workers and create an unwarranted labor scarcity. Federal regulations concerning hiring the handicapped also should be considered. The decision not to hire should be made on an individual basis and after careful consideration has been given to the following questions:

1. Will the applicant's hearing be further damaged to the handicapping degree by exposure to the noise? If it will be, then he should not be hired for that job unless adequate hearing protection can be provided.
2. In his hearing loss now at the point at which a small degree of further loss will place him in the handicapped classification and make him a compensation

problem? If the answer is yes, the individual should be so advised and not hired for this own good and that of the industrial employer.

3. Is the employee so highly skilled that he is essential to the job under consideration, and is the risk of further hearing damage unavoidable because of his vocation, such as that of a drop-forge operator or a chipper? In such a case one must take a calculated risk, since any employee who has experience in this line of work almost certainly has some hearing loss. It would be only good sense to hire him, but he should be provided with the best possible protection for this work.
4. Is the hearing loss progressive in nature? No matter what we determine to be satisfactory criteria for hiring personnel with sensorineural hearing loss, we should not place people with progressive sensorineural hearing impairment in noisy environments.

The otologist must recognize that persons with nerve deafness have not been shown to be more sensitive to further noise damage in most cases. In fact, under certain conditions, there is some indication that they may even be less sensitive. If all applicants with sensorineural hearing loss were prevented from working, no experienced workers would be available in numerous noisy professions where hearing loss is routine. Such applicants should be hired if the otologist can assure them and the employer that an adequate and effective hearing conservation program exists, and that the employee's hearing will not sustain additional damage as a result of excess noise on the job. In addition, local, state, and federal legislation and regulation should protect the employer from liability for any hearing loss not caused within his company. Preemployment audiograms are essential, of course. Moreover, an applicant's hearing level must meet the demands of his job and not create a safety hazard because of impaired hearing or as a results of attenuated hearing with hearing protectors. In some cases, making appropriate judgement requires that the physician acquire familiarity with the conditions of the specific job. The otologist should not hesitate to consult with the occupational physician or safety person responsible for health and safety in the specific industry involved.

5.2. Preventing Hearing Loss is Essential

All otologists and industrial physicians must recognize the importance of preventing deafness in our industries. It is the physician's responsibility to assure the population's well-being by educating himself on the subtleties of the problem and by coordinating the expertise of other physicians, technicians, attorneys, legislators, labor leaders, and others interested in preserving the *quality of life* of the industrial worker.

6. SPECIAL PROBLEMS IN OCCUPATIONAL HEARING LOSS

Most of this book and all current legislation deal with traditional concepts and problems of occupational hearing loss. This is certainly appropriate since they account for the greatest number of cases. However, the need for consideration of the problems of certain occupational groups with special requirements has become increasingly apparent. More research is needed to provide answers to their very important questions. Underwater construction personnel and musicians are illustrative examples of unusual problems encountered in these frontiers of occupational hearing conservation. These are covered in Chapters 28 and 30.

32
Establishing a Hearing Conservation Program

Robert T. Sataloff
Drexel University College of Medicine, American Institute for Voice and Ear Research, and Graduate Hospital, Philadelphia, Pennsylvania, USA

Joseph Sataloff
Thomas Jefferson University, Philadelphia, Pennsylvania, USA

Careful planning and attention to practical aspects of establishing a hearing conservation program are essential for success. A well-planned program based on medical concern for employee health and rich in educational groundwork is likely to succeed. Less comprehensive dedication often meets with resistance, failure, and continued hearing loss within an industry. Successful programs have at least the following characteristics:

1. Are mandatory and have complete management support.
2. Provide more than one type of hearing protector with some free selection available.
3. Allow for trial periods with free exchange of protectors.
4. Allow a break-in time for employees with problems.
5. Supervisory personnel wear protection, preferably muffs, for high visibility of their compliance with the program.
6. Have the cooperation of labor leaders and representatives.
7. Have a person designated with the responsibility for running the program.
8. Education, promotion, and encouragement are constant ingredients.
9. Monitoring audiometry is kept up-to-date and the results are used for constant evaluation of the hearing conservation program.
10. Hearing protection is properly fitted ... and worn!
11. Expert consultants are readily available for analysis and management of problems.
12. *Everyone* is included in the program, not just workers exposed to noise.

1. EDUCATIONAL PROGRAM

Education of management and labor is critical to development and acceptance of a hearing conservation program. Films and literature are available to assure that the necessary information is imparted correctly. Much of the material covered in this book may be simplified in a short lecture or film and summarized in a manual or in-plant program guide that should be provided as part of a program. Such a manual should contain:

1. The company's hearing conservation program policy
2. A glossary
3. A review of anatomy of the ear and types of hearing loss
4. Basic information regarding noise measurement and control
5. Engineering control and hearing protection information
6. Audiometric testing
7. The otologic history and record keeping procedures
8. Audiogram evaluation procedure and data-processing procedures

9. Hearing test equipment
10. Sample forms
11. Legal requirements and information on hearing impairment
12. Reference information and specific information about the company's hearing consultant
13. Additional information (an appendix) with miscellaneous information such as sample audiograms, information on different types of forms and equipment problems, etc.

At the heart of the hearing conservation program's success is personal contact and guidance from experts with vast clinical experience and personal medical concern for the workers. Although this chapter reviews briefly some of their information that may be included in the mechanical part of the educational process, explanation by and attention from physicians and audiologists with vast experience are invaluable. Films, manuals, and books on occupational hearing loss are helpful adjuncts, but are no substitute for personal attention and flexible expertise.

2. HEARING CONSERVATION PROGRAM POLICY

Each industry must develop its own hearing conservation program policy. This will be tailored to fit the needs of the industry. Attention must be paid to noise levels, climate, physical work requirements, and laws of the jurisdiction under which the industry falls. The policy will include specific provisions for education, testing, compliance, and consequences of a worker's refusal to comply. The OSHA Hearing Conservation Amendment may serve as a good model for hearing conservation program policies, and it establishes minimum requirements. In addition to OSHA guidelines, preemployment criteria for audiogram screening often pose special concerns.

In view of the compensation aspects of occupational hearing loss, it is advisable for all plants to have some criteria for screening employees when they are being hired. In many states, it is illegal to refuse to hire because of a handicap such as hearing loss. It is possible and advisable, however, not to employ an individual to perform a job that requires acute hearing, such as a telephone operator or a secretary. In view of this and other aspects, the following flexible criteria are suggested:

Give a baseline audiometric examination to all new employees prior to their starting date. Whenever an individual is applying for any employment and the hearing test shows the following levels, care should be taken to secure additional documentation of the probable cause and effect of the abnormal hearing thresholds.

1. Hearing levels in the speech frequencies of 500, 1000, and 2000 exceed an average of 21 dB in both ears (ANSI 1969).
2. An individual's audiogram shows a 40-dB loss or more at all frequencies in one ear and normal hearing in the other.
3. An individual has a descending high-frequency curve of hearing equal to or exceeding 30, 40, and 50 dB at 3000, 4000, and 6000 Hz, respectively, and any level at 8000 Hz, in one or both ears.
4. The arithmetic total for the frequencies 500, 1000, 2000, 3000, and 4000 Hz in at least one ear exceeds 125.
5. An individual gives a history of having had mastoid or other surgery in either ear and has a hearing loss of over 25 dB in each of 500, 1000, and 2000 Hz.

It is suggested, when an applicant's hearing reflects any of the above, that he or she be instructed to visit an otologist of his choice, at his expense, as a condition of employment. The otologist's otologic and audiologic report should be returned to the employer and then forwarded, with his copy of the applicant's audiogram to the hearing consultant. Only the employer's knowledge of the job situation can determine whether the applicant should be hired. The hearing consultant will return comments and recommendations as to frequency of retest for such an applicant. It is important to advise the employee that a condition of his employment will always be his agreement to wear hearing protectors when so advised.

3. GLOSSARY

The language of medicine and hearing conservation is foreign to most industrial personnel. In addition to recommending the acquisition of a good text on occupational hearing loss for the person responsible for a hearing conservation program, providing a few definitions at the beginning is helpful. We have found the following useful:

1. *A-Weighted Sound Level* (L_A)—The intensity of a sound, as measured through the A-weighting network of a sound level meter. The A-filter ignores many low frequency sounds reported in dBA.
2. *Acoustic Nerve*—VIII cranial nerve made up of two divisions. The cochlear division conducts hearing signals from the internal ear (cochlea) to the brain. The vestibular division conducts balance signals.
3. *Acoustics*—Science of sound; branch of physics.
4. *Acoustic Trauma*—An injury to the ear caused by a sudden and intense acoustic stimulus that results in some degree of temporary or permanent hearing loss.
5. *Acuity*—Sensitivity of hearing.
6. *Air-Bone Gap*—The difference in decibels between the hearing threshold levels for air conduction and for bone conduction.
7. *Air Conduction*—The path by which sound travels through the external and middle ear to the inner ear.
8. *Ambient Noise*—The all-encompassing noise associated with a given environment, usually a composite of sounds from many sources near and far.
9. *American Academy of Otolaryngology-Head and Neck Surgery* (AAO-HNS)—The final authority on matters pertaining to the ear and hearing measurement procedures. Membership in this group is a good indication of a physician's dedication to his/her profession and his/her qualifications. Members must meet rigid academic and practical requirements before being certified.
10. *ANSI*—American National Standards Institute.
11. *ASHLA*—American Speech-Language Hearing Association, better known as ASHA.
12. *Attenuate*—To reduce in amount.
13. *Audiogram*—The written record of the results of a hearing test. Those used in industry must contain information required by state or federal agencies.
14. *Audiologist*—A professional, specializing in the study and rehabilitation of hearing, who is certified by the American Speech-Language-Hearing Association or licensed by a state board of examiners.
15. *Audiometer*—An instrument designed to test hearing acuity.
16. *Auditory Fatigue*—The temporary increase in the threshold of audibility resulting from a previous auditory stimulus.

17. *BEL*—A unit used in describing a logarithm scale to the base 10. By itself, the term is relatively meaningless without statement of the phenomenon being measured and the starting point of the scale.
18. *Bench Readout*—A record showing the actual performance of an audiometer. This record should be presented after each electronic calibration.
19. *Binaural*—Pertains to the use of two ears.
20. *Biological Calibration*—An audiometric calibration check done on individuals with stable thresholds.
21. *Bone Conduction*—The pathway through the bones of the head by which sound reaches the inner ear.
22. *Calibrate*—To check an audiometer for uniformity and standard of accuracy and to bring it into compliance with expected performance.
23. *CAOHC*—Council for Accreditation in Occupational Hearing Conservation.
24. *Central Hearing Loss*—Impairment of hearing that occurs when there is damage to the auditory pathways of the brain.
25. *Cerumen*—Wax found in the external auditory canal.
26. *Circumaural*—Completely covering the external ear (pinna), as an earmuff.
27. *Cochlea*—A spirally coiled, tapered bony tube of about $2\frac{3}{4}$ turns located within the internal ear. It contains the receptor organs essential to hearing.
28. *Complex Noise*—Sound that covers a broad range of the frequency spectrum. The type of noise found in most industries.
29. *Conductive Hearing Loss*—An impairment of hearing due to failure of a vibration to be transmitted to the inner ear.
30. *Continuous Noise*—Sound that has little variation in on-time.
31. *Criterion Sound Level*—A sound level of 90 decibels.
32. *Crossover* (Lateralization, Shadow Hearing, Cross-Hearing)—The phenomenon in which sounds presented to one ear are coupled around or through the head and are heard in the other ear.
33. *Cycle*—A series of events that recur regularly and usually lead back to the starting point.
34. *Deafness: Noise Induced*—Deafness due to sudden-impact noise (explosion).
35. *Decibel*—One tenth of a bel. Based on a logarithmic scale. Unit of measure of sound pressure.
36. *Diplacusis*—Distortion of pitch.
37. *Discrimination*—The ability to distinguish words with similar vowel sounds but different consonants.
38. *Dosimeter*—See Noise Dosimeter.
39. *Eardrum* (Tympanic Membrane)—Separates the outer ear from the middle ear.
40. *Environmental Protection Agency* (EPA)—One of three primary regulatory agencies responsible for overseeing the working conditions of employees in the railroad and maritime industries.
41. *External Acoustic Meatus*—The ear canal, about 40 mm in length.
42. *Federal Railroad Administration* (FRA)—One of three primary regulatory agencies responsible for overseeing the working conditions of employees in the railroad and maritime industries. In the regulation of noise exposure, FRA shares joint responsibility with OSHA.
43. *Frequency*—(cps, Hertz)—The number of cycles that occur per unit of time. Frequency is a physical phenomenon that corresponds to the perception of pitch.
44. *Hair Cells*—The sensory receptor for hearing. Located within the organ of Corti.

45. *Hearing Loss: Noise Induced*—Diminution of hearing due to prolonged habitual exposure to high intensity noise.
46. *Hearing Threshold Level* (HTL)—Softest hearing level at which a tone is heard in a specified number of trials.
47. *Hertz* (Hz)—The term used to measure frequency (formerly cycles per second).
48. *Impulse Noise*—Noise of a transient nature such as that due to impact or explosive bursts.
49. *Intensity*—Amount of sound energy generated by a sound source. Corresponds to the cycle acoustical lack of perception of loudness.
50. *Interrupted Noise*—Continuous noise that is periodic, such a from a machine that runs for a period of time and is off for a period of time.
51. *Localization*—Determination of the apparent direction of a sound.
52. *Malingering*—The willful misrepresentation of threshold (or other) responses during an audiometric check or other test.
53. *Masking*—Use of specific, calibrated sound in one ear to prevent cross-over in order to be certain that the other ear is being tested accurately.
54. *Mastoid Bone*—Part of the temporal bone of the skull located behind the ear. It is on this process that the tuning fork and audiometer bone oscillator may be used for diagnostic tests.
55. *Maximum Power Output* (MPO)—The maximum sound pressure available on an audiometer for any given frequency.
56. *Medical Pathology*—A disorder or disease. For purposes of this discussion, a condition or disease affecting the ear that should be evaluated and treated by a physician specialist.
57. *Meniere's Disease*—A disease characterized by sensorineural hearing loss, ear sounds (tinnitus), vertigo, fluctuation and fullness.
58. *Mixed Hearing Loss*—Impairment of hearing that is due to combined conductive and sensorineural hearing losses in the same ear.
59. *Monaural Hearing*—Hearing with one ear.
60. *Neurotologist*—An otologist who further subspecializes in disorders of the inner ear and ear-brain interface.
61. *Noise*—Unwanted sound.
62. *Noise Dose*—The ratio, expressed as a percentage, of (1) the time integral, over a stated time or event of the 0.6 power of the measured SLOW exponential time-averaged, squared A-weighted sound pressure and (2) the product of the criterion duration (8 hours) and the 0.6 power of the squared sound pressure corresponding to the criterion sound level (90 dBA).
63. *Noise Dosimeter*—An instrument that integrates a function of sound pressure over time in such a manner that it directly indicates a noise dose.
64. *Noise Reduction Coefficient* (NRC)—A single number used to express the overall noise-absorbing efficiency of acoustic materials.
65. *Noise Reduction Rating* (NRR)—The most convenient method by which to estimate the adequacy of hearing protectors attenuation.
66. *NIOSH*—National Institute of Occupational Safety and Health.
67. *Occupational Safety and Health Administration* (OSHA)—One of the three primary regulatory agencies responsible for overseeing the working conditions of employees in the railroad and maritime industries.
68. *Octave*—The interval between two sounds having a basic frequency ratio of 2:1, for example.

69. *Octave-Band Analyzer* (OBA)—A device that allows measurement of sound intensity at various specific frequencies.
70. *Organ of Corti*—An aggregation of nerve cells lying in the cochlea which pick up vibrations and transmit them to the brain, where they are interpreted as sound. The switchboard of the hearing mechanism serving as a transducer from physical energy to electrical energy.
71. *Ossicle*—One of three small bones located within the middle-ear cavity. They are the malleus, incus, and stapes.
72. *Otitis Media*—Inflammation, infection of the middle ear.
73. *Otolaryngologist*—A physician specializing in diagnosis and treatment of disorders of the ear, nose, and throat.
74. *Otologist*—A medical doctor who specializes in treating the ear.
75. *Otosclerosis*—New bone formation in the bony covering of the inner ear that may cause immobility of third bony ossicle (stapes), producing conductive hearing loss. It may also produce sensorineural hearing loss in some cases.
76. *Otoscope*—An instrument used to view the ear canal and tympanic membrane.
77. *Overall Noise*—Measurement of the total noise in decibels without frequency breakdown.
78. *Paracusis of Willis*—The ability to hear in a noise environment as well as or better than a normal-hearing person. Characteristic of a conductive hearing loss.
79. *Permanent Threshold Shift* (PTS)—A permanent increase in the threshold of audibility for an ear at a specified frequency above a previously established reference level.
80. *Potential Harmful Noise*—Noise of sufficient intensity and appropriate frequencies to pose the threat of damaging the ear. Noise generally exceeding 90 dBA.
81. *Presbycusis*—Natural loss of hearing sensitivity that results from the physiological changes that occur with age.
82. *Psychogenic Deafness*—Deafness originating in or produced by the mental reaction of the individual but not due to impairment of the organ of hearing.
83. *Pure Tone*—The simplest and purest sound, electronically produced (not existing in nature), described by a simple sinusoidal function.
84. *Recruitment*—An abnormal increase in loudness compared to the actual increase in intensity.
85. *Representative Exposure*—Measurement of an employee's noise-dose of an 8-hour time-weighed average sound level that the employers deem representative of the exposures of other employees in the workplace.
86. *Rinne Test*—A diagnostic bone conduction test with a tuning fork used to distinguish between conductive and sensorineural hearing losses.
87. *Sensitivity*—Biological variation in an individual's reaction to harmful noise.
88. *Sensorineural Hearing Loss* (SNHL)—The impairment of hearing that occurs when there is damage to the cochlea or the cochlear nerve.
89. *Sociocusis*—Nonwork noise-induced hearing loss.
90. *Sound Level*—Ten times the common logarithm of the ratio of the square of the measured A-weighted sound pressure to the square of the standard reference pressure of 20 micropascals. Unit: decibels (dB). For use with this regulation, SLOW time response, in accordance with ANSI SI. 4-1971 (R1976), is required.
91. *Sound-Level Meter*—An instrument that measures the intensity of sound.

92. *Sound Pressure*—Fluctuations in air pressure caused by a vibrating body.
93. *Temporary Threshold Shift* (TTS)—Hearing loss suffered as a result of a noise exposure, all or part of which is recovered in an arbitrary period of time away from noise (accounts for the necessity to recheck hearing acuity at least 16 h after last noise exposure).
94. *Threshold*—The lowest level of sound consistently heard by the patient.
95. *Time-Weighted Average Sound Level*—That sound level, which if constant over an 8-hour exposure, would result in the same noise dose as is measured.
96. *Tinnitus*—An otological condition in which sound is perceived without any external auditory stimulation. It may be a whistling, ringing, roaring, buzzing, etc.
97. *Transducer*—A device that changes one form of energy to another.
98. *Wave*—A complete cycle from positive to negative pressure.
99. *Wave-Band Analysis*—Measurements of the frequency components of sound.
100. *Weber Test*—A diagnostic bone conduction test with a tuning fork used to distinguish between conductive and sensorineural hearing losses.

The hearing conservation personnel in each plant should have readily available basic information on the anatomy of the ear and physiology of hearing. This should include at least a simplified version of the material covered in Chapters 2 and 3. In addition to a picture of the ear, this should include basic definitions of conductive and sensorineural hearing loss and a simple review of how we hear.

4. NOISE MEASUREMENT AND CONTROL

Location of work areas where there is a noise hazard and elimination of the hazards are important management responsibilities of the hearing conservation program. Evaluation of noise hazards requires a noise survey conducted by an expert. The effect of noise on conversation can be used as a general guide to help determine when noise analysis should be made. If a voice must be raised in order to be heard at a distance of 1 yd, an analysis should be ordered. If conversation is possible at 1 yd using a normal conversation level, there is probably no hazard. In case of doubt, measurements should be made. The person conducting the survey will submit a written report and designate those areas determined to be excessively noisy. Following such a survey, which will determine noise levels produced by existing equipment, each item of machinery or equipment that is either modified or added to various work areas will require a new noise level test if such changes might raise the level to an "excessive noise area." In the development of new equipment or processes, the problem of noise inherent in the equipment or process should be considered carefully. Whenever possible, noise controls should be built into the equipment or specified when equipment is ordered.

There are three primary means of controlling exposure to noise:

1. Engineering controls
2. Administrative controls
3. Personal protection

If reduction of noise and/or administrative controls are not practical, then personal hearing protection is necessary.

4.1. Engineering Controls and Hearing Protection

4.1.1. Engineering Controls

Those personnel working in the field of noise control will require some fundamental knowledge of acoustics along with a high degree of ingenuity and determination.

It is expected that many satisfactory solutions will be developed among the various plants to handle similar types of noise problems because of specific plant conditions. For these reasons, it would be most difficult to attempt to establish any standard solutions, as each situation must be tempered with the actual noise characteristics in each case.

1. When the noise levels in a working area exceed 90 dBA, engineering control measures must be investigated, evaluated, and, where feasible, utilized to reduce the worker's exposure. The NIOSH *Control Manual* and various association guides will prove useful in your engineering studies. Note: Use of hearing protection must be considered as a temporary control until the above studies are completed and controls implemented. If the time-weighed average exceeds limits, use of hearing protection must be continued.
2. Recognized control measures should be implemented to reduce overall noise levels, even though the worker's exposure continues to be greater than 90 dBA.
3. When the hearing conservation program is activated at your plant, it will be necessary for the engineering personnel to give attention to the following in reducing and eliminating noise wherever it is found:
   a. Plant planning (new or revised layouts)
   b. Substitution
      (1) Use of quieter equipment
      (2) Possible use of a quieter process
      (3) Use of quieter materials
   c. Modification of noise source
      (1) Reduce driving force on vibrating surface
      (2) Reduce response of vibrating surfaces
      (3) Reduce the area of vibrating surfaces
      (4) Use directionality of noise source
      (5) Reduce velocity of fluid flow
      (6) Reduce turbulence
   d. Modification of sound wave
      (1) Confine the sound wave
      (2) Absorb the sound wave
      (3) Use resonance phenomena (mufflers and resonators)
      (4) Use noise canceling technology
4. It is further expected that reasonable and proper specifications concerning noise levels will be applied in the purchase of new equipment.
5. As the engineer is deeply involved with equipment, its operating characteristics, and performance, the success of this program to a large degree will depend on this efforts and attitude in understanding and appreciating the merits and worth of a successful hearing conservation program.

4.1.2. Hearing Protection

Hearing protection has been discussed in detail elsewhere in this book. It is a mainstay of any hearing conservation program. Comprehensive information regarding different types of hearing protectors, ordering information for the beginning of a program, care of hearing protectors, and other related concerns should be included in the in-plant program guide.

Fitting Hearing Protectors. The results of the noise energy survey will serve as a guide for posting those departments having critical noise levels with proper signs. This information will also be used for determining the types of protectors suitable for each department. Usually one department is fitted at a time, starting with the one having the highest noise levels and largest number of exposed personnel. Approximately 15 minutes per person should be set aside for initial fitting. To save time in a large facility, the fitting team could go to the workers, using area office space as temporary headquarters. In fact, fitting "in the field" is sometimes preferable in that there is immediate realization and appreciation of the noise reduction. Time schedules should be worked out in advance so that department managers can plan for expected interruptions. The fitting team should carry or display only those protectors previously determined to be suitable for the department being fitted. Fishing tackle boxes are excellent for transporting necessary supplies.

The protectors should be attractively displayed with the name of each type printed on cards. Sizing tools or kits are available, or several sets of different sizes can be used for the fitting trials. Bowls of washing and rinsing solutions should be set up for sanitizing the devices after each trial. Use reusable speculums for the ear observations with an otoscope. Employees with infected, reddened, or impacted ear canals should be referred to a physician before fitting with inserts and should use muffs until cleared for fitting.

5. HEARING CONSERVATION EDUCATION

Hearing conservation should be promoted through employee orientation, job instruction, training, posters, exhibits of hearing protection devices, and safety talks by supervisors.

Signs should be posted at entrances to high-noise areas warning of the hazard and directing employees to wearing hearing protectors.

1. It is highly recommended that management, supervisors, and employees (in that order) be exposed to a comprehensive education program on the ear, its function, possibilities of damage, preventive measures, and corrective measures if damage exists or if unusual deterioration is detected during employment.
2. Whenever noise levels exceed the maximum advisable limits, any employee who may be exposed to such noise, either temporarily or in the regular course of his work, should be required to wear hearing protection devices.
3. In order to obtain the compulsory utilization of such devices, the program will require the very best levels of understanding of the problem and cooperation between employees, plant, division, and area management, central personnel, and the company's medical director. Therefore, the following education program is recommended to each of the company's plants.
    a. Schedule a special conference for plant management to explain applicable laws, regulations, company philosophy, and plans for noise control or hearing conservation. This conference should be led by a member of area or division management and/or central personnel. Consider the use of audiovisual materials and have someone available to answer questions. (Divisions initiating such a program might consider divisional meetings for this purpose.)
    b. Schedule a similar conference for first- and second-level supervisors with division or plant managers as discussion leaders. As part of the

educational program, all management and supervisory personnel should be fitted with hearing protection. These should be worn whenever they enter noisy area—even if only passing through. Muffs are especially effective because of their high visibility. Distribute available educational literature, a list of questions and answers, and announce plans.

c. Schedule a 45-min employee meeting (of all plant employees) with appropriate plant management in charge. Present management's philosophy of the program. Show one of the commercially available films or slide shows. Following the showing, outline the basic steps in the program, enforcement and replacement policies, and general descriptions of caring for and replacing protectors. Encourage employees to discuss questions with their supervisor, the nurse, hygiene or safety personnel. Distribute available educational literature. Announce the specific date of the program. The hearing test program should also be explained at this time. Question-and-answer sheets should be provided to supervisors a day or two before meetings. Have all attendees sign an attendance list to ascertain full employee compliance. These signed lists should be kept on file for possible review by state or federal inspectors.
d. Management should then be prepared to issue protective devices in areas with excessive noise, put signs up in designated noisy areas, and post-educational posters as available.

6. RECORD KEEPING

In addition to verbal instructions concerning the fitting and hygiene of the protectors, handouts with the same information should be given to each employee. A record, signed by the employee, should be kept of the type of hearing protector and size issued. Master records should also be kept for inventory and informational purposes. If fitting has been deferred for medical evaluation, record the diagnosis and prognosis of the physician and when it will be suitable for hearing protection to be fitted. Notes of "excuse" should be handled immediately by telephoning the physician inquiring as to when the patient can participate in the hearing conservation program. Additional record keeping entails the issuance of muff-type protectors and the signed acknowledgment by the employee that these are to be returned on termination of employment. Also, records should be kept of muffs loaned to the employee who may have left the protector at home. All infractions of rules governing use of the protectors as noted by safety or supervisory personnel should also be noted on the employee hearing testing record.

Generally, after the initial issuance, all replacements due to loss or damage of the original issuance should be at cost to the employee (moldable plugs excepted). All reissuances of size protectors should be preceded by a sizing procedure if several months have elapsed since the last fitting. All such changes should be logged.

7. ENFORCEMENT OF MANDATORY PROGRAM

1. The wearing of muffs or inserts may be resisted by some employees for several reasons. All will be important at the time the employee takes such a position. Since noise is seldom, if ever, painful and hearing loss is gradual, the employee may not understand the need for protection. All concerned should understand

that Federal Standards state, "Personal protective equipment shall be provided *and used*"—50-204.10(b). The same rule is contained in all state standards.

2. Education, therefore, is essential for a successful program. Knowledge helps to win understanding and acceptance. Take all possible steps to acquaint questioning employees with the facts. Yet, some may continue to rebel. If an employee fails to respond to local management's discussions, then an interview should be scheduled with higher level(s) of management and/or the corporate medical director.

   Should any deterioration in hearing threshold be detected, the employee should be given an opportunity to transfer to a less noisy area. If unacceptable, then termination should be discussed with higher levels of management.
3. Mandatory use of protectors should be a condition of future employment.

8. MANAGEMENT OF COMPLAINTS FROM THE WEARING OF HEARING PROTECTION

1. The outer canal should be checked by the nurse using an otoscope.
2. Cerumen (wax).
   a. If excessive soft wax is observed, tell the employee that a sudden feeling of becoming deaf will be due to the plug pressing against soft wax and forming a complete seal where before there was a small opening enabling sound to get through.
   b. If a hard plug of wax is seen, warn the employee that the ear may become painful due to the ear protector pressing on wax, causing it to press against the eardrum. The wax should be removed by a physician or trained physician extender.
3. Dermatitis/skin irritation of outer canals. Inspect the ears; if skin appears dry and scaling, ask the employee if "ears ever itch." The usual reply is "yes." Warn these employees that they have a dormant condition which may flare up into an eczema-type irritation. Tell them that if they have any trouble, they should see you early, rather than delay until the condition becomes severe.

   Counsel employees about using baby oil and about protecting skin from hair spray, which is very irritating and drying. They can do a lot to help themselves. If the condition does not respond in 2 or 3 days, refer to a physician since the condition may be fungal in origin and a medical doctor's care is necessary (ear specialist preferred).

   Any of these employees should be informed at the outset that they may develop problems. It should also be explained that you will try to help them, but if you are unable to do so, it will be necessary for them to seek treatment from a physician.

   It is much better that you call attention to potential trouble before it arises, rather than let it happen and then convince them that it is not due to the protectors, but to a preexisting condition.
4. Another problem you may encounter is red, irritated ear canals. This is usually due to the plugs being too small. Fit with a size larger.
5. Some employees with sensorineural deafness have ringing (tinnitus) in their ears. The protectors will make them more aware of the noise. This is a real problem and difficult to cope with it. Work with these employees individually and show empathy. Keep working with and encouraging these employees. Their problems are real and they do need your understanding.

If employees are encountered who simply refuse to wear protectors, whether the problem seems real or imaginary to you, try different types of protectors and counsel them.

Each time you counsel an employee or make any change or adjustment in type of protector, it should be noted on the employee's individual health record. This may seem insignificant, but could be important information at some later date. (See sample forms 7 and 9 for recording hearing protection data in Appendix V.)

9. FOLLOW-UP

Quick collapse of a well-planned program can be expected if there is no follow-up to initial efforts. This entails continuing education programs, strict enforcement policies, frequent inspection of the devices by medical and safety personnel, treating complaints as problems and working individually with employees with problems, following up on medical excuses, maintaining accurate records, and using audiometric results as a guide to the effectiveness of the program.

10. AUDIOMETRIC TESTING

The educational program and written manual should include basic education regarding audiometers, interpretation of audiograms, the importance of soundproof audiometric booths, audiometer calibration, and other issues of concern, as discussed in Chapters 6 and 7.

10.1. Calibration of Audiometers

10.1.1. Functional–Biological

The importance of calibration of audiometers in the industrial setting cannot be overstressed. The audiogram is a medical and legal record. Its validity is dependent on the use of an accurately calibrated audiometer.

The well-cared-for normal human ear is a very stable instrument in that it does not experience significant fluctuations in day-to-day sensitivity. This factor of biological stability can be used to advantage in checking the stability of the audiometer, an electrical instrument, in which there may be instability. The use of the human ear to check the output of the audiometer is called "biological calibration."

It is required that a biological calibration of the audiometer be conducted each day the audiometer is in use. It is important to remember that if the audiometer is found to be out of calibration, the last valid audiograms are those that were done up to the time of the *previous* calibration check. All audiograms done *from* the time of the previous calibration check to the detection of the abnormality should be repeated.

The good operator is an observant one and may detect a sudden change in the performance o the audiometer during routine testing procedures. For example, it may be noted that three or four successive audiograms all show losses of the same degree in one ear, or that on retests, previously normal ears now show losses that cannot be accounted for. These apparent changes or unlikely findings may be due to a change in the audiometer rather than in the hearing of the subjects. If this situation develops, immediately conduct a biological calibration check on several of your biological subjects. If these checks prove that the audiometer is faulty, stop testing and have the audiometer repaired.

All biological calibration subjects should have stable hearing and should not have noisy occupations at the plant. Office personnel usually are ideal subjects. It is recommended that at least three people who are reliably available be used. Keep a separate biological calibration sheet on each subject. As the tests accumulate on each sheet for each subject, the operator has a previous baseline threshold record against which to compare the current test results. A normal variation of ± 5 dB from the baseline results at any frequency is to be expected. If the difference is as much as 10 dB at any frequency in either ear, call in one or more of your other biological subjects to see whether that change persists in these subjects also. If it does, the audiometer should have a periodic calibration (to be discussed later). If the change does not persist in the second and/or third subject, consider the change to be in the first subject, not in the audiometer.

In addition to recording the thresholds, also record the date and the time of the calibration check, the serial number and reference standard of the audiometer, and the signature and printed spelling of the tester. Maintain all these records in a separate calibration folder. Recording the time of the test is important, especially if a change of 10 dB or more is noted. Subsequent tests on other biological subjects will show, according to the time entered, that immediate follow-up tests did, or did not, substantiate the detected "change."

Each tester should know his or her own thresholds and in an emergency do a "self-audiogram." This is more of an audiometer check rather than a true biological calibration. It can be used as a preliminary procedure if difficulty with the audiometer is suspected (as previously described), or if biological checks must be performed routinely. (See sample form 12 in Appendix V.)

10.2. Self-Listening Test of Audiometer Function

There are other checks of the audiometer that must be conducted and recorded. These checks can be done by the operator on his/her own ears. These "self-listening" tests include checking:

1. That all dials and switches have no free play.
2. That all plugs and jacks are in their proper sockets and *completely seated.*
3. That earphone cords are not developing breaks: with the 1000-Hz tone on at 60 dB in the right ear, bend and flex the earphone cord to check for breaks or static in the steady tone. Repeat for the left earphone. Broken cords can be replaced by the plant electrician.
4. That there *are* intensity (volume) changes of tone as the attenuator is turned up *and* down: check in only one earphone at 1000 Hz. Move the hearing level dial up in 5-dB steps over the full range of the dial and back down in 5-dB steps.
5. That the tones sound clear and properly pitched: it is difficult to detect small changes in the various frequencies (tones) of the audiometer. Some medical departments have a full range of tuning forks that can be used to check the audiometer frequency. The operator places on earphone on one ear, the opposite one on the temple to keep that ear open. He or she presents a 500-Hz tone to the covered ear and then gently strikes the C_2 fork and holds it to the uncovered ear and determines whether the two are equal in tone (not loudness). Similar procedures are carried out with the other forks and frequency settings of the audiometer.
6. That the tones cone on immediately and go off immediately with pressing and releasing of the tone presenter: there should not be a slow rise to loudness and a gradual decline of intensity as the presenter is pressed and released.

7. That there is no static in the earphones along with the tone or any clicking in the earphones as the presenter is activated: sometimes rapid rotation back and forth of the hearing level dial will polish the contacts and remove dust particles which may be causing the static.
8. That the tone presenter does not make an audible mechanical click when it is activated: this kind of difficulty usually requires replacement of the tone presenter switch.
9. That the earphone seals are not cracked and misshaped: replace with new cushions of the same size and with the same size opening over the earpiece receiver.
10. That there is a no cross-talk between earphones: place the earphone selector switch on "right ear." Unplug the right earphone jackplug. Listen in the left earphone as you sweep through each frequency at 60 dB. There should be *no tone* in the left earphone. If there is, you have cross-talk which must be corrected at the factory or local repair agency. Repeat the procedure with the opposite earphone.

Note that in terms 3 and 9 above, replacement of some components can be done by plant personnel without fear of decalibrating the audiometer, and that in 1, 3, and 7, some minor tightening and cleansing is permissible at the plant. Any other repairs should be done by qualified personnel only.

If the earphones have been damaged and need replacement, a complete electronic recalibration must be obtained. The audiometer cannot be calibrated without its earphones. Do not interchange earphones with other audiometers without having an electronic recalibration.

All of the above listening tests should be conducted with each biological check and a separate record kept of these checks. Sample form 13 (Appendix V) shows a suggested method of recording these procedures.

Following are a few further suggestions about the proper care and handling of your audiometer:

1. Don't slap the earphones together (cushion to cushion) or place the earphones on a flat surface with the cushions down. There is a strong possibility that a buildup of pressure against the earphone diaphragms will cause a rupturing of the diaphragms.
2. Don't plug the right earphone into the left earphone circuit of the audiometer (or vice versa). The right and left audiometer circuits are calibrated individually to their respective earphones.
3. Don't subject your audiometer to temperature extremes or sudden changes in temperature (such as taking it from a warm building and placing it in the trunk of a car in winter).
4. Keep the audiometer free from condensation as much as possible.
5. Keep the audiometer free from dust and other airborne contaminants as much as possible.
6. Unless battery operated, leave the audiometer on. That is, don't turn the power off after each test.
7. If you have several audiometers in use at your facility, identify the headphone sets with the serial number of the audiometer to which each belongs.
8. Maintain the proper tension on the earphone headband. The earphone cushions should make light contact with each other when the headband is held at top center.

10.3. Acoustic Calibration

Federal regulations state that the audiometer shall be subjected to an annual (periodic) calibration check or to a periodic calibration when a biological check uncovers threshold changes greater than ±5 dB at any frequency; distorted signals; attenuator or tone presenter transients; or other severe operating difficulties. We suggest, however, that a careful check be made (as previously outlined in this chapter) to ascertain the permanence of the problem before the audiometer is shipped out for repairs. Notify the calibration agency of the specific problem you are having with the audiometer. Delineate between problems and required routine calibration checks. Ship the audiometer in its original carton if possible.

The acoustic calibration conducted by the repair agency should include the following:

1. Set audiometer to 70-dB hearing threshold level and measure sound-pressure levels of test tones using an NBS-9A-type coupler, for both earphones and at all test frequencies.
2. At 1000 Hz for both earphones, measure the earphone decibel levels of the audiometer for 10-dB settings in the range 10–70 dB hearing threshold level. This measurement may be made acoustically with a 9A coupler or electrically at the earphone terminals.
3. Measure the test tone frequencies with the audiometer set at 70-dB hearing threshold level, for one earphone only.
4. In making the measurements in 1–3 above, the accuracy of the calibrating equipment should be sufficient to prove that the audiometer is within the tolerances permitted by ANSI S3.6-1989.
5. A careful listening test, more extensive than required in the biological calibration, should be made in order to ensure that the audiometer displays no evidence of distortion, unwanted sound, or other technical problems.
6. General function of the audiometer should be checked, particularly in the case of a self-recording audiometer.
7. All observed deviations from required performance should be corrected.

When work is completed, you should obtain from the repair agency a certificate of calibration and a copy of the actual output levels developed by the audiometer as it was subjected to the various tests. (The latter is sometimes called a "bench certificate" or a "readout.")

10.4. Exhaustive Calibration

Once every 2 years an exhaustive-calibration check of the audiometer should be performed. This is a deeper check of the audiometer function and must be in compliance with ANSI S3.6-1989 Specifications for Audiometers.

As was suggested in the previous section, a certificate of calibration and a listing of actual audiometer outputs should be obtained for the plant records upon completion of this work. Keep all certificates and calibration data with the biological calibration logbook.

New audiometers or those returned from repair or calibration service should immediately be subjected to a biological calibration and a listing check before being placed into service.

If a replacement audiometer is to be used while the plant audiometer is being serviced, it is necessary to conduct biological and listening checks before the equipment is placed into service. Record all information on temporary equipment as is done with permanent equipment. Separate calibration records should be maintained on temporary equipment.

10.5. Summary

1. Functional calibration is to be conducted each day the audiometer is in use.
2. The use of electronic calibrators is acceptable for biological calibrations but true biological checks should be conducted on a planned schedule.
3. Permanent records should be kept of all calibration data.
4. Acoustic and exhaustive calibration checks should be conducted on a regular basis or when the audiometer is found to be malfunctioning or out of calibration.

10.6. What to Tell the Subject Regarding the Hearing Check

The technician is not qualified to diagnose hearing loss or its cause. That is, of course, the responsibility of the otologist. Therefore, it is often difficult for the technician to know what to tell the subject following the hearing check.

As a general rule, unless the tester is an otologist, it is best not to give any information to the subject. However, the technician will find that many people are very concerned or curious about their hearing when they leave the test booth. If the subject asks the tester about this, the tester cannot refuse information, but can be tactful in answering so as not to instill apprehension or exaggerated concern in the subject's mind.

Most subjects will be satisfied, and some would actually prefer to wait and let the physician tell them about the results of the hearing check. However, some are quite persistent and want immediate answers. Remember not to be at all devious with these people. Remember that if the subject has a significant degree of hearing loss, the technician should simply reiterate what is actually already known. That is, if hearing is normal or within normal limits for speech purposes, tell the subject that the hearing is good. If the subject has a known severe hearing loss explain that you do not know what caused it, but that the hearing will not get worse due to noise with the use of hearing protection.

Remember, do not be pressured into making a statement you are not qualified to back up in a court of law.

11. SCHEDULING HEARING TESTS

11.1. New Applicants

Preemployment hearing tests should be conducted on *all* applicants for employment. This baseline test will be the determining factor in hiring decisions. That is, if the hearing test reveals a unilateral hearing loss, for example, and upon otologic examination an active otitis is diagnosed, the decision to hire the applicant may be delayed until the condition is corrected. A diagnosis and prognosis should be required on all moderate to severe, unilateral and bilateral hearing losses before the applicant is hired.

An applicant with skill and experience in a trade will, in all likelihood, have a hearing loss if high noise levels are associated with that trade. In time, this may not be a necessary relationship, but at present one can expect that some degree of loss is to be found in the new employee with past noise exposure. The skilled workers should not be turned away because of hearing loss, but care should be exercised to determine whether the hearing loss will be detrimental to personal safety and that of fellow workers.

When an applicant with thresholds greater than 25 dB at any frequency is hired, two or three threshold readings should be taken to serve as a basis for permanent records. Some companies refer applicants (at their own expense) to an outside source to establish an independent baseline audiogram and diagnosis.

In view of the possibility that the applicant may have an ear or hearing condition that will require some attention, it is a much better policy to conduct preemployment, and *not* merely preplacement audiograms. In the latter method, the applicant is hired and then tested prior to assignment to a critical noise area. In many instances the test is never done because the medical department is not notified of the transfer, and the company possibly has incurred an unnecessary medical responsibility.

If large numbers of workers are hired at one time, it may be difficult to obtain preemployment audiograms on all of them. However, thresholds should be obtained within 60 days on all those assigned to critical noise areas. All preemployment and preplacement audiograms should be sent to your consultant for evaluation and recommendations.

11.2. Current Employees

All employees working in critical noise areas should have a "baseline audiogram," a hearing test that is preceded by a period of at least 14 h of quiet. This should provide "rested ears" for this reference test. If scheduling does not allow for this, a reference test is permitted after the employee has been on the job provided hearing protection was worn prior to the hearing check.

Schedule several employees (working in 85 dBA or above) for hearing tests at the start of the shift. This can be done by department or by several departments. The latter suggestion is preferable because it doesn't severely handicap one department supervisor by having several workers missing at the start of the shift. Scheduling alphabetically is another possibility. The chance of pulling more than one worker from a single department is remote when using this method.

Perform the test as quickly as possible without sacrificing accuracy of the results or good relations with the employees. On the other hand, the hearing testing program should not be detrimental to the productivity of the plant. Supervisory cooperation is lessened when production is slowed.

Employees working in areas below critical noise levels can be tested at any time as scheduling allows. As the hearing conservation program is a medial program, *all* employees will have hearing tests.

Your consultant will advise you on the frequency of which repeat hearing tests will be conducted. In general, those in critical noise areas will be retested annually, and those not in critical noise will be retested every 4 years.

Repeat (annual) tests can be performed "from the job." The results of these tests are compared with the initial baseline audiograms, and the efficiency of the hearing protectors, among other things, is determined. All results of repeat tests should be sent to your consultant for further evaluation and recommendation.

If possible, all terminated employees should also have audiometric examinations, especially those with a history of noise exposure, loss of hearing, or who are retiring for medical reasons.

Retesting of all permanent and temporary employees should be scheduled. Annual hearing tests should be conducted on those working in designated noisy areas. Those transferring in and out of these areas should be retested within 60 days of assignment to such employment.

Employees returning from a long sick leave, or after sustaining a head injury, should be retested as soon as possible.

Test as many employees as possible who come to the medical department for special examination or treatment.

All audiometric examinations should be made in the manner described in the technician's audiometry manual. Sufficient time must be scheduled so that the examination can be proceed without waste or unnecessary interruption. Additional testing should be performed according to the suggestions of the consultant.

12. THE OTOLOGIC HISTORY

Obtaining a complete otologic history at the time of the initial audiometric examination of current and prospective employees is *essential.* A short questionnaire expedites this procedure. At minimum, the following information must be obtained. (See sample form 15 in Appendix V.)

12.1. Record-Keeping Requirements

Accurate records are vital to an audiometric program. Since these records reflect a profile of the employee's hearing acuity during his employment, they are important as legal documents and may be subpoenaed and examined by the courts and lawyers involved. Because records are written evidence, they can be used to support the claims of the employee, employer, or health personnel.

Records should be:

1. Designed to suit the needs of the company: Preplanning? On computer? Serial reading?
2. Kept simple and uniform—no erasures!
3. Reviewed periodically to get the full value of a hearing conservation program. Audiograms that are done per schedule, but not reviewed, defeat the purpose of the program!

The occupational health nurse or technician should keep the following records vital to the audiometric program:

1. Annual *sound level readings* of noise exposure at all work stations must be obtained and records retained indefinitely. As the work area changes, new sound level readings must be taken.
2. When daily noise exposure time exceeds the permissible limits and/or hearing protection is used to control the employee exposure, records of the specific *control methods* used must be maintained.
3. *Audiograms* must be permanently maintained. (See sample form 11 in Appendix V.) Each audiogram should include:
   a. Employee Social Security number
   b. Employee name
   c. Employee clock number
   d. Employee sex
   e. Employee date of birth
   f. Employee service date
   g. Employee test number

   h. The date and time of test
   i. Test frequency readings for all frequencies recorded on serial form (sample form 11, Appendix V)
   j. Employee department number
   k. Employee shift number
   l. Type of audiogram (i.e., baseline, recheck, or other)
   m. Employee job number
   n. Years on present job
   o. Noise level at job
   p. Last exposure to high-level noise
   q. If hearing protection is used, what type?
   r. The examiner's name, certification number and/or date of certification
   s. The examiner's signature
   t. Significant otologic medical history and otoscopic findings of the employee (see sample form 11 in Appendix V)
   u. Signature of employee (optional)
   v. Audiometer serial number
4. Other logs to maintain are:
   a. Sample form 14, Appendix V (optional)
   b. Otologic and otoscopic history update (sample form 15, Appendix V (mandatory))
5. If the state requires registration of technicians, a copy of the certificate of registration must be available at the test area.
6. A record of date, time, and place of all audiometric calibrations must be filed at the test area. *Calibration records must be permanently maintained.*
   a. Acoustical electronic calibration should be done at intervals of not more than 12 months, and more frequently if malfunctions occur.
   b. Biological calibration should be done each day the audiometer is in use. At least two, but preferably three, subjects with stable thresholds are needed. Only one subject need be tested per calibration check unless the threshold at any frequency varies by more than 5 dB for the subject being tested. In that case, the other biological subjects should be immediately checked to determine whether the "change" is in the audiometer or in the first subject tested. If the change appears in all the subjects, the audiometer will need to be recalibrated. (See sample form 12 in Appendix V.)
   c. Self-listening checks of audiometer function should be done in conjunction with biological calibration checks. (See sample form 13 in Appendix V.)
7. The sound level inside the test area must be measured and recorded a minimum of every 12 months, or sooner if significant changes affecting noise generation occur in the surrounding environment. Records must show:
   a. Date
   b. Model and type of sound-level meter used and serial number
   c. Name of person measuring noise levels
   d. Readings, records retained indefinitely
8. Plant sound survey. Records retained indefinitely.
9. Log of all persons who have been referred to otologists (along with physicians' reports, if any).

10. File of reports and information from hearing conservation consultant.
11. OSHA No. 200. Bureau of Labor Statistics Log and Summary of Occupational Injuries and Illnesses.

13. AUDIOGRAM EVALUATION PROCEDURES AND DATA PROCESSING PROCEDURES

Evaluating audiograms, particularly large numbers of audiograms, is complex and of critical importance. Methods developed by the Hearing Conservation and Noise Control Corporation, Bala-Cynwyd, Pa., are state-of-the-art. This company was the oldest and largest hearing conservation corporation in the world. Its audiogram evaluation procedures and data-processing procedures are reprinted for the first time in Appendix VI.

14. HEARING TEST EQUIPMENT

Hearing test equipment (hearing test booth and audiometer) will be required at each location maintaining an in-house hearing conservation program. The make and model number of equipment recommended will depend on the noise levels and physical characteristics of the area in which it is to be placed. Plant personnel should select alternate areas for equipment placement prior to an on-site visit by the hearing consultant. Generally, the space required for the test booth is 8 ft × 8 ft, with no less than 8 in. of overhead clearance. The area selected should be readily accessible to medical and production personnel and have as few extraneous distractions as possible. Ideally, this area should be used exclusively for hearing conservation. Consultation should be made with the hearing conservation expert to be sure that the equipment is placed in a scientifically acceptable position avoiding extraneous noise and vibration.

15. MOBILE SERVICES

As corporations continue to cutback on expenses, medical departments are usually among the first to feel the effects of these cutbacks in the form of staff downsizing. Often hearing testing starts to lag or in some cases falls seriously behind schedule. One solution is to hire a mobile testing service to bring the program back on schedule or to take it over completely.

Finding a quality is sometimes difficult. The following factors should be considered in the selection process:

1. How long has the service been in operation? Obviously, those with years of service should be considered first.
2. How many test units do they have and how many employees can be tested at one sitting?
3. What are the set-up fees and charges per audiogram?
4. Is there an extra charge for audiogram evaluation? How quickly are reports made and available?
5. Are “instant” reports available?
6. What are the qualifications of the on-board technicians?

7. Are employee educational programs also provided by the van personnel?
8. Will they provide audiometer calibration and test room readings for each day the van is on site? (They must meet the same requirements for audiometer calibration and background noise as is required for in-plant testing.)
9. Are local testing facilities available for conducting tests on new hires, returns from sick leave or lay-off at times when the test van is not on site? (These testing facilities also must meet state or federal requirements.)
10. Consider keeping the booth and audiometer for situations outlined in 9 above.

16. SAMPLE FORMS

Forms used for record keeping can be extremely helpful in simplifying the collection of accurate information and recognition of important changes. Sample forms developed over 40 years by the Hearing Conservation and Noise Control, Inc., are reprinted in Appendix V.

17. LEGAL REQUIREMENTS

The medical department plays an instrumental role in compliance with federal and state standards and legislation. Guidelines set forth in the Hearing Conservation Amendment (Chapter 34) should be followed. Cooperation among the medical and management teams is essential.

18. HEARING CONSULTATION AND REFERENCE MATERIALS

Additional information supplied at the beginning of a hearing conservation program should include comprehensive information on the services provided by the hearing consultant and ways to reach the consultant staff easily when problems arise. In addition, references to occupational hearing loss literature should be made available. Sample audiograms and examples of practical problems and solutions are also helpful.

19. SUMMARY

Establishment of an effective hearing conservation program requires substantial expertise from a hearing conservation consultant. Programs are designed to protect hearing, as well as to comply with the law. To be successful, the program should be especially adapted for each industrial situation, should include the entire employee population, should have an active educational component and should be well organized and efficient. This will minimize "downtime" for individual employees and maximize the reliability and validity of the data gathered. When a hearing conservation program is properly developed and presented, employee resistance is extremely rare.

33

Occupational Hearing Loss: Legislation and Compensation

Robert T. Sataloff
Drexel University College of Medicine, American Institute for Voice and Ear Research, and Graduate Hospital, Philadelphia, Pennsylvania, USA

Long-term exposure to high levels of industrial noise has caused hearing loss in millions of employees throughout the world. In the United States, the federal government showed its serious concern by including a noise standard in the Occupational Safety and Health Act of 1972. The standard makes it necessary for industries to reduce noise by every feasible means where employees are exposed to 90 dBA or more for an 8-hour workday. If the noise cannot be reduced adequately, a hearing conservation program has to be established. Criteria are discussed in Chapters 25 and 26. Although the original standard did not include complete guidelines for hearing conservation or a clear-cut definition of "feasible noise controls," every noisy plant was encouraged to comply with its objective of abolishing occupational hearing loss. Subsequent legislation clarified the intent and requirements of the law (1983). The Secretary of Labor's Noise Proposal Committee in 1973 recommended practical guidelines for industries to follow in developing effective hearing conservation programs. Some of these were:

1. Posting of noise hazard areas
2. Educating management and labor about hearing conservation
3. Performing noise measurements whenever indicated

4. Performing preemployment and routine audiograms
5. Using hearing conservationists who are properly trained or certified
6. Properly calibrating all equipment, including noise measurement devices and audiometers
7. Describing standard methods of performing hearing tests, recording techniques, using quiet test rooms, and maintaining records
8. Informing management and employees about their hearing levels
9. Referring employees or their audiometric records and other information for otologic consultation
10. Providing hearing protectors that can be used effectively and monitoring with routine audiometry
11. Interpreting all audiograms and obtaining diagnoses for those with abnormal hearing levels
12. Obtaining good otologic noise exposure histories on all employees

If noisy industries provide all these services properly, they will not only comply with present and future OSHA requirements, but they also will be assured that their employees will not develop occupational hearing loss.

In 1979 the Department of Labor regulated the elements of a hearing conservation program which included the above guidelines and directed OSHA inspectors to enforce industrial compliance.

1. ECONOMIC IMPACT

The primary thrust of OSHA is for all industries to reduce noise by engineering means. This approach is excellent, but it presents serious shortcomings if the concepts are applied too rigidly. It is common knowledge that technical capability for reducing noise in many of our present manufacturing processes, machinery, and even newly built machinery simply does not exist. Furthermore, the cost of such technical modifications is estimated in billions of dollars and would take many years to complete.

In order to eradicate occupational deafness as rapidly as possible, hearing conservation programs must be adopted immediately by all noisy plants regardless of OSHA regulations, fear of inspections, or lack of any compensation claims for occupational hearing loss. It is untenable to permit employees in noisy industries to sustain hearing loss that can be prevented by available means. Concurrently, noise reduction programs must be implemented until safe hearing levels are attained.

2. IMPAIRMENT DISABILITY AND HANDICAP

All professionals involved in occupational hearing loss should understand the terms "impairment," "disability," and "handicap." It is also important to recognize that definitions and their applications have been changing. Most widely accepted definitions (and changes) have been promulgated by the World Health Organization. Traditionally, impairment is a bodily dysfunction (such as hearing loss). Disability refers to the consequences of that impairment to most people in the population, such as the expected difficulties understanding speech or functioning in certain jobs. Handicap refers to the disadvantage experienced by an individual with regard to his or her specific normal functions and activities. Handicap (the individualized, personal effect of an impairment) is generally not taken into account in compensation law. Rather, the law addresses disability, the expected adverse effects on a population with specified impairment. However, recently, the World Health Organization has advocated a somewhat more sophisticated analysis of health conditions (diseases, disorders, and injuries) and contextual factors (1,2). Current concepts include considerations such as "performance qualifiers" that describe what an individual does in his or her current environment, and "capacity qualifiers" that describe an individual's ability to complete a task or an action. These concepts have not been incorporated into occupational hearing compensation law, yet. However, they may be in the future.

3. COMPENSATION FOR OCCUPATIONAL DEAFNESS

The emphasis of the OSHA noise regulation is on *prevention* of occupational deafness. Compensation for occupational hearing loss is quite another matter. This provides financial remuneration when adequate preventive measures have not been taken.

The noise levels specified in OSHA actually are guidelines similar to a maximum speed limit for driving. If the stated level is exceeded, it does not necessarily mean that occupational deafness will result, any more than exceeding the speed limit always lead to an accident. Slightly exceeding the noise dose occasionally for short periods of time is not really a concern. Furthermore, the noise limits are based on large statistical studies and cannot be applied to individual cases for the purposes of paying workers' compensation. However, the maximum noise level designated in OSHA should not be exceeded for long periods of time, because hearing damage is likely to occur, and claims for compensation may be justified.

4. ABUSES OF NOISE STANDARD

In all likelihood, the 90-dBA and other specified limits may be applied improperly. Individuals unfamiliar with noise measurement may omit "A" after dB and merely stipulate 90 dB, an inaccurate application. In compensation cases for occupational hearing losses, the abuses may be more serious, especially if an individual's deafness is attributed to noise exposure solely because his noise dose exceeded limits for short periods of time. The standard is based on the average effect on many workers or exposure to continuous steady noise for 8 h daily for a working lifetime. Information relating hearing loss to impact or intermittent noise for a lifetime of exposure still is inconclusive. The application of this knowledge to the diagnosis of an individual's hearing loss requires sophisticated judgment. Testimony by expert otologists and acousticians still is the best means of deciding the diagnosis in individual claims.

5. LEGISLATION FOR WORKERS' COMPENSATION

Before 1948, gradual partial hearing loss caused by individual noise was not included in state workers' compensation laws. When deafness was included, it applied chiefly to explosions and injuries, and the text specified as a condition for compensation that deafness must be total in one or both ears. (Neither of these conditions occurs in occupational deafness as presently defined.) Little or no mention was made of partial hearing loss caused by long exposure to occupational noise; lawmakers were unaware of the far-reaching consequences of hearing loss, as it did not seem to cause loss of wages or earning power.

Large numbers of claims for hearing loss did not occur until after judicial interpretations in New York and Wisconsin established the principle of payment of compensation for partial loss of hearing without loss of earnings—the first instances in the field of compensation where payment was made without direct economic loss.

6. CONCEPTS IN WORKER'S COMPENSATION

The original basic objective of workmen's compensation was to provide payment for loss of earnings and for medical costs of injury related to employment. Under this concept, the employer gave up his common-law defenses of assumption of risk, contributory negligence, and negligence of a fellow employee. In return, the employee gave up his right to sue for whatever he could collect. Specific limits were placed on the amounts of liability. In addition to payments for loss earnings and medical costs, specific awards were established for accidental dismemberment; fingers, arms, legs, and so on.

After the enactment of legislation for these purposes, provision was made for payment of compensation for occupational diseases. Under the occupational disease provisions, there were no scheduled awards for loss of function only. Claimants had to establish a date of injury and sustain a loss of earnings in order to be compensated. The approach to the problem of compensation for partial hearing loss is based on the assumption that partial deafness which develops over a period of time as a result of exposure to noise is an occupational disease. Deafness caused by a single incident such as an explosion generally is considered an accident.

There are exceptions to the applications of this concept, however. In the state of Georgia in 1962, the court ruled that an employee's hearing loss resulted from the cumulative effect of a succession of injuries caused by each daily noise insult on the ear. On the basis of this reasoning, the court felt that the hearing loss was compensable under the "accidental injury" provisions of the law: the notion of "gradual injury" applied to hearing loss. It is almost a unique application in the United States.

Beginning in New York in 1948 and extending into Wisconsin in 1951 and Missouri in 1959, a series of legal patterns was established that made partial hearing loss caused by occupational noise exposure compensable even though no wages were lost. Since then, most states as well as the federal government have made such loss compensable by passing specific laws defining partial hearing loss as an occupational disease.

In some states and in the railroad system, compensation for partial hearing loss can be obtained only by recourse to litigation in court. Juries have, at times, awarded large sums of money to workers suing employers in this manner.

The influx of claims created a major problem for industry and insurance carriers, because prior to this period occupational deafness was not considered compensable. Self-insured industry had accumulated no reserves, and insurance companies had collected no premiums from which to make payments (accrued liability). This major economic problem subsequently was resolved by administrative rulings and compensation regulations that stabilized the manner in which claims could be filed and the way hearing impairment should be calculated.

A good example of a law covering workers' compensation for hearing loss was passed by the state of North Carolina (copies can be obtained by writing to: Chairman, Worker's Compensation Commission, State of North Carolina, Raleigh, NC). One section of this chapter is a proposal for a model legislation. This proposal simplifies the method of compensating for hearing loss and emphasizes greater payment to employees with substantial hearing losses, but it reduces the payment for those cases in which there is little or no real hearing impairment.

Important concepts and their justification in the model legislation are as follows:

1. As used in this Act:
   a. "Noise-induced occupational hearing loss" means permanent bilateral loss of hearing acuity of the sensorineural type due to prolonged, habitual exposure to hazardous noise in employment. For purposes of this supplementary act, sudden hearing loss resulting from a single, short noise exposure, such as an explosion, shall not be considered an occupational disease but shall be considered as an injury by accident.
   b. "Sensorineural hearing loss" means a loss of hearing acuity due to damage to the inner ear, which can result from numerous causes, as distinguished from conductive hearing loss, which results from disease or injury involving the middle ear or outer ear or both and which is not caused by prolonged exposure to noise.
   c. "Prolonged exposure" means exposure to hazardous noise in employment for a period of at least 1 year.
   d. "Habitual exposure" means exposure to noise exceeding the allowable daily dose, at least 3 days each week for at least 40 weeks each year.
   e. "Hazardous noise" means noise that exceeds the permissible daily exposure to the corresponding noise level as shown in the following table:

Noise level (dBA)	Permissible daily exposure
90	8 h
95	4 h
100	2 h
105	1 h
110	30 min
115	15 min

   f. "Hearing threshold level" means the lowest decibel sound that may be heard on the audiometer 50% of the times presented during an audiometric testing.
2. a. For purposes of determining the degree of hearing loss for awarding compensation for noise-induced occupational hearing loss, the average hearing threshold for each ear shall be determined by adding the hearing thresholds for the four frequencies 500, 1000, 2000, and 3000 Hz and dividing that sum by four. The percentage of disability shall be calculated from the better ear. If more than one audiogram is introduced for evaluation of disability, only the audiogram showing the lowest (best) average threshold shall be used to determine disability.
   b. If the better ear has a hearing loss of 25 dB or less, as measured from 0 dB on an audiometer calibrated to ANSI S3.6–1989 American National Standard "Specifications for Audiometers," or 15 dB or less, as measured on an audiometer calibrated to ASA-Z 24.5–1951 "American Standard Specifications for Pure Tone Audiometers for Screening Purposes," the hearing loss shall not be compensable.

c. If the hearing loss in the better ear is 26 dB or more (ANSI) or 16 dB or more (ASA) measured from 0 dB, the percentage of disability is as shown in the following table:

Hearing threshold level (dB ANSI)	Percent hearing disability	Hearing threshold level (dB ASA)
More than 26 to 31	5	More than 16 to 21
More than 31 to 36	10	More than 21 to 26
More than 36 to 41	15	More than 26 to 31
More than 41 to 46	25	More than 31 to 36
More than 46 to 51	35	More than 36 to 41
More than 51 to 56	50	More than 41 to 46
More than 56 to 66	70	More than 46 to 56
More than 66 to 76	90	More than 56 to 66
Over 76	100	Over 66

The percent hearing disability values shown are empirical. The actual values will vary in each state to make the awards consistent with compensation regulations.

3. a. An employer shall be liable for the hearing loss of an employee to which his employment has contributed. If previous occupational hearing loss or hearing loss from nonoccupational causes is established by competent evidence, including the results of a preemployment audiogram, the employer shall not be liable for the hearing loss so established whether or not compensation has previously been paid or awarded, and he shall be liable only for the difference between the percentage of disability determined as of the date of disability, as herein defined, and the percentage of disability established by the preemployment audiogram.
   b. An employer may require an employee to undergo audiometer testing at the expense of the employer at the time of termination of employment. The employee must be notified in writing of this requirement and the penalty, as provided herein, for noncompliance with such requirement at or before the employee's termination date. In the event of refusal or failure by the employee to undergo audiometric testing within 60 days after receipt of written notice of the scheduling of such test by the employer, the employee shall be penalized by losing any right to compensation as granted by this act, unless failure is due to a legitimate reason as determined by the division.
   c. Any employee who undergoes audiometric testing at the direction of an employer may request, within 2 weeks of such testing, a copy and brief explanation of the results, which shall be provided to him within 2 weeks of said request.
   d. For the purpose of verifying the degree of hearing loss for awarding compensation, an employee may introduce audiometric testing results obtained within 30 days after employer testing at his own expense from any individual approved for performing hearing tests pursuant to section 7.
4. In any evaluation of occupational hearing loss, only hearing levels at frequencies of 500, 1000, 2000, and 3000 Hz shall be considered.
5. Hearing levels shall be determined at all times by using pure-tone air-conduction audiometric instruments calibrated in accordance with American

National Standard ANSI S3.6–1989 (R 1971) and performed in an environment as prescribed by American National Standard S1.1–1977 ("Criteria for Permissible Ambient Noise during Audiometric Testing"). To measure permanent hearing loss, hearing tests shall be performed after at least 16-hour absence from exposure to hazardous noise. The calibration of an audiometric instrument used to measure permanent hearing loss shall have been performed within 1 year of the time of the hearing examination, to assure that the audiometer is within the tolerances permitted by the ANSI standards.

6. All hearing tests shall be performed by a person at the level of a certified audiometric technician or above; an individual who meets the training requirements specified by the Intersociety Committee on Audiometric Technician Training (American Industrial Hygiene Association Journal *27*:303–304, May–June 1966) and the State Department of Health. If hearing loss is demonstrated, an employee shall be referred for audiologic evaluation by a certified audiologist holding a certificate of clinical competence issued by the American Speech and Hearing Association or its equivalent or a physician certified by the American Board of Otolaryngology.
7. No claims for compensation for occupational hearing loss shall be filled until after 10 full consecutive calendar weeks have elapsed since removal from exposure to hazardous noise in employment. The last day of each exposure shall be the date of disability.
8. No reduction in award for hearing loss shall be made if the ability of the employee to understand speech is improved by the use of a hearing aid, nor shall the employer be obligated to furnish such hearing aids, including accessories and replacement, in cases of occupational loss of hearing.
9. No compensation shall be payable for loss of hearing caused by hazardous noise after the effective date of this act if an employer can properly document that, despite warnings, an employee has failed to properly and effectively utilize suitable protective device or devices provided by the employer capable of diminishing lose of hearing due to occupational exposure to hazardous noise.

7. IMPAIRMENT AND DISABILITY

An essential part of a compensation act is the manner of calculating how much compensation an employee should receive for a specific amount of hearing loss. It is first necessary to distinguish between impairment and disability. Impairment is a medical concept and is a deviation from normal. Disability involves many nonmedical factors and includes a concept of loss of ability to earn a daily livelihood, "loss of living power" or reduction of the individual's enjoyment of daily living. Hearing impairment contributes to a disability, but many other factors are involved. Compensation is awarded for disability.

7.1. Calculating Hearing Impairment

The ultimate test in any formula for determining hearing disability is the ability to understand speech, but because speech audiometry has certain limitations for practical use, pure-tone audiometry is used. The most commonly used frequencies for calculating hearing impairment are 500, 1000, and 2000 Hz. Recently the AAO has recommended that 3000 Hz also be included. A so-called "low fence" has been determined, below which a hearing loss is considered insufficient to warrant compensation. There is a difference of opinion as to precisely where this low fence should be. The Committee on Hearing and Bio-Acoustics (CHABA) has recommended that the low fence be placed at 35 dB. The AAO is recommending that the low fence be maintained at 25 dB. Each state has it own method of paying disability and uses its own formula and provides a method for measuring and calculating binaural hearing impairment. The hearing level for each frequency is the number of decibels at which the listener's threshold of hearing lies above the standard audiometric 0 for that frequency. The hearing level for speech is a simple average of the hearing levels at the frequencies 500, 1000, 2000, and now 3000 Hz. The following is an example of how to calculate hearing impairment for compensation purposes (AAO guidelines, 1978):

1. The average of the hearing threshold levels at 500, 1000, 2000, and 3000 Hz should be calculated for each ear.
2. The percent impairment for each ear should be calculated by multiplying by 1.5% the amount by which the above average hearing threshold level exceeds 25 dB (low fence) up to a maximum of 100%, which is reached at 92 dB (high fence).
3. The hearing handicap, a binaural assessment, should then be calculated by multiplying the smaller percentage (better ear) by 5, adding this figure to the larger percentage (poorer ear), and dividing the total by 6.

Examples

Mild Hearing Loss				
	500 Hz	1000 Hz	2000 Hz	3000 Hz
Right ear	15	25	45	55
Left ear	20	30	50	60

AAO Method: 25-dB Fence

1. Right ear $\dfrac{15+25+45+55}{4} = \dfrac{140}{4} = 35\text{-dB average}$

2. Left ear $\dfrac{20+30+50+60}{4} = \dfrac{160}{4} = 40\text{-dB average}$

Monaural Impairment

3. Right ear $35 - 25 = 10\text{ dB} \times 1.5\% = 15\%$
4. Left ear $40 - 25 = 15\text{ dB} \times 1.5\% = 22.5\%$
5. Better ear $15 \times 5 = 75$
6. Poorer ear $22.5\% \times 1 = 25.5$
7. Total $97.5 \div 6 = 16.25\%$

Model Legislation Method used to calculate the above loss:

1. Right ear $\dfrac{15+25+45+55}{4} = \dfrac{140}{4} = 35\text{-dB average}$

2. Left ear $\frac{20+30+50+60}{4}=\frac{160}{4}=40\text{-dB average}$

3. Better ear threshold = 35 dB = 5%

Severe Hearing Loss

	500 Hz	1000 Hz	2000 Hz	3000 Hz
Right ear	80	90	100	110
Left ear	75	80	90	95

AAO Method: Average Hearing Test Level

1. Right ear $\frac{80+90+100+110}{4}=\frac{380}{4}=95\text{ dB(use 92 maximum)}$

2. Left ear $\frac{75+80+90+95}{4}+\frac{340}{8}=85\text{ dB}$

Monaural Impairment

3. Right ear $92-25=67\text{ dB}\times 1.5\%=100.5\%$ (use 100%)
4. Left ear $85-25=60\text{ dB}\times 1.5\%=90\%$
5. Better ear $90\times 5=450$
6. Poorer ear $100\times 1=100$
7. Total $550\div 6=91.7\%$

New Jersey method used to calculate the above loss:

1. Right ear $\frac{80+90+100+110}{4}=\frac{380}{4}=95\text{ dB(use 92 maximum)}$

2. Left ear $\frac{75+80+90+95}{4}+\frac{340}{4}=85\text{ dB}$

3. Better ear $>81\text{ dB}=100\%$

8. COMPENSATION VARIABILITY

The disparity in payment for occupational deafness in the various states and the federal government was demonstrated by Joseph Law, of the General Accounting Office, who investigated the problem in the naval shipyards. Mr. Law pointed out the government's concern about the growth of hearing impairment compensation under the Act. "Claims for hearing impairment compensation from federal civilian employees have steadily increased from 500 in 1969 to nearly 9000 in 1976, or about 36,000 in those 8 years, for an estimated total accumulative liability of about 185 million dollars ... about 80 percent of the claimants have received an award which averaged about $7,000.00." Mr. Law further pointed out the great disparity in payment for compensation between the states and the federal government (Fig. 33.1). This problem stills exists, as discussed in Chapter 35. There is also considerable differences between hearing loss awards in the United States and those available in Canada and England. A review of the Occupational

(A)

Program	Formula	Frequencies K/Hz	Begin. fence	Conversion factor	Binaural impair	After exclusion for age	Max. comp rate	Binaural (weeks)	Effective rate per binaural per cent	Compensation
N.C.	AAOO	0.5, 1, 2	25	1.5%	9%	—	2/3 ($219)	150	$219	$1,971
Me.	AAOO	0.5, 1, 2	25	1.5 §	9	3% *	2/3 (212)	200	283	212
Fed. Mil	AAOO	0.5, 1, 2								0
Fed. Civ.	Own	1, 2, 3	25	1.5	24	—	3/4 (salary)	200	375	9,000
Wisc.	Chaba	1, 2, 3	35	1.75	11	7	2/3 (85.50)	216 ‡	123	862
Cal.	Own †	0.5, 1, 2, 3	25	1.5	17	—	70	240	190	3,237
N.J.	Own	0.5, 1, 2, 3, 4	25	1.5	27	—	40	200	80	2,160
Ore.	Own	0.5, 1, 2, 3, 4, 6	25	1.5	24	—	70	192	134	3,226

(B)

Freq.	dB level (ANSI) Left ear	Right ear
500	20	30
1,000	30	35
2,000	40	45
3,000	50	55
4,000	60	65
6,000	40	45

(C)

100%
Impairment
Federal civilian
Chaba
AMA/AAOO
25
Average dB level
90
(0.5, 1, 2 K/Hz)
25
90
(1, 2, 3 K/Hz)

Figure 33.1 Comparison of selected compensation criteria for occupational hearing loss. (A) Formulas used by the selected jurisdictions (i.e., North Carolina, Maine, Federal Military, Federal Civilian, Wisconsin, California, New Jersey, and Oregon). (B) Hearing levels of typical hearing loss claim of a federal civilian worker. These levels were used in computing the compensation awarded by the various jurisdictions in A. (C) Graphical comparison of the computer binaural impairments in A. The CHABA formulas, using the frequencies of 1000, 2000, and 3000 Hz, but a beginning fence of 35 dB and a conversion factor of 1.75%, are much closer to the AAOO formula than the Federal Civilian formula.

Safety and Health Act and of compensation practices in the United States, Canada, and England is presented in Chapters 34 through 39.

BIBLIOGRAPHY

1. International Statistical Classification of Diseases and Related Health Problems, Revision, Volumes 1–3, Geneva World Health Organization, 1992–1994.
2. Towards a common language for functioning, disability and health. Geneva World Health Organization, 2002.

34
OSHA Noise Regulation

Robert T. Sataloff
Drexel University College of Medicine, American Institute for Voice and Ear Research, and Graduate Hospital, Philadelphia, Pennsylvania, USA

Joseph Sataloff
Thomas Jefferson University, Philadelphia, Pennsylvania, USA

The history of the OSHA noise regulation and the hearing conservation amendment is complex, as with most other important regulations and laws. Comparatively little valid and reliable scientific data were available on which to base a noise standard. Practical measures, politics, economics, and numerous other factors played important roles in determining the final regulation. It is not our purpose to provide a historical background.

Rather, we will highlight the most important features of OSHA's requirements to comply with government regulations and to prevent occupational hearing loss.

For complete details of the current law, the reader is referred to the *Federal Register*, Volume 48, Number 46, Tuesday, March 8, 1983. This contains Occupational Safety and Health Administration Regulation 29 CFR 1910, Occupational Noise Exposure, Hearing Conservation Amendment, final rule. This final rule became effective April 7, 1983. This issue of the *Federal Register* contains a wealth of information, including specific details required for hearing conservation programs, appendices on noise exposure computation, methods for estimating the adequacy of hearing protector attenuation, audiometric measuring instruments, audiometric test rooms, acoustic calibration of audiometers, calculations and application of age corrections to audiograms, monitoring noise levels, availability of referenced documents, and definitions. Copies may be obtained by contacting the Docket Officer, Docket Number OSH-11, Room S-6212, U.S. Department of Labor, 200 Constitution Avenue, NW., Washington, DC 20210. This chapter reprints the selected sections of the law that are of particular practical importance.

1. OCCUPATIONAL NOISE STANDARD

1.1. 1910.95 Occupational Noise Exposure

(a) Protection against the effects of noise exposure shall be provided when the sound levels exceed those shown in Table G.16 when measured on the A scale of a standard sound level meter at slow response. When noise levels are determined by octave band analysis, the equivalent A-weighted sound level may be determined as follows:

Equivalent sound level contours (Fig. G.9). Octave band sound pressure levels may be converted to the equivalent A-weighted sound level by plotting them on this graph and noting the A-weighted sound level corresponding to the point of highest penetration into the sound level contours. This equivalent A-weighted sound level, which may differ from the actual A-weighted sound level of the noise, is used to determine exposure limits from Table G.16.

(b)(1) When employees are subjected to sound exceeding those listed in Table G.16, feasible administrative or engineering controls shall be utilized. If such controls fail to reduce sound levels within the levels of Table G.16, personal protective equipment shall be provided and used to reduce sound levels within the levels of the table.

(2) If the variations in noise level involve maxima at intervals of 1 second or less, it is to be considered continuous.

Table G.16 Permissible Noise Exposures

Duration per day, hours	Sound level dBA response
8	90
6	92
4	95
3	97
2	100
1½	102
1	105
½	110
¼ or less.	115

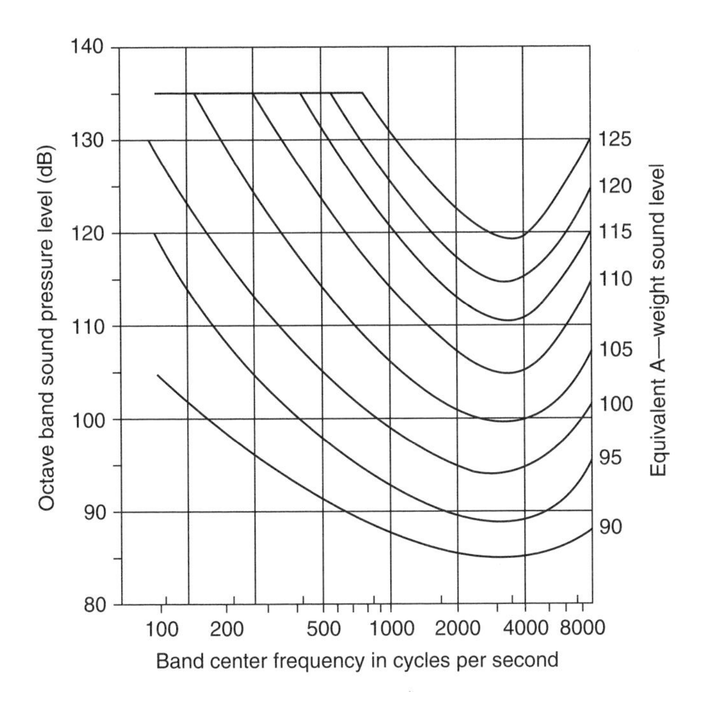

Figure G.9 Equivalent sound level contours.

When the daily noise exposure is composed of two or more periods of noise exposure of different levels, their combined effect should be considered, rather than the individual effect of each. If the sum of the following fractions: $C1/T1 + C2/T2 + \ldots + Cn/Tn$ exceeds unity, then the mixed exposure should be considered to exceed the limit value. Cn indicates the total time of exposure at a specified noise level, and Tn indicates the total time of exposure permitted at that level. Exposure to impulsive or impact noise should not exceed 140 dB peak sound pressure level.

(c) *Hearing Conservation Program.* (1) The employer shall administer a continuing, effective hearing conservation program, as described in paragraphs (c) through (o) of this section, whenever employees noise exposures equal or exceed an 8-hour time-weighted average sound level (TWA) of 85 decibels measured on the A scale (slow response) or, equivalently, a dose of fifty percent. For purposes of the hearing conservation program, employee noise exposures shall be computed in accordance with Appendix A and Table G.16a, and without regard to any attenuation provided by the use of personal protective equipment.

(2) For purposes of paragraphs (c) through (n) of this section, an 8-hour time-weighted average of 85 decibels or a dose of fifty percent shall also be referred to as the action level.

(d) *Monitoring.* (1) When information indicates that any employee's exposure may equal or exceed an 8-hour time-weighted average of 85 decibels, the employer shall develop and implement a monitoring program. (i) The sampling strategy shall be designed to identify employees for inclusion in the hearing conservation program and to enable the proper selection of hearing protectors.

(ii) Where circumstances such as high worker mobility, significant variations in sound level, or a significant component of impulse noise make area monitoring generally inappropriate, the employer shall use representative personal sampling to comply with the

Table G.16a

A-weighted sound level, L (decibel)	Reference duration, T (hour)
80	32
81	27.9
82	24.3
83	21.1
84	18.4
85	16
86	13.9
87	12.1
88	10.6
89	9.2
90	8
91	7.0
92	6.1
93	5.3
94	4.6
95	4
96	3.5
97	3.0
98	2.6
99	2.3
100	2
101	1.7
102	1.5
103	1.3
104	1.1
105	1
106	0.87
107	0.76
108	0.66
109	0.57
110	0.5
111	0.44
112	0.38
113	0.33
114	0.29
115	0.25
116	*0.22*
117	*0.19*
118	*0.16*
119	*0.14*
120	*0.125*
121	*0.11*
122	*0.095*
123	*0.082*
124	*0.072*
125	*0.063*
126	*0.054*
127	*0.047*
128	*0.041*
129	*0.036*
130	*0.031*

In the above table the reference duration, T, is computed by $T = 8/_2(L - 90)/S$ where L is the measured A-weighted sound level.

monitoring requirements of this paragraph unless the employer can show that area sampling produces equivalent results.

(2)(i) All continuous, intermittent and impulsive sound levels from 80 decibels to 130 decibels shall be integrated into the noise measurements.

(ii) Instruments used to measure employee noise exposure shall be calibrated to ensure measurement accuracy.

(3) Monitoring shall be repeated whenever a change in production, process, equipment or controls increases noise exposures to the extent that:

(i) Additional employees may be exposed at or above the action level; or

(ii) The attenuation provided by hearing protectors being used by employees may be rendered inadequate to meet the requirements of paragraph (j) of this section.

(e) *Employee notification.* The employer shall notify each employee exposed at or above an 8-hour time-weighted average of 85 decibels of the results of the monitoring.

(f) *Observation of monitoring.* The employer shall provide affected employees or their representatives with an opportunity to observe any noise measurements conducted pursuant to this section.

(g) *Audiometric testing program.* (1) The employer shall establish and maintain an audiometric testing program as provided in this paragraph by making audiometric testing available to all employees whose exposures equal or exceed an 8-hour time-weighted average of 85 decibels.

(2) The program shall be provided at no cost to employees.

(3) Audiometric tests shall be performed by a licensed or certified audiologist, otolaryngologist, or other physician, or by a technician who is certified by the Council of Accreditation in Occupational Hearing Conservation, or who has satisfactorily demonstrated competence in administering audiometric examinations, obtaining valid audiograms, and properly using, maintaining and checking calibration and proper functioning of the audiometers being used. A technician who operates microprocessor and audiometers does not need to be certified. A technician who performs audiometric tests must be responsible to an audiologist, otolaryngologist or physician.

(4) All audiograms obtained pursuant to this section shall meet the requirements of Appendix C: *Audiometric Measuring Instruments.*

(5) *Baseline Audiogram.* (i) Within 6 months of an employee's first exposure at or above the action level, the employer shall establish a valid baseline audiogram against which subsequent audiograms can be compared.

(ii) *Mobile test van exception.* Where mobile test vans are used to meet the audiometric testing obligation, the employer shall obtain a valid baseline audiogram within 1 year of an employee's first exposure at or above the action level. Where baseline audiograms are obtained more than 6 months after the employee's first exposure at or above the action level, employees shall wear hearing protectors for any period exceeding six months after the first exposure until the baseline audiogram is obtained.

(iii) Testing to establish a baseline audiogram shall be preceded by at least 14 h without exposure to workplace noise. Hearing protectors may be used as a substitute for the requirement that baseline audiograms be preceded by 14 h without exposure to workplace noise.

(iv) The employer shall notify employees of the need to avoid high levels of non-occupational noise exposure during the 14-hour period immediately preceding the audiometric examination.

(6) *Annual audiogram.* At least annually after obtaining the baseline audiogram, the employer shall obtain a new audiogram for each employee exposed at or above an 8-hour time-weighted average of 85 decibels.

(7) *Evaluation of audiogram.* (i) Each employee's annual audiogram shall be compared to that employee's baseline audiogram to determine if the audiogram is valid and if a standard threshold shift as defined in paragraph (g)(10) of this section has occurred. This comparison may be done by a technician.

(ii) If the annual audiogram shows that an employee has suffered a standard threshold shift, the employer may obtain a retest within 30 days and consider the results of the test as the annual audiogram.

(iii) The audiologist, otolaryngologist, or physician shall review problem audiograms and shall determine whether there is a need for further evaluation. The employer shall provide to the person performing this evaluation the following information:

(A) A copy of the requirements for hearing conservation as set forth in paragraphs (c) through (n) of this section:

(B) The baseline audiogram and most recent audiogram of the employee to be evaluated:

(C) Measurements of background sound pressure levels in the audiometric test room as required in Appendix D: *Audiometric Test Rooms.*

(D) Records of audiometer calibrations required by paragraph (h)(5) of this section.

(8) *Follow-up procedures.* (i) If a comparison of the annual audiogram to the baseline audiogram indicates a standard threshold shift as defined in paragraphs (g)(10) of this section has occurred, the employee shall be informed of this fact in writing, within 21 days of the determination.

(ii) Unless a physician determines that the standard threshold shift is not work related or aggravated by occupational noise exposure, the employer shall ensure that the following steps are taken when a standard threshold shift occurs:

(A) Employees not using hearing protectors shall be fitted with hearing protectors, trained in their use and care, and required to use them.

(B) Employees already using hearing protectors shall be refitted and retrained in the use of hearing protectors and provided with hearing protectors offering greater attenuation if necessary.

(C) The employee shall be referred for a clinical audiological evaluation or an otological examination, as appropriate, if additional testing is necessary if the employer suspects that a medical pathology of the ear is caused or aggravated by the wearing of hearing protectors.

(D) The employee is informed of the need for an otological examination if a medical pathology of the ear that is unrelated to the use of hearing protectors is suspected.

(iii) If subsequent audiometric testing of an employee whose exposure to noise is less than an 8-hour TWA of 90 decibels indicates that a standard threshold shift is not persistent, the employer:

(A) Shall inform the employee of the new audiometric interpretation; and

(B) May discontinue the required use of hearing protectors for that employee.

(9) *Revised baseline.* An annual audiogram may be substituted for the baseline audiogram when, in the judgment of the audiologist, otolaryngologist or physician who in evaluating the audiogram:

(i) The standard threshold shift revealed by the audiogram is persistent; or

(ii) The hearing threshold shown in the annual audiogram indicates significant improvement over the baseline audiogram.

(10) *Standard threshold shift.* (i) As used in this section, a standard threshold shift is a change in hearing threshold relative to the baseline audiogram of an average of 10 dB or more at 2000, 3000, and 4000 Hz in either ear.

(ii) In determining whether a standard threshold shift has occurred, allowance may be made for the contribution of aging (presbycusis) to the change in hearing level by correcting the annual audiogram according to the procedure described in Appendix F: *Calculations and Application of Age Corrections to Audiograms.*

(h) *Audiometric test requirements.* (1) Audiometric tests shall be pure tone, air conduction, hearing threshold examinations, with test frequencies including as a minimum 500, 1000, 2000, 3000, 4000, and 6000 Hz. Tests at each frequency shall be taken separately for each ear.

(2) Audiometric tests shall be conducted with audiometers (including microprocessor audiometers) that meet the specifications of, and are maintained and used in accordance with, American National Standard Specification for Audiometers, S3.6–1969.

(3) Pulsed-tone and self-recording audiometers, if used, shall meet the requirements specified in Appendix C: *Audiometric Measuring Instruments.*

(4) Audiometric examinations shall be administered in a room meeting the requirements listed in Appendix D: *Audiometric Test Rooms.*

(5) *Audiometer calibration.* (i) The functional operation of the audiometer shall be checked before each day's use by testing a person with known, stable hearing threshold, and by listening to the audiometer's output to make sure that the output is free from distorted or unwanted sounds. Deviations of 10 dB or greater require an acoustic calibration.

(ii) Audiometer calibration shall be checked acoustically at least annually in accordance with Appendix E: *Acoustic Calibration of Audiometers.* Test frequencies below 500 Hz and above 6000 Hz may be omitted from this check. Deviations of 15 dB or greater require an exhaustive calibration.

(iii) An exhaustive calibration shall be performed at least every two years in accordance with sections 4.1.2; 4.1.3; 4.1;, 4.3; 4.4.1; 4.4.2; 4.4.3; and 4.5 of the American Standard Specification for Audiometers, S3.6–1969. Test frequencies below 500 Hz and above 6000 Hz may be omitted from this calibration.

(i) *Hearing protectors.* (1) Employers shall make hearing protectors available to all employees exposed to an 8-hour time-weighted average of 85 dB or greater at no cost to the employees. Hearing protectors shall be replaced as necessary.

(2) Employers shall ensure that hearing protectors are worn:

(i) By an employee who is required by paragraph (b)(1) of this section to wear personal protective equipment; and

(ii) By an employee who is exposed to an 8-hour time-weighted average of 85 dB or greater, and who:

(A) Has not yet had a baseline audiogram established pursuant to paragraph (g)(5)(ii); or

(B) Has experienced a standard threshold shift.

(3) Employees shall be given the opportunity to select their hearing protectors from a variety of suitable hearing protectors provided by the employer.

(4) The employer shall provide training in the use and care of all hearing protectors to employees.

(5) The employer shall ensure proper initial fitting and supervise the correct use of hearing protectors.

(j) *Hearing protector attenuation.* (1) The employer shall evaluate hearing protector attenuation for the specific noise environments in which the protector will be used. The employer shall use one of the evaluation methods described in Appendix B: *Methods for Estimating the Adequacy of Hearing Protector Attenuation.*

(2) Hearing protectors must attenuate employee exposure at least to an 8-hour time-weighted average of 90 dB as required by paragraph (b) of this section.

(3) For employees who have experienced a standard threshold shift, hearing protectors must attenuate employee exposure to an 8-hour time-weighted average of 85 decibels or below.

(4) The adequacy of hearing protector attenuation shall be re-evaluated whenever employee noise exposures increase to the extent that the hearing protectors provided may no longer provide adequate attenuation. The employee shall provide more effective hearing protectors where necessary.

(k) *Training program.* (1) The employer shall institute a training program for all employees who are exposed to noise at or above an 8-hour time-weighted average of 85 decibels, and shall ensure employee participation in such program.

(2) The training program shall be repeated annually for each employee included n the hearing conservation program. Information provided in the training program shall be updated to be consistent with changes in protective equipment and work processes.

(3) The employer shall ensure that each employee is informed of the following:

(i) The effects of noise on hearing;

(ii) The purpose of hearing protectors, the advantages, disadvantages, and attenuation of various types, and instructions on selection, fitting, use, and care; and

(iii) The purpose of audiometric testing, and an explanation of the test procedures.

(l) *Access to information and training materials.* (1) The employer shall make available to affected employees or their representatives copies of this standard and shall also post a copy in the workplace.

(2) The employer shall provide to affected employees any informational materials pertaining to the standard that are supplied to the employer by the Assistant Secretary.

(3) The employer shall provide, upon request, all materials related to the employer's training and education program pertaining to this standard to the Assistant Secretary and the Director.

(m) *Recordkeeping*—(1) *Exposure measurements.* The employer shall maintain an accurate record of all employee exposure measurements required by paragraph (d) of this section.

(2) *Audiometric tests.* (i) The employer shall retain all employee audiometric test records obtained pursuant to paragraph (g) of this section:

(ii) This record shall include:

(A) Name and job classification of the employee:

(B) Date of the audiogram;

(C) The examiner's name;

(D) Date of the last acoustic or exhaustive calibration of the audiometer, and

(E) Employee's most recent noise exposure assessment.

(F) The employer shall maintain accurate records of the measurements of the background sound pressure levels in audiometric test rooms.

(3) *Record retention.* The employer shall retain records required in this paragraph (m) for at least the following periods.

(i) Noise exposure measurement records shall be retained for two years.

(ii) Audiometric test records shall be retained for the duration of the affected employee's employment.

(4) *Access to records.* All records required by this section shall be provided upon request to employees, former employees, representatives designated by the individual employee, and the Assistant Secretary. The provisions of 29 CFR 1910.20(a)–(e) and (g)–(i) apply to access to records under this section.

(5) *Transfer of records.* If the employer ceases to do business, the employer shall transfer to the successor employer all records required to be maintained by this section, and the successor employer shall retain them for the remainder of the period prescribed in paragraph (m)(3) of this section.

(n) *Appendices.* (1) Appendices A, B, C, D, and E to this section are incorporated as part of this section and the contents of these Appendices are mandatory.

(2) Appendices F and G to this section are informational and are not intended to create any additional obligations not otherwise imposed or to detract from any existing obligations.

(o) *Exemptions.* Paragraphs (c) through (n) of this section shall not apply to employers engaged in oil and gas well drilling and servicing operations.

(p) *Startup date.* Baseline audiograms required by paragraph (g) of this section shall be completed by March 1, 1984.

In addition to the above essential information, certain aspects of the appendices are particularly useful. For more complete information, the *Federal Register* should be consulted.

2. APPENDIX A: NOISE EXPOSURE COMPUTATION

This Appendix is Mandatory

2.1. Computation of Employee Noise Exposure

(1) Noise dose is computed using Table G-16a as follows:

(i) When the sound level, L, is constant over the entire work shift, the noise dose, D, is present, is given by: $D = 100\ C/T$ where C is the total length of the work day, in hours, and T is the reference duration corresponding to the measured sound level, L, as given in Table G-16a or by the formula shown as a footnote to that table.

(ii) When the workshift noise exposure is composed of two or more periods of noise at different levels, the total noise dose over the work day is given by:

$$D = 100\ (C_1/T_1 + C_2/T_2 + \cdots + C_n/T_n),$$

where C_n indicates the total time of exposure at a specific noise level, and T_n indicates the reference duration of that level as given by Table G-16a.

(2) The eight-hour time-weighted average sound level (TWA), in decibels, may be computed from the dose, in percent, by means of the formula: $TWA = 16.61 \log_{10} (D/100) \cdot 90$. For an eight-hour workshift with the noise level constant over the entire shift, the TWA is equal to the measured sound level.

(3) A table relating dose and TWA is given in Section II.

2.2. Conversion Between "Dose" and "8-Hour Time-Weighted Average" Sound Level

Compliance with paragraphs (c)–(r) of this regulation is determined by the amount of exposure to noise in the workplace. The amount of such exposure is usually measured with an audiodosimeter which gives a readout in terms of "dose." In order to better understand the requirements of the amendment, dosimeter readings can be converted to an "8-hour time-weighted average sound level" (TWA).

In order to convert the reading of a dosimeter into TWA, see Table A.1, below. This table applies to dosimeters that are set by the manufacturer to calculate dose or percent exposure according to the relationships in Table G.16a. So, for example, a dose of 91 percent over an eight hour day results in a TWA of 89.3 dB, and, a dose of 50 percent corresponds to a TWA of 85 dB.

If the dose as read on the dosimeter is less than or greater than the values found in Table A.1, the TWA may be calculated by using the formula: $TWA = 16.61 \log_{10} (D/100) + 90$ where TWA = 8-hour time-weighted average sound level and D-accumulated dose in percent exposure.

3. APPENDIX B: METHODS FOR ESTIMATING THE ADEQUACY OF HEARING PROTECTOR ATTENUATION

This Appendix is Mandatory

For employees who have experienced a significant threshold shift, hearing protector attenuation must be sufficient to reduce employee exposure to a TWA of 85 dB. Employers must select one of the following methods by which to estimate the adequacy of hearing protector attenuation.

The most convenient method is the Noise Reduction Rating (NRR) developed by the Environmental Protection Agency (EPA). According to EPA regulation, the NRR must be shown on the hearing protector package. The NRR is then related to an individual worker's noise environment in order to assess the adequacy of the attenuation of a given hearing protector. This Appendix describes four methods of using the NRR to determine whether a particular hearing protectors provides adequate protection within a given exposure environment. Selection among the four procedures is dependent upon the employer's noise measuring instruments.

Instead of using the NRR, employers may evaluate the adequacy of hearing protector attenuation by using one of three methods developed by the National Institute for Occupational Safety and Health (NIOSH), which are described in the "List of Personal Hearing Protectors and Attenuation Data," HEW Publication No. 76-120, 1975, pages 21–37. These methods are known as NIOSH methods 1, 2, and 3. The NRR described below is a simplification of NIOSH method 2. The most complex method is NIOSH method 1, which is probably the most accurate method since it uses the largest amount of spectral information from the individual employee's noise environment. As in the case of the NRR method described below, if one of the NIOSH methods is used, the selected method must be applied to an individual's noise environment to assess the adequacy of the attenuation. Employers should be careful to take a sufficient number of measurements in order to achieve a representative sample for each time segment.

NOTE: The employer must remember that calculated attenuation values reflect realistic values only to the extent that the protectors are properly fitted and worn.

When using the NRR to assess hearing protector adequacy, one of the following methods must be used:

(i) When using a dosimeter that is capable of C-weighted measurements:

(A) Obtain the employee's C-weighted dose for the entire workshift, and convert to TWA (see Appendix A, II).

Table A.1 Conversion from "Percent Noise Exposure" or "Dose" to "8-Hour Time-Weighted Average Sound Level" (TWA)

Dose or percent noise exposure	TWA
10	73.4
15	76.3
20	78.4
25	80.0
30	81.3
35	82.4
40	83.4
45	84.2
50	85.0
55	85.7
60	86.3
65	86.9
70	87.4
75	87.9
80	88.4
81	88.5
82	88.6
83	88.7
84	88.7
85	88.8
86	88.9
87	89.0
88	89.1
89	89.2
90	89.2
91	89.3
92	89.4
93	89.5
94	89.6
95	89.6
96	89.7
97	89.8
98	89.9
99	89.9
100	90.0
101	90.1
102	90.1
103	90.2
104	90.3
105	90.4
106	90.4
107	90.5
108	90.6
109	90.6
110	90.7
111	90.8
112	90.8
113	90.9
114	90.9

(Continued)

Table A.1 *Continued*

Dose or percent noise exposure	TWA
115	91.1
116	91.1
117	91.1
118	91.1
119	91.3
120	91.3
125	91.6
130	91.9
135	92.2
140	92.4
145	92.7
150	92.9
155	93.2
160	93.4
165	93.6
170	93.8
175	94.0
180	94.2
185	94.4
190	94.6
195	94.8
200	95.0
210	95.4
220	95.7
230	96.0
240	96.3
250	96.6
260	96.9
270	97.2
280	97.4
290	97.7
300	97.9
310	98.2
320	98.4
330	98.6
340	98.8
350	99.0
360	99.2
370	99.4
380	99.6
390	99.8
400	100.0
410	100.2
420	100.4
430	100.5
440	100.7
450	100.8
460	101.0
470	101.2
480	101.3

(*Continued*)

Table A.1 *Continued*

Dose or percent noise exposure	TWA
490	101.5
500	101.6
510	101.8
520	101.9
530	102.0
540	102.2
550	102.3
560	102.4
570	102.6
580	102.7
590	102.8
600	102.9
610	103.0
620	103.2
630	103.3
640	103.4
650	103.5
660	103.6
670	103.7
680	103.8
690	103.9
700	104.0
710	104.1
720	104.2
730	104.3
740	104.4
750	104.5
760	104.6
770	104.7
780	104.8
790	104.9
800	105.0
810	105.1
820	105.2
830	105.3
840	105.4
850	105.4
860	105.5
870	105.6
880	105.7
890	105.8
900	105.8
910	105.9
920	106.0
930	106.1
940	106.2
950	106.2
960	106.3
970	106.4
980	106.5
990	106.5
999	106.6

(B) Subtract the NRR from the C-weighted TWA to obtain the estimated A-weighted TWA under the ear protector.

(ii) When using a dosimeter that is not capable of C-weighted measurements, the following method may be used:

(A) Convert the A-weighted dose to TWA (see Appendix A).

(B) Subtract 7 dB from the NRR.

(C) Subtract the remainder from the A-weighted TWA to obtain the estimated A-weighted TWA under the ear protector.

(iii) When using a sound level meter set to the A-weighted network:

(A) Obtain the employee's A-weighted TWA.

(B) Subtract 7 dB from the NRR, and subtract the remainder from the A-weighted TWA to obtain the estimated A-weighted TWA under the ear protect.

(iv) When using a sound level meter set on the C-weighting network:

(A) Obtain a representative sample of the C-weighted sound levels in the employee's environment.

(B) Subtract the NRR from the C-weighted average sound level to obtain the estimated A-weighted TWA under the ear protector.

(v) When using area monitoring procedures and a sound level meter set to the A-weighting network:

(A) Obtain a representative sound level for the area in question.

(B) Subtract 7 dB from the NRR and subtract the remainder from the A-weighted sound level for that area.

(v) When using monitoring procedures and a sound level meter set to the C-weighted network:

(A) Obtain a representative sound level for the area in question.

(B) Subtract the NRR from the C-weighted sound level for that area.

4. APPENDIX C: AUDIOMETRIC MEASURING INSTRUMENTS

This Appendix is Mandatory

1. In the event that pulsed-tone audiometers are used, they shall have a tone on-time of at least 200 milliseconds.

2. Self-recording audiometers shall comply with the following requirements:

(A) The chart upon which the audiogram is traced shall have lines at positions corresponding to all multiples of 10 dB hearing level within the intensity range spanned by the audiometer. The lines shall be equally spaced and shall be separated by at least $\frac{1}{4}$ inch. Additional increments are optional. The audiogram pen tracings shall not exceed 2 dB in width.

(B) It shall be possible to set the stylus manually at the 10-dB increment lines for calibration purposes.

(C) The slewing rate for the audiometer attenuator shall not be more than 6 dB/sec except that an initial slewing rate greater than 6 dB/sec is permitted at the beginning of each new test frequency, but only until the second subject response.

(D) The audiometer shall remain at each required test frequency for 30 seconds ($\pm$3 seconds). The audiogram shall be clearly marked at each change of frequency and the actual frequency change of the audiometer shall not deviate from the frequency boundaries marked on the audiogram by more than $\pm$3 seconds.

(E) It must be possible at each test frequency to place a horizontal line segment parallel to the time axis on the audiogram, such that the audiometric tracing crosses the line

Table D.1 Maximum Allowable Octave-Band Sound Pressure Levels for Audiometric Test Rooms

Octave-band center frequency (Hz)	500	1000	2000	4000	8000
Sound pressure level (dB)	40	40	47	57	62

segment at least six times at the test frequency. At each test frequency the threshold shall be the average of the midpoints of the tracing excursions.

5. APPENDIX D: AUDIOMETRIC TEST ROOMS

This Appendix is Mandatory

Rooms used for audiometric testing shall not have background sound pressure levels exceeding those in Table D-1 when measured by equipment conforming at least to the Type 2 requirements of American National Standard Specification for Sound Level Meters, S1.4–1971 (R1976), and to the Class II requirements of American National Standard Specification for Octave, Half-Octave, and Third-Octave Band Filter Sets, S1.11–1971 (R1976).

6. APPENDIX E: ACOUSTIC CALIBRATION OF AUDIOMETERS

This Appendix is Mandatory

Audiometer calibration shall be checked acoustically, at least annually, according to the procedures described in this Appendix. The equipment necessary to perform these measurements is a sound level meter, octave-band filter set, and a National Bureau of Standards 9A coupler. In making these measurements, the accuracy of the calibrating equipment shall be sufficient to determine that the audiometer is within the tolerances permitted by American Standard Specification for Audiometers, S3.6–1969.

(1) Sound Pressure Output Check

A. Place the earphone coupler over the microphone of the sound level meter and place the earphone on the coupler.

B. Set the audiometer's hearing threshold level (HTL) dial to 70 dB.

C. Measure the sound pressure level of the tones at each test frequency from 500 Hz through 6000 Hz for each earphone.

D. At each frequency the readout on the sound level meter should correspond to the levels in Table E-1 or Table E-2, as appropriate, for the type of earphone, in the column entitled "sound level meter reading."

(2) Linearity Check

A. With the earphone in place, set the frequency to 1000 Hz and the HTL dial on the audiometer to 70 dB.

B. Measure the sound levels in the coupler at each 10-dB decrement from 70 dB to 10 dB, noting the sound level meter reading at each setting.

Table E.1 Reference Threshold Levels for Telephonics—TDH-39 Earphones

Frequency, Hz	Reference threshold level for TDH-39 earphones, dB	Sound level meter reading dB
500	11.5	81.5
1000	7	77
2000	9	79
3000	10	80
4000	9.5	79.5
6000	15.5	85.5

C. For each 10-dB decrement on the audiometer the sound level meter should indicate a corresponding 10 dB decrease.

D. This measurement may be made electrically with a voltmeter connected to the earphone terminals.

(3) Tolerances

When any of the measured sound levels deviate from the levels in Table E-1 or Table E-2 by ± 3 dB at any test frequency between 500 and 3000 Hz, 4 dB at 4000 Hz, or 5 dB at 6000 Hz, an exhaustive calibration is advised. An exhaustive calibration is required if the deviations are greater than 15 dB or greater at any test frequency.

7. APPENDIX F: CALCULATIONS AND APPLICATION OF AGE CORRECTIONS TO AUDIOGRAMS

This Appendix is Nonmandatory

In determining whether a standard threshold shift has occurred, allowance may be made for the contribution of aging to the change in hearing level by adjusting the most recent audiogram. If the employer chooses to adjust the audiogram, the employer shall follow the procedure described below. This procedure and the age correction tables were developed by the National Institute for Occupational Safety and Health in the criteria document entitled "Criteria for a Recommended Standard ... Occupational Exposure to Noise," (HSH-11001).

Table E.2 Reference Threshold Levels for Telephonics—TDH-49 Earphones

Frequency, Hz	Reference threshold level for TDH-49 earphones, dB	Sound level meter reading, dB
500	13.5	83.5
1000	7.5	77.5
2000	11	81.0
3000	9.5	79.5
4000	10.5	80.5
6000	13.5	83.5

For each audiometric test frequency:

(i) Determine from Tables F.1 or F.2 the age correction values for the employee by:

(A) Finding the age at which the most recent audiogram was taken and recording the corresponding values of age corrections at 1000 Hz through 6000 Hz;

Table F.1 Age Correction Values in Decibels for Males

	Audiometric test frequencies (Hz)				
Years	1000	2000	3000	4000	6000
20 or younger	5	3	4	5	8
21	5	3	4	5	8
22	5	3	4	5	8
23	5	3	4	6	9
24	5	3	5	6	9
25	5	3	5	7	10
26	5	4	5	7	10
27	5	4	6	7	11
28	6	4	6	8	11
29	6	4	6	8	12
30	6	4	6	9	12
31	6	4	7	9	13
32	6	5	7	10	14
33	6	5	7	10	14
34	6	5	8	11	15
35	7	5	8	11	15
36	7	5	9	12	16
37	7	6	9	12	17
38	7	6	9	13	17
39	7	6	10	14	18
40	7	6	10	14	19
41	7	6	10	14	20
42	8	6	11	16	20
43	8	7	12	16	21
44	8	7	12	17	22
45	8	7	13	19	24
46	8	8	13	19	24
47	8	8	14	19	24
48	9	8	14	20	25
49	9	9	15	21	26
50	9	9	16	22	27
51	9	9	16	23	28
52	9	10	17	24	29
53	9	10	18	25	30
54	10	10	18	26	31
55	10	11	19	27	32
56	10	11	20	28	34
57	10	11	21	29	35
58	10	12	22	31	36
59	11	12	22	32	37
60 or older	11	13	23	33	38

Table F.2 Age Correction Values in Decibels for Females

	Audiometric test frequencies (Hz)				
Years	1000	2000	3000	4000	6000
20 or younger	7	4	3	3	6
21	7	4	4	3	6
22	7	4	4	4	6
23	7	5	4	4	7
24	7	5	4	4	7
25	8	5	4	4	7
26	8	5	5	4	8
27	8	5	5	5	8
28	8	5	5	5	8
29	8	5	5	5	9
30	8	6	5	5	9
31	8	6	6	5	9
32	9	6	6	6	10
33	9	6	6	6	10
34	9	6	6	6	10
35	9	6	7	7	11
36	9	7	7	7	11
37	9	7	7	7	12
38	10	7	7	7	12
39	10	7	7	8	12
40	10	7	7	8	13
41	10	8	8	8	13
42	10	8	9	9	13
43	11	8	9	9	14
44	11	8	9	9	14
45	11	8	10	10	15
46	11	9	10	10	15
47	11	9	10	11	16
48	12	9	11	11	16
49	12	9	11	11	16
50	12	10	11	12	17
51	12	10	12	12	17
52	12	10	12	13	18
53	12	10	13	13	18
54	13	11	13	14	19
55	13	11	14	14	19
56	13	11	14	15	20
57	13	11	15	15	20
58	14	12	15	16	21
59	14	12	16	16	21
60 or older	14	12	16	17	22

(B) Finding the age at which the baseline audiogram was taken and recording the corresponding values of age corrections at 1000 Hz through 6000 Hz.

(ii) Subtract the values found in step (i)(A) from the value found in step (i)(B).

(iii) The differences calculated in step (ii) represented that portion of the change in hearing that may be due to aging.

Example: Employee is a 32-year-old male. The audiometric history for his right ear is shown in decibels below.

	Audiometric test frequency (Hz)				
Employee's age	1000	2000	3000	4000	6000
26	10	5	5	10	5
27*	0	0	0	5	5
28	0	0	0	10	5
29	5	0	5	15	5
30	0	5	10	20	10
31	5	10	20	15	15
32*	5	10	10	25	20

The audiogram at age 27 is considered the baseline since it shows the best hearing threshold levels. Asterisks have been used to identify the baseline and most recent audiogram. A threshold shift of 20 dB exists at 4000 Hz between the audiograms taken at ages 27 and 32.

(The threshold shift is computed by subtracting the hearing threshold at age 27, which was 5, from the hearing threshold at age 32, which is 25.) A retest audiogram has confirmed this shift. The contribution of aging to this change in hearing may be estimated in the following manner:

Go to Table F.1 and find the age correction values (in dB) for 4000 Hz at age 27 and age 32.

	Frequency (Hz)				
	1000	2000	3000	4000	6000
Age 32	6	5	7	10	14
Age 27	5	4	6	7	11
Difference	1	1	1	3	3

The difference represents the amount of hearing loss that may be attributed to aging in the time period between the baseline audiogram and the most recent audiogram. In this example, the difference at 4000 Hz is 3 dB. This value is subtracted from the hearing level at 4000 Hz, which in the most recent audiogram is 25 dB, yielding 22 dB after adjustment. Then the hearing threshold in the baseline audiogram at 4000 Hz (5) is subtracted from the adjusted annual audiogram hearing threshold at 4000 Hz (22). Thus the age-corrected threshold shift would be 17 dB (as opposed to a threshold shift of 20 dB without age correction).

8. APPENDIX G: MONITORING NOISE LEVELS NON-MANDATORY INFORMATIONAL APPENDIX

This appendix provide information to help employers comply with the noise monitoring obligations that are part of the hearing conservation amendment.

8.1. What Is the Purpose of Noise Monitoring?

This revised amendment requires that employees be placed in a hearing conservation program if they are exposed to average noise levels of 85 dB or greater during an 8 hour workday. In order to determine if exposures are at or above this level, it may be necessary to measure or monitor the actual noise levels in the workplace and to estimate the noise exposure or "dose" received by employees during the workday.

8.2. When Is it Necessary to Implement a Noise Monitoring Program?

It is not necessary for every employer to measure workplace noise. Noise monitoring or measuring must be conducted only when exposures are at or above 85 dB. Factors which suggest that noise exposures in the workplace may be at this level include employee complaints about the loudness of noise, indications that employees are losing their hearing, or noisy conditions which make normal conversation difficult. The employer should also consider any information available regarding noise emitted from specific machines. In addition, actual workplace noise measurements can suggest whether or not a monitoring program should be initiated.

8.3. How Is Noise Measured?

Basically, there are two different instruments to measure noise exposures: the sound level meter and the noise dosimeter. A sound level meter is a device that measures the intensity of sound at a given moment. Since sound level meters provide a measure of sound intensity at only one point in time, it is generally necessary to take a number of measurements at different times during the day to estimate noise exposure over a workday. If noise levels fluctuate, the amount of time noise remains at each of the various measured levels must be determined.

To estimate employee noise exposures with a sound level meter it is also generally necessary to take several measurements at different locations within the workplace. After appropriate sound level meter readings are obtained, people sometimes draw "maps" of the sound levels within different areas of the workplace. By using a sound level "map" and information on employee locations throughout the day, estimates of individual exposure levels can be developed. This measurement method is generally referred to as *area* noise monitoring.

A dosimeter is like a sound level meter except that it stores sound level measurements and integrates these measurements over time, providing an average noise exposure reading for a given period of time, such as an 8-hour workday. With a dosimeter, a microphone is attached to the employee's clothing and the exposure measurement is simply read at the end of the desired time period. A reader may be used to read-out the dosimeter's measurements. Since the dosimeter is worn by the employee, it measures noise levels in those locations in which the employee travels. A sound level meter can also be positioned within the immediate vicinity of the exposed worker to obtain an individual exposure estimate. Such procedures are generally referred to as a *personal* noise monitoring.

Area monitoring can be used to estimate noise exposure when the noise levels are relatively constant and employees are not mobile. In workplaces where employees move about in different areas or where the noise intensity tends to fluctuate over time, noise exposure is generally more accurately estimated by the personal monitoring approach.

In situations where personal monitoring is appropriate, proper positioning of the microphone is necessary to obtain accurate measurements. With a dosimeter, the microphone is generally located on the shoulder and remains in that position for the entire workday. With a sound level meter, the microphone is stationed near the employee's head, and the instrument is usually held by an individual who follows the employee as he or she moves about.

Manufacturer's instructions, contained in dosimeter and sound level meter operating manuals, should be followed for calibration and maintenance. To ensure accurate results, it is considered good professional practice to calibrate instruments before and after each use.

8.4. How Often Is it Necessary to Monitor Noise Levels?

The amendment requires that when there are significant changes in machinery or production processes that may result in increased noise levels, monitoring must be conducted to determine whether additional employees need to be included in the hearing conservation program. Many companies choose to remonitor periodically (once every year or two) to ensure that all exposed employees are included in their hearing conservation programs.

8.5. Where Can Equipment and Technical Advice Be Obtained?

Noise monitoring equipment may be either purchased or rented. Sound level meters cost about $500 to $1,000, while dosimeters range in price from about $750 to $1,500. Smaller companies may find it more economical to rent equipment rather than to purchase it. Names of equipment suppliers may be found in the telephone book (Yellow Pages) under headings such as: "Safety Equipment," "Industrial Hygiene," or "Engineers-Acoustical." In addition to providing information on obtaining noise monitoring equipment, many companies and individuals included under such listings can provide professional advice on how to conduct a valid noise monitoring program. Some audiological testing firms and industrial hygiene firms also provide noise monitoring services. Universities with audiology, industrial hygiene, or acoustical engineering departments may also provide information or may be able to help employers meet their obligations under this amendment.

Free, on-site assistance may be obtained from OSHA-supported state and private consultation organizations. These safety and health consultative entities generally give priority to the needs of small businesses. See the attached directory for a listing of organizations to contact for aid.

9. APPENDIX I: DEFINITIONS

These definitions apply to the following terms as used in paragraphs (cO through 9n) of 29 CFR 1910.95.

Action level—An 8-hour time-weighted average of 85 dB measured on the A-scale, slow response, or equivalently, a dose of fifty percent.

Audiogram—A chart, graph, or table resulting from an audiometric test showing an individual's hearing threshold levels as a function of frequency.

Audiologist—A professional, specializing in the study and rehabilitation of hearing, who is certified by the American Speech-Language-Hearing Association or licensed by a state board of examiners.

Baseline audiogram—The audiogram against which future audiograms are compared.

Criterion sound level—A sound level of 90 decibels.

Decibel (dB)—Unit of measurement of sound level.

Hertz (Hz)—Unit of measurement of frequency, numerically equal to cycles per second.

Medical pathology—A disorder or disease. For purposes of this regulation, a condition or disease affecting the ear, which should be treated by a physician specialist.

Noise dose—The ratio, expressed as a percentage, of (1) the time integral, over a stated time or event, of the 0.6 power of the measured SLOW exponential time-averaged, squared A-weighted sound pressure and (2) the product of the criterion duration (8 h) and the 0.6 power of the squared sound pressure corresponding to the criterion sound level (90 dB).

Noise dosimeter—An instrument that integrates a function of sound pressure over a period of time in such a manner that it directly indicates a noise dose.

Otolaryngologist—A physician specializing in diagnosis and treatment of disorders of the ear, nose and throat.

Representative exposure—Measurements of an employee's noise dose of 8-hour time-weighted average sound level that the employers deem to be representative of the exposures of other employees in the workplace.

Sound level—Ten times the common logarithm of the ratio of the squares of the measured A-weighted sound pressure to the square of the standard reference pressure of 20 micropascals. Unit: decibels (dB). For use with this regulation, SLOW time response, in accordance with ANSI S1.4–1971 (R1976), is required.

Sound level meter—An instrument for the measurement of sound level.

Time-weighted average sound level—That sound level, which is constant over an 8-hour exposure, would result in the same noise dose as is measured.

35

Formulae Differences in State and Federal Hearing Loss Compensation

Irvin Stander
Bureau of Workers' Compensation of Pennsylvania and Pennsylvania Bar Institute, Harrisburg, Pennsylvania; and Temple University, Philadelphia, Pennsylvania, USA

Robert T. Sataloff
Drexel University College of Medicine, American Institute for Voice and Ear Research, and Graduate Hospital, Philadelphia, Pennsylvania, USA

Many of the problems in hearing loss determination in compensation proceedings stem from the basic differences between medical and legal causation principles. The medical definition of cause and effect suggests scientific certainty, so that the alleged causative element must be one recognized as scientifically accurate. On the other hand, legal causation requires only that there be a cause-and-effect relationship within a reasonable degree of medical certainty, in others words—that the causal relationship between alleged work hazard and the resulting injury be one that is more likely than not. This is the dilemma that faces the doctor, the lawyer, and the adjudicator.

As we know, it is often difficult to quantify hearing loss on an absolutely accurate percentage scale. But such a determination is necessary to fix the amount of a financial allowance under the compensation laws. The measurement of hearing involves a complex analysis of the hearing level for a variety of pure tones and speech. The results of these measurements must then be related to an individual's ability to communicate effectively in a variety of listening situations.

However, when considering compensation awards we must try to relate the hearing impairment, as measured in decibels, to the projected handicap in percentages a person would have in ideal listening environments. This, of course, may be different from the real-world environment, but the real world requires us to make reasonable judgments on hearing impairment handicaps which can then be translated into specific loss compensation awards. In effect, we are compelled to use formulas or criteria for calculating hearing handicaps to arrive at impairment awards. These problems are not limited to determining the cause and extent of hearing impairments. They are present to the same degree in determining cause and extent relating to cardiovascular injuries, psychiatric impairment, and many other repetitive trauma injuries.

As would be expected, criteria or formulas for calculating hearing loss handicaps to arrive at disability awards vary extensively. The following is a brief overview of how the states and federal compensation agencies compare in their treatment of this problem:

1. Two states use the 1949 American Medical Association (AMA) formula.
2. Eighteen states use the 1959 formula adopted by the American Association of Ophthalmology and Otolaryngology (AAOO).
3. Two states and the Longshoremen's Act use the 1979 version of the AAOO formula (AAO formula).
4. The Federal Employees' Compensation Act uses the NIOSH standards coupled with the AMA formula.
5. One state uses the CHABA recommendation, differing as to audiometric frequencies used, and the low-fence provision.
6. Twenty-seven states, by far the majority, depend entirely on medical evidence, without specifying any particular formula or set of criteria.

Table 35.1 shows the formulas used by various states.

Before reviewing the formula differences in hearing loss determination in the several state and federal agencies, we shall survey the wide divergence of hearing loss benefits in these various jurisdictions. The formula provisions and the compensation rates of the various state and federal jurisdictions given in this chapter are subject to modification by their legislative bodies. The reader is urged to consult the current statutory provisions of the particular jurisdiction under consideration for the year of the injury. Following are the low and high benefits, as of January 1990:

For loss of hearing in one ear, claimant can get as much as $57,734 for the specific loss under the Federal Employees' Compensation Act and $33,832 under the Federal Longshoremen's Act. Among the states, the highest benefit for loss in one ear is awarded by Maryland at $56,500. On the low side for loss in one ear, compensation is awarded by Rhode Island at $5,400.

For loss of hearing in both ears, the specific loss benefits range from a high of $222,052 under the Federal Employees' Compensation Act and $130,124 under the Longshoremen's Act, with the highest state being Illinois at $172,560. On the low side are Rhode Island at $18,000.

The actual monetary award for a claimant in a particular jurisdiction is determined from the following factors, considered jointly: first, the number of weeks of compensation granted by that state for permanent or total hearing loss, or, where allowed, the number of weeks granted for various percentages of partial hearing loss, and second, the monetary value set for each week—usually on the basis of two-thirds of the claimant's average weekly wage, not in excess of a fixed maximum, generally set at 100% of the statewide average weekly wage for the particular year of the injury, and third, consideration of the formula, or other criteria, established by that state for the evaluation of hearing impairment.

Let us now consider the components of the various formulas, as well as the implications in the large group of states that depend entirely on medical evidence.

The original AMA method for determining hearing loss was first developed in 1942 and later modified in 1947. In consists of a weighted chart which used four frequencies: 512, 1024, 2048, and 4096 Hz. The 512- and 4096-Hz frequencies were valued at 15% each, the 1024-Hz frequency at 30%, and the 2048-Hz frequency at 40%. The ratio of hearing loss of the poorer ear to the better hearing ear was 1 to 5.

The next group to consider in depth the question of assessing hearing handicap was the 1958 Meeting of the Subcommittee on Noise of the American Academy of

Table 35.1 Occupational Hearing Loss Statutes

Formulas used by states for determining occupational hearing loss compensation

State	OHL compen-sation	Time limit to file	Separation from noise before filing	Minimum exposure in last employment	Hearing loss formula	Presbycusis deduction for aging	Deduction for pre-existing loss	Benefits (scheduled injury) One ear	Both	Choice of physician	Aural Rehabi-litation Provided?	Waiting period	Award for tinnitus (ringing noise)
AL	Y	1 y-q	NIS	NIS	ME	N	Y	$11,660	$35,860	Carrier	P	N	[a]
AK	Y	D-2 y	NIS	NIS	ME	N	N	[b]	[b]	Employee	P	N	Y
AZ	Y	1 y	NIS	NIS	59 AAOO	N	Y	$28,110	$84,006	Employee	Y	N	N
AR	Y	2 y	NIS	NIS	ME	N	N	$11,802	$44,398	Carrier	Y	N	P
CA	Y	D-1 y	NIS	NIS	79 AAOO	N	Y	[b]	[b]	Employee	P	N	Y
CO	Y	3-5 y	NIS	NIS	ME	N	Y	$9,991	$39,681	Carrier	Y	N	N
CT	Y	D-1 y	NIS	NIS	59 AAOO	N	N	$32,656	$97,968	Employee	N	N	P
DE	Y	D-y	NIS	NIS	ME	N	Y	$30,833	$71,944	Employee	Y	N	P
DC	Y	1 y	6 m	NIS	47 AMA	P	Y	$43,410	$166,964	Employee	Y	N	P
FL	Y	D-2 y	NIS	NIS	ME	N	Y	[c]	[c]	Carrier	Y	N	Y
GA	Y	1 y	6 m	90 d	59 AAOO	N	Y	$24,375	$48,750	Carrier	Y	6 m	N
HI	Y	D-1-2 y	NIS	NIS	59 AAOO	N	N	$26,998	$103,800	Employee	P	N	Y
ID	Y	1 y	NIS	NIS	ME	N	Y	[d]	$43,890	Carrier	Y	N	NIS
IL	Y	3 y	NIS	90 d	ME	N	Y	$43,140	$172,560	Employee	N	N	NIS
IN	Y[e]	2-3 y	NIS	NIS	ME	N	Y	$12,500	$39,500	Carrier	N	N	N
IA	Y	2 y	6 m	90 d	ME	Y	Y	$43,600	$152,600	Carrier	Y	N	Y
KS	Y	1 y	NIS	NIS	47 AMA	N	N	$10,980	$40,260	Carrier	Y	N	N
KY	Y	1-3	6 m	90 d	59 AAOO	N	Y	[d]	[d]	Employee	N	N	P
LA	Y	D-4 m	NIS	NIS	ME	Y	N	[f]	[f]	Carrier	Y	N	NIS
ME	Y	2 y	30 d	90 d	59 AAOO	Y	Y	$33,050	$88,200	Employee	Y	30 d	Y
MD	Y	2 y	6 m	90 d	59 AAOO	Y	Y	$56,500	$150,516	Employee	Y	6 m	NIS
MA	Y	D-1 y	NIS	NIS	ME	N	N	$20,297	$53,892	Employee	P	N	P
MI	N	D-4 m	NIS	NIS	ME	N	Y	[g]	[g]	Carrier	Y	N	N
MN	Y	D-3 y	NIS	NIS	ME	N	N	[c]	[c]	Employee	Y	N	P
MS	Y	D-2 y	NIS	NIS	ME	N	Y	$11,714	$43,929	Carrier	Y	N	P

(*Continued*)

Table 35.1 *Continued*

Formulas used by states for determining occupational hearing loss compensation

State	OHL compen-sation	Time limit to file	Separation from noise before filing	Minimum exposure in last employment	Hearing loss formula	Presbycusis deduction for aging	Deduction for pre-existing loss	Benefits (scheduled injury)		Choice of physician	Aural Rehabi-litation Provided?	Waiting period	Award for tinnitus (ringing noise)
								One ear	Both				
MO	Y	D-1 y	6 m	90 d	59 AAOO	Y	Y	$14,442	$53,051	Carrier	N	6 m	N
MT	Y	D-30 d	6 m	90 d	59 AAOO	Y	Y	[b]	[b]	Employee	Y	6 m	N
NE	Y	D-6 m	NIS	NIS	59 AAOO	N	Y	$23,400	[h]	Employee	N	N	Y
NV	Y	D-90 d	NIS	NIS	79 AAOO	N	Y	[b]	[b]	Employee	P	N	Y
NH	Y	2 y	NIS	NIS	59 AAOO	P	N	$25,200	$103,320	Carrier	Y	N	Y
NJ	Y	D-1/2 y	4 weeks	1 y 3 d 40 weeks	47 AMA (Plus 3,000 Hz)	P	N	$8,640	$50,400	Carrier	N	N	N
NM	Y	1 y	NIS	NIS	ME	NIS	Y	$15,682	$58,808	Carrier	NIS	N	N
NY	Y	90 d-2 y	3 m	90 d	59 AAOO	N	Y	$24,000	$60,000	Panel	P	6 m	Y
NC	Y	D-2 y	6 m	90 d	59 AAOO	N	Y	$39,200	$84,000	Carrier	Y	6 m	N
ND	Y	1 y	NIS	NIS	AMA	N	N	$20,850	$83,400	Employee	Y	5 d	N
OH	Y	6 m	NIS	NIS	ME	N	Y	$14,175	$70,875	Employee	Y	N	Y
OK	Y	D-3-18 m	NIS	NIS	AMA	N	N	$21,300	$63,900	Carrier	Y	N	Y
OR	Y	D-180 d To 5 y	NIS	NIS	500.6 K	Y	Y	$27,240	$87,168	Employee	Y	N	Y
PA	Y	3 y	NIS	NIS	AMA[i]	N	Y	$35,280	$152,880	Carrier	N	N	N
RI	Y	D-2 y	6 m	NIS	59 AAOO	P	Y	$5,400	$18,000	Employee	Y	6 m	N
SC	Y	D-2 y	NIS	NIS	ME	P	Y	$38,678	$79,773	Carrier	Y	N	Y
SD	Y	2 y	6 m	90 d	AMA	Y	Y	[j]	$61,200	Carrier	Y	N	N
TN	Y	1-3 y	NIS	NIS	ME	N	N	[j]	$77,250	Panel	N	N	P
TX	Y	6 m	6 m	NIS	59 AAOO	N	Y	[j]	$54,900	Carrier	Y	N	Y

UT	Y	1 y	6 m	NIS	ME	Y	Y	[j]	$32,500	Carrier	N	6 m	N
VT	Y	1 y	NIS	NIS	ME	N	Y	$37,804	$130,860	Employee	Y	N	Y
VA	Y	D-2 y	NIS	NIS	59 AAOO	N	Y	$26,700	$53,400	Panel	Y	N	Y
WA	Y	D-1 y	NIS	NIS	59 AAOO	N	Y	[k]	[k]	Employee	NIS	N	P
WW	Y	D-3 y	NIS	NIS	59 AAOO	N	N[b]	[b]	[b]	Employee	Y	N	P
WI	Y	NONE	14 d	90 d	CHABA	Y	Y	$6,624	$39,744	Employee	Y	2 m	N
WY	Y	D-1-3 y	NIS	NIS	ME	P	N	$12,400	$24,800	Employee	Y	N	N

[a]Alabama: Courts have the authority for establishing criteria to be used in determining the degree of hearing loss.
[b]Alaska, California, Montana, Nevada, and West Virginia: Ratings for compensation purposes are determined as a percentage of permanent total disability.
[c]Florida and Minnesota: Benefits paid according to degree of impairment and loss of earnings.
[d]Idaho and Kentucky: Weekly benefit amount is 55% of the state average weekly wage (SAWW).
[e]Indiana: Yes. If traumatic injury.
[f]Louisiana: No hearing loss schedule.
[g]Michigan: Hearing loss compensation based on wage loss.
[h]Nebraska: Loss of hearing in both ears constitutes permanent total disability.
[i]Pennsylvania: No award for under 10%.
[j]South Dakota, Tennessee, Texas, and Utah: Monaural loss is determined as a percentage of binaural loss.
[k]Washington: Benefits based on percentage of disability, or combination thereof, at the time of injury.
OHL, occupational hearing loss; P, possible; D, discovery rule; N, no; Y, yes; AMA, American Medical Association; AAOO, American Academy of Ophthalmology and Otolaryngology; NIS, not in statute, CHABA, Committee on Hearing Bioacoustics and Biomechanics (working group of the National Academy of Sciences); ANSI, American National Standards Institute.
Modified and updated from information in Herman AM, Anderson BE, Kerr TM (eds): State Workers' Compensation Laws, Washington, DC. U.S. Department of Labor, 1999, pp. 160–163.

Ophthalmology and Otolaryngology (AAOO). This conference included representatives from the areas of acoustics and bioacoustics, linguistics, otologists, audiologists, physicists, psychologists, speech pathologists, and many other specialists concerned with the problems of hearing.

The majority of this group believed that the ability to hear and repeat sentences correctly in quiet surroundings should be accepted as satisfactory hearing. The three frequencies they selected to the weighted equally were 500, 1000, and 2000 Hz, with a low fence of 25 dB. Hearing loss less than the low fence was considered satisfactory. In 1961, the American Medical Association Committee on Medical Ratings of Physical Impairments adopted the AAOO formula completely, thus replacing the AMA complex weighted formula of 1947.

In 1979 the AAOO formula was slightly revised by the American Academy of Otolaryngology and became known as the AAO formula. Under the AAO current formula, the percentage of hearing loss is calculated by taking the average, in decibels, of the hearing threshold levels in each ear for the frequencies of 500, 1000, 2000, and 3000 cps, or Hz. If such levels of hearing range 25 dB or less at these four frequencies, such losses are not considered significant. If the loss of hearing averages 92 dB (high fence), or more, at the four frequencies, then that constitutes total, or 100%, hearing loss. In 1984, using this formula as a basis, the AMA published its "Guide to the Evaluation of Permanent Impairment," and the formulas has now become the first officially recognized guide for the determination of hearing impairment. It is readily applied and can be used for determination of hearing impairment is one or both ears. It is also a formula that can be easily explained to a judge, a hearing official, or a jury that has the task of making the ultimate decision.

As noted in the agency formula summary above, several of the states have created their own formulas by making variations on the AMA or AAOO criteria, as follows:

1. Oregon uses an average of 500-, 1000-, 2000-, 4000-, and 6000-Hz frequencies with a low fence of 25 dB.
2. Wisconsin follows the recommendation of the CHABA Committee, which used an average of 1000, 2000, and 3000 Hz with low fence of 35 dB, and a better ear correction based on 4 to 1, instead of the customary 5 to 1.
3. California created a formula that uses an average of 500, 1000, 2000, and 3000 Hz with a low fence of 25 dB and the usual 5-to-1 better ear correction. This formula is the same as the 1979 AAO formula.
4. The Federal Employees' Compensation Act now (since September, 1984) follows the standard followed by the American Medical Association, which calculates the hearing loss for one ear (monaural) by testing at the 500-, 1000-, 2000-, and 3000-Hz frequency levels. The losses at each frequency are added up and averaged: and, using ANSI-ISO standards, the fence of 25 dB is deducted from the average loss. The result is multiplied by 1.5% to arrive at monaural hearing loss.
5. To determine the loss for both ears (binaural), the loss in each ear is calculated using the formula for monaural loss. The lesser loss is multiplied by 5, then added to the greater loss. The total is then divided by 6 to arrive at the binaural loss. As one can perceive, the loss in the ear which tested for better hearing will continue to be given five times the loss in the worse ear is determining binaural hearing loss for schedule award purposes.

Most formulas use a low fence of 25 dB and a high fence of 92 dB. The multiplying factor of 1.5% is calculated by subtracting the low fence from the high fence, which gives 67 dB, and applying that figure to the 100% total loss factor. Dividing 67 dB into 100% results approximately in the figure 1.5% dB used as the multiplier.

In most states there is no specified formula, and the compensation agency relies entirely on medical evidence for hearing loss assessment. This means that the determination of extent of hearing impairment is based solely on the physician's opinion. However, case decisions provide that the otologist must base his opinion on a recognized formula before it can be used to determine an award. This is usually the AAO formula. The state may also apply a special standard, for example, that the loss of hearing be for all practical intents and purposes.

The Council of State Governments drafted a Model Workers' Compensation Act in 1963 which gives some of the most explicit hearing loss standards and calculation instruction available. The following is a summary of these provisions:

1. Losses of hearing due to industrial noise for compensation purposes shall be confined to the frequencies of 500, 1000, and 2000 cps.
2. The percent of hearing loss shall be calculated as the average, in decibels, of the thresholds of hearing for the frequencies of 500, 1000, and 2000 cps. If the losses of hearing average 15 dB (ASA) or less in the three frequencies, such losses of hearing shall not then constitute any compensable hearing disability. If the losses of hearing average 83 dB or more in the three frequencies, then the same shall constitute and be total, or 100%, compensable hearing loss.
3. In measuring hearing impairment, the lowest measured losses in each of the three frequencies shall be added together and divided by 3 to determine the average decibel loss. For every decibel of loss exceeding 15 dB (ASA), an allowance of 1.5% shall be made, up to the maximum of 100%, which is reached at 82 dB.
4. In determining the binaural percentage of loss, the percentage of impairment in the better ear shall be multiplied by 5. The resulting figure shall be added to the percentage of impairment in the poorer ear and the sum of the two divided by 6.
5. There shall be deducted from the total average decibel loss, 0.5 dB for each year of the employee's age over 40 at the time of last exposure to industrial noise, to allow for nonoccupational causes.
6. No consideration shall be given to the question of whether or not the ability of an employee to understand speech is improved by the use of a hearing aid.

Although the specific numbers in the Model Act may be open to a revision since they date back to 1963 and were based on the ASA standard, the general method proposed is excellent, and deserves consideration by the states.

To give some idea of the relative results of hearing loss percentage determinations under the various formulas used by compensation agencies, here are the figures calculated in the case of an individual male applicant 65 years of age, with the following audiometric readings:

At hearing level of 500 Hz, the readings were right ear 20 and left 20. At 1000 Hz, the readings were right 30 and left 35. At 2000 Hz, the readings were both 45. At 3000 Hz, the readings were right 60 and left 65. At 4000 Hz, the readings were right 70 and left 75. At 6000 Hz, the readings were right 65 and left 70.

Following are the percentage of hearing loss determined under the various formulas applied to this case:

AAOO, 1959	10% loss
AAO, 1979	21% loss
NIOSH-FECA (Federal)	31% loss
Oregon (Oregon's formula includes 6000 Hz)	35% loss

The wide divergence in state compensation rates and benefits, as well as the disparity in the standards and criteria for determining hearing impairment, has led many critics of the state systems to propose the establishment of federal minimum standards in all aspects of the workers' compensation system.

36

Occupational Hearing Loss in the Railroad Industry

David J. Hickton
Burns, White and Hickton, Pittsburgh, Pennsylvania, USA

In January of 1991, the National Institutes of Health convened a symposium on hearing loss with special emphasis on occupational hearing loss and recreational noise exposures (1). At that conference, the Director of the National Institute of Occupational Safety and Health (NIOSH) predicted that the final census of occupational hearing loss claims in the railroad industry will exceed 100,000 (2). While this prediction appears excessive, the fact is that, as of January 1992, in excess of 60,000 claims have been presented by present and former railroad workers against their employers alleging some degree of hearing loss as a result of exposure to noise at work. This is contrasted with 1987 when less than 250 claims and lawsuits had been presented.

This problem has far-reaching consequences for all involved in the railroad industry. From the perspective of the railroads, mismanagement in handling this problem can lead to financial disaster and ultimately extinction. Further, efforts to comply with later-enacted government regulations can be distorted and used against even the most visionary company to misportray advanced hearing conservation efforts far exceeding the norm in American industry. Accordingly, the dimensions of the problem can serve to act as a disincentive for companies to do even more than what is required under current standards for fear of incurring untoward liability. This can increase the railroad's exposure to include possible regulatory liability.

From the standpoint of the railroad worker, it is self-evident that the continued financial stability of the railroads is desirable. Further, some railroad employers are skeptical of these claims and are discouraged from taking even ordinary safety measures with regard to noise. Simply put, many railroad employers view this type of claim as insignificant and not actionable due to the railroad's compliance with applicable governmental regulations. The existence of massive numbers of claims fuels this skepticism. Thus, unrestrained, the flow of railroad worker hearing loss claims could conceivably cause even greater numbers of hearing loss victims.

This chapter examines the unique nature of the occupational hearing loss problem in the railroad industry with reference to the legal standards, a profile of a typical Federal Employers' Liability Act (FELA) hearing loss claim, the regulatory scramble relating to the railroad industry, a survey of the medical and legal studies relating to noise and hearing loss in the railroad industry, the evolution of objective diagnostic criteria, and an objective assessment of the many unanswered questions presented today and the need for further research into these questions.

1. THE LEGAL STANDARDS

1.1. The Federal Employer's Liability Act

In virtually all of American industry (4), work-related injuries, including claims for hearing loss due to noise, are governed by a worker's compensation system. In the railroad industry, they are not governed by a worker's compensation system. Enacted by Congress in 1908, the Federal Employer's Liability Act (FELA), unlike most state worker's compensation statutes, requires the injured railroad worker to establish not only a job-related injury but that the injury was caused by his employer's negligence (5). In order to prevail, an injured railroad worker must establish the following elements by a preponderance of the evidence:

1. Breach of duty,
2. Foreseeability of harm,
3. Negligence, and
4. Causation (6).

Railroad liability is determined by whether the railroad's negligence played any part, even the slightest, in causing the employee's injury (7).

Also unique to FELA is a complete lack of guidance and limits on damage awards. Pursuant to most state worker's compensation statutes and under the federal employer's compensation administration, a mathematical formula establishes a threshold of loss and a mathematical calculation to assess damages (8). Under FELA, the injured railroad worker can recover damages for any or all of the following:

1. Pain and suffering,
2. Medical expenses,
3. Past wage loss, and
4. Future wage loss (9).

Recently, inventive plaintiffs' attorneys have begun to include psychological damages and hedonic damages to the list (10). Punitive damages are not permitted (11).

Under FELA, an employee's contributory negligence does not operate as a complete bar to recovery (12). However, the defense of contributory negligence is properly asserted to reduce any damage award by the percentage of negligence attributable to the employee (13). The defense of assumption of the risk is not proper in response to a FELA claim (14).

Under FELA, an injured employee is required to bring his claim within three years from the date when he knew he had a work-related injury (15). A physician's diagnosis is not required to commence the running of the statutes of limitations (16).

1.2. The Boiler Inspection Act

Enacted in 1911, the Boiler Inspection Act requires the railroad to ensure that any locomotive or appurtenances or parts thereof are in a proper and safe condition and that all parts and appurtenances have been inspected (18). Under the Boiler Inspection Act, absolute liability is imposed for any violation of the Act which is the proximate cause of the plaintiff's injury (19). The Boiler Inspection Act is expressly inapplicable to injuries where the locomotive in question is not in active service (20). Where the plaintiff's only alleged contact with the locomotives occurred in the railroad's shop facilities and involved repair work on engines or cars out of service, the Act and interpretive case law establish that a claim is not valid (21).

1.3. The Safety Appliance Act

The Safety Appliance Act was made law in 1893, and pertains to train braking and coupling systems and other safety appliances and parts necessary to make a moving train safe (23). Under this Act, the railroad is absolutely liable for any injury which occurs while a train is moving due to any malfunction of its safety appliances (24). In the context of the hearing loss litigation, this usually means an injury due to a malfunctioning horn or whistle which causes hearing loss.

1.4. Negligence *Per Se*

Workers claiming an occupational hearing loss may attempt to assert regulatory violations to establish negligence *per se* as evidence of the applicable standard of care or to preclude the raising of the comparative negligence defense (25).

Not infrequently, plaintiff's counsel will cite sound pressure level surveys which establish that noise levels in the subject workplace exceed those limits prescribed by the Occupational Safety and Health Administration (OSHA). The federal regulatory schemes do *not* prohibit noise levels in excess of those set forth therein but, rather, require that the individual worker not be exposed to excessively loud noises beyond a regulatorily established period of time. For example, the OSHA regulations do not prohibit noise levels exceeding 100 dB(A) in an industrial setting. Quite the contrary, noise exposure at a level of 100 dB(A) is permissible *so long as the employee is not exposed to noise at that level for a duration in excess of two hours in an eight-hour day.* Hence, the mere proof of noise exposure in excess of 90 dB(A) does not constitute proof of a regulatory violation. Similar analysis is applicable to the Locomotive Cab Noise Standard which, likewise, does not prohibit noise exposure in excess of 90 dB(A) but rather requires that exposures of greater than 90 dB(A) be limited in duration. In summary, the plaintiff has the burden of proof of establishing that he was exposed to noise levels exceeding those prescribed by the applicable regulation *and* that his exposure to such noise levels exceeded the permissible period of duration prescribed by the applicable regulation.

The negligence *per se* doctrine operates to engraft a particular legislative standard onto the general standard of care imposed by traditional tort law concepts, that standard of care to which an ordinarily prudent person would conform his conduct (26). To establish negligence, the jury under the negligence *per se* doctrine need not decide whether the defendant acted as an ordinarily prudent person would have acted under the circumstances; the panel must merely decide whether or not the relevant statute or regulation has been violated and, if it has, the defendant is deemed negligent as a matter of law (27). It is, therefore, common for plaintiff's counsel to argue that proof of a regulatory violation constitutes negligence *per se*, obviating further proof of the applicable standard of care and a breach thereof. But regulations promulgated under OSHA cannot furnish a basis for jury instruction on negligence *per se* (28). In a negligence action, OSHA regulations do provide evidence of the standard of care demanded of employers, but they neither create an implied cause of action nor establish negligence *per se* (29). This rule is consistent with 29 U.S.C. Section 653(b)(4) which provides that OSHA shall not be construed to supersede, diminish or affect the common law or the statutory duties or liabilities of employers with respect to injury to their employees. This is also consistent with cases which have held that a negligence *per se* rule is inconsistent with Section 653(b)(4) of the Occupational Safety and Health Act (30). Application of the negligence *per se* doctrine would affect the rights, duties or liabilities of the employer toward its employees in contravention of Section 653(b)(4) (31). Almost universally "in a negligent action, regulations promulgated under ... [OSHA] provide evidence of the standard of care exacted of employers ..." (32). Moreover, Section 53 of the FELA provides, in pertinent part: "[N]o ... employee who may be injured or skilled shall be held to have been guilty of contributory negligence in any case where the violation by [a] common carrier of any statute enacted for the safety for employees contributed to the injury or death of such employee" (33).

The matter of whether a worker's compensation scheme would be a "better" method for compensating the injured railroad employee has been and continued to be debated. Proponents of reform have for the last several years debated before Congress the repeal of FELA (34). Examples of severely injured workers who have lost cases and received no compensation are compared and contrasted with examples of workers with little or no injury who have become millionaires. Regardless of the merit of FELA as a means to deal with the single incident workplace injury, it cannot be debated that the existence of FELA with its requirement of proof of negligence and lack of a measurable standard of injury or damage has contributed greatly to the explosion of hearing loss claims alleging chronic exposure to noise in the railroad industry. If one accepts the notion that we will all experience some measure of hearing loss due to the normal aging process, it is conceivable that every single railroad worker can present a theoretical claim of occupational hearing loss due to the standard which exists under FELA. In that regard alone, it can be said that the existence of large numbers of hearing loss claims in the railroad industry may bear less upon the existence of noise and the presence or absence of an enlightened hearing conservation program in that industry and more to the existence of an antiquated legal standard which permits an unlimited number of claims.

2. THE TYPICAL FELA OCCUPATIONAL HEARING LOSS CLAIM

Hearing loss claims under FELA are in two general categories. First, there is the single incident claim where the employee sustained hearing loss due to a single noise exposure such as a horn, whistle blast, or explosion. The claim is presented under FELA, the Boiler Inspection Act, and the Safety Appliance Act. The claim is very straightforward and turns usually upon the railroad's conduct and any contributory fault of the employee.

The second broad category of railroad hearing loss involves claims of hearing loss due to chronic noise exposure over the course of the worker's employment with the railroad. These claims involve three principal crafts: train and enginemen, carmen, and maintenance-of-way workers. It is this type of claim by these workers which has accounted for the explosion of claims in the industry.

2.1. The Plaintiff Profile

The typical railroad occupational hearing loss plaintiff fits the following profile:

1. Age 50 or older,
2. Thirty years or more of service with the railroad,
3. Hunter,
4. Veteran (World War II, Korea, Vietnam),
5. No lost time from work,
6. No medical bills,
7. No prior medical treatment for hearing loss,
8. No or minimal complaints,
9. Claim is presented after mass screening of workers by union in conjunction with plaintiff law firm,
10. Plaintiff claims no awareness of hearing loss until within last three years, and
11. No use of hearing protection on job until 1980s.

In each individual case, a host of other relevant case-specific factors may be present including:

1. Family history of hearing loss,
2. History of viral infections,
3. History of recreational noise exposure,
4. History of noise exposure in other jobs,
5. History of systemic disorder relevant to hearing ability, and
6. History of blows to head.

2.2. Course of the Claim

Under FELA, concurrent jurisdiction in the state and federal courts is provided (35). The case is commenced by the filing of a Complaint in an appropriate jurisdiction, although many claimants have begun to file claims directly with their employer in the hopes of avoiding the need for a lawsuit.

Upon filing of the Complaint and an Answer on behalf of the railroad, the discovery process ensues. Generally, the plaintiff provides the following information to the defendant railroads so the railroad may evaluate the claim:

1. Medical authorization to review all prior medical records,
2. Plaintiff's expert ENT report,
3. Answers to various interrogatories relating to prior personal and work history, and
4. Independent medical examination by the ENT doctor selected by the defendant.

From the standpoint of the plaintiff's discovery as to the railroad, the plaintiff generally seeks the following information:

1. Noise exposure data relating to the locations where plaintiff worked,
2. The identify of plaintiff's supervisors and coworkers who may be deposed on history of complaints or noise levels,
3. Documents relating to the railroad's hearing conservation program, and
4. Documents relating to the railroad's general awareness of the association between excessive noise in the workplace and hearing loss.

Many plaintiff's attorneys present claims on behalf of railroad workers under FELA relying principally or entirely upon a speech presented to the medical and surgical subsection of the Association of American Railroads (AAR) in 1966 by Dr. Aram Glorig (36). Dr. Glorig, along with Dr. Joseph Sataloff, is recognized as one of the leading authorities in the field of occupational hearing loss. As will be seen later, Dr. Glorig was active in the initial work in this subject and continues in his 80s today to be involved in this field.

In 1966, Dr. Glorig was invited by the Association of American Railroads to speak to the AAR about hearing conservation. As of that date, Dr. Glorig had had no direct experience with railroads or railroad noise levels and had only seen a couple of Union Pacific trainmen who had been referred to his clinic due to hearing loss (37).

Dr. Glorig spoke in extremely strong terms to the AAR medical surgical group giving an impassioned plea for the institution of hearing conservation in railroad industry and, indeed, all American industry (38).

Careful reading of Dr. Glorig's remarks in 1966 and, indeed, in many depositions of Dr. Glorig taken on the subject, is revealing. Given Dr. Glorig's lack of specific knowledge as to railroad noise exposure, and the fact that the text of his speech before the AAR was virtually identical to an article he wrote earlier (39) as well as the presentation he made before Congress in connection with the enactment of the Hearing Conservation Amendment to OSHA in 1983 (40), it is clear that there is no basis to conclude that Dr. Glorig had special knowledge that the railroads were somehow out of step with the rest of American industry (41). Further, Dr. Glorig has acknowledged that much of American industry even today has yet to take steps on hearing conservation, and that it took the U.S. government and the U.S. military well over twenty years to heed Dr. Glorig's message (42). A critical review of Dr. Glorig's remarks and subsequent steps taken by the AAR to evaluate those remarks on a case-by-case and craft-by-craft basis throughout the 1970s, coupled with the fact that many railroads had enlightened and advanced hearing conservation programs in the 1960s and 1970s, and the fact that most railroads had a program during the early 1980s, has taken the initial "fizz" out of Dr. Glorig's famous speech rendering it interesting historically, but of questionable relevance.

Other plaintiffs attempt to prove railroad negligence through a broad discussion of the medical literature and government publications. One example of this is the citation to the National Safety Council Accident Prevention Manual, circa 1955 (43). This document dramatically shows the confusion among scientists with regard to the noise issue and with reference to other safety issues as well. A fair review of the National Safety Council Accident Prevention Manual references to workplace noise discloses many incorrect statements relating to damage risk criteria, noise engineering and hearing protection devices (44). Judged by what we know today, this reference is, at best, incorrect, and at work wholly inadequate given the mix of safety issues in a dynamic railroad work setting. Moreover, a review of the same manual discloses many recommendations that industrial safety equipment be manufactured of asbestos products (45). Clearly, today, this recommendation is unacceptable.

None of the foregoing is a criticism of the well-reasoned and well-intended advice of Dr. Glorig and others. The point remains that it is not possible to simply act as the continuum of knowledge unfolds on this or any other subject. Discriminating care must be employed to determine what steps must or should be taken given each circumstance.

Upon the conclusion of the discovery phase of the case, the plaintiff's case is ready for trial. In any case filed under FELA, the plaintiff is entitled to a trial by jury. The trial of an occupational hearing loss case can take as few as two days and as much as two weeks depending on the case-specific circumstances relating to the claim. If the plaintiff's claim is supported by noise engineering analysis and there is a detailed and involved medical history which must be considered, the case can be much more involved. On the other hand, if the plaintiff has a relatively clean medical profile and there is little contested about the nature and scope of the workplace, the case can turn almost entirely on the jury's determination of the negligence of the railroad and the contributory negligence of the plaintiff.

Given the lack of guidance with regard to damages under FELA, the question of what the plaintiff is entitled to is solely within the province of the jury. These awards range from nothing for the plaintiff to isolated reports of substantial awards.

3. MEDICAL/LEGAL STUDIES RELATING TO NOISE AND HEARING LOSS IN THE RAILROAD INDUSTRY

The volume of occupational hearing loss claims in the railroad industry has inspired tremendous research into the historical studies relating noise exposure and hearing loss in the railroad industry. While this research has unearthed a handful of remote works, it is generally accepted that the first real awareness of any association between noise exposure and hearing loss occurred after the conclusion of World War II (46). The Z24-X2 committee of the American Standards Association in 1947 reported the initial scientific study of the association between noise and hearing loss (47). This work was commenced as a result of the experience of vast numbers of World War II veterans who had sustained hearing losses during their service overseas (48).

With specific attention to the railroad industry, there had been a few isolated references in the medical literature to incidents of hearing loss among railroad workers (49). The references are merely anecdotal and at least one report hypothesizes causes of hearing loss other than noise (50).

In fact, a well-known study in Finland considered the early work in this area to be generally inconclusive:

> The results obtained varied greatly: according to some, workers with no special "occupational deafness" could be observed in engine crews, while others considered deafness marked enough to be a danger to safety ...
> Most investigators have arrived at the result that working on an engine is injurious to the hearing (citing Moos, Schwabach and others) ...
> Other investigators consider that this occupation is not in itself dangerous to the hearing and according to them hearing impairment and diseases of the ear are scarcely more frequent among train crews than in any other occupation. ... (51)

It is generally recognized that there have been three landmark studies of noise exposure and hearing loss which deal directly with the railroad worker. These are: The Wisconsin State Fair Study, the Kilmer Study or FRA/DOT Engine Cab Survey, and the Clark and Popelka Study of Hearing Levels of Union Pacific Trainmen.

3.1. The Wisconsin State Fair Study—1954

The initial work of the Z24-X2 committee led to a landmark survey of the hearing levels of the U.S. population throughout the U.S. Public Health Service in 1952 (52). Thereafter, the next significant study was known as the Wisconsin State Fair Study published in 1957 (53).

Due to an interest in potential hearing loss claims as a result of a state worker's compensation decision, the Wisconsin Manufacturer's Association commissioned and sponsored the Research Center of the Subcommittee on Noise in Industry of the Committee on Conservation of Hearing of the American Academy of Ophthalmology and Otolaryngology to conduct a massive survey of the hearing levels of participants at the 1954 Wisconsin State Fair. Ten audiometric test booths were constructed and more than 3500 hearing tests were administered to willing fairgoers (54). Physical examinations were administered and the test subjects completed a questionnaire with regard to personal history including work history, noise exposure information, and events or illnesses which could effect auditory thresholds (55).

The stated purpose of the study was to sample the hearing levels of an industrialized population to determine:

1. The effects of noise exposure on hearing loss,
2. To establish standards of normal hearing that would serve as a baseline from which to evaluate hearing loss,
3. To evaluate the effects of physiological and physical variables such as age and sex and pathological variables, such as ototoxic disease on hearing loss, and
4. To study the relations of pure tone thresholds to speech reception thresholds (56).

It is the first of these stated purposes which is directly pertinent to the analysis here. However, the guidance we gain from this landmark study on the other topics is also relevant to other issues discussed later.

At the conclusion of the testing, the audiograms were evaluated and classified. The Study identified six different categories of occupation. These are as follows:

Category 1: Drop hammer, riveting, and chipping on large metal plates
Category 2: Riveting and chipping on small castings

Category 3: Using other pneumatic tools (air-driven drills and screwdrivers, sandblasters, jack hammers) saws, planers, tools with steam or air hiss, tumblers

Category 4: Grinding, welding, using lathes, drills, bores, milling machines, furnaces, mixers.

Category 5: Maintenance (work such as that done by plumbers, electricians, carpenters, painters, guards, watchmen, stock room and tool crib employees, clipper and fork lift operators, timekeepers, inspectors, packers, loaders, hoist and conveyer operators)

Category 6: Construction, railroading, mining work in power plants, work of marine engineer, brewing (57)

These categories represent classes of occupations which reflect degrees of hearing losses with Category 1 representing the most hearing loss and Category 6 the least (58). Also, Category 5 was used to describe jobs for which noise levels would be low, aperiodic and insignificant in terms of exposure time (59). Category 6 was described as "waste basket" for jobs unassigned to any other category (60).

The inclusion of railroading as a Category 6 occupation in the Wisconsin State Fair Study may suggest that, based on the data in that study, railroading is a relatively risk-free occupation, although Category 6 was *also* a "waste basket" category for those occupations or classes of test subjects where there was insufficient data to draw any firm conclusion (61). As has been pointed out by Dr. Glorig, other occupations within Category 6 are now known *today* to involve high noise exposure (62). It is also true that the inclusion of riveting in Category 1, a task which was and is within the job responsibility of a certain class of railroad workers, would suggest that at least those workers who did riveting would be in a high risk category.

Nevertheless, the Wisconsin State Fair Study is universally accepted as a monumental work and it stands, even today, as a landmark in our evolving and continuing understanding of such subjects as aging and hearing mechanism, and the relationship between noise exposure and hearing loss as measured on an audiogram. Accordingly, in assessing the question of what the railroads knew or should have known about conditions within their workplace, as they set about to provide a safe place for their workers to work, the Wisconsin State Fair Study is consistently cited for the propositions that, as of that time, it was not generally accepted that railroading was an occupation which could lead to abnormal levels of hearing loss.

3.2. The Kilmer/FRA-DOT Assessment of Locomotive Crew-in-Cab Occupational Noise Exposure

In 1979, under sponsorship of the U.S. Department of Transportation and The Federal Railroad Administration (FRA), Roger Kilmer of the National Bureau of Standards conducted an extensive survey to investigate and assess diesel locomotive noise environments in terms of crew noise exposure (63). Recognizing that the railroad industry, unlike most other U.S. industries, was and is not subject to safety regulations of the Occupational Safety and Health Administration, but, instead, is subject to regulations of the FRA, Kilmer studied 18 test runs using 16 different locomotives on a variety of operational modes for varied trip lengths to determine whether locomotive crews were overexposed based upon the noise exposure criteria employed under OSHA (64).

The general conclusion of the Kilmer Study was that the noise exposure experienced by locomotive crews was within acceptable limits (65). Kilmer found this to be true due to the fact that high noise sources operate for short periods of time (66). Kilmer did find overexposure on onc run and did further recognize a potential for "over exposure" if criteria more stringent than those employed by OSHA were used (67).

The Kilmer Study has been generally accepted for the proposition that the typical railroad locomotive crew member (engineer, conductor, brakeman) is not likely to be subjected to sufficient workplace noise to cause noise-induced hearing loss.

3.3. Clark and Popelka Study

In the February, 1989 edition of *Laryngoscope*, Clark and Popelka published their work reporting their evaluation of the hearing sensitivity of 9427 railroad train crew members (68). Recognizing the results of Kilmer's work and drawing upon the hearing test data available from a comprehensive hearing conservation program at the Union Pacific Railroad, Clark and Popelka set out to determine whether substantial noise-induced hearing loss is caused by working in railroad locomotives (69).

Using multiple regression analysis, Clark and Popelka first classified the railroad workers by degree of loss and age (70). Then, these results were compared to a control population obtained from an International Standards Organizations data base of individuals from an industrialized country who were not exposed to noise (71).

A comparison of the data reflecting hearing levels of railroad trainmen to that of the control population showed little differences between the two samples (72). Further, a consideration of the effects of age and years of service among the trainmen shows no significant effects due to years of service (73). Based upon these results, Clark and Popelka conclude that railroad trainmen are not typically exposed to hazardous occupational noise and, further, that the analysis of Clark and Popelka validates the findings of the Kilmer Study (74).

Given its recent publication, the Clark and Popelka Study is currently under academic review. Thus far, two criticisms have been levied. First, the statistical analysis employed by Clark and Popelka has been challenged (75). Second, critics have challenged the selection of the control population (76). Clark and Popelka have squarely answered these criticisms (77).

In April, 1991, an article purporting to refute Clark and Popelka's work was published (78). However, a review of this work discloses a finding of "statistically significant but relatively small losses (2–7 dB for the different percentiles), in hearing sensitivity at frequencies above 2000 Hz and found in trainmen over the age of 45 years unscreened for sociocusis or exposure to gun noise when their hearing levels are compared to those of liked-aged unscreened males of the general United States population, as given in Annex B, ISO 1999" (79). Accordingly, the finding of this "rebuttal" does not refute but supports the work and conclusions of Clark and Popelka.

Thus, it would seem that the Clark and Popelka Study is valid and consistent with the Kilmer noise exposure study. Accordingly, taken together, the Kilmer and Clark and Popelka studies stand for the proposition that there is insufficient noise exposure experienced by railroad trainmen to reflect abnormal levels of hearing loss.

4. REGULATORY SCRAMBLE: OSHA V. FRA AND BEYOND

While compliance with government noise regulations is not an absolute defense in a FELA action (80), it is appropriate for the railroad defendant to prove or attempt to prove absence of negligence by citing compliance with appropriate government standards. Certainly, failure to comply with government standards may be decided by the jury to be evidence of negligence (81). From the standpoint of "what ought to be," it would seem self-evident that American railroads ought to be in the position to rely upon the appropriateness of government standards as the benchmark for hearing conservation efforts as well as the standard for negligence liability in any FELA trial (82).

Regrettably, what should be simple questions of what is the appropriate noise standard and what is its effective date have been and continue to be a source of much confusion. The railroads are principally regulated by the FRA which is an arm of the U.S. Department of Transportation. As will be seen below, the FRA enacted noise standards and regulations in 1980, and the railroads have, since that time, generally acceded to the FRA regulations and enforcement.

The evolving OSHA noise standard which was enacted in 1971, and was substantially refined in 1983, is quite different from the FRA standard as well as hearing conservation program requirements (83). The question of OSHA jurisdiction over the railroads was extensively litigated from 1974 through 1978. The question of OSHA jurisdiction was partially resolved by a policy statement issued by the FRA in 1978 (84). However, the controversy still remains to this day and it is *clear* only that railroad shops are within the jurisdiction of OSHA. Logically, it would follow that all other areas fall within the jurisdiction of the FRA. However, many other crafts and work tasks within the railroad industry are left in limbo as to whether OSHA or FRA regulations apply. It is most certain that this regulatory debate between OSHA and FRA has created tremendous confusion with regard to hearing conservation efforts in the railway industry and certainly has detracted from those efforts.

In addition, standards have been published by the International Standards Organization (85) which do not set specific damage risk criteria but which are often cited or miscited in FELA actions as applicable or at least relevant to the question of the railroad employer's conduct. These references present yet another and different benchmark for assessing railroad conduct and imposing railroad liability.

Still further, throughout the 1970s, there was tremendous interest by the Environmental Protection Agency (EPA) in enacting community noise as opposed to occupational noise standards (86). The misapplication of these EPA community noise standards to a consideration of an individual railroad worker's noise exposure is not infrequent. Moreover, the process by which the EPA conducted its analysis directly and through consultants led to yet another layer of regulatory personnel seeking access to railway property. Acting partly on their own initiative, and many times in concert with FRA, many railroads began in the 1970s and early 1980s to limit access to government regulators or investigators unless they were cleared or approved by the FRA or until the regulatory jurisdictional battle was resolved. This action by some railroads has been misportrayed as an effort to avoid government regulation entirely. In virtually any action where the issue of compliance with some noise standards is at issue, the railroad defendant is forced to explain its policies of access to government investigators during the 1970s and 1980s against the suggestion by plaintiffs that the railroad was acting in bad faith.

What follows is an examination of the content of each of the regulations which seems to have some relevance to the typical noise exposure of the railroad worker as well as more specific descriptions of the impact of this regulatory scramble.

4.1. The Evolving OSHA Standard

The Occupational Safety and Health Act was enacted by Congress in 1970 (87). The initial OSHA enactment contained no specific reference to a noise regulation but did empower the Secretary of Labor to enact appropriate regulations pursuant to OSHA (88).

In 1971, the Secretary of Labor, pursuant to OSHA, recognized the regulation contained in the 1968 Walsh-Healy Act which pertained to employers' contracts with the government in excess of $10,000 (89). Although the Walsh-Healy Act was immediately retracted upon enactment in 1968, the Secretary determined its regulatory provisions to be appropriate (90). OSHA provided the following guidance with respect to noise exposure over a typical work day (91):

90 dBA	8 h
95 dBA	4 h
100 dBA	2 h
105 dBA	1 h
110 dBA	1/2 h
115 dBA	1/4 h or less

Thus, as of 1971, OSHA endorsed the standard of 90 dBA for a time-weighted average over eight hours and the 5-dB trading ratio. The initial OSHA regulation contained no guidance or rulemaking with respect to a hearing conservation program other than a suggestion that employers were to take steps to use "feasible administrative or engineering controls or provide personal protection equipment" for employees who were in noise in excess of 90 dBA for eight hours a day (92).

Since 1971, a debate has continued between interested parties as to whether the OSHA standard of 90 dBA was appropriate. Many advocated that the standard be reduced to 85 dBA and others, including interested parties within the EPA and the American Counsel of Industrial Hygienists, sought a standard as low as 75 dBA (93). This led, in part, to the formation of the Inter-Industry Noise Committee organized by one of the principal editors of this text (Joseph Sataloff) with constituent membership of government, labor, academic, and industry leaders to survey all of the previous tests and studies and do such additional testing as was necessary to determine the appropriate level at which any damage will occur over an assumed work life of 30 years of greater.

The Inter-Industry Noise Study (94) which is still recognized as the finest of its kind both in terms of validity and constituent interests shows that there was no greater degree of hearing loss between samples of 82 dBA and 92 dBA with the exception of 6 to 9 dB at 3000, 4000, and 6000 Hz compared to a control population (95). Accordingly, the Inter-Industry Noise Study is often cited for the proposition that the OSHA standard of 90 dBA for eight hours is within the safe standard of noise exposure and the injury does not likely occur until the workers are exposed to 92 dBA over an eight-hour day (96).

In 1979, OSHA began the process of issuing a hearing conservation amendment to its initial regulations (97). This amendment was first enacted in 1981, later withdrawn, and ultimately enacted in 1983 with an effective date of March 1, 1984 (98). Pursuant to the terms of the OSHA Hearing Conservation Amendment, elaborate guidance was given to industrial employers for the first time by the government as to the components of a comprehensive hearing conservation program (99). The OSHA Hearing Conservation Amendment established an action level of 85 dBA over an eight-hour period and called for bi-annual audiograms and mandatory hearing protection for workers in such a noise exposure environment. The OSHA Hearing Conservation Amendment has over 20 sections with requirements on noise engineering, audiometric testing, noise exposure analysis, employee training, hearing protection provision, and recordkeeping, all of which serve to comprise portions of a comprehensive regulatory framework for protecting the hearing of the American industrial worker (100).

Most commentators recognize that the OSHA regulations were overdue upon enactment in 1983 (101). Few can debate that they stand as the first real enactment of any type that provided any guidance on the subject; that they were enacted with the broad participation of all interested segments including labor, management, government, and the medical community; and that the consequences to all interested parties would have been dramatically different had OSHA adopted a different standard.

4.2. Federal Railroad Administration Act—1970

Congress enacted the Federal Railroad Administration Act in 1970 providing for establishment of the Federal Railroad Administration to oversee the operations of the nation's railroads (102).

With reference to occupational noise exposure regulations, the Federal Railroad Administration adopted the Locomotive Cab Noise Standard in 1980 which provided a noise exposure regulation as follows (103):

87 dBA	12 h
90 dBA	8 h
95 dBA	4 h
100 dBA	2 h
105 dBA	1 h
110 dBA	1/2 h
115 dBA	1/4 h or less

Thus, the Federal Railroad Administration requirements on railroad noise exposure are roughly the same as the initial OSHA enactment. Some differences do exist. Under the FRA Standard, noise shall not exceed 115 dBA, while OSHA permits measures up to 140 dBA (104). Further, the Federal Railroad Administration has enacted no follow-up legislation or at any time called for or required a hearing conservation program in the railroad industry (105). If it can be established that the workers' exposure did not exceed the FRA standard, if the FRA controls, there is no regulatory violation (106). Presence or absence of a hearing conservation program is not relevant under the Federal Railroad Administration standards (107).

4.3. FRA v. OSHA Debate

Both FRA and OSHA have statutorily imposed duties to protect the safety and health of America's workers. But during the 1970s, the American railroad industry became immersed in a dispute between these two agencies regarding their respective jurisdiction prerogatives. The industry, because of interagency "turf fighting," found itself trying to "serve two masters" (108) and, consequently, not serving either very well.

The Federal Railroad Safety Act of 1970 (109) (enacted nine weeks prior to OSHA) was passed to "promote safety in all areas of railroad operations...." (110). The Occupational Safety and Health Act of 1970 (111) was enacted shortly thereafter and was passed in effort to "assure so far as possible every working man and woman in the nation safe and healthful working conditions...." (112). By its plain terms, the Occupational Safety and Health Act would also include the protection of the safety and health of American railroad workers. However, Congress was cognizant of the occupational safety and health protection functions provided by other federal agencies (e.g., FRA) and, consequently, provided that when another agency "exercises" its statutory authority for workplace safety and health conditions, OSHA is preempted from acting (113).

In 1971, OSHA issued regulations (enforced by the Department of Labor) which governed the permissible noise exposure for American workers (114). Again by its terms this would necessarily include American railroad workers (and, of course, since its enactment. OSHA had promulgated numerous other regulations regarding worker safety and health that embraced railroad workers). However, the following year the FRA published an advanced notice of proposed rulemaking indicating FRA's intention to issue specific standards regarding railroad employee safety and health (115). The FRA had historically addressed railroad worker safety in their role as chief enforcer of the Safety Appliance Act (116), the Hours of Service Act (117), the Signal Inspection Act (118), the Locomotive Inspection Act (119), and the Noise Control Act (120). The FRA terminated its rulemaking efforts in March of 1978 and, instead, issued a policy statement which purported to delineate the respective areas of OSHA and FRA jurisdictional authority for protecting railroad employees (121). By its plain terms, OSHA recognized the jurisdiction of other agencies including the FRA (122):

> Nothing in this chapter shall apply to working conditions of employees with respect to which other federal agencies ... exercise statutory authority to prescribe or enforce regulations affecting occupational safety and health.

Between 1970 and 1978 the American railroad industry received mixed signals from Washington regarding who the industry should look to for guidance and accountability. Although the FRA seemed to be primarily responsible for promoting the safety of railroad operations, the "exercising" of that responsibility (arguably sufficient to preempt OSHA) consisted primarily of an anemic and interminable rulemaking process which was eventually terminated; in the meantime, OSHA had acted apparently comfortable in the belief that the FRA's rulemaking process was not a sufficient enough "exercise" of power to divest it (OSHA) of the authority to regulate the operations of the railroads in the interests of worker safety and health.

During that period, the American railroad industry, uncertain over whether FRA or OSHA had authority over railroad worker safety and to what extent, turned to the courts for relief. They were not alone; other industries were also increasingly resorting to litigation in their own efforts to comply with apparently overlapping agency authority over worker safety and health (123).

In 1976 the Southern Railway Company, the Southern Pacific Transportation Company, and the Baltimore and Ohio Railroad Company challenged in three federal courts OSHA's authority to regulate the railroad industry (124). The railroads argued that since FRA had regulated railroad safety as far back as 1893 and because FRA, pursuant to the mandates of the Federal Railroad Act of 1970, had in 1975 published an advance notice of proposed rulemaking regarding railroad employee safety and health, the FRA has "exercised" sufficient power to preempt OSHA of authority to regulate the railroads. The courts rejected the preemption argument and concluded that the "exercise" contemplated by Congress was more than simply the initiation of or consideration of proposed rules. Hence, OSHA could retain control over certain areas of railroad worker safety until the FRA "exercised" its power in a more conclusive fashion.

The FRA "concluded" the exercise of its power in regulating the railroad industry for the time being when in 1978 it abruptly terminated its rulemaking efforts and in a policy statement issued that same year abdicated much of the responsibility for protecting the safety and health of railroad workers to OSHA (125). According to the policy statement, FRA has jurisdiction over all areas of railroad safety that are directly related to railroad operations, including those maintenance or repair duties performed by maintenance-of-way workers. The three areas of railroad operation that are regulated by FRA include: (a) tracks, road beds, and associated structures such as bridges; (b) equipment; and (c) human factors, such as hours of service, OSHA then regulates those shops, servicing areas, and other locations not directly related to operating railroads (126).

The FRA Locomotive Cab Noise Standard arguably governs noise levels encountered by railroad "train and engine" workers. The OSHA Hearing Conservation Amendment clearly limits noise exposure at locations not directly related to operating railroads. The amendment protects workers in the railroad industry traditionally described as "carmen." And in 1988, Congress ordered FRA to issue regulations for the safety of "maintenance-of-way employees." Those have yet to be issued (127).

4.4. Environmental Protection Agency (EPA)

The 1970s were punctuated by congressional efforts to respond to the public's heightened awareness of the potential health effects of exposure to deleterious levels of noise. The Environmental Protection Agency was created to establish and enforce environmental protection standards "consistent with national goals pursuant to the 1970 enactment of the National Environmental Policy Act" (128). The Noise Control Act of 1972 (129) (the Act) declared that "inadequately controlled noise presents a growing danger to the health and welfare of the Nation's population, particularly in urban areas" (130). The Act made it a policy of the United States to promote an environment for all Americans free from noise that jeopardizes their health and welfare (131). The Act purported to "establish a means for effective coordination of federal research and activities in noise control, to authorize the establishment of federal noise emission standards for products distributed in commerce, and to provide information to the public respecting the noise emission and noise reduction characteristics of such products" (132).

Congress made the Administrator or the Environmental Protection Agency responsible for coordinating the program of all federal agencies relating to noise research and noise control (133). The Administrator was directed to "develop and publish criteria with respect to noise which reflected the scientific knowledge most useful in indicating the kind and extent of all identifiable effects on the public health or welfare which may be expected from differing quantities and qualities of noise" (134).

Though the problems associated with deleterious levels of noise were painted with a wide brush by Congress in the Act, such problems within the railroad industry were presented narrowly for special attention. Congress specifically directed the Administrator to propose, after consultation with the Secretary of Transportation, noise emission regulations for surface carriers engaged in interstate commerce by railroad (135). The proposed regulations were to include noise emission standards setting such limits on noise emissions resulting from operation of the equipment and facilities of surface carriers engaged in interstate commerce by railroad which reflected the degree of noise reduction achievable through the application of the best available technology, taking into account the cost of compliance (136). The Secretary of Transportation was required, after consultation with the Administrator, to promulgate regulations to ensure compliance with all standards established by the Administrator (137).

The 1980 noise emission standards proposed and established by the Administrator and the regulations promulgated concurrently by the Secretary address what can fairly be described as "community noise" generated by the American railroad industry. There is little evidence to suggest that the Administrator or the Secretary sought at that time to regulate the type or amount of noise to which railroad workers are exposed within the scope of their employment (138), although an expansive reading of the Act seems to permit the Administrator such latitude. The Standards actually deal with permissible noise emissions 100 feet from locomotive and rail car operations. Nevertheless, plaintiffs commonly allege in their Complaints that the defendant railroad negligently exposed them to hazardous levels of noise in violation of standards established by the EPA. This is not prevalent in cases where the alleged sources of deleterious noise are coupling operations, retarders, or load cell test stands.

The noise emission standards for those operations established by the Administrator and enforced by the FRA in compliance with the Act regulate noise levels "only at receiving property locations" (139), that is, property that is "on or beyond the railroad facility boundary" (140). The "property" embraced by the standards means "any residential or commercial property that receives the sound from railroad facility operations but is not owned or operated by a railroad" (141). So while the noise emission standards do speak to specific noise sources, the measurement of said emissions for compliance purposes takes place off of railroad property away from the noise source. Nowhere in these standards or regulations are the requirements that the noise be measured "at the source." In fact, they require the contrary (142).

In 1972 Congress ordered the EPA to establish noise emission standards in the railroad industry. Thus far, the EPA has chosen primarily to control levels of noise on property on or beyond the railroad facility boundary. A cause of action was created for persons on that property aggrieved by the failure of the railroads to alleviate "community noise" as required by the law. While this effort to control "community noise" continues at the EPA, it appears that efforts to reduce "occupational noise" continue with more fervor at other agencies of the government.

4.5. International Standards Organization (ISO)

In 1975 the International Standards Organization (ISO) issued ISO Standard 1999:1975 "Acoustics-Assessment of Occupational Noise Exposure for Hearing Conservation Purposes" (143). This Standard was updated in 1990 (144). The ISO reference is *not* a noise exposure standard as the name suggests but, instead, is a resource document which uses a noise measurement technique (Leq), an energy limit of 85 to 90 dBA, and the 3-dB trading ratio (145). Using these criteria, ISO 1999 establishes predictable levels of hearing loss in a given population (146).

ISO 1999 is frequently mis-cited in FELA occupational hearing loss litigation. Some "experts" contend it is the enlightened European standard (147). This is false as many European countries employ the 90-dBA/5-dB trading ratio standard (148). Other commentators cite the ISO reference as a standard which has been exceeded to attempt to establish the negligence *per se* (149). This is patently untrue by the plain terms of the document (150). ISO 1999 is also employed as a means to calculate railroad noise levels particularly, it seems, where resort to OSHA/FRA measuring standards would yield an "underexposure" (151).

The significance of this regulatory overlap and confusion is that many railroads believed or were led to believe that compliance with the FRA regulations were both safe practice and all that was required. As the OSHA regulations have been issued, most railroads have made their standards stricter out of a sense of duty and upon the theory that the most conservative regulatory scheme is the safest one. As the regulations continue to shift, the question arises as to what standard a railroad is entitled to follow, or alternatively, whether a railroad has an obligation to independently assess the proper course through review of all literature and standards.

Clearly, as of today, no industry, including the railroad industry, can defend the absence of an appropriate hearing conservation program. However, what is necessary as a component of such a program or what is required from a regulatory standpoint are still topics of debate. It is in this context that railroads are subject to questions of judgment everyday in FELA hearing loss claims. Given all of the safety issues on a railroad, it is neither possible or sensible to simply take every step declared necessary by any or all commentators for the ultimate result could be a shut-down of operations. It is also frustrating to the railroad defendant and unfair to the industry to be deemed negligent where good faith compliance with FRA and OSHA regulations has been established.

5. THE EVOLUTION OF OBJECTIVE DIAGNOSTIC AND DAMAGE RISK CRITERIA

Without question, the central question in every FELA occupational hearing loss claim is: When did the worker/plaintiff develop permanent hearing loss? As a threshold matter, this question resolves the issue of the timeliness of the case as FELA provides that a claim must be presented within three years (152). Numerous courts have disposed of hearing loss claims presented under FELA on the basis that the worker knew, or should have known, of his hearing loss and the basis of his claim against his employer more than three years before the case was filed (153). To date, in excess of fifty reported decisions and many more by judges who have not written opinions but ruled summarily have been catalogued (154).

Of equal significance is the fact that the onset of hearing loss is the definitive fact with regard to whether the injury is occupationally related or caused by some other etiology. This owes to the fact that while occupationally induced hearing loss has been medically and legally classified as both a disease and an injury, it has always been recognized as a process which occurs immediately upon noise exposure to a maximum loss within 10 to 15 years (155). Accordingly, a railroad worker of 40 years seniority who maintains that he only within the last three years has had symptoms of hearing loss is either: (a) testifying untruthfully to "protect" the statute of limitations, (b) suffering from hearing loss due to another cause, or (c) experiencing a combination of causes for his hearing loss with no new injury or symptoms relating to the principally high-frequency loss component caused by noise.

This dynamic issue has been most controversial over the course of the railroad hearing loss litigation explosion of the last several years. A strong minority view that hearing loss injury continues has emerged. Some otolaryngology experts stress the insidious onset of occupational hearing loss (156). However insidious this injury/disease may be, it is most certainly observable if present, in fact, early in the worklife. Other experts insist the noise exposure has increased or the process of constant exposure has "spread" the loss into speech frequencies, presenting more symptoms of the injury (157). This is somewhat implausible due to improvements in equipment and the pervasive presence of hearing conservation measures taken in the last 15 years by some railroads.

This issue has festered in the railroad litigation, in part, because of a failure of the medical community to establish accepted diagnostic criteria for occupational hearing loss. Incredibly, in 1991 it is not uncommon for well-credentialed treating ENT doctors or hired ENT experts to make a diagnosis of occupational noise-induced hearing loss relying solely upon audiometric test results reflecting a "noise notch" and the workers' complaints of noise at work (158). Little investigation is conducted; no noise levels are examined and the doctor presumes sufficient noise at work if the audiogram suggests noise as the cause (159). The following deposition example is not atypical (160):

Q. . . . What were your essential (criteria for) development of occupational noise-induced hearing loss? Tick them off for us?

A. Exposure for 42 years and noise in his occupation. Obvious evidence of acoustic trauma on the metrics.

* * *

Q. What noise? What are you talking about? What are you talking about? What criteria?

A. If you go on the railroad you'll found (sic) out. You haven't heard the whistle?

Q. What whistles are blowing in the cabooses?

A. You hear the whistles and go through tunnels and you're on the caboose or wherever you are on the train.

* * *

Q. And you don't know what the decibel level of a caboose is?

A. No.

Q. You don't know the decibel level of a horn?

A. I know that if most of it is around 78 except when you add the two together that it's more than 78.

Q. As we sit here today, do you have a sufficient understanding that you can say that exposure to certain noise causes hearing loss where exposure to a noise less than that does not?

A. I think a certain exposure to a noise in certain individuals can cause a hearing loss.

Q. What level of noise?

A. Supposedly, according to the rules and regulations, it's 85 decibels for a certain length of time.

Q. What length of time?

A. It can be over two hours or within an eight hour day.

Under this flawed analysis, virtually *all* railroad workers will be claimants as all industrial workers including railroad workers can complain of the presence of noise at work and owing to presbycusis alone, all will experience, over time, some loss of hearing.

Other doctors give expert testimony that a railroad worker with profoundly asymmetrical hearing loss, a history of hunting, and a work history which includes hearing conservation measures (including hearing protection devices) are suffering from hearing loss exclusively caused by railroad noise (161).

Dr. Glorig, among others, has spoken clearly to the inadequacy of this "medicine" (162).

Q. . . . You believe that one of the problems in your profession, that being the ear, nose and throat doctors, is that many of the doctors are qualified to do surgery and other things but are not qualified to render opinions on noise exposure and hearing loss?

A. Absolutely, unfortunately true.

Q. And this has been a problem throughout your career?

A. Absolutely.

Q. This problem has been, really, the impetus behind the formulation of these eight criteria (American College of Occupational Medicine)?
A. That's correct.
Q. Because your view it is very dangerous that many doctors, by virtue of either circumstantial evidence or proximity in time, tie a loss to a noise exposure without doing a full diagnosis?
A. I've seen doctors call a one-eared loss with normal hearing in the other ear due to noise. That's a bunch of baloney. It cannot happen (163).

The examples are endless. Plaintiff's experts have offered testimony that: (a) noise exposure today can cause hearing loss 20 to 25 years later (164); (b) damage from noise causes retocochlear progressive diseases (165); (c) hearing "loss" is any deviation from audiometric zero (166); (d) claimant with average hearing level at all frequencies of 10 dB with no frequency worse than 20 dB is "permanently and seriously impaired" (167); (e) worker with 20 years hunting and three battlefield terms in World War II has *no* contribution from nonworkplace noise and has injury alleged to have occurred due to exposure to the office computer (168); (f) high blood pressure with family history of same alleged to be caused by chronic noise exposure (169); and (g) claimant with family history of mother, father, two uncles, and three brothers with *identical* hearing loss had *no* contribution from genetic history (171):

Q. You are not an ear doctor are you?
A. No, I'm not.

* * *

Q. You are not qualified to do audiometry, is that correct?
A. No, I'm not.
Q. You are not qualified to render the diagnosis of an ear doctor as to the cause of anyone's problem, is the correct?
A. I could give the diagnosis as to the cause of anyone's problem that — the same as (sic) anybody else who has an M.D.

* * *

Q. . . . Do you know what the characteristic pattern on an audiogram is for someone who has noise-induced hearing loss?
A. Yes.
Q. Can you describe?
A. They have loss in the high-frequency areas; loss of hearing. So when the noise gets high pitched, they lose the hearing. So they can't hear whistling?

* * *

Q. Do you have enough information to tell us what the characteristic pattern is for someone who has hearing loss by virtue of aging?
A. I don't know what that is.
Q. Okay. What is aging called by ear doctors, in terms of hearing loss? You don't know what that is?
A. But there is no hearing loss through aging. I don't know what you are talking about.

* * *

Q. Have you ever heard the term "presbycusis"?
A. I've heard of it.

Q. What does it mean to you?

A. Old age hearing. No such thing. Hearing loss (172).

Incredibly, this internal medicine specialist by virtue of liberal rules relating to expert testimony has recently testified in a burn accident and a diesel exposure case in addition to the above-referenced hearing loss claim (173).

In 1989, the American College of Occupational Medicine adopted eight criteria for the diagnosis of occupational hearing loss. The committee of experts developing the consensus criteria included Joseph Sataloff, Aram Glorig, and others. These criteria are (174):

1. It is always sensorineural affecting the hair cells in the inner ear.
2. It is almost always bilateral. Audiometric patterns are usually similar bilaterally.
3. It almost never produces a profound hearing loss. Usually low-frequency limits are about 40 dB and high-frequency limits about 75 dB.
4. Once the exposure to noise is discontinued there is no significant further progression of hearing loss as a result of the noise exposure.
5. Previous noise-induced hearing loss does not make the ear more sensitive to further noise exposure. As the hearing threshold increases, the rate of loss decreases.
6. The earliest damage to the inner ears reflects a loss at 3000, 4000, and 6000 Hz. There is always far more loss at 3000, 4000, and 6000 Hz than 500, 1000, and 2000 Hz. The greatest loss usually occurs at 4000 Hz. The higher and lower frequencies take longer to be affected than the 3000- to 6000-Hz range.
7. Given stable exposure conditions, losses at 3000, 4000, and 6000 Hz will usually reach a maximal level in about 10 to 15 years.
8. Continuous noise exposure over the years is more damaging than interrupted exposure to noise, which permits the ear to have a rest period (174).

Despite the credentials of the authors of the criteria, the need for them and their apparent logic, the eight criteria of the American College of Occupational Medicine have not yet been universally accepted in railroad hearing loss litigation (175). This is principally due to the fact that the criteria, if objectively employed, will eliminate many claims and they have been opposed aggressively by plaintiffs despite their scientific merit and authority.

The hope remains that the criteria adopted by the American College of Occupational Medicine will become uniformly accepted. Only objective medically recognized and accepted criteria can serve to eliminate the many meritless claims as well as expedite handling of the justified claims filed.

6 CURRENT ISSUES—FURTHER RESEARCH

One of the hidden potential benefits of a litigation explosion like the hearing loss claims in the railroad industry is the advancement of science which occurs due to the identification of unanswered questions, and the research which goes into those questions for the purpose of winning the case or cases at hand.

While many questions abound, it seems that three principal issues require resolution. The resolution of these questions would provide scientific benefits which transcend the current legal problem.

6.1. The Isolation of the Noise Component of a Given Hearing Loss

Under FELA, the defendant railroad is liable only for that damage which is caused by railroad negligence (176).

It is certainly likely that, given presbycusis and sociocusis alone, no one has a hearing loss which is due solely to noise at work. Indeed, OSHA and others have published an "age correction" table (included in this book) which attempts to account for expected hearing loss at a given age (177). The Wisconsin State Fair Survey includes a chart which shows the mean (average) hearing level/loss without regard to history or cause of the sample, clearly illustrating hearing deterioration with advancing age (178).

A comparison of this chart to any census of railroad hearing loss claimants would suggest hearing levels in the railroad workers "better than normal." Obviously, this comparison, alone, is an inadequate measure of "normalcy" or the relative contribution of various factors contributing to the hearing loss presented. Nevertheless, absent comparisons of this type, it is impossible for ENT doctors and certainly juries to parcel occupational noise exposure and other societal causes, let alone assess specific issues such as family history, viral infection, drug exposures, and other causes.

The medical community must resolve this problem through adequate review of clinical test results and case histories, and commitment to academic research, as well.

6.2. Is Noise-Induced Hearing Loss Progressive?

For years, it seems that the medical community has grappled with the question of whether long-term noise exposure causes a progressive injury. No definitive research

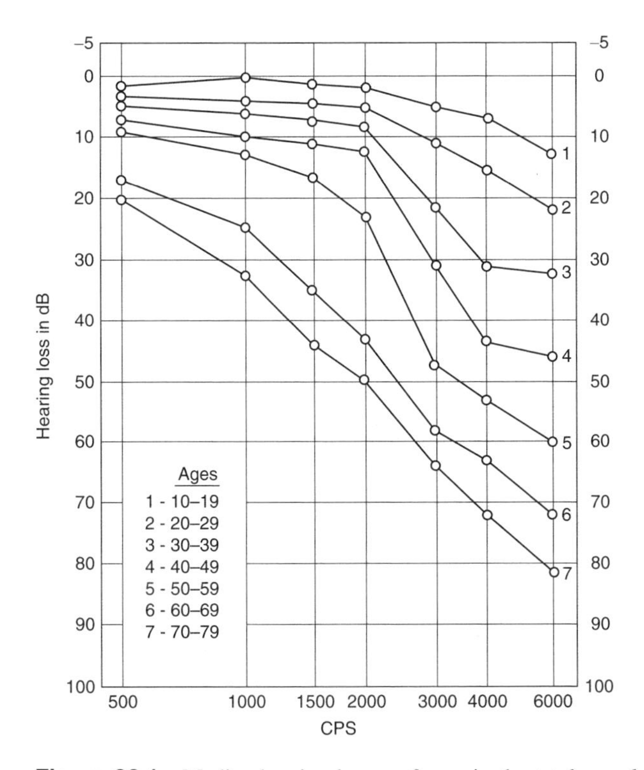

Figure 36.1 Median hearing losses of men in the total sample (left ear only).

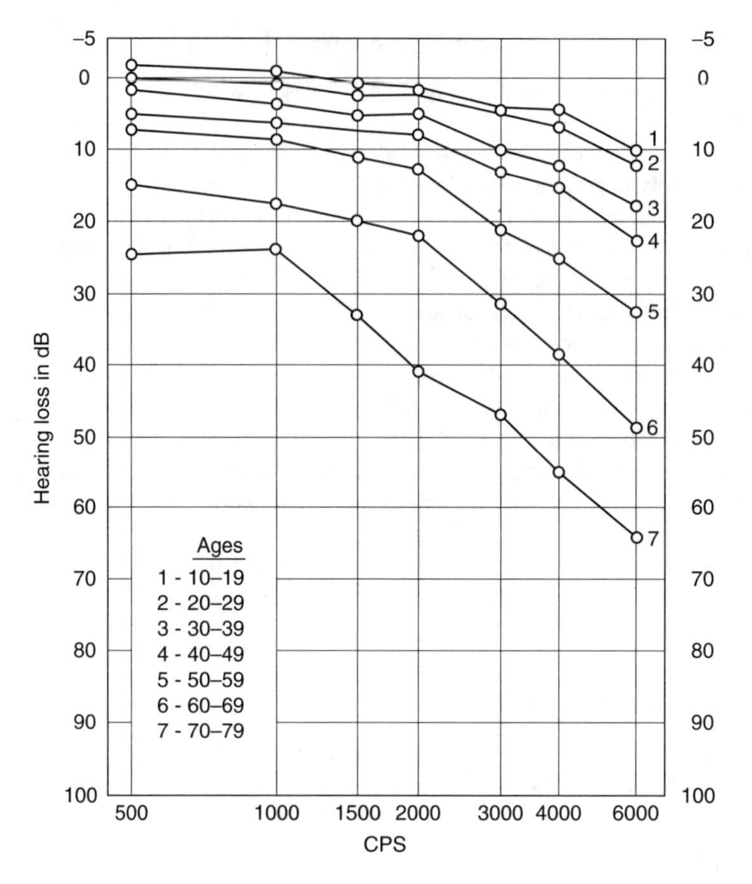

Figure 36.2 Median hearing losses of women in the total sample (left ear only).

exists to resolve the question. Owing to clinical evidence that injured persons generally experienced a stabilization of hearing levels upon removal from noise, and credible experts agree that noise-induced hearing loss is not a progressive phenomenon (179). The American College of Occupational Medicine Criteria endorse this concept (180).

In the context of FELA hearing loss litigation, it would seem that a worker's history of a progressive condition would constitute a meritless or defensible claim. Unfortunately, this is not so. Experts on behalf of plaintiffs frequently offer that the injury may be progressive or that a new injury has occurred (181):

Q. What about in the area of progression or non-progression, anything characteristic in terms of long-term noise-induced hearing loss?
A. It generally continues to progress.
Q. What about in the area of onset? Anything characteristic of long-term noise-induced hearing loss, as to its onset?
A. Generally, an insidious onset.

* * *

Q. When you say that one of the characteristics of it is progressive, what causes it to be progressive?
A. Continued exposure to the noise.
Q. What if an individual is removed from the noisy environment, does it continue to progress?

A. It may, yes.
Q. What causes it to continue to progress despite removal from the environment?
A. The damage to the inner ear hair cells is the area of the damage that is caused by noise. Once these hair cells are damaged, the supporting cells which are part of the inner ear beneath the hair cells will degenerate. As the supporting cells degenerate, the nerve fibers that go from the inner ear that subsequently end up in the cortex of the brain, go through the cranial nerves, will degenerate. If you take an individual away from noise, there will be no further damage to the hair cells, however, if the hair cells have been damaged over following years, these supporting cells may degenerate with a subsequent degeneration of the nerve.

Without regard to the imperative in the defense of the railroad hearing loss litigation, a scientific survey of the medical literature and clinical exams to prove or disprove the accepted truth that the injury is not progressive is necessary. Certainly this question would not only assist in the diagnosis but would also provide valuable insight into the nature of the injury.

6.3. Can Noise-Induced Hearing Loss Be Restored?

An accepted medical truism presented in railroad hearing loss cases is that noise-induced hearing loss is a permanent injury for which there is no cure or treatment. Recent advances in cochlear implants and hearing aid technology have provided some answers and hope for excellent treatment of hearing impairment. However, at the present time, there is no cure for occupational hearing loss.

From a lay, as opposed to a medical, standpoint, the explanation of hearing loss due to long-term noise exposure seems similar to orthopedic injury due to long-term exposure to heavy lifting or to the loss of eyesight that lawyers and other professions experience due to excessive reading and bad lighting. In the example of loss of a portion of eyesight, few would argue that being required to wear glasses restores eyesight. Yet, similarly, few would argue that any substantial disabling unrehabilitative injury has occurred with such diminution in visual acuity. The explanation for the difference in attitude regarding visual versus auditory acuity lies principally in societal impressions of hearing disability. However, in large part, we still do not fully understand, medically, the precise nature of the damage which occurs due to noise insult and how that injury may be restored. In simple terms, noise damage is often described by plaintiffs in litigation as similar to spinal cord injury. While this comparison is strained, even in that example tremendous scientific progress has been made. The prospect exists, hopefully in the near term, of restoring the use of limbs to spinal accident victims. If so, why cannot hair cell damage due to noise be restored?

These are challenges for the medical community which will not only end suffering but squarely solve medical/legal problems in the railroad and other industries.

7. CONCLUSION

When the dust settles, the occupational hearing loss problem may cost the railroad industry over $1 billion. While railroads may be noisy, they are not the noisiest industry in our country (182); and yet the financial impact of the cost upon the railroads will be disproportionately large.

Like most problems, the problem of hearing loss in the railroad industry is ultimately going to be solved by the medical, political, or business communities. The legal community plays a role but it cannot solve the problem alone. Unlike most problems of this nature, the legal system under FELA is not an entirely accurate barometer of the conduct of the defendant, the injury to the plaintiff, or the pervasiveness of the problem because vast numbers of FELA hearing loss claims would have no basis in liability or damages under state and federal compensation systems.

In the last several years, less than 100 of the pending FELA hearing loss claims have been tried to verdict. No pattern or guidance is apparent from these trials, as the results are widely varied. It seems that there have been a greater percentage of defendant's verdicts, however, several extremely large plaintiffs' verdicts occurred, as well. Many more cases have been settled. Again, no pattern has developed although approximately one-third of the cases settled or disposed of by courts prior to trial or settlement are time-barred. Further, the average settlement is modest as expected given the absence of any complaints, medical bills, or lost time from work.

Ultimately the final resolution of this large problem depends upon the objective and aggressive role of the ENT physician answering unanswered questions and applying medical appropriate diagnostic criteria in an objective and fair manner.

REFERENCES/NOTES

1. Noise and hearing loss: Consensus conference, *JAMA 23*:3185–3190 (1990).
2. Noise and hearing loss: Consensus conference, *JAMA 23*:3185–3190 (1990).
3. 45 U.S.C. §§51–60.
4. Only the railroad and maritime industries are excepted from workers' compensation laws. The FELA is the exclusive remedy for injured railroad employees superseding state workmen's compensation laws. *New York Central & Hudson River R. Co.* v. *Tonsellito*, 244 U.S. 360 (1917); the Jones Act of 1920, 45 U.S.C. §688, governs actions by maritime workers.
5. *Green* v. *River Terminal Railway Co.*, 763 F.2d 805, 808 (6th Cir. 1985) and cases cited therein. *See also Wilkerson* v. *McCarthy*, 366 U.S. 53 (1949).
6. *Practico* v. *Portland Terminal Co.*, 783 F.2d 255, 262 and 265 (1st Cir. 1985) (traditional negligence principles apply): Restatement of Torts (Second) §281.
7. *Rogers* v. *Missouri Pacific Railroad*, 352 U.S. 500 (1957).
8. Sataloff and Sataloff, eds., *Occupational Hearing Loss*, 3rd Ed., pp. 553–539 (1987).
9. *Chesapeake & Ohio Railway Co.* v. *Kelly*, 241 U.S. 485, 491, 36 S.Ct. 630, 632, 60 L.Ed. 1117 (1916) (manner of determining damages under FELA must be settled according to general principles of law as administered in the Federal courts). *See also Michigan Central RR* v. *Vreeland*, 227 U.S. 59 (1913) (pain and suffering); *Wagner* v. *Reading Co.*, 428 F.2d 289 (3rd Cir. 1970) (medical expenses); *Norfolk and Western Railway Co.* v. *Liepelt*, 444 U.S. 490 (1980) (past wages); *Pfeifer* v. *Jones & Laughlin Steel Corp.*, 678 F.2d 453 (3rd Cir. 1982), *cert. granted*, 103 S.Ct. 50, *vacated*, 103 S.Ct. 2541 (1983), *on remand*, 711 F.2d 570 (3rd Cir. 1983) (future wages).
10. In the matters of *Gardner, et al.* v. *Norfolk and Western Railway Co.*, CA No. 87C94-TS, *et seq.*, in the Circuit Court of Brooke County, West Virginia.
11. *Michigan Central Railway Co.* v. *Vreeland*, 277 U.S. 59, 71–72, 33 S.Ct. 192, 196, 57 L.Ed. 417 (1913); *Gulf Colorado and Santa Fe Railway Co.* v. *McGinnis*, 228 U.S. 173, 175–176, 33 S.Ct. 426, 427, 57 L.Ed. 785 (1913) (only compensatory damages are available in FELA actions). *See also Kozar* v. *Chesapeake & Ohio Railway*, 499 F.2d 1238, 1240–1243 (6th Cir. 1971).
12. 45 U.S.C. §53.
13. *Dale* v. *Baltimore & Ohio Railway Co.*, 000 Pa. 000, 552, A.2d 1037 (1989).
14. 45 U.S.C. §54.
15. 45 U.S.C. §56.

16. *Emmons* v. *Southern Pacific Transportation Co.*, 701 F.2d 1112, 1122 (5th Cir. 1983); *see also Urie* v. *Thompson*, 337 U.S. 163, 169 S.Ct. 1018, 93 L.Ed. 1282 (1949).
17. 45 U.S.C. §22 *et seq.* (Boiler Inspection Act) (§§22–34).
18. 45 U.S.C. §23.
19. 45 U.S.C. §23; *Green* v. *River Terminal Railway Co.*, 763 F.2d 805 (6th Cir. 1985); *Urie* v. *Thompson.*
20. 45 U.S.C. §23; *Garcia* v. *Burlington Northern R. Co.*, 818 F.2d 713 (10th Cir. 1987) (active service required).
21. 45 U.S.C. §23; *Steer* v. *Burlington Northern, Inc.*, 720 F.2d 975, 976–977 (8th Cir. 1983) (citing *Angell* v. *The Chesapeake and Ohio Railway Co.*) 618 F.2d 260, 262 (4th Cir. 1980); *see also Simpkins* v. *The Baltimore and Ohio Railway Co.*, 449 F.Supp. 613 (S.D. Ohio 1976).
22. 45 U.S.C. §1, *et seq.;* (Fed. Safety Appliance Act) [§§1–16].
23. 45 U.S.C. §§1–7.
24. 45 U.S.C. §§1–7. *St. Louis I.M. & S. R. Co.* v. *Taylor*, 210 U.S. 281, 28 S.Ct. 616, 52 L.Ed. 1061 (1908) (45 U.S.C. §§2–7 supplants common law rule of reasonable care on part of carrier as to providing appliances defined and specified therein, and imposes an absolute duty upon carrier to equip their cars with appliances complying with the standards established under 45 U.S.C. §§1–70); *Chicago B. & Q. R. Co.* v. *U.S.*, 220 U.S. 559, 31 S.Ct. 612, 55 L.Ed. 582 (1911) (duty under 45 U.S.C. §§1–7 is absolute and unqualified).
25. *Albrecht* v. *Baltimore and Ohio R. Co.*, 808 F.2d 329 (4th Cir. 1987); *Practico* v. *Portland Terminal Co.*, 783 F.2d 255 (1st Cir. 1985); *Bertholf* v. *Burlington Northern R.*, 402 F. Supp. 171 (E.D. Wa. 1975); *Moody* v. *Boston & Maine Corp.*, 921 F.2d 1 (1st Cir. 1990).
26. *Wendland* v. *Ridgefield Construction Services, Inc.*, 439 A.2d 954, 956 (Conn. 1981).
27. *Wendland* v. *Ridgefield Construction Services, Inc.*, 439 A.2d 956 (Conn. 1981).
28. *Wendland* v. *Ridgefield Construction Services, Inc.*, 439 A.2d 957 (Conn. 1981).
29. *Albrecht* v. *Baltimore and Ohio R. Co.*, 808 F.2d 329, 332 (4th Cir.) (1987).
30. *Albrecht* v. *Baltimore and Ohio R. Co.*, 808 F.2d 332 (4th Cir.) (1987) (citations omitted).
31. *See Albrecht*, 808 F.2d at 332–333; *see also Hebel* v. *Conrail*, 475 N.E.2d 652 (Ind. 1985) (violation of Occupational Safety and Health Administration regulations not admissible as evidence of negligence *per se* in order to impose strict or absolute liability in action against railroad under FELA, citing 653(b)(4); and *Wendland*, 439 A.2d at 957.
32. *Albrecht*, 808 F.2d at 332.
33. 45 U.S.C. 53; *see also Practico* v. *Portland Terminal Co.*, 783 F.2d 255 (1st Cir. 1985); and *Bertholf* v. *Burlington Northern R.*, 402 F.Supp. 171 (E.D. Wash. 1975).
34. *Hearings on the Federal Employer's Liability Act Before the Subcommittee on Transportation and Hazardous Materials of the House Commission on Energy and Commerce*, 101st Congress, 1st Session (1989). During the 101st Congress, bills were introduced repealing FELA and putting the railroad industry under the jurisdiction of the state worker's compensation system. S. 3214, 101st Congress, 2nd Session, 136 Cong. Rec. S15553 (1990); H.R. 5853, 101st Congress, 2nd Session, 136 Cong. Rec. H10104 (1990).
35. 45 U.S.C. §56.
36. Transcript of Proceedings, 46th Membership Meeting of American Association of Railroads, Medical and Surgical Officers, February 23–25 1966, San Francisco, California.
37. Transcript of Deposition, *Allen, et al.* v. *Norfolk and Western Railway Co.*, No. 89-L-1636, The Circuit Court of the Third Judicial Circuit, Madison County, Illinois, pp. 35, 97.
38. See Ref. 36.
39. A. Glorig, Noise in industry, *Amer. Ind. Hyg. Assoc., 14*(3) (1953).
40. 29 C.F.R. 1910.95.
41. Transcript of deposition of Aram Glorig; *Donald R. Allen* v. *Norfolk and Western Railway Co.*, No. 89-L01636, The Circuit Court Third Judicial Circuit, Madison County, Illinois, November 12, 1990, at p. 96.
42. Transcript of deposition of Aram Glorig; *Donald R. Allen* v. *Norfolk and Western Railway Co.*, No. 89-L01636, The Circuit Court Third Judicial Circuit, Madison County, Illinois, November 12, 1990, at p. 93.

43. National Safety Council, *Accident Prevention Manual*, 3rd Ed. (1955).
44. National Safety Council, *Accident Prevention Manual*, 3rd Ed., pp. 36–5, 36–6 (1955).
45. National Safety Council, *Accident Prevention Manual*, 3rd Ed. pp. 29–28, 29–31, 33–20, 36–35–7 (1955).
46. See Glorig deposition, Refs. 32 and 41, pp. 91–94 and 109–110, respectively.
47. A. Glorig, Noise: Past, present and future, Ear and Hearing, 1(1) (1980).
48. A. Glorig, Noise: Past, present and future, Ear and Hearing, 1(1) (1980).
49. Z. Abdruck, *Hearing Difficulties of Railway Personnel and Their Influence on Railway Operating Safety*, Verlag Von J. F. Bergmann, Weisbaden, Germany (in German) containing a dialogue between Moos and Pollnow (1882).
50. J. S. Lumio, Studies on hearing loss of railway engine employees in Finland, *Acta Otolaryngol.,37*:539–550 (1949).
51. J. S. Lumio, Studies on hearing loss of railway engine employees in Finland, *Acta Otolaryngol.,37*:539–550 (1949).
52. U.S. Department of Health, Education and Welfare Public Health Services, *Hearing Levels of Adults by Age and Sex: United States 1960–1962*, October, 1965.
53. A. Glorig et al., 1954 Wisconsin state fair hearing survey, *Amer. Acad. Ophthalmol. Otolaryngol.*, (1957).
54. A. Glorig et al., 1954 Wisconsin state fair hearing survey, *Amer. Acad. Ophthalmol. Otolaryngol.*, p. 5 (1957).
55. A. Glorig et al., 1954 Wisconsin state fair hearing survey, *Amer. Acad. Ophthalmol. Otolaryngol.*, (1957).
56. A. Glorig et al., 1954 Wisconsin state fair hearing survey, *Amer. Acad. Ophthalmol. Otolaryngol.*, p. 6 (1957).
57. A. Glorig et al., 1954 Wisconsin state fair hearing survey, *Amer. Acad. Ophthalmol. Otolaryngol.*, p. 6 (1957).
58. A. Glorig et al., 1954 Wisconsin state fair hearing survey, *Amer. Acad. Ophthalmol. Otolaryngol.*, (1957).
59. A. Glorig et al., 1954 Wisconsin state fair hearing survey, *Amer. Acad. Ophthalmol. Otolaryngol.*, (1957).
60. A. Glorig et al., 1954 Wisconsin state fair hearing survey, *Amer. Acad. Ophthalmol. Otolaryngol.*, (1957).
61. A. Glorig et al., 1954 Wisconsin state fair hearing survey, *Amer. Acad. Ophthalmol. Otolaryngol.*, (1957).
62. Transcript of deposition of Aram Glorig, *Allen* v. *Norfolk & Western Railway*, see Ref. 41, p. 104.
63. R. Kilmer, Assessment of locomotive crew in-cab occupational noise exposure, *U.S. Dept. Transport./Fed. Railway Admin.*, December (1980).
64. R. Kilmer, Assessment of locomotive crew in-cab occupational noise exposure, *U.S. Dept. Transport./Fed. Railway Admin.*, p. xiii (Executive Summary), December (1980).
65. R. Kilmer, Assessment of locomotive crew in-cab occupational noise exposure, *U.S. Dept. Transport./Fed. Railway Admin.*, p. 114, December (1980).
66. R. Kilmer, Assessment of locomotive crew in-cab occupational noise exposure, *U.S. Dept. Transport./Fed. Railway Admin.*, p. 112, December (1980).
67. R. Kilmer, Assessment of locomotive crew in-cab occupational noise exposure, *U.S. Dept. Transport./Fed. Railway Admin.*, December (1980).
68. W. W. Clark and G. R. Popelka, Hearing levels of railroad trainmen, *Laryngoscope, 99*:1151–1157 (1989).
69. W. W. Clark and G. R. Popelka, Hearing levels of railroad trainmen, *Laryngoscope, 99*, 3 (1989).
70. W. W. Clark and G. R. Popelka, Hearing levels of railroad trainmen, *Laryngoscope, 99*, 4 (1989).
71. W. W. Clark and G. R. Popelka, Hearing levels of railroad trainmen, *Laryngoscope, 99*, 8 (1989).

72. W. W. Clark and G. R. Popelka, Hearing levels of railroad trainmen, *Laryngoscope, 99*, 8 (1989).
73. W. W. Clark and G. R. Popelka, Hearing levels of railroad trainmen, *Laryngoscope, 99*, 9 (1989).
74. W. W. Clark and G. R. Popelka, Hearing levels of railroad trainmen, *Laryngoscope, 99*, 12 (1989).
75. K. D. Kryter, Letter to editor: Comments on *Hearing Levels of Railroad Trainmen*, W. W. Clark and G. R. Popelka, *Laryngoscope, 100*, 1134–1136 (1990).
76. K. D. Kryter, Letter to editor: Comments on *Hearing Levels of Railroad Trainmen*, W. W. Clark and G. R. Popelka, *Laryngoscope, 100*, 1134–1136 (1990).
77. W. W. Clark and G. R. Popelka, Letter to editor, *Laryngoscope, 100*, 1136–1138 (1990).
78. K. D. Kryter, Hearing loss from gun and railroad noise — Relating with ISO standard 1999. *J. Acoust. Soc. Am.*, 1991; 90(G):3180.
79. K. D. Kryter, Hearing loss from gun and railroad noise — Relating with ISO standard 1999. *J. Acoust. Soc. Am., 90(G)*, 3180 (1991).
80. *Hebel* v. *Conrail*, 475 N.E.2d 652 (Ind. 1985); *Hennessey* v. *Commonwealth Edison Co.*, 746 F. Supp. 495 (N.D. Ill. 1991); *Albrecht* v. *Baltimore and Ohio R. Co.*, 808 F.2d 329 (4th Cir. 1987). See also regulations under [OSHA] provide evidence of the standard of care exacted of employers.
81. *Wendland* v. *Ridgefield Const. Services*, 439 A.2d 954, 957 (Conn. 1981) (appropriate application of a violation of an OSHA regulation to a civil suit would be as evidence of the "standard of care" to be accepted or rejected by the jury); *National Marine Services, Inc.* v. *Gulf Oil Co.*, 433 F. Supp. 913, 919 (E.D. La. 1977), *aff'd* 608 F.2d 522 (5th Cir. 1979) (standard of care is the only appropriate evidentiary use of OSHA regulations).
82. In other settings, the defense of compliance with government standards is gaining acceptance. *See, for example, Lorenz* v. *Celotex Corp.*, 896 F.2d 148 (5th Cir. 1991); *Clarksville-Montgomery County School System* v. *U.S. Gypsum*, 925 F.2d 993 (6th Cir. 1991); *Hennessey* v. *Commonwealth Edison Co.*, 764 F. Supp. 495 (N.D. Ill. 1991).
83. *Compare* 29 C.F.R. §1910.95 and 49 C.F.R. §229.121.
84. 43 Fed. Reg. 10,583 (1978).
85. ISO 1999, *Acoustics — Determination of Occupational Noise Exposure and Estimation of Noise-Induced Impairment* (1990) and ISO 1999:1975.
86. 40 C.F.R. §201.10.
87. 29 C.F.R. §1910.95.
88. 29 U.S.C. §651(b)(3).
89. 41 U.S.C. §35(a–e).
90. 41 U.S.C. §39.
91. 29 C.F.R. 1910.95(b)(2).
92. 29 C.F.R. 1910.95(b)(1); *see also Forging Industry Association* v. *Secretary of Labor*, 748 F.2d 210, 212 (4th Cir. 1984), *on remand*, 773 F.2d 1436 (4th Cir. 1985).
93. Transcript of deposition of expert medical (ENT) witness, *Wayne Austin, et al.* v. *Norfolk and Western Railway Co.*, CA No. 87-C-166-TS, in the Circuit Court of Brooke County, West Virginia, July 10, 1989. PI-95; A. Suter et al., Noise and public policy, *Ear and Hearing*, Vol. 8, No. 4 (1987); A Suter, OSHA's hearing conservation amendment and the audiologist, *ASHA*, June 1984.
94. J. Sataloff et al., Inter-industry noise study, *J. Occup. Med., 20*(5) (1978).
95. J. Sataloff et al., Inter-industry noise study, *J. Occup. Med., 20*(5) (1978).
96. J. Sataloff et al., Inter-industry noise study, *J. Occup. Med., 20*(5) (1978).
97. Department of Labor/Occupational Safety and Health Administration, Bulletin 334, 1971 revision, 11-3-80 publication.
98. 29 C.F.R. 1910.95.
99. 29 C.F.R. 1910.95(c).
100. 29 C.F.R. 1910.95(c–p).
101. Transcript of deposition of Aram Glorig, *Donald Golike* v. *CSX Transportation, Inc.*, No. 88-L-172, in the Circuit Court of 20th Judicial District, St. Clair County, Illinois,

p. 93, 134 D. Lipscomb, Three little words, in *Hearing Conservation in Industry, Schools and the Military* (1988).
102. 49 U.S.C. §103.
103. 49 C.F.R. §229.121.
104. Compare 49 C.F.R. §229.121, *et seq.* with 29 C.F.R. §1910.95, *et seq.*
105. 49 C.F.R. §229.121.
106. 49 C.F.R. §229.121.
107. 49 C.F.R. §229.121.
108. C. A. Edwards, Safety and health regulation of the transportation industry: Can the industry serve two masters? *ICC Practitioners Journal.*
109. 45 U.S.C.A. §421, *et seq.*
110. 45 U.S.C.A. §421.
111. 29 U.S.C.A. §651, *et seq.*
112. 29 U.S.C.A. §651(b).
113. 29 U.S.C.A. §653(b)(1).
114. 29 C.F.R. §1910.95.
115. 40 Fed. Reg. 29153.
116. 45 U.S.C.A. §§1–14.
117. 45 U.S.C.A. §9.
118. 45 U.S.C.A. §61, *et seq.*
119. 45 U.S.C.A. §§22–34.
120. 42 U.S.C.A. §§4901, *et seq.*
121. 49 C.F.R. Part 221.
122. 29 U.S.C. §653(b)(1).
123. *Marshall* v. *Northwest Orient Airlines, Inc.*, 574 F.2d 119 (2nd Cir. 1978); *See Donovan* v. *Red Star Marine Services, Inc.*, 739 F.2d 774 (2nd Cir. 1984) for a contemporary illustration of this not wholly resolved jurisdictional issue.
124. *Southern Railway Co.* v. *OSHRC*, 539 F.2d 335 (4th Cir. 1976); *Southern Pacific Transportation Co.* v. *Usery*, 539 F.2d 386 (5th Cir. 1976); *Baltimore and O. R. Co.* v. *OSHRC*, 548 F.2d 1052 (D.C. Cir. 1976).
125. 49 C.F.R. Part 221.
126. 49 C.F.R. Part 221.
127. For an excellent comprehensive discussion of this topic, see M. Nowak, Occupational noise exposure of railroad workers—Which regulation applies?, *University of Pittsburgh Journal of Law and Commerce, 11*(1) (Fall 1991).
128. 42 U.S.C. §4321, *et seq.*
129. 42 U.S.C.S. §4901 *et seq.* (Noise Control Act of 1972.)
130. 42 U.S.C.S. §4901(a)(1).
131. 42 U.S.C.S. §4901(b).
132. 42 U.S.C.S. §4901(b).
133. 42 U.S.C.S. §4903(c)(1).
134. 42 U.S.C.S. §4904(a).
135. 42 U.S.C.S. §4916(a)(1).
136. 42 U.S.C.S. §4916(a)(1).
137. 42 U.S.C.S. §4916(b).
138. S. Rep. No. 92-1160, 92d. Cons., 2d Sess., reprinted in 1972 U.S. Code Cors: Ad. News, 4657.
139. 40 C.F.R. §201.1(v).
140. 40 C.F.R. §201.1(v).
141. 40 C.F.R. §201.1(w).
142. 40 C.F.R. §201.1(w).
143. ISO 1991, *Acoustic-Determination of Occupational Noise Exposure and Estimation of Noise-Induced Impairment.*
144. ISO 1991, *Acoustic-Determination of Occupational Noise Exposure and Estimation of Noise-Induced Impairment.* See Ref. 85.

145. ISO 1991, *Acoustic-Determination of Occupational Noise Exposure and Estimation of Noise-Induced Impairment.*
146. ISO 1991, *Acoustic-Determination of Occupational Noise Exposure and Estimation of Noise-Induced Impairment.*
147. ISO 1991, *Acoustic-Determination of Occupational Noise Exposure and Estimation of Noise-Induced Impairment.*
148. ISO 1991, *Acoustic-Determination of Occupational Noise Exposure and Estimation of Noise-Induced Impairment.*
149. ISO 1991, *Acoustic-Determination of Occupational Noise Exposure and Estimation of Noise-Induced Impairment.*
150. ISO 1991, *Acoustic-Determination of Occupational Noise Exposure and Estimation of Noise-Induced Impairment.*
151. Acoustical expert reports in the matters of *Austin, et al.* v. *Norfolk and Western Railway Co.*, CA No. 87-C-1665, in the Circuit Court of Brooke County, West Virginia.
152. 45 U.S.C. §56.
153. *In Re: Central Railroad Co. of New Jersey Wilmat Holdings, Inc.*, F.2d No. 91-5259, slip. op., 1991 WL 257696 (3d Cir. December 10, 1991); *Erickson* v. *Burlington Northern Railroad Co.*, No. CS-91-0044-JBH, slip. op. (E.D. Wash., October 18, 1991); *Turner* v. *Southern Railway Co.*, No. D073734, slip. op. (Ga. Super. Ct., October 16, 1991); *Snow* v. *Southern Railway Co.*, No. D-73736, slip. op. (Ga. Super. Ct., October 15, 1991); *Wimpey* v. *Norfolk Southern Railroad Co.*, No. D-73738, slip. op. (Ga. Super. Ct., October 11, 1991, *appeal pending*); *Lambert* v. *Norfolk and Western Railway Co.*, No. 1-90-0309, Judgment Order (S.D. W. Va. 1991); *Hicks* v. *CSX Transportation, Inc.*, No. R-91-403, slip. op. (D. Md., October 7, 1991); *Roesch* v. *Chicago and Northwestern Transportation Co.*, No. 90 C1792, slip. op., 1991 WL 183856 (N.D. Ill., September 16, 1991); *Clark* v. *Illinois Central Railroad Co.*, No. 89-CI-520, Order (McCracken Cir. 1991); *Spann* v. *Illinois Central Railroad Co.*, No. 99-CI-854, Order (McCracken Cir. 1991); *Dunn* v. *Norfolk and Western Railway Co.*, No. 89-0954, Opinion (W.D. Va. 1991); *Dillon* v. *Norfolk and Western Railway Co.*, No. 89-0954, Opinion (W.D. Va. 1991); *Robb* v. *CSX Transportation, Inc.*, No. D-73330, slip. op. (Ga. Super Ct., August 22, 1991); *Farley* v. *Norfolk and Western Railway, Co.*, No. 1:90-0902, slip op. (S.D. W. Va., August 15, 1991); *Estes* v. *Southern Railway Co.*, No. C-1-90-420, Order (S.D. Oh. 1991); *Van Buren* v. *Penn Central Corp. and Consolidated Rail Corp.*, No. 90-CU-73095, brief in Support of Defendant's Motion for Summary Judgment and Order Granting (1991); *Clark* v. *Illinois Central Railroad Co.*, No. 892-00107, Order (Mo. Cir. 1991); *McCoy* v. *Illinois Central Railroad Co.*, No. 892-00804, Order (Mo. Cir. 1991); *Goodwin* v. *Southern Railway Co.*, No. D-73735, slip. op. (Ga. Super. Ct., June 11, 1991); *Goodwin* v. *Southern Railway Co.*, No. D-73741, slip. op. (Ga. Super Ct., June 11, 1991); *Lowder* v. *Southern Railway Co.*, D-73740, slip. op. (Ga. Super. Ct., June 10, 1991); *Crisman* v. *Odeco, Inc.*, 932 F.2d 413 (5th Cir.) *cert. denied*, U.S., 112 S. Ct. 337 (1991); *Steele* v. *Southern Railway Co.*, No. D-71561, slip. op. (Ga. Super Ct., May 30, 1991); *Jackson* v. *Southern Railway Co.*, No. D-73739, slip op. (Ga. Super. Ct., May 28, 1991); *Taal* v. *Union Pacific Railroad Co.*, 106 Or. App. 488, 809 P.2d 104 (1991); *Meier* v. *Chicago and Northwestern Transportation Co.*, No. 90 C 2023, slip. op., 1991 WL 49620 (N.D. Ill., April 2, 1991); *In Re: Central Railroad Co. of New Jersey*, No. B-67-401, slip. op., 1991 WL 37886 (D.N.J., March 18, 1991), *aff'd, In Re: Central Railroad Co. of New Jersey Wilmat Holdings, Inc.*, F.2d No. 91-5259, slip. op. 1991 WL 257696 (3d Cir., December 10, 1991); *Richard* v. *Elgin, Joliet and Eastern Railway Co.*, 750 F. Supp. 372 (N.D. Ind. 1991); *Fries* v. *Chicago and Northwestern Transportation Co.*, 909 F.2d 1092 (7th Cir. 1990); *Kragel* v. *Long Island Railroad Co.*, 1990 WL 121532 (S.D. N.Y. 1990); *Van Zweden* v. *Southern Pacific Transportation Co.*, 741 F. Supp. 209 (D. Utah 1990); *McCoy* v. *Union Pacific Railroad Corp.*, P.2d, 102 Or. App. 620, 1990 WL 108767 (1990); *Albert* v. *Maine Central Railroad Co.*, 905 F.2d 541 (1st Cir. 1990); *Smith* v. *Cliff's Drilling Co.*, 562 So.2d 1030 (La. Ct. App. 1990) [Jones Act case]; *Crisman* v. *Odeco, Inc.*, 736 F. Supp. 712 (E.D. La. 1990); *Kestner* v. *Missouri Pacific*

Railroad Co., 785 S.W. 2d. 646 (Mo. Ct. App. 1990); *Turner* v. *Norfolk and Western Railway Co.*, 785 S.W. 2d 569 (Mo. Ct. App. 1990); *Aungst et al.* v. *Reading Co. and Consolidated Rail Corp.*, No. 87-0308, Memorandum (E.D. Pa. 1989); *Townley* v. *Norfolk and Western Railway Co.*, 887 F.2d 498 (4th Cir. 1989); *Bechtholdt* v. *Union Pacific Railroad Co.*, 722 F. Supp. 704 (E.D. Pa. July 31, 1989); *Fries* v. *Chicago and Northwestern Transportation Co.*, Decided and filed (E.D. Wis., April 18, 1989), *aff'd*, 909 F.2d 1092 (7th Cir. 1990); *Courtney* v. *Union Pacific Railroad Co.*, 713 F. Supp. 305 (E.D. Ark, 1989); *Reed* v. *Illinois Central Gulf*, No. 89-5039, Order (S.D. Ill. 1989); *McMeen* v. *Illinois Central Railroad Co.*, No. 89-L-46, Order (4th Cir. 1989); *Jones* v. *Maine Central Railroad Co.*, 690 F. Supp. 73 (D. Me. 1988); *Townley* v. *Norfolk and Western Railway Co.*, 690 F. Supp. 1513 (S.D. W.V. 1988); *Stokes* v. *Union Pacific Railroad Co.*, 687 F. Supp. 552 (D. Wyo. 1988); *Leach* v. *Southern Pacific Transportation Co.*, Memorandum Op. (N.D. Ca., May 13, 1988); *Campbell* v. *Consolidated Rail Corp.*, No. 87-2312, slip. op., 1988 WL 71283 (E.D. Pa., July 1, 1988); *Bailey* v. *Norfolk and Western Railway*, No. 1:85-0524, slip. op. (D. W. Va., May 5, 1986); *Lessee* v. *Union Pacific Railroad Co.*, 38 Wash. App. 802, 690 P.2d 596 (1984); *Travelers Insurance Co.* v. *Cardillo*, 225 F.2d 137 (2nd Cir. 1955); *Bealer* v. *Missouri Pacific Railroad Co.*, No. 91-3031 (5th Cir. 1991); *Singleton* v. *Consolidated Rail Corp.*, No. C-1-85-119 (S.D. Oh. 1986).

154. *In Re: Central Railroad Co. of New Jersey Wilmat Holdings, Inc.*, F.2d No. 91-5259, slip. op., 1991 WL 257696 (3d Cir. December 10, 1991); *Erickson* v. *Burlington Northern Railroad Co.*, No. CS-91-0044-JBH, slip. op. (E.D. Wash., October 18, 1991); *Turner* v. *Southern Railway Co.*, No. D-73734 slip. op. (Ga. Super. Ct., October 16, 1991); *Snow* v. *Southern Railway Co.*, No. D-73736, slip. op. (Ga. Super. Ct., October 15, 1991); *Wimpey* v. *Norfolk Southern Railway Co.*, No. D-73738, slip. op. (Ga. Super. Ct., October 11, 1991, *appeal pending*); *Lambert* v. *Norfolk and Western Railway Co.*, No. 1-90-0309, Judgment Order (S.D. W. Va. 1991); *Hicks* v. *CSX Transportation, Inc.*, No. R-91-403, slip. op. (D. Md., October 7, 1991); *Roesch* v. *Chicago and Northwestern Transportation Co.*, No. 90 C1792. slip. op., 1991 WL 183856 (N.D. Ill., September 16, 1991); *Clark* v. *Illinois Central Railroad Co.*, No. 89-CI-520, Order (McCracken Cir. 1991); *Spann* v. *Illinois Central Railroad Co.*, No. 99-CI-854, Order (McCracken Cir. 1991); *Dunn* v. *Norfolk and Western Railway Co.*, No. 89-0954, Opinion (W.D. Va. 1991); *Dillon* v. *Norfolk and Western Railway Co.*, No. 89-0954, Opinion (W.D. Va. 1991); *Robb* v. *CSX Transportation, Inc.*, No. D-73330, slip. op. (Ga. Super Ct., August 22, 1991); *Farley* v. *Norfolk and Western Railway, Co.*, No. 1:90-0902, slip op. (S.D. W. Va., August 15, 1991); *Estes* v. *Southern Railway Co.*, No. C-1-90-420, Order (S.D. Oh. 1991); *Van Buren* v. *Penn Central Corp. and Consolidated Rail Corp.*, No. 90-CU-73095, brief in Support of Defendant's Motion for Summary Judgment and Order Granting (1991); *Clark* v. *Illinois Central Railroad Co.*, No. 892-00107, Order (Mo. Cir. 1991); *McCoy* v. *Illinois Central Railroad Co.*, No. 892-00804, Order (Mo. Cir. 1991); *Goodwin* v. *Southern Railway Co.*, No. D-73735, slip. op. (Ga. Super. Ct., June 11, 1991); *Goodwin* v. *Southern Railway Co.*, No. D-73741, slip. op. (Ga. Super Ct., June 11, 1991); *Lowder* v. *Southern Railway Co.*, D-73740, slip. op. (Ga. Super. Ct., June 10, 1991); *Crisman* v. *Odeco, Inc.*, 932 F.2d 413 (5th Cir.) *cert. denied*, U.S., 112 S. Ct. 337 (1991); *Steele* v. *Southern Railway Co.*, No. D-71561, slip. op. (Ga. Super Ct., May 30, 1991); *Jackson* v. *Southern Railway Co.*, No. D-73739, slip op. (Ga. Super. Ct., May 28, 1991); *Taal* v. *Union Pacific Railroad Co.*, 106 Or. App. 488, 809 P.2d 104 (1991); *Meier* v. *Chicago and Northwestern Transportation Co.*, No. 90 C 2023, slip. op., 1991 WL 49620 (N.D. Ill., April 2, 1991); *In Re: Central Railroad Co. of New Jersey*, No. B-67-401, slip. op., 1991 WL 37886 (D.N.J., March 18, 1991), *aff'd, In Re: Central Railroad Co. of New Jersey Wilmat Holdings, Inc.*, F.2d No. 91-5259, slip. op. 1991 WL 257696 (3d Cir., December 10, 1991); *Richard* v. *Elgin, Joliet and Eastern Railway Co.*, 750 F. Supp. 372 (N.D. Ind. 1991); *Fries* v. *Chicago and Northwestern Transportation Co.*, 909 F.2d 1092 (7th Cir. 1990); *Kragel* v. *Long Island Railroad Co.*, 1990 WL 121532 (S.D. N.Y. 1990); *Van Zweden* v. *Southern Pacific Transportation Co.*, 741 F. Supp. 209 (D. Utah 1990); *McCoy* v. *Union Pacific Railroad Corp.*, P.2d, 102 Or. App. 620, 1990

WL 108767 (1990); *Albert* v. *Maine Central Railroad Co.*, 905 F.2d 541 (1st Cir. 1990); *Smith* v. *Cliff's Drilling Co.*, 562 So.2d 1030 (La. Ct. App. 1990) [Jones Act case]; *Crisman* v. *Odeco, Inc.*, 736 F. Supp. 712 (E.D. La. 1990); *Kestner* v. *Missouri Pacific Railroad Co.*, 785 S.W. 2d. 646 (Mo. Ct. App. 1990); *Turner* v. *Norfolk and Western Railway Co.*, 785 S.W. 2d 569 (Mo. Ct. App. 1990); *Aungst et al.* v. *Reading Co. and Consolidated Rail Corp.*, No. 87-0308, Memorandum (E.D. Pa. 1989); *Townley* v. *Norfolk and Western Railway Co.*, 887 F.2d 498 (4th Cir. 1989); *Bechtholdt* v. *Union Pacific Railroad Co.*, 722 F. Supp. 704 (E.D. Pa. July 31, 1989); *Fries* v. *Chicago and Northwestern Transportation Co.*, Decided and filed (E.D. Wis., April 18, 1989), *aff'd*, 909 F.2d., 1092 (7th Cir. 1990); *Courtney* v. *Union Pacific Railroad Co.*, 713 F. Supp. 305 (E.D. Ark, 1989); *Reed* v. *Illinois Central Gulf*, No. 89-5039, Order (S.D. Ill. 1989); *McMeen* v. *Illinois Central Railroad Co.*, No. 89-L-46, Order (4th Cir. 1989); *Jones* v. *Maine Central Railroad Co.*, 690 F. Supp. 73 (D. Me. 1988); *Townley* v. *Norfolk and Western Railway Co.*, 690 F. Supp. 1513 (S.D. W.V. 1988); *Stokes* v. *Union Pacific Railroad Co.*, 687 F. Supp. 552 (D. Wyo. 1988); *Leach* v. *Southern Pacific Transportation Co.*, Memorandum Op. (N.D. Ca., May 13, 1988); *Campbell* v. *Consolidated Rail Corp.*, No. 87-2312 slip. op., 1988 WL 71283 (E.D. Pa., July 1, 1988); *Bailey* v. *Norfolk and Western Railway*, No. 1:85-0524, slip. op. (D. W. Va., May 5, 1986); *Lessee* v. *Union Pacific Railroad Co.*, 38 Wash. App. 802, 690 P.2d 596 (1984); *Travelers Insurance Co.* v. *Cardillo*, 225 F.2d 137 (2nd Cir. 1955); *Bealer* v. *Missouri Pacific Railroad Co.*, No. 91-3031 (5th Cir. 1991); *Singleton* v. *Consolidated Rail Corp.*, No. C-1-85-119 (S.D. Oh. 1986).

155. Committee report: Occupational noise—induced hearing loss. *American College of Occupational Medicine Noise and Hearing Compensation Committee, J. Occup. Med., 31*(2) (1989).
156. Transcript of deposition of expert medical (ENT) witness, *Elmore et al.* v. *Norfolk and Western Railway Co.*, CA No. 87-C-170-W, in the Circuit Court of Brooke County, West Virginia.
157. Transcript of deposition of expert medical (ENT) witness, *Elmore et al.* v. *Norfolk and Western Railway Co.*, CA No. 87-C-170-W, in the Circuit Court of Brooke County, West Virginia.
158. Transcript of deposition of expert medical (ENT) witness, *Harold D. Mintz* v. *Southern Railway Co.*, (V 87-1200 December 17, 1990).
159. Transcript of deposition of expert medical (ENT) witness, *Harold D. Mintz* v. *Southern Railway Co.*, (V 87-1200 December 17, 1990).
160. Transcript of deposition of expert medical (ENT) witness, *Harold D. Mintz* v. *Southern Railway Co.*, (V 87-1200 December 17, 1990).
161. Testimony of expert medical (ENT) witness, *Lamblin* v. *Norfolk & Western Railway Co.*, CA No. 87-C-257-W, in the Circuit Court of Brooke County, West Virginia, Brooke Co. August 14, 1991.
162. Transcript of deposition of Aram Glorig, *Allen* v. *Norfolk & Western Railway Co.*, see Ref. 41, pp. 185–186.
163. Transcript of deposition of Aram Glorig, *Allen* v. *Norfolk & Western Railway Co.*, see Ref. 41, pp. 185–186.
164. Transcript of deposition expert medical (ENT) witness, *Congelli* v. *Conrail*, CD No. 81-1082, in the United States District Court for the Western District of New York (1985).
165. Transcript of deposition expert medical (ENT) witness, *Ratcliffe, et al.* v. *Norfolk and Western Railway Co.*, CA No. 87-C-167-W, in the Circuit Court of Brooke Country, West Virginia.
166. Transcript of deposition expert medical (ENT) witness, *Markum* v. *Conrail*, GD No. 86-2218, in the Court of Common Pleas of Allegheny County.
167. Transcript of deposition expert medical (ENT) witness, *Markum* v. *Conrail*, GD No. 86-2218, in the Court of Common Pleas of Allegheny County.
168. Transcript of deposition expert medical (ENT) witness, *Eric Bollinger* v. *Norfolk and Western Railway Co.*, No. 88-0360-R, the United States District Court in the Western District of Virginia.

169. Transcript of trial testimony of expert medical witness, *Tighe* v. *Conrail*, CA No. H-54192 Supreme Court of New York, Niagra County (1991).
170. Transcript of trial testimony of expert medical witness, *Tighe* v. *Conrail*, CA No. H-54192 Supreme Court of New York, Niagra County (1991), pp. 446–528.
171. See Ref. 170.
172. See Ref. 170.
173. Transcript of deposition of Erika Schwartz, M.D.; *Gail A. Duvall* v. *M. K. Rao, M.D., Anthony Golio, M.D., Damrong Hadsaiton, M.D. and Sheehan Memorial Hospital*, No. H-64663, the Supreme Court of the State of New York; *Richard Inglut* v. *Conrail*, No. H-50028, the Supreme Court of the State of New York.
174. Committee report: Occupational noise-induced hearing loss, American College of Occupational Medicine Noise and Hearing Compensation Senate Committee, *J. Occup. Med.* (1989).
175. See Refs. 162, 167, and 170.
176. *Dale* v. *Baltimore and Ohio Railway Co.*, Pa., 552 A.2d 1037 (1989).
177. 29 C.F.R. §1910.95 Appendix F.
178. A. Glorig, 1954 Wisconsin State Fair Survey, see Ref. 53, p. 28.
179. Transcript of deposition of Aram Glorig; *Donald Golike* v. *CSX Transportation, Inc.*, No. 88-L-172, in the Circuit Court of the 20th Judicial District of St. Clair County, Illinois, p. 40, 172–173. A. Glorig, An introduction to the industrial noise problem, *Illinois Medical Journal, 107*, No. 1, January 1955. Transcript of deposition of Joseph Sataloff, M.D.; *Wayne Austin, et al.* v. *Norfolk and Western Railway*, CA No. 87-C-166 TS, in the Circuit Court of Brooke County, West Virginia, p. 78–84.
180. See Ref. 175.
181. See Ref. 162.
182. Lipscomb, *Hearing Conservation in Industry School and the Military*, (1988), p. 47.

37

The United States: The Longshore and Harbor Workers' Compensation Act

Lawrence P. Postol
Seyfarth Shaw, LLP, Washington, DC, USA

When the 1972 Amendments to the Longshore Act were passed by Congress, if one were to ask employers or claimants about the importance of hearing loss claims, the response would have been uncontrolled laughter. Hearing loss claims were virtually unheard of, and at the preamendment maximum weekly compensation rate of $70, any award would have been, in any case, rather small. Now, only the claimants are laughing, although many of them cannot hear the laughter.

Due largely to ignorance, many employees in the maritime industry were exposed to noise in the workplace in the 1930s through the 1970s. The Occupational Safety and Health Administration (OSHA) began regulating noise exposure in the 1970s and early 1980s. Ironically, since shipyards are particularly noisy workplaces, a lot of attention was directed at them, and by the early 1980s, most shipyards had hearing surveillance-protection programs in place. Indeed, most shipyards now have model hearing protection programs. While longshoring work can at times be noisy, the noise is not as extreme or as constant as shipyards. In addition, the longshoremens unions have been resistant to any requirement that longshoremen wear hearing protection. Thus, noise exposure and hearing loss remain concerns for many stevedoring companies. Many of the exposed workers are now hard of hearing due to noise-induced hearing loss.

As employees became aware of their loss, its cause, and their rights under the Longshoreman's Act, many employers were, and some still are, flooded with hearing loss claims. Once one worker wins a recovery, word quickly spreads among the other employees as well as retirees. While few individual hearing loss claims exceeded $100,000, and most are in the $10,000–30,000 range, when 1000 former and present workers file their claims within a couple of years, the financial and administrative drain on an employer can be enormous and devastating. For employers who have already paid the piper (the claimants) and now take appropriate safety precautions, the tide of hearing loss claims has receded and only small waves can be expected in the future. However, for employers whose workers have not yet realized this potential goldmine, a tidal wave is out there, just waiting to engulf the unsuspecting employer. Moreover, since hearing protection is still not used by most (virtually all) longshoremen, employers can expect future claims from the longshoremen for the worsening of their hearing loss.

Although some employers and attorneys pay little attention to the defense of a hearing loss claim, a multitude of issues should be addressed. Is the hearing loss caused by the workplace noise? Is it permanent? Which test results are most reliable for measuring the claimant's hearing loss? Was the hearing loss impairment calculated under the correct formula? Who is the responsible employer and insurance carrier, a particularly important issue in the longshoring industry, where employees are hired out of a union hiring hall, and thus they can have a different employer each day? Was a timely claim filed and timely notice of injury given? Finally, what credit should be taken for prior compensation that has already been paid?

1. CAUSATION—OCCUPATIONALLY INDUCED HEARING LOSS

The workplace can cause hearing loss in essentially two forms: (a) an acoustic trauma injury in which a worker is exposed to an extremely loud noise, e.g., from an explosion, and (b) long-term exposure to noise. The type and appearance of a hearing loss from a trauma will, of course, depend on the nature of the trauma. For example, in which direction was the worker facing when the explosion occurred, and was the explosion outdoors or in an enclosed area? Etiology is usually not a problem with respect to a traumatically induced hearing loss since the loss appears immediately after the accident. Such is not the case with a hearing loss caused by continuous exposure to noise. A long-term, noise-induced hearing loss does, however, have a recognizable audiometric pattern that can usually be distinguished from other hearing loss, as discussed elsewhere in this book.

One of the primary concerns of employers in the hearing loss area is presbycusis—hearing loss due to age. Employers feel it is unfair for them to have to pay for such losses. However, the test results and losses from presbycusis may be similar to the loss from noise-induced hearing loss—bilateral and increasing at higher frequencies. Thus, it is often almost impossible to separate out and quantify the hearing loss from presbycusis and the hearing loss from noise exposure (although the latter, but not the former, usually has a recovery at 4000–8000 Hz). Moreover, under the aggravation doctrine, the employer is liable for a worker's total hearing loss, even if part of it is caused by presbycusis (*Prime v. Todd Shipyards Corp.*, 12 B.R.B.S. 190 [1980]; *Moore v. Newport News Shipbuilding & Dry Dock Co.*, 15 B.R.B.S. 28 [1982]; *Mowl v. Ingalls Shipbuilding*, 32 B.R.B.S. 51 (1998); *Newport News Shipbuilding & Dry Dock Co. v. Fishel*, 15 B.R.B.S. [CRT] 52

[4th Cir. 1982]). An employer can defeat such a hearing loss claim only if it can show that none of the worker's hearing loss was due to noise exposure at work.

There is a logical argument that the aggravation doctrine should not apply to scheduled injuries such as hearing loss. The aggravation doctrine is used because it is usually impossible to segregate out the effects of different medical impairments on a worker's economic disability—his loss of wages. With a scheduled injury, however, the medical impairment is the worker's disability. Thus, to the extent the medical experts can separate out the causes of the medical impairment, there is no need for the aggravation doctrine. For example, if an employer hires a man with a 25% hearing loss, and the employee ends up with a 31% hearing loss, the employer should pay only for the 6% occupationally caused loss and not the 31% total loss. Unfortunately, relying on the statutory language in section 8 (f) of the Act, the Fourth Circuit rejected this argument in *Newport News Shipbuilding & Dry Dock Co. v. Fishel*, 15 B.R.B.S. [CRT] 52 [4th Cir. 1982]. The 1984 Amendments to the Longshore Act, moreover, codified the *Fishel* decision, making clear employers are liable for a worker's entire hearing loss.

The inequity of an employer being held responsible for a worker's preexisting hearing loss, which is easily segregated from his occupationally caused hearing loss by a preemployment audiogram, was tempered by a modification to the second injury fund provision by the 1984 Amendments. Owing to a drafting defect, even an employer who qualified for second-injury fund relief under the old Act, was responsible for the *greater* of 104 weeks of compensation or the scheduled hearing loss disability attributable to the employer. The 1984 amendments corrected this error, to make the employer's liability for the lesser of the two (33 U.S.C. Section 908 [f] [1]). Thus, employers now must only pay for the portion of the compensation based on the occupationally caused hearing loss; the Special Fund will be liable for the portion of the compensation based on the preexisting hearing loss. However, the preexisting hearing loss must be documented by a preemployment audiogram. The Longshore regulations define a preemployment audiogram, for purposes of placing a claim under section 8(f), as one performed within 30 days of employment:

> In determining the amount of pre-employment hearing loss, an audiogram must be submitted which was performed prior to employment or within 30 days of the date of the first employment-related noise exposure. Audiograms performed after December 27, 1984 must comply with the standards described in paragraph (d) of this section. [20 C.F.R. 702.441 (c).]

Thus, if employers obtain preemployment audiograms, they can later obtain section 8(f) relief for the amount of hearing loss that existed when the worker was first hired. Unfortunately, an employer's assessment to pay for the Special Fund is based on the employer's use of the fund, and thus, the employer, via its assessment, usually has to pay in the area of 60% of the compensation for the preexisting disability (33 U.S.C. Section 944). Thus, the gross inequity and windfall to claimants continue in part.

Disputing causation is always difficult under the Longshore Act. The courts have held that a worker's mere testimony that noise at work was loud is enough to involve the statutory presumption that his injury (hearing loss) was caused by his work (*Damiano v. Global Terminal & Container Service*, 32 B.R.B.S. 261, 262 (1999); *Ramey v. Stevedoring Services of America* 31 B.R.B.S. (CRT 206, 210 (9th Cir. 1998)). Thus, the burden of proof is on the employer to show noise at work did not cause the worker's hearing loss, not even in part.

Nevertheless, if exposure data show noise levels are below 85 dB time-weighted average (TWA), then the employer can attempt to prove none of the workers' hearing loss was caused by noise. OSHA has opined 90 dB TWA is a safe level of noise exposure for which no noise protection is needed, and annual surveillance (hearing test) is not even needed if exposure is less than 85 dB TWA. Claimant's attorneys, however, argue individual susceptibility prevents a causation defense, even when noise surveys shows exposure below 85 dB TWA. Also, there is the problem of proving exposure levels from years where there were no noise test surveys, as well as the problem of workers who work over 8 h a day. The 85 dB TWA limit is based on 8 h a day. If a worker works more than 8 h in a day, the allowable level goes down. Thus, for example, for a 16-h day, the standard allowable 85 dB TWA level becomes only 80 dB TWA allowed.

If the employer through testimony and noise survey can show exposure levels were always below 85 dB TWA, then the employer has a chance for a causation defense. See, e.g., *Global Terminals & Container, Inc. v. Foreman*, 33 B.R.B.S. (CRT) 134 (3rd Cir. 1999); *Coffey v. Marine Terminals Corp.*, 34 B.R.B.S. (2000).

In short, if the employer exposed the worker to noise, and the employee has a hearing loss caused even in part by the workplace noise—as represented by a bilateral sensorineural hearing loss that is greater at the higher frequencies—then the employer will most likely be liable for the entire hearing loss. However, if the employer can document the level of the preexisting hearing loss, the Special Fund will be liable for the compensation attributable to the preexisting loss. Similarly, with appropriate noise survey data, employers can try to dispute causation, although it is usually an uphill battle.

2. DETERMINING THE EXTENT OF HEARING LOSS

One of the key factors in any hearing loss case should be the measurement and calculation of the extent of the hearing loss. Yet, employers often overlook this critical issue which can cost them thousands of dollars. Moreover, given the subjective nature of a patient's testing, it is critical that the hearing test be performed by a reliable tester who has excellent equipment and a quiet sound booth, and who conducts a complete audiogram which uses several internal tests to cross-check for reliability.

While the American Medical Association (AMA) formula is mandated under the new Longshore Amendments, in evaluating pre-1984 longshore cases and companion state claims, one should be aware that there are other impairment formulas in existence which can be utilized under some state worker's compensation statutes. Among these formulas are the now outdated Department of Labor/NIOSH formula, the ASHA formula, and the Ohio State formula.

The National Institute of Occupational Safety and Health (NIOSH), in its 1972 *Criteria for a Recommended Standard, Occupational Exposure to Noise*, recommended the use of the frequencies 1000, 2000, and 3000 Hz for measuring hearing loss. However, NIOSH did not recommend any particular formula that was to be utilized with these frequencies. Nevertheless, the Department of Labor, Office of Workers' Compensation Program (OWCP), adopted these frequencies for use in the AMA formula. Thus, the Department of Labor used the same 25-dB fence, 1.5% multiplication, and five times weight of the better ear, although it used the frequencies of 1000, 2000, and 3000 Hz. Since occupational noise-induced hearing loss affects the higher frequencies first and most severely, the Department of Labor formula gave a higher permanent disability loss than either the old AMA formula or the new (and higher compared to the old) AMA formula. As noted below, however, Department of Labor Administrative Law Judges were not bound by the OWCP Department of Labor formula, and frequently utilized the new AMA formula as a better measurement of hearing loss.

The Department of Labor contracted with researchers at Ohio State University to determine the most appropriate formula for measuring hearing loss. The Ohio State University investigation recommended that hearing impairment be measured at 500, 1000, 2000, 3000, and 4000 Hz, with a 15-dB low fence, a conversion factor so that a 70-dB threshold represented a 100% loss (1.81%), and with equal weight assigned to each ear in calculating the binaural hearing loss.

The American Speech–Language–Hearing Association (ASHA) also conducted a Task Force on the Definition of Hearing Handicap. The ASHA Task Force recommended a hearing loss formula utilizing 1000, 2000, 3000, and 4000 Hz, a 30-dB low fence, a conversion factor so that an 80-dB threshold represented a 100% loss (2.0%), and equal weight assigned to each ear in calculating the binaural hearing loss.

The dramatic effect the choice of hearing loss formula has on the amount of compensation a worker receives can be seen from the percentage hearing loss and the compensation listed below which would be awarded to a worker with the audiogram shown below, assuming the longshore 200 weeks' compensation for a binaural hearing loss and an average weekly wage of $450, giving a weekly compensation rate of $300.

	Left ear	Right ear
1000 Hz	30 dB	40 dB
2000 Hz	40 dB	45 dB
3000 Hz	55 dB	60 dB
4000 Hz	70 dB	75 dB
Formula	Impairment	Compensation
New AMA	20.63%	$12,378
DOL	26.66%	$15,996
ASHA	43.75%	$26,250
Ohio State	58.83%	$35,298

The choice of a compensation formula can obviously have a dramatic effect on the amount of compensation employers must pay their hearing-impaired work force. The 1984 Longshore Amendments resolved this previously troublesome problem for maritime workers' federal claims, since the 1984 Amendments mandate the use of the current AMA formula.

2.1. Longshoremen's and Harbor Workers' Compensation Act

The Longshoremen's and Harbor Workers' Compensation Act provides for a scheduled disability compensation for loss of hearing. Fifty-two weeks of compensation is provided for a monaural (one ear) hearing loss, and 200 weeks of compensation is provided for a binaural (both ears) hearing loss. Less than a total loss is prorated (33 U.S.C. Section 908 [c] [13] and [19]). If the hearing loss is due to noise exposure, the award should be made as a binaural loss, *Tarner v. Ingalls Shipbuilding, Inc.*, 26. B.R.B.S. (43)[1992]), but recent decisions indicate a monaural award can be given. *Garner v. Newport News Shipbuilding*, 25 B.R.B.S. (CRT) 122 (4th Cir. 1992). Unless a worker is totally disabled by a permanent scheduled disability, he can recover only the scheduled compensation irrespective of his actual wage loss (*Potomac Electric Power Co. v. Director*, 14 B.R.B.S. 363 [1980]). Conversely, a worker may recover his scheduled benefit even if he has no actual wage loss due to the scheduled injury (*Travelers Co. v. Cardillo*, 225 F.2d 137, 143 [2d Cir.], cert. *denied*, 350 U.S. 913 [1955]). A worker, however, cannot receive a scheduled benefit if he or she is already drawing total disability compensation (*Tisdale v. Owens-Corning Fiber Glass Co.*, 13 B.R.B.S. 167 [1983], *aff'd memo sub nom. Tisdale v. Director*, 698 F.2d 1233 [9th Cir. 1982], *cert. denied* 103 S. Ct. 2454 [1983]; *Mahar v. Todd Shipyards Corp.*, 13 B.R.B.S. 603 [1981]; *Rathke v. Lockheed Shipbuilding and Construction Co.*, 16 B.R.B.S. 77 [1984]).

The 1984 Longshore Amendments explicitly provide that hearing loss determinations "shall be made in accordance with the Guides for the Evaluation of Permanent Impairment as promulgated and modified from time to time by the American Medical Association" (33 U.S.C. Section 908 [c] [13] [E]).

2.2. Ensuring the Reliability of the Hearing Test

2.2.1. *Obtaining a Complete Audiogram*

The 1984 Longshore Amendments do not address directly the problem of obtaining a complete and reliable audiogram. Nevertheless, the Amendments do give presumptive (but nonconclusive) weight to an audiogram if it is performed by a *certified* audiologist or otolaryngologist, a copy of the audiogram was given to the employee at the time of testing, and no contrary audiogram was performed at the time:

> An audiogram shall be presumptive evidence of the amount of hearing loss sustained as of the date thereof, only if (i) such audiogram was administered by a licensed or certified audiologist or a physician who is certified in otolaryngology, (ii) such audiogram, with report thereon, was provided to the employee at the time it was administered, and (iii) no contrary audiogram made at that time is produced. [33 v. U.S.C. Section 908 (c) (13) (C).]

The Longshore regulations, however, allow a technician to perform the audiogram that is to be given presumptive weight if:

> An audiogram shall be presumptive evidence of the amount of hearing loss on the date administered if the following requirements are met: the audiogram was administered by a licensed or certified audiologist, by Board of Otolaryngology, or by a technician, under an audiologist's or physician's supervision, certified by the Council of Accreditation on Occupational Hearing Conservation, or by any other person considered qualified by a hearing conservation program authorized pursuant to 29 C.F.R. 1910.95 (g) (3) promulgated under the Occupational Safety and Health Act of 1970 (20 V.S.C. 667). Thus, either a professional or trained technician may conduct audiometric testing. However, to be acceptable under this subsection, a licensed or certified audiologist or otolaryngologist, as defined, must ultimately interpret and certify the results of the audiogram. The accompanying report may set forth the testing standards used and describe the method of evaluating the hearing loss as well as providing an evaluation of the reliability of the test results. [20 C.F.R. 702.441 (b) (1).]

In the preamble to the regulations, the justification for this provision is that the conference report allows for same:

> A new 702.441 specifies that audiograms can be administered by qualified technicians, as long as they are certified by an audiologist or otolaryngologist who, the regulations state, must be certified by a generally accepted professional hearing program. This clarification is based on the Conference Manager's Statement. [50 Fed. Ref. 383, 389 (1985).]

While the legislative history states that the regulations shall follow the hearing conservation program procedures, which in fact allow for testing by technicians, query whether the regulations exceed the scope of the statute that explicitly limits presumptive evidence testing to licensed or certified audiologists or otolaryngologists:

> The Senate recedes to the House. In requiring audiograms to be administered by certified audiologists or otolaryngologists, the conferees wish to assure that audiogram results are certified by competent medical personnel. In promulgating regulations under this section the conferees expect that the Department of Labor will incorporate audiometric testing procedures consistent with those required by hearing conservation programs pursuant to the Occupational Safety and Health Act. [H.R. Rep. No. 98-1027, 98th Cong., 2nd Sess. 27-28 (1984).]

The Longshore regulations provide a definition for defining a contrary audiogram in the "same time period," which would negate the presumptive weight of an audiogram. The regulations allow 30 days where there was noise exposure, otherwise, 6 months:

> No one produces a contrary audiogram of equal probative value (meaning one performed using the standards described herein) made at the same time. "Same time" means within thirty (30) days thereof where noise exposure continues or within six (6) months where exposure to excessive noise levels does not continue. Audiometric tests performed prior to the enactment of PubI.L.98-426 will be considered presumptively valid if the employer complied with procedures in this section for administering audiograms. [20 C.F.R. 702.441 (b) (3).]

The Longshore regulations also require calibration of equipment:

> In addition, the audiometer used for testing the individual's threshold of hearing must be calibrated according to current American National Standard. [20 C.F.R. 702.441 (d).]

Arguably, the term "audiogram" refers to a *complete* audiogram with all the consistency test procedures being performed, e.g., pure-tone air conduction testing, speech reception threshold (SRT), and speech discrimination testing. However, the provisions in subsection (c) is really a leftover provision from a larger section in the prolonged Senate bill which would have made the employer liable only for the extent of hearing loss it caused—hence the reference to an audiogram being presumptive evidence of the extent of hearing loss *at the time of testing*. The Senate amendment was otherwise struck in the House–Senate conference committee, and the reason for retaining subsection (c) without the Senate provision, and thus subsection (c)'s import, is not clear.

In any case, an employer should have employees tested by a certified audiologist or otolaryngologist (not by staff), and a complete and internally consistent audiogram should be obtained, with a copy to the employee. Either as a matter of statutory interpretation, or as a matter of the preponderance of the evidence, the employer should argue that such complete test results should be credited over incomplete audiograms performed by unskilled examiners.

3. COMPENSATING HEARING LOSS VICTIMS

Once the extent of hearing loss is determined, one must determine the appropriate compensation for the worker. This involves determination of the worker's average weekly wage. The employer must make sure that timely notice of the injury was given, and a timely claim for compensation was filed. Similarly, the subsequent exposure rule must be properly applied so as to determine the liable employer and carrier. Coordination of benefits must be made with past compensation awards by the same employer, other employers, and the state awards. Finally, appropriate penalties and interest must be paid.

3.1. Date of Injury and Retired Workers

For many years after the 1984 longshore amendments, there was a substantial issue of whether hearing loss for retirees were covered as a scheduled injury under section 8(c)(15), or a retiree occupational disease benefit under section 8(c)(23). The occupational disease benefit has a different average weekly wage, timeliness, and benefit provision. In *Bath Iron Works Corp v. Director*, 506 U.S. 153, 113 S.Ct. 692 (1993), the U.S. Supreme Court held that hearing loss is not an occupational disease, and thus is not covered by section 8(c)(23), because it has no latency periods. Noise exposure causes damages at the time of the exposure. Moreover, once the exposure ends, the damage is done. The hearing loss from exposure does not get worse with time, if the exposure has ended. This is unlike most occupational disease, where there is a 10- to 30-year latency period from the exposure to when the disease appear. Because hearing loss has no latency period, the Supreme Court held it does not fall under section 8(c)(23), which applies to occupational diseases from exposures which do *not* immediately cause impairment.

The *Bath Iron Works* case recognition of the lack of a latency period is important for dating the injury, and hence determining the average weekly wage. Since the injury is complete when the exposure ends, the date of injury is the date the employee last had noise exposure, which is usually the last date the claimant worked. Thus, for retirees, the date of injury and thus the average weekly wage is based on the last day the worker worked.

3.2. Responsible Employer and Insurer

The last noise exposure also sets the responsible employer and insurer who is liable for the hearing loss. The employer and insurer at risk at the time of last noise exposure before the *determinative* audiogram are responsible for the entire claim. The judge must decide which audiogram best measures the worker's hearing loss (the determinative audiogram), and then the employer and insurer who had coverage at the time of the last noise exposure before that determinative audiogram is liable for the hearing loss (*Good v. Ingalls Shipbuilding Inc.*, 26 BRBS 159, 163 (1992); *Ramey v. Stevedoring Services of America*, 134 F.3d 954 (9th Cir. 1998); *Benjamin v. Container Stevedoring Co.*, 34 BRBS (ALJ) 115 (1999)).

3.3. Timeliness of Notice of Injury and Filing of the Claim

Before the 1984 Amendments, some hearing loss claims could be barred by the 30-day notice of injury provision in section 12 and the 1-year statute of limitations in section 13 for filing a claim (33 U.S.C. Sections 12 and 13). Since a worker can notice his own hearing loss, and easily make the connection to the noise at work, some claimants were held to have been on notice of their occupational injury, and their claim was barred for failure to provide notice to the employer (*Canamore v. Todd Shipyards Corp.*, 13 B.R.B.S. 911, 194 [1981]; see also *Mattox v. Sun Shipbuilding & Dry Dock Company*, 15 B.R.B.S. 162 [1983] [denial of asbestosis claim]; *Janusziewicz v. Sun Shipbuilding*

& *Dry Dock Company*, 14 B.R.B.S. 705 [3rd Cir. 1982] [reverse and remand of occupational lung disease compensation award]; *Walker v. Sun Ship, Inc.*, 4 B.R.B.S. 1035 [3rd Cir. 1982] [untimely notice of occupational lung disease claim was not excused]).

The 1984 Amendments, however, virtually eliminate the possibility of a claim being excluded as untimely. After the 1984 Amendments, the prohibitions of sections 12 and 13 of the Act do not commence for a hearing loss until the employee receives an "audiogram, with accompanying report thereon, which indicates that the employee has suffered a loss of hearing":

> The time for filing a notice of injury, under section 12 of this Act, or a claim for compensation, under section 13 of this Act, shall not begin to run in connection with any claim for loss of hearing under this section, until the employee has received an audiogram, with the accompanying report thereon, which indicates that the employee has suffered a loss of hearing. [33 U.S.C. Section 908 (c) (13) (D).]

This provision applies, however, only to claims pending on or filed after the date of enactment of the 1984 Amendments, 33 U.S.C. Section 928 (a). Hence, claims that were already barred should not be revived. However, if the hearing loss manifests itself after enactment, or there is further noise exposure and hence a new injury, the 1984 Amendments apply to the "new" hearing loss (33 U.S.C. Section 928 [g]). The Department of Labor has interpreted this amendment to mean that once the audiogram and report have been received, the employee is subject to the thirty (30) day, one (1) year filing requirements set forth in sections 12 (a) and 13 (a) of the Act.

Finally, of course, the statute of limitation does not begin to run until the claimant not only has his or her audiogram, but also knows, or should have known, his or her hearing loss is work-related (*Macleod v. Bethlehem Steel Corp.*, 20 B.R.B.S. 234 [1988]).

3.4. Subsequent Exposure Rule, Coordination of Benefits, and Penalties and Interest

Oftentimes, when an employee's hearing is finally tested, the worker has been retired for many years, or he or she has had subsequent non-maritime jobs when he has had additional hearing loss. The employer is not liable for any hearing loss caused after the work exposure *if* the employer can prove the amount of the subsequent hearing loss. That is, the employee is entitled to compensation for his entire hearing loss, unless the employer can prove what the loss was when he left the workplace. Unless there was an audiogram at the time of retirement, it is usually impossible to separate out the hearing loss from the different causes.

If there is subsequent nonmaritime noise exposure after the worker leaves his maritime job, the last maritime employer will be liable for the total hearing loss, unless he can prove the extent of the loss when the worker left his maritime job (*Wyman v. Bath Iron Works Corp.*, 22 B.R.B.S. 302 [1989]; *Brown v. Bath Iron Works Corp.*, 22 B.R.B.S. 384 [1989]; *Labbe v. Bath Iron Works Corp.*, 24 B.R.B.S. 159 [1991]). Similarly, an employer is liable for the effects of presbycusis (*Ronne v. Jones Oregon Stevedoring Co.*, 22 B.R.B.S. 344 [1989]). Thus, it is usually beneficial for an employer to perform a termination-of-employment audiogram to establish the employee's hearing level when he leaves the employment, and, of course, to give a copy to the employee. Conversely, however, the last employer rule ends at the time of the audiogram for that level of loss, even if there is subsequent exposure (*Port of Portland v. Director*, 24 B.R.B.S. (CRT) 137 [9th Cir. 1991]).

An employee can, of course, recover under both the Longshore Act and any applicable state compensation act (*Sun Shipbuilding v. Pennsylvania*, 12 B.R.B.S. 828 [U.S. 1980]). If the state as a matter of statute or case law allows recovery under a more liberal hearing loss formula than the AMA formula, then the employee may pursue both his federal and state rights. Similarly, an initial employer may have paid compensation for an initial hearing loss or the same employer may have paid for an initial loss and then caused a further loss. The question of how much credit should be given for prior losses and compensation payments based on the initial loss must then be addressed.

Since wages generally increase over time, and hence compensation rates increase, employers generally would prefer a credit, based on the *percent* rating of the prior award, as opposed to the *dollars* of the compensation paid. Thus, if a 10% hearing loss increases to 25%, the employee would recover only for a 15% loss at his new wage rate. The Benefits Review Board, and at least one Court of Appeals, have held that the credit must be in terms of dollars, i.e., whatever compensation the claimant was previously paid (*Balzer v. General Dynamic Corp.*, 22 B.R.B.S. 447 [1989]; *Strachan Shipping Co. v. Nash*, 18 B.R.B.S. (CRT) 45 [5th Cir. 1986]). Other credit issues, however, remain unanswered. For example, what if a worker had a 10% hearing loss that was time barred, and he later is exposed to noise and his loss increases to a 15% disability; can he recover the full 15% or only the 5% increase?

Similar problems can arise when multiple employer/carriers are involved and with dual federal and state awards by the same employer. Unfortunately, there are no easy answers. Finally, employers often are aware that their employees have hearing losses, but are unaware that they are entitled to compensation. Nevertheless, if a notice of controversion is not filed within 14 days of the employer having knowledge that the employee has an occupationally caused hearing loss, the employer is liable for a 10% penalty (33 U.S.C. Section 914 [d] and [e]). Moreover, in any case, interest must be paid from the date of injury (manifestation), and the Benefits Review Board increased the interest rate from a fixed 6% rate to the applicable U.S. Treasury Bill rate for the time the payment was unpaid (*Grant v. Portland Stevedoring Co.*, 16 B.R.B.S. 267 [1984]).

4. OTHER AREAS OF INTEREST—OSHA, ERISA, AND HANDICAP DISCRIMINATION

Workers' compensation issues should never be considered in a vacuum. While all the potential tangential problems are beyond the scope of this discussion, a brief summary of the OSHA Hearing Conservation regulations (29 U.S.C. Section 651 et seq. and 29 C.F.R. Section 1910.95) is appropriate, as well as a reference to important considerations under the Employee Retirement Income Security Act (ERISA) (29 U.S.C. Section 10(H) and handicap discrimination statutes (e.g., 29 U.S.C. Sections 793 [a] and 794 [1976]).

The Hearing Conservation regulations limit the level and time period workers can be exposed to noise. Noise levels must be monitored, hearing protection and training provided, and audiometric testing must be made available to employees exposed to an 8-h TWA of 85 dB or above. The testing results must be preserved and the record available for the employee to view. Interesting, the OSHA regulations allow testing by a technician, but the Longshore Amendments do not for application of their presumptive weight provision (20 C.F.R. Section 702.441).

Most workers with occupational diseases (which have long latency periods and/or are a result of cumulative exposures) tend to be near retirement age. This is true in the case of hearing loss. The Supreme Court, in *Alessi v. Raybestos-Manhattan, Inc.*, 101 S.Ct.

1985 (1981), held that an employer can draft its ERISA pension plan so as to take a credit under its pension plan for any workers' compensation payments to the retired worker. Arguably, such a provision would be appropriate for even a scheduled injury. Obviously, such a contractual provision can save large sums of money.

In order to avoid or reduce the costs of occupational hearing loss claims, an employer may wish to refuse to hire or to retain employees with preexisting hearing losses. Suffice it to say that the federal and state handicap discrimination statutes may place limits on such selective processing of workers.

38
Occupational Hearing Loss in Canada

John T. Rutka
University of Toronto and University Health Network, Toronto, Ontario, Canada

Christine Johnston
Workplace Safety and Insurance Board of Ontario, Toronto, Ontario, Canada

1. INTRODUCTION

The United States and Canada have had significantly greater shared economies since the North American Free Trade Agreement (NAFTA) of January 1, 1994 commenced. As our natural resource and manufacturing based industries are similar it is no surprise that the demographics of occupational hearing loss are similar. However, the compensation processes (eligibility for claims, adjudication of appeals, and the actual awards, etc.) are rather different between the United States and Canada and vary substantially within Canada itself.

Compensation for work-related injuries in Canada falls under provincial jurisdiction and is administered by quasi-autonomous, provincially appointed, Workers' Compensation Boards in each of the ten provinces and the three federally administered territories (the Yukon, Northwest Territories, and Nunavut). The major exception continues to be the Armed Forces whose claims are handled by the Ministry of National Defense and the Department of Veterans' Affairs. Compensation for hearing loss caused in the workplace is arbitrated and paid through industry funded Workers' Compensation Boards. Appeal mechanisms are built into the system at numerous levels. Workers who take their claim through the compensation boards waive their rights and are ineligible for settlements in the courts of law.

Occupational hearing loss is the most pervasive of all specifically recognized occupational disease claims. Canada's most populous and industrialized province, Ontario with a population of 12 million as of 2004 had approximately 2300 new claims a year. (In 2004 more automobiles were manufactured in Ontario than in the acknowledged motor state of Michigan, for example). In British Columbia (BC), with a population of 4.5 million, where natural resource based industries (forestry and mining) reign supreme, there were approximately 2300 new claims, a remarkably similar number.

2. WORKERS' COMPENSATION BOARDS

Following World War I, Canada's Supreme Court Justices established an over-riding principle that industry was responsible for the protection (including prevention and health and safety) and compensation of its workers. This eventually led to the establishment of now longstanding Workers' Compensation Boards. Over the years the Boards have assumed other related roles from running treatment and rehabilitation programs to seeking third party opinions from designated assessment centers (DAC) or preferred providers. There remains considerable variation between provinces.

All Boards have statutory authority and are established by government but function as autonomous corporations. Most have a Chairman who is government appointed and are guided by a Board of Commissioners. Provinces such as Ontario and BC have both a president and a CEO in the same portfolio. The boards are inclusive in their membership consisting of government appointed lay people who represent a broad range of public interests from unions to management representatives. As stewards for workplace safety, the Board of Commissioners is expected to guide the Board and not to represent rational interest. The Boards also employ professionals in a wide range of areas, including statisticians, actuaries and many physicians. The most senior physician functions as the Medical Director for the province. The Medical Director is assisted by full-time physicians who often have particular areas of expertise (e.g., respirologists with an interest in asbestosis).

It is important to recognize that all the Boards are funded in their entirety by industry: one important difference between Canada and the United States. In Ontario, for example, workplace insurance coverage is compulsory for any business of industry named in Schedule 1, Regulation 175/98. Employers included in Schedule 1 must contribute to the insurance fund. The size of the premium is based on the Workplace Safety and Insurance Board's (WSIB) classification of the employer's business activity and the employer's total payroll. Funded by employers the WSIB is one of the top 10 disability insurers in North America. There is no minimum threshold of employees for registration purposes: once an employer hires a single employee, that business must register. Those who are self-employed can purchase optional insurance. Federal government workers are covered under the Government Employee Compensation Act (a federal statue administered by the provincial Workers' Compensation Boards) according to Schedule 2. (Schedule 2 employers include the railways, telephone companies within the legislative authority of the Parliament of Canada, telegraph companies, navigation companies, international bridges, the provincial government, including boards, commissions and Crown agencies, any airline with regularly scheduled international passenger service, municipalities, etc.). The employer's contribution varies from industry to industry and is based on industry classes. These classes are subdivided further into rate groups. In Ontario, the WSIB uses rate groups to set premium rates every year that are set at a level that aims to be financially responsible, fair and balances the needs of employers, workers and the workplace safety and insurance needs. The rate may be modified on an individual employer basis under the WSIB's safety incentive programs according to the employer claims experience and accident cost history. Any change to the Experience Rating Programs aims to ensure that good performers are rewarded while employers with poor health and safety records pay their fair share.

All Boards are required to be self-supporting and debt free. Awards are based on current premiums, calculated on an actuarial basis, and funds are set aside for future payments. Nevertheless, changes in government legislation, the rising cost of health care, the increasing impact of occupational disease, increasing life expectancy, and fluctuations in investment returns have brought issues of potential unfunded liability to the forefront.

In Canada, the Boards have the legal right to seek evidence and to subpoena witnesses, and they have built in appeal mechanisms through which the employer and the employee may challenge an award without cost. Should the appellate process fail to satisfy, there remains the possibility for the lengthy and expensive process of an independent judicial review where the costs are borne totally by the aggrieved party.

The enormity of Canada, approximately 8000 km from coast to coast, its historical and political development as a confederation of initially independent, somewhat autonomous former colonies of the British Crown, and its relatively sparse, unevenly distributed population (approximately 80% of Canada's population lives within 100 km of the U.S. border) has made it difficult to achieve a national consensus on how best to manage occupational hearing loss. One can appreciate why practices might differ from province to province.

Canada's three largest cities are Toronto, Montreal, and Vancouver. Other notable cities (over 500,000 people) include the Calgary–Edmonton corridor, Winnipeg, Ottawa, London (Ontario), Quebec City, and Halifax. These cities and their greater surrounding areas incorporate just over 80% of Canada's population. The population of Canada is multicultural in the urban centers and has two official languages, English and French, both enshrined in the Constitution and the Charter of Rights. Compensation services for occupational hearing loss in each province therefore tend to be centralized in a single major center such as Vancouver, Toronto, or Quebec City. Sector or affiliated branch offices are sometimes found in regional centers. The decentralization of many of the Boards' activities (for example, in Ontario throughout the 1990s) has been replaced again by a more centralized structure for particular areas such as occupational diseases adjudication. In the province of Ontario, the sector or regional offices usually deal with acute work-related injuries initially (e.g., lower back injury).

3. HEARING LOSS

Historically, hearing loss from acute trauma (such as head injury and explosions) has been compensable in Canada from the inception of the compensation acts. Hearing loss from chronic noise exposure has been compensated relatively more recently ranging from just before World War II in Saskatchewan to as late as 1975 in British Columbia. In 1947, Ontario passed the law making chronic exposure to noise levels potentially injurious to hearing compensable, with the first claim received 3 years later in 1950. By the mid-1960s most provinces had followed suit only to see an explosion in occupational hearing loss claims in the 1970s. The majority of claims appear to arise largely from well-unionized heavy industry such as mining, oil and gas, forestry, pulp and paper, automobile and automobile parts manufacturing. Trades such as construction also figure highly in the initiation of an occupational hearing loss claim. Although the incidence of work related hearing loss continues to grow, the average age of a hearing loss claimant in Ontario has increased to 61 from 55 years, which may reflect the efficacy of hearing conservation programs and the role of presbycusis in older workers.

The principle of compensating for hearing loss resulting from chronic noise exposure differs from the original charge of the Boards to compensate a worker for loss of earnings as a result of a work-related injury. Workers generally tended to receive compensation once they had been out of hazardous noise for 6 months, but were then compensated for life. In practice, this meant that hearing loss awards were an additional retirement pension. By the mid-1970s however, regulations were changed in most provinces such that compensation would be paid while the worker continued to work in noise with payments still continuing for life.

In Ontario, claims with an accident date before 1990 were eligible for a permanent disability (PD) award, which was calculated based on a percentage of net average earnings. Claims with an accident date on or after January 2, 1990 resulted in a permanent impairment award (non-economic Loss or NEL) in recognition of the fact that most workers continued in their employment despite their impairment and as a means to acknowledge that their quality of life had been affected. The NEL award is assessed using information from the medical file and with reference to the types of impairment in the American Medical Association's Guides to the *Evaluation of Permanent Impairment, 3rd Edition (Revised).* A lump sum award is provided when a specific weighted pure-tone threshold (minimum 26.25 dB bilaterally in Ontario) is reached according to the *AMA Guides.*

By contrast, few continuous retirement pensions are awarded now. The Boards, however, are still responsible for the medical treatment of injured workers so will pay for hearing aids, necessary rehabilitation and future health care benefits associated with the injury or disease of eligible workers for life.

4. ESTABLISHMENT OF ELIGIBILITY

The process by which a claim for occupational hearing loss can be initiated is common to all provinces, namely through workers or their union steward, their employer, or a physician. The claim is submitted to the appropriate Board's Claims Department. Workers are required to indicate physicians whose care they have been under and employers for whom they have worked in the past. Employers are contacted and are obliged to submit work records, noise dosimetry levels and exposure times. For the most part, the recording of noise levels have been part of industrial safety programs in industry since the 1970s. Every province has established data banks of noise levels associated with different companies and different jobs which help establish whether a worker has been reasonably exposed to noise levels potentially injurious to hearing.

The total burden of the claim, for purposes of simplicity, used to be placed squarely on the most recent employer. This often unfair practice for what in essence is a long latency disease has changed. Although the most recent exposure employer is named as the accident employer for the purpose of blame, the financial burden is spread amongst the appropriate rate group, which ultimately is derived from the type of industry it involves and its payroll.

Once a claim has been established medical evidence is sought by the Board's Claim Department. There is a statutory obligation on the part of physicians to provide the Board with copies of consultation records and hearing tests when requested.

5. ADJUDICATING A CLAIM

Adjudication continues to differ from province to province in the absence of national consensus. In all instances, when material has been collected regarding noise exposure and medical records, a Medical Officer, staff audiologist or adjudicator employed by the Board makes a preliminary evaluation of the claim. The next steps are to establish whether there is hearing loss, to quantify it accurately, to attribute it to a cause and to make therapeutic or rehabilitative recommendations. Most boards require the worker to be off work for at least 24–48 h before an audiometric evaluation.

The following presents a sampling of how various compensation boards in Canada from west to east adjudicate and assess occupational hearing loss claims. A summary can be found in Table 38.1.

In British Columbia, all adjudications and assessments (including audiometric testing) are preformed at their own facility in Vancouver if a differential diagnosis is needed or for assessment of a permanent impairment award. Legislation requires that employers (where employees are subjected to hazardous noise) submit results of annual testing of employees to the Board. The exposure criterion is 85 dBA for 8 h per day for 2 years or its equivalent. Eligibility for an award is determined from an average weighted audiometric threshold at 500, 1000, and 2000 Hz and must be a minimum of 28 dB in one ear to qualify. To qualify for hearing aids, the loss must be at least 30 dB at 3000 Hz bilaterally, or for unilateral hearing loss, 50 dB in the worst ear at 3000 Hz. The diagnostic evaluation also includes cortical evoked response audiometry. Hearing aids are provided in the claimant's home area in private practice clinics. Hearing aid evaluations are performed, aids are provided and aural rehabilitation is undertaken. The BC Board is unique in being the only Board in the country that undertakes its own evaluation for occupational hearing loss claims.

In Alberta, compensation for permanent impairment resulting from occupational hearing loss is based on the *Alberta Permanent Clinical Impairment Guide* of June 1, 1996. The exposure criterion is 85 dBA per day for 2 years or its equivalent. Similar to the computation of binaural hearing impairment from the AMA guides, the weighted pure tone loss for compensation purposes is based on an audiometric average at 500, 1000, 2000, and 3000 Hz rating both the poorer and better hearing ears. A presbycusis correction factor is not considered in the compensation process. The use of *Robinson's Tables*, however, may be employed.

In the first of the western provinces of Canada to be settled historically, Manitoba, the Board generally requires an otolaryngologic assessment. Where this is impractical in remote areas, a general practitioner's report may be acceptable. The exposure criterion is 85 dBA for 8 h per day for 2 years or its equivalent. Under circumstances where there are non-routine findings such as a conductive hearing loss or an asymmetric sensorineural loss, claimants are usually evaluated at the Health Sciences Centre in Winnipeg. A presbycusis correction factor of 2 dB per year is applied to all those over 60 years.

In Ontario, the WSIB rates the worker's permanent impairment using the prescribed rating schedule of the *AMA Guides, 3rd Edition.* The exposure criteria for occupational

Table 38.1 Provincial and Territorial Comparisons

Province	Exposure requirements	a. Entitlement criteria b. Minimum hearing loss level for hearing aid benefits	Presbycusis correction factor (or other)	Tinnitus compensation a. Allowed in claim b. Amount PI
British Columbia (BC)	85 dBA—8 h/day over 2 years	a. Average at 500, 1000, and 2000 Hz b. Hearing loss ≥30 dB/30 dB bilaterally for hearing loss benefits	Use of *Robinson's Tables*	a. Yes b. NA
Alberta (AB)	85 dBA—8 h/day over 2 years	a, b. Average at 500, 1000, 2000, and 3000 Hz ≥140 dB to be considered for hearing aids and/or pension award	Use of *Robinson's Tables*	a. Yes b. NA
Saskatchewan (SK)	85 dBA—8 h/day over 2 years	a. Average at 500, 1000, 2000, and 3000 Hz b. Hearing loss ≥35 dB bilaterally	No deduction	a. Yes b. 2%
Manitoba (MB)	85 dBA—8 h/day over 2 years	a. Average at 500, 1000, 2000, and 3000 Hz b. Hearing loss ≥26.25 dB bilaterally for benefits	2 dB/year after age 60	a. Yes b. 2%
Ontario (ON)	90 dBA—8 h/day over 5 years	a. Average at 500, 1000, 2000, and 3000 Hz b. Hearing loss ≥22.5 dB for benefits	0.5 dB/year after age 60	a. Yes (includes medical opinion and tinnitus match) b. 2%

Quebec (QB)	NA	a. NA b. Hearing loss $\geq$30 dB for benefits	NA	a. NA b. NA
New Brunswick (NB)	NA	a. Average at 500, 1000, 2000, and 3000 Hz b. Hearing loss $\geq$25 dB for benefits	NA	a. Yes b. NA
Prince Edward Island (PEI)	85 dBA—8 h/day over 2 years	a. Average at 500, 1000, 2000, and 3000 Hz b. Hearing loss must be $>$25 dB in one year for benefits	0.5 dB/year after age 60	a. Yes b. 2%
Nova Scotia (NS)	NA	a. b. Average at 500, 1000, 2000, and 3000 Hz $\geq$100 dB	2 dB/year after age 60	a. Yes b. NA
Newfoundland (NF)	85 dBA—8 h/day over 2 years	a. NA b. Hearing loss must be $\geq$25 dB for benefits	0.5 dB/year after age	a. Yes b. NA
Yukon	85 dBA—8 h/day over 5 years (but 2 must have been worked in the Yukon)	a. Average at 500, 1000, 2000, and 3000 Hz b. Hearing loss must be $>$25 dB for benefits	None accounted for	a. Yes b. 2%
Northwest Territories and Nunavut	90 dBA—8 h/day over 5 years	a, b. NA	NA	a, b. NA

NA—not available (at time of writing); OHL—occupational hearing loss; PI—permanent impairment.

hearing loss are two-pronged and include pervasive evidence of work-relatedness in claims for sensorineural hearing loss requiring a continuous exposure of 90 dBA of noise for 8 h per day for a minimum of 5 years or the equivalent and a pattern of hearing loss consistent with noise induced sensorineural hearing loss. A presbycusis correction factor of 0.5 dB/year over age 60 is applied to the weighted pure tone averages at 500, 1000, 2000, and 3000 Hz. To qualify for hearing aid benefits an individual must have at least a weighted loss of 22.5 dB in both ears. While stricter in eligibility for compensation, the corollary is that Ontarians receive benefits for their hearing loss at much lower threshold levels. Those claimants whose hearing loss is sufficient to result in a permanent impairment benefit as recognized by the *AMA Guides, 3rd Edition* (at or above 26.25/26.25 dB or 25/32.5 dB in the better/worse ear, respectively) are referred for an NEL assessment. Workers with an accepted claim for occupational hearing loss who return to work with the same accident employer are entitled to an NEL reassessment for any further deterioration. A new claim commences if the worker under the above circumstances begins with a new accident employer. The Ontario board, as routine matter of policy, requires the medical report of an otolaryngologist and an audiogram that has been performed by either a registered audiologist or otolaryngologist in the adjudication of occupational hearing loss claims. Contentious claims are sent to outside consultants with recognized expertise for an independent medical evaluation. The philosophy behind this approach prevents the WSIB from finding itself in a position of conflict if it were to make the assessment as well as the award itself. Finally, as individuals' susceptibility to noise varies, if the evidence of noise exposure does not meet the above exposure criteria, claims will be adjudicated on the merits and the justice of the case, having regard to the nature of the occupation, the extent of the exposure and any other factors peculiar to the individual case.

In the maritime province of New Brunswick awards for occupational hearing loss require a history of noise exposure reasonably considered to be deleterious to hearing (85–90 dBA) but with no specific duration or timeframe applied. A 25 dB weighted pure-tone loss at 500, 1000, 2000, and 3000 Hz is required for compensation.

Throughout Canada, all claims that reach NEL status are usually provided with a lump sum payment which recognizes in a monetary fashion an individual's impairment in their quality of life. If their loss prevents an individual from resuming equivalent work or should there be any future economic consequences then a permanent disability (PD) award is generally granted. At the time of the second edition of this book in 1993, the cost in pension payments in Ontario for occupational hearing loss alone was $1000 per year: the average age for a claim at the time was 55 years so anticipating a worker would live another 20 years this translates to $20,000 CAD. In 2004, the average cost per NEL award was significantly less than the award that was provided under the previous legislation. On the surface, it would appear that claimants have taken a drastic cut in their pension for their quality of life impairment. This, however, has been largely offset by compensation for the costs of hearing rehabilitation, which will be discussed later in this chapter.

6. TRAUMATIC HEARING LOSS

In general, acute traumatic hearing loss has always been more generously treated than chronic hearing loss from noise. This type of hearing loss is relatively rare and is usually accompanied by other injuries, the compensation for which usually overshadows that for the hearing loss.

For the purposes of definition, the WSIB for the province of Ontario defines traumatic hearing loss as one of sudden onset experienced after acute trauma such as

- exposure to noise from an explosion (acoustic trauma),
- direct physical trauma to the ear causing damage,
- trauma to the ear resulting in a severe concussion, or
- exposure to an ototoxic drug or chemical.

The WSIB assesses a worker's permanent impairment from traumatic hearing loss when maximum medical recovery (MMR) is reached. Since the healing time after trauma varies considerably (depending on the type of injury) medical confirmation of an MMR is needed. A period as long as 6 months to 1 year may be required to ensure that the MMR has been reached. A unilateral profound acute hearing loss is usually compensated as a 5% permanent impairment, and a bilateral profound acute loss as a 35% permanent impairment of the individual. In other provinces the awards may be different.

7. TINNITUS COMPENSATION

Compensation for unwanted head noise is not infrequently sought in Canadian occupational hearing loss claims. The inclusion criteria for an award varies slightly from province to province although most require an accepted occupational hearing loss claim and its persistence for over 2 years. In Ontario, the condition also requires its confirmation by an otolaryngologist, which is difficult for a subjective complaint, with facilities for the testing of tinnitus (i.e., tinnitus matching).

Most awards appear in the 2% permanent impairment range across Canada except under unusual circumstances (e.g., severe psychotraumatic disability as a result of the tinnitus). Where severe and bothersome tinnitus exists in the presence of a unilateral hearing loss it may impair an individual's speech discrimination: under these circumstances awards are often increased to a 3–5% permanent impairment level.

8. APPEALS MECHANISMS

In all provinces, either the employee or the employer may appeal awards. In Ontario, the first level of review of the same claim is by an appeals resolution officer. Should contention continue, a further level of appeal is available where each side can be represented by counsel if required. There is no recourse to the civil courts for injuries, real or perceived, when the formal Compensation Board route is pursued.

In Ontario, a semi-judicial Board called the Workplace Safety and Insurance Appeals Tribunal (WSIAT) that has been established by the Ministry of Labor independent of the Ontario WSIB as a final appeal for the employer or the employee. WSIAT decisions are binding on all parties and cannot be overturned under the same claim. The Ministry of Labor has established an office of the Workers' Adviser and an office of the Employer's Adviser to guide and if need be, assist the appealing parties.

9. REHABILITATION

The provision of hearing aids and what each provincial board will cover varies. Some provinces such as BC and Alberta have a co-payment option with a cap on the maximum amount funded, whereas others such as Ontario and New Brunswick have no ceiling and cover the full cost of hearing aids. Digital aids appear to be replacing analog hearing aids in conjunction with current changes in the hearing aid industry.

The Provincial Boards also pay for the upkeep and ongoing costs including batteries, repairs and replacements. Replacement time varies between provinces from every 3 years in Ontario to every 6 years in Quebec.

10. NOVEL PROGRAMS OF QUALITY ASSURANCE

Over the last 3 years, the cost for hearing aid provision has skyrocketed in all provinces.

In Ontario, costs of $25 million in 2001 increased to over $34 million in 2004 despite the annual number of claims staying relatively unchanged. Part of the cost increase has been the result of new technology in hearing aid manufacturing, the move to a 3-year replacement strategy and the inclusion of assistive listening devices (e.g., telephone amplifiers and FM systems).

In an attempt to ensure that both workers continue to receive benefit and that employers remain satisfied their premiums are well spent, the Ontario WSIB instituted in June 2004 an ambitious Program of Care for Noise Induced Hearing Loss that requires evidence-based care, fair and reasonable fees, improved provider relationships, and measured outcomes. The process included all the representatives and stakeholders for hearing health care (industry, unions, the Ontario Medical Association, the College of Audiology and Speech and Language Practitioners for the Province of Ontario, and the Association of Hearing Instrument Practitioners amongst others). Outside reviews of consensus recommendations arising from the Program of Care were sought in a process that is meant to be reiterative as issues, our scientific knowledge of hearing loss, advances in technology, etc. change. The introduction of a Client Orientated Scale of Improvement (COSI™)* to evaluate objectively the usefulness of a hearing aid is administered by the provider in two phases. In the first phase, the workers identify listening situations that they would like improved with new amplification. In the second phase, after the hearing aid is fitted, the change in hearing function for the identified listening situation is recorded. The prime goal of this program is accountability and the demonstration of individual benefit from hearing aids and other listening devices, not constraint of costs.

11. HEARING CONSERVATION

Risk tables for the provision of mandatory hearing protection vary from province to province, from 85 dBA for an 8 h continuous exposure over 2 years in British Columbia to 90 dBA for an 8 h continuous exposure over 5 years in Ontario. Most provinces have criteria closer to British Columbia's.

In Ontario, education for employers and their employees has become a prime goal in the prevention of occupational hearing loss. Guides to Noise Control and Hearing Conservation (NCHC) and an Audit Tool for industry is readily available to all stakeholders.

---

*Australian Hearing is the holder of the copyright and trademark in the COSI Questionnaire.

The Audit Tool has been created to:

- assess the completeness and quality of the program's components; and
- evaluate noise exposure and the audiometric data for both individuals and groups of employees exposed to hearing hazards.

Within the Audit Tool are sections on standards (company, management, and administrative), recognition (noise measurements, training, and content), the assessment of noise (monitoring audiometry and record keeping), and controlling noise (engineering and administrative controls and hearing protection). Further information concerning the NCHC, Audit Tools, Program of Care and other useful electronic links can be found at the WSIB website (www.wsib.on.ca).

12. CONCLUSIONS

Regulations concerning the prevention and assessment of occupational hearing loss continue to vary from province to province, which suggests the need for a national strategy to manage occupational hearing loss claims and their rehabilitation in future. The increased public awareness of occupational injuries and the increasingly high cost of compensation requires improved legislation and accountability that the dollars contributed by industry are wisely spent and that workers receive appropriate compensation for their work-related hearing impairments. We anticipate that the systems of compensation will continue to evolve. The introduction of novel programs of quality assurance and hearing conservation education will hopefully make the workplace safer and benefit all those with occupational hearing loss impairments.

ACKNOWLEDGMENTS

The help from all provincial medical officers and their respective occupational managers is gratefully acknowledged. Mr Daniel Hajioff's proof reading and editing of this chapter is also appreciated. The authors would also like to acknowledge the contribution of Dr Peter Alberti's chapter on *Occupational Hearing Loss in Canada* in a previous edition of this book. Dr Alberti's international reputation and interest in this aspect of industrial medicine has rightly resulted in his recognition as one of the founding fathers of occupational hearing loss in Canada.

39

Occupational Hearing Loss in the United Kingdom

F. Javed Uddin
Derbyshire Royal Infirmary, UK

Ann F. Dingle
Leighton Hospital, Crewe, UK

Jeremy F. Sharp
Derbyshire Royal Infirmary, UK

Liam M. Flood
James Cook University Hospital, Middlesbrough, UK

In the United Kingdom, legislation to protect those exposed to injurious noise at work and to compensate for a consequent hearing loss has been greatly influenced by economic considerations. The financial implications for industry, the highly concentrated nature of the problem geographically, and the desire to avoid swamping the limited resources of the audiologic services of the National Health Service (NHS) must be considered in framing legislation. The concept of a welfare state, introduced in 1948, has ultimately allowed a uniform, nationwide system of compensation. In addition, claims against employers by civil claims or by negotiated agreements between unions and insurance companies occur widely.

1. HISTORY

Until 1961, workers suffering a noise-induced hearing loss had little recourse at law unless they could prove common-law negligence on the part of the employer (1). However, the Factories Act (1961), although making no specific provisions relating to control of noise, required the employer to make and keep safe, so far as reasonably practical, the place of work (Section 29).

The Wilson Report (1963) considered that contemporary knowledge was insufficient to form a basis for legislative control but recommended further research and increased awareness of the hazards (2). The scientific basis for further legislation on noise was produced by the pioneering work reported by Burns and Robinson in 1970 (3). A joint investigation by the Medical Research Council and the National Physical Laboratory established the relationship between noise exposure and hearing loss and defined data regarding noise levels and duration of noise exposure with consequent hearing loss.

Accordingly, a Code of Practice was drawn up in 1972 by the Department of Employment. Both sides of industry, the Trades Union Congress (TUC) and the Confederation of British Industries (CBI), were represented on the Industrial Health Advisory Sub-Committee. Although the recommendations are voluntary, a breach of the code is admissible as evidence of negligence.

The Health and Safety at Work Act (1974) allows a civil claim and breach of statutory duty, in addition to a claim of negligence, for damages resulting from industrial noise. Few actions against employers resulted, however. Up to 1976, only 10 cases involving 11 plaintiffs were heard in the high court, and damages totaled no more than £51,000 (4). With increased public awareness, this picture of resigned acceptance is fast disappearing.

The National Insurance (Industrial Injuries) Act (1965) had established the principle of payment of industrial injury benefit for prescribed industrial diseases or for personal injury following an accident (e.g., acute blast injury to the ear). The Social Security Act (1975) finally included progressive noise-induced hearing loss in the list of prescribed diseases covered by the Industrial Injuries Scheme as Occupational Deafness (PD48).

The criteria of eligibility for compensation were initially very restrictive to avoid overburdening the NHS audiological service and aimed to keep the number of examinations within 10,000/year. In fact, the yearly total never reached more than one-third of this ceiling, and less than 3000 claimants proved eligible for benefits between 1975 and 1979. The yearly total of successful claims fell from 1367 in 1975 to 432 in 1979. Total benefits paid increased from £250,000 to £2,750,000 during this period. Only three occupations were covered (drop forging, shipbuilding, and heavy metalwork), with a minimum of 20 years employment, and claims were to be made within 1 year of leaving employment. Disability was calculated on a scale from 20% for a 50 dB loss to 100% for a 90 dB loss.

In 1979, the regulations were amended to cover a wider range of occupations. These included the use of pneumatic percussive tools, weaving machinery, spray guns, and nail cutting machines. In addition, 100% disability was considered to occur at 110 dB (Table 39.1). The regulations were further amended in 1983 and made eligible for compensation those workers employed in the immediate vicinity (50 ft) of the specified machinery, rather than solely the machine operator and his supervisors. Further to recommendations from the British Association of Otolaryngologists, the range of occupations covered by prescribed disease A 10 was increased to include the use of chain saws and high-speed wood-working machines (Table 39.2) (5). Both amendments led to marked increases in the number of claims made from the region of 3000 to 5000 in 1979 to

Table 39.1 Calculation of Disability

dB Hearing loss (averaged 1, 2, and 3 kHz)	Disability (%)
50–53	20
54–60	30
61–66	40
67–72	50
73–79	60
80–86	70
87–95	80
96–100	90
106	100

Calculation of average binaural hearing loss: average loss = 0.2 × [4 × (better ear) + 1 × (worse ear)].

approaching 12,000 in 1983. A review by the Industrial Injuries Advisory Council in 1982 also led to further relaxation of the criteria of eligibility for compensation (6). As most of the damage from industrial noise occurs in the first few years of exposure it was recommended that the minimum period of employment be dropped to 10 years. The requirement to apply within 1 year of ceasing work was felt unfair to those made redundant and thus was increased to 5 years. The somewhat high qualifying minimal loss of 50 dB has been retained. This is despite the "British Standard Method for Estimating the Risk of Hearing Handicap Due to Noise Exposure" (BS 5330:1976), which determined that a 3 dB hearing loss, averaged over 1, 2, and 3 kHz, caused significant handicap by impairing understanding of conversational speech with low levels of background noise.

2. DIMENSIONS OF THE PROBLEM IN THE UNITED KINGDOM

The pioneering research reported by Burns and Robinson in 1970 calculated that after a lifetime of exposure to 100 dB, 32% of industrial workers will exhibit auditory thresholds greater than 50 dB at 1, 2, and 3 kHz. At 90 dB exposure, 11% will be so afflicted, and even at the 80 dB level, 3% will suffer this handicap. It has been estimated that 600,000 Britons work in noise levels greater than 90 dB (A) over an 8-hour period (7).

Table 39.2 Occupations Covered by "Prescribed Disease A 10"

Any occupation involving:
The use of powered grinding tools on cast metal or on billets or blooms
The use of pneumatic percussive tools on metal
The use of pneumatic percussive tools for drilling rock in quarries, underground, or in mining coal
Work wholly or mainly in the immediate vicinity of a plant engaged in the forging of metal
The use of machines engaged in weaving fibers or high-speed false twisting of fibers
The use of machines engaged in cutting, shaping, or cleaning nails
The use of specific machines engaged in the working of wood and circular sawing machines
The use of chain saws in forestry

Two and a half million are exposed to noise levels in excess of 90 dB (A). More recently, it has been estimated that 11% of men and 6% of women in Britain nearly always need to raise their voices to be heard in the workplace, and that 3% of men and 2% of women encounter working conditions that leave them with ringing in their ears or temporary deafness on a daily basis (8).

The Social Security Act (1975) rendered 20,000 workers eligible for compensation, and the broadening of coverage provided by the subsequent legislation (1979) increased this figure more than sevenfold to 150,000. Estimates from the Health & Safety Executive's (HSE) Self-reported Work-related Illness (SWI) survey in 2003–2004 estimated that 81,000 people ever employed in Great Britain were suffering from work-related hearing problems in this year. The Department for Work and Pensions (DWP) estimate from a 10% sample of its records showed that 14,000 people were receiving disablement benefit for occupational deafness in 2001. The annual numbers of new awards of disablement benefit for occupational deafness show a clear downward trend in the number of awards each year in the years up to 1997–1998 inclusive and in 2000 when the annual number reached its lowest level at 225. The number rose slightly to 335 in 2003, but remained well below the levels seen in 1997 and earlier (9).

The problem is highly localized to the north of England and Scotland. The predominantly lighter industries of London account for less than 5% of all claims for compensation. In contrast, three-quarters of claims evaluated come from four regions: Tyneside and Teesside, 29.25%; Scotland, 28%; Yorkshire and Humberside, 10.25%; and West Midlands, 10.25%.

The industry groups with the highest average annual incidence rates of new cases qualifying for benefit (in 2002–2004) were extraction energy and water supply, manufacturing and construction. Occupations with high incidence rates included the armed forces and other laborers in making and processing and machine tool operatives (9).

3. COMPENSATION

Unless alleging negligence on the part of their employers, workers suffering from occupational deafness seek compensation through the National Insurance Scheme. If a worker is suffering from an acute onset of deafness due to an accident at work, even where there is no negligence or fault on the part of the employer, compensation is covered by the National Insurance (Industrial Injuries) Act (1965) and benefit is paid. However, the worker suffering a progressive hearing loss following prolonged noise exposure must satisfy strict criteria.

To qualify for disablement benefit, the worker must demonstrate a prolonged employment in a prescribed occupation with sufficient consequent hearing loss. A minimum of 10 years employment is required, and applications must be made within 5 years of leaving this work (5).

Occupational deafness (prescribed disease A 10) has been defined as "sensorineural hearing loss, amounting to at least 50 dB in each ear, being due in the case of at least one ear to occupational noise and being the average of hearing losses measured by audiometry over the 1, 2, and 3 kHz frequencies."

When initial application is made through the social security office, the claim is assessed to satisfy employment and audiological criteria.

1. *Employment criteria.* The insurance officers must confirm that the above employment conditions have been satisfied. However, should they disallow

this the claimant has the right of appeal to a Local Tribunal and ultimately to the Social Security Commissioner.

2. *Audiological criteria.* If these minimum requirements are met, the claimant undergoes audiometry and examination by a consultant otologist. Should this evaluation fail to satisfy the minimum audiometric requirements the insurance officer will disallow the claim and no further claim may be made for a period of 3 years. There is, however, a right of appeal, to be made within 10 days, to a Medical Board of two doctors and ultimately to a Medical Appeal Tribunal.

In the event of a satisfactory outcome to audiological assessment, the claimant is referred to a Medical Board, who will decide whether the claimant is indeed suffering from prescribed disease A 10 and will assess the extent and period of disablement. If they reject the claim, there is a right of appeal to the Medical Appeal Tribunal, whose decision is final. Further appeals (to the National Insurance Commissioner) may only be on a point of law and not on diagnosis.

The extent of disability is calculated on a scale from 20% to 100% (Table 39.3) and applies either for life or for a minimum period of 5 years, whatever the subsequent course of the disease. An allowance of disability due to presbycusis is made based on a 1% reduction for every year over 66 years in men and 75 years in women. This may not reduce the assessment reassessments. In practice, such allowance rarely influences the calculation of percentage disability. The requirement for early application following retirement from occupation implies that claimants are relatively young. Minor alterations in percentage disability are then usually canceled out by the requirement to round out the figure calculated to the nearest multiple of 10%. In the event of a conductive loss, the bone conduction levels are used to diagnose occupational deafness, i.e., there should be a 50 dB sensorineural loss in both ears (averaged 1, 2, and 3 kHz) due to noise in at least one of them. However, to calculate the degree of disability for compensation, the overall hearing losses are considered.

Tinnitus suffered as a result of noise exposure cannot be objectively measured and is rarely compensated. Where it is considered to cause a disability greater than that expected for the degree of noise-induced hearing loss, the examiner should include a report of the additional handicap suffered (10).

Benefit is paid as a weekly pension and will not compromise war pensions or any other national insurance benefits, such as a retirement pension or a sickness/invalidity benefit. However, should the claimant be so unfortunate as to have suffered other industrial diseases eligible for disablement benefit, the sum percentage disability may not exceed the 100% ceiling (see Table 39.3) (11).

Table 39.3 Industrial Disability Benefit Rates (2005)

Disability (%)	Weekly benefit paid (£)
100	123.80
90	111.42
80	99.04
70	86.66
60	74.28
50	61.90
40	49.52
30	37.14
20	24.76

The cost of medical rehabilitation of the hearing impaired is met by the NHS. Medical treatment and provision, replacement, and maintenance of the NHS hearing aids are free of charge.

4. CIVIL CLAIMS

The dual system of law in England allows the worker with noise-induced hearing loss to bring claims against both his employer and the State. Under statute law (that is, the Law of Acts), a worker may claim compensation from the State without proving negligence on the part of his employer. This is the basis of "no fault" compensation. Under common law, a worker may bring a civil claim against an employer but needs to prove negligence on the part of the employer. The date after which all employers are expected to be aware of the risks of noise, and to have taken measures for protection against noise, is 1963. Prior to this "date of guilty knowledge," employers cannot be held responsible for hearing loss due to noise.

The first civil claim against an employer was brought in 1969 and the claimant lost the case (12). Subsequently, in the 1970s and early 1980s, a small number of cases came to litigation. In the early 1980s, it became apparent that there was an enormous number of cases waiting to be tried and by 1984 the legal system was in danger of being swamped by over 20,000 outstanding claims.

One of the leading cases on industrial hearing loss is that of *Thompson v. Smiths Ship Repairers* (1984). This case brought to light matters regarding industrial compensation, health and safety at work, and employer's liability. It highlighted the number of pending cases, and while it was not a test case, the judge's ruling on the case suggested that similar disputes might be settled by agreement. This has become known as the Mustill judgment and as a result an agreement was set up between the General Municipal Boilermakers and Allied Trades Union (GMBATU) and the Iron Trades Mutual Insurance Company. This agreement was drawn up in January of 1984 and applied to members of the GMBATU bringing claims for hearing loss and mild or moderate tinnitus caused by excessive noise (more than 90 dB). A claimant is examined by one of an agreed panel of ENT specialists and hearing loss over 1, 2, and 3 kHz is recorded. The average hearing loss is calculated by the DHSS formula (Table 39.1) and the degree of disability is thus calculated. Compensation is determined accordingly.

The minimum average hearing loss *compensable* by this agreement is 25 dB. The time period insured is from 1963 to 1978. After 1978 employers do not concede negligence as hearing protection was being provided. In exceptional cases employees exposed to noise between 1953 and 1962 may be considered for compensation. In those cases where a claimant's employers are not insured by the Iron Trades Insurance Company, the company will endeavour to persuade other insurers to agree to the terms of the scheme.

Under this scheme, a large number of claims are processed and in the Newcastle and Sheffield area, where industry is at its most concentrated in the United Kingdom, in the region of 1000 claims per month are still being settled. A small number of claims may still be brought to court but the majority are now settled by out of court agreements.

5. GUIDELINES FOR THE DIAGNOSIS OF NOISE-INDUCED HEARING LOSS

Coles et al. (2000) have published guidelines to assist in the diagnosis of noise-induced hearing loss (NIHL) for medicolegal settings (13). Establishment of the diagnosis is

Table 39.4 Requirements (R) and Modifying Factors (MF)

R1	High-frequency impairment
R2	Noise exposure
a	100 dB (A)
b	90 dB (A) (susceptible individuals)
R3	Audiometric evidence of hearing loss
a	At least 10 dB difference
b	At least 20 dB difference
MF1	Clinical picture
MF2	Compatibility with age and noise exposure
MF3	Robinson's criteria
MF4	Complicated cases

based on the balance of probabilities, i.e., "more probable than not," in the presence of a reliably measurable hearing loss on audiometry. A factor to consider in the calculation is the age-associated hearing loss (AAHL) and various tables are available detailing the range of age-associated hearing data corrected for age and sex [ISO 7029 (1984)] (14). The remaining task is then to quantify the existence of a NIHL in the overall hearing impairment. A value of at least 10 dB at 4 kHz is considered significant.

The guidelines comprise three requirements (R1, R2(a,b), and R3(a,b)) and four modifying factors (MF1–4) as shown in Table 39.4. For a diagnosis of NIHL, either the combination of R1, R2a and R3a or R1, R2b and R3b needs to be fulfilled. The modifying factors then further strengthen or weaken the analysis.

Requirement R1 is the audiometric presence of a high-frequency sensorineural hearing loss and is taken as at least a 10 dB threshold increase at 3, 4, or 6 kHz when compared to 1 or 2 kHz (after correction for earphone type). Requirement R2 is the presence of a significant duration of loud noise exposure. R2a states that at least 50% of individuals exposed to the stated noise would suffer measurable hearing loss. Practically, this is taken as a daily 8-h continuous noise exposure of at least 85 dB (A) for a number of years leading to a cumulative exposure of at least 100 dB (A). R2b is applied to that minority of individuals who are more susceptible than average and the lower level of total noise exposure of 90 dB (A) is used. Requirement R3 relates to audiometric evidence of NIHL. This is in the form of either a downward "notch" in the 3–6 kHz range or a "bulge" down and to the left in this frequency range (Fig. 39.1). This bulge is typically seen rather than the characteristic notch due to the additional presence of an AAHL. R3a is met where the hearing threshold (notch or bulge) at 3, 4, or 6 kHz is at least 10 dB greater than at 1 or 2 kHz. R3b is to be used when R2b is only met for the noise exposure and states that the notch or bulge has to be at least 20 dB as opposed to 10 dB.

The modifying factors either strengthen or weaken the calculation. MF1 is based on a history of progressive auditory symptoms and after other causes of hearing impairment are accounted for. MF2 allows for checking the measured hearing impairment for compatibility with age, sex, and estimated total noise exposure (occupational and non-occupational) (15). MF3 is used with "borderline" cases of NIHL and compatibility is checked with Robinson's probability tests (16). These are a series of statistical tests to define criteria relating to the degree of conformity with the Burns and Robinson model of NIHL (3), asymmetry of hearing loss, and whether other diagnoses need to be considered that may have contributed to the measured hearing loss. MF4 is used for

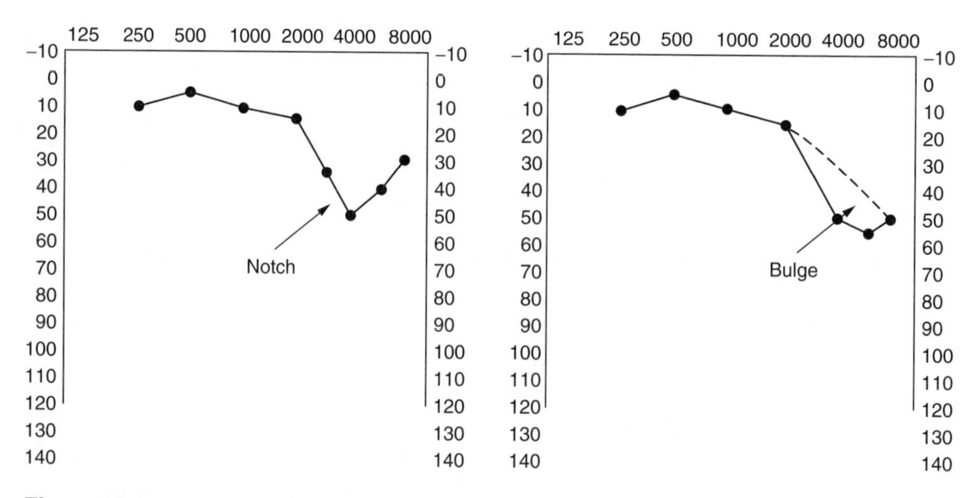

Figure 39.1 Example of an audiometric "notch" (left) and "bulge" (right).

complicated cases where there is marked asymmetry, other otological disorders, or a conductive hearing loss.

The calculation based on these guidelines is as follows (see, for example, Table 39.5). Firstly, the actual hearing threshold level (HTL) is measured (after correction for conductive hearing loss and earphone type). In the second step, the values at 1 and 8 kHz are taken as lower and upper "anchor points," respectively. Step 3 involves consulting sex- and age-adjusted tables for AAHL reference values. Step 4 calculates the differences between the values in step 3 and measured HTLs (step 2) at the two "anchor points," and step 5 then interpolates these "misfit" values for the intermediate frequencies (2, 3, 4, and 6 kHz). In step 6, the adjusted AAHL values are calculated by adding the interpolated "misfit" values (steps 4 and 5) to the AAHL reference values (step 3). In the final step the adjusted AAHL values (step 6) are compared to the actual measured HTLs (step 1), and thus provides an estimate of the probability of NIHL for that ear with a value of at least 10 dB taken as significant.

Table 39.5 Calculation of Possible Presence of NIHL (one ear only)

	Hearing threshold level (dB) at the various frequencies (kHz)					
	1	2	3	4	6	8
Step 1 (corrected measured HTL)	10	20	40	50	45	50
Step 2 ('anchor points')	10					50
Step 3 (AAHL tables)	9	16	23	34	39	46
Step 4 (anchor "misfits," 2 − 3)	+1					+4
Step 5 (interpolation)		+1	+2	+3	+4	
Step 6 (adjusted AAHL, (3 + (4 + 5)))	10	17	25	37	43	50
Step 7 (audiometric difference, 1 − 6)	0	+3	**+15**	**+13**	+2	0

Figures in bold face indicate a significant bulge of +15 dB at 3 kHz and of +13 dB at 4 kHz.

6. PROTECTION OF HEARING

The Noise at Work Regulations (1989) specifies the legal obligations of employers to prevent damage to the hearing of workers from exposure to noise at work (17). These requirements are in addition to the general obligations to protect worker's health specified in the Health and Safety at Work Act of 1974. The regulations were made with reference to a European Council directive on the approximations of the laws of the Member States relating to machinery. The Council directive requires that "existing national health and safety provisions providing protection against risks caused by machinery must be approximated to ensure free movement of machinery without lowering existing justified levels of protection in the Member States" (18). Machinery conforming to the European Community (EC) regulations will carry an EC Declaration of Conformity and will be labeled accordingly with the EC symbol. With regard to noise, the machinery must be designed to reduce noise to a minimum, although no specific levels are recommended. The U.K. Noise at Work Regulations apply to all workers covered by the Health and Safety at Work Act except crews of ships, aircraft, or hovercraft.

Three levels of noise are specified as action levels. The first action level is a daily personal noise exposure of 85 dB (A), the second action level is a daily personal noise exposure of 90 dB (A), the peak action level is a peak sound pressure of 200 Pa (140 dB re 20 μPa). The personal noise exposure or equivalent continuous sound level is calculated for fluctuating sound or varied length of exposure. Every employer is legally obliged to reduce noise exposure to the lowest level practicable. This includes modification of machinery where practical to reduce noise emission. When despite measures to control the noise at source, the levels lie between the first and second action level, ear protection should be provided at the request of the employee. When noise exposure is greater than the second action level or peak action level, ear protection must be provided and employers should ensure its use. Ear protectors should be suitable for the conditions where they will be used and should provide minimum protection of 5 dB (A) or the amount by which the exposure exceeds 90 dB (A) or 200 Pa. They should be readily available and of personal issue.

Where noise levels exceed the sound or peak action level, an employer is obliged to provide adequate information, instruction, and training of employees with regard to protection of hearing. Designated ear protection zones should be clearly marked and the wearing of ear protectors enforced in these areas. The Health and Safety Executive has limited powers to grant exemptions from the regulations in special circumstances. Exemptions apply when

1. There are substantial fluctuations in noise exposure from day to day. Control is exercised over the weekly exposure.
2. The use of ear protectors may increase danger outweighing the risks of hearing loss.
3. Where it is unpractical to wear ear protectors.

7. CRITICISMS

U.K. noise legislation has been criticized as being too strict in adopting criteria for selection of cases suitable for compensation, and too lenient in applying regulations to ensure hearing conservation.

Eligibility for compensation requires prolonged employment in a limited range of occupations known to be particularly hazardous. It does not attempt to refer to the noise levels experienced regardless of the nature of the work. A review of the workings of the Social Security Act (1975) has highlighted many problems (19). The Industrial Injuries Advisory Council in 1982 noted inconsistencies in assessment of claimants when reviewing the initial awards after a 5-year interval. Of 586 claims reassessed, 30.9% showed an audiometric improvement of more than 10 dB (reduced disability), and in no less than 13.3% the diagnosis was no longer satisfied (no disability)! In a disease that is usually progressive, it was also surprising to note that only 24.3% showed any deterioration of hearing. It was suggested that inaccurate audiometry was responsible (6); 3 kHz thresholds had frequently not been estimated, possibly because not all designs of audiometers allow this. Adequate calibration of audiometers was stressed. Limited bone conduction output of some audiometers prevents evaluation of bone conduction thresholds greater than 50–60 dB. As compensation is paid for a mean loss of 50 dB at 1, 2, and 3 kHz, this may involve estimating bone conduction thresholds up to 70 dB at 3 kHz. Although conceding that, after prolonged exposure to industrial noise some hearing loss is inevitable, the Council acknowledged the difficulties in excluding nonorganic deafness, whether artifactual, malingering, or psychogenic.

The peak daily personal noise exposure of 90 dB (A) specified by the noise regulations has been criticized as excessively high. As shown above, a far greater proportion of the work force is exposed to noise levels of 80–90 dB than to 90 dB and above. Although a smaller proportion will be handicapped as a result of this lesser exposure, the greater population implies a considerable scale of disablement nonetheless. An 85 dB limit would triple the number of workers protected (7). Indeed, it is the declared aim of the TUC to reduce this acceptable level to 80 dB (20). The General and Municipal Workers Union and the textile industry have been in the forefront of a TUC campaign to publicize the symptoms of occupational deafness through safety representatives, while advising claimants seeking disability benefit or damages from employers.

Noise legislation has been further criticized as ignoring the damaging effects of noise other than hearing loss. The effect on behavior, mood, efficiency, and accident rate should also be considered. A possible link between industrial noise exposure and hypertension has been suggested. A rise in diastolic pressure and total peripheral resistance is found and persists after cessation of noise exposure (21). Tinnitus, recruitment, and diplacusis or distortion are ignored in assessment for compensation. Settlements for hearing losses through civil litigation are somewhat low, ranging from £4000 to £7000 for a moderate loss requiring a hearing aid. Furthermore, there is no legal consideration of the effects of further noise exposure in leisure activities or military service. One in 10 of those exposed to high noise levels in discotheques are exposed to injurious noise in their work (22).

The place of monitoring audiometry is a debatable issue (23). Those in favor argue that audiometry will detect susceptible cases early, prove the efficiency of hearing conservation measures, increase public awareness of the hazard, and provide medicolegal protection for the employer by excluding previous NIHL. Those against argue the cost of equipment and testing, particularly for firms with factories that are widely scattered.

Little consideration has previously been given to noise abatement in machinery design. To educate engineers, the Institute of Acoustics has introduced a Diploma of Acoustics and Noise Control, following a part-time course for postgraduate students. Some U.K. universities provide similar courses for undergraduate students of engineering.

The noise regulations (1989) specify the need for reduction of noise at source and the Health and Safety Executive have published recommendations regarding reduction of noise emission levels by machinery.

8. SUMMARY

The U.K. worker suffering occupational deafness is recognized as having a disability and is consequently entitled to compensation through the National Insurance Scheme. In addition, workers may be eligible for compensation from their employers as a result of negligence. Guidelines are now available to assist in the diagnosis of noise-induced hearing loss in medicolegal settings.

REFERENCES

1. Kerse CS. The Law Relating to Noise. London: Oyez Publishing, 1975.
2. Report of the Committee of the Problem of Noise, Chairman Sir Alan Wilson. London: HMSO, 1963.
3. Burns W, Robinson DW. Hearing and Noise in Industry. London: HMSO, 1970.
4. Industrial Noise. The Conduct of the Reasonable and Prudent Employer, The Wolfson Unit for Noise and Vibration Control University of Southampton, Southampton, England, 1976.
5. If You Think Your Job Has Made You Deaf, Department of Health and Social Security Leaflet N1 207, London, 1991.
6. Occupational Deafness: Report by the Industrial Injuries Advisory Council. HMSO, London, 1982.
7. Editorial. Noise at work. Br Med J 1981; 283–A5S.
8. Jones JR, Hodgson JT, Osman J. Self-Reported Working Conditions. London: Health and Safety Executive HMSO, 1995.
9. Occupational Exposure to Noise and Hearing Difficulties in Great Britain. London: Health and Safety Executive HMSO, 2001.
10. The BAOL and BSA Method for Assessment of Hearing Disability, British Association of Otolaryngologists, 1983.
11. Social Security Benefit Rates, DHSS Leaflet N1 196, London, 1991.
12. Ping CW. Forensic audiology. J Laryngol Otol 1986; Suppl 11.
13. Coles RRA, Lutman ME, Buffin JT. Guidelines on the diagnosis of noise-induced hearing loss for medicolegal purposes. Clin Otolaryngol 2000; 25:264–273.
14. ISO 7029 Acoustics: Threshold of Hearing by Air Conduction as a Function of Age and Sex for Otologically Normal Persons, International Organization for Standardization, Geneva, 1984.
15. ISO 1999 Acoustics: Determination of Occupational Noise Exposure and Estimation of Noise-Induced Hearing Impairment, International Organization for Standardization, Geneva, 1990.
16. Robinson DW. The audiogram in hearing loss due to noise: a probability test to uncover other causation. Ann Occup Hyg 1985; 29:477–493.
17. Noise at Work, Health and Safety Executive HMSO, London, 1989.
18. Council Directive of 14th June 1989, Off J Eur Community, 1989; 32:9–32.
19. Henwood WR. Problems with occupational deafness as a prescribed disease. Health Trends, 1982; 74:76–78.
20. Editorial. Trade union action on noise-induced deafness. Lancet, 1979; 2:1067.
21. Andren L, Hansson L, Bjorkman M, Johnson A. Noise as a contributory factor in the development of elevated arterial pressure. Acta Med Scand 1980; 207:493–498.
22. Bickerdike J, Gregory A. An evaluation of hearing damage risk to attenders at discotheques, Leeds Polytechnic, Leeds, 1979.
23. Taylor W. In Steel PM, ed. Clinical Otolaryngology. London: Blackwell Scientific, 1979.

40

Tape Simulation of Hearing Loss

Robert T. Sataloff
Drexel University College of Medicine, American Institute for Voice and Ear Research, and Graduate Hospital, Philadelphia, Pennsylvania, USA

Joseph Sataloff
Thomas Jefferson University, Philadelphia, Pennsylvania, USA

Recently, attempts have been made to simulate the hearing of individuals through audio recordings. Those responsible for creating these tapes have put them forth as reasonable representations of the hearing of specific individuals. The tapes are manufactured on the basis of an individual's pure-tone audiogram, using electronic filtering of selected speech phrases. To the incompletely educated observer, the premise behind preparation of the tapes seems to make sense, and they even appear to be "scientifically" engineered. Consequently, these extremely inaccurate tapes could easily mislead a listener. However, closer scrutiny reveals that the tapes violate virtually all scientific guidelines and bear no predictable relationship to reality. Understanding why these tapes do not present even a reasonably valid approximation of the hearing abilities of the people on whose audiograms they were based, requires a working familiarity with several complex issues in otology and acoustics.

The author's (RTS) interest in and study of this subject were accelerated by the creation of a large number of tapes by specific group of individuals. This led to a collaborative investigation of the reliability and validity of the tapes by a team of specialists utilizing an otologist, physicist, engineer, physiological acoustician, and others, including Robert Thayer Sataloff, M.D., D.M.A.; Terrence A. Dear, M.S., M.E.; Kathleen K. Hodgdon, M.S.; K. Uno Ingard, Ph.D.; Merle Lawrence, Ph.D.; Paul L. Michael, Ph.D.; and John J. Portelli, M.S. The information presented in this chapter summarizes the insights and conclusions of this group. The specific tapes studied had avoidable, technical flaws such as use of one microphone, selection of biased speech samples ("the scarf was made of shiny silk"), simulated recordings in noise that did not include the normal increase in the speaker's intensity (Lombard effect), equipment errors, and other problems. Analysis of these factors and their consequences will be reviewed in this chapter. However, the study revealed that even if these shortcomings were corrected, fundamental scientific problems exist which make it impossible to create a recording, play it back to a group of listeners (such as judge and jury), and have even a modest assurance that what the listeners hear bears any relation at all to what is heard by the patient on whose audiogram the tapes were based. This chapter discusses some of the technical problems that illustrate how difficult it is to make high-quality tape recordings (and how easy it is for such tapes to be biased, intentionally or unintentionally), as well as the principles that dictate that even the finest tapes would be invalid.

Demonstration tapes are prepared on the basis of each individual patient's pure-tone audiogram. It is well known that routine industrial screening audiometry involves the presentation of pure tones to the listener subject in accordance with instrumentation specifications and procedures set forth in American National Standards Institute (ANSI) standards (1).

The purpose of this specific type of audiometry is to determine hearing sensitivity (2) only under test conditions specifically designed to exclude and eliminate a number of well-known factors and phenomena that aid humans in the hearing process. Factors and phenomena that are eliminated in pure-tone industrial screening audiometry include the Lombard effect, combination tones in the speech frequencies, and feedback and binaural enhancement, among others. This raises serious technical and scientific questions concerning the application of pure-tone audiometric test results to the production of audiotapes that purport to represent actual hearing abilities, particularly those that pertain to the hearing of human speech by individuals so tested.

As clearly pointed out by T. Miyakita and H. Miura in a peer-reviewed technical paper (3):

> Pre-employment and periodic follow-up audiometric examinations for noise exposed workers play a very important role in the early detection of high frequency hearing loss. However, pure-tone audiometry offers no direct information about the handicaps experienced by workers with noise induced hearing loss.

As noted by these authors, this is a fact concurred with D. H. Eldridge (4) and K. D. Kryter (5). The evidence clearly suggests that any attempts at portraying hearing handicaps experienced by workers using only their pure-tone audiometric test records would be inappropriate and inaccurate.

No one has come up with a satisfactory method of predicting speech discrimination or conversational hearing abilities from a pure-tone audiogram. There are many other psychological and physiological factors involved, including the adaptation of the individual to his hearing loss, the degree of recruitment, the dynamic range of hearing, the use of binaural hearing, and others. Recognizing these facts, and remembering that pure-tone audiometry measures simply the softest sound that can be heard at seven of the roughly 20,000 frequencies we can perceive, it is clear that the tapes are based on a false premise. It is not possible to take threshold information from seven sample frequencies and extrapolate a person's ability to hear. The ear and hearing are vastly more complicated than that. This drastic misuse of hearing data renders such demonstration tapes invalid from the outset.

Similarly, such demonstration tapes do not take into account the problems with trying to deduce broadband hearing acuity from simple pure-tone hearing tests. This problem involves a key technical principle upon which standard testing is based, termed *tonal representation.* By tonal representation, we refer to the technical fact that the fundamental principle of pure-tone audiometry is presentation of pure tones to one ear at a time, knowing that the response to each tone represents the listener's hearing sensitivity to a bandwidth of frequencies (in hertz) for which the frequency of the specified pure tone is central and generally representative. Each bandwidth is defined by appropriate ANSI Standards.

For example, the standard pure-tone frequency involved in industrial screening audiometry of 1000 Hz is characterized to represent a frequency bandwidth defined by $1000/\sqrt{2}$ Hz at the lower end of the band and $1000/\sqrt{2}$ at the upper end of the band.

In other words, the frequency bandwidth for which 1000 Hz, a pure-tone output by the ANSI standard audiometer system, is representative covers a frequency range of approximately 707 Hz to 1414 Hz. Note that this band of frequencies is an octave wide. Depending upon the quality of any filters used in the fabrication of audiotapes, the bandwidth of the octave filter centered at 1000 Hz might or might not span about the same frequency range.

The basic technical question then arises as to how the actual broadband hearing acuity, particularly in terms of response to information containing sound spectra that vary with time and level, could be deduced from the relatively simple hearing sensitivity response given to a standard pure-tone stimulus by the listener subject. One possibility that might tempt those who would presume to portray individual hearing acuity by making audiotapes based upon pure-tone responses is that the sound pressure levels within the frequency bands of interest and at each discrete frequency should be assumed equal in magnitude. Such assumptions are obviously and seriously flawed. If such a fallacious premise were assumed to have any validity, such an assumption would also result in similarly erroneous conclusions that narrow band analysis and even one-third octave band analysis—and all of the sophisticated instrumentation associated therewith—are irrelevant and unnecessary to understanding the important details of noise spectra. When it comes to hearing information contained in sound, for example human speech and musical spectra, it is well known that not only are the component details important for technical and scientific definition of the actual spectra, but that other phenomena such as lateralization, binaural enhancement, combination tones, and the Lombard effect—to name a few—are extremely important in determining individual hearing acuity and evaluating the hearing process efficiency for any individual.

Considering these technical facts, it is extremely difficult to understand how one might begin to assume, construct, synthesize or otherwise devise a scheme for producing an audiotape, particularly one based upon pure-tone responses, wherein it is claimed unequivocally that said audiotape defines and accurately portrays, or even approximates, the hearing handicap of the subject individual, particularly concerning the ability to hear human speech. Some of the questions that must be raised, where such claims may appear, include:

1. How are the sound pressure levels in the typical octave band, as defined above, distributed within the octave band to reflect the exact spectral response of the individual test subject that would be expected for other types of audiometric testing and what aspects of the measured individual response to the pure tone permit that definition and distribution?
2. Do the assumptions made for each pure tone vary or are they the same for each bandwidth for each pure tone, recognizing that the bandwidth for each pure tone varies substantially with respect to the number of discrete frequencies contained therein? For example, if we consider 500 Hz, another of the standard test frequencies, it is recognized that the bandwidth represented by 500 Hz covers a frequency range between approximately 354 Hz and 706 Hz. On the other hand, the bandwidth represented by the 1000-Hz center frequency covers the range from about 707 Hz to 1414 Hz.
3. What is the scientifically valid protocol for representing an individual's exact response to information containing sound pressure levels as a function of frequency, particularly their distribution within the bands represented by each pure tone, and how does one construct audiotape representation in such a way as to account for the inter- and intra-band variabilities using

only responses to pure tone presented in the context of industrial screening audiometry?

It appears that there is no scientifically valid methodology for application of pure tones to the development of unique and valid spectra, containing detailed sound pressure level versus frequency data, that would characterize the actual hearing ability of an individual, particularly that hearing pertaining to broadband noise in the speech frequency range.

Even if the tapes had been based on a scientifically valid premise, so many errors and biases were introduced during the creation of the specific tapes we studied that they would have had to be deemed invalid, in any case. While some of these errors might be avoidable, others are extremely difficult to eliminate. The recordings we studied were made through a single microphone, and designed to be played back to a person listening with two ears. Normally a person requires two ears to pick a sound out of background noise. If the person has only one hearing ear, the desired signal and background overlap, and sounds confusing. As recorded, these demonstration tapes placed the listener (a judge or juror, for example), in a position of someone with a one-sided hearing loss. He or she is forced to listen to the recording through the "one ear" of the microphone. Having two ears cannot undo this phenomenon once it has been recorded. Consequently, the recording technique itself makes it extremely difficult to separate speech from background noise. In addition, the taping process in this case did not adequately control the volume at which signals were recorded, the volume at which they are played back, the quality of the tape recorder on which they might be played to a jury, or the hearing level of each individual juror. All of these factors alter substantially the sound of the tapes. In fact, if several of the jurors have slight hearing losses as is common in people over 50, they will sound different to each juror. Moreover, there is no reason to believe that the perception of any of the jurors matches the hearing of the patient in question.

In addition, room acoustics create many additional reasons why a recorded presentation in a courtroom will not be realistic. Many of these considerations are summarized in a chapter by Nabelek (6). The numerous variables that affect such presentations include the dimensions of the room, reflection and absorption coefficients of all surfaces in the recording and playback room, and other factors. Extreme differences such as those found outdoors or in a gym or racketball court illustrate the problem. Such difference occur commonly among courtrooms that vary widely in age and material. The tapes will sound quite different in a large, old marble courtroom than in a modern courtroom with acoustically treated ceilings. In addition, different distances between the sound source and various listeners also substantially affect loudness (the inverse square law).

The tapes we studied were biased further by the selection of material, using sentences such as "the scarf is made of shiny silk." The producers of these tapes could hardly have found sentences that contain a higher percentage of high-frequency consonants. This proportion of high-frequency consonants (s, f, t, and z) is most unnatural in everyday speech. Naturally, electronic filtering of the higher frequencies makes such a sentence maximally difficult to understand, but such sentences are not representative of listening tasks encountered in the context of everyday conversation.

In addition to violating numerous scientific principles of otology, audiology, and acoustics, many technical errors were perpetrated in the creation of these tapes and in the production of many other tapes created by different groups. The technique that was used is not standard, standardized, or accepted by the scientific community. The recording and playback levels were not specified or standardized, and the Lombard effect (tendency to get louder in the presence of background noise) was artificially absent from the tapes.

Furthermore, there was no indication that the equipment on which the tapes was made was properly calibrated, the distortion of each piece of equipment was not specified, vocal intensity and the distance of the voice source from the microphone were not controlled, and the composition and intensity of background noise were neither specified nor controlled. The makers of the tape were also unclear regarding the use of white noise or pink noise to calibrate the equipment; and no calibration was done to ensure the integrity of the tapes after dubbing. Pink noise has noise-power-per-unit-frequency that is inversely proportional to frequency over a specified range. White noise has power-per-unit-frequency that is substantially independent of frequency over a specified range, although white noise need not be random. There was no indication that the recording and elective modifications were treated in accordance with decibels referenced to normal hearing threshold (dB HTL) at each frequency. Most standard electronic equipment is calibrated relative to 0.0002 microbar (dB SPL), and it is not appropriate for use in dealing with clinical hearing measures.

A particularly interesting problem was discovered through analysis by Dr. Paul Michael and his group using highly sophisticated equipment. The probable dynamic range of the tape recorder used to create the tapes is about 45 dB. If an individual with a hearing loss of 50 dB in a given band hears speech at 65 dB in that band, he will have a resultant sensation level of 15 dB. This suggests some intelligibility contribution from that band. However, if the tape recorder's dynamic range is 45 dB, the noise floor of the tape recorder will dominate at 20 dB (65–45), and the useful speech sensation level at 15 dB will be obscured in the tape recorder's noise floor. This electronic bias makes the speech on the tapes even more difficult to understand. This equipment artifact can afford the illusion that the individual isn't hearing anything at all, when, in fact, the instrumentation is actually limiting the intelligibility. Analysis of actual tapes indicated that this is not merely a theoretical consideration. Much useful information was lost because of this phenomenon and obscured in tape "hiss."

In addition, the clinical impressions created by the nonscientific tapes differ dramatically from information obtained through standard, accepted audiologic testing. Most of the tapes created the impression of a severe or profound, handicapping loss, even when based on audiograms of patients with mild or moderate hearing deficits. Discrimination scores performed by a certified audiologist in a standardized and scientifically accepted fashion were available for all of the people whose audiograms were used to make the tapes we studied. Consequently, clinical judgments regarding the expected hearing capabilities of the patients were possible. The discrimination scores ranged from 72% to 100%, indicating that all of the patients were able to hear quite well. This is consistent with vast clinical experience in patients with noise-induced hearing loss, although it is not consistent with the impression created by the tapes. The findings further emphasized that demonstration tapes cannot be construed as accurate representations of hearing in any individual case.

In summary, "demonstration tapes" are based on the false notions that the ability to hear speech can be extrapolated from a pure-tone audiogram. The techniques with which such recordings are made and filtered are not standard or accepted by the scientific community, and numerous technical and scientific errors are common in processing such tapes. Some are unavoidable. Laboratory analysis and clinical assessment have provided additional evidence that such tapes are invalid. They have no scientific merit, and are not suitable for use for medical or legal purposes.

REFERENCES

1. American National Standards Institute (ANSI) Standards References: J. B. Olishifski, and E. B. Harford, *Industrial Noise and Hearing Conservation*, National Safety Council, pp. 842–844 (1975). S1.6-1967 (R1971), Preferred frequencies and band numbers for acoustical measurements (agrees with ISO R266). S1.8-1969, Preferred reference quantities for acoustical levels. S1.11-1966 (R1971), Octave, half-octave, and third-octave band filter sets, specification for (IEC 225). S3.1-1960 (R1971), Background noise in audiometer rooms, Criteria for S3.2-1960 (R1971), Audiometers, specifications for (IEC 177). S3.7-1973, Coupler calibration of earphones, method for measurement of (revision and redesignation of Z24.9-1949). S4.1-1960 (R1972), Mechanically-recorded lateral frequency records, methods of calibration of (58 IRE 19.S1:IEEE Std 192-1958).
2. Miller M and Harris JD, Hearing testing in industry, in *Handbook of Noise Control, 2nd Ed.*, CM Haris, ed., McGraw-Hill, New York, pp. 10–11 (1979).
3. Miyakita T and Miura H, Monosyllable speech audiometry in noise expanded workers—Consonant and vowel confusion, *J. Sound Vibration, 127*(3):535–541 (1988).
4. Eldridge DH, The problems of criteria for noise exposure, in *Effects of Noise on Hearing*, DH Henderson, et al., eds., Raven Press, New York (1976).
5. Kryter KD, *The Effects of Noise on Man*, Academic Press, New York (1985).
6. Nabelek AK, Effects of room acoustics on speech perception through hearing aids by normal-hearing and hearing-impaired listeners, in *Acoustical Factors Affecting Hearing Aid Performance*, GA Studebaker and I. Hochberg, eds., University Park Press, Baltimore, pp. 25–46 (1980).

41

A Plaintiffs' Attorney's Perspective on Occupational Hearing Loss

Gregory J. Hannon
Hannon and Joyce, Philadelphia, Pennsylvania, USA

1. INTRODUCTION

It is widely acknowledged that there are millions of American workers who suffer from some degree of hearing loss resulting from exposure to noise while performing their jobs. It should be just as widely acknowledged that these workers deserve to recover awards which will reflect the damage they have sustained to their hearing.

For the past several years, this author has represented some of the thousands of railroad workers whose hearing has been or continues to be damaged by exposure to noise while working in the railroad industry. However, many of the topics discussed herein can be applied to the hearing loss sustained by workers in any occupation.

Workers in the railroad industry are able to recover damages for their hearing loss under a federal statute, the Federal Employers' Liability Act (FELA). The FELA does not contain some of the barriers to recovery which are encountered by workers covered by some State Workers' Compensation Acts or other legislation, as discussed in Chapters 33–35 and 37–39. The barrier erected by State Workers' Compensation Statutes which, in most states, only allow recovery in the so-called speech frequencies and ignore loss of hearing in the high frequencies, must be broken down. If workers have high-frequency hearing loss causally related to occupational noise and have consequent difficulty hearing in crowds, in multiparty conversations, over the telephone, while listening to the television, when attending church services, and when listening to the higher-pitched voices of their wives, children, and grandchildren, then they deserve to be awarded damages. It is not enough to acknowledge high-frequency hearing loss and its associated problems in the medical community but ignore it in the laws that provide recovery for occupational diseases.

The approach to the problem must be a realistic and candid one. The Plaintiff's Bar handling these cases should acknowledge the problems associated with occupational hearing loss cases and accept the reality that these are not life and death situations. On the other hand, the Defendant's Bar must also acknowledge that workers with high-frequency hearing loss have sustained physical damage to their ears and do experience genuine problems and suffer frustration, irritation, and loss of some of the enjoyments of life. All attorneys, whether representing plaintiffs or defendants, should push for legislation, such as the recovery afforded railroad workers under FELA, which allows recovery of reasonable damages for all workers who have any type of occupationally related hearing loss.

2. INITIAL ANALYSIS

The handling of occupational hearing loss cases should be prefaced by a learning process on the part of any attorney. It is necessary to read and understand the literature on the mechanics of noise and the medical literature on the effects that noise has on an individual's hearing. This knowledge of the agreed-upon concepts and disputed areas in the field of occupational hearing loss is essential in analyzing whether to pursue a potential occupational hearing loss case.

The presence of hearing loss, in and of itself, without any further scrutiny, is not enough to determine whether a worker has a viable case.

The following are a few tools used by plaintiff's attorneys which may help determine the viability of an occupational hearing loss case or ascertain the strengths and weaknesses of a particular case.

2.1. Preliminary Test

It can be helpful to perform a preliminary test. These preliminary tests should be performed by an individual who has been certified to administer preliminary audiograms. The test should be performed in an environment where outside noise is kept to a minimum in order to produce fairly accurate readings. If possible, the test should be administered after the worker is away from a noisy job for as long as practical or, if this is not possible, that the worker wear hearing protection at his job in order to avoid problems with temporary threshold shifts.

A preliminary test is not a full diagnostic audiogram. It must be reviewed by an audiologist and otolaryngologist in order to determine whether a full diagnostic audiogram is warranted.

However, even at this early stage, there are some positive signs that may be helpful to the lawyer's initial analysis. The following are some examples:

1. Approximate symmetry of the hearing thresholds in the right and left ears.
2. Hearing thresholds that show a hearing loss greater than 25 dB in any frequency and become worse in the high frequencies.

3. A severe hearing loss in the high frequencies possibly accompanied by low-frequency loss. In some test results there will be a recovery in the high frequencies at 6000 or 8000 Hz. The absence of recovery does not negate the presence of occupational hearing loss; it may simply be due to the type of exposure, i.e., impact noise, or the presence of an age effect factor.
4. A very severe or profound loss at every frequency in one ear, but a loss similar to (3) above in the other ear.

Variations or the converse of any of the above may indicate problems that should be examined more carefully after the diagnostic audiogram and discussion with an audiologist and otolaryngologist.

2.2. Initial Interview

The initial interview, in whatever form or style, should be thorough and can supply very important information to the attorney, consultant, audiologist, and otolaryngologist. Aside from basic information it should cover the following topics because they can play a role in the occupational hearing loss case:

1. The presence of tinnitus.
2. Wage loss resulting from hearing loss.
3. The specific ways in which the worker's hearing loss has interfered with his enjoyment of life activities and specific examples of the impact the occupational hearing loss has on the worker's social, work, and familiar settings.
4. Whether the individual is presently wearing hearing aids and, if so, the cost.
5. The number of years an individual has been in a noisy occupation.
6. The kind and duration of noise sources to which the individual has been exposed.
7. The existence of prior audiograms from any source and the date and place where the audiogram was performed.
8. The names of any other physicians who have examined the plaintiff for any medical problems related to that individual's ears.
9. Ototoxic medications taken by an individual.
10. Other significant medical problems.
11. The dates of employment and the individual's history of exposure to noise with any other employers who are not identified as defendants.
12. The individual's history of noise exposure while serving in the military.
13. The worker's hobbies indicating either the lack of exposure to loud and/or continuous noise or hobbies often associated with noise (e.g., hunting, target shooting, loud music, motorcycling, etc.).
14. The age of the worker.
15. Any family history of hearing loss.
16. The history of wearing hearing protection devices and, if so, when and under what circumstances.

It is crucial in this initial interview that the information be comprehensive, accurate, and candid.

2.3. Diagnostic Audiogram

A full diagnostic audiogram must be performed by an audiologist provided the audiologist and otolaryngologist have indicated that hearing loss does exist and the matter should be pursued. It is important that this audiogram include air and bone testing in order to identify possible conductive hearing loss versus sensorineural hearing loss. In occupational hearing loss, we are often dealing with high-frequency hearing loss and it would be helpful to have the speech discrimination test performed with background noise, in addition to the standard speech discrimination test.

The information from the interview and the results of the diagnostic audiogram should be blended together and discussed with an otolaryngologist. The development of the case would continue after this process has been completed and the medical specialist has rendered an opinion that an individual worker's hearing loss is occupationally related.

2.4. Legal Considerations

After the otolaryngologist has rendered the opinion that an individual's hearing loss is related to his occupational noise exposure there may be other legal considerations that come into play before a case is pursued. These considerations include the statute of limitations; where negligence is an issue in the case, the extent of the employer's knowledge of the dangers of exposing employees to loud noises; the existence of sound level surveys; and the establishment of hearing conservation measures and programs by the employer.

Completing this initial analysis before continuing to develop a case is important so that the client/worker and attorney can begin to understand the make-up of the case with all its pluses and minuses and the best way to pursue the case. It may also lead to the conclusion that there is no case to pursue.

3. CASE WORK-UP

Once the case is on its way toward some type of trial or hearing, the tools available to the attorney to gather information from the defendant should be used in a careful, thought-out manner. For example:

1. Are there specific documents in the possession of the defendant which you believe would be useful in developing the case?
2. What questions or interrogatories would lead to answers that would help develop more useful facts?
3. What deposition of other employees, supervisors, medical personnel, engineers, etc., will produce information that will bolster your position or lead to the production of useful documents?

If the plaintiff's deposition or statement is going to be taken by the defendant, the plaintiff must be thoroughly prepared. The plaintiff or claimant should know the areas that will be covered in the deposition. All of the information in the file should be reviewed with the individual prior to the deposition. The worker should be prepared to give details and examples of how the occupational hearing loss has affected his social, work, and family environment.

Demonstrative evidence such as models of the ear, audiogram blow-ups, audio tapes demonstrating the difficulties experienced by an individual with an occupational hearing loss, age effect charts, noise survey blow-ups, etc., can be very useful at the settlement stage or the hearing/trial of the matter. Demonstrative tools can be used to educate the defendant, jurors, judges, and referees about occupational hearing loss and its effects on an individual. They can also be helpful in trying to negotiate a settlement of the case.

In the early stages of developing the occupational hearing loss case, it will be necessary to secure the services of experts in the areas of the mechanics of noise, audiology, otolaryngology and, in selective cases, vocationalists, economists, psychologists, or psychiatrists. The psychologist or psychiatrist can explain the effects that occupational hearing loss can have on an individual. These include irritation, frustration, embarrassment, or depression, and the real impact on an individual's daily living in a social, work, or family environment. If the outcome of the case is going to hinge on a hearing or trial, a stepped-up and intense involvement of these experts in preparation for the trial will be extremely important.

4. SETTLEMENT

The settlement of hearing loss cases is a topic of much debate. There are, in the opinion of this writer, three basic hurdles that must be cleared initially by both sides before a more detailed healthy debate on the "value" of a case can take place. These hurdles are the following:

1. Hearing classifications. Taking the worker's hearing as a whole, low-frequency and high-frequency loss, the hearing loss is going to fall somewhere between very minimal and extremely profound. A method of classifying each individual's hearing loss somewhere in this spectrum has to be agreed upon by both parties. There may be loss in the high frequencies with no speech frequency loss. Hearing loss caused by noise usually begins in the high frequencies. However, if the exposure is severe over a long period of time, hearing loss may also develop in the low frequencies. There is medical literature that supports the position that some occupational hearing loss caused by noise over a period of time involves low frequencies as well as high frequencies. Therefore, the methodology used to establish the classifications (where an individual falls on the severity spectrum) may take into consideration low- and high-frequency threshold readings which, at a particular frequency, may be helpful to the plaintiff or the defendant. If a settlement is going to be achieved, compromises must be reached and a sincere effort must be made to find a middle ground that is going to result in a fair settlement to the plaintiff and the defendant.
2. Age effects. Age-related hearing loss, or what is commonly referred to as presbycusis, involves many of the same frequencies on an audiogram as hearing loss caused by exposure to occupational noise. Therefore, age is another hurdle that requires special handling before any detailed discussions on the case can begin. However age is handled in this initial stage, it has to be dealt with in a manner fair to all parties. The younger workers (for example, those individuals who are age 55 or younger) who have a moderate to severe loss, have a strong case where presbycusis is not an issue provided no other factors come into play. On the other hand, with older individuals (for example, those individuals who are 71 or older), age effect charts from an agreed-upon source should be taken into consideration by counsel. There are age effect factors other than presbycusis which should also be taken into consideration at this stage.
3. Establishing monetary values. General monetary values have to be established, taking into consideration the hearing loss classification and the individual's age.

These should be flexible values that can be increased or decreased by the absence or presence of other factors.

The factors that may play the most significant role in increasing monetary values are (a) the presence of tinnitus, (b) wage loss resulting from hearing loss, (c) psychological problems (i.e., depression, anxiety, frustration, etc.) resulting from occupational hearing loss and/or tinnitus, and (d) the necessity of hearing aids and their cost.

Factors that may play the most significant role in decreasing the monetary values are (a) other medical problems related to treatment of ear problems, (b) noise exposure without hearing protection outside the workplace (other employment, the military, noisy hobbies, etc.), and (c) the statute of limitation depending on the act or statute that governs the particular case.

Some of the other items in the initial or follow-up client interviews may also play a role in settlement discussions. Each case should be evaluated individually taking into consideration that there are many variations of the above factors and that individuals have different sensitivity levels that may cause variation in the degree of frustration, irritation, embarrassment, depression, and general loss of enjoyment of life caused by occupational hearing loss and tinnitus, which can accompany this loss.

5. CONCLUSION

The plaintiff's attorney has to have a solid educational foundation of the disciplines that play a role in occupational hearing loss. Careful and thorough analysis must be applied to each step in the handling of an occupational hearing loss case.

As is the case in all areas of law, preparation at all stages is the key to a successful and fair outcome. There are many positive and strong arguments that can be applied to the defenses raised, but they will only be available to those who take the time to learn this area of occupational disease law.

Where the Act (such as FELA) or cases governing a statute (such as Pennsylvania Workers' Compensation) realistically recognize occupational hearing loss and its effects on the worker, amicable resolutions of most of the cases is an attainable goal.

Where statutes absolutely preclude recovery because they fail to realistically recognize high-frequency hearing loss, tinnitus, and problems associated with occupational hearing loss, the effort should be made to adopt new statutes. The person hurt the most by statutes that preclude recovery is the individual worker who has permanently lost a portion of his hearing. Where recovery is allowed, the result must not only be monetary, but just as important, must also continue to lead to the establishment of hearing conservation program in order to provide a safe and healthy workplace for the American worker.

42

Workers' Compensation: Presenting Medical Evidence in Hearing Loss Cases

Irvin Stander
Bureau of Workers' Compensation of Pennsylvania and Pennsylvania Bar Institute, Harrisburg, Pennsylvania; and Temple University, Philadelphia, Pennsylvania, USA

One of the most important aspects in the proof and defense of occupational hearing loss claims is the medical evidence factor. The proper presentation of medical evidence in workers' compensation is a highly developed art, making great demands on the skill and ability of the legal practitioner. This is especially true in loss-of-hearing cases because of the following serious problems:

1. Was a claimant's hearing loss of occupational origin or one that would probably occur in most persons because of advancing age?
2. Where the complaint of hearing loss is claimed to follow a sudden noise, did the loss result from that noise, or was it present prior to the noise that occurred?

3. Was the claimant's physical condition responsible for hearing impairment that appears after long exposure to noise, and would it have occurred in any event because of his physical condition?
4. Where a slight hearing loss comes from occupational exposure, but a larger loss results from nonoccupational causes, how should we classify such hearing impairment?

These vexing questions tell us why a closed-claims survey has revealed that 86% of claims for repeated trauma cases have been sharply contested; most of the repetitive trauma claims have been for occupational hearing loss.

It has been aptly stated that the physician is the "gatekeeper" of our disability claims benefits system. He is needed to admit the claimant for benefits, and also to close the gates and terminate benefits. The presentation of medical evidence by the physician is therefore critical and deserves careful review from the viewpoint of the claimant's and defendant's attorneys, as well as from the perspective of the adjudicator (referee or judge).

Everyone in this field, whether a lawyer, industrial physician, otologist, safety engineer, hygienist, audiologist, nurse, or claims personnel, can benefit from a review of this subject because when each knows what it takes to present or defend a hearing loss claim, he can gain a better perspective of his respective responsibilities and functions in this litigious area of occupational injury law.

1. PREPARATION AND PRESENTATION OF CLAIMANT'S CASE—FUNCTIONS OF CLAIMANT'S ATTORNEY

The functions of the claimant's attorney are presentation of competent substantial evidence and persuasion of the adjudicator as the ultimate fact finder. The attorney does this by first conducting a careful investigation, comprehensive medical research, and intensive legal research.

1.1. Hearing Loss Cases—Interview of Client

The following items of evidence should be obtained from the claimant, especially with reference to a hearing loss claim:

1. Claimant's reasons why he believes his hearing loss is due to his work
2. Date of hearing loss onset; when claimant first connected it to his employment, and why
3. Date of claimant's last exposure to the noise hazard
4. Chronological description of the particular work factors that the claimant believes to be the cause of the hearing loss
5. Description of the machinery or devices believed to be the noise hazards
6. Whether any protective devices were used, with a description of the devices and the exact extent of their use
7. A full medical history of the current and previous ear or hearing problems, with all details of locations, names of witnesses, and all examining and treating physicians or hospitals
8. Full history of any prior claims affecting the ears or hearing
9. Originals or copies of any audiograms made by any physician or medical facility or at any time at the behest of the claimant

1.2. The Injury and Its Work Connection

In his effort to causally connect the claimant's hearing loss to his work environment, the claimant's attorney should conduct the following investigations:

1. Use available legal discovery methods to ascertain his client's work assignments; the types of machinery or devices producing the noise hazards; the length of time of such exposures; copies of any sound-level tests or surveys made by the employer; a copy of the report of the claimant's preemployment medical examination; and copies of any medical or dispensary records in the employer's possession.
2. Refer all investigative and medical material to a competent hearing specialist for his opinion. Such a hearing specialist (otolaryngologist) should, when forming his opinion, be satisfied that: (a) the claimant has the usual bilateral sensorineural hearing loss that accompanies such exposure; (b) the sound-level surveys of the working area reveal noise of sufficient intensity and type to cause hearing loss; (c) there has been sufficient exposure time; and (d) may local and systemic causes of hearing loss have been ruled out by a complete history, otologic examination, and hearing studies.

1.3. Presentation of Claimant's Medical Evidence

In preparing the claimant's medical evidence, his attorney should observe the following points:

1. In states in which formulas are used, the claimant's doctor must be prepared to establish the extent of the claimant's hearing loss so that his level of compensation can be validly determined. In other states, the physician may have to assert more subjective judgments. In Pennsylvania, for example, the claimant's doctor must be prepared to prove that the claimant has complete loss of hearing in one or both ears "for all practical intents and purposes." The physician's opinion on causation must be by a reasonable degree of medical certainty. When partial loss of hearing is claimed, accompanied by wage loss and a claim for benefits, or without wage loss—when a claim is made for suspension of benefits—the claimant's physician should be able to justify his determination of the extent of the hearing impairment by stating the method and formula used in its determination. He should also testify as to how the claimant's hearing impairment affects his ability to do his job and to perform the ordinary requirements of his daily life.
2. Some doctors are reluctant to testify, and the lawyer must explain that the doctor has a duty to the patient beyond treatment.
3. Explain to the doctor "medical causation" contrasted with "compensable (legal) causation."
4. Distinguish a sole "cause" from "aggravation of a preexisting condition," which may also be compensable under the law.
5. When an accident and health insurance form states "not related to employment," explain the difference between "medical causation" and "compensable (legal) causation" as it affects the entry on the accident and health form.
6. Some doctors "hate" court, call it a "waste of time," and hate cross-examination by lawyers. You can try to overcome this by careful preparation prior to the hearing.
7. The treating doctor is best, but if he is not available, use an expert for opinion evidence.

8. In preparing for opinion evidence, carefully review the claimant's facts, using your own notes or a copy of the transcript of testimony, and draft a carefully drawn hypothetical question for the expert based on the facts in the record or facts to be definitely presented.
9. Incorporate the treating doctor's records and hospital reports in the hypothetical.
10. Present your medical expert in court; the judge may base his opinion on the above factors and can also consult other doctors' reports and opinions, but he cannot rely entirely on their opinions.

1.4. Points on Rebuttal on Behalf of the Claimant

1. Carefully limit your cross-examination.
2. Don't try to get the defendant's doctor to admit he's wrong—this is generally a futile task.
3. Attack failure of defendant's doctor to review all available records.
4. Determine whether to bring back your expert for rebuttal.

2. PREPARATION AND PRESENTATION OF THE DEFENSE

2.1. Investigation of the Work Site

Many of the items discussed in connection with the preparation of the claimant's case apply equally to the defense of an occupational hearing loss claim, but following are some specific considerations in defense which should be carefully observed:

Assemble the following material from the employer as part of the claimant's work-site investigation:

1. Work assignment records, showing the position held, dates of assignment, description of duties, types of noise exposure and their duration, as well as the last date of exposure.
2. Description of the claimant's work sites, including diagrams showing the dimensions and layouts of the building, and all the machinery and other noise hazards at the times the claimant was on duty.
3. Description of any sound-level tests or surveys made in the claimant's work areas and their findings. If no tests or surveys have been made, ask the employer to have them made as soon as possible.
4. List of all safety precautions, acoustical engineering, and noise suppressors used to eliminate or reduce noise hazards.
5. Ascertain whether any other coemployees performing the same type of work had similar complaints about hearing loss.
6. Get copies of the report of the claimant's preemployment medical examination and all plant medical and dispensary records.

2.2. Investigations of Relevant Matters to Aid in Defense

1. Check to see whether preemployment audiograms are available. This will help your expert in determining whether the claimant's existing hearing loss was caused by his present job or preexisted this employment.

2. Investigate the outside activities of the claimant, such as using a chain saw, rock music exposure, military background, work with a jack hammer, hunting, etc.
3. Check the circumstances of any hearing tests, such as qualifications of the audiometric technician, period of time removed from noise, calibration of audiometer, use of soundproof booth, etc.
4. Remember that noise-induced hearing loss usually affects both ears equally.
5. Check for use of employer-provided ear protection as a factor affecting causation of hearing loss by present employment.
6. Check available medical records for existence of presbycusis (hearing loss caused by advancing age) or sociocusis (hearing loss caused by non-work-environmental factors).
7. Check carefully the qualifications of the claimant's physician being utilized in this case. Many may be competent treating physicians, but know little about diagnosing noise-induced hearing loss.

2.3. Points for Medical Investigations

1. Get all hospital records, laboratory reports, and charts.
2. Get all union welfare group or health insurance records for old health problems.
3. Get full reports and office records from the claimant's specialists.
4. Using authorizations, get opinions from the claimant's physicians.
5. Refer all investigative and medical material to a good hearing specialist who agrees to testify, if his opinion is favorable. Ask him very specific questions.
6. Research and develop medical articles on causes of hearing loss for use with your expert and in cross-examination of the opponent's expert.

2.4. Defense Trial Preparation

1. Assemble all original signed statements, group health records, personnel and payroll records, time cards, etc., for use at the trial.
2. Prepare the employer's witnesses for court appearance.
3. Meet personally with your medical expert and prepare for direct and cross-examination; discuss "soft spots" in the claimant's medical case; discuss the theory of defense and have the expert keep it well in mind.
4. Agree to deposition of the claimant's expert, but only if you get a copy of his written report well in advance.
5. Get a transcript of lay testimony, if possible, before the claimant's doctor testifies so that you can attack his hypothetical question, or at least evaluate it better.

2.5. Presenting the Defense

1. Meet your doctor in advance at his office to prepare for trial testimony and use of demonstrative evidence, such as charts, diagrams, hearing test results, etc.
2. Explore his weaknesses on cross-examination and try to deal with them.
3. Document what records and test results your doctor has reviewed.
4. Have your doctor review the transcript of the opposing doctor's testimony and develop specific areas of disagreement.
5. Develop your doctor's lack of bias and try to overcome the "treating-doctor mystique."

6. Review and explain to the court the "jargon" of medical terms and explain the standard of proof.

3. ADJUDICATOR'S POINT OF VIEW

3.1. Preparing Your Case

1. *Gather your records early*, preferably before you fill out the claimant's petition. This will avoid errors in dates and other important information.
2. *On medical research*, a suggested standard you should try to achieve: the lawyer must, for the purpose of this case, be as well informed as the doctor in *the intricacies of the particular disease involved in this case and the present state of medical research on the disease.*
3. *Notice of doctor's testimony*: Be sure to notify the adjudicator when you are presenting a doctor's testimony and try to arrange a definite time for his appearance.
4. *On legal research*: You must be familiar with the provisions of the act applicable to your case; all of the elements and issues involved in proof and defense; burdens of proof and presumptions in the act and the decided cases; the extent and duration of the benefits available; and the rules for medical benefits. For example, in connection with occupational hearing loss claims in Pennsylvania, the attorney should be cognizant of the following Pennsylvania statutory provisions and decided cases:
   a. Section 306(c) of the Act (77 P.S. Sect. 513) provides for specific loss allowances for "complete" loss of hearing in one ear of 60 weeks of compensation, or in both ears of 260 weeks, with a healing period of 10 weeks. (See below for case interpretation of "complete.")
   b. Section 108(n) of the Act (77 P.S. Sect. 27.1) relating to occupational diseases specifically *excludes* partial loss of hearing in one or both ears due to noise as an occupational disease.
   c. In *W.C.A.B.* and *Nissen* v. *Hartlieb*, 465 Pa. 249, 348 A.2d 746 (1976), the Supreme Court determined that "complete" loss of hearing is interpreted to mean of loss of hearing "for all practical intents and purposes."
   d. In *Hinkle* v. *H. J. Heinz Co.*, 462 Pa. 111, 337 A.2d 907 (1975), the Supreme Court determined that a claimant who has suffered partial loss of hearing without any loss of earnings is entitled to a suspension of compensation—which is a legal recognition that he has suffered a work-related injury and that his right to reopen the claim on proof of actual earning loss is preserved for a period of 500 weeks (10 years), instead of being limited to the present statutory period of 3 years for the presentation of such claim.
5. *Proving every allegation*: Be prepared to prove, and proceed with, every allegation in the claim petition or any other petition you have filed in your case.
6. *Role of adjudicator*: Remember that the adjudicator is the sole and ultimate arbiter of the facts and prepare your presentation with that in mind.
7. *Sufficient competent evidence*: When you prepare your case, you are seeking to present "sufficient competent evidence," and you should know the standards for

such evidence. Here are some definitions from the case law which you may find helpful:

a. "Sufficient competent evidence is such relevant evidence as a reasonable mind might accept as adequate to support a conclusion" (*W.C.A.B.* v. *Auto Express*, 346 A.2d 829 [1975]).
b. Here is a definition in the negative: "Reviewable as incompetent is only testimony which is so uncertain, equivocal, ambiguous, or contradictory as to make administrative findings of fact mere conjecture that fails to meet the test of substantiality" (*Brooks* v. *W.C.A.B. & Knight*, 392 A.2d 895 [1975]).
c. The courts have again reviewed *"capricious disregard of competent evidence"* by defining the converse in their rule for reversal where *no evidence* was presented to the referee to *support the prevailing party*, for whom the referee finds, nonetheless. This, they hold as a *"capricious disregard of competent evidence,"* which they define as a willful disregard of trustworthy, competent relevant evidence, of which one of ordinary intelligence could not possibly doubt as to its truth, quoted in *Russell* v. *W.C.A.B.*, 550 A.2d 1364 (1988).

3.2. Tools to Aid Preparation and Presentation

1. Use trial depositions and available discovery proceedings to assemble needed investigatory material.
2. Pretrial conferences can be used to explore the issues and discuss the order of presentation of the evidence, as well as the possibilities of an amicable resolution.
3. Assemble medical and hospital reports as early as possible, since they must be furnished to the opposite parties on request.
4. Subpoenas are available to enforce the attendance of witnesses and the production of books and other writings.
5. Hospital records are generally admissable as evidence of the medical and surgical matters and therein.

3.3. Questions of Proof of Medical Evidence

1. *Rules of evidence in compensation cases*: The rules of evidence are somewhat more relaxed in workmen's compensation than in other civil cases. In Pennsylvania, for example, the act provides that the referee is not bound by the common law or statutory rules of evidence in conducting any hearing, but that "all findings of fact shall be based upon sufficient, competent evidence to justify same."
2. *Evidence must support findings*: Although it is true that a workmen's compensation statute is generally construed with a fair degree of liberality toward claimants as remedial legislation, nevertheless, all proof must be carefully examined and irrelevant and incompetent testimony excluded in the fact-finding process, in order to lend support to the findings made by the adjudicator.
3. *Distinction between medical and legal causation*: Causation is both a medical and legal term, but the meanings are somewhat different. The medical definition of cause and effect calls for scientific certainty, so that the alleged

causative element must be one recognized scientifically. However, establishment of legal causation requires only that there be a cause and effect within a reasonable degree of medical certainty. This can be termed a quantitative standard.

4. *Medical proof in loss of hearing cases*: In some types of injury, the causal connection is obvious and medical proof may not be necessary, but in hearing loss cases, which are generally based on repetitive trauma, medical evidence will be required to establish the causal connection and extent of impairment by unequivocal medical testimony. This can be termed a quantitative standard.
5. *Expression of medical opinion analyzed*: (a) If medical testimony is required relating to causation, it must be unequivocal to support an award. *Haney* v. *Workmen's Compensation Appeal Board and Patterson Kelley Co., Inc.*, 65 Pa. Cmwlth. 461, 442 A.2d 1223 (1982). (b) An expression of medical opinion will satisfy the standard of unequivocal medical testimony if the expert testifies that in the expert's professional opinion there is a relationship, or that the expert thinks or believes there is a relationship. The testimony of the expert must be considered as a whole, and complete medical certainty is not required. Reservations relating to medical or scientific details do not affect the admissibility of the medical opinion so long as the expert does not recent the opinion or belief expressed. See *Lyons Transportation Lines* v. *Workmen's Compensation Appeal Board* (*Pogany*), 84 Pa Cmwlth. 546, 480 A.2d 358 (1984) and *Philadelphia College of Osteopathic Medicine* v. *Workmen's Compensation Appeal Board* (*Lucas*), 77 Pa. Cmwlth, 202, 465 A.2d 132 (1983). For a case in which the medical testimony was held to be equivocal and where there is an excellent discussion of the issue, see *Lewis* v. *Workmen's Compensation Appeal Board* (*Pittsburgh Board of Education*), 508 Pa. 360, 498 A.2d 800 (1985). (c) Many cases have considered the effect of various expressions of medical opinion. Reliance upon the effect of any specific expression may be misplaced. The standard is the expert's testimony as a whole, and not isolated expressions. *Michaelson* v. *Workmen's Compensation Appeal Board* (*R.R. Leininger & Son*), 560 A.2d 306 (1989). A physician need not use "magic words" such as "reasonable degree of medical certainty" or "fully recovered." *Williams* v. *Workmen's Compensation Appeal Board* (*Montgomery Ward*), 562 A.2d 437 (1989).

Although worker's compensation acts in many states do not specify any specific scientific formula for determination of hearing impairment, as many other states provide in their statutes, practitioners should be familiar with the generally accepted principles used by hearing specialists to fix percentages of hearing loss. The most widely accepted formula for assessing hearing loss is that adopted by the American Academy of Ophthalmology and Otolaryngology (AAOO) in 1959 and slightly revised in 1979 by the American Academy of Otolaryngology (AAO). These formulas are discussed elsewhere in this book.

In general, measurement of hearing loss is made by using an audiometer, which is an electroacoustic instrument so calibrated and standardized that it measures the hearing level or hearing threshold level (HTL) in the frequencies tested. For example, if the patient can hear a certain frequency (measured in cycles per second [cps] or hertz [Hz]) at audiometric zero decibels (which are units of sound-pressure level [dB]), he has no hearing loss, but if he cannot hear until that frequency tone is 30 dB louder than audiometric 0 dB, then he has a hearing loss (as defined by the HTL) at that frequency of 30 dB, and this is so recorded on the audiogram. It is important to ensure that hearing loss is permanent and not just temporary, and that the test is reliable and valid.

In examining or cross-examining a hearing specialist, one can insist that he justify his opinion on hearing impairment by reference to recognized scientific criteria, such as the current AAO formula. Since that is the formula, although imperfect, most frequently used by hearing specialists, you should be prepared to question him about the various factors covered by that formula—or any other formula he mentions in his medical report. The other formulas differ from the AAO formula mainly as to the audiometric frequencies used and the low fence, or exclusion, applied (e.g., New Jersey uses frequencies of 1000, 2000, and 3000 Hz and a low fence of 30 dB, and Wisconsin uses 1000, 2000, and 3000 Hz and a low fence of 35 dB).

3.4. Additional Factors to Be Considered in Hearing Impairment

In addition to the above-mentioned physical factors, consideration should also be given by the doctor and adjudicator to the following factors on the question of how the physical impairment affects the individual claimant:

1. Does the impairment produce inattention and slower responses of body and extremely movement?
2. Does the impairment affect the claimant's coordination? Is he awkward and/or clumsy? Does he confuse easily? Does he have problems with complying with instructions?
3. Does the impairment cause him to function with a depreciation of judgment and calculation where his skillful acts would depend on sound?
4. Does his impairment affect his security as a person? Does he tend to commit acts with error, embarrassment, or hesitation because of his loss of hearing?
5. Does his impairment subject him to more risks than a normal individual?

3.5. General Suggestions to Trial Attorneys

1. Prepare your medical witness with great care. His testimony is very important and can make or break your case.
2. If you are going to present a hypothetical question for your doctor to answer, prepare it carefully in writing, remembering that the "hypo" must be based on evidence presented or to be presented on the record.
3. Review your case from your opponent's viewpoint and then try to anticipate his moves by knowing your respective strengths and weaknesses.
4. Always keep in mind that the adjudicator is the ultimate finder of the facts and prepare fully to offer support for proper findings of fact at his level.
5. In proper cases when there is a sharp irreconcilable conflict in the medical evidence, try to persuade the adjudicator to exercise his power to appoint an impartial physician to examine the claimant, file a report, and be available for testimony.
6. Most important, attend every hearing, and if urgent reasons prevent your presence, be sure to notify the adjudicator, your opponent, and your own client, well in advance.

7. *To summarize*: Louis Nizer, the famous attorney, once wrote: "The only difference between a good lawyer and a bad lawyer is the effort they spend in careful preparation and presentation." The same case in the hands of a good lawyer can mean victory, and in the hands of a bad lawyer can mean defeat—a defeat mainly caused by failure to spend adequate time and effort in preparation and presentation. Similar time, effort, and preparation are required of a medical expert witness.

Remember that occupational noise exposure continues to pose one of the most pervasive threats to the quality of life. Human beings thrive on communication, and hearing loss detracts from our ability to maintain effective interpersonal communications.

Financial award for a handicap that occurs as a result of exposure to workplace noise is never as good as prevention of the handicap in the first place. However, when such loss does occur, the financial award should be fairly and equitably provided.

Appendix I
Anatomy of the Ear

This book will not review anatomy of the head and neck in detail. Most anatomy relevant to otolaryngologic practice is taught in medical school. However, many medical school curricula omit detailed discussion of special sensory structures including the ear. Consequently, a more complete discussion of otologic anatomy is provided here. Understanding the ear requires not only observation of gross anatomy, but also histologic study. There are two basic techniques for studying histology of the ear.

1. *Traditional temporal bone histology.* Decalcification, staining, section. Takes several months. Good for bones. Less useful for membranous structures. Unsatisfactory for studying hair cell loss.
2. *Surface preparation.* Microdissection after special staining. Excellent for non-bony structures. Takes only hours. Phase contrast microscope studies. Electron microscopy is also used to study the ear, of course.

The ear is divided into three parts differing in their gross anatomy, their histology, and their functions. The italized items in the following outline are most important structures. A good clinician should review them mentally as he/she examines an ear in order to recognize their normal conditions and understand the implications of their pathology. Familiarity with the non-underlined structures facilitates *special and functional understanding* of the major structures.

The External Ear

I. *Auricle*: Elastic cartilage, elastic fibers in perichondrium, skin with subcutaneous layer posteriorly only, a few hairs, sebaceous glands, and very few sweat glands. The helix should meet the cranium at a level not below a line drawn straight back from the lateral canthus of the eye. The axis of the ear should not be $>10^\circ$ off a perpendicular line dropped from the lateral canthal line.

II. *External Auditory Meatus*: S-shaped, curving inferomedially. Lateral portion, cartilaginous. Medial portion, bony. Thin skin, generally lacking papillae, firmly attached to perichondrium and periosteum.

   A. Cartilaginous portion: hairs. Extremely large sebaceous glands connected with hair follicles. Ceruminous glands are specialized coiled tubular apocrine sweat glands, surrounded by myoepithelial cells. Ducts open onto surface or into hair follicles.

B. Bony portion: small hairs and sebaceous glands on the upper wall only.

The Middle Ear

I. *Tympanic Cavity*
   A. Lateral wall: *tympanic membrane.*
   B. Medial wall: bony wall of the inner ear, facial nerve canal.
   C. Posterior wall: connects through tympanic antrum with *mastoid air cells.*
   D. Roof: tegmen tympani, *very thin bone* (also forms roof of mastoid air cells, antrum, and eustachian tube).
   E. Floor: bony covering of *jugular venous bulb.*
   F. Anterior wall: opening of *eustachian tube*, and bony covering of *internal carotid artery.*
   G. Contents: *ossicles*, tendons of tensor tympani and stapedius muscles, *chorda tympani nerve*, facial nerve canal.
   H. Mucosa with simple squamous epithelium except there is ciliated, cuboidal or columnar epithelium near the tympanic membrane and the auditory (eustachian) tube.

II. *Tympanic Membrane*—The "Eardrum"
   A. *Semi-transparent*, cone-shaped, with *malleus* attached to apex.
   B. Four layers: two layers of collagenous fibers and fibroblasts deficient in the anterior superior quadrant forming the flaccid Shrapnell's Membrane. Collagen fibers are radially arranged in the outer layer, circularly in the inner layer. Lined by skin laterally, by mucosa medially. Vessels and nerves reach the center of the tympanic membrane along the area of the manubrium of the malleus.

III. *Ossicles*
   A. *Malleus*: attaches to tympanic membrane at the *umbo.*
   B. *Incus*: connects the malleus and incus.
   C. *Stapes*: fits into oval window. Held in place by fibrous annular ligament. Fixed in the disease *otosclerosis.*
   D. The ossicles are supported by ligamentous attachments to bone.

IV. *Mucosa of the Middle Ear*
   There are five types of cells in the mucosa of the human middle ear, mastoid and eustachian tube. These cells are: (1) the nonciliated cell without secretory granules, (2) the nonciliated cell with secretory granules (including the goblet cell), (3) the ciliated cell, (4) the intermediate cell, and (5) the basal cell. The mastoid cavity has predominantly simple squamous or cuboidal epithelium, although ciliated cells can be found in some specimens. The posterior part of the mesotympanum and the epithympanum have taller epithelium, which is often ciliated. The promontory has secretory and nonsecretory columnar cells, occasional ciliated cells, and rarely goblet cells and glands. The middle ear mucosa has modified respiratory mucosa, and the nonciliated cell is the principal secretory cell of the middle ear. In its most active secretory phase, the nonciliated cell has the cytological characteristics of a goblet cell.

V. *Auditory Tube = Eustachian Tube = Pharyngotympanic Tube*
   A. One-third of the eustachian tube nearest the ear is bony. Two-thirds nearest the nasopharynx are cartilaginous. Elastic cartilage, except where bone meets cartilage (the isthmus), where there is hyaline cartilage. Cartilage runs mostly medially, but a superolateral ridge of cartilage

gives a shepherd's crook appearance in cross-section. There is fat adjacent to the open portion of the cartilaginous eustachian tube.

B. Mucous membrane, thin, low columnar ciliated in the bony part, but abundant pseudostratified ciliated tall columnar with mucous-secreting tubuloavelor glands and goblet cells at the pharyngeal end. Tubal tonsils (lymphoid tissue) may be found at the pharyngeal end.

The Labyrinth

A membranous labyrinth filled with endolymph is suspended in perilymph within a bony labyrinth. The whole inner ear is embedded in hard bone within which are large cavity called the vestibule, the spiraling cochlea, and three semi-circular canals. The inner ear has two divisions: *acoustic* (hearing) and *vestibular* (balance).

I. *The Membranous Labyrinth*
   A. All parts arise from a single otic vesicle of ectodermal origin. Cochlear duct and saccus endolymphaticus are derived from the saccule; semicircular ducts are derived from the utricle. Generally, squamocuboidal epithelium. All parts are interconnected. All parts are filled with endolymph.
   B. Components: Three semicircular ducts, utricle, ductus endolymphatics, saccus endolymphaticus, saccule, ductus reuniens, and the spiraling cochlear duct ending as the cecum cupulare.

II. *The Vestibule*: Through the oval window, the footplate of the stapes contacts the perilymph filling most of the vestibule. The vestibule contains the utricle, the saccule, five openings for semicircular canals, the opening for the vestibular aqueduct, the opening to the scala vestibuli of the cochlea, and the ductus reuniens leading to the cochlear duct.
   A. *Utricle*: Oblong membranous structure, five openings to the semicircular ducts and ampullae. Sensory functions at the macula. Simple squamous epithelium except around the macula. Gives off a duct medially which joins the duct from the saccule to form the endolymphatic duct.
   B. *Saccule*: Spherical. Anterior and inferior to the utricle. Contains a macula. Gives off ductus reuniens and a small duct that joins the duct from the utricle to form the endolymphatic duct.
   C. *Maculae*: Sensory structures within the utricle and saccule. Supporting cells and hair cells similar to those described under the vestibular system. Hairs embedded in gelatinous otolithic membrane which contains otoliths—3 to 5 micron stones of calcium carbonate and protein.
   D. Ductus endolymphaticus (endolymphatic duct) leads through the petrous portion of the temporal bone to the saccus endolymphaticus which lies between the layers of the meninges. Squamocuboidal epithelium in the duct changes to two types of tall columnar epithelium in the sac, suggesting specialization for resorption. Cellular debris is common in the sac, but not elsewhere in the labyrinth.

III. *The Vestibular System*
   A. Includes utricle (position of the head and linear acceleration) and saccule (function uncertain) and semicircular ducts.
   B. *Three semicircular ducts* called superior, posterior, and horizontal (or lateral). The nonampullated ends of the superior and posterior canals join, entering the utricle as the common crus (Hence, five openings

into the utricle instead of six). Sensory structures of the semicircular ducts: epithelium of the ampulla makes a transverse ridge called the *crista ampullaris.* It is covered with sensory epithelium and bounded on either end by the planum semilunatum. The hairs are embedded in the "gelatinous" cupula (no otoliths) which fits like a saddle over the crista.

C. *Hairs cells*
   1. Type I: Flask shaped, round basal nucleus surrounded by mitrochordia. Mitochondria also near surface. Enclosed in a chalice-like nerve terminal.
   2. Type II: Simple columnar. Nuclei at various levels, but usually higher than Type I. Numerous small nerve endings: granulated (efferents) and nongranulated (afferents).
   3. Features common to both types: 40–80, sterocilia, long microvili, probably nonmotile. One kinocilium—a modified cilium lacking a central pair of fibrils. Hairs are constricted at base, arranged in a regular hexagonal pattern on the cell surface, according to height and varying in size from 100 micron to 1 micron. The longest hairs are next to the kinocilium, which is at one side of the cell. Complicated specialized organelles differ between Types I and II. Nerve fibers leave the cells and transverse the crista becoming the nerves of the ampullae, which combine with each other and the nerves of the utricle and saccule to form the vestibular division of the eighth cranial nerve.

IV. *The Cochlea*: the acoustic system. Bony spiral anteromedial to the vestibule. Makes 2 and 3/4 turns around its axis, the modiolus.

A. *Modiolus*: Conical pillar of porous bone containing blood vessels, nerve bundles of the cochlear division of VIII, spiral ganglia. Its base is the deep end of the internal acoustic meatus.

B. *Spiral lamina* divides the cochlear spiral into upper and lower compartments meeting at the apex as the helicotrema. Spiral lamina itself is divided into two sections.
   1. *Osseous spiral lamina* (medial, with respect to the modiolus). Carries blood vessels, myelinated nerves. Spiral limbus at its lateral lip is the site of attachment for Reissner's membrane, and the area where the nerve fibers lose their myelin.
   2. *Membranous spiral lamina* or *basilar membrane* contains the *Organ of Corti.* Lateral attachment to the bony spiral is a thickening of periosteum called the spiral ligament (it is not a ligament).

C. From the area of the spiral ligament, arching up to the bony wall of the cochlea, are the spiral prominence and the *stria vascularis.* At the top of the stria, Reissner's membrane (the vestibular membrane) originates and courses diagonally down to the join the spiral osseous lamina the medial border of the spiral limbus.

D. *Reissner's membrane*: Differentiated cells probably for fluid and electrolyte transport. Divides the compartment above the spiral lamina in two. The upper part communicates with the perilymphatic space of the vestibule, and is called *scala vestibuli.* The lower division of the upper compartment is the *scala media.* It is filled with endolymph and connects via the ductus reuniens with the saccule. The compartment below the spiral lamina is the scala tympani, a perilymphatic space that meets the scale vestibuli at the helicotrema, and leads basally to the round window (covered by the secondary tympanic membrane, hence scala tympani) and to the cochlear aqueduct, which connects the perilymphatic space with the posterior cranial fossa.

E. *Stria vascularis*: There are at least two cell types. Basal cells with few mitochondria, but numerous ascending processes which separate marginal cells. Marginal cells, with smooth convex free surface, but basally, labyrinthine infoldings of plasmalemma containing many mitochondria. Abundant capillaries throughout.

F. Spiral prominence is continuous with epithelium of stria vascularis superiorly. Inferiorly, reflected from the wall onto the basilar membrane, forming the external spiral sulcus. Over the basilar membrane the cells become cuboidal cells of Claudius. In parts of the basal turns, cells of Boettcher come between the basilar membrane and cells of Claudius.

G. *The Organ of Corti*: The organ of hearing. Supporting cells and hair cells.
   1. Supporting cells: Tall, slender cells extending from basilar membrane to free surface. Upper surfaces are in contact with each other forming, with the upper surfaces of hair cells, the reticular lamina (reticular membrane). Pillar cells form the inner tunnel, through which nerves cross to the outer hair cells. Hensen's cells, Deiter's cells and the outer row of outer hair cells form the outer tunnel or space of Nuel. The arch formed by the pillars overlies the zona arcuata of the basilar membrane. Laterally, the membrane continues as the zona pectinata containing "auditory strings." Inner phalangeal cells support inner hair cells and some nerve fibers. Deiter's cells support outer hair cells, enclosing the nerve endings around the base of the hair cells, and giving off finger-like processes which surface about three cells away to join the reticular lamina. Border cells are also supporting cells.
   2. Hair cells: *Inner hair cells* forms a single row. Resemble vestibular Type I cells, somewhat. Similar hairs. No kinocilium, but basal body persists. Hairs arranged in a W-pattern. Mitochondria under the terminal web and at the base of the cells. Granular and nongranular nerve endings on base of cell. Inner hair cells receive 90–95% of all nerve endings. *Outer hair cells* generally forms three rows, but may be five rows near basal end. Highly specialized. Nerve endings as on inner hair cells. No kinocilium. More rows of hairs, and they are long toward the periphery and short in the center of the cells. Dense lipid-like inclusions, convoluted granular endoplasmic reticulum, and some mitochondria below terminal web. Mitochondria at base and along side parallel to plasmalemma. Outer hair cells receive only 5–10% of nerve endings.

H. Spiral limbus is the inner edge of scala media. Overhangs inner spiral sulcus. Auditory teeth of Huschke (collagenous fibers). Uniformly spaced between teeth on upper margin are interdental cells which secrete the tectorial membrane.

I. *Tectorial membrane*: Highly organized protein, fibrous, cellular structure. Overlies hairs like a cuticle. Hair marks detectable.

J. *Nerves*

1. Direct myelinated acoustic fibers radiate from spiral ganglia bipolar cells to nearest part of Corti's Organ. Spiral fibers are thicker and fewer than direct. Spiral fibers go radially toward Organ of Corti, but turn and follow a spiral course. Inner spiral bundle, spiral tunnel bundle, outer spiral bundle.
2. The simple processes from the spiral ganglion cells synapse in the cochlear nucleus. Axons from the cochlear nucleus cross the midline in the dorsal and intermediate acoustic striae, and the trapezoid body. Trapezoid body fibers continue to the superior olivary nucleus which receive ipsilateral and contralateral cochlear input. Cochlear and superior olivary axons ascend in the lateral lemniscus. Some synapse in the nuclei of the lateral lemniscus and cross through Crobst's commissure. Lateral lemniscus fibers eventually synapse in the inferior colliculus. Inferior colliculus axons course to the medial geniculate nucleus of the thalamus. Thalamic cells project homolaterally to the primary auditory cortex in the superior temporal gyrus (Brodman's area 41 and 42).

K. *Blood vessels of the cochlea*

1. Vestibulocochlear artery is the lowest part of first cochlear turn.
2. Cochlear artery proper penetrates modiolus. Its branches spiral to apex as the spiral tract. Supplies spiral ganglia and inner parts of the basilar membrane in arcades in the tympanic cover layer under the inner tunnel and the limbus. Stria vascularis and spiral ligament vessels do not communicate with vessels of the basilar membrane.
3. Vein distribution does not follow arterial distribution. Most of the veins traverse the scale tympani.
4. Rationale: Arteries are in the wall of the scala vestibuli and in the modiolus. Veins are in the scala tympanic. This distribution provides optimum damping of pulsations, protecting the ear from the sound of its own blood supply.

Appendix II
Otopathology

This appendix is not intended to be complete. It was designed to summarize the scope of otopathology and to serve as an aid for organizing further study of the field.

Most of these notes are in brief, outline form. However, certain diseases have been discussed in some detail. They have been selected either for their special interest because of their incidence, or they have been selected because of difficulty finding information about them in standard reference sources. In particular, an extensive discussion of diseases of the external ear is included here because it is generally not well covered in many standard texts.

Congenital Diseases of the Ear

This topic is so broad and interesting that even a modestly thorough discussion of it would more than double the size of this outline. Therefore, because it is such a specialized area, the topic is merely sketched to suggest its scope and present a few of the most common and relevant lesions. The reader is encouraged to consult other literature.

I. *Embryology of the Ear, in Brief*

A series of six tubercles located around the margins of the first pharyngeal or mandibular cleft develop into the auricle, and the cleft becomes the external auditory meatus. The middle ear is derived from the first and second branchial arches, for the most part. The malleus and incus are from the first branchial arch (Meckel's Cartilage), from which the mandible is also derived. Most of the stapes, except for the vestibular wall of the footplate, is derived from the second branchial arch (Reichert's Cartilage) as is the tympanic portion of the facial canal. The vestibular aspect of the footplate is from a differentiation of the otocyst called the lamina stapedialis which is at first continuous with the rest of the otic capsule, but later differentiates into the annular ligament.

Otic vesicles in the ectoderm form as invaginations of the otic placodes. The otic vesicles, or otocysts develop into a ventral component which gives rise to the saccule and the cochlear duct, and a dorsal part which forms the utricle, semicircular canals, and the endolymphatic duct. These comprise the membranous labyrinth. They are imbedded in mesenchyme with develops into the bony labyrinth.

II. *Malformations of the External and Middle Ear*
External and middle ear deformities may occur separately, but often they are associated.

A. *Primarily outer ear*: Aplasia (also middle ear, often), absence of part of an ear, asymmetry of the ears, polyotia, microtia, macrotia, Macacus ear (helix points up), Satyr ear (helix is pointed and retains the Darwinian tubercle), abnormal protrusion, drooping roll or dog ear, malposition or malrotation, adhesions, atresia of membranous or osseous canal, cleft lobule, fistula auris (defective closure of the first branchial cleft, supernumerary auricle, Darwinian tubercle (local thickening on the upper posterior border of the helix).

B. Middle ear anomalies, continued. In addition to the preceding list, note there can be complete agenesis of the middle ear.

Important ossicular malformations include congenital fixation of the stapes (failure of differentiation of the lamina stapedialis); ossicular absence; monopolar stapes.

Middle ear anomalies with unknown etiology, without associated anomalies, include entities such as abnormally positioned facial nerve, ossicular abnormalities, mal-positioned carotid artery, congenital absence of the oval of round window, persistent stapedial artery, internal carotid artery aneurysm, high jugular bulb, and congenital absence of the stapedial muscle and tendon. Middle ear anomalies with unknown etiology, but with associated **anomalies** include some ossicular malformations, congenital cholesteatoma (associated with branchial arch anomalies), and anomalies associated with Trisomy 18, Trisomy 13,15, and Reynauld's dysplasia.

There are numerous middle ear anomalies of genetic origin. Some have no associated anomalies such as otosclerosis.

C. Middle ear anomalies may be associated with branchial arch anomalies and others.

1. With branchial arch anomalies
   a. Atresia auris congenita
   b. Cleft palate, micrognathia, and glossoptosis
   c. Mandibulofacial dysostosis
   d. Mixed hearing loss, low-set malformed ear, and mental retardation
2. With skeletal anomalies
   a. Actuce phalosyndactylia
   b. Brevicollis
   c. Craniofacial dysostosis
   d. Craniometaphyseal dysplasia
   e. Dominant proximal Symphalangus and hearing loss
   f. Knuckle pads, leukonychia, and deafness
   g. Oral-facial-digital syndrome 11
   h. Osteitis deformans
   i. Osteogenesis imperfecta
   j. Osteopetrosis
   k. Otopalatodigital syndrome
3. With cardiac anomalies
   a. Congenital heart disease, deafness, and skeletal malformation

4. With renal anomalies
    a. Recessive renal, genital, and middle ear anomalies
5. With connective tissue anomalies
    a. Androplasia
    b. Hurler's syndrome
6. With ophthalmologic anomalies
    a. Congenital facial diplegia
    b. Duane's retraction syndrome
7. And with other abnormalities
    a. Chemodectoma of middle ear
    b. Histiocytosis

Inner ear anomalies can also be associated with prenatal infection (congenital syphilis and congenital rubella syndrome), and iatrogenic ototoxicity (thalidomide).

A. *Middle ear anomalies without associated anomalies*
    1. Aberrant facial nerve
    2. Anomalies of ossicles
    3. Anomalous internal carotid artery in middle ear
    4. Congenital absence of oval window
    5. Congenital absence of round window
    6. Congenital absence of stapedial muscle and tendon
    7. High jugular bulb
    8. Internal carotid artery aneurysm
    9. Persistent stapedial artery
B. *Middle ear anomalies with other associated anomalies*
    1. With branchial arch anomalies
        a. Congenital cholesteatoma of middle ear
        b. Ossicular malformations
    2. With chromosomal abnormalities
        a. Gonadal aplasia
        b. Trisomy 13–15 syndrome
        c. Trisomy 18 syndrome
C. *Anomalies of the inner ear*
    1. Five recognized groups:
        a. Michel—complete absence of the inner ear.
        b. Bing Siebenman—Osseous labyrinth is normal, basilar membrane is deformed, as are the Organ of Corti and the spiral ganglion and some of the membranous vestibular labyrinth.
        c. Mondini—Cochlea flattened from base to apex; lack of interscalar septum between middle and endolymphatic duct are dilated. Organ of Corti atrophic in some areas. Spiral ganglion may have reduced numbers of cells. Semicircular canals are usually normal.
        d. Scheibe—most common; osseous labyrinth normal. Reissner's membrane distended. Saccule may be dilated. Organ of Corti, stria vascularis and macula of the saccule are somewhat atrophic. Spiral ganglion is lacking.
        e. Michel and Mondini are present before birth, Bing Siebenman and Scheibe may occur after birth.
        Note: See below for Mondini–Alexander types.

D. Hereditary sensorineural deafness (see Chapter 13 and other topics below in this Appendix).

III. *Other Congenital Disease*

A. Congenital cholesteatoma—Originates from epidermis. Rests deposited in the temporal bone during development. Detected by damage to facial nerve or ossicles. Smooth, pearl-like surface, grossly and histologically like acquired cholesteatoma. See acquired middle ear diseases.

B. Congenital cysts and tumors—In addition to preauricular cysts and fistula, may see dermoid, hemangioma and lymphangioma, in the external ear.

C. Mondini–Alexander type—Anomalies of cochlear and vestibular divisions. Varying degrees of aplasia, usually incomplete spiral lamina. Dysplastic cochlear duct. Often anomalies of semicircular canals. Otic capsule boney abnormalities.

D. Manifestations of retrocochlear disease (see other literature and Chapter 10 in this book).

E. Otosclerosis—Hereditary, but not truly congenital.

F. Acquired congenital deafness—Thalidomide, chloroquine, other drugs, Rubella, encephalitis especially due to Chicken Pox virus, Erythroblastosis Fetalis, Anoxia, Hypoxia, Prematurity.

IV. *Diseases with Middle Ear Anomalies*

V. *Dominant Hereditary Sensorineural Deafness*

A. Hereditary sensorineural hearing loss without associated abnormalities

VI. *Hearing Loss and Inborn Errors of Metabolism*

A. Schaeffer's syndrome, TC syndrome, Waardenberg's syndrome

VII. *Hearing Loss in Renal Abnormalities*

A. Alport syndrome, Herrmann's syndrome

B. Muckle and Well's syndrome

VIII. *Hearing Loss and Ectodermal Abnormality*

A. von Recklinghausen's disease

IX. *Hearing Loss and Degenerative Neurological Disease*

A. Huntington's chorea

X. *Recessive Hereditary Sensorineural Hearing Loss*

A. Hereditary recessive sensorineural hearing loss without associated abnormality

B. Recessive hereditary hearing loss and inborn errors of metabolism

C. Albinism and hearing loss; Hurler's syndrome; Morquio's disease, Onychodystrophy, Tay-Sachs' disease, Wilson's disease

XI. *Hearing Loss and Degenerative Neurological Disease*

A. Freidereich's ataxia, Schilder's disease, Unverricht's epilepsy

XII. *Hearing Loss and Heart Disease*

A. Jervell's and Lange–Nielsen syndrome

B. Lewis's syndrome

XIII. *Hearing Loss in Endocrine Disease*

A. Pendred's syndrome

XIV. *Hearing Loss and Ophthalmologic Disease*

A. Alstrem's syndrome

B. Cokayne's syndrome

C. Norrie's disease

D. Usher's syndrome

Acquired Diseases of the Ear

External ear

I. *Diseases of Unknown Etiology*

A. *Malfunction of skin glands*

1. Seborrheic dermatitis: Erythematous, greasy, scaly eruption which flakes. Usually associated with seborrhea of scalp. Frequent secondary changes imposed by excoriations: crusting, inflamed and weeping. May be difficult to distinguish from contact dermatitis and streptococcal dermatitis.
2. Diffuse external otitis: discussed under gram negative organisms.
3. Asteatosis: dry ear. Deficiency of sebaceous secretion. May be iatrogenic. May occur with eczema or ichthyosis, and seen frequently in old age.
4. Infantile eczema
   a. Infantile intertrigo: small children. Lobule and postauricular sulcus. Bilateral and symmetrical. Erythematous and scaly with vesicular or vesico-pustular border. Itching. Secondary infection. Psoriasis often underlying. If due to Candida, more pinkish and glazed and more extensive.
   b. Infantile dermatitis: Irritation without involvement of the auricle of other exposed skin (although skin may be involved at times). No edema or adenopathy. Itching. Dry crusted debris at the entrance of the auditory canal. Soft, pale yellowish, moist fetid secretion fills the canal. Cultures often show corynebacteria (diphtherois).
5. Membranous external otitis: Otitis externa crouposa. Rare. Meatal skin swollen, gray, and wrinkled. May be seropurulent discharge. Marked discomfort. After 5–7 days, fibrinous membrane can separate as a complete cast of the meatus. Improves in 2–3 days. No consistent flora.
6. Chronic idiopathic myringitis: Usually myringitis is a manifestation of disease of the meatus, middle ear or mastoid. Persistent changes may include chronic infection, superficial ulceration or low-grade suppuration, all limited to the T.M. Cultures often show micrococci and gram negative organisms such as Aerobacter.
7. Hyperceruminosis: Usually apparent "wax formers" do not extrude wax easily because of anatomy of the meatus. May be due to rapid cerumen formation.

B. *Miscellaneous*

1. Lupus erythematosus: Chronic discoid: erythematous, circumscribed lesions with firm adherent scale; pigmentation and atrophy; ear is frequently involved, especially the lobule and concha. Systemic: diagnosis in usual manner.
2. Psoriasis: Superficial, pink lesions, with mica-like scale. Usually also on scalp, elbows, knees, and nails. Not as greasy as seborrhea and there is a greater tendency to form circumscribed plaques.
3. Lichen planus: Annular and linear patches of flat, lilac-colored, angulated papules. Dry and shiny. Scale. Trunk and flexor surfaces of the extremities. Mucous membranes in 25%.
4. Pemphigus—Lethal. Large flaccid bullae from noninflamed bases on the skin and mucous membranes. Occasionally seen on the ears.

II. *Infections*

A. *Bacteria*

1. Gram positive

a. Furunculosis—Hair infection. Staphylococcus.

b. Impetigo—Staphylococcus (staph), Streptococcus (strep), or both.

c. Pyoderma—Deep, boggy, frequently follow trauma. Usually staph.

d. Ecthyma—Infrequent in ear. Inflammatory. May lead to scarring and stenosis of canal. Mixed flora.

e. Cellulitis—Lymphangitis and adenopathy. Usually strep or staph.

f. Erysipelas—Hemolytic strep, occasionally staph pain, fever, anorexia. Warm, tender, smooth, elevated tense shining, red, indurated appearance with sharp line of demarcation. Rapid extension. May be vesicles and bullae.

2. Gram negative

a. Diffuse otitis externa—Swimmer's ear; hot, humid weather. Preinflammatory stage: initiated by swimming, diving, trauma; edema of stratum corneum interferes with formation of lipid cover; absence of wax; dry skin; itching. Acute inflammatory stage: (i) mild—pain on manipulation; edema, redness, odorless secretion, or exfoliated debris; some luster of drum. (ii) moderate—lumen partially occluded; moderate periauricular edema without adenopathy, obliteration of canal, gray or green seropurulent secretions, and exfoliated debris; small pustules appear. Infecting organism in all of the above is usually Gram negative, but may be fungal. Chronic inflammatory stage—thickening of the skin; dry, adherent debris; secretion may be present, gray-brown or greenish with a fetid odor. Usually Gram negative, rarely fungal.

b. Hemorrhagic or Bullous external otitis—No preceding URI. Sudden onset of severe pain. Later bloody discharge, without pain. Hemorrhagic bullae on meatal wall. TM is normal.

c. Otitis externa granulosis—Skin raw and coated with scanty, creamy pus, and granulations. Peculiar stalk nature of the granulations. Usually antecedent neglected otitis externa.

d. Perichondritis—May follow infection, trauma including surgery and temperature. Auricular cartilage becomes warm, soft, reddish, and thickened. Pus between the cartilage and perichondrium interferes with nourishment. Deformity is a common sequel. Cauliflower ear.

3. Acid fast

a. Tuberculosis (TB)—Lupus vulgaris, apple jelly nodules. Less common: TB of skin; TB of subcutaneous tissues.

b. Leprosy—Auricle becomes infiltrated, nodular, and enlarged.

B. *Fungi and yeasts*

1. Saprophytic

a. Aspergillaceae, mucoraceae, yeast-like fungi, dermatophytes, Actinomycetaceae and miscellaneous fungi—chronic or recurrent; tropical climates; itching and fullness, and there may be pain, deafness, and tinnitus in severe cases. Moist cerumen with exfoliated scales, sheets of epithelium. Canal may be line with structure like those on culture media.

b. *Aspergillus niger*—Skin of osseous meatus and TM covered velvety gray membrane with black spots.

2. Pathogenic

a. Superficial—*E. inguinal*, *T. gypseium*, *T. purpureum*, and *C. albicans*.

b. Deep—Blastomycosis, actinomycosis, and sporotrichosis. Should be thought of with granulomatous lesion of the canal.

C. *Virus*

1. Bullous myringitis—associated with URI. Acute vesicular or hemorrhagic bullae of the TM and adjacent canal. Sudden severe pain, relieved by rupturing blebs.
2. Herpes simplex—vesicles usually also on lips.
3. Herpes zoster—unilateral. Tense vesicles following course of cutaneous nerve.
4. Molluscum Contagiosum—usually children. One or more small, circumscribed raised, waxy nodules with central umbilication.
5. Variola and varicella—pustular vesicles as elsewhere.

D. *Protozoa*

1. Syphilis—primary, secondary, and tertiary.
2. Yaws—treponema pertenue. Clinically resembles syphilis, although not as serious. Indigenous to tropics.

E. *Animal parasites*

1. Pediculosis—edematous, red papules, may have central punctate from proboscis. Sometimes vesicles and bullae.
2. Scabies—itching, excoriated papules, and burrows. Ear lesions usually in children. Acarus scabiei.

III. *Neurogenic Eruptions*

A. Simple pruritus—predominantly middle aged women, tense, hyperactive. No gross lesions. Must rule out contact dermatitis, diabetes, jaundice, nephritis, bloody dyscrasias and dry skin, as well as other causes

B. *Neurodermatitis*

1. Localized—lichen chronicus simplex, excoriated, scaly, dry patches, more sharply circumscribed than in seborrhea. Neck, eyelids, anticubital and popliteal areas and ankles.
2. Disseminated—lesions as above, but wider distribution. Frequently psychogenic. Rule out atopic dermatitis.

C. Neurotic excoriations

D. Artifact dermatitis (malingering)—do not follow natural forms. Are usually discrete, inflamed, crusted, and ulcerated; different geometric appearances.

IV. *Allergic Dermatoses*

A. Contact dermatitis—suspect ear medicines, cosmetics, hair spray, eyeglass temples (nickel), telephone receivers, perfumes, and earrings.

B. Atopic dermatitis—associated with dermatitis of eyelids, face, neck, popliteal, and cubital fossae. History of other allergic disease such as asthma, hay fever, etc. Family history common.

C. Drug eruptions—any morphological form. Usually other parts of body also affected.

D. Infectious eczematoid dermatitis—adjacent focus of infection discharges and sensitizes skin leading to this condition.

E. Physical allergy—exposure to cold, heat, pressure, ultra violet (UV) rays, or visible rays may result in urticaria or crythema. Mechanism may be a true antigen antibody reaction against protein altered by the physical agent, or it may actually be another mechanism of response to trauma.

V. *Traumatic Lesions*
   A. *Contusions and lacerations*
      1. Traumatic
      2. Chondrodermatitis nodularis chronica helicis—harmless, common in telephonists, soldiers, nuns, and physicians. Hard, small immovable nodule surrounded by a hyperemic abscessed area on the helix.
   B. Surgical incisions
   C. *Hemorrhages*
      1. Hematoma—Between cartilage and perichondrium secondary to trauma; may result in cauliflower. Hemorrhages into skin of canal may occur secondary to base of skull and osseus meatus fracture.
      2. Vesicles and bullae—follow manipulation.
   D. Burns
   E. Frostbite—common site.
   F. *Radiation injury*
      1. Ultra violet
      2. Roentgen rays—auricle is said to be able to tolerate up to 600 R.
   G. Chemical—acids and alkalis. Medicines used in excessive strengths.

VI. *Senile Changes*
   A. Thinning of the epidermis, loss of elasticity, freckling and dryness, pruritus. Keratoses.
   B. Indolent ulcer, usually in one ear. Mild discomfort and pain, slight discharge, foul odor. Granulations in circumscribed area of osseous canal. Exposed bone in midst of ulcer. Differentiate from malignancy.

VII. *Vitamin Dyscrasias*
   A. Vitamin A—dry skin, scaly small follicular papules.
   B. Vitamin B—pellagra occasionally involving ears.
   C. Vitamin C—scurvy with subsequent acanthosis and hyperkeratosis.

VIII. *Endocrine Dyscrasias*
   A. Hypothyroidism—pale, sallow, greasy skin varying to cold, thick dry, and scaly skin.
   B. Hyperthyroidism—warm, moist thin, and translucent skin.
   C. Addison's disease—soft and bronzed skin, pigmentation associated with pigmentation elsewhere

IX. Foreign bodies—vegetable, animal, and mineral.

X. Otalgia (see other literature)

XI. *Tumors*
   A. *Benign cysts*
      1. Pseudocysts—fluid in tissue space without epithelial lining.
      2. Sebaceous cyst—common around the auricle, especially in the postauricular crease.
      3. Preauricular cysts—see congenital diseases.
      4. Incisional cyst or epidermal implantation cyst—due to implantation of squamous epithelium into subcutaneous tissue. Secretions are from sweat or sebaceous glands implanted.
      5. Dermoid cyst—see congenital diseases.

B. *Benign tumors*
   1. See congenital diseases.
   2. Papilloma or verrucosa wart. Rare in the external canal and even less common on the auricle.
   3. Keratosis obturans or cholesteatoma of the external auditory meatus is rare. Desquamating squamous epithelial mass is found deep in the canal. Frequently associated with bronchiectasis and sinusitis. Cause unknown. Pain. Sometimes deafness and otorrhea.
   4. Osteoma or hyperostosis
      a. Exostosis—most common tumor of the external canal. Much more common in males. Usually bilateral. Most common in saltwater swimmers. Covered by skin. Attached by broad base to osseous canal. Usually asymptomatic.
      b. Osteoma—single cancellous osteoma is relatively rare. Hearing loss and discomfort common. Will continue to grow.
   5. Adenoma—rare; glandular. Originate from sebaceous glands in the outer part of the canal. Small, soft, painless.
   6. Ceruminoma (hidralenoma, sweat gland tumor)—uncommon. Smooth, intraverted, polypoid swelling. Local recurrence and malignant change are common.
   7. Chondrodermatitis nodularis Chronicus helicis—discussed earlier.
   8. Chondroma—rare lesion of the auricle, usually associated with trauma. Ecchondromas are cartilaginous spicules in the external canal which continue to grow.
   9. Lipoma—extremely rare on the ear.
   10. Fibroma—very rare. Two types: soft and hard.
   11. Keloid—connective tissue hypertrophy, scar-like tumor. Commonly on lobule secondary to earring.
   12. Xanthoma—not a true neoplasm, but may resemble one. Diseases associated with a disturbance in cholesterol metabolism.
   13. Myoma—striated muscle, extrinsic muscles of the ear. Smooth muscle, arrectores pilorum or muscular coats of vessels. Extremely rare.
   14. Mixed salivary gland tumors—rare. Moveable masses tightly related to overlying skin. Ectodermal origin.
   15. Melanoma—fairly common. Usually intradermal (common mole).

C. *Malignant tumors*
   1. Incidence
      Approximately 90% of all skin cancers occur in the head and neck, and 6% of all skin cancers occur on the ear. The majority are on the helix and antihelix. Squamous cell is most common, then basal cell. Approximately 80–85% of aural cancers involve the auricle, 10–15% involve the canal, and 5–10% involve the middle ear.
   2. Malignant tumors of the auricle
      a. Squamous cell carcinoma—grows slowly, metastasises late. Small lesions on the helix or antihelix have excellent prognosis. Near the cartilaginous canal, much poorer.
      b. Endothelioma—perivascular origin, much slower than Epi.
      c. Basal cell—men more than women. Fifth or sixth decade. Locally malignant, but slow growing and does not usually metastasize. Rodent ulcer.

d. Cylindroma, adenocarcinoma, sarcoma, and malignant melanoma are rare on the auricle, but do occur.

3. Malignant tumors of the external auditory canal
Serious disease. Pain, otorrhea, bleeding, fullness, and decreased hearing may all be symptoms. Urgent to diagnosis early.
   a. Squamous cell—70% of all canal malignancies. First nodes are usually pretragal. Distant metastases are rare. Death is from local extension.
   b. Basal cell—rarer than on auricle, but prognosis worse because of late discovery and invasion of bonė and middle ear.
   c. Adenocarcinoma—least common. Arises from sweat or sebaceous glands.
   d. Cylindroma and adenoid cystic epithelium—not common. Is an adenocarcinoma arising from the ceruminous glands. Fifth decade of life. Slow growth. Pain becomes extremely severe. Bony destruction. Yellow firm, smooth masses in the canal. Distant metastasis late to pulmonary cervical nodes and kidneys.
   e. Sarcoma—Mesoblastic origin. Extremely rare, but in the ear one sees chrondrosarcoma, fibrosarcoma, squasarcoma, lymphosarcoma, Rhabdomyosarcoms, leiomyesarcoma, undifferentiated cylindrical cell type sarcoma. Prognosis relates to site and differentiation, but they metastasize early and the outlook is usually not good.
   f. Malignant melanoma—extremely rare. Prognosis is guarded as elsewhere with this lesion.
4. Addition to diseases of the auditory canal
Stenosis of ear canal—traumatic, postoperative, or congenital. Collapse of ear canal.

Diseases of the Tympanic Membrane

I. Many of the diseases of the external ear may involve the drum. See preceding discussion.

II. *Infection of Middle Ear Reflected in the Drum.*
   A. Acute suppurative otitis media—pain, fullness, congestion of vessels. Hyperemia; exudation; suppuration following rupture. Resolution with desquamation, closer of perforation.
   B. *Acute viral otitis media*
      1. Minimal findings, associated with blocked eustachian tube.
      2. Bullous myringitis—severe pain without fever or hearing loss. Epidemics. Multiple blebs in drum and canal.
   C. Acute necrotizing otitis media—bacterial, usually beta-hemolytic strep. Areas of poorest blood supply (kidney-shaped center of the pars tensa) affected first. True necrosis.
   D. Tuberculous otitis media usually secondary to primary TB spread through eustachian tube. Cloudy exudate, continued drainage. Pain is rare.
   E. Non-TB chronic otitis media—can result from persistent perforation, bone destruction, inflammation, cholesterol, granuloma, cholesteatoma, allergy, dermatitis. When behavior or response to treatment are atypical, must consider chloroma, lupus, malignant otitis externa or granulomatosis disease such as Wegener's of histiocytosis X.

III. Tympanosclerosis—benign; result of repeated infection. Not invasive. Tunica propria of the mucous membrane of the middle ear and TM becomes edematous and infiltrated with inflammatory cells during acute infections. Repeated recurrence leads to invasion by fibroblasts, organization of tunica propria into thick collagenous connective tissue. Hyalin degeneration results in smooth, white, slightly raised areas of dense tissue. Calcification may occur. In middle ear, ossicles may be fixed. Chalky, white patches are seen in the TM.

IV. Traumatic perforations—compression; instrumentation; burns or caustic agents; blasts 90% heal spontaneously, usually within 1 month.

V. Chronic granular myringitis—itching, otorrhea, and hearing loss. Replacement of the squamous layer by proliferating granulation tissue. Rare.

VI. Keratitis obturans—ear canal cholesteatoma. See external auditory canal discussion.

VII. Chronic adhesive otitis media—late sequel of chronic Otitis media, middle layers of drum destroyed.

VIII. Tympanic membrane in chronic otitis media—destruction of drum, partial (fibrous layers) or complete. Retraction perforation, cholesteatoma.

Diseases of Eustachian Tube

See other standard otolaryngologic literature.

Diseases of the Facial Nerve

Bell's palsy, facial palsy (inflammatory, traumatic, herpetic), and Ramsey–Hunt syndrome. Facial nerve tic, hemifacial spasm, facial nerve neuroma.

Diseases of the Middle Ear

I. *Middle Ear Effusions*
   A. Etiologies—adenoid hypertrophy, cleft palate, barotrauma, tumors (especially unilateral effusion in adult). High incidence of nasopharyngeal carcinoma, (especially in Cantonese), inflammation and antibiotics, allergy, iatrogenic (inadequate antibiotic therapy, radiation therapy, trauma especially to the torus during adenoidectomy), other causes (viral, immune, etc.).
   B. Classification of middle ear fluids—serous, mucoid, bloody, purulent.
   C. Symptoms—plugged up feeling, conductive hearing loss, autophony, often pain, often history of URI.
   D. Findings—Opaque or amber drum, creamy colored drum, retraction (or bulging) of drum, fluid level or bubbles, blue drum.
   E. Sequelae—atrophic tympanic membrane, ossicular erosion, tympanosclerosis, cholesterol granuloma and gland formation, chronic otitis media and cholesteatoma.

II. *Acute Otitis Media and Mastoiditis*
   A. *Types*
      1. Acute bacterial—usually acute middle ear infection. *H. influenza* (rare in patients >5 years old), pneumococci and beta-hemolytic streptococci are most common. *S. aureus* occasionally; also

pseudomonas aeruginosa. Self-limited diseases. Bulging drum. About 1–5% will become chronic, and may erode bone and lead to extradural or perisinus abscess or a subperiosteal abscess, leptomeningitis, sigmoid sinus thrombophlebitis, brain abscess, suppurative labyrinthitis, petrositis or facial nerve palsy.

2. Acute necrotic otitis media—chiefly in children severely ill with exanthem or other systemic infection. Rare. Usually beta-hemolytic strep. Necrosis or localized gangrene. Virulent because of impaired host resistance. Requires treatment.
3. Acute viral myringitis—normal hearing and absence of significant fever.

B. Malignant otitis externa—In diabetics, granuloma rises from floor of external. Canal at bony cartilaginous junction. Pain out of proportion. Poor response to prescription. Not a cancer. *Pseudomonas aeruginosa*. Severe destruction of middle ear and temporal bone. Gangrene. May lead to facial paralysis, death.

III. *Chronic Otitis Media and Mastoiditis*

A. Types—Allergic secretory; suppurative, benign mucosal type; tubercular; suppurative with Osteitis or osteomyelitis; suppurative with cholesteatoma.

B. Bacteriology—Usually (in order) *S. aureus* (32%), *B. proteus*, *Pseudomonas pyocanea*, *S. aureus* penicillin resistant, mixed, *E. coli*, *St. pyogenes*, *St. viridans* and pneumoniae (5%). No growth in 10.6%.

C. Location—Chronic infection in ciliated columnar areas is usually confined to the mucosa. In flat pavement epithelium of the attic and antrum, it is in association with erosion of bone and requires more vigorous attention.

D. Tubotympanic disease—Permanent perforation syndrome; persistent tubotympanic mucosal infection.

E. *Attic and antrum disease*

1. Occlusion of the attic floor by a chronic process limited to the anterior and posterior tympanic isthmuses.
2. Cholesteatoma: Congenital—see congenital disease.
   Primarily acquired—squamous epithelial cyst that arises in the pars flaccida and gradually invades the epitympanum, aditus and middle ear. Possibly an extension of rete pegs into unresolved mesenchyme in early life. Origin not visible by otoscopy. Conductive hearing loss due to immobility of destruction of ossicles. Wall (matrix) produces enzymes that result in absorption of bone (not proven). Thin layer of squamous epithelium forms the wall of the cyst. Center consists of desquamated keratinizing epithelium which undergoes degeneration. Cholesterol accumulates, giving rise to misnomer. Is really an epithelioma, or keratoma. Secondary acquired—repeated infections destroy fibrous layer of drum, usually in posterior superior quadrant. Perforations from disease. Negative pressure in ear that led to infection encourages invagination of squamous tissue when perforation heals. Negative pressure causes drum to retract in posterior superior gradrant, and desquamation of the squamous epithelium fills the retraction pocket. Histology the same as primary acquired. Another theory suggests that the secondary type results directly from metaplasia of the middle ear mucosa provoked by infection. Symptoms—foul, mild otorrhea, hearing loss variable, bleeding, earache, dizziness, headache suggestive of intracranial extension.

F. *Late middle ear sequelae*
   1. Adhesive otitis media
   2. Tympanosclerosis
   3. Cholesterol granuloma—stasis with mucosal edema, exudation, and hemorrhage, usually blood vessels to the ossicles and along the facial canal. Blue drum. Breakdown of red cells releases cholesterol crystals which act as foreign bodies and stimulate granulation tissue. Mucosa often undergoes metaplasia becoming ciliated secretory throughout the middle ear and mastoid. Granulation tissue walls off preexisting effusion producing cysts containing cholesterol crystals. Cholesterol clefts also in granulation tissue. Macrophages, round cells, and foreign body giant cells around clefts. Fibroblasts may invade and obliterate entire middle ear space.

G. Complications of suppurative otitis media and mastoiditis facial paralysis, labyrinthitis, suppurative labyrinthitis, petrositis (Gradenigo's syndrome), meningitis, lateral sinus thrombophlebitis, extradural abscess, subdural abscess, brain abscess, otitis hydrocephalus.

IV. Post surgical pathology—not covered in this discussion

V. Temporal bone fractures—not covered

VI. *Granulomas and Other Diseases of the Ear and Temporal Bone*

A. Tuberculous otitis media

B. Syphilis—any syphilitic lesions (gummas, etc.) may occur in the middle or inner ears. There is a syndrome of sudden deafness associated with late congenital syphilis in children or in later years. Path. is marked ectasia of the cochlear duct. Corti's organ may be absent, may be bone or connective tissue in the scala media. There is some evidence that this syndrome may be reversible if the diagnosis is made rapidly. Hearing loss in $\sim$35% of all lues cases, incidence higher in women.

C. Polyps of various types

D. *Histiocytosis X*
   1. Acute disseminated histiocytosis (Letterer–Siwe disease) before age of two. Lesions in soft tissue and bone.
   2. Chronic disseminated histiocytosis (Hand–Schuller–Christian disease). Children and young adults. Bone. Soft tissue. Juvenile xanthogranuloma is benign histiocytosis disease that may be isolated or occur in association with histiocytosis X.
   3. Eosinophilic granuloma—benign chronic disease. One or several bony lesions. May arise in soft tissue.

   Temporal bone lesion in histiocytosis X expands and presents as a granuloma in the external canal, may erode the mastoid cortex, tegmen, dura, labyrinth, lateral sinus, and zygoma. Secondary infection is common. Sheetlike layers of histiocytes. Occasional lymphocytes, plasma cells, and multinucleated giant cells.

E. Wegener's Granulomatosis—uncommon in the ear. Usually white male adults. Nasal sinus or ear symptoms. Appears severely ill out of proportion to usual nose or ear disease. Fever. Facial paralysis common. Tissue resembles closely periarteritis nodosa. Nonspecific granuloma with many round cells and some giant cells. Round cells are typically perivascular.

VII. *Tumors of the Middle Ear*

A. Glomus jugulare—glomus bodies (solid masses of epitheloid cells) exist normally along the course of Arnold's and Jacobson's nerves and in the

adventitia of the jugular bulb. Nerves from IX and blood supply through the tympanic branch of the ascending pharyngeal artery. Tumors enlarge concentrically. Five times more common in women. Familial disposition. Ten per cent are multicentric. Histopathology. shows large epitheloid cells with abundant acidophilic cytoplasm, often vacuolated. Nuclei are small and nonhyperchromatic without mitotic figures. Small alveolar.
Patterns separated by fibrous septa containing vessels. In some cases, cord-like deposition. Two kinds:

1. Glomus tympanicum—from Jacobson's nerve. Slow.
2. Glomus jugulare—from jugular bulb. Sometimes very rapid spread to adjacent and intracranial structures. The are nonchromaffin paragangliomas.

B. Aquamous cell—intractable external otitis, bleeding paralysis of facial nerve, severe pain especially at night. Prognosis is poor.
C. Fibrosarcoma—transitory paralysis of the facial nerve in attacks; prolonged pain.
D. Neurofibroma—can occur on any of the nerve structures.
E. Giant cell tumor—rare in the middle ear. Usually in long bones or TMJ, but does occur in ear.
F. Osteoma—relatively rare. Damage by expansion.
G. Rhabdomyosarcoms—most common temporal bone tumor in children <6 years. Painless otorrhea or bleeding.
Prognosis poor. Four types:

1. Pleomorphic—dense cellularity of spindle-shaped cells, Multiple nuclei. Striations.
2. Alveolar—connective tissue septa. Closely applied to the inner aspect of the trabecular longitudinal and cross striations.
3. Embryonal—Abundant eosinophilic cytoplasm. Long, tiny cells with bipolar processes. Enlarged around their single nuclei. Cross-striations in some.
4. "Tadpole cells" Botryoid—Gross description. Rest under mucous membrane. Histology-like embryonal type.

H. Salivary Gland Christoma—Rare. Histologically normal salivary tissue on the Middle ear cleft. It is usually associated with facial and ossicular chain congenital anomalies. Tumor usually covered by pseudostratified columnar epithelium.
I. Metastatic Tumors—in order of frequency: mammary cancer. Renal cell cancer and bronchogenic cancer. Melanoma also reported.
J. Tumors invading from adjacent structures—Meningioma, gliomas, neurilemmoma, cylindroma, dermoid cyst, tumors of the nasopharynx.

VIII. *Otosclerosis—extensive literature readily available, so only a brief discussion here*

A. Incidence—about 10% of the population.
Females: males are about 2:1; Hereditary; usually second third and fourth decade presentation with conductive hearing loss. Tinnitus often present. Pregnancy accelerates clinical findings.
B. Lesion may occur anywhere in the capsule, but symptoms only if stapes is fixed or if cochlea (especially the stria vascularis region) or vestibule are involved.
C. Two types: Otospongiotic—active; multinucleated osteclasts and osteoblasts, enlarged marrow spaces. No osteoclastic activity occurs in the

normal otic capsule bone. Otosclerotic—inactive; narrower marrow spaces with osteoblast on the walls. (Endochondral and endosteal layers.)

D. Conductive, sensorineural or mixed loss may develop.

IX. *Other Labyrinth Capsule Lesions (with generalized disease)*

A. Paget's disease—polyostotic or monostotic. Often familial. Continuous resorption and regeneration of bone. Deformities of the normal structures. Stapes fixation less frequent than in Otosclerosis.

B. Osteogenisis imperfecta—congenita and tarda.

C. Late congenital syphilis—often a Meniere's-like syndrome. Positive Hennebert's sign (positive fistula test with intact drum, especially with negative pressure). Often negative or equivocal serology with negative spinal fluid serology and positive FTA.

D. Radium-induced necrosis and carcinoma

E. Histiocytosis X.

Diseases of the Inner Ear

This set of notes outline only a few of the major ailments of the inner ear which are especially common and/or especially exciting in terms of future advancement in managing otologic disease.

I. *Diseases of the labyrinth*

A. Non-genetic sensorineural deafness in children—see other literature.

B. Sudden deafness—Etiology unclear but several factors appear related in some cases: virus, vascular, noise (special case), pressure changes. Symptoms: Deafness, tinnitis, vertigo. About one-third return to normal, one third return partially and one third remain totally deaf. See Chapter 11.

C. Functional deafness—diagnosed by special audiometric tests.

D. Presbycusis—old age deafness. High frequencies. Discrimination difficulties. According to Schuknect, four types: sensory, neural, metabolic, mechanical. Recruitment and diplacusis may be present.

E. Noise induced hearing loss—Exposure to blasts, gunfire or other loud noises, or chronic unprotected exposure to >90 dBA daily is likely to lead to bilateral, symmetrical hearing and diplacusis may be present. A major problem in American industry.

F. Ototoxic drugs—Partial list: streptomycin, dihydrostreptomycin, neomycin, gentamicin, cisplatin, chloramphenicol, kanamycin, vancomycin, polymixin B, ethacrynic acid, furosemide, salicylates, quinine, nitrogen mustard, tetanus antitoxin, thalidomide. Chemicals: carbon monoxide, mercury, tobacco, gold, lead, arsenic, aniline dyes, alcohol.

G. Labyrinthitis—Toxic, epidemic, viral, circumscribed labyrinthitis paralabyrinthitis, labyrinthitis fistula, and perilabyrinthitis), serous labyrinthitis, suppurative (commonly hematogentic), Labyrinthine ossification.

H. Tinnitus—Etiology unknown. Relation of character to lesion unreliable. occurs with a great many lesions. See Chapter 18.

I. Vertigo—Tabes dorsalis, pellagra, pernicious anemia, cerebral anoxemia, postural hypotension, paroxysmal atrial fibrillation, aortic stenosis, Adams–Stokes disease, metabolic disturbances, epilepsy, meningitis, encephalitis, syphilis, and head injury all produce non rotational dizziness, *not true vertigo*. See Chapter 19.

Vertigo—sensation of motion.

Meniere's syndrome—paroxysmal attacks of spinning vertigo lasting hours and accompanied by perceptive hearing loss, frequently fluctuating, usually more severe in one ear and with tinnitus and usually with a feeling of fullness. Audiogram is usually flat. Calorics are usually depressed bilaterally. Path: Endolymphatic hydrops.

Postural vertigo, Cupulolithiasis syndrome, Labyrinthine circumscribed labyrinthitis, serous labyrinthitis, Lermoyez's syndrome, otosclerosis inner ear syndrome, vesibular neuronitis, secondary endolymphatic hydrops.

II. Disorders of the central auditory nervous system: Brain stem, temporal lobe and other CNS lesions

III. *Otologic Manifestations of Retro-Cochlear Disease*
   - A. Congenital diseases
   - B. Infectious—Meningitis, Ramsay Hunt's syndrome, Encephalomyelitis, Neurosyphilis, brain abscess
   - C. *Neoplasms*
     1. Childhood—glioma, medulloblastoma, astrocytoma, and ependymoma.
     2. Adult—Subtentorial have especially definitive auditory and vestibular signs. Angle tumors may be meningioma, glioma, cholesteatoma, cyst, granuloma, metastatic, or aneurysm, but neuroma is most common.

        Acoustic neuroma: More than half of all tumors of the cerebellopontine angle. 9% of all intracranial tumors. Usually 30–40 year group. Usual site is Scarpa's ganglion. Bilateral in cases of von Recklinghausen's Disease. Early symptoms are tinnitus, hearing loss and dizziness and often a feeling of fullness. Later, weakness of other nerves. Sensorineural loss, poor discrimination, usually no recruitment or diplacusis, but tone decay present. Usually visible on x-ray.
   - D. Vascular lesions—Hypotension, hypertension, vertebrobasilar artery insufficiency, thrombosis of the posterior inferior cerebellar artery, thrombosis of the internal auditory artery
   - E. Traumatic lesions—cervical vertigo, skull fracture
   - F. Metabolic and drug induced—Hormonal effects, hypervitaminosis D or A. Vitamin Bl def, nicotinic acid def (Nystagmus in 20%), drugs
   - G. Demyelination and degenerative disease—multiple sclerosis, diffuse sclerosis, subacute combined degeneration, syringobulbia and syringomyelia, olivocerebellar atrophy, Friedreich's ataxia
   - H. Idiopathic disorders—Vestibular neuronitis, epilepsy
   - I. Psychogenic disorders

Miscellaneous Additions

Tumors not mentioned that occur in the temporal bone are Ewing's sarcoma, angiosarcoma, osteogenic sarcoma, and others.

Appendix III

Otosclerosis, Paget's Disease, and Osteogenesis Imperfecta Involving the Ear

Otosclerosis, Paget's disease, and osteogenesis imperfecta may present similar clinical pictures as far as their otologic manifestations are concerned. Consequently, some older literature suggests that they may be related or, in fact, that otosclerosis is actually one of the other two diseases. It now appears clear that these three diseases are distinct entities with some clinical similarities, but quite distinct histopathologic features. There are other diseases not covered in this short discussion whose presentation and pathology allows them to be confused with the diseases discussed. The basic features of these diseases are outlined in brief.

Otosclerosis

I. Incidence: About 10% of the population. Females:males are ~2:1. Hereditary, probably autosomal dominant. Presentation after puberty, usually in second, third, or fourth decade. Pregnancy accelerates clinical picture.

II. Presentation usually with conductive hearing loss. Tinnitus often present. Lesions may occur anywhere in the human otic capsule, but symptoms occur only if stapes is fixed or if cochlear (especially in area of stria vascularis) is involved, or the area of the vestibule.

III. Two types of otosclerosis can be found. (1) Otospongiotic-active: multinucleated osteoclasts and osteoblasts, enlarged marrow spaces. No osteoclastic activity occurs in the normal otic capsule. (2) Otosclerotic-inactive: narrower marrow spaces with osteoblasts on the walls.

IV. Conductive, sensorineural, or mixed hearing loss.

Paget's Disease

I. Incidence: Less common than otosclerosis; but there are more subclinical than apparent cases, so the disease is more common than it appears to be. Males:Females are 4:1. Hereditary, probably autosomal dominant previously considered sex-linked recessive. Usually presents after fourth decade.

II. Often found incidentally on X-ray. When clinical, usually presents with enlarged skull. Often, progressive mixed hearing loss (interestingly, Paget's original case was hard of hearing). Skull, spine, pelvis, femur, and tibia are most often involved. May be polyostotic or monostotic. When temporal bone is involved, lesions may involve otic capsule, bone of the tympanic cavity, and occasionally thc ossicles themselves.

III. The otic capsule lesions tend to involve the vascular periosteum and spare the endosteal layer and membranous labyrinth except in advanced lesions. The ossicles are usually fixed by impingement by deformed bone and spurs from their surrounding structures. The ossicles themselves may be involved and may be fixed by increased mass or, rarely, actual ankylosis. The external auditory canal may be stenosed, as by the internal canal (leading to pressure on the eighth nerve). Alkaline phophatase is elevated. Blood is shunted to the lesions. About 1% progress to osteogenic sarcoma.

IV. Conductive, sensorineural, or mixed hearing loss.

Osteogenesis Imperfecta

I. Incidence: 1 in 25,000 births. No sex prediliction. Autosomal dominant. A few reports of autosomal recessive families. Variable expressivity. Two types.

II. Osteogenesis imperfecta congenita is usually severe. Multiple fractures *in utero* or at birth. Still-born or low survival.

Osteogenesis imperfecta tarda is seen at later age. Relatively painless fractures with minor trauma. Long bones with slender shafts and abrupt widening near epiphyses. Small structure. Pectus excavatum and kyphoscoliosis are common. Hyperextensible joints and ligaments. Odontogenesis imperfecta, blue sclera. Hearing loss clinically identical to otosclerosis. Usually begins after puberty. Thirty-five per cent have hearing loss by age 40; 50% by age 60. More sensorineural incidence than in otosclerosis. May have bony lesion in footplate or atrophy of crura. Middle ear mucosa is hyperemic and stapes is fixed due to thick, friable bone. Some authors report bone similar to otosclerotic lesion in some cases. The two diseases have been known to co-exist. Van der Hoeve–DeKlein syndrome: blue sclera, bone fragility, and hearing loss.

III. Abnormalities of collagen fibers at the reticular stage. Calcium and phosphorus content are low, but enzyme activity is normal. Bone matrix is immature with large numbers of osteoblasts, but no apparent osteoblastic activity.

IV. Conductive, sensorineural, or mixed loss.

Appendix IV
Neurofibromatosis

Von Recklinghausen's disease (1882) occurs in three forms (overlap of forms also occurs).

I. Central forms, known as neurofibromatosis-II (NF-II)
   A. Multiple intracranial and intraspinal tumors
   B. Bilateral acoustic schannomas
II. Peripheral form (NF-I) (today, only NF-I is referred to as von Recklinghausen's disease)
   A. Cafe olé spots
III. Visceral form

NF-II

I. First report of bilateral acoustic tumors was Wishart in 1882 and Kanoblauch.
II. In 1940, Turner reported a family with bilateral acoustic tumors transmitted through generations.
III. No gross difference in appearance between unilateral and bilateral acoustic schwannomas. However, in NF-II, there may be multiple schwannomas involving an individual nerve, or involving numerous nerves, as well as multiple meningiomas. Schwannomas and meningiomas may be present in the same surgical specimen and indistinguishable grossly.

   Microscopically, NF-II schwannomas are histologically identical to sporadic acute unilateral acoustic schwannomas. In some cases they may show an intermediate pattern between meningioma and schwannoma (the cells of schwannomas and meningiomas originate from the neural crest). These schwannomas are different, however, from neurofibromas.
IV. *Recurrence*
   A. Tumor behavior can be different from unilateral acoustic schwannomas (old data).
   B. NF-II tumors may be more invasive and may infiltrate the cochlear nerve. This infiltration may be responsible for low success rate with tumor debulking.
   C. No real evidence that recurrence rate is higher, although there is a risk of new tumor on adjacent or residual cranial nerve. Tumor recurrence or recurrence has been reported infrequently, and usually after suboccipital/retrosigmoid approach; and these probably represent residual tumor.

V. Overall acoustic neuroma recurrence rate is ~0.5%. True recurrence rate with NF-II is unknown, but appears to be higher. The average time interval from removal to recurrence overall is ~10 years. Growth rate of recurrences range between 0.79 and 5.2 mm per year. Recurrence is most likely to occur in large tumor.

VI. NF-II effect primarily brain and spinal cord, although there may be other tumors.

VII. NF-II afflicts 1 in 35,000 or 40,000 individuals, but NF-1, which is usually diagnosed in infancy, does not become apparent in early adulthood or later.

- A. Accepted average tumor growth rate is 2.8 mm per year. House believes in one post op MRI at 5 years, at which time recurrent tumor would be 1.5 cm.
- B. NIH concensus conference diagnostic criteria for NF-II.
- C. Bilateral eighth nerve masses on MRI or CT.
- D. First-degree of relative with NF-II and either a unilateral eighth nerve tumor or two of the following: neurofibroma, meningioma, glioma, schwannoma, or posterior subcapsular lenticular opacity.

VIII. *Hearing Preservation Surgery Usually is Indicated*

- A. Growth pattern is unpredictable.
- B. Surgery is the mainstay of treatment, although radiation and chemotherapy have been used.
- C. Old concepts noted above have been abandoned, and hearing preservation is tried with all tumors less than ~2 cm. If hearing is lost on the first side, second tumor is observed until hearing begins to deteriorate. If hearing is preserved, second tumor is removed with attempted hearing preservation.
- D. If cochlear nerve can be preserved, cochlear implantation can be performed.

IX. *Hearing Preservation: Factors That Influence*

- A. Serviceable hearing 50 dB, 50% discrimination.
- B. Success in NF-II for saving hearing is the same as that in acoustic neuromas in general, 67%.
- C. Tumor located more medially in the IAC may be more favorable for middle fossa approach for hearing preservation, and may be less adherent to the cochlear nerve.
- D. Reduced caloric may indicate superior compartment tumor and be more favorable.
- E. Good preoperative ABR morphology may be more favorable.

Genetics

I. Autosomal dominant.

II. Deletion on the long arm of chromosome 22.

III. The NF-II is a tumor suppressor gene that is inactivated in NF-II tumors. The DNA sequence and protein product on which is acts had been determined.

IV. Penetrance is 100%, but expression is variable.

V. About 50% of NF-II cases are new mutations.

VI. Heterogeneous expression between families; there are two clinical subtypes of NF-II.

- A. *Gardner's subtype* (milder form)
  1. Onset of symptoms, usually hearing loss from bilateral vestibular schwannomas.

2. Few or other associated brain or spinal cord tumors.
3. Main age of onset is 27 years, only 12% have meningiomas.

B. *Wishart's subtype* (severe form)
1. Multiple intracranial and spinal tumors.
2. Developed at an early age with rapid progression of signs and symptoms.
3. Average age of presentation 17.4 years, >70% with meningiomas.

C. *Third subtype, called the Lee/Abbott subtype*
1. Childhood cataracts and early death due to cranial and spinal meningiomas and schwannomas.
2. Average age of presentation 14 years, >70% incidence of meningiomas.

Screening

I. Molecular testing peripheral blood sample.
II. Screening by MRI.
III. Screening should be performed early (under age 8) for at-risk family members; and gene carrier should be screened yearly.

Appendix V
Hearing Conservation Procedures

The forms included in this Appendix have been developed through years of practical experience with hearing conservation programs in industries of all types. They are published with the permission of Hearing Conservation Noise Control, Inc., Bala-Cynwyd, Pennsylvania, to illustrate some of the techniques used in the daily operation of a hearing conservation program. [*Note*: These forms were published in the second edition of *Occupational Hearing Loss* and are preserved here essentially unchanged. Their educational value remains, although HCNC no longer exists in the same form or location.]

SAMPLE FORMS

The forms contained in this section are shown as samples of forms used in various areas of hearing conservation programs. They are not essential to any client or to the HCNC system. The essential forms will be determined jointly by your company and HCNC and the other forms will serve as examples to be utilized at your discretion as your future needs dictate. These forms, if used, should be reviewed by and cleared with your company management and/or possibly your company's legal department:

- #1 Medical referral letter.
- #1A Audiological referral letter.
- #2 Diagnosis and prognosis report to be completed by the otologist.
- #3 Release of information and return to work.
- #4 Return to work report.
- #5 Release of information request.
- #6 VA release of information.
- #7 Record of counsel to employee concerning issuance of hearing protection.
- #8 Report of employees wearing hearing protection by department.
- #9 Composite report of number of employees checked and percentage wearing hearing protection by department.
- #10 Graph audiogram form for recording hearing thresholds. (Not appropriate for industry.)
- #11 Serial form for recording hearing thresholds (preferred).
- #12 Biological calibration log.
- #13 Self-listening test of audiometer function.
- #14 Log of hearing tests performed.
- #15 Otologic and otoscopic observation update.
- #16 OSHA standard change in hearing and hearing protection education letter.

FORMS FOR USE IN DATA PROCESSING

DP#1	Audiogram (see #11 above)
DP#2	Update Otologic History and Otoscopic Observation (see #15 above)
DP#3B	Change in Employee Information or Status (revised 4/90)
DP#5	Plant Audiometric Technician History
DP#10B	Audiogram Information Codes

OTHER FORMS USED IN AUDIOGRAM REVIEW

1. Retest Explanation
2. Request for Additional Information

#1 MEDICAL REFERRAL LETTER

For employees who should receive a medical referral as indicated by HCNC.

Date: / /

Dear ______________________:

The results of your hearing test indicate that you should have your hearing evaluated further by a medical specialist of your choice as soon as possible. Although this is not a work related problem, we would like you to help us follow up by having a report of this evaluation sent to us.

We are concerned about your hearing health and if you have any questions, please contact ______________________________.

Sincerely,

Complete and tear off for file Date: / /

I have been notified that I should see a medical specialist for my hearing.

EMPLOYEE SIGNATURE: ______________________________

REV. 7/85

#1A AUDIOLOGICAL REFERRAL LETTER

Date: / /

Dear ______________________:

The results of your pre-employment hearing test indicate that you should have your hearing re-evaluated. This examination, at this time, need only be a test similar to the one given at the plant. This is called an "air conduction test".

If you have already had your hearing tested by a doctor or an audiologist and this was done in recent months, please have a copy of the test sent to ______________________________.

We are concerned about your hearing health and we will gladly cooperate in any way possible.

Sincerely,

Complete and tear off for file Date: / /

I have been notified to have my hearing re-evaluated and/or to send a copy of the test to the plant.

EMPLOYEE SIGNATURE: __________________________________

REV. 7/85

#2 DIAGNOSIS AND PROGNOSIS REPORT TO BE COMPLETED BY THE OTOLOGIST

Date ....

RE: ....

Prior to consideration for employment, it will be necessary to have a report from an otologist (ear specialist) indicating:

a. Diagnosis; including degree of hearing impairment.

b. Etiology of the hearing loss.

c. Treatment proposed, if any.

d. Prognosis

e. Restrictions as to work environment and/or hearing protection.

f. Copy of otologist's audiogram.

A COPY OF THE AUDIOGRAM DONE AT .... IS ATTACHED.

<u>THIS INFORMATION IS TO BE OBTAINED AT THE EXPENSE OF THE APPLICANT.</u>

#3 RELEASE OF INFORMATION AND RETURN TO WORK

TO: Personal Physician of Employee of ....

When our employees are absent from work for medical reasons we ask that they provide our Company with a medical release from their physician prior to returning to work. This release and the other information requested on the form below is important to the Company so that it has written verification of the employee's medical release for work, date of release, and other information concerning the nature of the illness or injury that may place any temporary or permanent restrictions on the work activity to which an employee is assigned.

We appreciate your cooperation in this matter, and I am sure that our employee will appreciate it also.

I hereby authorize my physician to release all information concerning my health which is related to my absence from work and my return to work.

Date .... Signature of employee ....

FORM BELOW TO BE COMPLETED BY PHYSICIAN

TO: ....

Employee's Name: ....

Nature of Illness: ....

Date of First Treatment for the above condition: ....

Is the employee still under treatment? .... If yes, please list any medication that may affect the employee's mental altertness or physical reaction:

....

The above named employee is released to return to work on .... with/without restrictions. If with restrictions, please state the restrictions, whether permanent or temporary and, if temporary, the date the restrictions are removed:

....

Other remarks: ....

....

Physician's signature .... Date ....

#4 RETURN TO WORK REPORT

DATE: ------------------------------------

This is to certify that --

has now recovered sufficiently to be able to return to:

☐ Light

☐ Regular

working duties on -------------------------

Restrictions: --

Remarks: ---

#5 RELEASE OF INFORMATION REQUEST

TO: .... DATE: ....

....

.... FILE #: ....

You are hereby authorized to give to the ....
or any representative thereof, any and all information which may be requested regarding my physical condition and treatment rendered by you therefore, and to allow them or any physician appointed by them, to examine any X-ray pictures taken of me, and to inspect, review and make copies, including photostatic copies, of all medical records which you have regarding my condition or treatment.

WITNESS: .... DATE: ....

WITNESS: .... DATE: ....

#6 VA RELEASE OF INFORMATION

Form Approved
OMB No. 76–RO138

REQUEST FOR AND CONSENT TO RELEASE OF INFORMATION FROM CLAIMANT'S RECORDS

NOTE: The execution of this form does not authorize the release of information other than that specifically described below. The information requested on this form is solicited under Title 38, United States Code, and will authorize release of the information you specify. The information may also be disclosed outside the VA as permitted by law or as stated in the "Notices of Systems of VA Records" published in the Federal Register in accordance with the Privacy Act of 1974. Disclosure is voluntary. However, if the information is not furnished, we may not be able to comply with your request.

TO: Veterans Administration

NAME OF VETERAN *(Type or print)*

VA FILE NO. *(Include prefix)*

SOCIAL SECURITY NO.

NAME AND ADDRESS OF ORGANIZATION, AGENCY, OR INDIVIDUAL TO WHOM INFORMATION IS TO BE RELEASED

VETERAN'S REQUEST

I hereby request and authorize the Veterans Administration to release the following information, from the records identified above to the organization, agency, or individual named hereon:

INFORMATION REQUESTED *(Number each item requested and give the dates or approximate dates—period from and to—covered by each.)*

PURPOSES FOR WHICH THE INFORMATION IS TO BE USED

NOTE: Additional items of information desired may be listed on the reverse hereof.

DATE

SIGNATURE AND ADDRESS OF CLAIMANT, OR FIDUCIARY, IF CLAIMANT IS INCOMPETENT

VA FORM DEC 1975 **60-3288** EXISTING STOCKS OF VA FORM 07-3288, FEB 1974, WILL BE USED. ☆U.S. GOVERNMENT PRINTING OFFICE: 1979— 281-629-1190

#7 RECORD OF COUNSEL TO EMPLOYEE CONCERNING ISSUANCE OF HEARING PROTECTION

Date: ____________________

Employee Name: ____________________

Location: ____________________

Hearing Protection Issued: Yes ☐ No ☐

Date Issued

Type: muffs ☐ ____________

plugs ☐ ____________

molded ☐ ____________

combination ☐ ____________

other ☐ ____________

Hearing Protection Worn: Yes ☐ No ☐

____

Based on the results of your last audiometric test, a slight hearing deficiency was noted in the:

Left ☐ Right ☐ Both ☐ ear(s).

Continued exposure to noise without the aid of hearing protection could result in additional hearing deficiency. As a condition of your continued employment you must wear hearing protection when working in high noise level areas.

____________________ Company Representative

____________________ Employee

#8 REPORT OF EMPLOYEES WEARING HEARING PROTECTION BY DEPARTMENT

PLANT ________ DATE ________ DEPARTMENT ________ SHIFT ________ SIGNATURE OF PERSON MAKING REPORT ________

No. of EMPLOYEES IN DEPT. ________ NO. OF EMPLOYEES CHECKED ________ WEARING ________ NOT WEARING ________

	TYPE OF EAR PROTECTION								REASON GIVEN FOR NOT WEARING															ATTITUDE			ACTION TAKEN			
EMPLOYEES NAMES						Other	WEARING	NOT WEARING	1 Wk. Adjustment	Allowance for Plugs	Headache	Dizziness	Nausea	Pressure	Irritation	Roaring	Earache	Draining Ear	Difficulty Hearing	Does Not Like	Left At Home	Lost	Other	Good	Fair	Poor	Conference	Referred to Clinic	Referred to Mgmt.	REMARKS

#9 COMPOSITE REPORT OF NUMBER OF EMPLOYEES CHECKED AND PERCENTAGE WEARING HEARING PROTECTION BY DEPARTMENT

Department	Supervisor Responsible	Number of Employees	Date............		Date............		Date............		Date............	
			No. Checked	% Wearing	No. Checked	% Wearing	No. Checked	% Wearing	No. Checked	% Wearing

#10 GRAPH AUDIOGRAM FORM FOR RECORDING HEARING THRESHOLDS

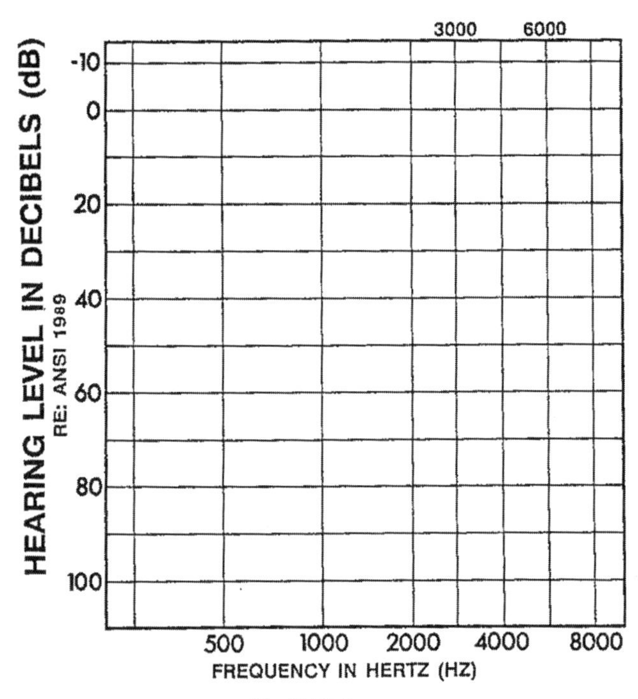

#11 AND DP#1 SERIAL FORM FOR RECORDING HEARING THRESHOLDS (PREFERRED)

Send Carbon Copy To: HCNC Data Center
225 Bala Ave., Suite 200
Bala-Cynwyd, PA 19004

SOC. SEC. # ___ ___ ___ - ___ ___ - ___ ___ ___ ___

CLOCK # ___ ___ ___ ___ ___ ___

COMPANY ____________ DIVISION ____________

LOCATION ____________

PLANT ____________

NAME ____________ (last, first, middle initial)

SEX ______ DATE OF BIRTH ___/___/___ (month, day, year) SERVICE DATE ___/___/___ (month, day, year)

Date	RIGHT Yes	RIGHT No	LEFT Yes	LEFT No
Is Eardrum Visible				
Is Perforation Present				
Is Drum Normal				
Is Other Present				

		RIGHT EAR							LEFT EAR																	
Enter test No.	Date and Time	500	1000	2000	3000	4000	6000	8000	500	1000	2000	3000	4000	6000	8000	Serial # Make Model	Dept. No.	Shift No.	Reas. for test code	Job No. or class.	Years on present job	Noise level code	Time lapse code	Hear. prot. code	TESTER Sign and print name	Tester cert no. or cert date

INITIAL OTOLOGIC HISTORY (Explain All "Yes" Answers)

UPDATE OTOLOGIC HISTORY (enter test number)

Audiometer Electronic Calibration

DATE ________	NO	YES	#	#	#	#	#	#	#	#	Test #	Date of Cal.
1. Ever had noises in ears?												
2. Ever had dizziness?												
3. Ever had fullness in ears?												
4. Ever had pain in ears?												
5. Have you had fluctuating hearing loss?												
6. Have you had sudden or rapid hearing loss?												
7. Have you had ear infections?												
8. Have you been to an ear specialist?												
9. Ear surgery recommended or performed?												
10. Have you had head injury or unconsciousness?												
11. Did you ever hunt or shoot?												
12. Do you have any noisy hobbies?												
13. Do you presently have other noisy job?												
14. Have you had mycins, quinine, excessive aspirin?												
15. Ever had hearing tested before?												
When? Where?	/	/										
16. Family history of hearing loss?												
17. Do you now have difficulty hearing?												
18. Ever had 1. Measles 2. Mumps 3. Chick. Pox 4. Scar. Fev. 5. Dipth.												
19. Yrs. Military Serv. Branch Job												
20. Have you ever had a prior noisy job?												

UPDATE OTOSCOPIC OBSERVATION

								a. Are ear canals obstructed?
								b. Are perforations present?
								c. Is other present?

UPDATE INSTRUCTIONS: Enter only "Yes" responses. Indicate with ✓ through number entered at top of column to show that all questions were asked. The purpose of the update is to discover new developments or worsening of old complaints. Use phrases such as "do you now have a new complaint of" or "since the last test" to elicit new information.

OTOLOGIC & OTOSCOPIC COMMENTS	Test Number	Question Number

#12 BIOLOGICAL CALIBRATION LOG

BIOLOGICAL CALIBRATION LOG

NAME:

AUDIOMETER Serial No. ____________
Make & Model ____________

	RIGHT EAR							LEFT EAR							
Date	500	1000	2000	3000	4000	6000	8000	500	1000	2000	3000	4000	6000	8000	Technician Signature

#13 SELF-LISTENING TEST OF AUDIOMETER FUNCTION

Date	Audiometer Serial Number	Make and Model	Jacks Seated	Cords O.K.	Headband Tension O.K.	Earphone Cushions O.K.	Dials Tight	Volume Increases and Decreases	Pitch O.K	Tone On-Off O.K.	Tone Presenter Inaudible	No Static	No Crosstalk	Tester Signature

#14 LOG OF HEARING TESTS PERFORMED

AUDIOGRAM NUMBER	EMPLOYEE NAME	LOCATION	TECHNICIAN	DATE	REMARKS

#15 AND DP2# OTOLOGIC AND OTOSCOPIC OBSERVATION UPDATE

COMPANY________________ LOCATION________________ PLANT________________

SOCIAL SECURITY # ________________ EMPLOYEE NAME ________________

UPDATE OTOLOGIC HISTORY	Enter Test Number									
"Do you now have a new complaint of"	#	#	#	#	#	#	#	#	#	#
1. Noises in ears?										
2. Dizziness?										
3. Fullness in ears?										
4. Pain in ears?										
5. Fluctuating hearing loss?										
6. Sudden or rapid hearing loss?										
7. Ear infections?										
"Since the last test"										
8. Have you been to an ear specialist?										
9. Was ear surgery recommended or performed?										
10. Have you had a head injury or unconsciousness?										
11. Have you been exposed to gunfire?										
12. Have you worked at noisy hobbies?										
13. Have you worked at another job that was noisy?										
14. Have you had myeins, quinine, excessive aspirin?										

UPDATE OTOSCOPIC OBSERVATION	Enter Test Number									
	#	#	#	#	#	#	#	#	#	#
a. Are ear canals obstructed?										
b. Are perforations present?										
c. Is other present?										

COMMENTS	test no.	question no.

UPDATE INSTRUCTIONS. 1. Leave boxes blank unless answer is "yes" (then enter "Y"). 2. If all boxes are left blank, put a checkmark through the test number at the top of the appropriate column to indicate that the questions were asked.
The purpose of the update is to discover new developments or worsening of old complaints. It is important therefore that questions are properly phrased as indicated.

IHCNC 1980 DP #2 [illegible]

#16 OSHA STANDARD CHANGE IN HEARING AND HEARING PROTECTION EDUCATION LETTER

SAMPLE combined employee notification and hearing protection education letter. This might simplify your in-house procedures. This is strictly for internal use and not for reporting to HCNC.

Please feel free to modify it in any way you think is necessary and/or check with your supervisor or corporate office before putting it into use.

SAMPLE

OSHA STANDARD CHANGE IN HEARING AND
HEARING PROTECTION EDUCATION LETTER

Date / /

Dear____________________________________

Your recent hearing test shows some changes as compared to your past hearing levels. We are advising you as stated below to help prevent further changes in your hearing. Please take similar precautions to protect your hearing when exposed to noise off the job.

Fitted______ Refitted ______ with hearing protection.

Type of hearing protection: Plugs_____ Muffs_____ Canal Caps ______
Other______

We have instructed you in the use and care of your hearing protection and require you to use them. Please contact ____________________
if you have any questions about your hearing health.

Sincerely,

Employee Name__________________ Employee Signature__________________ Date________

Employee SS#__________________

Witnessed by__________________________ Title__________________________ Date________

REV 4-83

To be completed in duplicate. File original in employee folder, give copy to employee.

DP#3B CHANGE IN EMPLOYEE INFORMATION OR STATUS

COMPANY: ______________ DATE: ______________

DIVISION: ______________ TECHNICIAN: ______________

PLANT: ______________ PHONE NO: ______________

LOCATION: ______________

CURRENT INFORMATION (as it appears on your reports)	CHANGE TO (as you want it to appear)	DELETE from File	DELETE from Test Sched
Name			
SSN (Mandatory)			
Name			
SSN (Mandatory)			
Name			
SSN (Mandatory)			
Name			
SSN (Mandatory)			
Name			
SSN (Mandatory)			
Name			
SSN (Mandatory)			
Name			
SSN (Mandatory)			
Name			
SSN (Mandatory)			
Name			
SSN (Mandatory)			
Name			
SSN (Mandatory)			

Delete from *File* removes Employee entirely; delete from *Schedule* permanently removes Employee Name from *Test Schedule*

Send to: HCNC Data Center, 225 BALA AVENUE, SUITE 200 • BALA-CYNWYD, PA 19004 DP#3B (Rev. 04/90)

DP#5 PLANT AUDIOMETRIC TECHNICIAN HISTORY

COMPANY ______________ DIVISION ______________ PLANT ______________

ADDRESS ______________ CITY ______________ STATE ______________ ZIP ________

YEAR IN WHICH HEARING TESTING COMMENCED AT THIS FACILITY ______________

IS HEARING TESTING DONE "IN-HOUSE"? YES ____ NO ____ WAS IT ALWAYS DONE "IN-HOUSE"? YES ____ NO ____

IF DONE "OUTSIDE", STARTING DATE __________

	NAME OF AUDIOMETRIC TECHNICIAN			DATE INITIALLY TRAINED		INITIAL COURSE BY HCNC?		DATE OF REFRESHER		REFRESHER BY HCNC		DO YOU HOLD A CAOHC*CERT?			DATE TERMINATED
	LAST	INITIALS	SS#	MONTH	YEAR	YES	NO	MON.	YEAR	YES	NO	YES	DATE ISSUED	NO	If applic.
1															
2															
3															
4															
5															
6															
7															
8															
9															
10															

INSTRUCTIONS: Please list ALL persons who have performed hearing tests since your testing program began. If any of these persons no longer are employed at your facility, indicate the date of their termination. If applicable, please indicate by * any persons trained but who do not perform tests.

HCNC 1980 DP#5 * Council for Accreditation in Occupational Hearing Conservation

DP#10B AUDIOGRAM INFORMATION CODES

REASON FOR TEST CODES:

Code	Description
1	Routine (annual; retest with physical exam etc; 1st test on current employee).
2	Pre-employment; Return from layoff; Pre-placement; Return from leave of absence; Rehire; Transfer within plant; Transfer from other plant in Company to this plant; Return from illness or surgery.
3	Summer; Co-op; Temporary help.
4	Termination; Layoff; Transfer away from plant.
5	Special request (By: Employee; employer; physician).
6	HCNC recommendation of 1 year after HCNC significant threshold shift or OSHA Standard Shift.
7	HCNC recommendation of less than 1 year.
8	Bone/air conduction test (this code only used by HCNC personnel performing the test).

SPECIAL REASON FOR TEST CODES:

Code	Description
A	Retesting within 30 days of last test (OSHA or PRESUMPTIVE OSHA STANDARD SHIFT).
B	Non-noise exposed personnel for possible exclusion from retest schedule.

REASON FOR TEST (DUPONT ONLY):

Code	Description
5	Before military leave
2	Return from military leave

HCNC DP#10B REV. 1986

NOISE LEVEL CODES: *

Code	Description
1	Less than 85 dBA
2	85 dBA or more
	Supplemental Codes (Rev. 1-1-87) **
4	Less than 80 dBA
5	80-84 dBA
6	85-89 dBA
7	90-94 dBA
8	95+ dBA
P	Potential for 85 dBA + ***
*	Unrecorded noise level data results in a one year retest
**	Mandatory for Dupont locations
***	Not to be used by Dupont locations

TIME LAPSE CODES:

Code	Description
1	Less than 1 hour
2	1 to 13 hours
3	14 hours or more

HEARING PROTECTION CODES:

Code	Description
1	Ear inserts
2	Muffs
3	Canal caps
4	Combination muffs and plugs
5	Issued but type not specified
6	Other
7	None

1. RETEST EXPLANATION

Increasing concern for compensation for hearing loss necessitates our recommending that new employees with substantial hearing losses be retested very soon after being hired.

The purpose of this recommendation is to verify the incoming hearing levels before the new employee has worked for any length of time. In certain instances we may suggest that a prospective employee be referred to a local otologist for a repeat audiogram, otologic evaluation and recommendations for employment. Your company policy should dictate whether this evaluation should be paid for by the company of by the prospective employee.

We suggest also that on all new hires and initial audiograms of current employees, all thresholds exceeding 25 dB be repeated and recorded. The second readings should be recorded above the first readings, similar to what is done for the recheck at 1000 Hz.

In the case of rehires, if you have previous thresholds and the incoming thresholds are worse by 10 dB or more repeat those thresholds and question and record the responses to noisy occupations and activities while not in your employ. If no previous hearing record exist, your recommendation to see an otologist should be tempered by the previous length of time and type of work involved. Send both preceding categories to us for our recommendations.

We are sensitive to the extra time involved but feel that current trends justify this type of verification.

....

2. REQUEST FOR ADDITIONAL INFORMATION

COMPANY .... ADDRESS ....

....

....

Medical Department:

We note that the otologic history obtained on the following indicates that medical attention for an existing hearing loss was obtained at some time in the past:

....

....

....

....

....

....

We recommend that where possible to do so, obtain a copy of all otologic and audiologic reports from the physician consulted for the hearing loss.

Upon receipt of those reports, please forward us a copy attached to the carbon copy of the audiogram you normally send us.

Thank you.

Audiogram Review Department
HCNC
Bala-Cynwyd, Pa

Appendix VI

Hearing Conservation Program Procedures and Reports

The material in this Appendix is reprinted with permission from the Reports Manual of Hearing Conservation Noise Control, Inc., Bala-Cynwyd, Pennsylvania. It describes procedures developed during more than 30 years of experience in providing hearing conservation services to industry and provides practical insight into the daily operations of a hearing conservation program. [*Note*: These reports were published in the second edition of *Occupational Hearing Loss* and are preserved here unchanged. Their educational value remains, although HCNC no longer exists in the same form or location.]

INTRODUCTION

Hearing Conservation Noise Control, Inc. (HCNC) is capable of processing audiometric data from several media. These include:

1. Paper records. Specifically audiogram form HCNC DP#1, which is designed for keying by our data entry personnel.
2. Magnetic tape and/or floppy diskettes. Exact data formats need to be discussed with our programming department.
3. Modem. Data transmitted to our office via telephone.

SHIPMENT OF AUDIOGRAMS TO THE CONSULTANT

Do not send original copies of the audiograms to the consultant

1. Send the carbon copies in batches of 50 or more once a month, whichever occurs first. If you will not be doing 50 hearing tests in a month's period you do not have to wait until 50 have been done before sending them to HCNC. Send them any time during the month to:

   HEARING CONSERVATION NOISE CONTROL, INC.
   Data Center
   225 Bala Avenue, Suite 200
   Bala-Cynwyd, PA 19004
   (215) 667-1711
   (215) 667-1776 (FAX)

2. After HCNC has entered the data into the computer, the copies are returned to you for matching-up with the original portion you kept at the plant. Please do not photocopy audiograms for transmission to HCNC.
3. Included in the pack of audiograms returned to you will be reports "A" and "B" (explained below). These reports will give you immediate feedback concerning the audiograms in that pack.

WHAT TO EXPECT

This section is concerned with how to interpret and use the information that HCNC will provide to you. Each of our six standard reports is included as well as a brief explanation of each one. The six reports are the Individual Audiogram Analysis, Employee Notification Letter, Follow-Up Report, Test Schedule, Year-to-Date Report of Employees with Changes, or Medical Referrals (and Notification of Such) Showing Date of Change, and the Quarterly/Year-to-Date Program Summary. All reports are printed on two-part carbonless paper, so that two copies are always available. Good recordkeeping practice dictates that all reports should be stamped with the date they were received at your office.

Employee Audiogram Analysis (Report A)

This report, individually printed for every audiogram reviewed, lists the employee's identifying information, the baseline, current, and next most recent test, the findings of the current test, and has a space for comments and a signature line.

1. Identifying information is exactly what was provided to us on the DP#1 (audiogram) form. Please scan over such things as the spelling of the employee's name, sex, date of birth, and social security number. If you find an error, please refer to File Maintenance Procedures, in the section entitled Special Situations.
2. The three tests mentioned above are printed out in descending order, with the baseline at the top, followed by the next most recent test, and finally, the current test. These are listed separately by ear, with the right ear above the left ear.
3. Following each test are the audiogram type codes, which describe the HCNC interpretation of hearing levels for that test. Next are listed the findings of the last two tests. Please refer to Appendix II for a description of the type codes and findings abbreviations.
4. The Comments section lists additional comments by HCNC regarding medical referrals. It can also be used for the plant technician, nurse, or physician comments.
5. The Signature line can be used either by plant personnel to indicate that the employee has been informed of the results of the analysis, or by the employee to indicate that he has received the results of his hearing test (either a copy of the analysis, or the notification letter).

Please refer to the sample on the next page for an example of the Individual Audiogram Analysis (Report A).

FOREST CORPORATION
DISPLAY DATA
SAMPLE IA 99999

RE: BELCHER, JOSEPHINE
000-00-0001

RE: BELCHER, JOSEPHINE
000-00-0001

EMPLOYEE'S AUDIOGRAM ANALYSIS 01/12/92

CLOCK#	DEPT	JOB#	BIRTHDATE	HIREDATE	SEX	NOISE	PROTECTION
401321	4047	447433	01/20/66	09/19/89	F	YES	INSERTS

RESULT OF ANALYSIS:
(RIGHT EAR) OSHA STANDARD THRESHOLD SHIFT

RETEST RECOMMENDATION BASED ON LAST AUDIOGRAM: 12 MONTH(S)
(01) MONTH(S) **

RIGHT KILOHERTZ

TEST#	TEST DATE	.5	1	2	3	4	6	8	TYPE	FINDINGS
01	01/27/90	00	00	00	00	00	20	10	A	
02	01/17/91	00	05	05	00	05	15	10	A	NO CHANGE
03	01/04/92	00	00	05	10	20	20	10	A	OSHA STS

LEFT KILOHERTZ

TEST#	TEST DATE	.5	1	2	3	4	6	8	TYPE	FINDINGS
01	01/27/90	05	05	10	05	10	20	15	A	
02	01/17/91	10	10	05	05	10	15	05	A	NO CHANGE
03	01/04/92	10	05	10	05	15	20	10	A	NO CHANGE

OTOLOGIC HISTORY

TINNITUS	N	EAR PAIN	N	EAR INFECTION	N	HEAD INJURY	N
VERTIGO	N	FLUCT. LOSS	N	EAR SPECIALIST	N	GUNFIRE EXP.	N
EAR FULLNESS	N	RAPID LOSS	N	EAR SURGERY	N		

** FOLLOW YOUR COMPANY POLICY CONCERNING SHORTER RETEST PERIODS (SHOWN IN PARENTHESES) FOR THRESHOLD SHIFTS.

CLIENT COMMENTS:

SIGNATURE: ______________________ DATE: __________

999-999-999-999 (C) HCNC, INC. 1984
REPORT -A-

FOREST CORPORATION
DISPLAY DATA
SAMPLE IA 99999

RE: EGGERT, PAULA
000-00-0003

RE: EGGERT, PAULA
000-00-0003

EMPLOYEE'S AUDIOGRAM ANALYSIS 12/23/91

CLOCK#	DEPT	JOB#	BIRTHDATE	HIREDATE	SEX	NOISE	PROTECTION
567890	3066	300600	06/16/47	12/02/89	F	YES	INSERTS

RESULT OF ANALYSIS: NO CHANGE

RETEST RECOMMENDATION BASED ON LAST AUDIOGRAM: 12 MONTH(S)

RIGHT KILOHERTZ

TEST#	TEST DATE	.5	1	2	3	4	6	8	TYPE	FINDINGS
01	12/04/89	10	05	00	10	45	70	65	E	
02	12/08/90	10	05	05	20	45	70	60	E	NO CHANGE
03	12/05/91	15	05	00	10	45	65	70	E	NO CHANGE

LEFT KILOHERTZ

TEST#	TEST DATE	.5	1	2	3	4	6	8	TYPE	FINDINGS
01	12/04/89	10	05	10	10	40	70	60	E	
02	12/08/90	10	15	10	20	40	65	65	D	NO CHANGE
03	12/05/91	10	10	10	20	45	65	60	D	NO CHANGE

OTOLOGIC HISTORY

TINNITUS	Y	EAR PAIN	N	EAR INFECTION	N	HEAD INJURY	N
VERTIGO	N	FLUCT. LOSS	N	EAR SPECIALIST	Y	GUNFIRE EXP.	Y
EAR FULLNESS	N	RAPID LOSS	N	EAR SURGERY	N		

CLIENT COMMENTS:

SIGNATURE: ______________________ DATE: __________

999-999-999-999 (C) HCNC, INC. 1984
REPORT -A-

FOREST CORPORATION
DISPLAY DATA
SAMPLE IA 99999

RE: PULLEY, JOSEPH
000-00-0002

RE: PULLEY, JOSEPH
000-00-0002

EMPLOYEE'S AUDIOGRAM ANALYSIS 12/21/91

CLOCK#	DEPT	JOB#	BIRTHDATE	HIREDATE	SEX	NOISE	PROTECTION
101011	3985	100200	06/12/51	11/01/88	M	YES	MUFF/PLUGS

RESULT OF ANALYSIS:
(RIGHT EAR) MEDICAL REFERRAL 1
(RIGHT EAR) HCNC SIGNIFICANT THRESHOLD SHIFT

RETEST RECOMMENDATION BASED ON LAST AUDIOGRAM: 01 MONTH(S)
(06) MONTH(S) **

RIGHT KILOHERTZ

TEST#	TEST DATE	.5	1	2	3	4	6	8	TYPE	FINDINGS	
01	11/01/88	05	00	05	05	10	10	05	A		
02	11/08/89	05	10	10	05	15	05	10	A		
03	12/10/91	40	30	20	15	10	10	15	F	HCNC SIG	MED-R

LEFT KILOHERTZ

TEST#	TEST DATE	.5	1	2	3	4	6	8	TYPE	FINDINGS
01	11/01/88	10	10	10	05	10	10	10	A	
02	11/08/89	15	10	15	05	10	10	05	A	
03	12/10/91	15	10	05	10	10	10	15	A	NO CHANGE

OTOLOGIC HISTORY

TINNITUS	Y	EAR PAIN	N	EAR INFECTION	N	HEAD INJURY	N
VERTIGO	Y	FLUCT. LOSS	Y	EAR SPECIALIST	N	GUNFIRE EXP.	N
EAR FULLNESS	Y	RAPID LOSS	N	EAR SURGERY	N		

1 CHANGE IN HEARING INVOLVING MOST IF NOT ALL TONES AND IS NOT RELATED TO NOISE EXPOSURE. WARRANTS OTOLOGIC EXAMINATION AND DIAGNOSIS. TRY TO OBTAIN A COPY OF THE EXAMINATION RESULTS FOR THE COMPANY FILE.

** FOLLOW YOUR COMPANY POLICY CONCERNING SHORTER RETEST PERIODS (SHOWN IN PARENTHESES) FOR THRESHOLD SHIFTS.

CLIENT COMMENTS:

SIGNATURE: ______________________ DATE: __________

REPORT -A-

Employee Notification Letter (Optional)

When included in your contract, the employee notification letters are printed on an individual basis at the same time as Report A. The letters are written in layman's terms and are designed to be handed out or mailed to the employee.

1. They describe the hearing levels, whether there have been any changes in hearing compared to the baseline test, and what further follow-up is recommended.
2. One copy of the letter should be kept in the employee's medical or personnel file. Document when the employee was given the letter, preferably by having him/her sign the plant copy of the letter or Report A.

```
FOREST CORPORATION
DISPLAY DATA
SAMPLE                 IA 99999

Dear Ms. BELCHER                                  SS# 000-00-0001

The results of your hearing check have been evaluated by our
consultant who has identified the following:

Your hearing in both ears is normal.

There has been no change in hearing in your left ear since the last check.

You have had a Standard Threshold Shift in your right ear.

Regulations require that you obtain an annual hearing examination. You will
be notified when the next hearing check is due.  Please wear your hearing
protection effectively.

Sincerely,                                        BELCHER, JOSEPHINE

FOREST CORPORATION                                DEPT. 4047

01/12/92                                          CLOCK # 401321
```

```
FOREST CORPORATION
DISPLAY DATA
SAMPLE                 IA 99999

Dear Ms. EGGERT                                   SS# 000-00-0003

The results of your hearing check have been evaluated by our consultant who
has identified the following:

Your hearing in both ears is satisfactory for hearing and understanding
conversation. There is however some hearing loss for the high-pitched tones
for which you were tested.

There has been no change in hearing in either ear since the last check.

Regulations require that you obtain an annual hearing examination.  You will
be notified when the next hearing check is due.  Please wear your hearing
protection effectively.

Sincerely,                                        EGGERT, PAULA

FOREST CORPORATION                                DEPT. 3066

12/23/91                                          CLOCK # 567890
```

```
FOREST CORPORATION
DISPLAY DATA
SAMPLE                IA 99999

Dear Mr. PULLEY                                   SS# 000-00-0002

The results of your hearing check have been evaluated by our consultant who
has identified the following:

Your hearing in the left ear is normal.

Your hearing in the right ear indicates that in some situations you may have
difficulty hearing or understanding conversation with this ear.

There has been no change in hearing in your left ear since the last check.

You have had a change in hearing at some tones in your right ear.

Regulations require that you obtain an annual hearing examination. You will
be notified when the next hearing check is due.  Please wear your hearing
protection effectively.

Because of the changes, you should see an ear doctor.

Sincerely,                                        PULLEY, JOSEPH

FOREST CORPORATION                                DEPT. 3985

12/21/87                                          CLOCK # 101011
```

Follow-Up Report (Report B)

This report, printed with every batch of audiograms reviewed, lists batch information, as well as those employees who need further attention.

1. Batch information lists at the top of the report the counts of audiograms received, audiograms processed, and audiograms returned for missing information. These counts should be used to verify that all audiograms submitted are accounted for on the report.

   Please note that audiograms which are already on file (and therefore not processed) are not included in these counts. They will have a note attached to them indicating that such is the case. Also, those audiograms which are returned because of missing information will also have a sheet attached, detailing what information was missing that prevented them from being processed. Those should be resubmitted once the missing information has been filled in.
2. Employees showing changes and/or medical referrals are listed next. Some identifying information, the finding(s), and the retest recommendations are printed. Please refer to Appendix II for a description of the abbreviations listed in the findings column. The retest recommendations are always listed in months. Follow your company policy regarding one-, two-, and six-month retest recommendations.
3. The Date-Notified column should be filled in by plant personnel when the employee has been notified that the condition listed on this report exists and/or that the recommendation for him/her to see a specialist has been made. Employees exhibiting an OSHA STS (for those plants not doing the optional 30-day retests), OSHA, CONF, and/or MED REF, must also be notified in writing of that condition. In either case, the plant has 21 days to

make this notification starting from the day the report of that finding is received at the plant from HCNC.

Once the date notified has been filled in, one copy of the Follow-Up Report should be sent to HCNC with your next batch of audiograms. The date notified will be entered into the employee's file, and that documentation will appear on the quarterly reports.

4. There is also a column marked New Base. This will alert you that HCNC has revised the baseline for that employee. The sequential test number to which the baseline was revised is listed. No further follow-up or notification is needed.

FOREST CORPORATION
DISPLAY DATA
SAMPLE IA 99999

FOLLOW-UP REPORT DATE OF REPORT: 12/21/91

AUDIOGRAMS RECEIVED	3
AUDIOGRAMS PROCESSED	3
AUDIOGRAMS RETURNED BECAUSE OF MISSING INFORMATION	0
AUDIOGRAMS RETURNED WITH NO ACTION	0

LIST OF EMPLOYEES SHOWING CHANGES AND/OR REQUIRING MEDICAL REFERRALS.

THESE PEOPLE WILL ALSO BE LISTED ON THE YEAR TO DATE REPORT. IF YOU WANT YOUR FOLLOW-UP DOCUMENTED ON THAT REPORT, PLEASE RECORD ON THIS LIST THE DATE THE EMPLOYEE WAS NOTIFIED AND SUBMIT IT WITH A FUTURE BATCH OF AUDIOGRAMS.

EMPLOYEE NAME	CLOCK#	DEPT#	JOB# NOISE	NEW BASE	FINDINGS	RETEST	DATE NOTIFIED
BELCHER, JOSEPHINE	401321	4047	447433		(R) OSHA STS	12(01)**	
000-00-0001				Y			
PULLEY, JOSEPH	101011	3985	100200		(R) MED. REF.	01	
000-00-0002				Y	(R) HCNC SIG	12(06)**	

** FOLLOW YOUR COMPANY POLICY CONCERNING SHORTER RETEST PERIODS (SHOWN IN PARENTHESES) FOR THRESHOLD SHIFTS.
999-999-999-999 (C) HCNC, INC. 1984
REPORT -B-

Test Schedule (Report C)

There are two versions of this report, the regular test schedule and the advance test schedule, which are always printed in combination once a month. They are usually mailed during the third week of the month.

1. The regular test schedule report lists the names of those employees who are due for a hearing test in the upcoming month, based on the date of their last test and the retest recommendation made at that time. As always, some identifying information is shown, as well as the date of their last test.

   In the column marked LATE, one asterisk will be printed for every month the employee is late for a test. Up to four asterisks will be printed. The employee will be carried on this report for up to twelve months unless we receive a current test or you ask us to delete him/her. See File Maintenance Procedures under the section marked Special Situations.

Only those tests not received before the reporting cut-off date for that month will appear as late. The cut-off date is fixed at five working days into the beginning of the next month. For example, tests due in April will be listed as late if they are not received by the fifth working day in May.

An employee who had an HCNC SIG on his/her last test will appear on the regular schedule one time exactly six months after the date of that test, then will drop off the schedule until a total of twelve months have passed.

2. The advance schedule lists those employees who will be due for a hearing test for the month following the regular schedule. No employees will be listed as late on this report. As with the regular schedule, an employee with an HCNC SIG will appear only once on this report.
3. Both schedules can be used to request file maintenance. See File Maintenance Procedures under the section entitled Special Situations for more information.

FOREST CORPORATION
DISPLAY DATA
SAMPLE IA 99999

TEST SCHEDULE REPORT DATE OF REPORT: 10/01/91

TESTS DUE NOVEMBER 1991

	CLOCK#	SHIFT	DEPT#	JOB # NOISE	DATE OF LAST TEST	LATE
ADAMS, STEVEN 223-45-6779	123400	1	0000	123000 N	11/23/87	
BARKER, EDWARD 231-16-0543	567800	3	9190	456000 Y	11/01/90	
DUNCAN, RAY R. 813-45-5678	432100	1	8670	321000 Y	09/06/90	**
WATKINS, ALBERT 242-22-1306	464800	1	34AT	4XA437 Y	11/06/90	

ADVANCE SCHEDULE FOR DECEMBER 1991

	CLOCK#	SHIFT	DEPT#	JOB # NOISE	DATE OF LAST TEST
ALLEN, MARK 265-75-3859	357800	1	0120	126000 Y	12/10/90
CARSON, BERNARD 207-88-9372	784600	1	9940	283000 Y	12/13/90
ELLIOT, DAVID 129-57-8473	184800	2	4840	949000 U	12/02/90
JONES, JOHN A. 384-56-4738	930300	1	2730	859000 Y	12/19/90

* ONE MONTH LATE FOR EACH ASTERISK (UP TO 4 ASTERISKS ARE PRINTED)

REPORT -C-

FOREST CORPORATION
DISPLAY DATA
SAMPLE IA 99999

YEAR TO DATE REPORT DATE OF REPORT: 12/31/91

EMPLOYEES WITH CHANGES, OR MEDICAL REFERRALS,
(AND NOTIFICATION OF SUCH), SHOWING DATE OF CHANGE

NAME	EAR	OSHA STS	OSHA CONF	OSHA REPL	HCNC SIG.	HCNC CONF	HCNC REPL	HCNC PROG	MED REF	NOISE (Y/N)	DATE EMP NOTIF.
ADAMS, JOE J. 123-45-6789	R	05/91	06/91			05/91				Y	06/91
	L				06/91			06/91	06/91		
AKERS, ADAM A. 987-57-4838	R	08/91		09/91						Y	
	L	09/91									
BAKER, BILL B. 321-45-9876	R				02/91			08/91		Y	
	L										
DAVIS, AL B. 123-65-3524	R									N	
	L				01/91		07/91				
SMITH, JOHN J. 107-22-3456	R	02/91	03/91		02/91	03/91			02/91	Y	03/91
	L	02/88	03/91		02/88	03/91					

999-999-999-999 (C) HCNC, INC. 1984
REPORT -D-

Quarterly/Year-to-Date Program Summary (Report E)

Report E is printed and mailed simultaneously with the Year-to-Date Report of Employees with Changes, or Medical Referrals, (and Notification of Such), Showing Date of Change (Report D), and as mentioned earlier, these two reports are designed to be used in combination to monitor the status of the Hearing Conservation Program. It lists audiogram counts, noise levels reported, and a numeric breakdown by quarter and year-to-date for both noise exposed and nonexposed employees of the findings listed on the preceding report.

1. The audiogram counts allow one to evaluate how the program is proceeding. The number of tests is listed, rather than the number of employees, since a particular employee may be tested several times during a quarter/year.
2. The noise levels reported can serve two functions. One, they can give a rough estimation of how many exposed vs. nonexposed employees were tested. Two, they can give an indication if the noise level codes on the audiogram are being filled in correctly.
3. The numeric breakdown of the findings which were listed by employee on Report D allow for statistical analysis of the effectiveness of the program. It also helps to identify whether proper employee notification is occurring.

FOREST CORPORATION
DISPLAY DATA
SAMPLE IA 99999

DATE OF REPORT: 12/31/91

QUARTERLY/YEAR-TO-DATE PROGRAM SUMMARY

	QTR	YTD	NUMBER OF EMPLOYEES BY NOISE LEVEL REPORTED ON LAST TEST	QTR	YTD
TESTS REVIEWED BY HCNC	29	82	NOISE EXPOSED	20	40
PAST DUE AS OF 10/01/91	2	4	NON-EXPOSED	5	26
			UNKNWN-EXPOSED	4	16

TOTALS OF EMPLOYEES WITH CHANGES OR MEDICAL REFERRALS BASED UPON STATUS OF ALL AUDIOGRAMS REVIEWED AT QUARTER-END AND YEAR-TO-DATE IT IS POSSIBLE FOR AN EMPLOYEE TO BE COUNTED IN MORE THAN ONE COLUMN IF THAT EMPLOYEE EXHIBITED MORE THAN ONE TYPE OF FINDING

	OSHA STS	OSHA CONF	OSHA REPL	HCNC SIG.	HCNC CONF	HCNC REPL	HCNC PROG	MED/AUD REF	#EMP NOTIF.
QTR:									
NOISE-EXPOSED	1	0	1	0	0	0	1	0	0
NON-EXPOSED	0	0	0	0	0	1	0	0	0
UNKWN-EXPOSED	0	0	0	0	0	0	0	0	0
QTR TOTAL	1	0	1	0	0	1	1	0	0
YTD:									
NOISE-EXPOSED	3	2	1	2	2	0	1	1	2
NON-EXPOSED	0	0	0	1	0	1	0	0	0
UNKWN-EXPOSED	0	0	0	0	0	0	0	0	0
YTD TOTAL	3	2	1	3	2	1	1	1	2

REPORT -E-

Index

A

D

E

F

G

H

I

J

K

L

M

N

O

S

T

U

V

W

X

Z